Physicochemical Hydrodynamics
Interfacial Phenomena

NATO ASI Series

Advanced Science Institutes Series

A series presenting the results of activities sponsored by the NATO Science Committee, which aims at the dissemination of advanced scientific and technological knowledge, with a view to strengthening links between scientific communities.

The series is published by an international board of publishers in conjunction with the NATO Scientific Affairs Division

A	**Life Sciences**	Plenum Publishing Corporation
B	**Physics**	New York and London
C	**Mathematical and Physical Sciences**	Kluwer Academic Publishers Dordrecht, Boston, and London
D	**Behavioral and Social Sciences**	
E	**Applied Sciences**	
F	**Computer and Systems Sciences**	Springer-Verlag
G	**Ecological Sciences**	Berlin, Heidelberg, New York, London,
H	**Cell Biology**	Paris, and Tokyo

Recent Volumes in this Series

Volume 170—Physics and Applications of Quantum Wells and Superlattices
　　　　edited by E. E. Mendez and K. von Klitzing

Volume 171—Atomic and Molecular Processes with Short Intense Laser Pulses
　　　　edited by André D. Bandrauk

Volume 172—Chemical Physics of Intercalation
　　　　edited by A. P. Legrand and S. Flandrois

Volume 173—Particle Physics: *Cargèse 1987*
　　　　edited by Maurice Lévy, Jean-Louis Basdevant, Maurice Jacob,
　　　　David Speiser, Jacques Weyers, and Raymond Gastmans

Volume 174—Physicochemical Hydrodynamics: Interfacial Phenomena
　　　　edited by Manuel G. Velarde

Volume 175—Superstrings
　　　　edited by Peter G. O. Freund and K. T. Mahanthappa

Volume 176—Nonlinear Evolution and Chaotic Phenomena
　　　　edited by Giovanni Gallavotti and Paul F. Zweifel

Series B: Physics

Physicochemical Hydrodynamics

Interfacial Phenomena

Edited by
Manuel G. Velarde
U.N.E.D.
Madrid, Spain

Plenum Press
New York and London
Published in cooperation with NATO Scientific Affairs Division

Proceedings of a NATO Advanced Study Institute on
Physicochemical Hydrodynamics: Interfacial Phenomena,
held July 1-11, 1986,
and of the Second EPS Liquid State Conference,
held July 12-15, 1986,
in La Rabida, Huelva, Spain

Library of Congress Cataloging in Publication Data

NATO Advanced Study Institute on Physicochemical Hydrodynamics: Interfacial Phenomena (1986: La Rabida, Huelva, Spain)
    Physicochemical hydrodynamics: interfacial phenomena / edited by Manuel G. Velarde.
      p.   cm.—(NATO ASI series. Series B, Physics; v. 174)
  "Proceedings of a NATO Advanced Study Institute on Physicochemical. Hydrodynamics: Interfacial Phenomena, held July 1-11, 1986, and of the Second EPS Liquid State Conference, held July 12-15, 1986, in La Rabida, Huelva, Spain"—T.p. verso.
  "Published in cooperation with NATO Scientific Affairs Division."
  Includes bibliographies and indexes.
  ISBN-13: 978-1-4612-8042-2   e-ISBN-13: 978-1-4613-0707-5
  DOI: 10.1007/978-1-4613-0707-5
  1. Surface chemistry—Congresses. 2. Hydrodynamics—Congresses. I. Velarde, Manuel G. (Manuel Garcia) II. EPS Liquid State Conference (2nd: 1986: Huelva, Spain) III. European Physical Society. IV. North Atlantic Treaty Organization. Scientific Affairs Division. V. Title. VI. Series.
    [DNLM: 1. Capillarity—congresses. 2. Surface-Active Agents—congresses. 3. Surface Tension—congresses. QC 183 N279p 1986]
QD506.A1N36   1986
541.3'453—dc19
DNLM/DLC
for Library of Congress
                          88-9899
                          CIP

© 1988 Plenum Press, New York
Softcover reprint of the hardcover 1st edition 1988

A Division of Plenum Publishing Corporation
233 Spring Street, New York, N.Y. 10013

All rights reserved

No part of this book may be reproduced, stored in a retrieval system,
or transmitted in any form or by any means, electronic, mechanical, photocopying,
microfilming, recording, or otherwise, without written permission from the Publisher

In memory of
Professor Ben Levich,
whose seminal work in the field of
physicochemical hydrodynamics opened
a new avenue of fascinating and
transdisciplinary research

PREFACE

This book contains lecture notes and invited contributions presented at the NATO Advanced Study Institute and EPS Liquid State Conference on **PHYSICOCHEMICAL HYDRODYNAMICS-PCH: INTERFACIAL PHENOMENA** that were held July 1-15, 1986, in **LA RABIDA** (Huelva) **SPAIN**. Although we are aware of the difficulty in organizing the contents due to the broad and multidisciplinary aspects of PCH-Interfacial Phenomena, we have tried to accomodate papers by topics and have not followed the order in the presentation at the meetings. There is also no distinction between the ASI notes and Conference papers. We have done our best to offer a coverage as complete as possible of the field. However, we had difficulties coming from the fact that some authors were so busy that either did not find time to submit their contribution or did not have time to write a comprehensive paper. We also had to cope with very late arrivals, postdeadline valuable contributions that we felt had to be included here.

Our gratitude goes to the NATO Scientific Affairs Division for its economic support and to the EPS Liquid State Committee for its sponsorship. Financial support also came from Asociacion Industrias Quimicas-Huelva (Spain), Caycit-Ministerio De Educacion Y Ciencia (Spain), Canon-España (Spain), Citibank-España (Spain), CNLS-Los Alamos Nat. Lab. (U.S.A.), CSIC (Spain), EPS, ERT (Spain), ESA, Fotonica (Spain), IBM-España (Spain), Junta De Andalucia (Spain), NATO, NSF (U.S.A.), ONR-London (U.S.A.), Petromed (Spain), Telefonica (Spain), UNED (Spain), UNED-Centro Asociado Huelva (Spain) and Universidad De Sevilla (Spain).

Many colleagues helped us in organizing the ASI and the Conference but four were particularly relevant: Prof. B. Nichols (Los Alamos), Prof. T. Riste (Kjeller), Prof. S. Bratos (Paris), and Prof. C. Vilchez, Director of UNED-Centro Asociado Huelva to whom we express our appreciation and gratitude.

Basil Nichols was of great help in many aspects and in particular in incorporating the cooperation of the Center for Nonlinear Studies (Prof. D. Campbell, Director) of Los Alamos National Laboratory, U.S.A.

Marian Martinez of CNLS-LANL and Goyi Alonso were keys to the successful development of the two meetings. Moreover, without the almost infinite patience, and efficiency of Goyi and the extraordinary assistance provided by Carlos Vilchez, I would not have been able to run the ASI and the Conference.

Finally, it is with great sadness that we see the publication of these Proceedings when Ben Levich is no longer with us. His seminal work opened new and fascinating avenues in the field.

Manuel G. Velarde,
Professor of Physics, U.N.E.D.,
Madrid (Spain),
ASI & Conference Chairperson.

CONTENTS

PART A. DROPS, BRIDGES, BUBBLES AND TWO-PHASE FLOW

A.1. The Deformation and Break-Up of Single Drops in Shear Fields 1
A. Acrivos

A.2. Detachment of Drops from Flat Solid Surfaces by a Simple Shear Flow 19
M.J. Mahé, M. Vignes-Adler, Ch. G. Jacquin, and P.M. Adler

A.3. Liquid Bridge Modeling of Floating Zone Processing 25
I. Martínez

A.4. Equilibrium Shapes and Free Vibrations of Liquid Captive Drops 53
A. Gañán and A. Barrero

A.5. Dynamics of Two-Dimensional Bubbles by the Lattice Gas Method 71
D. D'Humières, P. Lallemand and G. Searby

A.6. Effect of Interfacial Area on Flow Characteristics in Bubble Flow 87
G. Matsui, Y. Yamashita and T. Kumazawa

A.7. Effect of Surfactants on the Flow of Large Gas Bubbles in Capillary Tubes 97
D. Barthes-Biesel, N. Moulai-Mostefa and E. Meister

A.8. Motions of Stratified Gas-Liquid Interfaces Under Electromagnetic Fields in Horizontal Pipe Flow 115
J.-S. Chang

A.9.	Measurement of Interfacial Area and Geometry in Gas-Liquid Two-Phase Flow by an Ultrasonic Pulse Echo Technique E.C. Morala and J.-S. Chang	125
A.10.	Interfacial Transfers in Stratified Gas-Liquid Two-Phase Flow: A Critical Review J.-S. Chang	135

PART B. INTERFACIAL WAVES

B.1.	Surfactants and Capillary Waves: Experiments on Monolayers and on Microemulsions D. Langevin and J. Meunier	147
B.2.	High Frequency Capillary Waves on Liquid Surfaces: a Light Scattering Study J.C. Earnshaw and R.C. McGivern	163
B.3.	Some Surface Effects on Gravity Waves in a Viscous Fluid P.A. Tyvand	173
B.4.	Sustained Gravity-Capillary (Laplace) Waves at Interfaces and Marangoni-Benard Instability P.L. García-Ybarra and M.G. Velarde	179
B.5.	The Harmonic Oscillator Description of Longitudinal (Marangoni-Lucassen) Waves at Liquid Interfaces M.G. Velarde and X.-L. Chu	183

PART C. BUOYANCY-CAPILLARY CONVECTION WITH AND WITHOUT CHEMISTRY

C.1.	Experimental Investigations of the Effects of Surface Tension on Convection Caused by Heating from Below E.L. Koschmieder	189
C.2.	Influence of the Container Walls in Benard-Marangoni Convection: Experimental Approach P. Cerisier, R. Occelli, and J. Pantaloni	199
C.3.	Surface Tension Induced Convection in Presence of a Surface Tension Minimum J.C. Legros, M.C. Limbourg-Fontaine, and G. Petre	209

C.4.	Theories of Convective Instabilities Driven by Thermocapillary Forces S.H. Davis	227
C.5.	Some Problems in Marangoni Instability G. Lebon	253
C.6.	Interfacial Effects on the Onset of Convection in Horizontal Liquid Layers G.S.R. Sarma	271
C.7.	Experiments on Steady and Oscillatory Thermocapillary Convection in Space with Application to Crystal Growth D. Schwabe, R. Lamprecht, and A. Scharmann	291
C.8.	On Thermocapillary Flows in Containers with Differentially Heated Side Walls J.K. Platten and D. Villers	311
C.9.	Analysis of Flow Development Due to Marangoni Convection in a Mass Transfer System H.A. Dijkstra	337
C.10.	Two-Component Benard Steady Convection with Surface Adsorption X.-L. Chu, L.-Y. Chen and M.G. Velarde	343
C.11.	Interfacial Convection Driven by Surfactant Compounds at Liquid Interfaces. Characterization by a Solutal Marangoni Number E. Nakache and S. Raharimalala	359
C.12.	Surface Elasticity and Mechanical Instabilities at Liquid-Liquid Interfaces A. Sanfeld and A. Steinchen	367
C.13.	Volume or Surface Instabilities During Liquids Evaporation Under Reduced Pressure or/and Microwave Irradiation A. Steinchen-Sanfeld, M. Lallemant, P. Courville, P. Gillon, and G. Bertrand	387
C.14.	Basic Aspects of Tear Film Formation and Stability Frank J. Holly	401

C.15.	Nonequilibrium Phenomena in Microemulsions M.A. López Quintela, A. Fernández Nóvoa, and J. Quibén	411
C.16.	The General Phenomenon of Pattern Growth and Convection Driven by Chemical Reactions at Liquid Interfaces M.L. Kagan and D. Avnir	417
C.17.	Chemical Waves and Natural Convection S.C. Müller, T. Plesser, and B. Hess	423
C.18.	Kinetic Effects and Dynamics of Diffuse Interfaces P. Clavin	435
C.19.	A Turbulent Wrinkled Flame Simulation by Means of Cellular Automata R. Said and R. Borghi	469

PART D. INTERFACIAL INSTABILITIES, FINGERING, AND CRYSTAL GROWTH - RELATED PROBLEMS

D.1.	Instabilities and Surface Tension Y. Pomeau	487
D.2.	An Experimental Study of the Saffman-Taylor Instability P. Tabeling, G. Zocchi, and A. Libchaber	515
D.3.	Theory of the Needle Crystal M. Ben Amar and Y. Pomeau	527
D.4.	Dynamics of Unstable Interfaces N. Goldenfeld	547
D.5.	Convective and Interfacial Instabilities During Solidification S.R. Coriell, G.B. McFadden, and R.F. Sekerka	559
D.6.	Generalized Ginzburg-Landau Equations Applied to Instabilities in Systems Coupling Convection and Solidification T. Grauer and H. Haken	571
D.7.	Dendritic Pattern Formation Y. Saito, G. Goldbeck-Wood, and H. Muller-Krumbhaar	583

D.8.	Cellular Morphological Instabilities During Directional Solidification: Thin Sample Experiments S. de Cheveigné, C. Guthmann, and M.M. Lebrun	587
D.9.	Transport Processes During Directional Solidification and Crystal Growth: Scaling and Experimental Study D. Camel and J.J. Favier	595
D.10.	Molecular Beam Expitaxial Growth and Properties of Layered Semiconductor Structures for Advanced Photonic and Electronic Devices K. Ploog	619

PART E. LIQUID-VAPOR AND LIQUID-SOLID INTERFACES, CAPILLARITY, WETTING, SPREADING, AND RELATED EQUILIBRIA

E.1.	Synchroton X-Ray Studies of Liquid-Vapor Interfaces J. Als-Nielsen	639
E.2.	Statistical Mechanics of Interfaces B. Widom	657
E.3.	Utilization of the Second Gradient Theory in Continuum Mechanics to Study the Motion and Thermodynamics of Liquid-Vapor Interfaces H. Gouin	667
E.4.	The Surface Tension of Ionic Liquids M. Baus	683
E.5.	Landau Theory of Wetting Transitions E.H. Hauge	699
E.6.	The Dynamics of Wetting A.M. Cazabat and M.A. Cohen Stuart	709
E.7.	Spreading of Liquids on Solid Surfaces L. Leger	721
E.8.	The Effects of Polymers on Wetting by Nonvolatile Liquids A. Halperin	741

E.9.	Phase Equilibria of Fluids in Narrow Pores R. Evans and U.M.B. Marconi	751
E.10.	Orientational Order and Surface Phase Transitions M.M. Telo da Gama	753
E.11.	Wetting and Orientational Transitions in Nematic Liquid Crystals D.E. Sullivan and A.K. Sen	761
E.12.	A Model for Capillary Crystallization G. Navascués and P. Tarazona	775

PART F. MORE ON PHASE DIAGRAMS, MONOLAYERS, AGGREGATES, AND STRUCTURE OF INTERFACIAL LAYERS

F.1.	The Liquid-Solid Two-Phase Coexistence M. Baus	787
F.2.	Coexistence of Two-Phases: Long-Run Molecular Dynamics Computer Simulations J.J. Morales, F. Cuadros, and L.F. Rull	797
F.3.	Phase Separation of Binary Fluids Near a Critical Point Under Microgravity F. Perrot, P. Guenoun, and D. Beysens	811
F.4.	Influence of Sodium Sulfate on the Phase Diagram and Rheological Behaviour for a Sodium Dodecylbenzenesulafonate/Polyoxyethylene Fatty Alcohol/Water System J. Muñoz, C. Gallegos, V. Flores, and J.M. Pérez	833
F.5.	Statistical Thermodynamics of Pure and Mixed Amphiphilic Aggregates I. Szleifer, A. Ben-Shaul, and W.M. Gelbart	843
F.6.	Structural Studies in Langmuir Monolayers by Fluorescence Microscopy: a New Approach to the Phase Diagram of N-Pentadecanoic Acid at the Air-Water Interface F. Rondelez, J.F. Baret, K.A. Suresh, and C.M. Knobler	857

F.7.	ESR Study of Cu (II) in Polyacrylamide Networks S. Schlick	881
F.8.	Electric Potential and Charge Density Profiles in Inhomogeneous Interfacial Regions V.S. Vaidhyanathan	897
F.9.	Packing and Pairing of Ions Near a Charged Electrode T. Alts, B. D'Aguanno, P. Nielaba, and F. Forstmann	915
F.10.	Propagation of Sound Waves in the Presence of Weak Density Inhomogeneity: Application to Interfacial Phenomena and Phase Transitions in Fluids J. Maza, F. Miguelez, A. Veira, and F. Vidal	939

PART G. FLOW AND TRANSPORT IN EXCITABLE MEDIA. PATTERNS, CHAOS AND RELATED TOPICS

G.1.	Flow and Transport in Systems with Shape Change: Mass Transfer in Electrochemical Systems R.L. Sani, A.P. Peskin, and M.K. Maslanik	949
G.2.	Numerical Simulations of Buoyancy Driven Flows in Cylinders and Cavities for Vapour Crystal Growth P. Bontoux, F. Elie, C. Smutek, G.P. Extremet, A. Randriamampianina, E. Crespo, H. Branger, and B. Roux	963
G.3.	Onset of Oscillatory Convection in Horizontal Layers of Low-Prandtl-Number Melts H. Ben Hadid, B. Roux, A. Randriamampianina, E. Crespo, and P. Bontoux	997
G.4.	The Kuramoto-Sivashinsky Equation: Spatio-Temporal Chaos and Intermittencies for a Dynamical System B. Nicolaenko	1029
G.5.	Spatial Coherence and Temporal Chaos in Hydrodynamic Instabilities S. Ciliberto	1053
G.6.	Interfaces in Electrophoresis, Two-Phase Models, and Excitable Media P.C. Fife	1067

G.7. Field Induced Pattern Formation in Nematic Liquid
Crystals 1085
F. Sagués and M. San Miguel

G.8. Production of Ozone by High Frequency Surface
Discharge on a High Purity Alumina Ceramic
Dielectrics 1091
S. Masuda

PARTICIPANTS 1097

AUTHOR INDEX 1105

SUBJECT INDEX 1109

THE DEFORMATION AND BREAK-UP OF

SINGLE DROPS IN SHEAR FIELDS

Andreas Acrivos

Department of Chemical Engineering
Stanford University
Stanford, CA 94305-5025

ABSTRACT

A liquid drop of viscosity $\lambda\mu$ and volume V is placed in another fluid of viscosity μ with which it is immiscible. When the external fluid is sheared, the initially spherical drop will deform and may, in some cases, break into several fragments. It is desired to compute the deformation of this drop and the conditions under which it will break. This subject is of considerable fundamental interest in fluid mechanics as an example of a free-boundary problem and as a prototype for flow-induced deformation of a variety of flexible bodies such as red blood cells, macromolecules, flocs, elastic particles, etc. It is also closely related to dispersion processes in commercial blenders and mechanical emulsifiers where one is often required to estimate the maximum drop size of the dispersed phase for a given power input, or conversely to estimate the power that is required in order to insure that all the drops of the dispersed phase remain below a given maximum size.

When the particle Reynolds number is small, inertia effects can be neglected and the drop deformation is governed, in the absence of gravity, by two dimensionless groups in addition to the flow type: i) the viscosity ratio λ, and ii) the capillary number $G\mu a/\gamma$ where $a = (3V/4\pi)^{1/3}$ is the radius of the undeformed drop, G denotes the strength of the impressed shear field, and γ is the interfacial tension.

A theory will now be described for determining the drop deformation and the critical value of the capillary number at break-up, which makes use of a small deformation analysis (for $\lambda \geq 1$), slender-body theory (for $\lambda \ll 1$) and a full numerical solution of the relevant boundary value problem via an integral equation formulation. The theoretical predictions are shown to be in excellent agreement with the available experimental data.

INTRODUCTION

When a liquid drop of viscosity $\lambda\mu$ and volume V is suspended in another liquid of viscosity μ undergoing shear, the drop will deform and, under some conditions, it will break into two or more fragments if the strength of the applied shear field exceeds some critical value. This phenomenon of deformation and break-up, which has been discussed in two recent reviews,[1,2] plays an important role in a variety of physical processes of practical significance; for example, the rheology of emulsions, and the dispersion of one fluid into another with which it is immiscible. It is also of fundamental interest as a prototype free-boundary problem involving the unknown shape of the surface separating two phases and its evolution in response to changes in the applied shear field.

Let us imagine for simplicity that a = $(3V/4\pi)^{1/3}$, the radius of the undeformed drop, is much smaller than the macroscale of the applied shear flow and let us choose a moving coordinate system with origin O at the center of the drop. The undisturbed velocity $u_i^{(\infty)}(x)$ relative to the moving origin can then be expressed simply as

$$u_i^{(\infty)}(\vec{x}) = E_{ij}x_j + \frac{1}{2}\epsilon_{ijk}\Omega_j x_k \tag{1}$$

where E_{ij} and Ω_j are, respectively, the rate of strain tensor and the vorticity of the applied shear both evaluated at O, and $\vec{x}$ is the position coordinate measured from O. We also suppose that inertia effects are of negligible importance throughout the flow domain and that the drop is neutrally buoyant.

The motion is therefore governed by the classical incompressible creeping flow equations subject to the usual boundary conditions of continuity of velocity across S, the surface of the drop, plus the stress condition

$$[f_i]_s = \gamma n_i \frac{\partial n_j}{\partial x_j} \quad \text{on S} \tag{2}$$

where the symbol $[\ ]_s$ denotes the jump in the stress force f_i across S from the inside to the outside, n_i is the unit <u>outer</u> normal to S and γ is the interfacial tension which is assumed constant. Finally the kinematic condition becomes, at steady state,

$$u_j n_j = 0 \quad \text{on S.} \tag{3}$$

It is easy to show that, for a given type of shear flow, the system is governed by the two dimensionless groups: the viscosity ratio λ, and the capillary number $Ca = G\mu a/\gamma$, where G is the strength of the applied shear. We desire to compute the deformation of the drop as a function of λ and Ca, and in particular to determine Ca_c, the critical value of the capillary number, at which drop break-up will occur for a given flow configuration.

We begin by discussing the various theoretical studies on the subject.

THEORY (CREEPING FLOW)

It is important to appreciate at the outset that, in spite of the linearity of the creeping flow equations, the mathematical system is strongly non-linear owing to the <u>a priori</u> unknown shape of S and to the complicated expression for the curvature $\partial n_j/\partial x_j$ in the stress balance boundary condition, eq. (2). Thus, the mathematical analysis must be performed using approximate techniques which take advantage of the fact that, prior to break-up, the drop is often either almost spherical or long and slender. These two extreme cases will now be considered in some detail.

Small Deformation Analysis

In the absence of flow, i.e., when $Ca=0$, the drop is of course exactly spherical in shape owing to the action of surface tension. It is, therefore, logical to suppose that the shape will be almost spherical if the capillary number is sufficiently small. Thus, we seek to construct a solution in which each dependent variable of the system is expanded in a power series in Ca. We note parenthetically that the drop may also be almost spherical for any value of Ca if the viscosity ratio λ is sufficiently large provided, however, that the shear flow contains enough vorticity for the drop to spin almost as a rigid body. On the other hand, if placed in a pure straining flow, the drop will attain a steady shape which will depend primarily on the value of the capillary number Ca even if $\lambda^{-1} \ll 1$ although the time required to achieve this steady shape if the flow is started from rest will then be $O(\lambda)$. To avoid such complication, however, we shall discuss only the case $\lambda \sim O(1)$ and $Ca \ll 1$.

In principle, the analysis proceeds in a straightforward fashion starting with the creeping flow solution past a liquid drop of exactly spherical shape subject to all the boundary conditions except for the normal stress balance, i.e., the component of equation (2) along n_i which to a first approximation equals $m_i = x_i/a$. The new shape is then determined by expanding the curvature in the form

$$\frac{\partial n_j}{\partial x_j} = \frac{2}{a}\{1 + O(Ca)\} \qquad (4)$$

and then solving for the O(Ca) term through substitution in the normal stress balance where the left-hand side of eq. (2) has already been determined. The whole procedure is then repeated starting with the creeping flow solution for flow past a drop of "almost" spherical shape.

The general solution of the creeping flow equations in spherical coordinates is of course well-known[3] and hence, it is easy to obtain the leading term of the expansion described above for any choice of E_{ij} and Ω_j. Nevertheless, it is instructive at this point to anticipate the answer. Specifically, since, for any slightly deformed sphere, it is always possible to represent the distance $|\vec{x}|$ from any point on the surface to the origin O as a series of surface spherical harmonics, we let

$$\frac{|\vec{x}|}{a} = 1 + F_{ij}m_i m_j + \text{higher order harmonics} \qquad (5)$$

where m_i is the unit outer normal to a sphere. In view of the fact, however, that the coefficients F_{ij}, etc., multiplying these harmonics are obtained, to a first approximation, from the solution of the linear creeping flow equation, past a __spherical__ drop, it follows that they must be linear in E_{ij} with coefficients that are scalars and, therefore, all must vanish except for F_{ij} which must be scalar multiple of E_{ij} (note that to O(Ca), the drop deformation due to the action of the pure straining flow component of eq. (1) and that due to solid body rotation can be computed separately and then added, but of course an ambient flow with velocity equal to $1/2\ \epsilon_{ijk}\ \Omega_j\ x_k$ will not, if acting alone, deform the drop; hence, to leading order, Ω_j cannot enter into the expression for the shape of a slightly deformed drop). Thus, as originally pointed out by Taylor[4], the initially spherical drop first deforms into an ellipsoid with its principal axes coincident with those of E_{ij}.

To obtain the proportionality coefficinet between F_{ij} and E_{ij} it suffices then to select a convenient expression for E_{ij} and then solve the boundary value problem posed earlier. The simplest choice is given by the strain-rate of a uniaxial extensional flow

$$E_{ij} = G\{\delta_{i1}\delta_{j1} - \frac{1}{2}\delta_{i2}\delta_{j2} - \frac{1}{2}\delta_{i3}\delta_{j3}\} \qquad (6)$$

with extension along the 1-axis and a radically symmetric inflow into the origin along the x_2-x_3 plane. On performing the analysis we find that

$$\frac{|\vec{x}|}{a} = 1 + \frac{\mu a}{\gamma}\ \frac{19\lambda + 16}{8(\lambda+1)}\ E_{ij}m_i m_j + O(Ca^2) \qquad (7)$$

which, as mentioned above, applies for any choice of E_{ij}.

A convenient scalar measure of the deformation is the dimensionless deformation parameter D defined by Taylor as

$$D = \frac{L - B}{L + B} \qquad (8)$$

where L and B, the so-called half-length and half-breadth, are the largest and smallest distances from the interface to the center of the drop. In view of (7), we find that

$$D = \frac{19\lambda + 16}{16(\lambda+1)} \frac{\mu a}{\gamma} (E_{max} - E_{min}) + O(Ca^2) \qquad (9)$$

where E_{max} and E_{min} are, respectively, the largest and smallest principal strains, both of which are, of course, proportional to G. We remark, parenthetically, that Cox[5] has generalized somewhat the approach outlined above, in that he expanded his solution in powers of the small parameter D, rather than in powers of Ca, and thereby succeeded in treating simultaneously the two cases λ-O(1), Ca<<1 and Ca-O(1), λ^{-1}<< 1. A further discussion of this point is given by Rallison.[6]

As we shall show later on when comparing theory and experiments, eq. (9) remains remarkably accurate even for values of D which are, in many cases, as large as 1/2. On the other hand, such an analysis cannot predict the conditions for drop break-up without including at least some of the higher order terms in the expansion.

Unfortunately, to extend the small deformation theory to $O(Ca^2)$, it is necessary not only to obtain an expression for F_{ij} correct to $O(Ca)^2$, but also to include the higher order term $F_{ijk\ell}m_i m_j m_k m_\ell$ in eq. (5), the expression for the interface, since the fourth-order tensor $F_{ijk\ell}$ is itself $O(Ca^2)$. Although this increases the complexity of the analysis enormously, the algebraic manipulations could fortunately be performed on a computer using the symbolic manipulator REDUCE.[7] By evaluating the expression for D thereby derived[8] for various flow types and values of λ, it was found that: either D increased monotonically with Ca and reached a finite limit as Ca→∞, or, that the curve D vs. Ca had a turning point at a capillary number Ca_c. The latter was identified as being the critical capillary number for drop break-up, because it corresponds to the maximum value of Ca for which there exists a steady solution to the boundary value problem stated above that evolves continuously from the point D = Ca = 0. Also, solutions to the time-dependent equations showed that,[8] as expected, the drop would elongate indefinitely if Ca was increased beyond Ca_c.

We shall see later on that, in spite of its conceptual simplicity, the non-linear analysis just outlined gives values for Ca_c which are in surprisingly close agreement with the experimental results provide that λ is not too small. In retrospect this agreement is not very surprising because, just prior to their point of break-up, the drops are found to have a shape not too dissimilar from that of a sphere provided that λ is typically in excess of 0.1. Thus this approximate analysis, which properly accounts for all the physical factors that govern the break-up phenomenon, is able to accurately predict the critical capillary number even though, under some conditions, the drop shapes that are thereby calculated at the critical point are quite strange and unrealistic.[8]

Slender-Body Theory

Taylor[9] appears to have been the first to realize that relatively non-viscous drops ($\lambda \ll 1$) could become long and slender prior to break-up and that, for these cases, the critical capillary number could be calculated by adapting to creeping flows the well-known slender body analysis of aerodynamics.

To illustrate this procedure let us first consider an inviscid drop ($\lambda = 0$) which is placed symmetrically in an axisymmetric pure straining flow. We choose a cylindrical coordinate system with z being the distance along the axis of revolution and r the radial coordinate, and denote the equation for the surface of the drop by $r = \epsilon L R(z)$, where $R(z)$ is a dimensionless unknown $O(1)$ function, ϵ is the slenderness ratio and L is the half-length of the drop. Moreover, for this type of flow (c.f. eq. (6))

$$u_z^{(\infty)} = Gz \quad \text{and} \quad u_r^{(\infty)} = -\frac{Gr}{2} \tag{10}$$

In the absence of viscous stresses within the drop, the stresses exerted on the surface of the drop by the external flow, which are $O(\mu G)$, must be balanced by the $O(\gamma/\epsilon L)$ stress due to surface tension, hence $\epsilon = O(\gamma/\mu G L)$ which on account of the volume conservation requirement $\epsilon^2 L^3 = O(a^3)$ gives that

$$\frac{L}{a} = O(Ca)^2 \tag{11}$$

where $a = (3V/4\pi)^{1/3}$ is again the radius of the undeformed drop.

To determine the numerical value of the coefficient on the right-hand side of eq. (11), it is of course necessary to obtain an expression for $R(z)$ via the solution of the boundary-value problem. This can be achieved in the following manner.

First of all, it is clear from physical considerations that, due to the slenderness of the drop, the axial velocity component will equal everywhere that of the undisturbed flow with an error which will vanish asymptotically as $\epsilon \to 0$. Thus

$$u_z = Gz + o(1) \quad \text{as} \quad \epsilon \to 0.$$

On the other hand, in view of the kinematic condition (3) which, to leading order in ϵ becomes,

$$u_r = \epsilon \, L \, Gz \, \frac{dR}{dz} \quad \text{at} \quad r = \epsilon \, L \, R , \tag{12}$$

it is clear that the disturbance velocity in the radial direction must be of the same order as that due to the undisturbed flow, i.e. $O(\epsilon)$ on the surface of the drop. Such a disturbance flow, which does not contribute to the axial velocity, can be represented by a line source along the axis within the interval $-L \leq z \leq L$, and therefore

$$u_r = -\frac{Gr}{2} + \frac{G \epsilon^2 L^2}{r} R \left(\frac{R}{2} + z \frac{dR}{dz} \right) \tag{13}$$

where the strength of the line source has been chosen in such a way that eq. (13) satisfies eq. (12). Surprisingly then, we have been able to obtain the complete expression for the velocity profile in terms of the unknown shape of the drop as represented by the function $R(z)$, using strictly kinematic arguments, i.e., without having recourse to the equations of motion which, to this order, merely state that, as expected, the pressure outside the drop equals everywhere its undisturbed value. To obtain the expression for $R(z)$ it is now necessary to apply the normal stress balance which, to this order, becomes

$$2\mu \left(\frac{\partial u_r}{\partial r} \right)_{r=\epsilon LR} + p_o = \frac{\gamma}{\epsilon \, L \, R}$$

where p_o is the unknown pressure within the drop relative to that of the undisturbed flow. On substituting eq. (13) into the above and setting $\epsilon = \gamma/G\mu L$, we then arrive at

$$z \frac{dR}{dz} - \nu R = -\frac{1}{2} , \quad \nu = \frac{p_o}{2G\mu} \tag{14}$$

with boundary conditions $R(\pm L) = 0$. Thus as first shown by Buckmaster[10],

$$R(z) = \frac{1}{2\nu} \left\{ 1 - \left| \frac{z}{L} \right|^{\nu} \right\} \tag{15}$$

where, however, ν is still unknown.

Buckmaster[10] imposed the additional condition that $R(z)$ had to be analytic about $z = 0$ and thereby concluded that ν had to be an even integer. Although, as we shall see, this is the correct result, the assumption of analyticity, although plausible, is not entirely convincing. Another approach is to argue that the only asymptotic solution having physical significance is that which, with decreasing Ca, connects smoothly with the solution described earlier for a slightly deformed drop. Since the latter is analytic at $z = 0$, it seems reasonable to suppose that, again, only even integer values of ν should be allowed. Unfortunately, such an argument would exclude all values of ν since the drop shape given by eq. (15) has pointed ends, irrespective of ν, whereas the ends of a slightly deformed spherical drop are, of course, rounded. A more elaborate analysis is, therefore, required which takes into account the fact that the simplifications leading to eq. (13) apply only for $|z| > O(\epsilon L)$. Thus, the analysis has to be performed within the framework of the method of matched asymptotic expansions which, when applied to the present problem and after a considerable amount of mathematical manipulation, leads to the result[11] that matching between the various solutions can be achieved only if ν equals an even integer. Thus, a countably infinite number of asymptotic solutions has been found for $Ca \to \infty$, but all of them have been shown to be linearly unstable[11] except for that corresponding to $\nu = 2$. The only permissible asymptotic drop shape which is also linearly stable is therefore given by

$$R = \frac{1}{4} \{1 - \frac{x^2}{L^2}\} \qquad (16)$$

which, in conjunction with the volume conservation requirement

$$\frac{4\pi}{3} a^3 = 2\pi\epsilon^2 L^2 \int_0^L R^2 \, dz \qquad (17)$$

and the expression

$$\epsilon = \gamma/G\mu L \qquad (18)$$

derived earlier, gives that

$$\frac{L}{a} \to 20 \, (Ca)^2 \qquad (19)$$

It remains to be shown, however, that this asymptotic solution joins smooth with that branch of the deformation curve which originates from the spherical state $L = a$ at $Ca = 0$. We postpone a discussion of this point until later on and consider instead the case where λ, although small, is non-zero.

It is immediately apparent from eq. (19) that, according to the slender-body analysis just described, a truly inviscid drop when placed in an axisymmetric pure straining flow can elongate indefinitely with increasing Ca and, in principle at least, always attain a steady shape at a given Ca. This is not too surprising because the drop shape evolves so as to maintain a balance between two stresses, specifically: a) the $O(\mu G)$ stress of the external flow field which tends to compress the drop cross-section and thereby increase L; and b) the $O(\gamma/\epsilon L)$

stress due to the surface tension forces which opposes this tendency. Thus, with increasing Ca, it is always possible to create a drop having a local cross-sectional radius small enough to insure the existence of a steady shape. On the other hand, when ϵ^2 becomes comparable to λ, the pressure variation, Δp, within the drop reaches values which are comparable in magnitude to the other two stresses mentioned above, thereby creating the possibility of a force imbalance on the interface. The reason for this is that, within a drop which is long and slender, the flow is quasi-unidirectional, and hence, under creeping flow conditions, the pressure gradient along the axis of the drop, dp/dz, is balanced by the single viscous term $\lambda\mu(\partial^2 u_r/\partial r^2)$. The latter is, however, $O(\lambda\mu G/\epsilon^2 L)$ and <u>positive</u>, as can be shown by simple kinematic arguments, and hence Δp, the difference between the pressure at any point z and its value at the mid-plane $z = 0$, is a monotonically increasing function of z whose order of magnitude is $\lambda\mu G/\epsilon^2$. Thus, the term $\gamma/\epsilon LR$ in the normal stress boundary condition must now balance the sum of

$$2\mu\left(\frac{\partial u_r}{\partial r}\right)_{r=\epsilon LR} \quad \text{and} \quad p_o + \Delta p$$

both of which are monotonically increasing functions of z. It is therefore not surprising that, for $\lambda \neq 0$, a steady shape will not exist if the strength of the applied shear is increased beyond a critical value.

We see then that the equation for the drop shape R(z) will now contain the parameter λ/ϵ^2 which, in view of eq. (12), also equals the product $(G\mu a\lambda^{1/6}/\gamma)^2 (L\lambda^{1/3}/a)^2$. Therefore on account of the volume conservation requirement eq. (17)

$$\frac{L\lambda^{1/3}}{a} = \left(\frac{G\mu a}{\gamma}\lambda^{1/6}\right)^2 \times \text{function of } ((\lambda/\epsilon^2)) \qquad (20)$$

which rearranges into

$$\frac{L}{a}\lambda^{1/3} = F\left(\frac{G\mu a}{\gamma}\lambda^{1/6}\right) \qquad (21)$$

with

$$F \to 20 \left(\frac{G\mu a}{\gamma}\lambda^{1/6}\right)^2 \quad \text{as} \quad \frac{G\mu a}{\gamma}\lambda^{1/6} \to 0.$$

To obtain the expression for F in eq. (21) it is necessary to solve the normal stress balance equation and thereby determine R(z). As before, a continuous spectrum of solutions is found which reduces to a countably infinite set by requiring that, as explained earlier, R(z) be analytic at $z = 0$. Finally, by restricting the class of solutions to

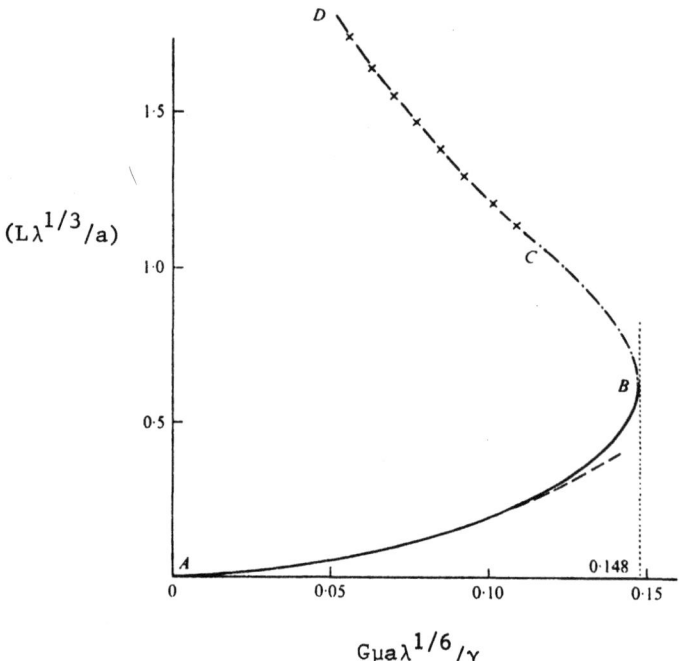

Fig. 1. The first two branches of the deformation curve for a drop with finite viscosity and zero inertial effects. ———, stable drop deformation curve; – – –, reference curve for zero-viscosity case; —·—·—·—, — + — + —, unstable steady-state solutions; B, point of breakup; C, point at which the two branches join. (Acrivos and Lo[11], by permission: Cambridge University Press (C.U.P.))

those which are linearly stable, we arrive at[11]

$$R(z) = \frac{1}{8} \left[1 + \left(1 - \frac{64\lambda}{\epsilon^2} \right)^{1/2} \right] \left(1 - \frac{z^2}{L^2} \right) \qquad (22)$$

a result also given by Buckmaster[12]. The above, in conjunction with equations (17) and (18), then leads to the deformation relation

$$\frac{G\mu a}{\gamma} \lambda^{1/6} = \frac{1}{(20)^{1/2}} \frac{(L\lambda^{1/3}/a)^{1/2}}{1 + \frac{4}{5} \left(\frac{L\lambda^{1/3}}{a} \right)^3} \qquad (23)$$

which was first stated by Taylor[9] without proof!

As can easily be seen from Figure 1, the deformation relation given by eq. (23) has a turning point at $G\mu a \lambda^{1/6}/\gamma = 5^{1/2}/3 \cdot 2^{7/8} = 0.148$ and $\frac{L\lambda^{1/3}}{a} = 4^{-1/3} = 0.63$. Hence, according to our analysis, this asymptotic branch of the deformation curve cannot extend beyond a critical shear rate $G_c = 0.148\gamma/\mu a \lambda^{1/6}$ which corresponds to the critical point for drop break-up.

The same type of analysis has been carried out for hyperbolic flows[13] (the two-dimensional analogue of the axisymmetric pure straining motion discussed above) for which

$$E_{ij} = G(\delta_{i1}\delta_{j1} - \delta_{i2}\delta_{j2}) \;,\; \Omega_j = 0$$

Surprisingly, even though the flow is not axisymmetric and, therefore, the cross-section of the drop is no longer circular, the criterion for drop break-up was found to be

$$G_c = 0.145 \; \gamma/\mu a \lambda^{1/6} \tag{24}$$

which is essentially identical to that for axisymmetric extensional flow.

On the other hand, several complications arise when the impressed flow is the simple shear

$$u_i^{(\infty)} = G \; \delta_{i2} \; x_2 \tag{25}$$

and $\lambda \ll 1$. First of all the analysis shows[14] that $(L\lambda^{1/3}/a)$ becomes here a function of $(G\mu a\lambda^{2/3}/\gamma)$ and that, although steady shapes exist for all values of this parameter, these become linearly unstable if $(G\mu a\lambda^{2/3}/\gamma)$ exceeds 0.055. In addition, this critical point can be reached only if G is increased very slowly and that a drop may break even when $(G\mu a\lambda^{2/3}/\gamma) < 0.055$ if G is given a jump increase from one subcritical value to another. Thus, the criterion given above for the break-up of slender drops in a simple shear flow is not a very useful one.

<u>Numerical Solutions</u>

With the advent of large scale computers, it has become feasible in recent years to study a large variety of fluid mechanical phenomena via the finite-difference numerical solution of the equation of motion. Although the problem at hand involves of course the additional complication of an <u>a priori</u> unknown surface geometry, a recently developed technique[15], which has been successfully applied in similar cases[16], renders such a computation feasible for values of the Reynolds number which are not necessarily small. Nevertheless, under creeping flow conditions, it appears to be more reasonable to employ a method of solution which takes advantage of the linearity of the Stokes equations. We note that the purpose of such numerical solutions is not only to bridge the gap between the two cases treated above analytically, i.e. drops that are slightly deformed or which are long and slender, but also to ascertain whether the deformation curve given by slender-body theory, e.g., equation (23), indeed connects smoothly, as Ca is reduced, with the corresponding expression as obtained from the small deformation analysis, i.e. eq. (7).

It is well known by now that, in creeping flow, the velocity $u_i(\vec{x})$ at any point $\vec{x}$ can be represented as a linear combination of two surface integrals along the closed interface S, one of which involves the unknown surface velocity and the other the unknown surface stress.[17] On making use of the appropriate expressions for $u_i(\vec{x})$, where $\vec{x}$ lies, successively, within the domain enclosed by S, exterior to that domain, and on S, and applying the condition of continuity of velocity plus the stress condition eq. (2), we obtain that[18], for any fixed point $\vec{x}$ on S,

$$\frac{1}{2}(1+\lambda)u_i(\vec{x}) + (1-\lambda)\int_S K_{ijk}(\vec{s}) u_j(\vec{y}) n_k(\vec{y}) dS_y$$

$$= u_i^{(\infty)}(\vec{x}) - \frac{\gamma}{8\pi\mu}\int_S J_{ij}(\vec{s}) n_j(\vec{y}) \frac{\partial n_k}{\partial y_k} dS_y \qquad (26)$$

where $\vec{y}$ is the variable point of integration on the interface S and

$$K_{ijk} = -\frac{3}{4\pi}\frac{s_i s_j s_k}{|\vec{s}|^5}, \quad J_{ij}(\vec{s}) = \frac{\delta_{ij}}{|\vec{s}|} + \frac{s_i s_j}{|\vec{s}|^3}, \quad \vec{s} = \vec{x} - \vec{y}.$$

For a given interface shape, eq. (26) is an integral equation of the second kind for the unknown interface velocity $u_i(\vec{x})$ whose component

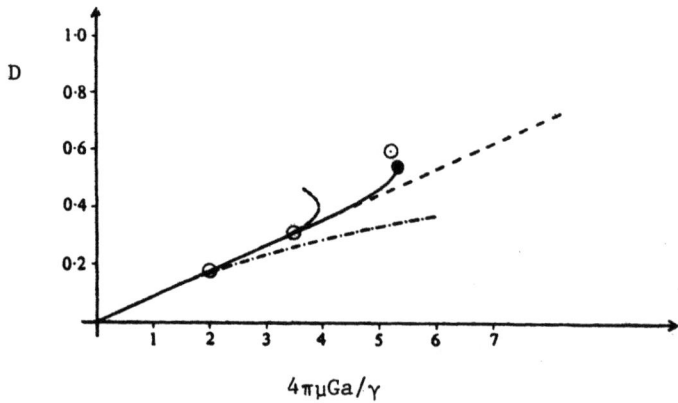

Fig. 2. Comparison of theories and experiments for the deformation D in a simple shear flow with $\lambda = 1$, ———, numerical results[19]; •, burst; – – –, linear theory[4]; - - - -, quadratic theory[8]; – · – · – · –, theory with inclusion of vorticity[19]; •, experiments[21]. (Rallison[19], by permission C.U.P.).

along the unit normal n_i determines the evolution of this shape as a function of time. The shape corresponding to steady state is then reached when the kinematic condition, eq. (3), is also satisfied.

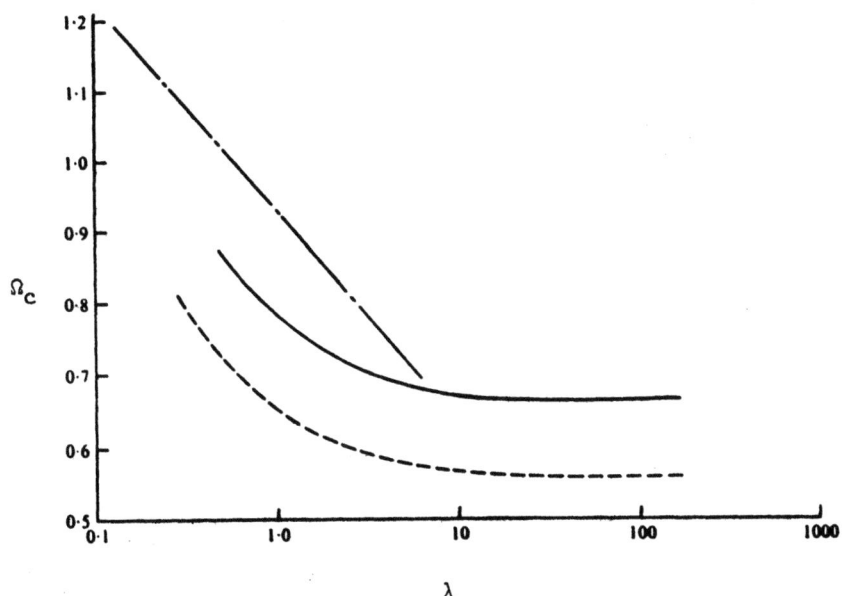

Fig. 3. Critical value of $\Omega = 4\pi\mu Ga/\gamma$ as a function of λ.
———, numerical solution[18]; – – –, quadratic theory[8]; - - - - -, slender body theory[9,11]. (Rallison and Acrivos[18], by permission C.U.P.)

Evidently, the numerical procedure based on eq. (26) simplifies considerably when $\lambda = 1$, a case which has been studied in some detail by Rallison for a variety of shear flows.[19] Numerical solutions for a wide range of λ had been given earlier for axisymmetric pure straining flows.[18].

Shown in Figures 2 and 3 are the results of such computations. It is evident that for $\lambda \geq 10^{-1}$ the small deformation analysis[8] accurately predicts the point of break-up as given by the more accurate numerical solutions and that, at least for axisymmetric pure straining flows, the slender body theory[9,11] is applicable for $\lambda < 10^{-2}$ and that it connects smoothly with that branch of the deformation curve which originates at $L/a = 1$ as $Ca \to 0$.

EXPERIMENTAL RESULTS

The earlier experiments on the subject were reported by Taylor[20] who created for this purpose the so-called "four-roll mill", a versatile instrument which, in principle, is capable of generating a variety of planar shear flows although, in Taylor's case, it was used only to produce a hyperbolic flow. Taylor[20] also studied drop deformation and break-up in simple shear flows which he produced in a "parallel-band" apparatus. Additional experiments in hyperbolic and in simple shear flows were subsequently performed over a wider range of parameters.[21-23] Also by employing an orthogonal rheometer, Hakimi and Schowalter[24] were able to generate flows with a variable vorticity-to-strain-rate ratio, but unfortunately their data were limited to only one viscosity ratio ($\lambda = 0.09$) and to small deformations ($D \leq 0.2$).

The most extensive and reliable experimental results currently available are undoubtedly those recently reported by Bentley and Leal[25] and by Stone, Bentley and Leal[26], who were able to employ the "four-roll mill" to its fullest potential through the development of an image processing system, which was then used in conjunction with a video camera and a computer to maintain the drop at the stagnation point of the impressed shear flow. As a result, these authors succeeded in generating the family of planar shear flows whose velocity gradient $\vec{\nabla}\vec{u}$ is given by

$$\vec{\nabla}\vec{u} = \frac{G}{2} \begin{bmatrix} 1+\alpha & 1-\alpha & 0 \\ -1+\alpha & -1-\alpha & 0 \\ 0 & 0 & 0 \end{bmatrix} \qquad (27)$$

where the single parameter α is a measure of the relative strength of the strain rate and the vorticity in the flow. In particular, $\alpha = +1$ for pure-straining (hyperbolic) flow, $\alpha = 0$ for simple shear flow, and $\alpha = -1$ for solid body rotation.

Shown in Figures 4-10 is a comparison between the experimental results[25] and the theoretical predictions. Clearly there is excellent agreement with the so-called $O(\epsilon^2)$ small deformation analysis when $\lambda \geq 10^{-2}$ and with slender body theory when $\lambda \leq 10^{-3}$. In the latter case, the earlier theory for hyperbolic flow[13] was modified for $0 < \alpha < 1$ by using an effective strain rate $G\alpha^{1/2}$ which takes into account the fact that long slender drops would be expected to align with the exit streamline of the flow field. Thus, the criterion for break-up of long slender drops given by eq. (24) generalizes to[25,27]

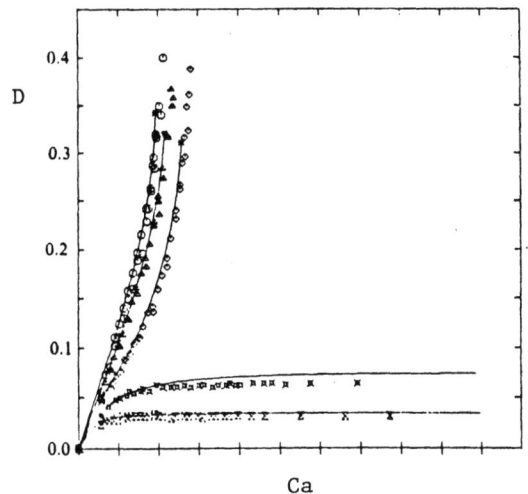

FIGURE 4.

Drop deformation curves for $\alpha = 1.0, 0.8, 0.6, 0.4$ and 0.2 for $\lambda = 57.0$.
⬠, $\alpha = 1.0$; △, $\alpha = 0.8$;
◇, $\alpha = 0.6$; ⊠, $\alpha = 0.4$;
✗, $\alpha = 0.2$; ———
quadratic theory[8]. (Bentley and Leal[25], by permission C.U.P.).

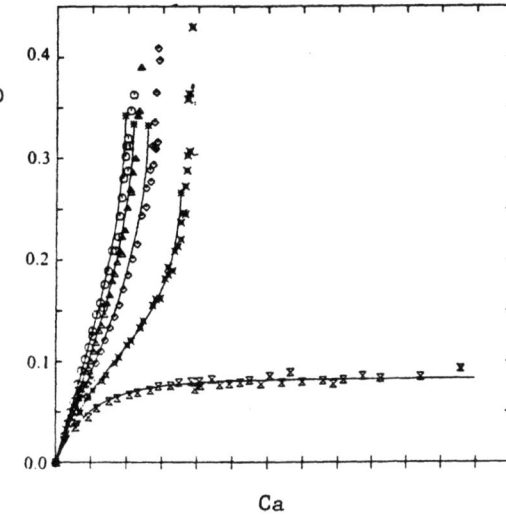

FIGURE 5.

Drop deformation curves for $\lambda = 26.0$. Legend as in Fig. 4.

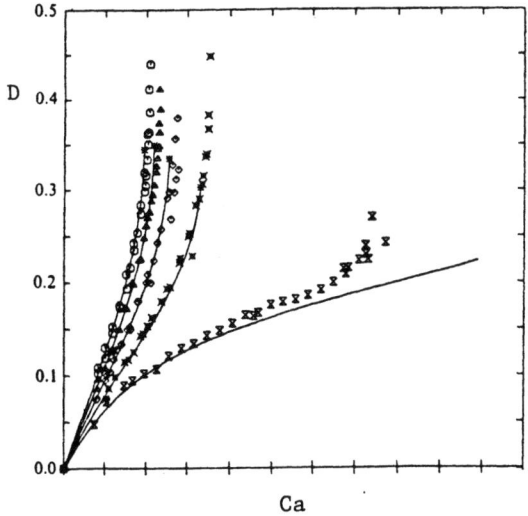

FIGURE 6.

Drop deformation curves for $\lambda = 14.0$. Legend as in Fig. 4.

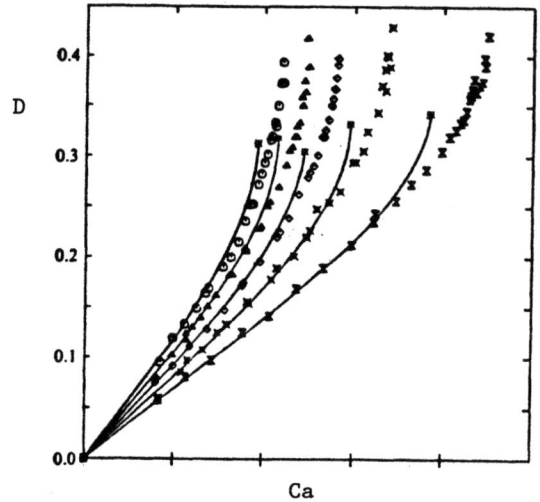

FIGURE 7.

Drop deformation curves for $\lambda = 2.80$. Legend as in Fig. 4.

FIGURE 8.

Drop deformation curves for $\lambda = 0.12$. Legend as in Fig. 4.

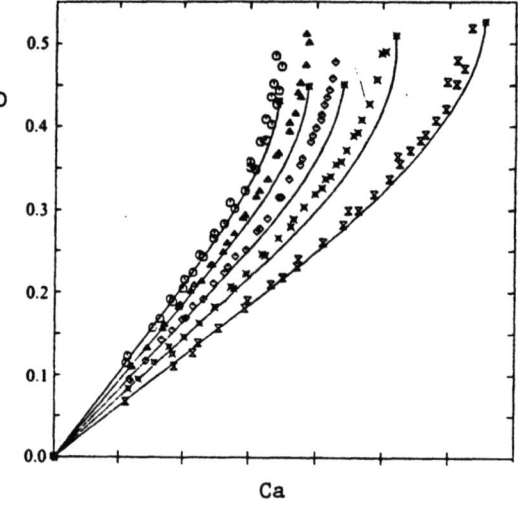

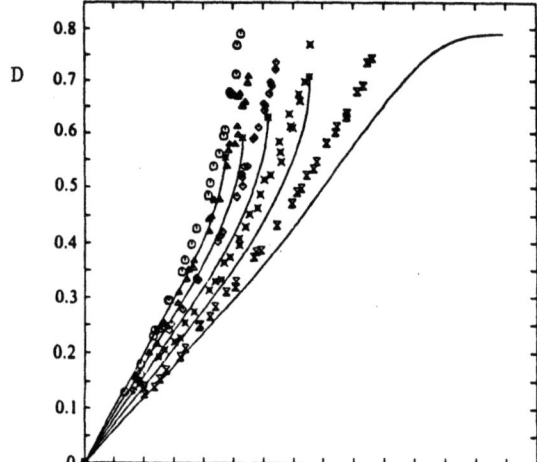

FIGURE 9.

Drop deformation curves for $\lambda = 1.1 \times 10^{-2}$. Legend as in Fig. 4.

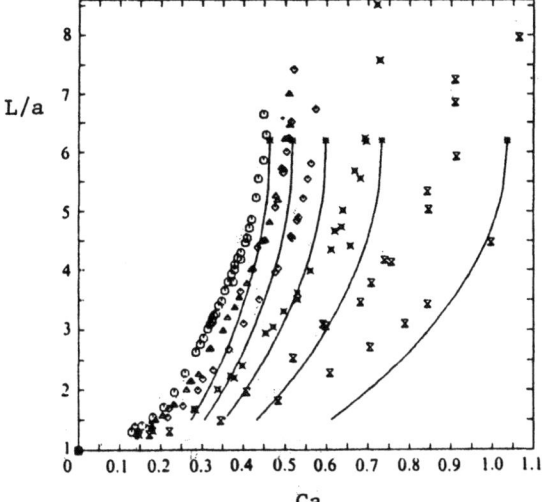

FIGURE 10.

L/a vs. Ca for
$\lambda = 1.1 \times 10^{-3}$.
Legend as in Fig. 4;
———— slender body theory, eq. (28).

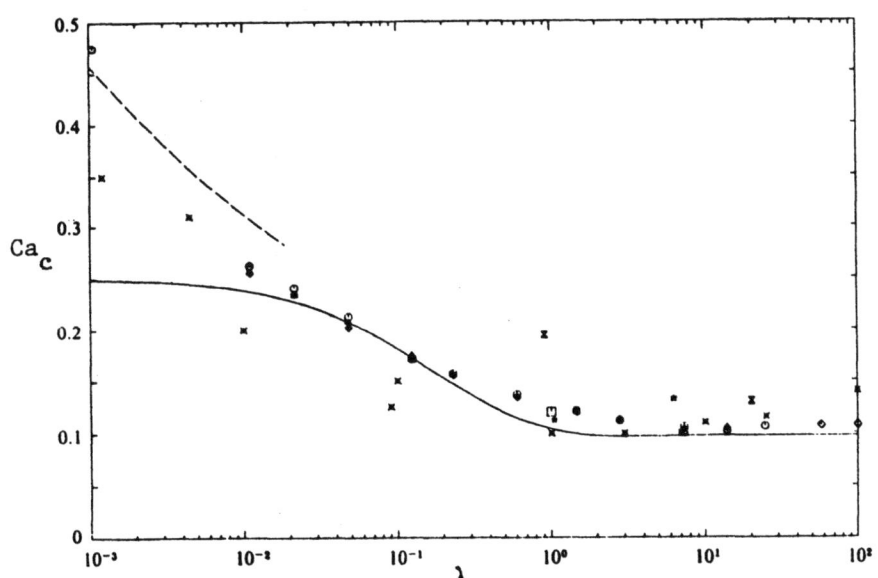

Fig. 11. Critical capillary number for burst as a function of λ for $\alpha = 1.0$, ⊙ , △ , ◇ , ref.[25]; ⊠ ref.[23]; ✷ ref.[21]; ⊠ ref.[20]; ▢ numerical results[18]; ———— quadratic small-deformation theory[8]; - - - - slender-body theory, eq. (28). (Bentley and Leal[25], by permission C.U.P.).

$$G_c = 0.145 \ \gamma/\mu a \lambda^{1/6} \alpha^{1/2} \qquad 0 < \alpha \leq 1 \qquad (28)$$

A summary of the experimentally determined critical values of the capillary number is given in Figure 11.

In view of this excellent agreement between the experimental and numerical results on the one hand, and the analytical predictions of the small deformation and slender body theories on the other, it can safely be concluded therefore that the latter can be used with confidence to predict the deformation and break-up of small drops freely suspended in general linear shear flow under conditions of small Reynolds numbers and negligible gravitational effects.

Acknowledgements

This review was supported in part by the National Science Foundation under grant 81-21713. It is my pleasure to acknowledge my past collaborations with N.A. Frankel, L.G. Leal, D. Barthes-Biesel, G.R. Youngren, T.S. Lo, J.F. Brady, J.M. Rallison and E.J. Hinch.

References

1. A. Acrivos, Ann. N.Y. Acad. Sci. 404, 1 (1983).
2. J.M. Rallison, Ann. Rev. Fluid Mech. 16, 45 (1984).
3. J. Happel, and H. Brenner, Low Reynolds Number Hydrodynamics, Prentice-Hall, Inc., 1965.
4. G.I. Taylor, Proc. Roy. Soc. Lond. A138, 41 (1932).
5. R.G. Cox, R.G., J. Fluid Mech. 37, 601 (1969).
6. J.M. Rallison, J. Fluid Mech. 98, 625 (1980).
7. D. Barthes-Biesel, and A. Acrivos, J. Comp. Phys. 12, 403 (1973).
8. D. Barthes-Biesel, and A. Acrivos, J. Fluid Mech. 61, 1 (1973).
9. G.I. Taylor, Proc. 11th Intl. Congr. Appl. Mech., Munich, pp. 790 (1964).
10. J.D. Buckmaster, J. Fluid Mech. 55, 385 (1972).
11. A. Acrivos, and T.S. Lo, J. Fluid Mech. 86, 641 (1978).
12. J.D. Buckmaster, J. Appl. Mech. E40, 18 (1973).
13. E.J. Hinch and A. Acrivos, J. Fluid Mech. 91, 401 (1979).
14. E.J. Hinch and A. Acrivos, J. Fluid Mech. 98, 305 (1980).
15. G. Ryskin and L.G. Leal, J. Fluid Mech. 148, 1 (1984).
16. G. Ryskin and L.G. Leal, J. Fluid Mech. 148, 37 (1984).
17. O.A. Ladyzhenskaya, "The Mathematical Theory of Viscous Incompressible Flow," Gordon and Breach, 1969.
18. J.M. Rallison and A. Acrivos, J. Fluid Mech. 89, 191 (1978).
19. J.M. Rallison, J. Fluid Mech. 109, 465 (1981).
20. G.I. Taylor, Proc. Roy. Soc. Lond. A146, 501 (1934).
21. F.D. Rumscheidt, and S.G. Mason, J. Coll. Sci. 16, 238 (1961).
22. S. Torza, R.G. Cox, and S.G. Mason, J. Coll. Int. Sci. 38, 395 (1972).
23. H.P. Grace, Engng. Found. 3rd. Res. Conf. Mixing, Andover, NH (1971).
24. F.S. Hakimi and W.R. Schowalter, J. Fluid Mech. 98, 635 (1980).
25. B.J. Bentley and L.G. Leal, J. Fluid Mech. 167, 241 (1986).
26. H.A. Stone, B.J. Bentley, and L.G. Leal, J. Fluid Mech. (to appear).
27. D.V. Khakhar and J.M. Ottino, J. Fluid Mech. 166, 265 (1986).

DETACHMENT OF DROPS FROM FLAT SOLID SURFACES

BY A SIMPLE SHEAR FLOW

M.J. Mahé[*], M. Vignes-Adler[*], Ch.G. Jacquin[**], and P.M. Adler[*]

[*]Laboratoire d'Aérothermique du CNRS, 92190 Meudon, France
[**]Institut Français du Pétrole, 92506 Rueil-Malmaison, France

INTRODUCTION

The interaction between liquids and solid walls is not well understood yet though this may have a large number of industrial applications (glass industries). This is certainly due to a lack of reliable experimental results which are difficult to obtain.

The specific purpose of the present work is to analyse the behaviour of a small alkane droplet immersed in water and deposited on a solid wall. The droplet is then detached from the solid wall by a water stream created along the wall. This is a three phases system where water-oil-solid interact together. The droplet sticks to the wall due to the existence of physico-chemical forces (we are deliberately vague on the exact nature of these forces). The adhesion of the droplet to the wall is measured by the value of the hydrodynamic stress necessary to its detachment. Some recent experiments are presented in this communication : they will be completed by additional ones in a forthcoming publication.

EXPERIMENTAL

Basically, the experimental device consists of two parts : the test section where the droplet is formed and deposited and the water loop necessary to create a well-controlled flow.

The water loop is briefly sketched in Figure 1. The water is forced by gravity into the test section between two constant level vessels : it is recirculated from the lower to the upper vessel by means of an auxiliary pump. The key point of this experiment is the cleanliness of the system. Only glass, teflon, Perpex, Viton and stainless steel are used ; each of these components was cleaned according to a specific procedure[1]. For instance, glass was degreased with an organic solvent, cleaned with boiling sulfuric acid, and then rinsed several times with ultra pure hot water, whose resistivity is about 18 M.Ω.cm. The components were then assembled ; gloves were used ; no grease, glue or surface active agents of any kinds were employed.

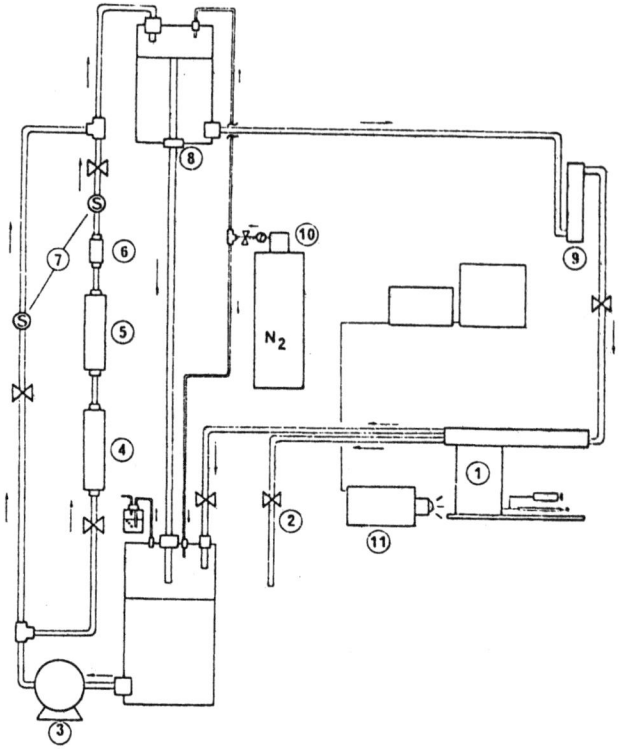

Fig. 1. Experimental set-up. (1) : test section (2) : outlet for decane. (3) : recirculation pump. (4) : activated coal. (5) : ion exchanger resin. (6) : microfilter. (7) : conductivity probes. (8) : constant level tank. (9) : flow meter. (10) : nitrogen supply. (11) : video camera.

The whole experiment is maintained under a small pressure of U-nitrogen. As a result, water is degased and does not contain any carbonic gas ; note that the free surfaces in the constant level vessels are also protected from contamination by U-nitrogen. Moreover, water is constantly partly recirculated through a cartridge of activated coal, an ion exchanger resin, and a microfilter whose cut-off is 0.2 µm. For the experiments which will be described below, water resistivity was always larger than 10 M.Ω.cm. The surface tension of water was measured and equal to 73 $\pm$ 0.1 mN.m^{-1} at 18.5°C.

The alkane used in this first series of experiments was decane ; the chemical product was carefully purified, by agitation on alumina, decantation ; it was then washed with pure water. Its interfacial tension with water was equal to 50.7 mN.m^{-1} at 23°C.

The decane droplets, saturated with water, were formed with an injection device, mainly composed by a syringe connected at its tip to a glass capillary. The tip of the capillary, whose diameter is about 20 µm, is located at the bottom of a rectangular vessel. The droplet may be detached from the capillary by a sudden mechanical vibration which is externally induced. Its rise velocity was measured, between two marks distant of 10 cm. Gravity drives the droplet to the glass ; at this moment, there is no flow in the water loop. A detailed view of the test section is given in Figure 2.

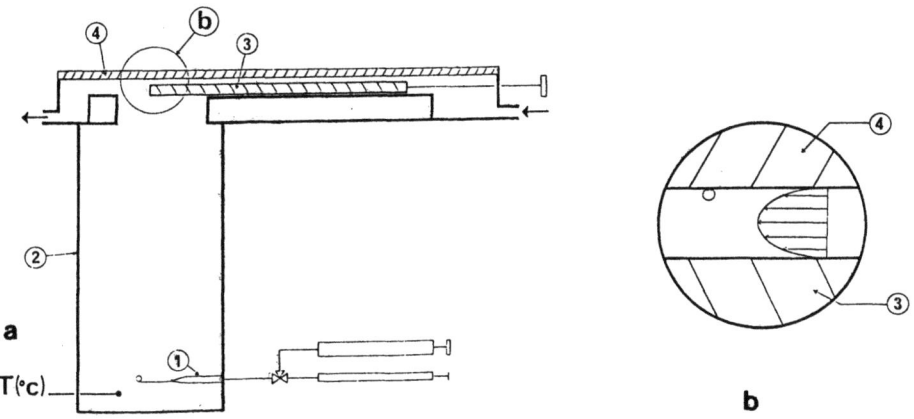

Fig. 2. Test section. (a) : the droplet is formed at the tip of a capillary (1) ; when it moves up under the action of gravity, its settling velocity may be measured in the vessel (2) ; as the lower wall (3) is opened, the drop will collide the upper wall (4). (b) : the lower wall (3) is closed ; it forms together with the glass plate (4) a rectangular channel. After a time interval Δt, a Poiseuille flow is generated in this channel.

Then, the vessel may be closed slowly, in order to form a rectangular channel whose upper wall is the glass. The transversal dimensions of this channel are 2 x 20 mm. Note that it is a float glass whose roughness is about 20 -30 Å. A well-defined Poiseuille flow may be generated in this channel with the help of the constant level tanks. Since the diameter $\emptyset$ of the droplets is often small with respect to the height of the channel, the relevant quantity is the shear rate γc at the wall, for which the droplet is detached. It was systematically measured as well as the time interval Δt between deposition and detachment.

RESULTS

Some preliminary results will be presented

Let us start with the rise velocity. It is commonly believed that the relative position of the data between the Stokes and Hadamard-Rybczynski reflects the degree of contamination by surface active agents[2]. Hence, such a measurement provides a scale for the hydrodynamic cleanliness of our experiments. Results are given in Figure 3. Clearly, intermediate data were obtained ; note however the large error bars for the small droplets, which are due mostly to a rough estimation of the diameter. Hence, in this first set of experiments, the system cannot be claimed to be totally clean in view of these data.

The critical shear rate γc, necessary to detach the droplet, was measured as a function of the diameter $\emptyset$ and of the time interval Δt. Reproducibility of these results is the touching stone of these experiments, and a lot of troubles were met : even minute amounts of impurities were found to dramatically change γc. This particular sensitivity is certainly due to the polar character of water. The results displayed in Figure 4 were obtained in a reproducible manner ; the whole set-up was disassembled

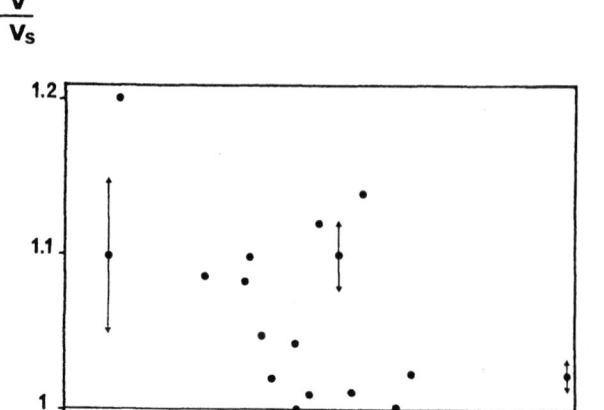

Fig. 3. Ratio of the measured rise velocity V and of the rise velocity of a solid sphere V_S as a function of the diameter $\emptyset$. Inertial effects are taken into account in V_S as indicated in [2]; for a fluid sphere, the ratio V/V_s is equal to 1.21; the order of magnitude of the experimental errors is indicated by the arrows.

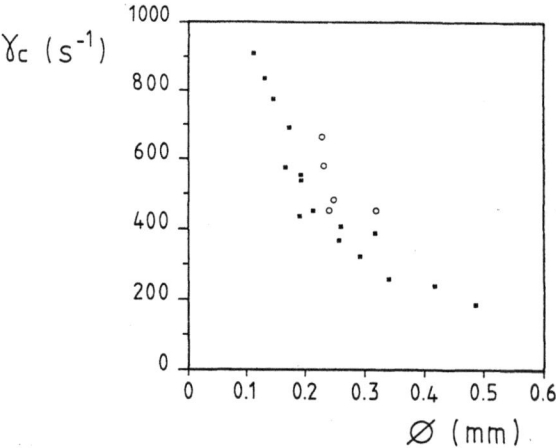

Fig. 4. The critical shear rate γ_c as a function of the drop diameter $\emptyset$. The time interval Δt is equal to 22'. Conditions are : ■ (water resistivity 10 MΩ.cm), O (water resistivity ranging from 1.5 to 1.8 MΩ.cm).

three times, cleaned again, and the same value of γ_c was measured on the same piece of glass, and on various samples of the same glass sheet.

γ_c is found to be a decreasing function of $\emptyset$, as it is expected on physical grounds, since the hydrodynamic forces are larger for a large droplet. γ_c seems to be of the general form $\emptyset^{-n}$, where the exponent depends upon Δt.

An other detail may be worth mentionning. An experiment was conducted as previously, but the nitrogen was stopped, and the various cleaning devices, activated coal..., were bypassed. The resistivity of water decreased and reached a value of 1 $M\Omega.cm$; γ_c was measured and was found significantly larger than previously (cf. Fig. 4). Hence, all the experimental features are important such as a controlled atmosphere. It is also a general conclusion of this series of measurements that the cleaner the system, the lower the value of γ_c ; any impurity tends to increase γ_c ; in this sense, the curves which are displayed in Figure 4, are the lowest ones that we could obtain.

The time interval Δt between deposition and detachment was systematically studied for a given diameter and displayed in Figure 5. In some earlier experiments, γ_c was still found to vary after a few days. Hence, the equilibrium between the droplet, water and the solid wall is very long to reach. The major issue in this respect is to know, whether or not, there is a water film between the decane droplet and the wall ; the thinning of such a film could thus be responsible of the slow variations of γ_c with Δt given in Figure 5. Some visual observations were made from the top of the test section ; after some time (typically $\Delta t \sim 15'$) for clean systems, and immediately for contamined ones, a line was visible around the droplet ; however, it was found impossible to decide whether it was a contact line or the contour of the water film.

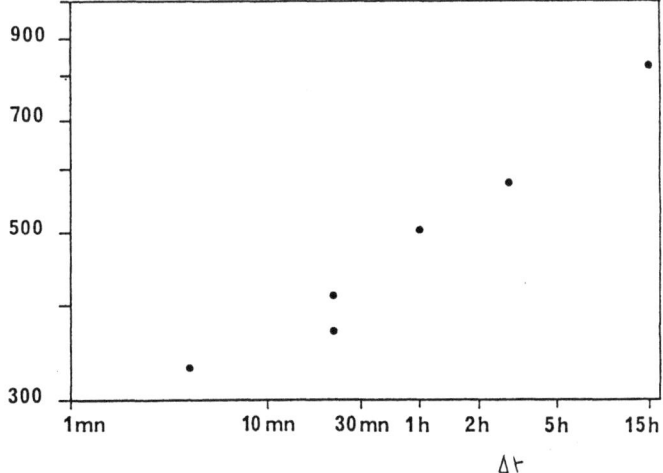

Fig. 5. The critical shear rate γ_c as a function of the time interval Δt. The drop diameter $\emptyset$ is equal to 0.26 mm.

23

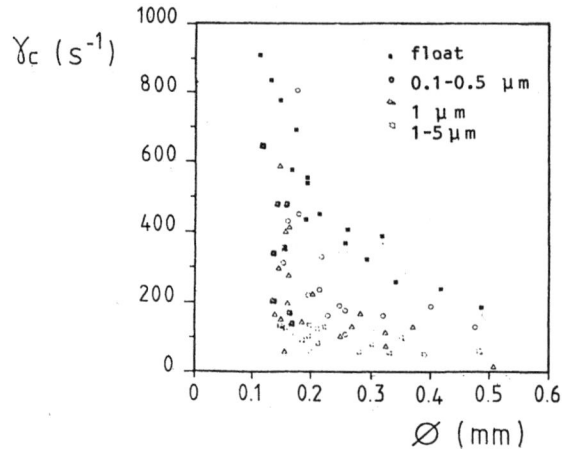

Fig. 6. The critical shear rate γ_c as a function of the time interval Δt for various roughnesses.

Eventually experiments were made with rough surfaces which were obtained either by a chemical corrosion (with vapor of hydrofluoric acid) or by a mechanical corrosion (with corundum particles) : three roughnesses have been obtained 0.1 - 0.5 µm, 1 µm and a few µm. Results are displayed on Figure 6.

There is some dispersion due to the random character of the geometrical irregularities. Nevertheless the critical shear rates are always smaller than the ones obtained for float wall ; the rougher is the glass, the lower the shear rate. The opposite effect was expected ; it was thought that an hypothetical contact angle hysteresis would have been an increasing function of the surface roughness ; the critical shear rate would have thus behaved in the same manner. The existence of a stable thin film between the decane droplet and the wall is probably confirmed by the present result.

ACKNOWLEDGEMENTS

This work was supported in part by a grand from the Institut Français du Pétrole. M.J. Mahé was supported by a E.N.S.P.M. fellowship.

REFERENCES

1. Mahé, M.M., Vignes-Adler, M., and Adler, P.M., in preparation.
2. Clift, R., Grace, J.R., and Weber, M.E., "Bubbles, drops and particles". Academic Press, New York (1978).

LIQUID BRIDGE MODELING OF FLOATING ZONE PROCESSING

I. Martínez

Universidad Politécnica de Madrid
E.T.S.I.Aeronáuticos, 28040-Madrid, Spain

ABSTRACT

A simplified fluidmechanical interpretation of some silicon growth experiments performed aboard Spacelab-1 is given, using the capillary liquid bridge theory, which is discussed in depth. A model is developed to simulate the outer shape during the floating zone process performed, and ideas are given for a more complete thermal simulation. These models would help experimentalists in crystal growth and zone refining to predict unstable configurations, avoid bridge disruption and achieve the desired shape at every stage.

INTRODUCTION

A lot of the thermal processing techniques used in present-day materials technology are still full of empirism, and thus, clues to enhance their performances are most difficult to draw. The lack of a suitable theory that could set guidelines to further progress is due to poor understanding of the complex thermodynamic, physico-chemical and fluidmechanical phenomena that interplay.

In particular, the floating zone technique of rod refining and single-crystal growth has lately caught a lot of attention because of its outstanding importance in the manufacture of semiconductor materials of the highest purity for the expanding microelectronics and optoelectronics industries, when dislocation, impurities and other defects must be kept down to several parts in 10^{12} to reach integration levels higher than the present 10^9 junctions per cm^2.

In this context, this paper presents a model that serves to explain some singular behaviour found during experimentation with melting rods of silicon in a microgravity environment, in particular predicting the evolution of the free molten interface. Equilibrium shapes and the stability of free interfaces and fusion- and solidification-fronts are key points to a controlled handling of molten zones.

To put this developments on a better perspective, a preliminary review of the classical theory of thermodynamics of interfaces is presented, what will show up the need to perform experiments in reduced gravity to get large curved interfaces and make the experimental analysis easier.

A detailed study of static stability and energy evolution of axisymmetric liquid bridges is further developed as the most simple model of the complex real problem, considering several configurations of major interest where important results have been found.

In order to test the analytical and numerical models available, a wide-range experimental program is being pursuit by this research team. On the one hand stays experimentation in a simulated microgravity environment [1] using Plateau's neutral buoyancy technique. On the other hand stays experimentation in real microgravity platforms: Spacelab-1 in 1983 [2], parabolic flights in NASA aircraft KC-135 and German sounding rocket TEXUS-10 in 1984, and TEXUS-12 [3] and Spacelab-D1 [4-7] in 1985. Results from these trials are also commented.

INTERFACES

Microgravity relevance

Most material systems appear as an ensemble of a few homogeneous bulk phases (solid, liquid or gaseous) with overall properties independent of the interface configuration. However, there are instances where interface effects play a dominant role, namely when volumetric forces become negligible ($\Delta\rho g L \ll \sigma/L$ for lengths L of interest) as in microgravity, and when the natural scale of the system (volume/interface area) is small. The latter refers to a finely divided heterogeneous mixture, as in colloidal systems (emulsions, foams, bubbling liquids, mists, etc), that are not considered here, despite its well-known importance.

Let us deepen in the analysis of why the microgravity environment is relevant to fluid science. Fluid science is an evolving subject that can be simply thought of as the conservation laws of physics applied to a local region of material (described by some constitutive equations) in local thermodynamic equilibrium, limited in space and time by appropriate initial and boundary conditions. At the atomic level, gravity is a weak volumetric force (10^{40} times smaller than the electrostatic one in the hydrogen atom and 10^{35} times in the molecule of hydrogen), but accumulates and becomes dominant when large masses are present, unless balanced by another body force. The main advantage that microgravity offers to fluid science is therefore the possibility of achieving large quiescent interfaces of uniform mean curvature, free from gradients of hydrostatic pressure, and showing high sensitivity to weak forces.

Experiments on Earth show many examples of capillary phenomena, but only in reduced dimensions (some millimeters) best characterized by the combination $\sqrt{\sigma/(\Delta\rho g)}$, known as capillary length, where σ is surface tension, $\Delta\rho$ is the density difference across the interface and g is gravity. It is important to notice that the material properties σ, $\Delta\rho$ and the contact angle θ (to be introduced later) are all independent of gravity. The spherical shape of small drops and bubbles, the colourful soap films, the varicosity and breaking of small jets, curved edges of all free surfaces, capillary rise and capillary pumping, retention of liquid in porous media, seepage on welding, displacement of a liquid by a more wetting one, flotation of ores by wetting agents, nucleation of drops and bubbles, and so on, are common examples.

The simplest system to be analyzed in microgravity is a large curved quiescent liquid/vapour interface of a pure substance with no external forces applied. From this basic configuration, extensions are made to include the effect of small additional perturbations: a constant residual acceleration, g-jitter, electrostatic and acoustic fields, and so on, although other possible perturbations as thermal, aerodynamic and electromagnetic fields are not contemplated here.

Under normal operating conditions, a neat division can be established between solid interfaces (where a solid and a fluid meet) and fluid interfaces (also called fluid/fluid interfaces) that may be liquid/liquid or liquid/gas. Their difference in mobility (solids retain their non-equilibrium shapes) renders their study widely apart, mainly because the diagnostic means are so distinct.

In what follows, the presence of a well defined fluid interface (that is, a heterogeneous system, and not a colloidal one) is always taken for granted, but in most instances a solid interface is also present, Thus, another useful classification is: interfaces without contact line and interfaces with contact line (further divided in: contact line without a solid and contact line with a solid, the latter being the most interesting). Several configurations of interest are presented in Fig. 1.

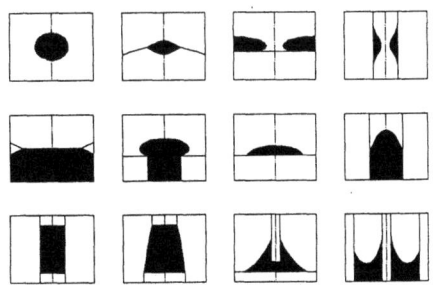

Fig. 1. Some axisymmetric fluid interface configurations of interest.

Another interesting classification from the point of view of materials science is: liquid-vapour, liquid-gas and liquid-liquid (immiscibles), for fluid interfaces, and liquid-solid, solid-vapour, solid-gas, solid-solid (grain boundaries) and solid-solid (distinct solids), for solid interfaces.

When two interfaces meet, a triple line exists that, if not anchored to a well defined edge, poses many problems both in practice (for positioning and handling) and in theory (for modeling the virtual displacements in stability studies).

Another remarkable thing to notice is that most of the methods to measure interfacial tension relay on the effect of gravity upon capillary systems, which rules them out in microgravity configurations as well as when the liquids are neutrally buoyant or near critical points.

But let us come back to the main point of characterizing interfaces. An interface is the region of high gradients in physicochemical properties where two volumetric phases meet. Under normal conditions, its thickness is only several molecular diameters wide. But one may wonder why do interfaces exist at all?, how is it in the interior?, and, from the practical point of view, how to treat them macroscopically. A simple one-dimensional model of phase transitions (where interfaces appear) gives some insight to the first point.

Let a given number (large) of molecules be constrained to lay on a line, each one separated by its neighbours by a kind of spring, as illustrated in Fig. 2a. When the particles are widely separated, the spring model should approach the ideal gas pressure/volume relationship $F=k/d$, where F is force, d distance (spring length), and k a constant. When distances between particles are reduced (Fig. 2b) a kind of Lennard-Jones potential should be added that for one dimension gives $F=k/d-a/d^2+b/d^4$. This spring model is represented in Fig. 2f and reminds of van der Waals equation of state.

When the overall length (volume occupied by all the molecules) is reduced, first a proportional uniform reduction of all separations takes place (Fig. 2b), but if overall squeezing continues, spring span ceases to be uniform; some of them keep a length d_2 and some others adopt a length d_1 (Fig. 2c). Further overall distance reduction only increases the population of d_2 springs at the expenses of d_2 ones (Fig. 2d) until all of them are d_1 type and again a uniform shortening is brought by subsequent overall distance reduction.

Of course, this is just a qualitative model and do not explains the tendence to coalescence and other important observed behaviour of real interfaces.

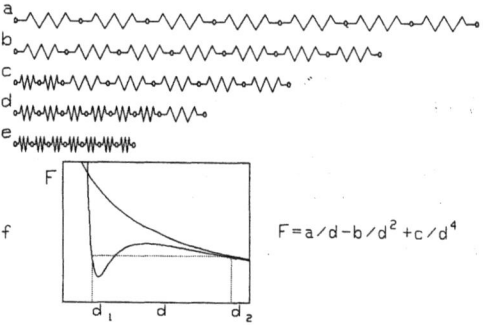

Fig. 2. A van-der-Waals-type of model for the occurrence of interfaces.

Fluid/fluid interfaces have great mobility, but solid/fluid interfaces lack it, their characteristic times are much longer and consequently they are normally under non-equilibrium conditions. Diffusion through interfaces and adsorption at them are vast subjects of research not dealt with here.

It is seen thence that interfaces appear as a consequence of thermodynamic instability, but why should interfaces have thermodynamic properties of their own and not be simply a geometrical surface separating two uniform different phases?

Microscopic and macroscopic views

Far from critical points the interface thickness is microscopic ($\sim 10^{-9}$m), but if a macroscopic patch is considered ($>10^{-7}$m of side), the continuum model can be applied to microscopic distances along the normal to the interface, and thermodynamic functions drawn as in Fig. 3. The macroscopic approach is applied to get rid of these finer details and retain only what shows up at the larger scale. The traditional way (Gibbs 1878) is to define a geometrical interface according to a certain criterion and extrapolate nearby bulk properties up to it (Fig. 3), assigning to this geometrical interface the excess value of the real physical properties and the extrapolated model. The usual criterion is to choose the interface at a location such that the excess density of one component vanishes. All interfaces have excess energy; besides, normal mixtures usually show excess concentration of solutes, but the contrary is the rule for electrolytic solutions. These excess properties are to be modeled by appropriate constitutive equations.

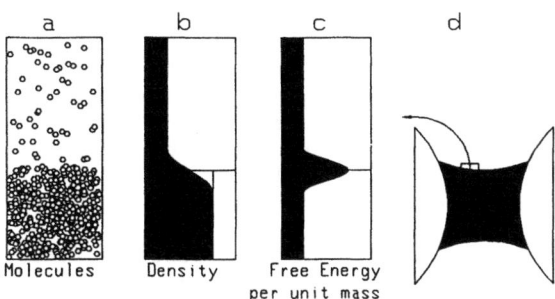

Fig. 3. Microscopic view of a liquid/vapor interface.

Because interfaces have excess energy, they tend to adopt a minimum area compatible with other constrains. To extend an interface under isothermal conditions a work $dW = \sigma \cdot dA$ must be done. For a pure substance σ only depends on temperature (pressure and temperature are coupled under phase transition), and has $d\sigma/dT<0$ and $d^2\sigma/dT^2>0$. The internal energy per unit area, $U/A = \sigma - T \cdot d\sigma/dT$, varies in a similar way as the enthalpy of vaporization $h_{fg}(T)$, as shown in Fig. 4 for the case of water.

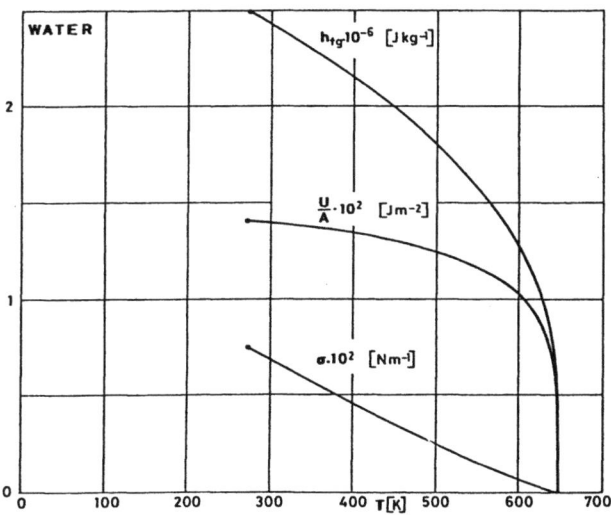

Fig. 4. The variation of surface energy U/A with temperature T is similar to the variation of the enthalpy of vaporization.

Between any two atoms or molecules, there is intermolecular attraction. Even in absence of net electrical charges, the electrons of each atom or molecule see and react to the motion of the electrons of all neighbouring molecules. The electrons arrange their motion accordingly, thus lowering the energy of the atomic entity. This correlated electron motion is the origin of van der Waals attraction and of surface tension. A molecule looses the more potential energy the more molecules of the same species surround it, that is, there exists an energy of vaporization. A molecule moving from the liquid phase into the gas phase needs to gain this vaporization energy. It gains about half that energy (see Fig. 4) if it cannot join the vapour and instead constitutes part of the surface, where van der Waals attraction of one half of the molecules is still acting. This excess in energy of the surface molecules relative to the bulk cause the surface tension.

This statement on the surface tension of a liquid and the energy of vaporization apply in like manner to the interface tension between two liquids and the energy of mixing. Molecules of the same species generally exhibit a stronger van der Waals attraction than molecules of different species. It needs energy to exchange a molecule 1 within a drop of molecules 1 for a molecule 2. It needs loosely half this energy, if the two molecules are just brought to the interface between liquids 1 and 2. The molecules loose only about one half of the stronger attraction by the molecules of their own species.

Also, from investigations into the maximum possible undercooling rates one can conclude that the interface energy between a solid and its melt amounts to about one half of the energy of fusion. Similar relations hold between the mixing energy of two liquids exhibiting a miscibility gap and their interface tension. The interface tension vanishes at the critical point of the miscibility diagram. When the critical point is approached, one of the coexisting phases spreads along the surface of the liquid mixture and of the vessel.

Thermodynamics of interfaces

Consider a closed system composed of two fluid phases 1 and 2 with an interface s in between, in absence of any external force field. For the equilibrium of the whole system (phase 1, interface s and phase 2) the total entropy must be a maximum

$$S = S_1 + S_s + S_2 = \text{maximum} \tag{1}$$

subject to the constrains of constant volume V, energy U and amount of substance n_i

$$V = V_1 + 0 + V_2 = \text{constant} \tag{2}$$
$$U = U_1 + U_s + U_2 = \text{constant} \tag{3}$$
$$n_i = n_{i1} + n_{is} + n_{i2} = \text{constant for every specie i} \tag{4}$$

Substitution in Eq. (1) of $dS=(1/T)dU+(p/T)dV-\Sigma(\mu_i/T)dn_i$ for bulk phases and $dS=(1/T)dU-(\sigma/T)dA-\Sigma(\mu_i/T)dn_i$ for the interfacial phase, where μ_i is the chemical potential for the i-th specie, with Eqs. (2-4) yields

$$T_1 = T_s = T_2 \tag{5}$$
$$p_1 = p_2 + \sigma dA_s/dV_1 \tag{6}$$
$$\mu_{i1} = \mu_{is} = \mu_{i2} \quad \text{for every specie i} \tag{7}$$

where conditions of independence for the variations of U_1, U_2, n_{i1}, n_{i2} and V_1 (not those of V_2 and A_s, which depends on that of V_1), have been used.

Analytical geometry readily shows (first found by Gauss) that dA_s/dV_1 is the mean curvature C of the interface (the sum of the inverses of the principal radii of curvature, also equal to the divergence of the normal vector changed of sign). It is seen thence that thermodynamic equilibrium requires constant temperature in all phases, a pressure jump across the interface proportional to local mean curvature (first stated by Young in 1804 and Laplace in 1805), and constant chemical potential throughout.

Introducing the surface concentration of component i, $\Gamma_i = dn_{is}/dA$, the Gibbs-Duhem equation for the interface phase (for unit area) is

$$0 = s_s dT_s + d\sigma + \Sigma \Gamma_i d\mu_{is} \tag{8}$$

which, in the limit of isothermal dilute solutions (chemical potential proportional to the logarithm of concentrations c) yields

$$\Gamma_i = \frac{-1}{RT} \frac{\partial \sigma}{\partial \ln c_i} \qquad \text{(for dilute solutions)} \tag{9}$$

showing that, under isothermal conditions, the variation of surface tension with solutal concentration (proportional to μ_i) is opposed to adsorption, thus, in normal solutions whose components tend to be adsorbed at the interface, the interface tension diminish, whereas in electrolytic solutions with an interface depleted of solutes, the surface tension increases.

However, under normal circumstances the diffusive process that tends to equalize μ_i has a characteristic time much longer than the mechanical process involved in adjusting the pressure field, as happens with thermal processes. Thus, from now ones. Thus, from now on, the influence of the concentration field will be neglected. In other words, equilibrium at T and V constant is reached when the Helmholtz free energy F=U-TS has a minimum. As F=σA, equilibrium implies minimum area, as already mentioned and, if there are several interfaces present, F=$\Sigma \sigma_i A_i$ must be a minimum. For a configuration as in Fig. 3d, because the area of solid interface is always constant, it suffices that the effective area

$$A_{eff} = A_{12} - A_{13} \cos\theta \tag{10}$$

be a minimum, for constant volume of bulk phases. If gravitational potential energy and rotation kinetic energy are included, the free energy F of the system is

$$F = \Sigma \sigma_i A_i + mgz_{com} - (1/2)I\omega^2 \tag{11}$$

where m is mass, g gravity, z_{com} center-of-mass height, I moment of inertia and ω rotation rate. Equilibrium requires dF=0 (with constant V) and stability requires $d^2F>0$ (with constant V).

SHAPES AND STABILITY

In the previous paragraphs a thermodynamic approach was followed to arrive at the formulation of equilibrium and stability conditions of an interface. Now, a purely mechanical presentation is made to give a different perspective.

Isothermal fluid static configurations satisfy the mechanical equilibrium condition

$$\nabla(p+U)=0 \qquad \text{at the bulk} \tag{12}$$

where ∇ stands for gradient, p is pressure and U the potential function of a possibly acting body force field. The boundary conditions are (Young-Laplace formulation)

$$p^+ - p^- = \sigma \cdot C \quad \text{at an interface} \tag{13}$$

and

$$\sin\theta_1/\sigma_{23} = \sin\theta_2/\sigma_{13} = \sin\theta_3/\sigma_{12} \quad \text{at a triple line} \tag{14}$$

where $p^+ - p^-$ is the pressure jump across the interface, σ is interface tension, C is local mean curvature and θ is fluid contact angle (Fig. 5).

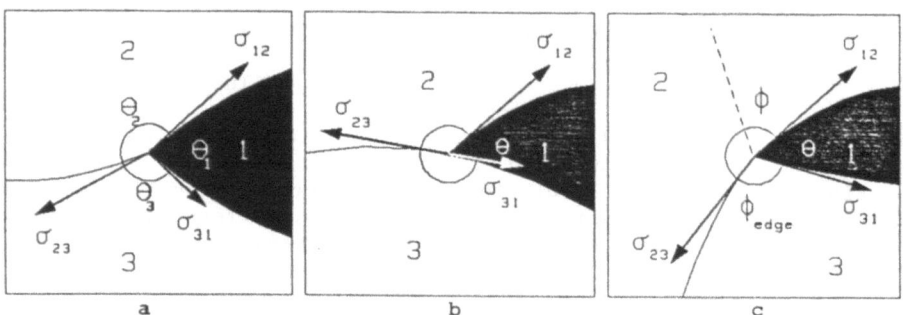

Fig. 5. Three typical contact-line configurations: **a)** at a three-fluid confluence, **b)** over a smooth solid, and **c)** anchored to a singularity (canthotaxis).

At thermodynamic equilibrium σ only depends on temperature for given substances. Although different liquid/liquid interfaces show wide variations in ρ in liquid/gas interfaces it is only marginally sensitive to the nature of the gas. Similarly, contact angles also depend on temperature and composition of the three phases meeting, but, if one of them is gaseous, it shows little sensitivity to the gas composition.

Canthotaxis

A degenerate case of Eq. (14) happens when one of the phases is solid (see Fig. 5b). In such a case it takes the form

$$\sigma \cdot \cos\theta = \sigma_{23} - \sigma_{13} \tag{15}$$

Furthermore, if the solid surface presents a singularity at the three phase line, contact angle is irrelevant in the range of values

$$\theta < \phi < \pi - \phi_s + \theta \tag{16}$$

On the other hand, the local mean curvature of a surface is simply the divergence of the normal vector changed of sign

$$C = -\nabla \cdot \vec{n} \tag{17}$$

which can be found expanded in different coordinate systems (rectangular, cylindrical or spherical) in most textbooks on analytical geometry. In any case, the problem of finding equilibrium shapes reduces to solving the second order partial differential equation (13) with the pressure jump given by Eq. (12) at the interface, plus one of the boundary conditions (14)-(16), plus the condition of known fluid volume or feeding pressure (of course, it is quite different to consider stability at constant fluid volume that at constant feeding pressure).

Thus, the capillary problem is well established, its formulation has been known for nearly two centuries, but direct solutions in general cases are only available since the arrival of computers. Before that, only planar, cylindrical, spherical and catenoidal surfaces were manageable interface shapes (and they mostly apply to U=0 in Eq. (12)).

Although it is so easy to obtain non-axisymmetric real interfaces in practice with a wire frame and a soap solution, it seems they have not deserved too much theoretical attention in the past, no doubt because its analytical difficulty. In what follows, only axisymmetric interfaces are considered.

It is customary to first treat interfaces in the absence of fields and then add the effect of gravity, solid-body rotation, electric fields, and so on. Concerning presentation of results, several approaches are followed. For instance, when gravity is present, some authors use the capillary length $\sqrt{\sigma/(\rho g)}$ and others prefer some geometrical length (radius of the support disc, for example) to nondimensionalize; whereas the former may seem sounder on Earth, the latter is preferable to clearly show the influence of gravity.

Liquid bridges

In the last decade, boosted by the availability of microgravity platforms, many studies on analytical and numerical modeling and experiments with liquid bridges have been performed. We shall limit here to problems associated with shape stability. A summary of results is shown in Figs. 6-9, where the stability limits for equal discs, unequal discs, equal discs with a Bond number and cylindrical columns in isorotation are presented.

The nominal configuration for most of the experiments in space platforms consists of a long cylindrical liquid bridge with its borders anchored to the sharp edges of two solid coaxial discs, and held by interfacial tension forces. Different mechanical stimuli are then applied through the supporting discs (stretching, vibration, rotation) and the liquid bridge

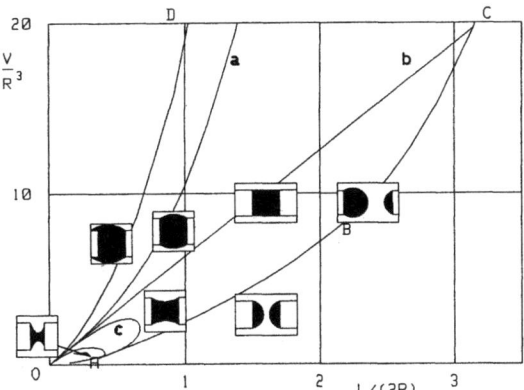

Fig 6. Stability limits for a floating zone of liquid volume V, held between equal discs of radius R a distance L apart. Curves a, b and c correspond to spherical, cylindrical and catenoidal shapes, respectively (in between are unduloids and outside nodoid-shapes). OA corresponds to minimum volume due to edge detachment, AB to symmetric instability (symmetric breakage respect to the mid plane between discs), BC to asymmetric breaking, and OD to maximum practical liquid volume (limit by edge overflow).

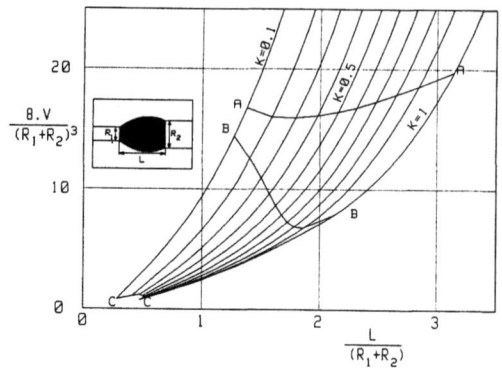

Fig 7. Stability limits for a floating zone of volume V, between unequal discs of radius R_1 and R_2 a distance L apart. The minimum stable volume for several disc ratio is shown. Curves AA, BB and CC correspond to bridges with minimum undulation, local cylindricity at the larger disc and catenoidal shapes, respectively.

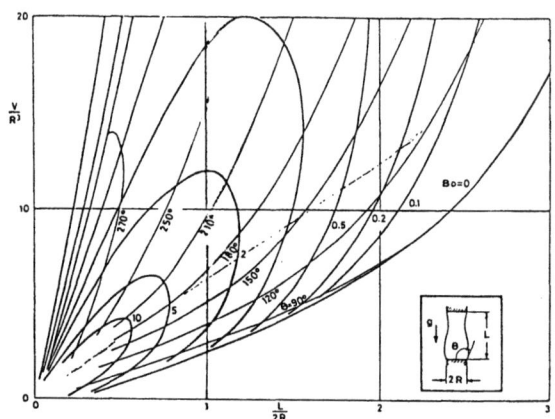

Fig 8. Stability limits for a floating zone of liquid volume V, held between equal discs of radius R a distance L apart, in the presence of an axial gravity field of acceleration g ($Bo = \rho g R^2/\sigma$) [8].

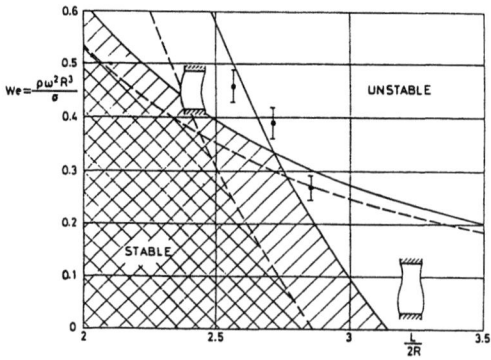

Fig 9. Critical rotation rates ω (solid lines) for cylindrical liquid bridges. Short bridges develop a C-mode deformation whereas bridges more slender than $L/(2R) = \frac{\pi}{2}$ break in an amphora-like mode. The stable area is reduced (shown cross-hatched) if an axial gravity is present as during Spacelab-D1 trials. The three experimental points correspond to the three amphora-mode breakages in that flight.

response is analyzed (mainly the outer shape deformation). Aside of its own scientific relevance in fluid mechanics this work has many potential spinoff applications, as in the modeling of the floating zone technique used in materials science for the processing of semiconductors and high melting point materials, what is treated later in the paper.

The first microgravity experiments in this project were performed in Spacelab-1 (Dec-83) with only partial success because of uncontrolled liquid spreading beyond the edges of the supporting discs, which were protruding only 0.5 mm from their base plate. In spite of that, the resourceful crew managed to improvise new working discs and achieved large (although not cylindrical liquid columns, as the one shown in Fig. 10a, which corresponds to an isorotation sequence at 7 rpm with the column showing the long awaited C-mode deformation (like a skipping rope).

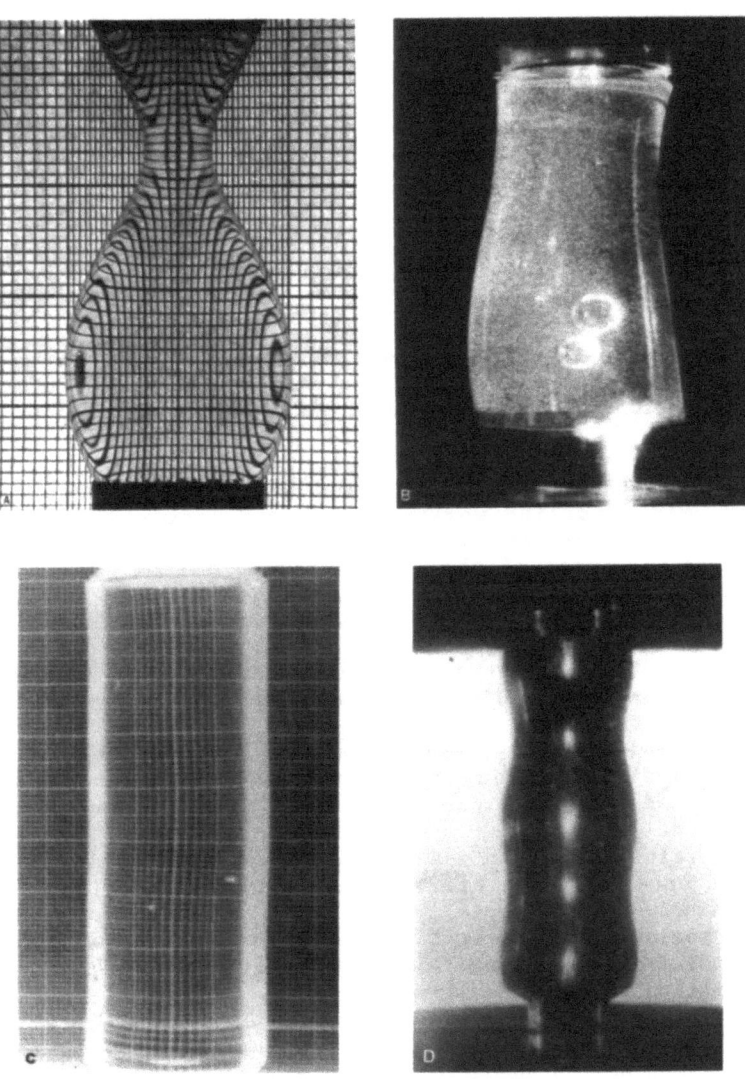

Fig. 10. Liquid columns obtained in different platforms: **a)** prior to rupture in a neutrally buoyant tank on the ground, **b)** C-mode deformation at 7 rpm in Spacelab-1, **c)** nearly perfect cylindrical zone in TEXUS-12, **d)** vibration at 1.1 Hz in Spacelab-D1.

35

In order to check the new disc design for the following Spacelab experiments, a series of parabolic flights with NASA KC-135 aircraft were flown in Dec-84, although the microgravity level achieved was too poor to handle large liquid masses.

Much better results were obtained on the TEXUS-12 rocket flight in May-85 (Fig. 11 shows the procedure followed), where perfect cylindrical liquid columns 82 mm long by 30 mm in diameter as the one shown in Fig. 10c were established (in a record time of about one minute). An earlier attempt in May-84 had been a complete failure due to equipment malfunction. A second campaign of parabolic flights were carried out in Jun-85 and again the ambient noise was too high to check Spacelab operations other than the necessary crew familiarization and training. These relatively cheap early trials may also serve to detect some materials incompatibility, as occurred indeed for other experiments, thus giving time to work out remedies and improving the reliability of multiuser facilities as the Fluid Physics Module used here.

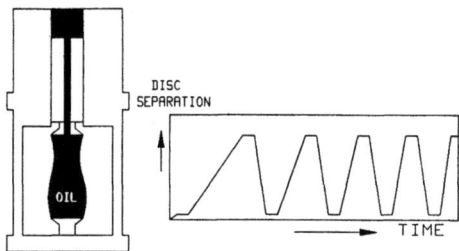

Fig. 11. Experiment in TEXUS-12 rocket. Movement of the upper disc assembly creates liquid columns as the one in Fig. 10c. Separation speed was increased until bridge disruption.

In November 1985, the sequence of Spacelab-1 trials were repeated aboard Spacelab-D1 with modified working discs and liquid injection system (Fig. 12) and this time everything run as foreseen. Payload Specialist Dr. Furrer was in charge of the experiment in flight, assisted by the investigator on ground when voice and video link was available. The experiment proceed as expected, with the following deviations:

- It was soon realized that the g-jitter effect was much more pronounced than Spacelab-1 and TEXUS-12 experience had shown. Long columns were continuously trembling.
- This noisy ambient stressed the operator, who on the first trial followed a slightly scarce filling law that caused the first disruption of the column soon after the nominal working length L=95 mm was reached.
- Experiment reinitialization was achieved thanks to the skillful operator (and not once but five times), demonstrating the importance of a well-trained Payload Specialist on board. During a real-time video link it was agreed to switch on background illumination to better follow the free surface oscillations (at expenses of a poorer tracer visualization by meridian plane lighting).
- The small rotation of the discs at 3 rpm, intended for overall viewing, was switched off to diminish ambient noise (work had to be stopped several times due to uncontrolled vibrations or Shuttle maneuvers).
- Axiall oscillation of one disc at several frequencies was exercised and liquid response was as foreseen though departures from a cylinder were larger near the filling discs due to a residual acceleration. A second breaking took place just after the axial oscillation exercise due to a Shuttle maneuver.

- Further breakings occurred at L=95 mm with ω=12 rpm, 100 mm with 10 rpm, and 90 mm with 13 rpm, all in an amphora-like mode. The last breaking showed an initial C-mode deformation (like a skipping rope), but ambient noise may have changed the final breaking mode.
- The last C-mode rotation trial, at L=75 mm with ω=16 rpm, much less sensitive to ambient noise, was not performed due to lack of time, in spite of the generous time extensions granted to this experiment.

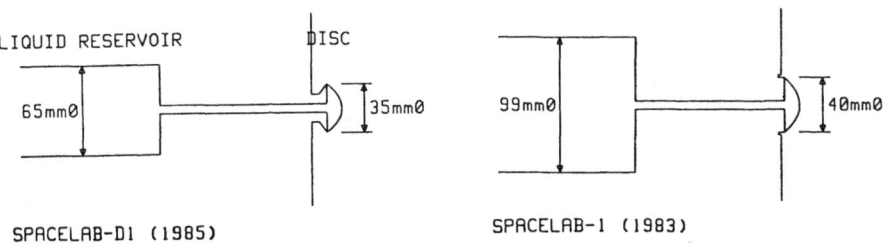

Fig. 12. Changes introduced in the working discs and the injection system.

The breaking of long cylindrical bridges has also been investigated by numerical simulation using a one-dimensional slice model [9]. The time evolution of the different energy terms (Fig. 13) gives a clear picture of the process. If some tracers are assumed to lay on the free surface, their respective paths may be traced with the model and can be seen in Fig. 14.

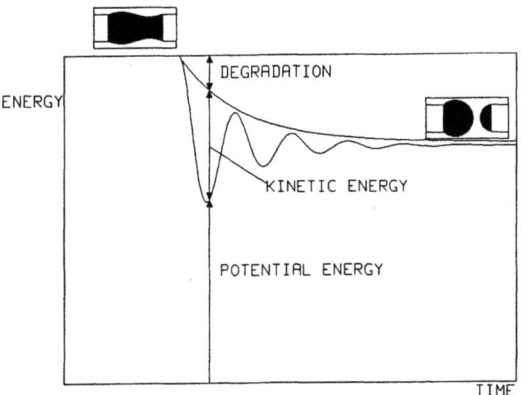

Fig. 13. Evolution of the free energy in a breaking bridge (an initially cylindrical column near the maximum stable length is slightly deformed to start the process.

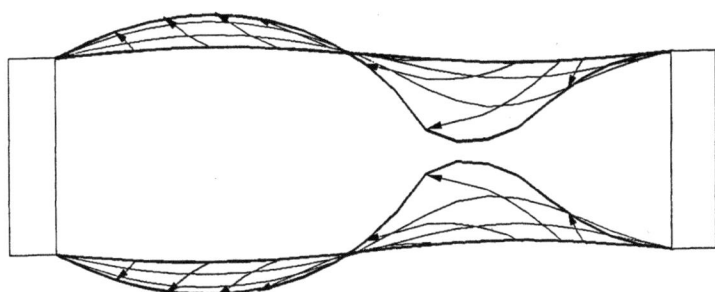

Fig. 14. Trajectory of surface tracers during the breaking of a bridge.

The most important result is the accurate prediction of the volume fraction of the two drops formed after bridge disruption, as shown in Ref. [5] for the experiments aboard Spacelab-D1 (a small free satellite drop is always observed in practice, but its volume is negligible).

CRYSTAL GROWTH

Introduction

Crystal growth is the aggregation of otherwise dispersed particles to an already existing solid lattice. This process demands some mobility of the particles, thence the disperse phase, although it may be a hot solid (recrystalization), is usually a fluid. Crystal growth from a gaseous phase is sometimes practiced (vapor growth techniques), as well as crystal growth from a cool liquid (solution growth techniques), but by far the largest quantity of materials is obtained by crystal growth from the melt.

This last technique is further classified according to the configuration of the three-phase contact line where the solid/melt interface meets the surrounding medium, that, although sometimes consisting of a solid wall or a molten salt, it is normally an inert gaseous atmosphere. Besides, three different configurations are usually distinguished: pulling a rod out of a molten bath (Czochralski technique), pulling a rod through a furnace (floating zone technique) or pushing a melt through a die (the most important of these being edge-defined film-fed growth). The latter seems to be the more versatile and is catching the market of semiconductor material production (mainly for solar cells), but the floating zone technique, the one dealt here, presents the advantage of the melt not being contaminated by the die or crucible materials and yields the purest materials.

It is important to realize that for crystal growth to occur the system needs to be in a non-equilibrium state, i.e. with a temperature (or concentration, or electric field, etc) gradient to provide the driving force, which must be very small to avoid instabilities at the solidification front (particularly constitutional supercooling). The decrease in entropy due to the ordering upon crystallization implies that heat must be removed from the sample.

Container-free floating-zone growth is a very common technique to get high purity crystals. The main advantage is that any contamination by the crucible material is avoided. Under gravity conditions, this method is used when the surface tension of the materials suffices to compensate the hydrostatic pressure in the melt. High vapor pressure of at least one of the components might prohibit this method, however.

To establish a floating zone, a relative motion between the sample rod to be processed and the furnace is imposed (Fig. 15), causing impurities to be swept to the rod ends by the difference in concentration at the melting and solidification fronts.

The interplay of thermodynamical, physicochemical and fluid-mechanical phenomena in this process is so entangled, that this technology is still full of empirism. Trying to better understand the underlying physics, a great effort is being devoted to the detailed study of more simple specific phenomena, both theoretically and experimentally. In particular, because one of the more conspicuous forces present is that of gravity, a proliferation of microgravity experiments has taken place since the availability of space platforms. Although some of them aim at obtaining new exotic materials, the majority are concerned with more fundamental research, using well known materials, as for instance Experiment 1-ES-321 "Crystal growth of a silicon rod" performed aboard Spacelab-1 in Dec-83, which is the one to be analyzed here. But before that, let us have a look at the other important application of floating zones: materials purification.

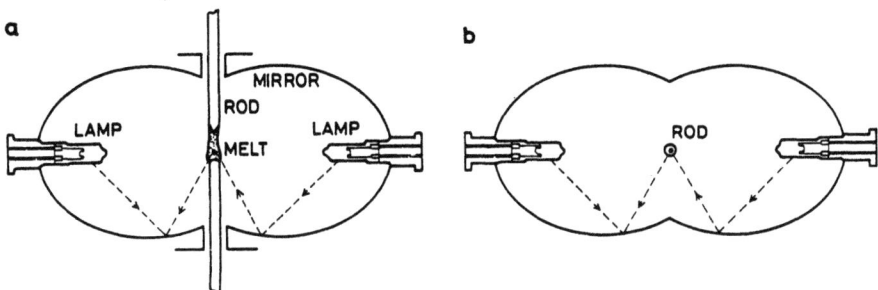

Fig.15. Double ellipsoidal mirror furnace used in the crystal growth of silicon rods aboard Spacelab-1. **a)** side view, **b)** plant view.

<u>Zone refining</u>

Although the main interest of recent floating zone work is on growing single crystals, its first application [10] was in the purification of germanium for the (at that time) emerging solid-state industry. The floating-zone technique was developed from the well-known molten-zone technique, where the melt was supported on a crucible instead of by surface tension forces. It has been applied to low melting point materials (organic solids with $T_f \sim 300°C$), medium temperatures (metals of $600 < T_f < 1000°C$) and high melting point materials (some metals, its oxides and some other inorganic substances with $1000 < T_f < 2000°C$). When the molten-zone technique was used, the crucible was a borosilicate glass or fused silica or platinum, according to the temperature ranges above mentioned. Sometimes the crucible was open to avoid problems of expansion and to have better access (to take out impurities).

The great advantage of crucible-free processing by the floating-zone technique is to get rid of thermal and solutal contamination, obtaining the purest man-made materials, with less than 1 ppm of impurities, that are needed not only for research but in the everdemanding microelectronics industry of today. Unfortunately, because of a large energy consumption and a small production rate, the cost of materials processed with this technique (less than 1 ppm impurities) rises to the order od 200 to 2000 $/kg.

To establish a simple model of the zone refining process, the following simplifications are introduced:

-One-dimensional geometry (Fig. 16) instead of the real shape. All the important effects of shape stability are thus forgotten.

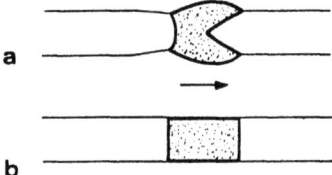

Fig. 16. Meridian section of a real floating zone, **a)**, and the model, **b)**.

-Negligible velocities for all interface fronts and internal motion in the liquid (due to Marangoni effects, diffusion, shrinking, etc).

Some other simplifications will be introduced later in the analysis, but let us see straight-on why impurities tend to segregate when the floating zone is made to travel.

A typical phase-equilibrium diagram for a roughly pure substance is shown in Fig. 17a, with the important example for silicon-aluminium in Fig. 17b. At constant pressure (the value is irrelevant for condense phases) the crucial point is that the solidus (SS' start of fusion) and liquidus (LL' start of solidification) lines are distinct.

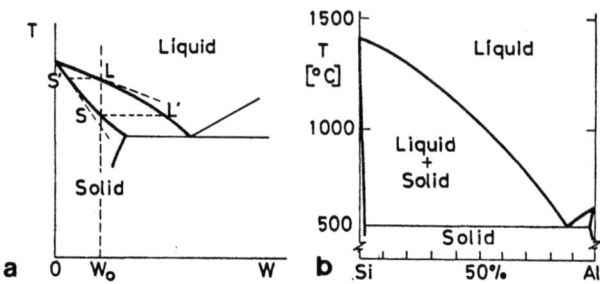

Fig.17. **a)** Typical phase diagram for a binary mixture of high purity: (SS' solidus line, LL' liquidus line). **b)** Phase diagram for silicon-aluminium.

There are certain substances for which the slopes SS' and LL' are positive (as for phosphor in gallium arsenide), but the segregation due to the different affinity of every component in the different phase takes place anyway. The extreme case where the solidus line is vertical (the solid phase does not dissolve impurities, as in salt water) is a trivial case without interest here. Thus, when a liquid mixture with an impurity concentration W_o starts to solidify (point L) the solid formed has a different concentration (point S'). Normally the solid phase is more pure and the remaining liquid (holding most of the impurities) is discarded. The model for zone refining is as follows: The initial condition is a solid rod of length Z_N with an impurity concentration profile $W_{s2}(Z)$, then a length (from the left) $Z=L(0)$ is melt, and afterwards the molten zone is made to travel to the right (forced by a relative movement of rod and heating device). At a generic instant we will have resolidified material from 0 to Z_1, the melt from Z_1 to $Z_1+L(Z_1)$, and unprocessed material as left by the previous pass (Fig. 18).

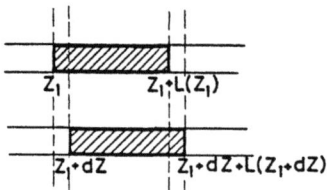

Fig. 18. Two successive instants in a zone refining process.

The following simplifications are introduced now:
    1) Diffusivities inside the solid are neglected and thus composition cannot vary with time (except when resolidified)

2) Diffusivities inside the liquid are large, thus, in a first approximation liquid composition is uniform
3) Thermodynamic equilibrium conditions apply to the solidification front Z_1 (and not to the melting front). That is, liquid and solid concentration of impurities are as in L and S' (Fig. 17a). For low impurity processing lines SS' and LL' may be approximated by their tangents at the origin W=0, thus

$$W_{s1}(Z_1) = k \cdot W_L(Z_1) \tag{18}$$

Where k is call the equilibrium segregation constant. A mass balance for impurities yields (see Fig. 3):

$$[W_L(Z_1)-W_{s1}(Z_1)]dZ + [W_L(Z_1+dZ)-W_L(Z_1)](L(Z_1)-dZ) + \int_{Z_1+L(Z_1)}^{Z_1+dZ+L(Z_1+dZ)}[W_{s2}(Z)-W_L(Z_1)]dZ = 0 \tag{19}$$

What is wanted is to know the final concentration profile $W_{s1}(Z_1)$, that depends on the initial concentration profile $W_{s2}(Z_1)$, the molten zone law $L(Z_1)$ and the segregation constant k.

Equation (19) has analytical solutions in certain cases, as when W_{s2}=constant. If, in addition, $L(Z_1)=Z_N-Z_1$ (all the rest of the rod is molten) the solution is

$$\frac{W_{s1}(Z_N)}{W_{s2}} = k\left(\frac{Z_N}{Z_N-Z_1}\right)^{1-k} \tag{20}$$

The above case serves to model the final solidification of a molten zone when the heating is removed (Fig. 19a). However, it presents a singular point at $Z_1=Z_N$, showing that when the molten length shrinks to zero its impurity contents diverges because the solid takes less than the liquid offers (the model assumes small concentrations and ceases to be valid).

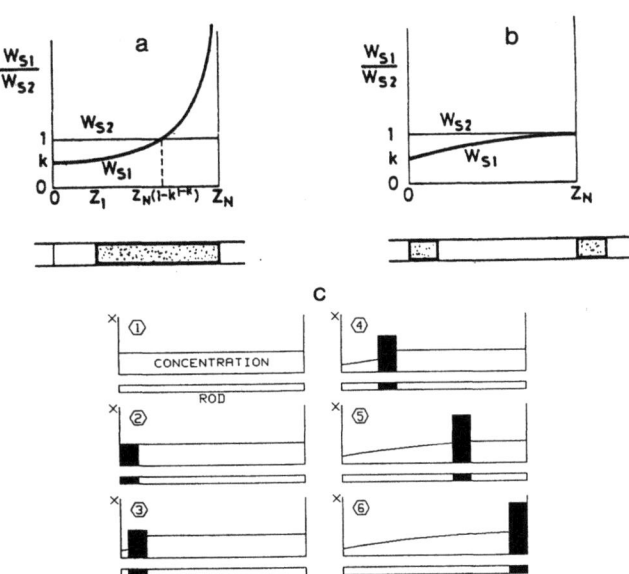

Fig. 19. Impurity concentration profile in a rod with an uniform initial concentration, after one processing pass. a) the full rod is molten and unidirectionally solidified, b) a constant length is molten and made to travel to the right, c) different time stages for case b).

Another interesting analytical solution with W_{s2}=constant is obtained when $L(Z_1)$=constant also (see Fig. 19b), yielding

The last solution serves to model the quasi-steady-state of the traveling zone. Note that, to represent a complete pass, this model must be combined with the former, at the right end, but if this part of the rod is to be discarded, the solution would be as in Fig. 19b.

In a more general case of $W_{s2}(Z)$, as for the second and following passes, recurs to numerical computations is necessary. From computer simulation with the above models, the following results can be obtained:

- The longest the molten zone, the best (limited in practice by shape instability)
- The initial rod should be as long as possible (at least 10 times the molten length) to avoid that the accumulated impurities at the end re-disolve
- For the above reason, every pass should be slightly shorter than the previous one
- Also, the length of the molten zone should be reduced from pass to pass
- The purest material lies at the starting end
- The efficiency of passes decreases because the adverse concentration gradient forces some re-mixing
- After so many passes as molten-zone lengths has the initial rod (roughly) there is little benefit in re-processing.

Although all kinetic effects have been neglected, it implies that the pulling law for the molten zone has to be very small in practice (typically ~1 mm/min). If not, diffusion will be unable to homogenize the liquid phase (an effective segregation coefficient velocity-dependent could still be used) and finally the solidification front would destabilize and dendritic structures would develop.

Floating zone growth in Spacelab-1

The analysis presented here concerns the growth of two silicon crystals aboard Spacelab-1 (1983) in a so called Mirror Heating Facility [11], sketched in Fig. 15, developed explicitly for zone growth experiments in Spacelab-1, and tested on earth by numerous experiments yielding high quality crystals of different electronic materials (e.g. Si, CdTe, GaSb, InP) [12-14]. The radiation of two 400 W halogen lamps is focused on the sample by two adjoint ellipsoidal mirrors. Since the zone is heated by incoherent light, additional forces such us those existing for instance in RF heating are avoided. Thus, this method is specially suited to investigate the behaviour of free liquid bridges.

The furnace is moved over the sample, thus making the molten zone to travel through prefabricated silicon rods from bottom to top (Fig. 20). The special shape shown was chosen to achieve dislocation-free growth by starting the crystallization in the lower part of the thin neck. A detailed performance description is given in [15]. Figure 21 shows crystals grown using that furnace. In such radiation heated silicon growth experiments, the solid/liquid interface at the feed rod (upper interface in Fig. 20) tends to be cone-shaped, whereas the growing interface (bottom) is slightly convex [12-16].

Touching of the apex of the cone with the growing crystal has to be avoided in order to maintain monocrystalline structure. Even worse, if both solid ends are being counterrotated, the touching may cause mechanical problems with the drives, and the melt may sputter over the furnace walls. That means that radiation power has to be high enough to exceed a minimal external length of the molten zone. But, on the other hand, an upper limit of this length also has to be attended in order to prevent liquid bridge disruption by capillary forces.

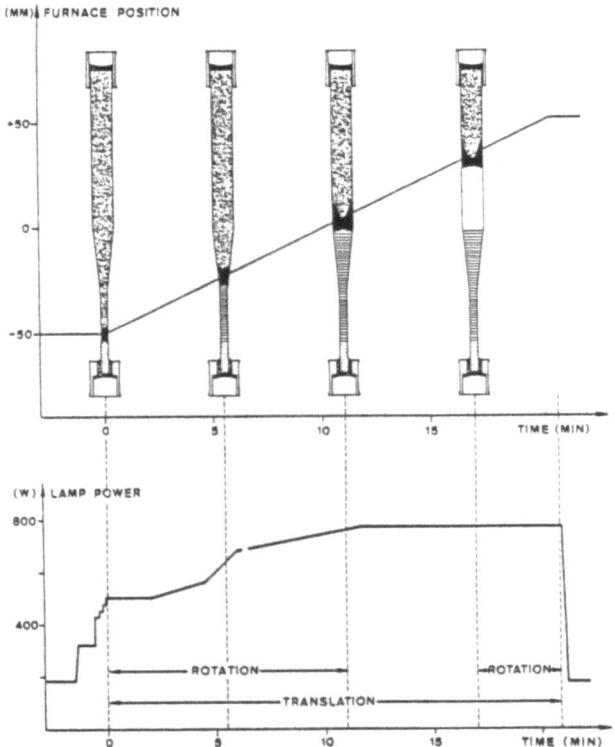

Fig. 20. Pulling and heating laws for Exp. 1-ES-321 aboard Spacelab-1.

There is a second critical problem aggravating floating zone growth under microgravity conditions. Surek and Chalmers [17] have shown that in silicon the angle between the meniscus of the melt and the growth direction (for constant cross-section growth) has to be $11°$, i.e. the molten zone should bulge slightly outward of the growing interface. On the earth, the interaction between gravity and surface tension always cause the melt to broaden at the lower end so that the interface adopts an S-shape (see Fig. 22).

In space the situation is different. After establishing the liquid zone, its volume is, due to an 8% decrease in volume of Si upon melting, too small to form a cylindrical bridge, let alone an outward bulging surface. As a consequence, the diameter of the growing crystal will initially decrease until the melt attains a meniscus angle of $11°$ (see Fig. 23). This diameter reduction tends to be less pronounced and would eventually recover, since in steady state growth the amount of material melting must equal the amount crystallizing.

But this transient decrease in diameter may affect the stability of the zone. The margin of stable zone lengths thus becomes even smaller in microgravity than on earth, and a very skillful experimentation is required to steer clear from zone disruption on the one hand and of solid/liquid interfaces touching on the other hand. This problem may be overcome in future by bringing the two solid ends (after a molten bridge is established) closer.

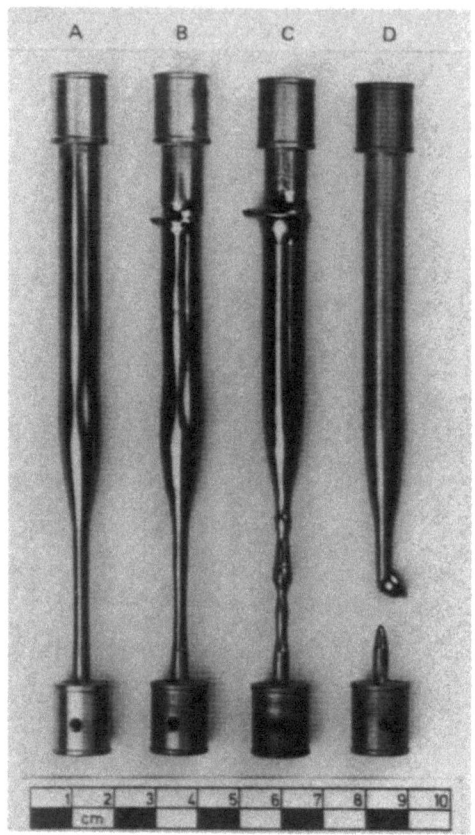

Fig. 21. Silicon rods: **A)** unprocessed, **B)** reference processed on ground, **C)** sample successfully processed in flight, **D)** sample processed in flight until bridge disruption occurred.

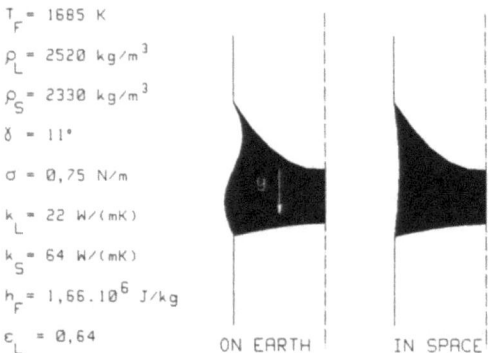

Fig. 22. Physical constants of silicon, and sketch of molten shapes (meridian section) when pulled on Earth and in space.

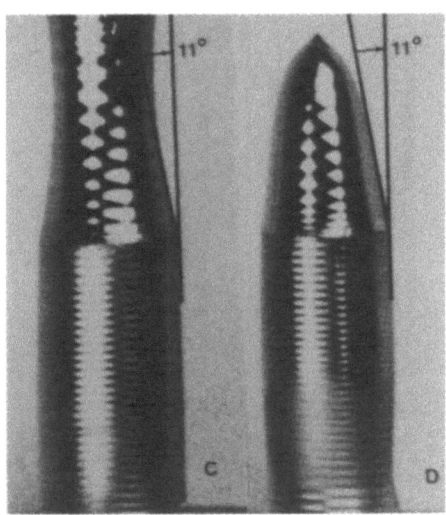

Fig. 23. Magnified view of the neck region for the two crystals processed in flight (Fig. 3C and 3D) showing the growth angle at start.

Rotation of the end supports is normally imposed to uniformize the heating. For instance, the furnace used here produced a two-lobed non-axisymmetrical radiation pattern. Fortunately, the rotation rates in use were so low that the influence on outer shape and stability limit seems negligible. Most of the sample, however, was grown without rotation, suffering the asymmetry of the radiation pattern.

Analysis of floating zone growth with the liquid bridge model

Two floating zone experiments were initiated during the Spacelab-1 mission. The first one took its scheduled course: a molten zone could be established in the neck and made to travel all the way to the end of the sample (Fig. 21C). There was video recording aboard and, during most of the time, TV transmission to the ground. The sequence of outer shapes of the melt was gathered from the video tapes and correlated with the crystals grown from these melts (Fig. 24). The irregularly shaped neck is the result of the difficulties encountered in adjusting by hand the proper zone length because of the meniscus problem mentioned above.

The second run had to be terminated shortly after starting of the zone travel because of rupture of the molten bridge (Fig. 22D shows the two pieces of that sample). Apparently, the zone length had become too large and the crystal diameter had decreased to such a degree that disruption near the lower interface occurred. The same occurrence had already taken place during a similar experiment in a sounding rocket [18].

A detailed view of the necks of the samples processed on board is presented in Fig. 23. The rotation rims appearing on the surface show that the samples were being rotated during growth. Crystal perfection evaluation, as well as a comparison of the earth grown and space grown crystals, concerning crystal structure, defect concentration and dopant distribution are reported elsewhere [15-16].

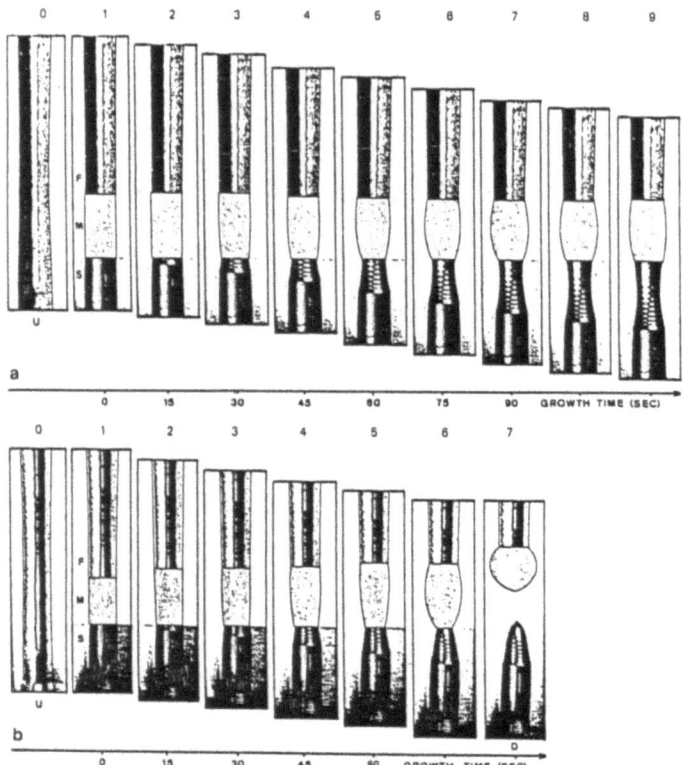

Fig. 24. Sequence of molten zones (depicted from video pictures). a) successful growth (Fig. 3C), b) bridge disruption in the second trial. U unmolten crystal, S seed crystal, M melt, F feed crystal. Note that shape D in b) immediately follows its previous picture; all others are 15 seconds apart.

What information can be gained by the experimenter with floating zones from the fluid-mechanical theory of liquid bridges? For instance, were the molten bridges equilibrium capillary menisci?, how far were they from capillary instability?. In all, what lessons from liquid bridge analysis can be advantageously used in crystal growth?

The fluid-mechanical behaviour of silicon floating zones can easily be explained by the simple liquid bridge theory of Laplace [19-26], provided the following simplifications are introduced:

    a) Axisymmetric geometry.
    b) Quasistatic configurations.
    c) Constant material properties.
    d) Negligible gravity, g-jitter and centrifugal effects.
    e) Deep melting (molten core).
    f) Free surface anchorage at both ends (seems to be quite realistic).

Under these hypothesis, the questions posed above may be answered as follows. Molten free surfaces at equilibrium should be part of a constant mean curvature surface (Plateau surface). They can be determined from three nondimensional parameters as, for instance, end diameter ratio H (see Fig. 25 for nomenclature), length divided by mean diameter Λ, and straight volume V (or neck to bulging ratio, what is easier to measure).

Although an explicit expression may be developed for small deviations from the cylinder, complicated graphical interpolation or involved computing are required in a general case. It should be noted, in any case, that in order to be able to detect deviations from equilibrium in real situations with small bridges (some millimeters in diameter), an accuracy of microns would be required.

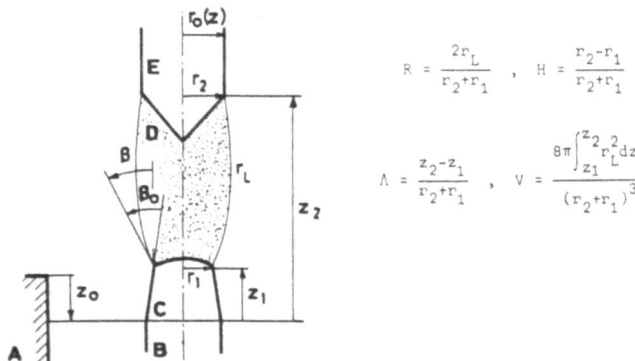

$$R = \frac{2r_L}{r_2+r_1} \quad , \quad H = \frac{r_2-r_1}{r_2+r_1}$$

$$\Lambda = \frac{z_2-z_1}{r_2+r_1} \quad , \quad V = \frac{8\pi \int_{z_1}^{z_2} r_L^2 dz}{(r_2+r_1)^3}$$

Fig. 25. Geometry and nomenclature used in the liquid bridge modeling of the floating zone growth. A furnace reference, B seed, C crystal grown, D melt, E feeding rod. All dimensions are scaled with seed radius $r_0(0)$, and the nondimensional parameters introduced are shown.

Once the three nondimensional parameters that determine the shape are measured, it is easy to look on the stability diagram for liquid bridges anchored to unequal discs [23] and see (Fig. 26) how far the stability limit is, with respect to the parameters mentioned above. However, from a

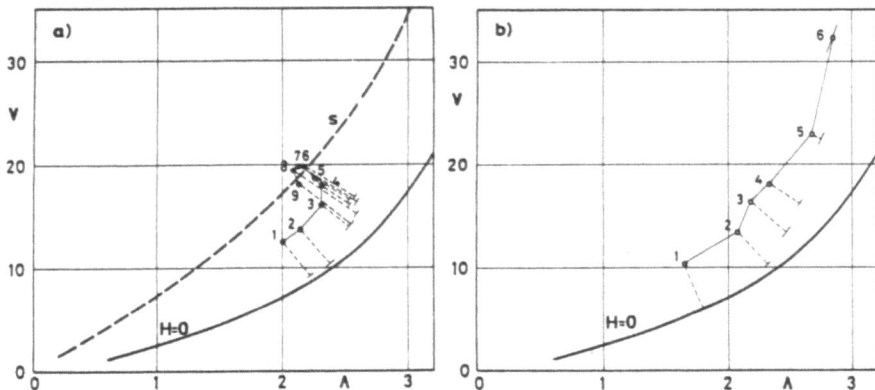

Fig. 26. Capillary stability diagram for a molten bridge. The points correspond to shapes in Fig. 5. Dashed lines indicate the stability margin (minimum distance to the stability boundary in the slenderness-volume diagram).
  a) Stable growth. It is seen that after shapes 3 and 4 the zone gets away from the unstable limit, approaching a point in curve s, which is the loci of cylindrical growth configurations for silicon (a barrel-shaped molten zone with 11° of bulging that moves along a cylindrical rod).
  b) Growth with bridge disruption. It is seen how the last shape in Fig. 21b was unstable (because of the smallness of the sample, 3.8 mm of diameter, the breaking is instantaneous).

practical point of view, what matters is the stability limits with respect to the directly controlled inputs: furnace power and pulling laws, which is a much more involved task.

The main lesson from liquid bridge theory is that the sequence of molten shapes during growth can be predicted as function of the position of the fronts and, if the small difference in density is accounted for, the volume of immersed solid tips. Thus, if a thermal analysis of furnace and rod could model these three parameter as functions of the heating and pulling laws, the behaviour of the floating zone should be entirely predictable and, consequently programed to a better advantage.

The geometry used is sketched in Fig. 25. All dimensions are scaled with the radius of the first resolidified rod section. If, for the sake of clarity, the difference between the density of liquid and solid silicon (the latter 8% larger) is neglected, the sequence of molten shapes during growth $r_L(z,t)$ is determined by knowing the initial rod shape $r_o(z)$ and the movement of the fronts: $z_1(t)$ and $z_2(t)$.

The procedure is as follows. From initial conditions, the liquid free surface is known. From that, resolidification should proceed with a constant receding angle of $\beta_o=11°$ from the liquid free surface, thus, diminishing the solidification radius and displacing some volume v_{dis} from the initial rod towards the molten bridge (ahead of the solidification front). The liquid bridge then tends to get a barrel shape and thus, if it does not gets destabilized in the way, the angle at the border will increase restoring the solidification front radius to that of the original rod. In mathematical form:

$$dz_1/dt = f_1(t) \tag{22}$$
$$dz_2/dt = f_2(t) \tag{23}$$
$$dr_1/dt = (dz_1/dt)\cdot\tan(\beta-\beta_o) \tag{24}$$
$$dv_{dis}/dt = \pi[r_o^2(z_1)-r_1^2]\cdot(dz_1/dt) \tag{25}$$

that can easily be integrated with a standard Runge-Kutta routine, once the initial conditions $z_1=0$, $z_2=z_2(0)$, $r_1=r_1(0)=r_o(0)$, $v_{dis}=0$ are fixed, and β is expressed as a function of the variables, as explained below.

To this aim, a further simplification is now introduced: the liquid bridge shape is assume not to deviate too much from a cylinder so that the handy linear approximation:

$$R^2(Z,H,\Lambda,V) = 1 + \frac{V-2\pi\Lambda}{2\pi\Lambda}\frac{\cos Z - \cos\Lambda}{\frac{\sin\Lambda}{\Lambda}-\cos\Lambda} + 2H\frac{\sin Z}{\sin\Lambda} + H^2\frac{\frac{\sin\Lambda}{\Lambda}-\cos Z}{\frac{\sin\Lambda}{\Lambda}-\cos\Lambda} \tag{26}$$

is valid. Here, a local scale is used for the shape such that it goes from $-\Lambda$ to $+\Lambda$, with $R(-\Lambda)=1-H$, $R(+\Lambda)=1+H$, Λ being the slenderness and $V=\int \pi R^2 dz$ the straight volume between ends (including immersed solid tips). These local parameters are related to the growth ones by

$$H = (r_o(z_2)-r_1)/(r_o(z_2)+r_1) \tag{27}$$
$$\Lambda = (z_2-z_1)/(r_o(z_2)+r_1) \tag{28}$$
$$V = 8\cdot(v_{dis} + v_o(z_2) - v_o(z_1))/(r_o(z_2)+r_1)^3 \tag{29}$$

where $v_o(z)$ is the volume of the initial rod $r_o(z)$ from 0 to z. Furthermore, $\beta_o=11°$ and, within the linear approximation above, β is directly given by

$$\beta = \arctan \frac{\frac{2H}{\tan\Lambda} + \frac{\frac{V}{2\pi\Lambda}-1-H^2}{1/\Lambda - 1/\tan\Lambda}}{2(1-H)} \tag{30}$$

For a given furnace and initial rod, the value $z_2(t=0)$ and the two functions $f_1(t)$ and $f_2(t)$ only depend on the heating and pulling laws imposed. Further work will be needed to build this thermal model, but the simple procedure detailed above allows one to gain much insight in the behaviour of floating zones processing. To that aim, simple trials can be run and the effects observed. Figure 27 presents the results of such a trial: the distance between solidification and melting fronts has been assumed constant and its influence on the stability of growing studied (similarly, other laws such as maintaining the molten volume, or the free surface area, etc., can be exercised). The same approach here applied to floating zones [27] has also been successfuly applied to the solidification of molten drops [28].

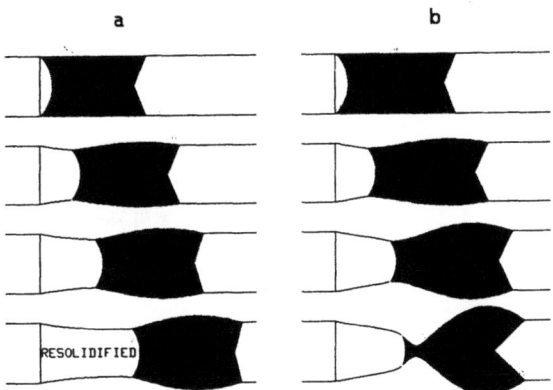

Fig. 27. Example of predicted evolution of the solidification radius, r_L, with distance from the seed, z, during floating zone processing of an initially cylindrical silicon rod. In this example, the external length of the molten bridge, z_2-z_1, is assumed constant with time. Note that if $z_2-z_1=3$ the crystal grows trying to recover the initial shape, whereas if $z_2-z_1=4$ the growing gets unstable and the bridge breaks.

The general procedure is implemented in a computer program that works as follows:

1. Read data for material properties (solidification angle, liquid and solid density, melting temperature, specific heats, enthalpy of fusion, thermal conductivities, etc).
2. Read initial rod data (meridian section profile for an axisymmetric rod).
3. Read experiment data (pulling and heating laws as functions of time t).
4. Compute (loop) molten bridge base parameters (position of solid/melt fronts and solid volume immersed inside the stright volume, as functions of time) either from real experiments or from a thermal model (still lacking).
5. Compute liquid bridge base parameters $H(t)$, $\Lambda(t)$ and $V(t)$ from their definitions (Fig. 25).
6. Compute liquid bridge shape from base parameters by using a liquid bridge model (as Eq. 26).
7. Show predicted molten-zone evolution for the data given. Both, a rod-fixed reference frame or a furnace-fixed one may be displayed.
8. If real experiment sequence is available, compare molten shapes with predictions. In fact, this is the case here, and Fig. 28 shows the comparison for shape 6 in Fig. 24b.

The above procedure may be used advantageously to study receding angles in other similar configurations as the one shown in Fig. 29.

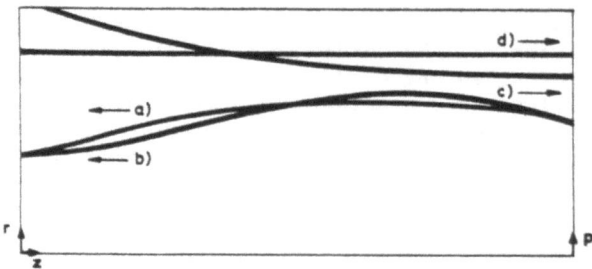

Fig. 28 Experimental **a)** and theoretically predicted **b)** shapes corresponding to stage 6 in Fig. 24b. Melt shape (after a video recording of the actual growth) appears more flatten than predicted. Its pressure profile **c)** should be uniform as for the theoretical shape **d** (assuming constant surface tension) and it is clearly not.

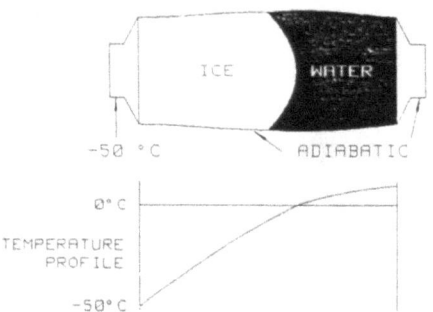

Fig. 29. An experiment proposal to study the receding angle and movement near the three-phase-line in a water column configuration.

Conclusions

A model is presented that explains the observed behaviour of molten bridges when a silicon rod is processed by the floating-zone technique under microgravity conditions. The analysis is focused in molten shape evolution, and assumes constant surface tension, no force field applied, and a given advancing angle at the solidification edge. The main results may be summarized as follows:

1. The stable evolution that occurred in the first trial of Exp. 321 aboard Spacelab is accurately predicted by the model (Fig. 26a).
2. The unstable evolution that occurred in the second trial of Exp. 321 is accounted for and accurately predicted by the model (Fig. 26b).
3. Uncertainty in shape discrimination, due to smallness of the sample, prevents a finer analysis of disturbances upon the equilibrium configurations.
4. Real bridges appear to be more flattened than expected (Fig. 28). Sizable departures of a constant pressure profile from observed shapes (pressure diminish towards the neck) need further explanation (maybe nonuniform surface tension due to temperature gradients).
5. From the history described above, the sequence of shapes may be computed, thus, the next step should be to model the thermal problem applicable to a particular type of furnace, to be able to predict the sequence of shapes directly from the heating and pulling laws.

Acknowledgement: This work is being sponsored by the Spanish Commission on Space Research.

REFERENCES

1. J. MESEGUER, L.A. MAYO, J.C. LLORENTE and A. FERNANDEZ, J. Crystal Growth 73, p 609, 1985.

2. I. MARTINEZ, ESA SP-222, p 31, 1984.

3. I. MARTINEZ and A. SANZ, ESA Journal Vol. 9, p 323, 1985.

4. I. DA RIVA and I. MARTINEZ, Naturwissenschaften 73, p 343, 1986.

5. J. MESEGUER, A. SANZ and J. LOPEZ, J. Crystal Growth (in press), 1986.

6. I. MARTINEZ and J. MESEGUER, Norderney Symposium on Spacelab-D1 results, DFVLR (FRG) (in press), 1986.

7. I. MARTINEZ, IAF-86-272 (to be published in Acta Astronautica), 1986.

8. D. LANGEBEIN, in Materials Sciences in Space, Springer-Verlag, 1986.

9. J. MESEGUER, J. Fluid Mech. 130, p 123, 1983.

10. W.G. PFANN, J. Metals 4, 1952.

11. A. EYER, R. NITSHE and H. ZIMMERMANN, J. Crystal Growth 47, p 219, 1979.

12. A. EYER, B.D. KOLBESEN and R. NITSHE, J. Crystal Growth 57, p 145, 1982

13. R. SCHONHOLZ, R. DIAN and R. NITSHE, J. Crystal Growth 72, p 72, 1985.

14. G. NAGEL and K.W. BENZ, Adv. in Space Research 4.5, p 23, 1984.

15. A. EYER, H. LEISTE and R. NITSHE, ESA SP-222, p 173, 1984.

16. A. EYER, H. LEISTE and R. NITSHE, J. Crystal Growth (submitted).

17. T. SUREK and B. CHALMERS, J. Crystal Growth 29, p 1, 1975.

18. A. EYER, H. LEISTE and R. NITSHE, J. Crystal Growth 71, p 173, 1985.

19. I. MARTINEZ, COSPAR Space Research XVIII, p 519, 1978.

20. I. DA RIVA and I. MARTINEZ, ESA SP-142, p 67, 1979.

21. I. MARTINEZ and D. RIVAS, Acta Astronautica 9, p 339, 1982.

22. A. SANZ and I. MARTINEZ, J. Colloid Interf. Sci. 93, p 235, 1983

23. I. MARTINEZ, ESA SP-191, p 267, 1983.

24. J. MESEGUER, J. Crystal Growth 67, p 141, 1984.

25. J. MESEGUER, ESA SP-222, p 297, 1984.

26. I. MARTINEZ and J.M. PERALES, J. Crystal Growth (in press), 1986.

27. I. MARTINEZ and A. EYER, J. Crystal Growth 75, p535, 1986.

28. A. SANZ, J. Crystal Growth 74, p 642, 1986.

EQUILIBRIUM SHAPES AND FREE VIBRATIONS OF LIQUID CAPTIVE DROPS

A. Gañán and A. Barrero

Escuela Técnica Superior de Ingenieros Industriales
Universidad de Sevilla, Sevilla 41012, Spain

INTRODUCTION

The equilibrium shape and small free oscillations of the liquid meniscus have been the subject of many theoretical and experimental studies. Renewed interest on this kind of problems has arisen from the possibility of producing single crystals of high quality and refining molten materials. Experiments on crystal growth have already been performed during Skylab and Spacelab missions.

The knowledge of the mechanical behaviour of the molten zone is of interest in order to obtain pure cristals of high quality. Free vibrations of liquid drops have been extensively investigated in the case of isolated spheres[1,2]. The effect of rotation on the natural frequencies of the oscillations of a liquid drop has been recently considered by Busse[3].

Vibrations of a liquid drop in partial contact with solid supports, assuming zero gravity and no rotation, small interface deformations and negligible viscous effects, have been studied for both cylindrical bridges and drops [4,5].

Here, following a method partially inspired in previous works[4,5], we analize the free vibrations of liquid captive drops (bridges or drops) or bubbles surrounded by an outer inmiscible liquid or gas, under the influence of both gravitational and isorotational fields. To solve the problem, we apply a generalized integral transformation to a modified Jacobi equation for the normal modes and arrive at a system of infinite algebric equations which may be truncated giving rise to an eingenvalue problem which yields the normal modes and natural frequencies ω_n.

For $\omega_n = 0$, which corresponds to the bifurcation case, our results reproduce the stability limits for cylindrical bridges and drops previously given by Rayleigh[2], Pitts[6], and Michael and Williams[7].

FORMULATION OF THE PROBLEM

We have considered the small amplitude free vibrations of a liquid captive drop or bubble (pendent, sessile or liquid bridge) surrounded by an outer liquid or gas under the influence of both gravitational and iso-rotational fields; a sketch of the considered problems are given in Fig.1. In a co-rotating system of reference the Navier-Stokes equations for the inviscid motion of both inner and outer incompressible fluids read

$$\nabla \cdot \vec{v}^j = 0, \tag{1}$$

$$\frac{\partial \vec{v}^j}{\partial t} + \vec{v}^j \cdot \nabla \vec{v}^j = -\frac{1}{\rho^j} \nabla p^j + \frac{1}{2} \nabla (\vec{\Omega} \wedge \vec{x})^2 - 2\vec{\Omega} \wedge \vec{v}^j, \tag{2}$$

where superscript $j \equiv o, i$ refers to outer or inner fluids respectively; for convenience we will use cylindrical coordinates $\vec{x}(r, \phi, z)$ and the angular velocity $\vec{\Omega}$ will be taken along the z axis. Equations (1) and (2) have been written in non-dimensional form using a characteristic length L that is either the capillary length $l_c = |\sigma/g(\tilde{\rho}^i - \tilde{\rho}^o)|^{1/2}$ for liquid drops, or the bottom radius R_o for bridges; in adittion we have used as characteristic time $|(\tilde{\rho}^i - \tilde{\rho}^o) L^3/\sigma|^{1/2}$ and $(\tilde{\rho}^i - \tilde{\rho}^o)$ as reference density.

The interface position $f(\vec{x},t) \equiv F(\phi,z,t) - r = 0$ satifies the equation

$$\frac{\partial f}{\partial t} + \vec{v} \cdot \nabla f = 0, \tag{3}$$

and the pressure jump accross the interface is balanced by the surface tension

$$p^i - p^o = \nabla \cdot \vec{n}, \tag{4}$$

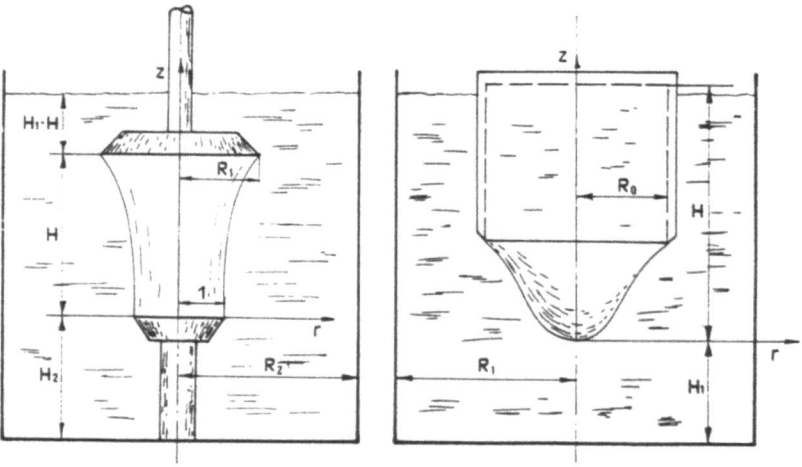

Fig. 1. Liquid bridge and drop configurations. All lengths have been non-dimensionalized using a characteristic length L that is either the base radius R_o for bridges or the capillary length $|\sigma/g(\tilde{\rho}^i - \tilde{\rho}^o)|^{1/2}$ for drops.

where $\vec{n}$ is the unit normal on the interface. Equations (1)-(4) must be solved subject to the following conditions:

a) The normal component of the velocity in both fluids must 1) vanishes at the solid surfaces, and 2) be equal at the interface;

b) The liquid volume is held constant

$$V = \int_0^{2\pi} \int_0^H F^2(\phi,z,t)\, dz\, d\phi. \tag{5}$$

LINEAR ANALYSIS

In the following, we will study small oscillations around axisymmetric equilibrium shapes and look for a linearized solution to the considered problem (1)-(5) in the form:

$$F(\phi,z,t) = F_e(z) + \sum_{m=0}^{\infty} \sum_{n=1}^{\infty} \eta_n(z) \exp[i(\omega_n t + m\phi)], \tag{6}$$

$$p^j(r,\phi,z,t) = p_e^j(r,z) + \sum_{m=0}^{\infty} \sum_{n=1}^{\infty} \rho^j \phi^j(r,z) \exp[i(\omega_n t + m\phi)], \tag{7}$$

$$\begin{Bmatrix} v_r^j(r,\phi,z,t) \\ v_\phi^j(r,\phi,z,t) \\ v_z^j(r,\phi,z,t) \end{Bmatrix} = 0 + \sum_{m=0}^{\infty} \sum_{n=1}^{\infty} \begin{Bmatrix} u^j(r,z) \\ v^j(r,z) \\ w^j(r,z) \end{Bmatrix} \exp[i(\omega_n t + m\phi)], \tag{8}$$

where F is the interface radius and subscript e refers to the equilibrium conditions; clearly, the perturbation quantities $|\eta_n|$ and ϕ^j are much smaller than F_e and $|p_e^j|$ respectively.

Introducing expansions (6)-(8) into equations (1)-(4) one arrives at the lowest order (equilibrium) problem:

$$\Omega^2 F_e^2/2 + (\pi^i - \pi^o) - Bz = F_{ezz}(1 + F_{ez}^2)^{-3/2} - F_e(1 + F_{ez}^2)^{-1/2}, \tag{9}$$

together with the equilibrium pressure field in both fluids

$$p_e^i = \pi^i + \rho^i F_e^2 \Omega^2/2 - B \rho^i z, \tag{10}$$

$$p_e^o = \pi^o + \rho^o F_e^2 \Omega^2/2 - B \rho^o z, \tag{11}$$

Bond number B is 1 for drops ($L = l_c$) or $[g R_0^2 (\tilde{\rho}^i - \tilde{\rho}^o)/\sigma]$ for bridges ($L = R_0$). Equation (9) must be solved subject to the following boundary conditions

$$F_e(0) = 0,\ F_e(H) = R_0 \quad \text{for liquid drops}, \tag{12}$$

55

$$F_e(0) = 1, \quad F_e(H) = R \quad \text{for liquid bridges};\tag{13}$$

furthermore, condition (5) reads to the lowest order

$$V = 2\pi \int_0^H F_e^2(z)\, dz.\tag{14}$$

Notice that three conditions are needed since the eingenvalue $(\pi^1 - \pi^0)$ is unknown.

The equation for the interface radial disturbance η_n (dynamical problem) becomes

$$F_e^{-1}\left[F_e(1+F_{ez}^2)^{-3/2}\,\eta_{nz}\right]_z + \eta_n(1-m^2)\,F_e^{-1}(1+F_{ez}^2)^{-1/2} =$$
$$\left[\rho^i \phi^i - \rho^o \phi^o\right]_{r=F_e(z)} - \Omega^2 F_e\,\eta_n = 0.\tag{15}$$

The pressure fields for both fluids are given by the equations[8]

$$\nabla^2 \phi^j = \left(\frac{2\Omega}{\omega_n}\right)^2 \phi^j_{zz},\tag{16}$$

the velocity fields satisfying momentum Navier-Stokes equations read

$$u^j = \frac{i(\omega_n \phi_r + 2m\Omega\phi/r)}{\omega_n^2 - 4\Omega^2},\tag{17a}$$

$$v^j = \frac{2\Omega\phi_r + m\omega_n\phi/r}{\omega_n^2 - 4\Omega^2},\tag{17b}$$

$$w^j = i\,\phi_z/\omega_n.\tag{17c}$$

Equations (15) and (16) must be solved with the conditions

$$\eta_n(0) = 0, \quad \eta_n(H) = 0,\tag{18}$$

$$u^j = 0 \text{ on solid surfaces such as } r = \text{const},\tag{19a}$$

and/or

$$w^j = 0 \text{ on solid surfaces such as } z = \text{const};\tag{19b}$$

condition (3) reads

$$i \omega_n \eta_n = \left[-u^i + v^i \frac{F_{e\phi}}{F_e} + w^i F_{ez}\right]_{r=F_e(z)} = \left[-u^o + v^o \frac{F_{e\phi}}{F_e} + w^o F_{ez}\right]_{r=F_e(z)} \quad (20)$$

Finally condition (5) for the volume disturbance becomes

$$\int_0^H \eta_n F_e \, dz = 0. \quad (21)$$

Equations (9)–(14) may be numerically integrated to obtain the axysimmetric equilibrium shape $F_e(z)$. Solutions of Equation (16) satisfying conditions (19) may be written as

$$\rho^j \phi^j = P_o^j + \rho^j \omega_n^2 \sum_{k=0}^{\infty} A_k^h R^j (\lambda_k^i r) Z^j (\mu_k^j z), \quad (22)$$

notice that A_k^j, ω_n and P_o^j are unknown and must be obtained from the analysis; functionals R and Z have different forms depending on the considered problem:

Liquid Bridges

$$\left. \begin{aligned} R^i(x) &= I_m(x), \quad Z^j = \cos x \\ R^o(x) &= I_m(x) - \frac{I_m(\lambda_k^o R_2)}{K_m(\lambda_k^o R_2)} K_m(x) \end{aligned} \right\} \text{ for } \omega_n^2 > 4\Omega^2, \quad (23)$$

and

$$\left. \begin{aligned} R^i(x) &= J_m(x), \quad Z^j = \cos x \\ R^o(x) &= J_m(x) - \frac{J_m(\lambda_k^o R_2)}{Y_m(\lambda_k^o R_2)} Y_m(x) \end{aligned} \right\} \text{ for } \omega_n^2 < 4\Omega^2, \quad (24)$$

Drops

$$\left. \begin{aligned} Z^i &= \exp(-\mu_k^i z) - \exp\left[\mu_k^i (z - 2H)\right] \\ Z^o &= \exp(\mu_k^o z) - \exp\left[-\mu_k^o (z - 2H_1)\right] \end{aligned} \right\} \text{ for } \omega_n^2 > 4\Omega^2, \quad (25)$$

$$R^j(x) = J_m(x)$$

$$z^i = \cos \mu_k^i z - \frac{\sin \mu_k^i H}{\cos \mu_k^i H} \sin \mu_k^i z \qquad \qquad (26)$$

$$z^o = \cos \mu_k^o z - \frac{\sin \mu_k^o H_1}{\cos \mu_k^o H_1} \sin \mu_k^o z$$

for $\omega_n^2 < 4\Omega^2$,

where J_m, Y_m, I_m and K_m are the first and second kind of Bessel and modified Bessel functions respectively; λ_k^j are related to μ_k^j by the expression

$$\lambda_k^j = \mu_k^j \left[1 - 4\Omega^2/\omega_n^2\right]^{1/2} . \qquad (27)$$

For bridges one finds

$$\mu_k^i = \frac{(k+1)\pi}{H} \quad , \quad \mu_k^o = \frac{(k+1/2)\pi}{H_1 + H_2} \quad ; \qquad (28)$$

for liquid drops, the conditions

$$\partial J_m (\lambda_k^j r)/\partial r = 0 \quad \text{at} \quad r = F_e(H) \quad \text{or} \quad r = R_1 \qquad (29)$$

yield the values of λ_k^j.

Equations (15), (16), (17) and (20) together with conditions (18), (19) and (21) and relations (23)-(29) lead to an eingenvalue problem whose solution determines the eingenvalues ω_n and the eingenvectors (P_o^j, A_k^j); P_o^j being zero for $m > 0$. We notice that the case $\omega_n^2 = 4\Omega^2$ is analized in a forthcoming paper.

SOLUTION OF THE EINGENVALUE PROBLEM

For numerical integration, it is proved convenient to introduce an intrinsic coordinate (as shown in Fig. 2) along the undisturbed equilibrium shape.

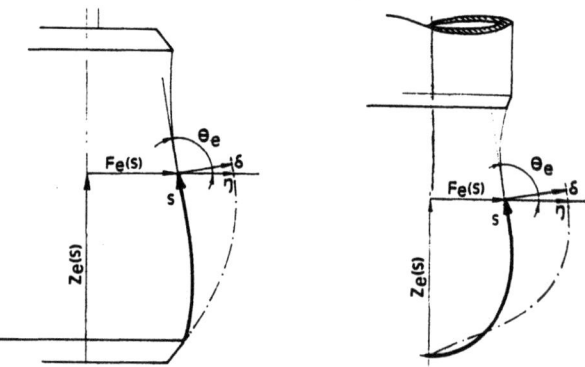

Fig. 2. Intrinsic coordinate s along the equilibrium shape. The sketch shows both radial and normal disturbance η and δ respectively.

In the new coordinate equation (15) may be written as

$$\Psi(\delta_n) = \rho^i \phi^i - \rho^o \phi^o , \qquad (30)$$

where the operator Ψ reads

$$\Psi \equiv \frac{d^2}{ds^2} + \frac{\cos\theta_e}{F_e}\frac{d}{ds} + B\cos\theta_e + \left(\frac{d\theta_e}{ds}\right)^2 + \frac{\sin^2\theta_e}{F_e^2} + \Omega^2 F_e \sin\theta_e - \frac{m^2}{F_e^2} , \qquad (31)$$

and normal mode $\delta_n(s)$ is defined as

$$\delta_n(s) = \eta_n[z_e(s)]\sin\theta_e(s) ; \qquad (32)$$

the boundary conditions are (see Fig. 3)

a) Liquid bridges

$$\delta_n(0) = 0 \quad \text{and} \quad \delta_n(s_o) = 0 \qquad (33a)$$

b) Drops

$$\delta_n(s_o) = 0 , \quad \left. d\delta_n/ds \right|_{s=0} = 0 , \quad \text{for } m = 0 \qquad (33b)$$

or

$$\delta_n(0) = 0 , \quad \delta_n(s_o) = 0 , \quad \text{for } m \neq 0 \qquad (33c)$$

s_o being the total length of the interface profile. The kinematic condition at the interface and the volume conservation become respectively

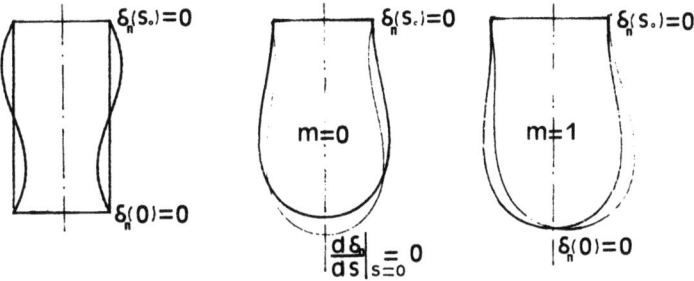

Fig. 3. Boundary conditions for bridge and drop deformations. Notice the difference in the boundary conditions for symmetric and non-symmetric drop deformation.

$$i\omega_n \delta_n = \frac{dz_e}{ds} u_I^j - \frac{dF_e}{ds} w_I^j , \qquad (34)$$

$$\int_0^{s_o} F_e(s) \delta_n(s) \, ds = 0 , \qquad (35)$$

u_I^j and w_I^j are the radial and vertical components of the disturbance velocity at the interface.

It may be easily shown (linear problem) that the solution of equation (30) with conditions (33) may be expanded as

$$\delta_n = (P_o^i - P_o^o) \left[\delta_p + \frac{\delta_p(s_o)}{\delta_H(s_o)} \delta_H \right] + \sum_{q=1}^{\infty} C_q \delta_q^* , \qquad (36)$$

where $\delta_p(s)$, $\delta_H(s)$, and $\delta_q^*(s)$ are solutions of the following problems:

$$\Psi(\delta_p) = 1 \text{ and } \Psi(\delta_H) = 0 , \qquad (37)$$

with the conditions

$$\delta_p(0) = \delta_H(0) = 0 \text{ (for liquid bridges and drops } m \neq 0) , \qquad (38)$$

or

$$d\delta_p/ds|_0 = d\delta_H/ds|_0 = 0 \quad \text{(for drops } m = 0) , \qquad (39)$$

and the Sturm-Liouville problem

$$\Psi(\delta_q^*) = g(\nu_q) \delta_q^* = \left[(1 - \nu_q^2) \left[B + \left(\frac{d\Phi_e}{ds} \bigg|_{s=0} \right)^2 \right] + (1 - \nu_q) \left(\frac{\sin \Phi_e}{F_e} \bigg|_{s=0} \right)^2 \right] \delta_q^* \qquad (40)$$

subject to the conditions

$$\delta_q^*(0) = 0 \quad \text{or} \quad d\delta_q^*/ds \bigg|_{s=0} = 0 , \quad \text{and} \quad \delta_q^*(s_o) = 0 \qquad (41)$$

Notice that the eingenfunctions δ_q^* are orthogonals in the interval $(0, s_o)$ and they may correspond to harmonic functions of the interface shape; in particular, it is easy to prove that for spherical or cylindrical shapes δ_q^* become spherical harmonics P_n or circular functions respectively.

Using orthogonal δ_q^* functions we carry out a generalized integral transformation (see the Appendix) of both equation (30) and conditions (34) and arrive at the following homogeneous system of infinite algebric equations

$$g(\nu_q) C_q D_q - \omega_n^2 \sum_{k=1}^{\infty} (\rho^i A_k^i E_{kq}^i - \rho^o A_k^o E_{kq}^o) = 0, \qquad (42)$$

$$(1 - 4\Omega^2/\omega_n^2) \left[G_q (P_o^i - P_o^o)/\delta_H(s_o) + C_q D_q \right] - \sum_{k=1}^{\infty} A_k^j S_{kq}^j = 0 \qquad (43)$$

$$S_{kq}^j = \left[\lambda_k^j H_{kq}^j + 2 m\Omega M_{kq}^j/\omega_n - \mu_k^j N_{kq}^j (1 - 4\Omega^2/\omega_n^2) \right] = 0. \qquad (44)$$

Introducing now expression (36) into condition (35) we arrive at

$$\frac{P_o^i - P_o^o}{\delta_H(s_o)} T + \sum_{q=1}^{\infty} C_q Q_q = 0, \qquad (45)$$

$$T = \int_0^{s_o} \left[\delta_H(s_o) \delta_p(s) - \delta_p(s_o) \delta_H(s) \right] F_e(s) ds ; \qquad (46)$$

expressions for the constants D_q, E_{kq}^j, G_q, H_{kq}^j, M_{kq}^j, N_{kq}^j, T, and Q_q are given in the Appendix. After some manipulation system (42)–(44) finally leads to

$$\omega_n^2 \sum_{k=1}^{\infty} \rho^i A_k^i \left[T E_{kq}^i - G_q \sum_{\alpha=1}^{\infty} \frac{g(\nu_q) Q_\alpha E_{k\alpha}^i}{\dot{g}(\nu_\alpha) D_\alpha} \right]$$

$$- \rho^o A_k^o \left[T E_{kq}^o - G_q \sum_{\alpha=1}^{\infty} \frac{g(\nu_q) Q_\alpha E_{k\alpha}^o}{\dot{g}(\nu_\alpha) D_\alpha} \right]$$

$$- g(\nu_q) T \sum_{k=1}^{\infty} A_k^j S_{kq}^j = 0. \qquad (47)$$

A simpler form of system (47) may be obtained for $m \neq 0$

$$\omega_n^2 \sum_{k=1}^{\infty} (\rho^i A_k^i E_{kq}^i - \rho^o A_k^o E_{kq}^o) - g(\nu_q) \sum_{k=1}^{\infty} A_k^j S_{kq}^j = 0. \qquad (48)$$

The eingenvalue problems (47) or (48) yield the natural frequencies ω_n and the corresponding eingenvector A_k^i and A_k^o, which determine the normal modes of the oscillations of the interface.

On the other hand, it is possible to obtain the stability criteria for the equilibrium shapes from an analysis of systems (47) or (48).

DISCUSSION OF THE RESULTS

First we have performed an analysis on the stability criteria. As it is known, when $\omega_n = 0$ (secular instability) a bifurcation gives rise to a new equilibrium shape family or to the breaking of the meniscus[9]. Then, for $m = 0$ it may be analitically shown that the vanishing of (46)

$$T = \int_0^{s_0} [\delta_H(s_0) \delta_p(s) - \delta_p(s_0) \delta_H(s)] F_e(s) \, ds = 0 \qquad (49)$$

is the required condition for the secular stability criteria ($\omega_n = 0$).

For $m \neq 0$ it is easy to see that

$$g(\nu_q) = 0 \quad \text{or} \quad \nu_q = 1 \qquad (50)$$

is the required condition for $\omega_n = 0$ [see expression (48)].

Fig. 4 shows the stability limits, obtained from (49) and (50), in the map of drop equilibrium shapes. Notice that drops whose profiles lay in the region bounded by $m = 1$ (o o o) and $m = 0$ (x x x) lines (on the left of $m = 1$) are stable. Curves of constant drop volume have also been plotted; then for a given volume and tube radius one may find the corresponding drop profile. It is of interest to notice that the crossing point of $m = 1$ and $m = 0$ curves corresponds to the maximum drop volume ($V = 18.963$) that one may achieve in a given gravitational field [10].

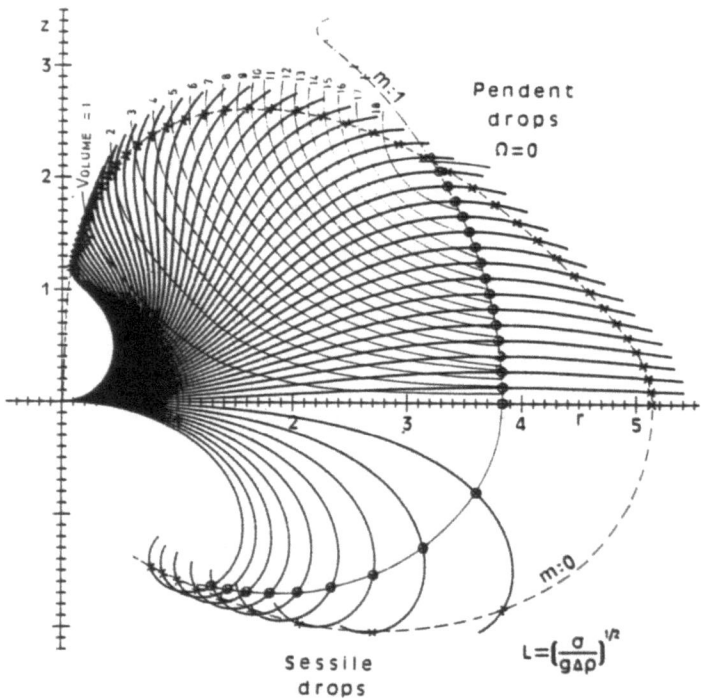

Fig. 4. Drop equilibrium shapes. The stability limits for $m = 0$ (x x x) and $m = 1$ (o o o) are also plotted. Profiles lying in the region bounded by $m = 1$ and $m = 0$ lines (on the left of $m = 1$) are stable. Curves of constant drop volume are also shown in the figure. Notice that all quantities have been non-dimensionalized using the capillary length.

Curves of maximum stable drop volume for a given Bond number is also given in Figure 5; notice that for practical purpose we have redefined the Bond number using the base radius of the drop as characteristic length. As it may be seen from Fig. 5 there exists a certain Bond number B^*, which depends on Ω, that for $B < B^*$ the stability limit is determined by the axisymmetric neutral mode ($m = 0$) while for $B > B^*$ the stability limit is controlled by a non-axisymmetric mode ($m = 1$).

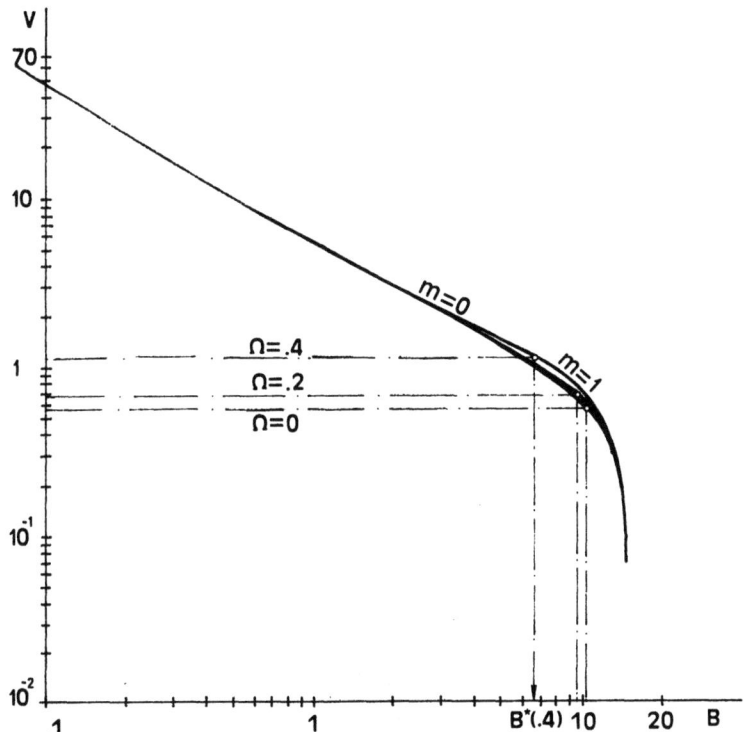

Fig. 5. Maximum drop volume vs. Bond number. Both volume and Bond number have been non-dimensionalized using the tube radius.

Secondly we have performed a set of numerical calculations for both drops and bridges. Natural frequencies ($n = 1, 2, 3$, and 4) for a drop without rotation and the corresponding normal modes for both axisymmetric and non-axisymmetric ($m = 1$) modes are given in Figs. 6 and 7. Figures 8 and 9 show the same kind of results for bridges.

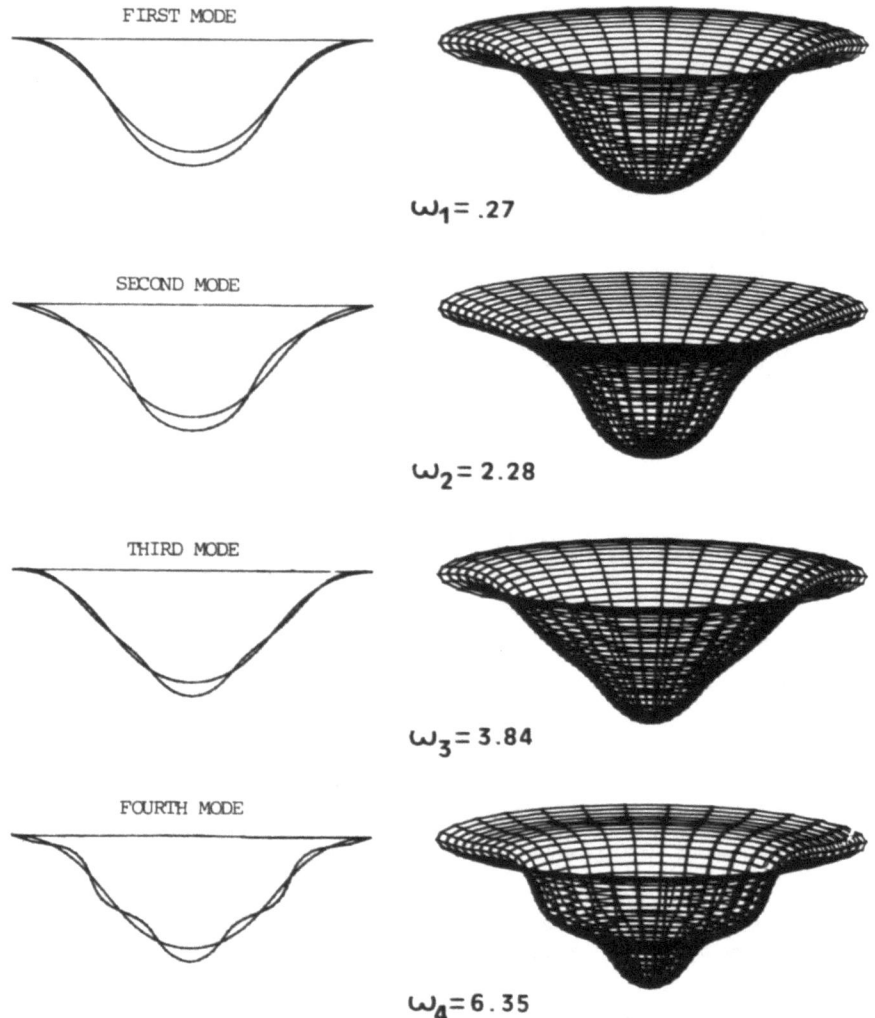

Fig. 6. Normal modes and natural frequencies for n = 1,2,3 and 4 of a pendent drop of volume V = 18.91, hanging from a tube of radius R_0 = 3.25 for Bond number B = 1. Axisymmetric case (m = 0).

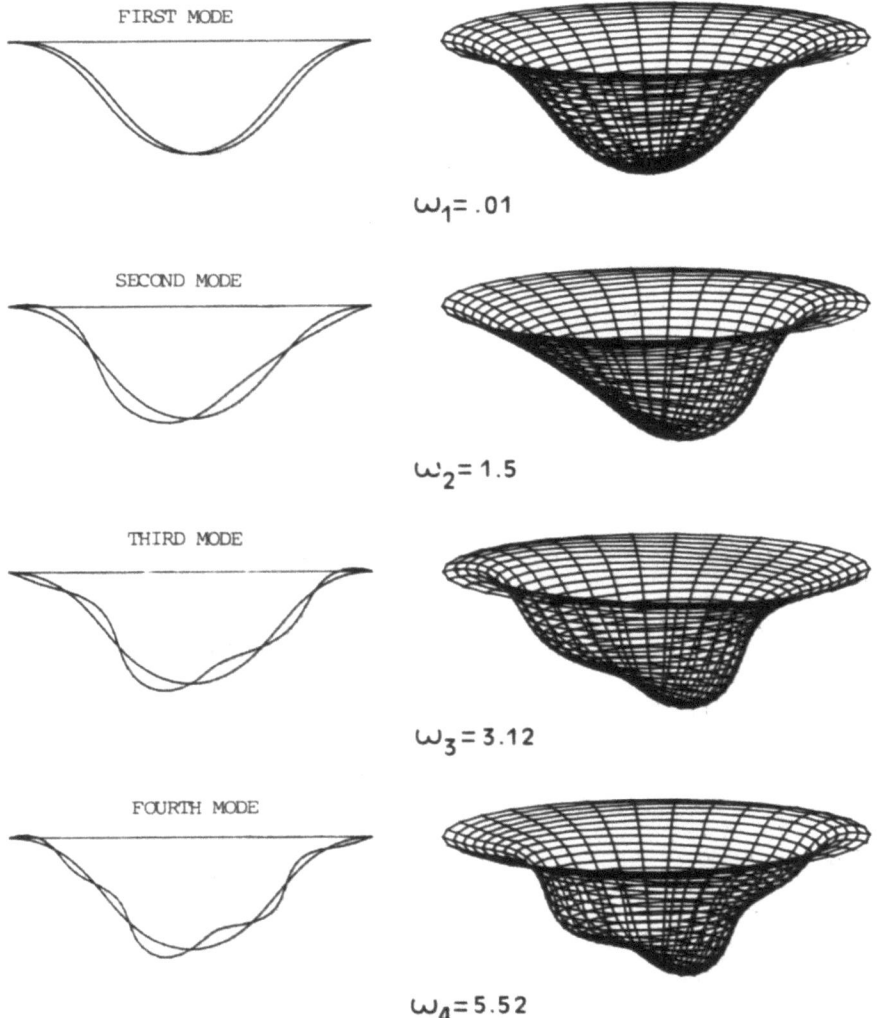

Fig. 7. Normal modes and natural frequencies for n=1,2,3 and 4 of a pendent drop of volume V=18.91, hanging from a tube of radius R_o=3.25 for Bond number B=1. Non-axisymmetric case. (m=1).

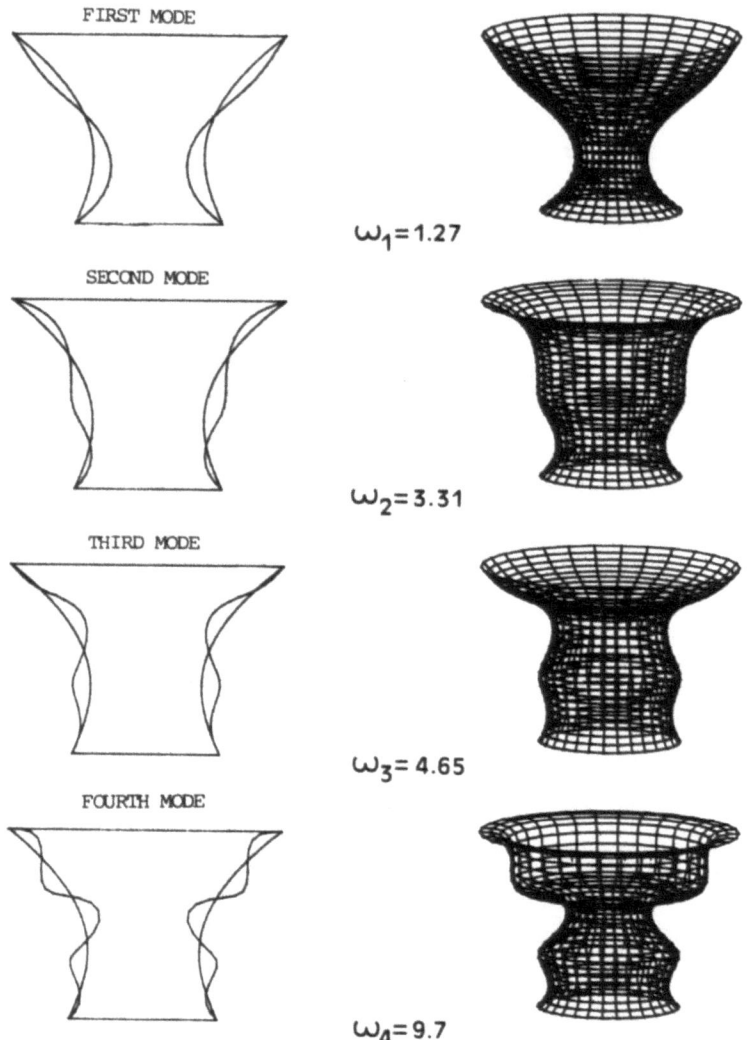

Fig. 8. Normal axisymmetric modes for n = 1, 2, 3, and 4 of a liquid bridge of volume V = 2.9, height H_o = 2.57, upper disk radius R_1 = 1.82 and Bond number B = .25.

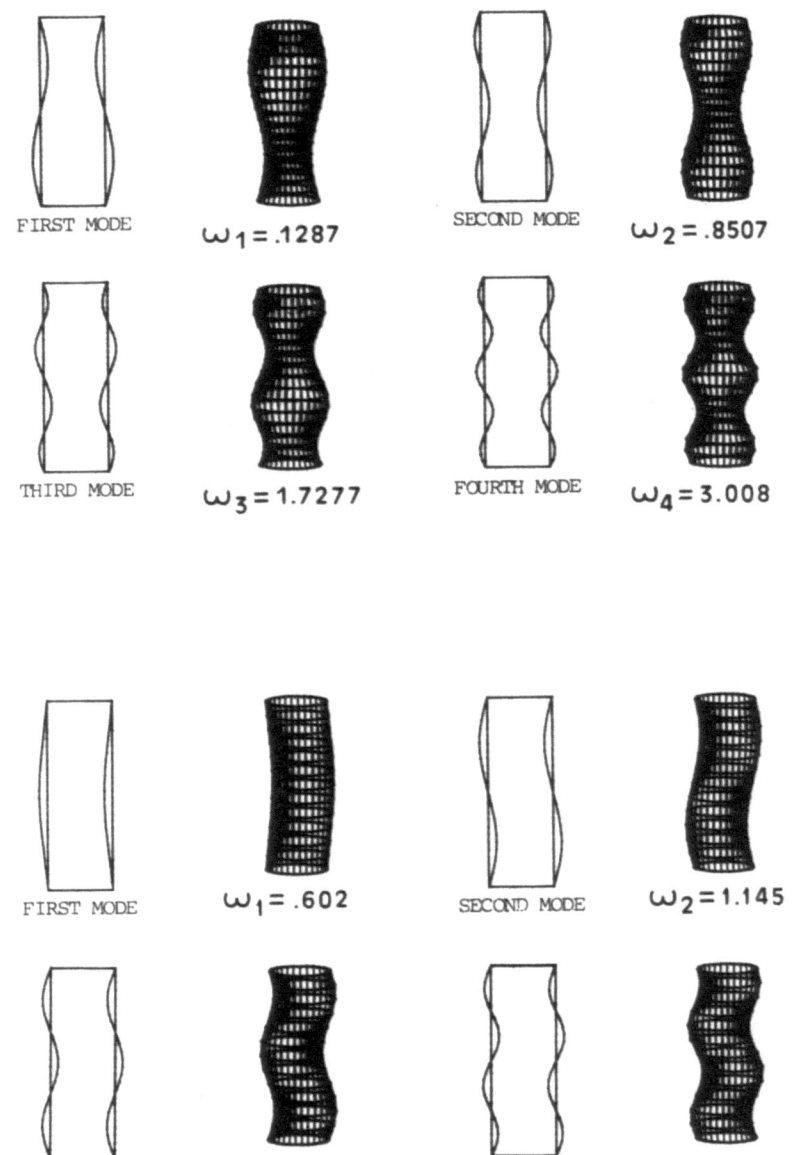

Fig. 9. Normal modes (m=0 and m=1) and natural frequencies of a cylindrical bridge; the cylinder slenderness is equal to 6. Notice that the calculated frequencies for m=0 agree with those reported by Sanz[4].

APPENDIX

Multiplying equation (30) and conditions (34) by $F_e \, \delta_q^* \, ds$, integrating them along the interval $(0, s_o)$, and taking into account both (37) and the orthogonality condition of δ_q^* functions

$$\int_0^{s_o} F_e \, \delta_q^* \, \delta_k^* \, ds = 0 \quad \text{if} \quad g \neq k, \tag{A1}$$

we arrive at equations (42) and (43); where the constants D_q, E_k^j, G_q, H_{kq}^j, M_{kq}^j, N_{kq}^j and Q_q are defined in the following form:

$$D_q = \int_0^{s_o} F_e \, \delta_q^{*2} \, ds, \tag{A2}$$

$$E_{kq}^j = \int_0^{s_o} R^j(\lambda_k^j F_e) \, Z^j(\mu_k^i z_e) \, F_e \, \delta_q^* \, ds, \tag{A3}$$

$$G_q = \int_0^{s_o} \left[\delta_H(s_o) \, \delta_p - \delta_p(s_o) \, \delta_H \right] F_e \, \delta_q^* \, ds, \tag{A4}$$

$$H_{kq}^j = \int_0^{s_o} \frac{d}{dr} \left[R^j(\lambda_k^j r) \right]_{r=F_e} Z^j(\mu_k^j z_e) \, \frac{dz_e}{dr} \, \frac{F_e}{\lambda_k^j} \, \delta_q^* \, ds, \tag{A5}$$

$$M_{kq}^j = \int_0^{s_o} R^j(\lambda_k^j F_e) \, Z^j(\mu_k^j z_e) \, \frac{dz_e}{ds} \, F_e \, \delta_q^* \, ds, \tag{A6}$$

$$N_{kq}^j = \int_0^{s_o} \frac{d}{dz} \left[Z^j(\mu_k^j z) \right]_{z=z_e} R^j(\lambda_k^j F_e) \, \frac{dz_e}{ds} \, F_e \, \delta_q^* \, ds, \tag{A7}$$

and

$$Q_q = \int_0^{s_o} F_e \, \delta_q^* \, ds. \tag{A8}$$

ACKNOWLEDGEMENTS

This research was performed under the auspicies of the Consejería de Educación de la Junta de Andalucía.

REFERENCES

1. KELVIN, LORD 1890, "Mathematical and Physical Papers", $\underline{3}$, 384, Clay.
2. RAYLEIGH, LORD 1894, "The Theory of Sound". Macmillan.
3. BUSSE, F.H. 1984, J. Fluid Mech., $\underline{142}$, 1.
4. SANZ, A. 1985, J. Fluid Mech., $\underline{156}$, 101.
5. STRANI, M. and SABETTA, F. 1984, J. Fluid Mech., $\underline{141}$, 233.
6. PITTS, E. 1974, J. Fluid Mech., $\underline{63}$, 487.
7. MICHAEL, D.H. and WILLIAMS, P.G. 1976, Proc. R. Soc. Lon. $\underline{A. 351}$, 117.
8. GREENSPAN, H.P. 1980, "Theory of Rotating Fluids" Cambridge U.P., New York.
9. BROWN, R.A. and SCRIVEN L.E. 1980, Phil. Trans. R. Soc. Lond. $\underline{A 297}$, 51.
10. BOUCHER, E.A. and EVANS, M.J.B. 1975, Proc. R. Soc. Lon. $\underline{A.346}$, 349.

DYNAMICS OF TWO-DIMENSIONAL BUBBLES BY THE LATTICE GAS METHOD

D. d'Humières and P. Lallemand
Laboratoire de Physique de l'Ecole Normale
Supérieure, 24 Rue Lhomond
75231 Paris Cedex 05, France

G. Searby
Laboratoire de recherche sur la Combustion
Centre de St Jérôme
13397 Marseille Cedex 13, France

Lattice gases, introduced recently as a way to calculate viscous flows at moderate Mach and Reynolds number, are currently the object of intense research. Among the variants of the initial proposal of Frisch, Hasslacher and Pomeau (1), the case of lattice gas mixtures has proven to provide an efficient means of studying hydrodynamical flows with free boundaries (2-4).

In this paper we first give some basic informations on lattice gases then we discuss in some detail a particular lattice gas that may describe fluid mixtures. We present the results of computer experiments designed to study the basic properties of this lattice gas mixture. Finally, we show the behaviour of two-dimensional bubbles as an illustration of the capabilities of lattice gases in the context of hydrodynamical flows with free boundaries. Questions raised by the current work will be addressed at the end.

DEFINITION OF LATTICE GASES

Simulation of hydrodynamical flows is usually performed starting from the Navier-Stokes equations that relate the time and space derivatives of macroscopic quantities: density, local velocity, temperature, concentration,... . These equations have general features which can be obtained simply by considering the conservation laws governing the basic microscopic events taking place in the fluid. The properties of the material that is used in a particular situation are included in an equation of state and in the values of some coefficients that appear in the constitutive equations. For newtonian fluids in which there are no temporal or spatial correlations, the equations of motion can be put in an adimensionalized form involving a few numbers like the Mach number, the Reynolds number, the Rayleigh number, etc.... . This forms the basis of the commonly used similarity laws, and makes it possible to perform calculations without any reference to the microscopic nature of the system that is considered. Similarly, some numerical methods are also used in which a set of particles are considered such as in the particle in cell technique. However there is no direct relationship between the particles used in these kinds of calculations and the real particles of the fluid under study.

Lattice gases result from a different approach. The detailed nature of the fluid is taken into account and a set of atoms or molecules interacting with short range forces is considered. In principle one can calculate the exact positions and velocities of all the particles present in the fluid by solving a large set of Newton's equations. Obviously practical considerations make this approach completely out of reach of even the largest computer one can think of, but in many cases a very small sample can suffice to provide significant results. The corresponding techniques are known as molecular dynamics (5) and usually physical studies are made with between a few hundred and a few thousand particles in order to get equations of state or transport coefficients. When macroscopic behaviours are looked for then larger numbers of particles are required. With roughly 10^5 particles, eddies behind a cylinder can be observed (6). However the necessary computer time becomes completely out of reach of most laboratories when complicated flows are under study. Note that a large part of the work in statistical mechanics has been devoted to develop theories allowing one to make shortcuts between two body interactions and macroscopic properties.

An alternate way to study fluids at the microscopic level has been introduced by Pomeau et al. (7) with a very simplified model of a gas in which a set of point particles move at constant speed and in a synchronized way along the links of a square lattice. A further simplifying assumption limits the occupation number of each directional link of the lattice to 0 or 1 particle giving them **boolean** character. The dynamics of the system can be seen as a succession of free flights from one vertex of the lattice (we shall often use the name site) to one of its four neighbours, and of collision steps in which particles may change their direction of motion. These collisions insure that the medium is viscous. The rules for these collisions can obviously not be chosen arbitrarily if one wishes to model a real system. Mass, energy and linear momentum must be conserved. When analyzed in detail using the general methods of statistical mechanics it is found that the system behaves as a perfect Fermi gas and that its large scale dynamics follow some sort of Navier-Stokes equation. However the symmetry of the underlying microscopic world leads to an unacceptable lack of isotropy for the nonlinear and viscous terms.

Recently, Frisch, Hasslacher and Pomeau (1) have proposed another type of lattice, obtained by paving the plane with equilateral triangles, and introduced the possibility of having particles with 0 velocity. We shall call these particles centers or rest particles and will show that they can play a very useful role. Fig. 1 shows the lattice and the basic rules from which collisions are constructed.

The microscopic description of the system is done as follows: at each site, located at position **r**, 7 numbers equal to 0 or 1 (7 bits of computer memory) are required to describe the system without ambiguity (they correspond to a maximum of 6 moving particles and 1 rest particle). These numbers $n_i(\mathbf{r}, t)$ evolve at discrete values of time (0 , τ , 2τ , ...) according to

$$n_i(\mathbf{r} + \mathbf{c}_i\tau, t+\tau) = n_i(\mathbf{r}, t) + \mathbf{C}(\{n_i(\mathbf{r}, t+\tau)\})$$

where $\mathbf{c}_i$ is the one of the 7 possible velocities, indicated in Fig. 1, and **C** is the collision operator.

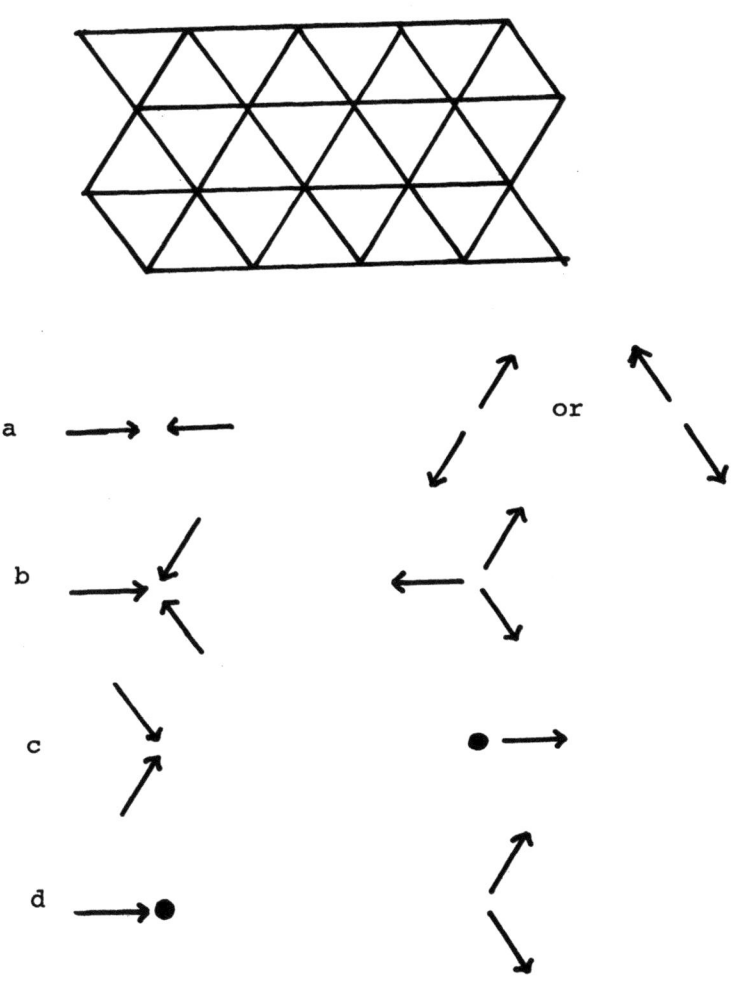

Fig.1 Geometry of the lattice and collision rules. Case of moving particles: a) two body collision, b) three body collision. Case involving rest particles: c) creation of centers, d) destruction of centers.

The collision operator can be either deterministic or include some randomness when a given precollision situation leads to several equivalent output conditions. In both cases **C** relates a 7-bit input word to another 7-bit word. For instance **C** (1000001) = 0100010 in the case of the elementary process that destroys a center as shown in Fig. 1. For the 7 bits model, it can be shown that out of the 2^7 possible precollision situations, 76 cases are active in the sense that they can produce a different postcollision situation, but with the same number of particles and the same total linear momentum. This corresponds to the conservation laws for mass and linear momentum in elementary collision events. Note that as there is only one non zero value for the particle speed, consideration of energy conservation brings nothing new. The elementary process in which 2 moving particles give rise to one center and one moving particle requires, however, the introduction of an internal energy equal to $mc^2/2$ for the rest particles.

The macroscopic properties of the lattice can be obtained (8) using the standard methods of statistical mechanics, as in the kinetic theory of gases. Local averages are defined for the density and the particle flux

$$\rho = \Sigma\, n_i$$

$$\mathbf{j} = \rho\, \mathbf{u} = \Sigma\, n_i\, \mathbf{c}_i$$

At thermodynamic equilibrium, the average population N_i is given by a Fermi-Dirac distribution

$$N_i = 1 / (1 + \exp (Q_i))$$

where $Q_i = a + b\, \mathbf{c}_i . \mathbf{u}$

If we call d_i the equilibrium population for $u = 0$, then for small **u**,

$$N_i = d_i + d_i(1-d_i)\, b_0\, \mathbf{c}_i . \mathbf{u} + \alpha_2\, u^2 +$$
$$b_0^2\, d_i\, (1 - d_i)(1 - 2 d_i)\, (\mathbf{c}_i . \mathbf{u})^2/2 + O(u^3)$$

where $b_0 = -2\, \Sigma\, d_i / \Sigma\, d_i(1-d_i)\, c_i^2$.

The macroscopic behaviour of the system can be obtained either by suitable linking of these local equilibrium properties, or by first writing a Boltzmann equation and then solving it using the Chapman-Enskog development. This leads to the continuity equation and the momentum equation. In terms of cartesian coordinates ($\alpha=1,2$)

$$\partial_t\, \rho + \text{div}\, \rho\mathbf{u} = 0$$

$$\partial_t\, \rho u_\alpha + \sum_{\beta\gamma\varepsilon} \partial_\beta (T^{\alpha\beta\gamma\varepsilon} u_\gamma u_\varepsilon) = - \partial_\alpha P + \sum_{\beta\gamma\varepsilon} \partial_\beta (S^{\alpha\beta\gamma\varepsilon} \partial_\gamma \rho u_\varepsilon)$$

The pressure P is given by

$$P = 1/2 \sum_i d_i\, c_i^2$$

The nonlinear term in the momentum equation involves the fourth order completely symmetric tensor:

$$T^{\alpha\beta\gamma\varepsilon} = 1/2 \ \beta_0^2 \sum d_i \ (1 - d_i)(1 - 2 d_i) \ c_{i\alpha} c_{i\beta} c_{i\gamma} c_{i\varepsilon}$$

This tensor which was anisotropic for the square lattice is now isotropic for the new triangular lattice. It can be seen that the momentum equation has essentially the form of the Navier-Stokes equation, except for the presence of terms of $O(u^3)$, and for the presence of a density dependent factor in the nonlinear term on the left-hand side.

The equation of state is given by $P = 3/7 \ d$, where d is the density per link ($d = \rho / 7$). The speed of sound is $\sqrt{3/7}$ in terms of microscopic units (l: length of lattice links, τ: propagation time). This value is close to 1, the maximum speed of the problem ("speed of light") and thus the present lattice would not be suitable to study supersonic flows. In addition the tensor T involves a density dependent term:

$$g(d) = 7/12 \ (1 - 2 \ d)/(1 - d)$$

and so situations involving large variations of the density (high subsonic and transsonic cases) may not be accurately modeled by the present lattice gas. This problem can be partly resolved by using a model with more than one particle at rest, and collision rules that allow a larger equilibrium value for the density of centers than that of moving particles.

The term $g(d) \neq 1$ appearing in the tensor T destroys the galilean invariance of the lattice gas model. Physically the macroscopic momentum is convected at speed $g(d)u$, different from the macroscopic flow velocity. Apparent invariance of the flow velocity can be restored by keeping the same definition of the flow velocity, but by looking at the momentum fluxes on a new time scale $t' = t/g(d)$. The apparent viscosity of the system is also renormalized, $v'= v/g(d)$. The validity of this renormalisation is probably limited to fairly low Mach numbers, although the corresponding errors have not yet been clearly elucidated.

The viscosity coefficients can be calculated either from a Chapman-Enskog analysis of the Boltzmann equation corresponding to this gas(8), or by a careful analysis of a steady state Couette flow(9).

In summary lattice gases provide a simple and efficient way to calculate two-dimensional flows at moderate Mach numbers and for Reynolds numbers given by

$$Re = u \ l \ /v' \ = \ u \ l \ g(d)/v \ .$$

Measured in microscopic units (10), the medium dependent term $g(d)/v$ reaches a maximum value of 3.39 for $d = 0.3$ as shown in Fig. 2 (top curve). The fact that this term is equal to 0 for $d = 0.5$ is due to the boolean character of the particles. For $d > 0.5$, the same results are found if all quantities depending upon the number of particles are replaced by a corresponding number of "holes". Note that if the number of active collisions used is reduced (medium curve), the maximum Reynolds number is reduced significantly. This is even more important if the centers are not used as shown in the lower curve of Fig. 2.

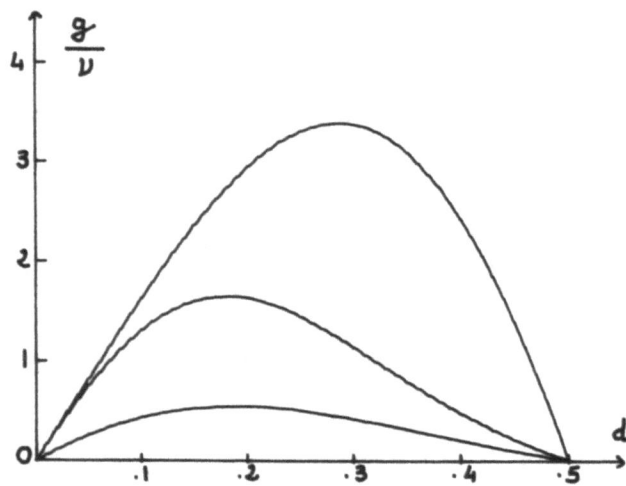

Fig.2 Value of the model dependent factor occurring in the Reynolds number. Top curve: model with centers and all possible collisions. Medium curve: model with centers and a reduced set of collisions. Lower curve: model with no centers.

Boundaries with obstacles can be introduced with no difficulty. The geometric boundary is first decomposed into a series of contiguous links of the lattice. On the particular sites located on the boundary, particles can be either reflected (this corresponds to the no-slip condition) or specularly reflected (which corresponds to the slip condition).

Several groups are currently using this lattice gas to perform flow simulations (11). As an example, we present here some results concerning tests of galilean invariance for the particular case of the Kelvin-Helmholtz instability.

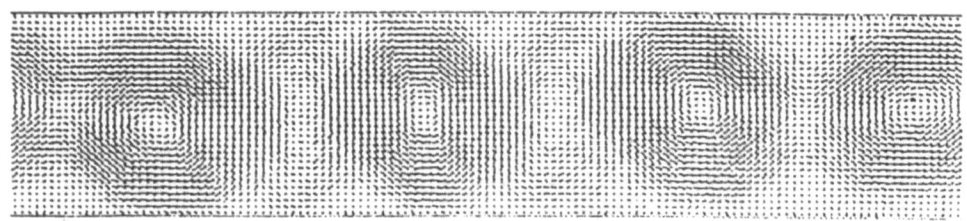

Fig.3 Map of the of flux of particles corresponding to the development of the Kelvin-Helmholtz instability, at time t = 4000. See text for numerical details.

Two parallel plates, of length 1024 and with the stick condition, are used to make a pipe, of size 1024 x 256 sites. Periodic conditions are assumed in the longer direction. The pipe is initially filled with two regions of fluid. They are of equal size and have the same mean density. The upper region has velocity $+ v + \delta$, the other has velocity $-v + \delta$. Due to the velocity gradient in the shear layer, the interface becomes unstable, and after a finite time eddies develop, as can be seen unstable, and after a finite time eddies develop, as can be seen in Fig. 3 for the case v = 0.128, and δ = 0. The exact development of the instability depends upon the microscopic distribution chosen at t=0. To produce reproducible results, the initial interface is not chosen flat, but is given a small periodic distortion. In this work the mean density per link is chosen equal to 0.30, and v = 0.128 in microscopic units.

Several experiments are performed for different values of δ. Then for the same time t, the flow distribution calculated with two different values of δ are compared to determine the mean difference between the particle fluxes as a function of an overall displacement. For real flows where galilean invariance applies one would get a minimum error for a displacement δt. Now, here advection of momentum takes place with velocity vg(d), so that one would expect that the minimum error occurs for δg(d)t. Fig. 4 shows the value of the displacement between two

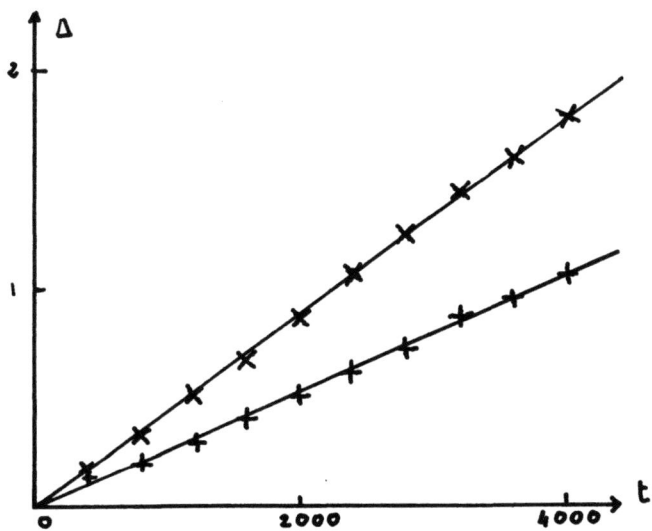

Fig.4 Time dependence of the displacement of the flows corresponding to different translational velocities with respect to the lattice. Crosses are obtained in numerical experiments, solid lines are theoretical predictions.

flows as a function of time together with the prediction δg(d)t. It can be seen that there a very good fit for δ = 0.128, and δ = 0.215. The error is found to remain roughly constant in time and about equal to the error between two flows that correspond to the same macroscopic conditions, but to different microscopic distributions. This results shows that restored galilean invariance holds in this particular case for a uniform translation of the fluids with respect to the lattice at speeds up to 0.25.

The extension of the model to three dimensional situations is related to the possibility of finding a system with a symmetry such that the tensors T and S are isotropic. It can be seen that these tensors have the same symmetry as that involved in the equations of elasticity. Now it is known that for all point groups in three dimensions a totally symmetric fourth rank tensor cannot be represented with less than three coefficients; instead of two for isotropic systems. This would preclude the extension of simple lattice gases to three dimensional problems. However by complicating slightly the problem, two solutions to this question have been proposed by d'Humières et al. (12).

LATTICE GAS MIXTURES

Lattice gas mixtures have been introduced by Clavin et al. (2) in order to study a variety of problems involving the coupling of hydrodynamical flows to either simple diffusion phenomena or reaction-diffusion problems. In the second case they showed that free boundaries can exist, and this provides a way to simulate hydrodynamical flows with free boundaries.

The mixture model consists in using the lattice gas presented in the first part of this paper, but in which each particle can be of one of two types, or "colors", (A or B). The state of a site is thus described by seven ternary bits (0, A, B) which are implemented as 7 + 7 binary bits. As in the first part of this paper, the first seven bits (n_i) describe the presence of a particle in direction i. The remaining seven bits (t_i) describe the type of the particle (0 = A, 1 = B) in the direction i. For an empty direction, the type specification is not relevant. The free flight step is exactly the same as in the original model of Frisch et al., the type attribute being translated with each particle.

Collisions are more complicated as there are now 3^7 different possible states. The rules for bits n_i are the same as in the first part of this paper, the additional rules for bits t_i can be of several kinds. If the number of particles of each type is conserved in collisions, then the lattice gas will be adapted to the simulation of diffusion phenomena. If the collision processes involve changes in the number of particles of each type, with a majority rule, like A + A + B $\rightarrow$ A + A + A
 then there will exist a sharp interface between regions filled with either pure A or pure B. At equilibrium the thickness of the interface will be as small as one mean free path (a few lattice units for typical conditions). Some preliminary results for this lattice gas mixture have been presented elsewhere (3). We shall just give here the main results.

Let us define the following macroscopic quantities:

$\rho = \Sigma n_i$, $\rho u = \Sigma n_i c_i$ total density and flux

$C = \Sigma n_i t_i / \rho$ concentration of B.

Non Reactive Case

These quantities satisfy the continuity equation and a

$$\partial_t \rho + \text{div } \rho \mathbf{u} = 0$$

$$\partial_t \rho u_\alpha + \partial_\beta (\rho g(\rho) u_\alpha u_\beta) = - \partial_\alpha P + \partial_\beta (\eta(\rho) \partial_\beta u_\alpha)$$

$$\partial_t \rho C + \partial_\alpha \rho C u_\alpha = \partial_\alpha D(\rho) \rho \partial_\alpha C$$

The first two equations are identical to those of the original Frisch et al. model, as the "color" of particles does not influence their dynamics. The last equation describes diffusion phenomena. It can be used in simple situations to determine the value of the diffusion coefficient $D(\rho)$. At low density the Chapman-Enskog analysis may be used to calculate $D(\rho)$. At higher densities, some hypothesis must be made concerning the absence of correlations between successive collisions in order to factorize higher order terms in the collision operator. Furthermore the large number of different collisions that may occur precludes closed form expressions for $D(\rho)$. Measurements of $D(\rho)$ can be performed by studying either the relaxation of an initial sinusoidal concentration distribution or the time evolution of a sharp flat interface between two regions filled with pure A or pure B. Both methods lead to the same numerical values for D, as shown in Fig. 5,

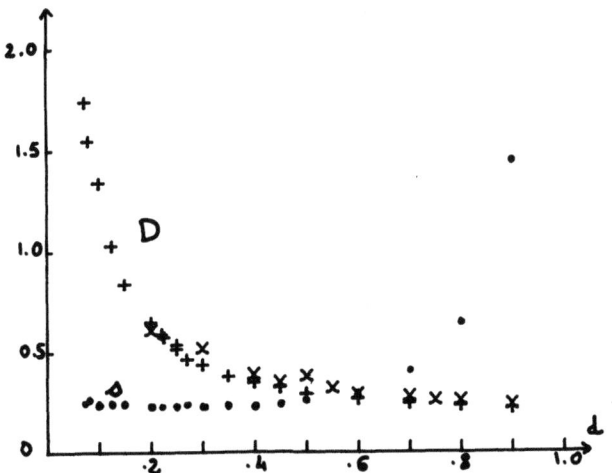

Fig.5 Density dependence of the diffusion coefficient for the lattice gas mixture. + correspond to standard diffusion experiments. x correspond to the relaxation of a bubble. correspond to the Schmidt number.

where the value of the Schmidt number $s = \eta / \rho D$ has been plotted. As was indicated in the first part of this paper the 7 bit model is invariant by changing particles into holes: the viscosity is an even function of d- 0.5. However the new 14-bit model is not invariant in the exchange of particles A and B. This allows one to choose the Schmidt number in a small range. Further possibilities of changing s are given by reducing the number of collisions affecting ν or those affecting D.

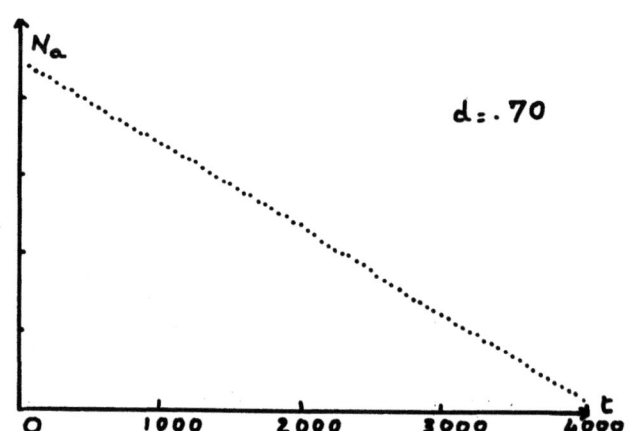

Fig. 6 Time dependence of the number of particles contained in an initially circular bubble.

Reactive Case

As indicated in the introduction of the mixture case, a sharp interface can be obtained using "type-exchange" reactions based on a majority rule.

For the simplest case, we consider 3-body collisions of the type:

$$A + A + B \underset{k_a^-}{\overset{k_a^+}{\rightleftarrows}} A + A + A$$

$$A + B + B \underset{k_b^-}{\overset{k_b^+}{\rightleftarrows}} B + B + B$$

These events can be called autocatalytic reactions.

A macroscopic analysis of the corresponding model leads to a diffusion-reaction equation

$$\partial_t C + \mathbf{u} \cdot \text{grad } C = D \Delta C + \varpi(C)$$

where C is the concentration of one of the species. The reaction rate $\varpi(C)$ is given by $k_a^+ C^2(1-C) + k_b^-(1-C)^3 - k_a^- C^3 - k_b^+ C(1-C)^2$ taking into account the boolean character of the particles. If we assume symmetry in the reaction rates, $k_a^+ = k_b^+ = k^+$ and

$k_a^- = k_b^- = k^-$, then ϖ is an odd function of C-0.5. When $\chi = k^- / k^+ < 1/3$, ϖ has three roots: C_1, C_2 and $1/2$. $C_1 = 0$ for $\chi = 0$. This means that for $\chi = 0$, ($k^- = 0$), the system will exhibit segregation of particles into regions that are pure A ($C_1 = 0$) or pure B ($C_2 = 0$). Note that similar results apply when higher order collisions are considered:

$$A + nB \underset{0}{\overset{k^n}{\rightleftarrows}} (n+1) B$$

$$nA + B \underset{0}{\overset{k^{-n}}{\rightleftarrows}} (n+1) A$$

where $n > 1$.

For steady state situations, and in the absence of velocity gradients for the mean flow (no stretching term present) or density difference between both sides fo the interface, the structure of the interface will be given by

$$D \, \partial^2 C / \partial x^2 + \varpi (C) = 0,$$

where x is the normal coordinate to the interface.

The preceding equation means that an initial distortion of the interface of the form $x = x^0 \cos kr$, will relax as $\exp - (D k^2 t)$, and that an initial circular bubble of radius R_0 will shrink with a value of the radius equal to $R(t)^2 - R_0^2 = -2Dt$.

To produce a stable bubble of radius R_0, we may either take a slight difference in the reaction rates k_a^+ and k_b^+, or apply a small pressure difference between the inside and the outside, as for Laplace law in the case where the surface tension is positive.

APPLICATIONS

The preceding model can be used to calculate a number of flows, using the concentration as a passive scalar. Examples corresponding to jets or the Kelvin-Helmholtz instability have been presented elsewhere (2, 11). Here we shall concentrate on bubbles.

Nonbuoyant Case

We consider a 2d mesh of typical size 512 x 512 lattice sites, with either periodic boundary conditions or rigid boundaries. As initial conditions, we fill the mesh with a fluid at rest and of uniform density. The exact microscopic distribution of particles is obtained from the equilibrium distribution function by using a suitable random number generator. A circular region of the lattice is filled with pure A, the rest with pure B. Fig. 6 shows the time dependence of the number of A particles. It decreases linearly with time. From such data an effective diffusion coefficient can be obtained. The values are found to be slightly larger than those obtained in the purely diffusive problem. This comes from the fact that a substantial number of the collisions occurring in the diffusion case have been replaced by reactive collisions.

Buoyant Case

We can apply to the gas some sort of gravity field, the strength of which is different for the two types of particles. This can be done in the following way. For a fraction ε of the lattice sites, a new collision rule is applied: the center, if filled, is transformed into a particle moving along the gravity field **g**. If **g** is parallel to one of 6 directions of the lattice links, this procedure is equivalent to applying a gravity field $g=\varepsilon(1-d)/7$. If g is parallel to the bissectrix of two adjacent lattice links, then $g=\sqrt{3}\varepsilon(1-d)14$. The (1-d) term is again due to the boolean character of the particles.

In a separate study, we first verified that this value of g is correct by measuring the stratification of the corresponding atmosphere. An approximately exponential variation of the density was found, corresponding to an isentropic atmosphere with a variable gravity. This procedure can then be applied to only one of the fluids: for instance that contained in the bubble.

As examples of results that can be obtained, we now present two cases:

Free buoyant bubble

We place a circular bubble filled with particles A in the 512 x 512 lattice considered above. Fig. 7 shows

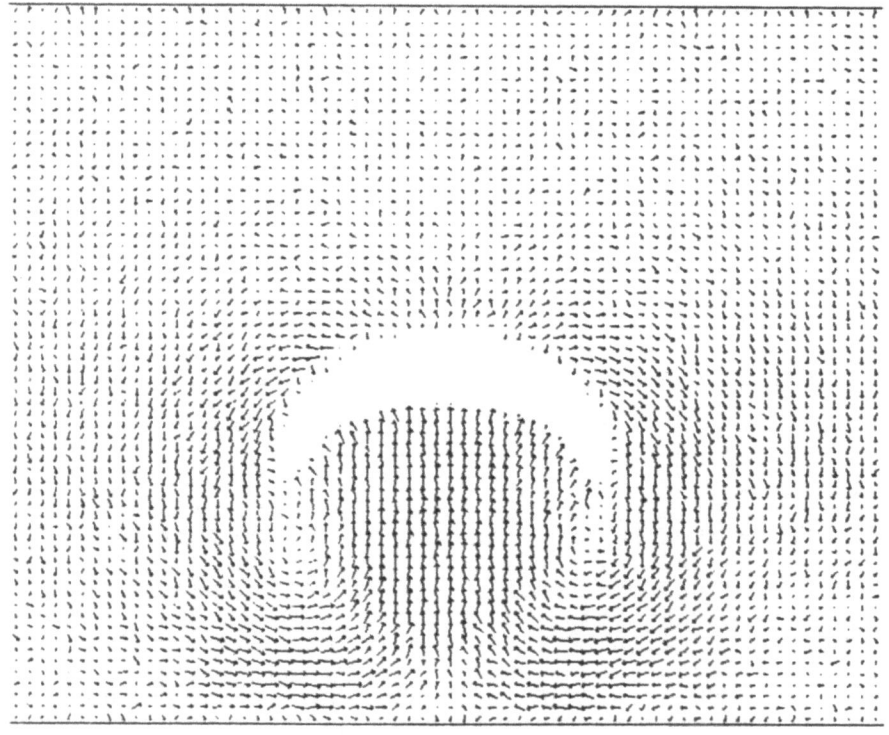

Fig.7   Map of B particle fluxes at time t = 1000.

the flux pattern of particles B around the bubble at time t=1000 One can clearly see the effect of the back flow on the lower part of the bubble. Fig.8 shows a time series of the shape of the interface between the two fluids. The deformation is characteristic of bubbles in situations where the ratio of lift forces to capillarity forces is large, as is well known for bubbles at large Eötvös numbers.

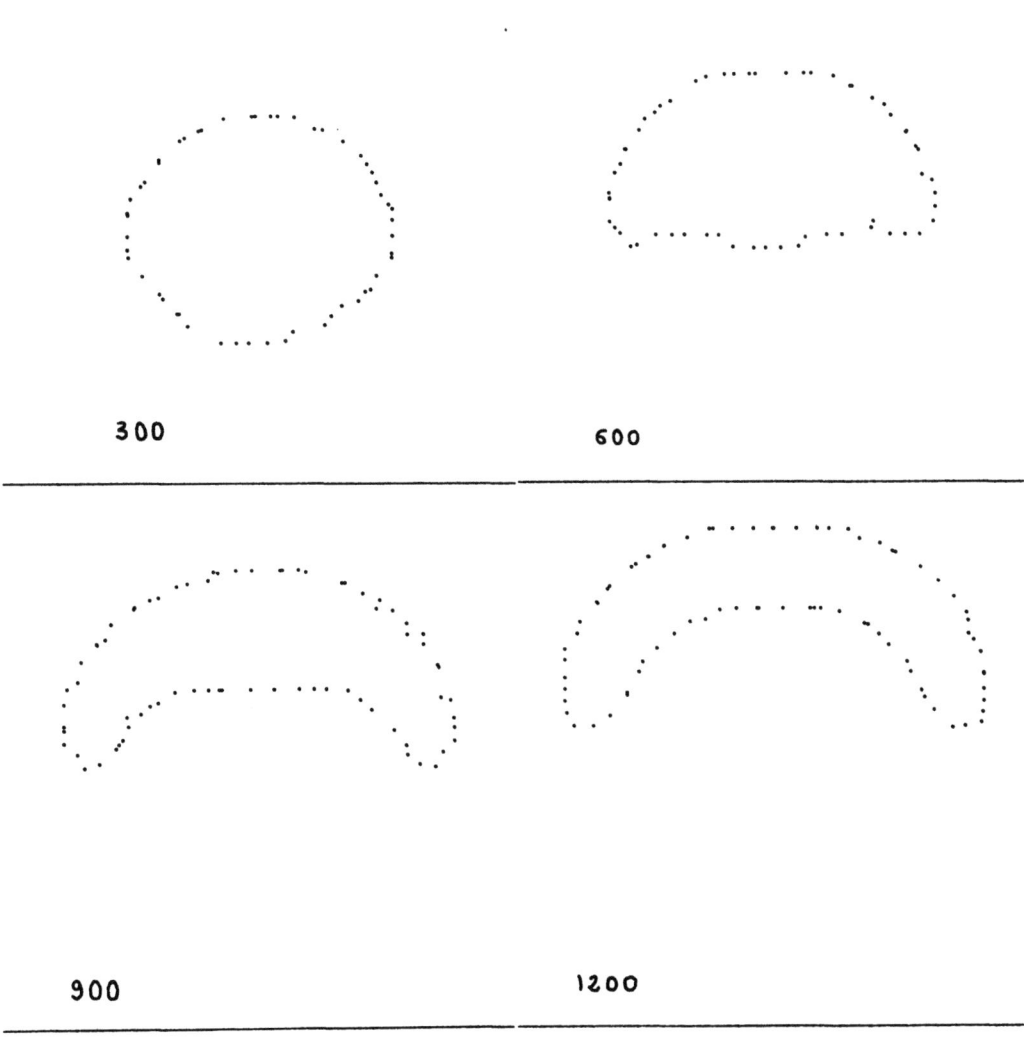

Fig.8   Series of shapes of a buoyant bubble located in free space at various instants.

Buoyant bubble in a channel
We now place a circular bubble of particles A in a 1024 X 256 channel. The sides of the channel are made of rigid boundaries set with the no-slip condition. Initially both fluids are at rest. We then apply a gravity field to the gas inside the

bubble in a direction parallel to the axis of the channel. As can be seen in Fig. 9, we find that the front part of the bubble has an approximately constant shape and that its back part becomes flat. The location of the front apex of the bubble is found to vary approximately as the square of the time.

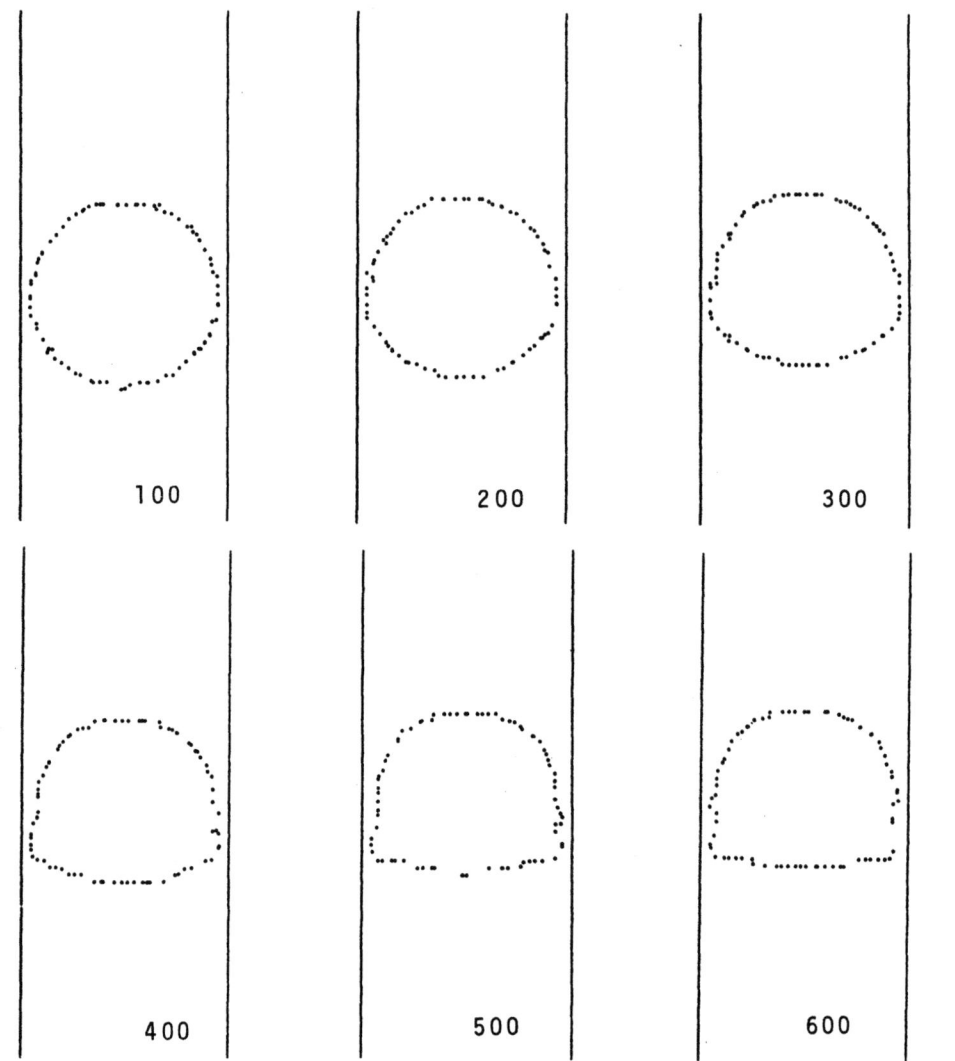

Fig.9 Series of shapes of a buoyant bubble located in a channel with rigid boundaries.

These examples show the ease with which hydrodynamical flows with free boundaries can be calculated.

QUESTIONS FOR FUTURE EXTENSIONS OF LATTICE GASES

The present work can be extended in several directions, however some questions must be answered first.

In order to study the nonlinear transport of concentration fluctuations, the present model must be modified as the convective terms for the momentum fluxes and the concentration are different in Eq.1. One way to do this is to use different collision rules for the motion of particles, as indicated in the first part of this paper. This allows one to obtain a model in which $g(\rho) = 1$ for a given value of ρ, and $dg(\rho)/d\rho = 0$ for that particular value of ρ.

A more difficult point is linked to the zero value of the surface tension for the present model. This is related to the fact that our model does not include any immiscibility condition. One might solve that problem by applying some book-keeping technique in which the number of each type of particles, which is not conserved in elementary collisions, would be conserved on average over a portion of space of the order of the square of the mean free path.

Other types of extension of the present model consider particles with different velocities. For instance, we can consider the lattice gas mixture made of two sets of particles:
a- mass 2, velocity 1/2, linear momentum 1, pressure $3d_a/14$
b- mass 1, velocity 1 , linear momentum 1, pressure $3d_b/7$.

Two regions filled with either a or b, will be in equilibrium if the pressure is the same, so that $d_a = 2\ d_b$. Collision rules between a and b particles can be of elastic type, as in the first part of this paper, since they have the same value for the linear momentum, or reactions can take place. Now conservation of mass leads to reactions of the type

a + b → b + b + b (b acting as a catalyzer).

Energy is not conserved in such a reaction, so that one may consider the model as a way to study exothermic chemical reactions, such as found in flames. Some results (2) have already been obtained with this model. It has been found that an initially flat interface separating regions of pure A and pure B is unstable. This is probably a manifestation of the Darrieus-Landau hydrodynamic instability produced by gas expansion through the reaction zone (14).

An other case of interest is provided by the following case:
a- mass 1, velocity 2, linear momentum 2, pressure $12d_a/7$,
b- mass 4, velocity 1, linear momentum 4, pressure $12d_b/7$.

Equilibrium will be obtained for the same number of particles, but as the inertia of the gases are different, instability of initially flat shock fronts (15) might be studied, even though the model is not strictly applicable to situations where shocks occur.

CONCLUSION

In this paper, we have given a brief description of what a lattice gas is, and have indicated their basic properties. We then extended the original hexagonal model of Frisch, Hasslacher and Pomeau, to the case where "color" is added to the description of each moving particle. This allows one to represent either diffusion phenomena, or reaction diffusion phenomena. When such questions, as those raised at the end, are resolved, then the technique of lattice gases may become a very useful tool for the simulation of complex flows.

REFERENCES

1- U. Frisch, B. Hasslacher and Y. Pomeau, Phys. Rev. Letters **56**, 1505 (1986).
2- P. Clavin, P. Lallemand, Y. Pomeau and G. Searby, J. Fluid Mech. 1987
3- D. d'Humières, P. Clavin, P. Lallemand and Y. Pomeau, Compt. Rend. Acad. Sci. Paris **II 303**, 1169 (1986).
4- D. d'Humières and P. Lallemand, Proc. of Conference on Supercomputers, Paris, 2-6 Feb. 1987, to be published by North Holland Co.
5- J.P. Boon and S. Yip, Molecular hydrodynamics, Mc Graw Hill, (1980)
6- D.C. Rapaport and E. Clementi, Phys. Rev. Letters **57**, 695 (1986).
7- J. Hardy, O. de Pazzis and Y. Pomeau, Phys. Rev. **A13**, 1949 (1976).
8- J.P. Rivet and U. Frisch, Comp. Rend. Acad. Sci. Paris, **II 302**, 267 (1986).
9- M. Hénon, to be published in J. Stat. Phys. (1987).
10- D. d'Humières and P. Lallemand, Physica **140A**, 327 (1986).
11- Proc. of Santa Fe Conference on large physical systems, 27-29 Oct. 1986, to be published in J. Stat. Phys. (1987).
12- D. d'Humières, U. Frisch and P. Lallemand, Europhysics Letters **2**, 291 (1986).
13- C. Borges and S. Zaleski, to be published in Ref. 11.
14- G. Darrieus, "Propagation d'un front de flamme", unpublished work presented at La Technique Moderne in 1938, and at Le Congrès de Mécanique Appliquée, Paris 1945.
L. Landau, Acta Physicochemica, USSR, **19**, 77 (1944).
15- Non Steady Flame Propagation, edited by G.H. Markstein, Pergamon Press, 1964.

EFFECT OF INTERFACIAL AREA ON

FLOW CHARACTERISTICS IN BUBBLE FLOW

Goichi Matsui, Yutaka Yamashita, and Toshio Kumazawa

Institute of Engineering Mechanics
University of Tsukuba
Sakura, Ibaraki, Japan

INTRODUCTION

Bubble flow is one of familiar patterns of two-phase flow which we encounter or use frequently in various energy conversion systems and industrial plant such as nuclear reactors, boilers, chemical plant, and air-lift pumps. But bubble flow exhibits inherent properties in the internal flow structures namely distributions of gas phase, phase velocities, and their fluctuations depending on the flow conditions and/or complex interactions between bubbles and liquid or a channel wall. Therefore, in order to improve the efficiency of the systems or apparatus using bubble flows and to guarantee their safe opertions, it is essential to know not only the global flow characteristics but also the local flow charcteristics. However, the mechanism of bubble flows has not been perfectly clear yet.

Baced on peculiar configurations of local void fraction or a gas-phase profile in a cross-section of a channel, Sekoguchi et al.[1] have classified bubble flow into three fundamental categories: sliding bubble flow, coring bubble flow, and intermediate bubble flow. They have indicated that the gas-phase profiles of bubble flow depend on the mixing condition and the flowrate.

Serizawa et al.[2], Inoue et al.[3], Theofanous & Sullivan[4], and Lance & Bataille[5] have shown experimental results that turbulence in flow is increased if gas is injected to liquid flow as bubbles, but Serizawa et al.[2] have found a special case that turbulence in bubble flow is weakened comparing with that in all liquid flow. Matsui and Yamashita[6] have also shown similar experimental results on these conflicting properties of bubble flows. In order to understand such characteristics of bubble flow, further theoretical and experimental work will be necessary.

To investigate the behavior of interfacial area between both phases may throw light on the problems of internal flow characteristics because the internal flow structure has a close relation to the behavior and distribution of the interfacial area. Thus, the problems associated with the effect of the interfacial area on the internal structure of bubble flow are important and current topics connected closely with the improvement of efficiency of the systems and plant using bubble flows. This paper deals with the internal flow characteristics for the interfacial area namely the bubble size and the number of bubbles per unit volume under the same flowrate condition and the same average void fraction in a channel. The bubble size was controlled by changing a mixing condition and by adding a surfactant to liquid. By combining the two means, four kinds of experimental cases are obtained.

Experiments were carried out for upward bubble flow in a square channel using nitrogen gas and water as working fluids. Then we could see three fundamental regimes of bubble flow mentioned above. From these experimental results, flow characteristics of structure peculiar to bubble flow were found.

EXPERIMENTAL APPARATUS AND INSTRUMENTATION

A schematic diagram of the experimental apparatus is shown in Fig.1. The channel system is a single closed loop and consists of a downcomer, a horizontal section, a pump, a turbine-type flowmeter, a gas-liquid mixer, a riser section, and a gas-liquid separator along the flow direction. The riser section includes the vertical test section with quick shutoff valves at both ends. The channel downstream of the mixer has a square cross-section of 30x30 mm^2 and is made of acrylic resin to allow visual observation and the measurement using laser beams. In the gas-liquid mixer, a sintered metal ring with 100 μm-mesh is used to make bubbles in flowing liquid by injecting gas into the liquid through the ring.

Nitrogen gas and water were used as working fluids. The nitrogen gas is injected into the flowing water through the mixer at the bottom part of the riser section, and upward bubble flow is produced. The gas is separated in the gas-liquid separator and is released into the atmosphere.

The water flowrate is measured by a turbine-type flowmeter, and the gas flowrate is measured by a float-type flowmeter but corrected for the exit pressure of the flowmeter. The average void fraction in the measurement section is obtained from the gas volume trapped between two quick shutoff valves.

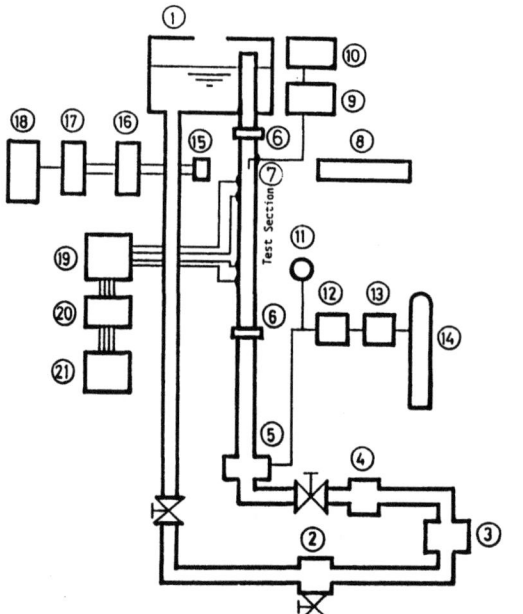

(1) Gas-liquid separator
(2) Drain
(3) Pump
(4) Turbine-type flow meter
(5) Mixer
(6) Quick shutoff valve
(7) Conductance probe
(8) Laser Doppler anemometer
(9) Counter processor
(10) Microcomputer
(11) Pressure gauge
(12) Float-type flow meter
(13) Pressure regulator
(14) Bomb
(15) Photomultiplier
(16) Counter processor
(17) Interface
(18) Microcomputer
(19) Amplifier
(20) Low-pass filter
(21) A/D converter and Minicomputer

Fig. 1. Schematic diagram of experimental apparatus.

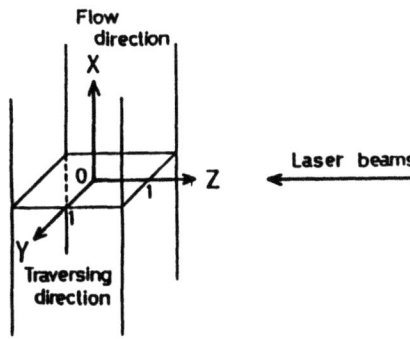

Fig.2. Coordinate system in the measurement cross-section (30×30 mm^2).

We employed a laser Doppler anemometer (LDA) and a double-sensor conductance probe to measure the internal characteristics of bubble flow at the cross-section about 1500 mm downstream of the mixer. Water velocity was measured by the former and bubble velocity, bubble length and local void fraction by the latter, respectively.

The coordinate system in the measurement cross-section is shown in Fig.2. The measuring volume intersected by laser beams and the probe are traversed every 1 mm from the center of channel to the wall in the y-direction.

The LDA employed has 3-beam two-component Ar-ion optics, two photomultipliers, and two counter processors. Digital data of detected signals are stored in a micro-computer through interfaces from the processors. The discrimination of two-direction burst signals of a particle was made based on the idea of overlap of two burst signals. The average values and the fluctuations of water velocities were obtained by processing the data stored.

The double-sensor conductance probe employed has two point-electrode with the space of about 3 mm and detected go to a counter processor and the digital data is stored in a micro-computer. The average values and fluctuations of bubble velocity, bubble length and local void fraction were obtained by processing the data stored. The spacing between two-electrodes of the probe used was measured using a microscope.

The size of bubbles was controlled by changing the mixing condition and by adding a surfactant to water. The Tween 20 (Polyoxyethylene sorbitan monolaurate) was used as the surfactant. By adding the Tween 20 to water, the resultant value of surface tension of water was reduced by about half. Thus, by using two kinds of mixing methods and a surfactant, four sets of experimental cases were investigated as follows:
(1) case that the 100 μm-mesh sintered metal ring is not used effectively, that is equivalent to the mixing method using several small holes.
(2) case that the surfactant is added to water under the case (1).
(3) case that the 100 μm-mesh sintered metal ring is used effectively.
(4) case that the surfactant is added to water under the case (3)

Experiments were conducted under the flowrate conditions of j_L=0.56 and 0.93 m/s for the water superficial velocity and j_G=0.012, 0.023, and 0.069 m/s for the gas superficial velocity.

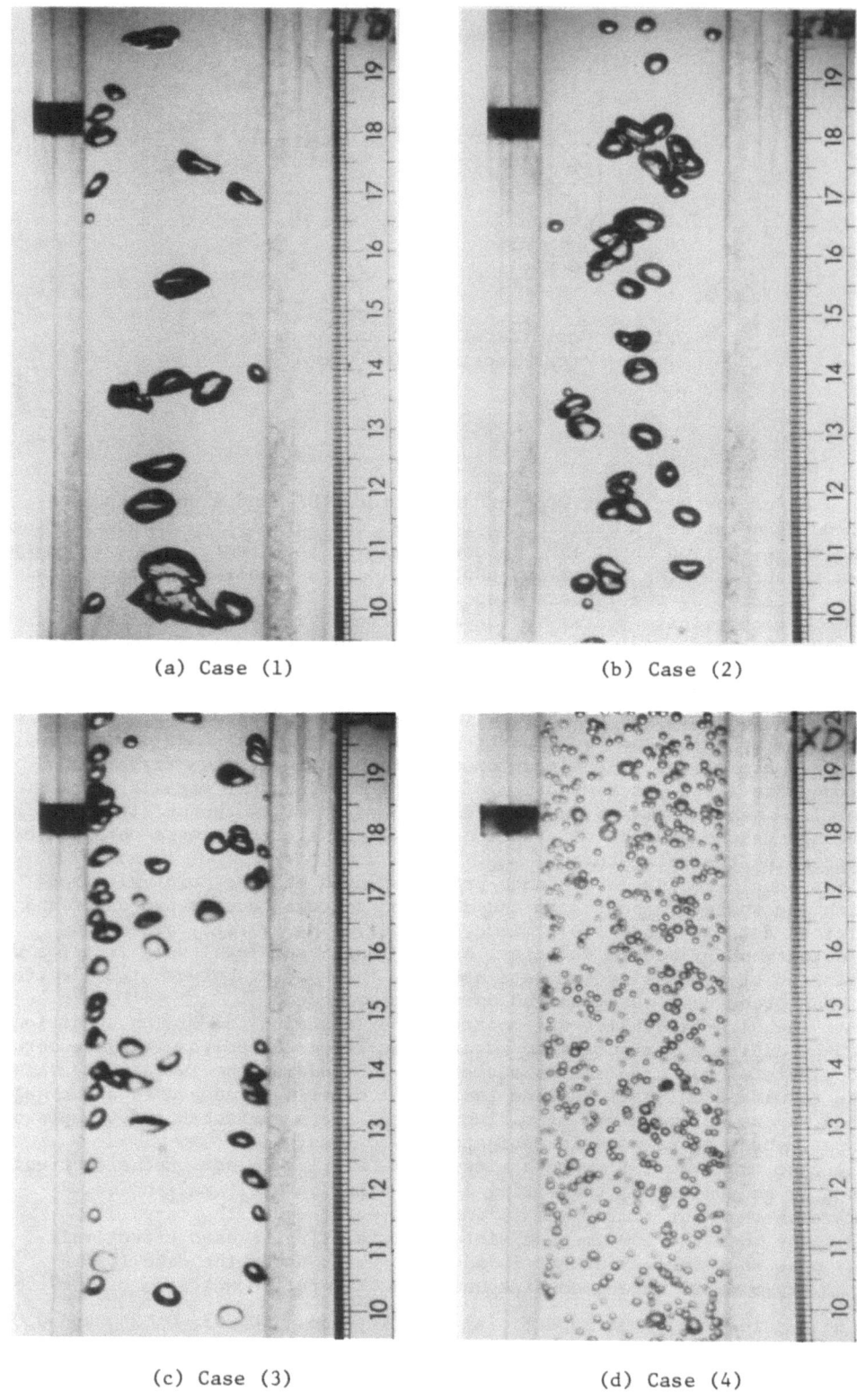

(a) Case (1) (b) Case (2)

(c) Case (3) (d) Case (4)

Fig. 3. Pictures of flow for the four experimental cases.

EXPERIMENTAL RESULTS AND DISCUSSIONS

Typical experimental results for four experimental cases under the flowrate conditions or the water superficial velocity of j_L=0.93 m/s and the gas superficial velocity of j_G=0.023 m/s and the average void fraction of $<\alpha>$=0.017 in the measurement section are shown in Figs.3 to 9. The pictures of flow for the four cases are shown in Fig.3 and the distribution of local void fraction or gas-phase distribution, in Fig.4. The cases (1) and (2) present that the bulk of bubbles flows in the center part of channel, that is called coring bubble flow, but the case (3) shows that the bulk of bubbles flows along the channel wall, that is called sliding bubble flow. The case (4) shows that bubbles flow uniformly over the cross-section, that is called intermidiate bubble flow. The pictures indicate that the reduction in bubble size under the conditions of the same average void fraction results in the change of bubble shape from the shape like an ellipsoid to the spherical one and an increase in the number of bubbles. The distributions of average bubble length shown in Fig.5 indicate that the bubble size becomes small in order of the case number. The gas-phase profile tends to flatten and bubbles tend to become small in order of the cases (1), (2), and (4). Then the interfacial area is found to increase in the same order. Assuming that all bubbles in the unit length of the channel along the flow are spherical with the same diameter, the diameter and the interfacial area of bubbles can be calculated using the average void fraction in the channel and the number of bubbles appeared in the pictures shown in Fig.3. The numerical results are listed in Table 1. The total interfacial area is normalized by the total surface area of the channel volume corresponding to the unit length of the channel. The bubble diameter is about twice the mean bubble length. These results are reasonable, supposing the length of spherical bubbles is detected at the same probability at all points on the bubbles.

The profiles of local average water velocity, the local average rise velocity of bubbles, the fluctuation intensities of water velocity, and the Reynolds stress of water-phase flow are shown in Figs. 6 to 9, respectively. The dash-dotted or solid line in the figures denotes the case for a single-phase flow of water alone. The flow of case (3) exhibits very peculiar characteristics comparing with the remains. The experimental results by Sekoguchi et al.[1] and Serizawa et al.[2] suggest that a sliding bubble flow appears under the special flowrate conditions. Thus, the effect of the interfacial area is discussed about the cases (1), (2), and (4).

The experimental results show that under the same flowrate conditions and the same average void fraction, the increase in interfacial area for the case (1) to (2) leads to a decrease in turbulence and an increase in water and bubble velocities, but the extreme increase in the interfacial area, which is corresponding to the case (4), induces a more decrease in turbulence and a uniform gas-phase distribution but does not lead to a further increase in both velocities. These results suggest that under the conditions of the same flowrate and the constant void fraction, the internal energy loss based on the interaction between bubbles and main flow may be reduced with the increase in the interfacial area but the extreme increase in the interfacial area may not necessarily induce a more reduction of energy loss although there may be any effect of secondary flow and that we can produce a bubble flow with the desired flow characteristics by controlling the interfacial area even if the way is limited.

CONCLUSIONS

The internal characteristics of bubble flow were investigated experimentally in upward nitrogen gas-water flow in a square channel using a laser Doppler anemometer and a double-sensor conductance probe under the conditions of the same flowrate and the same average void fraction. The interfacial

x direction	○	●	▲	▲
y direction	□	■	▽	▼
case number	(1)	(2)	(3)	(4)

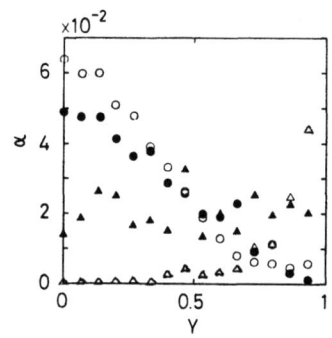

Fig. 4. Distribution of local void fraction, α.

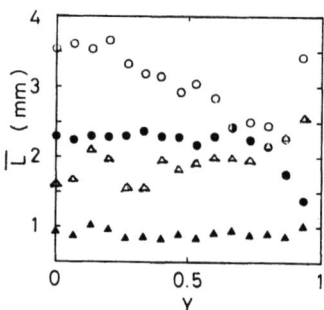

Fig. 5. Distribution of bubble length, L.

Table 1. Bubble number density, bubble diameter and interfacial area.

cases	bubble number density (/cm^3)	bubble diameter (mm)	normalized interfacial area
(1)	0.21	5.4	0.057
(2)	0.42	4.3	0.072
(3)	0.59	3.8	0.081
(4)	5.0	1.9	0.164

x direction	○	●	△	▲	——•——
y direction	□	■	▽	▼	——•——
case number	(1)	(2)	(3)	(4)	single phase

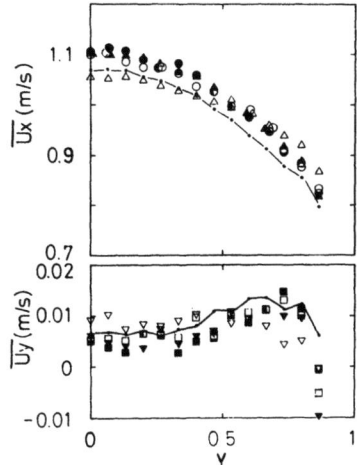

Fig. 6. Profiles of water velocities, $\overline{U_x}$ and $\overline{U_y}$

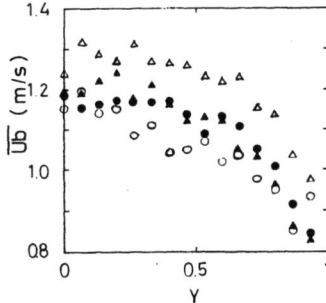

Fig. 7. Distribution of bubble rise velocity, $\overline{U_b}$

x direction	○	●	△	▲	—•—
y direction	□	■	▽	▼	—•—
case number	(1)	(2)	(3)	(4)	single phase

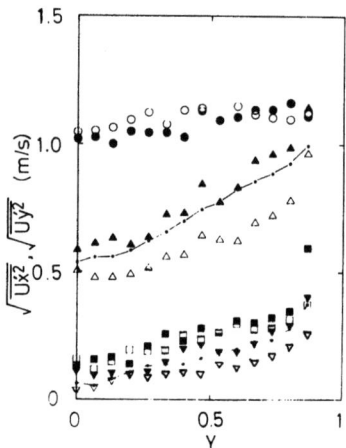

Fig. 8. Fluctuation intensity of Water velocities, $(\overline{U'^2_x})^{1/2}$ and $(\overline{U'^2_y})^{1/2}$

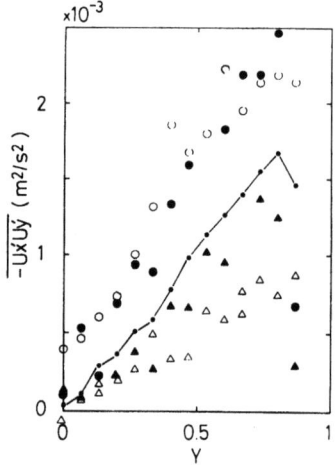

Fig. 9. Reynolds stress, $-\overline{U'_x U'_y}$

area is controlled by changing a mixing condition and adding a surfactant to water.

Main experimental results are as follows.

(i) The increase in the interfacial area under the condition of the constant average void fraction increases the number density of bubbles reducing in bubble size and flattens the gas-phase distribution, in order of the cases (1), (2), and (4). And the increase in the interfacial area (the case (1) to (2)) leads to a decrease in turbulence and an increase in water and bubble velocities, but the extreme increase in the interfacial area (the (4)) brings about a more decrease in turbulence but does not lead to a further increase in both velocities.

(ii) It is suggested that the characteristics of bubble flow can be changed into the desired ones by controlling the interfacial area.

As there may be some effect of secondary flow, a further study is necessary.

REFERENCES

1. K. Sekoguchi, M. Nakasatomi, H. Sato, andO. Tanaka, Heat transfer of vertical air-water bubble flow, Trans. JSME (B), 46-402:78 (1980).
2. A. Serizawa, I. Kataoka, and I. Michiyoshi, Turbulence structure in air-water bubbly flow, (ii) Local properties, Int. J. Multiphase Flow, 2:235 (1975).
3. A. Inoue, S. Aoki, T. Koga, and H. Yaegashi, Void fraction bubble and liquid velocity profiles in a vertical pipe, Trans. JSME (B), 42-360: 2521 (1976).
4. T. G. Theofanous and J. Sullivan, Turbulence in two-phase dispersed flows, J. Fluid Mech., 116:343 (1982).
5. M. Lance and J. Bataille, Turbulence in the liquid phase of a bubbly air-water flow, in : "Advances in two-phase flow and heat transfer," S. Kakac and M. Ishii, ed., Matinus Nijhoff Pub., Boston (1983).
6. G. Matsui and Y. Yamashita, Effect of bubble size on internal structure of bubbly-liquid flow in vertical square channel, Proc. Third Int. Symp. on Applications of Laser Anemometry to Fluid Mechanics, Lisbon, Portugal (1986).

EFFECT OF SURFACTANTS ON THE FLOW OF LARGE GAS BUBBLES

IN CAPILLARY TUBES

D. Barthes-Biesel, N. Moulai-Mostefa, and E. Meister

Université de Technologie de Compiègne
Division de Biomécanique, UA CNRS 858
BP 233, 60206 Compiègne Cedex, France

1. INTRODUCTION

There are many situations in nature and in industry where large gas bubbles are present in tube flow. Such may be case in human microcirculation after treatment with a faulty extracorporeal circulation machine or after a too rapid decompression of deep sea divers. The same problems are encountered in industry with the flow of foams, or with the displacement a liquid phase by a gas phase (e.g. secondary oil recovery). One situation of interest deals with the case of an horizontal tube, in which a large gas slug is present and where the flow is imposed either by a given pressure gradient or by a given flow rate. This problem was investigated in detail by Bretherton (1961) who solved it completely for very small values of the capillary number:

$C = \mu V/\gamma(0)$

where μ is the liquid phase viscosity, V the mean flow velocity and $\gamma(0)$ the surface tension between the liquid and the gas. He was able to calculate the thickness h of the liquid film around the bubble, as well as the bubble velocity U and the additional pressure drop ΔP. This last quantity was found to be independent of the bubble size. Bretherton ran experiments, measured the film thickness and found a good agreement with his theoretical predictions for values of C ranging from 10^{-4} to 10^{-2}. However he did not measure ΔP.

At about the same time, Marchessault and Mason (1960) conducted a short experimental study of the flow of large gas bubbles in horizontal capillary tubes, where they measured both the film thickness and the pressure drop. They found that the experimental value of ΔP was larger than the one predicted by Bretherton, and that it was proportionnal to the bubble length. They explained this result by postulating that the interface of the bubble behaved as if it were rigid, but they did not investigate the cause of rigidification.

Recently, Delaunay (1984) has conducted a thorough experimental investigation of the flow of large gas bubbles in small capillary tubes (diameter 300 µm). He designed an experiment where he could measure simultaneously the bubble length, the average film thickness, the mean flow rate and the additionnal pressure drop. He found for air-water systems, that ΔP first increased almost linearly with the bubble length, until it reached a plateau value, which was one to two orders of magnitude higher than the value predicted by Bretherton (see Figure 2 for example). These phenomena were enhanced when a surfactant (Tween 20) was added to the water. Delaunay interpreted these results by postulating that the rear part of the bubble interface was rigid over a length Lo, whereas the front part of the interface was a classical free-slip surface. The length Lo was estimated from the experimental data and correlated with the capillary number. It turned out that this simple model could indeed be used to analyze the data at least in an approximate sense. The main problem of Delaunay's work lies in the uncertainty in the determination of Lo, and in its phenomenological aspects.

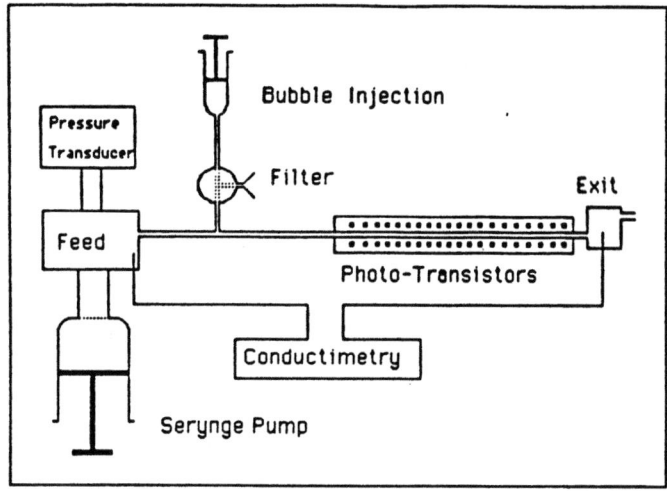

Figure 1. Experimental apparatus.

Since the presence of surfactants appeared to have a strong influence on the hydrodynamics of the system, it was decided to conduct systematic study of these phenomena by working with a single type of surface active molecules for which the physio-chemical thermodynamics were well known. Sodium Dodecyl Sulfate (SDS) was thus selected and, in this paper, we present some preliminary results on its effect on the flow of large gas bubbles in capillary tubes. Delaunay's experimental set-up is used to measure the essential hydrodynamic quantities for air-water systems.

We also present a mechanical model of the situation where the physio-chemistry of the interface is explicitely taken into account. It appears that the model can reproduce some of the experimental results.

In paragraph 2, the experimental method is described, and the main results are given. In paragraph 3, the hydrodynamics of the system are presented. A solution of the problem is given in paragraph 4 for the case where the bubble is very long and the capillary number very small. In paragraph 5 the experimental results are compared to the theoretical predictions.

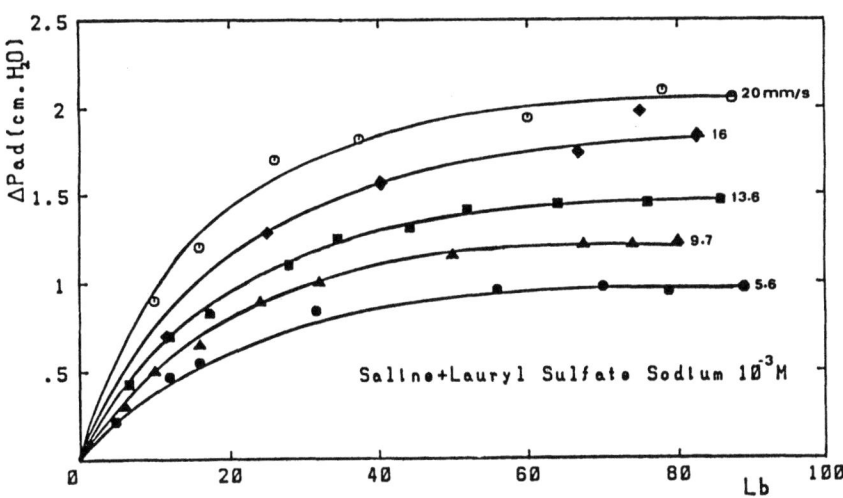

Figure 2. Additionnal pressure drop as a function of bubble length for different velocities.

2. EXPERIMENTAL RESULTS

2.1 Experimental procedure

The original experimental set-up of Delaunay has been improved, and is represented on Figure 1. The main part consists of a precision bore glass capillary tube (Glasswerk) with an internal diameter of 1000 µm, and lengths 25 or 50 cm. The capillary tube is mounted horizontally between a feed reservoir and a constant level exit reservoir. The water is pushed in the tube by means of serynge pump (Precidor), driven by a d.c. stepping motor, fitted with precision serynges. The system can operate either under a constant flow rate, or under a constant pressure head. A filtered air bubble (.2 µm filter) is injected manually in the tube by means of a precision serynge. Bubbles of different sizes are obtained by varying the injected air volume.

Couples of photo-diodes, photo-transistors are mounted along the tube in order to monitor the bubble length and the bubble velocity as it flows. Two platinum electrodes, located respectively in the feed and exit reservoirs, are connected to a conductimeter (Tacussel CDN 6), in order to measure the electrical resistance of the fluid in the tube. The pressure in the feed reservoir is obtained by means of a piezo-resistif pressure transducer (Foxbor 1800, 50" water).

The pressure signal, the electrical resistance, and the signals of the photo-transistors are all sent to an Apple IIE microprocessor. The acquisition of the pressure and resistance signals begins when the bubble arrives in front of the first optical detector. At the end of each run, the program computes the bubble velocity, the bubble length, the additional pressure drop and the film thickness as indicated by Delaunay.

Before each set of experiments, the system is washed with sulfochromic acid and copiously rinced with distilled water. The flowing fluid is distilled water with a given concentration of S.D.S. and 40 g/liter of NaCℓ for conductivity measurements.

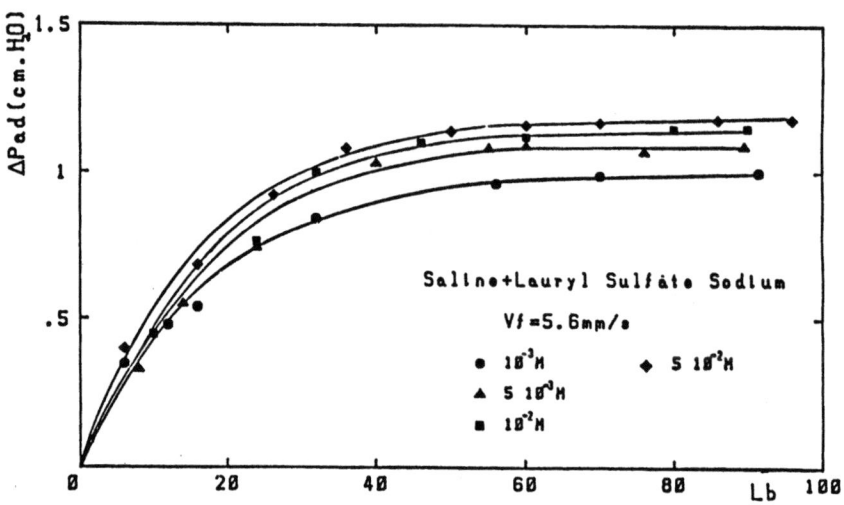

Figure 3. Effect of surfactant concentration on the pressure drop.

2.2 Results

The capillary number varies between $4\ 10^{-5}$ and $4\ 10^{-4}$. Experiments ran at a constant flow rate indicate that, for this range of capillary numbers, the difference between the mean flow velocity and the bubble velocity is of the order of the experimental errors. Consequently, the bubble velocity U is used to evaluate the flow rate in the case where the system is run under a constant pressure head.

Typical results for the pressure drop are shown on Figure 2. For a given concentration of S.D.S. (here 10^{-3} moles/ℓ), the additional pressure drop increases with bubble length until it reaches a plateau value. This behaviour is observed for different flow rates, and thus for different capillary numbers. It should be noted that this pressure-length dependency is completely in contradiction with Bretherton's theory which predicts a constant pressure drop for all bubbles, irrespective of their length. Furthermore the maximum ΔP is some 50 times larger than Bretherton's value.

The influence of the bulk concentration c of surfactant is then assessed, as shown on Figure 3. It appears that for values of c varying between 10^{-3} moles/ℓ to $5\ 10^{-2}$ moles/ℓ, the pressure drop increases with c. This range of variation of c corresponds to sharp variations of the surface tension between air and the contaminated water (see Figure 4). For $5\ 10^{-2}$ moles/ℓ, the critical micellar concentration has certainly been reached, and the corresponding interfacial phenomena become different.

A correlation showing the adimensionalized maximum pressure as a function of the capillary number is given in Figure 5, where it appears that for given physio-chemical conditions, the two quantities are correlated by the following relation :

$$\frac{\Delta P}{\mu U / a} = 130\ C^{-0.46}, \tag{2.1}$$

for a bulk concentration of S.D.S. of 10^{-3} moles/ℓ.

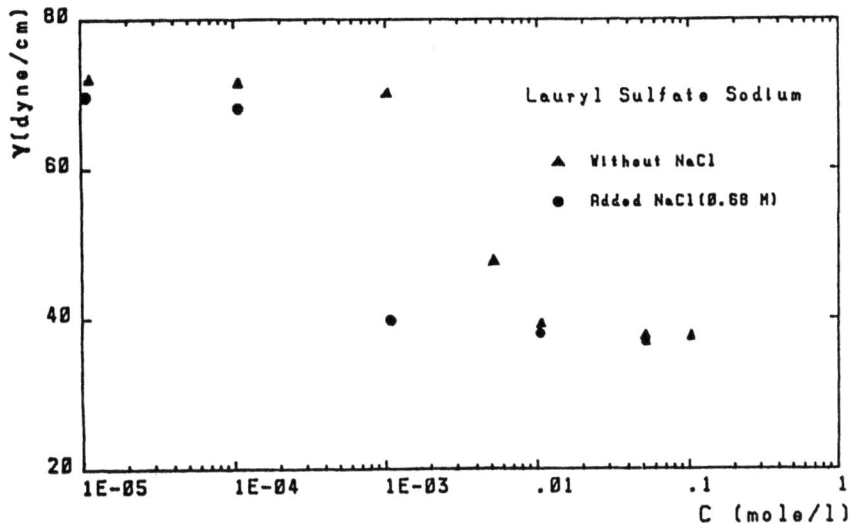

Figure 4. Variation of surface tension with bulk concentration of S.D.S.

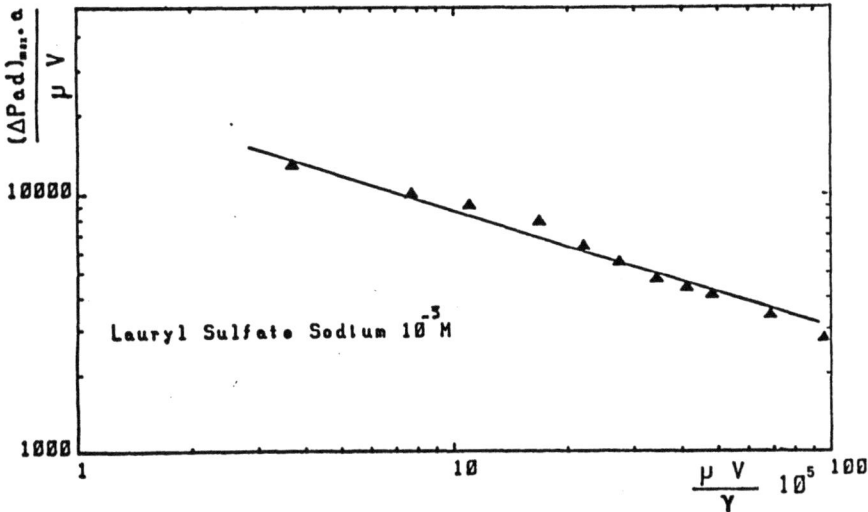

Figure 5. Correlation between the maximum pressure drop and the capillary number for given physio-chemical conditions.

3. EQUATIONS OF MOTION OF A GAS BUBBLE IN A CAPILLARY TUBE

Consider a capillary tube of radius a, in which is flowing a S.D.S. water solution of molar concentration c. The mean flow velocity is denoted V. A large air bubble, of length $L_b a$ flows in the tube (Figure 6). The surface tension between air and water is denoted γ. Non dimensional quantities are used throughout : lengths are scaled with a, velocities with U, the bubble velocity, stresses with γ(0)/a, and surface tension with γ(0), where γ(0) is the clean interface tension.

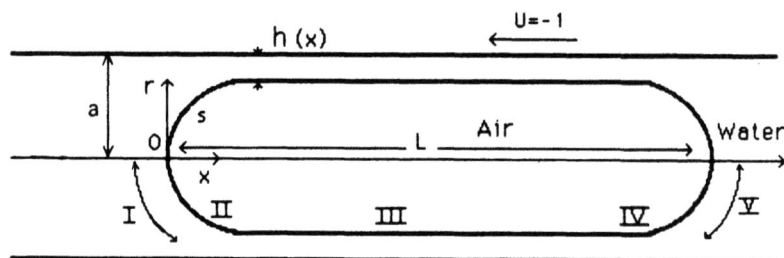

Figure 6. Schema of the bubble.

3.1 Equations of motion

The equations of motion are referred to a reference frame moving with the bubble. Owing to the geometry of the problem, cylindrical coordinates are used where x and r denote respectively the axial and radial coordinates. Thus in absence of inertia effects, the equations of motion are given by :

$$v_{,r} + u_{,x} + v/r = 0, \qquad (3.1)$$

$$c^{-1} p_{,x} = \frac{1}{r} u_{,r} + u_{,rr} + u_{,xx}, \qquad (3.2)$$

$$c^{-1} p_{,r} = v_{,rr} + \frac{1}{r} v_{,r} - \frac{v}{r^2} + v_{,xx}, \qquad (3.3)$$

where a comma denotes partial derivation and where p(x, r), v(x, r), u(x, r) represent respectively the pressure and the radial and axial components of velocity.

The associated boundary conditions are the following :

- no slip at the tube wall :

$$u(x, 1) = -1 \quad , \quad v(x, 1) = 0, \qquad (3.4)$$

- regularity conditions at the bubble ends :

$$u_{,r}(x, 0) = 0 \quad , \quad v(x, 0) = 0, \tag{3.5}$$

- stationarity of the interface :

$$v(x, H) - H'(x) u(x, H) = 0, \quad x \in [0, L_b], \tag{3.6}$$

where H measures the thickness of the film around the bubble, the interface of which is given by :

$$r = H(x), \tag{3.7}$$

$H'(x)$ denotes the derivative of H with respect to x.

- dynamic equilibrium of the interface :

$$\mathbf{n} \cdot \sigma \cdot \mathbf{n} = C_v \, \gamma(\Gamma), \tag{3.8}$$

$$\tau \cdot \sigma \cdot \mathbf{n} = - \tau \cdot \nabla \gamma, \tag{3.9}$$

where σ is the viscous stress tensor in the liquid, C_v is the total curvature of the interface, τ and $\mathbf{n}$ are respectively the unit tangent and normal vectors to the interface. For a Newtonian liquid, the normal and tangential stresses are respectively given by :

$$\mathbf{n}.\sigma.\mathbf{n} = -p + 2C \left[u_{,x} \, n^2_x + v_{,r} \, n^2_r + (v_{,x} + u_{,r}) \, n_x \, n_r \right], \tag{3.10}$$

$$\tau.\sigma.\mathbf{n} = C \left[2(u_{,x} - v_{,r}) \, n_x \, n_r + (v_{,x} + u_{,r})(n^2_r - n^2_x) \right], \tag{3.11}$$

where the internal bubble pressure is taken to be zero.

3.2 Interface behaviour

Equation (3.9) implies that there is a variation of surface tension along the interface. This will be due to the simultaneous effects of adsorption of tensio-active molecules at the interface and of convection of the latters due to the motion. It is thus necessary to investigate in detail the interfacial phenomena. Levich's (1962) classical analysis will be used.

It is first assumed that the surface tension varies linearly with the surface concentration Γ^* of surfactant. Correspondingly, in dimensional form, γ is given by :

$$\frac{\gamma(\Gamma^*) - \gamma(0)}{\gamma(0)} = - K \, \Gamma^*/\Gamma_\infty^*. \tag{3.12}$$

If the surface tension and the surface concentration Γ^* are non dimensionalized respectively with $\gamma(0)$ and with the maximum concentration Γ_∞^*, corresponding to the critical micellar concentration, (3.12) becomes :

$$\gamma(\Gamma) = 1 - K\Gamma, \tag{3.13}$$

where

$$\Gamma = \Gamma^*/\Gamma_\infty^* \tag{3.14}$$

The balance of surface active material at the interface is given in dimensional form by :

$$-j_n + \frac{d}{ds}(\Gamma^* u_s) - D_s \frac{d^2\Gamma^*}{ds^2} = 0, \qquad (3.16)$$

where s is the arclength along the interface, u_s is the velocity at the interface, D_s is the surface diffusion coefficient and j_n is the flux of surfactant normal to the interface. Following Levich, it is then assumed that the flux to the interface is controlled by adsorption/desorption. This will be the case if the concentration of surfactant in the solution in the vicinity of the interface is equal to the bulk concentration. This condition is assumed to be fulfilled here, since, owing to the bubble motion, there is a constant influx of fresh solution to the bubble. Consequently, according to Davies and Rideal (1963) and to Levich, j_n is given by the Langmuir isotherm :

$$j_n = \beta c (1 - \Gamma^*/\Gamma_\infty^*) - \alpha \Gamma^*, \qquad (3.17)$$

where βc is the rate of adsorption (proportional to the bulk concentration) and α the rate of desorption.

Combining (3.16) and (3.17), and using non dimensional quantities, we obtain :

$$\Gamma = \frac{k}{1+k} - \frac{k'}{1+k} \frac{d}{ds}(\Gamma u_s), \qquad (3.18)$$

where surface diffusion is assumed to be negligible with respect to the other interfacial phenomena. The parameters k' and k are defined by :

$$k' = \frac{U}{a\alpha}, \quad k = \frac{\beta c}{\Gamma_\infty^* \alpha}. \qquad (3.19)$$

It follows that k' measures the ratio of desorption time to convection time, whereas k measures the ratio of desorption time to adsorption time. Then, the combination of (3.9), (3.12) and (3.18) yields :

$$\tau \cdot \sigma \cdot n = - C N_s \frac{k+1}{k} \frac{d^2}{ds^2}(\Gamma u_s), \qquad (3.20)$$

where :

$$N_s = K \frac{k}{(k+1)^2} \frac{\gamma(0)}{\mu a \alpha}. \qquad (3.21)$$

Consequently, the parameter N_s depends only on the physio-chemistry of the system, while the hydrodynamics are measured by C.

4. SOLUTION FOR SMALL CAPILLARY NUMBERS

The motion of the bubble is thus completely determined by equations (3.1) to (3.9) and (3.20). Since the position of the interface is a priori unknown, the problem is non linear and no general analytical solution is presently available. It is of interest, though, to examine the case of small capillary numbers and large bubbles, since there exists then a thin film region, where the lubrication simplifications can be made.

Correspondingly, the solution will be sought in the case where:

$$C \ll 1. \tag{4.1}$$

Consequently, in this case, the bubble may be decomposed in five zones as shown on Figure 6. Regions I and V correspond to the downstream and upstream central parts of the bubble. They are almost semispherical, and thus have a curvature roughly equal to 2. Region III is a thin film region where the curvature of the interface is nearly unity. Regions II and IV are two thin regions where the curvature evolves from 1 to 2, and where viscous efforts will thus be important.

In principle, the full solution can be obtained by means of the method of matched asymptotic expansions, as shown by Park and Homsy (1984) in the case of two-phase displacement in an Hele-Shaw cell. Since the full development is long, we shall not go into all the analytical details of the solution, but shall rather concentrate on the identification of the main physical phenomena in each region.

4.1 Region I

From (3.2) and (3.3), it appears that to first order, in capillary number, the pressure is constant and equal to P_I. Consequently, the shape of region I is obtained from (3.8) and is given by the classical capillary statics solution:

$$H(x) = \frac{2\gamma}{P_I} [1 - (\frac{P_I}{2\gamma} x - 1)^2]^{1/2}, \tag{4.2}$$

where $r = H(x)$ is the equation of the interface. P_I is obtained after matching of the solutions in regions I and II. However, from previous work (Bretherton, Park and Homsy), it may be anticipated that $P_I/2\gamma$ is equal to 2 and thus that region I is spherical with a radius equal to 1.

In this preliminary study, it is assumed that there is an accumulation of surfactant molecules in region I, due to convective effects. The interface of I thus is "rigidified". This hypothesis is suggested by the similar problem of a drop sedimenting in a quiescent liquid, where it is well known that the downstream part of the drop interface can become a "no-slip" surface. Consequently, we assume that in I:

$$u_s = 0. \tag{4.3}$$

4.2 Simplified equations for film regions

Such regions are very thin, and it is thus useful to rescale the coordinates x, r and the velocity v, in order to take advantage of the geometry. Correspondingly, the following variables are now defined:

$$\rho = \frac{1-r}{R}, \quad h = \frac{1-H}{R}, \quad \xi = \frac{x-x_0}{L}, \quad w = -\frac{v}{V}, \quad (4.4)$$

where R, L and V are scaling factors to be determined, and where x_0 represents a shift of origin.

It will be further assumed that in those regions, the following condition is met:

$$\frac{k'}{k+1} \ll 1, \quad (4.5)$$

This will be the case if k' is such smaller than unity (slow flow) or if k is much larger than unity (large desorption times). Then (3.20) simplifies:

$$\tau \cdot \sigma \cdot n = -CN_s \frac{d^2 u_s}{ds^2} + O(CN_s k'/k). \quad (4.6)$$

It will be verified a posteriori that the conditions (4.1) and (4.5) are indeed fulfilled by the particular experimental system under investigation.

From Bretherton's work, it may also be anticipated that the proper choice for the scales R and V is:

$$R = C^{2/3}, \quad V = C^{1/3}. \quad (4.7)$$

In terms of the new variables, and to leading order, the problem equations become:

$$w_{,\rho} + C^{1/3} L^{-1} u_{,\xi} = 0, \quad (4.8)$$

$$p_{,\xi} = C^{-1/3} L u_{,\rho\rho}, \quad (4.9)$$

$$p_{,\rho} = 0. \quad (4.10)$$

The corresponding boundary conditions are:

- no slip at the wall:

$$u(\xi, 0) = -1, \quad w(\xi, 0) = 0. \quad (4.11)$$

- the regularity condition will be replaced by a matching condition with the neighbour regions.

- interface stationarity:

$$w(\xi, h) + L^{-1} C^{1/3} h' u(\xi, h) = 0. \quad (4.12)$$

- normal stress balance :

$$-p + 2\, C^{2/3}\, w_{,\rho}\, (\xi, h) = \gamma(\Gamma)\, (1 + Ca^{2/3}\, L^{-2}\, h''). \qquad (4.13)$$

- tangential stress balance :

$$u_{,\rho}\, (\xi, h) = L^{-2}\, N_s\, C^{2/3}\, \frac{d^2 u_s}{ds^2}, \qquad (4.14)$$

where to first order :

$$u_s(\xi) = u(\xi, h), \quad \text{and} \quad s = \xi. \qquad (4.15)$$

Equation (4.14) has been derived by Hirasaki and Lawson (1984) who considered a similar problem. They solved it in the case of a constant film thickness and found that both the front and the rear of the bubble were symmetrically rigidified, a result which does not seem very realistic in view of the convective motion.

The scaling L along the axial direction must now be specified independently for each film region.

4.3 Region II

In this region, the curvature of the interface undergoes an order 1 change, and thus viscous effects must be important and must balance the pressure variations. The two terms of (4.12) then have the same order of magnitude and consequently the proper length scale is :

$$L_{II} = C^{1/3}. \qquad (4.16)$$

It follows that (4.9) becomes :

$$p_{,\xi} = u_{,\rho\rho}, \qquad (4.17)$$

to which the following boundary condition is associated :

$$u(\xi, 0) = -1. \qquad (4.18)$$

It is clear that to be able to match u_s in region I and II, it must be equal to zero in II :

$$u(\xi, h) = u_s = 0. \qquad (4.19)$$

The solution to the flow field in region II is then very similar to the one given by Bretherton and by Park and Homsy. The only difference is due to condition (4.19) which replaces the zero tangential stress condition at the interface. Consequently, one finds that the film thickness is given by :

$$h''' = \frac{6(h - 2Q)}{\gamma h^3}, \qquad (4.20)$$

where Q is a measure of the flow rate in the film, and is defined by :

$$Q = -\int_0^h u\, d\rho. \qquad (4.21)$$

Matching of regions I and II yields that P_I/γ is indeed equal to 2, that the origin of region II is at $x_0 = 1$, and that the value of Q cannot yet be determined. Furthermore, the pressure drop in region II is of order $C^{2/3}$.

4.4 Region III

Region III consists of the main part of the bubble, which is surrounded by a slowly varying thin film. In III, there is a surface velocity gradient due to the presence of a non uniformity in surfactant distribution. Then, viscous forces in the film arise because of the retarding effect of surfactants at the bubble surface. This will be the case if both terms of equation (4.14) are of the same order of magnitude, and thus the proper scaling of region III is :

$$L_{III} = C^{1/3} N_s^{1/2}. \tag{4.22}$$

If $L_{III} \gg 1$, the origin of region III is also located at $x_0 = 1$. Then, the equations in region III are :

$$w,_\rho = 0,$$

$$u,_{\rho\rho} = 0 \quad , \quad p = p(\xi), \tag{4.23}$$

with the associated boundary conditions :

$$u(\xi, 0) = -1 \quad , \quad w(\xi, 0) = 0, \tag{4.24}$$

$$u,_\rho(\xi, h) = \frac{d^2 u_s}{ds^2} = \frac{d^2 u_s}{d\xi^2}, \tag{4.25}$$

$$-p = \gamma(\Gamma) = \gamma(\xi). \tag{4.26}$$

This set of equation is solved as follows. From (4.23), (4.24) and (4.25), the expression of u is obtained :

$$u = \frac{d^2 u_s}{d\xi^2} \rho - 1. \tag{4.27}$$

Then, using (4.21) :

$$h = \frac{2Q}{1-u_s}. \tag{4.28}$$

Eliminating h between (4.28) and (4.27), evaluated at $\rho = h$, leads to :

$$2Q \frac{d^2 u_s}{d\xi^2} + u^2_s - 1 = 0. \tag{4.29}$$

The change of variable :

$$\zeta = \xi/\sqrt{2Q}, \tag{4.30}$$

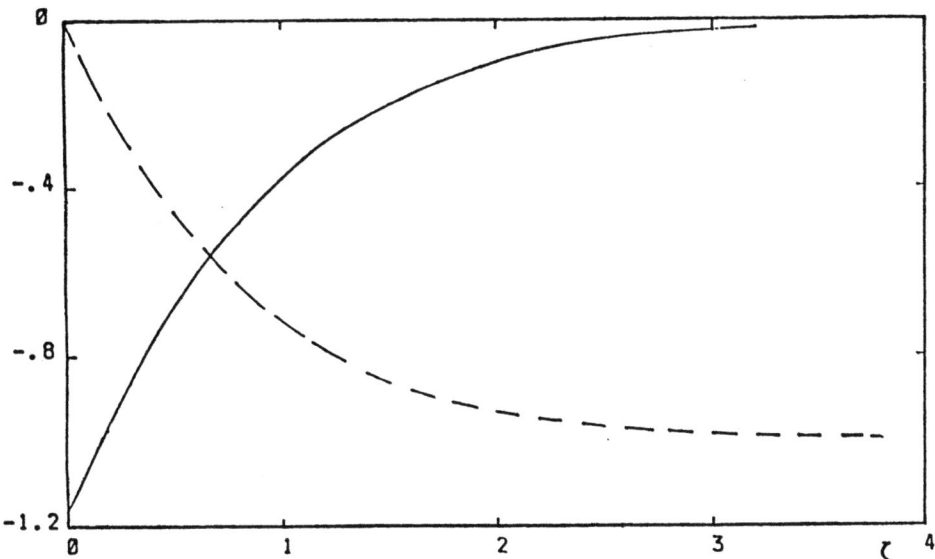

Figure 7. Numerical values of u_s (dashed line) and of $du_s/d\zeta$ (full line) as a function of ζ.

leads to a canonical form of (4.29) :

$$\frac{d^2 u_s}{d\zeta^2} + u^2_s - 1 = 0. \tag{4.31}$$

This equation was integrated numerically by means of the routine ODE of BLAISE package, with the boundary conditions :

$u_s = 0$ for $\zeta = 0$,

$$\frac{du_s}{d\zeta} = 0 \quad \text{for } u_s = -1. \tag{4.32}$$

The condition (4.32) means that for large values of ζ, a free slip interface is found. The graph of the solution of (4.31) is shown on Figure 7. The pressure in the film is obtained from (4.13). Thus, to first order, replacing $\gamma(\xi)$ by its value given by (3.14) and (3.15), p becomes :

$$p(\xi) = -\gamma_e - N_s^{1/2} C^{2/3} \frac{1}{\sqrt{2Q}} \left(\frac{du_s}{d\zeta}\right)_{\zeta=\xi/\sqrt{2Q}}. \tag{4.33}$$

where γ_e denotes the equilibrium surface tension obtained from (3.13) and (3.18). It appears that for large values of ξ (thus for long bubbles), the total pressure drop across the film region reaches a maximum value given by

$$\Delta P_{III}^{max} = 2 N_s^{1/2} C^{2/3} / \sqrt{6Q}. \tag{4.34}$$

109

4.5 Region IV

The same reasoning as for region II, leads to the same longitudinal scaling factor, and thus to the same basic equations (4.17) and (4.18). The only difference between II and IV resides in the interfacial boundary condition which becomes here (after matching of u_s between IV and III):

$$u_{,\rho} = \frac{1}{2Q}\left(\frac{d^2 u_s}{d\zeta^2}\right)_{\zeta_0} = \frac{1}{2Q}(1 - u_s^2)_{\zeta_0} \qquad (4.35)$$

where the right-hand-side of (4.35) is constant and is evaluated for:

$$\zeta_0 = \frac{L_b - 2}{L_{III}\sqrt{2Q}}. \qquad (4.36)$$

Thus integrating the motion equations in a fashion similar to the one used in section 4.3, we obtain the following expression for the film thickness:

$$\frac{d^3 h}{\delta\xi^3} = \frac{3\left[h - Q - \dfrac{(1 - u_s^2)}{4Q} h^2\right]}{\gamma h^3}. \qquad (4.37)$$

The change of variables:

$$t = \xi/Q\gamma^{1/3}, \quad z = h/Q, \qquad (4.38)$$

leads to a canonical form of (4.37):

$$\frac{d^3 z}{dt^3} = \frac{3\left[z - 1 - (1 - u_s^2)\, z^2/4\right]}{z^3}. \qquad (4.39)$$

The asymptotic behaviour de (4.39) for large negative values of t, may be obtained after matching with region III:

$$h \to 2Q/(1 - u_s) \quad \text{as} \quad \xi \to -\infty, \qquad (4.40)$$

which leads to the following asymptotic expression for z:

$$z = \frac{2}{1 - u_s} + \exp\left[(-3 u_s)^{1/3} \left(\frac{1 - u_s}{2}\right) t\right]. \qquad (4.41)$$

when $t \to \infty$, z is asymptotic to a parabola:

$$z = Kt^2 + K't + K'', \qquad (4.42)$$

where the value of the coefficients K, K' and K" are obtained after integrating numerically (4.39) subject to boundary condition (4.41). For matching with V, the important coefficient is K, which is found to vary but slightly with u_s, as shown on table 4.1.

Table 4.1 : Variations of K with u_s

$-u_s$	.1	.3	.5	.7	.9	1.0
K	.626	.668	.703	.702	.682	.669

4.6 Region V

The equations in Region V are essentially the same as in region I. Consequently, the shape of the interface is given by (4.2) where P_I is replaced by P_V. The matching of the solutions for the interfacial profiles, obtained in IV and V leads to :

$$\frac{P_V}{2\gamma} = 1 , \qquad (4.43)$$

$$Q = 2 K \gamma^{-2/3}. \qquad (4.44)$$

It follows from (4.43) that the forward part of the bubble is hemispherical with a radius equal to the tube radius. The value of Q thus completes the solution to the full problem.

The pressure drop across region IV is again $O(C^{2/3})$, and consequently the main contribution to the overall bubble pressure drop ΔP arises from region III :

$$\Delta P = \frac{2}{\sqrt{2Q}} N_s^{1/2} C^{2/3} \left[\left(\frac{du_s}{d\zeta}\right)_{\zeta 0} + \frac{2}{\sqrt{3}} \right] . \qquad (4.45)$$

For very long bubbles, ΔP obviously reaches a maximum value given by :

$$\Delta P^{max} = \frac{4}{\sqrt{6Q}} N_s^{1/2} C^{2/3} . \qquad (4.46)$$

This result is coherent with the experimental findings (see figure 2).

5. DISCUSSION

It has been found experimentally that, for fixed physio-chemical conditions, the maximum pressure drop was correlated to the capillary number by (2.1). Using the same non-dimensional quantities as in section 4, this correlation becomes :

$$\Delta P^{max} = 130 \ C^{0.54} , \qquad (5.1)$$

whereas the theory predicts that ΔP^{max} is proportionnal to $C^{2/3}$. The discrepancy is slight and may attributed to experimental errors and to the difficulty there is to determine the powers of C with any precision at low values. Consequently, it is felt that there is a reasonable

Table 5.1 : Values of N_s, obtained from experimental data

U cm/s	C 10^4	ΔP^{max} cm H_2O	$N_s^{1/2}$	L_{III}
2	2.9	1.5	298	20.3
1.6	2.3	1.4	324	18.7
1.3	1.9	1.1	289	17.6
.8	1.4	.9	290	15.9
.56	.8	.7	328	13.2

agreement between theory and experiments and that the parameter N_s may be computed from the experimental data :

$$N_s = \left(\frac{\Delta P^{max} \, C^{-2/3}}{1.15}\right)^2. \tag{5.2}$$

The values of N_s, obtained for five capillary numbers are shown on Table 5.1, where it appears that, indeed, $N_s^{1/2}$ is almost constant, with a mean value of 306 and a standard deviation of 17. The corresponding length scale of region III may be obtained from (4.22), as shown on Table 5.1, where it has been computed with the mean value of N_s. As predicted by the theory, L_{III} increases with the capillary number, which means that the plateau of pressure drop is reached later for longer bubbles. This is coherent with the experimental results of Figure 2.

Using now this value for N_s, it is possible to deduce from (4.33) and from Figure 7, the variation of the additionnal pressure drop with bubble length. The comparison between the experimental and theoretical curves is shown on Figure 8. It appears that the model reproduces correctly the experimental curve for low values of the capillary number, but that the prediction is not as satisfactory for higher values of C. A reason for this may be attributed to the fact that hypothesis (4.5) might not be fulfilled for large bubble velocities.

It is quite difficult to estimate the proper values for the adsorption and desorption kinetics for ionized surfactants, and electrolytic solutions. Following Rideal and Davies (1963), the rate of desorption is given by :

$$\alpha = 2 \; 10^4 \exp\left(\frac{ze \, \Psi - W}{RT}\right), \tag{5.3}$$

with

$$W = 12 \times 700 \; cal/mole. \tag{5.4}$$

Here z, e, Ψ, R, T represent respectively the valency of the ion, the electric charge, the surface electric potential, the gas constant, the temperature. If Ψ is computed by means of the Gouy equation, then α is equal roughly to 0.1 s^{-1}. Similarly the adsorption rate is given by (3.17) where :

$$\beta = 2.3 \; 10^{18} \; cm/s. \tag{5.5}$$

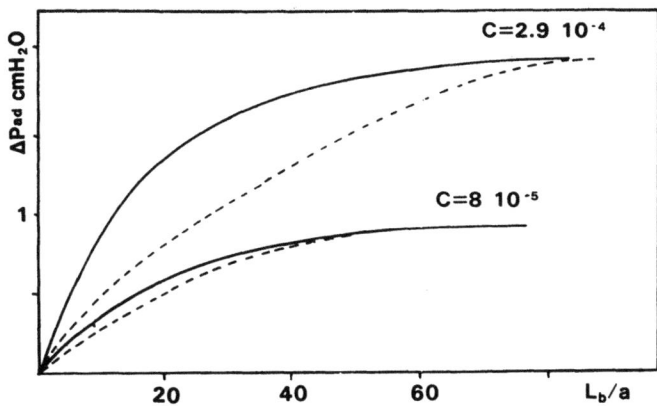

Figure 8. Comparison between the experimental pressure drop curves (full lines), and the model predictions (dashed lines).

Thus

$$\frac{k'}{k+1} = .43 \quad \text{for} \quad C = 8\ 10^{-5}$$
$$= 1.7 \quad \text{for} \quad C = 2.9\ 10^{-4} \tag{5.6}$$

Consequently condition (4.5) is approximatively fulfilled for the lowest value of C, but not for the larger one. This may explain why the theoretical curve is not closer to the experimental one in the later case.

Furthermore the value of N_s, computed directly from Rideal and Davies data is of the order of $3\ 10^3$, a large value, but notably smaller than the one obtained experimentally. This discrepancy may be explained in two ways. First of all, the experiments are very difficult to make and such low levels of additional pressure drop are hard to measure with precision. It is well possible that the experimental values of ΔP are overevaluated, and thus lead to too high values of N_s. Second, the data on adsorbed films, found in the litterature, usually deal with static situations where the deviations from equilibrium are small (an hypothesis used here). However it is clear that our system is rather more complex, since the film is in motion, and since there are parts of the interface where the displacement from equilibrium is important (at the rear), and where compressive forces are exerted on the adsorbed monolayer. Thus the physio-chemical parameters used in (5.3) to (5.6) may not be relevant in this situation.

However, the present system has the great advantage of showing the interplay of physio-chemical and hydrodynamic forces. The physio-chemistry may be controlled by using specific surfactants, and different phases (e.g. oil-water). The analysis of the system is quite complex to perform. A solution has been presented in a particular case, but obviously, it would be of interest to try and obtain other solutions where, for example, condition (4.5) is not imposed.

REFERENCES

1. F.P. Bretherton, The motion of long bubbles in tubes. J. Fluid Mech., 10, 166-188 (1961).
2. Blaise, User's manual. INRIA Report n° 56.
3. J.T. Davies, E.K. Rideal, Interfacial phenomena. Academic Press (1963).
4. M. Delaunay, Perte de charge additionnelle produite par des bulles de gaz dans un écoulement capillaire. Thèse de Doctorat. Université de Technologie de Compiègne.
5. G.J. Hirasaki, J.B. Lawson, Mechanisms of foam flow in porous media. SPE Preprint 012129 (1984).
6. V.G. Levich, Physiochemical hydrodynamics. Prentice Hall. (1962).
7. R.N. Marchessault, S.G. Mason, Flow of entrapped bubbles through a capillary. Ind. Eng. Chemistry, 52(1), 79-84 (1960).
8. C.W. Park, Homsy G.M., Two-phase displacement in Hele-Shaw cells : theory. J. Fluid Mech., 139, 291-308 (1984).

MOTIONS OF STRATIFIED GAS-LIQUID INTERFACES UNDER ELECTROMAGNETIC FIELDS IN HORIZONTAL PIPE FLOW

Jen-Shih Chang
Department of Engineering Physics, McMaster University
Hamilton, Ontario, Canada L8S 4M1

ABSTRACT

Stratified gas-liquid two-phase electromagnetic hydrodynamics in pipe flow has been studied experimentally and numerically. Experimental studies are conducted by using 1.27 and 1.9×10^{-2} [m] I.D. horizontal tubes under the air-water two-phase flow for the range of the gas superfacial velocity from 10^{-2} to 10 [m/s], the liquid superfacial velocity from 10^{-2} to 2×10 [m/s], the applied magnetic field of 0, 2 [T], and the applied voltage from 0 to 20 [kv]. Experimental results are analyzed by a volume averaged two-fluid one dimensional model. The results show that the effect of the applied electric field is influenced between the stratified smooth to wavy and stratified wavy to intermittent flow transition boundary. No significant magnetic field effect has been observed in present range of experiment.

INTRODUCTION

Two phase flow dynamics under the electromagnetic field has become increasingly important in the design, operation and control of many electric apparatus cooling systems. The effects of the electromagnetic field on the quantities of engineering interest such as heat transfer rate, flow rate, and pressure drop are poorly understood at the present stage. One significant difficulty in evaluating such parameters arises from the flow pattern (Fig. 1). In horizontal flow, the transitions between flow patterns were generally controlled by gas and liquid superfacial velocities, and several flow pattern maps were suggested [1-5]. In this paper, experimentally obtained flow pattern transition boundaries and time averaged void fraction for a horizontal air-water two-phase flow under an electromagnetic field are presented and compared with one-dimension, two-fluid models. Experimental studies are conducted by using 1.27 and 1.9×10^{-2} [m] I.D. horizontal tubes under the air-water two-phase flow for the range of the gas superfacial velocity from 10^{-2} to 10 [m/s], the liquid superfacial velocity from 10^{-2} to 2×10 [m/s], the applied voltage from 0 to 20[kv] and applied magnetic field of 0.2 [T]. Experimental results are analyzed by a volume averaged two-fluid one dimensional model.

CONSERVATION EQUATIONS UNDER ELECTROMAGNETIC FIELD

Conversion Equation

Under an electromagnetic field, the mass, momentum and energy conservation equations in fluid flow can be expressed as follows

(i) Mass conservation

$$\frac{\partial \rho}{\partial t} + \nabla \cdot (\rho \underline{U}) = 0 \qquad (2.1)$$

(ii) Momentum conservation

$$\rho \frac{\partial \underline{U}}{\partial t} + \rho(\underline{U} \cdot \nabla)\underline{U} = -\rho g \beta (T - T_s) - \nabla P + \underline{F}_{EB} + \nabla \cdot [v + \varepsilon_D) \nabla \underline{U}] \qquad (2.2)$$

(iii) Energy conservation

$$\frac{\partial T}{\partial t} + \underline{U} \cdot \nabla T = \frac{k}{\rho C_p} \nabla^2 T + Q_{EB} \qquad (2.3)$$

where ρ is the density, U is the flow velocity, β is the coefficient of thermal expansion of the fluid, k is the thermal conductivity, T is the temperature, P is the pressure, v and ε_D are viscosity and eddy viscosity, respectively. T_s is the reference temperature, C_p is the specific heat, $\underline{F}_{EB}$ and Q_{EB} is the momentum and energy change due to the presence of electric and magnetic fields, respectively.

Thermodynamics Under Electromagnetic Fields

In the absence of the electromagnetic field, the thermodynamic state of a fluid element is completely specified when two of its independent state variables are specified, e.g. when its density ρ and temperature T are given. The thermodynamical behaviour then can be expressed by a single characteristic function

$$A = A(\rho, T) \qquad (2.4a)$$

where A is Helmholtz free energy per unit mass. Under the electromagnetic field, additional two independent "mechanical" state variables become important. If we choose the magnetic flux density B and the dielectric flux density D as the two independent electromagnetic state variables, the Helmholtz free energy becomes

$$A = A(\rho, T, B, D,) \qquad (2.4b)$$

From the detailed study of Chu [6], pressure can be expressed by

$$P = \rho RT + \frac{1}{2}(\underline{B} \cdot \underline{H} + \underline{D} \cdot \underline{E}) - \nabla \left[\frac{1}{2} \rho H^2 \cdot \left(\frac{\partial \mu}{\partial \rho}\right)_T + \frac{1}{2} \rho E^2 \left(\frac{\partial \varepsilon}{\partial \rho}\right)_T \right] \qquad (2.5)$$

where μ is the permeability, ε is the dielectric constant and $\underline{B} = \mu \underline{H}$ and $\underline{D} = \varepsilon \underline{E}$. Thus the pressure in the fluid consists of three components.

(1) a "mechanical" part ρRT of the perfect gas
(2) an electromagnetic pressure 1/2 ($\underline{B} \cdot \underline{H} + \underline{D} \cdot \underline{E}$) which is recognized as the Maxwell's stress tensor without the magnetostruction
(3) the term $\nabla(1/2\ \rho H^2\ (\partial \mu/\partial \rho)_T + 1/2\ \rho E^2 (\partial \varepsilon/\partial \rho)_T)$ is the pressure due to electromagnetostruction effects given in the classical theory of electromagnetics.

The origin of these forces are coming from the free charge, the electrodipole and the magnetic dipole interactions. Therefore, momentum changes due to the electromagnetic field, $\underline{F}_{EB} = \nabla P_{EB}$ are

$$\underline{F}_{EB} = \rho_{ie} \underline{E} + \underline{J} \times \underline{B} - \frac{1}{2} E^2 \nabla \varepsilon - \frac{1}{2} H^2 \nabla \mu + \nabla \left[\frac{1}{2} \rho E^2 \cdot \left(\frac{\partial \varepsilon}{\partial \rho_T}\right) + \frac{1}{2} \rho H^2 \left(\frac{\partial \mu}{\partial \rho_T}\right) \right] \qquad (2.6)$$

Thus the momentum in the fluid consists of six components:

(1) term $\rho_{ie}\underline{E}$ is the momentum change due to the space charges.
(2) term $\underline{J} \times \underline{B}$ is the momentum change due to the charge particle motion.
(3) term $1/2 E^2 \nabla \varepsilon$ is the momentum change due to the dielectric property changes.
(4) therm $1/2 H^2 \nabla \mu$ is the momentum change due to the fluid permeability changes.
(5) term [   ] is the momentum change due to the electromagnetostrictions.

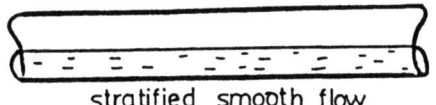
stratified smooth flow

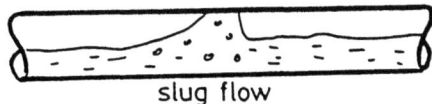
slug flow

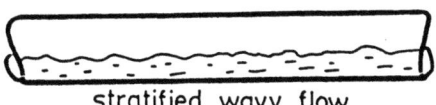
stratified wavy flow

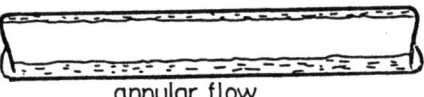
annular flow

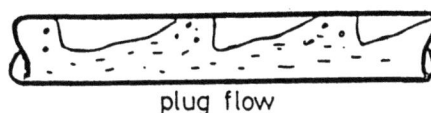
plug flow

Figure 1 The different types of flow regimes developed in a horizontal cocurrent air-water two-phase flow.

The additional energy terms due to the electromagnetic field can be obtained by similar procedures [6] and yield

$$Q_{EB} = \underline{J} \cdot (\underline{E} + \underline{U} \times \underline{B}) + \nabla \cdot [(\underline{E} + \underline{U} \times \underline{B}) \times (\underline{H} - \underline{U} \times \underline{D})] + \underline{E} \cdot \frac{d}{dt}\left(\frac{\underline{D}}{\rho}\right) + \underline{H} \cdot \frac{d}{dt}\left(\frac{\underline{B}}{\rho}\right) \quad (2.7)$$

SEPARATED FLOW MODEL OF STRATIFIED GAS-LIQUID FLOW UNDER ELECTROMAGNETIC FIELDS

<u>Cross-Sectional Averaged Equations</u>

In the present model we assume all flow regimes to be perturbations from the horizontal stratified flow regime [3,4]. The primary property of this regime is the liquid level. This is found by equating the pressure drops in air and water and considering the momentum balance. Consider the cross-sectional planes shown in Figure 2. D is the pipe diameter, h_L is the height of the liquid in the pipe, A_G and A_L are the gas and liquid crosssectional areas respectively, S_G is the gas perimeter, S_L is the liquid perimeter and S_i is the length of the interface between the two. In the second diagram U_G and U_L are the gas and liquid velocities, α is the angle of the pipe with the horizontal, E is the electric field, H is the magnetic field and γ is the angle of the of the electric or magnetic field with the pipe.

Since the gas-fluid system is in equilibrium the total change in momentum for each system must be zero. This gives the following equations for the fluid and gas phases respectively,

$$-A_L \frac{dp}{dx} - \tau_{WL} S_L + \tau_i S_i + \rho_L A_L g \sin\alpha - S_i F_{EB} \cos\gamma = 0 \quad (3.1)$$

$$-A_G \frac{dp}{dx} - \tau_{WG} S_G - \tau_i S_i + \rho_G A_G g \sin\alpha + S_i F_{EB} \cos\gamma = 0 \quad (3.2)$$

where dp/dx is the change in pressure along the pipe, τ_{WL} is the shear stress of the wall on the liquid, τ_{WG} is the shear stress of the wall on the gas, τ_i is the interfacial shear stress, ρ_L and ρ_G are the liquid and gas densities, F_{EB} is the electromagnetic force from (2.6) and g is the gravitational acceleration.

The shear stresses are evaluated in the usual manner, which is

$$\tau_{WL} = \frac{f_L \rho_L U_L^2 L}{2}, \quad \tau_{WG} = f_G \frac{\rho_G U_G^2}{2}, \quad \tau_i = f_i \frac{\rho_G (U_G - U_L)^2}{2} \tag{3.3}$$

The liquid and gas friction factors (f_L and f_G) are evaluated as suggested by [4] as follows

$$f_L = C_L \left(\frac{D_L U_L}{v_L} \right)^{-n}; \quad f_G = C_G \left(\frac{D_G U_G}{v_G} \right)^{-m} \tag{3.4}$$

where D_L, D_G are the hydraulic diameters, and v_L, v_G are the liquid and gas viscosities respectively. C_G, C_L, n and m are constants whose value depends on whether the flow is laminar or turbulent. The same values used by Taitel and Dukler [4], are $C_L = C_G = 0.46$, $n = m = 0.2$ for turbulent flow, and $C_L = C_G = 1.6$, $n = m = 1.0$ for laminar flow. The hydraulic diameters are

$$D_L = \frac{4A_L}{S_L}; \quad D_G = \frac{4A_G}{S_i + S_G} \tag{3.5}$$

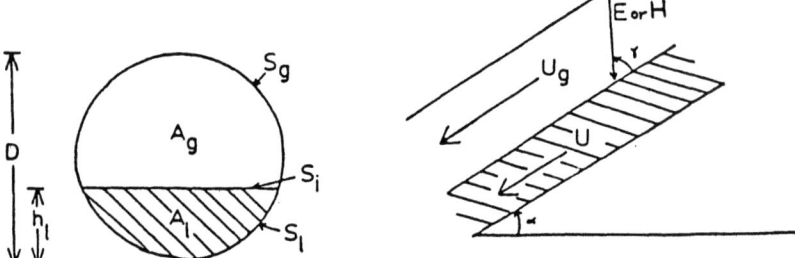

Figure 2   Gas-liquid system in equilibrium stratified flow.

The shear stress due to stratified smooth flow has been shown to be approximately equal to the gas shear stress, so $f_i = f_G$. It is useful to transform the equations to dimensionless form. The normalizing variables are: D for length, D^2 for area, and the superficial velocities U_G^S and U_L^S for the gas and liquid velocities. By combining equations (3.1) to (3.5), we obtain

$$X^2 \left[(\tilde{U}_L \tilde{D}_L)^{-n} \tilde{U}_L^2 \frac{\tilde{S}_L}{\tilde{A}_L} \right] - \left[(\tilde{U}_G \tilde{D}_G)^{-m} \tilde{U}_G^2 \left(\frac{\tilde{S}_G}{\tilde{A}_G} + \frac{\tilde{S}_i}{\tilde{A}_L} + \frac{\tilde{S}_i}{\tilde{A}_G} \right) \right]$$

$$- 4Y - 4\tilde{S}_i \left(\frac{1}{\tilde{A}_L} + \frac{1}{\tilde{A}_G} \right) Z = 0 \tag{3.6}$$

$$X^2 = \frac{\dfrac{4C_L}{D} \left(\dfrac{U_L^S D}{v_L} \right)^{-m} \dfrac{\rho_L (U_L^S)^2}{2}}{\dfrac{4C_G}{D} \left(\dfrac{U_G^S D}{v_G} \right)^{-m} \dfrac{\rho_G (U_G^S)^2}{2}} = \frac{|(dp/dx)_L^S|}{|(dp/dx)_G^S|} \tag{3.7}$$

$$Y = \frac{(\rho_L - \rho_G)g\sin\alpha}{\dfrac{4C_G}{D}\left(\dfrac{U_G^S D}{\nu_G}\right)^{-m}\dfrac{\rho_G(U_G^S)^2}{2}} = \frac{(\rho_L - \rho_G)g\sin\alpha}{|(dp/dx)_G^S|} \qquad (3.8)$$

$$Z = \frac{2\,DF_{EB}\cos\gamma}{\dfrac{4C_G}{D}\left(\dfrac{U_G^S D}{\nu_G}\right)^{-m}\dfrac{\rho_G(U_G^S)^2}{2}} = \frac{2\,DF_{EB}\cos\gamma}{|(dp/dx)_G^S|} \qquad (3.9)$$

The dimensionless quantities are denoted by a tilde ($\sim$). The dimensionless variables in equation (3.6) can be calculated with the dimensionless water level $\widetilde{h}_L = h_L/D$ only, as follows,

$$\widetilde{A}_L = 0.25\left[\pi - \cos^{-1}(2\widetilde{h}_L - 1) + (2\widetilde{h}_L - 1)\sqrt{1 - (2\widetilde{h}_L - 1)^2}\right];$$

$$\widetilde{A}_G = 0.25\left[\cos^{-1}(2\widetilde{h}_L - 1) - (2\widetilde{h}_L - 1)\sqrt{1 - (2\widetilde{h}_L - 1)^2}\right]; \qquad (3.10)$$

$$\widetilde{S}_i = \sqrt{1 - (2\widetilde{h}_L - 1)^2};\; \widetilde{U}_L = \widetilde{A}/\widetilde{A}_L;\; \widetilde{U}_G = \widetilde{A}/\widetilde{A}_G$$

$$\widetilde{S}_L = \pi - \cos^{-1}(2\widetilde{h}_L - 1);\; \widetilde{S}_G = \cos^{-1}(2\widetilde{h}_L - 1)$$

If we applied the electric or magnetic field perpendicular to the horizontal pipes, we obtain $\gamma = 90°$, $z = 0$ and $Y = 0$, and h_L can be unambiguously calculated with a knowledge of the flow rates alone.

<u>Flow Regime Transition Criteria</u>

(i)   <u>Transition from the stratified to the intermittent or annular flow regimes</u>

The intermittent flow regime covers the plug (elongated bubble), and slug flow regimes. Flow in these regimes starts as stratified flow.

As the liquid flow rate is increased, the waves will tend to reach the top of the pipe forming a bridge. For plug and slug flow, a complete bridge is formed.

Transition criteria from the stratified to the intermittent flow regimes can be modified from the work of Brunner and Chang [2], which is based on the Kelvin-Helmholtz instability analysis, and the results can be expressed as follows:

$$f_{EB}^2\left[\frac{1}{2}\frac{\widetilde{U}_G\, d\widetilde{A}_L/d\widetilde{h}_L}{\widetilde{A}_G}\right] \geq 1;\quad f_{EB} = \frac{U_G^S}{\sqrt{D}}\left[\frac{\rho_G}{(\rho_L - \rho_G)g - F_{EB}}\right]^{1/2} \qquad (3.11)$$

where f_{EB} is an electromagnetic Froude number of the form, and dA_L/dh_L can easily be found from the expression for A_L. It should be noted that if the electric field goes to zero, then the expression reduces to the formula obtained by Taitel and Dukler [4].

(ii)   <u>Transition from the stratified smooth to stratified wavy flow</u>

Stratified wavy flow occurs when the gas velocity is great enough to cause waves, but lower than that needed to cause the rapid wave growth characteristic of intermittent or annular flow. This happens when the force of the wind is great enough to overcome the viscous dissipation of the wave. By modification of the model presented by Taitel and Dukler [4], we obtain

$$\widetilde{U}_G > \left[\frac{4v(\rho_L - \rho_G)g^2 - 2F_{EB}\widetilde{U}_L^3(U_L^S)^3}{\beta \rho_G \widetilde{U}_L(U_L^S)(U_G^S)^2} \right]^{1/2} \qquad (3.12)$$

(iii)   Interfacial electromagnetic forces

The cause of above interfacial phenomena may be due to the differences of the dielectric and magnetic property of gas and liquids for present magnetic and electric fields direction, and the most important forces in the interfaces due to the electromagnetic fields are

$$F_{EBI} = \frac{\varepsilon_g (\varepsilon_\ell - \varepsilon_g)^2}{\varepsilon_\ell (\varepsilon_\ell + \varepsilon_g)} E_g^2 + \frac{\mu_g (\mu_\ell - \mu_g)^2}{\mu_\ell (\mu_\ell + \mu_g)} H_g^2 \qquad (3.13)$$

$$= \frac{\varepsilon_\ell (\varepsilon_\ell - \varepsilon_g)^2}{\varepsilon_g (\varepsilon_\ell + \varepsilon_g)} E_\ell^2 + \frac{\mu_\ell (\mu_\ell - \mu_g)^2}{\mu_g (\mu_\ell + \mu_g)} H_\ell^2$$

where E_g, E_ℓ, H_g, H_ℓ are electromagnetic field near interfaces for E and H perpendicular to the gas-liquid interfaces.

Numerical Results

By using the method employed by Chang et al. [1,3], the numerical results for the stratified smooth to slug transition using equations (3.6) to (3.12) to predict the flow transition under an electromagnetic field for a 0.0127 [m] and 0.019 [m] ID pipe are shown in Figures 3 nd 4, respectively. It should be noted that the result without applied field is equivalent to the results obtained by Taitel and Dukler [4]. The applied voltage causes the transition from stratified to slug flow to occur at lower gas and water flows. The effect increases as the voltage is increased, being more pronounced at low gas flows. This occurs because at low gas flow, the liquid level is near the top electrode, thus giving a large force, since, $F_{EBI} \propto [D-h_\ell]^{-2}$ for the conductive tap waters and gas [2,3]. Here, we assumed that the conductive component of F_{EBI} can be neglected due to the electrode arrangements and below corona onset applied voltages. Figures 3 and 4 also show show a significant tube diameter effect due to the E_{EBI} differences.

However, if we calculated the effect of magnetic field in equation (3.13), the order of magnitude of magnetic field component is much less than the both electric field and the gravity terms.

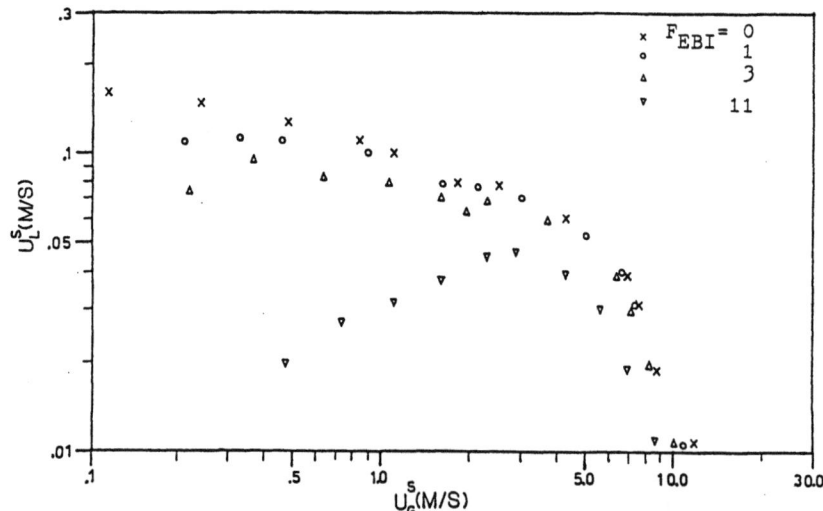

Figure 3   Predicted stratified-intermittent transition for a 0.0127m ID horizontal pipe for various EMHD numbers F_{EBI}.

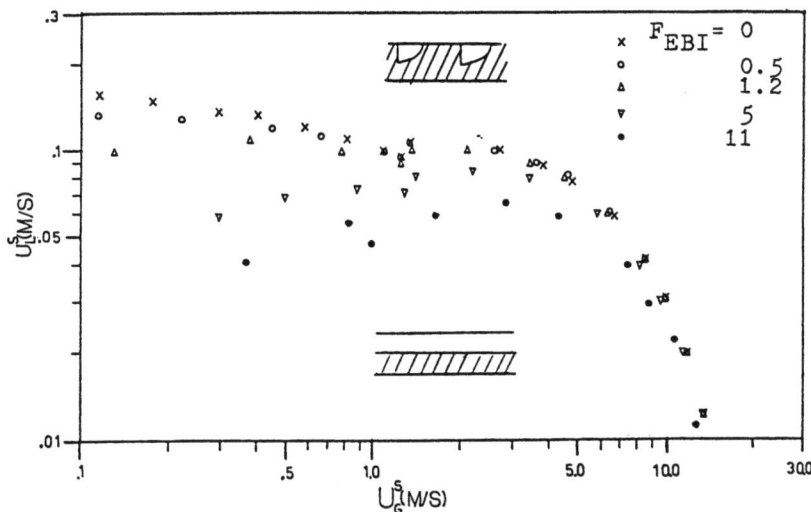

Figure 4 Predicted stratified-intermittent transition for 0.019m ID horizontal pipe for various F_{EBI}.

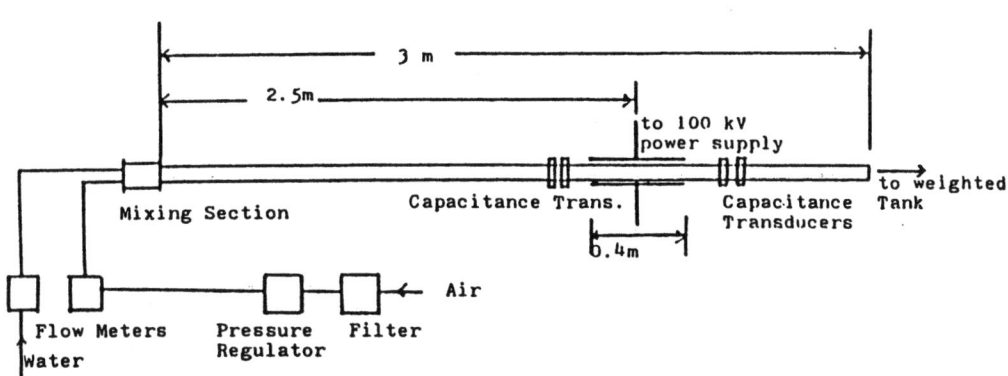

Figure 5 Experimental apparatus.

EXPERIMENTAL APPARATUS

Experimental set up is similar to that used by Brunner et al. [2,3], except 1.27 or 1.19 cm ID, 3 m long, 0.32 cm thick acrylic pipes were used in the present case, as shown in Figure 5. Compressed air and water were taken from the laboratory supply via filter and flowmeters and mixed in T-junctions. The electrodes were two flat pieces of aluminum 46 cm long, 9 cm wide and rounded at the edges to reduce corona effects. The time averaged void fraction was determined from ring type capacitance transducers [7] located end of the electrodes (Fig. 5). For the applied magnetic field experiments, the parallel electrodes was replaced by two electromagnets which capable to generate up to 0.2 [T] of magnetic field.

RESULTS AND DISCUSSION

The effect of applied voltage on the flow regime transitions for 1.9×10^{-10} [m] and 1.27×10^{-2} [m] I.D. tubes are shown in Figures 6 and 7, respectively. Figures 6 and 7 show that the non-monotonic effect of applied electric field on the stratified smooth to wavy flow

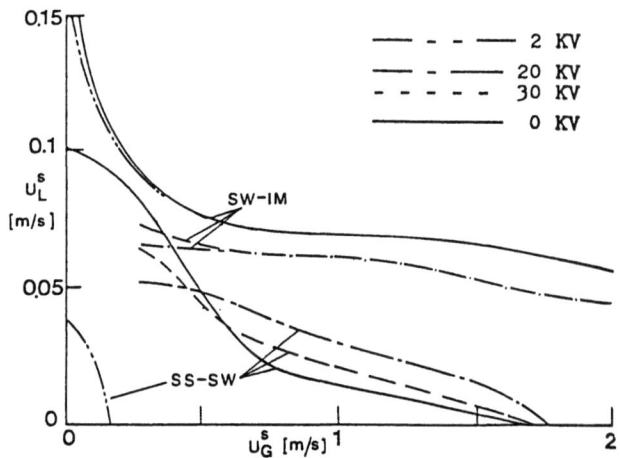

Figure 6 The effect of applied voltage on the flow regime transition boundaries for 0.019 I.D. tubes.

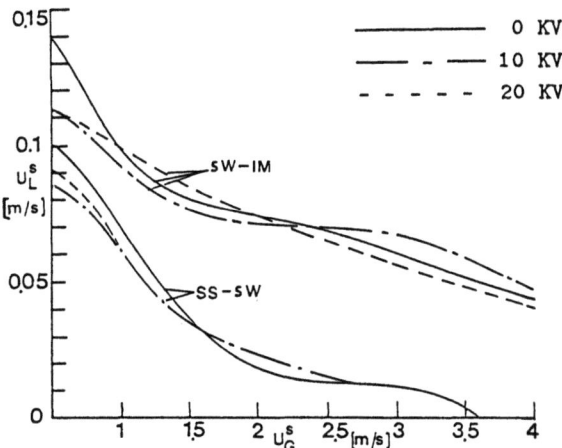

Figure 7 The effect of applied voltage on the flow regime transition boundaries for 0.0127 I.D. tubes.

boundary. The flow regime boundary shifted towards smaller superfacial velocities of gas and liquid areas when small electric field are applied and then shifted back towards larger superfacial velocity area when electric field increases. Since time averaged void fraction or time averaged liquid level in this case, is not influenced by apppplied field we may conclude that this nonmonotonic dependence origin from the surface wave. Multi-dimensional analyses are required for detail investigations.

For the stratified wavy to intermittent transition, flow regime transition boundary shifted toward smaller superfacial velocity area when applied field increases as shown in Figures 6 and 7. This observation agrees qualitatively with the Kelvin-Helmholtz instability analyses of Brunner et al. [2,3]. Figures 6 and 7 show that a significant influence of tube size effect on the transition boundaries, where the boundaries shifted toward smaller superfacial velocities of gas and liquid when tube size increases. For the applied magnetic field experiments, no significant influence has been observed for the present range of experiments.

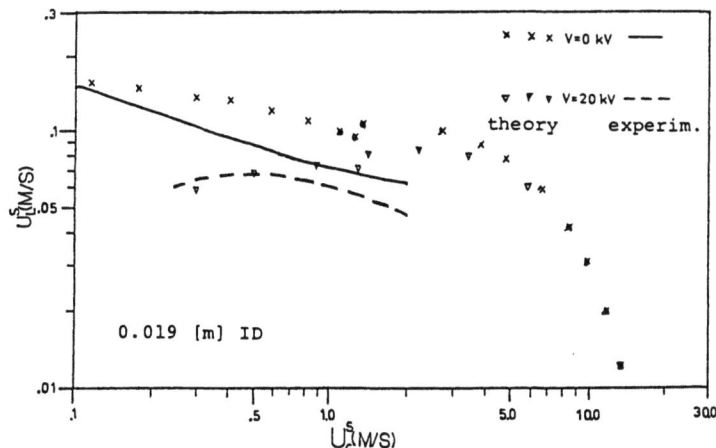

Figure 8 Comparison of flow pattern predictions and experimental obsevations for various voltages.

Typical experimental results are compared in Figure 8 with those obtained by present predictions. Figure 8 shows that present results agree both qualitatively and quantitatively with experimental results, except that the effect of applied field by the current prediction method overestimated EHD force effects for the smaller tube diameters. This may be due to the building up of space charges or the inadequate estimations of the constitutive relationships used in the present models. Nevertheless, the effect of applied electric field in a flow regime map was qualitatively confirmed.

Acknowledgement

The author wishes to thank K. Brunner, L. Lightstone, S.T. Revankar and P.T. Wan for valuable discussions. This work is supported by the Natural Science and Engineering Council of Canada.

References

1. J.S. Chang, S.T. Revankar, R. Raman and F.B.P. Tran, "Application of EHD Technique to a Nuclear Power Plant Emergency Core Cooling System", Trans. IEEE/IAS, IA-21, 715-722, 1985.

2. K. Brunner and J.S. Chang, "Flow Regime Transition under Electric Fields in Horizontal Two-phase Flow", Proc. 15th IEEE/IAS Conf., pp. 1052-1058, 1980.

3. K. Brunner, P.T. Wan and J.S. Chang, "Flow Pattern Maps for Horizontal Gas Liquid Two-Phase Flow under D.C. Electric Field", Electrostatics, Inst. Phys. Conf. Ser. No. 66, pp. 215-220, 1983.

4. Y. Taitel and A.E. Dukler, "A Model for Prediction for Near Horizontal Regime Transition in Gas-Liquid Two-Phase", AICh.E.J. 22, 47-54, 1976.

5. J.M. Mandhane, G.A. Gregory and K. Aziz, "A Flow Pattern Map for Gas Liquid Flow in Horizontal Pipes", Int. J. Multi-Phase Flow 1, 537-553, 1974.

6. B.T. Chu, "Thermodynamics of Electrically Conducting Fluids", Phys. Fluids 5, 473-484, 1959.

7. J.S. Chang, R. Girard, R. Raman and B.F.P. Tran, "Measurement of Void Fraction in Vertical Gas-Liquid Two-Phase Flow by Ring Capacitance Transducers", Mass Flow Measurements 1984", T.R. Hedrick and R.M. Reimer Eds., ASME Press, New York, vol. FED-17, pp. 93-100, 1984.

MEASUREMENT OF INTERFACIAL AREA AND GEOMETRY IN GAS-LIQUID TWO-PHASE FLOW BY AN ULTRASONIC PULSE ECHO TECHNIQUE

E.C. Morala and J.S. Chang

Department of Engineering Physics
McMaster University
Hamilton, Ontario, Canada L8S and 4M1

ABSTRACT

An ultrasonic pulse-echo instrumentation system is developed for two-phase flow applications. The technique is capable of locating instantaneously the location of a moving gas-liquid interface target by measuring the round trip time of flight of the echo of a pulsed ultrasonic signal reflected upon by the interface. Digital methods are employed and fast data acquisition rates are achieved by using transient memory schemes. Data retrieval and analysis is also enhanced by using a microprocessor.

The technique was applied to horizontal air-water pipe flow system in order to measure the interfacial area and geometry; and to characterize the various flow regimes developed. The results are in agreement with some two-phase flow correlations as well as with previous experimental works using different techniques.

1. INTRODUCTION

An ultrasonic pulse-echo instrumentation system was specially designed to measure instantaneously the location of a moving liquid-gas interface. This instrument has been demonstrated to be capable of measuring two-phase flow parameters such as interfacial area, geometry, void fractions and the location of bubbles for both horizontal pipe flow and vertically injected gas flow systems. For the case of the horizontal pipe flow systems, the instantaneous liquid level measurement was plotted against time, to characterize the different flow regimes developed, i.e., stratified smooth, stratified wavy, plug and slug flows. Since the contact type ultrasonic transducer was mounted from the outside wall of a metal pipe, this is yet the only available two-phase flow instrumentation technique, that is non-intrusive, which is capable of identifying the different flow regime patterns without having to construct a visible or transparent test section. The experimental results demonstrated that ultrasonic pulse echo technique offers a feasible instrumentation alternative for two-phase flow applications. Digital schemes were employed to facilitate data recording and analysis by a microcomputer. The location of the gas-liquid interface was determined to a maximum resolution of 0.075 mm. The sampling speed of the device is variable and its setting is only restricted to the round trip time-of-flight of the echo pulse from the transducer to the liquid-gas interface. In this work, an interfacial geometry and areas for a stratified smooth and wavey, slug and plug flow pattern are also presented.

2. PRINCIPLE OF ULTRASONIC PROPAGATION

Ultrasonic waves can be propagated either longitudinally or transversely. In longitudinal waves (sometimes called compressional waves), vibrations of the particles in the material take place in the direction of motion of sound. For the transverse waves (sometimes called shear waves), vibrations of the particles in the material occur normal to the direction of wave motion. Shear waves can generally be passed through only solids because liquids and gases do not support shear stresses.

For applications related to this study, longitudinal waves is of prime concern. The ultrasonic fields generated in a longitudinal propagation depend on the dimension of the ultrasonic transducer and also on the wavelength. This field is divided into distinct regions, namely the near field and the far field as seen in Figure 1a. The near field boundary ends at a distance, ℓ, from the transducer given by

$$\ell = \frac{a^2}{\lambda} \tag{2.1}$$

where a is the radius of the transducer, λ is the wavelength ($\lambda = f/c$), f is the resonant frequency of the transducer, and c is the sound speed over the medium of transmission.

The far field has a divergence angle, α, given by:

$$\sin 2\alpha = 0.5 \frac{\lambda}{a} \tag{2.2}$$

As indicated by Filipczynski et al. (1966), the best directional characteristics of longitudinal waves is achieved when $\lambda << 2a$.

When a wave is incident normally to the boundary between two media, both reflection and transmission occur. Part of the wave energy is reflected back to the incident wave and the rest is propagated in the other medium (see Figure 1b). The reflection and transmission coefficients R and T, respectively are defined

$$R = \frac{I_r}{I_i} \left[\frac{\rho_2 c_2 - \rho_1 c_1}{\rho_2 c_2 + \rho_1 c_1} \right]^2 \tag{2.3}$$

$$T = \frac{I_t}{I_i} = \frac{4\rho_1 c_1 \rho_2 c_2}{(\rho_1 c_1 + \rho_2 c_2)^2} \tag{2.4}$$

where I_i, I_r, I_t represent the intensities of the incident, reflected and transmitted waves, respectively, ρ is the density and c is the sound speed. The product ρc is also called the characteristic impedance. Values of R and T for different combinations of media are well documented (Krautkramer, 1969) in various textbooks. As seen from equation (2.3), a mismatch in characteristic impedances will result in a high reflection coefficient. For the case of liquid-gas interface, almost 99% of the incident wave is reflected back. This phenomenon is the basis for the application of ultrasonics in two-phase flow system. The location of a liquid-gas interface is determined by measuring the time of flight of the reflected wave relative to the incident wave.

Ultrasonic waves travel in straight lines. However, a phenomenon called diffraction, which is characteristic of all types of wave motion, can occur when the ultrasonic wave comes in contact with an obstacle. A change in their direction of motion or bending may occur. Figure 1c illustrates the effects of an obstacle to a wave motion in relation to its size and the wavelength. If the wavelength is very small compared to the size of the obstacle, the area behind the obstacle will not propagate the wave and a reduction of the intensity is effected.

Ultrasonic intensity diminishes with distance in the following manner:

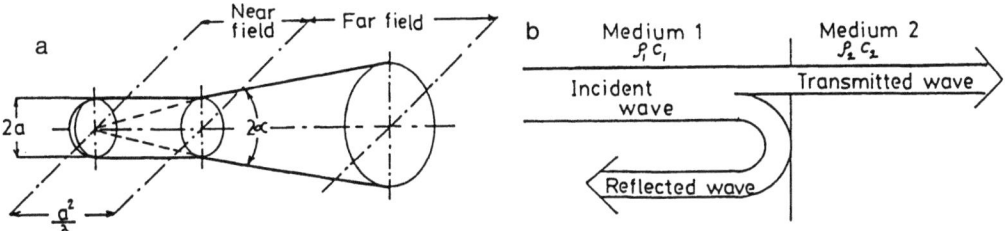

Figure 1 (a) The approximate shape of the fields in an ultrasonic beam for large values of $2a/\lambda$.
(b) Reflection and transmission of an ultrasonic signal at the interface between two media.

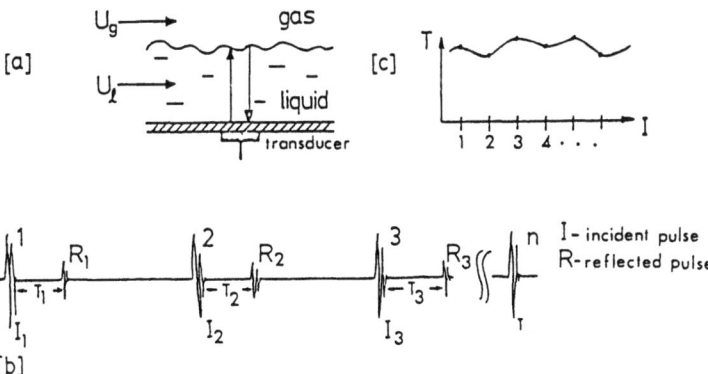

Figure 2 Ultrasonic waveforms obtained in a typical wavy interface.

$$I = I_o \exp(-a\ell) \quad (2.5)$$

I_o is the intensity of $\ell = 0$, I is the intensity at a distance ℓ, and a is the absorption coefficient. The absorption coefficient is also known as attenuation coefficient and is defined as

$$a = a_r + a_s \quad (2.6)$$

where a_r is the real absorption and a_s is the attenuation due to scattering and they are defined as

$$a_r = c_1 f \quad (2.7)$$

$$a_s = c_2 d^3 f^4 \quad (2.8)$$

where c_1 and c_2 are constants, f is the frequency and d is the grain size diameter.

3. EXPERIMENTAL APPARATUS AND PRINCIPLE OF OPERATION

3.1 Ultrasonic Diagnostic System

The application of ultrasonic pulse-echo technique to two-phase flow diagnostics has not yet really been well explored to date, since the interface is not stationary but rather is dynamically altered by the flow structure.

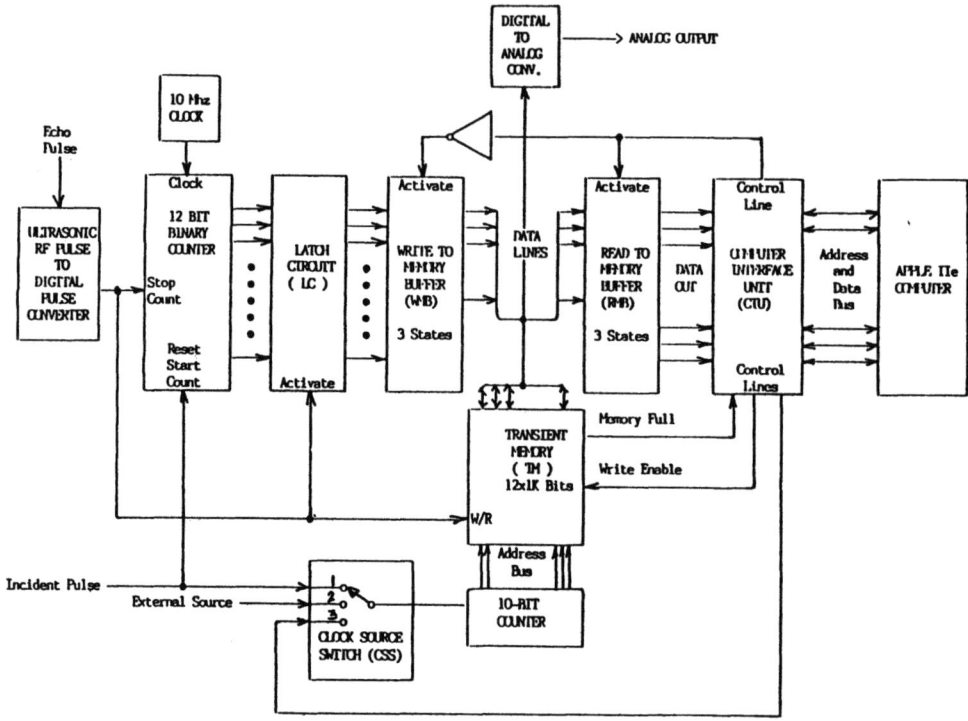

Figure 3 Block diagram of instrumentation design to dynamically measure the ultrasonic echo pulse time of flight.

To demonstrate the signal processing involved, consider the wavy flow of Figure 2a and the corresponding ultrasonic waveform obtained in Figure 2b. The incident pulse is periodically generated at a rate longer than the echo pulse flight times, $T_1, T_2, T_3, ..., T_n$, and fast enough so as to detect the moving interface. The individual values of T are plotted against incident number, I_n, as seen from Figure 2c. If the sound velocity, c, is known for the liquid medium in question, then the instantaneous liquid level, S(t), is known by the following:

$$S(t) = 1/2\, c\, T \qquad (3.1)$$

The block diagram of the specially designed circuit is shown in Figure 3.

In the following discussions, the abbreviations used are:

CIU — Computer Interface Unit
CSS — Clock Source Switch
DAC — Digital to Analog Converter
LC — Latch Circuit
RMB — Read Memory Buffer
TM — Transient Memory
WMB — Write Memory Buffer

The Panametrics 5052UA ultrasonic analyzer is used as our source for the incident pulse and the echo pulse. The incident trigger pulse is then used to start the 12-Bit Binary Counter. At the instant the echo-pulse is detected, it is converted to digital logic pulse and then in turn used to stop the counter and activate the latch circuit (LC). At this same instant, the binary count is then latched out through a three state Write Memory Buffer (WMB) and is written to the $12 \times 1k$ Transient Memory (TM). The address of the TM storage is generated by a 10 bit counter whose clock is synchronized by the incident trigger pulse via a manual or software controlled Clock Source Switch (CSS). This synchronization is necessary so that, for every occurrence of an incident trigger, the address bus of the TM is incremented, thereby data corresponding to the time of flight of the echo-pulse are stored sequentially in the transient memory. Each memory element is equivalent, in real time, to the period of the incident trigger. There is no restriction as to the setting of the period of the incident trigger except that it should be longer than the time of flight of the echo pulse.

As soon as the transient memory storage is full, it will send a signal to the Computer Interface Unit (CIU) to perform all of the following actions concurrently:

(1) Deactivate the WMB and place it in the high impedance state, thereby disallowing any more data transfer from the LC to the TM.

(2) The write enable signal to the TM is released, thereby disallowing any writing of data to the TM, and allowing the reading of data from the TM.

(3) Activate the three-state Read Memory Buffer (RMB) to enable the data transfer from TM to CIU, and consequently, by software control, from CIU to the Apple IIe microcomputer.

The control lines of the CIU generate, by software control, all the necessary signals to perform the above processes. It also provides the signals to properly transfer data from the TM to the Apple IIe computer. When data is ready to be transferred to the Apple IIe computer, the CSS is switched to position 3 (see Figure 3). This action allows the computer program to have control over which data storage element of the TM is to be read by the computer.

Another purpose of the CIU is to store 12 bits of data from the TM to two successive 8 bit data in the memory storage of the computer. Twelve data bits were chosen for the TM in order to provide for:

(a) Finer time resolution of the echo pulse time of flight. For our instrument, this translates to 0.075 mm water thickness resolution.

(b) Increased range of the position of the gas phase interface to be detected. For our instrument, this translates to 307 mm maximum range.

The CSS can also be switched to position 2 (Figure 3). This allows input of external source clock which provides for reading the data of the TM via a DAC. The analog output of the DAC is then connected to an oscilloscope for quick viewing of the waveform.

Once data are transferred to the computer, it is then an easy task to write programs to process these data. In our case, data are merely transferred to a disk storage device and consequently off-line analyses are performed. To achieve high speed data transfer, the I/O operation is written in machine language.

3.2 Horizontal Pipe Experiment

In the horizontal pipe experiment of Figure 4, air and water of known flow rates flow through a 2.0 cm I.D. horizontal pipe. A piezoelectric type transducer (contact type, 1/4", 2.25 MHz, Panametrics Model A5011) was mounted from the outside wall of the pipe, at a section where the flow is fully developed. The test section was actually made of aluminum to demonstrate the applicability of this technique to metallic pipes. The rest of the pipe was made of glass in order to visually observe the type of flow regimes developed. With this flow geometry, the different flow regimes observed were stratified smooth, stratified wavy, plug and slug flows. The mechanics of these flow regimes are thoroughly discussed in the works of Taitel and Dukler (1976). The different types of flow regimes developed in a horizontal two-phase flow geometry are shown in Figure 5.

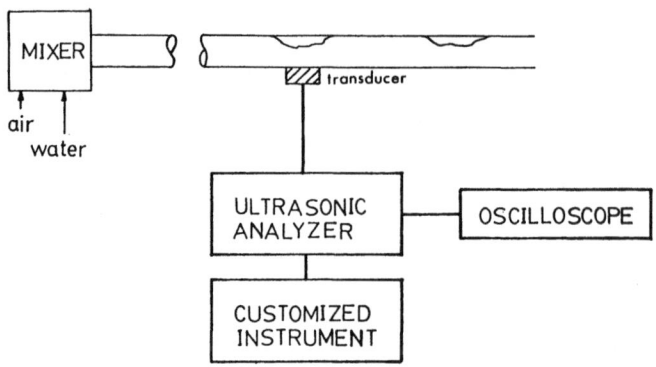

Figure 4 Experimental set-up of the horizontal pipe air-water two-phase flow system.

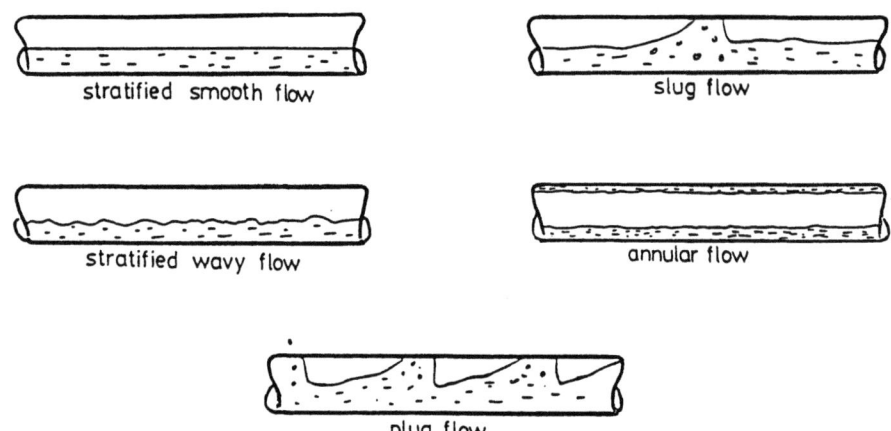

Figure 5 The different types of flow regimes developed in a horizontal cocurrent air-water two-phase flow.

During horizontal, cocurrent gas-liquid flow in pipes, a variety of flow patterns can exist. The flow patterns developed depend upon the manner by which the liquid and gas is distributed along the pipe. In the case of stratified smooth flow, the liquid phase flows at the bottom of the pipe and the gas phase at the top, and the intersurface between them is smooth. As the liquid rate is increased, the liquid level rises and a wave is formed which grows rapidly, tending to block the flow. Wavy flow occurs under conditions where the velocity of the gas is sufficient to cause waves to form but slower than that needed for the rapid wave growth towards the top wall. Plug flow is characterized when elongated gas bubbles form at the top part of the tube, and separated by sections of continuous liquid flowing slowly downstream. In slug flow, the liquid slugs are separated by gas pockets moving violently downstream. At still higher gas rates, there is insufficient liquid flowing to maintain or form the liquid bridge, and the liquid is swept up and around the pipe to form an annulus.

4. EXPERIMENTAL RESULTS FOR HORIZONTAL TWO-PHASE FLOW

4.1 Calibration Results

In order to calibrate the system, a horizontal pipe partially filled with a known amount

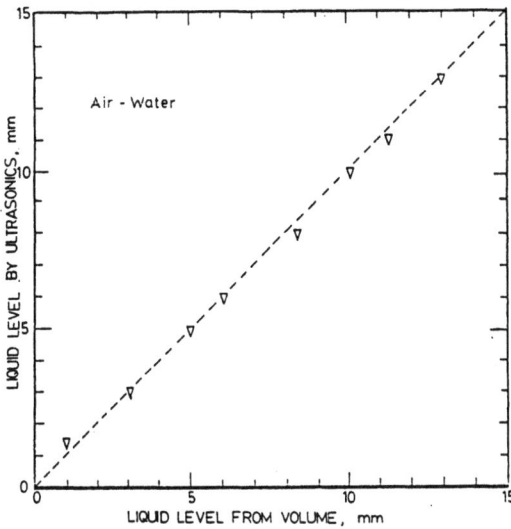

Figure 6 Comparison between measured liquid level in a horizontal pipe by ultrasonic pulse-echo technique versus the liquid level obtained by volume measurement.

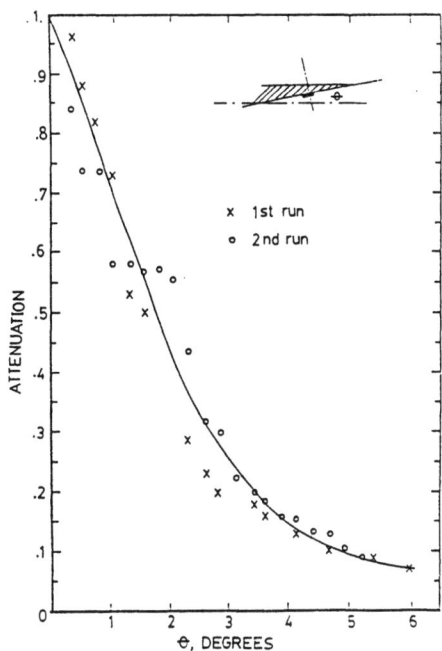

Figure 8 The attenuation of ultrasonic pulse echo with respect to the degree of inclination of the pipe.

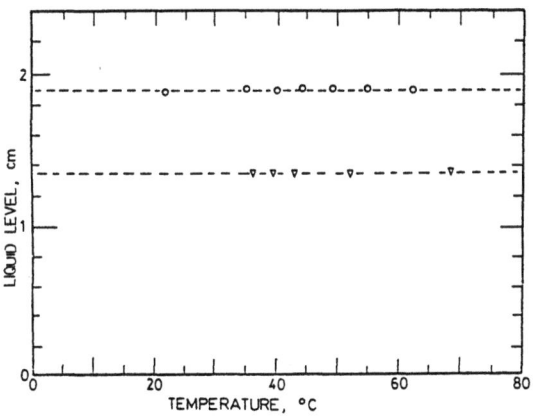

Figure 7 The effect of liquid temperature to the liquid level measurement using ultrasonic pulse echo technique.

of water (static) was used, and the liquid level (or film thickness) was measured. One measurement is taken from the known volume of water and the other by ultrasonic pulse echo method. The result is shown in Figure 6 which is in excellent agreement.

The effect of liquid temperature to ultrasonic film thickness measurement is shown in Figure 7, which indicates no temperature effects from 20°C to 80°C, where the effect of temperature and pressure to the sound velocity of water reproduced from Weast (1984).

Figure 8 shows the attenuation of the ultrasonic pulse-echo as a function of the angular displacement of the target interface. This information is useful in determining the maximum divergence of the ultrasonic beam, as well as in setting the pulse level detector to effectively focus the transducer.

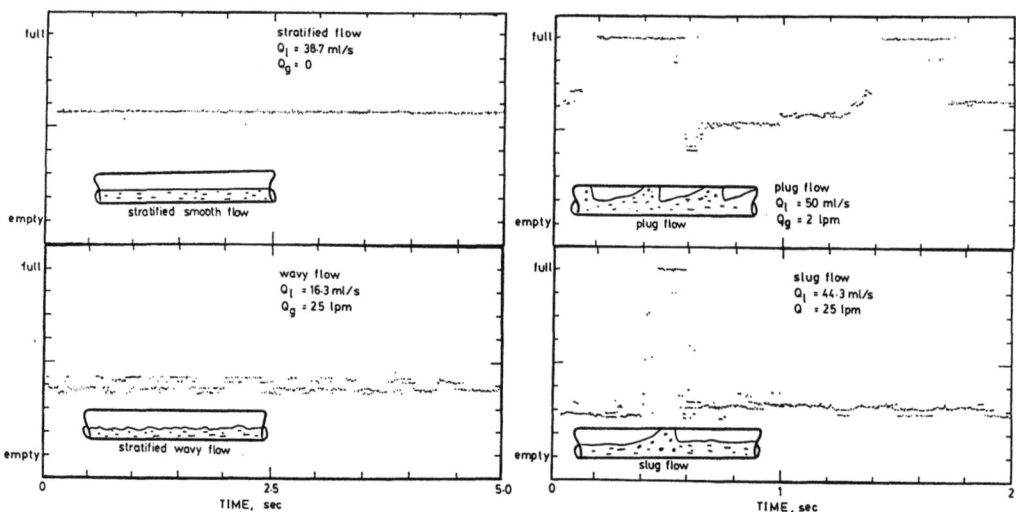

Figure 9 Typical ultrasonic signal as seen from the oscilloscope for different types of horizontal flow regimes.

4.2 Flow Regime Characterization

Flow regime characterization using ultrasonics was reported by Chang et al. (1982). However, their work yielded only qualitative results. The instrumentation design as discussed in Section 3.1 is implemented to characterize the flow regimes in a horizontal two-phase flow system and extends the work of Chang et al. Since this instrument measures the instantaneous liquid level over a period of time, the direct plot of a collection of 1024 data points will indicate the different flow patterns developed. Figure 9 shows the various waveforms for stratified, wavy, plug and slug flows, respectively. Multiple points occurring on the same time scale are due to the limited resolution of the dot matrix printer used. As seen from these plots, it is an easy task to identify the different flow patterns.

4.3 Bubble Speed and Interfacial Area

Ultrasonic pulse-echo technique may be used to measure approximately the bubble speed and interfacial area in a horizontal two-phase flow system. Figure 10 is a schematic diagram showing the suggested experimental setup to accomplish such a purpose. This is simply achieved by duplicating the instrumentation requirements. Two ultrasonic transducers are also used and mounted at a fixed distance, S, parallel to the flow of the bubble. The two transducers are excited at the same time by a single trigger source. Two output display waveforms are then produced. These two waveforms are basically the same except of course, the downstream side instrument will show a phase lag, ΔT. The bubble speed, U_B, is then known by,

$$U_B = \frac{S}{\Delta T} \tag{4.1}$$

To measure approximately the interfacial area, consider the instantaneous liquid level output of Figure 11a and the corresponding coordinate transformation of Figure 11b of a typical plug flow. Then, the incremental area, dA, is known as

$$dA = 2 \times dz \tag{4.2}$$

whereby

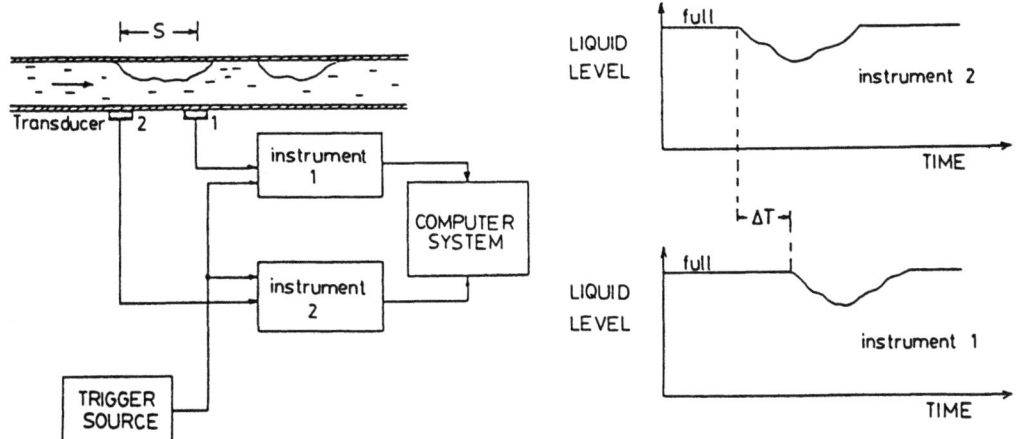

Figure 10   Proposed experimental set-up to measure bubble speed and interfacial area using ultrasonic technique in a horizontal two-phase flow system.

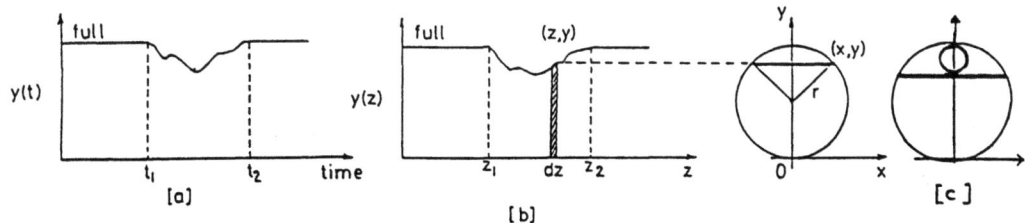

Figure 11   Projected interfacial size and shape for the proposed experimental set-up.

$$x = (2\,r\,y - y^2)^{1/2} \tag{4.3}$$

$$y = y(t); \quad z = U_B t; \quad dz = U_B\,dt.$$

Therefore, the interfacial area, A, is obtained by integrating Equation (4.2) over the limits between two points when the water level is full. That is,

$$A_B = 2U_B \int_{t=t_1}^{t=t_2} \left(2\,r\,y(t) - y(t)^2\right)^{1/2} dt \tag{4.4}$$

since the discrete values of y(t) are actually acquired by the instrumentation system, then the interfacial area per plug, A_B, may also be calculated using,

$$A_B = 2U_B \sum_{i=1}^{n} \left(2\,r\,y_i - y_i^2 \Delta t\right)^{1/2} \tag{4.5}$$

where i = 1 is that point in time corresponding to integration limit t_1 of equation (4.4), i = n is that point in time that corresponds to limit t_2, and Δt is the period of the ultrasonic trigger applied to the instrumentation system.

Alternatively, Simpson's rule may also be applied to the integrand of equation (4.4) to calculate the interfacial area.

Figure 12 Experimental results of the time-averaged liquid level and interfacial area for various volumetric flow ratios and flow regimes.

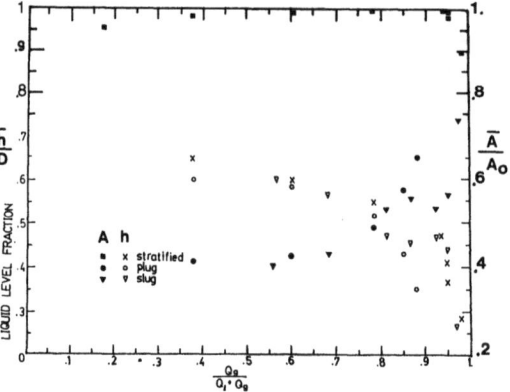

For the model assumed cylindrical cross-sectional bubble shape as shown in Figure 11c, we obtain

$$A_B = \pi U_B \sum_{i=1}^{n} \left(2r - y_i \Delta t\right) \tag{4.6}$$

We may assume that the maximum and minimum interfacial area for slug or plug flow are equations (4.6) and (4.5), respectively.

For the stratified smooth and wavy flow, the instantaneous interfacial area become equation (4.3), and the instantaneous interfacial area for annular flow may be assumed to be $\pi(2r-y)$. Typical time averaged interfacial area and liquid level is shown in Figure 12. By applying the polynomial regression model, result in Figure 12 can be approximated by

$$\overline{h}/D = 0.673 + 0.124 \beta - 0.442 \beta^2 \tag{4.7}$$

having a variance of 0.063, and normalized time average interfacial area. $\overline{A}/A_0$ becomes

$$\overline{A}/A_o = 0.327 - 0.124 \beta + 0.442 \beta^2 \tag{4.8}$$

where A_o is the πD, D is the tube diameter, β is the $Q_g/(Q_g + Q_\ell)$, and Q is the flow rate.

ACKNOWLEDGEMENT

The authors wish to express their appreciation to G.A. Irons and L. Matikainen for valuable discussions and comments. This work was supported in part by the Natural Sciences and Engineering Research Council of Canada.

REFERENCES

Chang, J.S., Ichikawa, Y. and Irons, G.A., 1982, Flow regime characterization and liquid film thickness measurement in horizontal gas-liquid two-phase flow, "Measurements in Two-Phase Flows", T.R. Heidrick and B.R. Patel, eds., pp. 7-12.

Filipczynski, L., Pawlowski, Z. and Wehr, J., 1966, "Ultrasonic Methods of Testing Materials", Butterworths, London.

Krautkramer, J. and H., 1969, "Ultrasonic Testing of Materials", Springer-Verlag, New York.

Matikainen, L., Irons, G.A., Morala, E.C. and Chang, J.S., 1986, Ultrasonic system for the detection of transient liquid/gas interfaces using the pulse-echo technique, Rev. Sci. Instrum., 57:1661-1666.

Taitel, Y. and Dukler, A.E., 1966, A model for predicting flow regime transitions in horizontal and near horizontal gas-liquid flow, AIChE J., 22(1):47-55.

Weast, R.C., 1984, "CRC Handbook of Chemistry and Physics", 64th edition, CRC Press.

INTERFACIAL TRANSFERS IN STRATIFIED GAS-LIQUID TWO-PHASE FLOW: A CRITICAL REVIEW

J.S. Chang
Department of Engineering Physics, McMaster University
Hamilton, Ontario, Canada L8S 4M1

ABSTRACT

Interfacial heat mass and momentum transfers in stratified gas liquid two-phase flow is critically reviewed. The significance of interfacial velocity, geometry and areas on the interfacial transfers are discussed and applied to the stratified gas-liquid two-phase flow predictions.

INTRODUCTION

The elucidation of two-phase flow dynamics has become increasingly important for the design, operation and control of many processes and power systems. This research, however, is at an early stage of development in spite of considerable activity, and we still cannot analyze many apparently simple but important phenomena. This is because the analysis is made immensely more difficult than in single phase flow by the presence of interfaces in the flow field, and the need to describe heat, mass and momentum coupling between the two phases. One of the major impediments to progress has been the difficulty in making good measurements of the velocities of each phase and the interfacial areas and shapes in two-phase flow (Young and Hsu 1982). Therefore, the development of advanced techniques to measure the above parameters is of high priority in the current research field. At the same time the development of constitutive equations for modelling relevant two-phase flow situations is of great engineering interest. Such two-phase flow modelling should recognize the evolution of interfaces in making predictions of such quantities of engineering interest as averaged pressure drop, flow rate and heat transfer. This is because the shape and motion of interfaces profoundly affects the heat, mass and momentum transfer between phases.

The ultimate objective of the present work is to survey the published literature and recommend suitable interfacial parameters correlations in the stratified two-phase flow. The recommended correlations should be designed to provide data of a character applicable to 1-D and multi-dimensional "multi-fluid" models. These should ideally be expected to predict the dynamics of each phase more accurately than "mixture" models (Boure 1978, Drew et al. 1979, Hancox et al. 1980, Carver 1982). Since, most of interfacial transfers in a 1-D model can be rewritten as (Ishii, 1982),

Interfacial transfer ~ (Interfacial Area) x (driving Force)

$$\sim A_i \, F_{DF}$$

First order geometrical effect can be expressed by

$$A_{i1} = \frac{\text{Interfacial Area}}{\text{Mixture Volume}} = \frac{1}{L_s}$$

and the driving forces are

1. Local turbulence
2. Interface Motion
3. Potential (temp. difference, relative velocity, etc.)
4. Geometrical Factors (α_g, A_i, etc.)

Therefore, interfacial parameters need to be considered for interfacial transfers for various flow regime is shown in Table 1. In this work, stratified two-phase flow is chosen to demonstrate a critical review of the interfacial transfers in two-phase flow.

INTERFACIAL PARAMETERS FOR STRATIFIED FLOW

Two-phase stratified laminar motion has been observed in a variety of situations most notably in pipeline transport. However, there is some confusion as to the transition criterion for laminar to turbulent stratified flow. Various authors (Grennel and Epstein 1962, Govier and Omer 1962, Chisholm 1968) have demonstrated that the Reynolds number, based on superficial velocity, should be less than 1000 for laminar flow regimes. Spriggs (1973) considered analyzing the two-phase situation using Hank's criteria (1963) and found a poor correlation between theoretical and experimental results. He proposed this was due to stabilizing forces in one phase which act to dissipate disturbances in the other before they grow into turbulent patches. In general, it has been assumed that, as with single-phase flow, transition to turbulence occurs when the Reynolds number exceeds about 2000. Here, the Reynolds number is based on average film velocity and hydraulic diameter of the flow. This paper is concerned with the effect of following parameters in the interfacial parameters in stratified flow:

1. laminar and turbulent flow
2. fully developed ($U_y = 0$) or underdeveloping flow ($U_y \neq 0$)
3. surface waves and liquid levels (or void fractions)
4. interfacial mass transfers
5. variable property effects

Here, the range of the flow velocity in terms of superfacial and local velocities are shown in Figure 1.

Interfacial heat and mass transfer

Interfacial heat transfer in a stratified smooth flow generally used a solid interface approximation (zero interfacial velocity) in parallel channel flow. The heat transfer coefficient from gas to liquid phases yield

$$Nu = 5 + 0.015 \, Re_f^a Pr_w^b \quad \text{(turbulent flow)} \tag{1}$$

$$a = 0.80 - 0.24/(4 + Pr_w)$$

$$b = 1/3 + 0.5 \exp(-0.6 \, Pr_w)$$

$$Nu = 1.86 \, (Re_f Pr_f \frac{D}{L})^{1/3} (\frac{\mu_f}{\mu_w})^{0.14} \quad \text{(laminar flow)} \tag{2}$$

in a fully developed flow (Sleicher and Rouse 1975). However, the interfacial velocity is finite even under a very low phase velocity. Therefore, an application of equations (1) and (2) must be reconsidered. If we considered stratified flow as shown in Figure 2, the energy conservation equations becomes

TABLE 1

The parameters need to be considered for interfacial transfers.

	Flow Regime	Flow Velocity	Interfacial Shape	Interfacial Area or Void Fraction
Dispensed Flow	Droplet Bubbly	Laminar Flow with & without Wake	Surface deformations	Mean Diameters and number density
Separated Flow	Stratified Annular Inverted Annular	laminar flow laminar flow with wake or turbulent	Surface waves	liquid film thickness or levels
Inermittent Flow	Slug Churn Liquid Slug	turbulent	Slug or churn length and shapes	Slug length and radius

$$\underline{U}\cdot\nabla T - \nabla\cdot[k(T)\,\nabla T] = 0 \tag{3}$$

For the condition without interfacial mass transfers (evaporations or condensation) and vertical velocity components (underdeveloping flow or surface wave), the heat transfer occurs only by the conduction phenomena.

Therefore, we obtain (Chang and Laframboise, 1978),

$$Nu_i = \left(\frac{dt}{dH}\frac{dH}{d\eta}\right)_i \tag{4}$$

$$H = \int F(t)\,dt$$

where $\eta = y/L$, $t = T/T_s$, and $f(t) = k(T)/k(T_s)$.

If we approximate $f(t) = t^\beta$, we obtain

$$Nu_i = \frac{(T_w/T_s)^{-\beta}[1-(T_w/T_s)^{1+\beta}]}{(1-\alpha_g)(1+w)} \tag{5}$$

$$= [1-(T_w/T_s)]/(1-\alpha_g) \quad (\text{for } \beta = 0)$$

The stratified smooth flow becomes wavy flow due to the momentum change caused by interfacial mass transfers (evaporation or condensation), or by the interfacial drags with following conditions,

$$U_g \geq \left[\frac{4g\,\mu_\ell(\rho_\ell-\rho_g)}{s\,\rho_\ell\rho_g U_\ell}\right]^{\frac{1}{2}} \tag{6}$$

where s is a sheltering coefficient ($\simeq 0.01$). With a presence of surface wave, the interfacial heat transfers will be enhanced by following mechanisms,

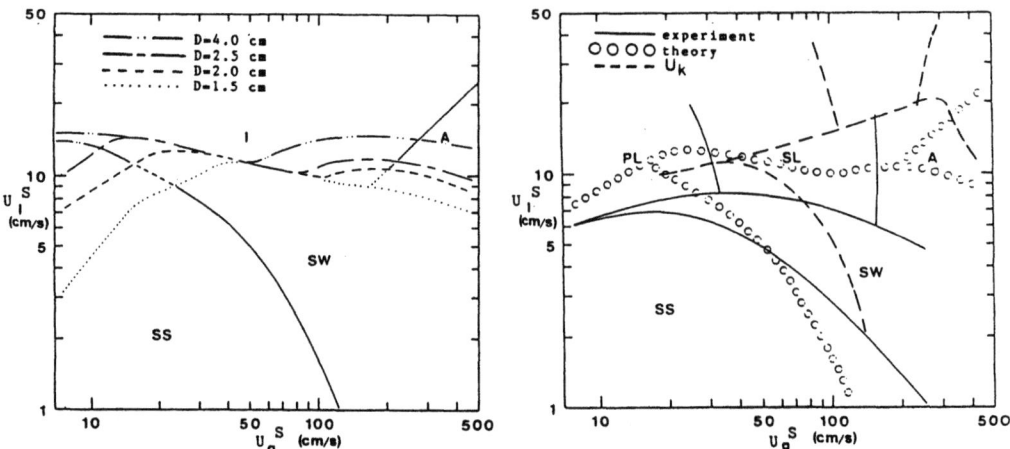

Fig. 1 Stratified flow boundary based on superfacial and local velocities for various tube diameters. SS = stratified smooth, SW = stratified wavy, I = slug or plug, A = Annulur flow. Here local velocity was defined from $U_k = U_k^s/\alpha_k$ (Lightstone and Chang 1986).

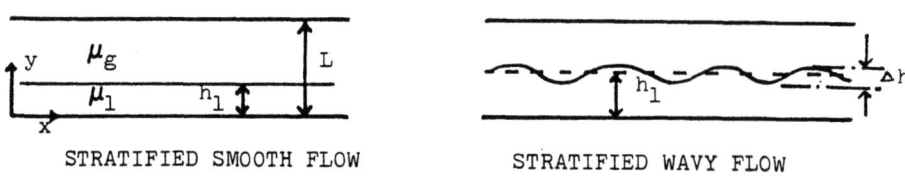

Fig. 2 Schematics of stratified flow.

Table 2 Constitutive equations for interfacial heat transfers

References	Processes	Equations	Applicable Ranges	Comments
Thomas R. M. (1979)	Condensation in horizontal channel	$h = 0.25 \, [K]/\, Re^{3/4} Pr^{1/2} \, [Kw/°C\, m^2]$	$u_\ell = 0$	Only steam is flowing
Segev et al (1981)	Counter-current flow in inclined channel	$Nu = 1.16 \times 10^{-3} Re_v^{0.28} Re_\ell^{0.87} Pr^{0.05}$ $Nu = 8.5 \times 10^{-4} Re_v^{0.25} Re_\ell^{0.85} Pr^{0.5}$	$T_{\ell in} = 23°C$ $0.1 \leq W_v \leq 0.32$ $0.21 \leq W \leq 0.43$ [kg/s]	Horizontal wavy interface 17-45° inclination
Ardron & So (1984)	Rewetting	liquid to interface $\lambda_{\ell-i} = 2\times 10^5 \alpha_v \alpha_\ell$ for $h_\ell \geq h_{\ell-i}$ $h_{\ell-i} = 0$ $h_\ell < h_{\ell-i}$ vapour to interface $\lambda_{v-i} = [(W_i k_v)/(AD_v C_v)] \, [0.023\, Re_{iv}^{0.8} Pr_v^{0.4}]$ W_i - interface width $Nu_{v-i} = 0.023 Re_{iv}^{0.8} Pr_v^{0.4}$	$2000 < Re_{iv} > 2000$	Rama-Model Experiment with heated & unheated tube

(i) increasing heat transfer surface
(ii) generation of velocity component perpendicular to the interface (e.g. U_y in Figure 2).
(iii) enhance local mixtures and temperature gradients.

For the condition with interfacial mass transfers, the existing published results are summarized in Table 2.

Interfacial momentum transfer

The interfacial shear stresses are often evaluated in the conventional manner using the form suggested by Blasius (1948)

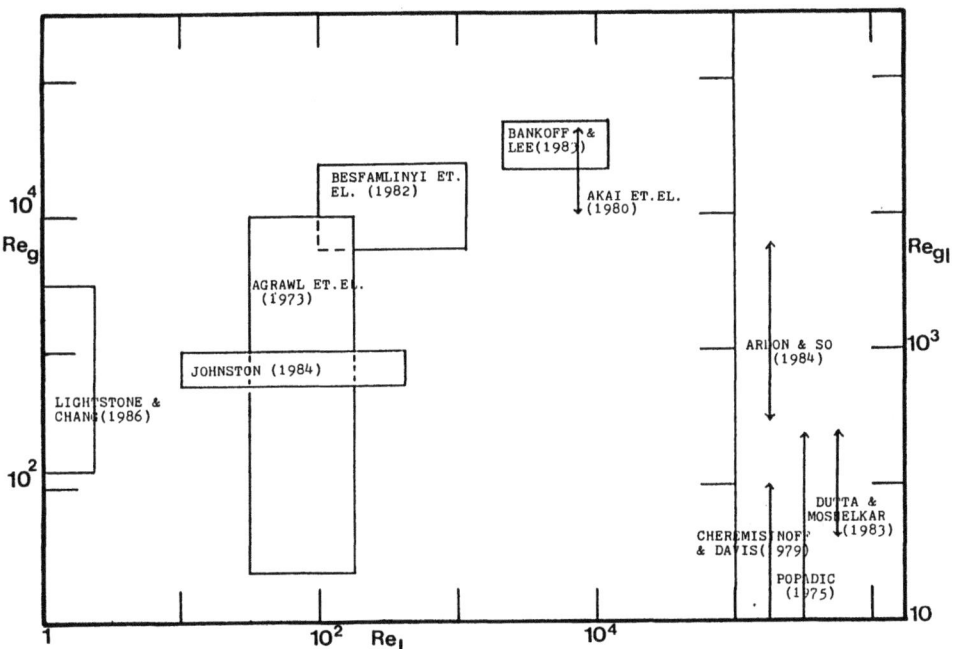

Fig. 3   Applicable ranges of interfacial momentum transfer constitutive equations based on the Reynolds numbers for stratified flow.

$$\tau_i = f_i \rho_g \frac{(u_g - u_\ell)^2}{2} \qquad (7)$$

with the interfacial friction factors evaluated from (Gazley, 1963)

$$f_i = f_g = C_g [\rho_g (D_g u_g) / \mu_g]^{-x_g} \qquad (8)$$

While, strictly speaking, this assumption is only valid for stratified smooth flow. The error incurred if the interface is wavy. This is verified a posteriori by direct comparison with experimental results.

In evaluation of the constants in the Blasius equation, it is assumed that the results from single-phase analysis may be applied directly. For the case of steady laminar flow the variables C and X are 16.0 and 1.0, respectively, corresponding to Poiseuille flow. For steady turbulent flow, these parameters have the values 0.046 and 0.2, respectively, corresponding to the well known Blasius equation for smooth pipe flow. In this analysis the transition to turbulent motion is assumed to occur in a given phase when that fluid's Reynolds number exceeds 2000. Here, again, the possible effect of surface waves is ignored. Recent experiment works by Lightstone and Chang (1984), Morala and Chang (1986) show that the Blasius type approximation with 1-D separated flow model agree relatively well with experimental values for laminar-turbulent or turbulent-turbulent combinations without surface waves.

For the pressure driven stratified incompressible laminar flow in a fully developed channel or pipe flow as shown in Figure 2. Several theoretical analyses have been conducted (Ranger and Davis, 1978). If we assumed that the differences between pressure drops in each phase is relatively small, we obtain, in our notations,

$$\tau_i = \frac{L^2}{2} \frac{dp}{dx} \left[\frac{U_\ell(2a_\ell - 1) + a_\ell^2(\mu_g - \mu_\ell)}{\mu_e + a_\ell(\mu_g - \mu_\ell)} \right] \quad (9)$$

for the channel flow with finite void fractions, $a_g = (1-a_\ell)$, where $U_i = U_g = U_\ell$, $\tau_g = \tau_\ell = \tau_i$ or $\mu_\ell (\partial U_{x\ell}/\partial y)_i = \mu_g (\partial U_{xg}/\partial y)$ are also assumed in the interface.

For a wavy interface, Akai et al. (1981) proposed that the wavy interface can be treated plainly as a turbulent flow over a rough solid wall channel as follows.

$$\frac{1}{\sqrt{f}} = 4.06 \ln\left(\frac{1-h_\ell}{\Delta h}\right) + 3.36 \quad (10)$$

Interfacial shear stress correlations determined by various authors are summarized in Figure 3 and Table 3. The results show that the effect of surface wave may be less influenced Blasius type equation as demonstrated by Shoham and Taital (1984) for the incompressible flow for small diameter pipes. However, recent experimental work by Sutradhar and Chang (1985) show that surfacewave becomes important source of interfacial frictions when the gas phases is the compressible flow. Recent 1-D separated flow model of Morala and Chang (1986) and the multi-dimension model of Aoki et al. (1981) also demonstrated that the effect of surface wave may also be significant even in the incompressible flow in the larger diameter pipes as shown in Figure 4. Therefore, the role of interfacial geometry must be investigated in detail.

<u>Interfacial area and geometry</u>

Interfacial area inside pipes without surface waves can be rewritten in terms of liquid levels h_ℓ or void fractions a_g as follows:

$$A_i = D[1-(2h_\ell/D-1)^2]^{\frac{1}{2}} \quad (11)$$

$$a_g = h_\ell^2[\pi/2 - \sin^{-1}(1-D/h_\ell)] - (h_\ell - D)(2h_\ell D - D^2)^{\frac{1}{2}}$$

With the presence of surface waves, equation (11) can be applied, if we can replace h_ℓ by a time averaged liquid level h_ℓ. Determination of the liquid level has been conducted by twin impedance probes (Davices, 1978) or by the ring type capacitance probes (Ouzu et al., 1973). However, the twin impedance probe is an intrusive device and the ring type capacitance probes can not be applied to the local measurements. Recently, Chang et al. (1980, 1986) developed ultrasonic pulse echo techniques to determine instantaneous liquid levels. Studies of interfacial areas and the geometries in stratified flow will be one of the most important goals of the ultrasonic techniques, since as we discussed in the previous section, interfacial heat, mass and momentum transfer in the stratified flow may be significantly influenced by interfacial geometry than the time averaged interfacial areas.

Table 3 Constitutive equations for interfacial momentum transfers.

References	Processes	Equations	Applicable Ranges	Comments
Akai et al (1980)	Co-current Air-Hg flow	$C_{fi} = 0.03$	$10^4 \leq (Re_v)_{eq} \leq 5 \times 10^4$ $Re_\ell = 8.04 \times 10^3$	wavy interface
Sinai (1983)	Co-current gas-liquid flow	$C_{fi} = 0.08$		Theory based on single phase Charnack equation
Taitel & Dukler (1975)	Flow regime modelling	$\tau_i = \rho_g f_i (U_g - U_\ell)^2/2$ $f_i = f_g = C_g \{\rho_g (D_g U_g)/\mu_g\}^{-x_g}$ $C_g = 16.0, x_g = 1.0$ (Laminar flow) $C_g = 0.046, x_g = 0.2$ (turbulent flow) $D_g = 4 A_g/(S_g + S_i)$		Experimental confirmation by Lightstone & Chang (1984) : Best fitting of experimental result to two-fluid model code TY
Shoham & Taitel (1984)	Co-current gas-liquid flow	$\tau_i = 0.0142 \, \rho_g U_g^2/2$ (turbulent-turbulent)		Modification of Cohen & Hanratty's experimental results (1968)
Cheremisinoff & Davis (1979)	Co-current gas-liquid flow	$f_i = 0.008 + 2Re_\ell/10^{-5}$ (turbulent-turbulent)	$Re_\ell \leq 100$	Modification of Miya et al.'s experimental results
Ishii (1982)		$\tau_i = f_i \rho_g (U_g - U_\ell) |U_g - U_\ell|/2$ $f_i = 0.005[1 + 300(1 - \alpha_g)/4]$	low entrainment	
Chan & Banerjee (1981)	Rewetting model in horizontal tubes	$F_i = (1/2) f_i U_g |U_\ell| \rho_g \, a_g/\alpha_g A_o$ $f_i = 0.005$		Based on Wallis's experiment (1969)
Popadic (1975)	Wavy film	$f = 24/Re_o$; $Re_o = [12n/(2n+1)](q/g\delta)(\rho_g \delta/k)^{1/n}$ δ – mean film thickness, k – consistency index, n – power law constant, $q = u_s \delta + (\rho_g/k)^{1/n} [n/(2n+1)] \delta^{2+1/n}$ – flowrate per unit width	$5 \leq Re_o \leq 200$	Experiment
Bassols, Juhel, & Rousseau (1982)		$\tau_i = (2/\pi) 10^{-2} \sqrt{\alpha(1-\alpha)} \, [1 + 75(1-\alpha)] \, (\rho_v/D)(U_v - U_\ell)^2$ $f_i = (2/\pi) 10^{-2} \sqrt{\alpha(1-\alpha)} \, [1 + 75(1-\alpha)] \, (\rho_v/D)$		
Ellis, & Gay (1959)		$f_i = 0.374 Re_g^{-0.111} (y_o/h_g)^{0.0862}$ y_o – wave roughness height h_g – gas phase thickness	$Re_g = h_g U_g \rho_g/\mu_g$	An alternate to this equation giving a better fit to the data is equation Grovier & Aziz (1977)
Grovier & Aziz (1977)		$Y_o = 2.84 Re_g^{-0.838} (y_o U_i^*/\mu)^{0.892}$ $U_i^* = \sqrt{\tau_i g/\rho}$: friction velocity		

(continued)

References	Processes	Equations	Applicable Ranges	Comments
Johnston (1984)	Air-water	$\tau_i = \rho_g f_i [U_g + U_\ell]^2 / 2$	$G_\ell = 2.11\text{-}30.00 \times 10^{-3}$ m^3/s $G_g = 0.9.00 \times 10^{-4}$ m^3/s	Experiment d = 0.057-0.127m
Agrawal, Gregory & Govier (1973)	Air-oil	$\tau_i = (0.804\, Re_g^{-0.285})^2 (\rho_g U_g^2 / g)$	P = 88.93-91kPa U_g = 0.0113-6.1m/s U_ℓ = 0.002, 0.004, 0.006m/s	Experiment d = 25.75mm
Ardron & So (1984)	Rewetting	$\tau_i = (-1)[(4W_i \rho_v f_i)/8A](U_v - U_\ell)\|U_v - U_\ell\|$ $f_i = \begin{cases} 0.184\, Re_{iv} & Re_{iv} > 2000 \\ 64/Re_{iv} & Re_{iv} < 2000 \end{cases}$ $Re_{iv} = (U_v - U_\ell) D_v / \nu_v$	2000 < Re_{iv} > 2000	Rams-Model Experiment with W_i - interface width heated and unheated tube
Besfamilinyi, et al. (1982)	Isothermal	$C_{fi} = 1.78 \times 10^{-3}\, Re_\ell^{0.32}$ and with the appearance of equal waves at the surface of the film: $C_{fi} = 1.07 \times 10^{-3}\, Re_\ell^{0.45}$	Re_ℓ = 100 to 1100 Re_g = 6500 to 23000	Experiment in rectangular chann L = 2.3m h = 19mm w = 165mm
Bankoff & Lee (1983)	Flooding – steam/saturated water flow	$f_i = 0.012 + 2.694 \times 10^{-4} (Re_\ell/1000)^{1.534} (Re_v - Re_v^*)/1000$ where $Re_v^* = 1.837 \times 10^5 Re_\ell^{-0.184}$	$1800 \leq Re_\ell \leq 11000$ $18600 \leq Re_v \leq 49300$	Experiment P = 1atm H = 0.076 m L = 1.46 m
Dutta & Moshelkar (1983)	Wavy film	$f = 24/Re$ $Re = Re_o + Re_s$ $Re_s = 12qu_s/g^2$ $u_s = \beta \rho g \delta$:slip velocity β = slip constant	$5.5 \leq Re \leq 200$	Re_s – part effect of slip

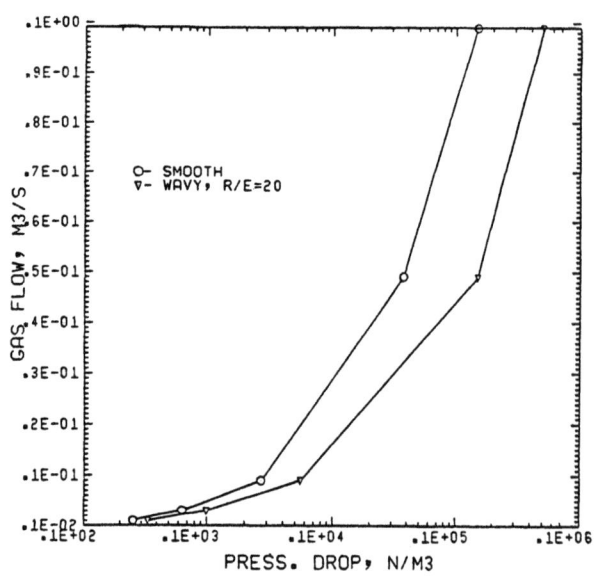

Fig. 4 Pressure drop predictions with and without interfacial momentum transfers for liquid flow rate 5×10^{-5} [m^3/s] I.D. 2.54×10^{-5} [m] tubes (Morala and Chang 1986).

Interfacial velocity

Most researchers assumed $U_i = U_{gi} = U_{\ell i}$ at the gas-liquid interface, however, no rigid justifications have been given. For the stratified smooth laminar flow, analytical solution has been given by various authors, based on this assumption with $\tau_i = \tau_{gi} = \tau_{\ell i}$. For the channel flow, as shown in Figure 2, the interfacial velocity becomes (Ranger and Davis, 1978).

$$U_i = -\frac{L^2}{2}\frac{dp}{dx}\left[\frac{\alpha_g(1-\alpha_g)}{\mu_\ell + (1-\alpha_g)(\mu_g-\mu_\ell)}\right] \quad (12)$$

The interfacial velocity becomes zero when $\mu_\ell >> (1-\alpha_g)$ or $\alpha_g = 1$. As we discussed in previous section, $U_i = 0$ are assumed for most of the wavy interfaces or turbulent flow.

CONCLUDING REMARKS

The significances of interfacial parameters in two-phase flow modelling have been discussed. Available published literature on determining experimentally or analytically, the interfacial parameters have been reviewed and presented in tabular forms for a stratified flow regime, and following the concluding remarks are obtained:

1. The study on interfacial parameters be continued in two systematic approaches as present work. First, a review study on the values or expressions of interfacial parameters. They should be analyzed and categorized according to the methods and experimental conditions. Second, a study of constitutive equations, derived from the need in developing certain models for particular cases. This should lead to a systematic list of specific requirements on the interfacial parameters. The overall study can be completed by combining the two approaches.

2. Development of custom made constitutive equations for each computer code must be conducted.

3. Development of new two-phase flow diagnostic technique must be conducted for the interfacial parameters.

4. Studies of interfacial area and geometry effects on the interfacial transfers must be conducted for stratified flow regimes.

5. Flow regime transition boundaries and the transition boundary between laminar and turbulent flow under the two-phase system must be studied.

ACKNOWLEDGEMENT

Author wishes to express his appreciation to D. Groenveld, B. Donevski and E.C. Morala for valuable comments and discussions. Work supported by the Natural Sciences and Engineering Council of Canada and AECL.

REFERENCES

Agrawal, S.S., Gregory, G.A., and Govier, G.W., 1973, "An Analysis of Horizontal Stratified Two-phase Flow in Pipes", Can. J. Chem. Eng., Vol. 51, pp. 280-286.

Bassols, S., Juhel, D., Rousseau, J.C., 1982, "A Synthesis of qualificaiton of the advanced reactor safety code CATHARE", European Two-Phase Flow Group Meeting, Centre d' Etudes Nucleaires de Grenoble — Equipe Mixte EDF/CEA/FRA — Service des Transferts Thermiques.

Besfamil'nyi, P.V., Leont'ev, A.I., and Tsiklauri, G.V., 1982, "The Hydraulic Resistance of a Horizontal Rectangular Channel with Stratified Two-Phase Flow", Teploenergetika, Vol. 29, No. 8, pp. 65-67.

Boure, J.A., 1978, "Constitutive Equations for Two Phase Flows", J.J. Ginoux, Ed., 'Two-Phase Flows and Heat Transfer with Applications to Nuclear Reactor Design Problems', Chapter 9, pp. 157-178, Hemisphere Publishing Co., Washington.

Chan, A.M.C. and Banerjee, 1981, "Refilling and Rewetting of a Hot Horizontal Tube III", ASME J. Heat Trans., Vol. 103, pp. 653-659.

Chang, J.S., Ichikawa, Y. and Irons, G.A. 1982, Measurement in Poly-Phase Flow, 1982, Henrich and Patel Eds., ASME Press, pp. 7-12.

Cheremisinoff, N.P. and Davis, 1979, "Stratified Turbulent-Turbulent Gas-Liquid Flow", AIChE J., Vol. 25, 48.

Dutta, A. and Mashelkar, R.A., 1983, "Interpretation of Drag Reduction Phenomenon in Wavy Films of Polymer Solutions", AIChE J., Vol. 29, No. 3, pp. 519-521.

Drew, D., Cheng, L. and Lahey, R.T., 1979, "The Analysis of Virtual Mass Effects in Two-Phase Flow", $\underline{5}$, p. 233-242.

Ellix, S.R.M., and Gay, B., 1959, "The Parallel Flow of Two Fluid Streams-Interfacial Chear and Fluid-Fluid Interaction", Trans. Instn. Chem. Engrs., Vol. 37, pp. 206-213.

Fujii, T., Honda, H., Oda, K., Kato, Y., Kawano, S., 1981, "Condensation from Steam Flow in a Horizontal Pipe", Jap. Soc. Mech. Eng./Ronbunshu, Vol. 47B, No. 421, pp. 1861-70.

Govier, G.W., and Aziz, R., 1977, "The Flow of Complex Mixtures in Pipes", Van Nostrand Reinhold Co., New York.

Hancox, W.T., Ferch, R.L., Liu, W., S. Nieman, R.E., 1980, "One Dimensional Models for Transient Gas-Liquid Flows in Ducts", Int. J. Multiphase Flow. Vol. 6, pp. 25-40.

Inoue, K., Kutsuna, H., Nakanishi, S., Yoshida, M., Kotaki, T., 1982, "Filmwise Condensation Heat Transfer in a Parallel Channel", Jap. Soc. Mech. Eng./Ronbunshu, Vol. 48B, No. 426, pp. 308-317.

Ishii, M., 1982, "Two-Phase Flow Formulations", Lecture presented at CENG (France).

Johnston, A.J., 1984, "An Investigation inot the Interfacial Shear Stress Contribution in Two-Phase Stratified Flow", Int. J. Multiphase Flow, Vol. 10, No. 3, pp. 371-383.

Lightstone, L. and Chang, J.S., 1986, "Gas-Liquid Two-Phase Flow in Horizontal Dividing and Nondividing Flow" (to be published).

Morala, E.C. and Chang, J.S., 1986, (In this issue) Determination of Interfacial Geometry and Areas by Ultrasonic Techniques.

Papadic, V.O., 1975, "Drag Reduction in Film Flow", AIChE J., Vol. 21, p. 610.

Segev, A., Collier, R.P., 1981, "Application of Battelle's Mechanistic Model to Lower Plenum Refill", NRC Report NUREG/CR-2030 (BMI-2077), Battelle Columbus Laboratories, NTIS, March.

Shoji, M., Takagi, N., 1982, "Studies of Critical Filmwise Boiling in a Horizontal Surface", Jap. Soc. Mech. Eng./Ronbushu, Vol. 48B, No. 435, pp. 2324-33.

Sinai, Y.L., 1983, "A Charnock-Based Estimate of Interfacial Resistance and Roughness for Internal, Fully-developed Stratified, Two Phase Horizontal Flow", Int. J. Multiphase Flow, Vol. 9, No. 1, pp. 13-19.

Shoham, O. and Taitel, Y., 1984, "Stratified Turbulent-Turbulent Gas-Liquid Flow in Horizontal and Inclined Pipes", AIChE J., Vol. 30, No. 2, pp. 377-385.

Sleicher, C.A. and Rouse, M.W., 1975. Int. J. Heat and Mass Trans., vol. 18, pp. 677-683.

So, C.B. and Ardron, K.H., 1984, "Numerical Simulation of the Rewetting of a Horizontal Tube Using a Two-Flid Model", Inter. Workshop on Fundamental Aspects of Post-Dryout Heat Transfer, Salt Lake City, Utah, U.S.A., April 2-4.

Suthradhar, S.C. and Chang, J.S., 1985, "The Role of Interfacial Frictions on the Shock Wave Propagation in Stratified Gas-Liquid Flow Inside Pressure Tubes", Proc. Specialists Meeting on Small Break LOCA Analyses in LWRs, Vol. 2, pp. 243-254. (In Press, 1986 J. Nucl. Eng. Design).

Taitel, Y., Dukler, A.E., 1976, "A Model for Predicting Flow Regime Transitions in Horizontal and Near Horizontal Gas Liquid Flow", AIChE, Vol. 22, p. 47.

Thomas, R.M., 1979, "Condensation of Steam on Water in Turbulent Motion", Int. J. Multiphase Flow, Vol. 5, p. 1-15.

Young, M.W. and Hsu, Y.-Y, 1982, "Model Development Experimental Programs as Part of the NRC Reactor Safety Research", Nuclear Safety, 23, No. 4, p. 407-415.

Symbols

D	pipe diameters
f	friction factor
k	thermal diffusivity
h	level
L	channel width or pipe length
Nu	Nusselt number
p	pressure
Pr	Pecolet number
Re	Reynolds number
v	velocity
x,y,z	coordinate
α	volume fraction
μ	viscosity
τ	shear stress

Subscripts

f	fluid
g	gas
i	at interface
ℓ	liquid
s	saturation
w	at wall surface

SURFACTANTS AND CAPILLARY WAVES:

EXPERIMENTS ON MONOLAYERS AND ON MICROEMULSIONS

Dominique Langevin and Jacques Meunier

Laboratoire de Spectroscopie Hertzienne de l'E.N.S.
24, rue Lhomond
75231 Paris Cedex 05 - France

I. INTRODUCTION

When a surfactant molecule is dissolved in water (or oil), because of its amphiphilic character it spontaneously adsorbs at the air-water (or oil-water) interface. As a consequence, the interfacial tension γ is decreased [1]. Another consequence is the appearance of a resistance to compression and to shear at the interface. This is characterized by two new surface properties, the two dimensional compression and shear elastic moduli $\in$ and S [2].

The surfactant layer also creates a resistance to bending [3]. This corresponds to an energy cost which is usually negligible, excepted when the interfacial tension is close to zero. This situation happens for instance at an oil-water interface when a convenient surfactant is used.

During the motion of the interface, an energy dissipation is produced into the surfactant layer. This is due for instance to molecular reorientations whithin the film or to exchanges with the bulk phases. These processes can be formally taken into account by introducing several two dimensional surface viscosities.

All these surface parameters cause noticeable changes in the propagation of waves at the interface. It was known to the ancient Greeks that thin oil layers could damp the waves on the sea. More recently it has been suggested that high surface elasticities and viscosities enhance foam stability [1]. It is also known that surfactant layers would shift surface instability thresholds to high values [4]. Up to now, there are however very little experimental studies in relation with these problems. The case of bending elasticity is only relevant for systems where γ is ultralow or zero like microemulsions or biological membranes. It is related to important problems like membrane fusion [5].

In this paper we will present several experimental techniques that allow to measure these surface parameters. The surface light scattering technique allows to study the temporal evolution of thermally excited surface waves of wavelengths λ_c in the range 100 μ < λ_c < 1mm. In another type of technique, the spatial evolution of electrically induced capillary waves in the range 1mm < λ_c < 10cm can be measured. In the systems that we have studied, the bending elasticity begins to affect the capillary waves only for $\lambda_c \lesssim 100$ Å. To have access to such small wavelengths, we have used optical reflectivity techniques which give informations about the integrated amplitude of the thermally excited capillary waves over all wavelengths.

In the following, we will begin by introducing some physicochemical concepts of surfactant action. We will then describe the experimental techniques used and recall the theoretical background necessary for the interpretation of the measurements. We will illustrate the potentials of the techniques by results obtained recently in our laboratory.

II. PHYSICAL CHEMISTRY OF SURFACTANT ACTION

1/ Water-surfactant mixtures

Surfactant molecules have both a polar hydrophilic part (frequently an ion) and a non polar hydrophobic part (frequently a long aliphatic chain). When surfactant molecules are dissolved in water, they spontaneously adsorb at the air-water interface, thus partitioning between the bulk solution and the interface. At very low surfactant concentration C, the surfactant layer at the interface is dilute (Fig. 1), and the surface tension of the solution close to that of pure water γ_w. When C increases, the layer becomes more compact. The surface concentration or number of adsorbed molecules per unit area is given by the Gibbs equation [1] :

$$\Gamma = - \frac{1}{kT} \frac{d\gamma}{d\ln C}$$

Frequently, above a concentration C_S (Szyszkowski concentration) the curve γ versus $\ln C$ has a roughly linear part, indicating that the layer approaches a maximum surface concentration $\Gamma \sim \Gamma_\infty$ (Fig. 1). When C becomes typically one or two orders of magnitude larger than C_S, aggregates of surfactant molecules called micelles begin to form in the bulk. The surface tension γ satu-

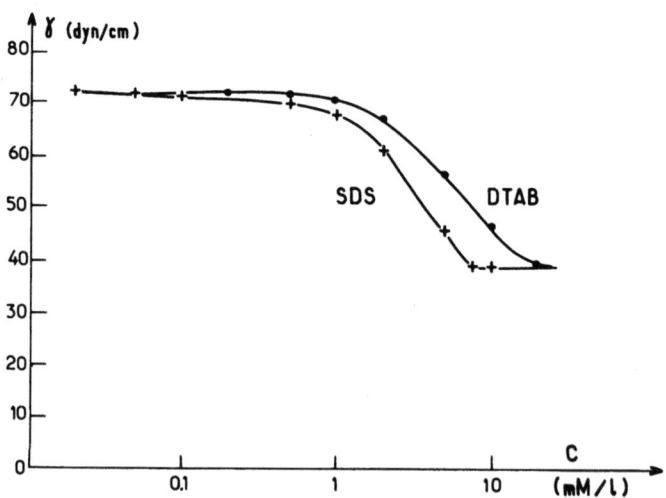

Fig. 1. Surface tension of aqueous surfactant solutions versus surfactant bulk concentration ; ○ : DTAB solutions ; + : SDS solutions.

rates to a constant value, because all the surfactant added only serves to form new micelles (Fig. 1).

When large surfactant concentrations are reached, liquid crystalline or gel phases are formed. We will not be concerned by these cases here. Let us also mention that some surfactants do not form micelles in water, but above a "critical aggregation concentration" directly give liquid crystalline phases. This happens for instance with the surfactants that have two aliphatic chains like the phospholipid molecules, which form lamellar liquid crystals in water. This has been explained in terms of simple geometrical considerations ([6]) : single chain ionic surfactants can pack and form curved layers because the polar heads are more cumbersome than the aliphatic chains ; double chain surfactants form flat layers because the double chain is about as cumbersome as the polar head (Fig. 2).

2/ Water-oil-surfactant mixtures

When the surfactant is more soluble in water than in oil, the evolution with surfactant concentration is very similar to that of water-surfactant mixtures. A surfactant monolayer adsorbs at the oil-water interface and is in equilibrium with surfactant monomers in the bulk. Then micelles form in the water phase above a critical micellar concentration. The micelles are swollen to a certain amount by the oil. If C is further increased, organized phases are formed.

When the surfactant is more soluble in oil, the micelles are formed in the oil, and excepted from this difference a similar evolution with C is observed.

When the surfactant is as soluble in oil than in water, the surfactant layers that form in the bulk phases have a tendency to be flat. We come back here to similar geometrical considerations than in the former paragraph (Fig. 2). The solubility of the surfactant in oil (or water) is indeed rela-

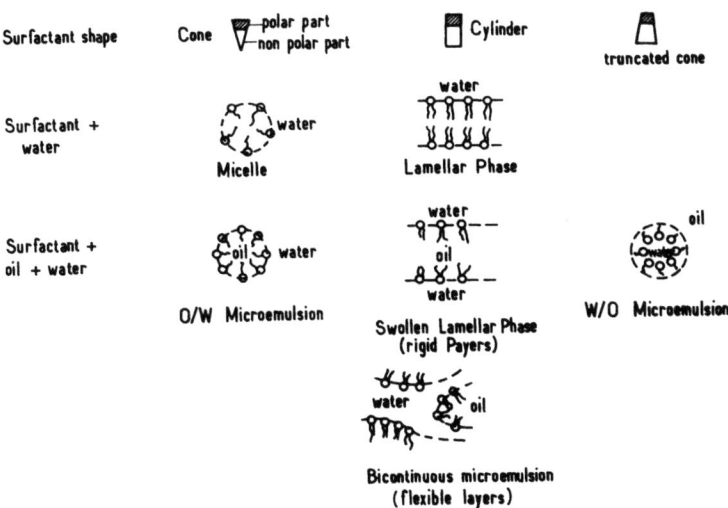

Fig. 2. Self organization of surfactant molecules according to their schematic shape.

ted to the interactions between non polar (or polar) parts of two neighbouring surfactant molecules, and therefore to the mean distance between non polar and polar parts in the surfactant film. This was at the origin of the success of the Bancroft rule which predicted the type of emulsion to be formed from surfactant solubility properties : oil in water (O/W) for water soluble surfactants and water in oil (W/O) for oil soluble surfactants.

3/ Water-oil-surfactant-alcohol mixtures

Typical oil-water interfacial tensions are of the order of 50 mN/m. With surfactants, the tensions drop to around a few mN/m. These values of γ are too high to allow for spontaneous emulsification. But when a cosurfactant is used, γ can decrease to about 10^{-2} mN/m and microemulsions spontaneously form, because the surface energy cost of droplets formation can be compensated by the gain in entropy of mixing ([7]). Let us mention that cosurfactant is not always necessary : some well choosen surfactants are very efficient in reducing the oil-water interfacial tension to below 10^{-2} dyn/cm by themselves, for instance the so called aerosol OT. But a common property of all the surfactant layers forming microemulsions is their flexibility ([8]). This is probably associated to the disorder introduced by alcohol molecules in the surfactant films (Fig. 3). When there is no alcohol in the film, a similar degree of disorder is obtained when the surfactant molecules have two chains of different lengths (like aerosol OT).

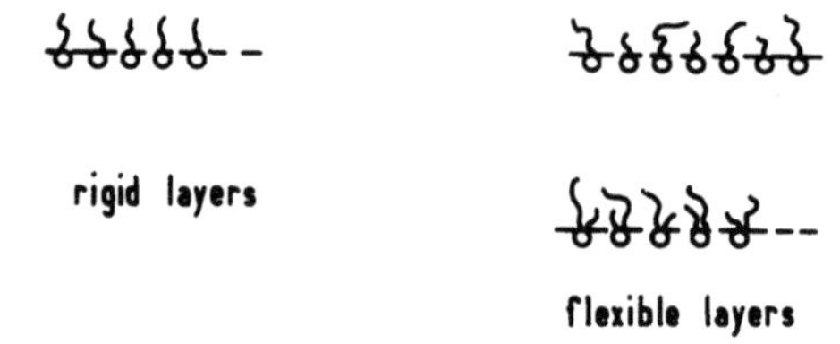

Fig. 3. Disorder and flexibility in surfactant layers.

As mentioned above, the problem of thermodynamic stability of microemulsions can be solved by considering only the free energy of formation of the microscopic oil-water interface and the entropy of mixing of oil and water. However the phase equilibria obtained in this simple way is never observed in practice. Theory predicts that oil in water (O/W) microemulsion should coexist with water in oil (W/O) microemulsion. Again from geometrical considerations, one type of curvature is always favoured. This can be taken into account by simply adding to the surface energy term a curvature term ([8])

$$F_S = \gamma \Delta A + \frac{1}{2} K (C - C_0)^2 \Delta A \qquad (1)$$

where K is the curvature elastic modulus, C the curvature of the surfactant film and C_0 its spontaneous curvature. The case $C_0 \neq 0$ leads to either W/O or O/W microemulsions depending on the sign of C_0. The case $C_0 \sim 0$ is more interesting : the surfactant film has a tendency to be flat but it is subject to thermal fluctuations. The film loses its orientational order beyond a "persistence length" ξ_K ([8])

$$\xi_K = a \exp 2\pi K /kT \qquad a = \text{molecular length} \qquad (2)$$

At the difference of droplets structures formed when C_0 is large and where only one component (oil or water) is continuous and like in a lamellar structure, the medium is continuous both in water and in oil (bicontinuous). But the surfactant layers are randomly oriented : the medium is sponge-like (Fig. 2).

The curvature properties of the layer depend on the scale ξ over which the deformation is produced : it is easier to bend a rough surface than a flat one. This leads to a renormalization of the elastic modulus

$$K(\xi) = K_0 + \frac{kT}{2\pi} \ln \frac{\xi}{a}$$

where K_0 is the modulus of the bare surface.

Even now, if the surface tension term is dropped out from eq.(1) microemulsion stability can result from the compensation of elastic energy (which decreases when ξ increases) and dispersion entropy (which also decreases and leads to an increasing free energy). The final result is that the characteristic dispersion size is ξ_K ([9]).

Let us now give some order of magnitudes. For typical flat surfactant layers as those formed by phospholipid molecules, the measured K values are in the range 10-100 kT. As a consequence, ξ_K is very large, and is macroscopic. This means that the layers are flat over large distances, i.e. that the dispersion is a lamellar liquid crystal (Fig. 2). When K is smaller and of the order kT, ξ_K is of order 100 Å and the layers are no longer macroscopically flat. This corresponds well to the sponge-like image given for the microemulsions in the inversion region between O/W and W/O structures ($C_0 \sim 0$).

The surface tension γ of the <u>macroscopic</u> oil-water interfaces covered by the same layers is of course not zero, but very small. Such interfaces are also very rough.

III. TWO DIMENSIONAL HYDRODYNAMICS

Let us now come back to the surfactant layer adsorbed at the free surface of water. It can be viewed as a two dimensional system. Its equilibrium position will be taken as plane Z=0. If $\vec{u}(\vec{r},t)$ is the displacement of a point $\vec{r}(x,y)$ of the surface at time t, the elastic energy per unit area associated to such a displacement is ([10]), if $\vec{u}$ is a variable of x and t only (wave propagating along Ox) :

$$F = \frac{1}{2} \epsilon \left(\frac{\partial u_x}{\partial x} \right)^2 + \frac{1}{2} S \left(\frac{\partial u_y}{\partial x} \right)^2 + \frac{1}{2} \gamma \left(\frac{\partial u_z}{\partial x} \right)^2$$

ϵ and S are the dilational and shear moduli respectively and γ the surface tension. For a deformation mode of wave vector $\vec{q}$:

$$\vec{u}(x,t) = \vec{u}_q(t) e^{iqx}$$

the elastic energy per unit area is :

$$F_q = \frac{1}{2}\epsilon q^2 u_x^2 + \frac{1}{2} S q^2 u_y^2 + \frac{1}{2}\gamma q^2 u_z^2$$

For each $\vec{q}$, three types of surface waves can exist : compression waves related to u_x, shear waves related to u_y and capillary waves related to u_z. All these waves can be thermally excited. Their mean square amplitude is deduced from the equipartition theorem ([10]) :

$$\langle u_x^2\rangle = \frac{kT}{\epsilon q^2} \qquad \langle u_y^2\rangle = \frac{kT}{Sq^2} \qquad \langle u_z^2\rangle = \frac{kT}{\gamma q^2} \qquad (3)$$

Because there are also dissipation processes during the three types of motion a dissipation function has to be introduced ([10]) (with $u_{ij}=\partial u_i/\partial x_j$) :

$$\psi_q = \frac{1}{2}\kappa q^2 \left(\frac{\partial u_{xx}}{\partial t}\right)^2 + \frac{1}{2}\eta_s q^2 \left(\frac{\partial u_{yx}}{\partial t}\right)^2 + \frac{1}{2}\mu q^2 \left(\frac{\partial u_{zx}}{\partial t}\right)^2$$

where κ, η_s and μ are the dilational, shear and transverse surface viscosities. These viscosities can be also (and more properly) defined as excess surface properties ([11])([12]).

All these parameters play important roles in the propagation of surface waves. The problem can be solved by writing the hydrodynamic equations in the bulk aqueous solution and taking as limit conditions at the air-water interface :

$$\sigma_{iz}(z=0) = \eta\left(\frac{\partial v_i}{\partial z} + \frac{\partial v_z}{\partial x_i}\right) - p\delta iz = P_i = \frac{\partial}{\partial z}\left[\frac{\partial F_q}{\partial u_{iz}} + \frac{\partial \psi}{\partial(\partial u_{iz}/\partial t)}\right]$$

which represents the equilibrium between the stress tensor components σ_{iz} at the surface (Z=0) and the force exerted on the surface per unit area P_i. $\vec{v}$ is the fluid velocity, η its viscosity and p the pressure in the fluid.

One thus find that shear surface waves are entirely decoupled from compression and capillary waves. The common dispersion equation of compression and capillary waves writes in reduced frequency units :

$$D(S) = S^2\left[(1+S)^2+y-\sqrt{1+2S}\right] + (\alpha y+\beta S)\left[S^2\sqrt{1+2S} + (y+\delta S)(\sqrt{1+2S}-1)\right] + \delta S^3 \qquad (4)$$

where $S=i\omega\rho/2\eta q^2$ and $y=\gamma\rho/4\eta^2 q$ is a dimensionless parameter, ratio of capillary over viscous forces ; ρ is the fluid density. α, β and δ are dimensionless viscoelastic parameters ; α is the ratio of dilational modulus over surface tension $\alpha=\epsilon/\gamma$; β and δ are reduced surface viscosities $\beta=Kq/2\eta$ and $\delta=\mu q/2\eta$.

When the surfactant layer is very dilute, the frequency of capillary waves ω_C is larger than the frequency of longitudinal or compression waves ω_L. When the layer becomes more compact, i.e. around the Szyszkowski concentration C_S, ω_L and ω_C becomes equal. The coupling between capillary waves and longitudinal waves is maximum for this particular concentration ([12]). As a result, the damping of the capillary waves is maximum (about three times the damping on pure water, for typical wavelengths $\lambda_C \sim 100\,\mu$). The frequency ω_C is also maximum around C_S. Let us note that the simple expression of ω_C for pure fluids $\omega_C = \sqrt{\gamma q^3/\rho}$ is unable to explain this behaviour since it predicts that γ should be a decreasing function of the surfactant bulk concentration C.

We have represented on Fig. 4, the frequencies of capillary waves as measured from surface ligth scattering versus reduced elasticity for different reduced dilational viscosities. The resonance $\omega_C = \omega_L$ occurs around $\alpha \sim 0.1$ ($\in \sim \gamma/10$) and is best seen when β is small (small surface viscosity).

The energy dissipation can be explicitely calculated when the only effective process is the adsorption-dissolution of surfactant molecules around the surface. This leads to ([12]) :

$$\in(\omega) = \varepsilon_0 \frac{1 + \Omega}{1 + 2\Omega + 2\Omega^2}$$

$$\kappa(\omega) = \frac{\varepsilon_0 \Omega/\omega}{1 + 2\Omega + 2\Omega^2} \tag{5}$$

$$\gamma(\omega) = \gamma_0 \qquad \mu(\omega) = 0$$

where $\in_0 = - d\gamma/d\ln\Gamma$ and $\Omega = \sqrt{\frac{D}{2\omega}} \frac{dc}{d\Gamma}$. D is the diffusion coefficient of the surfactant molecules.

When the relation between the surface and bulk surfactant concentrations Γ and c is known, it is also possible to relate $\in$, κ and C. A useful expression was proposed by Frumkins, in which the surface layer is treated like a regular mixture of water and surfactant :

$$\gamma = \gamma_w - kT \, \Gamma_\infty \left[\ln\left(1 - \frac{\Gamma}{\Gamma_\infty}\right) + \frac{H}{kT} \left(\frac{\Gamma}{\Gamma_\infty}\right)^2 \right] \tag{6}$$

$$\frac{\Gamma/\Gamma_\infty}{1 - \Gamma/\Gamma_\infty} = \frac{C}{C_S} \exp\left(\frac{2H}{kT} \frac{\Gamma}{\Gamma_\infty}\right)$$

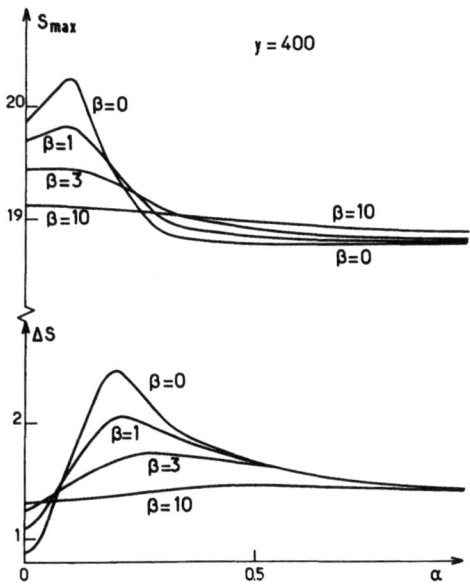

Fig. 4. Reduced peak frequency S_{max} and width ΔS of the spectra of the light scattered by a surfactant solution versus reduced dilational elasticity α for different reduced dilational viscosities β.

Γ_∞ is the maximum surface concentration and H the non-ideality parameter. These expressions fits very well the existing experimental data ([13]).

When the surfactant concentration is large enough and micelles or lamellae present in the bulk phase, supplementary exchange processes between the micelles or lamellae and the surface layer have to be taken into account ([15]). Let us also recall, that molecular reorientations in the surfactant layer can also contribute to ϵ and κ as well as to γ and μ. This has already been observed in insoluble layers, where this dissipation process is the only allowed ([10]).

All these results can be generalized for surfactant layers at oil-water interfaces. The only important difference is that because of the similar densities and viscosities of the coexisting fluid phases, the resonance phenomenon at $\omega_C = \omega_L$ is washed out. This comes from the fact that the boundary conditions at the interface impose that the fluid velocity is almost vertical during the propagation of capillary waves. It then follows that there is negligible coupling between capillary and longitudinal waves ([12]).

IV. EXPERIMENTAL TECHNIQUES

1/ Surface light scattering

The technique has been already widely described ([16]). Let us briefly recall that one measures the spectrum of the scattered light, which is identical to the spectrum of the thermally induced vertical displacement of the interface $u_z(\vec{r},t)$, provided that the optical detection is purely heterodyne. This spectrum is from the fluctuation dissipation theorem equal to :

$$P(\omega) = \frac{kT}{\pi\omega} \text{Im}\, \chi(\omega)$$

where $\chi(\omega)$ is the response function to an external pressure disturbance $\pi(\omega)$: $u_z(\omega) = \chi(\omega)\,\pi(\omega)$. $\chi(\omega)$ can be calculated from the hydrodynamic equations subject to limit conditions at the interface as explained in §III. The result for the air-water interface covered by a surfactant layer is the following :

$$P(\omega) = \frac{kT}{\pi\omega}\frac{\rho}{4\eta^2 q^3}\,\text{Im}\left\{\frac{S^2 + (\alpha\gamma + \beta S)(\sqrt{1+2S} - 1)}{D(S)}\right\}_{S = i\omega\rho/2\eta q^2} \tag{7}$$

where $D(S)$ is the dispersion equation in reduced units as introduced in § III (eq.4).

The horizontal displacement $u_x(\vec{r},t)$ also produces a slight polarizability variation at the interface level. This gives rise to some light scattering. But the corresponding intensity is completely negligible compared to those of $u_z(\vec{r},t)$ ([17]).

$P(\omega)$ is approximately lorentzian with a peak frequency ω_q and a width $\Delta\omega_q$. The variations of ω_q and $\Delta\omega_q$ with the viscoelastic coefficients ϵ and κ are represented in reduced units in Fig. 4, for a typical q value in the experiments. It can be noted that the relative frequency variations are small (less than about 10%). The relative width variations are larger, but the corresponding measurements are less precise because of instrumental broadening problems ([18]). It follows that the accuracy on the determinations of ϵ, κ and μ is poor excepted in the region $\alpha \sim 0.1$ and $\beta < 10$, i.e. when :

$1 < \epsilon < 50$ mN m^{-1}

$10^{-5} < \kappa < 10^{-3}$ mN m^{-1} s^{-1}

μ is easier to determine because it gives large modifications of the spectrum. However, experimentally, it is found negligible most of the time, excepted for some condensed insoluble layers ([10])([19]).

The measurements at oil-water interfaces are also feasible. They are even easier because the interfacial tension γ is usually smaller. The mean square amplitude of u_z is then larger (eq.3) and the scattered intensity is also larger. Because of the decoupling between capillary and longitudinal waves, the spectrum depends essentially only on γ and μ. Its rigourous expression is extremely complicated and can be found in ref. 16.

For the systems of interest in this paper, the oil-water interfacial tension is ultralow $\gamma \lesssim 10^{-1}$ mN/m, and μ is negligibly small. The spectrum reduces to a lorentzian line, centered at zero frequency and of half width

$$\Delta\omega_q = \gamma q/2(\eta + \eta')$$

where η and η' are the bulk viscosities of the coexisting phases.

Examples of experimental spectra are given on Fig. 5 and 6.

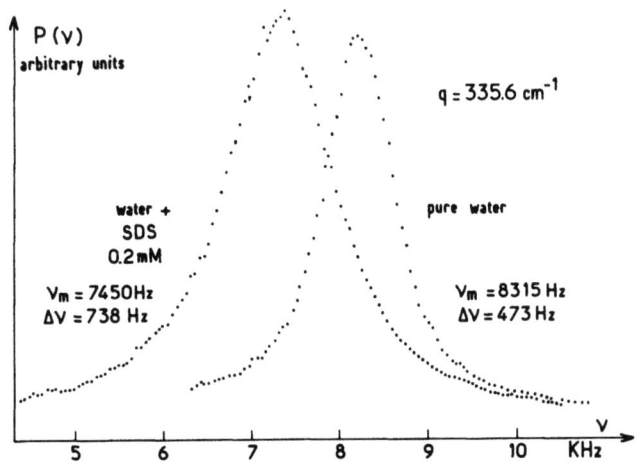

Fig. 5. Experimental spectra of the light scattered by the free surface of pure water and of a surfactant solution.

2/ Waves excited by electrocapillarity

This technique has been proposed and developed by A. Miyano and coworkers ([20]). The principle is very simple : a sharp metal edge close to the water surface is submitted to a sinusoidal voltage of frequency ω. Because of the difference in dielectric constant between air and water the surface is distorted and a wave of frequency 2ω is generated at the surface. The wave is detected by the deflection of a laser beam sent normally to the unperturbed surface. By translation of a mirror which controls the point of incidence of the beam on the surface, one can follow the propagation of the wave. An example of the signal obtained after detection is shown on Fig. 7. The spatial wavelength λ_c and damping $\Delta\lambda_c$ are related to frequency by ([13]) :

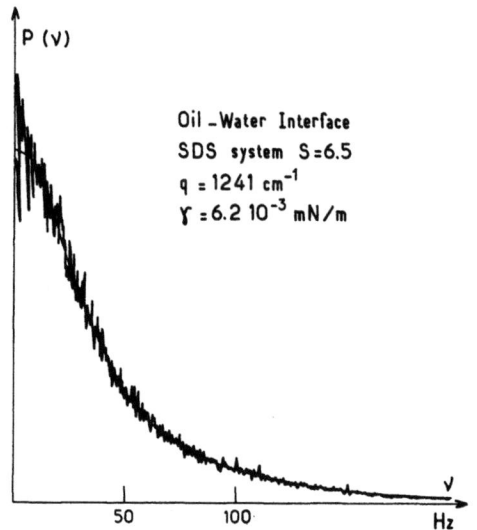

Fig. 6. Experimental spectra of the light scattered by an oil-water interface covered by a surfactant layer.

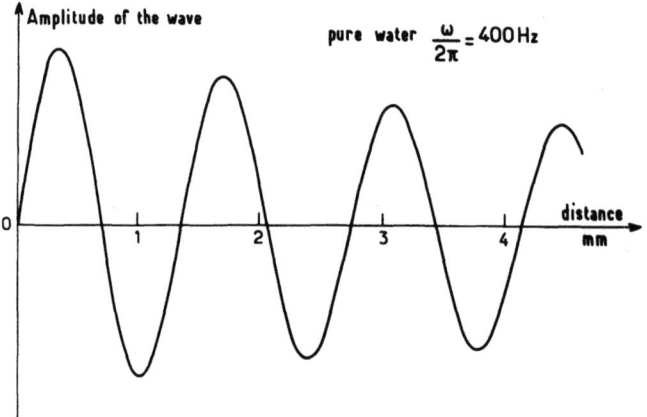

Fig. 7. Experimental signal from the set-up of §IV-2, obtained on the free surface of water ; $\omega/2\pi = 400$ Hz.

$$\lambda_c = 2\pi \sqrt[3]{\frac{\gamma}{\omega^2 \rho}}$$

$$\Delta\lambda_c = \frac{3\pi}{2} \frac{\gamma}{\eta\omega}$$
(8)

for a pure fluid. When a surfactant film is present, the roots of the dispersion equation (4) have to be used. Instead of looking for a complex solution in ω, q being real, as in §1, it is here necessary to take ω real and to find a complex q ($2\pi/q = \lambda_c + i\Delta\lambda_c$).

3/ <u>Reflectivity of a liquid interface. Case of normal incidence</u>

The reflectivity of a sharp interface between two fluids has been calculated by Fresnel. For normal incidence, it reduces to :

$$R_F = \left(\frac{n_1 - n_2}{n_1 + n_2} \right)^2$$

where n_1 and n_2 are the refractive indices of the two fluids.

In practice, the interface is never sharp, either because there is an intrinsic density profile across the interface, or because its is rough due to thermal fluctuations. As we have seen earlier, the surface roughness is due to the superposition of surface deformation modes of wave vector q. All the components with q values smaller than $k = 2\pi/\lambda$, λ being the light wavelength, will behave optically as diffraction gratings. All the components with q values larger than $2\pi/\lambda$ do not scatter light and behave optically as diffuse interfaces. It is then impossible in practice to make a difference between an intrinsic density profile and roughness of small periodicity. Physically it is even not clear how the intrinsic profile fits together with the roughness Fourier components of largest wave vectors $q_{max} \sim \pi/L$, L being the interface thickness. Some authors ([21]) even consider that there is no intrinsic profile and that the structure of the interfacial zone is entirely determined by thermal fluctuations. This is supported by recent X-rays reflectivity experiments on the free surface of several pure fluids ([22]).

In the case of light, λ is generally very large compared to L, so that the reflectivity R does not differ significantly from R_F. However, in the case where the interfacial tension between the coexisting phases is very low, L becomes large and can be deduced from the ratio R/R_F. It can be shown indeed, that the reflectivity loss due to roughness is ([23]) :

$$\frac{R}{R_F} = e^{-4k^2 <u_z^2>}$$

where

$$<u_z(\vec{r},t)^2> = \sum_q <u_q(t)^2> \sim \int_0^{q_{max}} \frac{kT}{\gamma q^2 + \Delta\rho g} \frac{q dq}{2\pi} = \frac{kT}{4\pi\gamma} \ln(1 + q_{max}^2 \ell_c^2)$$

$\Delta\rho = \rho_1 - \rho_2$ being the density difference between the coexisting phases, and ℓ_c the capillary length $\ell_c = \sqrt{\gamma/\Delta\rho g}$. The term $\Delta\rho g$ is usually negligible for capillary waves (eq.3) but has to be introduced here because in the high wavelength limit gravity waves are also thermally excited ; their contribution to $<u_z^2>$ is important because if they were not present $<u_z^2>$ would be infinite.

The reflectivity loss due to an intrinsic profile is less straightforward to calculate because it depends on the shape of the profile. Assuming that the profile is well represented by an error function :

$$\rho(z) = \frac{1}{2}\left\{(\rho_1 + \rho_2) + \frac{1}{\sqrt{\pi}}(\rho_1 - \rho_2)\int_0^{z/L_p\sqrt{2}} e^{-t^2/2}\,dt\right\}$$

This is the case close to a critical point where L_p is large (about 2 times the correlation length in the bulk) and where the profile has been calculated first by van der Waals and later on by Fisk and Widom. The forms found do not allow to calculate easily R, but lead to results very similar to that of the error function profile. It is of course more questionable to use the error function profile far from T_c where no mean field theory is applicable even for oil-water systems of very small γ.

Then :

$$\frac{R}{R_F} = e^{-4k^2L_p^2}$$

The total reflectivity loss is finally :

$$\frac{R}{R_F} = e^{-4k^2L^2}$$

with $L^2 = L_p^2 + \langle u_z^2 \rangle$

4/ Ellipsometry

Even for thin interfaces (L ∿ few Å), it is possible to gain information about interfacial structure by working at the Brewster incidence. If the interface were flat and sharp, then according to Fresnel, the reflection coefficient for light polarized in the plane of incidence would be zero : $R_\parallel = 0$. The phase shift between light reflected polarized in the plane of incidence $\rho_\parallel$ and the one polarized in the plane perpendicular $\rho_\perp$ would change discontinuously from zero to $\pm \pi$ at the Brewster angle. The interface being neither flat nor sharp, the phase shift vary more smoothly with incidence angle and is equal to $\pm \pi/2$ at the Brewster angle ; $R_\parallel$ is small but non zero. The reflected light is therefore elliptically polarized, and one can measure the amplitude and sign of :

$$\sqrt{\frac{R_\parallel}{R_\perp}} = \frac{\pi}{\lambda}\frac{\sqrt{\epsilon_1 + \epsilon_2}}{(\epsilon_1 - \epsilon_2)}\eta \qquad (9)$$

where ϵ_1 and ϵ_2 are the dielectric constant of the two coexisting phases. As in the case of normal reflectivity, η is the contribution of two terms ([24]), one due to the intrinsic profile and calculated by Drude :

$$\eta_p = \int_{-\infty}^{+\infty}\frac{(\epsilon(z) - \epsilon_1)(\epsilon(z) - \epsilon_2)\,dz}{\epsilon(z)} \qquad (10)$$

the second to roughness :

$$\eta_R = -\frac{3}{4}\frac{kT}{\pi\gamma}\frac{(\epsilon_1 - \epsilon_2)^2}{\epsilon_1 + \epsilon_2}q_{max} \qquad (11)$$

As η_R is proportional to q_{max}, the knowledge of the exact value of q_{max} becomes necessary. When a surfactant layer is present at the surface this is in fact easy, because eq.(3) can be further generalized to include besides gravity term, the curvature terms :

$$<u_z(t)^2> = \frac{kT}{\Delta\rho g + \gamma q^2 + Kq^4}$$

K being the curvature elastic modulus (§ II.3).

It follows that :

$$q_{max} = \frac{\pi}{2}\sqrt{\frac{\gamma}{K}} \tag{12}$$

V. EXPERIMENTAL DATA

1/ Water surfactant mixtures

We have studied aqueous solutions of two surfactants that have the same hydrophobic tail : sodium dodecyl sulfate (SDS) and dodecyl trimethylammonium bromide (DTAB) ([14]). The corresponding critical micellar concentration are 8mM for SDS and 15 mM for DTAB (Fig. 1).

The maximum damping of surface fluctuations as studied from surface light surface light scattering (§ IV.1) occurs well below the c.m.c. : about one tenth of it, for both SDS and DTAB solutions. Below this concentration, the viscoelastic parameters are very small, and above it, they increase rapidly (Table I). Close to the c.m.c. and above they are too large to be measured with accuracy.

Further information can be gained by analyzing the data using the adsorption-redissolution model of § III.1 and eq.(5) and (6). The parameters a, Γ_∞ and H have been calculated by fitting "static" surface tension measurements performed with a Wilhelmy plate on the same samples and shown on Fig.1 with eq.(6). It was found in this way, that the relaxation frequency ω_0 is always smaller than the frequency of the light scattering experiments for

Table I. Viscoelastic coefficients of SDS and DTAB aqueous solutions.

C (mM)	ϵ (mNm^{-1})		κ (mNm^{-1}s^{-1}) × 10^4	
	SDS	DTAB	SDS	DTAB
0.2	18	1.2	1.6	0.025
0.5	46	1.6	0.6	0.44
1	37	1.6	9.3	0.15
2	75	11	17	0.38
5	170	29	26	8
10	390	64	39	11
20	-	390	-	60

SDS solutions, for which eq.(5) reduces to :

$$\epsilon = \epsilon_0 = -\frac{d\gamma}{d\ln\Gamma} \qquad \kappa = 0$$

The measured values are however larger than these predictions. This means that, as in condensed insoluble layers ([10]) dissipation associated to reorientation processes in the layers play an important role.

The behaviour of DTAB layers is quite different. First, ω_0 becomes smaller than the light scattering frequencies before the c.m.c., and relaxation by the adsorption-dissolution process is then observable. Second, the measured values of ϵ and κ are in good agreement with the calculated ones from eq.(5). This means that the relaxation due to intralayer processes is now negligible.

These results illustrate the important role of the polar head of the surfactant molecules as far as two dimensional viscoelasticity is concerned.

Experiments at smaller frequencies with the device described in IV.2 are in progress on these systems.

2/ Water-oil-alcohol surfactant mixtures

We have studied more complicated mixtures made with the same surfactants : water + salt (47 wt%), toluene (47 wt%), SDS or DTAB (2 wt%), butanol (4 wt%) ([25]). The salt used is sodium chloride with SDS and sodium bromide with DTAB. It allows to adjust the spontaneous curvature of the surfactant film : at low salinities S (expressed in wt% in water), the surfactant is more soluble in water and O/W microemulsions are formed. At large salinities the surfactant is more soluble in oil : repulsions between the polar heads are screened, the surfactant tails become more bulky than the heads and the spontaneous curvature is reversed : W/O microemulsions are formed. The sequence of phase quilibria is the following :

$S < S_1$ O/W microemulsion in equilibrium with excess oil
$S_1 < S < S_2$ bicontinuous microemulsion in equilibrium with excess oil and water
$S > S_2$ W/O microemulsion in equilibrium with excess water

For SDS, $S_1 = 5.4$, $S_2 = 7.4$ and for DTAB, $S_1 = 2$, $S_2 = 5$.

The structures of these microemulsions have been determined by light ([25])([26]), X-rays ([26])([27]) and neutron ([28]) scattering. The characteristic sizes : droplets radii R for droplets microemulsions, oil and water domains size ξ for bicontinuous microemulsions, vary continuously with salinity and are maxima in the three phase domain where ξ is roughly constant. This can be explained rather simply if one assume that the area Σ occupied by a surfactant molecule in the film is constant. Indeed, if ϕ_o and ϕ_w are the oil and water volume fractions in the microemulsion phase and if the volume fraction of surfactant and alcohol are neglected :

$$R = \frac{3\phi_o}{C\Sigma} \qquad \text{oil droplets}$$

$$R = \frac{3\phi_w}{C\Sigma} \qquad \text{water droplets}$$

$$\xi = \frac{6\phi_o\phi_w}{C\Sigma} \qquad \text{bicontinuous structure}$$

where C is the surfactant concentration (expressed in number of molecules per unit volume). When S increases, the spontaneous curvature of the film C_0 decreases because of the screening of repulsion between the heads ; R increases as well as ϕ_0. In the bicontinuous microemulsions, $\phi_0 \sim \phi_w \sim 1/2$ so that $\phi_0\phi_w$ is roughly constant. At still larger S, the curvature is reversed and ϕ_w and R decrease now both with S.

An interesting comparison can be made between SDS and DTAB microemulsions. The sizes are larger in the first system. For instance in the bicontinuous region ξ = 240 Å for SDS and ξ = 100 Å for DTAB. If $\xi = \xi_k$ this means that K is smaller for DTAB than for SDS films. This is in accordance with the behaviour already observed with other viscoelastic coefficients in § V.1. It can of course be argued that the films of interest here contain alcohol. But the alcohol is the same as well as the relative proportion (3 alcohol molecules per surfactant molecule in the layers [25][26]).

The curvature elasticity has only been measured up to now for SDS layers. Normal reflectivity measurements have indicated that the macroscopic interfaces were very thick [23]. The thickness is maximum is the three phase domain (L $\sim$ 600 Å) and is entirely due to thermal roughness. The role of the intrinsic profile, if it exist, is beyond experimental accuracy. In these systems, the ellipsometric signal at the Brewster angle has an important contribution due to roughness, because of the very low interfacial tensions between coexisting phases. We will recall here only the simplest results obtained for the oil-water interfaces with a single surfactant film adsorbed at the interface [24]. By substracting the (small) contribution of the film to the measured value of η (eq.10-12), γ being measured independently by surface light scattering [25], the curvature elastic modulus has been deduced. κ do not vary very much with salinity and is about $(3 \pm 1)10^{-21}$ S.I in the intermediate salinities (three phase region). This is in reasonable agreement with the de Gennes prediction $\xi = \xi_k = a \exp 2\pi K/kT$ (eq.2). Let us recall that ξ = 240 Å in this system from which one gets : $K = 2.10^{-21}$ S.I. Experiments are in progress with the DTAB system to further test this point. Let us note that we have neglected here the renormalization problems that might significantly alter these simple comparisons [29][30].

Conclusions

We have shown how viscoelasticity of surfactant layers can be measured with several different optical techniques. Viscoelasticity play important roles in many different phenomena : foam and emulsion stability, control of interfacial instabilities, phase behavior of surfactant systems. We have shown that two model surfactants having the same hydrophobic chain, but different hydrophilic heads, can exhibit quite different viscoelastic behavior at the free surface of aqueous solutions. A memory of the difference is apparently conserved when a short chain alcohol is added to the film in order to provoke spontaneous emulsification of the aqueous surfactant solutions with an oil.

Acknowledgements

The work presented here has been performed in collaboration with A.M. Cazabat, D. Chatenay, A. Pouchelon, C. Otero, R. Ober, C. Stenvot and V. Thominet.

References

1. A. Adamson, Physical Chemistry of Surfaces, Wiley (1976).
2. V.G. Levich, Physico-Chemical Hydrodynamics, Prentice Hall (1962).
3. E. Evans and R. Shalak, Mechanics and Thermodynamics of Biomembranes, CRC (1980).
4. L.E. Scriven and C.V. Sternling, J. Fluid Mech. $\underline{19}$, 321 (1964).
5. W. Helfrich, Z. Naturforsch. $\underline{33a}$, 305 (1978).
6. J. Israelachvili, D.J. Mitchell, B.W. Ninham, J. Chem. Soc., Far. Trans. II, $\underline{72}$, 1525 (1976).
7. For a recent review, see D. Langevin, Physica Scripta $\underline{34}$ (1986).
8. P.G. de Gennes and C. Taupin, J. Phys. Chem. $\underline{86}$, 2294 (1982).
9. S. Safran, D. Roux, M.E. Cates, D. Andelman, Phys. Rev. Lett. $\underline{57}$, 491 (1986).
10. D. Langevin, J. Coll. Int. Sci. $\underline{80}$, 412 (1981).
11. F.C. Goodrich, J. Phys. Chem. $\underline{66}$, 1858 (1962).
    H. Brenner, these proceedings.
12. E.H. Lucassen and J. Lucassen, Adv. Coll. Int. Sci. $\underline{2}$, 347 (1969).
13. E.H. Lucassen, Progress in Surface and Membrane Sci. $\underline{10}$, 253 (1976).
14. V. Thominet, C. Stenvot and D. Langevin, submitted for publication.
15. M. van den Tempel, J. Fluid. Mech. $\underline{2}$, 205 (1977).
16. For a recent review, see D. Langevin, J. Meunier and D. Chatenay in "Surfactant in Solution", ed. K.L. Mittal and B. Lindman, Plenum 1984.
17. M.A. Bouchiat and D. Langevin, J. Coll. Int. Sci. $\underline{63}$, 193 (1978).
18. D. Langevin, Coll. and Surf., in press.
19. J.F. Crilly and J.C. Earnshaw, Biophys. J. $\underline{41}$, 197 (1983).
20. C.H. Sohl, K. Miyano and J.B. Ketterson, Rev. Sci. Instrum. $\underline{49}$, 1464 (1978).
21. F.l. Buff, R.A. Lovett and F.H. Stillinger, Phys. Rev. Lett. $\underline{15}$, 621 (1963).
22. J. Als Nielsen, these proceedings.
23. J. Meunier and D. Langevin, J. Phys. Lett. $\underline{43}$, L-185 (1982).
24. J. Meunier, J. Phys. Lett. $\underline{46}$, L-1005 (1985).
25. A.M. Cazabat, D. Chatenay, D. Langevin and J. Meunier, Adv. Coll. Int. Sci. $\underline{16}$, 175 (1982).
26. C. Otero, R. Ober and D. Langevin, manuscript in preparation.
27. L. Auvray, J.P. Cotton, R. Ober and C. Taupin, J. Phys. $\underline{45}$, 913 (1984).
28. A. de Geyer and J. Tabony, Chem. Phys. Lett. $\underline{113}$, 83 (1985).
29. J. Meunier, in preparation.
30. S. Safran, private communication.

HIGH FREQUENCY CAPILLARY WAVES ON LIQUID SURFACES:

A LIGHT SCATTERING STUDY

J.C. Earnshaw and R.C. McGivern

Department of Pure and Applied Physics
The Queen's University of Belfast
Belfast BT7 1NN, Northern Ireland

INTRODUCTION

Capillary waves have long been used as a probe of such liquid properties as surface tension and viscosity. Certain specifically surface effects can affect the propagation of such waves. It will be shown that waves of small wavelength and high frequency should be most susceptible to these effects. Thermally excited capillary waves scatter light, permitting experimental observations at rather high frequencies. This paper reports a study of high frequency capillary waves on various fluids for which surface effects have been reported or might plausibly be expected.

THEORETICAL BACKGROUND

A capillary wave upon a liquid surface constitutes a periodic perturbation of the surface. If the equilibrium surface is taken as the x-y plane with the x axis in the direction of propagation of the wave, the perturbation can be written as

$$\zeta = \zeta_0 e^{iqx} e^{i\omega t} e^{mz} \tag{1}$$

where q and ω are the surface wavenumber ($q = 2\pi/\Lambda$) and frequency respectively. Experimentally we define a real q and observe a complex frequency ($\omega = \omega_0 + i\Gamma$). For the free surface of a pure liquid the frequency ω_0 and damping constant Γ permit the liquid viscosity (η) and surface tension (γ) to be determined uniquely. In more complex situations (eg a monomolecular layer on the surface) the capillary waves are affected by further properties of the fluid, in particular the surface dilational modulus ε (= $d\gamma/d\ln A$, where A = area per molecule).

The propagation of capillary waves upon a liquid-air interface is governed by the dispersion equation[1]

$$D(\omega) = [\varepsilon q^2 + i\omega\eta(m + q)][\gamma q^2 + i\omega\eta(m + q) - (\omega^2\rho/q)] - [i\omega\eta(m - q)]^2 = 0 \tag{2}$$

This equation can readily be extended to a liquid-liquid interface[1]. The surface disturbance penetrates into the bulk fluid, the decay

distance being m^{-1}, where

$$m = \sqrt{q^2 + \frac{i\omega\rho}{\eta}} \quad , \quad \text{Re}(m) > 0. \tag{3}$$

From eqn 2 the capillary waves are seen to be coupled to longitudinal surface waves (governed by ε): under most circumstances this leads to an increase in Γ, accompanied by a smaller change in ω_0 (Fig 1). The roots of eqn 2 for fixed q may be complex conjugate, corresponding to propagating waves, or pure imaginary, for overdamped waves. Our experiments involved the propagating regime.

The thermally excited capillary waves on a liquid surface, while of microscopic amplitude, scatter light by virtue of the consequent variations in refractive index. The spectrum of the scattered light[2] reflects the temporal evolution of the waves, being the power spectrum of waves of a particular q:

$$P(q,\omega) = \frac{kT}{\pi\omega} \text{Im} \left\{ \frac{i\omega\eta(m + q) + \varepsilon q^2}{D(\omega)} \right\} \tag{4}$$

For propagating waves, the spectrum comprises a doublet ($\pm \omega_0$) symmetrically disposed about the laser frequency. The spectrum can be measured using time domain or frequency domain analysis. In practice we have worked in the time domain, using photon correlation to measure the Fourier transform of $P(\omega)$.

The spectrum given by eqn 4, whilst a complicated function of ω, is

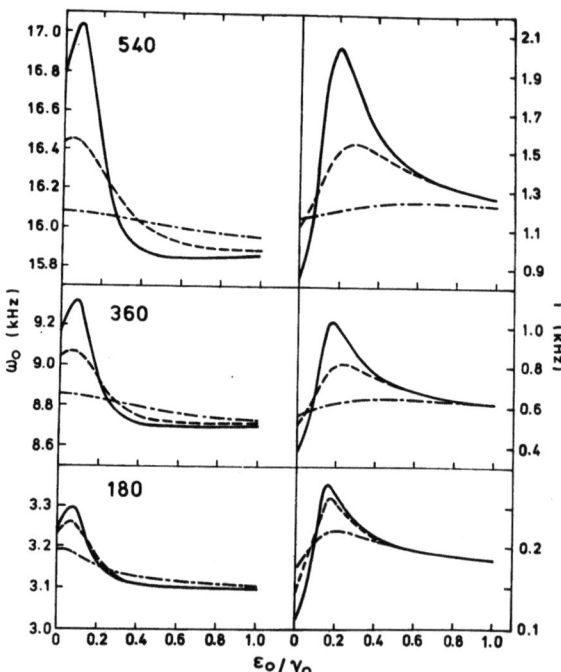

Fig 1 The variations of ω with ε for three different values of q (in cm^{-1}). Three different values of ε' are used: 0 mN.s/m (_____), 10^{-4} mN.s/m (_ _ _ _) and 5×10^{-4} mN.s/m (__ _ __).

approximately Lorentzian, centred upon ω_0 and of half width Γ. The deviations from an exact Lorentzian form can be satisfactorily accounted for by inclusion of a phase term in the Fourier transform

$$G(t) = B + A \cos(\omega_0 t + \phi) \exp(-\Gamma t) \tag{5}$$

used to fit the observed correlation functions to extract ω_0 and Γ.

Various specifically surface processes may affect the capillary waves. Here we will concentrate upon two such effects.

Firstly, dissipation of energy may occur within the surface layers. Following the usual rheological convention this can be incorporated into the above formalism by allowing the relevant elastic moduli to be complex. Here this leads to[3]

$$\gamma = \gamma_0 + i\omega\gamma'$$
$$\varepsilon = \varepsilon_0 + i\omega\varepsilon' \tag{6}$$

(note that there is no uniform notation in this field). While a complex ε is well established in studies of amphiphilic monolayers or surfactant solutions[4], the only clear evidence for a complex γ derives from studies of bilayer lipid membranes[5]. In principle both moduli may be complex for any given fluid interface. Thus to four specifically interfacial properties (γ_0, γ', ε_0, ε') may affect capillary waves. Indeed certain anomalies observed for waves on the free surface of water have been explained[6] in terms of a non-zero value of the transverse shear surface viscosity γ'. A surface excess quantity such as γ' might arise from changes, close to the surface, in the peculiarly structured nature of water. Such considerations motivated a study of aqueous solutions of certain alcohols, thought to enhance the inherent structure of water.

A second surface-associated process, familiar from studies of surfactant solutions, is diffusive relaxation between bulk and surface concentrations. An excess of the surfactant molecules is absorbed at the surface of such a solution. Should this absorbed layer be rapidly removed, the instantaneous surface concentration will be that of the bulk solution. The surface tension in this situation (the true dynamic tension[7]) will exceed the equilibrium value (γ_e). The surface adsorption, and thus the tension, will relax towards the equilibrium value over a period of time. Various processes may control this relaxation[4], but here we will restrict ourselves to diffusive motion of solute molecules. A capillary wave of frequency ω_0 will be associated with periodic compression and rarefaction of the surface layers, involving perturbation of the local adsorption from the equilibrium value. The instantaneous tension will thus vary about γ_e, its average value being in general unchanged. However the dilational modulus will be changed, the exact effect depending upon the relaxation mechanism involved[4]. For diffusion, it has been shown[1] that

$$\varepsilon_0 = \frac{d\sigma}{d\ln A} \frac{1 + a}{1 + 2a + a^2}$$
$$\varepsilon' = \frac{d\sigma}{d\ln A} \frac{a}{\omega(1 + 2a + a^2)} \tag{7}$$

where the differential is the equilibrium modulus and

$$a = \sqrt{1/\tau\omega_0},$$

τ being the diffusive relaxation time of the surface layers. Such a frequency dependent elastic modulus would modify the capillary wave dispersion behaviour of eqn 2.

EXPERIMENTAL

Our experimental arrangement has been described elsewhere in detail[8]. Briefly, the liquid surface is illuminated by light from an Ar^+ laser operated in TEM_{00} mode. Scattered light is mixed at the photomultiplier with a beam of the original laser light, unshifted in frequency. The spectrum of the detector output is as eqn 4. In practice instrumental effects broaden this spectrum: for a Gaussian beam profile the instrumental lineshape will be a Gaussian of standard deviation β. The observed correlation functions will thus be of the form

$$G(t) = B + A \cos(\omega_0 t + \phi) \exp(-\Gamma t - \beta^2 t^2/4) \qquad (8)$$

Deconvolution is particularly simple in the time domain. An objective function of this form fits our observations very well. The ω_0 and Γ values thus found at a particular q can be used to determine the physical properties of the system (γ_0, η). ω is substituted into the dispersion equation (eqn 2), which is then solved for γ_0 and η. In this procedure other surface properties ($\varepsilon_0, \varepsilon', \gamma'$) are assumed zero.

A particular advantage of our experimental system[8] is the ability to observe capillary waves over about an order of magnitude in q – $190 \lesssim q \lesssim 2000$ cm^{-1}. This exceptional range permits phenomena to be studied over a wide range of frequencies.

RESULTS

Surface Excess Effects

We have examined various liquids. Here we show data for the free surface of water at 19.45° C. Fig 2 shows the observed ω_0 and Γ values compared to the functional variations computed from the dispersion equation (eqn 2) using accepted values of γ_0 and η. The observed propagation of capillary waves upon clean water agrees very well with theoretical prediction. From (ω_0, Γ) we can extract values of γ_0 and η: averaged over all q the observed values are $\bar{\gamma}_0 = 72.45 \pm 0.15$ mN/m and $\bar{\eta} = 1.013 \pm 0.006$ mPa.s, compared with the accepted values of 72.83 mN/m and 1.016 mPa.s.

Experiments over a range of q values, such as that summarized in Fig 2, take several hours to perform. Control observations over such periods were made to rule out time-dependent changes in the system. Before commencement of light scattering observations, the surface of the water was aspirated to ensure the absence of any surface-active contamination. The consequent perturbation of the liquid died away over some seconds: light scattering was not possible during this time. Observations were made (at a single q value, 706.8 cm^{-1}) at intervals from 30 s to 2 hrs on five separate samples of water. The $\bar{\omega}_0$ and $\bar{\Gamma}$ values found did not change over this period and, at all times, were consistent with theoretical expectation. As the postulated surface viscosity of water[6] primarily derived from observed Γ values, Γ is shown as a function of time in Fig 3. These data are consistent with the value expected if the wave energy is only dissipated via the viscosity of the fluid.

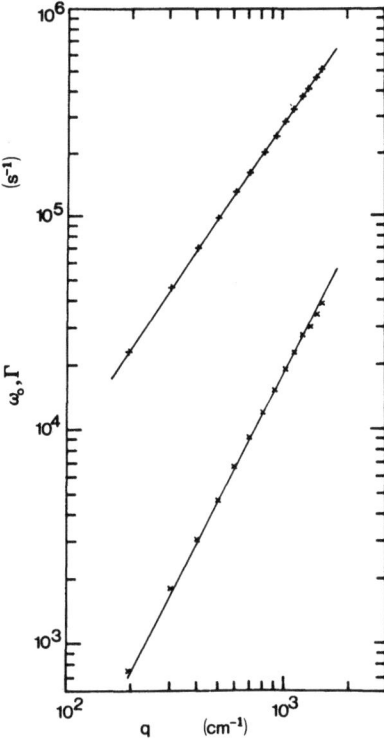

Fig 2   Observed ω_o (+) and Γ (x) values for water at 19.45°C, compared to the theoretical dispersion behaviour.

The data from these refined and extensive experiments thus offer no support for the hypothesized surface viscosity of water. We also conclude from these and other studies[8] that the instrumental effects in our experiments are essentially correctly allowed for in eqn 8. Any errors would cause the measured Γ values to deviate from expectation.

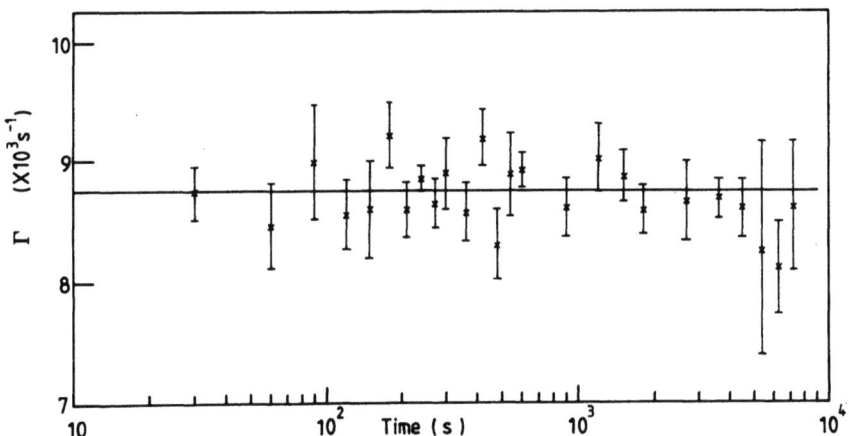

Fig 3   Variation with time of Γ observed for water at 20.5°C. The line indicates the value expected in the absence of surface viscosity effects.

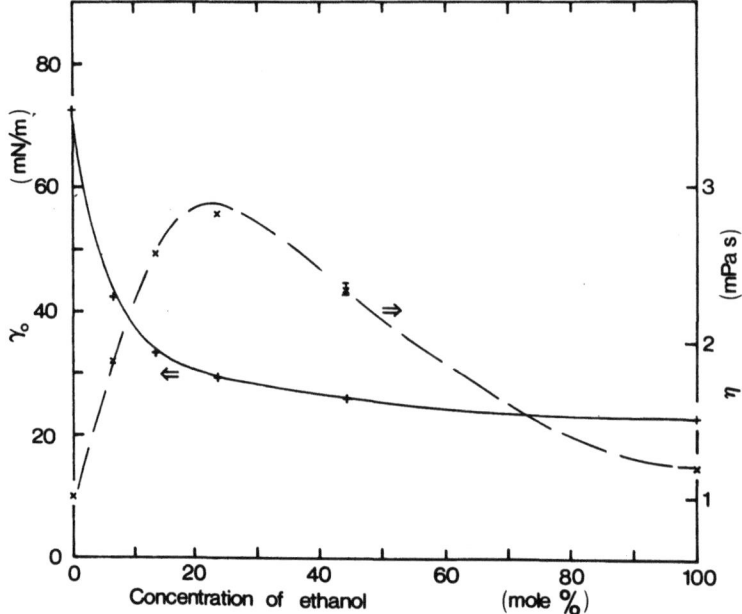

Fig 4 Tension and viscosity values observed for aqueous solutions of ethanol. The lines represent the accepted variations at 20°C.

Various materials modify the structure of water. For example certain ions are known as 'structure-makers' or 'structure-breakers' but the structure involved is not that intrinsic to water[9]. However, the properties of some aqueous alcohol solutions are consistent with the enhancement of some aspects of the inherent structure of water[10,11]. In all cases studied[8] the observed variations with q of ω_o and Γ were entirely in accord with theoretical predictions, based upon the accepted properties (γ_o, η) of the solutions. Fig 4 shows $\bar{\gamma}_o$ and $\bar{\eta}$, derived from the light scattering surface studies and averaged over all q, for various aqueous solutions of ethanol. The agreement with accepted variations[12,13] is clearly excellent. No specifically surface dissipative effects need be invoked.

It still seems reasonable to speculate that surface excess effects (eg viscosities) must exist for a liquid, such as water, which is highly structured in the bulk. However the magnitude of these effects is clearly currently undetectable. The surface layer will likely be of molecular dimensions, while the capillary waves studied penetrate rather far into the bulk: in the present work $[Re(m)]^{-1} \gtrsim 3$ μm. Experiments at very large q (ie very small m^{-1}) might be more affected by the hypothesized surface viscosities.

Surface Activity

Alcohols are known to be surface-active. The dynamic surface tensions (γ_d) of aqueous solutions of several alcohols have been inferred from nucleation studies[14]. The variation with alcohol content of the difference of γ_d from the equilibrium tension (γ_e), shown in Fig 5, closely resembles the behaviour of the surface excess adsorption[15].

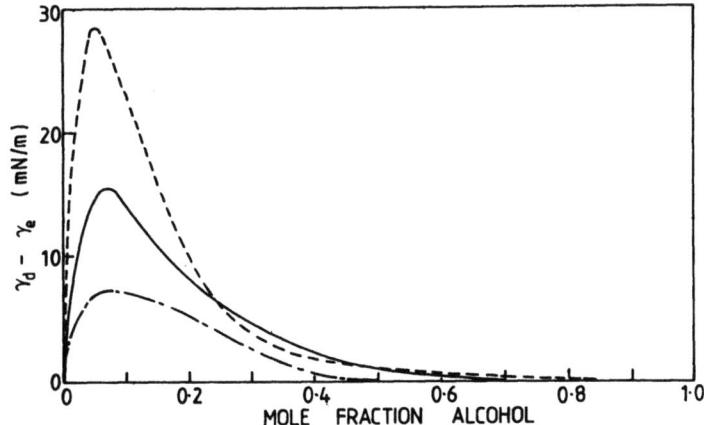

Fig 5 The variations of ($\gamma_d - \gamma_e$) with concentration of aqueous solutions of varius alcohols[14]: methanol (___ _ ___), ethanol (_____) and n-propanol (_ _ _ _).

As explained earlier, the behaviour of high frequency capillary waves will be determined by the equilibrium tension. However the elastic modulus ε will be frequency dependent (eqn 7), ε_o rising towards its equilibrium value for frequencies below the inverse diffusional relaxation time τ^{-1}.

Both ε_o and ε' affect the capillary waves studied[1]. In particular ε_o changes the wave damping Γ substantially, producing a maximum. This

Table 1 Observed tension and viscosity values for aqueous ethanol solution (6.84 mole%) at 20.0°C.

q cm^{-1}	ω_o × 10^5 s^{-1}	γ_o mN/m	η mPa.s
203.1	0.188	41.83	1.91
304.7	0.347	42.62	1.89
406.2	0.529	42.01	1.83
507.8	0.739	42.31	1.86
616.1	0.984	42.27	1.82
718.8	1.236	42.32	1.86
821.5	1.508	42.55	1.92
924.1	1.790	42.34	1.88
1036.3	2.126	42.77	1.95
1140.0	2.465	43.31	1.90
1254.5	2.826	43.64	1.92
1454.0	3.490	42.55	1.87
1567.9	3.867	42.18	1.95

Table 2 Observed tension and viscosity values for aqueous n-propanol solution (5.97 mole%) at 19.0°C.

q cm^{-1}	ω × 10^5 s^{-1}	γ_o mN/m	η mPa.s
244.2	0.181	29.49	2.14
336.3	0.332	30.04	2.21
448.4	0.508	29.79	2.05
560.5	0.703	29.71	2.17
672.6	0.934	30.68	2.19
784.7	1.163	30.32	2.18
896.8	1.424	30.66	2.13
1008.9	1.676	30.15	2.12
1121.0	1.967	30.86	2.27

maximum is reduced by the effects of ε' (see Fig 1), which saturate at large values of ε'.

Unfortunately the <u>equilibrium</u> modulus ($d\gamma/d\ln A$), which sets the scale of the frequency variations of ε_o and ε' (eqn 7), is essentially unknown for the solutions considered here. However, by analogy with results for various surfactants in dilute solution, for which this quantity exceeds the surface pressure (π), we would expect the equilibrium modulus to be quite large for the present samples. Thus the maximum value of ε_o should be of large magnitude, leading to observable changes in capillary wave propagation, particularly Γ, as ω approaches τ^{-1}.

The variations of tension and viscosity with frequency observed for the 6.84 mole% ethanol solution (close to the maximum surface excess concentration[13]) are given in Table 1. Neither quantity shows any observable frequency dependence, all data agreeing adequately with the accepted values (γ_o = 42.82 mN/m, η = 1.877 mPa.s). This suggests that $\tau^{-1} > 10^2(\omega_o)_{max} \sim 10^{-8}$ s (unless the effects of increasing ε_o and decreasing ε' cancel fortuitously, Fig 1). This would imply, given a diffusion coefficient for ethanol in water[16] $D \sim 10^{-5}$ cm^2/s, an adsorbed layer $\lesssim$ 4.5 nm, compatible with the monomolecular layer suggested by Guggenheim and Adam[15].

The surface adsorption of n-propanol is stronger than that of ethanol (Fig 5). We have studied n-propanol in aqueous solution close to the peak of adsorption (5.97 mole%). In this case also (Table 2) the light scattering data yield values of γ_o and η which do not vary with frequency and which agree well with accepted values. These data imply a limit on τ^{-1} of the same order of magnitude as that for ethanol.

DISCUSSION

For none of the fluids studied was any evidence found for specifically surface effects. While surface excess viscosities might be expected to exist in some cases, their influence upon capillary waves will be more noticeable when the depth of penetration of the wave motion (for which m^{-1}) into the bulk fluid is considerably closer to the thickness of the surface layers than in the present study (for which m^{-1} > 3 μm).

For surface-active solutes such as ethanol, relaxation processes such as diffusive exchange between surface and bulk solution should affect the capillary waves. Choice of a system with τ^{-1} nearer to the experimentally accessible ω_o values (or extension of the range of ω_o) should permit observation of these effects. Our present results are compatible with diffusive exchange involving a monomolecular layer of adsorbed alcohol.

ACKNOWLEDGEMENTS

The work was supported by the Science and Engineering Research Council. RCM acknowledges support as a British Gas Research Scholar. We thank Prof C M Sorenson for communicating unpublished results.

REFERENCES

1. E.H. Lucassen-Reynders and J. Lucassen, Properties of capillary waves, Adv. Coll. Interf. Sci. 2:347 (1969).
2. D. Langevin, J. Meunier and D. Chatenay, Light scattering by liquid surfaces, in: 'Surfactants in Solution', K.L. Mittal and B. Lindman, ed., Plenum, New York (1984).
3. F.C. Goodrich, The mathematical theory of capillarity. II, Proc. Roy. Soc. Lond. A. 260:490 (1961).
4. M. van den Tempel and E.H. Lucassen-Reynders, Relaxation processes at fluid interfaces, Adv. Coll. Interf. Sci. 18:281 (1983).
5. G.E. Crawford and J.C. Earnshaw, Phase transitions in monoglyceride bilayers. A light scattering study, Biophys. J. 49:869 (1986).
6. J.C. Earnshaw, Surface viscosity of water, Nature. 292:138 (1981).
7. R. Defay and I. Prigogine, 'Surface Tension and Adsorption', Longmans, London (1966).
8. J.C. Earnshaw and R.C. McGivern, Photon correlation spectroscopy of thermal fluctuations of liquid surfaces, J. Phys. D. to be published.
9. F. Franks, 'Water', Royal Society of Chemistry, London (1984).
10. B.L. Halfpap and C.M. Sorenson, The viscosity of supercooled aqueous solutions of ethanol and hydrazine, J. Chem. Phys. 77:466 (1982).
11. T.M. Bender and R. Pecora, A dynamic light scattering study of the tert-butyl alcohol-water system, J. Phys. Chem. 90:1700 (1986).
12. 'Handbook of Chemistry and Physics', 55th ed., CRC Publishing Co., Cleveland, Ohio (1974).
13. C.M. Sorenson, private communication.
14. D.H. Rasmussen, Dynamic surface tension and classical nucleation theory, J. Chem. Phys. 85:2272 (1986).
15. E.A. Guggenheim and N.K. Adam, The thermodynamics of adsorption at the surface of solutions, Proc. Roy. Soc. Lond. A. 139:218 (1933).
16. R.C. Reid and T.K. Sherwood, 'The Properties of Gases and Liquids', McGraw-Hill, New York (1966).

SOME SURFACE EFFECTS ON GRAVITY WAVES IN A VISCOUS FLUID

Peder A. Tyvand

Department of Physics and Meteorology
Agricultural University of Norway
N-1432 Aas-NLH, Norway

The complex dispersion relation for gravity waves in a viscous fluid of infinite depth was computed by Chandrasekhar[1]. He assumed a stress-free surface, which may be expressed by

$$\frac{\partial u}{\partial y} + \frac{\partial v}{\partial x} = 0, \quad y=0 \tag{1}$$

valid in linear theory. Two-dimensional motion is considered, where x is a horizontal coordinate along the undisturbed fluid surface, and y is a vertical coordinate opposite gravity. g denotes gravitational acceleration, while u and v are horizontal and vertical velocity components.

In cases where the surface is covered by a thin film which is almost immobile horizontally, a more appropriate condition may be the no-slip condition for the tangential motion

$$u = 0, \quad y = 0 \tag{2}$$

instead of condition (1). Condition (2) is valid in linear theory only.

We have calculated[2] the complex dispersion relation with condition (2). Both Lamb[3] and Landau & Lifshitz[4] have treated the long-wave limit of this dispersion relation.

We will now give a short review of the results. In the long-wave limit[3,4] nothing unexpected happens: Actually, the dispersion relation for the no-slip surface (assuming no rigidity or tension in the film) is equivalent to that at a stress-free surface with a rigid bottom located at a depth given by

$$h = 0.1053 \, \lambda \tag{3}$$

where λ is the wavelength. This is shown by a comparison with the complex dispersion relation for gravity waves in a viscous fluid layer, given by Wehausen & Laitone[5].

For wave numbers $k(=2\pi/\lambda)$ satisfying

$$0 < \frac{\nu^{2/3}}{g^{1/3}}k < 0.1632 \tag{4}$$

(ν denoting kinematic viscosity) the results are qualitatively as in the long-wave limit: Immobilization of the surface horizontally leads to an increased damping and reduced phase and group velocities, compared to a stress-free surface. However, as we approach the branch point given by $k=k_b$;

$$\frac{\nu^{2/3}}{g^{1/3}}k_b = 1.1729 \tag{5}$$

we find opposite effects in certain intervals.

Let us first note the result that k_b is reduced by 2% compared to the stress-free condition[1]. So the immobilization reduces the wave number range of travelling waves, but only slightly.

However, when

$$0.1632 < \frac{\nu^{2/3}}{g^{1/3}}k < 0.8304 \tag{6}$$

the damping is smaller for an immobile surface than for a stress-free surface. The maximal reduction is relatively large, exceeding 20%. This seems like a paradox: How is it possible that prevention of tangential motion can reduce the wave damping? Probably the paradox is resolved by noting what happens above the branch points, i.e. when $k>k_b$: Then the no-slip condition tends to "freeze" the surface shape, so it **reduces** the decay compared with the stress-free condition. However, this argument applies only to the physically dominating creeping mode[1]. For the rapidly decaying viscous mode[1], the situation is opposite: The decay is promoted by immobilization.

We also find unexpected effects on the phase velocity: When

$$0.5063 < \frac{\nu^{2/3}}{g^{1/3}}k < 1.0886 \tag{7}$$

the phase velocity is increased by the tangential immobilization. It never increases by more than 10%, and this is too small to cause any increased group velocity: We find that energy propagation is never enhanced by a tangential immobilization of the surface.

We have also taken surface tension T and flexural rigidity D of the film into account. A relevant dimensionless group in problems involving gravity, viscosity and surface tension is the Morton number;

$$M = \frac{g\rho^3 \nu^4}{T^3} = \frac{g\mu^4}{\rho T^3} \tag{8}$$

where ρ is the density of the fluid and μ is the dynamic viscosity. In an earlier paper[6] we found a criterion

$$M < 0.044789 \tag{9}$$

for travelling waves to exist with group velocity exceeding phase velocity. When we have a no-slip condition for the tangential motion, this criterion is changed to

$$M < 0.058406 \tag{10}$$

The last result assumes that the flexural rigidity of the surface film is zero. When it is nonzero, we have a dimensionless group which is analogous to the Morton number:

$$N = \frac{\rho^3 \nu^8}{gD^3} = \frac{\mu^8}{\rho^5 gD^3} \tag{11}$$

When T=0, we find the basic result that travelling waves exist for all wave numbers when

$$N < 64.23 \tag{12}$$

The physical reason for this possible absence of creeping motion when the waves are very short, is the strong restoring force due to the flexural rigidity.

When the thickness of the fluid layer is finite, the calculation of the dispersion relation is more complicated. The long-wave or small-viscosity limit has been studied since the previous century, and is given in Wehausen & Laitone[5]. Here we will only give the solution for the limit of creeping motion:

For a stress-free surface, the absence of inertia gives a local decay of an initially disturbed surface, governed by the factor $\exp(-\sigma t)$ where the decay rate σ is given by

$$\sigma = \frac{g}{2\nu k} \frac{\sinh 2kh - 2kh}{\cosh 2kh + 2k^2h^2 + 1} \tag{13}$$

valid when the bottom plane is rigid (no-slip condition):

$$\frac{\partial v}{\partial y} = v = 0, \quad y = -h \tag{14}$$

We also have a formula for the case of a horizontal bottom without shear stress (free-slip condition);

$$\sigma = \frac{g}{2\nu k} \frac{\cosh 2kh - 1}{\sinh 2kh - 2kh} \tag{15}$$

with the boundary conditions:

$$\frac{\partial^2 v}{\partial y^2} = v = 0, \quad y = -h \tag{16}$$

These formulas assume the stress-free condition (1) for the deformed upper surface. We have derived the corresponding formulas for a tangentially immobile surface, governed by eq. (2): For the no-slip bottom condition (14) we find:

$$\sigma = \frac{g}{2\nu k} \frac{\cosh 2kh - 2k^2 h^2 - 1}{\sinh 2kh + 2kh} \tag{17}$$

For the free-slip bottom condition (16) we find:

$$\sigma = \frac{g}{2\nu k} \frac{\sinh 2kh + 2kh}{\cosh 2kh + 1} \tag{18}$$

Eq. (13) gives a maximal value of σ at $kh = 2.1195$. Eq. (17) gives a maximal value of σ at $kh = 2.4475$. On the other hand, when there is free slip at the bottom, we will have stronger decay the longer the waves, as long as the creeping motion approximation is valid. Therefore eqs. (15) and (18) give no maximal values.

In all four cases represented by eqs. (13), (15), (17) & (18) we find the short-wave limit[3]

$$\sigma = \frac{g}{2\nu k} \tag{19}$$

So the decay of a free surface by creeping motion is not influenced by an immobilization, unless there is a bottom closer than one wavelength from the surface. But such a bottom may have strong effects, which is seen from eqs. (15)-(18).

These results for free surface creeping motion have been applied[7] to explain some features of the viscous buckling phenomenon investigated experimentally by Suleiman & Munson[8]. They have a thin layer of very viscous fluid above a slightly viscous, heavier fluid. The upper fluid is confined between two concentric vertical cylinders with a gap b. The inner cylinder is at rest and the outer cylinder has velocity U. Suleiman and Munson[8] found the criterion for buckling:

$$\frac{\mu U}{b(\gamma_b T_t)^{1/2}} > 4.1 \qquad (20)$$

Here γ_b is the specific weight of the bottom layer. T_t is the total surface tension of the viscous layer, which is the tension at the free surface plus the interfacial tension.

In order to find a theoretical argument for this criterion, we make some simplifying assumptions. We assume that the only influence of the lower layer is to give a hydrostatic response to the interfacial deflection. Then we neglect the lower layer altogether and replace it by the free-slip bottom condition (16). Furthermore, we assume that the basic Couette flow is not modified by the surface deformations. This is a valid assumption within linear theory, and leads to the superposition of two independent effects: (I) Horizontal convection with the basic Couette flow. (II) Creeping motion decay of the surface disturbance, influenced by gravity and the surface tension T.

The horizontal convection has one important effect in this context: Kinematic contraction. This means that the distance between the crests of a wavy disturbance changes with time due to gradients in the basic flow. We will now give a general theory of kinematic contraction. Let u_i (i=1,2) be an arbitrary horizontal flow field, and let λ_i be an instantaneous wave length vector, representing an arbitrary Fourier component of a white noise disturbance. The time derivative of λ_i is given by:

$$\frac{\partial \lambda_i}{\partial t} = \lambda_j \frac{\partial u_i}{\partial x_j} \qquad (21)$$

Let λ denote the absolute value of λ_i. We introduce the unit vector $n_i = \lambda_i/\lambda$ and find:

$$\lambda^{-1} \frac{\partial \lambda}{\partial t} = \lambda_i \lambda_j \frac{\partial u_i}{\partial x_j}/\lambda^2 = \frac{1}{2} n_i n_j \left(\frac{\partial u_i}{\partial x_j} + \frac{\partial u_j}{\partial x_i}\right) = n_i n_j \tau_{ij}/2\mu \qquad (22)$$

The relative rate of kinematic contraction is given by $(-\lambda^{-1} \partial \lambda/\partial t)$.

In eq. (22) τ_{ij} is the viscous stress tensor for the two-dimensional flow. It follows that maximal kinematic contraction always corresponds to maximal compressive viscous stress. The relative rate of kinematic contraction has its maximum

$$(-\lambda^{-1} \partial \lambda/\partial t)_{max} = U/2b \qquad (23)$$

for our Couette flow problem.

A necessary but not sufficient criterion for instability is that the wave slope increases with time;

$$\left|\frac{d\zeta}{\zeta}\right| < \left|\frac{d\lambda}{\lambda}\right| \qquad (24)$$

where ζ is the amplitude of the surface deflection. On the left hand side we insert the minimum decay rate for large Bond number

$$\sigma_{min} = (\rho g T)^{1/2}/\mu \qquad (25)$$

This is valid for both surface boundary conditions (1) and (2). Invoking eq. (23), we finally arrive at a necessary condition for buckling instability:

$$\mu U/b(\rho g T)^{1/2} > 2 \qquad (26)$$

This is in reasonable accordance with the experimental criterion (20). A more general version of criterion (26) is this: The maximal compressive viscous stress must exceed $2(\rho g T)^{\frac{1}{2}}$ in order to get buckling instability.

REFERENCES

1. S. Chandrasekhar 1955. The character of the equilibrium of an incompressible heavy viscous fluid of variable density. Proc. Camb. Phil. Soc. 57, 415-425.

2. P.A. Tyvand & K.M. Gjerde 1985. Effects of an elastic surface film on gravity waves in a viscous fluid. Manuscript.

3. H. Lamb 1932. Hydrodynamics. Cambridge Univ. Press, p. 632.

4. L.D. Landau & E.M. Lifshitz 1959. Fluid Mechanics. Pergamon Press, p. 244.

5. J.V. Wehausen & E.V. Laitone 1960. Surface waves. In: Encyclopedia of Physics, Vol. 9, Springer-Verlag.

6. P.A. Tyvand 1984. A note on gravity waves in a viscous liquid with surface tension. J. Appl. Math. Phys. (ZAMP) 35, 592- 597.

7. P.A. Tyvand 1984. Free surface creeping motion related to a buckling phenomenon. Phys. Fluids 27, 2199-2201. (Erratum: Phys. Fluids 28, 1214 (1985)).

8. S.M. Suleiman & B.R. Munson 1981. Viscous buckling of thin fluid layers. Phys. Fluids 24, 1-5.

SUSTAINED GRAVITY-CAPILLARY (LAPLACE) WAVES AT INTERFACES AND MARANGONI-BENARD INSTABILITY

P. L. Garcia-Ybarra and M. G. Velarde

Depto. Física Fundamental, U.N.E.D.
Apartado 60.141, Madrid 28.071. Spain

1. INTRODUCTION

When the equilibrium surface of a liquid layer open to the ambient air is disturbed the interface behaves like a membrane, it gets deformed, and then forces tending to return it to the original equilibrium state appear in the liquid. Capillary forces tend to reduce the increased surface and, moreover, if the liquid is in a gravitational field the disturbance gives rise to gravitational forces that tend to return the interface to is original level shape. Because of inertia, however, the liquid particles overshoot their original equilibrium position and in turn are affected by these forces. As a consequence transverse waves appear on the air-liquid interface that are called capillary-gravity waves. [1-3] These waves are eventually damped out by the viscosity of the liquid. For a frequency of oscillation Ω and kinematic viscosity ν, the viscous penetration length of the wave is of order $(\nu/\Omega)^{1/2}$. Thus for deep enough layers this penetration length is rather small with respect to the layer's depth. This is the case we shall consider here limiting our analysis to a high enough frequency range such that $\Omega \gg \nu/h^2$, where h is the depth of the liquid. For simplicity we shall let h go to infinity.

If the constraint leading to the interfacial disturbance is maintained, in particular, for a standard liquid heating the layer from above or cooling it from below, an interfacial instability may develop due to surface tension inhomogeneities thus leading to sustained oscillations. In this article we show that such a Marangoni-Bénard instability does indeed arise in a single-or two-component liquid layer subjected to temperature and/or concentration gradients. We provide the oscillation frequency and the critical Marangoni number as functions of the Prandtl and the capillary (deformation) numbers. The limiting values of these two quantities are also given for microgravity conditions and so our predictions are expected to be relevant to fluid dynamics (crystal growth, etc.) experiments conducted aboard spacecrafts where the gravitational acceleration may be as low as four or six orders of magnitude the value on Earth.

2. EQUATIONS AND RESULTS

Small, infinitesimally large disturbances in the liquid layer obey the continuity equation, the Fourier heat equation and the (linearized) Navier-Stokes equation. At the air-liquid interface we have

$$(\partial w/\partial t) = - (1/\rho)(\partial p/\partial z) + \nu (\partial^2 w/\partial z^2) \tag{1}$$

where we restrict consideration to a two-dimensional problem. t and z denote time and vertical coordinate and w, ρ and p are vertical velocity, density and pressure, respectively. w is indeed the liquid velocity and if we assume the *kinematic* condition $w = \partial \zeta /\partial t$ we have a direct relationship between a point at the *geometrical* interface and a *liquid* point at every time t. Moreover, at the interface there is a dynamic balance for the normal and tangential components of the stress tensor. We have

$$p - \rho g \zeta + \sigma \partial^2 \zeta /\partial x^2 = 2 \rho \nu (\partial w/\partial z) \tag{2}$$

and

$$\rho \nu (\partial w/\partial x) = (\partial \sigma/\partial T)[(\partial T/\partial x) - \beta (\partial \zeta/\partial x)] \tag{3}$$

where x accounts for the horizontal coordinate, g is the gravitational acceleration, σ is the surface tension, T denotes temperature and β is the temperature gradient induced across the liquid layer (β is defined positive when the layer is heated from below, i.e., from the liquid side). $\zeta(x,t)$ is the deformation of the interface.

Thus in the high frequency limit ($\Omega \gg 1$), using (2), (3) and the kinematic condition, Eq. (1) becomes

$$d\zeta/dt^2 + k^2 [4\nu - k\beta (\partial\sigma/\partial T)(2\kappa)^{1/2}/\rho\Omega^{3/2}](d\zeta/dt) +$$
$$+ [(gk+\sigma k^3/\rho)+(\partial\sigma/\partial T)\beta k^3(\kappa/2\Omega^{1/2}/\rho] \zeta = 0 \tag{4}$$

where we have assumed that

$$\zeta = - (A/\Omega) \cos(kx + \Omega t) \tag{5}$$

and similar Fourier mode expressions for the remaining disturbances. κ is the thermometric conductivity.

Eq. (4) is the harmonic oscillator equation obeyed by a point at the air-liquid interface. The damping coefficient may be positive, negative or vanishing according to the sign and values given to β and $(\partial\sigma/\partial T)$.

Using the capillary length as the space scale, $l = (\sigma/\rho g)^{1/2}$, Eq. (4) takes on a dimensionless, universal form

$$\frac{d^2\zeta}{d\tau^2} + [4a^2 + \frac{M}{\sqrt{2P^3\omega^3}}]\frac{d\zeta}{d\tau} + [\frac{a(1+a^2)}{PC} - \frac{a^2 M}{\sqrt{2P^3\omega}}]\zeta = 0 \tag{6}$$

with $\zeta = \xi/l$, $\tau = t\nu/l^2$, $a = lk$, $\omega = \Omega/\nu k^2$, $M = -(\partial\sigma/\partial T)\beta l^2/\nu\kappa$, $P = \nu/\kappa$ and $C = \rho\nu\kappa/\sigma l$. Further simplification of Eq. (6) can be achieved by definitely restricting consideration to the high frequency limit.

We see that in order to have vanishing damping it suffices to set $M = 0 (\omega^{3/2})$, with $P = 0(1)$. We can also set $C = 0 (1/\omega^2)$, which is a reasonable assumption provided we are far from a critical point or if the liquid layer is not too thin. Then the second term in the coefficient of ζ can be neglected and Eq. (6) reduces to

$$\frac{d^2\zeta}{d\tau^2} + [4a^2 + M/(2P^3\omega^3)^{1/2}]\frac{d\zeta}{d\tau} + [\frac{a(1+a^2)}{PC}]\zeta = 0 \qquad (7)$$

This equation is the simplest (high frequency) harmonic oscillator approximation to the oscillatory interfacial motion of the open surface in a Bénard layer. It contains, however, all the relevant physics. Indeed the damping coefficient vanishes whenever we set $M = -4a^2(2P^3\omega^3)^{1/2}$ thus allowing a free oscillation of (dimensionless) frequency given by $\omega^2 = (1+a^2)/a^3 PC$.

3. CONCLUSION

As a result of the preceding analysis we find that the minimal value of the Marangoni number needed to sustain the oscillatory motion is

$$M_c = -7.93 \, (P/C)^{3/4} \qquad (8)$$

with a frequency

$$\omega_c = 6\sqrt{5}/PC, \text{ i.e., } (\Omega/2\pi) = 0.12 \, (g/l)^{1/2} \qquad (9)$$

and a wave number

$$a_c = \sqrt{5}/5, \text{ i.e., } \lambda_c = 14.04 \, l \, (cm) \qquad (10)$$

where it does indeed appear that the oscillatory interfacial instability is to be expected for *negative* Marangoni numbers only. For standard liquids this means that the heating is from above or the layer is cooled from below. However, there are nostandard liquids like some high alcohol solutions, dodecylammonium (DAC), some binary alloys, and some liquid crystals that have a minimum in their variation of surface tension with temperature[1-6] for which the predicted oscillatory instability can be seen when heating the liquid layer from below.

Notice that under isothermal conditions with, however, an imposed concentration gradient all the results obtained in this section are still valid. It suffices to replace temperature by concentration, one diffusivity (heat) by the other (Fick's law) and, obviously, the Marangoni number (Ma) by the solutal Marangoni (elasticity) number. Moreover, the simultaneous action of both heat and mass diffusion with, however, no coupling or crosstransport phenomena like the Soret or Dufour effects merely amounts to an appropriately weighted combination of both Marangoni numbers. The numerical estimates of both wavenumber and oscillation frequency remain unaltered. Thus the result found that a Marangoni effect can sustain capillary-gravity waves overtaking the damping action of the viscosity is a consequence of the variation of the surface tension. Further details about this problem can be found in Ref. 7.

ACKNOWLEDGMENTS

This research has been sponsored by CAICYT (Spain).

REFERENCES

1. R. VOCHTEN and G. PETRE, J. *Colloid Interface Sci..* **42**, 320 (1973).
2. J. C. LEGROS, M. C. LIMBOURG-FONTAINE, and G. PETRE, *Acta Astronaut.* **11**, 143 (1984).
3. K. MOTOMURA, S.-I. IWANAGA, Y. HAYAMI, S. URYU, AND R. MATUURA, J. *Colloid Interface Sci.*, **80**, 32 (1981).
4. P. J. DESRE and J. C. JOUD, *Acta Astronaut* **8**, 407 (1981).
5. J. J. BIKERMAN, *Physical Surfaces* (Academic, New York, 1970), Chap. 2.
6. M. G. J. GANNON and T. E. FABER, *Philos. Mag.* A. **37**, 117 (1978).
7. P. L. GARCIA-YBARRA and M. G. VELARDE, *Phys. Fluids* **30**, 1649 (1987).

THE HARMONIC OSCILLATOR DESCRIPTION OF LONGITUDINAL (MARANGONI-LUCASSEN) WAVES AT LIQUID INTERFACES

M. G. Velarde and X.-L. Chu

Depto. Física Fundamental, U.N.E.D.
Apartado 60.141, Madrid 28.071
Spain

1. INTRODUCTION AND DISTURBANCE EQUATIONS

Let us consider two liquid layers at rest with an interface between them located at $z = 0$; z is the vertical coordinate say. Let η, ν ($\eta = \rho\nu$), D and ρ denote the corresponding dynamic viscosity, kinematic viscosity, mass diffusivity and density in each liquid. We shall denote with subscript "one" the lower liquid. Let us assume that a surface active component (surfactant) is distributed in each bulk phase with a given volume gradient and that it may be adsorbed at the interface according to Langmuir's law [1]. Then if we consider a disturbance at the interface that may eventually be amplified thus leading to interfacial instability the expected evolution of such disturbance is governed by equations valid on each side of the interface and at the interface itself that for the simplest two-dimensional problem in dimensionless form are [1-5]

$$\text{div } \mathbf{v}_1 = \text{div } \mathbf{v}_2 = 0 \tag{1}$$

$$(\partial \mathbf{v}_1 /\partial t) + \text{grad } p_1 - \nabla^2 \mathbf{v}_1 = N_\rho (\partial \mathbf{v}_2 /\partial t) + \text{grad } p_2 - N_\eta \nabla^2 \mathbf{v}_2 = 0 \tag{2}$$

and

$$(\partial C_1 /\partial t) - w_1 - S^{-1} \nabla^2 C_1 = (\partial C_2 /\partial t) - w_2 - S^{-1} N_D \nabla^2 C_2 = 0 \tag{3}$$

where $\mathbf{v}_i$ ($i = 1,2$) = (u_i, w_i) with u and w the horizontal and vertical velocity components of the disturbance velocity field. p denotes pressure. $S = \nu / D$ (Schmidt number). $N_D = D_2/D_1$, $N_\eta = \eta_2/\eta_1$. $N_\rho = \rho_2/\rho_1$. x accounts for the horizontal coordinate. C is the volume concentration of the surfactant.

Let $E = - (\partial \sigma /\partial C_1) \beta_1 l^2 / \eta_1 D_1$ be the elasticity (solutal) Marangoni number with σ the liquid-liquid interfacial tension. β is the volume concentration gradient of the surfactant. l is a characteristic length that may very well be the capillary length but need not to be so. The capillary length is given by $l^2 = \sigma_0 / (\rho_1 - \rho_2) g$ with σ_0 a reference value. $B = (\rho_1 - \rho_2) g l^2 / \sigma_0$ is the Bond number. Then when surface deformation and surface accumulation of the surfactant can be neglected the disturbance evolution equations (1) - (3) for longitudinal motions obey the following boundary conditions (b.c.) at $z = 0$ [1]:

$$w_1 = w_2 = 0 \tag{4}$$

$$(\partial w_1 / \partial z) = (\partial w_2 / \partial z) = 0 \tag{5}$$

$$(E/S)a^2 C_1 - N_\eta (\partial^2 w_2 / \partial z^2) + (\partial^2 w_1 / \partial t^2) = 0 \tag{6}$$

$$\Gamma S (\partial C_1 / \partial t) = \partial (C_2 - C_1) / \partial z \tag{7}$$

$$C_1 = N C_2 \tag{8}$$

where Γ is the Langmuir adsorption number [1], i.e., the slope of the Langmuir adsorption law at the interface per unit length l (in the two-dimensional problem). N is the ratio of Γ_1 to Γ_2 when Γ is evaluated from each side of the interface.

2. SOLUTIONS

Due to the linearity of the problem we may seek solutions of the form

$$w_1 = B_1 (-e^{az} + \exp m_1 z) \tag{9}$$

$$w_2 = B_2 (-e^{-az} + \exp(-m_2 z)) \tag{10}$$

$$p_1 = B_1 (\lambda/a) e^{az} \tag{11}$$

$$p_2 = - B_2 N_\rho (\lambda/a) e^{-az} \tag{12}$$

$$C_1 = R_1 \exp(q_1 z) - (B_1/\lambda) e^{az} + (B_1 S/\lambda(S-1)) \exp(m_1 z) \tag{13}$$

and

$$C_2 = R_2 \exp(-q_2 z) - (B_2/\lambda) e^{-az} + (B_2 S N_D^{-1}/\lambda(S N_D^{-1} - N_\rho N_\eta^{-1})) \exp(-m_2 z) \tag{14}$$

where a denotes a Fourier mode and λ a complex time constant whose imaginary part is a dimensionless frequency. $m_1^2 \equiv \lambda + a^2$, $m_1^2 \equiv N_\rho N_\eta^{-1} \lambda + a^2$, $q_1^2 \equiv S\lambda + a^2$ and $q_2^2 \equiv S\lambda N_D^{-1} + a^2$. The quantities B_1, B_2, R_1 and R_2 are the unknown disturbance amplitudes left undetermined in a linear theory.

Now let us take the time derivative of Eq. (7). We have

$$\Gamma S (\partial C_1 / \partial t^2) = \partial^2 (C_1 - C_2) / \partial z \, \partial t \tag{15}$$

On the other hand taking the z-derivative in Eq. (3) we can estimate the right hand side of Eq. (15). However we must estimate terms like $\partial \nabla^2 C_i$ (i = 1,2)$/\partial z$. For such purpose we use the relationships given earlier. For instance, using (5) we get

$$B_2 = - (m_1 - a) B_1 / (m_2 - a) \tag{16}$$

while using (6) and (7)

$$R_1 = - (B_1/\lambda(S-1)) - (\lambda S/Ea^2)(1 + N_\rho(m_1 - a)/(m_2 - a)) B_1 \tag{17}$$

and using (8)

$$R_2 = \frac{(m_1-a)N_\rho N_\eta^{-1}}{(m_2-a)\lambda(SN_D^{-1}-N_\rho N_\eta^{-1})} B_1 - \frac{\lambda S}{NEa^2}(1+N_\rho \frac{m_1-a}{m_2-a}) B_1 \qquad (18)$$

Now with $\lambda = i\omega$, we search for purely oscillatory disturbances. Using (3), (16), (18), after some lengthy, albeit straightforward calculus, Eq. (15) becomes

$$(1+N_\rho^{1/2}N_\eta^{1/2})\Gamma S^{1/2}\frac{d^2C_1}{dt^2} + \sqrt{2\omega}[1+N_\rho^{1/2}N_\eta^{1/2}](1+\frac{1}{NN_D^{1/2}})+(\frac{Ea^2}{S^2\omega^2}\Pi_1)]\frac{dC_1}{dt}$$

$$+ a\omega[(1+N_\eta)(1+\frac{1}{NN_D^{1/2}})+\frac{Ea^2}{S^2\omega^2}\Pi_2]C_1 = 0 \qquad (19)$$

with

$$\Pi_1 = (N_\rho N_D/N_\eta)^{1/2} - 1 \qquad (20)$$

and

$$\Pi_2 = N_D^{1/2} - 1 \qquad (21)$$

3. RESULTS AND CONCLUSION

Equation (19) is the simplest harmonic oscillator description of the surfactant oscillations along the interface. It is an equation for longitudinal interfacial oscillations whose damping coefficient can be zero with a suitable of the elasticity Marangoni number.

When the damping coefficient is set to zero we have the following two relationships

$$(1+\sqrt{N_\rho N_\eta})(1+\frac{1}{NN_D^{1/2}})+\frac{a^2 E\Pi_1}{S^2\omega^2} = 0 \qquad (22)$$

and

$$a[(1+N_\eta)(1+\frac{1}{NN_D})+\frac{Ea^2}{S^2\omega^2}\Pi_2] = \omega(1+N_\rho^{1/2}N_\eta^{1/2})\Gamma S \qquad (23)$$

To be satisfied, Eq. (22) demands that

$$E\Pi_1 < 0 \qquad (24.a)$$

e.g. $\qquad \text{sgn}(E) = -\text{sgn}(N_\rho^{1/2}N_D^{1/2}/N_\eta^{1/2} - 1) \qquad (24.b)$

that together with sgn(E) = sgn(β_1) yields the following consequence: To have oscillatory behavior we must have

$$(D_f/D_t) > (v_f/v_t) \qquad (25)$$

where "f" and "t" stand for "from" and "to", a way of indicating how the surfactant is being transported *from* and *to* the volume. Condition (25) is a condition for overstability in Marangoni convection obtained by earlier authors [6,8]. Here it appears as a necessary, albeit not sufficient condition for overstability.

Also years ago Lucassen [3] introduced a complex elasticity modulus, ε, which is related to our elasticity Marangoni number by the following relationship

$$E = - \frac{\varepsilon \, l^2 \, (k^2 + i\Omega/v_1)^{1/2}}{\eta_1 \, D_1} \approx - \frac{\varepsilon \, l^2 \, (i\Omega/v_1)^{1/2}}{\eta_1 \, D_1} \qquad (26)$$

with $a = k\,l$ and $\omega = \Omega\, l^2/v_1$. The quantity k can be assumed to be smaller than the inverse of the viscous penetration length. Then using these new variables, Eq. (22), reduces to

$$\varepsilon m k^2 + i\eta_1 \Omega m^2 (v_1/D_1) \left[\frac{[1+(\rho_2 \eta_2/\rho_1 \eta_1)^{1/2}][1+ N(D_1/D_2)^{1/2}]}{(v_1 D_1/v_2 D_2)^{1/2} - 1} \right] = 0 \qquad (27)$$

$m^2 = k^2 \, i\Omega/v_1 \approx i\Omega/v_1$. Eq. (27) is a generalization of the particular case discussed by Lucassen [3] (see also Eq. (40.b) in Ref. [5]).

Then using both Eqs. (22) and (23) we get

$$E_c = - \left[\frac{(N_D^{1/2} + N_\eta)(N_\rho^{1/2} N_\eta^{-1/2} - 1)}{S \, \Gamma \, \Pi_1 \, \Pi_2} \right]^2 \frac{(1 + N^{-1} N_D^{-1/2}) \, S^2}{\Pi_1} (1 + N_\rho^{1/2} N_\eta^{1/2}) \qquad (28)$$

and the dispersion relation

$$\omega_c = a_c \left[\frac{(N_D^{1/2} + N_\eta)(N_\rho^{1/2} N_\eta^{-1/2} - 1)}{S^{1/2} \, \Gamma \, \Pi_1 \, \Pi_2} \right]^2 \qquad (29)$$

The value E_c is the minimal value of the elasticity Marangoni number needed to sustain longitudinal interfacial convective oscillations of frequency ω_c. Using the fact that both ω and a must be positive numbers we get from (29) that the following relationship must be satisfied

$$(v_f/v_t) < 1 \qquad (30)$$

in order to have oscillations.

Thus putting together Eqs. (25) and (30) we have the conditions needed to sustain the *longitudinal* waves [3]. Lucassen, however, only considered naturally damped motions but he clearly emphasized that these interfacial motions were drastically different from gravity-capillary waves [1,2,3,9,10]. The latter are rather transverse motions originated in the geometric deformation of the interface and Laplace law. Thus they may appear in ideal liquids while the Marangoni-Lucassen waves necessarily demand dissipation in viscous liquids and surface tension variation. With strong enough dissipation i.e. for Marangoni numbers larger than E_c these waves can be sustained along the interface even if it is not deformed. Worded differently, at $E = E_c$ we have the threshold for Marangoni-Bénard overstability under solutal /surfactant gradients [1,2,6,7,11].

ACKNOWLEDGMENTS

This research has been sponsored in its early stage by the Stiftung Volkswagenwerk and subsequently by the CAICYT (Spain). X.-L. Chu acknowledges a predoctoral fellowship from the Spanish General Dicterorate for Science Policy.

REFERENCES

1. C.A. MILLER and P. NEOGI, *Interfacial phenomena*, Marcel Dekker, New York, 1985.
2. B. G. LEVICH, *Physicochemical Hydrodynamics*, Prentice-Hall, Englewood Cliffs, NJ, 1962.
3. J. LUCASSEN, Trans Faraday Soc. **64**, 2221 (1968).
4. E. H. LUCASSEN-REYNDERS and J. LUCASSEN, Adv. Colloid Interface Sci. **2**, 347 (1969).
5. R. S. HANSEN and J. AHMAD, Prog. Surface Membrane Sci. **4**, 1 (1971).
6. C. V. STERNLING and L. E. SCRIVEN, A. I. Ch. E. J. **5**, 514 (1959).
7. M. HENNENBERG, P. M. BISCH, M. VIGNES-ADLER and A. SANFELD, J. Colloid Interface Sci. **69**, 128 (1979).
8. H. LINDE, P. SCHWARTZ and H. WILKE, in *Dynamics and Instability of Fluid Interfaces* (T. S. Sørensen, editor), Springer-Verlag, New York, p. 75.
9. P. L. GARCIA-YBARRA and M. G. VELARDE, Phys. Fluids **30**, 1649 (1987).
10. M. G. VELARDE, P. L. GARCIA-YBARRA and J. L. CASTILLO, Physicochem. Hydrodyn. **9**, 387 (1987).
11. For a recent review see, for instance, J. C. LEGROS, A. SANFELD and M. G. VELARDE, in *Fluid Sciences and Materials Science in Space* (H. U. Walter, editor), Springer-Verlag, Berlín, 1987, pp. 83-140.

EXPERIMENTAL INVESTIGATIONS OF THE EFFECTS OF SURFACE TENSION

ON CONVECTION CAUSED BY HEATING FROM BELOW

E. L. Koschmieder

College of Engineering
University of Texas at Austin
Austin, Texas   78712 U.S.A.

INTRODUCTION

We shall in the following discuss the essential results of various experiments which deal with surface tension driven convection. We assume that the reader is familiar with the results of linear theory of buoyancy driven convection, as originally developed by Rayleigh (1916), and the results of the linear theories of Pearson (1958) and Nield (1964), describing surface tension driven convection. We begin logically with Benard's (1900) experiments, which were only 56 years later recognized as being a case of surface tension driven convection. Since then a few more experiments have been made which have, as their starting point, the consequences of surface tension in mind. These experiments are few and far between. Until now a systematic series of experiments which would have tried to elucidate critical features of surface tension driven convection has not been made. The reader will therefore not find a complete and satisfying picture of surface tension driven convection in the notes presented in the following. We will, in general, discuss experimental technique only cursorily, but focus on the results of the experiments.

THE EXPERIMENTS

Benard's Work

Benard set an example for virtually all subsequent experiments on convection in shallow horizontal fluid layers by striving for uniform heating from below and by trying to eliminate the bothersome influences of the lateral confinement of the fluid. The apparatus which he used is shown in figure 1.

The bottom brass plate of the apparatus was heated by steam from boiling water, hence at a temperature of 100°C. This arrangement approximates uniform heating very well, but has the disadvantage of fixing the bottom temperature at 100°C. The influences of the lateral wall were minimized by working with fluid layers of very small depth (between 0.5 mm and 1 mm) in an apparatus of substantial horizontal extend, namely of 20 cm diameter. The so-called aspect ratio of the fluid layer, the ratio of the horizontal extend divided by the fluid depth, was then about 200 or more approximating a fluid layer with negligible walls, in other words an infinite fluid layer. The fluid used was molten spermaceti. Spermaceti is

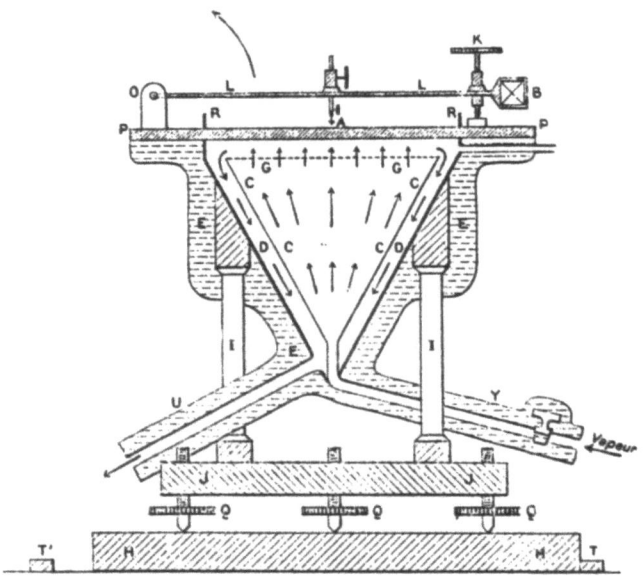

Fig. 1   Sketch of the apparatus used by Benard. The gadget on the top of the plate P-P served to measure the fluid depth. During the experiments this gadget was swung back about the pin marked 0. After Benard (1901).

rigid at room temperature but melts at 46°C. This probably implies that the viscosity of the liquid spermaceti varies substantially with temperature, and that, at least around 50°C, the fluid is rather viscous. Benard seems to have chosen this fluid over several others that he tested because of the high viscosity (which creates stable patterns) and the low volatility.

Crucial in connection with the topic of this paper was Benard's choice of the upper boundary of the fluid, which was ambient air. Benard realized that this was not an ideal arrangement, because his choice introduces, as he put it, an asymmetry into the setup; the fluid having a rigid boundary at the bottom and an apparently free boundary at the top. However, free on top means or implies that surface tension is negligible. Furthermore, the ambient air on top of the fluid makes it difficult to impose a uniform and time-independent temperature at the surface of the fluid, because the air will circulate when heated from below by the warm fluid. Air on top of the fluid makes, furthermore, the upper surface of the layer a very poor thermal conductor. But Benard had little choice with the upper medium if he wanted to observe the convective motions visually. Finally we note that the ambient air on the surface of the fluid fixed the temperature gradient across the fluid layer. Assuming that the room temperature was about 20°C there was then a temperature difference of 80°C across the layer. That means that the temperature gradient was of order of 1000°C/cm, which is extraordinarily large, in fact about two orders of magnitude larger than the temperature gradients usually employed in modern convection experiments. This means that second order non-Boussinesq effects, effects which we try to avoid in modern experiments, such as the variation of viscosity with temperature, have probably played a role in the formation of the patterns in Benard's experiments.

Visualization of the fluid motions was accomplished by Benard by suspended particles in the fluid (aluminum or graphite powder), or by

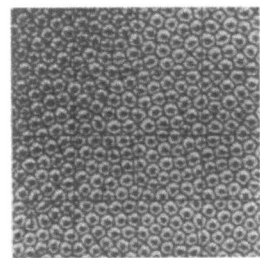

Fig. 2. Cellular pattern observed by Benard in 1 mm deep spermaceti. Visualization with graphite. Natural size. After Benard (1900).

various optical means, either interferometric or with the shadowgraph method.

The principal result of Benard's experiments was the discovery of the hexagonal convection cells, which are now commonly referred to as Benard cells. The observation of these cells seems to have fascinated Benard. There are many pictures of hexagonal cells in his papers, some with very regular cells, most of which are graphs based on photographs which apparently could not be focussed correctly. We reproduce here one photograph (figure 2) in which the pattern is not so regular, but which probably represents the usual appearance of the cell patterns.

A closer microscopic investigation of the cells revealed that the circulation of the fluid in a hexagonal cell is upwards in the center and downwards along the rim. Optical studies also showed that the surface of the fluid is depressed over the center of the cells and elevated at the cell rim, in particular the corners. The depression of the surface is of order 1 micron, for spermaceti about 1 mm deep. The results concerning characteristics of the cells are summarized schematically in figure 3. Although Benard investigated the circulation in the interior of the cells, his results were only qualitative, because there was at this time no technique available which would have permitted to make quantitative measurements of the velocity in the cells.

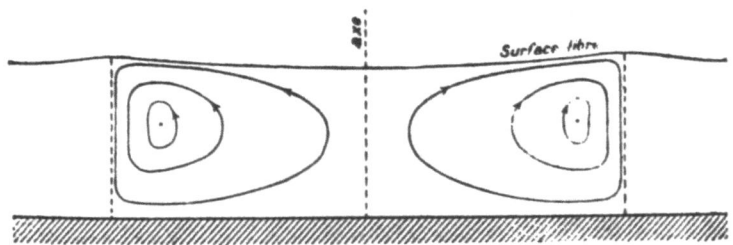

Fig. 3. Circulation in Benard cells. The curvature of the top surface is exaggerated a 100 times. After Benard (1900).

Benard made, furthermore, a great effort to determine a characteristic measure of the size of the hexagonal cells. This can be accomplished by counting the number of the cells covering a given area. This method is of good accuracy (about 1%) if the cells are regular hexagons. As characteristic dimension Benard choose the distance between two adjacent cell centers. He found that the ratio of the distance λ between the cell centers and the depth d of the fluid is constant for different fluid depths, at least in a first approximation. That means that cells of different depth are geometrically similar, as we expect it now from linear theory of convective motions. However Benard noted a number of exceptions from the rule λ/d = const, which exceptions have their origin in nonlinear conditions of the experiments. We will not pursue these problems here.

Linear theory of convection, whether it is buoyancy driven or surface tension driven, provides besides the form and the size of the cells another characteristic feature which conceivably might have appeared in Benard's experiments, namely the critical temperature difference required for the onset of convection. Benard did not notice the existence of such a critical temperature difference. This is probably due to the fact that he did not apply a gradually increasing temperature difference to the fluid layer but instead jumped immediately to the highly supercritical temperature difference of 80°C between the bottom plate and the ambient air. This then caused first a transient regime of comparatively short duration in which a critical temperature difference could not be observed. There was the possibility to observe the existence of the critical temperature difference in experiments in which the fluid gradually cooled down. But Benard was content with the observation that the pattern disappeared at a certain ΔT, before the liquid reached the melting point. Much later Benard (1930) expressed considerable skepticism about the concept of a critical temerature difference, stating that he had observed convection at temperature differences 10^{-4} to 10^{-5} times smaller than Rayleigh's critical ΔT. Rayleigh's ΔT_c is, however, irrelevant for Benard's experiments, since Rayleigh studied buoyancy driven convection, while Benard observed (unknowingly) surface tension driven convection.

There is no question about the quality of Benard's experiments. This was pioneering work which ended with an unambiguous, clearly reproducible result, the Benard cells. As far as the interpretation of his experiments is concerned, Benard was in error when he believed that the cells he observed originate from heating from below, implying thereby that the cause was buoyancy. He shared this misconception with all others working on Benard convection until 1956, when it was realized that Benard's experiments were the first case of surface tension driven convection.

Block's Paper

Block's (1956) one page article on surface tension as the cause of Benard cells in a liquid film was entirely qualitative, but changed the perception of the cause of the Benard cells completely. In his paper Block first referred to Benard's experiments, then on four lines to Rayleigh's and others analyses, mentioned then the verification of the critical nature of the onset of convection by Schmidt and Milverton (1935), in order to state that nevertheless "Benard cells, specifically defined here as the hexagonal cells in films (thinner than about 1 mm) with a free surface, may not be due to convective instability". In order to prove his point he established Benard cells and then covered the fluid with a silicone monolayer. Block observed that "wherever and as soon as the monolayer passed over the cells the surface deformation disappeared and the flow stopped". Since the temperature gradient or the corresponding (unstable) density distribution cannot vary so rapidly he concluded that the monolayer on top of the fluid must have affected the cause of the formation of the cells, that the cause was at the surface of the fluid, and that the cause was surface tension gradients.

Another significant observation dealt with the conditions for onset of convection. Block, who does not describe the heating nor gives the value of a temperature difference, noted that he had observed Benard cells in a fluid 50 microns (5×10^{-3} cm) deep. Since the depth of the fluid goes into the Rayleigh number R with the third power it follows that, (with a conventional fluid), one would need temperature differences of many thousand degrees in order to reach the critical Rayleigh number R_c for a layer that thin. Since instability obviously occurred although the applied temperature difference was much smaller than the critical ΔT, it follows that the critical Rayleigh number cannot be valid for such thin films. Finally, Block made another experiment in which a fluid layer with a "free" surface was cooled from below and hence should have been convectively stable. Nevertheless, he observed then cellular patterns, "the more regular patterns having the appearance of Benard cells". This is another indication that something is amiss with the explanation of convection under a supposedly "free" surface. Taking all observations together they make a strong point against the explanation of the Benard cells in thin layers with Rayleigh's theory. However in view of the qualitative nature of Block's observations one wonders whether his work would have had the impact it made if the same points had not been made theoretically only two years later by Pearson (1958).

Post Pearson-Nield Experiments

The first convection experiment after the clarification of the importance of surface tension for the Benard cells was made by Koschmieder (1967). The original motive for this experiment was the desire to produce regular hexagonal cell patterns. The lack of regularity of the cells in previous experiments seemed to be associated with lack of control of the conditions on top of the fluid. The thermal conditions were brought under control by a uniformly cooled glass lid placed close to the fluid surface. Motions of the air on top of the fluid were thereby suppressed and a uniform temperature established. The lid had to be made of glass in order to be able to observe the fluid flow. The fluid was silicone oil with depths ranging from about 4 mm to about 7 mm.

The onset of convection in these experiments began surprisingly with the formation of a pattern of circular concentric rolls in the circular container. Such circular concentric rolls in circular containers appear also when the lid is in contact with the fluid, when thereby surface tension is eliminated (buoyancy driven Rayleigh-Benard convection). This had been observed earlier by Koschmieder (1964). The size or the nondimensional wavelength of the circular rolls in the air-surface experiment differed however significantly from the wavelength of the rolls in Rayleigh-Benard convection. In the case of a 4.28 mm deep layer under an air surface the wavelength λ(for one roll) was 1.5 ± 0.016, while the critical wavelength (for one roll) in Rayleigh-Benard convection with either a rigid upper boundary is 1.008 or with a free upper surface is $\lambda = 1.171$ (for one roll). According to Nield (1964) (Table 1 therein) surface tension driven convection has a maximal possible (critical) wavelength $\lambda = 1.576$ (one roll) if the insulation on top of the fluid is perfect. Thus the observed wavelength of the air surface experiment of Koschmieder agreed very well with the theoretical prediction of Nield and differed significantly from the wavelength to be expected in the rigid-free case in Rayleigh-Benard convection.

When the formation of the pattern of circular concentric rolls was completed by the formation of the innermost circular cell, the cellular rings broke up and formed very regular hexagonal cells in very regular

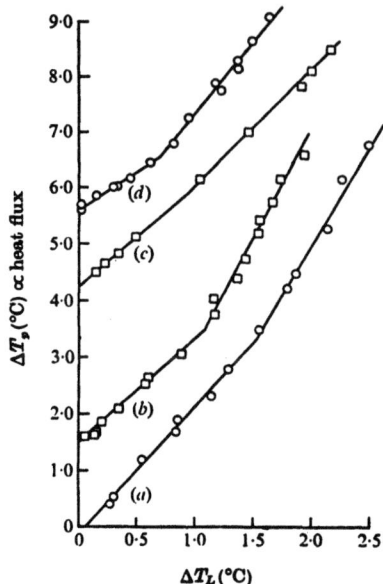

Fig. 4. Heat transfer through layers of 50 cs silicone oil. (a) 3.81, (b) 4.62, (c) 5.46, and (d) 6.55 mm deep. Curves (b) - (d) are displaced upwards in order to separate the points. After Palmer and Berg (1971).

arrangements. For example a set of 6 pairs of circular rolls with one inner circular cell formed 1-6-12-18-24-29-33 hexagonal cells. The wavelength of the hexagonal cells is the same as the wavelength of the circular rolls. It was thus shown that regular hexagonal patterns can indeed be produced with adequate control of the upper boundary conditions, and that the wavelength of the surface tension driven convection agrees with the value of λ predicted by Nield.

The next experiment to be done on surface tension driven convection was made by Palmer and Berg (1971). The aim of this experiment was the verification of the critical temperature difference (or critical Marangoni number M_c). The technique employed for this purpose was the same as used originally by Schmidt and Milverton (1935) namely the determination of the break in the heat-transfer curve which accompanies the onset of convection. The apparatus of Palmer and Berg consisted essentially of two blocks of aluminum, the cooled top block being 1.5 to 3 mm above the fluid surface. The bottom plate was heated electrically. The fluids used were three silicone oils of different viscosity, with thicknesses ranging from 1.5 to 10.3 mm. The heat flux through the oil was measured, an example of one of their curves is shown in figure 4. From the break of the heat flux curve follows the critical temperature difference. There was apparently no visual verification of the onset of convection, the type of convective flow is not mentioned at all. From the critical temperature difference follows the experimental critical Marangoni number M_c which was found to be, with moderate accuracy, in agreement with the expected theoretical values. Studying the data on the different figures one wonders whether there is a break of the heat transfer on the curve for the 1.52 mm deep layer. One also wonders about the inconsistent slopes of the different heat transfer curves, either before or after the critical temperature difference. For one and the same fluid the slopes of either the subcritical or the supercritical part of the curves should be the same. But, in all the existence of the critical temperature difference for surface tension driven convection had been verified and the values of ΔT agreed with Nield's theory.

A similar study has been made by Pantaloni, Bailleux, Salan and Velarde (1979). In this paper attention is focussed in particular on the relation $M/M_c + R/R_c = 1$. This formula was derived by Nield (1964) and shows the increasing importance of surface tension driven convection in comparison with buoyancy driven convection, when the depth of the fluid layer is decreased. In the absence of buoyancy driven convection (R = 0, or g = 0) the onset of convection takes place at M_c, as predicted by Pearson (1958). The experiments of Pantaloni et al. are inconclusive because they did not measure the value of dS/dT for the 500 cs silicone oil they used. Actually they determine a value of dS/dT assuming the validity of Nield's formula. But their value of dS/dT (0.027 dyne/cm°K) differs significantly from the value of dS/dT used by Palmer and Berg (1971), (dS/dT = 0.068 dyne/cm°K), and similar values used by Koschmieder (1967). Anyway, one can not determine the value of dS/dT and, at the same time, verify the validity of Nield's formula from the same data.

An interesting optical study of the deformation of the upper (free) surface of the hexagonal cells in Benard convection has been made by Cerisier, Jamond, Pantaloni and Charmet (1984). This seems to be the first investigation of this topic 84 years after Benard observed the depression of the surface at the center of the hexagons. As the study shows, the deformation of the surface is of order of a micron (10^{-3} mm), and is a function of the depth of the fluid as well as of the value of the supercriticality of the conditions. As is shown in figure 8 of Cerisier et al. the profile of the surface, for one and the same fluid depth, (e.g. 1.75 mm), is concave and becomes more concave with increased ε, while the profile is convex for a deeper layer (3.26 mm) and becomes more convex with increased supercriticality. Benard, who had worked with shallow layers, had noticed the concave deformation. The convex deformation of the surface indicates an increase of the influence of buoyancy. Buoyancy driven cells should have an elevated surface over the cell center, as has been expected on theoretical grounds. More about these experiments will be said in the article of Cerisier in this volume.

The most recent experiment on surface tension driven convection has been made by Koschmieder and Biggerstaff (1986). The aim of this investigation was the verification of the results of Pearson's (1958) theory. In order to do so one has to emphasize the importance of surface tension. This can be done by working with fluid layers of small depth. Since the Rayleigh number R, which characterizes buoyancy driven convection, is proportional to the third power of the depth it requires very large temperature differences in order to reach the critical Rayleigh number R_c if the fluid depth is, say, 1 mm. On the other hand, since the Marangoni number M, which characterizes surface tension driven convection, is proportional to the first power of the depth it will require much smaller temperature differences than in the buoyancy driven case, in order to reach the critical Marangoni number M_c, when the fluid depth is 1 mm. The onset of convection in a 1 mm deep layer should therefore take place close to M_c, at values of R far below R_c. Fluid layers of 1 mm depth or less were typical of Benard's experiments. Koschmieder and Biggerstaff's (1986) experiments amount, in many respects, to a modern repetition of Benard's experiments.

In Koschmieder and Biggerstaff's experiments the onset of convection was studied with carefully controlled conditions on top of the fluid layer, with very slowly and steadily increased temperature differences, and on a uniformly heated copper block. Under these circumstances a subcritical ($\Delta T < \Delta T_c$) pattern emerged in the fluid before the hexagonal cell pattern appeared near the critical Marangoni number, (if the temperature difference

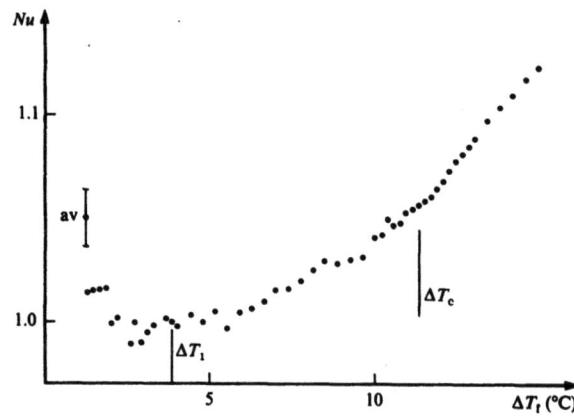

Fig. 5 Heat Transfer in a 0.93 mm deep layer of 100 cs silicone oil. After Koschmieder and Biggerstaff (1986).

was increased sufficiently). The thinner the fluid layer was, the earlier the subcritical pattern formed, a very unexpected feature. The subcritical pattern can be observed easily visually, but is also apparent in the gradual increase of the subcritical heat flux below ΔT_c on the heat transfer curves, see for example figure 5. Heat flux in figure 5 is given in terms of the Nusselt number Nu, which is the ratio of the actual heat transfer to the heat transfer by thermal conduction only. In the absence of motion heat is transferred by thermal conduction only and the Nusselt number is equal to one, up to the onset of convection. As shown in figure 5 the heat transfer in these convection experiments increased steadily for subcritical temperature differences. That means motions must occur in the fluid and their intensity must increase with increased ΔT. The temperature difference marked ΔT_1 in figure 5 indicates the temperature difference at which the subcritical pattern could first be seen. At ΔT_c the subcritical pattern changes spontaneously to a hexagonal cell pattern. Since the hexagons transfer more heat, the heat transfer curve has a break at ΔT_c. With supercritical ΔT very regular pattern of hexagonal cells form, see for example figure 6.

The consternation about the appearance of the subcritical pattern has ended when Koschmieder (1986) found that subcritical motions occur also in thin fluid layers in buoyancy driven convection, that means when the fluid is in contact with a rigid lid. The cause for the subcritical motion in buoyancy driven convection appears to be a non-Boussinesq effect, namely the variation of viscosity with temperature. Since the temperature differences required to reach either the critical Rayleigh number or the critical Marangoni number are between 20° C to 40° C in our experiments with thin fluid layers, we encounter effects not considered in Pearson's or Rayleigh's theories, which use the Boussinesq assumption. We have now learned that it is not feasible to verify unambiguously Pearson's theory in the laboratory. We have also learned that non-Boussinesq effects, in addition to surface tension effects, must have played a role in Benard's experiments, in which the temperature differences and the variation of viscosity were much larger than in the experiments of Koschmieder and Biggerstaff (1986).

<u>Zero-g Experiments</u>

Since it is so difficult to obtain good experimental results on surface tension driven convection, the question arises how one could do better and verify Pearson's theory. This ought to be possible in the zero-g environment on board of spacecrafts. As a matter of fact two attempts to do just that have been made on board of the Apollo 14 and Apollo 17 moon flights, (Grodzka and Bannister, 1972, 1975). In spite of admirable efforts of the astronauts to make these experiments a success, the results are inconclusive, because of difficulties to keep the surface of the fluid plane. In the attempts to correct this problem the fluid was stirred while it was heated. One does not know whether the resulting patterns were not just a consequence of the initial motions in the fluid layer. Other experiments in space are in preparation and will hopefully lead to conclusive results.

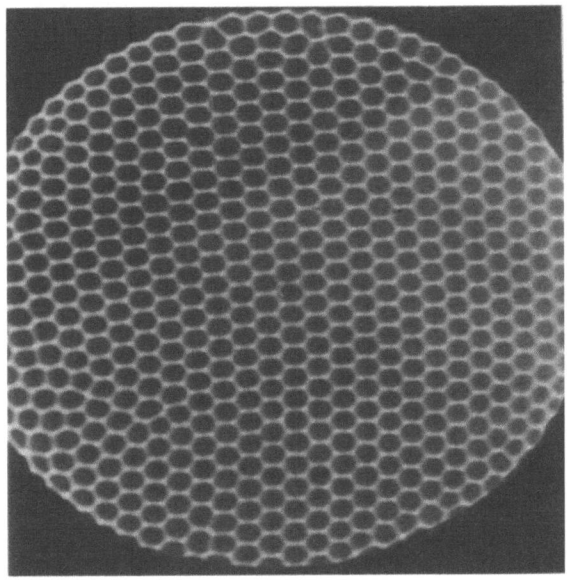

Fig. 6. Shadowgraph of a hexagonal convection pattern in a 100 cs silicone oil layer, 1.35 mm deep at $\Delta T = 1.19 \Delta T_c$. Diameter of field of view is 10 cm. After Koschmieder and Biggerstaff (1986).

CONCLUSION

In spite of its long history the knowledge about surface tension driven convection is far from complete. We do, most of all, not understand the principle which leads to the selection of the hexagonal pattern in surface tension driven convection. Contrary to buoyancy driven i.e. Rayleigh-Benard convection, where the planform of the motion is ambiguous in theory and experiment, in surface tension driven convection the cellular form seems to be absolutely certain, namely hexagonal. Nevertheless there is no clear-cut explanation for the preference for hexagons. One hopes that theoretical and experimental progress will solve this 86 year old problem in the near future.

REFERENCES

Benard, M., 1900, Rev. Gen. Sci. Pure Appl., 1:1261, 1309.
Benard, M., 1901, Ann. Chim. Phys., 23:62.
Benard, M., 1930, Proc. 3rd. Int. Congr. Appl. Mech., 1:120.
Block, M. J., 1956, Nature, 178:650.
Cerisier, P., Jamond, C., Pantaloni, J., and Charmet, J. C., 1974, J. Physique, 45:405.
Grodzka, P. G., and Bannister, T. C., 1972, Science, 176:506.
Grodzka, P. G., and Bannister, T. C., 1975, Science, 187:165.
Koschmieder, E. L., 1964, Beitr. Phys. Atmos, 39:1.
Koschmieder, E. L., 1967, J. Fluid Mech., 30:9.
Koschmieder, E. L., and Biggerstaff, M. I., 1986, J. Fluid Mech., 167:49.
Koschmieder, E. L., 1986, to be published.
Nield, D. A., 1964, J. Fluid Mech., 19:34.
Palmer, H. J., and Berg, J. C., 1971, J. Fluid Mech., 47:779.
Pantaloni, J., Bailleux, R., Salan, J., and Velarde, M. G., J. Non-Equilib. Thermodyn., 4:201.
Pearson, J. R. A., 1958, J. Fluid Mech., 4:489.
Rayleigh, Lord, 1916, Phil. Mag., 32:529.
Schmidt, R. J., and Milverton, S. W., 1935, Proc. Roy. Soc., A 152:586.

INFLUENCE OF THE CONTAINER WALLS IN BENARD-MARANGONI CONVECTION:

EXPERIMENTAL APPROACH

P. Cerisier, R. Occelli and J. Pantaloni

Systèmes Energétiques, Université de Provence
13397 Marseille Cedex 13, France

In account of its theoretical as experimental complexity the Bénard-Marangoni instability has been less studied than Rayleigh-Bénard's one. To our knowledge the only study concerning small aspect ratio containers is due to Rosenblat et al.[1]. They considered very small circular or square cylindrical vessels but in the particular case where the Rayleigh number R is zero (no buoyancy forces). They calculated the critical Marangoni number M and the wavenumber but they did not described the geometrical nature of the convective structure. The complete theories, that is to say the theories which take into account both buoyancy forces and surface tension forces, use more of less implicitely two hypotheses : (i) the liquid layer is of infinite extension and (ii) the convective structure is perfectly regular, without any defect[2]. As the experiments are necesseraly performed in finite containers the question is to know what is the influence of walls on the structural disorder and on the selected wavelength in the permanent and in the transient regimes. The problem of the selection of the structure will not be examined here. It is now admitted, from theoretical analysis and experimental studies, that when M is very different from zero the structure is hexagonal, but when M is zero it is a roll structure[3]. The experiments described in this paper correspond to cases when M is far from zero and where the aspect ratio is such as there is room for, at least, several hexagonal cells.

DISORDER AND WAVELENGTH

A qualitative description of the disorder in the hexagonal pattern of the BM convection has been made elsewhere[4]. It has been shown that the basic defect is made of a pentagon (P5) associated to a heptagon (P7). The simplest and the more natural way to characterize the disorder is therefore to calcultate the disorder density d_D which is defined by :

$$d_D = (n(P5) + n(P7)) / N$$

n(P5), n(P7) being respectively the numbers of P5 and P7 and N being the whole number of cells. An algorithm which provides d_D from the digitalization of a photograph of the convective structure has been written by R. Occelli[5]. This notion of disorder density is very fruitful, but it cannot be used in all cases. Indeed experiment shows that for a small amount of cells it can happen that no defects exist although the pattern cannot be considered as perfect : the hexagons (P6) are not regular, the pattern is more or less distorted. So another disorder function, has been defined as follows. It is independent of N and is based only on lattice distortions:

$$F_D = \sum_{j=1}^{n} \frac{1}{n_j} \sum_{i=1}^{n_j} | \ln \frac{l_{ij}}{\bar{l}} |$$

where n_j is the coordination number of the jth cell (6 for a P6, 5 for a P5 and 7 for a P7), l_{ij} is the distance from the centre of the jth cell to the centre of its $ij_i{}^{th}$ neighbour. Finally $\bar{l}$ is the average of all the l_{ij} but only the P6 are considered ($\bar{l}$ is related to the wavelength λ by $\bar{l} = 2\lambda/\sqrt{3}$). From a practical point of view, the difference between λ obtained from all P6 and λ obtained from all cells (P5, P6 and P7) is smaller than the fluctuations of λ. The measurement of λ of the hexagonal lattice is limited to cases where the disorder is not very important ($d_D < 0.5$). Beyond this value, speaking of a hexagonal lattice has no great meaning. Finally let us remark that all the marginals cells (i.e. the cells in contact with the walls) are not taken into consideration for the calculation of d_D, F_D and λ because, as it will be shown below, their shape is sometimes very different from the hexagonal type.

EXPERIMENTAL APPARATUS

The liquid to study is a silicon oil 47V100 which has a viscosity of 1 Stoke at 25°C. It is contained in a circular cylindrical vessel of 17 cm of diameter (Fig. 1 (A)). The walls are made of pyrex glass. The bottom is optically flat and made of copper. This vessel is surrounded by an outer guard ring (B) of the same oil and heated by the same heating device (C). The latter is built with a multilayered sandwich of copper plates and electrical resistors. The temperature of the upper surface of the device is

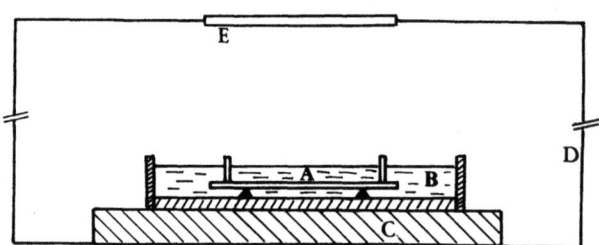

Fig. 1 Schematic of the apparatus. (A) liquid layer ; (B) outer guard ring ; (C) heating device ; (D) outer box ; (F) window

uniform and constant. The whole is set in a box (D) with a window (E) for the observation. The convectives cells are visualized using small aluminium flakes or by the shadowgraph technique. The temperatures of the lower and the upper liquid surfaces are measured with two thermocouples. Each of them is soldered to a small copper disc of 5 mm of diameter. The precision on the temperature is ±0.1°C and on the depth of the layer it is ±0.05 mm.

BASIC CONSIDERATIONS

From all the photographs of hexagonal convective structures published by various authors[6] and from all our numerous observations, it can be concluded : *in permanent regime the marginal cells are always perpendicular to the container walls* (Fig. 2). Such a result also exists in RB convection[7]. *So each wall, at least locally, imposes its direction upon the structure. The presence of defects in the pattern allows to obtain a cellular arrangement compatible with boundary conditions i.e. to reconcile two uncompatible pattern directions* (Fig. 3). This is similar to microcrystalline structures in which randomly oriented microcrystals are separated by disordered grain boundaries.

From this geometrical point of view the square, the rectangular, the pentagonal and the circular vessels, for instance, are uncompatible with

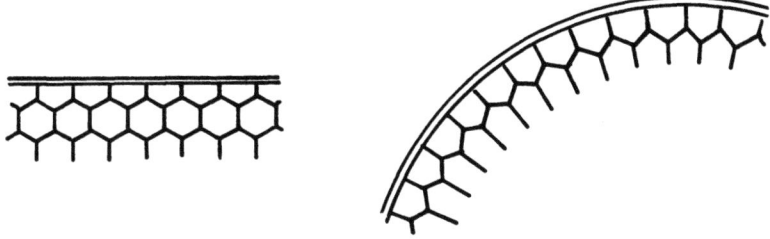

Fig. 2   Schematic position and shape of cells along a wall of a container in permanent regime

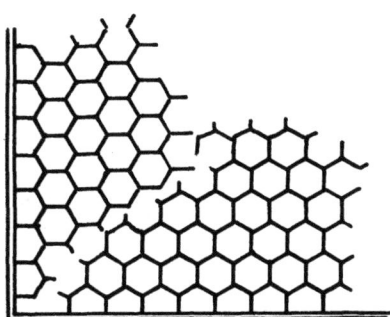

Fig. 3   Example of two uncompatible patterns induced by two perpendicular walls

the hexagonal pattern. Some distortions in the lattice are necessary because at least two uncompatible lattices are imposed by the walls (Fig. 2 - 3). Besides Rivier et al.[8] showed from Euler's theorem that a hexagonal arrangement in a circular container must contain at least five pentagons.

On the contrary the regular hexagonal and the equilateral triangle vessels are compatible with the hexagonal pattern. No distortion in the lattice due to the walls is expected if a certain arrangement of the cells in the angles is hydrodynamically possible and stable (Fig. 4). This is generally more or less well verified in the hexagonal cylindrical container.

EXPERIMENTAL RESULTS

Permanent Regime

   Aspect Ratio. To compare vessels with various shapes but with a similar amount of cells, the aspect ratio is defined by $\Gamma = \sqrt{S}/d$ where S is the surface area of the liquid and d its depth. To determine the influence of the aspect ratio Γ, for a fixed distance to the threshold $\varepsilon = 0.05$, seven regular hexagonal boxes have been used with values of Γ 13.6 - 19.3 - 30.7 - 42.0 - 52.6 - 60.3 - 82.3 respectively and in which the numbers of cells is respectively 21 - 57 - 110 - 159 - 192 - 252 - 530 (the marginal cells are excluded). The results are shown in Fig. 5. The results for $\Gamma = 180$ are taken in ref[5]. It clearly appears that two states of disorder can be distinguished. When Γ is small the structure is strongly influenced by the walls. There are no defects ($d_D = 0$) but some distortion of the lattice exists ($F_D \neq 0$). Then when Γ increases d_D and F_D also increase and have a maximum for $\Gamma \simeq 53$ and decrease to a constant value reached when Γ is greater than 70. So beyond this value the layer can be considered as being infinitely extended. The walls are without notable influence on the central part of the pattern, at least for a hexagonal pattern and as far as d_D and F_D are concerned. For $\Gamma \simeq 53$ there is a competition between the strains induced by the walls and the domains of hexagonal patterns with different orientations created in several places. The variation of λ as a function of Γ displays a different behaviour from d_D or F_D but the existence of two different states of disorder is confirmed. When $\Gamma > 70$ the selected wavelength is constant and corresponds to the value predicted by the linear theory ($\lambda = 3.01$) taking into account the experimental uncertainties.

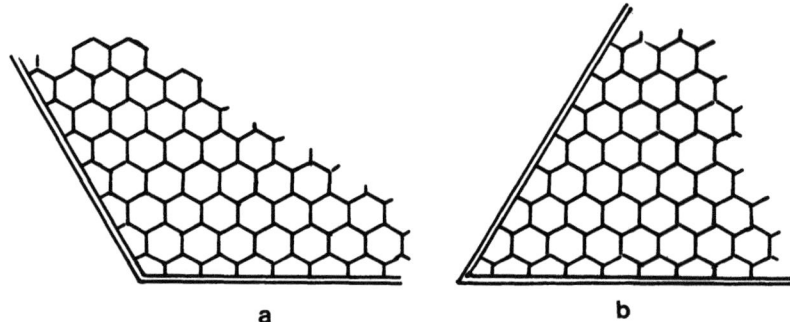

Fig. 4 Arrangement of the cells in a corner (a) of 120° ; (b) of 60°

Vessel shape. Three different vessels (hexagonal V_H, circular V_C and square V_S) are used at the same distance to the threshold and they have similar aspect ratios ($\Gamma = 85$). The corresponding values of d_D and F_D are respectively 3.7-6.1-8.2 % and 0.028-0.027-0.035. As expected d_D is minimum for V_H and maximum for V_S. But it is perhaps more suprising at first view to observe that F_D has about the same value for V_H and V_C. Not only the defects but also the strains induced by the walls must be taken into consideration. In V_C the strains induced by the walls are more uniformly distributed, so the distortion in the lattice is small. In V_H the strains induced by the angles are localized and promote strong distortions. So some sort of compensation (more defects but less strains in V_C, less defects but more strains in V_H) could explain that F_D is about the same in both containers. In V_S the disorder function is significantly higher, in agreement with previous comments.

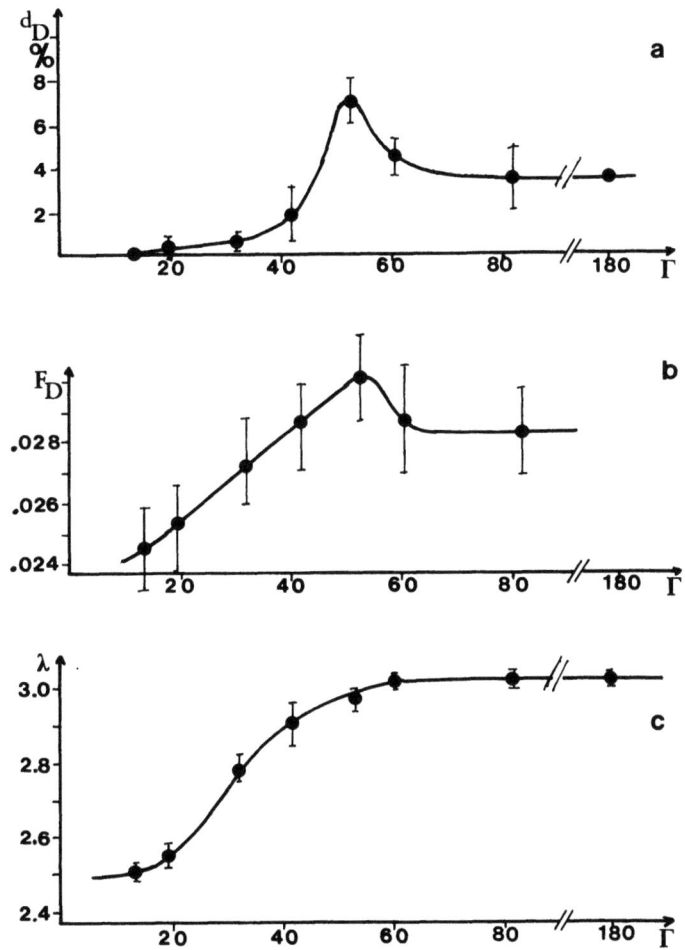

Fig. 5 Influence of the aspect ratio Γ of a hexagonal vessel.
(a) disorder density ; (b) disorder function ; (c) wavelength
The bars show the standard deviation

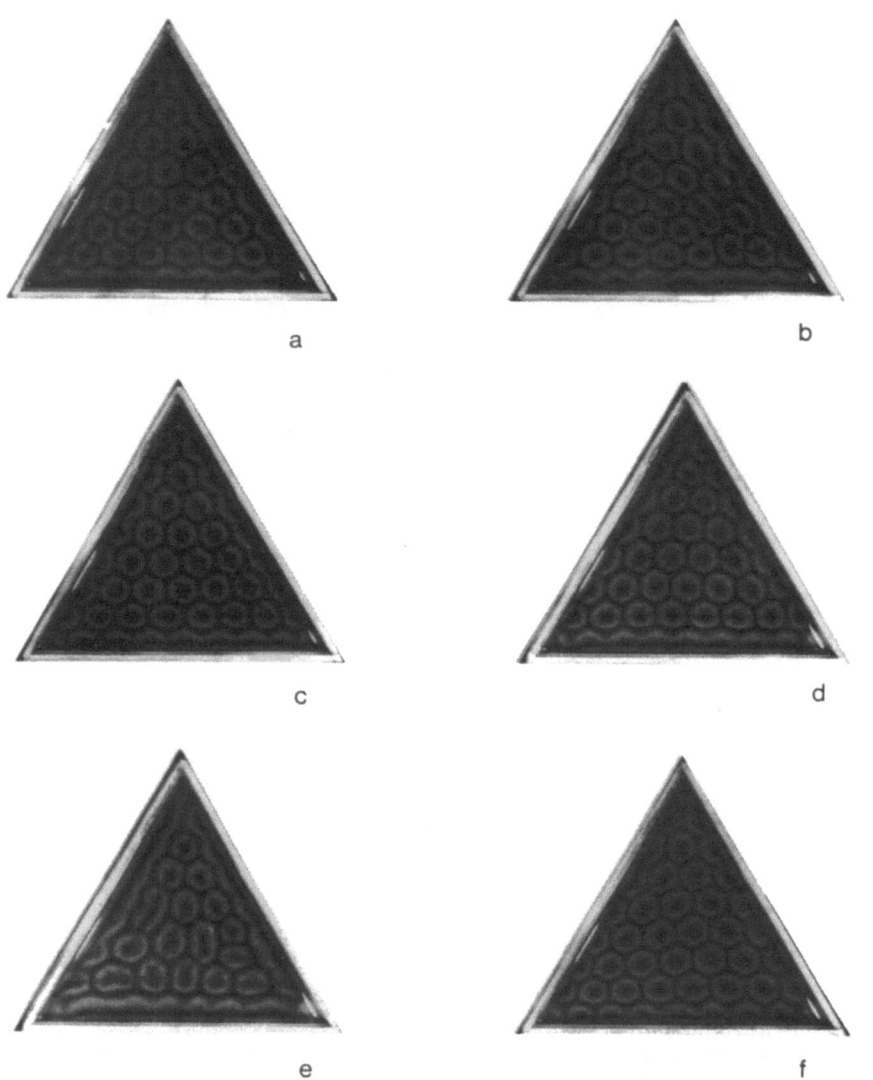

Fig. 6 Hexagonal pattern in an equilateral triangular cylindrical vessel (aspect ratio $\Gamma = 19$). Succession of ordered and disordered structures (a) $t = 4^H 56'$; (b) $t=31^H 28'$; (c) $t=32^H 15'$; (d) $t=71^H 53'$; (e) $t = 75^H 43'$; (f) $t=87^H 57'$

In the limit of experimental uncertainties the same value of the non-dimensional wavelength $\lambda = 3.01$ is found in the three vessels at $\varepsilon = 0.05$. This value is in agreement with the value $\lambda = 2.99$ provided for $\varepsilon = 0$ by the linear theory. These results stress the importance of the vessel shape on the disorder ; but if the geometrical compatibility of the vessel shape with the hexagonal structure is necessary, it is not sufficient to obtain or to keep a regular pattern as it is suggested by the following experiment.

An equilateral triangle cylindrical (V_T) vessel is used (Fig. 6). The distance to the threshold is $\varepsilon = 1$ and the aspect ratio is 19. So 21 complete hexagonal cells in the central part of the container and 24 marginal cells on the edges can exist in this container. A perfect hexagonal structure can be forced using a thermal technique described elsewhere[9]. After a few hours an ordered structure made of 21 imperfect hexagons is observed ($d_D = 0$ but $F_D \neq 0$). This structure is identical to the one observed when appearing spontaneously (Fig. 6a). But it is not stable : after a few hours, the ordered structure progressively disappears and transforms into a disordered structure ($d_D = 0$) (Fig. 6b). After a few hours the ordered structure appears again and so on (Fig. 6c-f). As the central structure evolves, modifications of the convective state along the walls are observed

Fig. 7 Break-up of induced rolls in a square cylindrical vessel (aspect ratio $\Gamma = 11$; $d = 1,34$ cm ; $\varepsilon = 3$). (a) $t = 2'30"$; (b) $t = 18'07"$; (c) $t = 38'$; (d) $t = 2$ hours.

(marginal cells or rolls or irregular rolls). This phenomenon is only observed when the aspect ratio is small. In a large V_T ($\Gamma = 85$) a fluctuating disordered structure, similar to those observe in V_C, V_H or V_S exits.

Transient Regime

The influence of the walls is also exhibited in a small aspect ratio vessel when initial conditions, different from the stable regime, are imposed to the convective structure. Two examples of such a behaviour are described here.

Square cell structure. Using the thermal technique[9] four pairs of rolls are forced in a square container ($\Gamma = 11$) (Fig. 7a). The depth layer is 1.34 cm. The roll wavelength is that provided by the linear theory. But in Bénard-Marangoni convection and for $M \neq 0$ the roll structure is not stable[3]. So the rolls are progressively broken. If the aspect ratio was large, and also the number of rolls as a consequence, (let us say 32 pairs of rolls), a spontaneous break-up of rolls in the central part of the container would iniate hexagonal cells[10]. Here in a box with a small aspect ratio the rolls are stabilized by the walls and only a cutting of rolls is possible, so 16 square cells are created (Fig. 7b-d). This structure is metastable : it can be observed during several hours or days according to the value of the layer depth because three quarters of the cells are leant against the walls and only four cells constitute the central part of the structure. Let us note that in this experiment the depth layer is 1.34 cm so that means that buoyancy forces are much more important than surface tension forces. The life time of transient rolls is large. When the depth layer is only a few millimeters the apparition of hexagonal cells is considerably more rapid.

"Viscoelastic" Behaviour. A hexagonal structure is imposed in a rectangular vessel having a small aspect ratio ($\Gamma = 28.5$) as shown in Fig. 8a with an imposed wavelength a little bit greater than the final value found in preliminar experiments. The orientation of the lattice is such that two walls of the container are uncompatible with it because the marginal cells (in Fig. 8a cells along the "horizontal" direction) are not perpendicular to them. Fastly they become perpendicular to the walls inducing a first strain in the lattice. The marginal cells along the other two walls (in Fig. 8a cells along the "vertical" direction) are perpendicular to them but they are also unstable because of their small size. They progressively disappear for the benefit of the neighbouring cells which transform into long marginal cells (Fig. 8b). But these latters are too long in comparison to the central cells (and also in comparison to the final selected wavelength). So the length of the marginal cells tend to decrease exercing a second strain on the central lattice. As a consequence the central cells tend to become longer (Fig. 8c) : first the average wavelength increases (Fig. 9) instead of decreasing as expected. Then, on account of the strong strain existing in the pattern, a "tearing" of the lattice takes place : a dislocation appears close to a wall (Fig. 8c) and propagates fastly along the Oy direction (Fig. 8d) : the average wavelength decreases sharply (Fig 9) Then the structure can relaxe : the average wavelength decreases to the final value (Fig. 9). The variation of λ as a function of time is very similar to the creeping as a function of time in a medium (Fig.9). The same experiment achieved in a large aspect ratio container ($\Gamma = 65$) provides, as expected, a classical relaxation of λ as a function of time. It is noteworthy to see that the initial marginal cells are rapidly getting perpendicular to the walls but the disorder is appearing in the cental part of the pattern independently of the edges.

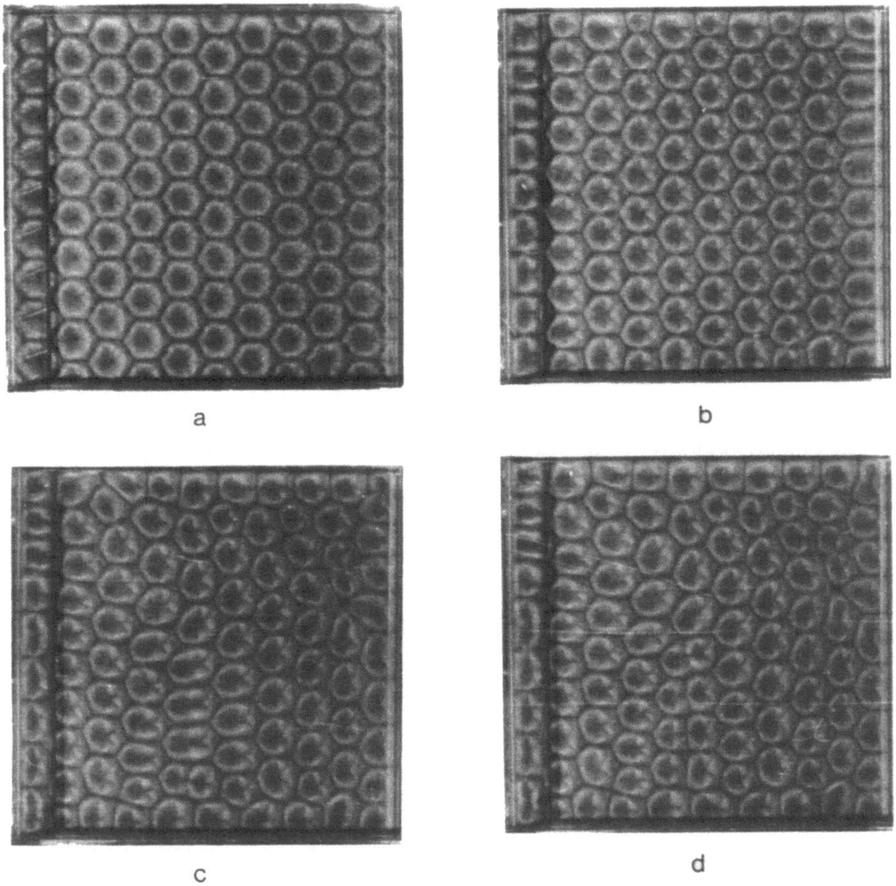

Fig. 8 Transient regime of an imposed hexagonal pattern in a square box
(aspect ratio Γ = 28.5 ; d = 0.878 cm ; ε = 2.2) (a) t = 10" ;
(b) t = 42'30" ; (c) t = 4^H02'10" ; (d) t = 4^H32'25")

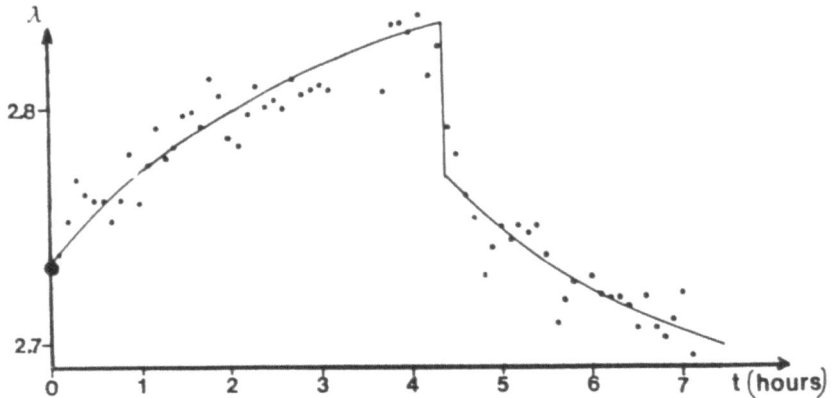

Fig. 9 Average wavelength as a function of time in a hexagonal pattern
imposed in a square cylindrical vessel (aspect ratio Γ = 28.5 ;
d = 0.878 cm ; ε = 2.2)   imposed wavelength ;   measured wavelength

CONCLUSION

From a few examples described above the influence of the walls has been exhibited. It would be interesting to complete this study with other experimental studies such as the influences of wetting angle, of the thermal conductivity and of the mean temperature of the walls, etc... A theoretical analysis would be very helpful but the phenomena are strongly non linear and the boundary conditions are difficult to formulate.

REFERENCES

1. S. Rosenblat, S.H. Davis, and G.M. Homsy, Non linear Marangoni convection in bounded layers-Part 1. Circular cylindrical containers, J. Fluid Mech., 120 : 91 (1982) - Part 2. Rectangular cylindrical containers, J. Fluid Mech., 120 : 123 (1982)

2. S. Chandrasekhar, "Hydrodynamic and Hydromagnetic Stability" Clarendon, Oxford (1961)
   P. Cerisier, R. Occelli and J. Pantaloni, Velocity and Temperature fields in Bénard-Marangoni instability. Phys. Chem. Hydr. (to appear)

3. A. Schlüter, D. Lortz and F. Busse, On the stability of the finite amplitude convection, J. Fluid Mech., 23 : 129 (1965)
   A. Cloot and G. Lebon, A non linear stability analysis of the Bénard-Marangoni problem, J. Fluid Mech., 145 : 447 (1984)
   P. Cerisier, C. Perez-Garcia, R. Occelli and J. Pantaloni, Cellular structures in Bénard-Marangoni instability, Phys. Chem. Hydr., 6 : 653 (1985)

4. J. Pantaloni and P. Cerisier, Structure defects in Bénard-Marangoni instability, in "Cellular structures in Instabilities", Lectures Notes in Physics, J.E. Wesfreid and S. Zaleski ed. Springer Verlag (1984)

5. R. Occelli, E. Guazzelli and J. Pantaloni, Order in convective structures, J. Phys. Lett., 4 : L-597 (1983)

6. M. Velarde and M. Normand, Convection, in "Scientific American", 243, 1 : 92 (1980)

7. J. Bergé and M. Dubois, in "Dynamical critical phenomena", Enz C.P. ed., Lecture Notes in Physics, 104 : 288 (1985)

8. N. Rivier, R. Occelli, J. Pantaloni and A. Lissovsky, Structure of Bénard convection cells, phyllotaxy and crystallography, J. Phys., 45 : 49 (1984)

9. P. Cerisier, C. Perez-Garcia, C. Jamond and J. Pantaloni, Phys. Lett 112 A, 8 : 366 (1985)

10. P. Cerisier, C. Jamond and J. Pantaloni, Stability of roll and hexagonal patterns in Bénard-Marangoni convection, (submitted)

SURFACE TENSION INDUCED CONVECTION IN

PRESENCE OF A SURFACE TENSION MINIMUM

J.C. Legros, M.C. Limbourg-Fontaine, and G. Petre

Chem. Phys. Dept. (CP 165)
Free University Brussels
50, Ave. F.D. Roosevelt
1050 - Brussels Belgium

INTRODUCTION

Considerable efforts are accomplished to process materials under reduced gravity conditions. This environment offers the possibility to handle liquid phase without any contact with a container, and leads to the possibility to avoid any contamination due to interactions with the sample holder. It offers also the possibility to handle large stable liquid floating zones.

The liquid/gas interfaces, always present in these containerless processes are submitted to capillary forces, which become of predominant importance in the mass and heat transfer in the liquid phases and in the exchanges through the interfaces. Inhomogeneities in composition or/and temperature modify the interface tension and can induce motions: this convection, generally called Marangoni convection, can become oscillatory if the constraints exceed a given level. These oscillations disturb deeply the crystal growth processes. The relation between non--steady flows and the formation of striations is established. These periodic variations of composition and of morphological defect density limit the quality of the produced crystals . Different experimental techniques are developed to avoid these striations. In presence of a surface tension extremum these oscillations could possibly appear for larger temperature differences. The analysis of the causes producing oscillatory convection under normal gravity and microgravity conditions must be well understood to take advantage of this new process possibility.

The experimental conditions under which the crystals are grown correspond to large temperature gradients (i.e. large Marangoni numbers) and consequently the non-linear contributions to the phenomena cannot be neglected.

Up to now, the analytical treatment of the conservation equations is limited to the creeping regime, in that case the non-linear terms are neglected. On the other hand, numerical simulations starting with the non-linear equations and simple boundary conditions are restricted to small Marangoni numbers due to numerical instabilities appearing when increasing the convective velocities (decreasing the boundary layer thickness). These types of approach are not yet directly related to the practical problems encountered when growing crystals.

Unfortunately there exists little experimental informations on the surface tension induced flows and the transport processes within the fluid for configurations and conditions that would be applicable to space processing. The goal of our experimental approach was to provide data on Marangoni convection under microgravity conditions and thus without any effect of the additional gravity forces always present on earth, even if the Marangoni effect can be dominant (e.g. in thin layers).

Experiments were performed during the flights of the high altitude sounding rockets Texus 8 and Texus 9 and during the D_1 mission of Spacelab.

Because the induced flows are sensitive to the configuration geometry and to the thermal boundary conditions, we choose to perform these experiments in a rectangular observation volume in which the dimensions of the liquid phase are equal to 3 cm x 1 cm x 1 cm. These dimensions allow a comparison with realistic Bridgman configurations.

A temperature difference is imposed at two opposite lateral walls. The generated temperature gradient, parallel to the liquid/gas interface, corresponds to Marangoni numbers of the order of 10^6. We are not able to use more important thermal constraints (as during crystal growth processes) because of the use of transparent aqueous solutions which allow to determine the velocity profiles inside the liquid bulk phase.

We choose to study a $6 \; 10^{-3}$ molal aqueous n-heptanol solution, which presents a surface tension minimum as a function of the temperature, the Marangoni number is vanishing. The use of this particular system allows us to study the behaviour of interfaces with increasing or decreasing surface tension by only changing the mean temperature.

The following paragraphs are first devoted to the description of the preparation of the STEM (Surface Tension Minimum) experiment as well on earth than during short duration microgravity experiments. Finally the results obtained during the D_1 mission are given and discussed.

THERMOCAPILLARY CONVECTION IN AQUEOUS N-HEPTANOL SOLUTION

The goal of our D_1 - WL-FPMOS experiment was to study the convective behaviour of an aqueous n-heptanol solution with a non isothermal free surface. This system has the particular property to present at equilibrium a surface tension minimum around 40° C [1], thus at this temperature T min, the Marangoni number Ma = $(\partial \sigma / \partial T . \Delta T L)/\rho \nu \kappa$ is vanishing.

During preliminary tests on earth, using large surface (18 cm diameter) in order to avoid any meniscus effects, it was verified that the direction of the surface motions could be reversed by varying the mean temperature around which the temperature gradient is established [2].

We wanted to determine the structure of the remaining convection without the perturbating effects of the gravity induced convection, when imposing at two opposite lateral boundaries of the liquid/gas interface two temperatures respectively higher and lower than the temperature T min, and the effects of this minimum on the slowering of the velocities. We are also interested to know if the oscillations only appear for larger thermal gradients.

TABLE 1

Aspect ratio A = H/L : 0.33

Mean value of residual g taken : $10^{-3} g^0 = 1$ cm.s^{-2}

Mean value of $\dfrac{\partial \sigma}{\partial T}$ between 35 and 70º C : 0.086 mN m^{-1} K^{-1}

Bond number Bo = $\rho g L^2 / \sigma$: 0.26

Marangoni number Ma = $\dfrac{\partial \sigma}{\partial T} \dfrac{\Delta T L^3}{\rho \nu \kappa}$: 800 000

Grashof number Gr = $\dfrac{\beta g \Delta T L^3}{\nu^2}$: 2 000

Ratio Ma / Gr : 400

Reynolds number (using the marangoni velocity as scaling factor[16])

$$Re = \dfrac{\partial \sigma}{\partial T} \dfrac{\Delta T H}{\nu^2 \rho}$$: 3 000

Re$_\sigma$ A^2 = 327 ≫ 1 ⟶ boundary layer flow regime

Thickness of this velocity boundary layer: 0.055 cm.

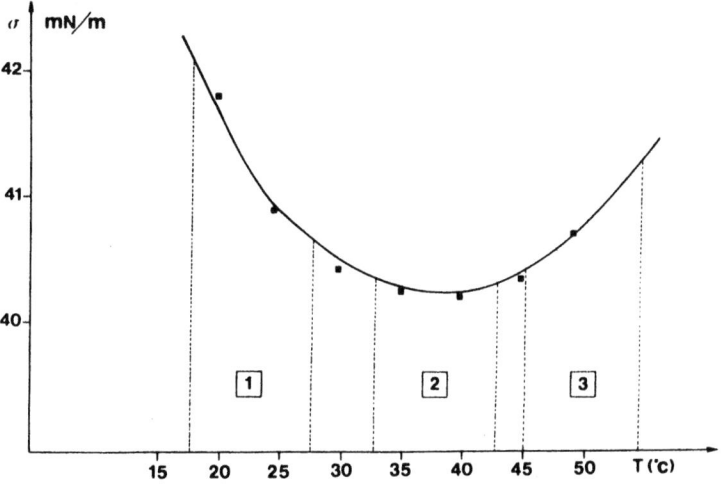

Fig. 1. Surface Tension of a 6.3 10^{-3} molal n-heptanol aqueous solution as a function of the temperature. (Wilhelmy - plate method measurements).

Theoretical and numerical approaches

In the case of the n-heptanol solution, three typical cases can be identified (cf. Fig. 1):

1. the surface tension is decreasing with temperature as in usual liquids. as in usual liquids. The surface forces tend to induce interface motions from hot to cold (zone 1),

2. the surface tension gradient changes sign somewhere along the interface. The amplitudes of the surface forces change their directions (zone 2),

3. in the third zone represented in fig. 1, the surface tension is increasing with temperature, this has to lead to motions in the unusual direction in the surface, from cold to hot.

Napolitano et al.[3] have discussed some theoretical aspects of the Marangoni convection for systems with non-linear surface tension variations, including the existence of a minimum. This study is restricted to the creeping regime ($Re_\sigma \ll 1$), existing when the largest driving speed is much smaller than the smallest diffusion speed. In that case an analytical evaluation of the convective flow pattern is possible.

The general solution is a linear combination of the unit solutions corresponding to the purely buoyant case, to the purely even Marangoni flow and to the purely odd Marangoni case. This type of approach yields the results that if T min is in the center of the interface, an opening of the interface is observed.

Another approach of this problem was made by numerical simulations[4,5] using a two-dimensional finite difference technique and simple boundary conditions. The non-linear dependence of the surface tension as a function of temperature, approximated by the simple second degree equation $\sigma = AT^2 + BT + C$ yields two modified Marangoni numbers:

$$M_1 = -\left[B + 2A \frac{T_1 + T_2}{2} \right] \frac{\Delta T H}{\kappa \nu \rho}$$

$$M_2 = \frac{2A \Delta T^2 H}{\kappa \nu \rho}$$

For the linear case, $A = 0$, M_2 is vanishing and M_1 is restoring the classical Marangoni number. All the possibilities, including different aspect ratios, gravity and various positions of the minimum along the interface have been investigated for small thermal constraints. If the Marangoni number and/or the Grashof number are increased, numerical instabilities appear. The oscillatory behaviours observed during ground tests was not reproduced by this attempt. This limit its applicability to situations which are not ot direct interest in the frame of space processings.

Reference ground test

We choose to study the behaviour of an interface of 3×1 cm^2 over a liquid layer 1 cm thick. These dimensions can be compared with experimental Bridgman configuration.

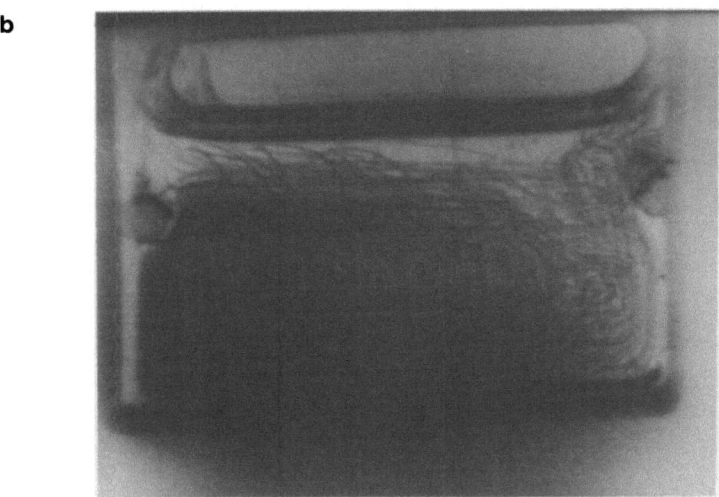

Fig. 2. Marangoni -and buoyancy- induced convective cell for different heights of n-heptanol solution. A is the aspect ratio. (a) A=0.1, (b) A=0.5, (c) A=0.3. Case (c) is shown between Figures 3 and 4.

Fig. 3.　　　　　　　Fig. 2(c).　　　　　　　Fig. 4.

Fig. 3. Velocity components measured by L.D.A. at different depths of the n-heptanol solution for the case A=0.3. The measurements are performed along the vertical symmetry axis of the cell. The corresponding convective pattern is represented by the trajectories of tracers. The hot side of the picture is at $60°$ C and the cold side is at $45°$ C.

Fig. 4. Numerical simulation of the convection induced under normal gravity conditions. The surface tension of the fluid is increasing with the temperature. Again this corresponds to the case A=0.3.

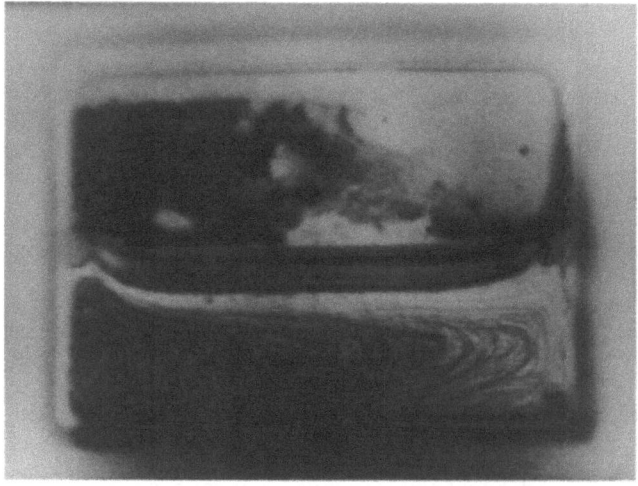

Fig. 5. Buoyancy induced convection in contaminated n-heptanol solution. The velocity is vanishing at the surface. The surface tension induced convection disappears.

The observation volume is constituted in all the experiments reported in the paper by a stainless steel frame, closed by two optical Pyrex windows. Its volume is $3 \times 2 \times 1$ cm^3.

Partly filled with a liquid layer, whose thickness is around 1 cm, this leads to an aspect ratio for this phase $A = H/L = 0.33$. The temperature difference is imposed by monitoring the temperatures of two opposite lateral metallic walls.

To be able to control the shape of the interface during ground tests and to create a liquid/gas interface parallel to the bottom of cell at reduced gravity, antiwetting coatings of Teflon were used on the metallic frames and on part of the windows (Fluomicron Coatings N.V., Antwerpen, Belgium). Generally, the depth of the liquid at the center of the interface was 8.6 mm. If the imposed temperatures, T_1 and T_2, are lower than T min, the surface tension is decreasing with the temperature.

This situation corresponds to the usual case[6,7]: the gravity induced and the surface tension induced convections reinforce each other. In our geometry under these conditions, we observed a single convective cell with a maximum of velocity in the surface.

When the temperatures T_1 and T_2 are both higher than T min, the surface tension is an increasing function of temperature. We have studied in detail this unusual situation. The convective patterns have been observed, using latex tracer particles.

The velocity field have been studied by laser doppler anenometry (LDA) with imposed temperatures at the cold and the hot side respectivel equal to 45 and 60° C.

When the thickness of the fluid in the center of the meniscus is 8.6 mm, two superposed contrarotative cells are observed: near the bottom of the cell there exist buoyancy induced motions and near the interface, a surface tension driven connective cell is present with motions in the surface from cold to hot side. The relative importance of the two cells is depending on the aspect ratios (see Fig. 2).

The paths of 90 μm latex particles reported on these figures arise directly from video recordings using the persistent properties of the camera.

In the interface is intentionally slightly contaminated (in contact during 1 second with a previously handled screwdriver) the extension of the surface tension induced convection is supressed (see Fig. 5).

Under the same thermal conditions and in the same experimental cell with a 8.6 mm thick layer some LDA results are reported on Fig. 3.

It gives the horizontal component of the velocity at different depths at the center of the cell. These data can also be compared on Fig. 4 with the shape of the velocity distribution obtained by numerical simulation (see parag. 2.1)[9].

Informations on the temperature field have also been obtained using diferential interferometry[10].

Following the light path of He-Ne laser beam, this interferometer is constituted by a beam expander, a polarizer, a Wollaston prisme, with a 30' angle deviation, a specially corrected lens with 85 mm focal length, the observation volume with optical windows, a second identical objective lens, a second 30' angle deviation Wollaston prism, a $\lambda/2$ plate and a polarizer. This interferometer is similar to the interferometer developed by MBB-ERNO and used in our Texus 9 experiment (see parag. 2.3.).

We observed that the interferograms change cyclically as a function of time. At an imposed temperature difference of 60-54° C, oscillations were observed in only a reduced part of the bulk phase (near the hot side). This behaviour was confirmed by LDA measurements[8].

If the thermal constraint is increased up to 65°-45° C, the whole bulk phase and interface velocities oscillate. This behaviour can be reproduced when the tickness of the liquid layer is increased to obtain a flat interface (the lateral limits of the liquid phase remain attached at the antiwetting Teflon barriers). But is necessary in that case to impose a larger temperature difference (69.1 - 45° C) in order to recover a similar oscillatory behaviour.

This indicates that the shape of the interface or/and the aspect ratio plays a role in this phenomena. These oscillations disappear abruptly when the surface is intentionally contaminated. These oscillations are thus linked with surface tension induced convection. quantitative analysis of this oscillatory behaviour is presently progressing.

Texus experiments

The sounding rocket flights of the Texus programme can provide six minutes of high quality microgravity. Even if this time is too short to obtain a convective steady state in the case of our experiment, the measurements that we performed yield very interesting informations. Our first experiment was performed during the Texus 8 flight in 1983[11]. The hardware of the experiment has been designed and built by MBB-ERNO.

Its complete description has already been given elsewhere[12]. The trajectories of 90 µm diameter laxer tracer particles were recorded on a 16 mm film, 20 frames per second.

The temperatures of the lateral walls were electronically monitored using Pt 100 thermosensors. During the experiment the temperatures remained constant (+/- 0.1° C), 46° C at the cold wall and 66º C at the hot wall, i.e. both higher than T min.

The obtained results can be summarized as follow:

1. During the filling under microgravity conditions, the creeping of the liquid up along the stainless steel walls was stopped by two small Teflon blocks and by grooves coated with Teflon in the Pyrex windows.

    This allows to create a rather flat liquid/gas interface.

2. The expected unusual direction for the fluid flow at the surface, from the cold to the hot side was observed.

3. The convective motions are initiated at the interface near the hot wall and developed slowly until a single cell occupies the whole observation volume after 6 minutes of the experiment.

4. At a given time, the velocities at the interface are position dependent. they are higher near the hot wall, decreasing with a slowing down $v/t = -0.0136 \cdot 10^{-3}$ m s^{-2}.

5. The velocities near the interface are higher than in the bulk.

From this experiment, we can deduce that in the used range of temperature, the surface tension is increasing with the temperature, indirectly indicating that a surface tension minimum is present at a lower temperature, and thus is still existing under such non equilibrium convective conditions. On May 1984 a second experiment has been performed during the Texus 9 flight[13]. A differential interferometer was added to the hardware in order to have informations on the temperature field inside the liquid phase.

The imposed temperature were 32 and 50° C, these temperatures are respectively lower and higher than T min. Latex tracer particles could also be followed in the dark parts of the interferograms.

The obtained results are the following:

1. The partial filling under microgravity leads to a flat interface.
2. Marangoni convection is initiated at the wall and a convective cell is growing.
3. The movements in the surface are from cold to hot region, in the hot two-third of the interface, the cold part remain nearly at rest.
4. The interferogram contained less fringes than in the ground tests and are not oscillating.
5. The opening of the interface, i.e. motions from the 40° C interface region toward two opposite lateral walls at 32 and 50° C was not recorded, as expected from theoritical arguments and from numerical simulations[3,4,5].

These Texus experiments were too short to achieve a steady state, the surface velocities were still increasing at the end of microgravity period.

Parabolic flights

The following steps in the preparation of our D_1 experiment were accomplished during parabolic flights of the specially adapted KC KC 135 aircraft of NASA.

On December 1984 and July 1985, experiments were performed during 4 flights with 20-25 s short periods of reduced gravity[14].

The real STEM hardware was tested in the engineering model of the FPM. The observation volume of our hardware was partly filled and some improvements get the filling better and more reliable. The experiment procedure was reviewed by the payload specialists and modified in order to get it faster.

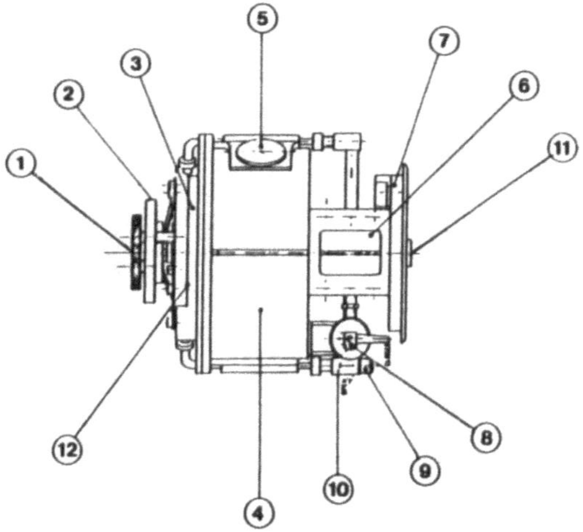

Fig. 6. STEM : The experimental device.

1. Transport screws.
2. Piston flange and piston springs.
3. n-heptanol solution container.
4. Cylindrical water reservoir.
5. Hole for water filling.
6. Pyrex windows.
7. Fixing disk with Minco heater and mechanical thermal switch.
8. Valve.
9. Bleed screw.
10. Connector for the thermistors.
11. Connector for electrical heater and thermistors for thermal regulation.
12. Membrane.

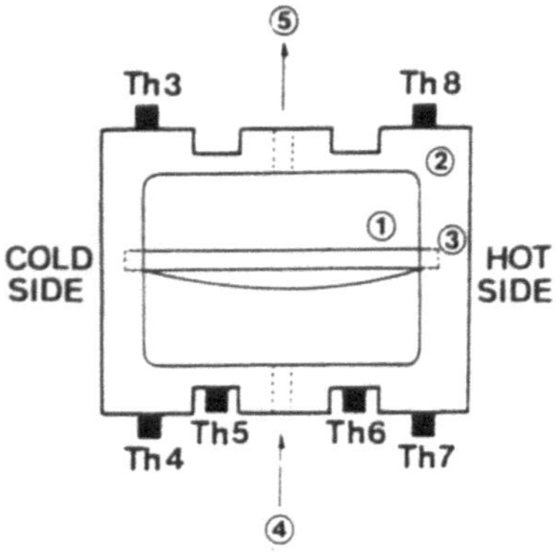

Fig. 7. The observation volume with the positions of thermistors.

1. Pyrex window,
2. Stainless steel frame,
3. Teflon-coated grooves,
4. inlet of fluid,
5. Outlet of gas.

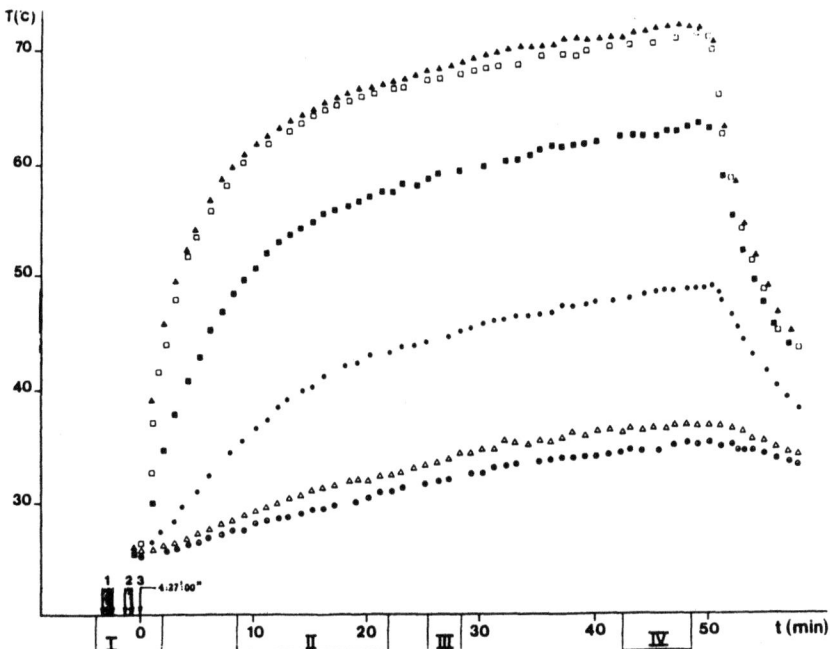

Fig. 8. Temperatures at thermistors 3 , 4 , 5 , 6 , 7 , 8 , as a function of time and time schedule of the experiment.

1. Filling of the cell
   from MET 4/04 : 24 : 50 to 04 : 25 : 18.

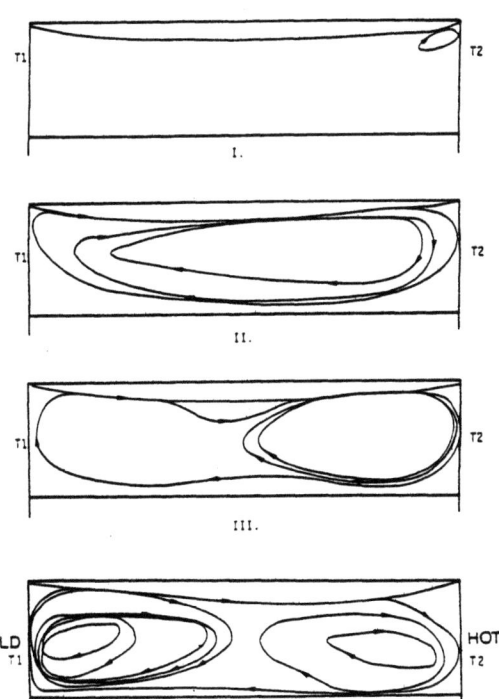

Fig. 9. Convective cells as observed on the video tape during the sequences I, II, III and IV. T_1 is the temperature at the cold side and T_2 the temperature at the hot side of the cell.

D_1 MISSION RESULTS

The experimental hardware used during our D_1-WL-FPM 05 experiment has been described in detail elsewhere[15]. This apparatus was developed and manufactured in collaboration with the Belgian Space Aeronomy Intitute (Brussels). During the mission, this cell was integrated in the Fluid Physics Module (FPM).

At the beginning of the experiment, the liquid was introduced in the observation volume by the payload specialist Dr. R. Furrer by the use of the longitudinal displacement mechanism of the FPM to move the piston of the STEM cell (see Fig. 6).

The streamlines of the induced flow are visualized by the paths of latex particles. These trajectories had to be recorded on a 16 mm film with the Vinten camera B of FPM and on a video tape. Unfortunately, as for some other experiments on FPM, this Vinten camera fails during our experiment and we have no film recording.

On the other hand the VITR video recordings are of very poor quality and represent only partly the very accurate work performed by Dr. R. Furrer and decrease highly the accuracy of the obtained experimental results.

The visualization of our tracer particles has been optimized during ground tests on the EM of FPM by the use of light diffuser sheets.

But it appears during the real experiment during the mission that the light intensity was significantly lower on the FM than on the EM. If the diffusing raster was used, the remaining light intensity was too low for the use of the video camera. This contributes also to the low contrast of the recorded data.

The available video recordings of tracer trajectories are during the four following sequences given in Mission Elapsed Time (MET):

Sequence 1   from   04 : 23 : 57   to   04 : 29 : 35
Sequence 2   from   04 : 35 : 24   to   04 : 49 : 24
Sequence 3   from   04 : 53 : 04   to   04 : 54 : 24
Sequence 4   from   05 : 11 : 12   to   05 : 16 : 32

The filling of the cell took place between 04:24:50 and 04:25:12, during the first recording sequence.

The liquid level was adjusted in order to have a suitable interface shape rather flat but allowing to follow tracer particles in the interface. The creepinf of the liquid along the walls in Pyrex and in stainless steel was successfully stopped by the Teflon coating of Fluomicron Coatings NV.

The smaller thickness of the layer in the center was 8.6 mm.

The heating phase started on 04:27:00 and stopped on 05:17:00. Six thermistors whose locations are con Fig. 7 allowed to follow the temperatures. The recorded values of the six given temperatures are reported on Fig. 8.

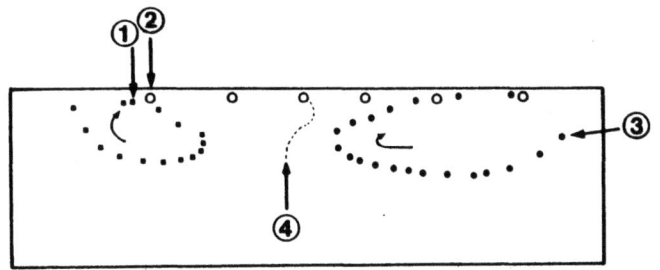

Fig. 10. Positions of four particles flowing during sequence IV of the VITR recording. The arrows indicate the starting position.

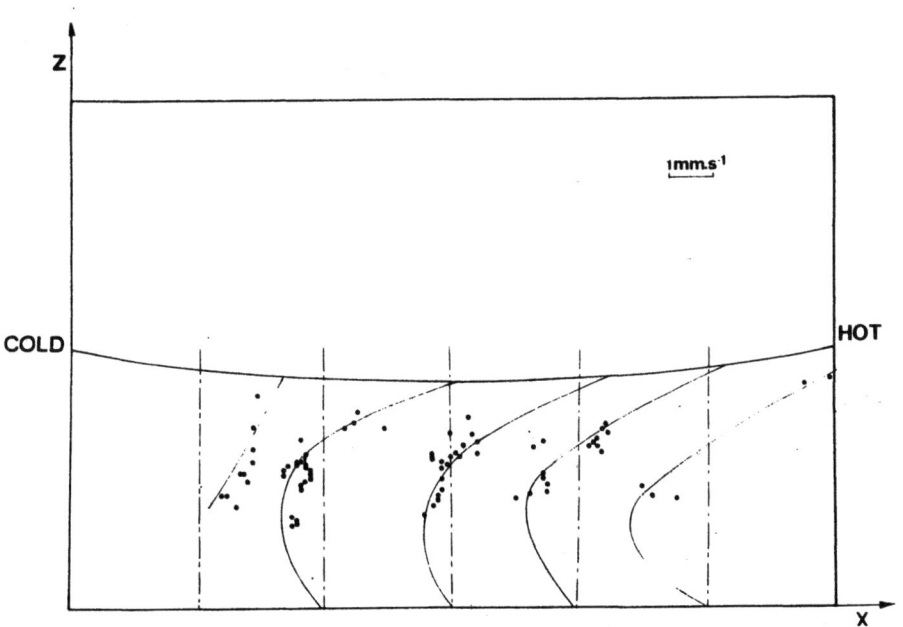

Fig. 11. Mean velocities in different zones as a function of depth in the liquid (sequence IV).

221

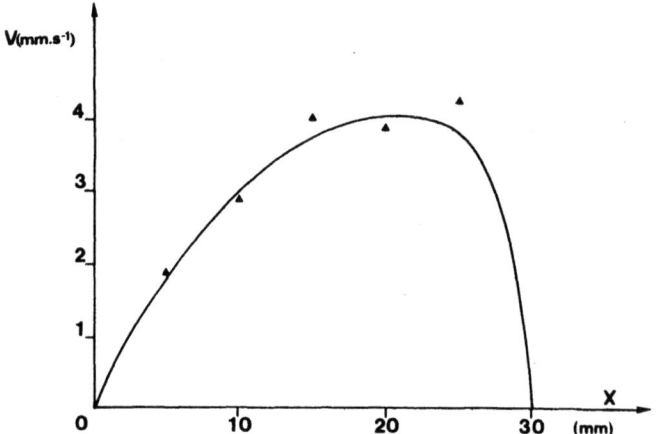

Fig. 12. Extrapolated velocities in the interface as a function of their positions along the interface (sequence IV).

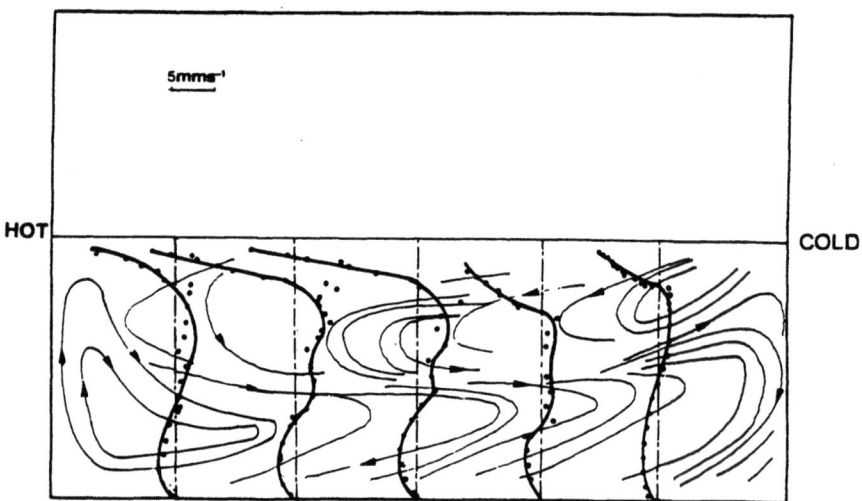

Fig. 13. LDA velocity measurements along five vertical axis in mid plane of the cell.
The imposed temperature difference is 71-35° C.

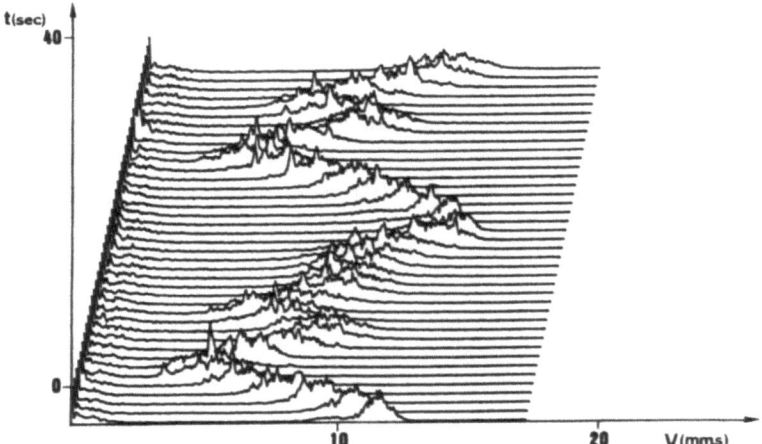

Fig. 14. Oscillatory velocity behaviour recorded from LDA measurements with a temperature difference of 71-35° C.

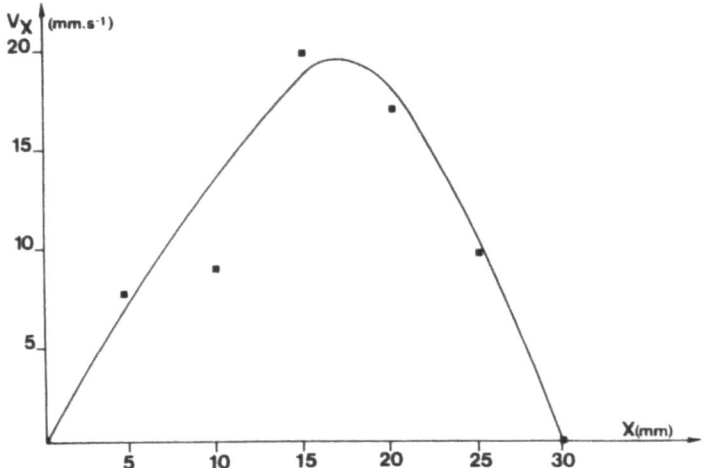

Fig. 15. Interface velocity measurements on earth by LDA as a function of the position in the interface.
The imposed temperature difference is 71-35° C.

The total heating phase of the experiment was 50 minutes long.

During this time the temperature of the cold side of the observation volume increased only from 26 to 37° C by the use of the high thermal capacity of the heat reservoir filled with 300 g water.

15 minutes after the beginning of the heating phase a 36º C temperature difference is established and will remain nearly constant up to the end.

These experimental conditions for the aqueous n-heptanol solution under the microgravity of Spacelab can be represented by dimensionless numbers reported in Table 1[16].

The 4 flow patterns represented on Fig. 9 are representative respectively of the convection during the 4 video sequences.

At time 04:27:30 (30 seconds after heating) a thermocapillary induced convective cell develop near the hot wall of the cell, as during the Texus experiments. The fluid moved in the surface from the cold to the hot side.

This unique convective cell will develop during the three first sequences and occupies the whole volume. let us remark that during the short third sequence, closed streamlines appeared in the hot part of the cell.

The fourth sequence shows the formation of two convective cells, rotating in the same direction, these cells are included in the large initial convective motion as shown on the sketch of Fig. 9. We are presently performing a numerical study in order to simulate this unexpected phenomena imposing the same surface velocity to a liquid phase with identical geometry.

From the determination of the horizontal velocities at the center of the interface as a function of time, it can be seen that it took to be at a steady state a rather long time, of the order of 30 minutes.

We have represented on Fig. 10 a typical step of the data analysis of the vido tape. The positions of 4 particles are reported during sequence 4, the time intervals between each points is 1 second. The starting points are identified by arrows. This figure is interesting because it shows clearly the two circulation loops and the large movement along the interface. The fourth particle trajectory is puzzling. During a short time this particle path that we can follow with difficulty only during a short time, seems to indicate the existence of a third cell flowing in contradirection with the two other ones.

From all the informations obtained similarly it is possible to evaluate velocities at different locations and different times. On Fig. 11, we have reported mean velocities, evaluated in different zones in the bulk phase during sequence 4. From the cold wall, zone 1 is extend from 2.5 to 7.5 mm, zone 2 from 7.5 to 12.5 mm, zone 3 from 12.5 to 17.5 mm, zone 4 from 17.5 to 22.5 mm and zone 5 from 22.5 to 27.5 mm.

The extrapolated velocities in the surface (in Fig. 11) are given on Fig. 12 as a function of their positions along the interface.

Some results of reference ground tests performed after the mission, with an imposed temperature difference equal to 71-35º C (corresponding to the conditions of sequence 4) are described herebelow. On Fig. 13, we give mean velocities evaluated by LDA along five vertical axis positionned at the center of each zones as defined for Fig. 11. Some characteristic streamlines arising from pictures obtained by the same way than for Fig.2 have been drawn in order to visualize the cellular structure. The dispersion of the evaluated velocities at different points is large due to the fact that oscillation behaviour is present as shown on Fig. 14 obtained by LDA using short sampling time. In the case of the data of Fig. 13, the sampling time is larger than for Fig. 14 but does not allow to obtain constant values.

The interface velocity values coming from Fig. 13 are reported on Fig. 15 as a function of their horizontal coordinate. This last figure has to be compared with Fig. 12.

Comparison with references ground tests shows that the velocitie in the interface obtained during microgravity experiments are smaller. these was also obtained during our Texus 8 and Texus 9 experiments.

ACKNOWLEDGEMENTS

The authors are very greatfull to dr. R. Furrer, Payload Specealist of the D_1 mission, who performed their experiments and participated to discussion around evaluation of the obtained results. they are also deeply endebted to Professor A. Jaumotte for his constant support and interest in our experiments.

This research is financially supported mainly by a "Action de recherche Concertée" contract (SPPS) and a "Fonds National de la Recherche Fondamentale Collective" contract.

All along this work, we have also received fundings from "La Loterie Nationale", the "Banque Nationale", the "Fonds National de la Recherche Scientifique", the "Fondation Universitaire A. et D. van Buuren" the "Commission de la Recherche de l'U.L.B.".

We should like to thank Dr. A. Gonfalone and Dr. D. Frimout from ESTEC for many helpfull discussions.

REFERENCES

1. Vochten,R., Pétré, G., J. Colloïd Interface Sci., <u>42</u>, 329 (1973).
2. Pétré, G., Azouni, M.A., J. Colloïd Interface Sci., <u>98</u>, 261 (1984).
3. Napolitano, L.G., Golia, C., Viviani, A., L'aerotecnica, <u>63</u>, 29 (1984).
4. Legros, J.C., Pétré, G., Limbourd, M.C., Villers, D., Platten, J.K., 5th Elgra meeting proceedings in Elgra News, <u>6</u>, 41 (1984).
5. Villers, D., Platten, J.K., Phys. Chem. Hydrodyn, <u>16</u>, 435 (1985).

6. Ostrach, S., Materials Processing in Space, A Workshop, Indian Academy of Sciences Bangalore (1982) p. 81.

7. Schwabe, D., Lamprecht, R., Scharmann, A., Naturwissenschaften 73, 360 (1986).

8. Limbourg, M.C., Submitted to PCH (1986).

9. Legros, J.C., Limbourg, M.C., Pétré, G., Proceedings of the symposium on Fluid Dynamics and Space, VKI - ESA, Brussels (1986).

10. Limbourg, M.C., Legros, J.C., to be published.

11. Limbourg, M.C., Pétré, G., legros, J.C., Phys. Chem. Hydrodyn, 6, 301 (1985).

12. Legros, J.C., Limbourg, M.C., Pétré, G., Acta Astronautica 11, 143 (1984).

13. Legros, J.C., Pétré, G., Limbourg, M.C., Adv. Space Res, 4, 37 (1984).

14. Legros, J.C., Limbourg, M.C., ESTEC Working paper 1457, 28 (1986).

15. Limbourg, M.C., Pétré, G., legros, J.C., Van Ramsbeeck, E., Acta Astronautica 13, 197 (1986).

16. Ostrach, S., see ref. [6] p. 73.

THEORIES OF CONVECTIVE INSTABILITIES DRIVEN BY THERMOCAPILLARY FORCES

Stephen H. Davis

Department of Engineering Sciences and Applied Mathematics
Northwestern University
Evanston, IL 60201 USA

1. INTRODUCTION

In this paper we discuss several classes of convective instabilities that are driven by variations in surface tension on interfaces between immiscible fluids. We begin by introducing the underlying mechanical effects and then turn to the predictions of the theories and the mechanisms of the instabilities.

An interface S between two immiscible fluids possesses localized properties, the most prominent of which is the interfacial (or surface) tension σ. This represents the magnitude of the force per unit length normal to a cut in the interface (see Levich 1962).

The surface tension σ usually depends on the scalar fields in the system (e.g. the electrical field, the temperature field), as well as on the concentration of foreign materials on the interface (Levich 1962). In the present paper we focus on a single such field, the temperature T. Thus, we consider only thermocapillarity and pose an equation of state

$$\sigma = \sigma(T) \ . \qquad (1.1)$$

Surface tension enters the description of the dynamics of the system through the force balance at the interface S.

On the one hand, the jump in normal stress at S balances surface tension times twice the mean curvature H of S. In the absence of viscosity, this is the Laplace relation which states that the pressure is larger on the concave side of S by an amount $|2H\sigma(T)|$. Thus, thermocapillarity can alter the capillary pressure jump or give it variations from point to point, depending on T.

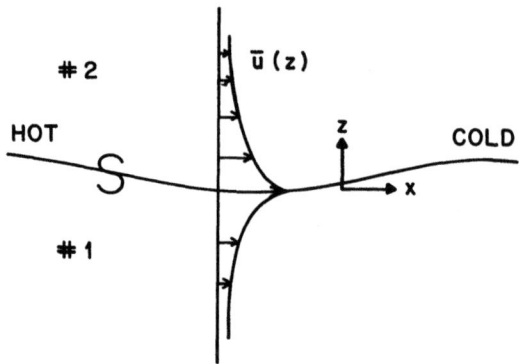

Fig. 1. Sketch of two-fluid system in which a temperature gradient is imposed along the interface S between two immiscible fluids #1 and #2.

On the other hand, there is a jump in shear stress at S balanced by the surface-tension gradient. If we represent the equation of state (1.1) by a linear law

$$\sigma = \sigma_0 - \gamma(T-T_0) \;, \tag{1.2}$$

then the surface-tension gradients on S are proportional to γ; temperature gradients along S induce shear stresses on S that result in fluid motion. This is shown in Figure 1, where a temperature gradient is <u>imposed</u> along S. For common liquids, we have $\gamma = -\frac{d\sigma}{dT} > 0$ so that there is surface flow from the hot end toward the cold end. Since the bulk fluids are viscous, they are dragged along; bulk-fluid motion results from interfacial temperature gradients. This is called the <u>thermocapillary</u> effect (Levich 1962).

Thermocapillary effects can dominate the dynamics in the containerless processing of crystals, the behavior of weld pools, the rupture of thin films, the movement of contact lines, and the propagation of flames over liquid fuels. In many situations, the transport of heat across interfaces can be dramatically increased through the presence of additional mixing processes triggered by instabilities.

Clearly, the presence of an interface with variable surface tension can modify already known instabilities. However, since surface tension

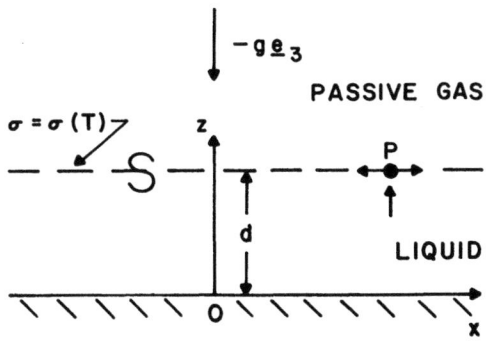

Fig. 2. Sketch of the one-layer system.

is a localized force at the interface, thermocapillarity can drive new instabilities. In this article we aim to describe their mechanisms in detail and define highly simplified systems that highlight these. Further details and generalizations are given in the survey by Davis (1987).

The simplified cases we examine involve a simple geometry — the <u>one-layer system</u> — in which there is a liquid layer whose lower boundary is a rigid plate and whose upper boundary is a <u>planar interface</u> with a passive gas (having negligible viscosity and density), as shown in Figure 2. In practice an interface may be subject to a temperature gradient $[\bar{\nabla T}]_{EXT}$ with arbitrary orientation. We let (e_1, e_2, e_3) be unit vectors in the (x,y,z) directions. If

$$e_1 \cdot [\bar{\nabla T}]_{EXT} = e_2 \cdot [\bar{\nabla T}]_{EXT} = 0 \;, \tag{1.3}$$

then $[\bar{\nabla T}]_{EXT}$ is imposed normal to S, the planar position of the interface, and there exists a purely static basic state. Instabilities of this state lead first to steady Marangoni convection, as identified by Pearson (1958). If

$$e_3 \cdot [\bar{\nabla T}]_{EXT} = 0 \;, \tag{1.4}$$

then $[\bar{\nabla T}]_{EXT}$ is imposed along S, and no static states exist. Thermocapillarity can drive a steady shear flow whose instabilities are often time-periodic hydrothermal waves as identified by Smith and Davis

(1983). We simplify the discussion here by taking S to be planar and ignoring all body forces. Davis (1987) discusses some of the instabilities present when these restriction are relaxed.

2. FORMULATION FOR NON-DEFORMABLE INTERFACES AND ZERO BODY FORCES

Consider a liquid layer bounded below by a rigid plane and above by a passive gas. The plane lies at z = 0 and the mean position of the interface lies at z = d, as shown in Figure 2. The liquid is a Newtonian fluid with constant values of the viscosity μ, the density ρ, the specific heat c_p, the thermal conductivity k, and the volume expansion coefficient α; $\kappa = k/\rho c_p$ is the thermal diffusivity, and $\nu = \mu/\rho$ is the kinematic viscosity. The surface tension σ of the interface varies with temperature T of the liquid as given in eqn. (1.2). Body forces are absent.

In order to simplify the discussion, we take the surface tension σ so large that the interface remains planar. When one suppresses the surface deflection, one eliminates certain instabilities while one modifies others; we discuss here only those survivors whose presence do not depend on the existence of interface corrugations. See Davis (1987) for a full discussion of the more general case.

The system is subject to an <u>imposed temperature gradient</u> $[\overline{\nabla T}]_{EXT}$,

$$[\overline{\nabla T}]_{EXT} = - b_{\parallel} e_1 - b_{\perp} e_3 \qquad (2.1)$$

where e_i are the unit vectors in the (x,y,z) directions. Thus $|b_{\parallel}|$ and $|b_{\perp}|$ are the magnitudes of horizontal and vertical temperature gradients, respectively. Let us define

$$b = (b_{\parallel}^2 + b_{\perp}^2)^{1/2} \ . \qquad (2.2)$$

We consider the fully three-dimensional system and scale all distances on the average liquid depth d. The velocity vector $v = (u,v,w)$, pressure p, temperature difference $T - T_0$, surface tension σ and time t are referred to scales $V_M = \gamma bd/\mu$, $\mu V_M/d = \gamma b$, bd, σ_0 and $\mu/\gamma b$, respectively. As a result, there arise the following dimensionless groups:

$$M = \frac{\gamma b d^2}{\kappa \mu} \ , \quad P = \frac{\nu}{\kappa} \ . \qquad (2.3a,b)$$

Here M is the Marangoni number, and P is the Prandtl number. In most of what follows, either $b_\parallel = 0$ or $b_\perp = 0$, so that the ratio $b_\parallel/b_\perp$ is not an additional parameter.

The governing equations for the liquid layer are the Navier-Stokes, the energy, and the continuity equations:

$$P^{-1}M(\frac{\partial \mathbf{v}}{\partial t} + \mathbf{v}\cdot\nabla\mathbf{v}) = -\nabla p + \nabla^2 \mathbf{v} \quad , \tag{2.4a}$$

$$M(\frac{\partial T}{\partial t} + \mathbf{v}\cdot\nabla T) = \nabla^2 T \quad , \tag{2.4b}$$

$$\nabla\cdot\mathbf{v} = 0 \quad . \tag{2.4c}$$

The interface S lies at $z = 1$. The kinematic condition is

$$w = 0 \quad \text{on} \quad z = 1 \quad , \tag{2.5a}$$

and the stress conditions are

$$\mathbf{n}\cdot\underline{\underline{T}}\cdot\mathbf{n} = 0 \quad \text{on} \quad z = 1 \quad , \tag{2.5b}$$

and for $\alpha = 1, 2$,

$$\mathbf{t}^{(\alpha)}\cdot\underline{\underline{T}}\cdot\mathbf{n} = -\mathbf{t}^{(\alpha)}\cdot\nabla T \quad \text{on} \quad z = 1 \quad , \tag{2.5c}$$

where $\underline{\underline{T}}$ is the stress tensor of the liquid. Here $\mathbf{n}$ is the unit normal to S pointing out of the liquid, and $\mathbf{t}^{(\alpha)}$, $\alpha = 1,2$, are orthonormal tangent vectors to S. The velocity scale V_M, the "Marangoni velocity scale," represents a balance between surface-tension gradients on the interface and the shear stresses generated by them, as captured by eqn. (2.5c).

3. $[\nabla\bar{T}]_{EXT} \perp S$: MARANGONI INSTABILITY

3.1. Basic State

We consider the situation in which, from eqn. (2.1), $b_\parallel = 0$ and $b_\perp > 0$, so that the liquid layer is heated from below normal to the interface.

Suppose that the rigid plate of Figure 2 is a perfect heat conductor fixed at temperature T_B, $\hat{T} = T_B$ at $z = 0$, while the interface is cooled by air currents according to the law $-k\nabla\hat{T}\cdot\mathbf{n} = h(\hat{T}-T_{AIR}) + Q_0$, where h is the unit thermal surface conductance, Q_0 is an imposed heat flux to the environment, and carets denote <u>dimensional</u> quantities. If we nondimensionalize as in Section 2, identifying T_0 with T_T (to be defined in a moment), then we have the two thermal boundary conditions as follows:

$$T = \frac{T_B - T_T}{b_\perp d} \quad \text{at} \quad z = 0 \tag{3.1}$$

and

$$\nabla T \cdot \mathbf{n} + B\left[T + \left(\frac{T_T - T_{AIR}}{b_\perp d}\right)\right] + Q = 0 \quad \text{at} \quad z = 1, \tag{3.2}$$

where

$$B = \frac{hd}{k} \quad \text{and} \quad Q = \frac{Q_0}{b_\perp k}. \tag{3.3}$$

We seek a <u>basic state</u> in which the fluid is static,

$$\overline{\mathbf{v}} = 0, \tag{3.4a}$$

the heat transfer is purely conductive,

$$\overline{T} = 1 - z, \tag{3.4b}$$

and the pressure is constant

$$\overline{p} = 0. \tag{3.4c}$$

Thus, we select T_T to be the interface temperature in the basic state, and we have that

$$T_B - T_T = b_\perp d, \quad T_T - T_{AIR} = (b_\perp d)B^{-1}, \quad Q = 0. \tag{3.5}$$

The stability problem is obtained by disturbing all quantities $\mathbf{v}$, T, and p. The governing system for this nonlinear problem has the following non-dimensional groups:

$$M^{(\perp)}, P, B, \varepsilon, \text{ and } (k_1, k_2). \tag{3.6}$$

Here we have inserted a superscript "$\perp$" to remind us that the imposed temperature gradient is normal to S (in the basic state), and that $b = b_\perp$ in the definition of $M^{(\perp)}$. In addition, ε measures the amplitude of a disturbance (at some initial time), and (k_1, k_2) are the wave numbers (or, more generally, the horizontal scales) of the disturbance.

3.2 Linear Stability Theory

In the limit $\varepsilon \to 0$, the disturbance equations and boundary conditions can be linearized about the basic state. This theory was first examined and explained by Pearson (1958). He seeks neutral conditions, which correspond to the onset of steady <u>Marangoni convection</u>. His curves are sketched in Figure 3. The result, as shown by the solid curves in Figure 3, is that pure conduction is unstable if $M^{(\perp)} > M_L^{(\perp)}$ when the lower plate is a fixed temperature boundary, where

$$M_L^{(\perp)} \approx 79.6 \quad , \quad k_L \approx 1.99 \quad \text{for} \quad B = 0 \quad . \tag{3.7}$$

Figure 3 also shows Pearson's result for the case when the lower plate is a fixed heat flux boundary. Here the dashed curves apply and

$$M_L^{(\perp)} \approx 48 \quad , \quad k_L = 0 \quad \text{for} \quad B = 0 \quad . \tag{3.8}$$

Notice that an increase in B results in a stabilization of the basic state. In the above, k is the overall wave number,

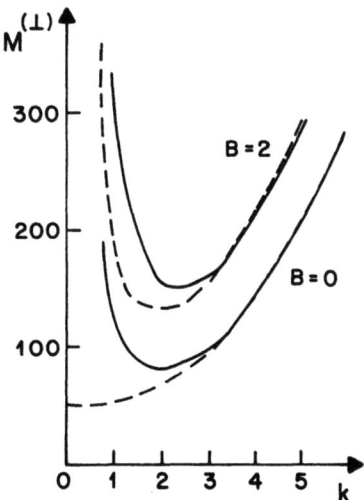

Fig. 3. Critical Marangoni number versus overall wave number for Marangoni convection as calculated by Pearson (1958) for the lower plate as a fixed-temperature surface (solid curves) or as a fixed heat-flux surface (dashed curves).

$$k = (k_1^2 + k_2^2)^{1/2} \quad . \tag{3.9}$$

In experiments it is convenient to impose a temperature across the gap between two horizontal plates that enclose the liquid layer bounded by a thin air gap present to help control the thermal environment of the interface. Thus, the two-layer problem is of practical importance. This problem has been considered by Zeren and Reynolds (1972) and then corrected and extended by Ferm and Wollkind (1982).

3.3 Physical Mechanisms

The instability identified by linear theory can be explained following Pearson (1958). If the plate is heated, as shown in Figure 2, a conductive temperature profile is achieved in which $b_z > 0$.

"MARANGONI INSTABILITY"

Assume that a disturbance creates a hot spot (compared with its neighbors) at a point P on S. If the surface tension decreases with temperature ($\gamma > 0$), there is a net surface traction away from P. Since the fluid is viscous, subsurface fluid is dragged away from P. By conservation of mass, an upflow beneath P is created. This rising fluid, since it comes from below, has been warmed by the conductive profile $\bar{T}$; it can maintain the heat excess at P if $\gamma b_z d$ is large enough, i.e. if either the temperature gradient is steep or if the surface tension is highly temperature sensitive. The fluid is recycled by cooling along the interface and falling at a distance away from P. Thus, if $M^{(\perp)}$ is large enough, a steady, finite amplitude convection can be maintained.

The above described convection is most easily generated if $B = 0$. When B is raised from zero, the fluid rising from below P loses some of its heat to the atmosphere, leaving less available to generate surface-tension gradients. As shown in Figure 3, increasing B gives rise to larger values of $M_L^{(\perp)}$ necessary to maintain the convection. From eqn. (3.2) in the limit $B \to \infty$ the interface is isothermal, surface-tension gradients are absent, and so $M_L^{(\perp)} \to \infty$; Marangoni convection is absent.

3.4 Energy Stability Theory

Energy theory is a variational formulation whose output is a criterion for stability against disturbances of arbitrary amplitude ε.

Davis (1969) formulates the theory and finds a critical value $M_E^{(\perp)}$ of $M^{(\perp)}$ for which $M^{(\perp)} < M_E^{(\perp)}$ is a sufficient condition for stability. $M_E^{(\perp)}$ is a function of B but is independent of P. Below the energy-theory value, stability is guaranteed; above the linear-theory value instability is guaranteed (Pearson 1958). Between the curves, the basic state may be stable for ε small but unstable for ε large enough. Where or whether subcritical instabilities exist in the range $M_E^{(\perp)} < M^{(\perp)} < M_L^{(\perp)}$ is a question that can only be answered by bifurcation theory or by direct numerical simulation.

3.5 Bifurcation Theory

The post-instability behavior of the system can be examined by bifurcation theory using perturbations in powers of ε near $M^{(\perp)} = M_L^{(\perp)}$.

Scanlon and Segel (1967) pose the one-layer model in the absence of gravity. For simplicity they let $P \to \infty$ and allow the layer to be infinitely deep. They consider the interaction of two disturbances that allow a competition between (two-dimensional) rolls and (three-dimensional) hexagonal convection and obtain a pair of amplitude equations governing the selection. These have the form

$$\frac{dA}{dt} = \sigma A - \beta AB - A(c_1 A^2 + c_2 B^2) \tag{3.10a}$$

$$\frac{dB}{dt} = \sigma B - \frac{1}{4} \beta A^2 - B[\frac{1}{2} c_2 A^2 + (4c_1 - c_2) B^2] \tag{3.10b}$$

where σ is the linear theory growth rate proportional to $M^{(\perp)} - M_L^{(\perp)}$, and the c_i and β are positive (computable) coefficients. The A and B are the time-dependent amplitudes in w_1, the linearized vertical component of velocity,

$$w_1 = \{A(t)\cos \frac{1}{2} \sqrt{3} \, kx \cos \frac{1}{2} ky + B(t)\cos ky\} f(z) \quad , \tag{3.11}$$

and $f(z)$ is a (computable) eigenfunction of the linearized theory. Here, if $A = 0$, $B \neq 0$, there is roll-cell convection; if $A = \pm 2B$, then hexagonal convection is present.

Scanlon and Segel (1967) find that in the range near $M^{(\perp)} = M_L^{(\perp)}$, where validity is expected, the only stable convective state has hexagonal planform; this convection is characterized by upflow in cell centers. As $M^{(\perp)}$ is increased from zero, the conductive state is stable until $\delta M = 0$, where

$$\delta M = (M^{(\perp)} - M_L^{(\perp)})/M_L^{(\perp)} \quad . \tag{3.12}$$

However, hexagonal convection is stable from $\delta M = -0.023$ onward. Thus, as $M^{(\perp)}$ is increased, conduction will remain stable until the system jumps to convection for $-0.023 < \delta M < 0$. This stable subcritical convection creates a dynamic hysteresis behavior (Scanlon and Segel 1967) in the response of the system.

The magnitudes of A and B are formally $O(\varepsilon)$. If β were zero, the equilibria of system (3.10) would have $A, B = O(\sigma^{1/2})$, so that we would have $\varepsilon = O(\sigma^{1/2})$. Now, if $\beta \neq 0$, then strictly speaking, one should retain both the quadratic and cubic terms in system (3.10) only if they are comparable, i.e. if $\beta\epsilon^2 \sim \epsilon^3$. Thus, only if $\beta = O(\epsilon)$ can higher powers in A and B be neglected. Since Scanlon and Segel find for β a numerical value independent of ϵ, their results may be only suggestive.

Why can one consider only the two disturbances defined in eqn. (3.11)? Any horizontal structure $\exp[ik_1 x + ik_2 y]$ with $k_1^2 + k_2^2 = k^2$ is allowable. Kraska and Sani (1979) attempt to use such a wider class, which includes the pair of eqn. (3.11). They fail to obtain closure; the more disturbances that they include, the more that are required. However, their analysis may be in error, given that the adjoint operator that they present is incorrect.

Cloot and Lebon (1984) consider the nonlinear problem with $B = 0$ or 1, $P = 7$, 70 or 500. The full range of horizontal planforms is included, and powers series representations in ϵ are examined. They find that hexagons are preferred near $M^{(\perp)} = M_L^{(\perp)}$ and extend to subcritical values of $M^{(\perp)}$. For $P = 7$, they begin at $\delta M = -0.003$, again showing the narrow range of subcritical convection. Furthermore, Cloot and Lebon (1984) examine supercritical convection and determine when such steady states are stable against disturbances of various wave lengths and planforms and find the stable range to be $k < k_L$. However, the comment made above about the analysis of Scanlon and Segel (1967) applies here as well. Formally, both quadratic and cubic terms can only be retained simultaneously if $M^{(\perp)} = O(\epsilon)$.

In all of the above, the convecting layer is unbounded laterally, so that the horizontal wave vector (k_1, k_2) forms a continuous spectrum. The linear theory determines the wave-vector magnitude $k = k_L$ at $M^{(\perp)} = M_L^{(\perp)}$ but not its phase. Thus, the planform is undetermined. Further, at $M^{(\perp)} > M_L^{(\perp)}$, a whole range of k corresponds to modes that grow according to linear theory. The above theories aim at obtaining the preferred planform (hexagons) and wave number through nonlinear selection. An alternative approach is to focus on a modified problem, i.e. a convection layer with lateral boundaries. These boundaries break the continuous spectrum and allow analyses of one or a few competing modes, presumably present when the container width "a" (scaled on layer depth) is not too large.

Rosenblat et al. (1982 a,b) consider such situations when the container is circular and rectangular. In each case they simplify the problem by posing "slippery sidewalls," thus allowing solution of the linear stability problem by separation of variables. P is arbitrary and now "a" is present as well.

We focus here on the cylindrical case where the linearized problem has a vertical velocity component that can be represented as follows:

$$w_1 = A_m(t) f(z) \cos m\theta \, g_m(r) \tag{3.13}$$

where $f(z)$ is a similar structure function as that in eqn. (3.11), except that the present layer has finite depth, g_m is an appropriate Bessel function, and m is an integer giving the azimuthal variations. Figure 4 shows the neutral curve for the problem; note that except for a small range of radii a (1.7 < a < 2.0), the most "dangerous" mode is nonaxisymmetric.

For most aspect ratios the weakly nonlinear theory focuses on a single mode gives an amplitude equation of the form

$$\frac{dA_m}{dt} = \sigma_m A_m - c_m A_m^3 \,, \quad m \neq 0 \,, \tag{3.14}$$

$c_m > 0$, corresponding to supercritical, pure mode m convection. Here σ_m is the linear theory growth rate and c_m is a computable Landau constant. In the range 1.7 < a < 2.0., there is pure axisymmetric convection with

$$\frac{dA_0}{dt} = \sigma_0 A_0 - \beta_0 A_0^2 - c_0 A_0^3 \,. \tag{3.15}$$

Both upflow and downflow at the container center are possible in each case. When the aspect ratios have special values as shown in Figure 4, linear theory is ambiguous since $M^{(\perp)}$ is equal for two modes and a nonlinear theory must be used to make a pattern selection. Thus, when "a" is fixed near one of the crossovers and $M^{(\perp)}$ is increased through $M_L^{(\perp)}$, there can be a sequence of transitions by primary, then secondary or tertiary bifurcations. Besides giving information on such sequences at these special values of "a", these sequences can mirror the transitions far from these special points, which would be widely separated in $M^{(\perp)}$ and otherwise inaccessible to analysis. Rosenblat et al. (1982a) analyze the double eigenvalues in neighborhoods of points A and B of Figure 4 using an asymptotic simplification of a series-truncation representation as an aid to the calculation of the coefficients.

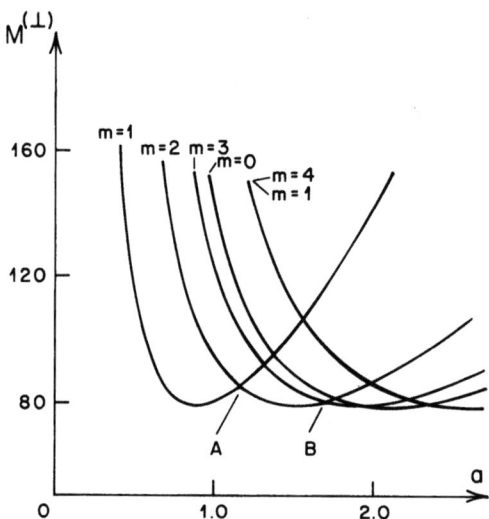

Fig. 4. Critical curves of Marangoni number versus box radius for Marangoni instability in a circular cylinder with "slippery" walls, (B = 0; from Rosenblat et al. (1982a)). The envelope of minima is the neutral curve. The points A and B indicate double eigenvalues where modes of two azimuthal wave numbers m have the same $M_L^{(\perp)}$.

Near point A, modes m = 1 and m = 2 interact. If the linearized vertical velocity component w_1 is represented as

$$w_1 = \{A_1(t)g_1(r)\cos\theta + A_2(t)g_2(r)\cos 2\theta\}f(z) , \qquad (3.16)$$

then the amplitudes near $a_A \approx 1.20$ satisfy

$$\frac{dA_1}{dt} = (\sigma_1+\Delta)A_1 + c_1 A_1 A_2 - A_1(c_2 A_1^2 + c_3 A_2^2) , \qquad (3.17a)$$

$$\frac{dA_2}{dt} = \sigma_2 A_2 - c_4 A_1^2 - A_2(c_5 A_1^2 + c_6 A_2^2) . \qquad (3.17b)$$

Here the c_i are (computable) coefficients, σ_i are the growth rates of the pure linear-theory modes, both of which vanish at $a = a_A$, $M^{(\perp)} = M_L^{(\perp)}$, and Δ is a splitting parameter whose sign shifts consideration to either $a < a_A$ or $a > a_A$. The two individual bifurcation points are at $\delta M = 0$ and $\delta M = \Delta$.

For $a < a_A$, as $M^{(\perp)}$ is increased through $M_L^{(\perp)}$, a mixed m = 1-2 mode bifurcates, followed by a transition to a pure m = 2 mode at $\delta M > \Delta$, beyond the second critical value of linear theory. In contrast, linear theory predicts that there would be only an m = 1 mode at this value of "a". For $a > a_A$, as $M^{(\perp)}$ is increased through $M_L^{(\perp)}$, a pure m = 2 mode bifurcates. This can persist, or alternatively a <u>time-periodic</u> (Hopf) secondary bifurcation can occur, leading to stable oscillatory convection at $\delta M < \Delta$.

When $a \approx a_B$, the modes m = 2 and m = 0 compete. Here various dynamic hysteresis and jump phenomena are predicted, and the predicted behavior is quite complex. Clearly, if such predicted phenomena do occur in experiment, there is a very rich catalogue of behavior left to be examined.

4. $[\nabla \bar{T}]_{EXT}$ ∥ S: HYDROTHERMAL INSTABILITIES

4.1 Basic State

We consider the situation in which, from eqn. (2.1), $b_∥ \neq 0$ and $b_⊥ = 0$, so that the liquid layer is heated horizontally, tangent to the interface.

Suppose that the rigid plate of Figure 2 is a zero-heat-flux surface. The interface is cooled by air currents according to law (3.2), where we exert an axial temperature gradient by imposing the following condition the air temperature (in dimensional form):

$$T_{AIR}(x) = -b_\parallel x \quad . \tag{4.1}$$

The imposed axial temperature field induces an axial surface-tension gradient that drives motion in the layer. We consider two basic states, as defined by Smith and Davis (1983).

First, there is the <u>linear-flow basic state</u> in which
$$\overline{v} = (\overline{u}(z),0,0) \quad , \quad \overline{u}(z) = z \quad , \tag{4.2a}$$
the temperature is
$$\overline{T} = -x + \overline{\theta}(z) \quad , \quad \overline{\theta}(z) = \frac{1}{6} M^{(\parallel)}(1-z^3) \quad , \tag{4.2b}$$
the pressure gradient is
$$\overline{p}_x = 0 \quad , \tag{4.2c}$$
and
$$Q = \frac{1}{2} M^{(\parallel)} \quad . \tag{4.2d}$$

Here the velocity profile is linear and the temperature field consists of the impressed axial nondimensional profile (-x) plus the <u>flow-induced vertical structure</u> $\overline{\theta}(z)$, a balance between horizontal convection and vertical conduction; these are shown in Figure 5a. The linear-flow basic state is an <u>exact solution</u> of the governing system.

Second, there is the <u>return-flow basic state</u>, in which the interface is flat,
$$\overline{v} = (\overline{u}(z),0,0) \quad , \quad \overline{u}(z) = \frac{3}{4} z^2 - \frac{1}{2} z \quad , \tag{4.3a}$$

the temperature is
$$\overline{T} = -x + \overline{\theta}(z) \quad , \quad \overline{\theta}(z) = -\frac{1}{48} M^{(\parallel)}(3z^4 - 4z^3 + 1) \quad , \tag{4.3b}$$
the pressure gradient is
$$\overline{p}_x = \frac{3}{2} \quad , \tag{4.3c}$$
and
$$Q = 0 \quad . \tag{4.3d}$$

Here the velocity profile is quadratic, and the temperature field consists of the impressed axial profile (-x) plus the <u>flow-induced vertical structure</u> $\overline{\theta}(z)$; these are shown in Figure 5b. The return-flow profile basic state is an <u>exact solution</u> of the governing system.

The linear-flow state simulates an <u>open system</u>. For example, if the plate at z = 0 had finite extent, then near the center of the plate the solution (4.2) might hold while the fluid would exit the channel (at large positive x), recirculate below the plate, say, and then return through the open end (at large negative x).

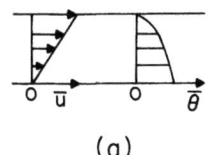

(a)

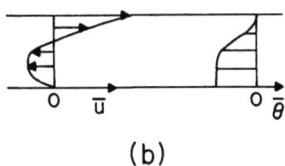

(b)

Fig. 5. Sketches of the velocity and flow-induced temperature profiles for the (a) linear-flow basic state and (b) return-flow basic state.

The return-flow state has a nonzero pressure gradient, obtained by setting to zero the flow rate across any vertical section; it simulates the flow in a <u>closed system</u> (say, a long slot). Here the endwalls create the pressure gradient causing the near-surface flow to return near the bottom of the layer. The solution (4.3) represents the "core flow" away from the ends, where the flows turn. Sen and Davis (1982) use matched asymptotic expansions to obtain this core flow, the endwall-region corrections, and the matching of these. The approximation they use there is that $P = O(1)$, and $M^{(II)} = O(1)$ as $A \to 0$. Here A is the aspect ratio of the slot. Note that large numerical values of $M^{(I)}$ are allowed here.

Smith and Davis (1983) have identified two distinct classes of instabilities associated with the above-defined basic states. There are convective instabilities that do not depend strongly on the deflection of the interface. There are surface-wave instabilities that depend intrinsically on the interaction of the flow and surface deflection. We discuss only the former of these. See Davis (1987) for a broader and deeper discussion.

4.2 <u>Linear Theory</u>: <u>Linear Flow</u>

In this case we follow Smith and Davis (1983), and examine the basic state for instability using normal modes for any quantity $\phi(x,y,z,t)$ of the form

$$\phi'(z)\exp\{i(k_1 x + k_2 y) + \sigma t\} \qquad (4.4a)$$

where

$$\sigma = \nu + i\omega \tag{4.4b}$$

is the complex growth rate.

In the present case we use the following parameters to characterize the system: $M^{(II)}$, P, B, k_1 and k_2. Since these are linear theories, it follows that $\varepsilon \to 0$. Smith and Davis (1983) find that increasing B delays the instability, as in the case of Marangoni convection. Hence, we discuss here only the limit $B \to 0$.

The critical conditions $M_L^{(I)}$ versus P are obtained by minimization over (k_1, k_2). Figure 6 shows that the neutral curve consists of three parts: when $P > 1.60$, there is stationary convection in the form of longitudinal rolls ($k_1 = 0$); when $0.60 < P < 1.60$, there are two-dimensional ($k_2 = 0$) waves that travel downstream; when $P < 0.60$ there are oblique waves that are nearly longitudinal rolls ($k_1 \approx 0$) traveling nearly crossstream, but with a component of phase speed directed opposite to the surface flow. The largest value of $M_L^{(I)} = 21.3$ occurs at $P = 0.60$.

4.2.1 <u>Flow-Induced Marangoni Convection</u>. The static layer of Section 3 has symmetry for every (k_1, k_2), pair and so the planform in Marangoni convection is undetermined by linear theory. The addition of shear in the basic state breaks this symmetry in favor of stationary longitudinal rolls when $P > 1.60$ and $B = 0$. Smith and Davis (1983) find that $k_1 = 0$ is the only possible stationary mode.

The interpretation of this instability stems from the flow-induced vertical temperature distribution $\bar{\theta}(z)$ given in eqn. (4.2c). This form makes the layer "heated from below." Let us write the linearized disturbance equations for longitudinal rolls in normal-mode form:

$$\left(L - \sigma M^{(II)} P^{-1}\right) U' = M^{(II)} P^{-1} \frac{d\bar{u}}{dz} W' \tag{4.5a}$$

$$\left(L - \sigma M^{(II)} P^{-1}\right) L W' = 0 \tag{4.5b}$$

$$\left(L - \sigma M^{(I)}\right) T' = M^{(II)} \bar{T}_x U' + M^{(I)} \bar{T}_z W' \tag{4.5c}$$

where

$$L = \frac{d^2}{dz^2} - k_2^2 \ . \tag{4.5d}$$

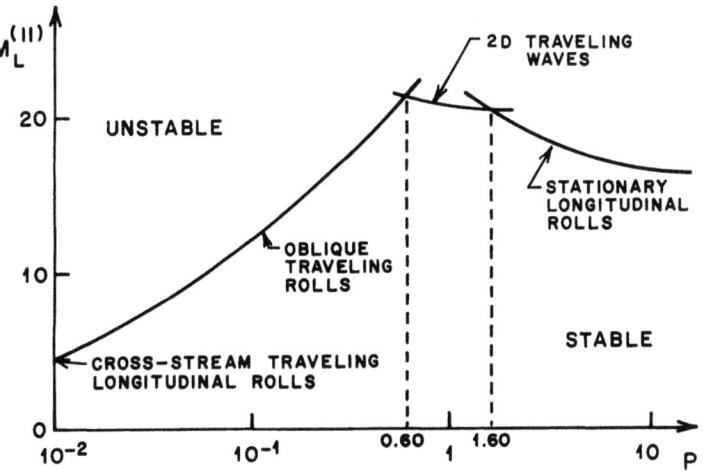

Fig. 6. Critical curves (B=0) for the linear-flow basic state (from Smith and Davis 1983).

For $P \to \infty$, $U' \to 0$, eqns. (4.5) reduce to the equivalent of those of Pearson (1958), except that in nondimensional form the magnitude of the vertical temperature gradient in eqn. (4.5c) is not unity. In the Pearson case (Figure 3 above), we have the equivalent of $M_L^{(I)} \approx 48$. If we estimate that the profile $\frac{1}{6} M^{(II)}(1-z^3)$ has the effective average gradient $\frac{1}{6} M^{(II)}$, then Pearson's result, applied in the present case, should give

$$\frac{1}{6} [M_L^{(II)}]^2 \approx 48 \qquad (4.6a)$$

or

$$M_L^{(II)} \approx 17 \ . \qquad (4.6b)$$

This is in excellent agreement with the computed value for $P \to \infty$ of Smith and Davis (1983), namely $M_L^{(II)} \approx 15.5$.

Now, when P is finite but $P \gg 1$, we see from eqn. (4.5a) that

$$U' \sim -M^{(II)} P^{-1} \frac{d\bar{u}}{dz} W' \qquad (4.7)$$

where $\frac{d\bar{u}}{dz} = 1$. When $W' > 0$, there is upflow beneath a surface hot line (along the x-direction). From relation (4.7), it follows that $U' < 0$ and $U' = O(P^{-1})$. This in turn produces a convective cooling, as seen from eqn. (4.5c). Here we have $\bar{T}_x = -1$, $\bar{T}_z < 0$, so that the destabilizing term $M^{(II)} \bar{T}_z W'$ is opposed by the stabilizing term $M^{(II)} \bar{T}_x U'$. Thus, as shown in Figure 6, $M_L^{(II)}$ increases as P is decreased from infinity. Smith and Davis (1983) argue further that the stabilization as P decreases is accompanied by an increase of k_{2_L}.

"FLOW-INDUCED MARANGONI INSTABILITY"

Our discussion on the mechanism for this instability paraphrases that of Smith and Davis (1983). Assume that a disturbance creates a hot line L (compared to its neighbors) in the flow direction on S. If surface tension decreases with temperature ($\gamma > 0$), there is a net surface traction away from L in the cross-stream directions. Since the fluid is viscous, subsurface fluid is dragged away from L. By conservation of mass, an upflow beneath L is created. This rising fluid, since it comes from below, has been warmed by the flow-induced vertical profile and reinforces the heat excess of the hot line, promoting sustained Marangoni convection. When $P \to \infty$ the fluid rises with zero downstream velocity perturbation, but when $P < \infty$ this perturbation is non zero. The rising fluid moves into a region where the basic-state speed is higher; it moves the particle upstream from a cooler downstream location, cooling the hot line and opposing Marangoni convection. As P decreases from infinity, it strengthens its opposition, leading to larger values of $M_L^{(II)}$.

4.2.2 <u>Hydrothermal Instability</u>. For $P \to 0$ the preferred modes are longitudinal rolls ($k_1 = 0$) that propagate in the cross-flow directions and have $M_L^{(II)} \sim P^{1/2}$. As P is increased to $P = 0.60$, the axes of the rolls rotate slightly into oblique waves that propagate in directions pointing against the flow up to $7.5°$ to the cross-stream direction. Thus, for $P < 0.60$, they are almost transversely propagating. They arise from an instability that involves a transfer of energy from the imposed horizontal (axial) temperature gradient to the disturbances through perturbations in the horizontal velocity field.

Smith (1986) shows for small P that the inertia dominates the viscous forces. Equation (4.5a) shows that $U' < 0$ and is of magnitude $M^{(II)}$ when $W' > 0$; the perturbation induced is an upstream component of velocity almost oppositely phased with T'. In eqn. (4.5c) the term $\bar{T}_x U'$ dominates the term $M^{(II)} \bar{T}_z W'$.

"Hydrothermal Instability at Small P"

The mechanism can be described following Smith (1986). Assume that a disturbance creates a hot line L (compared with its neighbors) in the flow direction on S. If the surface tension decreases with temperature ($\gamma > 0$), there is a net surface traction away from L in the cross-stream directions. Since the fluid is viscous, subsurface fluid is dragged away from L. By conservation of mass, an (almost in-phase) upflow beneath L is created. Since the fluid rises into a region where the basic-state speed is higher, it has an inertially driven (out-of-phase) negative downstream velocity component. This upstream moving particle has been cooled by the imposed axial temperature gradient and so lowers the heat excess in the hot line. In turn the cooling weakens the upflow, but the upstream velocity continues to increase, since there is still upflow. The fluid inertia causes the decreasing temperature at L to overshoot and become negative. The hot line L is now a cool line and the process reverses through the thermocapillary effect. An increase in $M^{(II)}$ strengthens the downstream velocity perturbation, leading to a maintained convective state through the extraction of energy from the

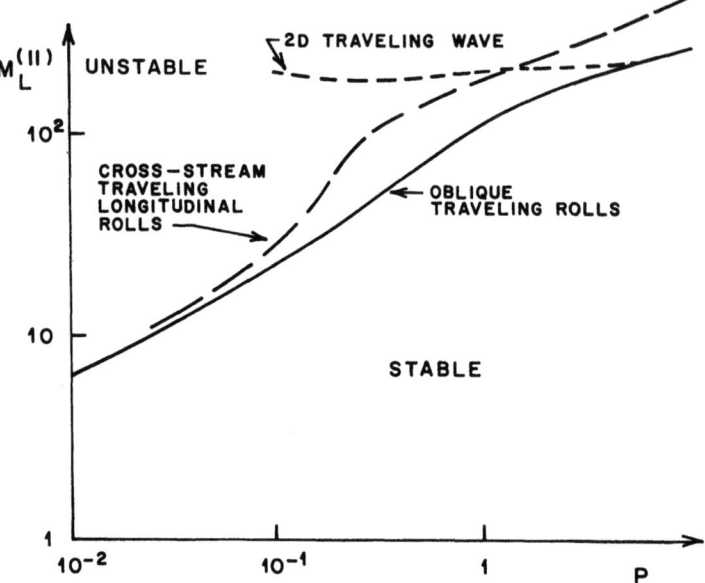

Fig. 7. Critical curves (B=0) for the return-flow basic state (from Smith and Davis 1983).

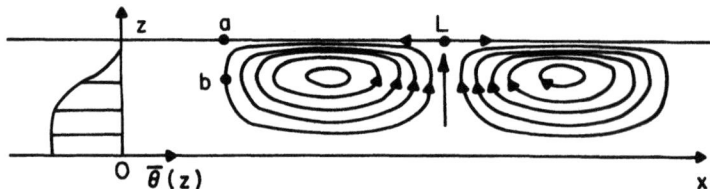

Fig. 8. Sketch of the streamlines and flow-induced vertical temperature profiles for the one-layer system with the return-flow basic state.

imposed axial temperature gradient. The flow-induced vertical temperature gradient makes the layer "heated from below" on a static basis, but it is the mean horizontal velocity distribution that has the major effect on P. As P is increased from zero, viscous effects increase and so a larger $M_L^{(II)}$ is required to maintain the heat excess of the hot line. This is associated with an increased k_L.

For $0.60 < P < 1.60$, two-dimensional hydrothermal waves become the preferred mode. This instability is related to the "Hydrothermal Instability at Large P" discussed below.

4.3 Linear Theory: Return Flow

The critical conditions $M_L^{(II)}$ versus P are obtained by minimization over (k_1, k_2). Figure 7 shows a smooth neutral curve that for $P \to 0$ has $M_L^{(II)} \sim P^{1/2}$ and $k_1 \to 0$; it is a longitudinal roll propagating crossstream. As P is increased, $M_L^{(II)}$ increases and the rolls have axes that rotate into oblique waves; for $P \to \infty$ the waves are nearly two-dimensional ($k_2 \ll 1$), propagating in the directions $7.90°$ from the upstream direction and we have
$$M_L^{(II)} \to 398.$$

4.3.1 Flow-Induced Marangoni Convection.

The flow-induced vertical distribution $\bar{\theta}(z)$ given in eqn. (4.3c) corresponds to the layer being "cooled from below." Thus, no stationary convective instability is present.

4.3.2 Hydrothermal Instability.

We saw earlier for the linear flow for $P \ll 1$ that instability arises from a transfer of energy to the disturbances from the imposed horizontal temperature field through a horizontal-convection mechanism. For small P, the return flow is susceptible to the same instability. Now, however, the flow-induced vertical gradient opposes this, so that a slightly larger value of $M^{(II)}$ is required here to sustain the instability than was the case for the linear-flow profile. More importantly, the instability is retarded here by the alteration of the velocity profile (Smith 1986), since $|\frac{d\bar{u}}{dz}|$ is smaller here than it was in the linear-flow case.

For large values of P energy transfer from the vertical flow-induced temperature distribution to the disturbances through vertical convection becomes the dominant mechanism of instability. For $P \to \infty$, let us consider a hot line L oriented cross stream to the flow along the y-axis. Figure 8 shows the instantaneous flow lines; these are centered near the interface. The hot line induces thermocapillary stresses that cause flow upstream and downstream as shown. A vertical upflow beneath L is induced by conservation of mass. This upflow cools L as a result of the flow-induced vertical profile shown. The warm fluid leaves L on S moves upstream to point a and descends toward point b. As is shown in equation (4.5c), it is most effective in heating a point b near the maximum of $\bar{\theta}_z$. Calculations (Smith 1986) show that the temperature at point b can be nearly 20 times that at L. Vertical conduction then strongly heats point a, and the hot spot effectively moves upstream, consistent with the results of Smith and Davis (1983). We can summarize this argument, taken from Smith (1986), as follows:

"Hydrothermal Instability at Large P"

Assume that a disturbance creates a hot line L (compared with its neighbors) cross stream to the flow direction on S. If the surface tension decreases with temperature ($\gamma > 0$), there is a net surface traction away from L in the upstream and downstream directions. Since the fluid is viscous, subsurface fluid is dragged away from L. By conservation of mass an upflow beneath L is created. If the original disturbance has closed streamlines, the downward flow leaving S upstream of L causes intense subsurface heating near the maximum of the gradient of the flow-induced vertical temperature profile. This elevated temperature internal to the layer produces a large conductive heating of S upstream of L. Thus, L effectively moves upstream. The heating upstream of L is proportional to $M^{(\parallel)}$, so that instability is maintained for large enough $M^{(\parallel)}$. The energy for the instability comes from the flow induced vertical temperature field, which, on a static basis, is stabilizing.

By the above mechanism, $\bar{\theta}_z$ determines the value of $M_L^{(\parallel)}$ and the distance upstream that the hot line travels. Both of these should decrease with $\bar{\theta}_z$ and so in the linear-flow case in which $\bar{\theta}_z < 0$, $M_L^{(\parallel)}$ is quite small and the hot line is induced to move downstream, consistent with the preferred mode for $0.60 < P < 1.60$.

Xu and Davis (1984) consider a return-flow profile for a circular-cylindrical geometry in which an axial temperature gradient is imposed.

They find convective instabilities both qualitatively and quantitatively similar to those discussed above.

4.4 Nonlinear Theory

There has been no energy theory analysis for systems having $[\nabla \bar{T}]_{EXT} \parallel S$.

The only bifurcation theory is that of Smith (1985). He considers convective instabilities of the return-flow basic state. These hydrothermal instabilities are studied with $B = 0$ and $10^{-3} < P < 10$. By the linear theory of Section 4.3, there are always two oblique waves that propagate with components in the direction opposite to the surface flow; one propagates left (cross stream) and one propagates right. If we write the linear theory component of velocity w_1 as

$$w_1 = A_L(t) f_L(z) e^{i(k_1 x + k_2 y)} + A_R(t) f_R(z) e^{i(k_1 x - k_2 y)}, \qquad (4.8)$$

then $(k_1, \pm k_2)$ are the critical values of the wave numbers, and the $f(z)$ are the (computable) eigenfunctions. Smith (1985) finds that

$$\frac{dA_L}{dt} = \sigma A_L - A_L(c_1 A_L^2 + c_2 A_R^2) \qquad (4.9a)$$

$$\frac{dA_R}{dt} = \sigma A_R - A_R(c_2 A_L^2 + c_1 A_R^2) \qquad (4.9b)$$

where the c_i are (computable) complex coefficients and σ is the linear-theory growth rate of the modes. By analyzing system (4.9), Smith finds that (a) the instabilities are always <u>supercritical</u> and (b) the only stable supercritical states are the pure modes $A_L = 0$, $A_R \neq 0$ or $A_L \neq 0$, $A_R = 0$. All mixed modes, including the cross-stream standing-wave case, are unstable.

Thus, the nonlinear theory fixes the relative phases of the linearized modes by determining the selected linear combination of the left- and right- propagating modes. In the Smith (1985) analysis only pure modes are stable in the laterally unbounded layer. If lateral boundaries were present, it might happen that the preferred mode would be waves standing in the cross-stream direction but propagating against the surface flow.

4.5 Discussion

When $[\nabla \bar{T}]_{EXT} \parallel S$, surface-tension gradients drive shear flows. These shear flows induce vertical temperature profiles whose structure controls the instability characteristics of the system.

The flow-induced temperature distribution for the linear-flow state is "heated from below," giving rise to flow-induced steady Marangoni convection at large P in the form of longitudinal rolls. This instability is absent in the return-flow state, since there the flow-induced temperature distribution is "cooled from below."

The imposed tangential temperature gradient is the source of energy for hydrothermal instabilities when P is small. For $P \to 0$ these are longitudinal waves that propagate cross stream. As P is increased, the axes of the rolls rotate and then have a phase velocity with a component directed opposite to the free-surface flow. In the return-flow case, the axes rotate nearly $90°$ as $P \to \infty$, and these waves derive their energy from the "statically stable" vertical distribution of temperature. In the linear-flow case, the axes rotate only slightly and give way to two-dimensional waves at P near unity; these waves derive their energy from the vertical thermal structure. At high P, the flow-induced Marangoni convection preempts the hydrothermal waves.

The flow-induced vertical temperature field also serves as the gross measure of instability. For example, in the "statically unstable" linear-flow case we have $M_L^{(II)} \approx 15.5$ for $P \to \infty$, where in the "statically stable" return-flow case we have $M_L^{(II)} \approx 398$ for $P \to \infty$.

5. CONCLUSIONS

In this article we have focused on convective thermocapillary instabilities in planar layers in which the interface is non-deformable and where body forces are absent. Clearly, the relaxation of such restrictions allows the family of instabilities to be greatly enlarged; see Davis (1987) for certain elaborations.

There are a number of excellent sources of information on other features of variable surface-tension effects. Among these are Sternling and Scriven (1959), Levich (1962), Levich and Krylov (1969), Kenning (1968), Sørensen (1979), Velarde and Castillo (1981), Normand et al.(1977).

Acknowledgment

This work was supported by the National Science Foundation, Fluid Mechanics Program.

References

Cloot, A. and Lebon, G. 1984. A nonlinear stability analysis of the Benard-Marangoni problem. J. Fluid Mech. 145:447-469.

Davis, S. H. 1969. Buoyancy-surface tension instability by the method of energy. J. Fluid Mech. 39:347-359.

Davis, S. H. 1987. Thermocapillary instabilities, Ann. Rev. Fluid Mech. 19 (to appear).

Ferm, E. N. and Wollkind, D. J. 1982. Onset of Rayleigh-Benard-Marangoni instability: Comparison between theory and experiment. J. Non-Equilb. Thermodyn. 7:169-190.

Kenning, D. B. R. 1968. Two-phase flow with nonuniform surface tension. Appl. Mech. Rev. 21:1101-1111.

Kraska, J. R. and Sani, R. L. 1979. Finite amplitude Benard-Rayleigh convection. Int. J. Heat Mass Trans. 22:535-546.

Levich, V. G. 1962. Physicochemical Hydrodynamics, Prentice-Hall, Englewood Cliffs, NJ.

Levich, V. G. and Krylov, V. S. 1969. Surface-tension-driven phenomena. Ann. Rev. Fluid Mech. 1:293-316.

Pearson, J. R. A. 1958. On convection cells induced by surface tension. J. Fluid Mech. 4:489-500.

Rosenblat, S., Davis, S. H. and Homsy, G. M. 1982. Nonlinear Marangoni convection in bounded layers. Part 1. Circular cylindrical containers. J. Fluid Mech. 120:91-122.

Rosenblat, S., Homsy, G. M. and Davis, S. H. 1982. Nonlinear Marangoni convection in bounded layers. Part 2. Rectangular cylindrical containers. J. Fluid Mech. 120:123-138.

Scanlon, J. W. and Segel, L. A. 1967. Finite amplitude cellular convection induced by surface tension. J. Fluid Mech. 30:149-162.

Scriven, L. E. and Sternling, C. V. 1964. On cellular convection driven by surface tension gradients: effect of mean surface tension and viscosity. J. Fluid Mech. 19:321-340.

Sen, A. K. and Davis, S. H. 1982. Steady thermocapillary flows in two dimensional slots. J. Fluid Mech. 121:163-184.

Smith, M. K. and Davis, S. H. 1983. Instabilities of dynamic thermocapillary liquid layers. Part 1. Convective instabilities. J. Fluid Mech. 132:119-144.

Smith, M. K. 1985. The nonlinear stability of thermocapillary shear layers. Bull. Am. Phys. Soc. 30:1732.

Smith, M. K. 1986. Instability mechanisms in dynamic thermocapillary liquid layers. Phys. of Fluids. In press.

Sørensen, T. S. 1979. Dynamics and Instability of Fluid Interfaces. Lecture Notes in Physics. Vol. 105. Springer-Verlag, Berlin.

Sternling, C. V. and Scriven, L. E. 1959. Interfacial turbulence: hydrodynamic instability and the Marangoni effect. A.I.Ch.E.J. 5:514-523.

Velarde, M. G. and Castillo, J. L. 1981. Convection Transport and Instability Phenomena. Eds. J. Zierep and H. Oertel, Jr. Braun Verlag, Karlsruhe.

Xu, J.-J. and Davis. S. H. 1984. Convective thermocapillary instabilities in liquid bridges. Phys. Fluids 27:1102-1107.

Zeren, R. W. and Reynolds, W. C. 1972. Thermal instabilities in two-fluid horizontal layers. J. Fluid Mech. 53:305-327.

SOME PROBLEMS IN MARANGONI INSTABILITY

G. Lebon

Liège University
Institute of Physics, B5
B-4000 Liège, Belgium

INTRODUCTION

The purpose of this note is to study some aspects of Marangoni problem in a zero-gravity environment. In most of the papers concerned with Marangoni instability, the following hypotheses are made:

1. The upper surface is flat and not deformed.
2. The viscosity of the fluid is temperature-independent.
3. The Fourier law of heat conduction is taken for granted.
4. The disturbances are infinitesimally small and a linear analysis is justified.
5. The surface tension is a linearly decreasing function of the temperature.

In the present work, one examines the consequences resulting from the relaxation of these various assumptions. In particular, one studies the coupled influence of a deformable upper surface and a temperature dependent shear viscosity on the onset of convection (Cloot and Lebon, 1985). Moreover, it is generally admitted that heat conduction in a fluid is governed by Fourier's law relating linearly the heat flux to the temperature gradient. But in some special circumstances (Straughan and Franchi, 1984; Lebon and Cloot, 1984), it is adequate to replace Fourier's law by the so-named Maxwell-Cattaneo relation, which involves in addition the time derivative of the heat flux. The consequences of using Maxwell-Cattaneo instead of Fourier law on Marangoni instability are discussed. All the above mentioned problems are solved by using the normal mode technique.

Clearly, this approach is unappropriate for treating nonlinear situations, like these occurring when the surface tension is a quadratic function of the temperature. This has motivated the development of a nonlinear analysis (Cloot and Lebon, 1984). It is based on Gorkov-Malkus-Veronis iterative method and is firstly applied to the classical problem of a surface tension decreasing linearly with the temperature. The band of allowed steady convective solutions is determined as a function of the wave-number and the Marangoni number. The shape of the convective cells is also derived. As last application, the method is applied to determine the form of the stable convective cells when the surface tension is a quadratic function of the temperature (Cloot and Lebon, 1986) : this situation is typical of long chain dilute alcohols (Vochten et al., 1973).

2. THE PHYSICAL PROBLEM AND THE BASIC EQUATIONS

Consider a thin fluid layer of thickness d, extending laterally to infinity, and subject to a temperature drop ΔT between the lower and upper boundaries. The reference state is the rest state with solutions

$$\mathbf{v}_{ref} = 0 \qquad \frac{\partial T_{ref}}{\partial z} = -\frac{\Delta T}{d} = \text{constant}, \qquad (2.1)$$

the z-direction is normal to the limiting faces. Let $\mathbf{u}$ and θ be the velocity and temperature disturbances respectively:

$$\mathbf{u} = \mathbf{v} - \mathbf{v}_{ref} \qquad \theta = T - T_{ref} \ .$$

For a Boussinesquian fluid, the disturbances satisfy the following net of balance equations:

$$\nabla \cdot \mathbf{u} = 0 \qquad \text{(mass balance)}, \qquad (2.2)$$

$$P^{-1} (\frac{\partial}{\partial t} + \mathbf{u} \cdot \nabla) \mathbf{u} = -\nabla p + \nabla^2 \mathbf{u} \qquad \text{(momentum balance)}, \qquad (2.3)$$

$$(\frac{\partial}{\partial t} + \mathbf{u} \cdot \nabla) \theta = \nabla^2 \theta + w \qquad \text{(energy balance)}, \qquad (2.4)$$

where $\nabla = (\frac{\partial}{\partial x}, \frac{\partial}{\partial y}, \frac{\partial}{\partial z})$, $\nabla^2 = \nabla_1^2 + \partial^2/\partial z^2$, $\nabla_1^2 = \partial^2/\partial x^2 + \partial^2/\partial y^2$.

The above equations are written in non-dimensional form with the coordinates x, y, z, the time t, the temperature T and the pressure p scaled by d, d^2/κ, ΔT, $\rho \kappa \nu / d^2$ respectively, κ is the heat diffusivity, ρ the density, ν the kinematic viscosity, $P = \nu/\kappa$ the Prandtl number and w the z-component of the velocity.

At the boundaries, one has:

at $z = 0$ (lower rigid surface): $w = \partial w/\partial z = 0$, $\theta = 0$, $\qquad (2.5)$

at $z = 1$ (upper free surface): $w = 0$, $\partial \theta/\partial z = 0$, $\qquad (2.6)$

$$\partial^2 w/\partial z^2 = \text{Ma} \ \nabla_1^2 \theta. \qquad (2.7)$$

Equations (2.5) and (2.6) express respectively that the lower boundary is perfectly heat conducting and the upper one adiabatically insulated, relation (2.7) is the boundary condition for a free undeformed surface with a temperature dependent surface tension $\xi(T)$; Ma is the dimensionless Marangoni number defined by

$$\text{Ma} = - \frac{(\partial \xi / \partial T) \ \Delta T \ d}{\rho \nu \kappa} \ . \qquad (2.8)$$

It is generally admitted that the surface tension is a monotically linearly decreasing function of the temperature. Such a behavior is typical of a large class of fluids like water, silicone oil, water-benzene solutions, etc... In exceptional cases, one may find systems like some alloys, molten salts or liquid crystals with a surface tension growing linearly with the temperature. In the first case, Ma is positive; in the second, Ma is negative. Moreover, in some systems, like in aqueous long-chain alcohol solutions, the surface tension behaves like a quadratic function of the temperature.

For further purpose, it is useful to recall the main results of the normal mode technique.

3. THE NORMAL MODE TECHNIQUE

For small disturbances, the non-linear terms in (2.3) and (2.4) may be dropped and **u** and θ may be expanded in the form

$$(\mathbf{u}, \theta) = [U(z), V(z), W(z), \Theta(z)] \exp[i\mathbf{k}\cdot\mathbf{x} + \sigma t] , \qquad (3.1)$$

k is the horizontal wave vector, σ is the stability growth parameter, U, V, W, Θ the amplitudes of the perturbed quantities, **x** the position vector in the horizontal plane.

At marginal stability (Reσ = 0) and within the hypothesis of exchange of stability (Imσ = 0), one obtains after substitution of (3.1) in (2.3) and (2.4) :

$$(D^2 - k^2)^2 W = 0 , \qquad (3.2)$$

$$(D^2 - k^2)\Theta = -W , \qquad (3.3)$$

where D stands for d/dz and k^2 for **k**·**k**.

In terms of the amplitudes, the boundary conditions read as

at $z = 0$: $W = DW = 0$, $\Theta = 0$, $\qquad (3.4)$

at $z = 1$: $W = 0$, $D\Theta = 0$, $\qquad (3.5)$

$$D^2 W = -Ma\, k^2\, \Theta . \qquad (3.6)$$

The problem can be solved by means of a Rayleigh-Ritz variational method (Lebon and Perez-Garcia, 1980). Non trivial solutions exist only for a given dependence of Ma versus k, the marginal stability curve. By taking the minimum of this curve, one obtains the critical Marangoni number Ma^c and the critical wave number k^c corresponding to the onset of flowing : these values are

$$Ma^c = 79.61 , \qquad k^c = 1.99 .$$

4. SURFACE DEFORMATION AND A TEMPERATURE-DEPENDENT VISCOSITY

Assume that the viscosity of the fluid layer is a linear function of the temperature

$$\nu = \nu_o [1 + \frac{(\partial_T \nu)}{\nu_o} (T - T_o)] , \qquad (4.1)$$

with T_o an arbitrary reference temperature, and that the upper layer is deformed, with $\eta(x,y,t)$ the deflexion above the mean height d (see fig. 1) :

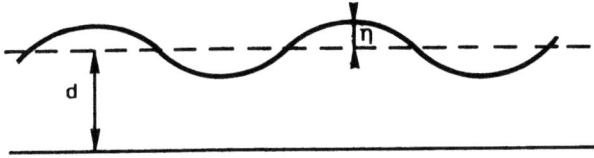

Fig. 1 - Surface deflexion

In dimensionless notation, the relevant linearized equations for the velocity and temperature disturbances are

$$P^{-1} \frac{\partial}{\partial t}(\nabla^2 w) = \nabla^4 w - R_\nu (2\nabla^2 \partial w/\partial z + z \nabla^4 w) \quad , \qquad (4.2)$$

$$\frac{\partial}{\partial t}\theta - w = \nabla^2 \theta \quad , \qquad (4.3)$$

with at $z = 0$: $w = \partial w/\partial z = \theta = 0$, (4.4)

at $z = 1 + \eta$: $w = \partial \eta/\partial t$, (4.5)

$$\rho \nabla_1^2 w - \partial^2 w/\partial z^2 = - Ma \nabla_1^2 (\theta - \eta) \quad , \qquad (4.6)$$

$$P^{-1} \frac{\partial}{\partial t}(\partial w/\partial z) - \nabla^2 (\partial w/\partial z) - 2\frac{\partial}{\partial z}(\nabla_1^2 w) + Cr^{-1}\nabla_1 \eta =$$
$$R_\nu (\nabla_1^2 w - \partial^2 w/\partial z^2 - \frac{\partial}{\partial z}\nabla^2 w) \quad , \qquad (4.7)$$

$$\partial \theta/\partial z = 0 \quad . \qquad (4.8)$$

The new dimensionless numbers appearing in (4.2) and (4.7) are the viscosity number R_ν and the crispation number Cr respectively defined by

$$R_\nu = \frac{(\partial_T \nu)_o \Delta T}{\nu_o} \quad , \qquad Cr = \frac{\nu_o \kappa}{\xi_o d} \quad ; \qquad (4.9)$$

Cr is related to the degree of deformability of the surface and reduces to zero for an infinite surface tension; typical values of Cr are

for water   : $Cr = 4.10^{-6}$,
for mercury : $Cr = 3.10^{-5}$,
for glycerin : $Cr = 4.10^{-3}$,

R_ν accounts for the effects arizing from the temperature dependent-viscosity, since the variation of the kinematic viscosity with respect to the temperature is opposite for gases ($\partial_T \nu > 0$) and for liquids ($\partial_T \nu < 0$) , R_ν may be invariably positive or negative. Characteristic values for R_ν are

for water : $-0.3 < R_\nu < -0.2$ ($20°C < T < 80°C$) ,
for oil   : $-0.75 < R_\nu < -0.35$ ($20°C < T < 80°C$) ,
for air   : $R_\nu = 0.37$ ($0°C < T < 50°C$) .

For infinitesimally small perturbations, we seek solutions of the form

$$(w, \theta, \eta) = [W(z), \Theta(z), E] \exp(i\mathbf{k}.\mathbf{x} + \sigma t) \quad . \qquad (4.10)$$

It was shown numerically by Vidal and Acrivos (1966) that the principle of exchange of stability holds for a layer with constant viscosity, bounded by an upper undeformable surface. This seems to remain true for a deformable surface as long as a single component fluid is considered (Castillo and Verlarde, 1982).

At marginal stability ($\sigma = 0$), the governing equations for the amplitudes W and Θ are

$$(D^2 - k^2)^2 W = R_\nu [2 (D^2 - k^2) DW + z (D^2 - k^2)^2 W] \quad , \tag{4.11}$$

$$(D^2 - k^2) \Theta = - W \quad . \tag{4.12}$$

At the boundaries, the equations are the same as (3.4) and (3.5) with (3.6) replaced by

$$Ma^{-1} D^2 W + k^{-2} Cr [3 k^2 DW - D^3 W + R_\nu (D^2 W + D^3 W - k^2 DW)]$$
$$= - k^2 \Theta \quad . \tag{4.13}$$

The surface deflection has been eliminated by combining (4.6) and (4.7). After that the velocity and the temperature fields are determined, the amplitude E of the deflection is easily obtained from equation (4.7) and is given by

$$E = \frac{Cr}{k^4} [D^3 W - 3 k^2 DW - R_\nu (D^2 W + D^3 W - k^2 DW)] \quad . \tag{4.14}$$

It is seen that the curves Ma(k) present two minima (see fig. 2) : one of them is located around $k^c = 2$ to which corresponds a finite value of Ma^c, the other is located at the origin $k^c = Ma^c = 0$. It is a well-known property (Smith, 1966) that, in absence of gravity effects, there exists a singularity in the short wave number limit $k \to 0$, in the sense that disturbances with a zero wave-number are unconditionally unstable. It follows from the present analysis that this anomaly is not removed by the introduction of a temperature-dependent viscosity. It can only be eliminated by allowing for the presence of gravity waves (Smith, 1966).

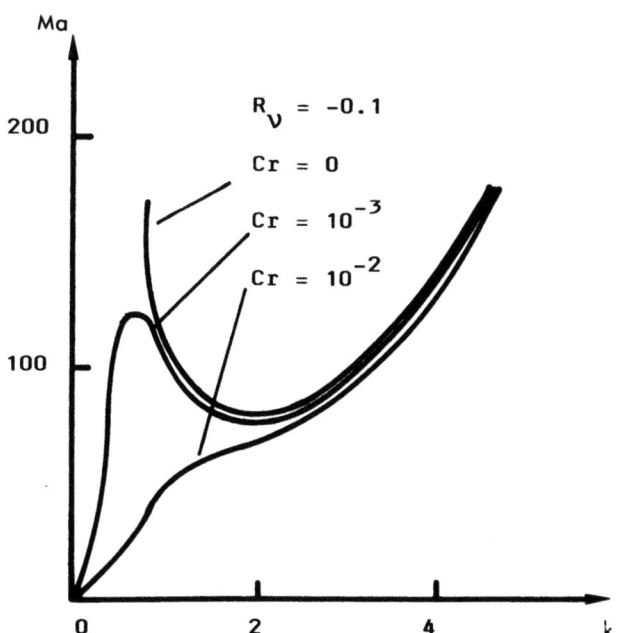

Fig. 2 - Marginal stability curves Ma versus k for various values of Cr, at fixed R_ν.

Table 1 : Critical values of Ma and k for various assigned values of Cr and R_ν (h = 0)

R_ν		Cr = 0	10^{-5}	10^{-4}	10^{-3}
-0.5	$Ma^c =$	69.85	69.83	69.63	67.95
	$k^c =$	1.95	1.95	1.95	1.9
-0.3	$Ma^c =$	72.97	72.96	72.81	71.20
	$k^c =$	1.95	1.95	1.95	1.9
-0.1	$Ma^c =$	77.06	77.04	76.93	75.70
	$k^c =$	1.99	1.99	1.99	1.95
0	$Ma^c =$	79.61	79.60	79.50	78.48
	$k^c =$	1.99	1.99	1.99	1.95
0.1	$Ma^c =$	82.64	82.63	82.55	81.76
	$k^c =$	2	2	2	2
0.3	$Ma^c =$	90.88	90.87	90.86	90.61
	$k^c =$	2.05	2.05	2.05	2
0.5	$Ma^c =$	104.6	104.6	104.6	103.2
	$k^c =$	2.05	2.05	2.05	2.05

Table 1 indicates that the critical Marangoni number decreases with increasing crispation number, at fixed values of R_ν. As a consequence, weak surface deflections are destabilizing, at least with respect to disturbances of not too small wave-numbers (k > 1). Such a result is in accord with the conclusions of Scriven and Sternling (1964) and Davis and Homsy (1980). This result is physically easily interpreted: increase of Cr from zero to a finite value means that the mean surface tension changes from infinite to smaller values and consequently, less energy is required to generate convection. It is also noticed that the critical wave-number is practically not affected by the surface deflection.

We now fix Cr and allow R_ν to vary. Clealy viscosity plays an important role on the onset of convection; for instance, going from $R_\nu = 0$ to $R_\nu = 0.5$, one changes the critical Marangoni number by more than 30 %. It is seen in Table 1 that stability is increased when R_ν grows.

Informations about the profile of the relief of the upper surface is provided by the ratio E/DW. It is calculated that at the convective threshold, the quantity E/DW is positive : E/DW = 2.24 for Cr = 10^{-5} and E/DW = 2.48 for Cr = 10^{-3}. This means that, where the fluid is rising (DW < 0), the surface is depressed (E < 0) and where the fluid is falling (DW > 0), the surface is raised (E > 0). For $R_\nu = 0$, we recover Scriven and Sternling's result that if the instability is driven by surface-tension gradients, the free surface over an upwelling current is always depressed; this result is confirmed experimentally by Linde (1981) and Cerisier et al. (1984). The present analysis shows that Scriven and Sternling's conclusion is unchanged when the viscosity is allowed to be temperature-dependent.

5. MARANGONI INSTABILITY IN A MAXWELL-CATTANEO FLUID

It is interesting to analyze to which extent the results about Marangoni instability are modified when the Maxwell-Cattaneo constitutive equation

$$\tau (\dot{\mathbf{q}} + \mathbf{S}.\mathbf{q}) = -\mathbf{q} - \lambda \nabla T \tag{5.1}$$

is employed instead of the classical Fourier law

$$\mathbf{q} = -\lambda \nabla T. \tag{5.2}$$

In (5.1), an upper dot designates the material time derivative, τ is a constant relaxation time, λ the heat conductivity and $\mathbf{S}$ the skew-symmetric spin tensor; in cartesian coordinates:

$$S_{ij} = \frac{1}{2} \left(\frac{\partial v_i}{\partial x_j} - \frac{\partial v_j}{\partial x_i} \right).$$

The l.h.s. of (5.1) must not be confused with Jaumann's derivative $D_j \mathbf{q}$ expressed by

$$D_j \mathbf{q} = \dot{\mathbf{q}} - \mathbf{S}.\mathbf{q}. \tag{5.3}$$

By selecting (5.1) instead of (5.2), one avoids the unpleasant property of propagation of heat disturbances with an infinite velocity; indeed, by substituting (5.1) in the energy equation, one obtains a hyperbolic equation while by using Fourier's law, one should have a parabolic equation (e.g. Lebon and Boukary, 1985). Moreover, it has been shown recently (Boukary and Lebon, 1986) that expression (5.1), in contrast with Jaumann's derivative, conciliates the results of the kinetic theory of gases and the principle of objectivity of continuum mechanics.

Our objective is to explore the consequences resulting from the substitution of Fourier law by Maxwell-Cattaneo expression for the simplest Marangoni problem : no surface deformation, a constant viscosity. At marginal stability, the basic equations are given by the set (3.2)-(3.6) at the exception of (3.3) to be replaced by

$$(D^2 - k^2)(\Theta + Ca\, W) = -W, \tag{5.4}$$

Ca is the so-called Cattaneo dimensionless number defined by (Lebon-Cloot, 1984)

$$Ca = \frac{\tau \kappa}{2d^2}. \tag{5.5}$$

The critical Marangoni number at which stationary convection occurs depends on the parameters k and Ca and is obtained from

$$Ma = \frac{8 k^2 \cosh k\, (k - \sinh k \cosh k)}{(k^3 \cosh k - \sinh^3 k) - 4\, Ca\, k^2 \sinh k\, (\sinh^2 k - k^2)}. \tag{5.6}$$

In the limiting case Ca = 0, one recovers Pearson's result (1958) established for a classical Fourier fluid. In (5.6), the quantities $(k - \sinh k \cosh k)$ and $(k^3 \cosh k - \sinh^3 k)$ are negative while $(\sinh^2 k - k^2)$ is positive. It follows that at marginal stability, Ma is necessarily positive. By selecting the Jaumann derivative, one should have obtained a plus sign in front of the quantity Ca in (5.6), allowing for negative Ma values and the occurence of steady convection by heating from above.

As the critical Marangoni number is smaller than its classical counterpart, we see that Maxwell-Cattaneo fluid is less stable than the Fourier one. It must however be realized that the corrections are minute as the values of Ca range from 10^{-6} to 10^{-8}.

Although exchange of stability has been demonstrated for a classical Fourier fluid, it is no longer guaranteed in a Maxwell-Cattaneo fluid. Hence, it is necessary to examine the behavior of the system under oscillatory convection. This is performed by putting $\sigma = i\sigma_1$, where σ_1 is real, in the balance equations. The relevant equations are solved by using a Galerkin method; the following results for σ_1^2 and Ma are obtained:

$$\sigma_1^2 = \frac{910 \left(\frac{23}{1260} + L\ Ca\right) \left(\frac{113}{1820} M + NK/P\right) - \frac{23}{630} MNCa}{113\ C^2 \left[(K/P)\ L - \frac{23}{1260} M\right]}, \qquad (5.7)$$

$$Ma = 2\left[\left(\frac{23}{1260} + L\ Ca\right)\left(MN - \frac{113}{1820}\sigma_1^2\ K/P\right) + \frac{23}{630}\ Ca\ \sigma_1^2 \left(\frac{113}{1820} M + KP/P\right)\right] /$$

$$a^2 \left[\left(\frac{23}{1260} + L\ Ca\right)^2 + \left(\frac{23}{630}\ Ca\ \sigma_1\right)^2\right], \qquad (5.8)$$

K, L, M, N are polynomial functions of k^2. The analysis simplifies considerably in the two limiting cases of a zero and an infinite Prandtl number.

For $P = 0$, σ_1^2 is unconditionally positive while Ma is a negative quantity with order of magnitude Ma $\sim -10^{14}$ for Ca $= 10^{-8}$. Theoretically, surstability is possible by heating from above but practically, the large values of Ma prevent us for observing this mechanism of instability. Similar results are obtained for $P = \infty$: overstability can only occur in the heated from above case, with very large values for Ma. It clearly appears that in Maxwell-Cattaneo fluid, oscillatory convection does not play an important role.

6. A NONLINEAR FORMALISM

In stability problems, two problems are of crucial interest. Firstly, to study the behavior of the fluid beyond the instability point predicted by the linear theory, secondly to determine the nature of the convective cells.

Gaining informations about these questions require a nonlinear approach. The first nonlinear theory of Marangoni's problem is that of Scanlon & Segel (1967). Their model, however, is very rough: the layer is assumed to be infinitely deep, and the Prandtl number is infinite. Scanlon & Segel use a successive approximation technique and predict the emergence of stable steady hexagonal cells at the onset of convection. Another nonlinear approach is due to Kraska & Sani (1979), whose results were not very convincing. To our knowledge, the only other nonlinear analysis is that proposed recently by Rosenblat et al. (1982). These authors study Marangoni convection in cylindrical and rectangular containers of finite extent. Their technique consists of expanding the field variables in series of eigenfunctions of the linear stability problem with time-dependent amplitudes. They conclude that hexagonal cells cannot appear in small-sized containers. It must, however, be observed that these results have been reached under rather simplifying restrictions and in particular by assuming that the no-slip condition is replaced by a zero tangential vorticity along the sidewall boundaries.

In this section, a different formalism is proposed. It consists of using an iterative scheme suggested by Gorkov (1957) and Malkus and Veronis (1958).

This technique was applied with success to a pure Bénard problem without surface tension effects by Schlüter, Lortz and Busse (1965).

Some simplifying assumptions are introduced : the Boussinesq approximation is applied, the top free surface of the fluid is flat and non-deformable, non-inertial effects, like rotation, are omitted and the surface tension depends linearly on the temperature.

The Gorkov-Malkus-Veronis method is based on an expansion of the field quantities $\mathbf{u}$, θ as well as Marangoni's number in powers of a small parameter ε :

$$\mathbf{u} = \sum_{i=1}^{N} \varepsilon^i \mathbf{u}^{(i)} \quad , \quad \theta = \sum_{i=1}^{N} \varepsilon^i \theta^{(i)} \quad , \quad Ma = M^{(o)} + \sum_{i=1}^{N} \varepsilon^i M^i \qquad (6.1)$$

$M^o(k)$ is the marginal value of Ma derived from the linear analysis, $\mathbf{u}$ and θ represent the steady solutions.

Introducing the developments (6.1) in the set of field equations and equating the different powers of ε yields a hierarchy of inhomogeneous differential equations. Since the flows considered have relatively small amplitudes, we have limited the analysis to the second order of approximation which is justified since ε remains very small ($10^{-7} < \varepsilon < 10^{-1}$).

The preferred mode of convection is determined by examining the stability of the steady solutions with respect to disturbances of infinitesimal size, denoted by $\tilde{\mathbf{u}}$ and $\tilde{\theta}$. Without loss of generality, we can express their time-dependence as

$$(\tilde{\mathbf{u}}, \tilde{\theta}) \sim \exp(\sigma t).$$

When the expansions (6.1) are inserted in the linearized perturbed equations, one obtains terms that are proportional to the various powers of ε. This suggests to develop similarly $\tilde{\mathbf{u}}$, $\tilde{\theta}$ and σ in power series of ε. At each order of approximation, the domain of stability of the steady solution is obtained by imposing that the real part of σ is negative.

In this note, we shall not enter into the details of the formalism. A complete description can be found in a recent work by Cloot and Lebon (1984). We shall here briefly discuss the main steps of the method.

6.1. The steady solutions

In the linear approximation, the solution $w^{(1)}$, $\theta^{(1)}$ is simply given by

$$(w^{(1)}, \theta^{(1)}) = [W^{(1)}(z), \Theta^{(1)}(z)] \phi(x,y) \qquad (6.2)$$

where $W^{(1)}$ and $\Theta^{(1)}$ have been determined earlier in section 2, while $\phi(x,y)$ satisfies the relation

$$\nabla_1^2 \phi + k^2 \phi = 0 \quad , \qquad (6.3)$$

whose solutions are

$$\phi(x,y) = \sum_{n=-N}^{+N} C_n \exp i(\mathbf{k}_n \cdot \mathbf{x}) \quad , \qquad (6.4)$$

with $|\mathbf{k}_n|^2 = k^2$, whatever the value of n and C_n, an arbitrary complex quantity. Solutions (6.4) corresponding to $N = 1,2,3$ represent two dimensional rolls, rectangles and hexagons respectively.

We now derive the second order approximation. The set of steady inhomogeneous equations in power ε^2 generated by inserting (6.1) in (1.3)-(1.4) is given by

$$\nabla^4 w^{(2)} = P^{-1}[\nabla_1^2(\mathbf{u}^{(1)} \cdot \nabla u^{(1)}) - \frac{\partial^2}{\partial x \partial z}(\mathbf{u}^{(1)} \cdot \nabla u^{(1)}) - \frac{\partial^2}{\partial y \partial z}(\mathbf{u}^{(1)} \cdot \nabla v^{(1)})] ,$$

(6.5)

$$\nabla^2 \theta^{(2)} + w^{(2)} = \mathbf{u}^{(1)} \cdot \nabla \theta^{(1)} .$$

(6.6)

The equations governing the behaviour of $u^{(2)}$ and $v^{(2)}$ are of no interest at this stage. The boundary conditions keep their form (2.5) - (2.6), with superscript (2) on each variable, with the exception of (2.7), replaced by

$$\frac{\partial^2}{\partial z^2} w^{(2)} - M^{(o)} \nabla_1^2 \theta^{(2)})_1 = M^{(1)} (\nabla^2 \theta^{(1)})_1 ,$$

(6.7)

subscript 1 means that the corresponding quantities are evaluated at the upper surface $z = 1$.

In order to obtain non-trivial solutions, Fredholm's existence theorem must be satisfied.

If $U = [w(x,y,z), \theta(x,y,z), \theta(x,y,z = 1)]$ denotes a solution of the set (6.5) - (6.7), the solvability condition requires that the adjoint solution U^* of the linear problem be normal to the inhomogeneous part of (6.5) - (6.7). Explicitly, the existence condition yields the following results:

$$C_{-N} = \ldots C_{-1} = C_1 = \ldots C_N = \pm 1\sqrt{2N} ,$$

(6.8)

$$M^{(1)} = 0 \text{ for } N = 1 \text{ (rolls) and } N = 2 \text{ (rectangles)} ,$$

(6.9)

$$M^{(1)} \neq 0 \text{ for } N = 3 \text{ (hexagons)} .$$

(6.10)

The solutions corresponding to $N > 3$ have not been observed experimentally and therefore are not analysed in the present work.

After the solvability condition has been established, it remains to calculate the second-order solution $w^{(2)}$ and $\theta^{(2)}$. They are of the form

$$(w^{(2)}, \theta^{(2)}) = (w_H^{(2)}, \theta_H^{(2)}) + (w_p^{(2)}, \theta_p^{(2)}) ,$$

(6.11)

where $(w_H^{(2)}, \theta_H^{(2)})$ are the solutions of the second-order homogeneous problem and are formally identical with the first-order solutions $w^{(1)}$ and $\theta^{(1)}$; $(w_p^{(2)}, \theta_p^{(2)})$ are particular solutions and are assumed to consist of two parts:

$$(w_p^{(2)}, \theta_p^{(2)}) = \sum_{m,1}(w_{m1}^{(2)}, \theta_{m1}^{(2)}) \phi_m \phi_1 + \sum_m (\bar{w}_m^{(2)}, \bar{\theta}_m^{(2)}) \phi_m ,$$

(6.12)

$(w_{m1}^{(2)}, \theta_{m1}^{(2)})$ are solutions of the non-homogeneous set (6.5), (6.6) and the homogeneous left-hand side of (6.7); $(\bar{w}_m^{(2)}, \bar{\theta}_m^{(2)})$ are solutions of the homogeneous parts of (6.5) and (6.6) and the boundary condition (6.7). The values of $M^{(2)}$ are obtained from the solvability condition of the third-order solutions. Despite the restrictions imposed by the solvability conditions, we are still faced with an infinite number of steady solutions, namely the set of all solutions consisting of rolls, rectangular and hexagonal cells. To remove this degeneracy, we shall examine the stability of the solutions with respect to infinitesimally small perturbations.

6.2. Stability of the steady solutions

In a first step, we consider a small disturbance with wavenumber $\tilde{k}$ identical with wavenumber k of the basic steady solution.

Let $\tilde{u}$ $(\tilde{u},\tilde{v},\tilde{w})$ and $\tilde{\theta}$ be the small-amplitude perturbations of the steady solution u (u,v,w) and θ with a time-dependence of the form $\exp(\sigma t)$. The disturbances $\tilde{u}$ and $\tilde{\theta}$ obey a set of equations that are readily derived by linearizing (1.2)-(1.4). The boundary conditions are still given by (1.5)-(1.7) with a tilda on every field variable. When the series expansions (6.1) for u and θ are used, one obtains equations with terms proportional to the powers of ε. This is why it is appropriate to develop similarly the growth rate σ and the disturbances $\tilde{u}$ and $\tilde{\theta}$ in terms of the same parameter ε:

$$\sigma = \sigma^{(0)} + \sum_{i=1} \varepsilon^{i} \sigma^{(i)} , \qquad (6.13)$$

$$\tilde{u} = \tilde{u}^{(0)} + \sum_{i=1} \varepsilon^{i} \tilde{u}^{(i)} , \qquad \tilde{\theta} = \tilde{\theta}^{(0)} + \sum_{i=1} \varepsilon^{i} \tilde{\theta}^{(i)} . \qquad (6.14)$$

The coefficients of each power of ε generated by inserting (6.13) and (6.14) in the equations satisfied by $\tilde{u}$ and $\tilde{\theta}$ must vanish identically. As a result, we are faced with a sequence of linear inhomogeneous equations setting up an eigenvalue problem for the growth rates $\sigma^{(i)}$, with $i = 0, 1, 2, \ldots$ We now discuss briefly the results.

Up to the second order, one finds

$$\sigma_{HH} = \varepsilon \sigma_{HH}^{(1)} + \varepsilon^2 \sigma_{HH}^{(2)} ,$$
$$\sigma_{RR} = 0 + \varepsilon^2 \sigma_{RR}^{(2)} , \qquad (6.15)$$

with real numerical values for the growth rates $\sigma_{AB}^{(i)}$. The subscript HH(RR) means that one examines stability of hexagons (rolls) with respect to disturbances taking themselves the form of hexagons (rolls). The domain of stability is obtained by requiring that σ_{HH} and σ_{RR} are negative. At this point of analysis, it is not possible to predict which structure, either rolls or hexagons is preferred because for both planforms, one can find ε values such that

$$\sigma_{HH} < 0 , \quad \sigma_{RR} < 0.$$

A definite answer can only be obtained by including a larger class of disturbances.

With that in mind, we introduce disturbances which are of different nature than the reference steady solutions: it is now found that, whatever ε,

$$\sigma_{RH} = \varepsilon \sigma_{RH}^{(1)} > 0 , \qquad (6.17)$$

$$\sigma_{HR} = \varepsilon \sigma_{HR}^{(1)} < 0 , \qquad (6.18)$$

from which follows that rolls are not stable.

To be sure that hexagons are stable, it remains to examine their stability with respect to disturbances of different wave numbers ($\tilde{k} \neq k$). Calculations show that there exist regions where

$$\sigma_{HH} = \sigma_{HH}^{o} + \varepsilon\, \sigma_{HH}^{(1)} + \varepsilon^{2}\, \sigma_{HH}^{(2)} < 0 ,$$
$$\sigma_{HR} = \sigma_{HR}^{o} + \varepsilon\, \sigma_{HR}^{(1)} + \varepsilon^{2}\, \sigma_{HR}^{(2)} < 0 ,$$
(6.19)

for every mode. Inequalities (6.19) determine a new domain of allowable values of ε. The actual zone of stability results from the intersection of the three stability domains defined by inequalities (6.15), (6.18) and (6.19). The numerical calculations, represented on fig. 3, show the existence of a narrow domain of stable hexagonal cells, characterized by wave-number larger than the critical wave-number $k^{c} \approx 2$ (see the horizontally dashed area).

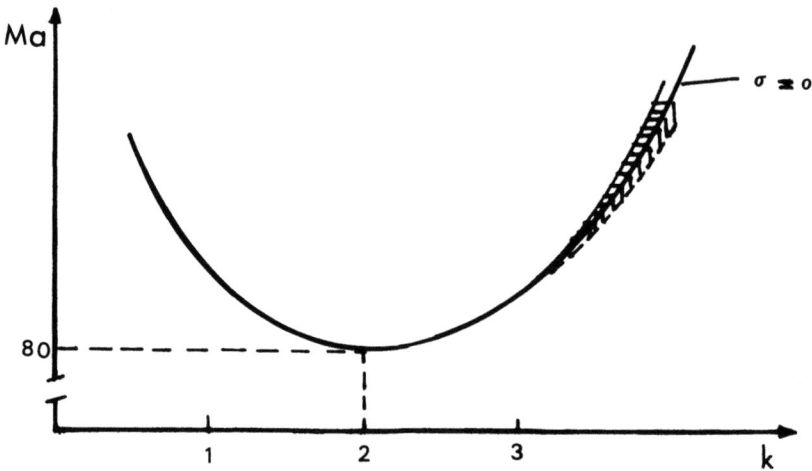

Fig. 3 - Regions of supercritical ▦ and subcritical hexagons ▥

In absence of gravity, the region of stable hexagons is very narrow. As shown by Cloot and Lebon (1984), the stability area enlarges when gravity effects are included and may cover half the area above the critical marginal curve. Another important result is worth mentioning. Figure 3 displays the presence of a extremely thin subcritical band where $Ma < Ma^{o}(k)$, i.e. where the linear theory predicts a state of rest.

It should be noticed that the existence of a subcritical instability in Marangoni's problem was also displayed by the energy method (Nield, 1964; Lebon-Garcia, 1980; Lebon-Cloot, 1982). But the latter predicts a much wider subcritical band : a factor 10 is found between the present and the energy treatments. This is not surprising if it is recalled that the energy method yields only sufficient conditions of stability and therefore leads to critical values which are much more smaller than the actual values.

7. THE ROLE OF A NONLINEAR TEMPERATURE DEPENDENCE OF THE SURFACE TENSION

Up to now, it is assumed that the surface tension is a linear function of the temperature. As stated earlier, this is surely a good model for many fluids but is not appropriate for describing surface tension effects in long chain dilute alcohols. Indeed, as shown by Vochten et al. (1973), a better model is provided by a parabolic function of the form

$$\xi(T) = \xi_m + \frac{b}{2} (T - T_m)^2 \qquad (7.1)$$

where ξ_m is the surface tension corresponding to the minimum of the curve $\xi - T$ (see fig. 4), b stands for

$$b = (\partial^2 \xi / \partial T^2)_m > 0 \;,$$

and is assumed to be a positive quantity, for a n-heptanol water solution, b is of the order of 10^{-6} N/m K^2 while ξ_m takes values ranging from 3.10^{-2} to 7.10^{-2} N/m.

Our objective is to determine whether there exist stable steady solutions in a Boussinesquian fluid layer when the motion is generated by a temperature gradient applied normally to the free surface. The governing equations are given by (2.2) - (2.7) but the kinematic boundary condition at the upper surface (2.7) must be substituted by (Cloot and Lebon, 1986),

$$\text{at } z = 1 : \partial^2 w / \partial z^2 = - M [(\nabla \theta)^2 + \theta \nabla_1^2 \theta + f \nabla_1^2 \theta] \;, \qquad (7.2)$$

where f is given by

$$f = \frac{T_1 - T_m}{\Delta T} \;, \qquad (7.3)$$

with T_1 the temperature at the upper face, ΔT the temperature difference between the limiting faces. The quantity M appearing in (7.2) is a new dimensionless number, defined by

$$M = \frac{(\partial^2 \xi / \partial T^2) (\Delta T)^2 d}{\rho \nu \chi} \quad (> 0) \;, \qquad (7.4)$$

M is the Marangoni number of second order. It plays a role comparable to Ma in the classical Marangoni problem and is related to the inverse of the radius of curvature of the function $\xi(T)$. M takes values between 100 and 1000 for aqueous alcohols when ΔT is of the order of a few degrees and d about one centimeter.

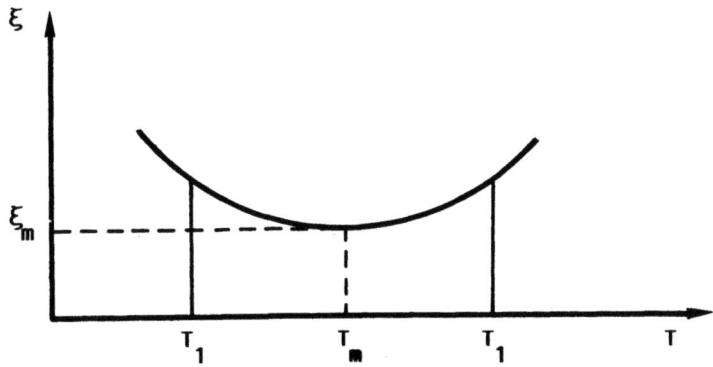

Fig. 4 - Surface tension versus temperature

For large values of f, it is justified to drop in (7.2) the nonlinear terms, but for f 1, the nonlinear contributions are dominant. For that range of f values, we shall apply the technique developed in the previous section. Like in § 6, the problem is solved in two steps. Firstly, one calculates the steady solutions of the eigenvalue problem set up by equations (2.2) - (2.6) and (7.2). Unfortunately, there exists an infinite number of mathematical admissible; one examines the stability of each solution with respect to infinitesimally small superimposed disturbances.

In view to simplify the mathematical resolution and since we are primarily concerned with the surface effects, we shall neglect the nonlinear contributions $\mathbf{u}.\nabla\mathbf{u}$ and $\mathbf{u}.\nabla\theta$ in the balance laws of momentum and energy respectively. The latter write then as

$$\nabla^4 w_{ss} = 0 \quad (0 < z < 1), \tag{7.5}$$

$$\nabla^2 \theta_{ss} + w_{ss} = 0 \quad (0 < z < 1), \tag{7.6}$$

where the subscripts ss refer to the steady solution.

As said above, equations (7.5) and (7.6) together with the appropriate boundary conditions has an infinity of solutions. This means that any particular cellular pattern, like rolls, squares, hexagons,... is mathematically admissible. However, observations show a tendency toward a single well defined cellular structure. In order to determine the preferred form of convection, one superposes infinitesimally small disturbances. The stability of solutions consisting of rolls, rectangles and hexagons has been investigated; other planforms, like pentagons, octogons, etc are not considered because no experimental evidence of their existence has ever been displayed.

The perturbations are assumed to be given by

$$\tilde{\mathbf{u}} = \mathbf{u}'(x,y,z) \exp(\sigma t), \tag{7.7}$$

$$\tilde{\theta} = \theta'(x,y,z) \exp(\sigma t), \tag{7.8}$$

where $\mathbf{u}'$ (u',v',w') and θ' are their amplitudes, σ is a real parameter whose sign determines the stability of the steady solutions.

The disturbances θ' and $\mathbf{u}'$ obey the linearized equations

$$\sigma\theta' + \mathbf{u}_{ss}.\nabla\theta' + \mathbf{u}'.\nabla\theta_{ss} = w' + \nabla^2\theta', \tag{7.9}$$

$$Pr^{-1}[\sigma\nabla^2 w' + \nabla_1^2(\mathbf{u}_{ss}.\nabla w' + \mathbf{u}'.\nabla w_{ss}) - \frac{\partial^2}{\partial x \partial z}(\mathbf{u}_{ss}.\nabla u' + \mathbf{u}'.\nabla u_{ss})$$

$$- \frac{\partial^2}{\partial y \partial z}(\mathbf{u}_{ss}.\nabla v' + \mathbf{u}'.\nabla v_{ss})] = \nabla^4 w', \tag{7.10}$$

plus two similar equations for the u', v' components which are of no use for our later developments. The boundary condition at the upper boundary involving the second order Marangoni number reads as

$$z = 1 : \frac{\partial^2}{\partial z^2} w' = -M\nabla_1.[(\theta_{ss} + f)\nabla_1\theta' + \theta'\nabla_1\theta_{ss}]. \tag{7.11}$$

To avoid costly and lenghly calculations, the Prandtl number is supposed infinite. From calculation and experimental observations, it is expected that $Pr = \infty$ is a reasonable hypothesis in the description of fluids with $Pr > 5$.

It is found (Cloot and Lebon, 1986) that only hexagonal cells can be stable. The regions in the M-k plane corresponding to negative values of the stability parameter σ, i.e. to stable hexagons, are represented by dashed areas on fig. 5.

It follows from fig. 5, that hexagonal planform is unconditionnally stable for values of M smaller than 800 for f = 0.1.

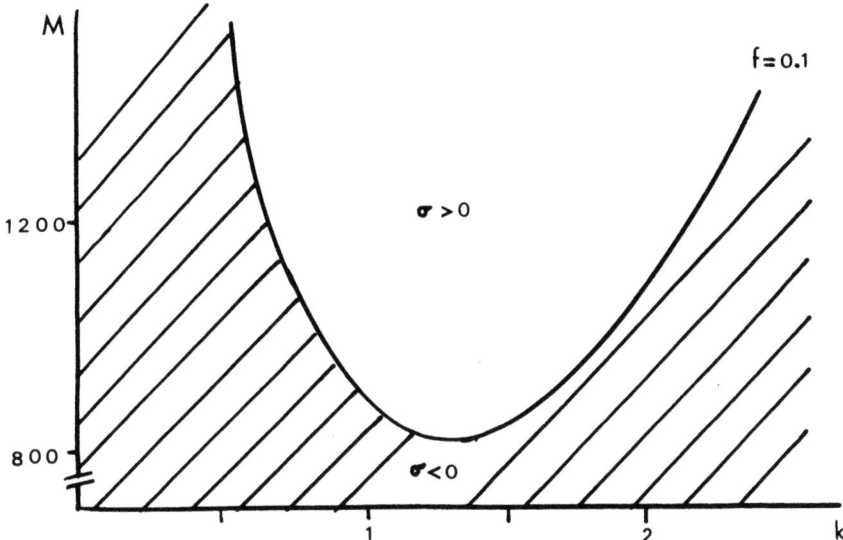

Fig. 5 - Regions of stable ▨ and unstable ☐ hexagons at a fixed value of f (f = 0.1)

8. CONCLUSIONS

Several questions concerned with Marangoni's instability are surveyed. More specifically, one has examined what roles do play in determining stability some features like a surface deformation, a temperature-dependent viscosity and a surface tension depending nonlinearly on the temperature. The Gorkov-Malkus-Veronis iterative method is also revisited in order to study the behaviour of the flow beyond the point of linear instability.

The main results which are obtained can be summarized as follows :

1. instability can be driven, even in absence of buoyancy effects, by surface tension gradients,

2. the role played by a surface deformation on the onset of convection is generally minute and therefore can be neglected in most applications,

3. the results may change appreciably by allowing the viscosity to depend on the temperature,

4. only hexagonal convective cells seem to be stable if the surface tension is a quadratic function of the temperature,

5. the same conclusion is reached for a Boussinesquian fluid with a surface tension depending linearly on the temperature. This result is obtained by performing a nonlinear approach à la Gorkov-Malkus-Veronis,

6. a narrow band of subcritical instability is displayed.

Of course, we are fully aware that the above modelling leaves open many questions. Among them, it may be asked what happens when the layer, instead of horizontal infinite extent, is confined in finite boxes. What are the consequences of application of a horizontal temperature gradient instead of a vertical one ? This problem is of crucial importance in the study of crystal growth mechanisms. Another question is related to the principle of exchange of stability. Although it has been taken for granted in the present work, it should be demonstrated in each individual case. Up to now, Marangoni's instability in two-component mixtures or in electrolytes has only received partial treatments (Castillo and Velarde, 1982). It should also be worth to examine the role of surface tension driven convection in nonnewtonian fluids, like polymers. Of course, this problem is made difficult by its full nonlinear character. More generally, a full theory is still lacking. In that respect, the actual progresses in the bifurcation theory opens promising ways towards this objective.

Interesting themes of research in Marangoni instability are also provided by the hydrodynamical theory of fluctuations. Similarities between second-order equilibrium phase transition and non-equilibrium instabilities have been widely emphasized in the literature (for a review see Jou, 1982). Although an impressive lot of works has been devoted to the Rayleigh-Bénard problem, a few rare and recent papers are concerned with surface-tension driven instabilities.

REFERENCES

BOUKARY, S. and LEBON, G., 1986, A comparative analysis of binary fluid mixtures by extended thermodynamics and the kinetic theory, Physica, 1986.

BUSSE, F., 1978, Non-linear properties of thermal convection, Rep. Prog. Phys., 41 : 1929.

CASTILLO, J. and VELARDE, M., 1982, Buoyancy-thermocapillary instability : the role of interfacial deformation in one and two-component fluid layers heated from below and above, J. Fluid Mech., 125 : 463.

CERISIER, P., JAMOND, C., PANTALONI, J. and CHARMET, J.C., 1984, Déformation de la surface libre en convection de Bénard-Marangoni, J. de Physique, 45 : 405.

CLOOT, A. and LEBON, G., 1984, A non-linear stability problem of the Bénard-Marangoni problem, J. Fluid Mech., 145 : 447.

CLOOT, A. and LEBON, G., 1985, Marangoni instability in a fluid layer with variable viscosity and free interface, in microgravity, PCH Phys.- Chem. Hydr., 6 : 453.

CLOOT, A. and LEBON, G., 1986, Marangoni convection induced by a nonlinear temperature-dependent surface tension, J. de Physique, 47 : 23.

DAVIS, S.H. and HOMSY, G., 1980, Energy stability theory for free surface problems : buoyancy thermocapillary layers, J. Fluid Mech., 198 : 527.

GORKOV, L., 1957, Steady convection in a plane liquid layer near the critical point, Sov. Phys. JETP, 6 : 311.

JOU, D., 1982, in Stability of Thermodynamic Systems, Casas-Vazquez, J. and Lebon, G., Eds., Springer Lect. Notes in Phys., 164 : 138.

KRASKA, J. and SANI, R., 1979, Finite amplitude Bénard-Rayleigh convection, Int. J. Heat Mass Transfer, 22 : 535.

LEBON, G. and PEREZ-GARCIA, C., 1980, Study of surface tension effects in thermal convection by variational methods, Bull. Cl. Sci. Acad. Roy. Belgique, 66 : 520.

LEBON, G. and PEREZ-GARCIA, C., 1981, Convective instability of a micropolar layer by the method of energy, Int. J. Engng. Sci., 9 : 1321.

LEBON, G. and CLOOT, A., 1982, Buoyancy and surface-tension driven instabilities in presence of negative Rayleigh and Marangoni numbers, Acta Mech., 43 : 141.

LEBON, G., 1982, in Stability of Thermodynamics Systems, Casas-Vazquez, J. and Lebon, G., eds., Springer Lect. Notes in Phys., 164 : 41.

LEBON, G. and CLOOT, A., 1984, Bénard-Marangoni instability in a Maxwell-Cattaneo fluid, Phys. Lett., 105A : 361.

LEBON, G. and BOUKARY, M.S., 1985, Is the principle of objectivity in violation with the kinetic theory, Phys. Lett., 107A : 295.

LINDE, H., 1981, In Proc.Euromec. Coll., 138 : 117, Karlsruhe Univ. Press, Karlsruhe.

MALKUS, W. and VERONIS, G., 1958, Finite amplitude cellular convection, J. Fluid Mech., 4 : 225.

NIELD, D., 1964, Surface tension and buoyancy effects in cellular convection, J.Fluid Mech., 19 : 341.

PEARSON, J.R.A., 1958, On convection cells induced by surface tension, J. Fluid Mech, 4 : 189.

ROSENBLAT, S., DAVIS, S.H. and HOMSY, G., 1982, Nonlinear Marangoni convection in bounded layers, J. Fluid Mech., 120 : 91 and 120 : 123.

SCANLON, J. and SEGEL, L., 1967, Finite amplitude cellular convection induced by surface tension, J. Fluid Mech., 30 : 149.

SCHLUTER, A., LORTZ, D. and BUSSE, F., 1965, On the stability of steady finite amplitude convection, J. Fluid Mech., 23 : 129.

SCRIVEN, L. and STERLING, C., 1964, On cellular convection driven by surface-tension and surface viscosity, J. Fluid Mech., 21 : 321.

SMITH, K.A., 1966, On convective instability induced by surface-tension and surface viscosity, J. of Fluid Mech., 24 : 401.

STRAUGHAN, B. and FRANCHI, F., 1984, Bénard convection and the Cattaneo law of heat conduction, Proc. Roy. Soc. Edinburgh.

VIDAL, A. and ACRIVOS, A., 1966, Nature of the neutral state in surface-tension driven convection, Phys. Fluids, 9 : 615.

VOCHTEN, R., PETRE, G. and DEFAY, R., 1973, Study of the heat of reversible adsorption at the air-solution interface, J. Colloid. Sci., 42 : 310.

INTERFACIAL EFFECTS ON THE ONSET OF CONVECTION

IN HORIZONTAL LIQUID LAYERS

Gabbita Sundara Rama Sarma

Institute for Theoretical Fluid Mechanics
German Aerospace Research Establishment (DFVLR)
Bunsenstr. 10, D-3400 Göttingen, Federal Republic of Germany

1. INTRODUCTION

Horizontal liquid layers are prone to convective instability when subjected to appropriate vertical gradients of temperature and/ or concentration. Whenever fluid-fluid interfaces are present the two potentially destabilizing mechanisms inducing convective instability in the configuration are buoyancy and surface tension. The onset of convection in the configuration depends also on the boundary conditions imposed, i.e., on the physical properties of the adjoining media, more so, if one of the driving mechanisms is itself of interfacial origin. The two destabilizing mechanisms in the configuration in general reinforce each other, i.e., the adverse (temperature/concentration) gradient necessary to induce instability when both the mechanisms are present is smaller than that required when only one of them is involved. This well known result due to Nield[1], however, turns out to be critically dependent on the detailed characterization of the two-fluid interface and the associated boundary conditions.

Most of the work on this classical configuration specified the interface as a perfectly planar boundary. Relaxing this quasi-rigidity (and more so by including surface viscosity and elasticity)[2,3] of the interface can radically alter the stability characteristics of the configuration. Allowing duly for interfacial disturbances can considerably affect even the interaction between the two destabilizing mechanisms under appropriate circumstances some of which may be relevant to a low-gravity environment[4]. Depending on the other boundary conditions and the relevant parameter ranges even a possible opposing tendency in this interaction[5] can be demonstrated. Such a possible counteraction between inherently *stabilizing* agencies such as Coriolis and Lorentz fields is well known[6]. But here the ambivalence of intrinsically *destabilizing* mechanisms is brought about by the interfacial conditions. The physical explanation for this a priori unexpected result is that the two mechanisms tend to induce flows of opposite sense directly under the two-fluid interface. Domains of dominance for the two mechanisms can indeed be distinguished[5].

The similarities between the cellular morphologies in crystal growth[7,8] and in the Bénard configuration have long been recognized but there is as yet no unified theoretical framework encompassing the two problems. But even here we find departures from the classical Bénard case when the upper boundary is formed by the solidification front of a horizontal, molten liquid layer below[9]. Effects of interfacial waviness and imperfect thermal boundary conditions play an important role also in this context.

Although the Bénard configuration has been studied extensively for nearly a century there are still unresolved and new aspects of mathematical and physical interest on the relevant stability problems[10,11,12]. The configuration serves as a convenient and tractable model for theoretical and experimental investigations and has potential applications in several fields. The currently active microgravity research programs[7,8,13,14] have also given a fresh impetus to this area. References 10 - 14 survey the developments in the field from different points of view and will be cited for appropriate details in the following. The present contribution deals with a rather specific aspect of the interfacial characterization and its effect on the onset and patterns of convection. Since the detailed reports on the pertinent analyses and experiments are to be found in the literature to be cited, we shall restrict ourselves here to a summary of the relevant results in § 3 and § 4 and proceed to some comparative observations in § 5.

2. NOMENCLATURE

2.1. *Symbols*

B	Magnetic induction field (B_0: magnitude of B-field)
K	Thermal conductivity
c_0	Specific heat
d	Mean thickness of the liquid layer
g	Acceleration due to gravity (g_0: terrestrial value)
k_x, k_y	Disturbance wave numbers in x,y (horizontal) directions
q	Heat transfer coefficient at the two-fluid interface
β	Coefficient of thermal expansion
γ	Electrical conductivity
ΔT	Temperature difference ($T_0 - T_1$)
κ	Thermal diffusivity $K/(\rho c_0)$
λ	Disturbance wavelength $2\pi/\sqrt{k_x^2 + k_y^2}$

μ   Dynamic viscosity
ν   Kinematic viscosity
ρ   Density
σ   Interfacial energy at the two-fluid interface
Ω   Angular speed of rotation

Subscripts: 0, 1 denote bottom and top of liquid layer
          : f, m denote frozen and molten phases.

2.2. Dimensionless perturbation amplitudes

W : Vertical component of velocity; X : Vertical component of current density; Z : Vertical component of vorticity; ε : Departure from criticality; ζ : Interfacial deflection; Θ : Temperature.

2.3. Dimensionless parameters

Bo = $\rho g d^2/\sigma$ (Bond); Cr = $\kappa\mu/\sigma d$ (Crispation/Capillary); $D_1 = d_f/d_m$ (Thickness ratio: frozen/ molten layer); $D_0 = d_b/d_m$ (Thickness ratio: bottom block/ molten layer); Fl = (W''' - 3 a^2 W')/W' (Flow indicator: Eq. 19); $K_0 = K_m/K_s$ (Ratio of thermal conductivities: melt/ bottom solid block); Ma = $d|\Delta T(d\sigma/dT)|/\kappa\mu$ (Marangoni); Nu = qd/K (Nusselt/Biot); Q = $B_0^2 d^2\gamma/\mu$ (Chandrasekhar); Ra = $g\beta\Delta T d^3/\nu\kappa$ (Rayleigh); Ta = $2\Omega d^2/\nu$ (Taylor); $a = 2\pi d/\lambda$ (Disturbance wave number); ξ : Interfacial curvature effect (Eq. 25); Π : Non-Boussinesq property variation.

3. CONVECTIVE INSTABILITY UNDER A TWO-FLUID INTERFACE

3.1. Formulation of the problem

Here we consider the basic problem of thermal instability driven both by gravity and capillarity in a horizontal liquid layer under various boundary conditions. In view of the practical interest in controlling instability in the configuration we include the effects of uniform rotation about a vertical axis and a uniform vertical magnetic field[4,5]. The configuration consists of an infinite, horizontal, imcompressible liquid layer of mean thickness d rotating about a vertical axis at a constant angular speed Ω. The lower and upper horizontal boundaries are nominally at constant temperatures T_0, T_1 respectively ($T_0 > T_1$) while a uniform magnetic induction field of strength B_0 may also be applied transverse to the layer. The layer is bounded at least on one side by an ambient gas and is subjected to a variety of thermal and electromagnetic boundary conditions. Fig. 1 is a schematic representation of the configuration and the boundary conditions (b.c.) **1, 2, 3** (circled numbers in Figures) considered.

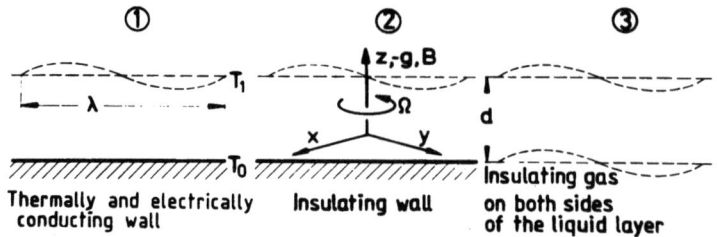

Figure 1. Basic Bénard-Marangoni configuration under b.c. **1, 2** and **3**.

The conditions for the onset of convection in this configuration can be formulated as a problem of[6] stability of the initial, linear temperature profile across the layer. An earlier asymptotic investigation[15] indicates that the incipient instability in the configuration is likely to be in the stationary mode even if oscillatory modes are admitted by the system under some circumstances. Within the standard framework of the normal modes procedure[6] the special features associated with the boundary conditions at a wavy two-fluid interface[2,3,4,5] are of critical importance in the present case. In contrast to earlier investigations, we specifically allow for non-zero interfacial deflection, i.e., a *finite* surface tension σ in addition to a finite variation of σ with temperature in formulating the pertinent stress and thermal balance conditions. In dimensionless terms the aforesaid stability problem can be stated[4,5] as follows with $DF \equiv F'$ denoting d.(dF/dz) for any F(z), multiple primes denoting multiple derivatives.

$$(D^2-a^2)^2 W - Ta.DZ - Q.D^2W = Ra.a^2\theta \tag{1}$$

$$(D^2 - a^2)\theta + W = 0 \tag{2}$$

$$(D^2 - a^2) Z + Ta.DW + Q.DX = 0 \tag{3}$$

$$(D^2 - a^2) X + DZ = 0 \tag{4}$$

The boundary conditions(b.c.) are for cases **1, 2, and 3** :

$$W(1) = 0 = DZ(1) = X(1) \tag{5}$$

$$-Nu.W''(1) + Ma.a^2 D\theta(1) = 0 \tag{6}$$

$$Nu.Cr. \{-W'''(1) + 3a^2 W'(1)\} + a^2(Bo + a^2)\{D\theta(1) + Nu.\theta(1)\} = 0 \tag{7}$$

and for Nu = 0 (6) and (7) are to be replaced by

$$D\Theta(1) = 0 \tag{8}$$

$$(Bo + a^2)\{W''(1) + Ma.a^2\Theta(1)\} \\ + Ma.Cr. \{-W'''(1) + 3a^2 W'(1)\} = 0 \tag{9}$$

whereas at the lower boundary (individually different) for
case **1**

$$W(0) = 0 = DW(0) = Z(0) = \Theta(0) = DX(0) \tag{10}$$

case **2**

$$W(0) = 0 = DW(0) = Z(0) = D\Theta(0) = X(0) \tag{11}$$

and for case **3** the same conditions as (5) - (9) hold at z = 0.

Equations (1) - (4) describe, respectively, the transport of momentum, energy, and current density in a stationary perturbed state of the basic configuration. The boundary conditions for cases **1, 2, 3** are specifications of the thermal and electromagnetic characteristics of the adjoining media in each case. The ambient gas at z = d in cases **1** and **2** and at z = 0, d in case **3** is an electrically perfect insulator. Moreover, at the interface the heat transfer characteristics are specified (for simplicity)[2,4] via a constant heat transfer coefficient q. In cases **1** and **2** the layer is bounded at z = 0 by a solid wall which is considered a thermally and electrically perfect conductor in case **1** and insulator in case **2**. In contrast to earlier treatments with a flat two-fluid interface, the boundary conditions are applied here at a disturbed wavy interface with an a priori unknown amplitude ζ. The normal and tangential stress balance conditions at the two-fluid interface after Taylor expansion about the mean (undisturbed) interface lead[2,3,4,5] to the following relations between the dimensionless amplitudes W, Θ and ζ.

Normal stress balance:

$$Cr.\{W'''(1) - 3 a^2 W'(1)\} = a^2 (Bo + a^2)\zeta \tag{12}$$

Tangential stress balance:

$$a^2 Ma.\{\Theta(1) - \zeta\} + W''(1) + a^2 W(1) = 0 \tag{13}$$

$$DZ(1) = 0 \tag{14}$$

The thermal boundary condition at the wavy interface also yields a relation between θ and ζ.

$$D\theta(1) + Nu.\theta(1) = Nu.\zeta \tag{15}$$

Equations (12), (13), (15) have been used to eliminate ζ from the other b.c.

In view of the large number of parameters involved we must restrict our attention to some "special" situations with representative orders of magnitude for the above parameters. Results for various special cases have been reported earlier[4,5] and discussed in the light of other results in the literature. In illustrating the significance of the interfacial curvature (Cr ≠ 0) and of gravity waves (Bo ≠ 0) in contrast to the cases Cr = Bo = 0, the small values of Cr = $10^{-3}, 10^{-4}$; Bo = 0.05, 0.01 have been found[2,3,4,5] particularly instructive. For the chosen values of Cr the corresponding Bo-values can be estimated for different classes of substances and various g-levels since it turns out that the dimensionless ratio Bo/Cr = $gd^3/\kappa\nu$ = Ra/(βΔT) is the key parameter determining the long wave stability characteristics of the system. The values chosen for Bo = 0.01, 0.05 with Cr = 10^{-3}, 10^{-4} may also be regarded as representative estimates for experiments in an earth laboratory on thin liquid layers (d ~ mm) and experiments in an orbital laboratory with thicker layers (d ~ cm, g ~ $10^{-4} g_0$, the nominal gravity level quoted for Spacelab[14]). The results to be discussed are based on general analytical solutions and similar data can readily be generated for other "representative" parameter values.

3.2. *Neutral stability characteristics*

The dependence of the stability characteristics on Nu, Ta and Q has been discussed in detail elsewhere[4] with special reference to small Ra demonstrating the importance of the long wave modes (a → 0) for the present configuration. Asymptotic analysis shows that Ma ∝ a^2, as a → 0, for Bo = 0, Cr ≠ 0, whereas Ma → Ma_1 ≠ 0 when Bo ≠ 0 and Cr ≠ 0. Figs. 2, 3 show some typical neutral stability curves for the three boundary conditions 1, 2 and 3. B.c. 2 is qualitatively between 1 and 3 sharing some of the features of 1 and 3 in different parameter ranges. We restrict ourselves here to the least stable situation[4,5], Nu = 0.

3.3. *The flat and curved interface*

Figs. 2 and 3 show that the flat interface curves (Cr = 0) differ drastically from those of the wavy interface (Bo ≠ 0, Cr ≠ 0). The latter show the possibility of *two local* minima in Ma (e.g. for b.c. 1, Cr = 10^{-4}, Ta = Q = 0.1, at a → 0 and a ≃ 2). The important observation for the later discussion is that in contrast to the case[2] Bo = 0, Cr ≠ 0 there *does exist* a non-

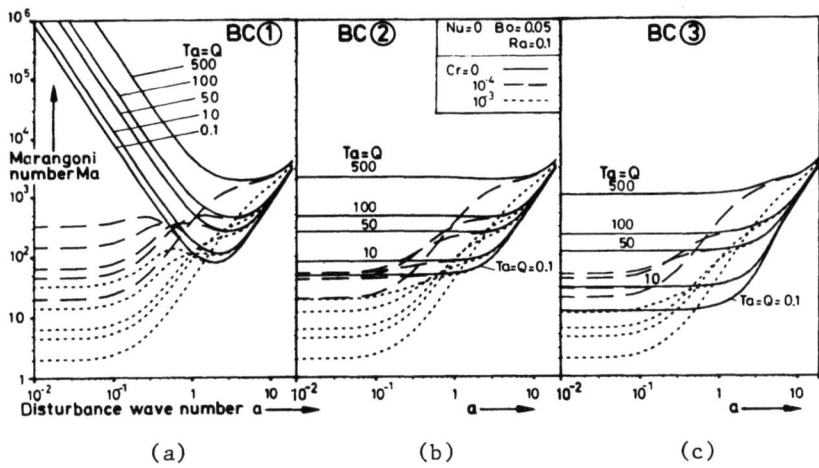

Figure 2. Neutral curves for thermocapillary convective instability in a horizontal liquid layer under b.c. 1, 2 and 3.

zero critical Marangoni number $Ma = Ma_c$ when *both* Bo and $Cr^{3,4}$ are nonzero. This Ma_c is given by the absolute minimum in Ma along the respective neutral stability curves for given Ta, Q, Cr, Nu, Bo and Ra. The corresponding critical wave number a_c may tend to zero in certain parameter ranges (cf. Fig. 2(a) b.c. 1 at large Ta, Q and Figs. 2(b),(c), 3(a)). Thus at nonzero gravity the convective instability sets in only above a finite threshold value Ma_c even when the two-fluid interface is wavy[2] with critical wavelengths λ_c being finite or infinite depending on the parameter ranges and the boundary conditions[4].

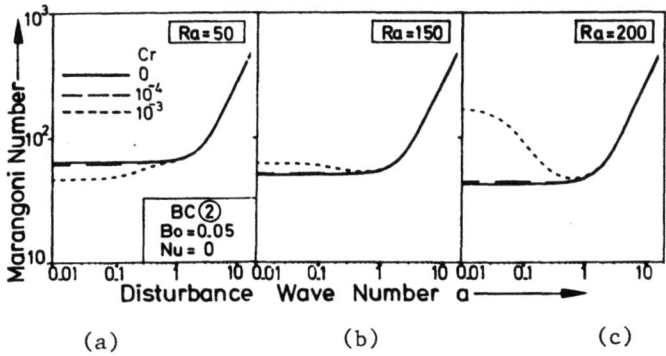

Figure 3. Neutral curves for thermocapillary convective instability in a horizontal liquid layer under b.c. 2.

In Fig. 2 we also see that Ma_c increases with Ta and Q in general, indicating a stabilizing tendency of the Coriolis and Lorentz forces. However, a monotonic increase in Ma_c is limited to a certain range of Ta and Q if the fluid interface is wavy (cf. b.c. 2, 3 ; $Cr = 10^{-3}$, 10^{-4}; Ta = Q = 50 vs. 100) and

Ma_c may decrease with Ta and Q. Thus, the anticipated stabilizing action of these agencies has also to be optimized within an appropriate parameter range especially for thin liquid layers[4]. Since the dependence of the stability characteristics on Ta and Q has been reported in detail elsewhere[4,5] we shall use only the case of negligible Coriolis and Lorentz effects (Ta = 0.1 = Q) and take Nu = 0 for purposes of the following discussion.

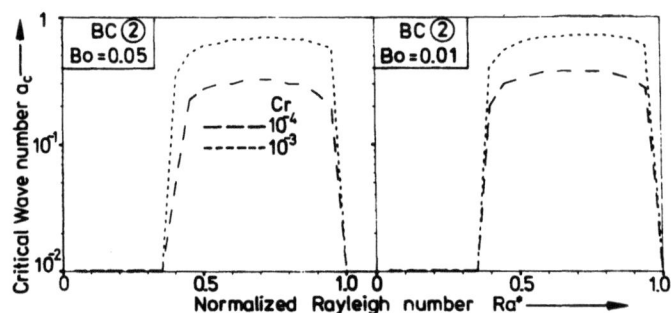

Figure 4. Variation of critical wave number a_c with normalized Rayleigh number for a horizontal liquid layer under b.c. 2.

Fig. 3 illustrates the effect of buoyancy (Ra) on the neutral curves with b.c. 2 as a case intermediate between b.c. 1 and 3. We observe that Ma_c at Ra = 150 can be *higher* than that at Ra = 50 or at Ra = 0.1 (Fig. 2(b)) for Cr ≠ 0. Further, the stable domain at Cr ≠ 0 can be larger than at Cr = 0 (cf. Fig. 3(b),(c)). Moreover, the critical wavelength λ_c can become finite even under b.c. 2 as Ra increases above a threshold and buoyancy takes hold (cf. Fig. 3(b),(c), Cr = 10^{-3}). The full range of variation of a_c with Rayleigh number under b.c. 2 is illustrated in Fig. 4 where $Ra^* = Ra/Ra_0$, with $Ra_0 = Ra_c$ for Ma = 0. Since Ma_c can occur at $a_c = 0$ in certain cases an asymptotic analysis was carried out[4] in the limit $a \to 0$. Thus, we get the following limits Ma_1 along the respective neutral stability curves for Nu = 0, Ta = 0 = Q, Ra ~ 0 (1).

B.c. 1 : $Ma_1 = f_1(Bo/Cr)$ (16)
where $f_1(Bo/Cr) = (2/3)Bo/Cr$

B.c. 2 : $Ma_1 = (960-3Ra)/f_2(Ra,Bo/Cr)$ (17)
where $f_2(Ra,Bo/Cr) = [20+(1440-12Ra)Cr/Bo]$

B.c. 3 : $Ma_1 = 12(1-Ra/120)/f_3(Ra,Bo/Cr)$ (18)
where $f_3(Ra,Bo/Cr) = [1-Ra\, Ma_1(Cr/Bo)^2]$

The above results show the role of the ratio Bo/Cr in determining the long wave stability characteristics of the configuration when both Bo and Cr are nonzero. We note that this ratio is independent of not only ΔT but also σ and hence its influence is not limited by the size of σ. As Bo/Cr → 0, Ma_1 → 0 and at *zero gravity* there is strictly *no critical* Ma below which the liquid layer is stable against arbitrary long wave disturbances[2]. In general, at smaller Bo/Cr the long wave modes become more important[3,4]. Numerical results in Figs. 2, 3(a) agree well with the asymptotic results in the limit $a \to 0$.

3.4. *The Bénard-Marangoni interaction*

Based on the general information gathered from the neutral stability curves such as Figs. 2, 3 we now discuss in the following only the variation of the *critical Marangoni number* Ma_c with respect to Ra under different boundary conditions, in order to study the coupling between the two potentially destabilizing agencies due to thermocapillarity and bouyancy.

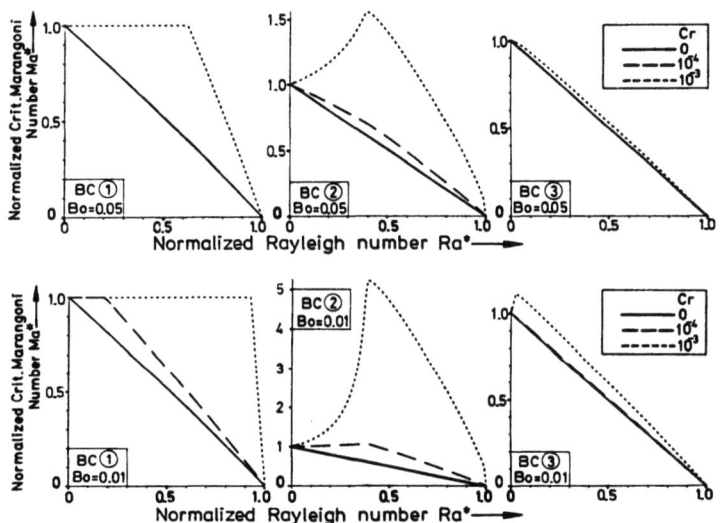

Figure 5. Variation of Ma^* with Ra^* under b.c. **1, 2** and **3**.

For this purpose we normalize Ma_c(Ra ≠ 0) and Ra respectively[4,5,12] in terms of Ma_0 (Ma_c at Ra = 0) and Ra_0 (Ra_c at Ma = 0) and set $Ra/Ra_0 = Ra^*$, and $Ma_c/Ma_0 = Ma^*$. Fig. 5 illustrates the correlation between Ma^* and Ra^* at Ta = 0.1 = Q, i.e., the onset of the pure Bénard-Marangoni convective instability (under negligible Coriolis and Lorentz fields) when we take both the effects of gravity waves (Bo $\neq$ 0) and of finite interfacial tension (Cr $\neq$ 0) at the two-fluid interface into account. We notice that under b.c. **1, 2**, and **3** a linear correlation, i.e., a "tight coupling" in the sense of mutual support between the two destabilizing agencies[1], obtains only for the flat interface (continuous curves with Cr = 0). At Cr $\leq 10^{-4}$ the above correlation holds approximately for Bo = 0.05 under b.c. **1** and **3** but not for Bo = 0.01.

The drastic departure from the strictly monotonic correlation between Ra^* and Ma^* is to be clearly seen under b.c. **2** at Cr = 10^{-3} for Ra^* less than some threshold value ($\simeq$ 0.39), wherein we find that Ma^* in fact *increases* with Ra^*. Ma^* stays well above unity for Cr = 10^{-3} over a considerable range of Ra^* before decreasing ultimately to zero as $Ra^* \to 1$ with nonzero a_c (cf. Fig. 4) in the "downhill" part. Similar behavior is also to be seen for b.c. **3** at Bo = 0.01. Thus, under b.c. **2** and **3**, there can be a regime in which an increase in buoyancy up to a certain threshold Ra_t can indeed *delay* the onset of convective instability beyond the critical gradients required at zero gravity. The threshold Ra_t is higher for lower Bo/Cr. Although too small to be seen on the plot for b.c. **1** there exists a similar regime near $Ra^* \to 0$. But the more obvious finding in the case of b.c. **1** is that bouyancy need not *reduce* the critical Marangoni number until Ra exceeds a certain threshold value. At the other extreme as $Ra^* \to 1$, i.e., when buoyancy dominates, our numerical results agree with the predictions of the asymptotic energy stability analyses[16,17] for b.c. **1** in the limit of small Cr which indicate a possible *stabilizing* tendency of the interfacial curvature. This is even more pronounced under b.c. **2** at larger Ra^* (cf. Fig. 3(b),(c), Cr = 0 vs. Cr = 10^{-3}, 10^{-4}). Thus, the present results spanning the entire range $0 \leq Ma^*, Ra^* \leq 1$ clearly indicate the dual role of the wavy interface. Whereas the layer tends to be destabilized by Cr $\neq$ 0 in the long wave limit, it is restabilized by Bo $\neq$ 0 (Ra $\neq$ 0). The quantitative significance of nonzero Bo and Cr is far greater in the low-buoyancy limit, say, in a low-gravity environment (real or simulated) and in thin liquid layers[4,5].

This ambivalence in the buoyancy-capillarity interaction may partly be a plausible explanation for the quicker onset of convection in the Apollo demonstration experiments[18]. Whereas in the ground experiment onset of convection in the *same thin* layer could have been delayed somewhat by the counteracting buoyancy effect, there could be no such benefit from buoyancy at g ~ $10^{-8} g_0$ aboard Apollo. At very small Bo/Cr = $gd^3/\kappa\nu$, as the asymptotic results show, Ma_c is also small. Thus, for given layer thickness, quicker onset of convection than on ground is indeed likely at such low gravity levels. Side walls and practical lower bounds on observing incipient convection could lead to higher Ma_c and finite cell size. Since both the ground and space experiments were also presumably well in the supercritical regime quantitative theoretical comparisons do not seem tenable here. But physically, even in that regime it is likely that evolution of the finite cells observed took longer in the ground experiment than in space due to the counteraction contribution of buoyancy in the former. Furthermore, the undercurrents associated with the side wall effects, nonuniform thickness of the layer in space[18], together with some heat transfer to the environment (Nu ≠ 0), might also have changed the pertinent a_c and Ma_c to begin with.

3.5. The counteraction mechanism

The possible counteraction between the two inherently destabilizing agencies of capillarity and buoyancy can be associated with the opposite flows[2,5] induced in the immediate vicinity of the two-fluid interface where the thermocapillary tractions operate. In view of the kinematic condition for a stationary, neutral mode, W(1) vanishes at the fluid interface. Therefore, the fluid immediately below moves towards the interface if W'(1) is negative and away from it if W'(1) is positive. Consequently, the flow is upward below the troughs of an interfacial wave and downward below the crests if ζ and W'(1) are of the same sign.

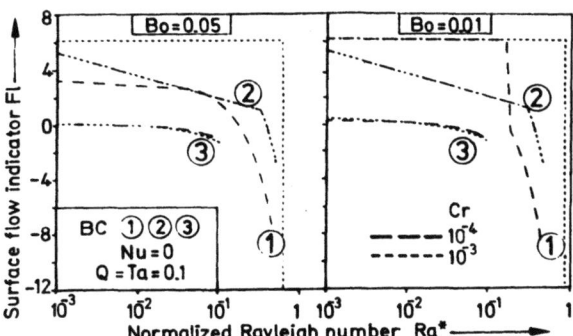

Figure 6. Variation of surface flow indicator Fl with Ra^* under b.c. **1, 2** and **3**.

In Fig. 6 we show the values of the surface flow indicator Fl (cf. Eq. 12) given by

$$Fl = [W'''(1) - 3 a^2 W'(1)] / W'(1) \tag{19}$$

under the three boundary conditions. We find that Fl is positive as $Ra^* \to 0$ and goes through zero to negative values as Ra^* exceeds a certain finite threshold value. The trend is clearly visible for b.c. 1 and 2 but is present numerically in b.c. 3 as well. The Fl-sign reversal point in b.c. 2 and 3 is close to the change over of λ_c to finite values. But the results for b.c. 1, wherein $a_c \simeq 2.0$ even for $Cr = 10^{-4}$, $Bo = 0.05$ as $Ra^* \to 0$ while Fl changes sign at $Ra^* \simeq 0.2$ also with $a_c \simeq 2.0$, show that the dominant regime of the respective driving mechanism: capillarity or buoyancy, is to be identified with the flow reversal, rather than the wavelength.

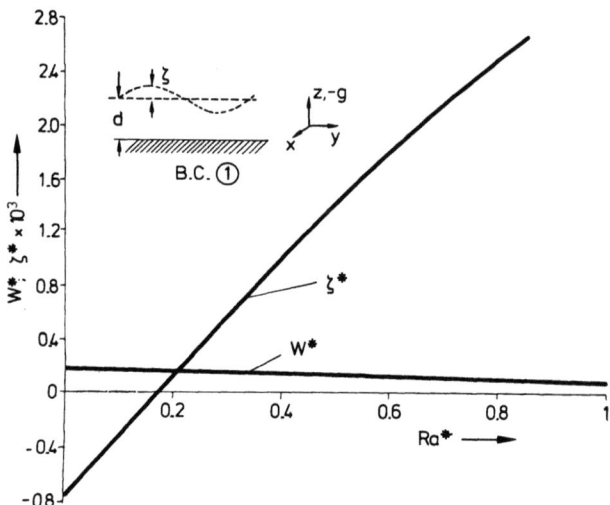

Figure 7. Variation of vertical velocity amplitude W^* below the interface ($z/d = 0.995$) and deflection amplitude ζ^* of a critical disturbance wave ($\lambda_c \simeq \pi d$) with bouyancy (Nu = 0, Bo = 0.05, Cr = 10^{-4}).

This point is clearly demonstrated in Fig. 7 wherein $W^* = W$ at $z/d = 0.995$ and ζ^* are plotted as functions of Ra^* for b.c. 1 at Cr = 10^{-4}, Bo = 0.05, Nu = 0. We see that in fact the flow just below the wavy interface is upward ($W^* > 0$) under a trough ($\zeta^* < 0$) as $Ra^* \to 0$ and under a crest ($\zeta^* > 0$) as Ra^* approaches 1. We see from the asymptotic result (Eq. 16) that $Ma_1 = 333.3$ for the above Bo, Cr (for finite Ra). Hence $Ma_c \simeq 80$ with $a_c \simeq 2$ (cf. Fig. 2(a))[1,12] over the entire range $0 \leq Ra^* \leq 1$. This shows unambiguously that the two mechanisms dominate in distinct domains while *the pertinent critical wavelength remains finite*. Thus, upward flows under crests and troughs respecively are indeed associated with buoyancy dominated and surface-tension dominated regimes, as can be concluded also from several earlier experiments[14,19,20]. The exact value of Ra^* at $\zeta^* = 0$ for b.c. 1 checks well with that of Cerisier et al[20,21] (although their parameter values are different). This agreement is however not fortuitous since the case in Fig. 7 corresponds precisely to the *finite critical wavelength* case considered wherein the *finite deflections* could be observed[20,21] in the experiments as well. The results in § 3.2 - § 3.4 encompass *all disturbance modes* including the long wave mode possible for low Bo/Cr. In the latter case observation of (slightly supercritical) interfacial deflections and the flows thereunder may be more difficult than in the finite λ_c case, which (as per our asymptotic results) occurs for larger Bo/Cr. Under the experimental conditions of Cerisier et al[20] Bo/Cr is large enough so that only the mode close to that of Nield[1] needs to be considered. Here the influence of Bo and Cr is marginal and hence the above mentioned agreement is in order. In our analysis we include the long-wave mode ($\lambda_c \to \infty$, $Ma_c = Ma_1 \neq 0$), discovered by Smith[3], for which the influence of Bo and Cr is significant. This basic aspect also leads to the departures shown in Fig. 5 from the well known "tight coupling" of the destabilizing mechanisms[1]. Besides, λ_c can be finite even under b.c. 2 and 3 with some heat transfer to the ambient gas[4]. Therefore, the sign reversal of Fl is rightfully to be attributed to the increasing influence of buoyancy over that of surface tension.

4. CONVECTIVE INSTABILITY UNDER A SOLIDIFICATION INTERFACE

4.1 *Formulation of the problem*

In contrast to the Bénard-Marangoni problem in § 3 we consider here a horizontal layer of a single-component melt underneath its solidification interface. The entire molten-frozen bi-layer is bounded above and below by solid blocks. Carefully controlled experiments were carried out by Dietsche[22] by enclosing cyclohexane liquid between two copper blocks maintained at temperatures $T_1 < T_m < T_0$ straddling its melting point T_m. The molten-frozen bi-layer emerges as a steady conductive state which loses its stability to a steady convective state in the melt after a critical threshold Ra_c is crossed. In the supercritical convective regime roll-like and hexagonal structures were identified in certain parameter (Ra, D_1) ranges with an intermediate transition region[22]. These structures, in turn, leave their traces on the solidification front as well. The configuration serves as a convenient model for controlled experimental study of the admittedly more complicated structural changes arising in alloy solidification[23]. The model is also amenable to a somewhat simplified (mono-diffusive instead of double-diffusive situation) theoretical investigation. A weakly nonlinear stability analysis[9] to second order in amplitude ε and $D_1 \sim O(\varepsilon) \ll 1$, making use of the results of Busse[10] for $D_1 = 0$, showed the possible emergence of the two types of structures. It is of some practical interest to take the finite thermal conductivity of the lower boundary block into account. Firstly, metallic melts (for which the model study is ultimately intended) are far better conductors than the containing walls. Moreover, even with less conducting model liquids (such as cyclohexane, silicone oils etc.) optical studies of structural evolution can be carried out[11,12,19,20] with transparent boundaries of glass or plexiglas (whose thermal conductivity is only a few times that of the liquid enclosed). A linear stability analysis was carried out for this more general case. The problem differs from that of § 3 in the boundary condition due to phase change at the upper interface and due to finite thermal conductivity of the lower boundary. The problem of steady onset of convective instability in the melt layer can be stated as follows.

$$(D^2-a^2)^2 \, W = Ra.a^2\theta \qquad (20)$$

$$(D^2 - a^2)\theta + W = 0 \qquad (21)$$

The boundary conditions are:

$$W(0) = 0 = DW(0) = Nu_0\theta(0) - D\theta(0) \qquad (22a, b, c)$$

$$W(1) = 0 = DW(1) = Nu_1\theta(1) + D\theta(1) \qquad (23a, b, c)$$

Here, the melt layer mean thickness, d_m, and $\Delta T = (T_0 - T_m)$, are used for scaling lengths and temperature respectively. The above problem differs from that of § 3 in b.c. (22c, 23c) and also from the one with the usual Robin- boundary conditions since the parameters $Nu_0 = a.K.\coth(a.D_0)$ and $Nu_1 = a.\coth(a.D_1)$ here *depend on the disturbance wave number a* due to the coupling of thermal disturbances in the various layers. We may note also that the layer thickness ratio $D_1 = (K_f/K_m) (T_m - T_1)/(T_0 - T_m)$. Just as in the problem of § 3 the phase change b.c. here holds at an a priori unknown location of a wavy interface whose amplitude ζ has to be found from the complete solution to the problem and is given by the additional relation:

$$(D^2-a^2)^2 \, W(1)/ \, Ra.a^2 \, (= \theta(1)) = \zeta \qquad (24)$$

4.2. Neutral stability characteristics

In Figs. 8(a), 8(b), 9 and 10 we illustrate the steady convective stability characteristics of the configuration (shown as inset in Fig. 10) for different ranges of the ratios D_0, D_1, and K_0. The neutral stability curves in Figs. 8(a), (b) are similar to those for the standard Benard-case[6] and indicate a minimum Ra_c for the onset of convection at a finite critical wavelength λ_c. In view of the finite thermal conductivity at the upper and lower boundaries of the melt layer the pertinent Ra_c is, in general, lower than that for the standard case with perfectly conducting boundaries (cf. Fig. 9). Even when the lower boundary block is perfectly conducting ($K_0 \to 0$) $Ra_c \simeq 1493$ with $a_c \simeq 2.8$ as against $Ra_c \simeq 1708$ with $a_c = 3.1$ for the standard case[6]. The variation with D_0 is not significant and shows a slight decrease in Ra_c. As D_0 increases from 0.1 to 10.0, Ra_c decreases by less than 1% for $0.01 \leq K \leq 100$. The more important variation in Ra_c is with respect to the ratios of frozen/molten layer thickness D_1 and of thermal conductivity K_0 illustrated in Fig. 9. Fig. 10 shows the steady decrease in a_c with increasing K_0 indicating an increase in the size of the emerging structures as the lower boundary becomes more and more insulating. In the limit of perfectly insulating boundaries there is a single cell, that is,

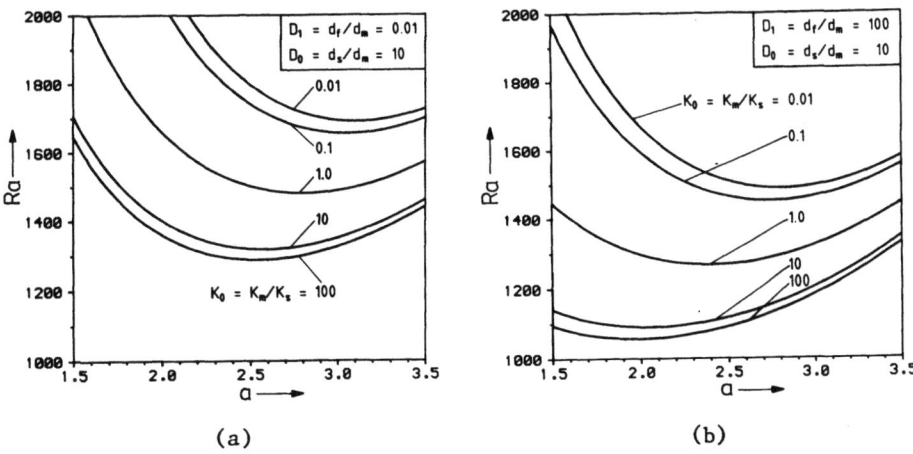

(a)   (b)

Figure 8. Neutral curves for convective instability in a horizontal melt layer under its solidification front
(a) $D_1 = 0.01$, $D_0 = 10$;   (b) $D_1 = 100$, $D_0 = 10$.

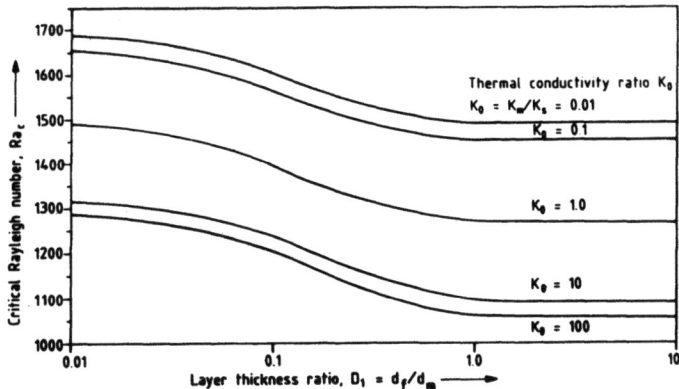

Figure 9. Variation of critical Rayleigh number with layer thickness and thermal conductivity of bottom boundary.

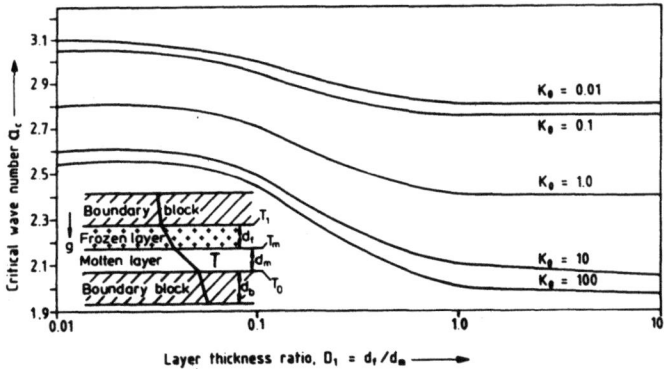

Figure 10. Variation of critical wave number with layer thickness and thermal conductivity of bottom boundary.

285

but for the influence of the side walls. Thus, in general, the cellular patterns arising in a convective melt layer and their traces on the frozen layer above are critically dependent on the overall b.c. on either side and with increasing K_0, in particular, there is an increase in λ_c. A "saturation effect" on Ra_c and a_c with respect to large D_1 is also to be seen in Figs. 9, 10 suggesting a trend towards structural stabilization as the melt layer becomes thin (cf. § 5). Here we considered only the steady convective mode, assuming the validity of the principle of exchange of stabilities[6]. The principle can be shown to hold when there is negligible change of density during fusion. Experiments of Dietsche[22] also show that unsteady modes occur at very high $Ra \gg Ra_c$.

5. DISCUSSION OF ANALOGIES

Now we comment briefly on the similarities between the results of the weakly nonlinear theories on the selection of cellular patterns in the respective configurations. For the classical Bénard case (Nu_0, $Nu_1 \to \infty$, $D_1 \to 0$ in § 4) Busse et al[10] showed that in the supercritical regime only rolls are stable under the strict Boussinesq approximation whereas hexagons can exist[24] when there is either non-Boussinesq property variation (e.g. viscosity variation with temperature) or some other asymmetry. Fig. 11 shows

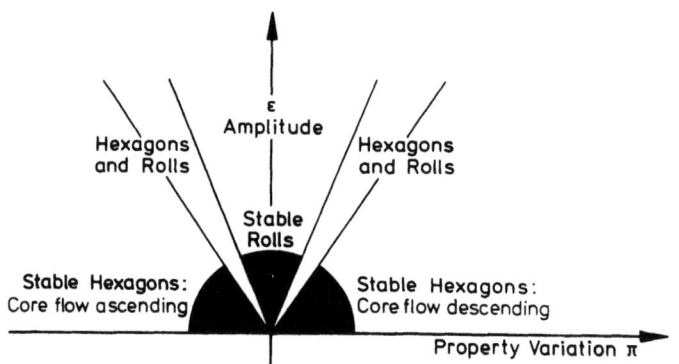

Figure 11. Schematic representation of stability domains for rolls and hexagons in the Benard configuration.

schematically the domains of stability for rolls and hexagons for the standard case[10] with repect to perturbation amplitude ε (a measure of the departure from criticality) and a thermophysical property variation parameter $\Pi \sim d\beta/dT$, $d\mu/dT$ etc. or a combination thereof[24,25] provided they are all equally small deviations from the Boussinesq approximation. We see in Fig. 11 that hexagons are possible for all ε if Π is large enough. The upward core flow (near cell centers) is typical of liquids (with $d\mu/dT < 0$) whereas the opposite is true of gases. The results reviewed in § 3 show however, that the upper liquid-gas interface is respectively convex or concave upward for liquids at the cell centers depending on whether bouyancy or surface tension is more dominant. This finding is in accord with experimental observations[11,12,19,20]. A similar selection mechanism due to imperfect thermal boundary conditions can also be demonstrated[24]. For instance, in the extreme limit of an insulating top and bottom wall (Nu_1, $N_0 \to 0$) Busse and Riahi[26] took advantage of the analytical solution for the linear stability problem to show that indeed square cells are preferred to rolls which should result for the same Boussinesq fluid when both walls are perfectly conducting. Similarly, Davis and Segel[25] considered the effects of finite curvature ($\zeta \neq 0$) at the upper two-fluid ("free") interface, *neglecting surface tension* in contrast to the case in § 3. Here again the a priori unknown deflection amplitude ζ is given (in our notation, using κ/d as the velocity scale) by:

$$\zeta = \xi(W''' - 3\ a^2 W')/a^2 \tag{25}$$

Here ξ is proportional to $\beta \Delta T$ which is retained only in the buoyancy term under the Boussinesq approximation. Thus, the proper allowance for an *interfacial disturbance* (even without surface tension) leads to this parasitic term equivalent to a non-Boussinesq effect and thereby favours structures other than rolls[10,20,24]. One may therefore expect that hexagonal/ polygonal shapes are favoured when the effects of surface tension (σ and $d\sigma/dT \neq 0$) are significant. The problem in § 3 accounts for these effects and in view of the similarity between Eqs. (12), (24) and (25) we see that the analogous parameters are ξ, $1/Ra$ and $Cr/(Bo + a^2)$. In the problem of § 4 the experimental results show[9,22] that the $Ra - D_1$ correlation with the structural patterns depends on the total bi-layer thickness ($d_f + d_m$) as well. Since $\Delta T = (T_0 - T_m)$ was held fixed at $\simeq 1°C$, here the thermophysical property variations for cyclohexane can indeed be neglected. This indicates that it is most likely the *nonzero deflection* of the wavy solidification front that influences the pattern selection. Upward flows due to buoyancy under the crests of a wavy solidification front may also tend to melt the front above, thereby accentuating the local corrugations. At somewhat high $Ra > Ra_c$ roll-like structures were found as $D_1 \to 0$ while hexagonal patterns were found at lower Ra with $D_1 > 1$. This can be understood from the schematic diagram in Fig. 11 and is in accord with the results of Busse[24] and Davis and Segel[25] since the strict Boussinesq case is approached in the former limit ($D_1 \to 0$) and the effects of interfacial curvature (non-Boussinesq surrogate) should increase as the melt layer becomes thinner than the frozen layer ($D_1 > 1$). In view of the expressions for ζ in Eqs. (12), (24) and (25) we can compare the effects of a wavy interface on the pattern selection in a convective horizontal layer with a liquid-gas interface above and that with a solidification front above. When the ratios $Cr/(Bo + a^2)$ and $1/Ra$ are relatively more significant and D_1 is sufficiently large ("saturation regime" in Figs. 9, 10) we can anticipate a hexagonal pattern. The well known occurrence of such patterns in thin liquid films[11,12,19,20] and in the experiments of Dietsche on the frozen/molten bi-layer of cyclohexane with $K_0 \to 0$ conform to this expectation (cf. Ra(rolls) >> Ra(hexagons); D_1(hexagons) >> D_1(rolls))[9,22].

For want of a complete nonlinear theory for highly supercritical convective regimes in horizontal liquid layers there seems to be some advantage in studying plausible analogies[23] such as those considered here for comparing the effects of interfacial and thermal boundary conditions. Such analogies may be useful preliminaries for undertaking any massive numerical efforts for tackling the general problem. It would also be of interest to conduct careful experiments on the configuration of § 4 with a transparent bottom block to allow visualization of flow patterns and their transitions.

6. REFERENCES

1. D. A. Nield, Surface tension and buoyancy effects in cellular convection, *J F M*, 19:341 (1964).
2. L. E. Scriven and C. V. Sternling, On cellular convection driven by surface-tension gradients: Effects of mean surface tension and surface viscosity, *J F M*, 19:321 (1964).
3. K. A. Smith, On convective instability induced by surface-tension gradients, *J F M*, 24:401 (1966).
4. G. S. R. Sarma, Effects of interfacial curvature and gravity waves on the onset of thermocapillary convective instability in a rotating liquid layer subjected to a transverse magnetic field, *P C H*, 6:283 (1985).
5. G. S. R. Sarma, Interaction of the surface tension and buoyancy mechanisms in a horizontal liquid layer, Paper No. AIAA-85-0986, presented at the AIAA 20th Thermophysics Conference, June 19-21, 1985, Williamsburg, Va.
6. S. Chandrasekhar, "Hydrodynamic and Hydromagnetic Stability," Oxford University Press, London, (1961).
7. R. Jansen and P. R. Sahm, Solidification under microgravity, *Mater. Sci. Engg.*, 65:199, (1984)
8. Y. Malmejac et al, "Challenges and Perspectives of Microgravity Research in Space," ESA BR-05, European Space Agency, Paris, (1981).
9. S. H. Davis, U. Müller, and C. Dietsche, Pattern selection in single-component systems coupling Benard convection and solidification, *J F M*, 144:133 (1984).
10. F. H. Busse, Non-linear properties of thermal convection, *Rep. Prog. Phys.*, 41:1929 (1978).

11. E. L. Koschmieder, Bénard convection, *Adv. Chem. Phys.*, 26:177 (1974).

12. C. Normand, Y. Pomeau, and M. G. Velarde, Convective instability: A physicist's approach, *Rev. Mod. Phys.*, 49:581 (1977).

13. J. Zierep and H. Oertel jr.(Editors), "Convective Transport and Instability Phenomena," Verlag G.Braun, Karlruhe (1982).

14. S. Ostrach, Low-gravity fluid flows, *Ann. Rev. Fluid Mech.*, 14:313, Annnual Reviews Inc., Palo Alto, California, (1982).

15. G. S. R. Sarma, On oscillatory modes of thermocapillary instability in a liquid layer rotating about a transverse axis, *P C H*, 2:143, (1981).

16. S. H. Davis and G. M. Homsy, Energy stability theory for free-surface problems: buoyancy-thermocapillary layers, *J F M*, 98:527, (1980).

17. J. L. Castillo and M. G. Velarde, Buoyancy-thermocapillary instability: the role of interfacial deformation in one- and two- component fluid layers heated from below or above, *J F M*, 125:463 (1982).

18. P. G. Grodzka and T. C. Bannister, Heat flow and convection experiments aboard Apollo 17, *Science,* 187:165, (1975).

19. J. Pantaloni, R. Bailleux, J. Salan, and M. G. Velarde, Rayleigh-Bénard-Marangoni instability: New experimental results, *J. Non- Equilib. Thermodyn.*, 4:201 (1979).

20. P. Cerisier, C. Peres-Garcia, R. Occeli, and J. Pantaloni, Cellular structures in Bénard-Marangoni instability, *P C H,* 6:653 (1985).

21. C. Perez-Garcia et al, Linear analysis of surface deflection in Bénard-Marangoni instability, *J. Phys.*, 46:2047 (1985).

22. C. Dietsche, Einfluß der Bénard-Konvektion auf Gefrierflächen, Research Report KfK 3724, Kernforschungszentrum Karlsruhe, accepted as Doctoral Thesis by the University of Karlsruhe (1984).

23. D. T. J. Hurle, On similarities between the theories of morphological instability of a growing binary alloy crystal and Rayleigh-Bénard convective instability, *J C G*, 72:738 (1985).

24. F. H. Busse, The stability of finite amplitude cellular convection and its relation to an extremum principle, *J F M*, 30:625, (1967).

25. S. H. Davis and L. A. Segel, Effects of surface curvature and property variation on cellular convection, *Phys. Fluids,* 11:470 (1968).

26. F. H. Busse and N. Riahi, Nonlinear convection in a layer with nearly insulating boundaries, *J F M*, 96:243 (1980).

ACKNOWLEDGEMENTS

The work reported here received partial support through a Collaborative Research Grant No. 419/84 from the NATO- Scientific Affairs Division, Brussels, Belgium. The author expresses his sincere thanks to the Project Coordinator, Professor R. Narayanan, University of Florida, Gainesville, Florida.

EXPERIMENTS ON STEADY AND OSCILLATORY THERMOCAPILLARY CONVECTION IN SPACE

WITH APPLICATION TO CRYSTAL GROWTH

D. Schwabe, R. Lamprecht, and A. Scharmann

I. Physikalisches Institut der Justus-Liebig-Universität
Giessen, Heinrich-Buff-Ring 16, D-6300 Giessen, FRG

INTRODUCTION

The various types of liquid motion (convection) due to inhomogeneities of the interfacial tension in free liquid surfaces are called Marangoni effects. The inhomogeneities of the interfacial tension can be of thermal or chemical origin. Marangoni effects are a common phenomenon for all liquids with a free surface (interface between liquid and gas or between liquid and liquid). Being surface effects, Marangoni effects are of higher importance in small volumina or near the free surface. Marangoni effects are flow phenomena which are independent of gravity. The investigation of Marangoni effects in an earth laboratory meets principle difficulties because they are invariably coupled with buoyant convection under normal gravity. Therefore experiments on Marangoni effects under microgravity in space are useful to investigate the pure surface tension driven flow.

What is ment by the term "thermocapillary convection"? The surface tension σ is temperature dependent. The surface tension of cold liquid is higher than the surface tension of hot liquid. Therefore, if we produce a temperature gradient in the free surface (or free interface) of a liquid, we have a surface tension gradient in this free surface which dragges the surface from hot regions to cold regions. The surface motion is transferred to liquid layers beneath the surface because of the liquid's viscosity. If the temperature gradient is intentionally and experimentally directed along the free surface, the fluid flow is called thermocapillary flow. This is in contrast to the situation where the temperature gradient is experimentally directed perpendicularily to the free surface. The latter case gives rise to the well known Bénard-Marangoni instability with the hexagonal convection cells when exceeding a critical temperature gradient (Marangoni number).

We report here on two experiments under microgravity. The first studies thermocapillary convection in an open cavity. The second is a study of the transition from steady to oscillatory thermocapillary flow in floating zones under microgravity. Both experiments are somewhat related to situations in crystal growth from the melt. The first experiment is related to crystal growth and zoning a material in an open boat, the second simulates the half of a zone in floating zone crystal growth.

THERMOCAPILLARY CONVECTION IN AN OPEN CAVITY

This experiment was conducted in the double rack "Prozesskammer" during the spacelabmission D1 in november 1985. The experiment chamber was built by the I. Physikalisches Institut der Justus-Liebig-Universität Giessen, the diagnostic unit and experiment command unit was built by MBB-ERNO in Bremen.

Experimental

The experiment consists of a quartz boat (cuvette) with two inserted heaters (fig. 1) with a gap between the heaters for the fluid sample. There are some special points to mention about the experiment because it was done in spacelab under microgravity. For instance, how to fill an open cavity under microgravity? This is not a trivial problem because a liquid will creep out through every capillary slit under microgravity. We have chosen the following solution: As liquid we use the paraffine tetracosane ($C_{24}H_{50}$) with a melting point of 50.9°C. This we filled into the boat in our laboratory and solidified it. The solid sample was than molten under microgravity. All rims which are in contact with the fluid were made as sharp as possible and coated with NUFLON (trademark of Fluomicron), which is not wetted by the paraffine. The NUFLON acted as antispreading barrier against capillary outflow.

To align the temperature gradient parallel to the free surface we used a double-walled heatshield made from plates of quartz glass which sourrounded the fluid sample. The glass plates are in contact with both heaters. So in the heat shields we have a linear temperature gradient parallel to the free surface.

The fluid sample is a 20 mm • 20 mm • 20 mm volume with free upper surface. We have tracer particles in the fluid (ceramic eccospheres from Emerson and Cuming) and illuminate a central vertical section with a light band. This central vertical section is observed with a 16 mm motion picture camera through the side wall of the cuvette.

A cross section of the experiment cell is given in figure 2. The outer shell is made from aluminium with four windows made from quartz glass. The experiment cell is made vacuum-tight because of safety reasons.

The diagnostic unit consists of a laser and some optics to produce a light band for illumination of the central vertical section of the fluid sample. Another part of the diagnostic unit is the 16 mm motion picture camera to observe the fluid motion indicated by the tracer particles in the light band.

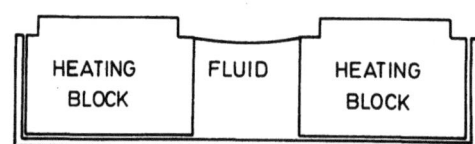

Fig. 1. Sketch of the principle of the experiment. A fluid sample is contained in a cuvette made from quartz glass between two heaters made from aluminium. A temperature difference between the two heaters generates thermocapillary convection in the free upper surface of the fluid.

Table 1. The physical data of n-tetracosane ($C_{24}H_{50}$) at 90°C

melting point	T_s	50.9	°C
density	ρ	0.753	$g \cdot cm^{-3}$
	$\partial\rho/\partial T$	$-6.6 \cdot 10^{-4}$	$g \cdot cm^{-3} \cdot K^{-1}$
thermal diffusivity	χ	$6.588 \cdot 10^{-4}$	$cm^2 \cdot s^{-1}$
dynamic viscosity	η	$2.44 \cdot 10^{-2}$	$g \cdot cm^{-1} \cdot s^{-1}$
kinematic viscosity	ν	$3.24 \cdot 10^{-2}$	$cm^2 \cdot s^{-1}$
Prandtl number	Pr	49.1	
surface tension	σ	26.0	$dyn \cdot cm^{-1}$
temperature dependence of surface tension	$\partial\sigma/\partial T$	$6.7 \cdot 10^{-2}$	$dyn \cdot cm^{-1} \cdot K^{-1}$

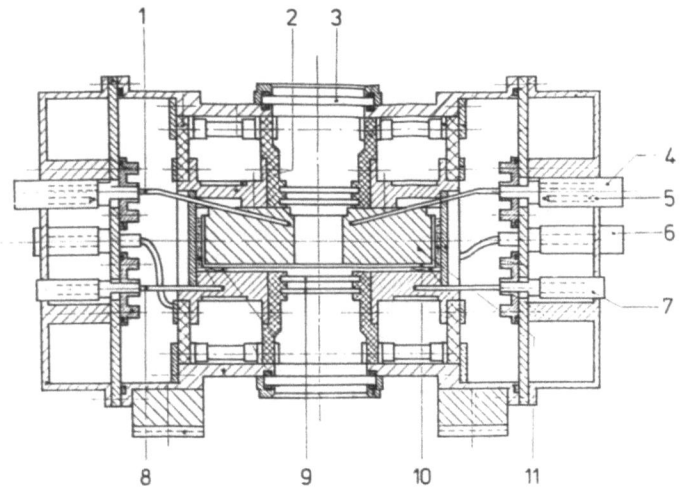

Fig. 2. Cross section through the experiment cell

    1 - thermocouple for the heating block
    2 - resistant heating wire
    3 - outer quartz glass window
    4 - plug for thermocouple for the heating block
    5 - PT 100 resistance thermocouple for calibration
    6 - plug for heater current
    7 - plug for thermocouple for the heater
    8 - thermocouple for heater
    9 - double walled quartz glass for thermal insulation
    10 - quartz glass cuvette (boat)
    11 - heating block

The experiment command unit was programmed for a semi-automatic run of the experiment. After melting the sample, both heaters are at 60°C for approximately one hour. Than the right heater was held at 60°C and the left heater was risen in 5°C-steps to 120°C. Than the right heater was brought to 90°C and than risen in 10°C-steps to 110°C. After each temperature step there was a thermalisation time of approximately 15 minutes before each take of 45 seconds with the motion picture camera. Thus the experiment was programmed to observe the convection for $\Delta T = 0°C, 5°C, \ldots 55°C, 60°C$ with the right heater at 60°C, and for $\Delta T = 30°C, 20°C,$ and $10°C$ with the left heater at 120°C.

Experimental results and discussion

The tetracosane was molten as preprogrammed. With both heaters at 60°C one can see a clear and transparent melt without bubbles and with the well illuminated tracer particles. But the tracers did not move in contrary to the expectation of a faint movement towards the middle of the liquid surface which is the coldest spot when both heaters are at 60°C. Some tracers sticked to the liquid surface.

After this the temperature difference between the two heaters was increased. Up to the temperature difference of $\Delta T = 55°C$ there is no motion of the tracers in contrary to the expectation of flow from hot to cold in the free surface. Only for the temperature difference of $\Delta T = 60°C$ there is strong motion in the expected way (figure 3).

How to explain this suppression of Marangoni convection for smaller Marangoni numbers (smaller ΔT)? One of the most difficult problems in experiments on Marangoni convection is to have a clean surface, so to speak to have not dirt on the free surface. In this respect our experiment failed. The tetracosane was not clean enough. We had some dirt on the free surface and it was even enough dirt to build up a dirt skin which acted like a solid layer. This skin suppressed thermocapillary flow totally in the beginning of the experiment. Only for the temperature difference of $\Delta T = 60°C$ this solid skin ruptured, thermocapillary flow could start, and the flow compressed the dirt in front of the cold wall (and the dirt was mixed into the bulk liquid). With free and clean surface strong thermocapillary flow was observed under microgravity (figure 3).

Such a stable solid skin on the free surface has never been observed in experiments on ground. On ground there is always some fluid motion due to buoyancy which hinders the formation of a "solid" and complete skin of dirt. In the microgravity experiment we gave one hour time, with both heaters at 60°C, for the built-up of the skin by diffusion of dirt from the bulk to the surface. Once such a solid skin has been formed on the free surface under microgravity, it is only a disturbance in connection with the surface tension force that can rupture the skin. As long as the surface skin is completely covering the surface, surface tension forces are completely suppressed. The disturbance must be a surface disturbance, therefore.

Figure 3 shows a photograph of the streamlines under microgravity with $\Delta T = 60°C$ between the two heaters. In figure 4 a sketch of this photograph is given. The photograph has been taken directly from the screen during projection of the 16 mm motion picture film. The photograph shows that the surface is bent and that the cuvette is no longer filled up to the rim. In the middle of the liquid zone the free surface is 1,8 mm below the rims. The antiwetting barrier at the hot heating block has failed and some tetracosane has leaked out there. Under microgravity (fig. 3 and fig. 4) we have one big convection cell which occupies almost the whole volume. The centre of the cell is situated at $x = 12$ mm and $z = 6$ mm (the origin of the coordinate system is in the left upper corner with the z-coordinate pointing down).

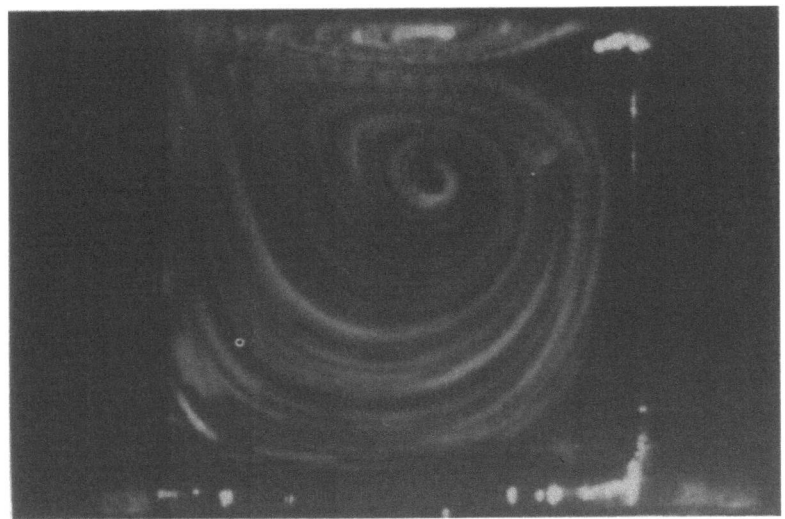

Fig. 3. Photograph of the streamlines of thermocapillary convection under microgravity at a temperature difference between the heaters of 60°C. Exposure time of 10 s.

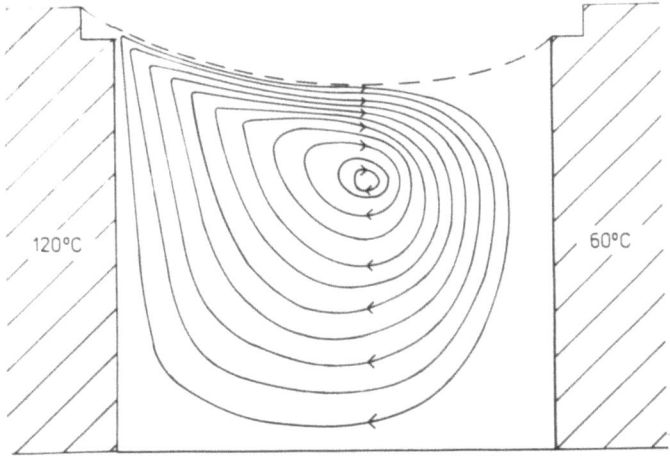

Fig. 4. Sketch of the streamlines of thermocapillary convection under microgravity at a temperature difference of the heaters of 60°C.

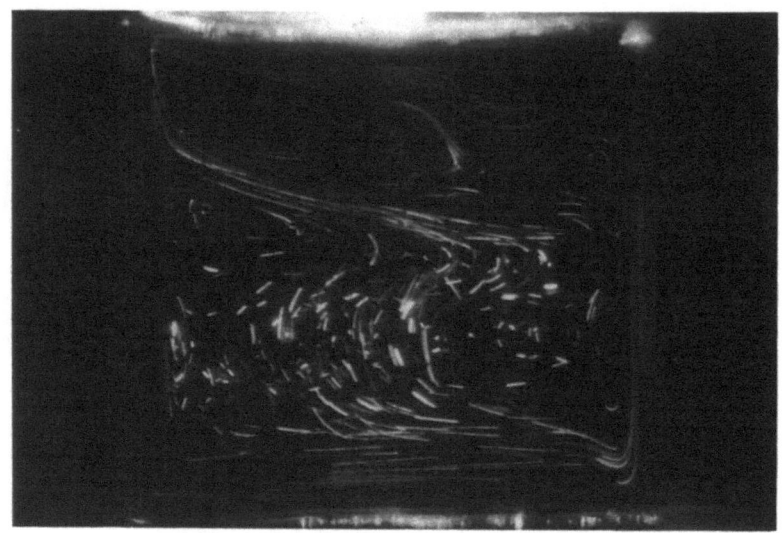

Fig. 5. Photograph of the streamlines of coupled thermocapillary and buoyant convection under normal gravity at a temperature difference of 60°C between the heaters.

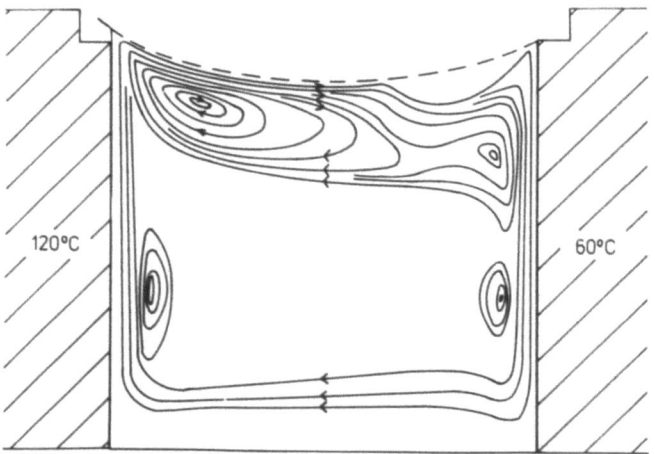

Fig. 6. Sketch of the streamlines of coupled thermocapillary and buoyant convection under normal gravity at a temperature difference of 60°C between the heaters.

The motion is from hot to cold in the free surface and the back flow in the bulk fluid is from cold to hot, as expected for thermocapillary convection. The flow is fairly twodimensional, which can be seen from the tracer particles staying in the light band. The free surface is not totally clean: in front of the colder heating block the dirt-skin is compressed and the flow dives down before reaching the cold wall. Therefore the convection cell is strongly asymmetric.

Figure 5 and 6 show the coupling of thermocapillary and buoyant forces in the 1-g reference experiment. The photograph is taken directly during the experiment. The reference experiment has been done under the same conditions as the microgravity experiment, e.g. the bending of the free surface and a dirt-skin in front of the cold heating block have been simulated. Comparing figures 5 and 6 with figures 3 and 4, one sees that the flow under normal gravity is much more complex than the flow under microgravity. The fluid volume is occupied by more than one convection roll. In the upper part we have two vortex centres near the free surface. One near the hot wall at $x = 3,5$ mm and $z = 2,5$ mm and one near the cold wall at $x = 18$ mm and $z = 5$ mm. Both vortex centres are nearer to the free surface than the one under microgravity because it is hot fluid circulating near the free surface which cannot dive down under normal gravity. The fluid circulation near the free surface is mainly driven by surface tension forces. The flow very near the hot wall and very near the cold wall circulating along the bottom of the cuvette is due to buoyancy forces.

The Marangoni number $Ma = -\partial\sigma/\partial T \cdot \Delta T \cdot L \cdot \eta^{-1} \cdot \chi^{-1}$, with $\partial\sigma/\partial T$ = temperature dependence of surface tension σ, ΔT = temperature difference between the two heating blocks, L = length of the liquid zone, η = dynamic viscosity and χ = thermal diffusivity, is of the order of $4.6 \cdot 10^5$ in both the experiments under 1-g and under μ-g. The Grashof number $Gr = \alpha \cdot g \cdot \Delta T \cdot H^3 \cdot \nu^{-2}$, with α = thermal volume expansion coefficient, g = value of gravitational acceleration, H = liquid height in the cuvette, ν = kinematic viscosity, is of the order of $3.3 \cdot 10^5$ for normal earth gravity and of the order of $5.3 \cdot 10^2$ for reduced gravity in the spacelab. From the modified Bond number $Bo = Gr \cdot Ma^{-1}$ one can read, that buoyancy and thermocapillary forces are of the same order of magnitude under normal earth gravity, whereas thermocapillary forces are dominating under microgravity. This is fully confirmed by observation (fig. 3 - 6).

Figure 7 shows a velocity profile of v_x for normal earth gravity and for microgravity. Both profiles show typical features of the surface tension driven flow, namely the highest stream velocity in the free surface and a steep decrease of flow velocity towards the bulk fluid. The surface flow from hot to cold occupies only a few millimeters of the fluid depth.

There are distinct differences between the microgravity- and the normal earth gravity-experiment. Under microgravity the maximum flow speed in the free surface reaches 20 mm·s^{-1} whereas it is only 10 mm·s^{-1} under normal gravity. Under microgravity the back flow penetrates deeply into the bulk fluid. The maximum back flow velocity is - 4 mm·s^{-1} at $z = 13$ mm. Under normal earth gravity, both, the surface flow and its back flow are nearer to the free surface. The vortex centre lies at $z = 4,2$ mm and the back flow velocity has its maximum at $z = 6$ mm with - 3 mm·s^{-1}. This confinement of the surface flow and its back flow is due to the fact that the fluid in this vortex is relatively hot, and therefore stabilized on top of the sample by buoyancy forces. In the normal gravity experiment there is a region between $z = 10$ mm and $z = 15$ mm with almost no motion in x-direction. Near the bottom of the cuvette there is again circulation in the normal earth gravity-experiment due to the buoyancy driven flow which circulates up near the hot wall and down near the cold wall in boundary layers.

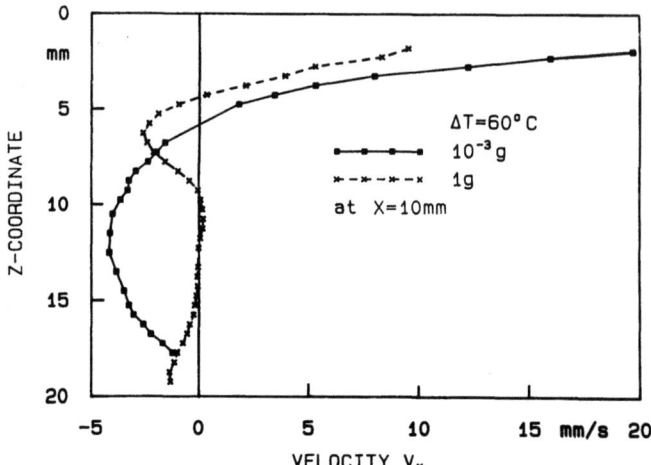

Fig. 7. Velocity profiles of v_x at $x = 10$ mm and at a $\Delta T = 60°C$ between the two heaters for μ-g and 1-g.

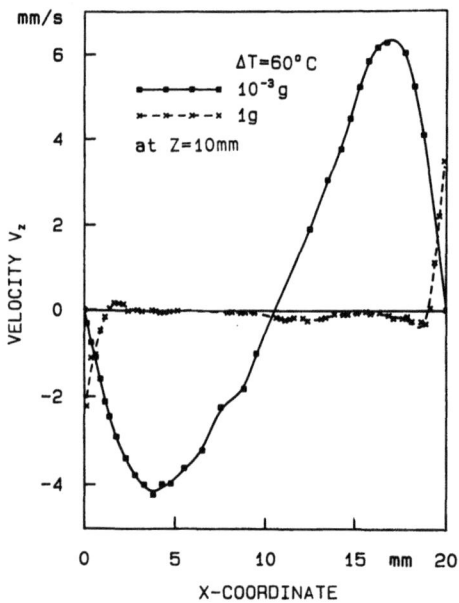

Fig. 8. Velocity profiles of v_z at $z = 10$ mm and at a $\Delta T = 60°C$ between the two heaters for μ-g and 1-g.

Fig. 8 shows the velocity profile of v_z for normal earth gravity and for microgravity at $z = 10$ mm. The differences between normal gravity and microgravity are enormous. Because v_z has been taken at $z = 10$ mm we see that thermocapillary convection penetrates deeply into the volume under microgravity whereas it is confined to the surface region under normal gravity. Under normal gravity we have only upflow near the hot wall and downflow near the cold wall in a boundary layer. This boundary layer is so thin that we could not measure the drop of v_z to zero (because of the non slip condition at the wall). Under normal gravity and at $z = 10$ mm the liquid is at rest in almost the whole volume except in the boundary layers. In contrary to this, under microgravity the surface tension driven flow convects the whole liquid volume. Heat and mass transport by surface tension driven convection seem to be more effective under microgravity than under normal earth gravity.

The description of earlier work and its relation to crystal growth from the melt can be found in references[1-3].

CRYSTAL GROWTH OF SILICON UNDER MICROGRAVITY

In this paragraph we review the work done by other authors on crystal growth of silicon under microgravity. Silicon growth was done by the floating zone method and by recrystallising a molten silicon sphere. In both methods the melt has a free surface with strong temperature gradient along it and therefore thermocapillary convection is likely to occur in the melt. The original idea of crystal growth under microgravity was to grow the crystal under more steady conditions than on earth. On earth one has buoyant convection in the melt which is in the most cases time dependent. Time dependent flow in the melt gives time dependent growth conditions for the crystal which leads to the occurance of striations in doped material. The striations are made visible by slicing the crystal with its growth direction in the plane and by polishing and etching the slice. Striations are then visible as faint lines perpendicular to the growth direction. Thus it was hoped to grow striation free crystals under microgravity because of no time dependent buoyant flow under microgravity.

It would have been very difficult to discuss from interface shape whether we had surface tension driven convection playing any role during growth of silicon in these experiments if there were not the phenomenon of striations. Eyer and coworkers grew silicon under microgravity[4,5] and found the same type of striations in crystals grown under microgravity as in crystals grown on earth. It was concluded that both crystals suffered the same type of time dependent growth conditions. Since gravity driven flow has been eliminated in the experiment under microgravity, the result suggests that in both crystals - the one grown under microgravity and the one grown under normal earth gravity - the same type of time dependent flow has caused the striations, namely time dependent thermocapillary convection.

Eyer and Leiste[6] and Croell and coworkers[7] conducted one more nice experiment on ground[6] and under microgravity[7] to reveal the importance of the free melt surface for the origin of the striation pattern; they covered the melt with an inert surface coating of SiO_2. In these crystals no striations were present. The melt surface was not free and thermocapillary convection was supressed.

Kölker recrystallized a silicon sphere under microgravity during the spacelab-1 mission[8] and spacelab-D1 mission[9]. In both cases he found striations in the crystals grown under microgravity.

As a result of all silicon growth experiments we have: there must be unsteady flow under microgravity if the surface is free. We claim this flow to be oscillatory or turbulent thermocapillary convection.

Striations in the silicon crystals are an indirect proof for unsteady flow in the melt from which the crystals are grown. We have conducted a microgravity experiment in which we studied oscillatory thermocapillary convection directly[10].

THE TRANSITION FROM STEADY TO OSCILLATORY THERMOCAPILLARY CONVECTION IN FLOATING ZONES UNDER MICROGRAVITY

The laminar flow state of thermocapillary convection shows a transition to a time dependent (oscillatory) state when exceeding a certain critical Marangoni number Ma^c of the order of 10^4. This has been discovered in simulated floating zone experiments (liquid cylindrical column with free cylindrical surface, suspended by surface tension forces between two cylindrical rods which are differentially heated to impose an axial temperature gradient along the free surface)[11-14].

The objective of the described experiment was, to measure under microgravity (μ-g $\cong 10^{-4}$ g) this critical Marangoni number Ma^c in a floating zone; to observe the critical parameters of the transition from laminar to oscillatory flow. This transition is an interesting but complex hydrodynamic problem and microgravity is needed to reduce coupling of thermocapillary forces with buoyancy forces. Moreover, it is of some importance in floating zone crystal growth and other growth techniques from the melt, especially for growth under microgravity[1]. In crystal growth time independent growth conditions are favourable because time dependent conditions can give rise to dopant inhomogeneities (compositional striations). In order to avoid such striations in μ-g float-zone crystal growth the Marangoni number (given by the growth conditions) must be lower than the critical number Ma^c for transition to oscillatory flow. The experiment yielded this number for a special situation. Further efforts in ground based and microgravity experiments are needed to extrapolate this knowledge for use in real growth systems. For this to happen, more features of the oscillatory thermocapillary flow state should be known to allow comparison with indirect observations in crystal growth (striation frequencies, striation amplitude, local variation of striation amplitude). Results concerning the oscillation frequencies under microgravity are reported, therefore.

Experimental

The μ-g-Experiment was performed on SPAS-01 during its free-flying phase on the 22th of June 1983 (space shuttle flight No 7; Challenger). The satellite SPAS-01 experienced very low residual gravity (of the order of 10^{-5} g) during the time the experiment was active. Two experiment cells with zone lengths l = 4.00 mm and l = 4.79 mm where operated at the same time. In both cells the solid $NaNO_3$-sample was molten to establish a floating zone with radius a = 3.00 mm, which is suspended between two cylindrical graphite rods of the same radius. After a thermalisation of 30 minutes the temperature difference ΔT between the two graphite rods was increased "linearily" in time to pass the transition point (steady $\rightarrow$ oscillatory) and afterwards decreased (oscillatory $\rightarrow$ steady). The rate of heating up was of the order of $+ 3.6 \cdot 10^{-3}$ K$\cdot$s^{-1} and cooling down was with $- 7.2 \cdot 10^{-3}$ K$\cdot$s^{-1}. Actually the temperature changes were in steps: every 100 s the temperature difference was increased by 0.36 K during heating up and decreased by 0.36 K every 50 s during cooling down. A thermocouple TZ was introduced through the upper rod into the liquid zone to register the temperature oscillations generated by oscillatory flow.

Table 2. Physical properties of liquid $NaNO_3$ at 320°C

melting point	T_m	306.8	°C
density	ρ	1.903	$g \cdot cm^{-3}$
coefficient of expansion	$\partial\rho/\partial T$	$-3.80 \cdot 10^{-4}$	K^{-1}
surface tension	σ	119.7	$dyn \cdot cm^{-1}$
temperature coefficient of surface tension	$\partial\sigma/\partial T$	-5.5	$dyn \cdot cm^{-1} \cdot K^{-1}$
dynamic viscosity	η	$2.82 \cdot 10^{-2}$	$g \cdot s^{-1} \cdot cm^{-1}$
kinematic viscosity	ν	$1.48 \cdot 10^{-2}$	$cm^2 \cdot s^{-1}$
thermal conductivity	λ	$5.71 \cdot 10^4$	$erg \cdot s^{-1} \cdot cm^{-1} \cdot K^{-1}$
thermal diffusivity	χ	$1.67 \cdot 10^{-3}$	$cm^2 \cdot s^{-1}$
specific heat	c_p	$1.80 \cdot 10^7$	$erg \cdot g^{-1} \cdot K^{-1}$
Prandtl number	Pr	8.9	
factor in Marangoni number used for $\bar{T} = 325°C$ near the transition point	$\|\partial\sigma/\partial T\|\eta^{-1}\chi^{-1}$	1211	$cm^{-1} \cdot K^{-1}$

The signals have been recorded on a tape. The temperature oscillations monitored by TZ are expected to be mainly in the 0.5 Hz-region and have been sampled with 20 Hz. The experiment cells have been developed and built by the I. Physikalisches Institut der Justus-Liebig-Universität Giessen. The MAUS-container, structure, power-supply and control, experiment command unit and data recording have been developed by ERNO-Raumfahrttechnik (Bremen).

Idential 1-g reference experiments at normal earth gravity (even the $NaNO_3$-sample was the same) have been performed before (30.12.1982) and after (02.02.1984) the flight experiment.

Most of the experimental technique is described in ref. 14. Details and results of other microgravity experiments in which we could show the existence of the oscillatory flow state under μ-g and measured Ma^c are given in ref. 15-17.

Results and Discussion

Both experiments performed well. During the heating-up phase of 99.75 min we observed first the steady state for approximately 50 min and then the oscillatory state for 50 min. During the cooling down phase we observed oscillations for approximately 30 min and no oscillations for 20 min.

Figure 9a shows the signal from the thermocouple in the zone (TZ) near the transition point for the zone with length $l = 4.79$ mm. The numbers on the abscissa give the time in seconds $\cdot 10^{-3}$ after start of the experiment. At $7.80 \cdot 10^3$ s no oscillations are visible. The flow is steady. A temperature step by 0.36 K in ΔT occurs at $7.82 \cdot 10^3$ s and, the fluid system needs approximately 10 s to come to a new equilibrium temperature which is constant for the next 90 s. At $7.90 \cdot 10^3$ s, at the end of this ΔT-plateau, very faint temperature oscillations are visible. The next ΔT-step occurs at $7.915 \cdot 10^3$ s. During this ΔT-plateau between $7.92 \cdot 10^3$ s and $8.01 \cdot 10^3$ s the oscillation amplitude grows and needs approximately 50 s to reach a steady maximum value. From these data one can read two important time

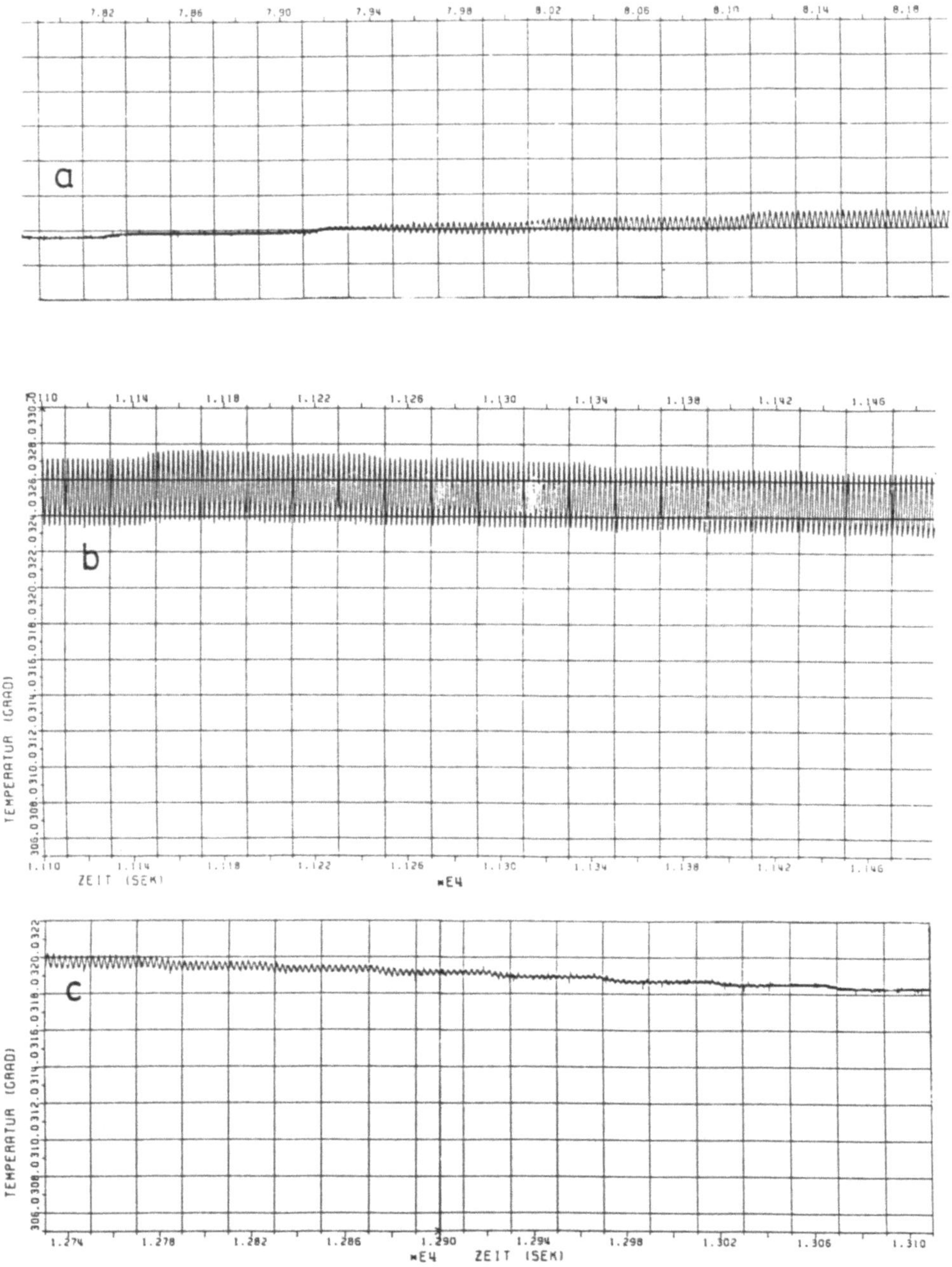

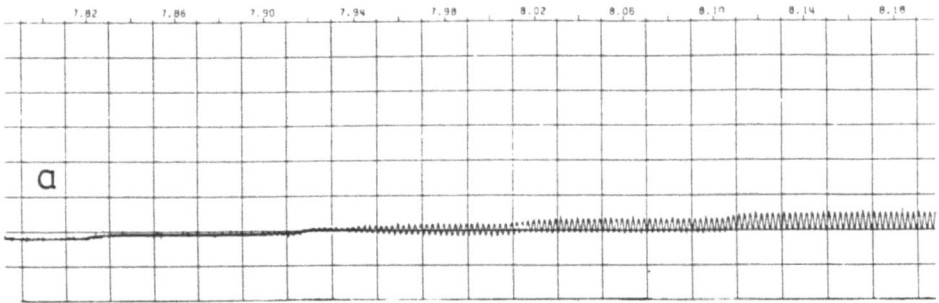

Fig. 9. Temperature oscillations under μ-g in the zone with l = 4.79 mm;
a. - onset at approx. $7.90 \cdot 10^3$ s when ΔT is 16.73 K; b. - maximum oscillation amplitude at the highest Marangoni number (Ma = 16625, ΔT = 28.66 K); c. - fading of oscillations during decrease of Ma.

constants of the experiment: 10 s to reach equilibrium after a ΔT-step of 0.36 K, and 50 s to develop from the steady to an oscillatory state. Once the oscillatory state has developed, further ΔT-steps (e.g. at $8.01 \cdot 10^3$ s and $8.11 \cdot 10^3$ s) increase the oscillation amplitude, but steady amplitude is reached in approximately 10 s. The peak-peak oscillation amplitude is 1.0 K at $8.14 \cdot 10^3$ s where the Marangoni number is ∿ 5% above the critical.

Figure 9b shows the temperature oscillations at maximum ΔT and the point when heating up is reversed to cooling down. The peak-peak temperature oscillation amplitude reaches 3.88 K which is approximately 13% of the applied ΔT.

Figure 9c shows the fading of oscillations during decrease of ΔT. The convection is back in the steady state approximately at $13.04 \cdot 10^3$ s. For the sake of brevity of the paper most of the gained direct TZ-data of the μ-g-experiment are not shown. The TZ-data of the 1-g reference experiments are not presented at all; they show features comparable to those of the μ-g-experiments and are of the same quality.

The temperature signal from the thermocouple in the zone with l = 4.00 mm is given in figures 10a - 10d for selected times. The onset of oscillations occurs in the ΔT-plateau between $8.30 \cdot 10^3$ s and $8.40 \cdot 10^3$ s. The frequency spectra of this zone are very interesting which is shown and discussed in more detail later. Obviously the onset of oscillations is not with a single frequency (main frequency) but with a modulation of the amplitude (beat frequency). This beat vanishes near $8.54 \cdot 10^3$ s (figure 10a), occurs again at higher Marangoni numbers around $9.40 \cdot 10^3$ s to vanish very abruptly at $9.54 \cdot 10^3$ s (figure 10b). The maximum peak-peak temperature oscillation amplitude at maximum Marangoni number at $11.18 \cdot 10^3$ s (figure 10c) is 5.5 K which is 18% of the applied temperature difference. Decreasing the temperature difference (and the Marangoni number) one expects the beat-frequencies to occur again. This is partly true, because the beat occurs just at the fading of the oscillations in figure 10d. But the frequency spectra in the heating up cycle differ from those in a cooling down cycle. Therefore hysteresis effects must be discussed. The Fourieranalysis in case of a beat frequency shows the main frequency and their harmonics to be split into fine peaks (fig. 11), whereas in case of no beat we only find a sharp peak of the main frequency and some of its harmonics (fig. 12).

The critical Marangoni numbers are evaluated from the oscillation signal by plotting the peak-peak oscillation amplitude over the Marangoni number and extrapolating towards zero amplitude (figures 13a, b; figures 14a, b, c). We observe a significant difference in the critical Marangoni numbers during increase of Ma (↑) and decrease of Ma (↓). The same holds for the development of the oscillation amplitude with increasing/decreasing Ma: the amplitude rises steeper when increasing Ma which indicates that the onset of oscillations is hindered. This is a hysteresis effect which can occur for many reasons. Comparing the two 1-g reference experiments with the zone with length l = 4.79 mm (figures 14b and 14c) we seem to have a good reproducibility of the measurements; the reproducibility is good enough to think of the hysteresis as being real. Hysteresis effects affect the transition point.

Table 3 summarizes the values of the measured critical Marangoni numbers. The absolute error in Ma^c is ± 250 as estimated from the precision of the temperature- and lengths-measurements. The main error comes from the need of absolute temperature measurements and uncertainties in extrapolating the amplitude towards zero. Comparing the μ-g and 1-g experiments directly, the estimated error in Ma^c is ± 150 only, because in this case only relative temperature measurements are needed. In this error analysis possible variations of the surface tension properties (changes in atmosphere, surface tension-

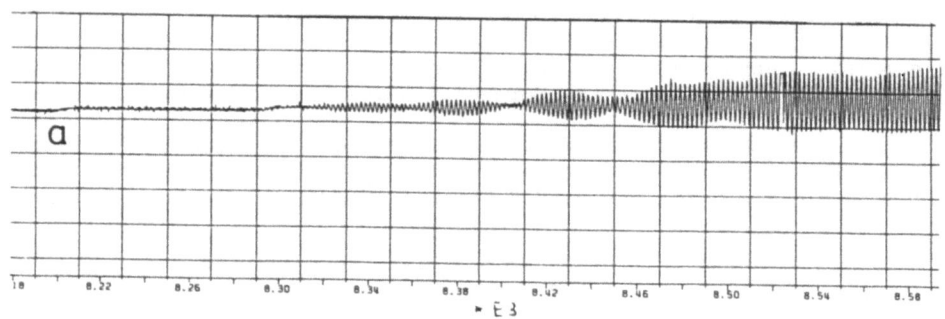
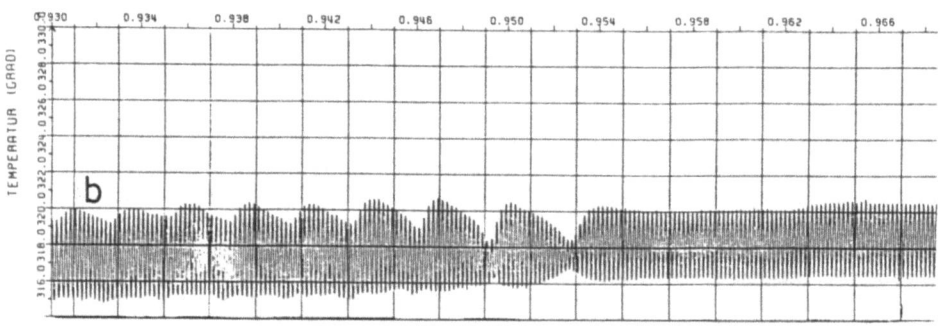
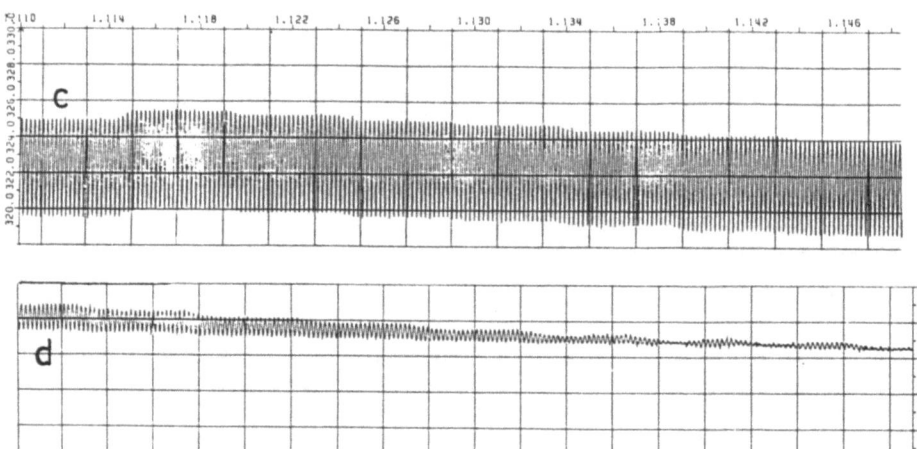

Fig. 10. Temperature oscillations under μ-g in the zone with l = 4.00 mm;
a. - onset of oscillations with beat frequency which vanishes very soon; b. - strong beat at higher Marangoni number; c. - maximum oscillation amplitude at maximum Ma; d. - fading of oscillations with beat frequency.

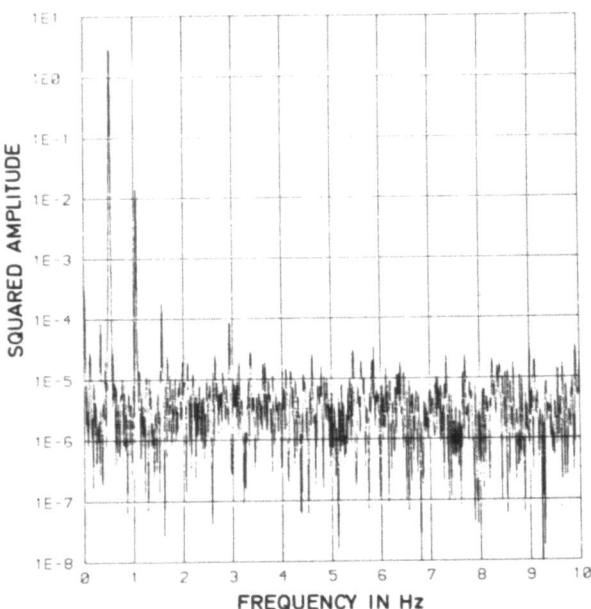

Fig. 11. Fourieranalysis of the temperature oscillations with beat under µ-g of the zone with length l = 4.00 mm. Time interval analysed 09342 s - 09437 s.

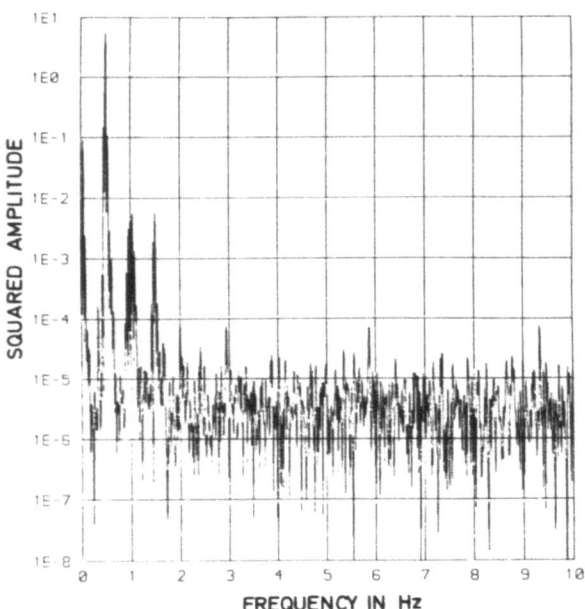

Fig. 12. Fourieranalysis of the temperature oscillations (without beat) under µ-g of the zone with l = 4.00 mm. Time interval analysed 09532 s - 09627 s.

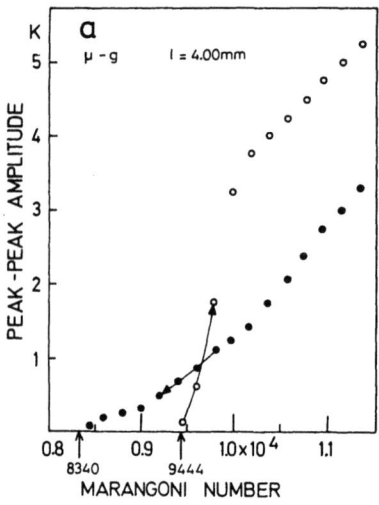
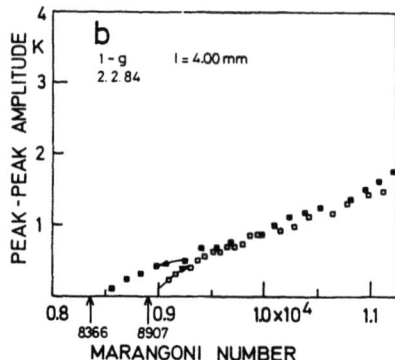

Fig. 13. Oscillation amplitudes during increase and decrease of Ma in the zone with l = 4.00 mm; a.- microgravity; b.- 1-g reference experiment.

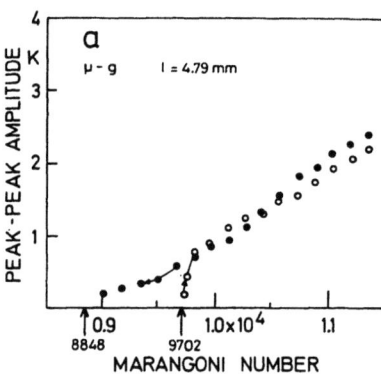
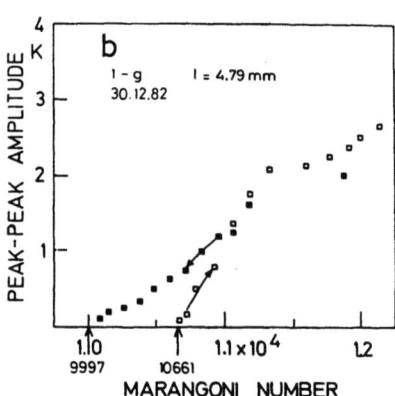
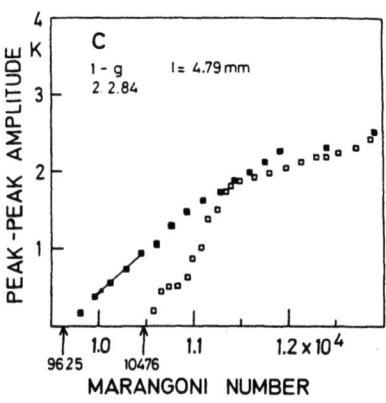

Fig. 14. Oscillation amplitudes during increase and decrease of Ma in the zone with l = 4.79 mm; a.- microgravity; b.- normal gravity before flight; c.- normal gravity after flight.

Table 3. Critical Marangoni numbers Ma^c during increase of ΔT (↑) and decrease of ΔT (↓) of the μ-g and 1-g experiments

Zone Length l	Critical Marangoni Number Ma^c					
	30.12.82 1-g		02.02.84 1-g		22.06.83 μ-g	
	↑	↓	↑	↓	↑	↓
4.79 mm	10660	9990	10470	9620	9700	8840
4.00 mm			8900	8360	9440	8340

active impurities, dirt- or dust-particles incorporated into the liquid zone during flight) are not taken into account.

The critical Marangoni numbers measured in this work compare well with those measured earlier[14]. It has been anticipated by the authors that a smaller Ma^c will be measured under μ-g, especially for longer zones. The main argument was, that the thermocapillary forces are directed in oposite direction to buoyancy forces in the used heating-from-above-situation. This argument is strong for long zones and, in the zone with l = 4.79 mm Ma (μ-g) is smaller than Ma (1-g). For the zone with l = 4.00 mm this is not true and further work is under preparation to understand this.

Figures 15-17 show the frequencies of the temperature oscillations measured in the various experiments with the same sample but under different gravity-conditions and at different times. The main frequencies are taken from the TZ-data by measuring the time needed for 20 oscillations at the end of each ΔT-plateau. The error in this frequency evaluation is given approximately by the size of the crosses or circles used in the figures. Figure 15 shows the most interesting frequency-data under μ-g. The oscillations start with main frequency $f_1 = 0.492$ Hz and soon jump to $f_2 = 0.475$ Z. During this time a beat frequency (larger broken circle; the arrows mark the time interval of significant beat frequency amplitude) $f_b \approx 0.024$ Hz is observed. We therefore conclude that during this time the two oscillations with f_1 and f_2 occur with comparable amplitude, the beat frequency being the difference frequency $f_b \approx f_1 - f_2$. After this jump to a lower frequency there is a steep increase of f with Ma until Ma $\approx 1.17 \cdot 10^4$, where the jump-back to higher frequency occurs from $f_2 = 0.488$ Hz to $f_1 = 0.518$. This transition is again accompanied by the occurence of a beat frequency $f_b \approx 0.036$ Hz. After this jump to the higher frequency f_1 the frequency decreases with increasing Marangoni number. This is an unexpected behaviour which is reproduced in the cooling down cycle. There the frequency increases with decreasing Marangoni number until Ma $\approx 0.93 \cdot 10^4$ is reached, where the transition from $f_1 \approx 0.510$ Hz to $f_2 \approx 0.476$ Hz occurs. This transition is accompanied by the existence of a beat frequency $f_b \approx 0.021$ Hz (compare to fig. 10). We could not observe such frequency jumps in the corresponding 1-g reference experiment. But the frequencies of the μ-g-experiment and the 1-g-experiment are close enough to conclude that they have the same origin, which can only be oscillatory thermocapillary convection.

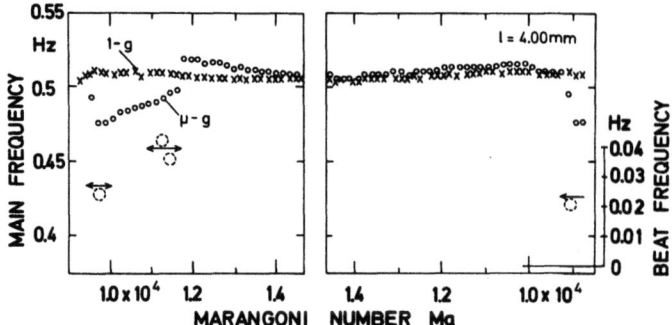

Fig. 15. Frequency dependence on Marangoni number under μ-g and 1-g for zone length l = 4.00 mm during heating up (left) and cooling down (right)

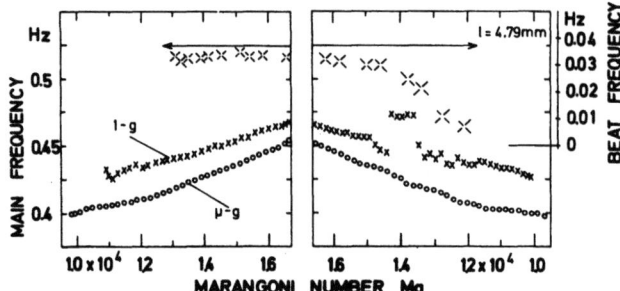

Fig. 16. Frequency dependence on Marangoni number under μ-g and 1-g for zone length l = 4.79 mm during heating up (left) and cooling down (right)

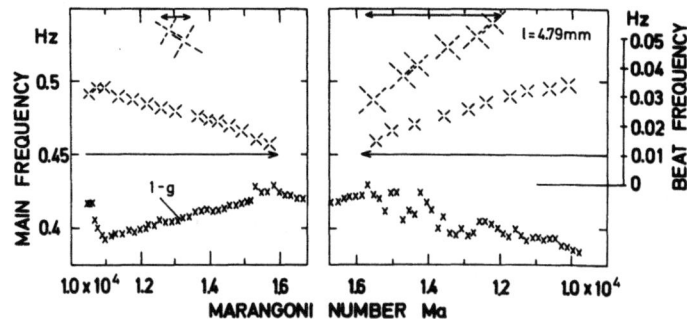

Fig. 17. Frequency dependence on Marangoni number under 1-g for zone length l = 4.79 mm (second 1-g reference experiment after flight, to be compared with fig. 16)

The frequency dependence in the larger zone with l = 4.79 mm under μ-g is rather normal, showing an increase in f during increase in Ma and decrease in f during decrease in Ma (fig. 16). In this case the frequency jump up and down is most pronounced near Ma = $1.4 \cdot 10^4$ during the cooling down cycle in the 1-g reference experiment. The beat-frequency is observed when the Marangoni number exceeds $1.3 \cdot 10^4$. We also note a hysteresis ($\Delta Ma \approx 0.1 \cdot 10^4$) in the occurence of the beat frequence and a frequency dependence of f_b during cooling down.

Figure 17 shows the results of the second 1-g reference experiment with the zone with l = 4.79 mm. We observe again beat frequencies during almost the whole experiment time, except for Ma > $1.6 \cdot 10^4$. The beat frequency f_b decreases with increasing Ma in contrast to the first 1-g reference experiment, and a second beat frequency $f_b' = 2\, f_b$ has been observed (largest broken crosses in fig. 17). Comparing the 1-g reference experiments in figures 16 and 17 we must conclude, that the reproducibility in frequency data was not good because the main frequency was allover lower in the second 1-g reference experiment. A nonspeculative reason for this is not yet known.

CONCLUSION

In an experiment with an open cavity we have demonstrated thermocapillary convection under microgravity and under normal earth gravity. With a free and clean surface strong thermocapillary convection has been observed under microgravity. Thermocapillary convection under microgravity was totally suppressed by a "solid" dirt-skin on the free surface. The velocity along the free and clean surface from hot to cold was 20 mm $\cdot$ s^{-1} and the back flow penetrated deeply into the bulk fluid under microgravity. Under normal earth gravity the thermocapillary eddy was stabilized near the free surface by buoyancy forces. Heat and mass transport by thermocapillary convection seem to be more effective under microgravity than under normal earth gravity.

A review of silicon crystal growth experiments under microgravity shows that the growth conditions were not steady. It was concluded that oscillatory or turbulent thermocapillary convection in the melt is the reason for unsteady growth (growth striations).

In a microgravity experiment we measured the critical Marangoni numbers Ma^c for the transition from steady to oscillatory thermocapillary convection in two small $NaNO_3$ floating zones. We could compare the μ-g-measurements to laboratory experiments under 1-g because these small zones are stable on earth. The critical Marangoni numbers under μ-g and 1-g are of the same order of magnitude (10^4) with a tendency to smaller Ma^c under μ-g for longer zones.

The temperature oscillations (and flow oscillations) under μ-g are of significant amplitude and comparable to those observed under 1-g. Such flow oscillations could be the reason for the striations found in the silicon crystals.

A new feature of oscillatory thermocapillary convection has been found under μ-g and under 1-g, namely frequency jumps and beat frequencies. The beat frequencies are of special interest considering real growth systems like float-zone silicon, because oscillatory flows (and growth conditions) are detected indirectly via dopant striations in the grown crystal. It can be anticipated that the main oscillation frequencies of thermocapillary flow in silicon (if they occur) must be higher in frequency than in $NaNO_3$. It is not yet clear whether the growing crystal can respond significantly with marked striations to oscillation frequencies of the order of some Hz, and whether our methods to reveal such striations have enough spatial resolution. Beat frequences (if they occur) give us the possibility to identify oscillatory flow in float zone silicon even when the main frequency is to high to be detected.

ACKNOWLEDGEMENTS

The work was supported by the Federal Ministry for Research and Technology of the FRG under contract no 01 QV 470 and no 01 QV 148. This work would not have been possible without the help of many friends from the DFVLR, Köln and MBB-ERNO, Bremen, and the payload specialists on board of the space shuttle.

REFERENCES

1. D. Schwabe; J. Physico Chemical Hydrodynamics 2 (1981) 263-280
2. D. Schwabe and A. Scharmann; J. Crystal Growth 52 (1981) 435-449
3. R. Lamprecht, D. Schwabe, A. Scharmann, and E. Schultheiss; J. Crystal Growth 65 (1983) 143-152
4. A. Eyer, H. Leiste, and R. Nitsche; ESA SP 222 (1984) 173-182
5. A. Eyer, H. Leiste, and R. Nitsche; J. Crystal Growth 71 (1985) 173-182
6. A. Eyer and H. Leiste; J. Crystal Growth 71 (1985) 249-252
7. A. Croell, W. Müller, and R. Nitsche; Floating Zone Growth of Surface Coated Silicon Under Microgravity, pending publication in J. Crystal Growth
8. H. Kölker; ESA SP 222 (1984) 169-172
9. H. Kölker; Crystalization of a Silicon Sphere in the Spacelab D1-Mission, private communication
10. D. Schwabe and A. Scharmann; ESA SP 222 (1984) 281-289
11. D. Schwabe, A. Scharmann; F. Preisser, and R. Oeder, J. Crystal Growth 43 (1978) 305-312
12. D. Schwabe and A. Scharmann; J. Crystal Growth 46 (1979) 125-131
13. Ch.-H. Chun and D. Schwabe; Marangoni Convection in Floating Zones in "Convective Transport and Instability Phenomena", Eds. J. Zierep and H. Oertel, Verlag G. Braun, Karlsruhe (1982) 297-317
14. F. Preisser, D. Schwabe, and A. Scharmann; J. Fluid Mech. 126 (1983) 545-567
15. D. Schwabe, F. Preisser, and A. Scharmann; Astronautica Acta 9 (1982) 265-273
16. D. Schwabe and A. Scharmann; ESA SP 191 (1983) 213-218
17. D. Schwabe and A. Scharmann; Z. Flugwiss. Weltraumforsch. 9 (1985) 21-28

ON THERMOCAPILLARY FLOWS IN CONTAINERS WITH

DIFFERENTIALLY HEATED SIDE WALLS

J.K. Platten and D. Villers

University of Mons
Department of Thermodynamics
B-7000 Mons, Belgium

INTRODUCTION

The problem to be investigated in this paper is that of a liquid layer in a rectangular container heated from one side and cooled from the other side. The lower wall of the container is insulated. The upper surface of the liquid is open to the ambiancy (Fig. 1). In this container, flow is driven by density differences in the gravitational field and by surface tension gradients. In a microgravity environment, surface tension gradients are the only cause of motion. Indeed, since surface tension σ is temperature dependent, it is clear that a steady surface tension gradient exists along the upper free surface. Usually surface tension decreases with temperature ($\partial\sigma/\partial T \simeq -0.15$ mN.m^{-1}.K^{-1} for water), but we do not like to restrict ourself to this usual case. Rather, we will concentrate on the less known case of systems for which surface tension increases with temperature ($\partial\sigma/\partial T > 0$). Indeed, if one looks at the surface tension of e.g. the system $MnO-SiO_2$ in the temperature range 1300 -1600 °C, one observes that surface tension may decrease or increase with T depending on the composition of this particular silicate[1]. When the content in SiO_2 exceeds 30 %, then $\partial\sigma/\partial T > 0$. It seems to be a general rule that, in binary SiO_2 systems, the temperature coefficient of surface tension increases with the percentage of SiO_2, becoming thus positive at a given percent of SiO_2.

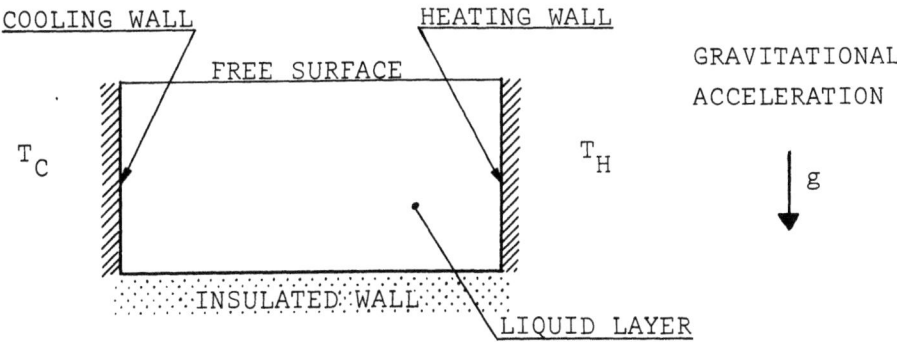

Fig. 1. Sketch of the experimental situation investigated.

Thus not only the system MnO-SiO$_2$ has a positive temperature coefficient of surface tension, but also the following systems : MgO-SiO$_2$; FeO-SiO$_2$; CaO-SiO$_2$; Li$_2$O-SiO$_2$.[2] Not only ceramics (in the temperature range 1000-1600 °C) have a positive surface tension coefficient, but it appears that this is also true for some liquid alloys, like Ag-Pb, Cu-Pb, Ag-Sn, Al-Sn, in the temperature range 600-1000 K.[3] This could be important for crystallisation experiments in microgravity. Thus we have at least two examples of materials of technological importance for which surface tension increases with temperature. One realizes immediately that it is not easy to study the flow field induced by surface tension gradients in the liquid phase of those materials in the neighbourhood of 1000 °C. Fortunately, dodecylammonium chloride (DAC) in water, has also a positive surface tension coefficient, but near room temperature[4]. This is also true for some aqueous solutions of long chain alcohols[5], say from C_5 to C_{10}, where in the notation C_i, the index i refers to the number of carbon atoms in the alcohol. The solubility of these alcohols decreases with the length of the carbon chain. For example, the solubility of n-heptanol in water is $\simeq$ 0.012 mole/liter, that of n-decanol only $\simeq$ 0.0002 mole/liter. Let us focus our attention on the system water-n heptanol at a concentration of 6.10^{-3} mole/liter. At low temperature (T < 40 °C) surface tension decreases with T, has a minimum near T $\simeq$ 40 °C and increases when T > 40 °C. The experimental data may be accurately fitted by the empirical law

$$\sigma = 45.79 - 0.29\ T + 3.80\ T^2 \qquad (1)$$

In Eq. (1), σ is expressed in mN.m^{-1}, T in °C and the concentration in heptanol is $6.3\ 10^{-3}$ mole/liter. One realizes the importance of these systems for an experimental study of the flow field in the case of a positive surface tension coefficient, since it will be sufficient to work near 50 °C. We may hope that such a study could result in a better understanding of the flow field developped in materials of technological importance in another temperature range.

There are several ways to attack the problem of thermocapillary convection : purely analytical work, numerical experiments, laboratory experiments and experiments in microgravity. Analytical work is generally restricted to simple geometry (e.g. an infinite cavity) with additional simplifying assumptions like no inertial terms, flat interface, constant value of $\partial\sigma/\partial T$,... The names of Davis, Sen, Smith[6-8], Homsy[9], Strani[10], Napolitano[11], Kirdyashkin[12] and others are associated with this kind of work. Numerical simulations of the flow field is based on the numerical solution of the complete Navier-Stokes equations, using e.g. finites differences[13] or the finite elements method[14]. Experiments may be done in microgravity (and some experiments have already be done by Legros and coworkers[15] during the Texus flights n° 8 and 10 and also during the D1-Spacelab mission) or in earth conditions. For example, Maekawa and coworkers[16] have done some experiments in silicone oil. We must also mention experimental results obtained by Kirdyashkin[12] in ethyl alcohol, but at the present time, the only experimental work on earth on systems for which $\partial\sigma/\partial T > 0$ is due to Limbourg-Fontaine et al.[15], at least to our knowledge. The central part of the present paper is devoted to a more extensive experimental study, by Laser-Doppler velocimetry, of the flow field in the system water-n-heptanol.

Our initial interest in this problem was aroused by the paper by Limbourg-Fontaine et al.[15] in which the Texus 8 experiment was described. They used a container of 3 cm length and 1 cm width. The depth of the liquid layer was typically 0.8 cm. The two side walls were maintained at T_C = 46 °C and T_H = 66 °C. The flow field was made visible by the use of rather large latex spheres (of 100 μm diameter) and the motion of the spheres was recorded by a camera. Thus an estimation of the velocity of a particular sphere can be given at different positions in the liquid. Their main result is that in the surface the motion is from cold wall to hot wall. Since in the tempera-

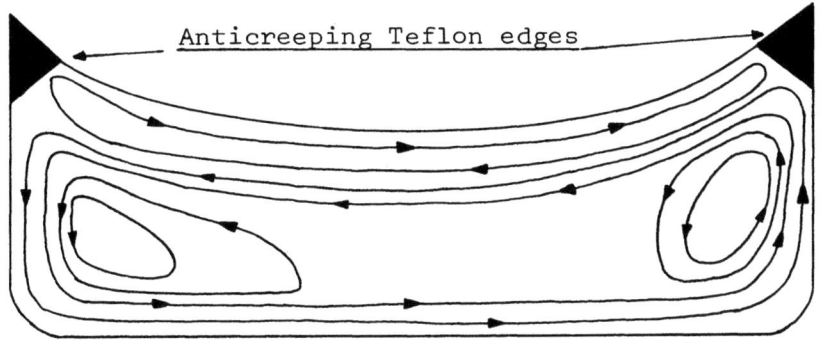

Fig. 2. Possible representation of convective cells in the system water-n-heptanol between 46°C and 66°C.

ture range 46 °C - 66 °C surface tension increases with T, the motion is thus from low surface tension to high surface tension region, as it should be. This experiment just demonstrates that in dynamical conditions $\partial\sigma/\partial T$ remains positive above 40 °C. On the other hand, this experiment will probably not serve for a quantitative comparison with theoretical predictions, since they did not operate in steady conditions. Indeed in the Texus 8 experiment, the duration of microgravity (μ-g) conditions is only 430 seconds and the filling of the observation cell was achieved 188 seconds after the lift off of the rocket. Thus at the best, they had only 250 seconds of μ-g conditions. Anyway the maximum surface velocity recorded was of the order of 0.8 mm.s^{-1}, and is thus probably too small. In the same paper[15] one may also find the results concerning the same experiment that was repeated on earth, in order to compare with the μ-g experiment. Fig. 7 of this paper[15] aims to a "schematical representation of the convective cells"; however we must confess that we do not understand what the lines drawn on the figure represent: probably trajectories of particular latex spheres at particular times. Anyway they cannot represent streamlines. The important facts in this tentative of vizualisation of the flow field are :
- the surface velocity is once again from cold wall to hot wall
- near the hot side wall, fluid raises, and goes down near the cold wall, inducing the usual thermogravitational convection cell
- probably due to the high temperature gradient imposed, secondary vortices are created inside this thermogravitational cell.

On the basis of the paper by Limbourg-Fontaine et al.[15] we propose on Fig. 2 a possible representation of the convective cells in the system water-n-heptanol ($\partial\sigma/\partial T > 0$) in the gravitational fields ; the main result to be confirmed, is the existence of two counterrotating cells : the upper cell is due to surface tension effects, the lower one to density differences. Therefore we undertook in the same cell a Laser-Doppler Velocimetry (LDV) study of thermocapillary convection. LDV is known to be a non intrusive technique since the velocity probe is optical, and there is no need to seed the flow with rather large latex spheres of 100 μm diameter, giving rise to important sedimentation problems. Also there is no need for calibration. LDV seems thus an appropriate technique for the study of thermocapillary flows. But before starting an experimental program, one has to know exactly the operating conditions and/or the important parameters to be varied and this need a theoretical support, developped in the next paragraph.

---

* The authors are very indebted to D^r. Legros and Mrs. Limbourg of the University of Brussels for the loan of their observation cells.

A THEORETICAL MODEL

The aim of this paragraph is a better understanding of the important parameters describing the flow field, rather than a precise determination of velocity and temperature profiles in the real conditions of the experiment described later in this paper. The main assumption is that we suppose that the heat sources are rejected *ad infinitum* (infinite horizontal layer) and that the upper free surface remains flat (see Fig. 3). In these conditions, the full Navier-Stokes equations have a steady solution under the form

$$V_z = 0 \; ; \; V_y = 0 \; ; \; V_x = V_x(z) \quad (2.a)$$

$$T = T(x,y) \quad (2.b)$$

solution of

$$\frac{\partial V_x}{\partial x} = 0 \quad (3.a)$$

$$\frac{1}{\rho_m} \frac{\partial p}{\partial x} = \nu \frac{\partial^2 V_x}{\partial z^2} \quad (3.b)$$

$$\frac{1}{\rho_m} \frac{\partial p}{\partial z} = - g \frac{\rho(T)}{\rho_m} \quad (3.c)$$

$$0 = - V_x \frac{\partial T}{\partial x} + \kappa \left(\frac{\partial^2 T}{\partial x^2} + \frac{\partial^2 T}{\partial z^2}\right) \quad (3.d)$$

Notations are standard; ρ_m is the mean (or reference) density. Eliminating the pressure between Eqs. (3.b) and (3.c) one gets

$$\nu \frac{\partial^3 V_x}{\partial z^3} = - \frac{g}{\rho_m} \frac{\partial \rho(T)}{\partial x} \quad (4)$$

We need an equation of state

$$\rho/\rho_m = 1 - \alpha (T - T_m) \quad (5)$$

$$\alpha = - \frac{1}{\rho_m} \frac{\partial \rho}{\partial T}$$

Eq. (4) becomes

$$\nu \frac{\partial^3 V_x}{\partial z^3} = \alpha g \frac{\partial T}{\partial x} \quad (6)$$

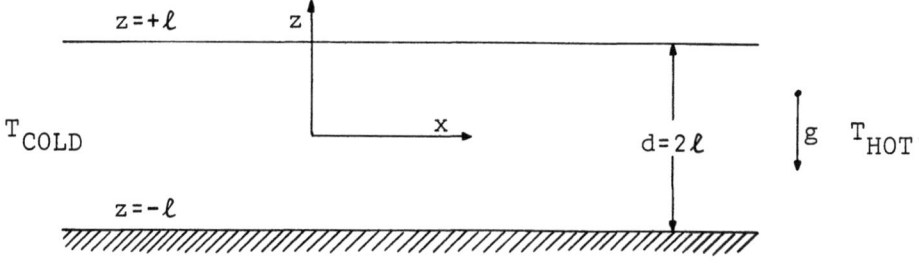

Fig. 3. Geometry and coordinate system for the theoretical model.

Since the ℓ.h.s. of Eq. (6) is a function of z only and the r.h.s. is not a function of z, the two sides must be equal to some constant K

$$\nu \cdot \frac{\partial^3 V_x}{\partial z^3} = K \qquad (7.a)$$

$$\alpha g \frac{\partial T}{\partial x} = K \qquad (7.b)$$

Integrating these two equations (7.a) and (7.b) one gets

$$V_x = \frac{K}{\nu} \left(\frac{z^3}{3!} + C_1 \frac{z^2}{2!} + C_2 z + C_3 \right) \qquad (8)$$

and

$$T(x,z) = \Theta(z) + \frac{K}{\alpha g} x \qquad (9)$$

Once known the three integration constants C_1, C_2 and C_3 in Eq. (8), the substitution of Eqs. (8) and (9) into Eq. (3.d) yields a differential equation for $\Theta(z)$. It is clear that $\Theta(z)$ must be a polynomial of the 5th degree in z, but since we are not interested in the temperature profile, the integration will not be performed here. The three integration constants involved in Eq. (8) are determined by the three following standard conditions:

$$V_x = 0 \quad \text{for} \quad z = -\ell \qquad (10.a)$$

$$\int_{-\ell}^{+\ell} V_x(z) \, dz = 0 \qquad (10.b)$$

$$\rho_m \nu \left(\frac{\partial V_x}{\partial z} \right) = \frac{\partial \sigma}{\partial T} \cdot \left(\frac{\partial T}{\partial x} \right) \quad \text{for } z = +\ell \qquad (10.c)$$

In Eq. (10.c) we suppose $\partial\sigma/\partial T$ to be a constant, but its sign is not *a priori* prescribed. We may thus with this approximation describe systems with negative (as usual) or positive temperature surface tension coefficients. We just impose that surface tension decreases or increases linearly with T. The model is thus unadequate to describe the system water-n-heptanol for which surface tension is a quadratic function of T. But on the other hand, if the temperature gradient is small enough, one may always approximate the surface tension variation by a linear law. With the conditions (10) we obtain

$$C_1 = \frac{3}{4} \frac{\partial\sigma/\partial T}{\rho_m \alpha g \ell} - \frac{\ell}{4} \qquad (11.a)$$

$$C_2 = \frac{1}{4} \frac{\partial\sigma/\partial T}{\rho_m \alpha g} - \frac{\ell^2}{4} \qquad (11.b)$$

$$C_3 = -\frac{1}{8} \frac{\partial\sigma/\partial T}{\rho_m \alpha g} \ell + \frac{\ell^3}{24} \qquad (11.c)$$

By substituting these integration constants into Eq. (8), and by using a dimensionless space coordinate $Z = z/\ell$ ($-1 \leq Z \leq +1$), one gets the velocity profile

$$V_x = \left(\frac{\partial T}{\partial x} \right)_{Z=1} \frac{g\alpha\ell^3}{\nu} \left[\frac{1}{6} Z^3 - \frac{(1-3k)}{8} Z^2 - \frac{(1-k)}{4} Z + \frac{(1-3k)}{24} \right] \qquad (12)$$

In this last expression, k is defined by

$$k = \frac{\partial\sigma/\partial T}{\rho_m g\alpha\ell^2} = -\frac{Ma}{Ra} \; ; \; Ma = \frac{-(\partial\sigma/\partial T) \Delta T \ell}{\rho_m \kappa \nu} \; ; \; Ra = \frac{g\alpha \Delta T \ell^3}{\kappa \nu} \qquad (13)$$

Thus it appears that k is minus the ratio of the Marangoni number to the Rayleigh number and that the shape of the velocity profile (i.e. the expression between brackets in Eq. (12)) does depend on k only. Therefore the shape of the velocity profile is independent of the temperature difference between cold and hot walls, which only appears in a multiplying factor defining a characteristic velocity U^*. Also there is no need to define ΔT, both in the Marangoni and Rayleigh numbers, since this ΔT cancel in their ratio. On the other hand, it would be a mistake to replace in Eq. (12) $(\partial T/\partial x)_{Z=1}$ by $(T_H - T_C)/L$ where L is the length of the container. Thus at the present time Eq. (12) cannot serve for a quantitative comparison with experimental velocity profiles. Only the shape of the curve, related to k has some importance for the time being. Let us now examine the different cases of Eq.(12).

<u>Limit k = 0</u>

That means clearly that the Marangoni number must be equal to zero, or in other words, as far as the boundary condition is concerned, that the following condition is fulfilled: $\partial V_x/\partial Z = 0$ at $Z = 1$. This is a so-called "free boundary". The velocity profile reduces to

$$V_x = (\frac{\partial T}{\partial x})_{Z=1} \frac{g\alpha\ell^3}{\nu} \frac{1}{24} [4Z^3 - 3Z^2 - 6Z + 1] \qquad (14)$$

and has roots at $Z_1 = -1$; $Z_2 = (7-\sqrt{33})/8 \simeq 0.157$ and $Z_3 = (7+\sqrt{33})/8 \simeq 1.593$ (i.e. Z_3 is outside the liquid layer). Also, V_x possesses extrema at $Z_4 = -1/2$ and $Z_5 = +1$. This velocity profile is sketched on Fig. 4. In the surface (Z=+1) the motion is from hot to cold wall, since $\partial T/\partial x > 0$ according to Fig. 3.

<u>Limit k → − ∞</u>

Since k is negative, that means that we are in the usual case $\partial\sigma/\partial T < 0$; moreover g = 0. This limit corresponds to the case where thermogravitation is not acting and thermocapillarity alone induces convection. The velocity profile reduces to:

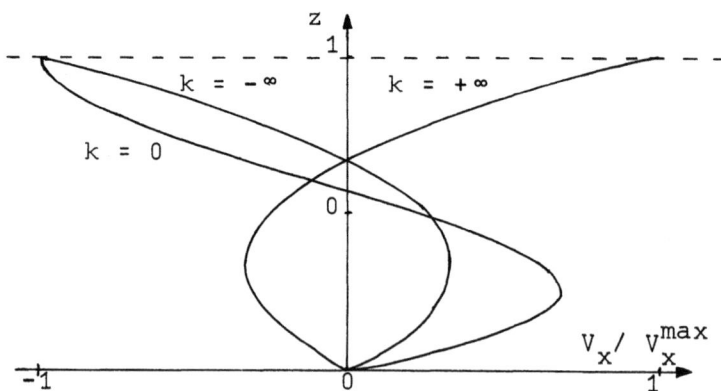

Fig. 4. Plot of equation (12) with the three limits: k = −∞ , 0 and +∞ .

$$V_x = \left(\frac{\partial T}{\partial x}\right)_{Z=1} \left(\frac{\partial \sigma}{\partial T}\right) \frac{\ell}{8\rho_m \nu} (3Z^2 + 2Z - 1) \quad (-1<Z<+1) \quad (15)$$

This is exactly the velocity profile that has been termed "the return flow solution" in the paper by e.g. Davis et al.[6-8] This velocity profile is also shown on Fig. 4 and has roots at $Z_1 = -1$ and $Z_2 = 1/3$ (instead of 0.157). V_x possesses an extremum at $Z_4 = -1/3$ (instead of -0.5). In the surface, $V_x < 0$, and according to Fig.3, the motion is from hot to cold wall.

Limit $k \to +\infty$

Since k is positive, this means that we are in the unusual case $\partial\sigma/\partial T > 0$. Also Ra=0. Eq. (15) is still valid, but the directions of motions are reversed, i.e. from cold to hot wall at the interface.

General Case

Let us now examine the general case given by Eq.(12). The surface velocity is given by:

$$V_x^{surface} = U^\star \left(\frac{3k-1}{6}\right) \quad (16)$$

While k is negative, surface tension forces act to produce surface motion in the same direction as the thermogravitational motion characterized by k = 0. Of course, when $k < 0$, $|V_x|$ in the surface is greater that when k = 0, since there are two causes of motion acting to produce both motion from hot to cold wall in the surface. For positive values of k ($\partial\sigma/\partial T > 0$, the unusual case), the surface velocity remains negative while $k < 1/3$. Since surface tension increases with temperature, the motion should be from cold to hot wall due to thermocapillarity alone, but taking into account the "dominating" gravitational forces (since k is small, $k < 1/3$), the motion in the surface remains from hot to cold wall, but the amplitude of the surface velocity is reduced compared to the case k = 0.

At the value k=1/3, surface tension and buoyancy forces taken separately would produce surface velocity of the same amplitude, but with a change of sign. As a result of the presence of both forces, the surface velocity is exactly zero: there is no surface motion and the flow is symmetric, as for an upper rigid boundary:

$$V_x = U^\star \left(\frac{Z^3 - Z}{6}\right) \quad \text{with } k = 1/3 \quad (17)$$

When $k > 1/3$, the result of the dominating surface tension forces compared to buoyancy, is to produce a positive surface velocity, i.e. motion from cold to hot wall. Thus in the range $1/3 < k < 1$, Marangoni and buoyancy convections compete and two counterrotating convective cells are created, according to Fig. 5, where the shape of the velocity profile has been plotted for different values of k. The velocity profile possesses extrema at values of Z which has already been called Z_4 and Z_5. One finds easily

$$Z_{4,5} = \frac{1 - 3k \pm \sqrt{9k^2 - 14k + 9}}{4} \quad (18)$$

Thus when k goes from 0 to $+\infty$, Z_5 varies from +1 to -1/3, whereas Z_4 goes from -1/2 to $-\infty$. For $k < 0$, Z_5 is outside the liquid layer and $-1/3 < Z_4 < -1/2$. There is a particular value of k such that $Z_4 = -1$ (and also $Z_5 = 0$), namely k = +1. At this value of k, there is only one cell with a clockwise rotation, i.e. a motion from cold to hot wall in the surface. Thus when $k > 1$, the lower cell, due to buoyancy forces, disappears. All these results are shown on Fig. 5.

In conclusion, the crude theoretical model reveals different regimes:
(i) one convective cell with an anticlockwise rotation, i.e. in the direction of the usual thermogravitational convection.
(ii) two superposed counterrotating convective cells when $1/3 < k < 1$.
(iii) one convective cell due to surface tension forces in the reverse direction of the thermogravitational motion when $k > 1$.

Therefore, these results suggest experiments at different values of k, thus experiments in which the depth of the liquid layer is changed, instead of the temperature gradient which should be kept constant, and not too high in order to avoid an instability of the basic velocity profile.

A NUMERICAL MODEL

The crude theoretical model presented in the previous paragraph has some shortcomings and therefore we cannot hope a quantitative comparison between experimental results obtained on the system water-n-heptanol and the predictions of the model. First of all, the model supposes an horizontal infinite layer with no vertical velocity component. Any real experiment is conducted in a container of finite size of given aspect ratio. In the "Space Experiments", described in paper[15], the aspect ratio is of the order of 4, and probably for technical reasons, the lenth of the apparatus could not be extended much beyond a few centimeters. If one wants to compare space experiments with earth experiments, it is reasonable to perform the experiments in the same apparatus. Thus we would like to have theoretical predictions of the flow field in a container with prescribed aspect ratio, taking into account the vertical velocity component near the two lateral boundaries. Also the crude theoretical model supposes that surface tension increases or decreases linearly with temperature. On the contrary, surface tension of the system water-n-heptanol is a quadratic function of T, as shown by Eq.(1). In particular the theoretical model of the previous paragraph is unable to describe the flow field near 40°C, i.e. near the minimum in the surface tension curve. Also the horizontal temperature gradient at the surface is an important parameter and there is no way to relate it to the temperature difference between cold and hot walls, since these walls (or heat sources) are rejected at infinity. For all these reasons, we would like to have numerical predictions of the flow field, thus to develop a numerical model which

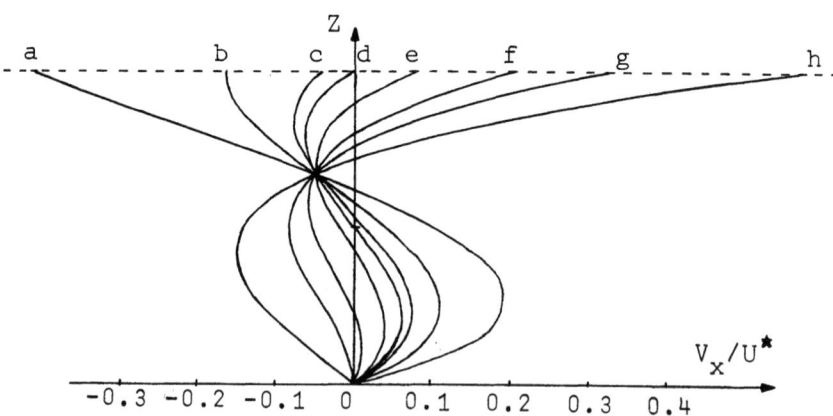

Fig. 5. Plot of function (12) for different values of k
(k = -0.5 (a), 0. (b), 0.25 (c), 1/3 (d), 0.5 (e), 0.75 (f), 1. (g), 1.5 (h)).

releases some of the unacceptable assumptions of the theoretical model. We shall go into a minimum of details here, since the model, and the main results, have already been given elsewhere[13]. Let us say that the full Navier-Stokes equations, including the nonlinear terms, have been numerically integrated, using standard finite differences for the space derivatives. For the time integration, in order to find a steady state solution starting from some prescribed initial conditions, we have used the standard ADI scheme. The main assumption that remains in the numerical model, is that the upper surface remains flat, even if a flat surface may not be compatible with the correct contact angles at the boundaries. Anyway this numerical model seems more appropriate since it takes into account the quadratic variation of surface tension with temperature. In fact this numerical model was developped before the crude theoretical model and therefore the results already published in paper[13] should be considered as preliminary results, since some of the predictions of the theoretical model were not checked by numerical experiments.

When integrating the full Navier-Stokes equations, it is very common to use dimensionless quantities for space, time, temperature, velocity or stream function, etc... When doing so, dimensionless numbers will appear in the starting partial differential equations, namely the Prandtl number and the Rayleigh number, or the Grashof number defined as the ratio of the Rayleigh to the Prandtl number. The boundary conditions are also modified, compared to the boundary conditions used in the theoretical model. First of all, we have now lateral side walls on which some conditions must be given. Evidently, they are at prescribed temperatures. Also, they are taken to be rigid (no slip boundary conditions, implying $V_x = V_z = 0$). The lower wall is also rigid and is taken insulating. At the upper free boundary, we need first a thermal boundary condition, that we write as:

$$\frac{\partial T}{\partial z} + Bi (T - T_{gas}) = 0 \qquad (19)$$

where Bi is the Biot number and T_{gas} is the temperature of the gas phase above the liquid layer. Most of the numerical experiments were performed with Bi=0. More drastic are the changes in the mechanical conditions at this upper free surface. We have now

$$\rho_m \nu (\frac{\partial V_x}{\partial z} + \frac{\partial V_z}{\partial x}) = \frac{\partial \sigma}{\partial T} \qquad \text{with} \quad \sigma = A T^2 + B T + C \qquad (20)$$

Thus in the nondimensional analysis, two different dimensionless groups may appear (say two Marangoni numbers M_1 and M_2) related to the two material constants A and B. These two Marangoni numbers are defined by (see paper[13] for more details)

$$M_1 = -(B + 2A(\frac{T_H + T_C}{2})) \frac{\Delta T \, S_\ell}{\kappa \, \rho_m \, \nu} \qquad (21)$$

$$M_2 = \frac{2A \, \Delta T^2 \, S_\ell}{\kappa \, \rho_m \, \nu} \qquad (22)$$

where S_ℓ is some length scale. When A = 0, M_1 becomes formally identical to the usual Marangoni number. Thus, we see here that ΔT appears explicitely both in the two Marangoni numbers and in the Rayleigh number, in contradistinction with the model developed in the previous paragraph where there was no need to define ΔT since only the ratio Ma/Ra was relevant. In this problem it is quite natural to take for the temperature scale $\Delta T = T_H - T_C$, i.e. the temperature difference between the side walls, instead of the temperature

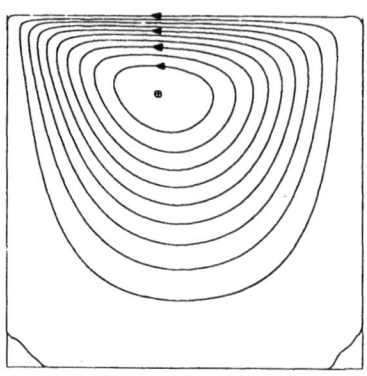

Fig. 6. Computed streamlines without gravity forces (Ra = 0) when $M_1 = 200$ and $M_2 = 0$.

difference between bottom and top (for example at the center of the cavity) as one should do in the standard Bénard problem where the liquid layer is heated from below. Also, the length scale will be taken as the distance between the lateral walls (by analogy with the temperature scale). Thus $S_\ell = L$, but we could without difficulty have chosen $S_\ell = d = 2\ell$, and the ratio of these two length scales is the aspect ratio of the container, which must appear somewhere in the computer program. In paper[13], we have thus taken $S_\ell = L$.

Let us now show some results that were obtained by means of these numerical experiments.

Results without Gravity Forces : (Ra = 0)

In the first case, M_2 is taken equal to zero to retrieve the classical case of a linear variation of surface tension with temperature. When $M_1 > 0$ ($\partial\sigma/\partial T < 0$), on the surface the fluid comes from the hot wall (low surface tension) and goes to the cold wall, and this motion induces convection in the bulk (Fig. 6). Let us observe that two little vortices are created in the lower corners by the motion in the bulk phase. On the contrary, when $M_1 = 0$ (and $M_2 \neq 0$), we are at a mean working temperature $(T_H + T_C)/2$ equal to the

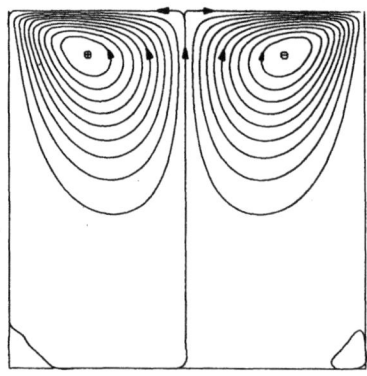

Fig. 7. Computed streamlines without gravity forces (Ra = 0) when $M_1 = 0$ and $M_2 = 1000$.

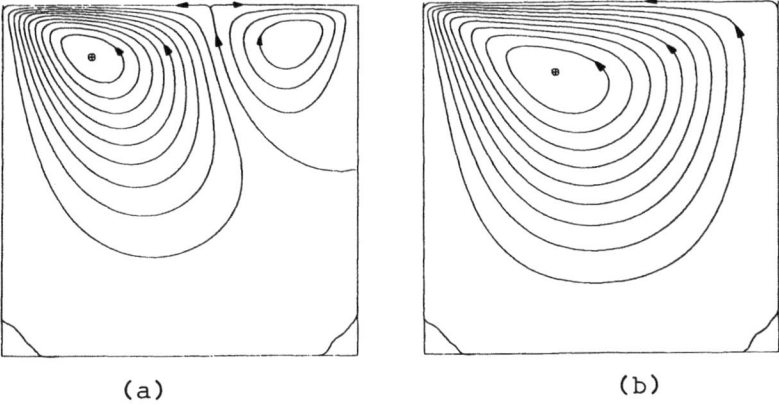

Fig. 8. Computed streamlines without gravity forces (Ra = 0) when M_2 = 1000 and M_1 = 100 (case(a)) or M_1 = 500 (case (b)).

temperature for which surface tension is minimum and the numerical value of M_2 is a measure of $\Delta T = T_H - T_C$. Once again the motion is from low surface tension region to high surface tension region and this induces two convective cells as displayed on Fig. 7. The flow pattern is perfectly symmetrical. If we now take the same value for M_2, but we increase M_1 in such a way that the dimensionless temperature at which the minimum in surface tension occurs, increases up to the temperature T_H of the right wall, the left cell, corresponding to a region of the surface where σ decreases with T, becomes more important and the null central streamline shifts towards the right wall (Fig. 8.a) and even disappears when M_1 is large enough (Fig. 8.b). We have also done numerical experiments in which the aspect ratio is changed, by increasing the depth of the layer. We have observed that the two upper convective cells remain almost unchanged (at constant values for M_1 and M_2) both in amplitude and in real size. Thus, the surface tension forces create convective cells which are almost indifferent to the exact position of the bottom of the container when it moves downwards. However the two small vortices in the corners and shown on Fig. 8.a are now combined into one single convective cell of very small amplitude (say one thousand of the velocity at the surface).

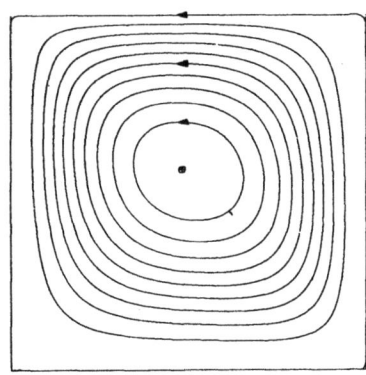

Fig. 9. Computed streamlines with conjugated buoyancy and surface tension forces Ra = 7000, M_1 = 250 and M_2 = 0.

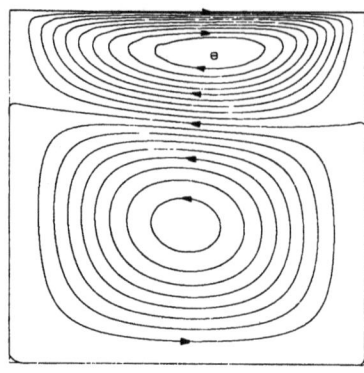

Fig. 10. Computed streamlines with two superposed cells when Ra = 700, M_1 = -200 and M_2 = 0.

Results with Gravity Forces : (Ra ≠ 0)

More relevant to the present work are the results when both gravity forces and surface tension forces act together to determine the flow field. When gravity forces act alone, the fluid rises near the hot wall and goes down near the cold wall (in the usual case when density decreases with increasing temperature). In Fig. 9, we show the case when surface tension and gravity forces act together and we suppose that surface tension decreases linearly with temperature (M_1 > 0 and M_2 = 0) pulling the fluid at the upper boundary from hot wall to cold wall. Therefore, the two effects cooperate and we observe only one convective cell. On the other hand, when the first Marangoni number M_1 is negative and M_2 = 0, (surface tension increases linearly with temperature), the surface tension forces pull near the upper boundary a fluid element from cold wall to hot wall. Therefore due to the two combined effects, two superposed counterrotating cells are created (Fig. 10), if $|M_1|$ is large enough. The relative amplitude of the maximum velocity in both cells depends on the particular values adopted for M_1 and Ra. We have also studied the effect of the aspect ratio on the convective flow pattern. For example, if we decrease the height of the container, the real size of the upper cell remains also the same (if the aspect ratio is not too

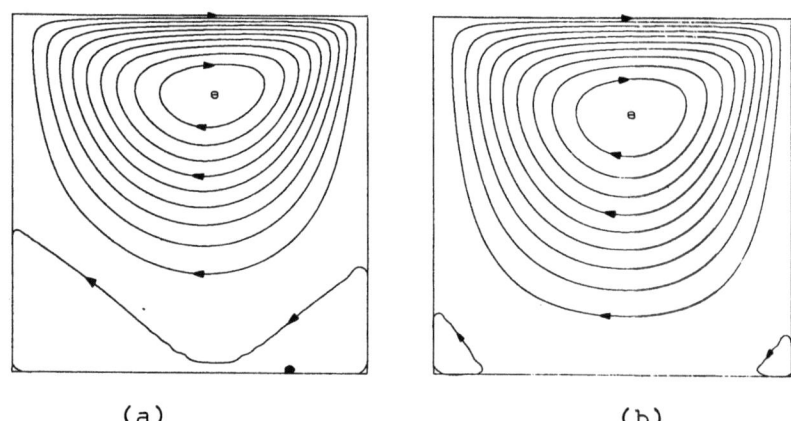

(a) (b)

Fig. 11. Computed streamlines when Ra = 700, M_2 = 0 and M_1 = -200 in cavities where $L/2\ell$ ≅ 1.33 (case (a)) and $L/2\ell$ = 2 (case (b)).

small), but the lower cell due to gravity forces finally breaks into two small vortices (Fig. 11.a and b). Thus the variation of the horizontal velocity component with Z, deduced from Fig. 10, 11.a and 11.b, ressembles successively to curves e,g and h of Fig. 5. Thus the results of the more realistic "numerical model" are in qualitative agreement with the results of the crude theoretical model. We have also verified that when $|M_1|$ is not large enough, only one cell is observed, with an anticlockwise rotation and a very small negative surface velocity. Additional details concerning computed steamlines and isotherms may be found in the original paper[13]. Since the present model seems more appropriate to real experiments in the system water-n heptanol, we may hope a quantitative comparison with experiments. With the crude theoretical model, only a qualitative agreement can be expected. However, as we shall see later, this is a too optimistic view of reality.

EXPERIMENTAL RESULTS

As already said, the experiments were conducted in the observation cells that were used by the Brussels group for their own observations, including the microgravity experiments. Thus a more detailed description may be found in the original paper[15]. Let us recall here that we have a rectangular container of length L = 3 cm., the width being 1 cm. In the experiments described below, the height of the liquid layer d = 2ℓ will be changed. A constant temperature difference of 5 °C was imposed between the side walls of the container. For the system water-n heptanol both T_C and T_H have to be greater than 40 °C, a temperature for which surface tension is minimum. We have taken T_C = 49.5 °C and T_H = 54.5 °C. Therefore in this temperature range we have $\partial\sigma/\partial T > 0$. The shape of the horizontal velocity profile on the vertical median of the cavity was recorded and we are interested in the variation of this profile when we change the height of the liquid. We have also done a comparison experiment in pure water.

The Technique

The velocities are measured by a Laser-Doppler Velocimetry (LDV) system from T.S.I., consisting of a Spectra Physics 15 mW He-Ne laser (λ = 632.3 nm), a beamsplitter (the spacing between the two beams of equal intensity after the splitting is 50 mm.), a focusing lens of focal length 120 mm. Rotating mounts provide convenient rotation of both the beam polarity and the beamsplitter which then rotates the direction of measurement. Sometimes a beamspacer is added in order to change the 50 mm. spacing between the two beams at the exit of the beamsplitter. At the intersection of the two beams, interference fringes are formed and the fringes spacing is uniquely determined by the wavelength of the laser light and the angle between the two beams, in other words, by the spacing between the two beams at the entrance in the lens and by the focal length of the lens. In the conditions of the present experiments, the fringes spacing is 1.7 µm (details concerning the LDV technique may be found e.g. in[17]). Particles moving with the fluid (either natural like in tap water, or sometimes added, like small latex spheres of 0.9 µm diameter when the working fluid is very clean) become source of scattered light when they pass in the optical probe which is the intersection of the two laser beams. The velocity of these particles, which is assumed also to be the velocity of the fluid, is determined by multiplying the fringes spacing by the frequency of the periodic component of the intensity of the scattered light. Thus what we determine is the velocity component perpendicular to the planes of the fringes. By rotating all the optical components, one rotates also these planes and one is able to measure all velocity components in a plane perpendicular to the optical axis. In the present experiment, we are able to measure all the components of the velocity vector in the z-x plane. The measurement of the third component (aligned parallel to the width) requires another optical arrangement. Thus the scattered light is

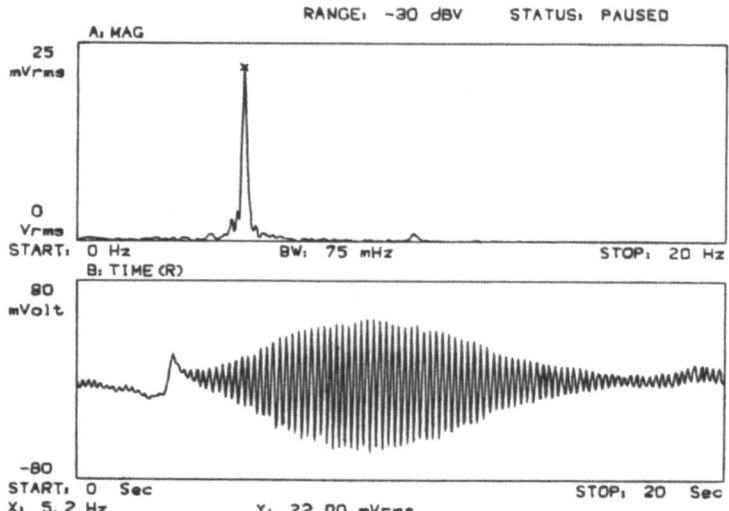

Fig. 12. Typical LDV signal and its Fourier transform.

collected by a receiving optics assembly and focused on a photomultiplier tube. The a.c. component of the electrical signal is next analysed in frequency. This is performed by taking the Fourier transform of the discretized signal, using of course the Fast Fourier Transform algorithm. This is usually done on 1024 sample points and with any computer, but in practice one asks to the computer to work in real time, at least if the frequency is not too high. There are now implementations of FFT on specialized minicomputers with continuous representation on the screen of the signal and its Fourier transform. The analysis of the signal was performed in the present case with the H.P. 3561A Dynamic Signal Analyser, which performs the FFT in real time up to frequencies of about 7.5 KHz. A typical example of the signal and its Fourier transform is given on Fig. 12 and shows the quality of the measurement even at frequencies as slow as $\simeq$ 5 Hz (or velocities $\simeq$ 10 µm/s). At the time the experiments were done, we did not yet have a frequency shifter system (Bragg cells) in order to resolve the directional ambiguity, since standard Laser-Doppler velocimeters cannot distinguish between forward and reverse flow. Therefore the directions of the velocity was determined by direct visual observation through the receiving optics assembly, with the help of a focusing eye-piece ; this is made possible thanks to the small velocities to be observed.

Marangoni Convection in Pure Water

First of all, let us quote Levich's book (see reference[18], p. 390) :
"Experiments with liquids having clean surfaces are relatively rare in actual practice. Usually, the liquid is covered with surface active substances to a greater or lesser degree. Such a film modifies the forces exerted on the surface of the fluid ... The presence of surface active substances, therefore, may lead to a significant change in the hydrodynamic regime".
This is particularly true with water, where the presence of traces of these surface active substances reduces drastically the surface tension. Therefore, to our knowledge, Marangoni convection in pure water has never been reported before. Thus cleanness of the surface is the key to a successful thermocapillary convection experiment. It is our experience that water from the tap, that has flown during several hours, does not contain these surface active substances. However, all the containers and accessories in contact with the

working fluid and specially the experimental cell, need a serious "cleaning" or "rinsing" before they can be used. This supposes a washing in sulfochromic acid and a rinsing by tap water during e.g. 12 hours. Sometimes a steam bath of 4 hours is sufficient. Also all the manipulations of the different containers and accessories must be performed wearing gloves. When the working fluid is not water alone, the other component must be distillated (say 2 or 3 times) and, as a matter of fact, the distillation apparatus must also be clean in the sense described before. When working in these conditions of cleanness, one may see Marangoni convection during a few hours (i.e. the duration of the determination of a velocity profile) if one has some luck. Fig. 13 shows the variation of the horizontal velocity component with Z, for a layer of $\simeq 7.8$ mm. height, when $T_C = 24.76$ °C and $T_H = 29.82$ °C. We say that this experiment with a clean surface (the experimental points are labelled with ●) demonstrates Marangoni convection in pure water. Indeed if the upper surface was a real "free" surface with no surface tension effect, then $Ma = 0$ and Eq. (14) would be applicable implying $(\partial V_x/\partial z) = 0$ at the surface. Now, this condition does not seem to be experimentally true. Also, a measure of the relative importance of the surface velocity is e.g. the ratio of the surface velocity to the maximum velocity in the return flow, and following Eq. (14) this ratio is

$$\frac{V_x(Z = 1)}{V_x(Z = 1/2)} = -1.46 \tag{23}$$

Experimentally this ratio is of the order of -3 ; thus the surface velocity is twice the surface velocity that would prevail if surface tension effects were ignored. Therefore, we believe that the present experiment demonstrates unambiguously the existence of Marangoni convection. As a matter of fact, when the liquid has stayed for a lot of time in the container, the surface must become unclean. Thus, once the velocity profile has been determined, say from top to bottom, one has to check again the surface velocity. A decrease of 10 % in the surface velocity is allowed between initial and final time of the experiment, otherwise the data are not validated and one has to clean the container again and restart a new set of velocity determinations. On the other hand, if one waits 12 hours the initially clean surface can been polluted to such a degree to produce a drastic reduction in the surface velocity (Fig. 13, where the corresponding points has been labeled +). This time the

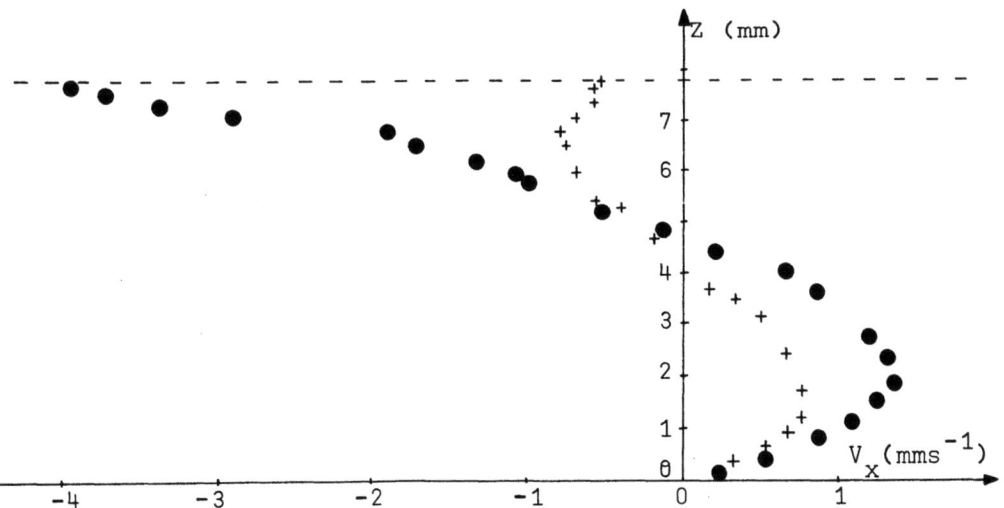

Fig. 13. Marangoni convection in pure water: clean surface(●) and polluted surface(+) after 12 hours.

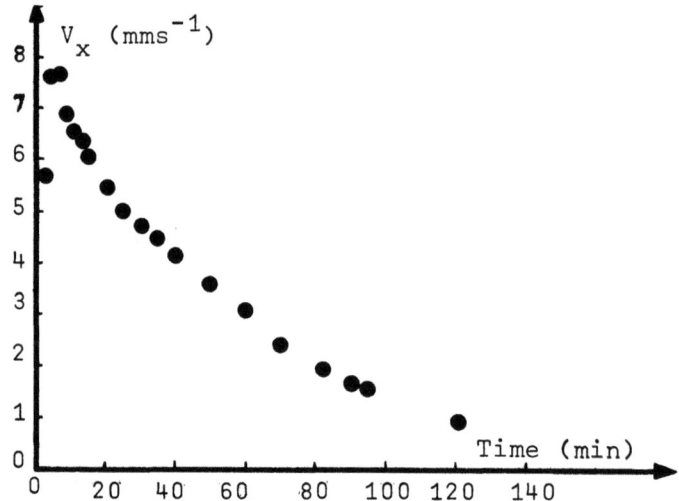

Fig. 14. Decrease of surface velocity due to surface active substances.

velocity is no longer maximum in the surface as it should be for a "free" surface without surface tension effects. The film formed by these surface active substances has a very significant effect in the viscous flow regime and the motion of the surface active substance, with its own viscosity, should be examined. In some sense, the fluid layer, instead of being bounded at its top by a rigid boundary (another layer of infinite viscosity) implying $V_x = 0$, or a gas phase of almost zero viscosity implying $\partial V_x/\partial z = 0$, is in reality bounded by a very thin layer of a substance of a priori unknown viscosity and this implies as boundary condition

$$\frac{\partial V_x}{\partial z} = B V_x \qquad (24)$$

Here B is a constant which contains the viscosities of the working fluid and of the film, as well as their thicknesses. In view of the curve obtained in "pure water" after 12 hours, it seems realistic to propose that the boundary condition (24) is fulfilled.

Sometimes one has no luck from the beginning, and the upper surface becomes covered with these undesired substances from the beginning of the experiment, implying that the surface velocity decreases immediately with time. Fig. 14 presents results of such an unlucky experiment : the surface velocity decreases almost exponentially with time. Therefore in any well conducted thermocapillary convection experiment, one has to wait a certain time in order to check if there is no decrease in the surface velocity, next to perform the measurements as quick as possible (it is our experience that this takes typically one hour) and finally to check again the surface velocity. This procedure for "earth experiments" is hard to follow in microgravity conditions at the present time. We shall come back to the implications of the microgravity procedure later, and show that it prevents a quantitative comparison with theoretical predictions.

Thermocapillary Convection When Surface Tension Increases with T

We present now several experiments that were realized with different heights of liquid in a solution water-n heptanol $6.3 \cdot 10^{-3}$ mole/liter. At this concentration and at a mean temperature of 52 °C, surface tension increases with temperature, approximatively $\partial \sigma/\partial T = + 0.1$ mN.m^{-1}.°C^{-1} according to Eq. (1).

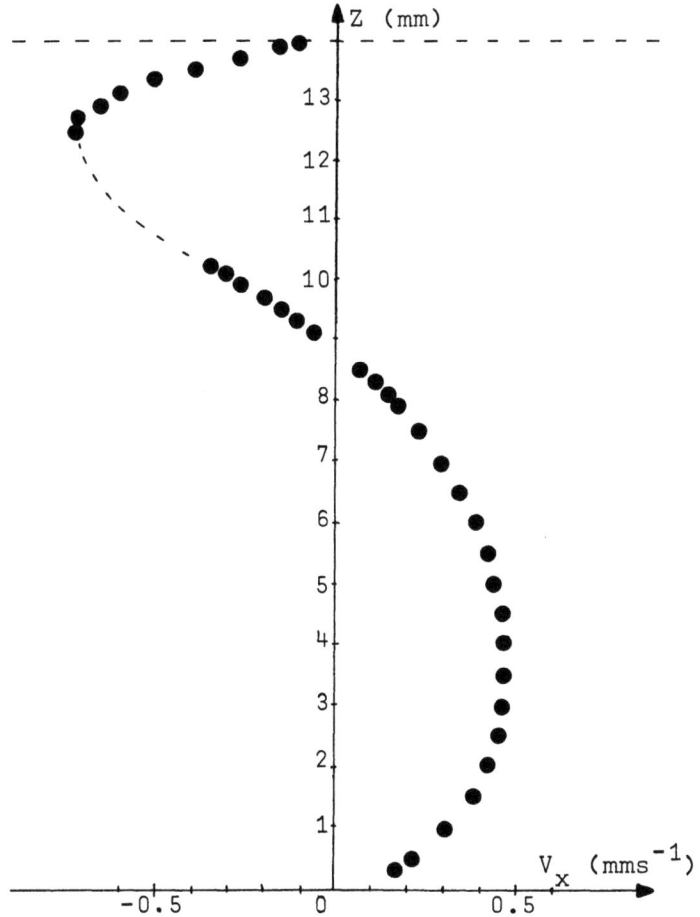

Fig. 15. Horizontal velocity profile on the vertical median of a 14 mm layer of water-n-heptanol solution.

a) $2\ell = 14$ mm : We observe one single convective cell with an upward motion near the hot wall, a downward motion at the cold wall, motion from cold to hot near the bottom and from hot to cold at the surface of the fluid. The measured horizontal velocity profile at the middle of the container is given on Fig. 15. This profile differs from the usual thermogravitational convection profile (Fig. 4, case k = 0) by a substantial reduction of the velocity at the interface, due to the unusual surface tension effect ; still the surface velocity is from hot to cold.

b) $2\ell = 10.2$ mm : With this slight decrease of the height of the fluid, surface tension forces overcome buoyancy and we observe a surface motion from cold to hot. (Fig. 16). This implies the coexistence of two counterrotating convective cells. In this case the upper cell (induced by surface tension effects) remains thin, say 1/6 of the total height, and the surface velocity is less than the ones in the core flow.

c) $2\ell = 7.4$ mm : The velocity profile given on Fig. 17 ressembles the previous case, except that now there is a substantial increase of the surface velocity which is now greater than anywhere in the bulk phase.

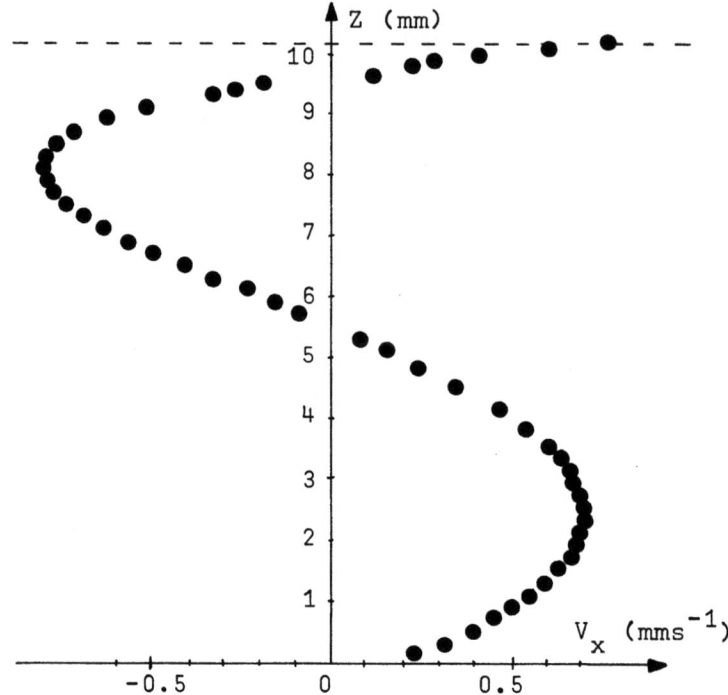

Fig. 16. Horizontal velocity profile on the vertical median of a 10.2 mm layer of water-n-heptanol solution.

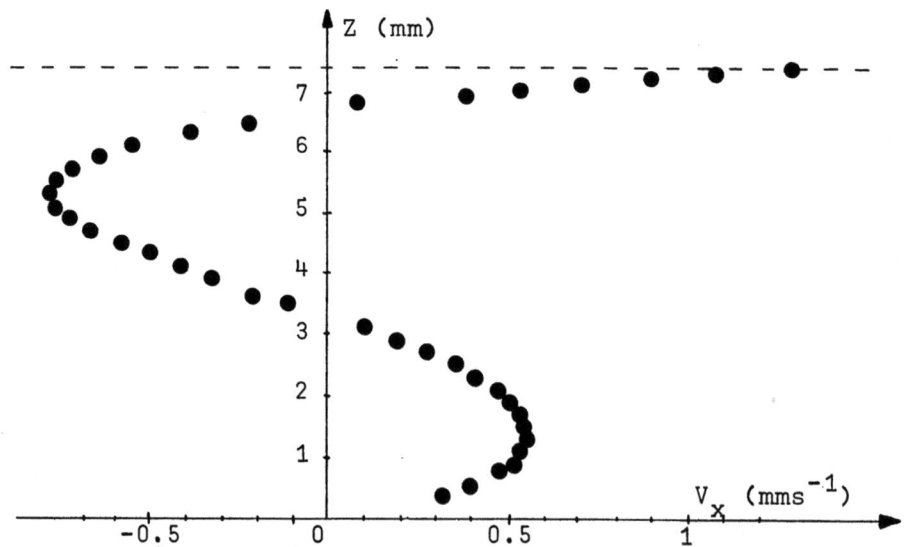

Fig. 17. Horizontal velocity profile on the vertical median of a 7.4 mm layer of water-n-heptanol solution.

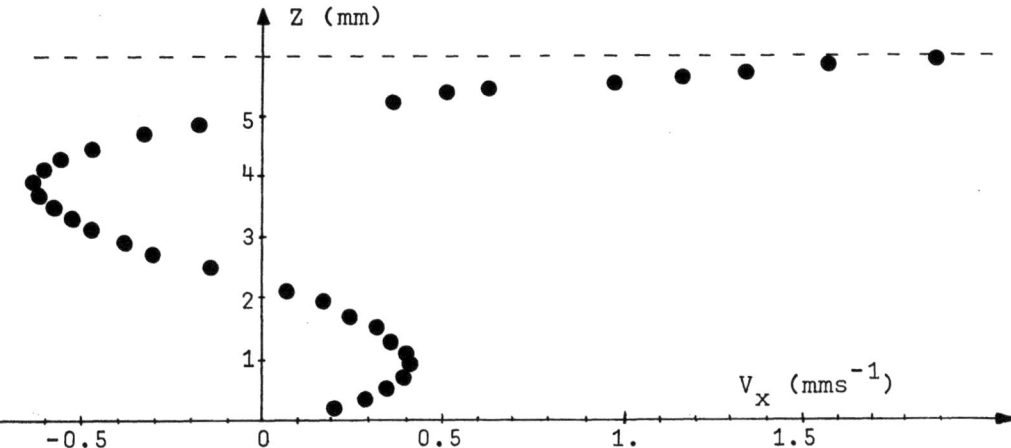

Fig. 18. Horizontal velocity profile on the vertical median of a 6 mm layer of water-n-heptanol solution.

d) $2\ell = 6$ mm: The velocity profile is displayed on Fig. 18 and once again corresponds to two counterrotating convective cells. However there is an important difference between the results of Fig. 16, 17 and 18: the points of zero velocity are shifted downwards. This means that the size of the convective cell due to surface tension gradients increases compared to that of the convective cell induced by density differences.

e) $2\ell = 4.9$ mm: By still lowering the height of the layer (or by increasing k), we observe only single convective cell induced by surface tension effects. The convective cell due to buoyancy has disappeared (Fig. 19). Nevertheless, we believe that the influence of gravity field can still be observed on Fig. 19. Indeed one may clearly observe near the bottom of the container that the experimental points verify more or less the condition $\partial V_x/\partial z \simeq 0$. Since the lower boundary is rigid, this observation can only be explained by the fact that $k = 1$ (Ma = Ra), as shown on Fig. 5, curve g.

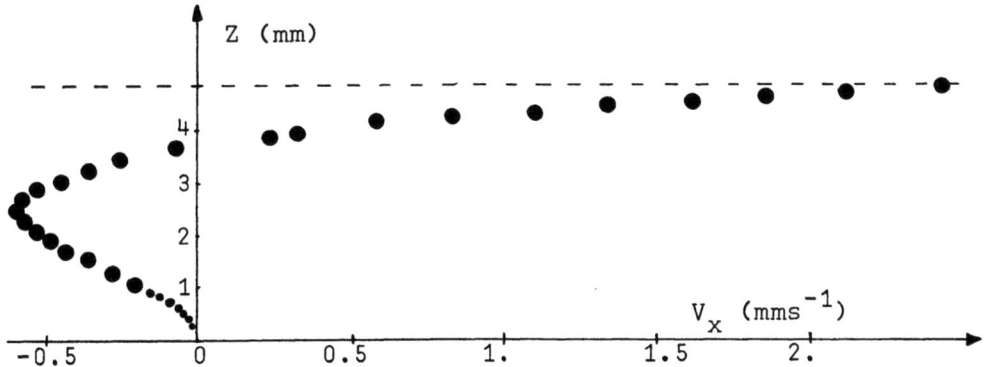

Fig. 19. Horizontal velocity profile on the vertical median of a 4.9 mm layer of water-n-heptanol solution.

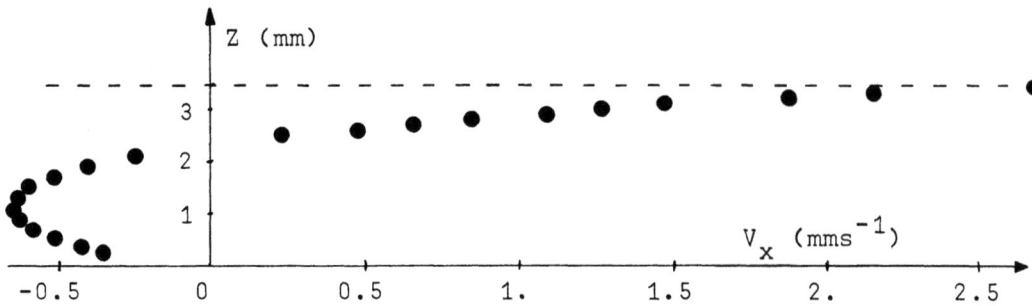

Fig. 20. Horizontal velocity profile on the vertical median of a 3.3 mm layer of water-n-heptanol solution.

f) $2\ell = 3.3$ mm : This time, the liquid layer is so thin, that we cannot discern any gravity effect on the horizontal velocity profile (Fig. 20). The only apparent cause of convection is the surface tension gradient and we have a motion from cold to hot wall at the top of the container, a downward motion near the hot wall, an upward motion near the cold wall (very unusual indeed!) Thus this type of motion is exactly the one that has been observed in microgravity conditions ; but on earth, we have the advantage to wait, if necessary, for a steady state, to restart a set of experiments if the surface becomes unclean for some reason, etc ...

We should also note that in the two last cases, two little vortices remain in the corners of the cavity, as sketched on Fig. 21. This is demonstrated by measuring V_x along the bottom of the container. Of course it is not easy to go very close to the two side walls since one of the laser beams will bump the lateral wall. Anyway, we have measured $V_x(x)$ 0.6 mm. above the bottom of the container, with $2\ell = 4.5$ mm. and the result is given on Fig. 22. Clearly near the two side walls the horizontal velocity is positive. However, the profile is not symmetric with respect to the center located at x = 15 cm. Also V_x is more or less constant in the central part of the cavity. These two small vortices could be due to the residual effect of gravity or they could be secondary vortices induced by the upper convective cell due to surface tension gradients.

Finally we have also measured in one case the variation of the horizontal velocity component along the surface (in fact as close as possible to the surface, and this means 0.1 mm. which is the dimension of the optical probe).

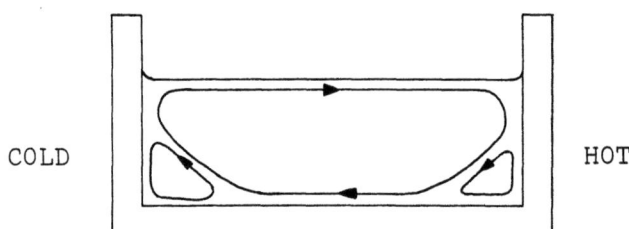

Fig. 21. Sketch of one convective cell with little vortices.

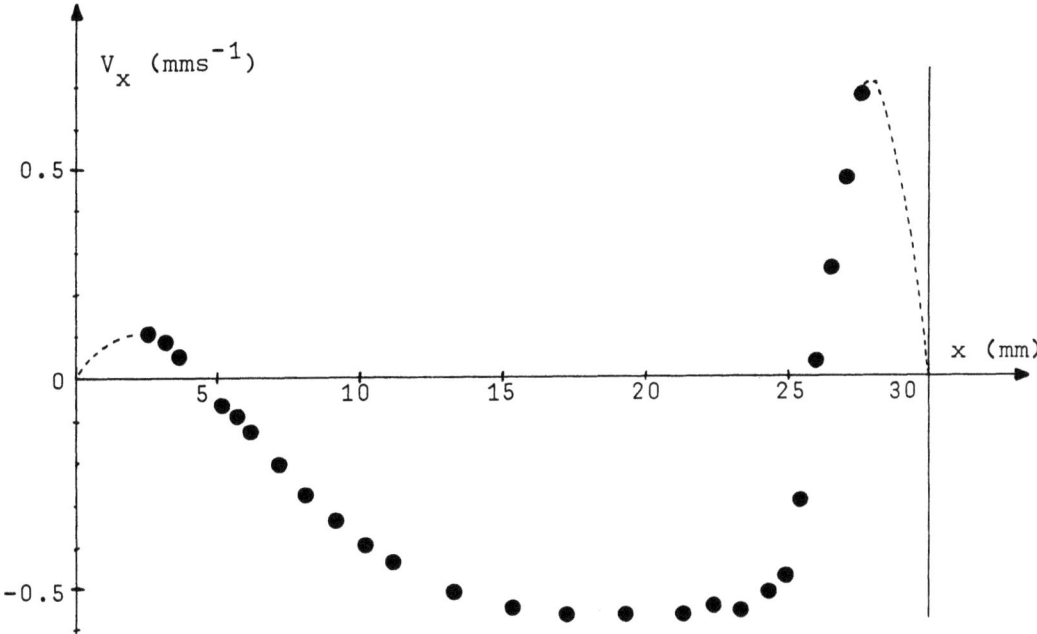

Fig. 22. Horizontal velocity profile 0.6 mm above the bottom of a 4.5 mm layer water-n-heptanol.

The result is given on Fig. 23 for a layer 8.7 mm thick. Two important points should be emphasized. First, one may observe that V_x is not a constant along the surface as imposed in the crude theoretical model. This is due to the fact that the aspect ratio is of the order of 3.5 only and also that the surface is not flat. Therefore quantitative agreement with the model is not expected. Also the profile is not symmetric maybe because the temperature gradient is not constant along the surface.

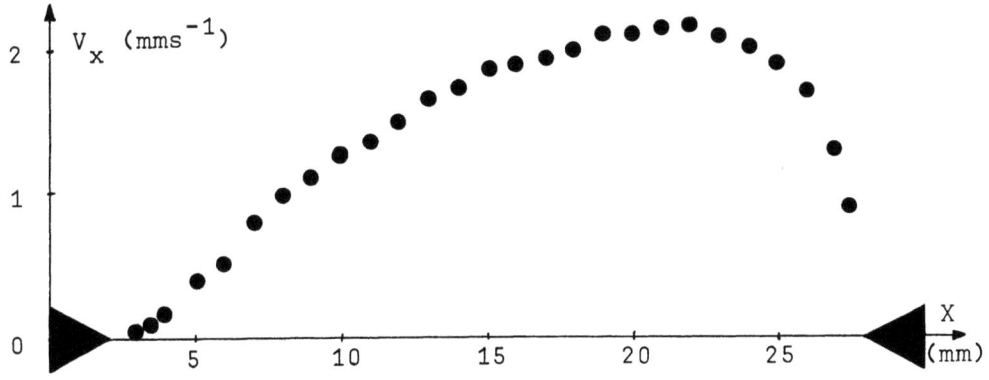

Fig. 23. Horizontal velocity profile along the surface (between the two anticreeping Teflon edges) in a 8.7 mm layer water-n-heptanol.

DISCUSSION OF THE EXPERIMENTAL RESULTS

The experiments that we have presented in the previous paragraph show a good qualitative agreement with theoretical or numerical predictions. The three foreseen different convective regimes are actually obtained. Firstly, for negative values of k ($\partial\sigma/\partial T < 0$, case of pure water), we obtain one single convective cell in the direction of the thermogravitational motion ; secondly, for the water-n heptanol solution (k > 0 or $\partial\sigma/\partial T > 0$) we successively observe one anticlockwise rotating convective cell for small k (or large values for the depth of the layer), next two superposed counterrotating cells when the liquid depth decreases and finally only one convective cell at large k with motion from cold to hot in the surface, and clearly this convective cell is due to surface tension effects. Is a quantitative comparison with theory possible ? We have tried to fit Eq. (12) through the experimental points : we have used a computer program that find the best value of k (that minimizes the total error) for each velocity profile supposed to obey the analytical law (12). We give in the table below the best value of k that we could obtain. On the other hand, knowing the properties of the liquid and the height of the layer, one can estimate the ratio - Ma/Ra, also reported in the table below. The results are also plotted on Fig. 24. Even if there are strong differences between computed values of - Ma/Ra and values of k determined using the experimental velocity profiles, the relation is almost linear between both variables.

Table

liquid	height 2 ℓ	best k	-Ma/Ra
solution	14.0 mm	0.227	0.453
solution	10.2 mm	0.402	0.884
solution	7.4 mm	0.497	1.62
solution	6.0 mm	0.599	2.47
solution	4.9 mm	0.92	3.70
solution	3.3 mm	3.24	8.15
pure water	7.8 mm	-1.50	-3.83

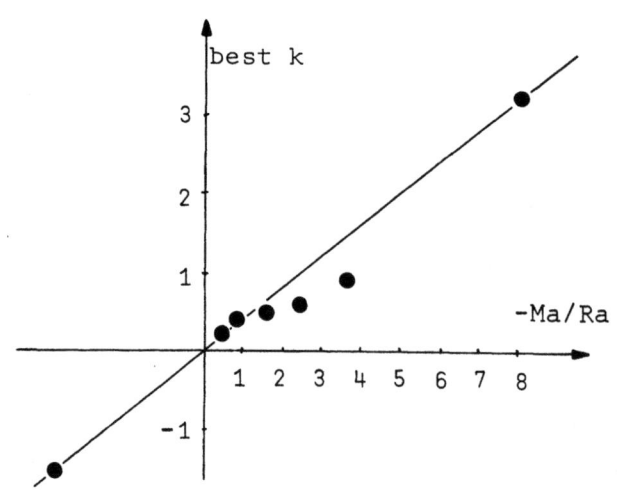

Fig. 24. Plot of best computed k versus the -Ma/Ra calculated from data.

Thus the quantitative comparison between experiments and an "elementary model" is not really successful for different reasons, some of them are discussed below.

When calculating the Marangoni number, there is a first difficulty to know the exact value of $\partial\sigma/\partial T$ at the measuring point. Indeed the surface tension temperature coefficient is deduced from Eq. (1), which is the "equilibrium" surface tension of the mixture. It is not sure that in dynamical conditions the surface tension obeys the same law. An even if Eq. (1) remains valid, we have to know the temperature at the measuring point, i.e. at the center of the cavity. We suppose that, at that point, the temperature is the mean temperature $(T_H + T_C)/2$. This is probably not true because the container is not laterally insulated and therefore the temperature at the measuring point could be smaller than 52 °C (the ambiancy is at 25 °C) and this could lower the calculated value of Ma. Also the apparatus was not designed to measure the temperature profile. The elementary model is also unadequate to correctly describe the present experiment, or in other words, the present apparatus is not well designed to verify the theoretical predictions deduced from Eq. (12). Indeed the model supposes that the aspect ratio is infinity, instead of ranging from 2 to 10 in the experiments. Also in the model V_x is not a function of x, which is experimentally not true due to the presence of lateral walls (see Fig. 23). For the same reason, $V_z = 0$ in the model ; once again this is not true and upward and downward motions modify the horizontal velocity profile. Finally the model is a two-dimensional model, but the real flow is truly three-dimensional, with $V_y \neq 0$. This nonzero third velocity component has not been measured in the present optical arrangement, but its existence is due to an existing temperature gradient in the third direction, itself due to a bad thermal insulation. The existence of three-dimensional flow is indirectly demonstrated by observing that

$$\int_{-\ell}^{+\ell} V_x \, dz \neq 0 \tag{25}$$

for any of the experimental profiles displayed in the above figures. One should verify that

$$\iint V_x(x,y,z) \, dz \, dy = 0 \tag{26}$$

but this remains to be done. Finally the model supposes a linear variation of surface tension with temperature, whereas the true variation is quadratic. Thus we need another model or a new experimental cell (and even both). It was not possible to change anything to the observation cell that has been used for the microgravity experiments and that shall still be used in future flights. At the present time we build an observation cell 15 cm large instead of 3 cm in which the temperature profile can be determined by thermocouples. On the other hand, one could hope that the numerical model described in a previous paragraph could give a better quantitative agreement with the experiments. As we already said, the numerical calculations were intended for a comparison with the space experiments. One has to know that in the space experiments the observation cell is usually stored in the rocket a long time (days or weeks) before the launch and therefore the cleanness of the surface is not assured. Let us once again quote Levich's book[18], p. 391 :

"... at high Reynolds numbers, the presence of surface active substances does not influence the liquid flow. ... At high liquid velocities, the surface active substances will be entirely swept away ("blown off").".

Therefore, in the space experiments, in order to avoid a velocity reduction by the presence of surface active substances, one asks high liquid velocities, i.e. high temperature differences, as large as 20 °C. In these conditions, the two Marangoni numbers M_1 and M_2 are rather large. When trying to apply the numerical model to the space experiments, we have found

$$Gr \simeq 0$$

$$M_1 \simeq -1\ 10^6$$

$$M_2 \simeq 1.2\ 10^6$$

where Gr is the Grashof number. For the earth experiment described in the paper by Limbourg et al[15], we have found for the corresponding numbers

$$Gr \simeq 5\ 10^6$$

$$M_1 \simeq -0.57\ 10^6$$

$$M_2 \simeq 0.63\ 10^6$$

At such high values for the dimensionless numbers, the numerical model (using of course simple algorithms for space and time discretization) shows numerical instabilities and fails to give any useful information. At the present time, the program works up to Marangoni numbers of the order of 10^4 and typically this corresponds to a temperature difference of $\simeq 0.2$ °C. Therefore no quantitative comparison with the space experiments is possible at the present time. Even worse, no quantitative comparison is possible today between analytical calculations, numerical simulations, microgravity and earth experiments for at least one of the following reasons :
- no steady state achieved in the time of the experiment
- aspect ratio not appropriate
- temperature difference not appropriate

Therefore, we do construct a new observation cell, in which the aspect ratio may be changed (up to $\simeq 75$) in order to compare with the simple analytical model. Also, the temperature profile shall be measured. Due to a correct insulation, it shall be possible to impose a very small temperature gradient, in order to compare with the numerical model which predicts results only up to $\Delta T \simeq 0.2$ °C. Simultaneously, the numerical algorithms should be improved in order to be able to compare with experiments at large ΔT.

Finally, we would like to question the usefulness of the space experiments at the present time in the steady state study of thermocapillary convection. Is it really necessary to have a value of Ma/Ra as large as 10^3 or 10^4 ? Certainly not, since the velocity profile does no longer change when this ratio exceeds more or less 5 ; and in order to have this value one has just in normal gravity to do experiments on layers a few millimeters thick. Since in the actual arrangement the optical probe in LDV is 90 μm thick (and the thickness could be reduced up to $\simeq 30$ μm by the use of a beam expander system) there is no practical difficulty to determine velocity profiles in layers of say one millimeter thick. If one wants to do space experiments, one needs a theoretical support that justifies the absolute necessity of large values of k. For example, the basic velocity profile given by Eq. (12) must at some high value of the Reynolds number become unstable, leading finally the system to a new convective regime, for example an oscillatory flow. It could be (but for the time being we have no theoretical support) that the transition to oscillatory flows in systems for which surface tension increases with T, occurs at such large values of k, that they could be obtained only in reduced gravity. Without this theoretical support, we propose once again earth experiments and we would like to end this paper by giving the first result concerning oscillatory flows. In the search for oscillatory flows, if any in normal gravity, we have to increase the temperature difference between cold and hot walls. The cold wall was maintained at a constant temperature of about 45 °C ($T_C = 44.8$ °C). The temperature of the hot wall was varied. For each temperature gradient, we record the horizontal velocity in the surface at the center of the container. This was done on a layer 8.8 mm thick. As long as the temperature of the hot wall was below $\simeq 62$ °C, we always observe steady convection. As a matter of fact, the amplitude of V_x increases with ΔT. When $T_H = 59.4$ °C (or $\Delta T = 14.6$ °C), the $V_x \simeq 7.2$ mm/s when $T_H = 61.35$ °C (or $\Delta T = 16.55$ °C), then the horizontal velocity increases up to 8.2 mm/s. It seems that there is a well defined critical value of ΔT for which oscillatory flow starts, namely $T_H \simeq 62.5$ °C (or ΔT ≃ 17.5 °C). When $T_H \simeq 65$ °C, the flow remains oscillatory with a rather large amplitude ; typically the velocity oscillates between 7.6 mm/s and 12.7 mm/s. The period of oscillations is close to 20 sec. When ΔT is increased further, (say T_H = 70 °C), then we observe very irregular oscillations, and eventually the flow becomes steady after a few hours. These preliminary results show that oscillatory flow may be obtained in normal gravity and there i   d for an extensive study of these oscillations. However, we propose to start with the study of the hydrodynamic stability theory of the basic velocity profile in order to have an idea of the range of the Reynolds number for which instability is expected and also of the frequency of this oscillatory flow, at least in the first stage of the transition, or close to the critical point. Such a study should reveal if large values of k has to be attained before observing an oscillatory flow, but in view of preliminary experimental finding on a layer 8.8 mm thick, we believe that instability is expected even for small k. This study is left for future work.

As we have tentatively shown in this paragraph when trying to compare experimental results and theoretical predictions, we are only at the beginning of a whole story and a lot of things remains to be done.

ACKNOWLEDGMENTS

This research has been supported by the FONDS DE LA RECHERCHE FONDAMENTALE COLLECTIVE(Belgium) under Grant # 2.4520.83.

REFERENCES

1. T.B. King, The surface tension and structure of silicate slags, J. Soc. Glass Technol. 35, 241 (1951)
2. J.F. Elliott, M. Gleiser and V. Ramakrishna, "Thermochemestry for Steelmaking", (Vol. 2, Thermodynamic and Transport Properties) Addison-Wesly Pub. Co. (1960)
3. P.J. Desre and J.C. Joud, Surface tension temperature coefficient of liquid alloys and definition of a "zero-Marangoni number alloy" for cristallisation experiment in microgravity environment, Acta Astronautica, 8, 407 (1981)
4. K. Motomura, S.I. Iwagana, Y. Hayami, S. Urgu and R. Matuura, Thermodynamics studies on adsorption at interfaces - IV. Dodecylammonium chloride at water-air interface, J. Colloid Interface Sci., 80, 32 (1981)
5. R. Vochten, PhD Thesis, Thermodynamisch studie van de reversiebele molaire adsorptie-warmte aan grensvlakken voor niet ionogene tensiden, University of Ghent, Belgium (1976)
   R. Vochten and G. Petré, Study of the heat of irreversible adsorption at the air-solution interface, J. Colloid Sci., 42, 320 (1973)
6. A.K. Sen and S.H. Davis, Steady thermocapillary flows in two-dimensional slots, J. Fluid Mech., 121, 163 (1982)
7. M.K. Smith and S.H. Davis, The instability of sheared liquid layers, J. Fluid Mech. 121, 187 (1982).
8. M.K. Smith and S.H. Davis, Instabilities of dynamic thermocapillary liquid layers, J. Fluid Mech. 132, 119 (1983) and 132, 145 (1983)
9. G.M. Homsy and E. Meiburg, The effect of surface contamination on thermocapillary flow in a two-dimensional slot, J. Fluid Mech., 139, 443 (1984)
10. M. Strani, R. Piva and G. Graziani, Thermocapillary convection in a rectangular cavity : asymptotic theory and numerical simulation, J. Fluid Mech., 130, 347 (1983)

11. L.G. Napolitano and C. Golia, Influence of surface tension minimum on surface and buoyancy driven flows in Stokes regime, in "Proceedings of the 4th European Sympodium on Material Science under Microgravity", ESA SP-191, 229 (1983)

    L.G. Napolitano, C. Golia and A. Viviani, Numerical simulation of unsteady thermal Marangoni flows, in "Proceedings of the 5th European Symposium on Material Science under Microgravity", ESA SP-222, 251 (1984)

12. A.G. Kirdyashkin, Thermogravitational and thermocapillary flows in a horizontal liquid layer under the conditions of a horizontal temperature gradient, Int. J. Heat Mass Transfer, $\underline{27(8)}$, 1205 (1984)

13. D. Villers and J.K. Platten, Marangoni convection in systems presenting a minimum in surface tension, PhysicoChemical Hydrodynamics, $\underline{6(4)}$, 435 (1985)

14. C. Cuvelier and J.M. Driessen, Thermocapillary free boundaries in crystal growth, J. Fluid Mech., $\underline{169}$, 1 (1986)

15. M.C. Limbourg-Fontaine, G. Petré and J.C. Legros, Texus 8 experiment : effects of a surface tension minimum on thermocapillary convection, PhysicoChemical Hydrodynamics, $\underline{6(3)}$, 301 (1985)

16. T. Maekawa, I. Tanasawa, J. Ochiai, K. Kuwahara, M. Morioka and S. Enya, Two-dimensional Marangoni and buoyancy convection related to crystal growth techniques in space, in "XXV COSPAR", 1 (1984)

    and by the same authors in a different order, plus K. Sezaki Experimental study of Marangoni convection, in "Proceedings of the 5th European Symposium on Material Science under Microgravity", ESA SP-222, 291 (1984)

17. J. Wiedemann, "Laser-Doppler Anemometrie", Springer-Verlag (1984)

18. V.G. Levich, "Physicochemical Hydrodynamics", Prentice-Hall (1962)

ANALYSIS OF FLOW DEVELOPMENT DUE TO MARANGONI CONVECTION IN A MASS TRANSFER SYSTEM

Henk A. Dijkstra

Mathematical Institute
University of Groningen
The Netherlands

INTRODUCTION

Surface tension gradients influence the mass transfer rate of purification processes, like distillation and adsorption, performed e.g. in a packed column. The Marangoni-effect is responsible for minute convective movements at the gas-liquid interface which increase the mass transfercoefficient. Moreover surface tension gradients can increase the contact area between gas and liquid. From measurements[1] it was found that the number of transfer units, the degree of performance of the transfer process, can be increased by a factor up to 3 by surface tension effects.

To study the development of the convective movements the following experimental setup was chosen (Figure 1). Acetone evaporates at the gas-liquid interface and is then absorbed by active carbon. For this system the parameter pair (Ma,Ra), explained below, has only positive values. If this pair assumes sufficiently high values, experimentally, convective cells (roll cells) develop after several minutes. To study the system at Ra = 0 an experiment in space was performed during the D1 Spacelab flight[2].

A theoretical analysis of this experimental system is presented in this paper. The analysis is comprised of three parts:
(i)   the mass transfer analysis of the motionless system
(ii) the linear stability analysis of the motionless state
(iii) the development from disturbances to roll cells, a numerical approach.

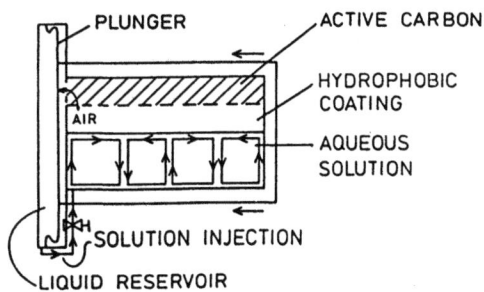

Fig. 1. A sketch of the experimental Marangoni convection container, dimensions 10*3.5*1 cm, as used during the D1 Spacelab mission.

ANALYSIS

A mathematical model of the system together with part (i) and (ii) of the analysis was presented elsewhere[3]. The boundary initial value problem consists of the two mass balance equations, one for the liquid and one for the acetone, and the Navier Stokes equation. The sidewalls are assumed either stress free or rigid, the bottom wall is rigid and the interface is nondeformable and flat. The concentration distribution c at t = 0 is assumed to be homogeneous.

The system is characterised by the following parameters:

$Sc = \frac{\mu}{D\rho}$, the Schmidt number,

$Bi = \frac{mH}{RD}$, the Biot number,

$A = \frac{L}{H}$, the aspect ratio of the container,

$Ra = \frac{-\frac{\partial \rho}{\partial c} H^3 g c_o}{\mu D}$, the Rayleigh number and

$Ma = \frac{-\frac{\partial \sigma}{\partial c} H c_o}{\mu D}$, the Marangoni number.

In these equations μ, D, ρ, σ and g are the dynamic viscosity, diffusivity, density, surface tension and gravitational acceleration, respectively. The symbols m and R represent the distributioncoefficient of the acetone and the resistance of the gasphase to mass transfer from the liquid phase, respectively. Moreover L and H are the length and height of the container.

Initial Flow Development: Linear Theory

Using penetration theory[4] or film theory to describe the mass transfer at the interface, the concentration distribution for the motionless liquid can be found analytically. A thin boundary layer, whose thickness increases with $\sqrt{t}$, develops at the interface.

The linear stability[3] of this motionless evaporating state was investigated for the case of stress free sidewalls and Ra = 0. Assume that for a short time $\bar{t}$ the fluid stays motionless and that at $t = \bar{t}$ a disturbance is present. Then the critical Marangoni number and the critical wavelength depend on $\bar{t}$ and Bi.

For Ma exceeding the critical Marangoni number, an unstable mode grows exponentially and although not proven analytically, the numerical results below indicate that no oscillatory behaviour occurs.

For the case of the experiment in Space (Sc = 800, Bi = 142, A = 4, Ra = 0 and Ma = 10^8) the results indicate that for small $\bar{t}$, the fastest growing wavelengths are very small. With increasing $\bar{t}$ these preferred wavelenghts increase. The experimental $\bar{t}$ so depends on the accuracy of a (velocity) measuring device.

Summarising, the results, from linear theory of the model, indicate that small cells will develop near the interface and grow exponentially in time. This picture is only valid for times close to $\bar{t}$.

Flow Development: Nonlinear Theory

The analysis of the stages of development where non-linear effects become important is up to now limited to a numerical one. The numerical

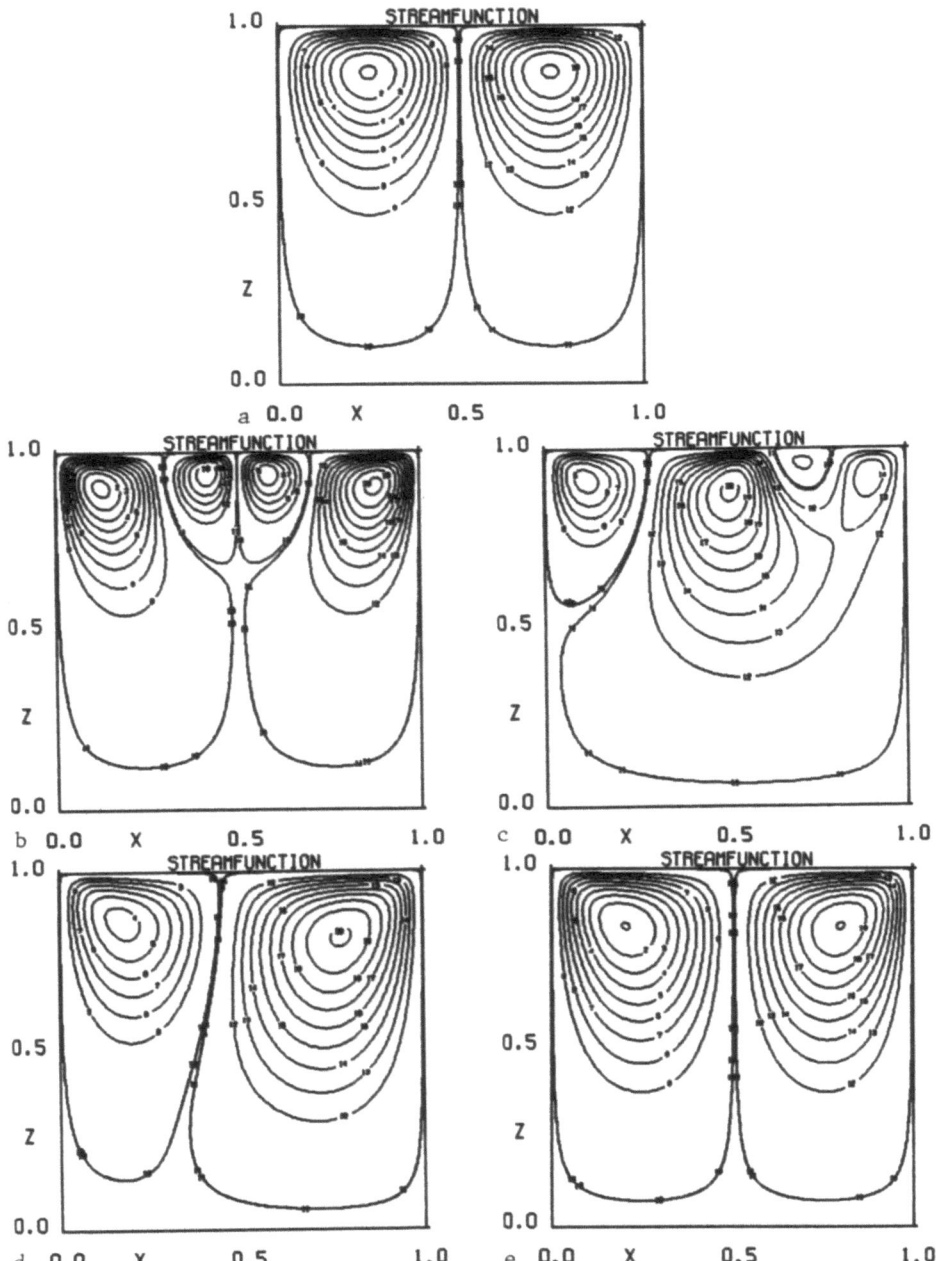

Fig. 2. Contour plots of the streamfunction ψ at different times t (in seconds). The quantity ψ_{max} is the gridmaximum of ψ. The ith contour represents values $(11 - i)*\psi/(10*\psi_{max})$, $i = 1,\ldots,9$. Contour 10 has the value $\psi/(100*\psi_{max})$ and the value of contour i, $i = 11,\ldots,20$ is the negative of that of $21 - i$.
a: $\psi_{max} = .45 \; 10^{-6}$, t = 25.   b: $\psi_{max} = .27 \; 10^{-1}$, t = 225.
c: $\psi_{max} = .32 \; 10^{-1}$, t = 425.   d: $\psi_{max} = .41 \; 10^{-1}$, t = 625.
e: $\psi_{max} = .34 \; 10^{-1}$, t = ∞.

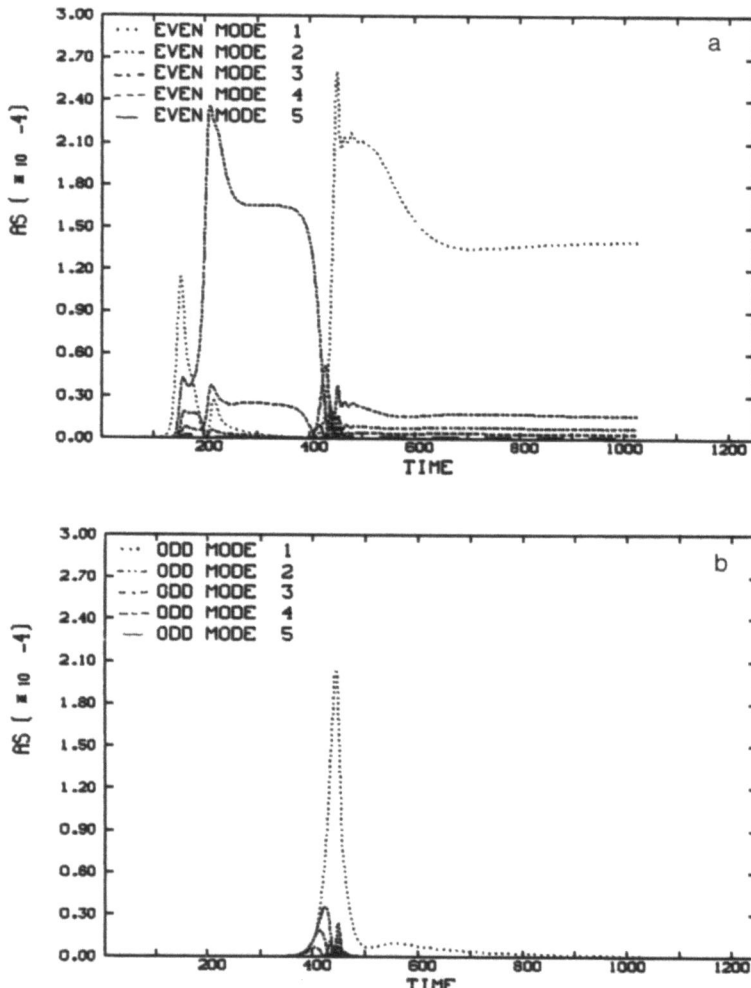

Fig. 3. Time development of Fourier-modes of the interface concentration. The squared amplitudes AS of the modes are plotted as a function of time (in seconds). Figure 3a: even modes; Figure 3b: odd modes.

methods used here are of finite difference type, second order in space and time[5]. An example of results obtained will now be given:

The boundary condition for the concentration at the bottom wall is changed[3] from $\frac{\partial c}{\partial z} = 0$ to $c = 1$ to obtain nontrivial steady solutions. This does not change the results of the linear analysis if $\tilde{t}$ is small. The parameters in the example are: $Sc = 7.9$, $Ma = 10^4$, $Ra = 0$, $Bi = 1.4$ and $A = 1$. The initial condition at $t = \tilde{t}$ consists of a sum of an (unstable) Fourier component (an eigenfunction of the linear system) and the motionless state solution.

In figure 2a the initial conditions are shown for the streamfunction. Four pictures of the streamfunction at different times in the unsteady simulation are shown in the figures 2b to 2e. Subcells appear (2b), the symmetry axis at $x = 1/2$ disappears (2c) and finally a two cell solution appears as a steady state (2e).

A description of this development can be given by plotting the squared amplitudes of the Fourier (x-)modes of the interfacial concentration (Figure 3a and 3b). The initial even mode with wavelength $\lambda = 1$ grows exponentially until even modes with wavelength $\lambda = 1/2, 1/4,...$ are growing due to (weak) nonlinear interactions. The even mode with wavelength $\lambda = 1/2$ becomes the dominant mode for a certain time range.

The nonlinearity of the concentration distribution from the motionless state is an essential factor for the appearance of the subcells, because these subcells do not appear when this distribution is taken to be linear[5] (the classical case).

Odd modes are growing, the symmetry axis vanishes, due to (strong nonlinear) interactions of the modes. During a small time interval the dynamics of the system is complex, resulting in a selection of the original mode as the dominant one. The final steady state is a multimode solution which also is the selected solution when the simulation is started with a sum of a random disturbance and the motionless solution[5].

An explanation of the development is the following. The subcells appear because the apparent aspect ratio is larger than A in the beginning of the simulation; the disturbance has only a relative large amplitude in the concentration boundary layer. Because the thickness of this layer becomes larger with time the apparent aspect ratio becomes smaller and the selected wavelength will be larger. This can be seen by the fact that the quasi steady solution ($300 \leq t \leq 400$) looses stability with respect to odd modes, symmetry is lost and the original wavelength becomes dominant.

REFERENCES

1. W. B. Patberg, A. Koers, W. D. E. Steenge and A. A. H. Drinkenburg, Effectiveness of Mass Transfer in a Packed Distillation Column in Relation to Surface Tension Gradients, Chem. Eng. Science 38:917 (1983).
2. J. H. Lichtenbelt, A. A. H. Drinkenburg and H. A. Dijkstra, Marangoni Convection and Mass Transfer from the Liquid to the Gas Phase Under Microgravity Conditions, Naturwissenschaften 73:356 (1986).
3. H. A. Dijkstra and A. I. van de Vooren, Initial Flow Development Due to Marangoni Convection in a Mass Transfer System, Int. J. Heat Mass Transfer, 28:2315 (1985).
4. R. B. Bird, W. E. Steward and E. N. Ligthfoot, "Transport Phenomena", John Wiley & Sons, New York.
5. H. A. Dijkstra, in preparation.

TWO-COMPONENT BÉNARD STEADY CONVECTION WITH SURFACE ADSORPTION

X. -L. Chu[1,2], L.-Y. Chen[1] and M. G. Velarde[2]

[1] Physics Department, Huazhong University of Science and Technology (HUST), Wuhan, P. r. China
[2] Depto. Física Fundamental, UNED, Apartado 60.141, Madrid 28071 (Spain)

1. INTRODUCTION

The Bénard-Marangoni instability of a two-component liquid layer open to air has received great attention in the last twenty years [1-2]. A series of studies made by Velarde and collaborators [3-8] analyzed a two-component liquid model, where buoyancy, surface tension, surface deformation, heat and mass transfer as well as Soret effect in the volume were all taken into account. For steady transitions, the sufficient condition for instability can be approximated by the simple relation

$$M/M_c + E/E_c + R/R_c = 0 \qquad (I.1)$$

where M, E and R are Marangoni, elasticity (solutal) Marangoni and Rayleigh numbers respectively. M_c is the critical value taken at $E = R = 0$ and, correspondingly, are E_c and R_c. Three different mechanisms, described by M, E and R, interact with each other, and affect the stability of the system. When $E > 0$, one has, $(\partial R/\partial E)_M < 0$, which means that the larger the elasticity number, the lower the stability of the fluid. The elasticity (solutal) Marangoni number plays a destabilizing role. However, no surface adsorption has been considered in these papers. For a multi-component liquid with surfactant solute, the equilibrium solute concentration in the surface is, usually, much higher than in the bulk [9], thus leading to solute accumulation at the open surface as the exchange of solute between surface and bulk are through adsorption-desorption process. Generally, this adsorption-desorption process is much faster than the diffusion process in the bulk and the whole system is controlled by the slow diffusion processes. However, there are cases where slow surface processes have been observed [9-12]. The following mechanisms are believed to be able to considerably reduce surface process rates

1. Adsorption barriers for dipolar molecules
2. Reorientation
3. Complex formation and chemical reactions
4. Formation of the surface structure with anisotropic structure as in a liquid crystal.

If, one or more of these mechanisms affect surface processes in a Bénard fluid layer so that they become so slow that take place on the scale of the diffusion process in the bulk, we must not ignore the solute adsorption and eventual accumulation. One expects that the stability of the system may be strongly affected by surface adsorption. Experiments have shown taht even a slight surface adsorption [13] can greatly affect the stability of the system.

On the other hand, other authors [14-16] have investigated the stability of the Bénard fluid with insoluble surfactant floating on the surface. Palmer and Berg [15] studied a binary liquid layer with surface-active material accumulation at surface by adsorption. They remarked the stabilizing effect of the solutal Marangoni number. It follows from their work that $(\partial R/\partial E)_M < 0$ and $(\partial R/\partial E)_R < 0$, which indicates that the system will become more and more stable as E increases. In their model, the solute was considered uniformly distributed in the bulk as in the steady reference state. This is, however, not valid in general as a solute concentration profile may originate from a temperature gradient (Soret effect) [2].

Figures (1a) and (1b) illustrate the destabilizing role of the solutal Marangoni effect in the adsorption-free case and the stabilizing role in the uniform bulk distribution case. Fig. (1c) shows the concentration profile used in this paper. We take both surface adsorption accumulation and bulk nonuniform distribution into consideration. One may notice that the solute profiles in Fig. (1a) and (1b) are just two special cases of Fig. (1c). The latter reduces to (a) or (b) by taking vanishing surface adsorption or uniform bulk distribution.

In the present paper, we review the effects of surface solute accumulation on the stability of a two-component Bénard liquid layer. We shall consider the case of an infinitely extended layer as well as for illustration, in a more realistic situation, the case of a cylindrically bounded liquid layer though we shall not study the appearance of cellular patterns.

2. THE SURFACE ADSORPTION PROBLEM

Let us consider a shallow two-component Bénard liquid layer, thickness d, with surface adsorption and accumulation. The system is bounded at the bottom by a rigid plate and it is open to the ambient air. We define a surface region by separating a very thin layer near the surface (see Fig. (2)), and assume all the surface properties to be only related to this region. We define the surface solute concentration: $\Gamma = n^*/A$, where $n^* = n - A \int_{d-\delta}^{d} C \, dz$, A is the surface area, n the total solute in moles in the surface region, with C the bulk concentration in the surface region without adsorption, and δ the thickness of the surface region which loosely is chosen to be the minimum thickness including all the solute adsorption and accumulation in this surface region. Consequently n^* is the

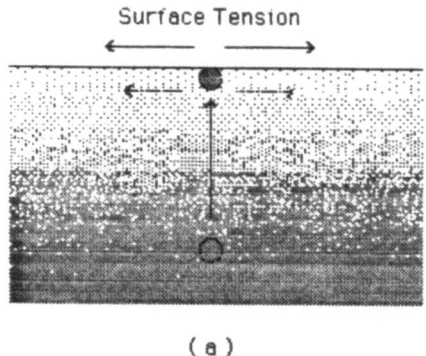

Fig.1. Three possible solute or surfactant distributions

(a) Leading to a destabilizing Marangoni effect

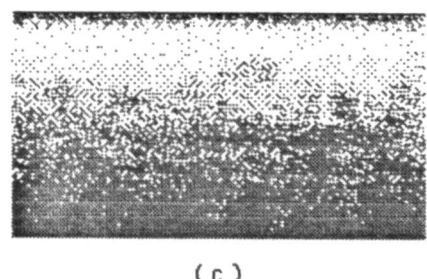

(b) Leading to a stabilizing Marangoni effect

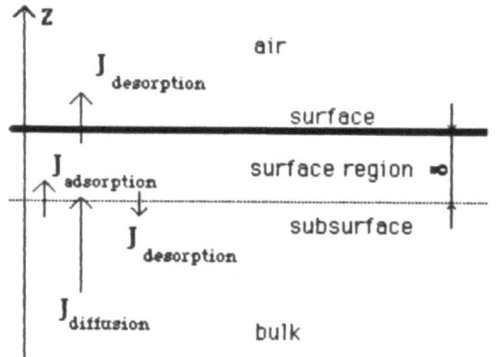

(c) Actual distribution used in this text

Fig.2. Transport phenomena near the open surface

excess-solute in the surface region, and Γ is the surface excess concentration of solute, simply called surface excess in the rest of the paper. Usually δ is of the order of molecular diameters, thus $\delta/d \ll 1$. Therefore we can treat is a two-dimensional surface. Assume that the solute transfer from the volume to the surface takes place in two steps. First, the solute diffuses to the boundary of the surface region (called subsurface here after) and then, by adsorption, it gets into the surface region, as shown in Fig. (3). Thus we have for the surface mass balance equation

$$\partial \Gamma/\partial t + \nabla \Sigma \cdot (\Gamma V_\Sigma - D_\Sigma \nabla_\Sigma \Gamma) = J_{ads} - J_{des1} - J_{des2} \qquad (2.1)$$

where J_{ads}, J_{des1} and J_{des2} are solute adsorption, desorption to liquid and desorption to air fluxes, respectively. In the linear approximation, they can be expressed as

$$J_{ads} = a_1 C(d) \qquad (2.2)$$

$$J_{des1} = b_1 \Gamma \qquad (2.3)$$

and $\quad J_{des2} = b_2 \Gamma \qquad (2.4)$

with $C(d)$ the bulk concentration near the surface and a_1, b_1 and b_2 the equilibrium constants for the surface. We take $a_1 \geq 0$, $b_1 \geq 0$ and $b_2 \geq 0$. Beacause the thickness d has been chosen such that all the solute accumulation is included in the surface region, there is no solute accumulation in the subsurface, which gives another mass balance equation in the subsurface

$$J_{diff} = J_{ads} - J_{des1} \qquad (2.5)$$

where $J_{diff} = -D (\partial C/\partial z)_{z=d}$ accounts for the mass diffusion near the surface. We now assume that the surface tension σ is a function of the solute excess Γ and temperature T. To a first approximation

$$d\sigma = (\partial \sigma/\partial \Gamma)_0 d\Gamma + (\partial \sigma/\partial T)_0 dT \qquad (2.6)$$

where the subscript "0" denotes a reference state to be defined later on.

3. EQUATIONS

To focus our attention on the surface adsorption accumulation and its influence on the stability of the system, here we present a very simple model, from which, by a straightforward analysis, we can obtain a clear physical picture about the dual effect of the surface tension traction.

We neglect thermal effects and assume the whole system to be in thermal equilibrium. The density of the solute ρ is a function of the bulk solute concentration C; we take $\partial\sigma/\partial C < 0$. A solute mass flux in the boundaries-in the bottom, e.g., by a porous plate and on the surface by desorption-maintains a linear concentration profile in steady state, with concentration higher at bottom and lower near the surface. In the simplest case we have the equations [17-22].

$$\text{div } \mathbf{U} = 0 \tag{3.1}$$

$$\partial \mathbf{U}/\partial t + \mathbf{U}\nabla\mathbf{U} = -(1/\rho_0)\text{ grad } P + (\rho/\rho_0)g\mathbf{k} + \nu\nabla^2\mathbf{U} \tag{3.2}$$

$$\partial C/\partial t + \mathbf{U}\nabla C = D\nabla^2 C \tag{3.3}$$

together with boundary conditions (b.c.) at the bottom

$$\mathbf{U}|_{z=0} = 0 \tag{3.4}$$

$$(\partial C/\partial z)_{z=0} = \text{const. (or } C|_{z=0} = \text{const.)} \tag{3.5}$$

and at the open surface

$$U_z|_{z=d} = 0 \tag{3.6}$$

$$\rho_0 \nu (\nabla \mathbf{U} + (\nabla \mathbf{U})^T)\mathbf{k}|_{z=d} = \nabla_\Sigma \sigma \tag{3.7}$$

$$- D (\partial C/\partial z)_{z=0} = a_1 C(d) - b_1 \Gamma \tag{3.8}$$

$$\partial \Gamma/\partial t + \nabla_\Sigma (\Gamma \mathbf{U}_\Sigma - D_\Sigma \nabla_\Sigma \Gamma) = a_1 C(d) - b_1 \Gamma - b_2 \Gamma \tag{3.9}$$

A steady motionless solution is $\mathbf{U}^r = 0$, $C^r = C^r(0) - z\ \Delta C/d$, $J^r_{\text{diff}} = J^r_{\text{des2}}$, $\Gamma^r = \Delta C D/b_2 d$ and $\sigma^r = $ const. This is our reference state. Perturbations of this state can be Fourier analyzed and in dimensionless form the equations become

$$S^{-1}\lambda(\partial^2/\partial z^2 - a^2) w = -R a^2 C + (\partial^2/\partial z^2 - a^2)^2 w \tag{3.10}$$

$$\lambda C = (\partial^2/\partial z^2 - a^2) w \tag{3.11}$$

together with b.c. at $z = 0$

$$w = \partial w/\partial z = 0 \qquad (3.12)$$

$$\partial w/\partial z = 0 \qquad (3.13)$$

and at $z = 1$

$$w = 0 \qquad (3.14)$$

$$\partial^2 w/\partial z^2 = -E a^2 \gamma \qquad (3.15)$$

$$\lambda \gamma = B_2 C - (B_1 + B_2)\gamma - P_s a^2 \gamma + A_1 \qquad (3.16)$$

Here a is the Fourier wavenumber, λ is the time constant that determines the stability. We restrict consideration to the stationary transition, i.e., the stability boundaries obtained by taking $\lambda = 0$. We have introduced the following dimensionless groups [19-22]. $S = \nu/D$ (Schmidt number), $R = g\alpha\Delta C d^3/D\nu$ (Rayleigh number), $E = (-\partial s/\partial \Gamma)_r \Gamma^r a_1 d^2/D^2 \rho \nu$ (Elasticity solutal Marangoni number), $A_1 = a_1 d/D$ (Adsorption number), $B_1 = b_1 d^2/D$ (Desorption number), $P_s = D_\Sigma/D$ (Surface diffusion number), and Surface excess number $B_2 = \Delta C d/\Gamma^r$. The surface excess number B_2, is chosen to be directly proportional to the concentration difference ΔC and inversely proportional to the surface excess Γ^r, so that we can easily treat some limit cases, e.g., vanishing surface accumulation ($B_2 \to \infty$) or uniform bulk solute distribution ($B_2 \to 0$).

Fig. 3 gives several typical neutral stability curves. It clearly appears that, the curve suddenly breaks at $a = a^*$, and develops into two branches. The branch to the left of a^* has a minimum, whereas the branch to the right of a^*, that, aparently, has never been reported, has a maximum. This behavior is characteristic of an agent playing dual effects on the stability. We identify the left branch (with minimum) to be the destabilizing branch, and the right one (with maximum) the stabilizing one. By optimizing these neutral stability curves, we get the threshold values, which are ploted in Fig. 4. When ($B_2 \to \infty$) (vanishing surface adsorption), our result agrees with earlier predictions, e.g., $(\partial R/\partial E) < 0$, which shows E as a destabilizing agent. As B_2 decreases, due to the stabilizing role of

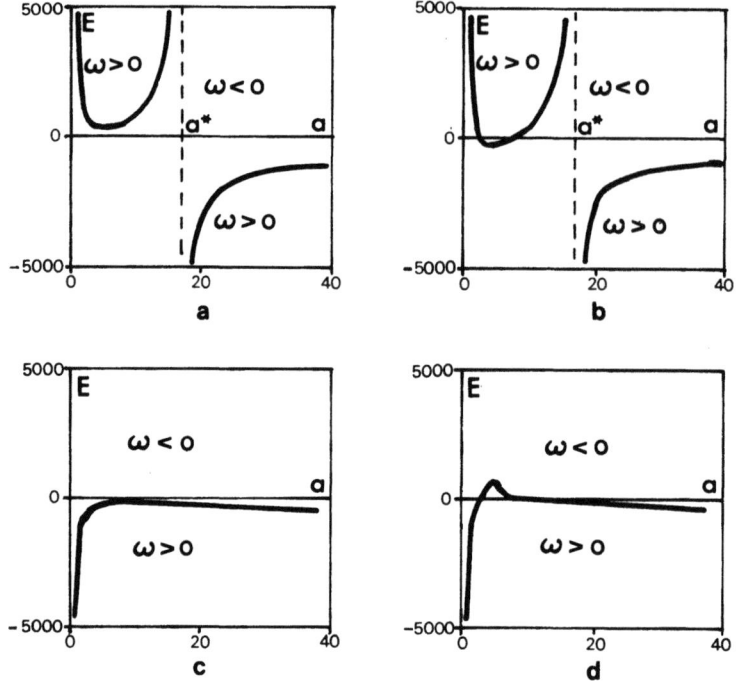

Fig.3. Solutal convection. Neutral stability lines providing the elasticity number versus wavenumber at different R and B values.
(a) R=100, B_2=500 (b) R=1000, B_2=500 (c) R=100, B_2=5 (d) R=1000, B_2=5.
In all cases A_1=10, P_s=1 and B_1=1.

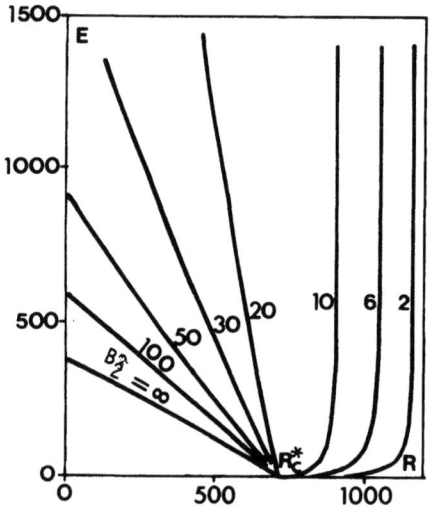

Fig.4. Thermosolutal convection
Neutral stability lines
as we vary the surface
excess number B_2.

surface accumulation, the stable region becomes larger and larger, though E still plays a destabilizing role. When B_2 is less than a critical value B_{2c}, the slope becomes positive and $(\partial R/\partial E) > 0$, thus showing the stabilization induced by the Marangoni effect. This result agrees with the results reported by Palmer and Berg [15-17].

An heuristic interpretation of the dual effect of E on the stability can be given. When $B_2 \to \infty$ (vanishing surface adsorption), a liquid bubble with higher concentration in the bottom is driven to the surface region, causing a local increase of solute concentration, the local surface tension diminishes. In turn, the surface tension will pull the liquid out and promote more fluid flow into the surface region from bottom. This is a positive feedback loop, and the solutal Marangoni force plays a destabilizing effect. With the increase of surface adsorption, the variation of surface tension by this mechanism becomes weaker and weaker and the system becomes more stable. Further increase of surface adsorption, reverse the situation discussed above. That is, the concentration at the point of ascending liquid may be lower than that of its surrounding. Then a negative feedback loop appears and the solutal Marangoni effect now is a stabilizing factor.

4. BOUNDED CYLINDRICAL LAYER

In this Section we consider the same problem as above but for a liquid enclosed in a cylinder of height L and horizontal diameter 2r, with the aim of illustrating the role of the earlier discussed phenomena upon the onset of steady convection in a conventional set-up. We have also considered that the layer is heated from below as in the standard Bénard problem. As we shall not consider the development of cellular patterns [2-16-20], the assumption of initially axisymetric disturbances with a flat cylinder is simple enough for a first-order analysis. In this case the disturbance equations are

$$\text{div } \mathbf{u} = 0 \tag{4.1}$$

$$\partial \mathbf{u}/\partial t = -(1/\rho_0)\text{ grad } P + (\delta\rho/\rho_0)g\mathbf{k} + \nu\nabla^2\mathbf{u} \tag{4.2}$$

$$\partial\theta/\partial t - w(\Delta T/L) = \kappa\nabla^2\theta \tag{4.3}$$

$$\partial c/\partial t - w\ (\Delta C_1/L) = D\ \nabla^2 n + C^0_1\ C^0_2\ D'\ \nabla^2\theta \tag{4.4}$$

together with the b.c.

- at the bottom ($z = 0$)

$$\mathbf{u} = 0 \tag{4.5}$$

$$\theta = 0 \tag{4.6}$$

$$D\ (\partial c/\partial z) + C^0_1\ C^0_2\ D'\ (\partial\theta/\partial z) = 0 \tag{4.7}$$

- at the free surface for simplicity assumed undeformable ($Z = L$)

$$w = 0 \tag{4.8}$$

$$(\partial c/\partial z) = 0 \tag{4.9}$$

$$- D\ (\partial c/\partial z) + C^0_1\ C^0_2\ D'\ (\partial\theta/\partial z) = k_1 c - k_{-1}\gamma \tag{4.10}$$

$$\rho^0\nu\ (\partial\mathbf{u}_\Sigma/\partial z) = \text{grad}_\Sigma\sigma \tag{4.11}$$

$$(\partial\gamma/\partial z) + \Gamma^0\text{div}_\Sigma\mathbf{u}_\Sigma - D\ _\Sigma\nabla^2_\Sigma\gamma - k_1 c - k_{-1}\gamma \tag{4.12}$$

We also assume the lateral walls rigid, i.e., $\mathbf{u} = 0$ at the periphery, and the following two equations of state

$$\rho = \rho_0(1 - \alpha\theta + \beta n) \tag{4.13}$$

$$\sigma = \sigma_0(\partial\sigma/\partial T)\theta + (\partial\sigma/\partial C_1)\ c \tag{4.14}$$

For convenience in the cylindrical case we choose the following scales: unit length, r; unit time r^2/κ, with κ the thermal diffusivity (thermometric conductivity); velocity $\kappa L/r^2$; temperature ΔT; concentration (mass fraction) of, say, component "one", ΔC_1; surface concentration, $\Gamma^0\Delta C_1/C^0_1$; and pressure, $\rho_0\nu\kappa L/r^3$. Thus in dimensionless form the equations are (no confusion is expected although we use the same notation for dimensional and dimensionless quantities).

$$\text{div } \mathbf{u} = 0, \qquad \mathbf{u} = (u,v,w) \tag{4.15}$$

$$p^{-1}\ (\partial\mathbf{u}/\partial t) = -\text{ grad } p + R^*\theta\mathbf{k} + R^*S\ c\ \mathbf{k} + \nabla^2\mathbf{u} \tag{4.16}$$

$$(\partial\theta/\partial t) = \nabla^2\theta + w \tag{4.17}$$

$$(\partial c/\partial t) = r_D \nabla^2 (c - \theta) + w \tag{4.18}$$

together with the b.c.

- at $z = 0$

$$\mathbf{u} = \theta = \partial(c - \theta)/\partial z = 0 \tag{4.19}$$

- at $z = L/r = h$

$$w = \partial\theta/\partial z = 0 \tag{4.20}$$

$$\partial c/\partial z = A^*_1 (\gamma - c) \tag{4.21}$$

$$\partial \mathbf{u}_\Sigma/\partial z = -M^* \, \text{grad}_\Sigma \theta - E^* r_D \, \text{grad}_\Sigma \gamma \tag{4.22}$$

$$H^* (\partial\gamma/\partial t + N^* \, \text{div}_\Sigma \mathbf{u}_\Sigma - r^s{}_D \nabla^2 \gamma) = c - \gamma \tag{4.23}$$

where with slight variations with respect to previous notation we now use the following groups: $P = \nu/\kappa$; $r_D = D/\kappa$; $R^* = \alpha g r^4 \Delta T/\nu \kappa L$, $S = -\beta \Delta C_1/\alpha \Delta T$, $A^*_1 = K_1 r/D$, $H^* = \kappa \Gamma^0/k_1 C^0_1 r^2$, $N^* = L C^0_1/r \Delta C_1$, $h = L/r$, $r^s{}_D = D_s/\kappa$, $M^* = -(\partial\gamma/\partial T) r^2 \Delta T/\rho\nu\kappa L$ and $E^* = -(\partial\sigma/\partial\Gamma)\Gamma^0 r^2 \Delta C_1/\rho\nu D L C^0_1$. H^* accounts for the excess-solute accumulation at the air-liquid interace.

The problem is easily solved with a single trial function Galerkin approach. In cylindrical coordinates (ρ, φ, z) we use

$$u_\rho = z(3z - 2h) \cos n\varphi \, U(\rho) \tag{4.24}$$

$$u_\varphi = z(3z - 2h) \sin n\varphi \, V(\rho) \tag{4.25}$$

$$w = u_z = -z^2(z - h) \cos n\varphi \, W(\rho) \tag{4.26}$$

$$\theta = z(z - 2h) \cos n\varphi \, W(\rho) \tag{4.27}$$

$$c = 2zh \cos n\varphi \, W(\rho) \tag{4.28}$$

$$\gamma = 2h^2 \cos n\varphi \, W(\rho) \tag{4.29}$$

with

$$U(\rho) = -[J'_n(k^*\rho) - J'_n(k^*) \rho^{n+1}]/k^* J_n(k^*) \tag{4.30}$$

$$V(\rho) = n[J_n(k^*\rho)/\rho - J_n(k^*) \rho^{n+1}]/k^{*2} J_n(k^*) \tag{4.31}$$

$$W(\rho) = [J_n(k^*\rho)/J_n(k^*)] - \rho^n \tag{4.32}$$

Note that u_r, u_φ and w satisfy the continuity equation (4.1). It gives

$$U' + U/\rho + nV/\rho - W = 0 \tag{4.33}$$

Here as earlier a dash denotes a derivative with respect to the corresponding argument.

Then using the above indicated trial functions the standard Galerkin method yields the neutral stability curves. As their analytical expresions involve rather cumbersome, albeit elementary realations between all the parameters in the problem we shall report here the results found in table and figure form only. Note that we may use two groups of scales (for tall and flat cylinders, respectively). The other set of scales can be obtained by writting

$f = 1/h$, $kf = k^*$, $R = \alpha g L^3 T/\nu\kappa$,

$M = - (\partial\sigma/\partial T) L\Delta T/\rho\nu\kappa$, $E = - (\partial\sigma/\partial \Gamma)\Gamma^0 L\Delta N/\rho\nu D N^0_1$,

$H = \kappa\Gamma^0/k_1 N^0_1 L^2$, $N = N^0_1/\Delta N_1$ and $A_1 = k_1 L/D$

In Table 1 we provide the critical values for the solutal (elasticity) Marangoni number, E, when we set to zero the Rayleigh, R, and thermal Marangoni, M, numbers and the buoyancy ratio, S. Values of E are given as we vary the aspect-ratio f as well as the excess-solute number, H. The latter has a slightly stabilizing effect.

Fig. 5 depicts the neutral stability curves in the plane (E,H) as we vary the Rayleigh and thermal Marangoni numbers.

Figure 6 shows how the (E,R) neutral stability lines are affected by the values taken by the excess-adsorption parameter, H. We also show here the influence of the buoyancy ratio. It appears that H has no influence upon the critical values of the Rayleigh number whereas the buoyancy ratio does not alter the critical elasticity number.

Figure 7 depicts the neutral stability lines in the (M,E) plane as we vary the Rayleigh and the excess-solute numbers. Again the latter plays a stabilizing role whereas the former is indeed destabilizing when the layer is heated from below.

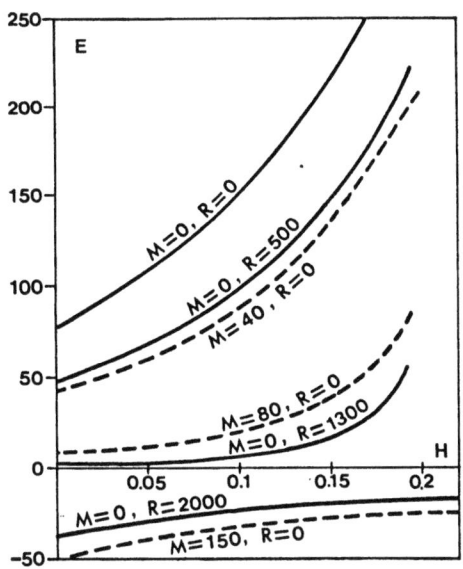

Fig.5. Thermosolutal convection with a free surface capable of solute accumulation. Neutral stability curves provide elasticity number(E) vs excess solute number(H) at different Rayleigh(R) and thermal Marangoni(M) numbers.

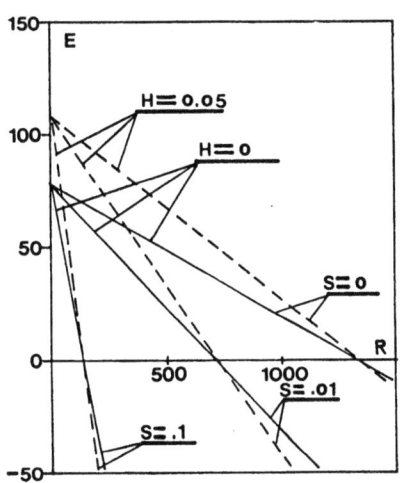

Fig.6. Thermosolutal convection with a free surface. Neutral lines as we vary the excess solute number(H) and the buoyancy ratio(S). Plotted is the elasticity Marangoni number(E) vs the Rayleigh number(R).

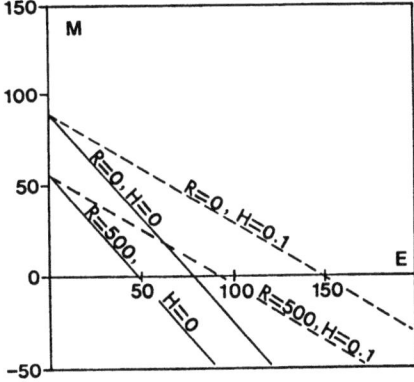

Fig.7. Thermosolutal convection with a free surface. Neutral lines using both Marangoni numbers as we vary the Rayleigh and excess-solute numbers(R and H, respectively).

ACKNOWLEDGMENTS

This research has been supported by CAICYT (Spain).

TABLE 1

Critical solutal (elasticity) Marangoni numbers as we vary the aspect-ratio of the cylinder, f, and the excess-solute number, H. j is the number of zeroes of the Bessel function $J_0(\rho)$. $R = M = S = 0$, $A1 = r_D = r^s_D = 0.01$ and $N = 10$.

	H = 0.0		H = .05		H = .1	
f	E	j	E	j	E	j
2.8	79.6	2	105.9	1	151.6	1
4	78.8	3	106.9	2	162.4	2
5.2	78.6	4	108.0	3	150.7	2
6.4	78.4	4	109.1	4	150.7	3
7.6	78.2	5	108.2	4	153.6	4
8.8	78.2	6	107.8	5	150.8	4
10	78.2	7	107.8	6	150.7	5

REFERENCES

1. L. E. Sriven and C. V. Sternling, *Nature* (London), **187** (1960), 186.
2. M. G. Velarde and J. L. Castillo, in *"Convective Transport and Instability Phenomena"*, edited by J. Zierep, Braun Verlag, Karlsruhe, 1981.
3. J. L. Castillo and M. G. Velarde, *Phy. Lett*, **A66** (1978), 489.
4. J. L. Castillo and M. G. Velarde, *J. Non-Equilib. Thermodyn*, **5**, (1980), 111.
5. J. L. Castillo and M. G. Velarde, *J. Fluid Mech*, **125** (1982), 463.
6. J. L. Castillo and M. G. Velarde, *J. Colloid Interface Sci*, **108** (1985), 264.

7. P. L. García-Ybarra and M. G. Velarde, *Phys. Fluids.* **30** (1987) 1649.
8. J. L. Castillo, P. L. García-Ybarra and M. G. Velarde, *Phys. Lett. A* **122** (1987), 107, *Phys. Fluids* **30** (1987) 2655.
9. J. T. Davies and E. K. Rideal, *"Interface Phenomena"*, Academic Press, 1963.
10. R. Defay and J. R. Hommelen, *J. Colloid Sci.*, **13** (1958), 553.
11. M. J. Schwuger and W. D. Hell, *Prog. Colloid Polymer Sci.*, **65**, (1978), 68.
12. C. Tsonopovlos, J. Newmen and J. M. Prausnitz, *Chem. Eng. Sci.*, **26** (1971), 817.
13. G. Pétré and P. Debelle, *Proc. 5th Int. Cong. Surface Active Subst.* (Barcelona), **B** (1969), 559.
14. R. Defay and G. Pétré, in *"Surface and Colloid Science"*, edited by E. Matijevic, **3** (1973), 27.
15. H. J. Palmer and J. C. Berg, *AIChE J.*, **19** (1973), 1082.
16. J. C. Berg and A. Crivos, *Chem. Eng. Sci.*, **20** (1965), 737.
17. J. J. Palmer and J. C. Berg, *J. Fluid Mech.*, **51** (1972), 385.
18. H. J. V. Tyrrell, *Diffusion and Heat Flow in Liquids*, Butterworths, London, 1961.
19. R. S. Schechter, M. G. Velarde and J. K. Platten, *Adv. Chem. Phys.*, **26** (1974), 265.
20. Ch. Normand, Y. Pomeau and M. G. Velarde, *Rev. Mod. Phys.*, **49**, (1977), 581.
21. Xiaolin Chu and Liaoyuen Chen, *Commu. Theo. Phys.* (Beijing) **6** (1986), 237.
22. Xiaolin Chu and Liaoyuen Chen, *Commu. Theo. Phys.* (Beijing) **6** (1986), 81.

INTERFACIAL CONVECTION DRIVEN BY SURFACTANT COMPOUNDS AT LIQUID

INTERFACES: CHARACTERIZATION BY A SOLUTAL MARANGONI NUMBER

E. Nakache and S. Raharimalala

Laboratoire de Chimie Physique
Université Pierre et Marie Curie
11 rue Pierre et Marie Curie, 75231 Paris Cedex 05, France

KEYWORDS/ABSTRACT : Marangoni effect ; liquid interfaces ; hydrodynamic instability, convection, solutal Marangoni number ; long chain surfactant compound ; acetone ; acetic acid.

We have studied the possibility for Marangoni effects to occur at an oil water interface containing a long chain surfactant compound which is known to damp interfacial convection. This was realized by choosing appropriately the organic solvent. In the range where the instability occurs a solutal Marangoni number recently determined on the basis of an experimental model has been evaluated. Its value is compared to those calculated for oil-water interfaces containing compounds which are known to induce interfacial convection (acetone, acetic acid).

1. INTRODUCTION

Marangoni interfacial instabilities created by the mass transfer of surface active solutes through a liquid-liquid interface enhance interfacial exchanges[1-4,16]. They occur as eruptions, cell convections or osccillations of drops which reduce the limiting diffusion layer allowing a better transfer through the interface. This is of great interest in liquid extraction, oil recovery or pharmaceutical industry.
    Among the investigators[5-9] attracted by this phenomenon, Haydon and Davies[2] were the first to point out the "kicking" of drops of water placed in contact with a solution of 4% acetone in toluene. Their interpretation of the phenomenon is based on a local increase of the concentration of acetone at the interface. This compound being moderately surface active, promotes a local change in the excess pressure and causes the drop to "kick". These workers observed also that a small quantity of a long chain surfactant molecule inhibits the movement. This was attributed to the presence of a strongly held monolayer at the interface which prevents acetone to affect the interfacial tension.
    Indeed this effect can be also interpreted in terms of an absorption desorption process according to a mechanism proposed recently[4] : A long chain surface active substance is strongly adsorbed at the interface where it forms a highly viscous monolayer which prevents the

interfacial convection to occur. On the contrary if a less surfactant compound is used, the desorption process which follows the absorption step may allow some clearing of the interface and the "kicking" may go on. If this interpretation is correct, one could find a long chain compound which should promote interfacial convection provided that the system allows its desorption.

In this paper we show that such a system exists. We investigate the desorption role in the occurence of the instability through a theoretical Marangoni number recently calculated[10] which takes into account particularly the variation of the interfacial density of the surfactant compound with its concentration.

2. MATERIALS

Drops of pure solvents were formed from an AGLA micromoter syringe. their diameter was 0.2 cm and they were hung 0.5 cm below the surface of a 20 mL solution of the surfactant dissolved in the other solvent. Before an experiment the two solvents were previously saturated by each other. All measurements were carried out at 24 ± 0.1°C in the absence of forced convection.

When it occured, the "kicking" of a drop was related to local contractions and expansions of the surface, visible owing to an emulsion which appears in some cases. These movements forced the drop to deviate from the vertical axis and thus to oscillate. The frequency of the pulsations was determined by taking the average number of sixty oscillations for three different drops, during the first minutes of their contact with the aqueous solution. Indeed two or three minutes afer the beginning of the movements the frequency decreased with time as the drop/solution system reached the equilibrium state.

The long chain surfactant used was dodecyltrimethylammonium bromide-symbolised by $C_{12}Br$-, a compound which is similar to the chloride salt -$C_{12}Cl$- used by Haydon[2] to inhibit the "kicking" phenomenon. This compound was bought from SIGMA laboratories and was cristallised twice in ethanol. Its superficial tension versus aqueous concentration curve did not show any minimun near the c.m.c., which is considered as a purity test. Acetic acid and acetone were pure MERCK compounds. Water was distillated. Pure toluene and nitroethane, respectively from MERCK and CARLO ERBA laboratories, were used as organic solvents. They were washed twice with distillated water before use. All the vessels were cleaned with sulfochromic acid then washed several times with distillated water before use.

When necessary the interfacial tension was measured at the equilibrium on a flat interface with a detaching method using a stirrup[12] instead of a ring. The distribution coefficients of $C_{12}Br$, acetic acid and acetone were determined respectively by conductimetry of the aqueous phase, acido-basic titration of the water phase and gas chromatography.

3. RESULTS AND DISCUSSION

3.1 "Kicking" experiments

We observed pulsations of nitroethane drops in contact with $C_{12}Br$ aqueous solutions, contrarily to what could be expected from Haydon's results (fig. 1). However, as this author reported, we did not see any oscillation if nitroethane was replaced by xylene (a solvent like toluene), and we got the kicking phenomenon with a drop of water hung in a solution of acetone or acetic acid in xylene.

Fig.(1) shows that when the instability of the drop occurs, the frequency of the movement increases with the initial concentration C_o from a threshold to a bifurcation of the curve which correspond to violent oscillations preceeding the detachment of the drop. As cationic compounds like $C_{12}Br$ are known to reduce the interfacial tension when their concentration is increased, we can deduce that the increase of the movement is related to an interfacial tension variation. This proves the presence of a Marangoni effect.

The question is now to look for a model which can account for this effect. Recently an experimental mechanism[4] and a theoretical model[10] have been proposed. We shall use them in order to point out the parameters which rule the phenomenon and to understand their relationship.

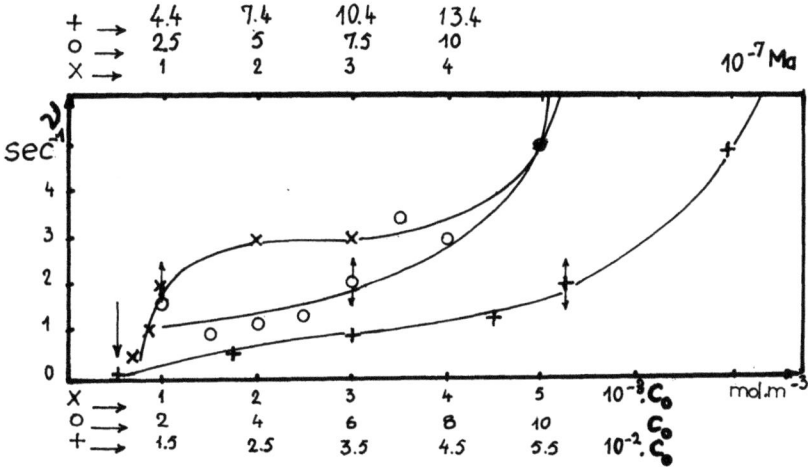

Figure 1. Influence of the initial concentration and Ma on the frequency of the movement. o $C_{12}Br$ in water/nitroethane drop ; + acetic acid in xylene/water drop ; x acetone in xylene/water drop.

3.2 Parameters governing the interfacial convection

The first obvious condition, necessary to get the instability, is that the system must be far from equilibrium. This is possible if the surface active solute is dissolved in the phase where it is the less soluble and if the distribution coefficient P between this phase and the other one is very small.

During the relaxation towards the equilibrium the transfer of the surface active solute may result in a hydrodynamical effect, which depends on transfer processes (diffusion, convection, adsorption-desorption) and on interfacial tension gradients. A coupling between these effects may promote an instability. According to Haydon and Davies[2] the arrival of an eddy of solution bringing more surfactant molecules at the interface promotes a local convective flux which may create a local interfacial tension gradient. According to Sternling and Scriven[1], if appropriate conditions of viscosity, diffusivity, surface activity and concentrations are fulfilled, this gradient is amplified and it generates an interfacial convection, causing the drop to kick. But if the absorbed layer is too condensed, i.e. if the interfacial density is too large, then the layer will resist the convection and the instability will be damped. The movement is sustained only if some clearing of the interface occurs[11]. This is the case for acetone and acetic acid which are poor surfactant solutes, easily desorbed. It results that the interfacial density Γ which governs the interfacial tension plays a major role in the "kicking" phenomenon. In order to evaluate this parameter at oil water interfaces it can be assumed[13] that the adsorption-desorption process is faster than the diffusion one. This induces that the adsorbed layer is always in equilibrium with two little zones located on either side of the interface. According to Gibbs the value of Γ is then related to the interfacial tension and to the equilibrium concentration of one of the phases.

3.3 The theoretical Marangoni number

The theoretical treatment proposed[10] on the basis of the experimental mechanism assumes an ideal situation where a finite amount of surfactant solute C_o is instaneously released at a distance L from a plane and motionless interface. This "puff" of surfactant arriving at the interface by a diffusion-convection process can figure the eddy arriving near the drop suggested by Haydon and Davies. The adsorption-desorption process at the interface is assumed to be faster than the diffusion-convection one. Therefore the local equilibrium condition is valid. The adsorption isotherm is supposed to be ideal, inducing that the interfacial density varies linearly with the concentration.

The treatment of this model adapted to the limiting condition of a spherical drop provides a solutal Marangoni number which characterises the interfacial convection with regards to diffusion :

$$Ma = \frac{RT}{D_1 \mu_1} \frac{\Gamma}{C_1} C_o d \qquad (1)$$

where d, Γ, D_1, μ_1 and C_o are respectively the diameter of the drop, the interfacial density, the bulk diffusion coefficient, the bulk dynamic viscosity and the initial concentration of the "puff". C_1 is a local concentration of phase 1 at time t in local equilibrium with phase 2 so that

$$P = \frac{C_1}{C_2} \qquad (2)$$

As an example it is interesting to notice that a compound which is strongly adsorbed at an oil-water interface gives a large Γ for a small C_1, leading to a great Marangoni number of the system.

3.4 Application to our experiments. Calculation of the experimental Ma

In order to apply relation (1) to our experiments we considered in a first approximation that C_o is the concentration of phase 1 just before its contact with the drop and that, just after the contact, owing to the assumption of local equilibrium, C_o is divided into two local concentrations, C_1 and C_2, related by

$$C_o = C_1 + C_2 \qquad (3)$$

Thus from relation (2) and (3) we can deduce

$$C_1 = C_o \frac{1 + P}{(P)} \qquad (4)$$

Γ was evaluated from γ versus log C_1 curves, by applying Gibbs relation

$$-\frac{d\gamma}{d \ln C_1} = RT\Gamma$$

where is the equilibrium interfacial tension between phase 1 and 2. The Γ versus C_1 curves (fig. 2) show an ideal behaviour in their major part, justifying the assumption of ideal isotherms. These curves also point out that the ratio Γ/C_1 which is related to the adsorbability of the solute and to the Marangoni number is decreasing from $C_{12}Br$ to acetone :

$C_{12}Br$ x/w > $C_{12}Br$ w/n > acetic acid x/w > acetone x/w

x, w and n are respectively xylene, water and nitroethane.

For the different solution/drop systems investigated Table I gives the constant parameters needed to calculate Ma. The diffusion coefficient of $C_{12}Br$ in water was calculated from conductimetric measurements[14].

Table 1

solution (1)/drop	P	μsolvent(1) poises	D(in solvent(1)) m^2/sec
acetone in xylene/water	0.50 ± 0.05	0.72 10^{-3}	2.6 10^{-9} (in C_6H_6)
acetic acid xylene/water	0.040 ± 0.003	"	2.1 10^{-9} (in C_6H_6)
$C_{12}Br$ in xylene/water	0.041 ± 0.03	"	2 10^{-10} ($C_{12}SO_4Na$)
$C_{12}Br$ in water/nitro-ethane	0.475 ± 0.005	10^{-3}	9.2 10^{-10}

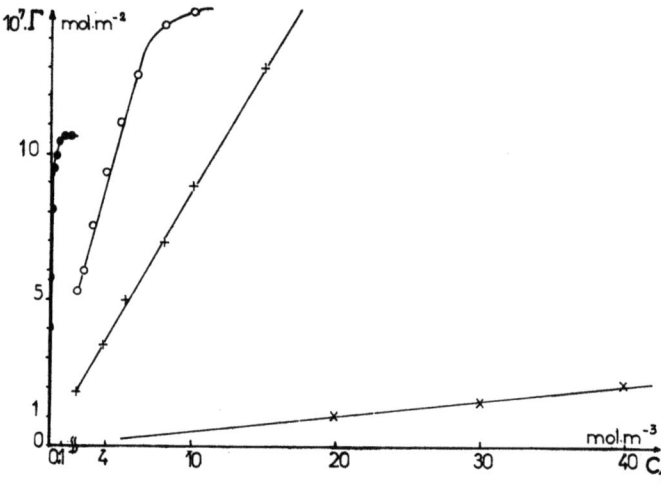

Figure 2. Variation of the interfacial density versus the equilibrium concentration in phase 1. ● $C_{12}Br$ in xylene/water ; ○ $C_{12}Br$ in water/nitroethane ; + acetic acid in xylene/water ; × acetone in xylene/water

For the other parameters some approximations have been made : the viscosities of solutions (1) were assumed to be those of their solvents. The acetone and acetic acid diffusivities were taken in benzene, which structure is similar to xylene. The diffusivity of $C_{12}Br$ in xylene was assumed to be that of sodium dodecyl sulfate in this solvent[15]. This compound has a C_{12} long chain like $C_{12}Br$ but its polar head is a sulfate ion instead of a bromide one.

In figure 3 the variation of Ma is plotted versus C_o in a range, where the adsorption isotherms are ideal. The different systems can be divided in instable ones, for which Ma is smaller than 10^8 and stable ones, beyond this Ma value. For a given C_o the adimensional number relative to $C_{12}Br$ in xylene/water interface is quite two orders of magnitude greater than the Ma of other systems. This is consistent with the adsorbability of $C_{12}Br$ in xylene which is also quite different from that of the other systems (fig 2). As a matter of fact convection is much higher than diffusion, and the surfactant solute is very rapidly transferred to the interface where it is strongly adsorbed. A viscous layer may appear very quickly, which may prevent the interfacial convection to go on. Reversely, in the other systems, the convection and the adsorbability are lower. The convective instability may then be sustained.

Arrows in fig. (3) and (1) indicate the threshold of the "kicking" of the drop. The corresponding critical Ma numbers remain in the range 10^6-10^7. For that values the corresponding concentration of the $C_{12}Br$ in xylene/water system is so small (about $10^{-3} mol.m^{-3}$) that the interfacial tension variation is negligible (about 0.1 mN/m). This is consistent with Davies and Haydon's observations concerning the onset of the interfacial convection only if this variation is at least a few mN/m.

According to the assumptions made for the experimental determination of Ma, we have not investigated further the Ma values corresponding to the bifurcation in the curves of figure (1) because the adsorption isotherms are no more ideal.

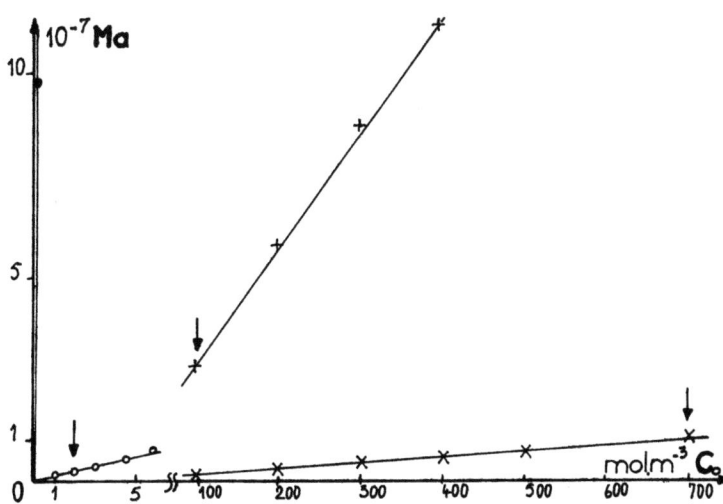

Figure 3. Variation of Ma with C_o. Symbols are the same as in figure 2).

4. CONCLUSION

The approach of the "kicking" of drops as a Marangoni effect characterized by a solutal Marangoni number has provided a "permitted" range in which the instability may occur. Although this result has to be confirmed by investigating more oil-water interfaces, it seems that a condition for a long chain surfactant like $C_{12}Br$ to promote instability is that the Marangoni number of the solution-drop system should be in the "permitted" range, the other condition being the minimum interfacial tension variation of a few mN/m pointed out by Davies and Haydon. This is of great importance for the prevision of spontaneous convective interfacial transfer through oil-water interfaces used in liquid-liquid extraction, oil recovery or pharmaceutical industries.

ACKNOWLEDGMENT The authors gratefully aknowledge Pr. M. Dupeyrat and Dr. M. Vignes-Adler for very stimulating discussions.

REFERENCES

1. C.V. Sternling, L.E. Scriven, A.I.Ch.E.J. 5:514 (1959)
2. J.T. Davies, Turbulence Phenomena, Academic Press New-York/London (1972) ; T.V. Davies, D.A. Haydon, Proc. Roy. Soc. A 243:492 (1958)
3. J. Zierep, H. Oertel Jr., Convective Transport and instability phenomena, Braun Verlag, Karstuhe (1982)
4. E. Nakache, M. Dupeyrat, M. Vignes-Adler, Faraday Discuss. Chem. Soc., 77:189 (1984)
5. D.A. Haydon, Proc. Roy. Soc., serie A, 243:483 (1958)
6. K. Durrani, C. Hanson, M.A. Hugues, Metallurgical Transactions B, 8B:169 (1977)
7. M.V. Ostrovsky, R.M. Ostrovsky, J. Coll. Int. Sc. 93:392 (1983)
8. E.S. Perez de Ortiz, H. Sawistowski, Chem. Eng. Sc. 28:2051 and 2063 (1973)
9. H. Sawistowski, in Recent advances in liquid extraction, C. Hanson Ed., Pergamon Press, Oxford 1975
10. M. Vignes-Adler Submitted for publication : M. Vignes-Adler, P.M. Adler, E. Nakache, M. Dupeyrat, Annals N.Y. Acad. Sc., 1985
11. M. Dupeyrat, E. Nakache, M. Vignes-Adler, in Chemical Instabilities, G. Nicolis and F. Baras Ed. NATO Advanced Science Institute, Series C D. Reidel, Dordrech, 1984
12. J. Guastalla, J. Chim. Phys., 53:470 (1956) ; Ibid. 68:822 (1971)
13. T.V. Davies, E.K. Rideal, "Interfacial Phenomena", Academic Press New-York/London (1963)
14. R.A. Robinson, R.H. Stokes, Electrolyte solution. Butterworth Scientific Publication 2nd edition (1959).
15. B. Lindman, P. Stilbs, M.E. Moseley, J. Int. Coll. Sc., 83, 569 (1981)
16. A. Sanfeld, A. Steinchen, "Chemical Instabilities", G. Nicolis and F. Baras Ed. D. Reidel Pub. Company p199 (1984)

SURFACE ELASTICITY AND MECHANICAL INSTABILITIES AT LIQUID-
LIQUID INTERFACES

A. Sanfeld and A. Steinchen

Chimie Physique (U.L.B.), Campus Plaine
Bd du Triomphe, B-1050 Brussels, Belgium

INTRODUCTION

At mechanical equilibrium the interface between two immiscible fluids is a transition region in which the density, the chemical composition and the related physical properties abruptly change[1]. An organized internal molecular layer leads to a lack of symmetry, and the pressure near the boundary becomes anisotropic in contrast to the isotropic pressure in the bulk phase. The presence of adsorbed molecules in the interface and of heteregeneous charged and polarized regions when fields are applied enlarges the anisotropic domain[2-5]. Furthermore the transition region is characterized by excess forces strongly depending on the physico-chemical properties. These forces may be described in terms of a renormalized functional interfacial tensions and of a so called-surface polarization[6]. Out of equilibrium, for example when motion takes place, the thickness of the transition layer interfacial enlarges. At a critical constraint leading to surface turbulence, mixed zones of the two liquids appear inducing emulsification and formation of micro and macro droplets[7]. A general description of the time evolution of this transition region requires the dynamical analysis of the behaviour of functional instead of local functions in the classical Newton-Cauchy equations. Conceptual difficulties occur from the contradictory concept of a pure mathematical and mechanical model involving discontinuities and also of a microscopic description based on statistical mechanics[8-10]. To avoid such problem, we adopt a model based on an extrapolated Cauchy-Newton two-dimensional mechanics using bulk phase tractions for describing the non autonomous character of the surface. During the motion of the interface, both mechanical and electrochemical quantities will locally change. The interdependence between the electrochemical stresses and the chemical composition leads to a strong coupling between the molecular dynamic properties and the hydrodynamic motion. When submitted to physico-chemical constraints, fluid interfaces at rest are able to convert electrochemical energies directly into mechanical energy.

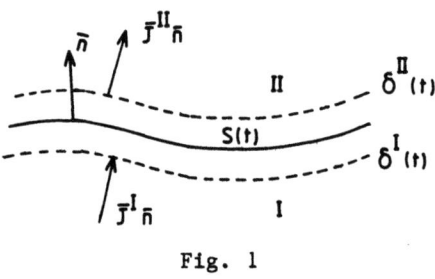

Fig. 1

In the absence of very long range forces as gravity, the role of the capillary stresses are considerably enhanced and important modifications of surface behaviour are expected. The early self-consistent studies on the fluid interfaces concerned the classical Gibbs model. For a moving and arbitrarily deformed interface, we may, for example, represent the densities of the bulk liquids I and II as Heaviside step-functions in the distance perpendicular to the surface. Considering (Fig.1) a moving surface element with area $S(t)$ depending on time t, we may define the position of the surface plane (unique position of the Gibbs dividing surface).

The two dashed line parallel to the surface line represent surfaces linked to specific solvent molecules near the surface. Therefore, the distances (t) generally vary with time. The unit normal to the surface is directed from liquid I to liquid II. Restricting matter transfers to diffusion processes relative to a solvent, the fluxes through the boundaries $-\delta^I$ (t) and $+\delta^{II}$ (t) are the surface integrals $\oint \bar{J} \overset{i}{\cdot} \bar{n}$ dS due to diffusion and not to convection, as those boundaries follow the solvent molecules. In addition to the components of the bulk liquids, other species may undergo interface diffusion ($\oint \bar{J}_s \bar{n}_s$ dS ; $\bar{n}_s$ is the inwardly directed unit normal to the boundary curve of the surface element) and also chemical reactions. This model may be related to a two-dimensional system characterized by singular surface quantities : sorption variables, viscosities, elasticities and external forces.

The transition from the quasi two-dimensional phase to an actual three-dimensional medium remains up to now a controversial open questions.

We will try, in the next section, to introduce more explicitely a consistent general formalism considering intrinsic surface quantities in addition to relaxation processes of the direct adjacent phases. To obtain analytical expressions, we restrict our study to a linear expansion of the perturbed equations.

THE USE OF THE DYNAMICAL SURFACE ELASTICITY AS A KEY PARAMETER FOR THE STABILITY ANALYSIS OF SURFACES

The classical analysis of stability of fluids consists to follow the time evolution of perturbations of velocity around a given reference state. In the case considered here, the reference state is motionless. The partial differential equation describing the evolution of the velocity perturbation in the bulk of the fluid is the Navier-Stokes equation if the considered fluid is Newtonian.

If the density of the fluid is concentration or temperature dependent, the momentum balance equation is coupled with the mass balances of each solute compounds and to the Fourier equation for heat transfer, like in the Bénard problem. In the Bousinesquian approximation, this composition or temperature dependence of the density appears only in the term accounting for the external force on the volume element. In the absence of gravity, this term vanishes and the Navier-Stokes equation may be solved independently of the mass balances and of the energy balance. This is no more the case even for systems without gravity when an electric force is applied on charged regions of the fluid like the regions near a charged interface (double layers). The solution of the Navier-Stokes equation has to verify boundary conditions one of which is the surface, momentum balance. In its general form this balance reads [11]

$$\Gamma \dot{\overline{v}}^s = \overline{\nabla}_s (\overline{\overline{\pi}}) + \overline{F}^s + \Delta_s (-\overline{\overline{P}} + \overline{\overline{T}}) \cdot \overline{n} \qquad (1)$$

where Γ is the total surface mass density, $\overline{\overline{\pi}}$ is the intrinsic surface stress tensor, F^s is the total surface intrinsic body force.

In the horizontal plane, Eq.(1) is the Marangoni condition while along the normal coordinate z, the same relation is the generalized Laplace-Kelvin condition. Along the horizontal coordinates, x,y, we assume for a two-dimensionnal Newtonian system

$$(\overline{\nabla}\overline{\overline{\pi}})_{\{{x \atop y}\}} = (\overline{\nabla}\sigma)_{\{{x \atop y}\}} - \eta_d \nabla_{\{{x \atop y}\}} v^s_{z,z} + \eta_s \nabla^2_s v^s_{\{{x \atop y}\}} \qquad (2)$$

where the phenomenological coefficients η_d and η_s are the intrinsic surface dilational and shear viscosities, and is the interfacial tension. An analogous equation may be written in the curvilinear coordinates [11]. Assuming the superposition of all contributions to the total force, we get

$$\overline{F}^s = \overline{\Gamma g} + \overline{F}_e^s + \overline{F}_m^s \qquad (3)$$

where $\overline{\Gamma g}$ is the surface weight, $\overline{F}_m^s$ is the excess chemical force due to very short range interactions and F_e is the electrical force.

The surface tension σ is thermodynamically defined by a mechanical contribution σ_M due to the surface composition and an electrical contribution σ_E due to the influence of the double layer[5]:

$$\sigma = \sigma_M - \sigma_E \qquad (4)$$

where

$$\sigma_E = \frac{1}{4\pi} \int \epsilon E^2 \ (\sim \sqrt{g^*})_{,1} \ \delta x^1 \qquad (5)$$

with g^* the determinant of the matrix of the space fundamental tensor and 1 is the coordinate curve of the field lines in the general curvilinear orthogonal coordinates. Usually authors neglect the inertial term $\Gamma \frac{d\overline{v}}{dt}$ because of the smallness of Γ and only few authors take into account the intrinsic surface viscosities. Without this two terms, the surface mass balance (1) reduces to the usual equation

$$\overline{\nabla}_s \sigma = -\overline{F}^s - \Delta_s (-\overline{\overline{P}} + \overline{\overline{T}}) \cdot \overline{n} \qquad (6)$$

The surface tension depends on the concentration of the adsorbed surface active solutes Γ_γ. At constant temperature, one assumes

$$d\sigma = -\sum_\gamma \alpha_\gamma d\Gamma_\gamma \qquad (7)$$

where

$$\alpha_\gamma = -\frac{\partial \sigma}{\partial \Gamma_\gamma} \qquad (8)$$

Maintaining constant the number of adsorbed molecules of each solute in the considered surface element of area equation (7) becomes

$$d\sigma = -\sum_{\gamma} \frac{\partial \sigma}{\partial \Gamma_{\gamma}} \Gamma_{\gamma} \frac{dA}{A} = \epsilon_G \frac{dA}{A} \qquad (9)$$

The quantity ϵ_G was first introduced by Gibbs to study the dilation of liquids films. It is usually called Gibbs surface elasticity. In the present approach, the local variations of surface are related to the longitudinal displacements by

$$\frac{\delta A}{A} = -D v_z^s /\omega \qquad (D \equiv \frac{d}{dz}) \qquad (10)$$

Where ω is the frequency of the perturbations. If the number of adsorbed molecules is maintained constant in each surface element during the motion, the local variation $\delta\sigma$ becomes

$$\delta\sigma = -\epsilon_G \frac{D v_z^s}{\omega} \qquad (11)$$

When there is interchange of surface active material between the surface layer and the bulk, or surface diffusion, it is still possible to define a surface elasticity relating the changes of σ to the change of A. This quantity however now depends on the complete description of the dynamic processes of mass transport [11,13]. According to all the relaxation processes including thermal effect, the perturbation of σ be written in terms of the variation of the surface area A and in terms of the fluctuation of the normal velocity at the surface. For a normal mode, it reads

$$\delta\sigma = -\epsilon_d(\omega,k) \frac{D\delta v_z^s}{\omega} + k\Psi(\omega,k) \frac{\delta v_z^s}{\omega} = -\sum_{\gamma} \alpha_{\gamma} \delta\Gamma_{\gamma} + \alpha_{th} \delta T^s \qquad (12)$$

where ε_d is the dynamical surface elasticity related to the longitudinal displacement $D\delta v_z^s$, Ψ is a phenomenological dynamical quantity related to the normal displacement and k is the wavenumber of the perturbations.

The phenomenological coefficients ε_d and Ψ are related to all the relaxation processes due to mass exchanges, chemical reactions and electrical effects.

The external forces acting on the surface (surface body forces and electrical stresses) also depend on the dynamic properties of the system : mass and charge transport, modification of the electrical interactions during the motion. As for the surface tension we may define two surface external perturbed forces δF_T and δF_N in terms of phenomenological quantities ε_F, ϕ_{xy}, ε_F and ϕ_z related to dynamical electrical contributions (diffuse layer, discrete ionic and dipolar interactions etc...),

$$\delta F_T = k^2 \varepsilon_F (\omega,k) \frac{Dv_z^s}{\omega} + k^3 \phi_{xy} (\omega,k) \frac{v_z^s}{\omega} \qquad (13)$$

$$\delta F_N = k^2 \sigma_F (\omega,k) \frac{v_z^s}{\omega} + \phi_z (\omega,k) \frac{Dv_z^s}{\omega} \qquad (14)$$

Fundamentally, the quantities δF_T and δF_N depend on the contributions of the surface forces and also on the bulk tractions (see 6 Eqs.241 and 242). Now comming to the general surface mass balances, one assumes

$$\dot{\Gamma}_\gamma = -\Gamma_\gamma (\bar{\nabla}.\bar{v}^s + a^*) - \bar{\nabla}_s.\bar{J}_\gamma^s - \Delta_s \{\bar{J}_\gamma^\beta\}.n + R_\gamma^s \qquad (15)$$

where $\bar{\nabla}_s.\bar{v}^s$ is the surface divergence of the surface velocity, a* is the change of the surface metric, R_γ^s is the source of surface chemical reactions, $\Delta_s \bar{J}_\gamma^\beta.\bar{n}$ accounts for the interchange of mass between the adjacent bulk phases and the surface, $\bar{J}_\gamma^s$ is the singular diffusion-migration flux on the surface

$$\bar{J}_\gamma^s = - D_\gamma^s (\bar{\nabla}_s \Gamma_\gamma + z_\gamma \Gamma_\gamma \bar{\nabla}_s \phi) \qquad (16)$$

with D_γ^s the surface diffusion coefficient. The sorption fluxes J_γ^β are related to the difference of electrochemical potentials between the surface and the sublayers, ϕ being the surface electrical potential and Z_γ the charge of the ion.

According to the process considered, the analytical form of the coefficients ε_d, ψ, ε_F, ϕ_{xy}, σ_F and ϕ_z may be obtained. They depend on the wavenumber, on the frequency of the perturbation and on the constraints in the steady state considered. For diffusional and chemical processes, fundamental expressions were derived by M. Hennenberg (see for example an exhaustive discussions in Ref [14]). An extension to electrical effects was given by P.M. Bisch [6].

The general formulation of the dispersion relation obtained as compatibility condition in a linear stability analysis may be written as the combination of two modes L and T

$$LT - C = 0 \qquad (17)$$

the condition $L=0$ corresponds to the dispersion relation for the pure longitudinal modes, while $T=0$ corresponds to the pure transversal modes. The term C accounts for the coupling between the two modes of the surface wave propagation. From the mathematical point of view L is the coefficient of the longitudinal displacement Dv_z^s in the tangential stress balance and T is the coefficient of the transversal displacement v_z^s in the normal stress balance. Taking into account the mass and momentum balances, one obtains the two stresses :

tangential stress

$$\{\omega \Gamma + k^2(\eta_d + \eta_s) + \sum_\beta \mu^\beta(q^\beta + k) + \frac{k^2}{\omega}(\varepsilon_d + \varepsilon_F)\} Dv_z^s$$
longitudinal displacements

$$+ \{\Delta_s \{\mu^\beta (q^\beta - k)\} + \frac{k^2}{\omega}(\psi + \psi_{xy})\} kv_z^s = 0 \qquad (18)$$
transversal displacements

normal stress
$$\{ \omega (\Gamma + \sum_\beta \rho^\beta/k) + \sum_\beta \mu^\beta (q^\beta + k) + \frac{k^2}{\omega}(\sigma^0 + \sigma_F) + \frac{g}{\omega}\Delta_s \{\rho^\beta\} \} k v_z^s$$

transversal displacements
$$+ \{ \Delta \{ \mu^\beta (q^\beta - k) \} + \frac{k\phi_z}{\omega} \} Dv_z^s = 0 \qquad (19)$$
longitudinal displacements

where σ^0 is the surface tension in the reference steady state, μ^β is the viscosity of the bulk phase β and $q^\beta = (k^2 + \frac{\omega}{\nu^\beta})^{1/2}$.

Each stress is thus a combination of two displacements. The elasticticity ε_d, and the additional contribution ε_F appear in the terms proportional to the longitudinal displacements of the tangential stress balance. The dynamic elasticity also includes dissipative contributions analogous to viscous effects[13-15], in contrast to the Gibbs elasticity which is a constant storage modulus. Both the storage modulus and the loss modulus depend throughly on the interfacial electrical forces. The quantity appears in the terms of the transversal displacements in the normal balance equation. When σ_F is independent from ω and k, the body forces only give a constant contribution to the interfacial tension. As discussed by Miller and Scriven[16] for wavelengths larger than the thickness of the diffuse layer, the electrical forces only give a negative contribution to the interfacial tension. For small wavelength σ_F however depends on k. It also depend on ω when the Boltzmann distribution in the diffuse layer is not instantaneously restored. In that case, a dissipative effect associated to the normal displacement, leads to an additional "viscous" resistance.

The model of Langevin et al.[17] for the propagation of surface waves includes a surface viscosity associated to normal displacements. As shown previously [18,19] such dissipative effects can be ascribed to the relaxation time of the electrical interaction during the motion.

Finally the quantities ($\phi_{xy} + \psi$) and ϕ_z contribute to the coupling between longitudinal and transversal surface motion. They may lead to a new behaviour of the surface waves propagation (see for example [6,20,21]).

DYNAMICAL SURFACE ELASTICITY FOR SPECIFIC CASES

i) Transfer of matter by diffusion processes

Let us assume that there is no accumulations of matter at a sublayer which is a limiting plane separating diffusion and sorption processes. However in the reference state, there is no accumulation of matter at the interface. For one surface active solute in ideal solutions, crossing the interface, and in the absence of electrical forces neglecting also the surface intrinsic body forces, the modes of the dispersion relation (17) become

$$L = \omega^2 \Gamma + \omega k^2 (\eta_d + \eta_s) + \omega \sum_\beta \mu^\beta (q^\beta + k) + k^2 \varepsilon(\omega,k) \tag{20}$$

$$T = \omega^2 (\Gamma + \sum_\beta \rho^\beta/k) + \omega \sum_\beta \mu^\beta (q^\beta + k) + k^2 \sigma + g \Delta_s \{\rho^\beta\} \tag{21}$$

$$C = \omega \Delta_s \{\mu^\beta (q^\beta - k)\} \cdot \omega \Delta_s \{\mu^\beta (q^\beta - k)\} + k^2 \psi(\omega,k) \tag{22}$$

where

$$\varepsilon(\omega,k) = \frac{\varepsilon_G \{1 + \frac{(\beta*D)}{\Gamma\omega} \Delta_s \{\frac{(D^\beta - k)}{(1 + D^\beta p^\beta/a)(p^\beta + q)}\}\}}{\{1 + \frac{k^2 D^s}{\omega} + \sum_\beta \frac{D^\beta p^\beta b}{(1 + D^\beta p^\beta/a)(a\omega)}\}} \tag{23}$$

$$\psi(\omega,k) = \frac{\varepsilon_G \{\frac{(\beta*D)}{\Gamma\omega} \sum_\beta \frac{(q^\beta + k)}{(1 + D^\beta p^\beta/a)q^\beta + (p^\beta)}\}}{\{1 + \frac{k^2 D^s}{\omega} + \sum_\beta \frac{D^\beta p^\beta b}{(1 + Dp^\beta/a)a\omega}\}} \tag{24}$$

with D^s the surface diffusion coefficient, $(\beta^* D)$ the unperturbed diffusion flux ($\beta^{*I} D^I = \beta^{*II} D^{II}$), a and b the adsorption and desorption kinetic constants respectively and $p = \{ k^2 + \omega/D \}^{1/2}$

The phenomenological quantities $\varepsilon(\omega,k)$ and $\psi(\omega,k)$ are thus related to the processes leading to the variation of the surface tension, i.e. mass exchange between the bulk phases and the surface. For vanishing fluxes in the steady state ($\beta^* D = 0$), the coupling term $\psi(\omega,k)$ vanishes and the dynamical surface elasticity reduces to

$$\varepsilon(\omega,k) = \frac{\varepsilon_G}{1 + \frac{k^2 D^s}{\omega} + \sum_\beta \frac{D^\beta p^\beta_b}{(1 + D^\beta p^\beta/a) \, a\omega}} \qquad (25)$$

As k and ω are complex quantities the wavenumber (the frequency) is a pure real number and the frequency (the wavenumber) is a complex imaginary quantity - the phase angle of ε is generally not a multiple of 2π, especially if we are looking at the response of the system to an oscillatory disturbance and

$$\varepsilon = \varepsilon_r + i \varepsilon_i = |\varepsilon| \exp i\theta \qquad (26)$$

The interface behaves thus like a visco-elastic medium as in the three dimensional theory of viscoelasticity, ε_r is a storage modulus and ε_i a loss modulus.

There are two kinds of dissipative processes at the interface. The first one is linked to the intrinsic viscosity and does not depend on the exchange of matter. The other one is solely due to the exchange and it leads to an important change in the rheological properties of the interface.

For the simple usual case where the diffusion is the determining step, $a \to \infty$ and $b \to \infty$ with $\frac{a}{b} = K$.

Equations (23) and (24) reduce then to

$$\varepsilon(\omega,k) = \frac{\varepsilon_G \{1 + \frac{\beta*D}{\Gamma\omega} \Delta_s \{\frac{p^\beta - k^\beta}{p^\beta + q^\beta}\}\}}{\{1 + \frac{k^2 D^s}{\omega} + \sum_\beta \frac{D^\beta p^\beta}{K\omega}\}} \qquad (27)$$

$$\psi(\omega,k) = \frac{\varepsilon_G \{\frac{\beta*D}{\Gamma\omega} \Delta_s \frac{q^\beta + k}{p^\beta + q^s}\}}{\{1 + \frac{k^2 D^s}{\omega} + \sum_\beta \frac{D^\beta p^\beta}{K\omega}\}} \qquad (28)$$

ii) Surface chemical reactions and sorption fluxes

That case was studied in an exhaustive way by W. Dalle Vedove et al.[22-24]. Let us consider pure longitudinal waves. Neglecting the surface mass density and the surface viscosity, the dispersion relation reads, for a system with a phase of high viscosity neighbouring a phase with a negligible viscosity

$$\mu(q + k) = -k^2 \frac{\varepsilon}{\omega}(\omega,k) \qquad (29)$$

For several species γ undergoing surface reactions and exchanging themselves with the neighbouring phases by linear adsorption processes, the normal mode expansion of the perturbed balance reads for uncharged species and for a flat interface

$$\omega \delta \Gamma_\gamma = \sum_\beta c_{\gamma\beta} \delta \Gamma_\beta - k^2 D_\gamma^s \delta \Gamma_\gamma + \Gamma_\gamma^0 Dv_z^s \qquad (30)$$

where $c_{\gamma\beta}$ are the chemical matrix elements related to the chemical sources and to the sorption fluxes. In the reference steady state, the surface concentrations satisfy the system of algebraic relations

$$(I_{\Gamma_\gamma} + J_\gamma^{ad})^0 = 0 \qquad (31)$$

with J_γ^{ad} and I_{T_γ} are respectively the sorption flux and the chemical source of γ.

The system of Eq.(30) for all species γ may be solved, we then get

$$\delta \Gamma_\gamma = \frac{\det L^{(\gamma)}}{\det L} D v_z^s \qquad (32)$$

where

$$L_{ij} = (\omega + k^2 D_j^s) \delta_{ij} - c_{ij} \qquad (33)$$

$$L_{ij}^{(\gamma)} = L_{ij} + \delta_{j\gamma} (\Gamma_i^0 - L_{ij}) \qquad (34)$$

Combining Eqs.(32) and (12) for systems undergoing pure longitudinal waves, we get for systems without heat fluxes,

$$\frac{\varepsilon}{\omega} = \sum_\gamma \alpha_\gamma \frac{\det L^{(\gamma)}}{\det L} \qquad (35)$$

In the absence of chemical surface reactions but in the presence of sorption processes, Eq.(35) reduces to

$$\frac{\varepsilon}{\omega} = \sum_\gamma \frac{\alpha_\gamma \Gamma_\gamma^0}{\omega + k^2 D_\gamma^s + \sum_\beta b_\gamma^\beta} \qquad (36)$$

where b_γ^β are the sorption constants from the surface to the bulk phases β.

The dispersion relation (29) with relation (35) leads to

$$\mu(q + k) = - k^2 \sum \alpha_\gamma \frac{\det L^{(\gamma)}}{\det L} \qquad (37)$$

The quantities $\frac{\alpha_\gamma}{\mu}$ are except for scaling factors, the Marangoni numbers for each component γ. For instabilities with exchange of stability, the solution of Eq.(29) leads to the instability condition

$$\left(\frac{\varepsilon}{\omega}\right)_{\omega = 0} < 0 \qquad (38)$$

For periodic regimes, the instability conditions are

$$\text{Re}\left(\frac{\varepsilon}{\omega}\right) < 0 \quad \text{and} \quad \text{Im}\left(\frac{\varepsilon}{\omega}\right) < 0 \qquad (39)$$

Several examples are studied by Dalle Vedove et al. (see for example[24,25]). From a general point of view, the linear theory of stability enlarges the first approaches due to Sanfeld and Steinchen[26,27] and extended later on by M. Hennenberg and T.S. Sørensen [28].

The general conclusion predicted are:

a) Mechano chemical instabilities result from a particular competition between chemical and convective processes. Both induce opposite local variations of the interfacial composition.

b) Surface chemical kinetics has to include a non linear step to induce instabilities.

c) Chemical auto (or cross) catalysis is an intrinsic cause for surface convection. Non auto (cross) catalytic steps are however also able to induce interfacial movements.

d) Chemical and mechanical processes are uncoupled for homogeneous perturbations. For infinite viscous media, the mechanical stability is guaranteed and the problem reduces itself to a pure chemical behaviour.

e) Surface diffusion stabilizes small wavelengths.

f) Linear sorption kinetics are unable to destabilize the surface.

g) For non autocatalytic schemes, increasing viscous coefficients contribute to stabilize the system. Unstable states occur for non viscous phases. On the other hand, for autocatalytic schemes, the system remains unstable for high viscous phases while non oscillatory regimes are stabilized for zero viscous coefficients).

h) More recently,[29] we investigated the role of the differences of surface activities of each component. They may be responsible for surface convection even in the case of intrinsically stable mechanisms.

Let us finally mention three interesting recent contributions on the chemo-Marangoni instability.

a) Sakata and Funada[30] studied the onset of instability driven by surface tension gradients in an isothermal gas-liquid reaction process. Including a deformation of free surface, effects of capillary and gravity waves were considered. These effects are important for the disturbance of small wavenumbers. They have shown the existence of two distinct mechanisms of instability, either with the essential effects of surface deformation or without them. For the former, the critical Marangoni number linearly depends on the ratio of the Weber number to the Crispation number and on the reaction parameter. For the latter, it depends on the reaction parameter and the diffusivity ratio.

b) Pismen[31] addresses the problem of formation of stationary patterns of convection and chemical activity due to a chemical autocatalytic reaction in a film with a free surface. The simple mechanism involves a single reactant supplied from the liquid phase.

He combines ideas of the catastrophe theory and long-scale expansion in order to derive a prototype equation that retains qualitative features of dynamics of the full non linear problem. Bifurcations to stationary patterns that occur in the prototype equation indicate the role of Marangoni convection in stabilizing unstable inhomogeneous states of a quiescent system. The prototype equation can produce as well more complex patterns of flow and chemical activity, in particular, turbulent behavior that has been observed is its truncated versions.

One of the most important result is the role of Marangoni convection in stabilizing unstable inhomogeneous solutions of the reaction-diffusion process. It is an important generalization of our linear approach.

c) As we have shown, transfers of surfactants through a fluid interface may induce Marangoni convection. Far from equilibrium and for finite perturbations, no analytical prediction is known up to now. M. Adler[32] has used a numerical method to solve the momentum and the matter balances in such conditions. She investigated the role of two dimensionless parameters : the partition coefficient related to the driving force and the Marangoni number which compares convective and diffusion transfers to surface inhomogeneities. The influence of the partition coefficient on the interfacial transfer by convection is discussed in detail. Far from equilibrium distribution the interfacial transfer is strongly increased by the Marangoni convection. This result is important for a large variety of problems and particularly for solvent extraction. In that case, however, other constraints may play a determinant role as mentioned above (for example, surface kinetics[33] and microemulsification[34]).

iii) Role of the electrical double layer

Let us briefly recall the results obtained for a charged interface between two immiscible Newtonian fluids[18]. The interface in a steady state at rest is submitted to electrical constraints : potential jump between the two phases and non-equilibrium surface charge density. The Boltzmann ionic distribution is maintained in the diffuse layer even in the perturbed state. The sorption of surface active ions is taken into account. For low-tension systems, we consider two surface dynamic behaviors : the slow (the frequency is sufficiently low as compared to the sorption step and to the surface diffusion) and the fast regime (the frequency of the wave motion is large enough so that the sorption and the surface diffusion become negligible). For these two limiting cases, only the negative electrical contribution of the total interfacial tension may be responsible for the onset of surface motions. For constant surface potential or for constant surface charge, the viscosity increases the maximum wavelength and decreases the fastest rate of growth. It thus stabilizes the system. The study of longitudinal waves for nearly insoluble monolayers shows that specifically adsorbed counterions contribute to increase the surface elasticity. The kinetics of sorption of counterions also leads to a compositional contribution to the dilational surface viscosity. Some qualitative agreements are shown with results obtained from induced surface waves and light-scattering techniques.

The same type of analysis was carried out for systems with larger relaxation times of the diffuse layers. When these relaxation times are of the same order of mangitude as the characteristic time of the hydrodynamic motion, dissipative effects in the double layer influence the hydrodynamic stability of the interface. In order to describe the competition between convective and diffusive transport in the diffuse layers, we restrict our analysis to ideally polarized systems (i.e. negligible fluxes through the interface), with small potential differences accross the diffuse layers. When the coupling between the transversal and the longitudinal motion is negligible, the only possibility of surface instability is the onset of transversal motion corresponding to negative or vanishing values of the total interfacial tension. The double layers contribute to the destabilization of the interface through their negative contribution to the surface tension. The transport of ions through the diffuse layers gives rise to a damping term that is responsible for a slowing down of the interfacial motion in both stable and unstable situations. For large values of the interfacial tension, the coupling term between the two modes of displacements may become important for systems with different viscosities and densities and for very slow diffusion processes. We predict the possibility of unstable states even for positive values of the surface tension. These values of the surface tension. These unstable regimes are due to a positive feedback between the normal deformation produced by the electrical stress originated by the longitudinal deformation and the longitudinal viscous stress originated by the normal deformation.

In Table 1, we summarize the stability criteria for polarized interfaces undergoing an electrical constraint (potential drop) for small surface charge

$$(\varepsilon^{II} C_\infty^{II})^{1/2} D^{II} > (\varepsilon^{I} C_\infty^{I})^{1/2} D^{I} \quad \text{stable}$$

$\rho^I \mu^I > \rho^{II} \mu^{II}$

$$(\varepsilon^{II} C_\infty^{II})^{1/2} D^{II} < (\varepsilon^{I} C_\infty^{I})^{1/2} D^{I} \quad \text{unstable}$$

$$\varepsilon^{II} C_\infty^{II} > \varepsilon^{I} C_\infty^{I} \quad \text{stable}$$

$\rho^I \mu^I > \rho^{II} \mu^{II}$ $\{ D^I = D^{II} \}$

$$\varepsilon^{II} C_\infty^{II} < \varepsilon^{I} C_\infty^{I} \quad \text{unstable}$$

The quantities ε^β are the dielectric constants for both bulk phases ($\beta = I, II$) and C^β are the concentrations.

For small and finite potential drop, the system becomes unstable if the phase where the potential drop is the largest is also the phase with the smallest $\rho\mu$.

Several examples are discussed in our previous papers[18,19,20]. Finally, let us mention unit operation approaches for the stability of fluid interfaces under microgravity conditions. Recent contributions are developed in this field by Velarde et al (see for example[36]).

Acknowledgments

The paper was financially supported by NATO and CONTRACT STI019 ECC Brussels.

REFERENCES

1. J.S. Rowlinson and B. Widom, "Molecular Theory of Capillarity Clarendom Press, Oxford, 1982.
2. A. Steinchen, R. Defay and, A. Sanfeld, J. Chem. Phys., 68, 83, (1971); ibid 69, 1241, (1971).
3. R. Defay, A. Sanfeld and A. Steinchen, J. Chem. Phys., 68, 518, (1971); 64, 1374, (1971).
4. R. Defay and A. Sanfeld, Electrochem. Acta, 12, 913, (1967); ibid J. Chem. Phys., 70, 895, (1973); ibid An. Quin. 70, 856, (1975).
5. A. Sanfeld, "Introduction to thermodynamics of charged and polarized layers", Wiley, 1968.
6. P. Bisch, Thesis, Université Libre de Bruxelles, 1980.
7. J.T. Davies, "Turbulence Phenomena" Academic Press, New York, 1972.
8. H. Brenner, P.C.H. Meeting, La Rabida (Spain), July, 1986.
9. Evans, P.C.H. Meeting, La Rabida (Spain), July, 1986.
10. M. Baus, P.C.H. Meeting, La Rabida (Spain), July, 1986.
11. T.S. Sorensen, Ed. in "Lecture Notes in Physics", 105, (1976).
12. W.J. Gibbs, Scientific papers (Ed. Dover), 1981.
13. M. van den Tempel, J. Non Newtonian Fluid Mech. 2, 105, (1977).
14. M. Hennenberg, Thesis, Université Libre de Bruxelles, 1980.
15. M.C. Maru, V. Mohan and D.T. Wasan, Chem. Eng. Sci., 34, 1283, (1979).
16. C.A. Miller and L.E. Scriven, J. Colloid Interf. Sci. 33, 371, (1970).
17. D. Langevin and C. Griesmar, J. Phys. D 13, 1189, (1980).
18. M. Prévost, P.M. Bisch and A. Sanfeld, J. Colloid Interf. Sci., 88, 353, (1982).

19. P.M. Bisch, A. Steinchen and A. Sanfeld, J. Colloid Interf. Sci., 95, 361, (1983).
20. P.M. Bisch, van Lansweerde-Gallez and A. Sanfeld, J. Colloid Interf. Sci. 71, 501, (1979); ibid 513.
21. D. Gallez, A. Sanfeld and P.M. Bish, P.C.H. 3, 1, (1982).
22. W. Dalle Vedove, P.M. Bisch and A. Sanfeld, J. Non Equil. Therm. 5, 35, (1980).
23. W. Dalle Vedove and A. Sanfeld, J. Colloid Interf. Sci. 84, 318, 1981; ibid 328.
24. W. Dalle Vedove, Thesis, Université Libre de Bruxelles, 1984.
25. A.R. Marquez Garcia, W. Dalle Vedove and A. Sanfeld, J. Chem. Soc. Far. Trans. 2, 77, 2303, (1981).
26. A. Sanfeld and A. Steinchen, Biophys. Chem. 3, 99, (1976).
27. A. Sanfeld, A. Steinchen, M. Hennenberg, P.M. Bisch, J. van Lamsweerde and W. Dalle Vedove, in "Lecture Notes in Physics", 105, 168, (1979).
28. T.S. Sørensen, M. Hennenberg, A. Steinchen and A. Sanfeld, J. Colloid Interf. Sci. 56, 191, (1976).
29. A. Sanfeld and A. Steinchen, Final Report ECC Contract STI019 JC Brussels, (1986).
30. M. Sakata and T. Funada, J. Phys. Soc. Japan, 50, 696, (1981).
31. L.M. Pismen, J. Colloid Interf. Sci. 102, 237, (1984).
32. M. Adler, Final Report ECC Contract STI019 JC Brussels, (1986).
33. G.J. Hanna and R.D. Noble, Chem. Rev. 85, 583, (1985).
34. M.K. Sharma, S.Y. Shiao, V.K. Bansal and D.O. Shah, ACS Symp. Series 272-87, 1983 (Ed. D.O. SHah), Washington D.C. (1985).
35. A. Sanfeld, M. Lin, A. Bois, Y. Panaiotov and J.F. Barret, Adv. Coll. Interf. Sci. 20, 101, (1984).
36. M.G. Velarde, A. Sanfeld and J.C. Legros, in "Orbit 2000", ESA PUBLICATION, Dordrecht 1987.

VOLUME OR SURFACE INSTABILITIES DURING LIQUIDS EVAPORATION UNDER REDUCED
PRESSURE OR/AND MICROWAVE IRRADIATION

Annie Steinchen-Sanfeld

Service de Chimie Physique II
Université Libre de Bruxelles (Belgium)

Michel Lallemant, Pascal Courville, Pascale Gillon, and
Gilles Bertrand

Laboratoire de Recherches sur la Réactivité des Solides
(U.A. 23), Faculté des Sciences Mirande, B.P. 138
21004 Dijon Cédex (France)

INTRODUCTION

Since a few years, Lallemant, Bertrand and al.[1-8] published several papers on evaporation and dehydration phenomena. Their investigations on evaporation of polar liquids were developped both in the experimental and in the theoretical field.

Experimentally they measured very accurately the temperatures in bulk phases and in the surface and their time evolution. They determined the evaporation flux in relation to the depression and/or microwave power constraints. The theoretical analysis[9-11] is principally an hydrodynamic stability analysis extending the classical Rayleigh-Bénard buoyancy instability criteria and the surface instability criteria (Marangoni and vapor recoil instabilities) to their experimental conditions.

EXPERIMENTAL SET-UP

The evaporating liquid fills a square quartz cell (b) crossing a rectangular microwave cavity (a) (Fig. 1) connected to a microwave source. The system is evacuated and the pressure P is fixed by equilibration of the vapor pressure with a condensor at fixed temperature (cold spot f). The liquid-vapor level in the evaporation cell is monitored by a system of photodiodes (connected to a Thomson TH 793 L monitor (g) governing the opening of the valve of the mercury piston (d). The level of the mercury is measured in function of time with a cathetometer (accuracy 1/50 mm). The rate V_o of evaporation is directly connected with the slope of the mercury level versus time plot.

Several thermocouples measure the temperatures in the vapor and in the liquid (outside of the field). Surface temperature is measured by infrared thermography. A video camera records bulk and interfacial motion.

The product studied is 95° ethyl alcohol (azeotropic mixture) provided by Carlo Herba.

The microwave field applied is supplied by a Mikrotron mark 3 source. The frequency chosen is 2450 MHz. To avoid field perturbation in the guide, the liquid interface is maintained a few millimeters outside the waveguide.

In previous experiments, a cylindrical evaporating tube was used. The observed phenomena and the sequence of structures observed were the same as in the square cell. This last was chosen to ensure an homogeneous microwave field inside the liquid and to avoid the effects of concentration of field due to the cylindrical lens.

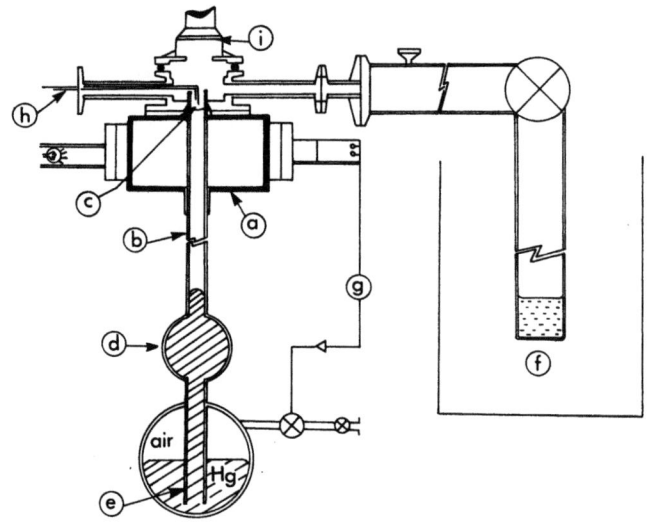

Fig. 1. Schematic view of the experimental apparatus

    a : waves guide RG 112/u
    b : evaporator ; square or cylindrical tube
    c : evaporative interface
    d : degasing bulb
    e : mercury container
    f : condenser
    g : system of maintaining the liquid level
    h : thermocouples
    i : radiothermometer or infra-red camera.

RESULTS

Fig. 2a, 2b gives the rate of evaporation of ethyl alcohol versus depressure or microwave power constraints. On these curves, we discriminate three main domains : the first in the vicinity of the equilibrium pressure is characterized by a quasi linear evolution of evaporation rate with pressure, the second shows a more rapid increase of the flux but a still monotonously increasing regime and, a third one corresponds to a dramatic rise of the flux simultaneously with quasi periodic "jumps" of the interface. Between domains 2 and 3, a domain of decreasing flux is observed corresponding to a competition between the endothermal surface evaporation process and the thermal flux (the so-called Smith-Topley effect).

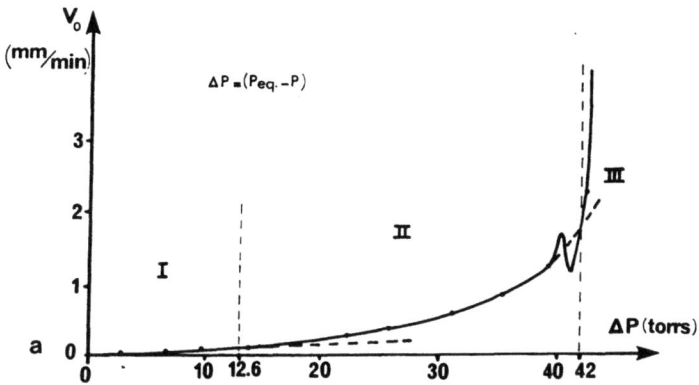

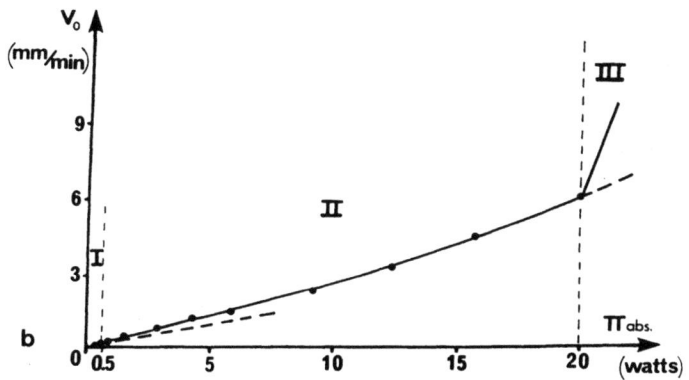

Fig. 2. Evaporation rate of ethyl alcohol
a. versus low pressure at $\pi_i = 0$
b. versus microwave irradiation for Peq
Three different modes of evaporation.

STABILITY ANALYSIS

In order to describe the transitions observed between the various regimes we will analyse the stability of the reference steady states before the onset of the various bifurcations. The first transition seems to correspond to the onset of convection in the bulk of the liquid by a mechanism similar to the classical buoyancy driven Rayleigh-Bénard instability. The peculiarity of the evaporating system however is that the temperature field is non linear in the liquid and that its extent varies in depth with the applied constraints. The second region corresponds to an evaporating flux enhanced by the bulk convection. The transition to the third region corresponds to surface effects, the most important of which being the differential vapor recoil.

First bifurcation : onset of convection

The first steady state the stability of which is analysed corresponds to the solution of the pure Fourier stationary equation without convection, but when the microwave power is on, there is an additional term in the energy balance due to the dipolar dissipation[12].
For a unidirectional thermal field normal to the interface in the coordinate system moving with the interface, it reads :

$$D_{th}^l D^2 T + V_o DT + \frac{\omega_e E_o^2 \varepsilon''}{8 \pi \rho^l C_p^l} = 0 \qquad (1)$$

with D_{th}^l the thermal diffusivity in the liquid
V_o the evaporation rate
E_o the intensity of the electric field in the liquid
ρ^l the liquid density
C_p^l the heat capacity of the liquid
ω_e the angular frequency of microwave
ε'' the dielectric loss coefficient
D   derivative along the z coordinate normal to the interface.

This equation is only valid for pure adiabatic wall conditions. In the system considered here these conditions are practically fulfilled as was observed by I.R. thermography.

For not too large deviations from the reference temperature $T°$ (taken for example as the ambiant temperature), the dielectric loss ε'' may be linearized around its value at $T°$ i.e

$$\varepsilon''(T) = \varepsilon''_o \left[1 + \alpha_e (T - T°)\right] \qquad (2)$$

where $\alpha_e = \frac{\partial \varepsilon''}{\partial T}$ is the slope of ε'' versus T plot.

For ethyl alcohol azeotropic mixture, we have estimated this value according to the data available in ref. 13. The solution of equation 1 together with 2 and the boundary conditions $T = T_a$ for $z = -\infty$ and $T = T_s$ for $z = 0$ give respectively an exponential profile of T for the system without applied field like in directional crystal growth[14], and a damped cosine form profile for the system with a positive slope of the dielectric loss versus T curve under irradiation (fig. 3).

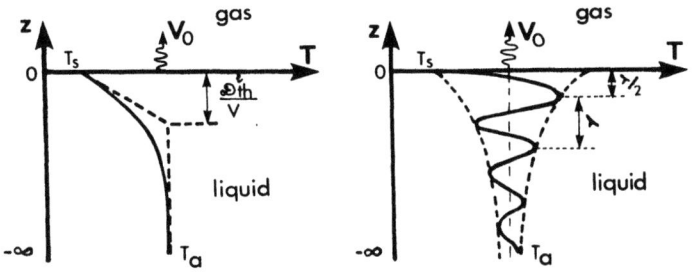

Fig. 3. Stationary temperature profiles.

Without field as well as for a system with $\alpha_e = 0$, one finds :

$$T(z) = T_a + (T_s - T_a) \exp\left[-\frac{V_o}{D_{th}^l} z\right] \quad (3)$$

In the presence of field for $\alpha_e > 0^*$

and $\left[\dfrac{V_o}{2 D_{th}^l}\right]^2 < \dfrac{\omega_e E_o^2 \varepsilon_o'' \alpha_e}{8 \pi D_{th}^l \rho^l C_p^l}$

$$T(z) = T_a + (T_s - T_a) \cos\left[\frac{2\pi z}{\lambda}\right] \exp\left[-\frac{V_o z}{2 D_{th}^l}\right] \quad (4)$$

with

$$\frac{2\pi}{\lambda} \cong \sqrt{\frac{\omega_e \varepsilon_o'' \alpha_e E_o^2}{8 \pi D_{th}^l \rho^l C_p^l}}$$

Both curves $T(z)$ are characterized by an adverse temperature gradient extending in a region near the interface whose depth : D_{th}^l/V_o for (3) or $\lambda/2$ for (4) depends on the constraints.

These length scales chosen for the definition of non dimensional variables are thus constraint dependent. The corresponding Rayleigh numbers with "mobile boundary" characterizing the stability of the liquid with regard to the onset of Bénard convection are respectively :

for low pressure constraint (index p)

$$Ra_p = \frac{\alpha g}{D_{th}^l \nu^l} (T_a - T_s)\left[\frac{D_{th}^l}{V_o}\right]^3 \quad (5)$$

---

* $\alpha_e > 0$ is observed for alcohol[13]. For water $\alpha_e < 0$ the behavior of the system could be totally different ; further experimental data are required to answer this point.

and for microwave power constraint (index Π)

$$Ra_\Pi = \frac{\alpha g}{D_{th}^1 \nu^1}(T_a - T_s)\left[\frac{\lambda^3}{4}\right] \qquad (6)$$

where α is the volume expansion coefficient
ν^1 the kinematic viscosity of the liquid
g the acceleration due to gravity.

These Rayleigh numbers vary non linearly with the applied constraint, T_s and Vo as well as T_s and λ being function of the respective constraints of pressure and microwave power. This non-classical Bénard problem is close to the problem of stability of the melt during unidirectional growth of crystals[14] and is also encountered in oceanic and atmospheric layers[15,16].

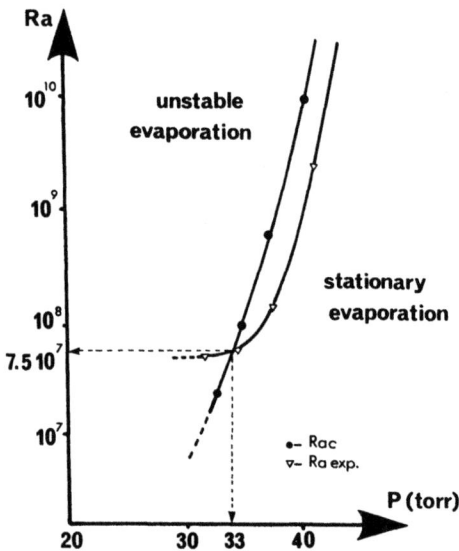

Fig. 4. Curves of experimental Rayleigh number and critical Rayleigh number versus the pressure for steady evaporation of the ethyl alcohol.

Moreover, the aspect ratio of the system is function of the constraint. The corresponding critical Rayleigh numbers have thus to be obtained for each value of the constraint. Catton[17] performed the numerical calculation of Ra_c for liquids confined in parallelipipedic containers with isothermal walls. In fig. 4 we have plotted the curves of Ra and Ra_c versus the constraint. To this purpose, we have assumed that the values of Ra_c calculated by Catton for isothermal walls were not too different from those obtained for adiabatic walls[18]. The two curves Ra and Ra_c cross each other for a pressure of 33 torrs (4 389 Pa), in good agreement with the first transition of evaporation regime observed experimentally on fig. 2.

Additionally to the measurements of evaporation rate, we have observed
the size of the convection cells from above by I.R video camera recording of
the evaporating surface and we have measured the velocity of latex particles
carried out by the fluid motion by optical video recording through the
lateral quartz walls of the square tube. For increasing microwave power
(or decreasing pressure) we see (fig. 5) that the convection rolls near the
bifurcation point are axisymmetric, then for larger deviation from the critical value, the convection cells become non stationary and their structure is
more or less antisymmetric.

A boundary layer sets in above the natural convection cells which in
turn becomes unstable for larger values of the constraint as seen in fig. 5d
in which hexagonal Marangoni cells are established over a depth of a few
tenth of millimeters corresponding to the thickness of the former thermal
boundary layer.

For a further increment of the constraint, hot spots appear fig. 5e
immediately followed by a violent destabilization of the interface due to
the vapor recoil mechanism (fig. 9).

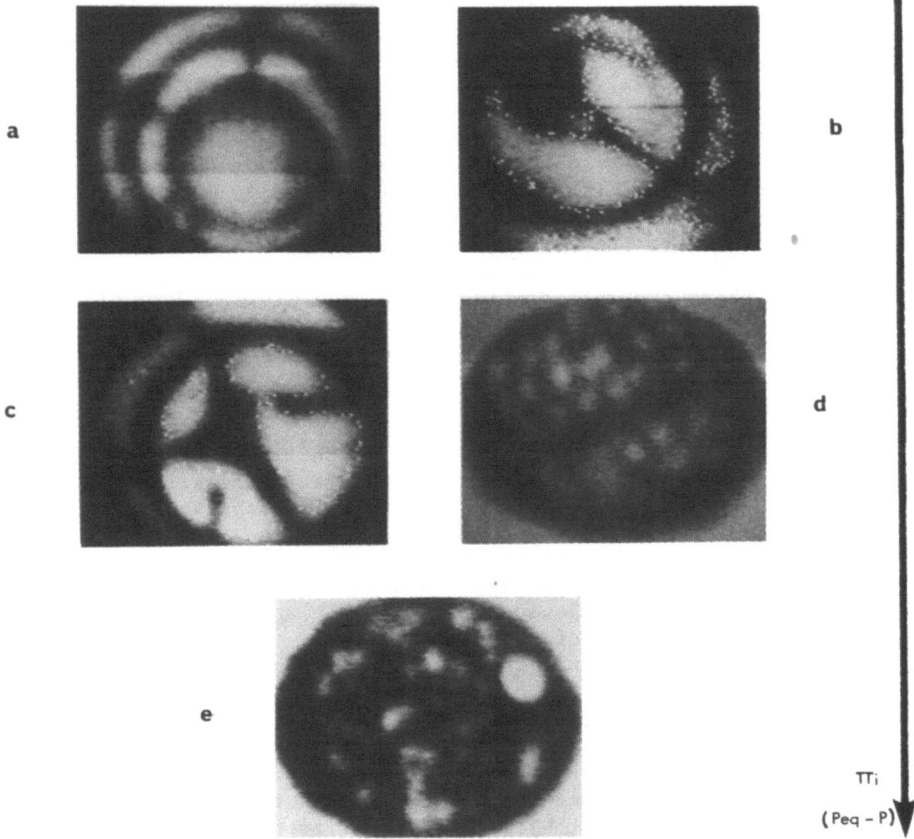

Fig. 5. Infra-red thermography of the surface showing
    (a)(b)(c) convection cells
    (d)   Marangoni cells
    (e)   Hot spot before the onset of the differential
          vapor recoil instability
for increasing constraint.

Second bifurcation : instability of the boundary layer

According to Palmer[19] two major coupled mechanisms are responsible for the dramatic increase of evaporation rate observed at low pressure in the experiments of Hickman[20] : The well known surface tension destabilizing mechanism (Marangoni instability) and the vapor recoil mechanism. In the experiments performed here, under low pressure and/or microwave irradiation the same behavior as in the experiments of Hickman[20] has been observed. In some of them, the onset of hexagonal Marangoni cells has been observed very clearly as shown in fig. 5d. These cells when they are observed, always appear for a slightly lower constraint as the dramatical increase of evaporation rate induced by the differential vapor recoil. Moreover, from the hot center of some of these cells, thermals may invade the bulk of the underlaying liquid. When the liquid is under microwave irradiation, these thermals give rise to a positive feedback in the local increment of temperature due to the positive slope of the dielectric loss coefficient with T. This feedback effect has already been observed by Bose and al.[13]. The onset of these thermals under microwave is immediately followed by a violent destabilisation of the interface by differential vapor recoil.

The mechanisms of amplification of the perturbations at the surface of the boundary layer are very clearly analysed by Palmer[19]. He distinguishes the following mechanisms : a local increase of the surface temperature will increase the local evaporation rate and decrease the local surface tension. The interface will undergo a subsequent increase of the normal force the result of which is a local depression allowing the departing vapor to shear the liquid surface. This motion will drag hot liquid up to the point of already higher temperature (auto-amplification). Similarly the surface tension gradient induced by the perturbation of temperature will induce the flow of hot liquid up to the surface amplifying by the way the disturbance.

Apart from these two main amplification mechanisms, Palmer also discussed the role a fluid inertia, of the viscous dissipation and of the moving boundary. He stressed on the competition between the vapor-recoil mechanism and the surface tension mechanism. This competition seems to be corroborated by some of our experiments.

The interpretation of our results lies in the frame of the stability analysis developped by Palmer[19] although we are still questionning on the validity of the precluded principle of exchange of stability.

This standard linear stability analysis gives a characteristic equation relating the critical Hickman number, characterizing the vapor recoil instability, to the non-dimensional wavenumber $\alpha = k\delta$ and to the other dimensionless groups at the condition of neutral stationary instability. Among the scaling factors appearing in all the dimensionless groups defined in this analysis we find the thickness δ of the boundary layer and the thermal gradient β in the boundary layer. These two quantities are very difficult to evaluate experimentally.

Typical values given by Palmer to describe experiments of low pressure evaporation of polyphenyl ether are $\delta = 0.1$ mm and $\beta = 2°C/cm$. These values are however not justified on the basis of experimental measurements of both quantities.

We have calculated the marginal stability curve (fig. 6) by the characteristic equation given by Palmer[19] (eq. 37).

With the dimensionless parameters defined as follows :

Hickman number : $N_H = \left[\dfrac{\partial V_o}{\partial T_s}\right] \dfrac{V_o^* \delta^* (T_a - T_s)}{D_{th}^l \sigma^*} \left[\dfrac{\rho^l}{\rho^v} - 1\right]$

Marangoni number : $N_{MA} = -\left[\dfrac{\partial \sigma}{\partial T_s}\right] \dfrac{\delta^* (T_a - T_s)}{D_{th}^l \mu^l}$

Bond number : $N_{BO} = \dfrac{\delta^{*2} g}{\sigma^*} (\rho^l - \rho^v)$

Reynolds number : $N_{RE} = \dfrac{V_o^* \delta^*}{\nu^l}$

Where μ and ν are respectively the dynamic and kinematic viscosity coefficient respectively, g is the gravitational acceleration and σ is the value of surface tension ; * denote the unperturbed state and l, v refer to liquid and vapor respectively.

For ethyl alcohol at the temperature of the interface in the neighborhood of the second bifurcation point in fig. 2 the values chosen for the physical parameters inserted in eq. 37, ref. 19 are reported in table I.

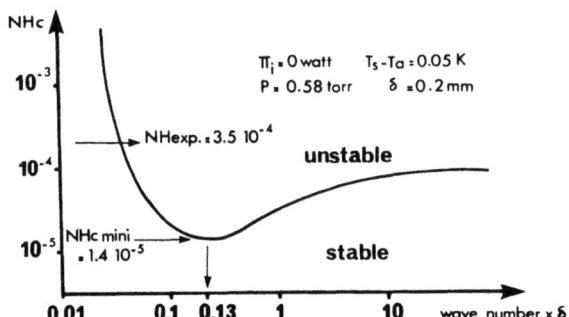

Fig. 6. Typical behavior of the critical Hickman number with the dimensionless wavenumber.

The values of β as well as δ being unknown in the present experiments, we have calculated the curves N_{Hc} versus the reduced wave number $\alpha = k\delta$ by eq. 37 ref 19 in which we have taken for δ the values 0.1 mm and 0.2 mm and for ΔT the values 0.1°C and 0.05°C. These curves were calculated for pres-

Table 1. Physical parameters of ethyl alcohol in the neighborhood of the 2nd bifurcation

$P = 77.3$ Pascals	$\rho^v = 1.7\ 10^{-6}$ g cm^{-3}
$T_s = 257$ K	$\rho^l = 0.83$ g cm^{-3}
$V_o = 2\ 10^{-3}$ cm sec^{-1}	$\mu^v = 7.4\ 10^{-5}$ poise
$\sigma = 25.6$ dynes cm^{-1}	$\mu^l = 2.7\ 10^{-2}$ poise
$(dV_o/dT) = -54$ cm s^{-1}.K^{-1}	$D^l_{th} = 10^{-3}$ cm^2 sec^{-1}
$(d\sigma/dT) = -0,084$ dyne cm^{-1} K^{-1}	

sures from 0.58 torr to 1.3 torrs. The minimum values of the critical Hickman number and the "experimental" Hickman numbers corresponding to the assigned δ and ΔT are plotted against pressure in fig. 7 in which $N_{Ma} = 2.9$. The critical Hickman and the "experimental" Hickman cross each other for $P \cong 0.59$ torr, pressure in good agreement with the observed change of regime in the evaporation rate. A noticeable fact is that whatever the values of δ and ΔT the intersection of both curves occurs for about the same pressure ($P_{int} = 0.59$ torr) and for the same value of the Hickman $N_{Hint} = 10^{-5}$.

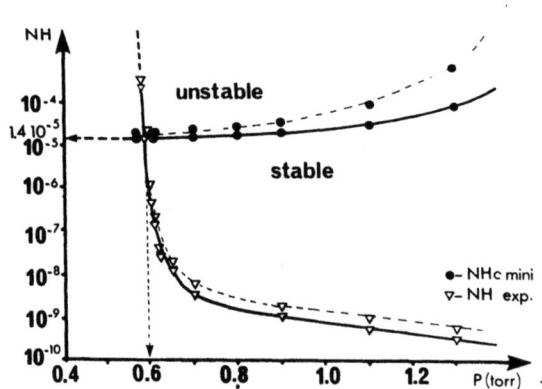

Fig. 7. Theoritical stability predictions for the evaporation of ethyl alcohol under low pressure in two cases :
$\delta = 0.1$ mm ; $T = 0.1°c$
$\delta = 0.2$ mm ; $T = 0.05°c$

As already stressed by Palmer, the coupling between the surface tension driven instability and the vapor recoil instability is increased when the Bond number is decreased. In the examples we have considered, the Bond numbers calculated for $\delta = 0.1$ mm and 0.2 mm were respectively $3 \cdot 10^{-3}$ and $1.3 \cdot 10^{-2}$. The Marangoni number was taken $= 2.9$. If we taken $N_{Ma} = 0$ in eq. 37 of ref. 19 the minimum of the curve N_{Hcrit} versus α is shifted towards the low values of α.

Conversely, if we calculate $N_{Ma\ crit}$ for $N_{Hcrit} = 0$ by eq. 37 ref. 19 we see that the minimum of $N_{Ma\ crit}$ is shifted towards the large values of α. The coupling has thus the effect to bring nearer the critical wavelengths for both instabilities.

The plot $N_{Ma}/N_{Ma}*$ versus N_H/N_H* (fig.8) shows that for increasing values of the Hickman number, the system first may exhibit convection driven by surface tension, then become stable and unstable again due to the effect of differential vapor recoil.

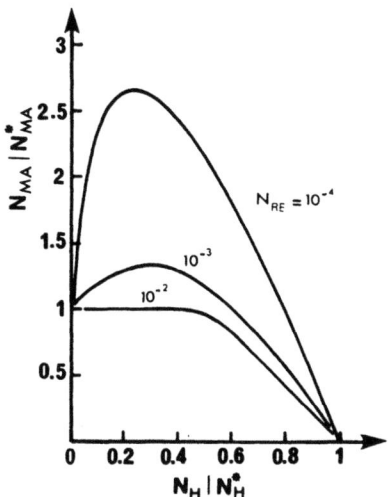

Fig. 8. The normalized critical Marangoni number versus the normalized critical Hickman number for different Reynolds numbers according to Palmer[19].

EXPERIMENTAL EVIDENCES

This behavior has been confirmed by our experiments. The pictures shown in fig. 5 show that for decreasing pressure or for increasing microwave power, the following sequence is observed : a first bifurcation point corresponding to the onset of convection by a buoyancy driven instability followed by a second bifurcation corresponding to the onset of convection in the thermal boundary layer by a surface tension driven instability as revealed by the small hexagonal cells observed fig. 5d. Increasing further the microwave power or decreasing the pressure, a quiescent domain is observed after which hot spots appear (fig. 5e) at the surface together with thermals followed by a violent destabilization of the interface by differential vapor recoil (fig. 9).

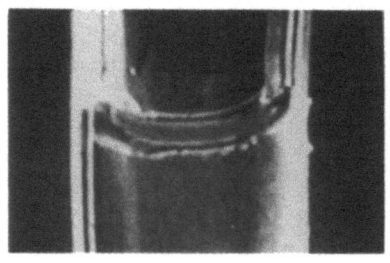
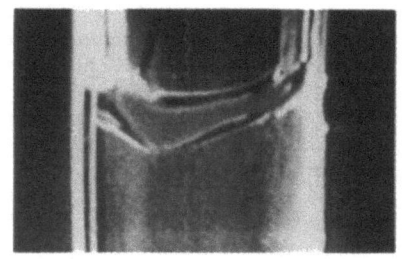

a b

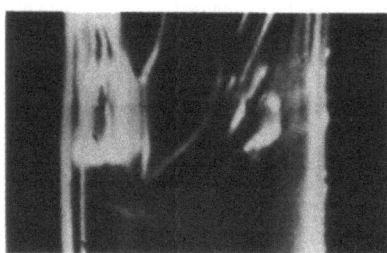

c d

Fig. 9. Differential vapor recoil instability.

For experiments performed near the equilibrium pressure under microwave irradiation, the domain of the Marangoni instability seems to be narrower as well as the stabilization zone. This seems to corroborate the prediction of Palmer that for increasing Reynolds number the preferred wavenumber for instability driven by surface tension is increased and thereby the degree of competition between Marangoni and vapor recoil mechanisms is decreased. The lifetime of the Marangoni cells is very short for high pressure and high microwave irradiation (high Reynolds number) and for a very small increase of microwave power the vapor recoil instability sets in.

CONCLUSION

The rate of evaporation of ethyl alcohol azeotropic mixture has been measured under low pressure and/or microwave irradiation. Simultaneously the motion induced by buoyancy, by surface tension or by differential vapor recoil have been recorded by optical or I.R. video camera. The surface temperature was measured by I.R emissivity measurements and the temperatures in both bulk phases were measured by thermocouples.

The evaporation rate shows three characteristic domains separated by two transition points corresponding i) to the onset of convection by buoyancy effect and ii) to the instability of the thermal boundary layer by coupled Marangoni and vapor recoil instability (Fig. 10).

The theoretical predictions of these bifurcation points by means of a linear stability analysis are in good agreement with the experimental results.

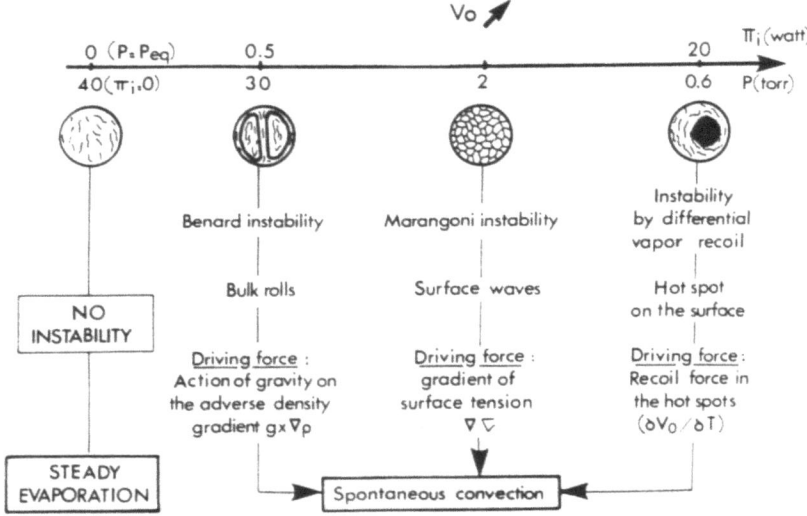

Fig. 10. Hydrodynamic instability of liquids under low pressure or/ and under microwave irradiations.

An important fact to notice when irradiating the material by microwave field is the specificity of behavior due to the dipolar nature of the system. The specific variation of the dielectric loss constant with the temperature gives rise to an additional feedback loop that is responsible for an autoamplification of the destabilizing effect of the vapor recoil mechanism for material with a positive value of $d\varepsilon''/dT$.

The oscillatory behavior of the hot spots appearing before the ejection of the surface by amplification of the vapor recoil seems to indicate that the marginal state of stability of the boundary layer is overstable. A theoretical analysis is now under investigation in this direction.

REFERENCES

1. G. Bertrand, M. Lallemant, A. Mokhlisse, G. Watelle, Variation anormale de la vitesse de décomposition d'un solide. I Cas des déshydratations d'hydrates salins, J. Inorg. nuclear Chem., 36, 1303:1309, (1974).
2. G. Bertrand, M. Lallemant, A. Mokhlisse, G. Watelle, Abnormal variation of the rate of decomposition of a solid. II A property common to interfacial endothermic reactions, J. Inorg. nuclear Chem., 40, 819:824, (1978).
3. G. Bertrand, M. Lallemant, A. Mokhlisse, G. Watelle, Characteristic features of the evaporation of liquids, Phys. Chem. Liq., 6, 215:224, (1977).
4. G. Bertrand, M. Lallemant, A. Mokhlisse, R. Prud'homme, Caractéristiques des réactions endothermiques d'interface : lois d'évolution en relation avec les états hydrodynamiques d'interface, C.R. Acad. Sc. Paris, Série II, 292, 571:576, (1981).
5. G. Bertrand, M. Lallemant, A. Mokhlisse, R. Prud'homme, Caractéristiques des réactions endothermiques d'interface : recherche des états thermodynamiques d'interface en relation avec l'écart à l'équilibre, C.R. Acad. Sc. Paris, Série II, 292, 49:52, (1981).

6. E. Abtal, M. Lallemant, G. Bertrand, Activation des liquides polaires sous champ microonde, I - Evaporation en régime interfacial stationnaire, J. Chim. Phys., 82, 381:389, (1985).
7. E. Abtal, M. Lallemant, G. Bertrand, Activation de l'évaporation des liquides polaires sous champ microonde, II - Du régime stationnaire au régime non stationnaire, J. Chim. Phys., 82, 392:399, (1985).
8. N. Roudergues, D.E.R. Etude expérimentale de l'interaction d'une évaporation endothermique avec le transfert de chaleur - Phénomènes instationnaires, Dijon (1984).
9. P. Gillon, Thesis Dissertation, Instabilités hydrodynamiques des liquides polaires sous pression réduite et/ou sous champ microonde, Dijon (1986).
10. M. Lallemant, P. Courville, G. Bertrand, CNRS FORMATION O.M.M., "Transfert de l'énergie des microondes dans les matériaux diélectriques", Thiais 1985.
11. P. Courville, Thesis, Dijon (1987) to be published.
12. C.F.F. Böttcher, P. Bordewijk, "Theory of dielectric polarization", vol. II, Elsevier Scientific Publishing Company, Oxford, (1978).
13. T.K. Bose, R. Chahine, C.A. Kyel, R.G. Bosisio, Computer based permittivity measurements and analysis of microwave power absorption instabilities, J. of Microwave Power, 19, 127:134, (1984).
14. D.T.J. Hurle, E. Jakeman, A.A. Wheeler, Effect of solutal convection on the morpholgical stability of a binary alloy, J. of Crystal Growth, 58, 163:179, (1982).
15. M. Coantic, Les couches mélangées et l'interface air-mer, in : "l'Océanologie spatiale", Cepadues-Editions, (1982).
16. K.B. Katsaros, The aqueous thermal boundary layer, Boundary Layer Meteorology, 18, 107:127, (1980).
17. I. Catton, Convection in a closed rectangular region : the onset of motion, Trans. of the A.S.M.E., Feb., 186:188, (1970).
18. M. Aza Azouni, Survey of thermoconvective inst. of confined fluids, J. non-equilibrium thermodyn., 4, 321-348, (1979).
19. H.J. Palmer, The hydrodynamic stability of rapidly evaporating liquids at reduced pressure, J. Fluid Mech., 75, 487:511, (1976).
20. K. Hickman, Surface behavior in the pot still, Ind. and Eng. Chem., 44, 1892:1902, (1952).

BASIC ASPECTS OF TEAR FILM FORMATION AND STABILITY

Frank J. Holly

Dry Eye Institute
P.O. Box 98069
Lubbock, TX 79499, USA

INTRODUCTION

The exposed surface of the eye is covered by a thin (less than 10 micrometers) fluid film surrounded by a fluid meniscus adjacent to the lids when the lids are open. This fluid film is called the tear film. The tear film provides the cornea, the most powerful refractive structure of the eye, with an optically smooth surface to ensure the projection of sharp visual images to the retina. The tear film also bathes the superficial epithelium of the cornea in an aqueous medium that appears to be indispensable to the well-being of the nonkeratinized epithelial cells.

The fluid part of the tear film consists of two layers.[1] The outermost layer consists of lipids from the meibomian glands of the eyelids and is only about 0.1 micrometer thick in the open eye. Underlying this lipid layer is the aqueous tear layer that forms 99% of the total thickness of the tear film. Aqueous tears are secreted by the main and accessory lacrimal glands.

The superficial epithelium is coated with a hydrated mucin layer that may be in a gel-like state. The thickness of this layer has not been well-characterized, and for several decades it was believed to be only 0.02-0.04 micrometers. More recent studies, however, appear to indicate that in the living eye, the mucus layer could have a thickness of the order of micrometers.[2] Most of this layer can be easily detached, hence the earlier much lower value (Figure 1).

PHYSICO-CHEMICAL MODEL OF THE TEAR FILM

Confined fluid film

The preocular tear film in the open eye thus basically is a fluid film confined by the lids and the adjacent meniscus. Such confined aqueous films are known to rupture immediately if the supporting solid surface becomes hydrophobic provided that the thickness of the film does not exceed 100 micrometers[3]. The hydrophilicity of the ocular surface appears to be due to the mucus layer that is not only intensely hydrophilic due to its high carbohydrate content but is also capable of trapping lipid contaminants and masking their hydrophilicity.

The "Black" Line

When the tear meniscus has a low volume and the eye is wide-open, an interesting phenomenon occurs. The tear film thins locally at the junction of the tear film and the meniscus. This local thinning can be made visible by staining the tear film with sodium fluorescein. Then the tear film and the meniscus become bright yellow-green while the local thinning appears as a dark blue line. The contour profile of the local thinning has been determined[4] and is shown in Figure 2. For films thicker than 100 micrometers, where gravity and thus hydrostatis pressure become important, such a configuration would be unstable, since under the convex surface the fluid pressure is positive while under the concave surface it is negative and the effect of the height difference between fluid levels would tend to increase the pressure difference.

Such "black" lines can be produced in small circular troughs with hydrophilic bottoms when the amount of liquid in the troughs is gradually decreased. As the total weight of liquid (W) approached the value necessary for an equilibrium meniscus, that is

$$W = 2R\pi\gamma \cos\theta$$

where R is the radius of the trough, γ the surface tension, and θ the receding contact angle of the meniscus with respect to the trough wall,

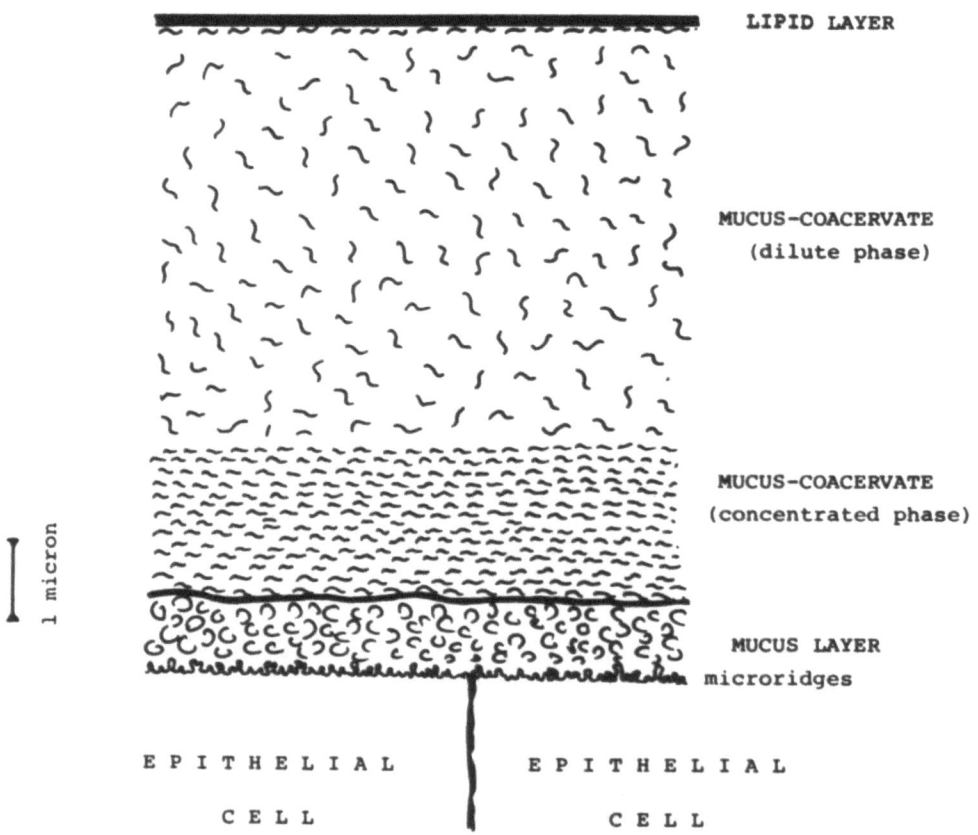

Figure 1. Schematic diagram of tear film structure drawn to scale.

the local thinning suddenly occurred adjacent to the meniscus, but no film rupture has taken place. Thus, it appears that the development of the "black line" is a direct result of insufficient quantity of liquid being present to form an equilibrium meniscus where the negative pressure resulting from the curvature and surface tension of the liquid surface is compensated by the height of the liquid column, i.e. the hydrostatic pressure in the meniscus.

Tear film formation

Periodically the upper eyelid moves downward and eliminates the tear film-air interface in a process known as blinking. As the distance between the lower and upper eyelids decreases, the superficial lipid layer is compressed similarly to a water-insoluble film in a Langmuir trough. It is energetically unfavorable for the lipid film to creep under the eyelid. With decreasing inter-lid distance, the thickness of the lipid layer increases as indicated by the appearance and increasing order of interference colors. When the lids are completely closed, the clearance is estimated to be no more than 10 micrometers. Such a thousand-fold compression would increase the lipid film thickness a thousand-fold to a value of 100 micrometers that can readily be contained between the lid edges. In the closed eye, the aqueous tear layer is present between the globe of the eye and the lid and serves as a lubricant (Figure 3).

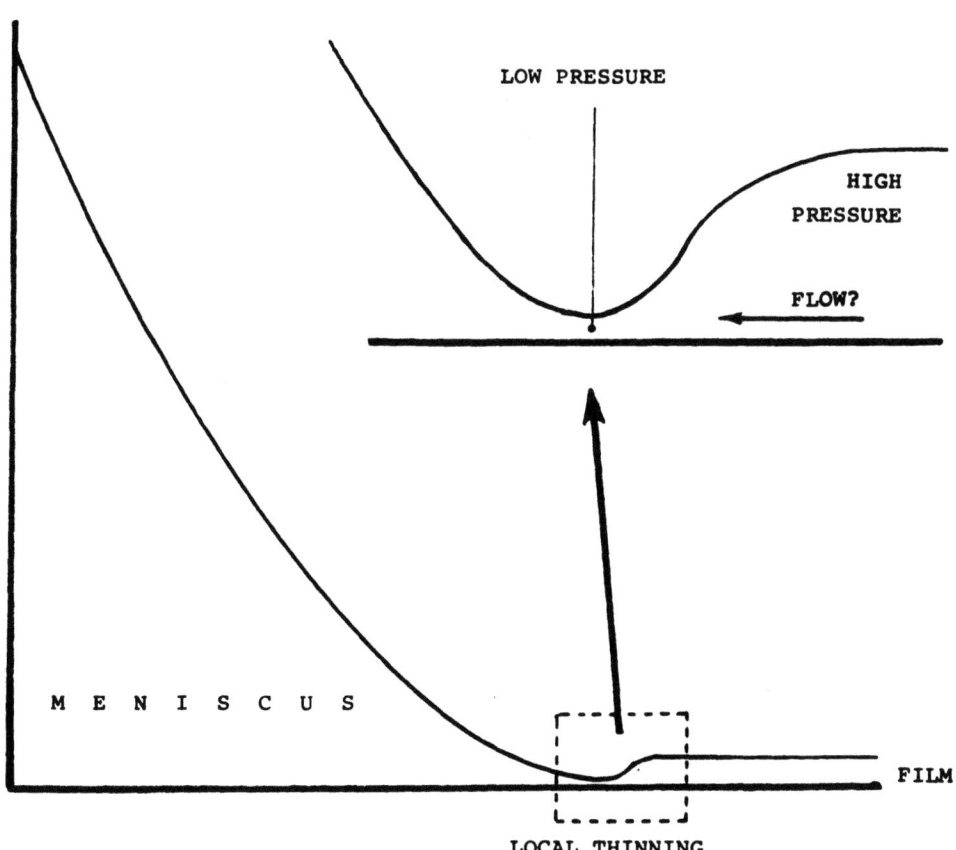

Figure 2. Contour profile of tear meniscus and adjacent "black line."

When the eyelid opens following a blink, the compressed lipid layer will spread first as a monolayer, then as a duplex film. Since the spreading velocity of the meibomian lipids is estimated to be about 25 cm/sec and since the higher velocity of the moving eyelid does not exceed 15 cm/sec, the motion of the eyelid is the limiting factor in the lipid spreading and thus a bare aqueous tear surface does not exist even temporarily in the eye.[3]

Tear Film Rupture

If blinking is prevented or voluntarily suppressed, after a certain time interval, the so-called tear film break-up time (BUT), the continuity of the tear film is compromised and "dry spots" of gradually increasing area begin to form. Originally it was believed that these local ruptures resulted from evaporation. Actually, under normal conditions, it takes about 10 minutes for the whole tear film to evaporate.[5] The depression-like appearance of these dry spots also suggests local nonwetting rather than spot-wise drying.

In the early seventies a tentative mechanism of tear film rupture was proposed based on qualitative surface chemical considerations.[3]. It was assumed and later experimentally demonstrated that even confined aqueous

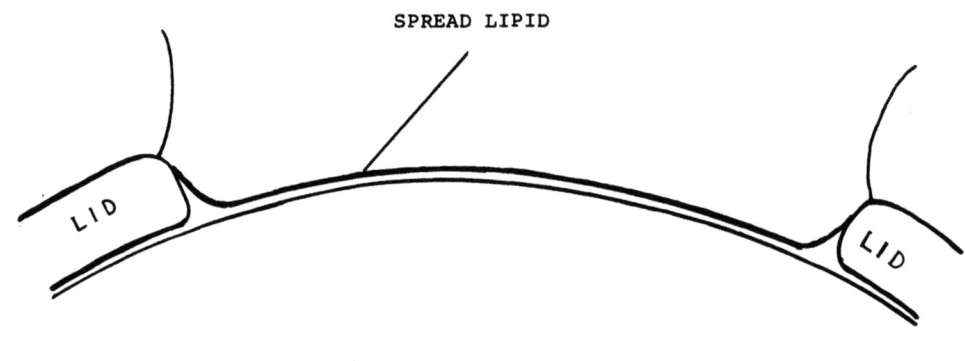

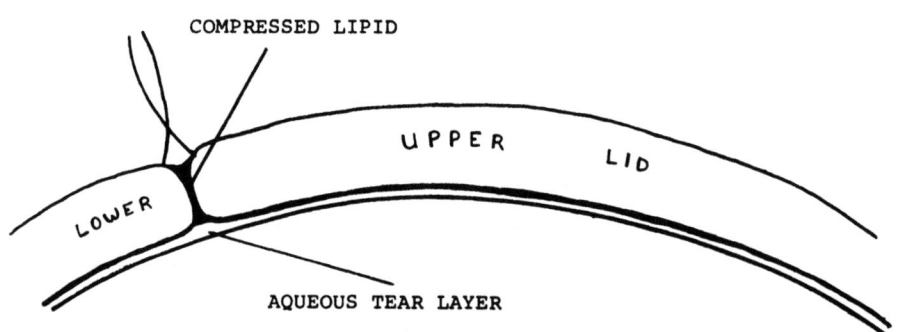

Figure 3. Schematic diagram of the effect of blinking on the tear film.

films will rupture if the supporting solid has a hydrophobic surface once the thickness of the film is less than 100 micrometers. Since the tear film is initially continuous following a blink, the epithelial surface must be originally hydrophilic. Then it also follows that when the tear film spontaneously ruptures over certain locations, then the hydrophilicity of the solid surface of those locations must have been compromised. Since the process of hydrophilic to hydrophobic conversion is energetically not favored while the solid is covered with an aqueous tear layer, some free-energy driven external process has to occur which causes the change. The most plausible process may be the migration (diffusion) of lipid molecules from the superficial lipid layer to the mucus layer.[6] The following factors would favor the process:

<u>The proximity of the lipid to the mucus</u>. Even if the thickness of the tear film had not been adversely affected, it does not exceed 10 micrometers. Diffusion over such a small distance is relatively rapid and the diffusion flux varies inversely as the second power of the distance. Hence, if there is a local thinning of the tear film decreasing its thickness ten-fold, the diffusion flux of the lipid will increase by a hundred-fold.

<u>Increased solubility of the lipid</u>. Lipids in general have low solubility in water. However, biopolymers such as certain proteins can associate with lipids increasing their solubility. Lipid-protein interaction also increases the film pressure of the lipid layer thereby decreasing the surface tension of the tear film. Such a complex formation has been observed with mucus glycoproteins; the interaction energy of this process could offset the increase in interfacial free-energy. In any case, the hydrophobicity of the lipid contamination would be masked initially by the mucus layer until the lipid content of the mucus layer would increase above a certain level that may be as high as 50 to 70% by weight.

The following factors would facilitate the local thinning of the tear film, thus the rate of lipid contamination resulting in diminished tear film break-up time:

<u>Finite surface tension gradient in the lipid layer</u>. If at a given location the local surface tension of the lipid film is lowered, e.g. by a higher concentration of polar lipids, lipid spreading accompanied by local thinning would occur. If a minute amount of highly polar skin lipids (sebum) is placed on the tear film surface, a large dry spot immediately forms. This surface tension gradient-driven local thinning is the direct result of a localized Marangoni flow. This phenomenon is used indirectly to dewet solid surfaces of complex configuration.

<u>The roughness of the epithelial surface</u>. The normal epithelial surface morphology consists of microridges and possibly some microvilli. The presence of such a "wrinkling" of the semi-fluid cell membranes must indicate that the water-cell interfacial tension is nearly zero and may even have been initially negative, causing microridge formation. The scale of this roughness is too small to interfere with the stability of the tear film. Furthermore, the possibly gel-like mucus layer appears to smooth out the ridges in addition to serving as a lipid trap.

More recently other, more sophisticated, mechanisms have also been offered that are based on parametric analysis employing basic physics, chemistry, and hydrodynamics. Since the temperature of the underlying tissue is higher near body temperature, 37°C, and the corneal surface temperature under average conditions is lower, near 34°C, the effect of this temperature difference across the tear film has been taken into account by Lin and Brenner.[7,8] They found that the Marangoni and

Hamaker coefficients are the most important parameters in defining tear film stability. They also found that the conditions in the tear film are favorable for the development of a convective Marangoni circulation in the tear film and came to the conclusion that this convection would not lead to tear film rupture. While surface tension and viscosity tend to stabilize the tear film, intermolecular forces due to coherent dipole-dipole interaction could be a causative factor in tear film rupture.

Velarde and co-workers[9] found that while the temperature gradient perpendicular to the tear film may be responsible for a Marangoni-Benard type convection, it will also assist in lowering the surface-active solute (lacrimal surfactant) concentration gradient which is needed for film instability. They succeeded in calculating the convection threshold values in terms of the capillary numbers characterizing the deformation of the film surface.

Ruckenstein and Sharma[10,11] have also analyzed the stability of the tear film in basic terms. They considered the stability of the underlying mucus layer separately and showed that a basic instability could arise in this layer from van der Waals interactions due to fluctuation in its thickness. According to them, the rupture of the mucus layer would expose the aqueous tear film to the possibly hydrophobic epithelial cell membranes resulting in the immediate rupture of the tear film.

An interesting but possibly untenable conclusion was proposed by Haberback, who envisioned sudden local permeability (or possibly hydraulic conductivity) increases in the surface epithelium due to the shedding of superficial cells resulting in the local drainage of the tear film appearing as dry spots.[12] Even though this hypothesis has not been formally refuted, approximate hydrodynamic approximate calculations suggest that the model may not be realistic.

<u>Water Wettability of the Corneal Epithelium</u>

In recent years, some criticism[13] has been directed toward earlier work that assigned a hydrophobic character to the demucinized corneal epithelium. The point has been raised that cell membranes are known to have negligibly small interfacial tension when in an aqueous medium. This is also true for normal corneal epithelium as shown by the existence of microridges. However, these microridges can be readily eliminated by topical application of surface-active substances that would tend to increase interfacial tension when adsorbed at the interface (surface-active preservatives, topical anesthetics, etc.). The microridges also disappear when the epithelium is exposed to air even if the air is saturated with water vapor. Clearly, a hydrophilic interface, at least initially, becomes a surface of relatively high free energy when exposed to a gaseous phase, where the possibility of forming hydrogen bonds across the interface is drastically reduced. We have shown[14] using hydrogel and other polymeric surfaces that when hydrophilic surfaces are exposed to air or to a condensed hydrophobic medium (e.g. nonpolar liquid hydrocarbon), a hydrophilic to hydrophobic conversion will take place at the surface (or interface) that will lower the energy of the boundary layer. The conversion is most likely due to orientational changes of the amphipathic surface polymeric chains which would require little energy. Hence, once the corneal epithelium is exposed to air, even if it is the air of a bubble used in the captive-bubble technique of wettability determinations, the epithelial surface will take on its most hydrophobic configuration possible.

In the two-step mechanism of tear film rupture,[11] it is assumed that once the mucus layer ruptures, the underlying epithelium exposed to the aqueous tear layer will have hydrophobic properties. This is quite unlikely since the hydrophilic-hydrophobic conversion is energetically not favored unless the cell membranes are exposed to a hydrophobic medium. On the other hand, if the composition of the epithelial surface is changed, e.g. by lipid adsorption, then the interface will become hydrophobic and film rupture will take place.

It is probable or even likely that the epithelial surface would remain wettable even in the absence of the mucus layer if two conditions were fulfilled: 1/ if the epithelium were not exposed to air, and 2/ if lipid contamination were completely prevented. However, unless the environment of the tear film could be maintained lipid-free, an impossible task, the mucus layer is needed to prolong the wetting time (break-up time), because of its pronounced masking activity of the lipid, until relatively high lipid concentrations in the mucus layer are achieved, which requires some time.

BIOCOMPATIBILITY OF CONTACT LENSES

The insertion of a contact lens into the eye imposes a considerable stress on the lacrimal system. It stands to reason that in order to have good visual acuity and maintain the integrity of the corneal epithelium, the contact lens in the eye should be covered with a continuous tear film and also have an intact tear layer under the lens (Figure 4).[15] The observations obtained by slow motion cinematography and biomicroscopy support this notion.

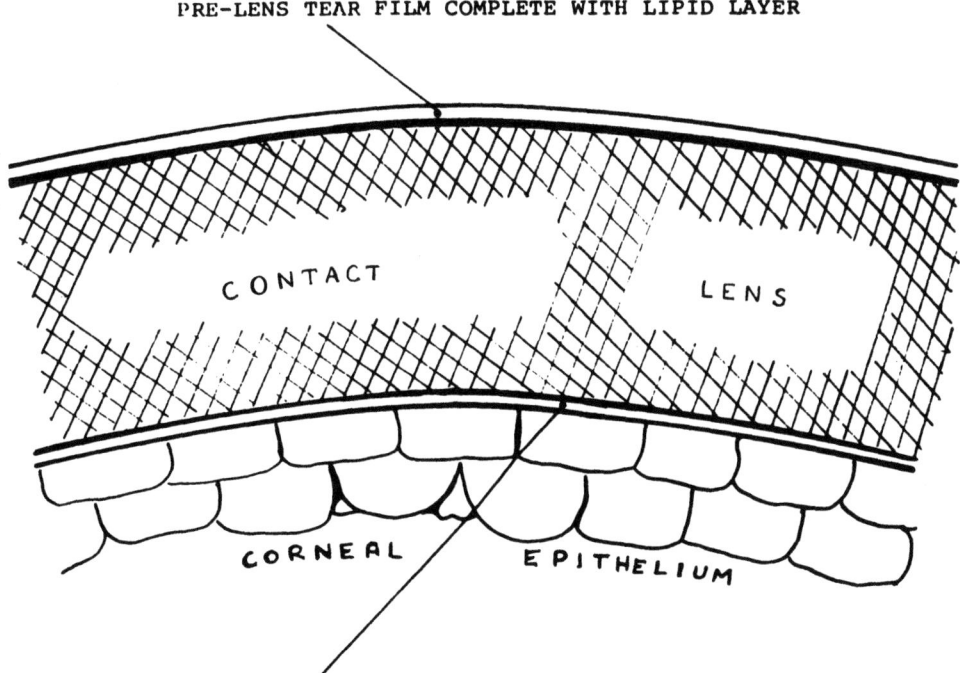

Figure 4. Schematic diagram of the cross-section of a contact lens in situ.

Tear film and contact lens

In the closed eye there must be a continuous aqueous tear layer surrounding the lens and acting as a lubricant; otherwise blinking would become painful and quite injurious to the tissue. Upon opening the lid, the superficial lipid layer would spread over the lens, dragging tears with it, much the same way as the lipid and tears spread over the ocular surface in the absence of a contact lens, and a continuous tear film would form. If the pre-lens tear film is thinner than the preocular tear film, the spreading speed may become somewhat lower but not drastically so.

In the open eye thus we have a tear film complete with the superficial lipid layer in front of the lens and a continuous aqueous tear layer without the lipid layer underneath the lens. This arrangement results in good visual acuity and allows some motion to the lens following each blink, which promotes the exchange of tears beneath the lens; a process referred to as "pumping." Even with pumping, the removal of contaminated mucus and redistribution of fresh mucin underneath the lens are hindered by the presence of the lens. However, due to the presence of the lens that acts as a lipid-impermeable barrier, the contamination rate of the mucus layer underneath the lens is drastically diminished compensating for the hindered "rejuvenation" of the mucus layer.

Black line around contact lenses

Depending on its edge design, the contact lens in the eye may have a significant tear meniscus around it. Thicker lenses, especially those with poorer edge design, may have local thinning ("black line") around the meniscus. This would indicate an unsatiated negative pressure that would tend to thin the post-lens tear layer as well as the pre-lens tear film. The sign and the magnitude of the disjoining pressure (the negative gradient of film free energy with respect to thickness)[16] will determine the fate of the post-lens tear layer. As long as both the posterior surface of the lens and the ocular surface underneath the lens are hydrophilic, it would take a considerable force or pressure (up to 140 atmospheres) to squeeze the aqueous layer out of the gap between the lens and the eye. Even if only one of the surfaces were to remain hydrophilic, the disjoining pressure would remain positive, resisting the thinning of the post-lens tear layer.[17]

Adhesion of contact lens to cornea

However, if for some reason, such as lipid-contamination and hydrophilic to hydrophobic transition, both the posterior lens surface and the ocular surface become hydrophobic, the disjoining pressure would become negative, making the thinning of the post-lens tear film energetically favorable. Under such conditions, the post-lens tear film would spontaneously thin and completely "drain" from under the lens. Since the ocular surface (and for hydrogel lenses the lens surface also) is pliable, intimate molecular contact could be established between the lens and the cornea, resulting in contact adhesion.[18] In such a case, a clean separation of the two solid surfaces usually is not possible unless the interfacial region can again be rehydrated, accumulating a weak boundary layer of water that inevitably leads to adhesive failure. If an adhered lens is forcibly removed from the eye, conditions favor a cohesive failure in the weaker bulk phase, i.e. epithelial damage would occur.

Tear film rupture over contact lenses

When the eye is kept open, the pre-lens tear film degenerates. Typical dry spot formation resulting in local dewetting of the lens can happen over certain lenses. However, the most typical degradation process involves rapid thinning over the lens followed by a receding fluid front usually toward the interior fornix, although de-wetting moving superiorly can also be observed. Characteristically, the break-up or wetting time over lens surfaces is considerably shorter than over the ocular surface; it seldom exceeds ten seconds.

If the pre-lens wetting time is shorter than the time interval between consecutive blinks, the lens surface will be exposed to the atmosphere. This facilitates lipid contamination as well as favoring hydrophilic to hydrophobic conversion of the lens surface. Even in cases of well-fitted and well-wetting lenses, this could occur as contact lens wearers are inclined to become "lazy blinkers."

Such premature de-wetting of contact lenses would result in deposit formation of tear film components over the anterior lens surface. One has to distinguish between the thermodynamically reversible adsorption of tear components and accumulative deposit formation. The former is self-limiting and usually results in favorable wetting properties (i.e. lowering of interfacial tension) and thus increases biocompatibility. However, if the adsorbed species is altered at the surface due to exposure to the atmosphere, e.g. protein denaturation or precipitation of chemically altered crystalloids, then the process is no longer self-limiting and deposit formation occurs that can obscure the lens, jeopardizing its transparency and interfering with lubrication, thus adversely affecting comfort. Such a despoliation of lenses is the most common cause of contact lens incompatibility. In certain acute cases, the rate of despoliation can be so high that the lens may become opaque in a matter of minutes.

Basic condition of lens compatibility

Basic considerations suggest[15] that one factor of over-riding importance determining biocompatibility of lenses is the magnitude of tear-lens interfacial tension. As long as it can be kept minimal, the adsorption of the proper species only will be ensured and the stability of either the pre-lens or the post-lens will not be compromised. Unfortunately this parameter, solid-water interfacial tension, cannot be directly measured. The capillary forces acting on the contact lens as a function of lens geometry have not been analyzed in detail either; therefore, they cannot be taken quantitatively into account in lens design.

Improvement of lens biocompatibility

There is considerable merit to the idea that the biocompatibility of lenses should be improved and ensured in a manner similar to the way nature does it for the ocular surface.[15] Hydrophilic polymers loosely adsorbed on the surface would maintain a low interfacial tension. If and when they lost hydrophilicity due to lipid contamination or other factors, they could be readily removed by the shear forces created by blinking and replaced with uncontaminated polymers.

CONCLUSIONS

The continued existence of an intact thin fluid film over the cornea and conjunctiva is imperative for good visual acuity and the well-being of exposed ocular tissues. Despite the complexity of the system, the basic physical and chemical properties of the components will determine the stability of the film and thus control its physiology.

If an external stress is placed on the lacrimal system by inclement atmospheric conditions such as turbulent, dry, and dusty weather or by the insertion of a contact lens, it may interfere with the functioning of the system sufficiently to cause dysfunction, resulting in pathological changes in the ocular surface. Understanding the underlying basic factors may lead to improved diagnostic and treatment modalities of lacrimal disorders.

REFERENCES

1. F.J. Holly, Physical chemistry of the normal and disordered tear film, Trans Ophthalmol. Soc. V. K. 104:374 (1985).
2. B.A. Nicholas, M.L. Chiappino, and C.R. Dawson, Demonstration of the mucus layer of the tear film by electron microscopy, Invest. Ophthalmol. Vis. Sci. 26:464 (1985).
3. F.J. Holly, Formation and rupture of the tear film, Exp. Eye Res. 15:515 (1973).
4. J.E. McDonald and S. Brubaker, Meniscus-induced thinning of the tear films, Am. J. Ophthalmol. 72:139 (1971).
5. S. Iwata, M.A. Lemp, F.J. Holly, and C.H. Dohlman, Evaporation of water from the precorneal tear film and cornea in the rabbit, Invest. Ophthalmol. 8:613 (1969).
6. F.J. Holly, On the wetting and drying of epithelial surfaces, in: "Wetting, Spreading, and Adhesion," J.E. Padday, ed., Academic Press, London (1977).
7. S.P. Lin and H. Brenner, Tear film rupture, J. Coll. Interface Sci. 89:226 (1982).
8. S.P. Lin and H. Brenner, Stability of the tear film, in: "The Preocular Tear Film: In Health, Disease, and Contact Lens Wear," F.J. Holly, ed., Dry Eye Institute, Lubbock, TX (1986).
9. M.G. Velarde, J. Murube del Castillo, J.L. Castillo, and M. Garcia, Marangoni-Benard convection and macrodynamics of tear flow, in: "The Preocular Tear Film: In Health, Disease, and Contact Lens Wear," F.J. Holly, ed., Dry Eye Institute, Lubbock, TX (1986).
10. A. Sharma and E. Ruckenstein, Mechanism of tear film rupture and formation of dry spots on the cornea, J. Coll. Interface Sci. 106:12 (1985).
11. E. Ruckenstein and A. Sharma, A surface chemical explanation of tear film break-up and its implications, in: "The Preocular Tear Film: In Health, Disease, and Contact Lens Wear," F.J. Holly, ed., Dry Eye Institute, Lubbock, TX (1986).
12. F.J. Haberich and B. Lingelbache, Kritische Ubersicht uber unsere Kenntnisse and Vorstellung eine Neuen Arbeithypothese uber die die Stabilitat des Praekornealen Tranenfilm (PKTF), Klin. Mbl. Augenheilk. 180:115 (1982).
13. J.M. Tiffany, Ocular surface chemistry, in: "Transactions of the International Society for Contact Lens Research," 3rd Sci. Mtg., E. Kenyon, ed., Cambridge, England (1984).
14. F.J. Holly and M.F. Refojo, Hydrogel-water interface, in: "Colloid and Interface Science," Vol. III, M. Kerker, ed., Academic Press, New York (1976).
15. F.J. Holly, Basic aspects of contact lens biocompatibility, Colloids and Surfaces 10:343 (1984).
16. B.V. Derjaguin, On the repulsive forces between charged colloid particles and on the theory of slow coagulation and stability of lyophobe sols, Trans. Faraday Soc. 36:203 (1940).
17. F.J. Holly, Biophysical aspects of epithelial adhesion to stroma and its clinical applications, Invest. Ophthalmol. Vis. Sci. 17:552 (1978).
18. P. Fanti and F.J. Holly, Silicone lens wear III. Physiology of poor tolerance, Cont. Intraocul. Lens Med. J. 6(2):III (1980).

NONEQUILIBRIUM PHENOMENA IN MICROEMULSIONS

M.A. López Quintela, A. Fernández Nóvoa and J. Quibén

Universidad de Santiago
Facultad de Química
Departamento de Química Fisica
E-Santiago de Compostela

Introduction

When a small amount of surfactant is dissolved in water, molecular aggregates are formed which can solubilize oils giving rise to clear solutions called microemulsions. It has long been known that most microemulsions exhibit critical phenomena which are in certain respects similar to those of multicomponent fluids. The similarities are that there exist scaling laws when these systems approach the critical point or cloud point, and that their dynamics exhibits critical slowing down(1-13). However the physics of these systems differs appreciably from the general properties of multicomponent fluids in that the critical exponents found for micelles and microemulsions seem to be non-universal(1,2,8,12) and in the existence of critical phenomena at temperatures which differ appreciably from the critical temperature, Tc(8). These facts, especially the latter, are the reasons for the special significance of microemulsions for the study of non-equilibrium and/or non-linear phenomena(14).
In this brief report we will refer to various non-equilibrium phenomena which can be observed in critical microemulsions.

Experimental

The microemulsions employed had the following composition by weight: water(0.6587), 2-butoxyethanol(C4E1)(0.3147), dodecane(0.0050) and KCl(0.0210).
The microemulsion separates into two phases when the critical temperature Tc = 298.46K at 1 atm is approached from below. This microemulsion seems to be near a tricritical point which adds interesting new aspects to the problems here considered(15).

Hydrodynamic instabilities

When the microemulsion (A in figure 1) is introduced in a closed Petri dish 12 cm in diameter and 1.5 cm deep, and is heated very slowly from beneath, the formation of spatial dissipative structures is observed, as can be seen in figure 2.

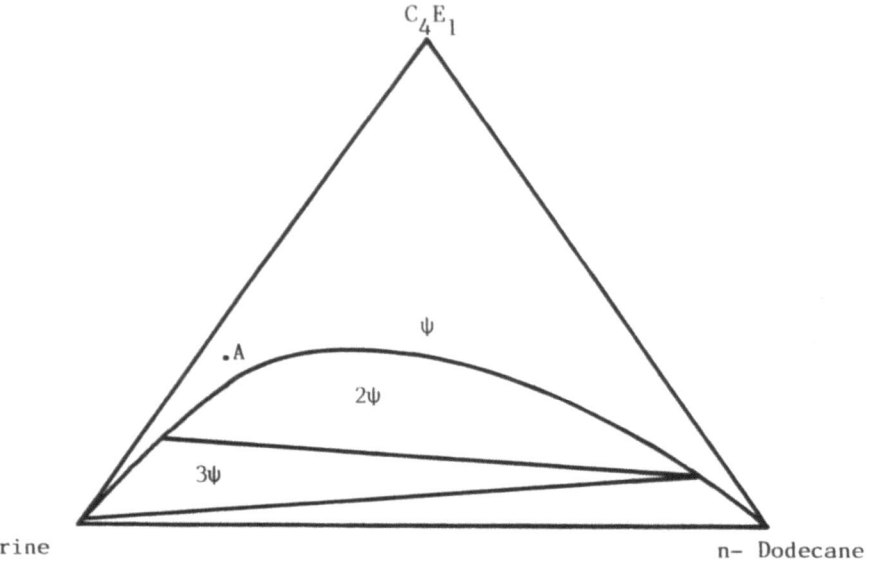

Fig. 1: Approximate phase diagram of the system water/C4E1/dodecane/KCl at T=298.15 K. A: the microemulsion employed in this study.

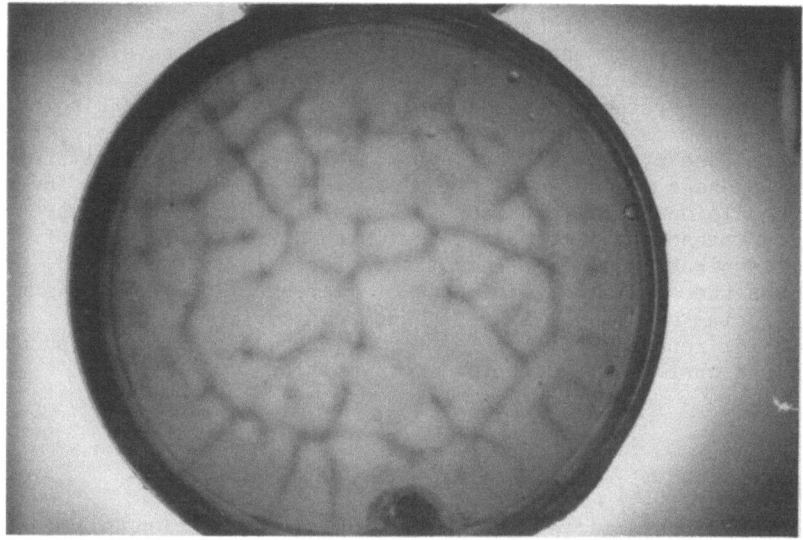

Figure 2. Typical spatial dissipative structures observed in a critical microemulsion in a closed Petri dish heated from below.

Patterns are observed as convection lines with stronger scattering than the rest of the fluid. It can also be seen that the fluid has a dextrorsum motion. In all the experiments carried out so far, the observed convection patterns have been transient: when the critical temperature Tc was reached, the fluid separated into two phases with subsequent disappearence of the patterns. A similar phenomenon can also be observed when the fluid is heated from above. When experiments were carried out with a similar microemulsion in which water had been replaced by a slightly basic solution of the indicator bromothymolblue, it was observed that the fluid forming convection lines had a lower pH than the rest of the fluid. Since in micellar solutions of ionic or polar surfactants(16) the negatively charged micelles' attraction of protons from the bulk phase gives them a lower pH than the latter, it may be concluded that the convection lines are predominantly composed of C4E1 and dodecane.

At one degree below the critical point we investigated the influence of the temperature difference between the top and the bottom of the Petri dish. The structures did not disappear when this difference was reduced, though the Rayleigh number was sub-critical(17-19). This implies that forces other than buoyancy are significant for the establishment of the observed hydrodynamic instabilities. One factor which might contribute to the instability of the microemulsion is interfacial tension. If we consider a microemulsion as a microheterogeneous solution, it is possible to define an interfacial tension between the droplets of the microemulsion and the bulk phase.

Such interfacial tensions may contribute to the formation of dissipative structures in the same way as surface tension contributes to Marangoni instabilities(17-25). Besides a temperature gradient, variation of the interfacial tension with temperature may also force the self-organization of the fluid. At this stage it is not easy to quantify the influence of this force, because it is difficult to relate it to the measured interfacial tensions between microemulsions and their excess phases. However it is possible to deduce by means of a theoretical model(26) that the Marangoni number for such an interfacial tension at one degree below the critical temperature should be supra-critical, due to the interfacial tension tending to zero on approaching the critical point(10,13)

Fluctuations

The great fluctuations which appear when the critical point is approached are able to cause important changes in the behaviour of the systems as compared to the same system far away from the critical temperature. The importance of stochastic fluctuations for the formation of dissipative structures has recently been analysed by various authors(27-29) who predict the amplification of the fluctuations in the neighbourhood of a bifurcation point to bring the system to a new far-from-equilibrium state. As far as we know, the influence of internal fluctuations on the onset of instabilities has not been observed experimentally, because of their small size: only the influence of external fluctuations has been investigated (30-35). Since in our system fluctuations are very large, we hope to be able to determine the influence of the internal fluctuations on the behaviour of the system.

Quenching

We have carried out some experiments in which the single phase microemulsion in the Petri dish was quenched to a two phase state by a rapid change in temperature. It was observed that the phase separation takes place via the formation of spatial dissipative structures(figure 3). These spatial structures are transient patterns which become more and more irregular until a chaotic structure is observed and two phases appear. Experiments are projected to see whether this occurs by well defined routes (deterministic chaos(36-39)).

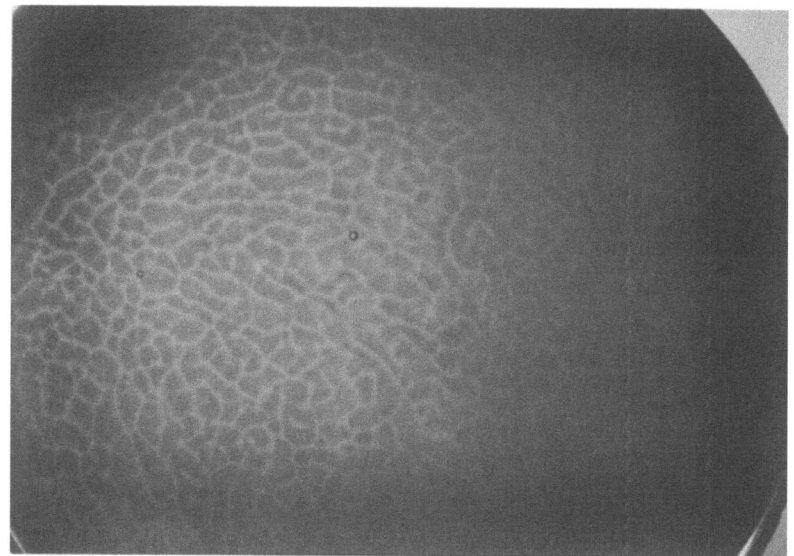

Figure 3. Spatial dissipative structures observed in a quenched microemulsion(T initial = 293 K, T final = 299 K).

Nonlinear relaxations of fluctuations

The experiments reported so far show that interesting non-linear phenomena are associated with the large, slow, non-Poisson fluctuations present in critical microemulsions. In order to get more information about these non-linear phenomena we have carried out some experiments involving the relaxation(40) of these fluctuations by pressure-jump method (with optical and conductometric detection), temperature-jump and stopped-flow techniques(41). The general result of these investigations is that far away from the critical point the fluctuations decay monotonically to equilibrium, whereas near the critical point they undergo an oscillating relaxation to equilibrium (figure 4).
This makes it evident that non-linear couplings are involved in the relaxation of fluctuations which, qualitatively at least, explains the appearance of dissipative structures in the quenching of microemulsions(41).
The findings described above clearly show the existence of very interesting non-linear phenomena in critical microemulsions. At present we are studying these phenomena in critical microemulsions by means of pressure-jump, temperature-jump, stopped-flow, calorimetric and quenching techniques and determining physical parameters such as density, viscosity and fractal dimension, the latter by means of diffusion-controlled reactions, in order to get more information for the interpretation of the

various general results reported here, a more detailed description of which will be published elsewhere.

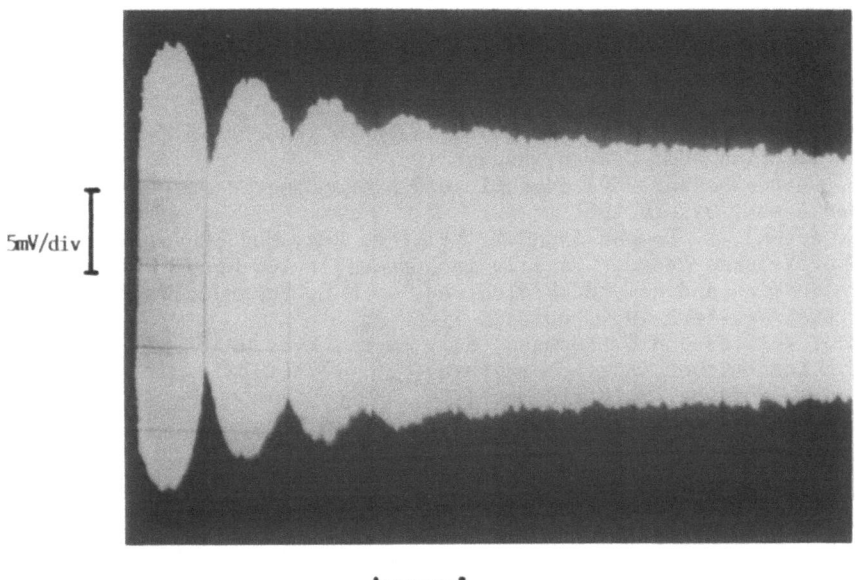

Figure 4. Typical oscillating behaviour for the relaxation of fluctuations in a critical microemulsion, observed by performance of a p-jump with conductometric detection(T= 298 K, ΔP= 150 bar , perturbation time = 1 sec).

Acknowledgement: We thank Prof. M.G. Velarde for his interest in our work. We also thank the Secretaría de Estado de Universidades e Investigación, the Xunta de Galicia and the Stiftung Volkswagenwerk for their financial support of different parts of this work.

References

1. J.S Huang and M.W. Kim, Phys. Rev. Lett., 47, 1462 (1981)
2. R. Dorshow, F. de Buzzaccarini. C.A. Bunton and D.F. Nicoli, Phys. Rev. Lett., 47, 1336 (1981)
3. M. Corti, V. Degiorgio and M. Zulauf, Phys. Rev. Lett., 23, 1617(1982)
4. M.W. Kim, J.S. Huang, Phys. Rev. B, 26, 2703 (1982)
5. G. Fourche, A.M. Bellocq and S. Brunetti, J. Colloid Interface Sci. 88, 302 (1982)
6. A.-M. Cazabat, D. Chatenay, D. Langevin and J. Meunier, Faraday Discuss. Chem. Soc., 76, 291 (1982)
7. M. Kotlarchyk, S.H. Chen and S. Huang, Phys. Rev. A, 28, 508 (1985)
8. C. Toprakcioglu, J.C. Dorte, B.H. Robinson and A. Howe, J. Chem. Soc. Faraday Trans. 1, 80, 413 (1984)
9. M.W. Kim, J. Bock and S. Huang, Phys. Rev. Lett., 54, 46 (1985)
10. D. Langevin in Physics of Amphiphiles: Micelles, Vesicles and Micro-

emulsions, ed. by V. Degiorgio, North-Holland, Amsterdam 1985, pp. 181
11. S.H. Chen and M. Kotlarchyk in Physics of Amphiphiles: Micelles, Vesicles and Microemulsions, ed. by V. Degiorgio, North-Holland, Amsterdam 1985, pp. 768
12. S. Huang and M.W. Kim in Physics of Amphiphiles: Micells, Vesicles and Microemulsions, ed. by V. Degiorgio, North-Holland, Amsterdam 1985, pp. 864
13. D. Guest and D. Langevin, J. Colloid Interface Sci. 112, 208 (1986)
14. H. Haken, Advanced Synergetics: Instability Hierachies of Selforganizing Systems and Devices, Springer-Verlag, Berlin 1983
15. M. Kahlweit, R. Strey P. Firman and D. Haase, Langmuir, 1,281(1985)
16. E. Dutkiewicz, M.A. López Quintela and W. Knoche in Fast Reactions in Solution DGM, Leuven, August 1982
17. S. Chandeasekhar, Hydrodynamic and Hydromagnetic Stability, Clarendon Press, Oxford 1961
18. C. Normad, Y. Pomeau and M.G. Velarde, Rev. Mod. Phys. 49, 581 (1977)
19. M.G. Velarde and J. Castillo in Nonequilibrium Cooperative Phenomena in Physics and Related Fields, ed. by M.G. Velarde, Nato Plenum Press, New York 1984, pp. 179
20. M.G. Velarde and C. Normand, Sci. Amer., 243, 92 (1980)
21. E.L Koschmieder, Adv. Chem. Phys., 26, 177 (1974)
22. J.R.A. Pearson, J. Fluid Mech., 4, 489 (1958)
23. L.E. Scriven and C.V. Sternling, J. Fluid Mech., 19, 321 (1964)
24. K.A. Smith, J. Fluid Mech., 24, 401 (1966)
25. M.G. Velarde in Dynamics and Instabilities of Fluid Interfaces, ed. by T.S. Sorensen, Springer-Verlag, Berlin 1979, pp. 260
26. M.A. López Quintela, to be published
27. R. Graham in Fluctuations, Instabilities and Phase Transitions, ed. by T. Riste, Plenum Press, New York 1975
28. G. Nicolis and C. Van der Broeck in Nonequilibrium Cooperative Phenomena in Physics and Related Fields, ed. by M.G. Velarde, Nato Plenum Press, New York 1984, pp. 473
29. G. Nicolis in Systems Far from Equilibrium, ed. by L. Garrido, Springer-Verlag, Berlin Heidelberg 1980, pp. 91
30. W. Horsthemke and R. Lefever, Noise-Induced Transitions, Springer-Verlag, Berlin 1983
31. J.P. Gollub, J.F. Steinman, Phys. Rev. Lett., 45, 5511 (1980)
32. P. De Kepper, W. Horsthemke, C.R. Acad. Sci. Paris Ser. C278, 251 (1978)
33. S. Kabashima, S. Kogure, T. Kawakubo, T. Okada, J. Appl. Phys., 50, 6296 (1979)
34. S. Kabashima, T. Kawakubo in Systems Far from Equilibrium, ed. by L. Garrido, Springer-Verlag, Berlin 1980, pp. 395
35. S. Kai, T. Kai, M. Takata, K. Hirakawa, J. Phys. Soc. Jpn., 47, 1379 (1979)
36. G. Ahlers and R. Behringer, Phys. Rev. Lett., 40, 712(1978)
37. F.H. Busse in Hydrodynamic Instabilities and the Transition to Turbulence, ed. by H.L. Swinney and J.P. Gollub, Springer-Verlag, Berlin 1981, pp. 97
38. M.G. Velarde in Nonlinear Phenomena at Phase Transitions and Instabilities, ed. by T. Riste, Plenum Press, New York 1981, pp. 205
39. P. Sergè, Y. Pomeau and C. Vidal, L'ordre dans le chaos: Vers une approche déterministe de la turbulence, Hermann, Paris 1984
40. H. Strehlow and W. Knoche, Fundamentals of Chemical Relaxation, Verlag Chemie, Weinheim 1977
41. M.A. López Quintela, A.F. Nóvoa and J. Quibén, to be published

THE GENERAL PHENOMENON OF PATTERN GROWTH AND CONVECTION DRIVEN BY

CHEMICAL REACTIONS AT LIQUID INTERFACES*

Michael L. Kagan and David Avnir

Department of Organic Chemistry, The Hebrew University
of Jerusalem, Jerusalem 91904, Israel

We have investigated the phenomenon of non-equilibrium pattern growth and onset of convections driven by chemical reactions at a horizontal liquid interface. The phenomenon is exceedingly general. Various possible mechanisms are discussed with the evidence pointing to time dependent hydrodynamic instabilities. Some of the possible mechanisms were tested experimentally and by simulations. Image analysis techniques were developed for quantitative treatment of pattern growth kinetics.

1. INTRODUCTION

It seems almost by definition, that homogeneity is unknown in Nature and that at all levels of scale it is possible to find patterning, from molecular scales[1] to networks of superclusters in the cosmos[2]. At each level the same questions are asked: "What are the mechanisms leading to the initial break in homogeneity ?" "What are the critical parameters ?" "What determines the limits of growth and the final form of the pattern ?"[3-10]
In chemistry, the recognition and explanation of patterns in time and space has come relatively recently[11]. Chemical oscillations, best typified by the stirred Belousov-Zhabotinskii reaction[12], represent patterns in time. Although such oscillations have been shown to exist in a large number of modifications[13] the actual number of unique oscillations is still limited[14]. Oscillating reactions have been shown to exhibit period doubling and chaotic behaviour in common with other non-linear phenomena[15]. When such reactions are carried out in unstirred conditions the solution becomes an excited medium and chemical waves are formed[16]. These are spatio-temperol patterns caused by the coupling of the complex reactions to diffusion[17]. Spatial patterns in purely chemical systems are best represented by Lisegang Rings[18], formed and fixed in a non-transportable medium by the reaction of a diffusing salt with an impregnated salt to give a discrete set of precipitation rings. Present explorations point to instabilities in the nucleation processes[19].

---

*Outline of Ph.D. Thesis, M.L. Kagan, The Hebrew University of Jerusalem, 1987 (in English, available upon request from D.A.).

Spatial patterns in purely chemical-diffusion systems have been widely predicted[20,21] but, apart from the appearance of stationary bands in the unstirred Belousov-Zhabotinskii reaction[22], none realized.

Whereas in chemistry the awareness of the significance and possibility of pattern growth has been a relatively recent development, in hydrodynamics, instabilities, patterning, and self-organization has been the norm rather than the exception[23,24]. The non-linear properties of fluid convection provide a classic example of homogeneity breaking. When the density gradient of a fluid becomes destabilized by either temperature or concentration changes, a critical point is reached at which the restraining forces on the fluid, i.e. viscosity, are overcome and a transition from molecular dissipation to bulk macroscopic dissipation occurs. The convections form predictable geometric patterns such as hexagonal cells or rolls as determined by the boundary conditions[25-26]. Surface properties of liquids are also responsible for patterning behaviour such as the thermo-capillary[27] and the chemo-capillary[28] effects. Another example of an instability is in the case of two immiscible solutions in contact with each other to form an unstable density gradient (Taylor-Rayleigh instability). This leads to patterning in the form of penetrating fingers[29].

Changes in the chemical composition of layers of multicomponent solutions can also lead to convections. During extraction processes (with or without reaction) interfaces can undergo rapid movement in the form of oscillations, waves and streamers[30,31]. Patterns evolve even in situations where layers of miscible solutions are in contact in an initially stable density gradient. In the case in which the solutions are of macropolymers, (for instance, a solution of dextran over dextran plus polyvinylpyrrolidone,)[32] fingers of the upper solution are seen to move rapidly downwards with a corresponding movement from the lower layer upwards. Such organized and rapid transfer of material has been compared to axon transportation[33]. In the simpler case of, say, sugar and salt, or heat and salt, the development of dynamic instabilities, in the form of fingers, are due to the differences in rates of diffusion and expansion coefficients of the solutions (double-diffusion)[34-35].

2. THE GENERAL PHENOMENON OF PATTERN GROWTH AND CONVECTIONS DRIVEN BY CHEMICAL REACTIONS AT LIQUID INTERFACES

When a chemical reaction occurs in a solution, it is usually accompanied by changes in densities as well as diffusion rates. We have investigated the consequences of such changes on the stability of reacting media and have found a remarkably wide and general phenomenon: It was originally observed by Möckel[36] that when a reaction takes place in a horizontal layer in an unstirred solution, the product layer increases in concentration homogeneously until at a critical level a pattern, in the form of lines, begins to grow and spread out within the layer[40]. Almost simultaneously, product material begins to descend (or ascend) down the lines in the form of sheets or down the nodes, at line intersections as fingers, until the entire solution was homogenized. Types of reactions were: photochemical in organic solutions[37]; photochemical in aqueous solutions[38]; gas/liquid[39]; liquid/membrane/liquid[40]. Patterns evolved under various boundary conditions with a depth limitation of 0.7 mm. Between this and about 10 mm, the patterns had a regular cellular structure whose average size was depth dependent. Above 1.5 cm there was no correlation.

It was found that the phenomenon occurred whether the reaction takes place from the bottom or the top of the solution, although the onset time and shape of the patterns differed. Reactions in a vertical plane caused immediate convections and no patterns[41].

We have thus revealed a very extensive phenomenon involving a variety of reactions under various physical conditions.

3. COMPUTERIZED IMAGE PROCESSING AND ANALYSIS

To obtain quantitative structural information and correlate it to various physical parameters of the experiment, we have developed computerized image processing and analysis techniques[42,43], and made use of high-resolution densitometry[44,45]. These techniques have also been used to clarify the type of mechanism that is responsible for this phenomenon. The former was used to measure the growth rate of the lines, the shapes of the patterns in their planform, the density of the lines, the build-up of the reaction layer, the wavelength of the first visible fluctuations and the break-up of the layer into the bulk. With computerized high resolution densitometry, we investigated reactions in thin layers under a microscope and detected the existence of a clear bifurcation point, i.e., the rapid growth of initial fluctuations in the local concentration to form a packet of stable states. We followed the transitory nature of the patterns as they faded away leading to a coalescent point as the system returned to homogeneity. Changes of onset time with concentration were also obtained by this method.

4. MECHANISTIC STUDIES (A) - SURVEY OF SOME POSSIBLE MECHANISMS

In our investigation of the mechanism (or mechanisms) responsible for this phenomenon we excluded the possibility of the following trivial causes: the sources of the energy and matter input, whether a light source, gas source or diffusion through a membrane, were homogeneous within the time scale of the reaction; the solution was homogeneous chemically and thermally before the reaction proceeded; external influences such as vibrations were excluded; pre-patterns by either mixing memory or evaporation were prevented by allowing the solution to come to rest for sufficiently long time and by covering the solution[46,47].

Non-trivial mechanisms that were investigated included: feedback loops in the reaction kinetics[48], by the formation of hot spots either by the exothermicity of the reaction or by preferential light absorption of one species and strong uphill diffusion (Soret effect) of another. The latter was excluded by replacing light reactions by gas/liquid reactions and the former by rapidly removing temperature inhomogeneities. Marangoni surface tension instabilities were excluded by changing drastically the properties of the interface without preventing the patterns, and finally by performing the reaction in a layer far from any boundary where surface tension effects are insignificant.

Pattern formation due to highly non-linear chemical kinetics were excluded by the observation that the appearance of patterns occurs in such 'simple' cases as first order photochromic reactions[49]. Non-linear concentration dependent diffusion rates, such as rapid diffusion at low concentrations of product (as occurs in aggregation processes) was excluded by the addition of anti-aggregants as well as amounts of product in the starting solution.

5. MECHANISTIC STUDIES (B) - SUGGESTED MECHANISMS

The elimination of a purely chemical mechanism to explain the phenomenon and the obvious importance of the fluid medium (patterns did not form in gels or below 0.7 mm) led us in search of a mechanism in which chemical reactions and hydrodynamic instabilities are coupled. We concentrated on the photochemical reaction to produce the soluble blue dye known as Turnbull's Blue:

$$(HCOOH)_2 + 4FeCl_3 \xrightarrow[40W]{360 \text{ nm}} 4FeCl_2 + 4HCl + 2CO_2$$

$$FeCl_2 + K_3Fe(CN)_6 \longrightarrow KFeFe(CN)_6 + 2KCl$$

[$FeCl_3$ - 0.001 M; Oxalic acid - 0.013 M, $K_3Fe(CN)_6$ - 0.005 M, the little CO_2 formed remains dissolved and no bubbles form during the course of the reaction].

The following changes in the fluid properties of the reacting solution were found: total volume increased during the course of the reaction; the product rose when the reaction vessel was held at an incline; the expansion coefficient of Turnbull's Blue was less than that of the starting materials; the temperature rise during the reaction was less than 0.2°C, far below the critical value for the onset of thermal convections; the patterns and convections occurred even in the presence of a very strong stabilizing temperature gradient; the diffusion rate of Turnbull's Blue was measured and found to be less than that of the starting materials. (Turnbull's Blue - 2.5×10^{-7} cm^2/sec; $FeCl_3$ - 1.1×10^{-5} cm^2/sec; $KFe(CN)_6$ - 9.4×10^{-6} cm^2/sec).[51]

We can therefore say that the product, in this particular reaction, forms a less dense layer, so that when irradiated from below, a lighter homogeneous layer is created, that begins to diffuse upwards. As the reaction continues the concentration gradient becomes steeper until reaching a critical value at which point convective motion begins, the layer breaks up and the product ascends in cells. This is a time-dependent Benard-Rayliegh problem[52,53] which is iso-thermal and does not involve the physical addition of new material to destabilize the system.

Irradiating from above produces a homogeneous layer that is lighter than the bulk, i.e. statically stable. In a typical double diffusion experiment, the two solutes exist as separate layers (reservoirs) where the concentration gradients across the interface are diffusion dependent only. In our case the product is formed by a photochemical reaction within the starting materials with a gradient determined by the Beer-Lambert Law of light absorption plus diffusion into the bulk. The starting materials are also diffusing back into the reaction zone. Since the product has a slower rate of diffusion and a lower expansion coefficient than either of the two principle starting materials ($FeCl_3$, $K_3Fe(CN)_6$; oxalic acid is in excess) a time dependent double diffusion mechanism for a three component system seems the most likely possible explanation of the phenomenon.

6. COMPUTER SIMULATIONS AND MODELLING

We have used recently developed methods for solving numerically time-dependent, non-linear, two dimensional problems[54]. These algorithms are being used to model the above two mechanisms by solving the full Navier-Stokes equation for a multicomponent system coupled to a chemical reaction and diffusion terms.

We have also investigated by this method solutions to reaction diffusion problems. We have shown that patterns can form from simple reaction in which the non-linearity exists in the diffusion term as a concentration dependent diffusion rate[55].

Figure 1   Patterns formed during the photoreduction of Fe^{III} (Section 5).

7. MICROGRAVITY EXPERIMENT

We have designed and are currently building, a microgravity experiment to be launched aboard the US Space Shuttle. This will greatly reduce and decouple the buoyancy effects from possible diffusion/reaction nonlinear couplings, and indicate whether the latter is a contributing factor in the convective pattern formation.

ACKNOWLEDGMENTS: Supported by the Volkswagen Foundation, by the Israel Space Agency through the National Council of R & D, and by the F. Haber Research Center for Molecular Dynamics, Jerusalem. The microgravity experiment is carried out in collaboration with the DFVLR (the German Space Agency). Parts of this study were carried out in collaboration with Drs. S. Peleg, R. Kosloff, W. Ross. We thank them for their help.

REFERENCES

1. D. Avnir, D. Farin and P. Pfeifer, Nature, 308:261 (1984).
2. J.O. Burns, Sci. Am., 255(1):30 (1986).
3. D'Arcy Thompsom, "On Growth and Form", J.T: Bonner, ed., Cambridge University Press, Cambridge (1969).
4. E. Schrodinger, "Wath is Life?", Cambridge University Press, Cambridge (1944).
5. N. Weiner, "Cybernetics ", J. Wiley, New York (1948).
6. P. Glansdorff and I. Prigogine, "Thermodynamic Theory of Structure, Stability and Fluctuations", J. Wiley, New York (1971).
7. M. Malacinski, ed., "Pattern Formation", Macmillan (1984).
8. J.G. Miller, "Living Systems", McGraw Hill, New York (1978).
9. A.T. Winfree, "The Geometry of Biological Time", Springer-Verlag, New York (1980).
10. H. Haken, "Synergetics", Springer-Verlag, New York (1983).
11. G. Nicolis, I. Prigogine, "Self-Organization in Non-Equilibrium Systems", J. Wiley, New York (1977).
12. J.J. Tyson, "The Belousov-Zhabotinsky Reaction", in: Lecture Notes in Biomathematics, S.A. Levin, ed.", Springer-Verlag, New York (1976).
13. I.R. Epstein, K. Kustin, P. de Kepper, M. Orban, Sci. Am. March, pp. 112-113 (1983).
14. G. Nicolis and J. Portnow, J. Chem. Rev., 73:365 (1973).

15. H.L. Swinney and J.C. Roux, in: "Non-Equilibrium Dynamics in Chemical Systems", C. Vidal, A. Pacault (eds), Springer-Verlag, New York (1984).
16. R.J. Field and M. Burger, eds., "Oscillations and Traveling Waves in Chemical Systems", J. Wiley, New York (1985). S.C. Müller, T. Plesser and B. Hess, Naturwiss, 73:165 (1986).
17. M. Herschkowitz-Kaufman, Comptes Rendus, 270C:1049 (1970). S.C. Müller, T. Plesser and B. Hess, Science 230:661 (1985).
18. K.H. Stern, Chem. Rev., 54:79 (1954). S. Kai and S.C. Müller, Science as Form 1:9 (1985).
19. S. Kai, S. Müller and J. Ross, J. Phys. Chem., 87:806 (1983).
20. I. Prigogine and R. Lefever, Adv. Chem. Phys., 39:1 (1978).
21. A.M. Turing, Phil. Trans. R. Soc. London, 237B:37 (1952).
22. B.J. Welsh, J. Gomaton, A.E. Burgess, Nature, 304:611 (1983).
23. Lord Rayleigh, Phil. Mag., 32:529 (1916).
24. J. Zierep and H. Oertel Jr., eds., "Convective Transport and Instability Phenomena", G. Braun, Karlsruhe (1982).
25. S. Chandrasekhar, "Hydrodynamic and Hydromagnetic Stability", Dover, New York (1981).
26. E. Moses and V. Steinberg in: "Patterns, Defects and Micro Structure", Proc. NATO Adv. Res. Workshop, Austin (1986).
27. L.E. Scriven and C.V. Sternling, J. Fluid Mech., 19:321 (1964).
28. J.C. Berg and A. Acrivos, Chem. Eng. Sci., 20:737 (1965).
29. D.M. Sharp, Physica D, 12:3 (1984).
30. M. Dupeyrat and E. Nakache, Bioelect. Bioenerg., 5:135 (1978).
31. A. Orell, J.W. Westwater, Chem. Eng. Sci., 16:127 (1961).
32. T.C. Laurent, B.N. Preston, W.D. Comper, G.C. Checkley, K. Edsman, L.O. Sundelof, J. Phys. Chem., 87:648 (1983).
33. W.D. Comper, B.N. Preston and L. Austin, Neurochem. Res., 8:943 (1983).
34. M.E. Stern, J. Fluid Mech., 35:209 (1969).
35. J.S. Turner, Ann. Rev. Fluid Mech., 17:11 (1985).
36. P. Mockel, Naturwiss., 66:575 (1979).
37. M. Kagan, A. Levi, D. Avnir, ibid., 69: 548 (1982).
38. D. Avnir, M. Kagan, A. Levi, ibid., 70:141 (1983).
39. D. Avnir, M. Kagan, ibid., 70:361 (1983).
40. D. Avnir and M. Kagan, Nature, 307:717 (1984).
41. Unpublished results.
42. M. Kagan, S. Peleg, E. Meisels, D. Avnir, in: "Modelling of Patterns in Space and Time", eds., W. Jager and J.D. Murrey, Lecture Notes in Biomath., 55:146 (1984).
43. M.L. Kagan, D. Avnir, S. Peleg, in preparation.
44. N. Stockbridge, W.N. Ross, Nature, 309:266 (1984).
45. D. Avnir, M.L. Kagan, W. Ross, submitted for publication.
46. J-C. Micheau, M. Gimenez, P. Brockmans, G. Dewel, Nature, 305:43 (1983).
47. J.C. Berg, M. Boudart, A. Acrivos, J. Fluid Mech., 24:721 (1966).
48. J. Ross and M. Flicker, J. Chem. Phys., 60:3458 (1974).
49. M. Gimenez, J-C. Micheau, Naturwiss., 70:90 (1983).
50. W.M. Riggs, L.E. Bricker, Ann. Chem., 38:897 (1966).
51. International Critical Tables, McGraw Hill (1927).
52. E.G. Muhler, R.S. Schecter, E.H. Wissler, Phys. Fluids, 11:1901 (1968).
53. B.S. Jhaveri and G.M. Homsy, J. Fluid, Mech., 114:251 (1982).
54. D. Kosloff and R. Kosloff, Comput. Phys., 52:35 (1983).
55. D. Avnir, M.L. Kagan, R. Kosloff and S. Peleg, in: "Non-Equilibrium Dynamics in Chemical Systems", C. Vidal and A. Pacault eds., Springer-Verlag, New York (1984).

CHEMICAL WAVES AND NATURAL CONVECTION

Stefan C. Müller, Theo Plesser and Benno Hess

Max-Planck-Institut für Ernährungsphysiologie
Rheinlanddamm 201, D-4600 Dortmund 1, FRG

1. INTRODUCTION

The nonlinear dynamics of chemical systems evolving sufficiently far from thermodynamic equilibrium leads to many remarkable phenomena. For instance, in homogeneous reactive solutions bistability and hysteresis, periodic oscillations or chaotic behaviour may occur [1]. Without stirring the coupling of complex reaction kinetics to diffusion can lead to the propagation of chemical waves, that is concentration gradients move through space, as best observed in thin excitable solution layers in a petri dish [2-4]. Spatial patterns in solution layers can also be stationary or transient [5-10]. For several systems it has been established that such structures are generated by reaction-convection coupling [11] without necessarily involving excitable or oscillatory kinetics [12].

A system of outstanding impact on chemical dynamics is the Belousov-Zhabotinskii (BZ) reaction, in which the oxidation and decarboxylation of malonic acid by bromate ions is catalyzed by metal ions [4,13-14]. Since its discovery this reaction has been in the focus of interest, because it displays a larger number of interesting temporal and spatial phenomena than any other known reaction. In particular, there has been systematic research on chemical waves such as kinematic waves [15], target patterns [2], spiral [3] or scroll waves [16]. Usually, these are indicated by red-to-blue redox transitions of the catalyst ferroin. Spatial structures other than waves have been reported in the BZ and closely related reactions and are frequently refered to as stationary patterns (sometimes also called "mosaic" patterns). In some reports the issue of the type of physical transport processes involved in the pattern formation remains open or at least ambiguous [5,17], in others the role of convection has been discussed to some extent [6,7]. However, efforts for a more detailed approach and explanation of the effects of reaction-convection coupling in the BZ reaction are still limited.

From the experimental point of view modern observation and image evaluation techniques lead to progress in the investigation of spatial patterns, and evidence for their application to waves has been given recently [18-20]. These video and computer based spectrophotometric techniques are also suitable to improve the experiments of convection induced effects, not only because of the high spatial resolution but also as a consequence of good intensity contrast due to selection of a specific wavelength of transmitted light [21].

In this paper several physico-chemical effects of hydrodynamic flow in the BZ reaction shall be described qualitatively. All observations are performed in reactive layers of a depth chosen large enough such that evaporative cooling of the liquid/gas interface can induce natural convection [22]. Structural details are emphasized that are sensitively detected with the two-dimensional spectrophotometer, some of which have not been reported previously. In the following section the preparation of the reactive media and the experimental setup are briefly described. The result section is subdivided according to various initial conditions used for the experiments provoking different spatial structures: Stationary patterns in quiescent, excitable media; superposition of these patterns on propagating waves; decomposition and reorganization of spiral waves; transient patterns appearing in an oscillating medium. A few additional measurements are reported that corroborate the role convection plays under the given experimental conditions.

2. MATERIALS AND METHODS

2.1. Sample preparation

Excitable or oscillatory BZ systems were obtained by preparing solutions with appropriate concentrations of sodium bromide, sodium bromate, malonic acid, and sulfuric acid. About 2 min after mixing the catalyst and indicator ferroin was added. Reagent grade chemicals were used throughout. All solutions were filtered with 0.44 μm Millipore filter. Concentrations are specified for each system in the corresponding figure captions.

A volume of the final mixture resulting in a layer thickness of 1.0 to 1.6 mm was placed in a siliconized, dust-free petri dish. The dish was constructed of an optically flat Pyrex glass plate of 68 mm diameter to which a cylinder of Pyrex glass of 12 mm height was glued. A glass plate was used as a cover, leaving an air gap of about 11 mm above the layer. The cover could be easily removed and repositioned during an experiment without mechanical disturbances.

Circular waves of chemical activity were initiated by immersing the tip of a hot platinum wire into the layer with a micromanipulator. Spiral shaped waves were produced by disrupting a small section of a wave front with a gentle blast of air ejected from a micropipet [19].

The ambient temperature was kept at 24°C within ± 1°C. No further thermostating of the liquid layer was undertaken. Local temperature measurements were performed with a thermocouple device to an accuracy of 0.05°C.

2.2. Image recording

The patterns evolving in the covered or open dish were investigated by digital spectrophotometry in two dimensions. The apparatus consists of optical components and a video camera mounted on a vibration-isolated table, and a fast, large memory computer for storage of the digitalized image data and further data processing. A detailed technical description is given in [21] and applications to various chemical patterns are reported in [23].

Specific properties are: the sample layer is illuminated from above with an expanded, parallel, and spatially homogeneous light beam emerging from a 150 W high-pressure mercury lamp. The wavelength of maximum ferroin absorption (490 nm) is selected by inserting an interference filter into the light path. For detection of the oxidized form of the catalyst, ferriin, a wavelength of 610 nm is selected. A mirror and a photolens system with closeup attachments serve for imaging a square section of the object plane of typically 10 x 10 mm^2 on the photoconductive target of the camera (Hamamatsu C-1000, vidicon tube N983).

The raster resolution is 512 x 512 pixels. An analog to digital converter of a video frame buffer converts the video signal into one out of 256 intensity levels. The grey level of one pixel is stored as one byte in the digital memory array of the buffer. The equipment is linked to a Perkin-Elmer 3230 computer. Acquisition and storage of the 2D data are feasible at a frequency of 30 frames per minute.

The stored image data obtain quantitative information about the intensity distribution in the liquid that can be transformed into concentration values of ferroin [23]. However, the convection induced effects considered here involve three-dimensional concentration distributions coupled with deformations of the liquid/gas interface. Since only 2D projections can be observed, no attempt of a quantitative analysis is undertaken.

3. RESULTS

The following experimental conditions will be distinguished that lead to different types of patterns related to natural convection:

(1) The solution is quiescent but excitable and the dish is left uncovered from the beginning of the experiment (after filling the solution into the dish); observations are carried out without any wave (section 3.1) and during the propagation of waves through the detection area (section 3.2).

(2) After a distinct pattern in an excitable medium has already evolved, the cover is removed and later put back again. This is a simple way to "switch" on and off natural convection (section 3.3).

(3) The solution is in an oscillating state and remains uncovered (section 3.4).

3.1. Stationary patterns

An example for the first case is shown in Fig. 1. The pattern was observed at 490 nm in an excitable BZ solution prepared with an initial ferroin concentration of 2.6 mM. No wave fronts travel through the detection area. About 2 min after placing the solution in the uncovered dish, the onset of spatial inhomogeneities in the ferroin distribution becomes detectable. The contrast between bright and dark areas increases in time and the initially rather diffuse structure evolves into a distinct stationary pattern of remarkable regularity (Fig. 1a, after about 3 min). This pattern slowly disappears when the dish is covered, and it fails to develop when the cover is kept on the dish from the start of the experiment. This is a strong indication of the importance of evaporative cooling evoking temperature gradients. These gradients, which amount to a few tenths of a degree per centimeter (see section 3.5), give rise to a convective structuration process driven by surface tension gradients [22,24].

The pattern in Fig. 1b was observed 25 s after that of Fig. 1a at the wavelength of 610 nm. While at 490 nm (Fig. 1a) the reduced form ferroin has its maximum absorption and the oxidized form ferriin almost does not absorb at all, at 610 nm the absorption of ferriin slightly surpasses that of ferroin, although it is still small. If one considers specific absorption properties only, the bright lines in Fig. 1a correspond to low ferroin concentration, which implies high ferriin concentration as a consequence of the conservation of the total amount of the redox couple. This should lead to a higher absorption of ferriin at 610 nm, which is in fact observed in Fig. 1b. In this image the dark lines correspond exactly to the bright ones in Fig. 1a.

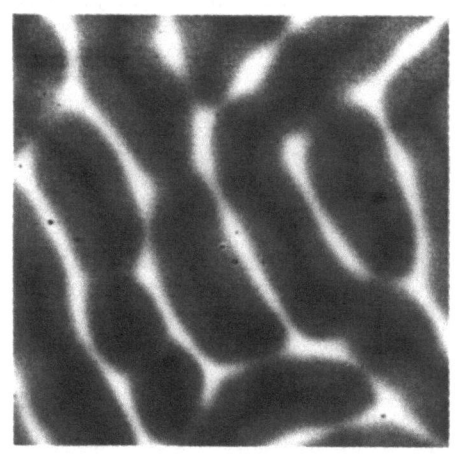
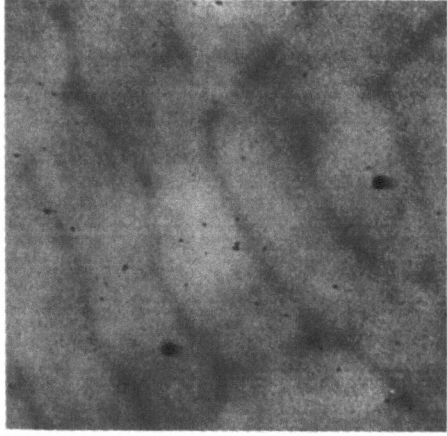

a b

├─────┤
2 mm

Fig. 1.

Stationary ("mosaic") pattern forming in an open 1.3-mm layer of an excitable solution, observed at 490 nm, the maximum absorption of reduced ferroin (a), and at 610 nm where absorption of oxidized ferriin is larger than that of ferroin (b). Initial concentrations: 71 mM NaBr, 324 mM $NaBrO_3$, 140 mM $CH_2(COOH)_2$, 356 mM H_2SO_4, 2.6 mM ferroin.

In terms of ferroin concentration, the contrast of intensities between bright and dark lines in Fig. 1a corresponds to differences in the order of 0.1 mM of the catalyst. Note that further structural details in the darker areas of Fig. 1a can not be detected in Fig. 1b because of much lower intensity contrast.

Comparison of the two images indicates that the stationary patterns mainly reflect the spatial distribution of chemical composition rather than optical effects caused by gradients in the refractive index of the layer. Such gradients are to be expected in any convecting layer due to inhomogeneities in temperature and chemical composition [24] and should influence the direction of the transmitted parallel beam, as shown, for instance, for a metabolyzing biochemical system [11]. It appears, however, that in the investigated system the refractive index gradients are small. Preliminary experiments with Schlieren type methods at almost non-absorbing wavelengths did not give conclusive evidence. Whether differences in the depth of the liquid layer due to surface deformations [24] are of significant influence, as suggested by previous experiments in chemical model systems [25,26], can not be answered at this stage of the investigations.

3.2. Coexistence of waves and stationary patterns

If wave fronts move into the detection area, at which the ferroin concentration almost vanishes, it becomes difficult to observe stationary and moving patterns simultaneously (Fig. 2). Since the intensity contrast between the front and

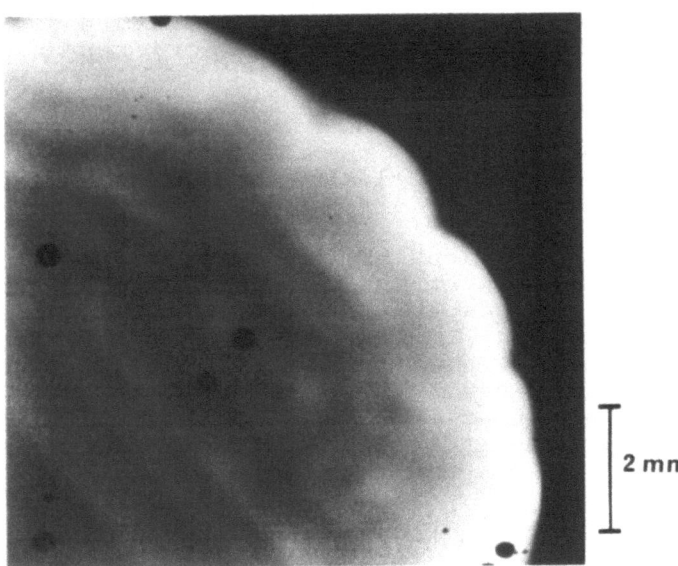

Fig. 2.
Chemical wave front and faintly visible stationary pattern in an open 1.3-mm layer with initial concentrations as in Fig. 1. The shape of the front is circular with small distortions on the scale of the stationary pattern.

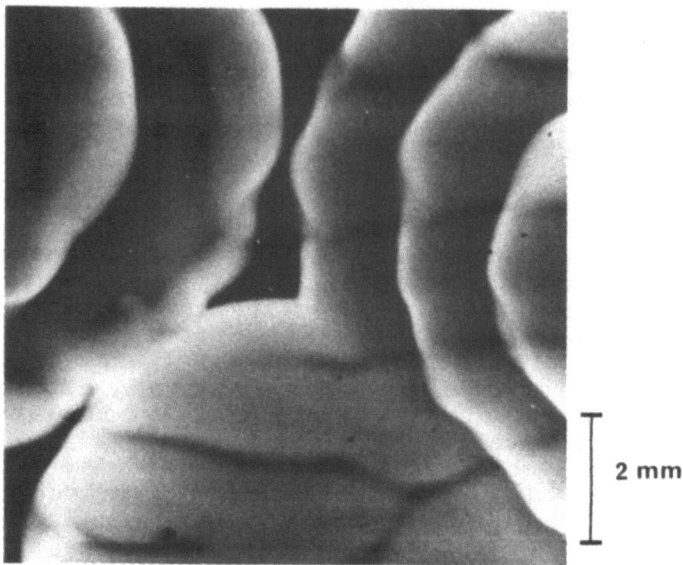

Fig. 3.
Multiple wave fronts in an open 1.6-mm layer of an excitable solution with initially 63 mM NaBr, 384 mM NaBrO$_3$, 125 mM CH$_2$(COOH)$_2$, 422 mM H$_2$SO$_4$, and 0.65 mM ferroin. This choice of concentrations allows simultaneous detection of fronts and dark lines indicating a network of convection cells.

back of the waves is much larger than that of the previously existing structure (by at least a factor of 5 for the given chemical mixture), the intensity of the incident beam has to be strongly reduced in order not to overexpose the camera target at the highly transparent fronts. In the picture of Fig. 2, taken again at 490 nm, a line network similar to that shown in Fig. 1a is only faintly visible as a background structure of the brighter, almost circular section, whereas it can not be detected at all in the remaining black part of the image.

The small distortions of the front in Fig. 2 are correlated with the line network. This correlation is demonstrated more clearly in systems with lower initial ferroin concentration. As shown in Fig. 3, where the initial mixture contained only 0.65 mM ferroin, the stationary line pattern and the superposed multiple waves can be well distinguished. Around the cross-over points of waves and lines the usually sharp fronts are somewhat blurred and slightly reduced in intensity (compare distortion in dye model systems [26]).

3.3. Decomposition and reorganization of waves

More pronounced effects on the shape of waves caused by convection are shown in Fig. 4. A spiral pattern with sharp fronts and regular geometric shape was initiated in an excitable medium with 3.4 mM ferroin concentration, and its evolution was first observed in a covered dish (Fig. 4a). Subsequent removal of the cover leads, after 90 s, to a strong di-

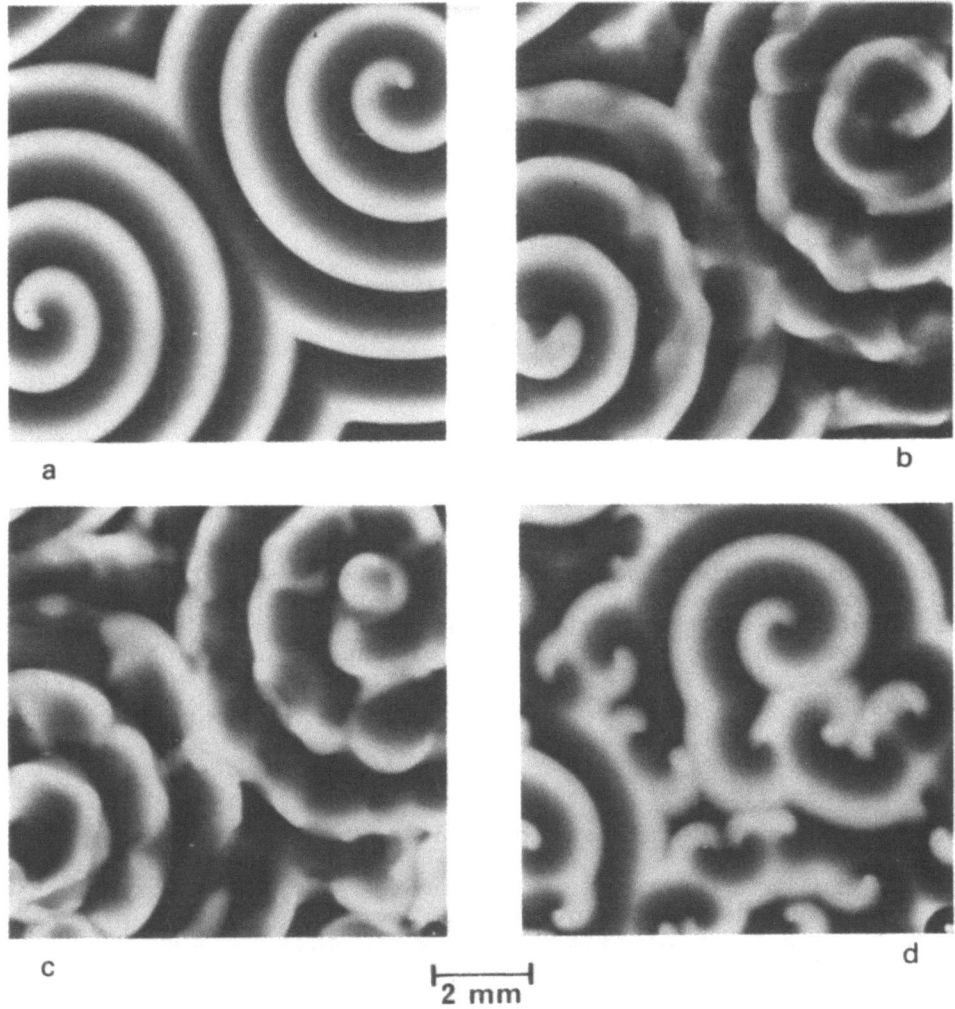

Fig. 4.
A pair of spiral waves in an 1.2-mm layer of an excitable solution (a) is strongly distorted 90 s (b) and decomposed 105 s (c) after removing the glass cover from the petri dish. 220 s after replacing the cover again, partial reorganization of the spiral pattern has taken place (d). Initial concentrations: 48 mM NaBr, 340 mM NaBrO$_3$, 95 mM CH$_2$(COOH)$_2$, 378 mM H$_2$SO$_4$, 3.5 mM ferroin.

stortion of the wave fronts (Fig. 4b) and, after 105 s, to a decomposition of the pair of spirals into a rather disorganized structure (Fig. 4c). The initial geometry remains barely recognizable. Note the spoke-forming tips which, during the subsequent propagation, span the pitch between successive spiral whorls. The distance between the small tips has the same scale as the structure in Fig. 1. When the dish is covered again, a partial reorganization into a more regular pattern consisting of many revolving spiral tips takes place.

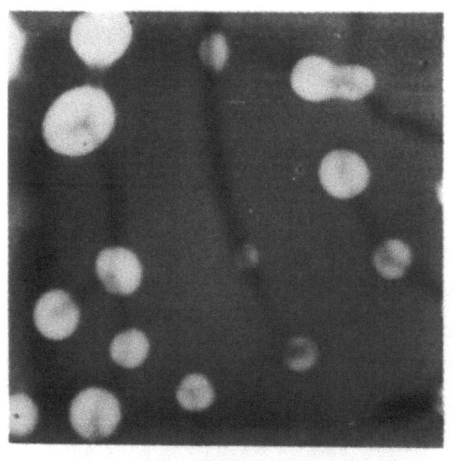 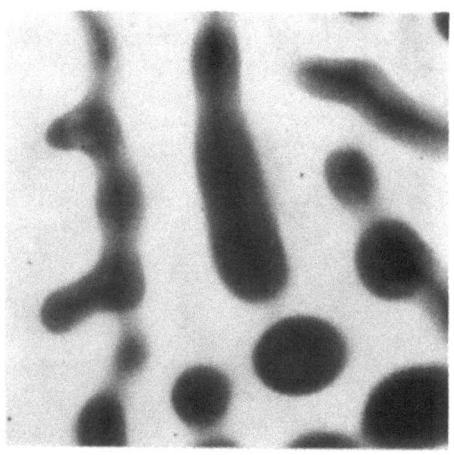

|—————|
2 mm

Fig. 5.
Transient "droplet" pattern forming during a bulk oscillation in an open convecting layer of 1.6-mm thickness. Image (a) was observed at the beginning of the oxidizing phase, image (b) shortly after the beginning of the reducing phase of the reaction cycle. Initial concentrations: 62 mM NaBr, 370 mM $NaBrO_3$, 123 mM $CH_2(COOH)_2$, 406 mM H_2SO_4, 0.96 mM ferroin.

3.4. Inhomogeneous transitions during oscillations

A quite different situation in the BZ reaction which produces another pattern of high complexity is depicted in Fig. 5. Due to a different choice of initial concentrations (essentially a reduction of ferroin concentration) the solution layer undergoes several bulk oscillation before resting in a quiescent state. With the surface left open to evaporative cooling a network of convection cells was allowed to evolve which is indicated by faint dark lines, analogous to those in Fig. 3. Under such conditions, the fast transition from the reduced to the oxidized state of ferroin, that is from dark to bright, does not take place homogeneously in space, but starts at sites lined up along the boundaries of the convection cells (Fig. 5a). Trident branching points of the lines are the prefered centers of excitation from which rapidly expanding circular waves emerge. Shortly afterwards the waves melt into each other, thus completing the redox transition.

Picture b of Fig. 5 was taken during the reversed phase of the oscillations during which the sytem slowly relaxes back to the reduced state. One observes that the areas close to the dark lines are more reduced (darker) than those farther away. To a certain degree, picture b is a complement of picture a, because dark areas are roughly replaced by bright ones and vice versa. Quite generally, the underlying convection pattern induces a phase shift of the oscillations, but processes seem to be involved that depend sensitively on the chemistry.

3.5 Further observations

Several observations corroborate that convective flow in the liquid layer is a major source for the above shown structural phenomena. (1) No stationary pattern or wave front decomposition are observed when the petri dish is covered. (2) The length scale of the stationary patterns is found to be of the order of the layer depth and varies accordingly in the investigated range. (3) Horizontal temperature gradients of up to 0.5 °C/cm from the center to the boudaries exist in the open dish, while covering the dish leads to the disappearence of such inhomogeneities. (4) Dust particles which may be present in the sample tend to move in an open layer, but not so under a cover. (5) There are indications that under oscillatory conditions the velocity of motion of dust particles varies periodically, a phenomenon to be investigated in more detail.

4. DISCUSSION AND CONCLUSIONS

The experiments presented here provide ample evidence that there exist a variety of patterns in excitable or oscillatory solutions of the BZ reaction, the occurrence of which is closely related to the exposure of a liquid/gas interface to evaporative cooling. It is known that under such conditions a liquid layer with a thickness larger than a critical value becomes unstable with respect to convective flow due to gradients in surface tension. The results suggest that this is in fact the case for all the investigated systems. The application of spectrophotometric techniques with high sensitivity and spatial resolution proves to be an efficient tool for the detection of structural details.

The stationary pattern in a quiescent excitable system (Fig. 1) suggests that convection induces local changes in the degree of oxidation which, however, remains small and does not reach the excitation level necessary to trigger a travelling wave front. The chemical processes causing this increase in ferriin concentration are possibly related to flow-dependent input and/or output of chemical compounds, for instance bromine produced by the reaction.

While the chemical processes occurring at the moving fronts in a hydrodynamically stable layer are discussed and at least partially understood by theoretical approaches on the basis of reaction-diffusion equations [27], the very nature of the lines in a convecting chemical system has not been clarified yet. It is an open question to which extent the bright lines in Fig. 1 or, correspondingly, the dark lines in Fig. 3 have to be attributed to absorption indicating concentration variations of chemical species or to refractive index gradients indicating variations in density.

The distortion (Fig. 2) and the decomposition of wave fronts (Fig. 4) support previously given evidence [17] showing that the interplay of natural convection with a reaction-diffusion pattern may give rise to order-disorder transitions, discussed and predicted on theoretical grounds for the case of controlled Bénard type convection [28]. Recently, image sequences similar to those shown in this paper have been treated by image analysis techniques with the goal of finding numerical criteria for the characterization of order and disorder [29].

In this context the question whether spatial chaos occurs is most interesting.

The experiments in a convecting layer of an oscillating solution (Fig. 5) focus again on the issue of excitability and its sensitive dependence on the local physico-chemical conditions. Nucleation for the transition from the reduced to the oxidized state of the catalyst takes place along the boundaries of convection cells. Those locations where such boundaries merge (trident branching points) are among the first to trigger a wave of chemical activity, shortly before the bulk oscillation starts. This suggests that the degree of local mixing associated with shear flow is of significant influence on the local chemistry.

Further insight into the coupling between reaction and convection is to be expected from experiments of Bénard-Rayleigh type. Image analysis techniques are at hand [30] that improve the perception and quantification of moving or stationary chemical gradients in such patterns.

ACKNOWLEDGEMENT

This work was in part supported by the Stiftung Volkswagenwerk, Hannover.

REFERENCES

1. C. Vidal and P. Hanusse, Inter. Rev. Phys. Chem. 5:1 (1986)
2. A.N. Zaikin and A.M. Zhabotinskii, Nature 225:535 (1970)
3. A. Winfree, Science 175:634 (1972)
4. "Oscillations and Traveling Waves in Chemical Systems" R.J. Field and M. Burger, eds., John Wiley N.Y. (1985)
5. A.M. Zhabotinskii and A.N. Zaikin, J. Theoret. Biol. 40:45 (1973)
6. M. Orban, J. Am. Chem. Soc. 102:4311 (1980)
7. K. Showalter, J. Chem. Phys. 73:3735 (1980)
8. P. Möckel, Naturwissenschaften 66:575 (1979)
9. D. Avnir and M. Kagan, Nature 307:717 (1984)
10. A. Boiteux and B. Hess, Ber. Bunsenges. Chem. Phys. 84:393 (1980)
11. S.C. Müller, Th. Plesser and B. Hess, Z. Naturforsch. 40c:588 (1985)
12. M. Gimenez, J.C. Micheau, P. Borckmans, G. Dewel, Nature 305:43 (1983)
13. A.M. Zhabotinskii, Dokl. (Proc.) Acad. Sci. USSR 157:3921 (1964)
14. J.J. Tyson, "The Belousov-Zhabotinskii Reaction", Lecture Notes in Biomathematics Vol. 10, Springer Berlin, Heidelberg (1976)
15. H.-G. Busse, J. Phys. Chem. 73:750 (1969)
16. B.J. Welsh, A.E. Burgess, Nature 304:611 (1983)
17. K.I. Agladze, V.I. Krinsky and, A.M. Pertsov, Nature 308:834 (1984)
18. P.M. Wood and J. Ross, J. Chem. Phys. 82:1924 (1985)
19. S.C. Müller, Th. Plesser and B. Hess, Science 230:661 (1985)
20. A. Pagola and C. Vidal, J. Phys. Chem. 91:501 (1987)

21. S.C. Müller, Th. Plesser and B. Hess, Anal. Biochem. 146:125 (1985)
22. J.K. Platten and J.C. Legros, "Convection in Liquids", Springer Berlin, Heidelberg (1984)
23. S.C. Müller, Th. Plesser and B. Hess, Naturwissenschaften 73:165 (1986)
24. J.C. Berg, A. Acrivos and M. Boudart, Adv. Chem. Eng. 6:61 (1966)
25. S.C. Müller and Th. Plesser, in: "Modelling of Patterns in Space and Time", Lecture Notes in Biomathematics Vol. 55, W. Jäger and J.D. Murray eds., Springer Berlin Heidelberg (1984)
26. S.C. Müller, Th. Plesser and B. Hess, Ber. Bunsenges. Phys. Chem. 83:654 (1985)
27. J.P. Keener and J.J. Tyson, Physica 21D:307 (1986)
28. D. Walgraef, in: "Non-Equilibrium Dynamics in Chemical Systems", C. Vidal and A. Pacault eds., Springer Series in Synergetics Vol. 27, Springer Berlin Heidelberg (1984)
29. M. Markus, S.C. Müller, Th. Plesser and B. Hess, in: "Chaos in Biological Systems", H. Degn, A. Holden and L.F. Olsen, eds., Plenum Press, New York (1987), in press
30. S.C. Müller, Th. Plesser and B. Hess, Biophys. Chem. 17: (1987), in press

KINETIC EFFECTS AND DYNAMICS OF DIFFUSE INTERFACES

Paul Clavin

Laboratoire de Recherche en Combustion
Univ. de Provence / CNRS
B. 252, Centre St Jérôme - 13397 Marseille cedex 13

1. INTRODUCTION

The purpose of this paper is to present a review of some recent results concerning the theoretical description of kinetic effects on the dynamics of interfaces. Results obtained from different models such as those for flame fronts in gas, premixed combustion in porous medium or gasless combustion and near equilibrium interfaces between two different phases (crystal growth) will be discussed comparatively. Special emphasis will be put on the nonlinear aspects of this kinetic effects.

The three next sections of the paper are devoted to the study of different models representative of local structures and dynamics of such interfaces.

In section 2 and 3, the attention is first focused to cases such as combustion fronts for which, kinetics effects are dominant in the local dynamics when the hydrodynamical mechanisms are neglected. In this context, a distinction is exhibited between cases for which a singular perturbation analysis in large values of the activation energy is proved to be useful (Lewis number close to unity) and cases for which a semi-phenomenological free boundary model is introduced (Large Lewis number).

By opposition, the case of fronts governed by reaction-diffusion equations but for which kinetic effects can be treated as perturbations, are presented in section 4. This is essentially the case of quasi equilibrium interfaces separating two coexisting phases. In this case, as for the Stefan problem, the dominant part of the dynamics is usually governed by purely diffusive phenomena. A particular interface model will be solved to point out how nonlinear kinetic effects can affect the front stucture. In this section a simple kinetic model based on two autocatalytic reactions will also be presented to describe free boundary problems in flows by using lattice gas algorithms /1/ which can be easily implemented on cellular automata (see the paper by d'Humière et al. in these proceedings).

The last section concerns a stability analysis of nonlinear patterns constituted by curved fronts in which the external noise at the tip is found to be determinant. The particular case of the needle case will be presented.

2. PREMIXED GAS FLAMES

Classical results of premixed gas flame are briefly recalled in this section. One begins with a basic presentation of the present state of the theoretical description of the dynamics of premixed gas laminar flames and of the role of kinetics effects by comparison with hydrodynamical phenomena. The simplest reaction-diffusion model to describe the kinetic effects is then investigated in the context of the classical asymptotic analysis in large values of the reduced activation energy /11/.

(a) Hydrodynamics versus kinetics and diffusion phenomena

The dynamics of a premixed gas flames is governed by two phenomena involving different length and time scales (for recent reviews see Refs. /4/):

i) reaction-diffusion phenomena controlling the inner structure of the interface and characterized by the flame speed $u_L \simeq (D/\tau_r)^{1/2}$ and the flame thickness $d \simeq (D\tau_r)^{1/2}$ which are based on a diffusion coefficient D, and a characteristic reaction rate $1/\tau_r$.

ii) hydrodynamical phenomena involving a characteristic length scale of the order of magnitude of the wavelength Λ of the wrinkled flame front which is usually much larger than d, ($d/\Lambda \ll 1$). These phenomena are developed in external zones (outside the flame thickness), and correspond to incompressible gas flows, they are induced by deflection of the streamlines across the flame sheet in where the density variation is localised.

The hydrodynamical mechanisms produce the Darrieus-Landau instability of the planar front characterized by a linear growth rate σ proportional to the absolute value of the wavenumber $k=2\pi/\Lambda$, $\sigma = a_1 u_L k$. The dimensionless coefficient a_1 is a positive number increasing with the gas expansion coefficient γ. When the reaction-diffusion phenomena are taken into account, the linear stability analysis of planar fronts yields

$$\sigma = a_1 u_L k - a_2 D k^2 + O(k^3) \qquad (1)$$

where the dimensionless coefficient a_2 depends on the diffusive and kinetics properties of the reactive gas mixture. When $a_2 > 0$, diffusion and kinetics stabilise the short wavelengths. In this case, the marginally stable wavelength $\Lambda^* = 2\pi(a_2/a_1)d$ is of order of magnitude of the flame thickness. This result cannot explained the existence of stable and smooth curved flame fronts whith a characteristic size much larger than the flame thickness as the ones observed for flames propagating in tubes. Following Zeldovich et al. /3/, such nonlinear patterns with $d/\Lambda \ll 1$ can be explained by the following nonlinear hydrodynmical mechanism: local perturbations growth effectively with a linear growth rate similar to Eq. (1) but they are stretched and swept along the mean shape of the front by the tangential component of the induced gas flow /2-3/. Similar mechanisms have been recently studied in other cases such as the Saffman-Taylor fingers, the Davies-Taylor bubbles and the needle crystal /2/ (see the last section of this paper). Moreover, thanks to the acceleration of gravity g, unconditionally stable planar flame fronts can also be obtained for flame propagating downwards at sufficiently low speed (small Froude number $Fr = u_l^2/gd$). Recent detailed analysis and experimental studies of the limits of stability of planar flame fronts propagating downwards are presented in Refs. /5/ and /6/ respectively.

Using the fact that the positive coefficient a_1 decreases to zero with the gas exansion coefficient γ, a bifurcation analysis of freely propagating flames has been carried out by Sivashinsky /7/ in the limit of small gas expansion $\gamma \to 0$ ($a_1 \simeq \gamma/2$) which corresponds to an unstable range of wave number shrinking to zero ($k \to 0$). When the dimensionless coefficient a^2 is positive, the result takes the form of a non linear equation for the evolution of the unstable front that can be written in the following form by using conveniently reduced variables:

$$\partial \alpha / \partial T = L(\alpha) + |\nabla \alpha|^2/2, \qquad (2-a)$$

where $X = \alpha(Y,Z,T)$ is the equation of the front and L is the linear operator corresponding to Eq.(1). Its Fourier transform representation L_k can be written in the following reduced form:

$$L_k = K - K^2. \qquad (2-b)$$

where $K = kd(a_2/a_1)$ is the reduced wave number and where the reduced time is defined by $T = (t/\tau_r)(a_1^2/a_2)$. For a flame propagating downwards in conditions close to the instability limit, it can be easily shown from the analysis of Pelcé Clavin /5/ that Eq. (2-a) is still valid in the limit $\gamma \to 0$ but with L defined by:

$$L_k = \epsilon - (K-1/2)^2 \qquad (2-c)$$

where the bifurcation parameter is $\epsilon = 1/4 - \gamma a_2/(2Fr.a_1^2)$. Notice that the non linear term of Eq. (2-a) is of a purely geometrical nature. It corresponds to the propagation of a tilted front with a given normal velocity. Eq. (2-a) does not contained the convection mechanism produced by the tangential components of the induced flow. Such effects do not appear either in the following order of the analysis in perturbation of the gas expansion coefficient /12/. They can only be obtained at higher orders.

(b) Description of the kinetics effects

The simplest model for the study of dynamics of flame fronts is based on the exothermic decomposition of a single limiting component. The temperature dependence of the reaction rate is represented by an Arrhenius term in which the activation energy is assumed to be much larger than the thermal energy. It is convenient to introduce the reduced mass fraction of the limiting component ψ and the reduced temperature θ defined in such a way that $\psi=1-\theta=1$ in the initial fresh mixture (unburnt gases) and $\psi=0$ in the burnt gases where in addition $\theta=1$ for adiabatic combustion. The two quantities θ and ψ satisfy similar conservation equations coupled by a reaction rate which can be written as $\omega(\theta)Y$ for an order one reaction. The main difference between the θ and ψ equations concerns the diffusion terms which involve two different diffusion coefficients. The Lewis number defined as the ratio of the heat diffusion coefficient to the species diffusion coefficient, $Le=(D_\theta/D_\psi)$, is close to unity in ordinary gas mixtures. In the planar case, hydrodynamical effects are almost trivial and thanks to the isobaric approximation, these two equations form a closed system describing the flame structure. In the non planar case, the change of density couple these two equations with the momentum equations and the equation of continuity essentially through the multidimensional convective terms. In the case $(d/\Lambda)\ll 1$, a multiscale analysis such as the one developed in Refs. /5/ and /8/ can be used to reduce the full problem of the flame dynamics to a free boundary problem in which the flame is considered as a surface of discontinuity for the density and the flows variables. The corresponding jump conditions as well as the local equation for the front evolution is obtained by the analysis of the local flame structure controlled by the diffusion-reaction phenomena i). The results of such analysis are presented in Refs./8/. It is instructive to mention the form of the so obtained modification of the normal flame speed u_n:

$$(u_n/u_L - 1) = m_2 \{ d/R + \tau_r (\vec{n}.\vec{\nabla}\vec{u}.\vec{n})_f \} \tag{3}$$

where m_2 is a dimensionless coefficient depending on the diffusive and reactive properties of the mixture, R is the mean curvature radius of the front, $(\vec{\nabla}\vec{u})_f$ is the value at the front of the rate of stress tensor associated with the upstream gas flow, and $\vec{n}$ is the unit vector normal to the front. At this stage of the analysis, the characteristics at the front of the upstream flow is considered as given but unknown quantities. The full solution is obtained in principle by solving the nonlinear free boundary problem ii) in the same spirit as in the linear hydrodynamical problem of Darrieus and Landau but with more complete boundary conditions at the front. The main difficulties lie in the nonlinearty of the hydrodynamical phenomena ii) and in the vorticity character of the flow in the burnt gases. No exact nonlinear solution has been yet obtained for steady states even for a simplified model $m_2=0$. In this respect, the situation is worth than for the Hele-Shaw or the needle crystal problems for which family of exact solutions are known in the limiting case of zero surface tension. Only approximate solutions have been carried out for flame propagation in tubes /3//7//9/.

The problem of determining such steady non linear patterns will not be considered in this paper. Only their stability property will be studied in the last section. The attention will be focused on the kinetics effects appearing in the phenomena i) and which control the coefficients a_2 and m_2 in Eqs. (1) and (3). In a first step, a qualitative description of these phenomena can be obtained with a simplified model of flame structure called the diffusive-thermal model in which the convective transport mechanism associated with the front wrinkling induced flow is neglected reducing the problem to the solution of a system of reaction-diffusion equations. The nonlinear part of the reaction rate $\omega(\theta)$ involves the reduced activation energy β (called the Zeldovich number). The solution for the flame structure has been successfully carried out in the singular limit of an infinitly stiff reaction rate characterized by an infinitly large value of the Zeldovich number, $\beta \to \infty$. When $\beta \gg 1$, the dominant order of $\omega(\theta)$ is proportional to $\exp\{\beta(\theta-1)\}$ and the dimensionless reaction-diffusion equations controlling the flame structure can be written as:

$$\partial \theta/\partial t - \Delta \theta = \omega(\theta)\psi, \qquad (4\text{-}a)$$

$$\partial \psi/\partial t - (1/Le)\Delta \psi = -\omega(\theta)\psi,$$

with

$$\omega(\theta) = (\beta^2/2Le) \exp\{\beta(\theta-1)\}, \qquad (4\text{-}b)$$

and with the following boundary conditions:

$$\text{fresh mixture,} \quad x=-\infty: \quad \theta=0, \quad \psi=1,$$
$$\text{burnt gases,} \quad x=+\infty: \quad \theta=1, \quad \psi=0. \qquad (5)$$

The flame thickness $d = (D\tau_r)^{1/2}$ and the characteristic time τ_r are the units of length and time. The coefficient $(\beta^2/2Le)$ is arbitrarily introduced in the definition of τ_r to make the dominant order of the laminar flame speed u_L exactly equal to $u_L = (D/\tau_r)^{1/2}$ (in the limit $\beta \to \infty$). The dominant order of the steady planar flame structure is thus described by Eqs. (4-a) with $\partial/\partial t \to \partial/\partial x$ and $\Delta \to \partial^2/\partial x^2$. By the way, notice that the form of Eqs. (4-a) points out the famous "cold boundary difficulty" : the initial condition ($x=-\infty$) is not in a steady state because the reaction rate is not exactly zero, $\omega \simeq \exp(-\beta)$. Following Zeldovich and Frank-Kamenetskii /10/, the relevant solution can be caught in the asymptotic limit $\beta \to \infty$ in which the reaction rate becomes singular i.e. $\omega(\theta) = 0$ everywhere excepted in a thin "boundary layer" located in the hot side of the temperature profile (see Fig.1). Because of the stiffness character of the reaction rate, the planar travelling wave solution obtained at the dominant order of the asymptotic analysis of Eqs (4), is of a different nature from the one selected by the Fisher equation. This asymptotic solution is of a nature similar to the case of an inflamation temperature model (see Refs. /14/ for a recent review of this problem).

The relative thickness of the thin reactive layer is of order of magnitude of $1/\beta$ in such a way that this layer is characterized by the following order of magnitude: $\partial/\partial\xi_n = O(\beta)$ with $1-\theta = O(1/\beta)$, $\varphi = O(1/\beta)$ where ξ_n is the coordinate normal to the reactive sheet. The analysis of Zeldovich and Frank-Kamenetskii can be extended to the non planar case as follows. When one considers the case

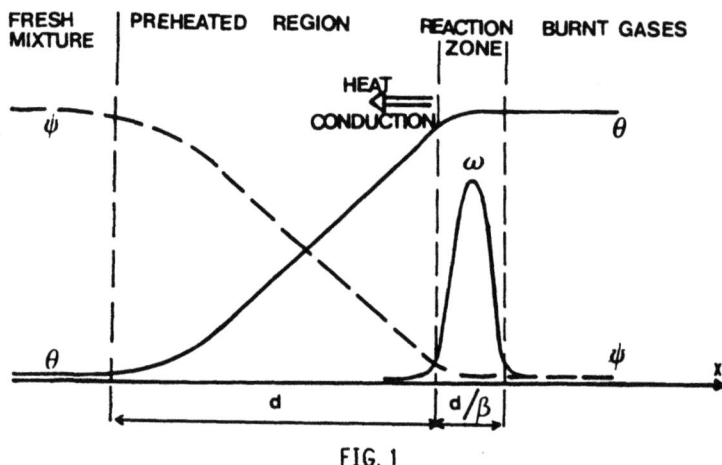

FIG. 1

$$Le = O(1) \qquad (6\text{-a})$$

and if one assumes that

$$\partial/\partial t = o(\beta^2), \qquad (6\text{-b})$$

the dominant order of Eqs. (4-a) yields

$$\partial^2\theta/\partial\xi_n^2 = (1/Le)\partial^2\varphi/\partial\xi_n^2 = \omega(\theta)\varphi. \qquad (7)$$

This equation is written in the moving frame of the reaction sheet in which the origin $\xi_n = 0$ is choosen. The reactant is completly consumed ($\varphi = 0$) at the downstream side of the reaction sheet where in addition one assumes that the dominant order of the heat flux is zero ($\partial\theta/\partial\xi_n = O(1/\beta)$). Using these boundary conditions, Eq. (7) can be integrated with (4-b) throughout the boundary layer. Then, matching conditions provide us with the dominant order of the jump conditions across the reactive sheet that the θ and ψ solutions must verified (see Ref. /11/ for more details)

$$[\partial\theta/\partial\xi_n + (1/Le)\partial\psi/\partial\xi_n]_-^+ = 0, \qquad (8\text{-}a)$$

$$[\partial\theta/\partial\xi_n]_-^+ = -\exp\{\beta(\theta_f - 1)/2\}, \qquad (8\text{-}b)$$

where θ_f is the temperature at the reactive sheet:

$$\theta(\xi_n = 0_+) = \theta(\xi_n = 0_-) = \theta_f. \qquad (8\text{-}c)$$

Superscript + and subscript - refer to downstream and upstream side respectively. Eq.(8-a) expresses that the total enthalpy flux is the same on both sides (the sum of the chemical and the thermal energy is conserved). Eq.(8-b) is a kinetic condition expressing the rate at which the chemical energy is released and evacuated by heat diffusion. This rate depends on the flame temperature θ_f. As a consequence of the high sensitivity of the reaction rate ($\beta \gg 1$), a small variation of the flame temperature produces an important modification of the local burning velocity.

In fact as clearly shown in Eq.(8-b), the asymptotic analysis is meaningful only when the relative modification of the local flame temperature is sufficiently small:

$$1 - \theta_f = O(1/\beta). \qquad (9)$$

By noticing that $\psi = 1-\theta$ is a solution of Eqs.(4-5) for Le = 1, it can be understood that a sufficient condition to insure the validity of Eq.(9) is:

$$Le - 1 = O(1/\beta) \qquad (10)$$

The distinguished limit (10) is the key point of the stability asymptotic analysis of the steady travelling wave solutions of Eqs. (4-5) /11/. As usual, the satbility analysis consists in determinig the linear growth rate σ of a initial perturbation of the front corresponding to a given wave number k. Because of the kinetic effects (see Eq. (8-b)), each order $(1/\beta)^{n+1}$ of the asymptotic expansion of θ_f is determined by the preceding order $(1/\beta)^n$ in the expansion of $\theta(\xi)$. When Eq.(10) is verified, the leading order $(1/\beta)^0$ of the Eq.(8-a) leads to a trivial result in the sense that θ and ψ satisfies the energy conservation (8-a) whatever be the growth rate σ. The leading order of σ can be determined only by considering the next order $(1/\beta)^1$ in the θ and ψ expansions. The two first orders $(1/\beta)^0$ and $(1/\beta)^1$ form a self consistent system. Eqs. (8) are verified at the order $(1/\beta)^1$ only for a particular value of σ which corresponds to the solution that we are looking for. In other words, the asymptotic analysis shows that in the thermal diffusive approximation, the kinetic effects enter at the leading order of the linear growth rate of flames. As shown by the result presented below (see Eq.(11)), the marginal stability limits will be found within the domain corresponding to (10). Thus, the cases that cannot be treated by the asymptotic analysis (Le-1=O(1)) correpond to strongly unstable flame fronts.

It is worthwhile to notice that at the two first orders in the asymptotic expansion in large values of β, the asymtotic analysis is equivalent to solving a free boundary problem which was considered originally /7/ and which can be formulated as follows: The temperature and concentration equations are Eqs. (4-a) with the boundary conditions (5) but with $\omega=0$ and with the jump conditions (8) at the free boundary. In this more phenomenological approach, no special attention is payed further to the limit $\beta\to\infty$. β and 1-Le are considered as finite and order one quantities in the jump conditions (8). One of the merit of the more rigorous asymptotic analysis is to point out the limits of validity of the corresponding results. In the next section we will present an instructive case for which these two approaches are not equivalent.

The results /7//11/ show that the planar front is stable for

$$-2 < \beta(Le-1) < +32/3. \tag{11}$$

The upper limit corresponds to a Hopf bifurcation which is not easily accessible in gases where the Lewis number is close to unity. In the planar case, a similar bifurcation appears at $\beta(Le-1)=4(1+3^{1/2})$ which is slightly larger than 32/3. This instability has been confirmed by numerical analysis /12//13/. Such mechanisms are more relevant for solid combustion phenomena notably to explain the spinning mode of combustion /15/. The lower stability limit of (11) corresponds to the change of sign of the dimensionless coefficient a_2 in Eq.(1) (with $\gamma=0 \Rightarrow a_1=0$) and is usually interpreted as the mechanism for the appearence of cellular structures on flame front/7/. This instability has been recently confirmed by direct numerical simulation of Eqs.(4a-b) in 2-D and unsteady case /18/. The corresponding bifurcation analysis yields the Kuramoto-Sivashinsky equation which has the same form as Eq.(2-a) but with a different linear operator L which can be written in a conveniently reduced manner as

$$L_k = K^2 - K^4. \tag{12}$$

The main interest of this equation is to describe nonlinear patterns presenting a chaotic behavior. Concerning the cellular threshold on planar flames propagating downwards, recent experiments /6/ show that the observed structures are more likely described by the mechanisms of Eq.(2-c) than those of Eq. (12). Similar asymptotic analysis of flame structure can be extended to the case with hydodynamics ($\gamma \neq 0$) of strectched and curved flame fronts under flow inhomogeneities. Such an analysis leads to derive equation (3) with relevant expression for the Markstein constant m_2.

3. GAS COMBUSTION IN POROUS MEDIA AND GASLESS COMBUSTION

By using a recent analysis of reverse combustion fronts /17/, we present in this section different cases corresponding to large Lewis numbers for which the classical asymptotic analysis cannot be used safely to describe the dynamics of fronts. A semi phenomenological approach is presented and is studied comparatively with direct numerical simulations of the problem.

(a) Formulation

A comprehensive theory describing the dynamics of reverse combustion fronts propagating in a combustible porous medium, supported by a forced oxidant flow has been recently developed /16/. The coal is ignited at one end of the seam and oxygen-containing gas is forced through the other end. The combustion front is thus drawn against the gas flow toward the gas injection end. The gas flow is governed by the Darcy law. Pyrolysis and combustion enhance the permeability of the porous coal vein in the burnt side. As a consequence hydrodynamics produces an instability mechanism of the front leading to fingering through a mechanism similar to the Saffman-Taylor instability /16//17/. Notice that the equation describing the development of such fingers cannot be obtained by a bifurcation analysis as Eq.(2-a) for unstable gas flames. Fingering will not be studied here where we will limit the attention to the inner structure of the front. A further simplification occurs when the fuel is devolatilized abundantly ahead of the reaction zone under a thermal equilibrium process in such a way that the limiting species is the oxidant sent with the forced gas flow. In this case, the limiting process is the combustion of the premixed mixture and not the devolatilization of the fuel. Thus, the problem is reduced to solve the combustion of a premixed gaseous mixture flowing through a porous medium. Heat diffusion through the solid porous material is the basic transport mechanism for the flame propagation. Molecular diffusion of the limiting component in the gas phase flowing through the pores is much less efficient and can be neglected in a first approximation. This limiting case corresponds to $Le = \infty$.

In the diffusive-thermal approximation, the simplest model for reverse combustion waves is similar to Eqs.(4-a) but with $Le = \infty$ and with the appearance of a ϵ coefficient in the time dependent term resulting from the difference of density between the gas and the solid phase. This coefficient ϵ is proportional to the ratio of the gas density to the solid density and is defined in such a way that $0<\epsilon<1$. A typical value is $\epsilon \simeq 10^{-2}$. In the referential frame moving with the steady planar travelling wave, the reduced governing equations yields :

$$\partial\theta/\partial t + \partial\theta/\partial x - \Delta\theta = \omega(\theta)\psi, \qquad (13\text{-a})$$

$$\epsilon\partial\psi/\partial t + \partial\psi/\partial x = -\omega(\theta)\psi, \qquad (13\text{-b})$$

with

$$\omega(\theta) = \beta \exp\{\beta(\theta-1)\}. \qquad (13\text{-c})$$

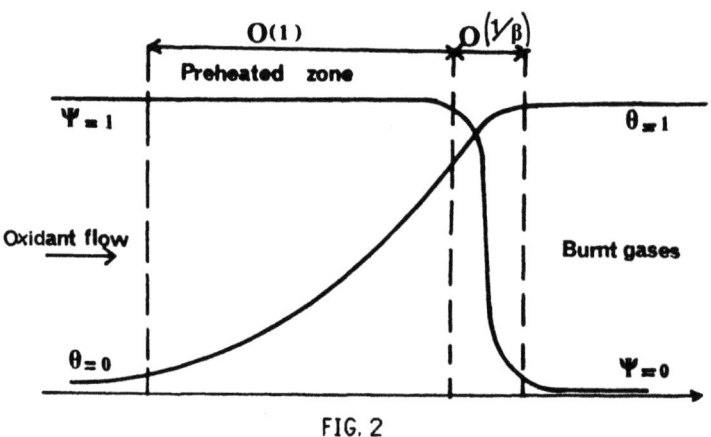

FIG. 2

The boundary conditions are given by Eq.(5). The coefficient β in the prefactor of the r.h.s. of Eq.(13-c) is introduced arbitrarily in such a way that the dominant order of the asymptotic expansion ($\beta \to \infty$) of the steady planar travelling wave solution is solution of Eqs.(13-a-b) with $\partial/\partial t=0$. The only difference with the gasless combustion model is the presence of the ϵ term. In particular, the steady planar travelling wave solutions are the same. The typical temperature and concentration profiles are plotted in Fig.2. Two cases are particularly interesting :

-i) $\epsilon=1$ is the classical gasless combustion model /15/ which corresponds to Eqs. (4-a) for $Le=\infty$.

-ii) $\epsilon=0$ corresponds to the quasisteady state approximation for the gas in reverse combustion fronts. This approximation is extensively used in these cases /16/.

(b) The asymptotic analysis

In the referential frame attached to the reaction zone ($\xi=x-\alpha$, $\eta=y$, $\zeta=z$), the dominant order (in the limit $\beta\to\infty$) of the equations governing the reactive boundary layer are obtained from Eqs.(13) by assuming

$$1-\theta = O(1/\beta), \quad \varphi = O(1), \tag{14-a}$$

$$\partial/\partial\xi = O(\beta), \quad \partial\theta/\partial t = o(\beta), \quad \epsilon\partial\varphi/\partial t = o(\beta), \tag{14-b}$$

to give:

$$(1-\epsilon\partial\alpha/\partial t)\partial\varphi/\partial\xi = (1+|\nabla\alpha|^2)\partial^2\theta/\partial\xi^2 = -\beta\varphi\exp\{\beta(\theta-1)\}. \tag{15-a}$$

Notice that for the gasless combustion model ($\epsilon=1$), Eqs. (15-a) can be written as:

$$u_n \partial\varphi/\partial\xi_n = \partial^2\theta/\partial\xi_n^2 = -\beta\varphi\exp\{\beta(\theta-1)\} \tag{15-b}$$

where u_n and ξ_n are the normal front speed and the coordinate in the direction normal to the reaction sheet respectively. In the same manner as for Eqs.(7), Eqs.(15) can be integrated through the reactive layer and matching conditions can be applied to provide us with jump conditions across the reactive sheet that the first orders of the asymptotic expansion of the θ and φ solutions must verified :

$$\varphi|_- = 1 \quad \text{and} \quad \varphi|_+ = 0. \tag{16-a}$$

When one limits the attention to the cases where the heat flux in the burnt side is small,

$$\partial\theta/\partial\xi|_+ = o(1), \tag{16-b}$$

the two other jump conditions are :

$$-[\partial\theta/\partial\xi]_-^+ = (1-\epsilon\partial\alpha/\partial t)/(1+|\nabla\alpha|^2) \tag{17-a}$$

which can be written in the case of gasless combustion ($\epsilon=1$) as:

$$-[\partial\theta/\partial\xi_n]_-^+ = u_n. \tag{17-b}$$

The second jump condition is :

$$-(1-\epsilon\partial\alpha/\partial t)[\partial\theta/\partial\xi]_-^+ = \exp\{\beta(\theta_f-1)\}, \tag{18-a}$$

to give for $\epsilon=1$:

$$-u_n[\partial\theta/\partial\xi_n]_-^+ = u_n^2 = \exp\{\beta(\theta_f-1)\}, \qquad (18\text{-b})$$

where (17-b) has been used.

Eqs.(17) express the conservation of energy. Notice that Eq.(17-b) is the same relation as in the Stefan problem for crystal growth modelisation /19/. By relating the normal burning velocity u_n to the flame temperature θ_f, Eqs.(18) describe kinetic effects. Eq.(16) points out that the only problem to solve in the outer regions is of a purely thermal nature. When the temperature profile is linearized around the steady planar solution $\bar{\theta}$ in the form

$$\theta = \bar{\theta}(\xi) + \theta'(\xi)\alpha(y,z,t) + \ldots, \qquad (19\text{-a})$$

the equation for the perturbation θ' in the outer regions yields

$$(\sigma + k^2)\theta' + d\theta'/d\xi - d^2\theta'/d\xi^2 = (\sigma + k^2)d\bar{\theta}/d\xi. \qquad (19\text{-b})$$

The solutions of Eq. (19-b) satisfying the boundary conditions (5) are:

$\xi<0:$ $\quad \theta' = \exp(\xi) + (\theta'_f - 1)\exp(\chi_+\xi)$ in the preheated zone, $\qquad (19\text{-c})$

$\xi>0:$ $\quad \theta' = \theta'_f \exp(\chi_-\xi)$ $\qquad$ in the burned side, $\qquad (19\text{-d})$

with $\quad \theta'_f \equiv \theta'(\xi=0)$. $\qquad (19\text{-e})$

and $\quad 2\chi_\pm \equiv 1 \pm (1+4\sigma+4k^2)^{1/2}$. $\qquad (19\text{-f})$

The asymptotic expansion of $\theta'(\xi)$ can be written as

$$\theta'(\xi) = \theta'_0(\xi) + \theta'_1(\xi)/\beta + O(1/\beta^2). \qquad (20)$$

As shown by Eqs.(18), and in order to avoid transcendental large terms, the asymptotic analysis of dynamical problems can be used in the limit $\beta\to\infty$ only if Eq. (9) is verified i.e. $\theta'_f = O(1/\beta) \Leftrightarrow \theta'_0(\xi=0) = 0$. Contrary to the usual premixed gas flame presented in the preceding section, and in the absence of a distinguished limit such as (10), Eqs. (17) for conservation of energy provides us with a non trivial result at the leading order θ'_0 of the asymptotic expansion. The so obtained result corresponds to a case where the kinetic effects are neglected ($\theta'_f=0$ at the leading order). In this perturbative approach, the kinetic relation (18) is only useful to determine $\theta'_1(\xi=0)$ which can be used afterwards to describe the kinetics effects which appears here at the next order of the asymptotic expansion. Thus, in the case Le=∞, the asymptotic expansion method in large values of β has to treat the kinetic effects on the front dynamics as a perturbation around a zero order solution. Such a procedure cannot be relevant to solve problems such as the one described by Eqs. (13) for which kinetics are expected to be dominant in the dynamics as soon as β is large enough. To see that we will compare later on the results of a direct numerical simulation of Eqs. (13) to the leading order of the asymptotic expansion and to the phenomenological free boundary problem defined in the preceding section (see comments just above Eq.(11)). This weakness of the asymptotic analysis for Le = 1 does not concern the steady and planar travelling waves in which constant and uniform transcendental terms can be easily absorbed in a convenient scaling.

The leading order of the asymptotic expansion is obtained by introducing Eqs. (19) with $\theta'_f=0$ in the jump conditions (17):

$$-2\epsilon\sigma = 1 - (1 + 4\sigma + 4k^2)^{1/2}. \qquad (21\text{-}a)$$

whose solutions are:

$$\epsilon = 0: \quad \sigma = -k^2 \qquad (21\text{-}b)$$

$$\epsilon \neq 0: \quad 2\epsilon^2\sigma = 1-\epsilon \pm \{(1-\epsilon)^2 + 4\epsilon^2 k^2\}^{1/2} \qquad (21\text{-}c)$$

The corresponding results are plotted in Figs. 3. Notice that for $\epsilon=1$ one obtains the leading order of the Mullins-Sekkerka instability of crystal growth. This is not surprising because in the approximation $\theta'_f = 0$, the gasless combustion model ($\epsilon=1$) represented by Eqs. (19) and jump conditions (17-b) and (18-b), is equivalent to the Stefan problem with a constant normal front speed u_n = Ct. Notice also that ϵ = 0 is a singular limit for the dynamical properties without kinetics. There is an unstable branch of solution in the range $0 < \epsilon < 1$ with a linear growth rate diverging as $1/\epsilon^2$ when $\epsilon \to 0$. Thus the result (21-b) which is obtained at the leading order of this asymptotic analysis ($\beta \to \infty$) of Eq. (13) with ϵ set equal to zero at the biginning /16/,is in this context questionable for real reverse combustion fronts ($\epsilon \simeq 10^{-2}$). We will see in the following that thanks to the kinetics effects, this result is in fact valid for small enough values of ϵ. In general, the results (21) are not representative of the dynamical properties of Eqs. (13) when $\beta \gg 1$.

(c) The free boundary model

A much better representation of the corresponding phenomena can be obtained with the phenomenological free boundary model represented by Eq. (13-a) but with ω =0, the boundary conditions (5) for θ and the jump conditions (17) and (18) in which β will be considered as a finite quantity of order unity. Such an approach has been extensively used during the last ten years in the Russian litterature for gasless combustion ($\epsilon=1$). Recent developments can be found in Refs/15/.

The results corresponding to the phenomenological free boundary model is obtained when Eqs. (19 c-d) are used with the jump conditions (17) and (18) to give:

$$-2\epsilon\sigma = 1 - (1+4\sigma+4k^2)^{1/2} - 4(\epsilon/\beta)\sigma(1+4\sigma+k^2)^{1/2}, \qquad (22\text{-a})$$

in which the perturbation of the combustion temperature has been given by Eq. (18):

$$\theta'_f = -2\epsilon\sigma/\beta. \qquad (22\text{-b})$$

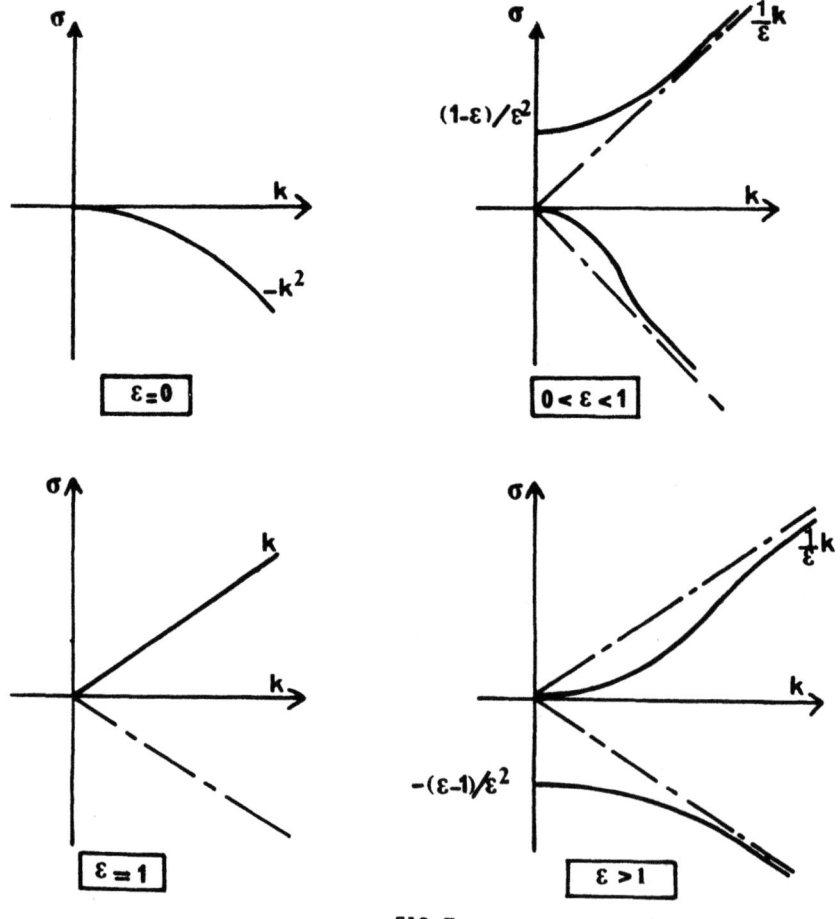

FIG. 3

By comparison with (21-a), the kinetic effects are easily identified as the last term of the r.h.s. of Eq. (22-a). The key point is that this last term introduces solutions of Eq.(22-a) corresponding to high frequencies ($\sigma=O(\beta/\epsilon)$) that cannot be caught by a perturbative expansion in powers of $1/\beta$ around the solution of Eq.(21-a). Questions concerning the pertinence of such solutions for the original system (13) are left open at this stage. This point will be discussed at the end of this section in the light of the results of numerical analysis. Let's first have a look to the roots of Eq.(22-a). To simplify, consider the planar case, k=0. Eq. (22-a) can be written as a cubic equation which is reduced to a quadratic equation in the planar case because of the invariance by translation (when k=0, $\sigma=0$ is a root). The corresponding results predict that the planar front is intrinsically stable for sufficiently small positive values of ϵ.

Stability domain: $\quad\quad 0 < \epsilon < \epsilon^*$. $\hfill$ (23-a)

A Hopf bifurcation appears for $\epsilon = \epsilon^*$ with ϵ^* given by

$$\epsilon^* = 8/(\beta - 4\beta^{-1}),\hfill (23\text{-b})$$

and with a high frequency $\omega^* \equiv -i\sigma$,

$$\omega^* = \beta(1 - \epsilon + 2\epsilon\beta^{-1})^{1/2}/4\epsilon \simeq \beta^2(1 - 4\beta^{-1})/32.\hfill (23\text{-c})$$

By standard bifurcation method this bifurcation can be proved to be supercritical. It is also worthwhile to mention that the third root σ which goes to zero with k→0 corresponds to the precedently questionnable solution (21-b) which is in fact regularly perturbed for small ϵ and large β. Thus, the quasi steady state approximation leading to (21-b) is proved to be justified in the domain (23-a). In the gasless model, $\epsilon=1$ and the Hopf bifurcation corresponds to a limiting value β^* for β. This bifurcation which is analogous to the one appearing in premixed gas flame for sufficiently high Lewis numbers (see the upper bound limit in (11)), has been used extensively to describe the spinning mode in solid combustion /15/.

In the absence of a working asymptotic method, the validity of the results obtained with the phenomenological free boundary model are questionnable for representing the solution of the original system of Eqs.(13) at least for two reasons. The first one is that one applies jump conditions with β considered as a finite quantity, through a reaction sheet which is known to have a zero thickness only in the limit $\beta\to\infty$. The second reason is that the predicted dynamical behavior of the front at the bifurcation, involves high frequencies which question the assumptions (14-b) used to derive the jump conditions. In fact, the quasi steady state assumption of the reaction layer of a premixed gas flame has also been questionned for sufficiently large Lewis number (Le > 1) by recent analysis /20/.

TABLE I

	Numerics	Theory (Eqs. 23)
$\beta = 10$	$\epsilon^* = 0.91$, $\omega^* = 1.6$	$\epsilon^* = 0.83$, $\omega^* = 1.75$
$\beta = 15$	$\epsilon^* = 0.59$, $\omega^* = 4.7$	$\epsilon^* = 0.54$, $\omega^* = 5.02$
$\beta = 30$	$\epsilon^* = 0.28$, $\omega^* = 25.7$	$\epsilon^* = 0.27$, $\omega^* = 24.02$

(d) Numerical analysis

In order to determine to what extend the results of the free boundary model are representative of the Eqs.(13), a direct numerical simulation of these Eq.(13) has been recently carried out by B. Denet/17/ for unsteady but planar cases. The characteristics of the numerical method can be found in Ref/17/. The results are plotted in Fig. 4 and Table I. The numerical results are in surprisingly good qualitative and quantitative agreements with the phenomenological model. The time dependent solution is found to converge toward a stable travelling wave for sufficiently small values of ϵ. The numerical solution is found to experience a Hopf bifurcation for critical values ϵ^* and with oscillatory frequencies ω^* which are plotted comparatively to the results (13) in Table I.

The marginal stability limits are plotted in Fig.4. Typical time dependent reaction rates as obtained from the numerical analysis are plotted in Figs. 5. The picture in Fig. 5-a presents the undamped oscillatory behavior at the instability threshold corresponding to

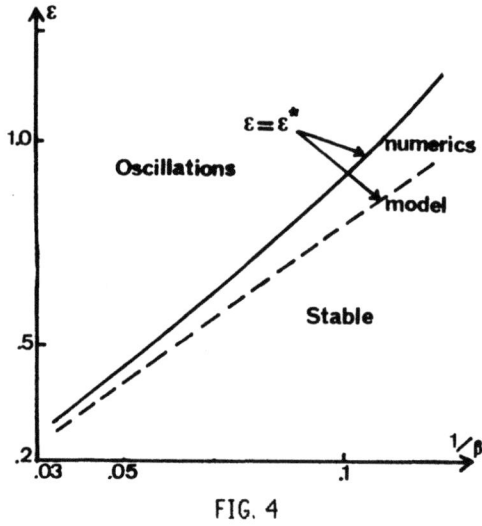

FIG. 4

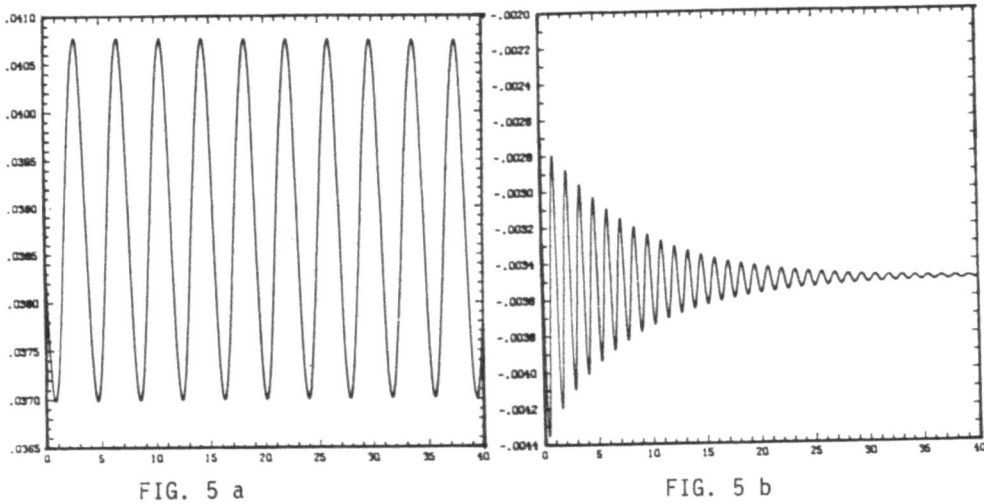

FIG. 5 a   FIG. 5 b

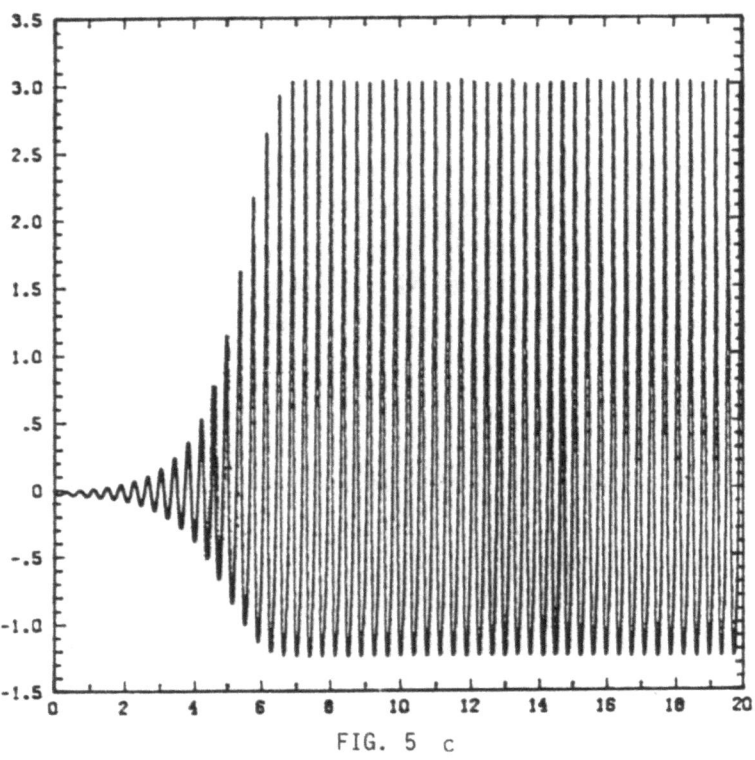

FIG. 5 c

$\beta=10$ and $\epsilon=0.91$. A dampted oscillatory behavior is shown in Fig. 5-b corresponding to a stable case $\beta=15$ and $\epsilon=0.58$. An example of unstable planar front is shown in Fig 5-c for $\beta=25$ and $\epsilon=0.36$. Notice that in all these examples, the high frequency is effectively picked up by the numerical method. But one must emphazise that this high frequency corresponds to the phenomena described by the phenomenological free boundary problem which assumes an inner reactive layer in quasi styeady state approximation. High frequency effects in the inner regions predicted in Ref. /20/ are not observable here and must not be important if they ever exist.

(e) Conclusions

In conclusion, for combustion process described by reaction-diffusion equations with high activation energy, the kinetics effects are dominant in the dynamical properties described by the diffusive-thermal approximation. In the case of large Lewis number, the classical assymptotic expansion do not work for describing the dynamics. The associated phenomenological free boundary model is proved to provide us with good results.

4. DIFFUSE INTERFACE MODEL FOR COEXISTING PHASES

(a) Free boundary model for solidification fronts

In the diffusion-limited regime, the motion of a solidification front propagating into a supercooled melt is mainly controlled by the rate at which the latent heat diffuses away from the fronts (see the paper of N. Goldenfeld in these proceedings). In a first approximation, the interface between the solid and the molten liquid can be considered at equilibrium. At the leading order, the temperature of the interface is the equilibrium temperature.

$$\Theta_f = 0 \tag{24}$$

The dimensionless temperature field Θ is governed by a diffusion equation

$$\partial\Theta/\partial t - D_T \Delta\Theta = 0, \tag{25-a}$$

with boundary conditions in the supercooled melt

$$x = -\infty: \quad \Theta = -\delta, \tag{25-b}$$

where δ is the undercooling and D_T is the thermal diffusivity which is assumed to be the same in the two phases. Two conditions must be used at the interface. The first one is the energy conservation (similar to Eq.(17-b)):

$$\dot{u}_n = -D_T[\partial\Theta/\partial\xi_n]_-^+, \tag{26}$$

and the second one describes the kinetics and surface tension effects and is written as:

$$(u_n/u_L) = (d_0/R) + \beta\Theta_f \tag{27}$$

where R is the radius of curvature and Θ_f the temperature at the front. u_L is a reference velocity, d_0 is the capillary length. The dimensionless coefficient β is a measure of the importance of the kinetic effects. Eq.(27) is known as the Gibbs-Thomson relation. Kinetics and surface tension effects are usually small in the sense that Θ_f as given by (27), is small compared to the undercooling δ. When these effects are neglected ($\beta=0$, $d_0=0$), Eq.(27) is equivalent to the equilibrium condition (24). Even in this case, non trivial solutions (Ivantsov solutions) do exist. This situation is similar to the one considered in the preceding section to obtain the result (21-a). Contrary to the combustion case, the corresponding solutions are here good candidates at least locally to work in perturbation even if small surface tension and/or kinetics effects will be proved to be important as for example to lift the degeneracy (see the lectures of Yves Pomeau, M. Ben Amar and Nigel Goldenfeld in these proceedings). We will not consider these problems in this paper. We will study here models of diffuse interfaces. The motivations for studying such models lie in the theoretical description of the structure of the interfaces and of their local behavior as described by the Gibbs-Thomson relation.

(b) Simulation of free boundaries in flows

It has been recently proved that lattice gas models with Booleans particules can provide a powerful method to study viscous flow at moderate Reynolds and Mach numbers (see the paper of Dominique d'Humières et al. in these proceedings). By introducing two different types of particules A and B that can react following kinetic schemes such as (28), an extension of this model has been recently proposed to similate flow problems with free boundaries /1/.

$$A + A + B \xrightarrow{k_1} A + A + A$$
$$B + B + A \xrightarrow{k_2} B + B + B \tag{28}$$

The corresponding macroscopic equation of evolution of the mass fraction φ of A is

$$D\varphi/Dt - D\Delta\varphi = \omega(\varphi), \tag{29-a}$$

with

$$\omega(\varphi) = k_1\varphi^2(1-\varphi) - k_2\varphi(1-\varphi)^2, \tag{29-b}$$

to give as a particular case

$$k_1 = k_2 = k \;\Rightarrow\; \omega = (k/2)\varphi(1-\varphi)(\varphi-1/2). \tag{29-c}$$

D/Dt is the Lagrangian derivative and D the binary diffusion coefficient. Notice that in the absence of convection, Eq.(29) derives from a Landau-Ginzburg potential:

with
$$\partial\varphi/\partial t = -\delta f/\delta\varphi, \tag{30-a}$$

$$f = \iiint \{(1/2)D|\nabla\varphi|^2 + W(\varphi)\}d^3r, \tag{30-b}$$

$$\omega = -\partial W/\partial\varphi. \tag{30-c}$$

When $k_1 = k_2$, A and B coexist (see Fig. 6-b) and Eqs.(30) and (29-c) admit a steady planar solution describing an equilibrium interface between pure phases A and B. As shown by Cahn and Hilliard /21/, the structure of such an interface is governed by:

$$D(d^2\varphi/d\xi_n^2) = \omega(\varphi). \tag{31-a}$$

The corresponding concentration profile can be easily computed from Eq.(31-a). The thickness of this interface is

$$d = (D/k)^{1/2}. \tag{31-b}$$

When such an interface is developed in a flow field whose the characteristic length scale Λ is much larger than d ($\Lambda \gg d$), a multiscale analysis based on Eq.(29-a) shows that the leading order of the local equation of evolution for the wrinkled front is /1/:

$$u_n/(Dk)^{1/2} = d/R \tag{32}$$

where u_n is the normal component of the front velocity relatively to the flow. Notice that this equation is similar to Eq.(3) for premixed gas flame excepted for the absence of stretch effects by the inhomogeneities of the flow. This difference is directly related to the symetry property of the φ profile solution of Eq.(31-a) /1/.

When $k_1 > k_2$ ($k_1 < k_2$), the phase B is more (less) stable than phase A (see Figs.6-a and 6-b). In this case, Eqs.(30) admit a steady travelling planar wave solution propagating the stable phase into the metastable one with a single wave speed related to $|k_1-k_2|$ /1/.

(c) Diffuse interface model for crystal growth

The purpose of this section is to present a model for the interface structure which includes kinetics and surface tension effect. The basic idea is to consider ψ of Eq.(30-a) as an order parameter of the phase transition in such a way that $\psi=0$ in the liquide phase and $\psi=1$ in the solid phase. Then, Eq.(30-a) is coupled to the temperature field by making the reaction rates k_1 and k_2 in Eq.(29-b) depending on temperature in such a way that at low (high) temperature, $\theta<0$ ($\theta>0$), the solid (liquid) phase becomes the most stable as described by Fig.6-a (Fig.6-c), the equilibrium temperature corresponding to $\theta=0$ (see Fig.6-b). A similar model with a linear temperature dependence of $k_{1,2}$ has been introduced recently in Ref./22/ (see also the lectures of G. Caginalp and P.C. Fife in these proceedings).

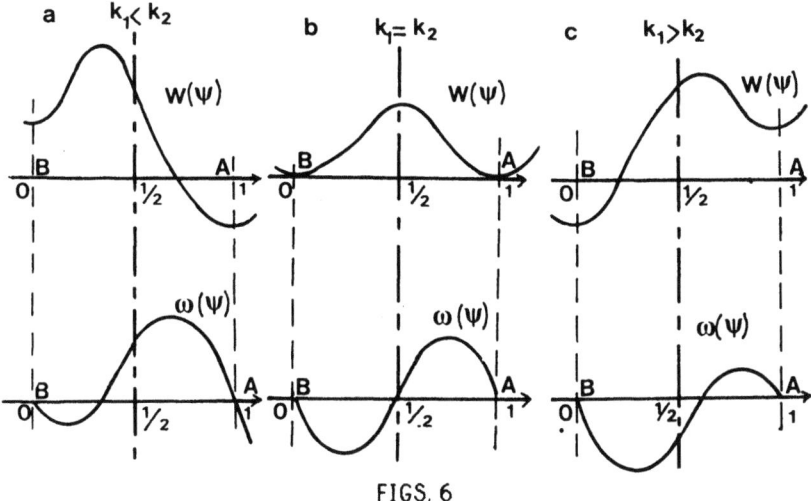

FIGS. 6

Two typical situations can be investigated :

- i) the crystal growth in an undercooled melt described by Fig.7-a in which the stable solid state propagates freely into the metastable liquid state

- ii) the directional solidification represented in Fig.7-b in which the stable solid state is forced to propagate into the stable liquid state.

Let's first formulate the problem by assuming that the interface can be described by the classical macroscopic conservation equations.

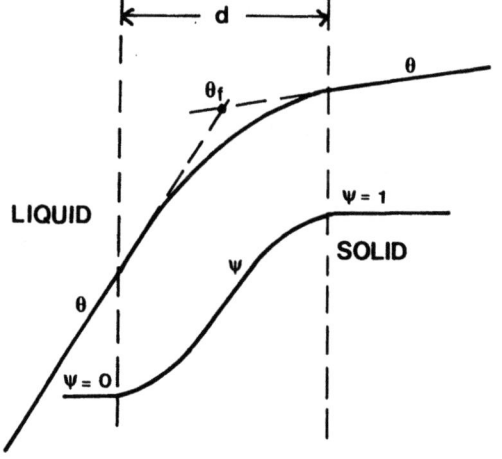

Fig. 7a

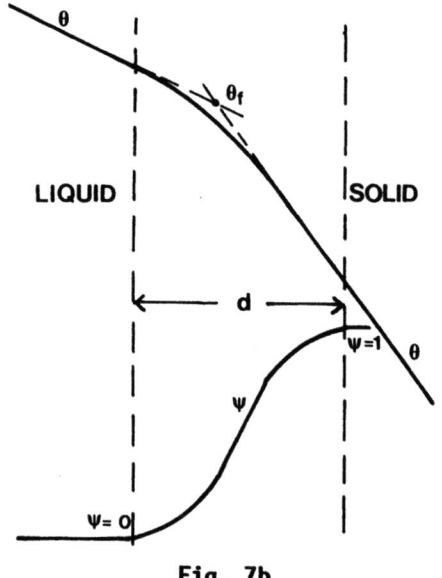

Fig. 7b

The conservation of the total energy can be written as:

$$\rho \, Dh/D\tau = -\vec{\nabla}\cdot\vec{q},\qquad(33\text{-a})$$

where $h(T,\varphi)$ is the enthalpy and q the heat flux. By assuming a simple form for h,

$$\partial h/\partial T = C_p = C^t, \qquad \partial h/\partial\varphi = -Q = C^t,\qquad(33\text{-b})$$

where C_p is the specific heat and Q the latent heat. Using the Fourier law, Eq.(33-a) can be written as:

$$D\Theta/D\tau - D\varphi/D\tau = D_T\,\Delta\Theta,\qquad(34\text{-a})$$

where Θ represents here the reduced temperature defined by $\Theta = C_p(T-T_e)/Q$ (T_e is the equilibrium temperature) and where $D_T = \lambda/\rho C_p$ is the thermal diffusivity. Eq.(34-a) is coupled to an equation similar to Eq.(29-a) but in which ω is now depending on Θ:

$$D\varphi/D\tau = D\,\Delta\varphi + \omega(\varphi,\Theta)/\tau_r.\qquad(34\text{-b})$$

ω is here the dimensionless reaction rate and τ_r the characteristic reaction time associated to one of the reaction rate k_1. The thickness of the interface is $d = (D\tau_r)^{1/2}$. Let u_L be the velocity of the interface, the characteristic length Λ for the variation of the temperature outside the interface is $\Lambda = D_T/u_L$. If one assumes that the order of magnitude of u_L is $u_L \sim (D/\tau_r)^{1/2}$, one obtains that:

$$d/\Lambda = O(1/Le),\qquad(35)$$

where Le is the Lewis number $Le = D_T/D$. Thus, in order to describe a thin interface one must assume that Le is a large number and one has to work in the limit $Le \to \infty$. In this sense, the Lewis number plays here a similar role to the reduced activation energy β in flame theory (see § 2 and § 3). For self consistency let's change the notation:

$$Le \equiv \beta.\qquad(36\text{-a})$$

As shown by Eq.(35), the relative temperature variation inside the interface is of order $1/\beta$, $\Delta\Theta = O(1/\beta)$. In order to force the temperature at the interface to stay stuck close to the equilibrium temperature of coexistence $\Theta=0$, it is natural to assume that the temperature dependence of ω is of the following form:

$$\omega(\varphi,\beta\Theta).\qquad(36\text{-b})$$

In the limit $\beta \to \infty$, Eqs.(36-a) and (36-b) define a distinguished limit which corresponds to a high sensitivity of the reaction rate ω. This presents a certain analogy with flame theory. This distinguished limit is the main difference with other works on this topic /22/ in which the dependence of ω on Θ is assumed to be linear and is treated as a perturbation. Notice that such a distinguished limit will allow to describe non linear kinetic effects. Kinetics effects are usually neglected in the Gibbs-Thompson relation (27). They could be of some importance in not yet explained nonlinear phenomena that are observed in the interface structure /23/.

When the thickness of the interface $d=\sqrt{D\tau_r}$, and the reaction time τ_r are used as unit of length and time respectively, the Eqs.(34) yields:

$$(1/\beta)\, D\theta/Dt - D\varphi/Dt = \Delta\theta \qquad (37\text{-a})$$

$$D\varphi/Dt = \Delta\varphi + \omega(\varphi,\theta) \qquad (37\text{-b})$$

where θ is a rescaled temperature defined by

$$\Theta = \theta/\beta \qquad (37\text{-c})$$

in such a way that the θ variation is of order unity in the interface. The dependence of ω on θ is assumed to be such that $\theta = 0$ is represented by Fig. 6-b and $\theta < 0$ ($\theta > 0$) corresponds to Fig.6-a (Fig.6-c). At the dominant order of the asymptotic analysis in large values of β ($\beta \to \infty$), the solutions of (37) obey a similar free boundary model as the one described by Eqs.(25-a) (26) and (27) but with a generalized Gibbs-Thomson relation including non linear kinetics effects. Notice that working in the similar manner than for flames /11/, the two first orders of the asymptotic expansion must be worked out in the external zone. The Stephan condition (26) is found to be verified at these two first orders when the position of the interface is defined by the intersection of the two temperature asymptotes in such a way that the interface temperature θ_f is unambigously defined (see Figs.7).

(d) Structure of the interface

Let's consider to simplify the steady planar case. According to Eqs.(37) the dominant order of the structure of the interface is given by:

$$-m\, d\varphi/d\xi = d^2\theta/d\xi^2, \qquad (38\text{-a})$$

$$m\, d\varphi/d\xi = d^2\varphi/d\xi^2 + \omega(\theta,\varphi), \qquad (38\text{-b})$$

where m is the reduced velocity of the front defined positively when the front propagates toward the liquid :

$$m = u_n/(D/\tau_r)^{1/2}. \tag{38-c}$$

According to Figs.7, the boundary conditions are :

$$\text{liquid phase, } \xi = -\infty : \varphi = 0, d\theta/d\xi = \theta'_{-\infty}, \tag{38-d}$$

$$\text{solid phase, } \xi = +\infty : \varphi = 1, d\theta/d\xi = \theta'_{+\infty} \tag{38-e}$$

A direct integration of (38-a) shows that the Stefan condition (26) is verified. The front speed is given by the difference of the slopes of the temperature asymptotes $m=\theta'_{-\infty}-\theta'_{+\infty}$
And one obtains :

$$\xi<0 : \quad -m \int^\xi \varphi d\xi + \xi \theta'_{-\infty} = \theta - \theta(\xi=0), \tag{39-a}$$

$$\xi>0 : \quad -m \int^\xi (\varphi-1)d\xi + \xi \theta'_{+\infty} = \theta - \theta(\xi=0). \tag{39-b}$$

In order to determine the interface temperature θ_f, one must solve Eq.(38-b). As an instructive example, we will solve this problem by using a particular model for $\omega(\theta,\varphi)$ inspired from the two triangles model introduced in Ref./14/ and in which $\omega(\varphi)$ is piecewise linear. Let's consider the model sketched in Figs.8 and where ω is defined by :

$$0<\varphi<\varphi_0(\theta) : \quad \omega(\varphi,\theta) = \varphi, \tag{40-a}$$

$$\varphi_0(\theta)<\varphi<1 : \quad \omega(\varphi,\theta) = (1-\varphi), \tag{40-b}$$

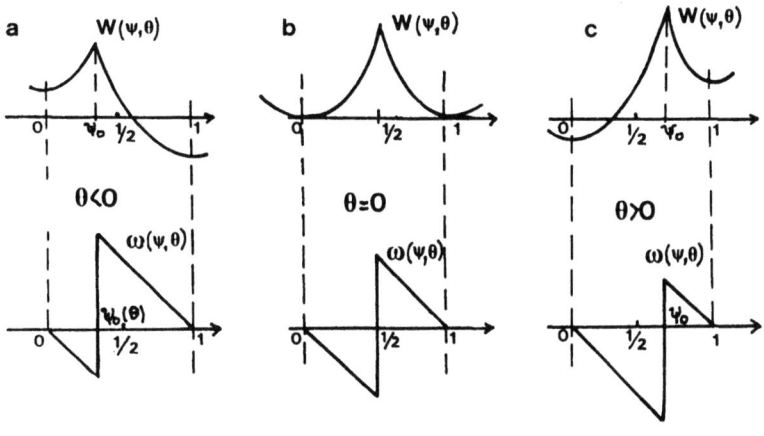

FIG. 8

where $\psi_0(\theta)$ varies with the temperature in such a manner that (see Figs.8) :

$$\theta < 0 : 0 < \psi_0(\theta) < 1/2, \qquad (41\text{-a})$$

$$\theta > 0 : 1/2 < \psi_0(\theta) < 1. \qquad (41\text{-b})$$

The given law $\psi_0(\theta)$ characterizes the phase transition process. Assuming that ψ_0 is a monotonously increasing function of θ, on can also use its inverse $\theta = \theta_0(\psi_0)$. Using Eqs.(40), a straightforward integration of Eq. (38-b) will provide the ψ profile interms of $\psi(\xi=0)$). The origin of the ξ-axis can be choosen arbirarily. Let assume for convenience that $\xi=0$ corresponds to the particular point of the flame structure where the relation

$$\psi = \psi_0(\theta), \qquad (42\text{-a})$$

is verified. Assuming continuity of the ψ derivative at the origin, one obtains :

$$m/(m^2+4)^{1/2} = -\psi_0 \qquad (42\text{-b})$$

where $\psi_0 = \psi(\xi=0)$ is the value of ψ where Eq.(42-a) is verified. Thus, $\theta(\xi=0) = \theta_0(\psi_0)$ is determined as well as the θ-profile as given by Eqs.(39). Now it is easy to determine the interface temperature θ_f defined as the value of θ at the intersection of the two asymptotes corresponding to Eqs.(39). This provides us with a non linear relation between the interface temperature θ_f and the velocity of the front m describing non linear kinetics effects in the Gibbs-Thomson relation. Notice that this relation is not intrinsic. It depends on a free parameter left in the problem which is $\theta'_{+\infty}$ for example. Thus a first result coming out from this model is that the kinetics effects depend in general on the external solutions. In the particular case where one considers that the heat flux in the solid phase is negligible (more precisely when $\theta'_{+\infty} = O(1/\beta)$), one obtains an intrinsic kinetic relation between m and θ_f which depends only on the law $\psi_0(\theta)$. This case is of the interest for the needle crystal. Assuming for simplification a linear law

$$\psi_0 = \alpha \theta \qquad (43\text{-a})$$

the kinetic relation takes the form

$$\theta_f = -2\psi_0(1-\psi_0)/(1+\psi_0) + \psi_0/\alpha \qquad (43\text{-b})$$

where $-\psi_0$ is a monotonously increasing function of m given by Eq.(42-b). When $\alpha > 1/2$, θ_f is positive and increases monotonously with m. This case is not very interesting because it corresponds to a metastable solid phase. More interesting is the case where $\alpha < 1/2$. For small increasing values of m, θ_f begins to decrease from zero. It reaches a negative minimum value for a critical value of m and then increases toward positive values.

This simple model exhibits a surprisingly rich family of solutions. For example it predicts in some cases, existence of intermediate phases of a kinetic nature. This happens when there are more than one point of the ψ-profile satisfying Eq.(42-a). This phenomena can be understood as follows. Because ψ is a monotonously increasing function of ξ, the temperature profile $\theta(\xi)$ can be consider as a function of ψ. When this function $\theta(\psi)$ is put in the "reaction term" ω of Eq.(38-b), the order parameter ψ is found to satisfy a classical reaction diffusion equation with an effective reaction rate $\Omega(\psi)$:

$$md\psi/d\xi = d^2\psi/d\xi^2 + \Omega(\psi). \qquad (44)$$

The so called "intermediate kinetic phases" correspond to the case where the effective potential associated with $\Omega(\psi)$ presents a local minimum for ψ between 0 and 1 as shown in Fig.9. In the above model, such situations are predicted to appear for sufficiently large values of the front speed and when the law $\psi_0(\theta)$ is sufficiently non linear as sketched by the S-shaped curve presented in Fig.10 or when the ψ-variation of the enthalpy of formation h in Eq.(33-a) presents a local maximum characteristic of an activation energy. When such intermediate phases appear, the structure of the interface can be strongly modified.

The model (40-41) is also useful to investigate the dynamical aspects as for example the intrinsic stability of the structure of the interface.

5. STABILITY OF CURVED FRONTS

Recently, Zeldovich et al /3/ presented a simple explanation of the anomalous stability of curved flame propagating in channels. The analysis is limited to the evaluation of the growth rate of an initially localised perturbation initiated close to the tip, stretched and carried along the front by the tangential flow. The attention is focused on stationary fronts of a size R much larger than λ_c, the most unstable wavelength of the planar front, and on initial perturbations of size λ_c anticipated to be the most dangerous ones. By using a Wentzel, Kramers and Brillouin (W.K.B.) method, the dominant order of the growth rate is shown to be proportional to the large number R/λ_c of the order of the Reynolds number defined as the ratio of the tube radius R and the flame thickness. A a result, a stability criterion is obtained, relating the Reynolds number to the amplitude of the external noise in the flow. Pierre Pelcé shows that similar ideas can be applied to other types of fronts /2/. The analysis of these different fronts shows that, as a general rule, the stability criterion relates non dimensional numbers (characterising the steady state solutions) to the amplitude of the initial perturbation at the tip. This section is deveoted to the results concerning the needle crystal which have been presented originally in Refs./2/. Contrary to the flame and the Saffman-Taylor finger, the dominant part of the growth rate of the side branching instability comes from the tail and not from the tip of the front.

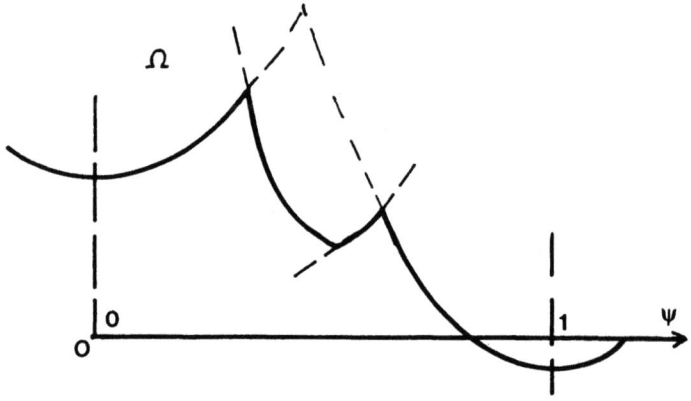

Fig. 9

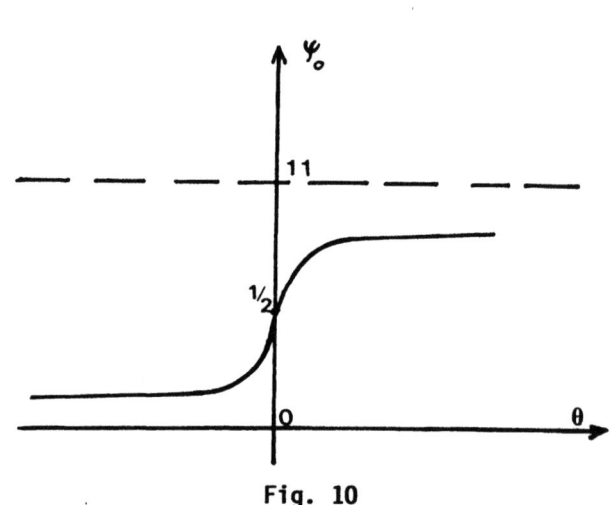

Fig. 10

Stationary shapes of needle crystal growing with constant velocity V have been obtained in the limit of small undercooling δ, including the effect of surface tension /24/. Each corresponding solution is labelled by an eigenvalue C of a nonlinear similarity integrodifferential equation. (See the papers of M. Ben Amar and Y. Pomeau in these proceedings). The stationary solution is not bounded by lateral walls, and side-branching is expected to be observed in real experiments /19/. We study here the dynamics of the sidebranching by evaluating the linear growth rate of a localized disturbance advected in the frame of the crystal by the non uniform tangential velocity $u_\tau = V \sin \theta$.

Following Ref./24/, the growth velocity V of an axisymetrical needle cristal is given by :

$$V = (4D/d_0)C^{-1}(\delta/\text{Log}\delta)^2 \qquad (45\text{-a})$$

where D is the heat diffusion coefficient and d_0 the capillary length. The characteristic length R of curvature at the tip satisfies the Ivantsov relation :

$$RV/D = -2\delta/\text{Log}\delta. \qquad (45\text{-b})$$

The linear growth rate is $\sigma = Vk\{1-\{Dd_0/V\}k^2\}$ and the corresponding most unstable wavelength on a planar front is $\lambda_c = 2\pi(3\,Dd_0/V)^{1/2}$. This implies that (R/λ_c) depends only on the eigenvalue C :

$$R/\lambda_c = C^{1/2}/2\pi\sqrt{3}. \qquad (45\text{-d})$$

As a result, the W.K.B. method is valid at the tip only for sufficiently large values of C. As the C-spectrum is not yet completly known, we have to assume the existence of such large values of C and focus our attention on them. Perturbations associated with the most unstable wavelength λ_c and starting in the vicinity of the tip (θ=0) are the most dangerous. The stretching of the wavelength is given by :

$$\lambda/\lambda_c = u_\tau(\ell)/u_\tau(\lambda_c), \qquad (46\text{-a})$$

or in a non dimensionnal form :

$$\lambda/R = U_\tau(s)/U'_\tau(s=0), \qquad (46\text{-b})$$

where the following limit has been used : $(\lambda_c/R) \ll 1$. u_τ is the stationary tangential velocity field expressed in the moving frame of the finger, ℓ is the coordinate along the front and s the dimensionless coordinate ℓ/R. U_τ is the dimensionaless stationary field and U'_τ is its s-derivative respectively. U_τ is defined by $U_\tau(\ell) = VU_\tau(s)$.

Three regions of the front can be identified (see Fig.11):

- The tip ($s\ll 1$, $\theta\simeq 0$) where the wavelength of the perturbation initially equal to λ_c, remains small compared to the local radius of curvature.

- An intermediate region ($s=O(1)$) in which, because of the stretching mechanism, the wavelength becomes of the order of the local radius of curvature.

- The tail ($s\gg 1$, $\theta\simeq \pi/2$) where the wavelength saturates to a value of order R which is very small compared to the local radius of curvature, and thus the W.K.B. can be used again.

In the first region, the growth rate is evaluated as

$$\Gamma_I = \int_{\lambda_c}^{\ell} (2\pi V/\lambda)[\cos\theta -(\lambda_c/\lambda)^2](d\ell/u_\tau), \qquad (47\text{-}a)$$

where the numerator of the integrand corresponds to the linear growth rate of the Mullins-Sekerka instability of a planar interface with a normal velocity $V\cos\theta$, in the quasisteady state approximation. When C is large, the shape of the crystal is close to the Ivantsov paraboloid whose radius of curvature at the tip is $\rho=(R/2)$ and one may introduce the small parameter $\epsilon=U_\tau(\lambda_c/\rho) \simeq (4\pi\sqrt{3})/\sqrt{C}$ where $U_\tau(s) = \sin\theta$ and $s=1/\rho$, ($U'_\tau(0)=1$). In the limit $\epsilon\to 0$ the dominant order of (47-a) is given by:

$$\Gamma_I = (2\pi/\epsilon)\int_1^\infty [(3x^2-1)/3x^4]dx = [4/(9\sqrt{3})]C^{1/2}. \qquad (47\text{-}b)$$

A dimensional analysis shows that the contribution of the intermediate region to the total growth rate is of order unity and is thus negligible compared to (47-b).

In the tail, the local linear growth rate σ becomes negative at a distance ℓ_f from the tip. The total growth rate decreases for $\ell>\ell_f$. The corresponding angle θ_f is given by expressing the equality of the local marginal wavelength and the actual stretched wavelength given by Eq.(46-b):

$$(\lambda_c \sin\theta_f)/\epsilon = \lambda_c/(3\cos\theta_f)^{-1/2} \qquad (48)$$

which leads to $\theta_f=(\pi/2-\epsilon^2/3)$. The W.K.B. method yields the following expression Γ_{III} for the contribution of the tail to the maximum value of the total growth rate:

$$\Gamma_{III} = \int_{\ell'}^{\ell_f} (2\pi V/\lambda)[\cos\theta -(\lambda_c/\sqrt{3}\lambda)^2](d\ell/u_\tau), \qquad (49\text{-}a)$$

where $\rho \ll \ell' \ll \ell_f$. When the dummy variable $y=U_\tau(s)$ is introduced, Eq.(49-a) yields:

$$\Gamma_{III} = U'_\zeta(s=0) \int_{U_\zeta(\lambda')}^{U_\zeta(\lambda)} 2\pi\{[\cos\theta(s(y)) - \epsilon^2/3y^2].[y^2 U'_\zeta(s=U_\zeta^{-1}(y))]\} dy, \qquad (49\text{-b})$$

where $U_\zeta(s_f) = \sin\theta_f = 1-\epsilon^4/18$. In Eq.(49-b), $U'_\zeta(s)$ is computed by using the equation for the Ivantsov paraboloid with a radius of curvature ρ. This leads to:

$$\Gamma_{III} = \int_{U_\zeta(s')}^{1-(\epsilon^4/18)} [(1-y^2)^{1/2} - \epsilon^2/3y^2]/[y^2(1-y^2)^2]^{-1} dy, \qquad (49\text{-d})$$

whose dominant order is:

$$\Gamma_{III} = 3/2\epsilon^2 = C/32\pi^2. \qquad (49\text{-e})$$

The maximum of the total growth rate is given by the contribution of the tail, $\Gamma_{III} \gg \Gamma_I$. A criterion for the stability of the needle crystal can be roughly obtained by setting that the maximum value of the amplitude of the perturbation becomes of the order of the radius of

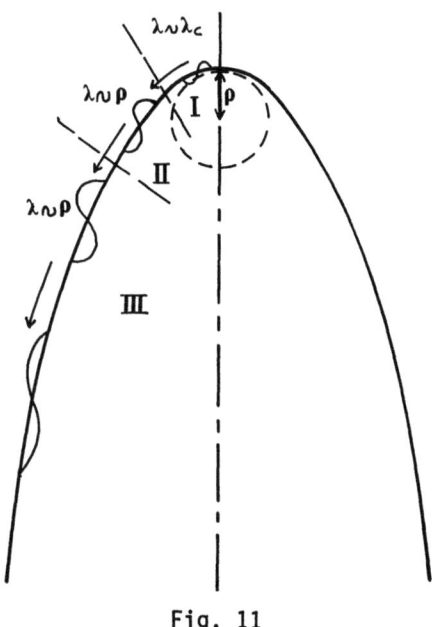

Fig. 11

curvature ρ. This criterion relates the amplitude A_i of the initial perturbation at the tip to the eigenvalue C:

$$C = -32\pi^2 \log(A_i/\rho). \qquad (50)$$

The physical nature of the external noise controlling the initial amplitude of the perturbation can be varied : thermal fluctuations, inhomogeneities of the upstream medium, residual or intrinsic turbulence... So, stability limits are predicted to be depending on the experimental conditions.

The fact that the side branching instability is always observed in experiments of dendritic growth could be explained by an actual value of C always exceeding the critical value given by Eq.(50) and associated with either unavoidable external noise such are thermal fluctuations or disorder induced by crystallisation (defects, dislocations...) or some intrinsic dynamical properties of the tip. Although Eq.(50) has been derived only for large values of C, one can expect that the corresponding result is still valid at least qualitatively for any value of C which is encountered in experiments.

ACKNOWLEDGEMENTS

The work reported in section 3 was performed in collaboration with J. Monteiller and P. Pelcé. The numerical analysis has been carried out by B. Denet with the help of P. Haldenwang. The results of section 5 are extracted from the thesis of P. Pelcé. I have also benefited from fruitful discussions with Y. Pomeau.

REFERENCES

1. P. Clavin, P. Lallemand, Y. Pomeau and G. Searby, J. Fluid. Mech. to appear (1987)
2. P. Pelcé, Thèse Université de Provence, Marseille (1986)
   P. Pelcé and P. Clavin, Europhys. Lett. to appear (1987)
3. Ya.B. Zeldovich, A.G. Istratov, N.I. Kidin and V.B. Librovich, Combust. Sci. Technol. 24 : 1 (1980)
4. G.I. Sivashinsky, Ann. Rev. Fluid. Mech. 15 : 179 (1983)
   P. Clavin, Prog. Energy Combust. Sci. 11 : 1 (1985)
   P. Clavin in "Combustion and Nonlinear Phenomena". Les Editions de Physique (1986)
5. P. Pelcé and P. Clavin, J. Fluid. Mech. 124 : 219 (1982)
6. J. Quinard, Thèse Université de Provence, Marseille (1984)
   J. Quinard, G. Searby and L. Boyer, Lecture Notes in Physics, 210 : 331 (1984) AIAA Progress Series 95 : 129 (1984)
7. G.I. Sivashinsky, Acta Astronautica, 4 : 1177 (1977)
8. P. Clavin and G. Joulin, J. Physique Lettres, 44 : L1 (1983)
   P. Clavin and F.A. Williams, J. Fluid Mech. 116 : 251 (1982)
9. P. Pelcé in "Combustion and Nonlinear Phenomena". Les Editions de Physique (1986)
10. Ya.B. Zeldovich and D.A. Frank-Kamenetskii, Acta Physicochimia URSS, IX-2-341 (1938)
11. G. Joulin and P. Clavin, Combust. Flame 35 : 139 (1979)

12. G. Sivashinsky and P. Clavin, J. Physique, to appear (1987)
13. N. Peters, J. Warnatz, Notes on Numerical Fluid Mechanics, Viewig vol.6 (1982)
    B. Larrouturou, Lectures in Applied Maths, vol.24 AMS : 415 (1986)
14. P. Clavin and A. Linan, NATO ASI Series B, 116 : 291 (1984)
    P. Clavin, Physica D to appear (1987)
15. G. Sivashinsky, SIAM Appl. Math, 40 : 432 (1981)
16. J.A. Britten, Ph. D. Thesis, Univ. Colorado (1984)
    J.A. Britten, W.B. Krantz, Combust. Flame, 60 : 125 (1985)
17. P. Clavin, B. Denet, J. Monteiller and P. Pelcé, J.M.T.A. (J. Theoretical & Applied Mech.) N° spécial : 173 (1986)
18. B. Denet, B. Larrouturou, Combust. Sci. Technol. submitted (1987)
19. J. Langer, Rev. Mod. Phys. 52 : 1 (1980)
20. D.S. Stewart, Combust. Flame, to appear (1986)
21. J.W. Cahn and J.E. Hilliard, J. Chem. Phys. 28 : 258 (1958), 31 : 688 (1959)
22. J.B. Collins, H. Levine, Phys. Rev. B 31-9 : 6119 (1985)
    G. Caginalp and P. Fife, These Proceedings and References cited therein.
23. J.M. Bilgram and P. Böni, NATO ASI Series C 120 : 351 (1984)
24. P. Pelcé and Y. Pomeau, Studies in App. Math. 74 : 245 (1986)

A TURBULENT WRINKLED FLAME SIMULATION BY MEANS
OF CELLULAR AUTOMATA

R. Said and R Borghi

Faculté des Sciences et Techniques
Université de Rouen
76130 Mont-Saint-Aignan
France

I. INTRODUCTION: LAMINAR FLAMES AND TURBULENT WRINKLED FLAMES

In a medium where fuel and oxidizer have been premixed, the combustion takes the form of a very thin flame front propagating towards the fresh mixture. The structure of such a front can be theoretically described for a simple chemistry, and numerical estimates are possible even if the chemistry is complex. [1]. It reveals, at ordinary pressures, a very small thickness separating the almost perfectly fresh mixture from the burned gases at the equilibrium temperature and concentrations; thicknesses of 0,3 mm are typical, and that allows very often to consider the flame as a propagating 2-D surface in the 3-dimensional space. The velocity of propagation of this surface can be evaluated; it depends on the diffusivity of the medium (heat and species diffusivity, often very close) and of a global time characterizing the chemistry. When the flame surface is deformed, due to an incoming flow, the burning velocity also depends on the curvature of the flame with respect to the incoming flow [2].

From a mere dimensional analysis, one can estimate, for the laminar premixed plane flame, the velocity of propagation (with respect to the fresh mixture) and the characteristic thickness,

$$u_L \sim (D/\tau_c)^{1/2} \qquad\qquad e_L \sim (D\tau_c)^{1/2}$$

where D is the thermal diffusivity, and τ_c the chemical time. When the flame is propagating in a turbulent premixed medium, the upstream flow field continuously displaces, stretchs, and curves the flame surface. If the smallest scales of the fluctuating velocity field, say η, the Kolmogorow microscale, is larger than the thickness e_L, the turbulent flame is then made up of laminar flamelets, and is called a wrinkled flame. When $\eta < e_L$, or when a stretching time of the turbulence (for instance the Kolmogorow time $\tau_K = (\epsilon/\nu)^{1/2}$, where ϵ is the disipation rate of the turbulent kinetic energy, and ν the viscosity) is too small with respect to τ_c, the turbulent flame structure is more complicated. A description of what happens is proposed in ref [3], and two new regimes, called wrinkled-thickned and thickned flame exist. Figure 1 shows the instantaneous profile of a flamelet in a turbulent wrinkled flame. Visualization is made possible by seeding the fresh gases with small particles, (disappearing at a given temperature) and after illumation with a laser sheet [4]. The prediction of the properties of turbulent flames is a complex problem. For low enough turbulent wrinkled flames ($u' \ll u_L$), where u' is a characteristic velocity of the turbulence), theory exists [5]. On the other hand, for thickned flames or thickned-wrinkled flames, some models have been proposed of theoretically unknown range of applicability but with real practical success ([6], [7], [8]).

We shall here study the wrinkled flame regime in order to fill the gap between existing models, namely the wrinkled flame regime with varying u'/u_L. An important phenomenon not considered in ref. [5] occurs in this regime. It is the formation of prockets due to the interaction of two flame fronts when the turbulent motions push them one against the other. Following Schelkin's analysis, one can see that pockets of burned gases in the fresh medium, or pockets of fresh mixture in burned gases could reach significant size when $u'/u_L > 1$, [3]. In order to take into account this

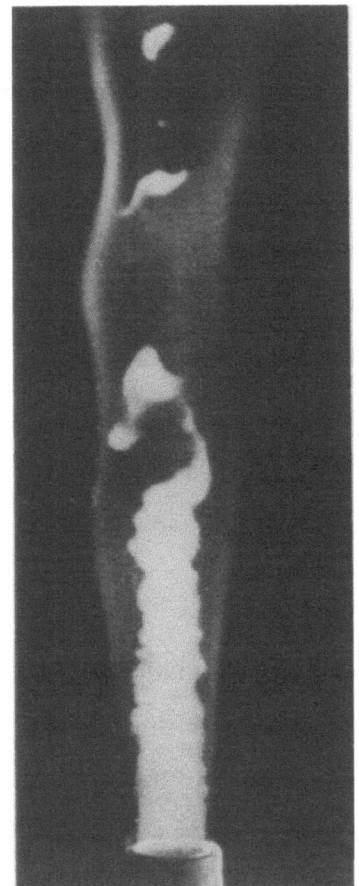

Figure 1. Turbulent wrinkled flame.

phenomenon we carry out a numerical simulation using "cellular automata" [9]. They easily describe the propagation and interaction phenomena but in order to account for turbulent motion, we have introduced within these cellular automata random walks of the cells, coupled to flame propagation and flame interaction algorithms. In real life turbulence is always so closely related to the heat release that combustion influences strongly the turbulent properties. However for simplicity in the simulation, we shall prescribe the latter independently of combustion; and moreover we shall assume it isotropic and homogeneous. Our purpose, however, is not to provide a new computation method to be applied for the prediction of practical flames. The phenomena involved in turbulent flames are so complicated that we only pretend to get insight about the relevant parameters, albeit to a qualitative level. Hopefully this will help in the search of more realistic models of turbulent combustion.

II. CELLULAR AUTOMATA

II. 1. FOUNDATIONS

The proposed simulation involves three steps, and can be developed on a line with only one-dimensional displacements, or in a plane or a three-dimensional space. We here develop only the one and two-dimensional models. Let us begin with a line. This line is divided into a number of adjacent sites; within each site there is one "fluid particle", which can either be burned (black) or fresh (white); incompletly burned particles are no allowed, because we assume always that the reactions are very fast ($\tau_c / \tau_t \ll 1$). Then:

i) The first step is a random displacement of all fluid particles, based on a random choice of a turbulent velocity v with a rms value, u', and zero mean.

ii) The second step is a combustion step between fluid particles. After the step 1, some sites are occupied by two (or more particles). In such a case we choose to "coalesce" all particles in only one, which is burned if one of the particles on the site is burned, fresh if all the particles are not burned. Similarly, we have to look at unoccupied sites. The choice is inade to fill this site with a burned particle if one neighbour is burned, or with an unburned, fresh particle if all the neighbours are unburned.

iii) The third step is a propagation step. It consists in the displacement, with a prescribed velocity, u_L, of the boundary between burned and fresh particles (this is a flame front), from the burned to the fresh side.

In the one-dimensional version, each of the above described steps is easy, In steps 2 and 3, we just look at two neighbours. For the two-dimensional version, each step needs an additional choice. First, we have to choose a lattice in order to represent the plane. We have chosen a square lattice, the simplest albeit with particular simetry properties. It may very well be that an hexagonal lattice, where each site is related to six other sites would be more appropriate. Then we have to define in a precise way the neighbours; with a square lattice, we can call neighbours only the four nearest sites (it is the von Neuman neighbourhood), or the eight closest sites (Moore's neighbourhood). Finally, the propagation step is not so simple, due to the flame front itself. In a square lattice, we have adopted the following procedure: each fresh site must get burned if there is at least one burned site in its neighborhoud.

There is a length scale in the lattice (say Δx); there is also a time scale Δt, i.e., the frequency with which the three steps are repeated. The random walks of each particle are taken independent of the other and thus the space correlation length is Δx. The velocities, v, are randomly chosen, each Δt, independent to the previous ones, and correlation time is

Δt. In order to account for turbulent motions we must have $u' \simeq \Delta x/\Delta t$. Then for a regular lattice the simulated turbulence is homogeneous and isotropic. We intend to vary u'/u_L from $u'/u_L \simeq 1$ to $u'/u_L \gg 1$. Note that the case $u' = 0$ can be described without step 1, and $u_L = 0$ without step 3. Obviously, each $u_L \Delta t$ is not necessarily commensurate with Δx, and so only discrete values of u'/u_L are used. There is no problem at each step if Δx is a multiple of $u_L \Delta t$. If $u_L \Delta t = \Delta x/n$ (n > 1) the step 3 is performed only every n steps.

II. 2. PHYSICAL INTERPRETATION

Step 1 is clearly the turbulent diffusion. Simulations of turbulent dispersion of fluids particles or solid (or liquid) particles have been proposed recently, with very similar Langragian methods [10, 11]. There are three crucial items: the simulation of the time and space velocity correlations, and the simulation of the probability density function p.d.f. for the components of the velocity. Here we just take a crude approximation: the time correlation is unity if $t < \Delta t$, and zero otherwise. The space correlation is unity if $x < \Delta x$, and zero otherwise. As we restrict to the case of homogenous and isotropic turbulence only one single random velocity is needed, the two components of the velocity are independent and with the same r.m.s. value. We have tried as global shape a gaussian or a tophat one and the results are rather similar. Although these drastic approximations could be improved, they are not too disturbing to account for turbulent diffusion: Following Taylor [12], we take the mean square displacement,

$$\overline{n^2}(t) = u'^2 \int_0^t \int_0^t R(\xi) \, d\xi \, dt' \text{ with } R(\xi) = \begin{cases} 1 & \text{if } t' < \Delta t \\ 0 & \text{if } t' \geq \Delta t \end{cases} \quad (1)$$

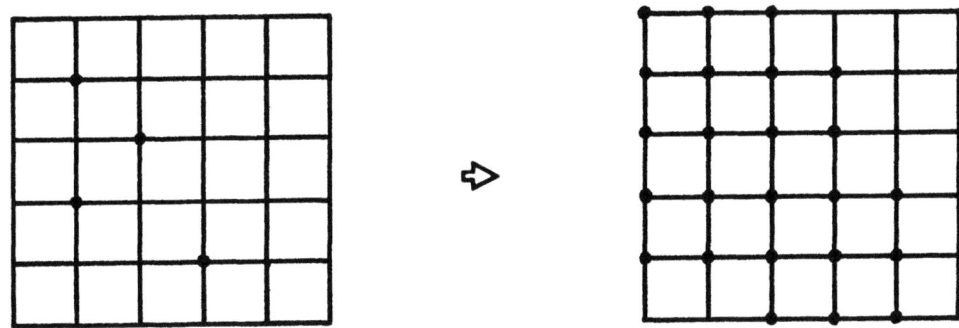

Figure 2. Propagation using Moore's neighbours (one-dimensional case)

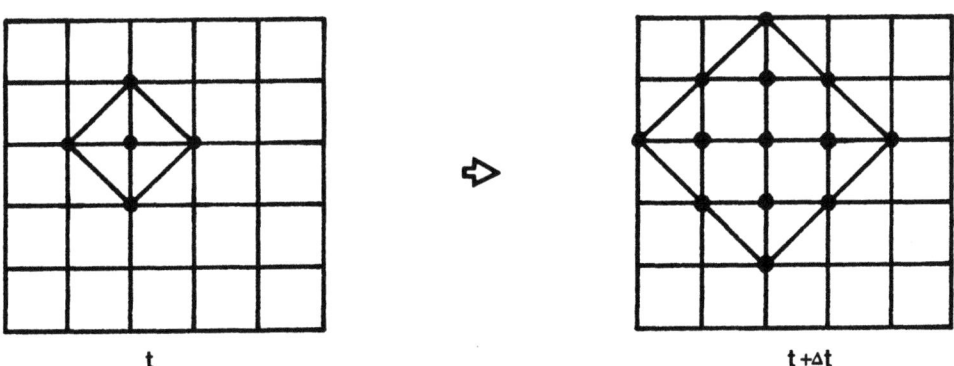

Propagation using Neuman's neighbours.

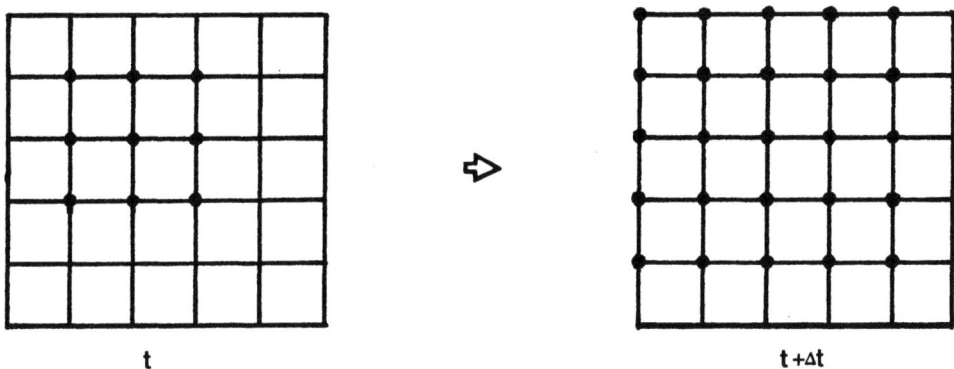

Propagation using Moore's neighbours.

Figure 3. Propagation in dimension two

where $\overline{n^2(t)} = u'^2 (t^2/2)$ if $t < \Delta t$ and $\overline{n^2(t)} = u'^2 t \Delta t$ if $t \geq \Delta t$ and the equivalent turbulent diffusion coefficient is $D_t = u'^2 \Delta t$ when $t \geq \Delta t$ [12]. In order to match with the known relation $D_t = \overline{u'^2} \tau_L$ we see that $\Delta t = \tau_L$ suffices. The Lagrangian time τ_L is usually taken proportianal to τ_t (the Eulerian one). Thus we shall identify Δt with τ_t.

Step 3 is clearly related to the laminar propagation of a flame. We have chosen a constant velocity of the flame, irrespective of the curvature of the front (without local extinctions). This is a first attempt; and extinctions could be taken into account, but the right account of the curvature is difficult for a square lattice. Therefore the simulation is restricted to the case where τ_c is always very small with respect to all turbulence scales. The flame propagation with our *automaton* does not correspond exactly to a normal propagation, the flame propagating from a point is not a circle, but a square, whose size depends on the type of neighbourhoud (fig. 3); with Neumann neighbourhood, the equivalent laminar propagation velocity is $(2/\pi)^{1/2}$, whereas with Moore's, it is $2/(\pi)^{1/2}$ if step 3 is done every time step. This shortcoming, however, is not crucial for our purpose.

The second step represents just a simple interpolation when we are concerned with the filling of an unoccupied site. But when two particles interact it represents a physical process. It is the volumetric burning of a fluid volume. We may expect that this phenomenon actually occurs when two flame fronts are close enough; it is necessary to avoid that the flame surface by unit of volume increases to infinity due to turbulent dispersion. The particular rule chosen for this step, that is that

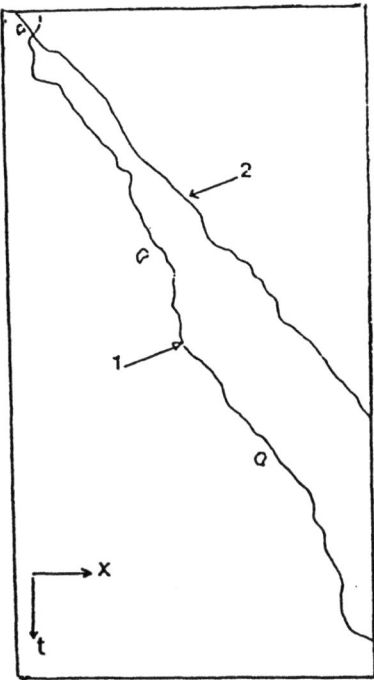

Figure 4. $U' = 1.15$

Propagation of a flame in one dimension

1: $U_L = 0.5$;   2: $U_L = 1$

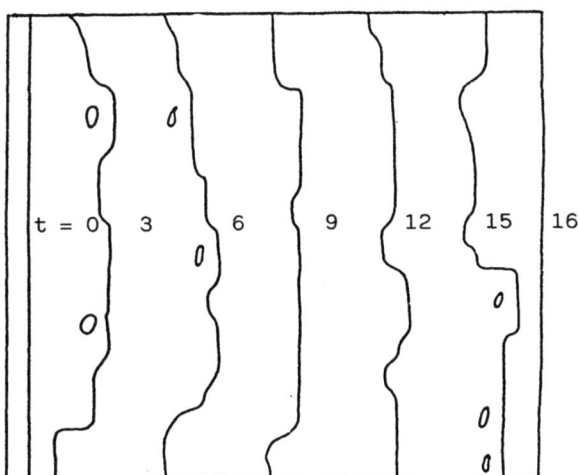

Figure 5. Propagation of a flame
in two dimensions using Moore neighbours

$U'/U_L = 1.15$: $U_L = 1$

one burned particule suffices always to ignite the other, is consistent with our hypothesis that τ_c in infinitely small.

II. 3. SOME WORKED SIMULATIONS

In fig. 4 is illustrated the case of a one-dimensional flame propagation for two values of u'/u_L. It clearly appears the influence of turbulence: the front is strongly corrugated, some pockets appear and the mean velocity of propagation is increased when u'/u_L increases.

Fig. 5 shows the flame propagation in a two-dimensional medium; at t-0 only a small zone of the plane is burned, here several pictures corresponding to flames at different times have been superimposed. These simulations have been made with 53 fluid particles and sites in dimension one, and 20 x 40 sites and fluid particles in two dimensions, on a Apple II microcomputer. For other simulations we have used a DPS 6 computer with a 170 x 170 lattice. As the lattice is bounded, we have assumed as boundary condition a reflection condition. Then we can follow the flame propagation only for a certain time interval which depends on u_L.

III. RESULTS AND DISCUSSION

III.1. Let us obtain quantitative results from the numerical simulations. We have first calculated u_T, the turbulent propagation velocity, given an initial distribution of fluid particles, with only one line of black sites (burned particles) on the left side, the flame develops with an irregular front; in addition, each realization is of course different we have to compute u_T as a mean value of the front velocity; the mean value has been taken here over the whole propagation of the same flame and for thirty realizations.

The mean flame thickness l_t, is also of interest. It has been definied here as the root mean square of the fluctuations of the flame front position with respect to its mean value, at each time step the average value being taken over many realisations. When pockets appear, each pocket front has been considered as an individual flame.

III.2 THE "TURBULENT FLAME VELOCITY"

A very old empirical relation [13] concerning u_T, the "turbulent flame velocity", is

$$u_T = u_L + cu', \text{ with } c \simeq 2 \qquad (2)$$

More recent experiments lead to a similar relationship with, however, c function of the (turbulence) Reynolds number [14]. A theoretical study of Clavin and Williams [5] predicts in the limit $u'/u_L \ll 1$ a quadratic dependence,

$$(u_T / u_L) = 1 + \alpha (u' / u_L)^2 \qquad (3)$$

without dependence of the turbulence scale in α. A recent numerical simulation by Ashurst and Ban [15] agrees with the linear dependance on u' and the dependance on l_t, but the correct value of c is unsettled.

It is obvious that if the turbulence is very large ($u'/u_L \gg 1$), if the Damkohler number τ_t/τ_c goes to infinity (the reaction is very fast with respect to the turbulence) and if the (turbulence) Reynolds number $Re_T = (u'l_t)/\nu$ (where l_t is the space integral scale of the turbulence and ν the viscosity) is very large, as usualy stated, an heuristic argument leads to

$$u_T = \beta u' \qquad (4)$$

with β constant. On the other hand, if we use the estimation of $u_L \alpha \sim (\nu/\tau_c)^{1/2}$ (with Schmidt and Lewis numbers close to unity), we obtain

$$Re_T = (u'/u_L)^2 \tau_t/\tau_c \qquad (5)$$

Note then that very large values of both Re_T and τ_t/τ_c are compatible with finite values of u'/u_L. Is such a case, β in (4) is expected to be a function of u'/u_L, without appearence of the scale l_t, because $Re_T \gg 1$. Equation (2), with c constant, or (3) could be acceptable. However, if Re_T is finite, that implies that both u'/u_L and τ_t/τ_c are finite (excluding the case where one of them is zero, wich is of no interest). In such a case β in (4) may depend on Re_T and τ_t/τ_c or, by ussing (5), perhaps to a formula like (2) with c depending on Re_T. In the light of this argument, formulas (2) or (3) are not contradictory. In addition, when τ_t/τ_c is finite but $u'/u_L \gg 1$, the relation (5) shows that $Re_T \gg 1$ is implied and we could expect that

$$(u_T/u') - f(\tau_t/\tau_c) \qquad (6)$$

such a relation has been numerically obtained by an assumed pdf method, as in ref. [7]. The particular simulation that we performed is the case $\tau_t/\tau_c \gg 1$ and u'/u_L finite.

Fig. 6 shows the curves u_T/u_L in terms of u'/u_L, as obtained from one-dimensional and two-dimensional simulations. The first result is clearly a linear relation, in agreement with (2) and in contradiction with (3). The slope of the curve, c, is found here of order unity in the one-dimensional case, and 1.8 in the two-dimensional case. This last figure is close to 2, the value often found in experiments.

It is obvious that the precision of our value is not high, the number of sample-simulations is not large enough. Nevertheless it is interesting that the value found in 2-D simulation is rather close to the experiments. This our cellular automata shares common features with real life. It was to be expected that one-dimensional simulations would not give quantitavely the same results as the two-dimensional. For instance, in percolation, where also lattices are used, it has been shown that the dimension of the space was an important parameter which is also the case in our problem here.

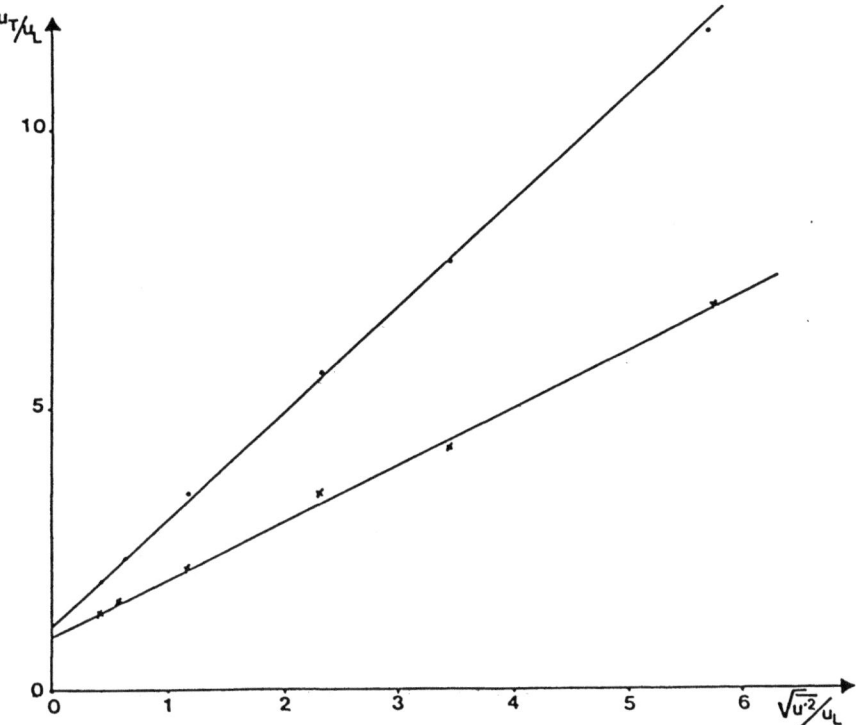

Figure 6. The turbulent flame speed
(●) for dimension one   (✗) for dimension two

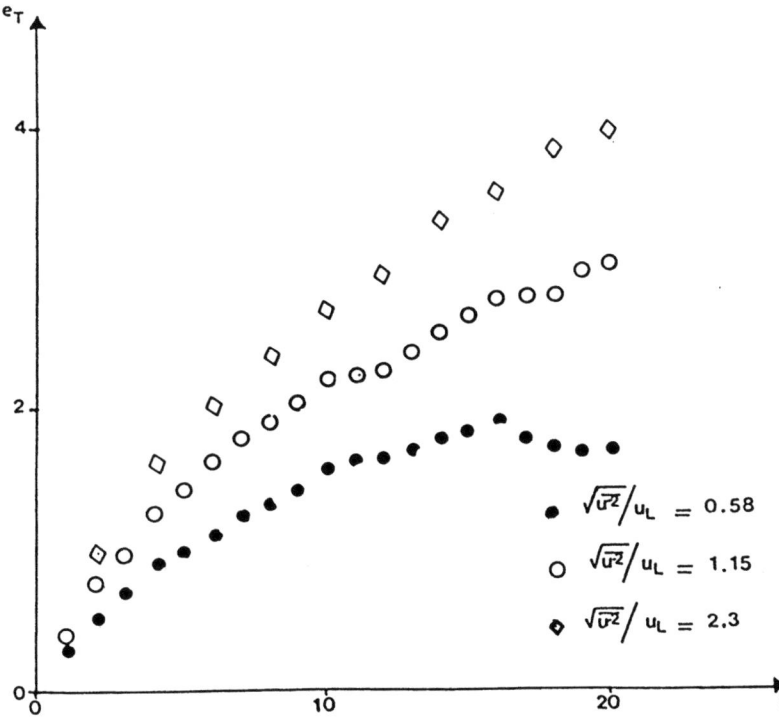

Figure 7. Turbulent flame thickness.

In fact, experimental values of u_T are very difficult to define. There are many ways to 'measure" them, each being approximately equivalent, but differing in a factor. In addition, in real flows, combustion influences turbulence, trough several different (and apposite in signe) phenomena. In our simulation, no such effect is taken into account. Ashurst and Barr |15| have tried to explain the low value (betwen 0.5 and 0.9) that they found by relating it to the effect of the Reynolds number. As we have said the Reynolds number dependence of c can occur only if τ_t/τ_c is not infinetely small which is not the case in our simulation, and presumably also in theirs (at least in their computation with a zero thickness flame).

Another interesting point is the linear dependence always found in all existing numerical simulations. An explanation to clarify the difference with the Clavin-Williams formula remains to be found ...

III.3 THE MEAN FLAME THICKNESS

An interesting question concerning the mean flame thicknss is just its existence with a value independent of time as the flame propagates. The turbulent flame speed actually shows such an independence, as found both experimentally and with our simulation.

The thickness of a laminiar flame e_L, is found constant, but if $u'/u_L \gg 1$, the thickness of the turbulent flame (definied as the r.m.s. value of the fluctuation of flame front position with respect to its mean value) could behave like $(n^2)^{1/2}$ (1). The turbulent flame speed has been the subject of many studies, experimental and theoretical; but that is not the case for the turbulent flame thickness. It is often said that is of the order of the turbulence length scale [13], but this statement has not been substantivated. Reference [16] is the only one with details. For wrinkled flames with pockets.

$$e_T \alpha \ (l_0 u/u') \ \text{Log} \ (1 + (u'/u_L)) \tag{7}$$

Figure 7 shows the results of our simulations. As time goes on e_T increases and apparently approaches an asymptotic value when u'/u_L is small enough. However, for the same propagation times such a limit is not obtained when u'/u_L becomes large.

III.4 SIMULATION OF AN OBLIQUE FLAME

Fig. 1 is a photograph of a stabilized flame. This flame does not propagate in the sense of our simulations. Actually, it is anchored at the edges of the tube, and it adapts itself to the velocity of the incoming flow in order that its normal velocity exactly compensates the velocity of the flow perpendicular to the mean flame position. It is possible simulate such a flame with the two-dimensional automaton. A given mean velocity u has to be prescribed for the fluid particles. In addition to the fluctuations, fluid particles have to be continuously released upstream with those at the edges being prescribed always burned. The results found are given in fig. 8, for two values of u_L. While fig. 1 corresponds approximately to $\overline{u}/u_L = 12$ and $u'/u_L = 0.03$ the results given in fig. 8 correspond to $\overline{u}/u_L = 6$ and $u'/u_L = 0.18$; $\overline{u}/u_L = 18$ and $u'/u_L = 0.18$ respectively.

IV. CONCLUSIONS

We have presented here a first attempt to simulate the turbulent propagation of flame fronts by means of "cellular automata". The rather simple models used here lead to results not far from what is

Figure 8. Simulation of a turbulent wrinkled flame

experimentally known of turbulent flames. Thus we have here a promising research avenue. The main improvement that we are now studying is the more realistic simulation of turbulence, including the possibility to take into account smoother spatial correlations for turbulent motions. This will permitis to get a full range of scales for the motions, and will lead to different sizes of curvature radii in the flame fronts.

REFERENCES

[1] J. WARNATZ. Flame propagation and ignition in hydrocarbon - air mixtures up to octane. -Proceedings of a Symposium on Numerical Simulation of Combustion Phenomena.- INRIA, Sophia Antipalis, FRANCE, May 1985.

[2] P. CLAVIN - Progress Energy Combustion Sciences, 11 - (1985) 1-59.

[3] R. BORGHI. On the structure and morphology of turbulent premisced flames - in Recente advances in the aeronautical sciences, (C. Bruns, C. Cascik, editors), Plenum Press 1984.

[4] J. P. DUMONT, R. BORGHI - Etude qualitative par tomographie laser de la structure des flames turbulentes - Combustion Science and Technology. 48 (1986) 107 - 128.

[5] P. CLAVIN, F. A. WILLIAMS - J. Fluid Mech, 9, (1979), 589.

[6] R. BORGHI - Models of turbulent combustion for Numerical prediction. - In Prediction methods for turbulent flows, W. Kolmann, editor. - Hemisphere, Guernsey (1980)

[7] also K.N.C. BRAY - Turbulent flow with premisced reactants p. 115 - 183, in Turbulent Reacting Flows - (P.A. Libby and F.A. Williams, editors. - Topics in Applied physics vol. 44 - Springer-Verlag, Heidelberg (1980).

[8] R. BORGHI, P. MOREAU, C. BONNIOT - Theoretical predictions of a high velocity premisced turbulent flame, in Physico chemical Hydrodynamics. D. B. Spalding, editor. - Advance Rub. Guernsey (1977) Vol. II. p. 827.

[9] N. H. PACKARD, S. WOLFRAM. (1985) 901 - 946.

[10] P. A. DURBIN. J. Fluid Mech. **100** (1980). 279 - 302

[11] J. K. DUKOWICZ. J. Comput. Phys, 35, (1980) 229 - 253.

[12] G. I. TAYLOR. Proc. London. Math. soc. - AZO (1922) 196.

[13] K. I. SCHELKIN. Combustion Hydrodynamics. Fizika goreniya i Vzryva, (1968) 455 - 468.

[14] R. G. ABDEL - GAYED, K. J. AL-KHIS-HALI, D. BRADLEY. Procs. Royal Soc. London, **A391** (1984) 393 - 414.

[15] W. T. ASHURST, P. K. BARR. (1983) 227 - 256.

[16] A. TALANTOV, V. M. ERMOLAEV, V. K. ZOTIN, E.A. PETROV in Combustion, Explosion and Schocks Waves - Vol. 5, pp. 73 - 75, (1969).

INSTABILITIES AND SURFACE TENSION

Yves Pomeau
Laboratoire de Physique de l'ENS,
24 rue Lhomond 75231, Paris
France

INTRODUCTION

I shall present below recent progress made in the study of the effect of surface tension on various instabilities. The attention will be on two well known physical phenomena: Saffman-Taylor fingers in long rectangular channels and needle crystals growing at the expanse of a supercooled melt. These two topics share the common property that under suitable assumptions(to be presented below) the fully nonlinear regime is completely described by nonlinear integrodifferential equations for functions of a single variable, with explicitely known solutions in the absence of surface tension. Thus it is natural to consider there surface tension as a small perturbation in the usual mathematical sense. As shown later on however it turns out that this perturbation expansion leads to rather nontrivial difficulties. Furthermore, for Saffman-Taylor fingers it makes sense too to study the opposite limit("large surface tension") that leads also to interesting mathematical structures.

This paper is thus divided in two main parts: section 2 is devoted to the determination of the shape of Saffman-Taylor fingers in the opposite limits of small and large surface tension. Section 3 gives some details about the formation of needle crystals, more being to be found in the papers by Muller-Krumbhaar and by Ben Amar at this conference.

I shall be concerned exclusively with the determination of patterns remaining steady in a moving frame. These patterns (when they exist) are eventually unstable, and so possibly without much experimental meaning. But I won't consider this sort of question (the interested reader should look at the review by the Chicago group [1] as well as at the paper by Paul Clavin at this conference).

SAFFMAN-TAYLOR FINGERS

In a porous medium or in Hele-Shaw cells($\simeq$2dimensional porous medium) the fluid velocity **u** is related to the pressure field p via Darcy's law:

$$\mathbf{u} = -k/\mu \, \nabla p$$

where k is the permeability of the medium depending on the geometry of the solid phase only and where μ is the shear viscosity of the fluid. Consider now two immiscible fluids (oil and water for instance) flowing in such a porous medium when the less viscous fluid pushes the more viscous one, a situation of interest for petroleum engineering. Under that circumstance a flat interface separating the two fluids is unstable because a small bump of the less viscous fluid in the more viscous one tends to grow(see fig. 1). By the point effect the pressure gradient is larger near the bump than on the flat surface: the pressure field is harmonic($\Delta p=0$) and in the electrostatic analogy the less viscous fluid is a better conductor. As a larger pressure gradient means a larger velocity by Darcy's relation cell. I will limit myself to a sketchy presentation, aiming to give a feeling of the sort of methods used in that field. Notations will be almost the same as in Mclean and Saffman.

As said before, the velocity field (a vector field in the plane) is given in the viscous fluid (i.e. outside of the finger) by Darcy's law:

$\mathbf{u} = -k/\mu \nabla p = \nabla \Phi$

to be supplemented by the incompressibility condition:

$\nabla \cdot \mathbf{u} = \Delta \Phi = 0$, where Φ is the harmonic velocity potential. The boundary conditions(b.c.) are:

- $\partial \Phi/\partial y = 0$ at $y = \pm a$ (no flow across lateral boudaries located at $y = \pm a$, x and y being the Cartesian coordinates in the plane of the cell as shown on fig.2. Notice that these coordinates do not move with the finger)

- $\partial \Phi/\partial n = U \sin\theta$, U being the velocity of the finger parallel to the axis of the cell and θ being such that $U\sin\theta$ is the component of finger velocity normal to the finger surface exactly equal to the same component of the fluid velocity on the finger surface by this last b.c.

- $p - p_0 = T/R$, T capillary constant. This is Laplace's relation for the capillary pressure jump across the finger surface, other contributions being not written out. As the fluid filling the finger is inviscid, the pressure inside this finger is uniform.

- From mass conservation $\Phi \simeq \lambda U x$ as $x \to \infty$, although $\Phi \to 0$ (an arbitrary constant actually) for $x \to -\infty$ and $\lambda a < |y| < a$.

The rest of the calculation will take advantage of the fact that Φ is harmonic and thus the real part of the analytic function $\Phi + i\Psi$, Ψ being the stream function.

Let us turn now to new quantities defined in the frame of reference of the finger and by using dimensionless quantities:

$x' = (x - Ut)/a$, $y' = y/a$, $R' = R/a$, $\Phi' = (\Phi - Ux)/(1-\lambda)Ua$ and

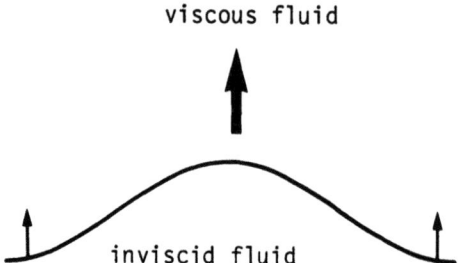

Fig.1. Physical mechanism of instability of a plane front of a less viscous liquid progressing in a more viscous one: arrows give local velocities. Near the bump the velocity grows (point effect in Electrostatics) thus pushing further growth of the bump with time.

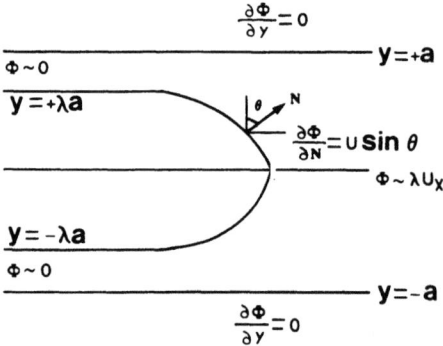

Fig.2. Boundary conditions for the Saffman-Taylor finger in a Hele-Shaw cell with the "physical" dimensions.

the bump grows bigger and bigger. As shown by Saffman and Taylor [2] the growth of this instability in long rectangular Hele-Shaw cells leads to the formation of steadily moving fingers. Below I will study those steady fingers. In the first subsection I will explain how to reduce the computation of the full 2d fluid flow to the solution of a set of two integrodifferential equations as shown by Mclean and Saffman [3]. Surface tension enters into those equations through a single dimensionless parameter denoted as κ.

I will then explain in the next subsection what to do in the large surface tension limit($\kappa \to \infty$); This makes appear a nonuniform expansion with an inner/outer matching that cannot be solved completely explicitly. The general structure of this limit has some resemblance with the well known Landau-Levich solution for the falling film [4].

In the last subsection I will consider the opposite limit (small surface tension or $\kappa \to 0$). This was a challenge for years. Without surface tension($\kappa=0$), Saffman and Taylor [2] have found a continuum of solutions indexed (for instance) by the relative width λ of the finger in the channel($0<\lambda<1$) , although experiments as well as the numerics done later showed that λ tends to 1/2 as κ tends to zero. Regular perturbation doe not explain this non trivial limit behavior. As shown in this last subsection this is related to transcendentally small terms , out of reach of ordinary perturbation theory.

The differential equations of Mclean and Saffman

Below I derive the equations of Mclean and Saffman giving the shape of symmetric steady advancing fingers of inviscid fluid displacing a viscous(Newtonian) fluid in an infinitely long Hele-Shaw $\Psi'=(\Psi-Uy)/(1-\lambda)Ua$.

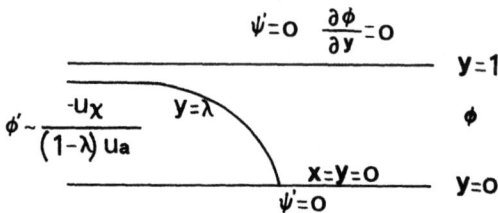

Fig.3. Boundary conditions as in Fig.2. with reduced quantities and in the moving frame of reference.

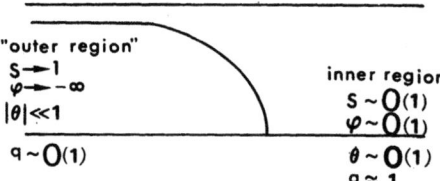

Fig.4. Inner and outer regions in the large surface tension limit. The finger tends to fill almost completely the cell, and a thin layer of viscous liquid remains on the lateral boundaries (outer region).

The corresponding b.c. are explained on fig. 3.

A relation between Φ and Ψ on the finger surface follows from the fact that they are two harmonically conjugate functions. Following the same method as Saffman and Taylor one uses Φ and Ψ as new rectangular coordinates instead of x and y, which defines a conformal change of coordinates. With those new coordinates the b.c. on the finger surface take a simple form. They become even simpler if one maps conformally the $w=\Phi+i\Psi$ plane into the σ-plane such that $\sigma = e^{-(w-\Phi_0)\pi} = s+it$, s and t real. The potential plane (= the region occupied by the viscous fluid) is mapped into the upper half σ-plane and the velocity vanishes everywhere on the real σ-line(t=0) but for 0<s<1, that is the image of the finger surface. The complex quantity $u'-iv'=q\,e^{-i\theta'}$ is an analytic function of σ as well as $-\ln q+i\theta'$ without singularities in the upper half σ-plane so that it is given there by a Cauchy integral calculated (for instance) on the real axis. Computing the limit value of this Cauchy integral on the real axis one gets the following Cauchy-Plemejl condition:

$$\pi \ln q(s) = -\fint_0^1 ds'\,[\theta'(s')-\pi]/(s'-s)$$

where $\fint$ is for Cauchy principal part.

A similar relation would hold along any steady flow line(that is not necessarily the one along the finger surface),because no relation has been used other than $\mathbf{u}=\nabla\Phi$ and $\nabla\cdot\mathbf{u}=0$. Thus it remains to take into account Laplace's condition for the capillary pressure jump on the finger surface.

The dimensionless curvature of the finger reads with the variables introduced before:

$1/R' = d\theta'/dS'$ (S'= arc length along the finger) $= \partial\theta'/\partial\Phi'\,d\Phi'/dS'$
$= q'd\theta'/d\Phi'$.

But $d\theta'/d\Phi' = d\theta'/ds \; ds/d\Phi$ and $ds/d\Phi = -1/\pi s$ on flow lines (i.e. at constant Ψ), as it results from the very definition of σ as a function of $\Phi+i\Psi$. Putting everything together in the pressure condition one gets after some algebra the sought pair of integrodifferential equations:

$$\kappa q s (q s \theta_s)_s = q - \cos\theta \tag{1.a}$$

$$-\pi \ln q(s) = -s \int_0^1 ds' \; \theta(s')/[s'(s'-s)] \tag{1.b}$$

where $\theta = \theta' - \pi$ and $\kappa = kT/[Ua^2(1-\lambda)^2]$.

The b.c. are:

$\theta(0) = q(1) = 0$; $q(0) = 1$ and $\theta(1) = -\pi/2$. Furthermore the relative width of the finger can be computed, once a solution of the above equations is known through the formula:

$$\ln(1-\lambda) = \int_0^1 ds' \; \theta(s')/s' \tag{2}$$

As announced, surface tension enters in the final formulae through the parameter κ only. Furthermore one may check that

$q_0 = (1-s)^{1/2}(1+\alpha s)^{-1/2}$ and $\theta_0 = \cos^{-1} q_0$ is a solution if surface tension is set to zero ($\kappa=0$). That was the one parameter family of solutions found by Saffman and Taylor. The parameter α is related to the width of the finger as $\alpha = (2\lambda-1)(\lambda-1)^{-2}$. As λ is arbitrary between 0 and 1, α may be anywhere between -1 and $+\infty$.

Let us end this with a series of remarks:

(i) This set of equations was solved numerically by Mclean and Saffman [3], by Vandenbroeck [5.a] and by Romero [5.b]. All concluded to the existence of a discrete set of solutions at a given κ and that $\lambda \to 1/2$ as $\kappa \to 0$ for any solution.

(ii) The regular expansion schema worked out by Mclean and Saffman near $\kappa = 0$ cannot explain this quantization of the solutions because as shown in last subsection below this depends on transcendentally small effects.

(iii) The general problem of evolution of the interface between viscous and inviscid fluid has been solved under quite general condition in 2d by B.Shraiman and D.Bensimon [6] but without surface tension. A very important feature of this solution is that it leads to singularities after a finite time. Thus surface tension is needed to prevent these singularities and to continue the evolution. Among all those solutions the steady finger is a special case.

S.Tanveer [7] has extended the above analysis to bubbles of inviscid fluid carried along with the viscous fluid. For that case too the zero surface tension limit is very singular.

(iv) The continuum of solutions found in the absence of surface tension is related to a nontrivial conformal invariance: the b.c. as well as the equations with $\kappa=0$ are invariant under the one parameter group of transformations

$$s \to s(1+\alpha)/(1+\alpha s).$$

As the conformal group in 2d is much larger than in 3d this leads one to speculate that the degeneracies of the 2d problem are absent in 3d and perhaps that there is no need to add surface tension (or like phenomena) to limit the number of possible shapes for steady fingers (advancing-for instance-in a cylindrical porous medium). Note however that another line of reasoning could lead to the opposite conclusion: without surface tension perturbations of a plane front with arbitrarily short wavelength are unstable so that fingers may be seen as resulting from the nonlinear superposition of unstable fluctuations and this does not depend on the dimensionality.

In the next two subsections I shall consider successively the limits $\kappa \to \infty$ and $\kappa \to 0$.

The "large" surface tension limit

This reports a work already published in the Studies of Applied

Mathematics [8]. The limit under consideration is obtained by taking formally $\kappa \to \infty$ in the integrodifferential equations of Mclean and Saffman. As κ is proportional to the capillary constant T, it would seem that this is equivalent to set surface tension equal to infinity. However this is not quite true because at the same time λ tends to 1 and κ is proportional to $(1-\lambda)^{-2}$. Let us consider however for the moment that $\kappa \to \infty$ is equivalent to a very large surface tension. Thus λ should tend to one in this limit because the finger increases as much as possible its radius of curvature to lower the capillary energy. As shown in figure 4 this leads quite naturally to the splitting of the problem into a thin lateral boundary layer remaining on the lateral sides of the cell and into a front part. These two regions will be called below (rather arbitrarily) outer ($s \simeq 0$) and inner (s finite) region respectively. The behaviour of θ and q in the two regions has to be made consistent by an appropriate matching. A general difficulty appears when one wants to study the equations (1.a and b) in their original form in the $\kappa \to \infty$ limit: no scaling of the variable s can absorb the κ dependance because of the formal dilatation invariance of any operator involving s. Thus applying a classical methods, one introduces the new variable $\varphi = \ln s$. In terms of this new variable the equations (1.a and b) become:

$$\pi \ln q(\varphi) = -\fint_{-\infty}^{0} d\varphi' \, \theta(\varphi')(e^{\varphi'-\varphi}-1)^{-1} \qquad (3.a)$$

$$\kappa q(q\theta_\varphi)_\varphi = q - \cos\theta \qquad (3.b)$$

with the b.c. $\theta(-\infty) = q(0) = 1$; $\theta(0) = -\pi/2$; $q(-\infty) = 1$. Notice that here as before the subscript are for derivatives.

Actually it will turn out to be more convenient to use the φ-derivative of the Plemejl-Cauchy condition that reads:

$$q_\varphi/q = -1/[2(e^{-\varphi}-1)] - 1/\pi \fint_{-\infty}^{0} d\varphi' \, \theta_\varphi \, [e^{\varphi'-\varphi}-1]^{-1} \qquad (3.c)$$

The outer region corresponds to s close to zero and thus to φ

very large and negative. From the above integral it appears that in this domain θ has to be small although q is of order 1 by the b.c. This, together with the form of the differential equation (3.b) leads one to propose the following scalings in this outer region: $\varphi^* = \varphi \, \kappa^{-1/3}$, $\theta^* = \theta \, \kappa^{1/3}$, where the starred quantities remain of order 1 in the $\kappa \to \infty$ limit, although, as already said q remains of order 1 in this domain. The integral equation reduces in the outer region to the simple relation:

$$q_{\varphi^*}/q = \theta^*/\pi$$

and the differential equation (3.b) becomes now at large κ:

$$q(q(q_{\varphi^*}/q)_{\varphi^*})_{\varphi^*} = (q-1)/\pi \tag{4}$$

This equation being of third order has three possible behaviors at $\varphi^* \to -\infty$. Only one can be retained because of the b.c. which reduces the number of physically admissible solutions to a one dimensional manifold corresponding actually to the translational invariance of solutions of (4) which is autonomous. But this freedom disappears once the other b.c. $q(0)=0$ is accounted for. Notice also that this equation can be put in a form rather similar to the one found by Landau and Levich for the falling film problem. Finally the θ^* variable is "enslaved" and is uniquely defined once the solution of the third order equation (4) is known. The scaling in the inner region (s and φ finite which corresponds to the tip of the finger) depends in part upon matching conditions with the outer range. This matching takes place near $\varphi^* \to 0$. In this range one has from the above equations for the outer problem:

$$q \simeq (4/3\pi)^2 \kappa^{-1/2} (-\varphi)^{3/2} \tag{5.a}$$

$$\text{and } \theta \simeq 3\pi/2\varphi \tag{5.b}$$

Indeed these relations define the asymptotics of the inner solution. This is consistent with the scaling $q \simeq \kappa^{-1/2} Q(\varphi)$, Q, φ and θ

being of order 1 in this domain. The differential equation becomes:

$Q(Q\theta_\varphi)_\varphi = -\cos\theta$, which yields the pendulum equation by introducing the variable Ω through the implicit change of variables $d\varphi/Q(\varphi)=d\Omega$:

$\theta_{\Omega\Omega} + \cos\theta = 0$.

This pendulum equation needs one constant of integration to be solved with the proper b.c. It appears that this constant (T.Dombre and V.Hakim, private communication) is fixed by matching at the next order with the outer solution. It remains to relate κ to the physically meaningful parameter λ. This is done by calculating the integral occuring in (2). The outer and inner contributions to this integral diverge separately but those divergences cancel out once the matching region is properly treated and one gets the following expansion for λ as a function of κ:

$\lambda \simeq 1 - b\,\kappa^{-1/2} + ...$

where b is specified by the higher order matching and is finite (T.Dombre and V.Hakim, private communication). Coming back now to the definition of κ one gets the result that KT/U is bounded from above to allow the existence of fingers. If U falls below some value proportional to surface tension the flat interface is stable and no steady finger may exist. A similar conclusion has been also reached by Boris Shraiman by a direct study of the problem in its original formulation(i.e. without reduction to integrodifferential equations posed on the finger surface).

Other solutions with mutiple bumps in the finger front may also exist (T.Dombre and V.Hakim, private communication) corresponding to more and more oscillations in the pendulum equation.

Fingers in the low surface tension limit ($\kappa \to 0$)

As already said this limit was a challenge because regular expansion near $\kappa=0$, starting from the continuum of explicitely known solutions at $\kappa=0$ does not show the quantization of those solutions and the simple property that λ tends to $1/2$ as κ tends to zero. Concerning this point we refer the interested reader to reference [3] as well as to Mclean's thesis (Caltech 1982). It appears there that the regular expansion poses by itself nontrivial questions. Below I will give a general idea of the manner in which one can perform this sort of expansion beyond any algebraic order [9]. The method we shall follow is basically the one exposed by Kruskal and Segur [10] in a geometric model for needle crystals and not every details of this will be explained.

The first important remark is that the Saffman-Taylor solution (q_0, θ_0) has a square root singularity located in the complex s-plane (i.e. in the σ-plane but we shall keep the notation s for real or complex variable) at $s=-1/\alpha$ which is not in the physical domain. As the small expansion parameter κ is in front of the highest derivative, the succesive terms of the expansion in powers of κ become more and more singular at increasing order near this singularity. On the other hand there is also the possibility that high frequency perturbations exist because again the small paramater κ is in front of the highest derivative.

These two remarks are at the basis of the method of Kruskal and Segur. Then the calculation proceeds as follows: first one transforms the equations in order to have the analytic continuation of their solution everywhere in the complex s-plane (actually another variable than s will be used), this will allow to reach the vicinity of the singularity at $s=-1/\alpha$. By appropriate scalings the behavior of the solution nearby this singularity will be reduced to a parameterless inner differential equation. Far from this singularity a WKB like analysis will show that beyond the regular perturbation they are also transcendentally small terms. The interest in those terms comes from the fact that they do not satisfy the boundary condition unless some supplementary relation is satisfied. This will lead to a quantization condition, unreachable by the regular perturbation analysis because this one satisfies the b.c. order by order. Finally the inner-outer matching near the singularity will allow to fix some prefactor of the WKB terms and the b.c. will impose that the imaginary part of this prefactor cancels, which is the sought solvability condition. As this is given by a nonlinear equation the prefactor cannot be computed in a closed form as a function of the only remaining free quantity that is α. However a second WKB expansion in the inner problem will allow to reach the "highly excited" modes in the set of possible shapes for the fingers. Owing to the relative complication of the whole analysis the most salient features only will be exhibited.

We shall do first the change of variable from s to r with $s = 1/ch^2 r$, the physical domain (i.e. the finger surface) corresponding to r real negative. With these new variables the b.c. at the tip is that $\theta(r)+\pi/2$ and $q(r)$ have to be odd functions of r and all the calculation will consist in trying to cancel the _even_ WKB contributions to those functions.

As announced one has to write first the complex extension of the equations of Mclean and Saffman. The Cauchy-Plemejl condition becomes:

$$\ln q(s) = -s/\pi \int_0^1 ds' \theta(s')/[s'(s'-s)] + i\theta(s)$$

As the integral is done on the real axis, the result remains smooth nearby the singularity. Let us denote it as $I(r)$. One has

$$\ln q(r) = I(r) + i\theta(r),$$

The functions $q(r)$ and $\theta(r)$ are singular near $r = i\pi + w_0$, with w_0 solution of $sh(w_0/2) = \alpha^{1/2}$ this being valid near $\lambda = 1/2$ or $\alpha = 0$, which is the domain of parameters we want to study. Let us derive now the equations for the inner problem in the r-variable. From the equation for w_0 one sees two different cases for the disposition of the singularities of the basic solution:

–If $\alpha > 0$ (equivalent to $\lambda > 1/2$) there is a pair of singularities at the same distance from the real axis and symmetrical with respect to $i\pi$.

–If α is negative (or λ less than $1/2$) the two singularities are on the imaginary axis at unequal distances from the real physical axis.

In the first case the two WKB contributions may cancel each other at the origin by a sort of destructive interference effect although this is no more possible in the second case. This explains why solutions exist at $\lambda > 1/2$ only. From the Cauchy-Plemejl relation one has near the singularity:

$$q \simeq q_0 e^{i(\theta - \theta_0)},$$

Inserting this into the differential equation and keeping the dominant terms one gets:

$$ik\omega(\omega q_\omega)_\omega = \alpha - \omega^2/4 + q^{-2} \qquad (6)$$

where $\omega = r - i\pi$ and where α is supposed to be small as well as ω. Thus it is convenient to scale everything in (6) to have an inner problem without any small or large quantity in it. This is achieved by scaling q as $\alpha^{-1/2}$, ω as $\alpha^{1/2}$ and κ as $\alpha^{3/2}$. Introducing furthemore A by $A = |\alpha|^{3/2}/\kappa$ one gets the sought inner equation (written with the same functions and variables as before to avoid proliferation of notations):

$$i\omega(\omega q_\omega)_\omega = A(\varepsilon - \omega^2 + 1/q^2) \qquad (7)$$

where ε is the sign of α. This is an equation posed in the complex plane and it needs some b.c. It has to match the Saffman-Taylor solution for ω with a large negative real part: $q \simeq -1/\omega$ as $\mathrm{Re}\,\omega \to -\infty$. The other b.c. comes from the condition that there is no transcendentally <u>large</u> contribution to $\theta(0)$ and $q(0)$ because the WKB analysis on the imaginary axis gives a priori two exponential factors: one is very large when computed at $r=0$ and is to be excluded by this b.c. for the inner problem although the other one is always there and will be considered below.

Following Kruskal and Segur let us look at the (possible) WKB terms in the solution of the inner equation. They are found by linearizing (7) around its dominant part at large ω that is by putting $q \simeq -1/\omega + q_1$ so that q_1 satisfies the homogeneous equation:

$$i\omega(\omega q_{1,\omega})_\omega = 2A\omega^3 q_1.$$

On the imaginary axis and for large ω's, q_1 is given by a WKB like expression:

$$q_1 \simeq C(A)(i\omega)^{-3/4} \exp[-\int^\omega (-2iA\omega')^{1/2} d\omega'] \qquad (8)$$

Indeed they are two possible choices of sign for the argument of the exponential in (8). One of them is excluded, as already said by the condition that no transcendentally large contribution may appear in the physical domain. This is checked by matching the WKB term written above with the one found in the outer region that includes the physical range too. Finally let us notice that the constant C(A) that appears in (8) depends on some knowledge of the nonlinearities in the inner region. One of the b.c. is that q is an odd function, which is equivalent to say that it is purely imaginary (if analytic) on the imaginary axis. This implies from the above formula that C(A) must be purely imaginary(recall that iω is real there), which is the sought quantization condition. From the direct numerical studies of Vandenbroeck [5.a] we know that there is an infinite set of values of A such that this quantization condition is satisfied. This leads one to think that the large "quantum" numbers in this set may be analysed by some version of the quasiclassical method.

These large quantum numbers are reached for A large (notice again that A large in the inner equation means that the factor of the highest derivative is small). So we want to solve (7) for A large and $\varepsilon=1$. Outside of the vicinity of the singularities at $\omega=\pm 1$, the solution of (7) is the superposition of a sum of terms obtained by a regular expansion in inverse powers of A and starting as $(1-\omega^2)^{-1/2}$ and of WKB contributions. This splitting is no more valid near the singularities at $\omega =+$ or -1 wherein one has to solve a new inner equation. This one follows from the remark that by putting $\omega=\pm 1+\eta$, the new variable η should be small in these two inner regions. Neglecting highest powers of η in (7) one has immediately:

$$iq_{\eta\eta} = A(2\eta + 1/q^2) \qquad (9)$$

where now the large parameter A may be eliminated by the scalings $q \to qA^{-1/7}$; $\eta \to \eta A^{+2/7}$, which yields a new parameterless inner equation formally identical to (9) but with A=1. Recall that this gives the behavior of q near the singularities at $\omega = \pm 1$. The WKB term written in equation (8) has to match the asymptotics of a WKB term in the asymptotics of the solution of (9). This is equivalent to say that C(A) may be taken as equal to the WKB part of the solution of (7), when computed along the horizontal path from $\omega = \pm 1$ to 0 [which fixes the lower bound in the integral (8)] because one has such a rapidly varying solution of (7) in the large A limit.

As the large parameter in (9) may be completely absorbed into the scalings at large A, this inner equation is relevant only for fixing a pure number in front of C(A) so that this function is proportional to

$$\exp \int_0^{\pm 1} d\omega' (-2iA)^{1/2} (\omega'^2 - 1)^{3/4} / \omega'.$$ Accounting for the superposition of the contributions of the two singularities at $\omega = \pm 1$ one gets that the phase of C(A) is the sum of a constant [depending as explained from the inner problem (9)] plus $\pi(A/2)^{1/2}$ which results from the phase of the exponential just written. As said before C(A) must be purely imaginary by the "solvability" condition, which is realised (at large A) if $A \simeq 2(N + \text{const}/\pi)^2$, where N is a (large) natural integer. This last condition is well verified even by the first few modes (N not so large) found numerically. This concludes our analysis of the Saffman-Taylor problem with surface tension.

STEADY NEEDLE CRYSTALS IN 2D

This section is mainly devoted to the derivation and study of the integral equation for the shape of a needle crystal of solid growing in an undercooled melt. Many simplifying assumptions will be made, some of them could be easily avoided and some not. More details on this may be found in the review article of J.Langer [11]. I shall assume that the growth is limited by the molecular diffusion of the latent heat generated by the thermodynamic phase change and I will also assume that the solid and its melt have the same heat diffusion coefficients (symmetrical model). This approximation becomes invalid if the growth is limited by impurity diffusion because the diffusion coefficient in the solid is negligible in that case. Moreover this neglects convection phenomena in the fluid although they are often important in real experiments. Moreover I shall restrict myself to 2 dimensional situations and look for solutions that are steady in a moving frame. It appears likely that they are unstable, although this instability may grow on the almost flat parts of the needle crystal (sidebranching) without changing too much the front part that will be studied here.

Thus the starting point will be the integrodifferential equation of Nash and Glicksman [12] giving the shape of a needle crystal growing at constant velocity. This equation depends on two dimensionless parameters (the undercooling Δ and the velocity u) but it is not obvious to understand the sort of relationship between those two numbers that follows from this. Looking at the small undercooling limit [13] one may get a more simple integral equation with a single parameter and it may be shown that this parameter is a sort of nonlinear eigenvalue and can thus have discrete values only.

This has been confirmed by some recent numerical studies [14] which have shown that this discrete set is actually empty and that steady solutions exist with anisotropy in the surface tension only.

Thus our starting point is the following integrodifferential equation [12]:

$$\Delta + K = u \int_0^\infty d\tau/(4\pi\tau) \int_R dx_1 \exp\{-[(x-x_1)^2 + (\xi - \xi_1 + u\tau)^2]/4\tau\} \quad (10)$$

where :

- Δ is the dimensionless undercooling measured in such a way that $\Delta=0$ at equilibrium and $\Delta=1$ if a flat interface between the solid and its melt moves at constant velocity, the generation of latent heat being just enough to heat the melt at the coexistence temperature.

- $\xi(x)$ is the Cartesian representation of the needle surface in its moving frame, the coordinates being such that the growth velocity is directed toward ξ positive and that x is perpendicular to this velocity.

- the integral on the r.h.s. is done on the whole real axis and ξ_1 is for $\xi(x_1)$.

- K is the curvature of the interface and is a function of x equals to $\xi_{xx}(1+\xi_x^2)^{-3/2}$.

- the velocity unit is D/d_0, d_0 being the capillary length proportional to surface tension and D the heat diffusion coefficient. Indeed the length unit is d_0.

Let us comment now upon the physical meaning of the various terms in equation (10). This equation may be seen as expressing the fact that on the interface the temperature is equal to the one given by the Gibbs-Thomson condition. This temperature field results from the addition of the undercooling Δ and of the perturbation generated by the latent heat as expressed by the integral on the r.h.s. One recognizes in the argument of this integral the Green's function of the diffusion kernel.

Indeed one may think of a numerical solution of (10). This has been the subject of recent investigations [ref.14, and Ben Amar's talk at this conference] and is by itself a nontrivial problem. Below I will insist upon an analytical approach. This will consist into a derivation of the limit form of (10) when Δ becomes very small, a rather realistic assumption for many experimental situations. The starting point will be the exact Ivantsov solution found by setting to zero the curvature term K in (10).

As shown in [13] this Ivantsov solution gives a parabola $\xi = -ax^2$, a>0 and related to the velocity u and the undercooling Δ by:

$$\Delta = u/\sqrt{\pi} \int_0^\infty dx\ (ax^2+u)^{-1} \exp{-x^2/4} \tag{11}$$

that becomes for Δ small:

$$u/a \simeq 4\Delta^2/\pi \tag{12}$$

As often noticed this Ivantsov solution does not fix the growth velocity of the needle crystal. It relates the curvature of the parabola at the tip to this velocity only. This is because there is not enough dimensionalizing parameters in the integral equation (10) if one omits the curvature term, or said in another fashion, without this term this equation is invariant under the transformation $u \to \lambda u$, $(x,\xi) \to (x/\lambda, \xi/\lambda)$, λ arbitrary. As this invariance is absent if the Gibbs-Thomson effect is included it is tempting to conclude that u is related in an unique fashion to Δ if (10) is satisfied, as it is written. However this requires to be checked.

Scaling the time as u in (10) and performing the τ-integration one obtains:

$$\Delta + \xi_{xx}(1+\xi_x^2)^{-3/2} = u/2\pi \int_R dx_1 \, F(xu,\xi u; x_1 u, \xi_1 u) \quad (13),$$

F being the function:

$$F(xu,\xi u; x_1 u, \xi_1 u) = \exp{-u(\xi-\xi_1)/2} \, K_0(u|\mathbf{r}-\mathbf{r}_1|/2),$$

where K_0 is the modified Bessel function of zero order and where

$|\mathbf{r}-\mathbf{r}_1| = [(x-x_1)^2 + (\xi-\xi_1)^2]^{1/2}$. Later on the argument of the function F will be abbreviated as xu to emphasize mainly the scaling dependance of F. Let us assume now that the Ivantsov relation (12) is still valid, at least dimensionnaly with surface tension included. This implies that quantities like u times some length, as some occur in the argument of F are small because they scale as u/a or as Δ^2 from (13) at small Δ. This gives the idea of simply taking the limit of small arguments for F in (13) to study this limit $\Delta \to 0$. But this does not work so simply since the result is a massively diverging integral on the r.h.s. of (11) because of its large distance behavior. This is basically because the Ivantsov relation (13) cannot be found by a simple scaling on the two sides of (10). In a sense the very problem of getting the small Δ limit of (10) is the same with or without the Gibbs-Thomson term.

But the long distance divergence found by taking without care the limit of small argument in F comes from the part of the needle crystal that is almost flat and thus where the curvature term becomes negligible. This gave to Pierre Pelcé and myself the idea of the following trick: suppose that u is known for a given Δ, thus there is a well definite Ivantsov parabola with the same paramaters (i.e. Δ

and u). For this Ivantsov parabola one has an identity derived from (13) without the curvature term:

$$\Delta = u/2\pi \int_R dx_1 \, F(xu, \xi^{Iv} u; x_1 u, \xi_1^{Iv} u),$$ where ξ^{Iv} is for the Ivantsov solution. Inserting this relation into the complete equation (13) one gets:

$$\xi_{xx}(1+\xi_x^2)^{-3/2} = u/2\pi \int_R dx_1 \, [F(xu, \xi u; x_1 u, \xi_1 u) - F(xu, \xi^{Iv} u; x_1 u, \xi_1^{Iv} u)]$$

so that in the new integral on the r.h.s. the large distance behaviour leads to a cancellation between the actual contribution of the needle crystal and the one of the Ivantsov parabola. As said before those two contributions cancel because the curvature becomes small far from the tip and the needle crystal tends to the Ivantsov parabola.

Performing the expansion of F for small arguments one gets the sought integral equation:

$$\xi_{xx}(1+\xi_x^2)^{-3/2} = -y/4\pi \int_R dx_1 \ln\{[(x-x_1)^2+(\xi-\xi_1)^2] / [(x-x_1)^2+(\xi^{Iv}-\xi^{Iv}_1)^2]\}, \quad (14)$$

In this last equation the quantity y is defined as $y=ua^{-2}$ although u and a are also related by the Ivantsov relation (11) and (12). So once y is known any physical parameter is defined. The equation (14) has an interesting mathematical structure because y is a nonlinear eigenvalue in the sense of Zeldovich. In differential equations eigenvalues appears whenever they are more b.c. than the order of the equation so that the only possibility to satisfy those b.c. is to fix the value of parameters in the equations. In the present case things are less simple because (14) is an integral equation and the number of free parameters in it is not obvious. Actually a careful analysis of the asymptotics of solutions of (14) shows that y is a nonlinear eigenvalue. Indeed this does not solve the problem completely. If one thinks of quantum bound states in a 1D potential for instance, no such bound state exists if the potential is everywhere positive and tends to zero at infinity, a fact not revealed at all by the asymptotics.

So let us sketch the analysis of the asymptotics of (14). We know that in the limit $x \to \infty$, ξ tends to the Ivantsov parabola: $\xi(x) \simeq -x^2 + \zeta(x)$, with $\zeta(x) \to 0$ as x tends to infinity. Let us try to estimate the behavior of the integral on the r.h.s. of (14) in this limit. If the integration variable is of the same order as x, the argument of the logarithm differs from 1 by a small quantity of order $\zeta(x)$. If this integration variable is of order 1 (i.e. if one considers the tip contribution to the large distance behavior of the final result), then any quantity with subscript 1 in the argument of the logarithm is small compared to the same quantity without subscript and the argument of the logarithm is still close to 1. Finally at large x one may replace the logarithm in (14) by the first term of its Taylor expansion near its argument 1:

$$\ln\{ [(x-x_1)^2 + (\xi-\xi_1)^2] / [(x-x_1)^2 + (\xi^{IV} - \xi^{IV}_1)^2] \} \simeq \Lambda(x, x_1),$$

with $\Lambda(x, x_1) = 2(x+x_1)^2 (\zeta - \zeta_1) / [(x^2 - x_1^2)(1 + (x+x_1)^2)]$.

Due to the denominator in this last expression the integral is dominated at large x by x_1 of order 1 (i.e. by the tip contribution). This gives:

$$\int_R dx_1 \, \Lambda(x, x_1) \simeq 2/x^2 \int_R dx_1 \, \zeta(x_1) .$$

If the integral on the r.h.s. converges at large x_1 as it has to be verified a posteriori, this $1/x^2$ dependance has to match a similar one coming from the curvature term. Far from the tip the needle crystal is almost parabolic so that the curvature there is simply the one of the parabola and it tends to zero as $1/x^3$ as it can be verified. Thus this cannot match the $1/x^2$ behavior of the r.h.s. unless the coefficient of this last term vanishes exactly. This gives the following condition:

$$\int_R dx_1 \, \zeta(x_1) = \int_R dx_1 \, [\xi(x_1) + x_1^2] = 0.$$

This can be seen as a sort of quantization condition as the one imposed to the bound states not to grow exponentially at large distance. The study of the next order term in the asymptotics of $\xi(x)$ is rather intricate and is done in ref.[13]. The final result is:

$$\xi(x) \simeq -x^2 + \ln|x|/[\pi y x^2] + ...$$

The next order terms may be computed order by order by matching the curvature and integral contributions in (14). This expansion defines a function with one degree of freedom (the parameter y) and this one should thus be fixed by the "quantization condition" derived before. This leads to the conclusion-basically correct- that y is quantized.

As said before numerical as well as other analytic studies have shown that with the present formulation there is no possible y, although an infinite discrete set of solutions is found when one adds the effect of an anisotropic surface tension in the Gibbs-Thomson condition which does not change the main points of the analysis presented here.

CONCLUSION

The formation of nonequilibrium patterns via interfacial instabilities gives a rather unique chance to carry the fully nonlinear analysis of these instabilities. It is thus noticeable that in the few cases presented here where this can be done nontrivial mathematics are involved. A major problem for future investigations is the inclusion in this picture of more complicated time dependance than implied by the assumption of steadiness in a moving frame .

ACKNOWLEDGEMENTS

This exposes a personal view of works done in collaboration with P.Pelcé, R.Combescot, T.Dombre, V.Hakim and A. Pumir. I have also benefited from conversations with M.Ben Amar and P.Clavin.

REFERENCES

[1] D.Bensimon,L.P.Kadanoff,S.Liang,B.Shraiman and C.Tang to appear in Rev. of Mod. Phys.

[2] P.G.Saffman and G.I.Taylor; Proc. Roy. Soc. A245,312 (1958)

[3] J.W.Mclean and P.G.Saffman; J. Fluid Mech. 102, 455 (1980)

[4] L.D.Landau and B.V.Levich; Acta Physicochem. USSR 17,42 (1942)

[5.a] J.M.Vandenbroeck; Phys. of Fluids 26, 2033 (1983)

[5.b] L.Romero,; PhD. Dissertation, Dpt of Math , Caltech (1982)

[6] B.Shraiman and D.Bensimon; Phys. Rev. A30, 2840 (1984)

[7] S.Tanveer "New solutions for steady bubbles in a Hele-Shaw cell", Caltech Preprint (July 1986)

[8] Y.Pomeau; Studies in Applied Math. 73, 75 (1985)

[9] T.Dombre, V.Hakim and Y.Pomeau; CRAS 302, 803 (1986); D.C. Hong and J.Langer; Phys.Rev. Lett. 56,2032 (1986); B.Shraiman; Phys.Rev. Lett.

56, 2028 (1986); R.Combescot, T.Dombre, V.Hakim Y.Pomeau and A.Pumir; Phys.Rev. Lett. 56, 2036 (1986)

[10] M.Kruskal and H. Segur; Aeronautical Res. Associates of Princeton Tech. Memo 85-25 (1985)

[11] J.Langer; Rev. of Mod. Phys. 52, 1 (1980)

[12] G.E.Nash and M.Glicksman; Acta Metall. 22, 1283 (1974)

[13] P.Pelcé and Y.Pomeau; Studies in Applied Math. 74, 245 (1986)

[14] D.Meiron; Phys.Rev. A33, 2704 (1986) ; D.Kessler, J.Koplik and H. Levine; Phys. Rev. A33, 3352 (1986).

AN EXPERIMENTAL STUDY OF THE SAFFMAN TAYLOR INSTABILITY

P.Tabeling

Groupe de Physique des Solides
24, rue Lhomond, 75005, Paris (France)

G.Zocchi and A.Libchaber

The James Franck Institute and E.Fermi Institute
University of Chicago 5640 S.Ellis Avenue
Chicago 60615 (U.S.A)

INTRODUCTION

When a gas is pushed into a viscous fluid through a Hele-Shaw cell, one observes, after a long transient, the formation of a finger which propagates steadily along the channel. This simple situation defines the Saffman Taylor[1] problem. In spite of its simplicity, the limiting sizes of those fingers at large velocities and their stability has not been understood for a long time. A particularly striking fact observed in the experiments is that fingers occupy about one-half of the channel width at large velocities. In the model of Saffman and Taylor[1], where surface tension is neglected, all finger sizes are allowed. Moreover, fingers of size 1/2 are found linearly unstable, which is also in direct conflict with experiment. The problem was then to understand how surface tension acts to select finger sizes and to ensure stability.
Answers to these questions came only recently [2,3]. Actually, all these studies assume the system to be two-dimensionnal. Experimentally, the interface is a three-dimensionnal meniscus which drains a film as it moves. The purpose of this paper is to describe an experimental study of this film, performed recently at the University of Chicago. We will see that the film modifies the sizes of the stationary fingers predicted by theory. We also study the stability of the fingers.

THE EXPERIMENTAL ARRANGEMENT

The Hele-Shaw cell consists of two long rectangular glass plates separated by two aluminium spacers which define the walls of the channel. The plates are 120 cm long, 10 cm wide and 1.27 cm thick. The channel terminates by two plexiglass pieces in which a large cavity has been machined. Those pieces are connected to a hydraulic circuit, which in turn is connected to a syphon. For large velocities (a few cm/s), we use a magnetic pump, driven by a D.C. motor, to pull the viscous fluid. In the low velocity range (< 1 cm/s), the viscous fluid is pulled by the action of gravitational forces. In both cases, the regulation of the velocity is better than 1%. The cell has a very uniform gap, as shown by interferometric measurements. The nonuniformities are located near the extremities of the cell and do not exceed 10^{-3} cm there. In the central part of the cell, the gap is uniform within $5 \: 10^{-4}$ cm.

Oils are chosen as the viscous fluid in most of the experiments because they wet glass. Experiments carried out with water showed that the advancing front is dramatically perturbed by droplets generated along the interface. When wetting fluids are used, the system is well defined. Various oils are used : vacuum pump oils, lubrificating oils and silicon oils. Their viscosity and surface tension are measured by classical methods.

Several cells are studied. The gap between the plates and the width of the channel are easily varied by changing the size and the location of the spacers. Three sets of spacers are used, corresponding to gaps of 800, 480 and 173 µm. It finally appeared that the quality of the experiment depends strongly on first avoiding pinning effects at the walls, then having a constant velocity all along the channel and a very uniform cell with regard to its depth.

3 - STUDY OF STATIONARY FINGERS

When the interface starts moving, small corrugations grow, compete together and finally we obtain a single finger which occupies a significant part of the channel and propagates steadily along it. An example of a steady finger is shown in Picture 1. The dark line which defines the interface is a meniscus of oil which scatters the light coming from below. Behind the tip, the interface is parallel to the fingers. One can obtain very long fingers, almost as long as the channel itself.

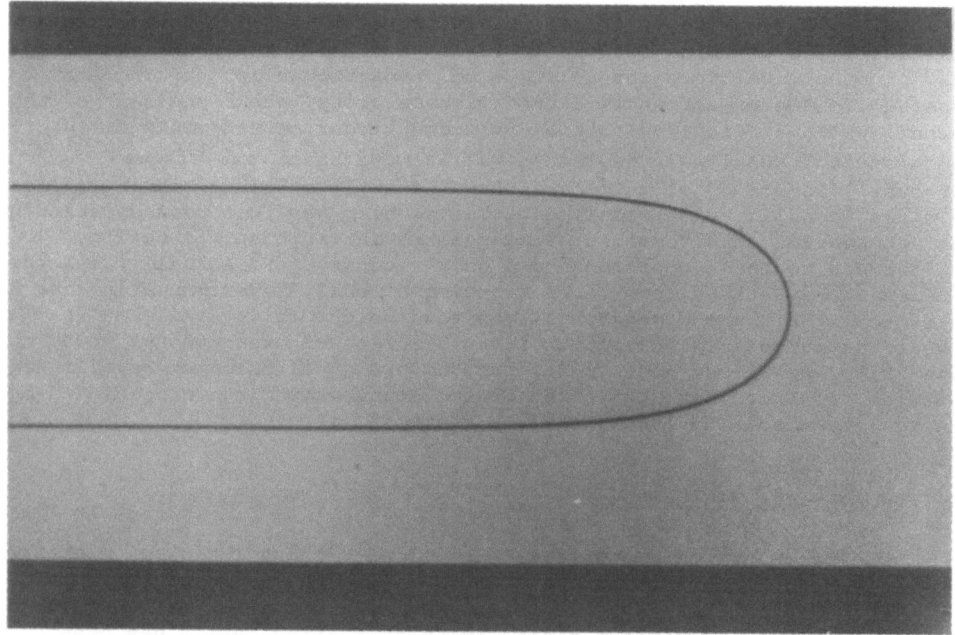

Picture 1. Stationary finger for $1/B \sim 5000$, with b=800 µm.

Fingers are defined by their velocity U, their relative size λ (compared to the width of the channel), and indeed their shape. In order to describe the various regimes of the system, we use dimensionless number[4]

$$1/B = 12 \ (w/b)^2 \mu U/T \ ,$$

where μ is the fluid viscosity, and T its surface tension against air.

In the very low velocity regime ($1/B < 50$), the finger shape is semi-circular (see Picture 2). There is a continuity between the finger at rest and the one moving very slowly since at zero velocity, the shape is also semi-circular. The relation between $1-\lambda$ and $1/B$ is linear in this regime; we observe a universal law

$$1-\lambda = .011/B,$$

which is valid at any value of the aspect ratio w/b, in the range $4 < 1/B < 50$. Smaller values of $1/B$ have not been studied.

The main results obtained at larger values of $1/B$ ($0 < 1/B < 250$) are summarized in Figure 1. Experiments are performed with a large aspect ratio w/b = 65, with two different oils, and the results are compared with those of Saffman and Taylor[1], obtained with an aspect ratio two times smaller. We obtain a result previously shown by Saffman and Taylor : for a given aspect ratio, the experimental results are universally described by a capillary number, such as $1/B$. However, each set of points lie on a curve which depends on the value of the aspect ratio. If we now look at the theoretical results of Mac Lean and Saffman[5], we find that they are in disagreement with the two experiments (see the continuous line on Fig.3). It is thus reasonable to suspect that another effect, not taken into account by theory, is involved in the experiments. The additionnal parameter of the experiment is related to the film drained by the meniscus(see below).

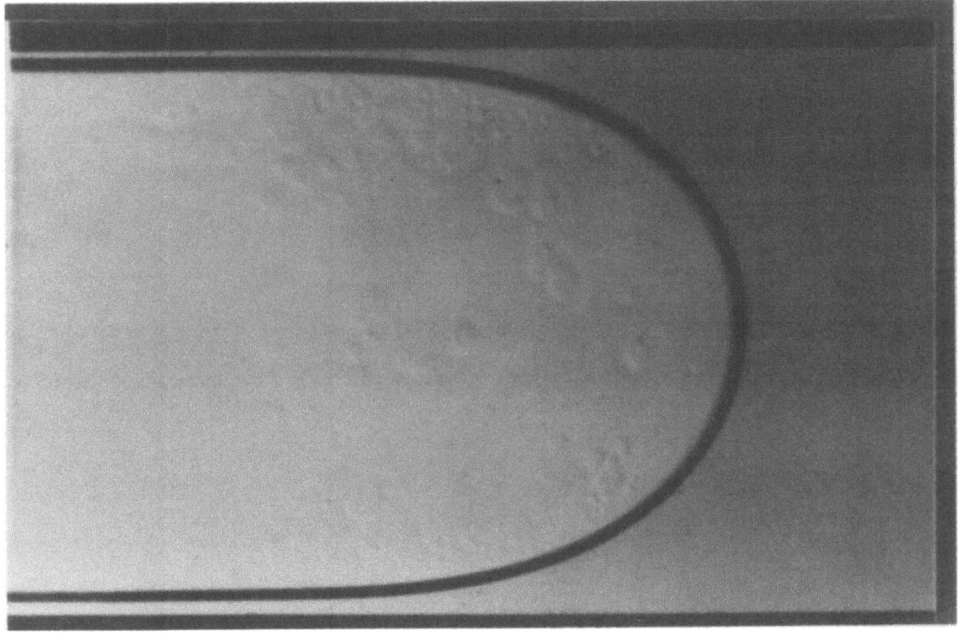

Picture 2. Finger in the very low velocity regime.

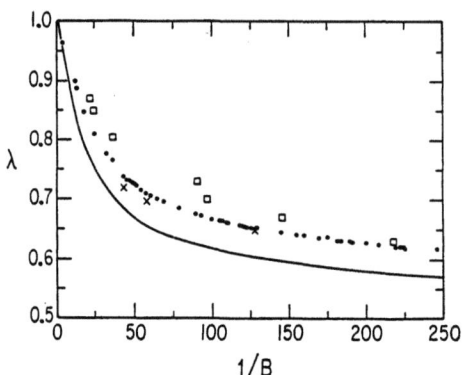

Figure 1. Relative size of the finger as a function of 1/B ; □ Saffman-Taylor[1] experiment with w/b = 31.8 ; ● and ✕ experiments with w/b = 65, with two Rhodorsil oils of different viscosities (μ = 98 cp and 480 cp) ; —— numerical results of Mc Lean and Saffman[2].

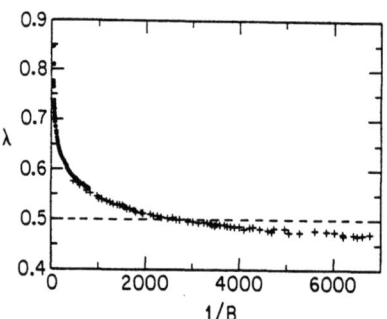

Figure 2. Variations of the finger sizes with 1/B for two oils (μ=98 cp and 480 cp) with w/b = 65.

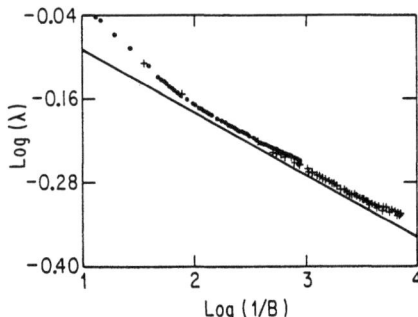

Figure 3. The same curve as in Fig.2 but plotted on a log-log scale.

The results obtained in the range of large values of 1/B are plotted on Fig.2. All the curves go below the value $\lambda=1/2$ as 1/B is increased.

On a log-log scale, we obtain that a power law of the form
$$\lambda \sim (1/B)^{-.08}$$
roughly fits the experimental data at large values of 1/B (See Fig.3).

These results are not in disagreement with those of Saffman and Taylor[1], when the experimental uncertainty is taken into account. In their case, the uncertainty is 5%, which allow values of λ smaller than 1/2, by an amount comparable with the one which we have observed for the same values of 1/B. The absence of plateau is in disagreement with the numerical results of McLean and Saffman[5] and also with more recent analytical work[2]. A plausible reason for such disagreements lies in the existence of the film drained by the meniscus of oil.

4 - STUDY OF THE FILM DRAINED BEHIND THE FINGER

The interface between air and oil is a meniscus which lives in a three-dimensionnal space. When the interface moves, the meniscus produces a film behind. The problem is analogous to the Landau-Levich[6] film drained by a vertical plate pulled out of a liquid. The transposition of this problem to the case of a meniscus moving through a Hele-Shaw cell has been performed by Bretherton[7]. According to him, the film thickness t follows a 2/3 power law in the form :
$$2 t/b = 0.643 \ (3 \ Ca)^{2/3},$$
where $Ca = \mu U/T$ is the capillary number. This law also applies for the case of curved interfaces, provided that the interface curvature is small compared to b. U must then be interpreted as the local velocity of the interface ; now, in the Saffman Taylor problem, since the motion along the interface is essentially directed normally, U can be replaced by the normal component U_n of the velocity at the interface.

The problem is that U_n is not constant along the interface ; then, the film thickness is not uniform and the finger sees a channel of variable gap. Intuitively, one feels that the problem may be strongly affected by the presence of the film.

The method for studying the film consists in bringing an expanded He-Ne laser beam almost normally to the cell and looking at the interferences fringes in the reflected light. Figure 4 shows the principle of the measurement. Since the index of oil and glass are similar, the

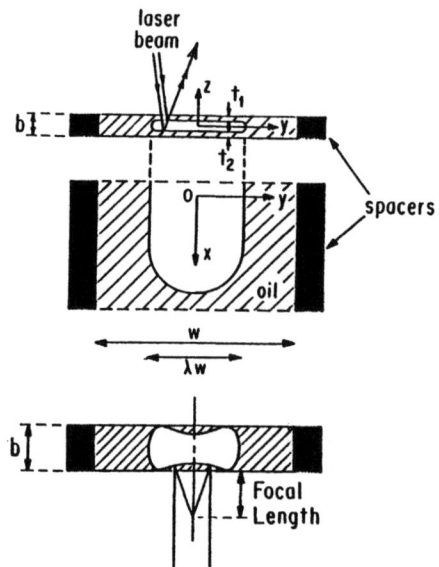

Figure 4. Principle of measurement of the film thickness (two methods).

reflection at the interface oil/glass is small. The reflection is important only at the oil/air interfaces of the top and bottom films. Thus between two consecutive fringes the gap between the two films has varied by a distance 1 (1 = wavelenght of the light used); assuming that the two films are alike, and the film thickness is zero at the sides of the finger, we finally deduce the absolute thickness t everywhere behind the finger. This method can be used when the film thickness is not large compared to 1. This condition is well satisfied when 1/B is smaller than 250. For larger values of this parameter, we use another method, which exploits the fact that the film can be used as a convergent mirror. By measuring its focal lenght, and assuming a particular shape for the film, the maximum thickness t_{max} is determined.

In the low velocity regime (1/B < 250), interferometric method is used and t can be plotted in function of the lateral coordinate y on Figure 5, for different values of the finger velocity. All the curves are self-similar and they fit very well a law of the form :

$$t = t_{max} \cos^{2/3}(\pi y/\lambda w),$$

where t_{max} depends on the finger velocity. Another expression for t can be obtained by using the normal velocity at the interface as the variable; We find :

$$t = t_{max} (U_n/U)^{2/3},$$

which is the local form of Bretherton law. Moreover, the dependance of t_{max} with the capillary number Ca is found in good agreement with Bretherton 2/3 power law for capillary numbers up to 10^{-2}.

When the velocity is increased, the film thickness tends to saturate at a value which depends on the aspect ratio w/b (see Figure 6). An experimental law in the form :

$$t_{max} = \kappa b \ (1-e^{-\gamma w/b})(1-e^{-\beta Ca^{2/3}}),$$

with $\kappa=.119$, $\gamma=.0038$, $\beta=8.58$ fits all the data. This law reduces to Bretherton law at small values of the capillary number.

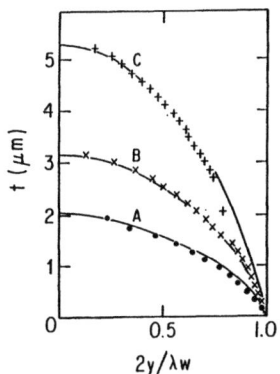

Figure 5. Variation of the film thickness with coordinate y, for three values of the velocity, with w/b = 65 ; A : U = .10 cm/s, B : U = .13 cm/s ; C : U = .265 cm/s.

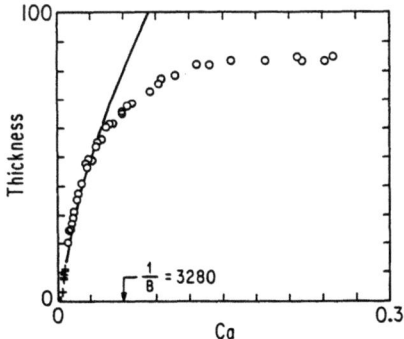

Figure 6. Variation of the film thickness with capillary number Ca.

5 - THE INFLUENCE OF THE FILM ON THE FINGER SIZE

The influence of the film on the finger size can be estimated by using a result of Park and Homsy[8], giving the expression for the pressure drop δP across the interface when a film is drained by the meniscus. Their result is

$$\delta P = T(\pi/4R + 7.80/b \ (\mu U_n/T)^{2/3}),$$

where R is the local radius of the finger. This expression can be used to define an effective surface tension T^* which roughly takes the effect of the film into account. We then define T^* by the expression :

$$T^* = T(\pi/4 + \alpha \lambda w/b \ (\mu U/T)^{2/3}),$$

in which α is a coefficient resultant from the average of $RU_n^{2/3}$ along the interface. With this T^*, we plot the experimental data and those of Saffman and Taylor. A good fit between theory and experiment is obtained[9] with $\alpha = 1.7$ (see Fig. 7). This result thus indicates that the previous disagreement between theory and experiment is mainly due to the existence of the film.

6- THE STABILITY OF THE FINGERS

When the finger velocity is increased, the interface becomes unstable in the way shown in Picture 3. A disturbance appears on one side of the finger, grows, recede backwards and further decreases. In a frame of reference moving with the finger, the disturbance looks like a wave-packet which travels backwards while in the laboratory frame, this packet looks stationary. The perturbations are stretched as they move along the interface, as the result of the velocity gradient they encounter. A particular feature of this instability is that it is localized ; when a disturbance grows on one side, it leaves the other side unperturbed. There is no well defined treshold value of 1/B for this instability but a narrow region above which the fingers loose stability. The location of this "treshold region" in the parameter space varies from one cell to the other. Instability is observed for 1/B around 7000 in the cell described above (see §2).

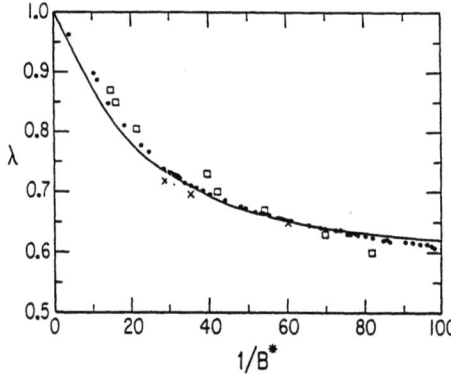

Figure 7. The same as Fig.1, but replacing T by an effective surface tension T^*

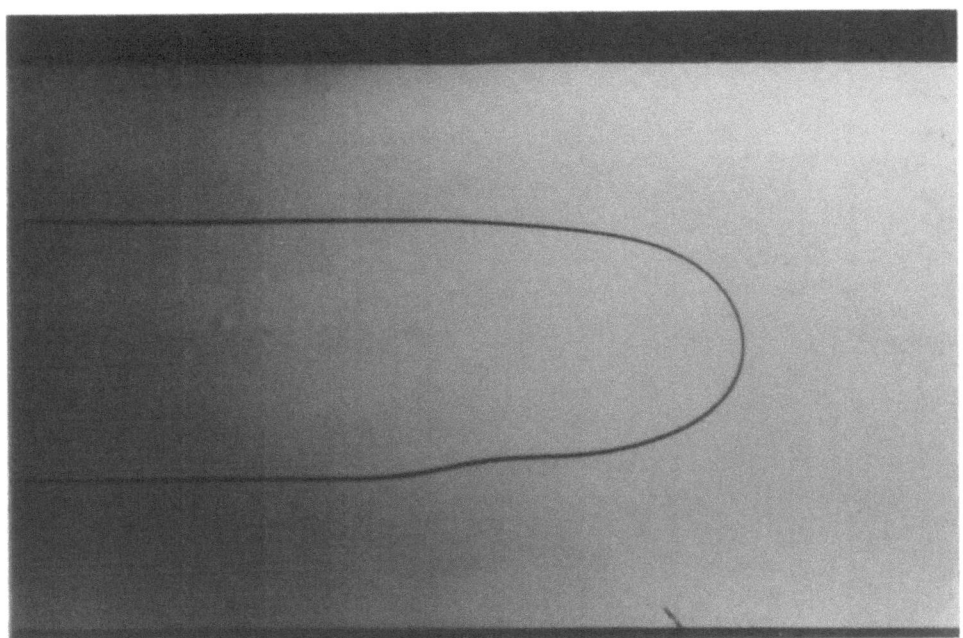

Picture 3. Asymmetric instability.

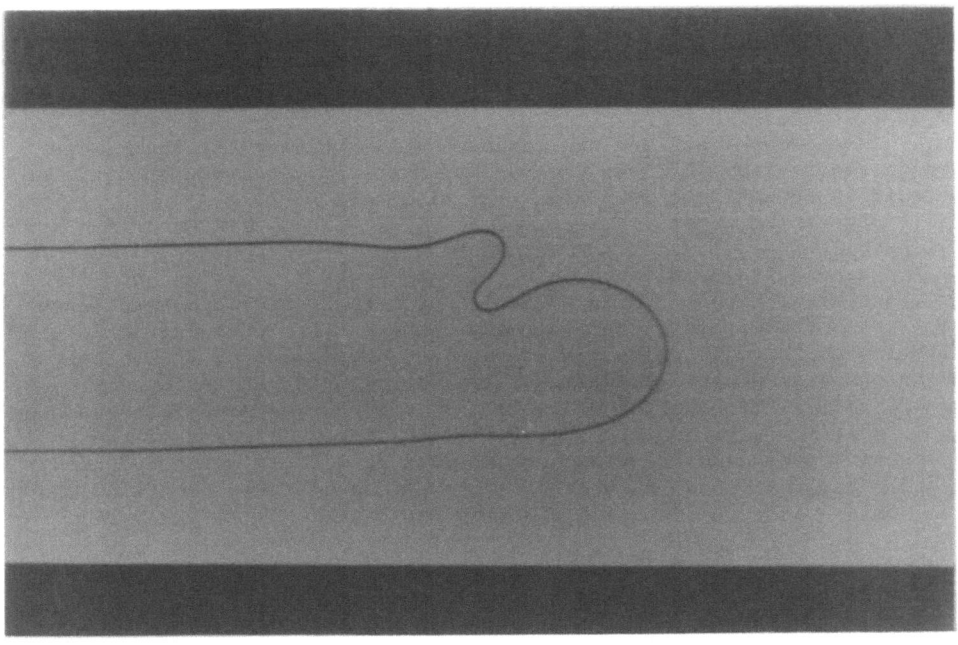

Picture 4. Side-branching instability.

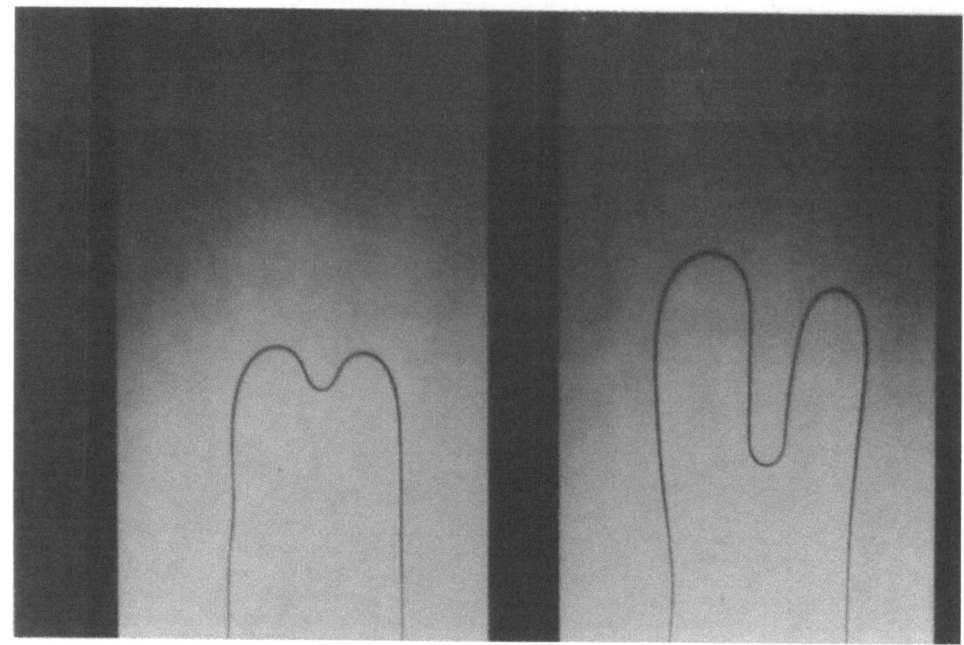

Picture 5 . Tip-splitting instability.

When 1/B is close to the treshold region, we usually observe one or two disturbances growing along the interface and then decaying for a complete course of the finger along the channel. The number of events increases with 1/B ; for a value of 1/B larger then 12000, the disturbances appear almost everywhere.

The growth rate and the maximum amplitude of the disturbances also increase with 1/B. When the maximum amplitude is large enough, the instability degenerates into side-branching (see Picture 4).

Symmetric modes are observed at large values of 1/B, in the form of tip-splitting, as shown in Picture 5.

Many features concerning the finger stability are in good agreement with theory [3]: the evolution of the asymmetric disturbances which are advected along the interface, the onset of the instability which is sensitive to noise. Also, a close comparison of the shapes of the unstable fingers with numerical results leads to an excellent agreement (see the paper by D.Bensimon[3] where such a comparison has been made). However, a crude estimate of the amplification rates of the asymmetric modes leads to values much lower than those given by theory. A probable explanation of such differences is due to the presence of the film, which tends to stabilize the fingers.

7 - CONCLUSION

The main result of this experiment is that the film plays an important role in the Saffman Taylor problem. This was suspected by Pitts (1980) and perhaps by Mc Lean & Saffman but no clear account of its importance has been reported previously.

Recently, much progress has been done on the problem of the selection of Saffman Taylor fingers at large velocities[2]. It would be interesting to see how the theoretical analysis is modified when the film is taken into account. Similarly, there is also no theoretical model for the effect of the film on the stability of the fingers. Other questions, such as the tip-splitting and side-branching instabilities, or the evolution of the film at large capillary numbers are also not understood theoretically.

Acknowledgements

We thank D.Bensimon, M.Jensen, L.Kadanoff and P.Pelce for very illuminating discussions. This work was supported by the National Science Fundation under Grant N. DMR-8316204, and also by an NSFINT-8412371 exchange award for P.Tabeling.

REFERENCES

1. Saffman,P.G., Taylor, G.I, 1958, Proc. Roy. Soc. A245,312
2. Numerical studies of stationary fingers have been performed by Mc Lean and Saffman, P.G., 1981,J.Fluid Mech,102,455, Vanden-Broek M.,1983, Phys. Fluids 26,2033 ; Analytical studies have been performed by Shraiman, B.I, 1986,Phys.Rev.Lett,56,2028, Hong D.C.,Langer J.S., 1986, Phys.Rev.Lett., 56,2032, Combescot R.,Dombre T.,Hakim V. Pomeau, 1986,56,2036.
3. Numerical studies of the stability of Saffman-Taylor fingers have been performed by Mc Lean and Saffman, 1981,J.Fluid.Mech,102,455, DeGregoria and Scwartz,L.W.,1985, Exxon preprints, Bensimon, D. 1986, Phys.Rev.A, 33,1302, Kessler,D.A.,Levine,H, 1985, Phys.Rev A,32,1930; The analogy between the stability of Saffman Taylor fingers and that of premixed flames has been pointed out by P.Pelce (private communication).
4. Tryggvason, G. and Aref, H. 1983, J.Fluid.Mech.,154,287.
5. Mc Lean,J.W. and Saffman P.G. 1981,J.Fluid.Mech.,102,455.
6. Landau,L. and Levich, B. 1942, Acta Phisicochimica USSR 17,42
7. Bretherton F.P.1961,J.Fluid Mech,10,166
8. Park, C.W., and Homsy,G.M 1984, J.Fluid Mech.139,291.
9. Tabeling, P. and Libchaber, A. 1986,Phys. Rev.A,33,794

THEORY OF THE NEEDLE CRYSTAL

M. Ben Amar and Y. Pomeau
Groupe de Physique des Solides
Ecole Normale Supérieure
24 rue Lhomond
75231 Paris cedex 05, France

INTRODUCTION

During the last two years, important improvements have been made in the understanding of hydrodynamical instabilities leading to formation of patterns as the Saffman-Taylor fingers, directionnal solidification and free dendrites. Among those instabilities, some lead to a continuous family of steady states as shown some time ago, under the condition that the effects of surface tension are neglected (Gibbs-Thomson effect). When they are included, they are usually believed to be responsible of the velocity selection observed in experiments.

This paper is devoted to the problem of a free dendrite growing in its undercooled melt. The temperature field, due to both generation of latent heat L at the interface and imposed uniform undercooling Δ far away from the tip, obeys the diffusion equation, if local thermodynamic equilibrium is assumed. We look for steady needle crystal growing at constant velocity u without change of shape. In the reference frame of the crystal, the shape is given by the solution of a rather complicated integro-differential equation established by Nash and Glicksman [1]. This one is valid as soon as we restrict ourselves to the symmetric model of solidification [2]. Besides the undercooling, measured by $\Delta = (T_M - T_\infty)c/L$, where T_M is the melting temperature, two physical parameters enter this equation; the Peclet number p and the capillary length d_0. Focusing to low undercooling, the Nash and Glicksman equation can be transformed into a non linear eigenvalue problem [3], which only depends of one parameter C. When the Gibbs Thomson effect is neglected, the profile of the needle-crystal is the Ivantsov paraboloid with an arbitrary growth velocity for a given undercooling Δ. Adding surface tension effects, one can expect to have the velocity selection, so we decided to study the Pelcé-Pomeau equation in two geometries: in 2 dimensions and 3 dimensions but with axisymmetry. Analytically, we can reach two limiting cases: vanishing or strong capillary effects. In the first case, one can imagine a perturbation expansion around the Ivantsov solution but on simplified models of solidification [4], it has been pointed out that weak surface tension effects act as a singular perturbation. In order to show this, we apply the method by Kruskal and Segur [5]. In particular, we shall prove that there is no needle crystal solution if one restricts to weak <u>and</u> isotropic surface tension effects. At large surface tension a uniform scaling shows the absence of solution with low velocity. Numerical analysis [6] confirms these predictions. It constitues the only mean of investigation of the problem for intermediate C values.

Either in 2 or 3 dimensions we never find any solution proving, once again the role of the anisotropic effects. When introduced via the surface tension coefficient, they are responsible of the velocity selection since they restore an infinite discrete set of solutions. Since it is out of our purpose to demonstrate it, we refer to our previous works [6,7] valid in 2d. Now, we extensively develop our analysis of the velocity selection considering the 2d and 3d case which gives results very similar to the 2d geometry despite its greater complexity.

NEEDLE-CRYSTAL GROWING IN AN UNDERCOOLED MELT

We consider the symmetric model of solidification at low undercooling, which gives a realistic description of heat diffusion. We look for steady state needle-crystal solutions of the temperature diffusion equation with free boundary conditions: that is a pure crystal moving without change in shape at constant velocity u along the y axis in its infinite undercooled melt. In this paper, we consider two geometrical situations for the needle-crystal.

THE 2 D MODEL OF SOLIDIFICATION

Using Green's funtion techniques, Nash and Gliksman [1] have established a non linear integro-differential equation for the profile $y=\xi(x)$ of the crystal. When surface tension effects are neglected, as proved by Pelcé and Pomeau [3], this equation restores the continuous family of Ivantsov parabolae (in 2d $\xi_{iv}(x)=-ax^2$) with an undetermined velocity u: only the product ua^{-1} is related to the undercooling Δ which is the physical driving force. Adding surface tension effects, one can expect to lift the degeneracy, that is to determine both the scale parameter (given by a) of the asymptotic Ivantsov parabola and the growth velocity u.

In the limit of low undercooling, Pelcé and Pomeau [3] have shown that the Nash and Glicksman equation can be transformed into a non linear eigenvalue (C) problem, free of any dimension quantity. As the unknown eigenvalue C is equal to ua^{-2}, by solving this equation, one can calculate both the velocity u and the needle crystal size, measured by a, using the relations:

$$u=(16/\pi^2 C)\Delta^4 [D/d_0] \qquad a=(4/\pi C)\Delta^2 d_0$$

D is the heat diffusion coefficient, d_0 the capillary length.
Let us write this integral equation, using a^{-1} and $[D/d_0]$ as length and velocity unit respectively:

$$\xi_{xx}(1+\xi_x^2)^{-3/2}=-(C/4\pi)\int_{-\infty}^{+\infty}dt\{\ln[1+(\xi(x)-\xi(t))^2(x-t)^{-2}]-\ln[1+(x+t)^2]\} \quad (1)$$

ξ_x (resp. ξ_{xx}) means the first (resp. the second) derivative relative to x. This equation does not take into account the kinetic effect [8] negligible at low undercooling. It is an eigenvalue problem, since three boundary conditions are imposed:

at $x \Rightarrow \pm\infty$ $\xi(x) \simeq -x^2$
at $x=0$ $\xi_x(0)=0$

Following a strategy applied by Vanden-Broeck for an analogous problem [9], we restrict the x variation to the half positive axis, in order to relax the last condition: so we allow a cusp at the tip. The full profile of the needle crystal will be derived by symmetry around the y axis and the velocity selection comes from the condition of vanishing of the cusp measured by $\xi_x(0)$ in term of C:

$$\xi_x(0)=\omega(C)$$

From an analytical point of view, we can reach the behaviour of ω in two limits: C going to zero or infinity which physically means strong or small surface tension effects. For intermediate C values, only numerical investigation can be performed, we give the results in the last part of the paper.

Velocity Selection At Low Capillary Effects

Since, at zero surface tension, the Ivantsov parabolae are solutions, one can expect for ξ to deviate weakly from $\xi_{iv}=-x^2$, as soon as weak capillary effects are involved. This corresponds to C going to infinity in eq. (1) and suggests a perturbation expansion in power of C^{-1} as:

$$\xi(x)=-x^2 + C^{-1}y_1(x) + \ldots + C^{-i}y_i(x) + \ldots \tag{2}$$

this expansion suggests three questions:
- existence of each y_i
- the so called quantization condition (see appendice B of [3]) of solvability of equation (1):
$$\int_0^\infty y_i(t)dt=0$$
- convergence of the series.

Each y_i is given by an inhomogeneous integral equation of Fredholm type which can be solved recursively by Fourier transformation. As an example, let us give here the first order:

$$-4\pi/(1+4x^2)^{3/2}=\int_{-\infty}^{+\infty} dt\, (x+t)(1+(x+t)^2)^{-1}[(y_1(x)-y_1(t))/(x-t)] \tag{3}$$

The Fourier transform of y_1 is given by:

$$y_1(p)=2e^{k/2}/ch(k/2)\,[(k/2)\,K_0(k/2) + th(k/2)\,K_1(k/2)] \quad \text{with } k=|p| \tag{4}$$

K_0 (resp. K_1) are the modified Bessel function of zero (resp. first) order.

On this example, one can notice that the two first requirements (y_1 exists and satisfies the solvability condition) are checked and that y_1 is an even function of x. These results are easily extended to each y_i. So this expansion, as it stands, gives no information on the velocity selection since it predicts a rigorously vanishing function for $\omega(C)$. In order to go further, as it is now well established that these problems [10] require a singular perturbation analysis, we must look for transcendental corrections which do not appear in (2) since they cannot be expanded in powers of C. We have adapted to this problem a method proposed by Kruskal and Segur [5], which is very efficient when non local equations are involved. This method performs an analytical continuation of (1) in the complex z-plane, selecting three important regions: plus infinity, an internal region around 1/2 (or -1/2) where both (1) and (3) show singularities and the part of the imaginary axis between 1/2 and zero. A path in the complex z plane relates these three regions, imposing asymptotic matching to local approximations of the analytical continuation of ξ. In particular, if we define $\Phi(z)$ as the deviation of ξ to the Ivantsov parabola:

$$\Phi(z) = \xi(z) + z^2$$

we know that $y_1(z)$ (see eq.3) is a valid approximation to $\Phi(z)$, when the real part of z goes to plus infinity. In regions of the complex z plane without singularity of Φ, or example on the imaginary axis between i/2 and zero, we can perform a linearization around $\xi_{iv}(z)$ both for the integral and the curvature term, in order to get transcendentally small terms formally absent from the regular expansion (2). In the internal region, a local equation for $\Phi(z)$, free of any parameter, takes into account the most singular terms in the regular expansion in C^{-1} and is a fair approximation of (1). Using a stretching transformation both for the function $\Phi(z)$ and the variable z, as in boundary layer methods, we define two exponents α, β such that:

$$\Phi(i/2+v) = C^\alpha F(C^\beta v) \qquad (5)$$

with $w = C^\beta v$ as the natural variable of F.

We fix $\alpha + 2\beta = 0$ in order to get the same order of magnitude for terms coming either from the Ivantsov solution ($\xi \sim v^2$ when $v \to \infty$, independent of C) or from the deviation Φ. As v should be small, by definition of an internal region, we deduce:

$$\beta > 0 \qquad \text{consequently} \quad \alpha < 0$$

For this reason, Φ remains small enough in this boundary layer to allow the linearization of the integral in (1). Once linearized, the analytical continuation of the integral here called I, is the sum of two terms, a local and a non local one:

$$I = -C(1+2iz)\Phi(z)/2(1+4z^2) - C/2\pi \int_{-\infty}^{+\infty} dt\, \Phi(t)(z+t)[(z-t)(1+(z+t)^2)]^{-1} \qquad (6)$$

Since the integration is performed on the real axis where $\Phi(t)$ is a regular function of t, the non local term cannot be singular for $z \simeq i/2$. So we neglect it and keep only the first one which is purely local and directly proportionnal to the singular function $\Phi(z)$, in the vicinity of $i/2$.

To obtain the scaling exponents α and β, we recall that asymptotic matching has to be performed between Φ and y_1 in their common domain of validity, that is for:

$$z = i/2 - i\tau + v \quad \text{with } v \to +\infty$$

τ is a necessary small quantity. Asymptotic analysis of the Fourier transform of (4) leads to:

$$y_1(z) = (iz)^{-3/2} \quad \text{and} \quad \Phi(z) = (iv)^{-3/2}/C$$
$$\text{Consequently:} \quad F(w) = (iw)^{-3/2} = (iv)^{-3/2} C^{-3\beta/2} \tag{7}$$

Joining (5) to (7), we deduce: $\alpha - 3\beta/2 = -1$ which gives:

$$\alpha = -4/7 \quad \text{and} \quad \beta = 2/7 \tag{8}$$

with $\alpha + 2\beta = 0$ and yields then a non linear similarity equation for F:

$$4(-2 + F'') = -F(4iw - 2iF')^{3/2} \tag{9}$$

F' (resp. F") means the first (resp. the second) derivative relative to the natural variable of F: w.

In all this analysis leading to (9), the most singular terms arising from the curvature in (1) have not been included. For example, in the denominator $(1+\xi'^2)^{-3/2}$, we have neglected F'^2, which is a priori more singular than F', because it was subdominant at large C. This restrict the domain of validity of our local equation to the outside of the internal region, the most useful one for the following of the demonstration. In the core of the internal region where F' as given by (9) diverges, we have established [7] an another local equation which has the right properties of asymptotic matching with (9).

In order to derive the sought derivative at zero, we need only odd terms of the Taylor expansion which represents Φ in the vicinity of zero. On the imaginary axis, these terms only contribute to the imaginary part of Φ, which greatly simplifies our analysis. Since F is restricted to a region of extension $C^{-2/7}$ around i/2, we need a new equation for Φ as soon as (z=it) is not in the immediate neighbourhood of i/2. The solution of the full problem (i.e. Φ) must match asymptotically F, so we examine now the asymptotic behaviour of F for large w=-it. Notice that, on the imaginary axis, equation (9) is transformed into:

$$4(2+F'') = F(4t+2F')^{3/2} \qquad (10a)$$

with purely real coefficients.
Let us study the asymptotic expansion of F when t goes to infinity:

$$F(-it) = t^{-3/2} \sum a_n t^{-n/2} = t^{-3/2} + 3/4 \, t^{-5} + \ldots \qquad (10b)$$

and define:

$$F(-it) = F_N(t) + g(t) \qquad (10c)$$

where F_N is the Nth truncation of F (10b). Notice that F_N is purely real, so we do not need to calculate it. The remaining, after truncation, $g(t)$ can be derived by the solution to an homogeneous linear equation, independent of N if subdominant powers of t are neglected.

$$g'' - 3/(2t) \, g' - 2t^{3/2} g = 0$$

Keeping only the decreasing eigenfunction, one gets:

$$\text{Im}(g(t)) = \text{Im}[F(w=-it)] = A \, t^{3/8} \exp-(2^{5/2}/7 \, t^{7/4}) \quad \text{when } t \to \infty$$
$$\text{and } \Phi(i/2 - iv) = A \, C^{-13/28} v^{3/8} \exp-(2^{5/2} C^{1/2} v^{7/4}/7) \qquad (11)$$

where A is a fixed non linear eigenvalue which can be found numerically. Now we can close the path in order to reach x=zero. Putting:

$$\chi(t) = \text{Im}(\Phi(it) \, C \qquad (12)$$

For $\chi(t)$, we obtain an homogeneous second order differential equation derived from the analytical continuation of equation (1) once linearized around the Ivantsov parabola. This equation is local since the integral term in (6) is purely real due to even parity of Φ when defined on the real axis.

$$\chi'' + 12t\chi'/(1-4t^2) - C(1-2t)(1-4t^2)^{1/2} \chi/2 = 0 \qquad (13)$$

Solving this equation in the WKB limit [11] valid at large C, we find:

$$\chi(t) = B \, (1-4t^2)^{3/8} (1+2t)^{1/4} [\exp(C^{1/2} S(t)) - \exp-(C^{1/2} S(t))] \qquad (14a)$$

$$S(t) = \int_0^t (1/2-u)^{1/2} (1-4u^2)^{1/4} \, du \qquad (14b)$$

By matching the two Φ approximations (5) and (12), we deduce B in term of A, from (11) and (14a):

$$B = A/2 \, C^{15/28} \exp{-C^{1/2} S(1/2)} \quad \text{with } S(1/2) \simeq 0.2176$$

So finally:

$$\Phi'(0) = \omega(C) = B(2/C)^{1/2} = A \, 2^{-1/2} C^{1/28} \exp{-C^{1/2} S(1/2)} \qquad (15)$$

As expected, in the limit of large C, $\omega(C)$ which measures the strength of the cusp at the tip has a transcendentally small amplitude which cannot be derived by ordinary perturbation expansion. It never vanishes (unless accidentally when A is zero) proving that with weak and isotropic Gibbs-Thomson effect, there is no needle crystal at low velocity. When we take into account anisotropic effects in the surface tension [8,12] we obtain an infinite discrete set of solutions. For a four fold anisotropy, we have established previously [7] that the C eigenvalues obey the following rule:

$$C^{1/2} T(\varepsilon) = \varphi_0 + k\pi$$

where $T(\varepsilon)$ is a function of the anisotropy coefficient [8,12] which enters the surface tension, φ_0 is a constant non linear phase shift and k a positive integer. We shall now investigate the 3d case, trying to emphasize the similarity with the previous 2d one, despite the greater complexity of each step of the analysis.

THE 3D NEEDLE-CRYSTAL WITH AXISYMMETRY

In the limit of small undercooling, Pelcé and Pomeau have applied the same asymptotic analysis than in 2d, in three dimensions with axisymmetry. They also derived a non linear eigenvalue (C) integro-differential equation for the profile ξ of the needle crystal as soon as axisymmetry is assumed. When this profile is described by $\xi(r)$, where r is the distance to the axis of symmetry, which also carries the growth velocity u, they obtain:

$$\xi_{rr}(1+\xi_r^2)^{-3/2} + \xi_r(1+\xi_r^2)^{-1/2}/r = C/\pi \int_0^{+\infty} r_1 dr_1 K[(4rr_1)^{1/2}/D(r,r_1)]/D(r,r_1)$$

$$-(\xi \leftrightarrow \xi^{iv}) \qquad (16)$$

.with $D(r,r_1) = \{ (\xi(r)-\xi(r_1))^2 + (r+r_1)^2 \}^{1/2}$

.K is the Jacobi complete elliptic integral of the first kind [13]

.the parenthesis, at the right, means that the contribution of the Ivantsov solution $\xi_{iv}(r) = -r^2$ must be subtracted to the integral.

.As in the 2d case, C is a nonlinear eigenvalue since we impose two conditions to the solution $\xi(r)$: vanishing first derivative at the tip, asymptotic behaviour given by the Ivantsov paraboloid. Once calculated, C determines both the velocity of the steady state needle crystal and the curvature (here called a) of the asymptotic Ivantsov paraboloid at the tip. We recall the values of these quantities in term of the undercooling Δ, since they are differents from those of the 2d case:

$$u = (2\Delta/\ln\Delta)^2/C \qquad a = -(2\Delta/\ln\Delta)/C \qquad (17)$$

A first important difference immediatly arises from the second term in the curvature (16) suggesting a singular behaviour of $\xi(r)$ at r=0, for any C value. In order to have a well posed problem, we slightly modify the curvature term introducing explicitly the first derivative at the tip $\xi_r(0)$, so that the l.h.s. of (16) is replaced by:

$$\xi_{rr}(1+\xi_r^2)^{-3/2} + (\xi_r - \xi_r(0))(1+\xi_r^2)^{-1/2}/r \qquad (18)$$

In that way, (16) can be solved either analytically or numerically for each C value, with a well behaved solution at r=0, making possible linearization if necessary.

Extended W K B Analysis Applied to The 3d Needle-Crystal

As previously, this one concerns large C values and a solution to (16) which weakly deviates from $\xi_{iv}(r)$. As in the 2d case, let us examine first the regular perturbation expansion:

$$\xi(r) = -r^2 + C^{-1}y_1(r) + C^{-2}y_2(r) + \ldots C^{-i}y_i(r)\ldots \qquad (19)$$

We know that this expansion is singular when all orders y_i have a vanishing first derivative at zero, due to transcendental corrections but these one are negligible (so hidden) when $y_1'(0) \neq 0$, as suggested by the numerical results of Kessler et al [14]. Consequently, let us look first at the parity of each function y_i, solution to a linear integral equation. We consider only y_1, given by the solution of the linear integral equation:

$$-(4+8r^2)/(1+4r^2)^{3/2} = C/\pi \int_0^{+\infty} dr_1\, r_1 E(k)\, t(r,r_1)\, \{y_1(r)-y_1(r_1)/(r-r_1)\} \quad (20)$$

- $E(k)$ is the complete elliptic integral of the second kind [13] with argument:

$$k = (4rr_1)^{1/2}\, \{(r^2-r_1^2)^2+(r+r_1)^2\}^{-1/2}$$

- $t(r,r_1) = (1+(r-r_1)^2)^{-1/2}\, (1+(r+r_1)^2)^{-1}$

After changing the sign of r, we take into account the following relation, valid for the elliptic integral E:

when $(r \Rightarrow -r)$, $E(k) \Rightarrow E(ik/k') = E(k)/k'$ (see 8.127 of [13])

with $\quad k' = \{(r^2-r_1^2)^2+(r-r_1)^2\}^{1/2}\, \{(r^2-r_1^2)^2+(r+r_1)^2\}^{-1/2} \quad (21)$

We find that $y_1(r)$ and $y_1(-r)$ obey the same equation, proving that y_1 is an even function of r. This argument can be extended to each order y_i of the regular expansion (19), since each y_i is calculated recursively by an integral equation having the same kernel as y_1. So, as in the 2d case, the perturbation expansion gives no information about the velocity selection and is in apparent contradiction with some numerical results [14].

Now, we can apply the Kruskal and Segur method in order to get transcendentally small terms in C^{-1} which are responsible to the velocity selection. Clearly (20) exhibits a singular behaviour in the vicinity of i/2 (and -i/2), as in the 2d case, but the integral, even linearized is more complicated than in 2d and we failed to inverse it explicitly in order to get y_1. Focusing to the internal region around i/2, we assume that we can perform a linearization of the integral term, the validity of this hypothesis will be checked once all the stretchings will be done. In the inner region, only the local contribution to the integral is singular. If Φ represents the deviation of ξ from the Ivantsov parabola, it is given by:

$$C\, \Phi(z)\, H(z)$$

where $H(z)$ is the analytical continuation of $H(r)$ defined on the real axis by:

$$H(r) = P.P \int_0^{\infty} dr_1\, r_1 E(k)\, t(r,r_1)\, (\pi(r-r_1))^{-1} \quad -ir/(1+4r^2) \quad (22)$$

In the 2d case, let us recall that we have found:

$$H(z) = -(1+2iz)/2(1+4z^2)$$

We prove hereafter that $H(z)$ has the same expression for the axisymmetric needle-crystal. The following part may be skipped at the first reading.

We give here some details concerning the calculation of the Cauchy integral in (22), which we call J, since it is not obvious at all. Very curiously, it is easier to find the final result from the original Nash and Glicksman equation[1], valid for every undercooling. When linearized around the Ivantsov paraboloid $\xi(x,y) = -ax^2 - by^2$, this equation [1] gives for J:

$$J = -P.P \int_0^\infty d\tau/(4\pi\tau)^{3/2} \int_{-\infty}^{+\infty} dx_1 \int_{-\infty}^{+\infty} dy_1 [(a(x_1^2-x^2)+b(y_1^2-y^2)+u\tau)/2\tau] \exp[-R(x_1,y_1,\tau)]$$

$$R(x_1,y_1,\tau) = [(x-x_1)^2 + (y-y_1)^2 + (a(x_1^2-x^2)+b(y_1^2-y^2)+u\tau)^2]/4\tau \qquad (23)$$

We introduce the same variables as in appendice A of [3], which are very efficient for deriving the relation between the undercooling and the Peclet number:

$$X = (x-x_1)/\sqrt{\tau} \qquad Y = (y-y_1)/\sqrt{\tau} \qquad T = \sqrt{\tau}(ax^2 + by^2 + u) \qquad (24)$$

and we express J in term of three integrals: J_1, J_2, J_3 as follow:

if $G(X,x,Y,y,T) = -[X^2 + Y^2 + (T - 2axX - 2byY)^2]/4$

$$J_1 = -(4\pi)^{-3/2} \int_0^{+\infty} dT \int_{-\infty}^{+\infty} dX \int_{-\infty}^{+\infty} dY \exp[G(X,x,Y,y,T)]$$

$$J_2 = 2ax(4\pi)^{-3/2} \int_0^\infty dT/T \int_{-\infty}^{+\infty} dX\, X \int_{-\infty}^{+\infty} dY \exp[G(X,x,Y,y,T)] \qquad (25)$$

The value of J_3 is deduced from J_2 by making the following change:

$a \Rightarrow b \qquad x \Leftrightarrow y \qquad X \Leftrightarrow Y$

By the same kind of symmetry argument as in [3], we deduce that J_1 is x and y independent, and can be easily computed at $x=y=0$, (for instance):

$$J_1 = -1/2$$

For J_2 we perform first the integration over X, taking into account formula (3.462,6) of [13]. Let us give the intermediate result:

$$J_2 = a^2x^2/\pi(1+4a^2x^2)^{3/2} \int_0^\infty dT/T \int_{-\infty}^{+\infty} dY \; (T-2byY) \exp-[Y^2/4+(T-2byY)^2/4(1+4a^2x^2)]$$

Since this new writing of J_2 looks as the first one (25), we reiterate the process, integrating first on Y, second on T. Finally we derive:

$$J_2 = 2a^2x^2/(1+4a^2x^2+4b^2y^2)$$

$$J_3 = 2b^2y^2/(1+4a^2x^2+4b^2y^2)$$

and:

$$J = -1/2(1+4a^2x^2+4b^2y^2) \tag{26}$$

As it has been pointed out in 2d [14,15], corresponding to b=0, this result is independent of the undercooling. Taking into account equations (22)+(26) we obtain for the axisymmetric case, with convenient length units:

$$H(r) = -(1+2ir)/2(1+4r^2) \tag{27}$$

So in the internal region near i/2, the integral term is reduced to the same expression as in the 2d case. Considering the curvature terms, only the first one, the most singular in the vicinity of i/2, has to be retained, so we can perform the same stretching transformation (5) as before. So we derive the same non linear differential equation (9) for F, defined in the same way as eq. (5). Let us now start with the last step of the Kruskal and Segur method involving the calculation of Φ between i/2 and zero. We find for:

$$\chi(t) = C \; \text{Im}(\Phi(it))$$

a second order differential equation which is purely local if one considers symmetry argument (21)

$$\chi'' + 12t\chi'/(1-4t^2) + (\chi'-\sigma)/t - C\chi(1-4t^2)^{1/2}(1-2t)/2 = 0 \tag{28}$$

σ is a constant which must be equal to $C\,\xi_r(0)$ (see eq.18) in order to make regular the 3d curvature term. This assumption has to be checked once the integration of (28) is achieved. One can notice that, contrary to the 2d case, this equation (28) is inhomogeneous, so we solve it by the method of variation of parameters, within the WKB approximation [11]. The eigenfunctions of the homogeneous equation are:

$$Z_\pm(t) = r(t)\,\exp(\pm C^{1/2} S(t))$$

with $\quad S'^2 = (1-4t^2)^{1/2}(1-2t)/2$

and $\quad r(t) = t^{-1/2}(1-4t^2)^{3/8}(1+2t)^{1/4}$ \hfill (29)

Since the Van Vleck factor is singular at t=0, WKB analysis is valid as soon as $t > C^{-1/2}$ [11]. In the vicinity of zero, a local equation can be established, if t < 1/2:

$$Z'' + Z'/t - CZ/2 = 0$$

The solutions of this last equation are the modified Bessel functions of zero order:

$$Z_+(t) = B_+\,I_0[(C/2)^{1/2} t] \quad\text{and}\quad Z_-(t) = B_-\,K_0[(C/2)^{1/2} t] \hfill (30)$$

As C is large, there is a consistent overlapping interval in t which allows asymptotic matching between the two approximations (29) and (30), so we deduce:

$$B_+ = (2C\pi^2)^{1/4} \quad\text{and}\quad B_- = (2C/\pi^2)^{1/4}$$

We need a particular solution of (28) vanishing at zero. This one can be written either in term of I_0 and K_0, in the vicinity of zero, or in term of Z_+ and Z_- as soon as $t > C^{-1/2}$.

$$\chi(t) = \sigma\,I_0[(C/2)^{1/2} t]\int_0^t dx\,K_0[(C/2)^{1/2} x] - \sigma\,K_0[(C/2)^{1/2} t]\int_0^t dx\,I_0[(C/2)^{1/2} x]$$

When $t \to 0$, $\chi(t)$ behaves as σt, which is consistent with the fact that σ represents the first derivative at t=0. On the imaginary axis, outside the internal region, $\chi(t)$ is given by:

$$\chi(t) = \sigma(2C)^{-1/2} \left\{ Z_+(t) \int_0^t dx \, (1-4x^2)^{-3/2} Z_-(x) - Z_-(t) \int_0^t dx (1-4x^2)^{-3/2} Z_+(x) \right\}$$

So we deduce:

$$\chi(1/2 - v) = \sigma \, 2^{5/4} \, \pi^{1/2} \, C^{-3/4} \, v^{3/8} \exp[C^{1/2} S(1/2)] \exp-[2^{5/2}/7v^{7/4}]$$

which is an asymptotic expansion, valid when $v \to 0$, and has to be compared to (11) to find the first derivative $\xi_r(0)$:

$$\xi_r(0) = A \, 2^{-5/4} \, \pi^{-1/2} C^{2/7} \exp-[C^{1/2} S(1/2)] \qquad (31)$$

where $S(1/2)$ has the same meaning as before (15), A is the non linear eigenvalue of the internal region, yet introduced in (11). We can have an estimate by direct numerical integration either of (1) or (16). Notice the difference in the C exponent : 2/7 instead of 1/28.

As expected, the cusp at the tip of the needle-crystal is a transcendentally small correction as a function of C. Within the framework of the model, we have proved that there is no needle crystal solution with axisymmetry when C is large. Let us now consider the other limit of a small C.

NEEDLE-CRYSTAL SOLUTIONS WITH STRONG SURFACE TENSION EFFECTS

For practical purpose, it is of interest to study the behaviour of $\omega(C)$, when C goes either to zero or infinity, since any possible change of the sign of $\omega(C)$ would give a proof of existence of an eigensolution of eq. (1) or (16), at some intermediate value of C. Moreover, checking numerically both analytical behaviours ensures us of the accuracy of our codes when C describes the full positive real axis. We can use them to get isolated intermediate eigenvalues, difficult to predict theoretically. When strong capillary effects are involved, one can expect the needle crystal profile to deviate enough from the Ivantsov solution, so prohibiting any linearization of (1) and (16). Nethertheless, far away from the tip, the profile must recover the Ivantsov parabola (or paraboloid). So we describe it, either in 2 or 3 dimensions, by the following stretching transformation:

$$\xi(t) = C^\alpha G(C^\beta t) \qquad (32\,a)$$

We impose:

$$\alpha + 2\beta = 0 \qquad (32\,b)$$

in order to match with the Ivantsov parabola at infinity. ($\xi \sim t^2$ without explicit C). Clearly, β is a positive exponent since the asymptotic behaviour is expected to be reached rather far away from the tip. We deduce the following inequality:

$$\alpha + \beta < 0 \qquad (32\,c)$$

which enables us to neglect 1 when compared to the first derivative. Let us focus now on the 2d case: taking into account the most important contributions in (1), we derive:

$$G'' \, |G'|^{-3} = -(2\pi)^{-1} \int_{-\infty}^{+\infty} dt \, \{ \ln|G(x)-G(t)| - \ln|x^2-t^2| \} \qquad (33)$$

This equation, free of any parameter, is obtained with:

$$\beta = 1/4 \quad \text{and} \quad \alpha = -1/2$$

The reader can check that (33) has the same asymptotic behaviour as (1), if one closely follows the demonstration given in appendice B of [3]. Clearly, the first derivative G' can never vanish, so it has the same negative sign as x describes the positive axis. Finally, we deduce:

$$\xi'(0) = -b^2 \, C^{-1/4} \qquad (34\,a)$$

In table (1) we have gathered some numerical results [6] in agreement with our predicted scaling when C is small (C < 5). It indicates that:

$$b^2 = 1.1 \pm 0.1$$

For the axisymmetric needle crystal, different exponents are expected due to the second term in the curvature (16) which dominates at small C. Looking at the integral, one can wonder how to approximate the elliptic integral K since two domains are available for the t integration variable (16):

i) t is different of r: taking into account (32 c), we notice that the argument of K is always roughly zero, so K is equivalent to $\pi/2$ over the whole integration domain.

ii) t is in the neighboorood of r, the argument of K is about one, and K exhibits a singular logarithmic behaviour.

In the first case, one can check that the integrand is C independent, in the second one, it behaves like $C^{-\beta}$, so we conclude that the neighboorood of the singularity dominates the integral. When replacing K by its limit around r=t, we obtain formally the same integral operator as in the 2d case (1), proving, once again, a general expression for the integral operator independent of the geometry. Finally, we find

if $\beta = 1/2$ and $\alpha = -1$

$$(G'(r)-G'(0))/r|G'(r)| = -(\pi)^{-1}\int_0^\infty dt\, \{\ln|G(r)-G(t)| - \ln|r^2-t^2|\} \tag{35}$$

and we deduce :

$$\xi'(0) = -d^2 C^{-1/2} \tag{34b}$$

with an estimated $d^2 = 2.36 \pm 0.5$

NUMERICAL INVESTIGATION OF THE EXISTENCE OF THE NEEDLE-CRYSTAL

We have proved that no needle crystal solution exists with either low or arge velocity (which corresponds respectively to large or small C values) and isotropic surface tension. With B. Moussallam, we have solved numerically both (1) and (16) in order to detect eventually an even number of eigenvalues C which cannot be reached analytically. The algorithm of our code has been extensively explained in [6] for the 2d case. We follow the same strategy for the 3d case when axisymmetry is assumed. Let us recall that we use an iteration process with a first order $\xi^{(0)}$ solution of an equation coming from the linearisation of (1) or (16) around the Ivantsov solution. As an example, let us give this equation for the 2d zero order profile:

$$-2(1+4x^2)^{-3/2}+C^{-1}\{\Phi''(1+4x^2)^{-3/2}-12x\Phi'(1+4x^2)^{-5/2}\} = \int_{-\infty}^{+\infty} K(x,t)(\Phi(x)-\Phi(t))dt \tag{36}$$

where K(x,t) is the kernel of the linear integral operator explicitly written

TABLE I. CUSP AMPLITUDE AT THE TIP OF THE NEEDLE-CRYSTAL
in 2 and 3 dimensions

C	0.1	0.5	1.0	2.0	3.0	5.0	10.0
ω(2d)	2.10	1.40	1.15	0.95	0.85	0.70	0.51
b^2 (34a)	1.18	1.17	1.15	1.13	1.12	1.05	0.9
ω(3d)		3.97	2.55	1.67	1.36	1.054	0.746
d^2 (34 b)		2.68 *	2.55*	2.36	2.36	2.36	2.36

* These values are given only for indications since they are poorly calculated by our code due to very slow convergence.

in (3), and Φ as usual represents the deviation from the Ivantsov parabola. This equation has been studied by Kessler et al. [14] and Barbieri et al.[16] to establish the velocity selection at low capillary effects. We use it, even at intermediate or low C values, since it represents a valid approximation of (1) when x goes to infinity. Consequently, our method ensures the convenient asymptotic behaviour predicted by Pelcé and Pomeau for ξ and Φ. When x or t is discretized, according to a linear scale with constant mesh for v=atan(x), (36) can be put in matrix form, so we need only a simple inversion matrix code to obtain a numerical estimate of $\xi^{(0)}$. Our iteration method turns out to be very efficient since we succeed to solve both (1) and (16) in a domain of C values ranging from 1 to about 5000.

For small C (which roughly corresponds to C < 5.), table (1) shows that we check the predicted scaling of the cusp at the tip. Since neither b nor d (34) vanish, we are sure that there is no needle crystal as soon as strong isotropic surface tension effects are involved.

For large C, the transcendental correction found for the cusp at the tip fits our numerical data as soon as C>200. As an example, in figure (1) we draw a curve representing:

$$\ln|\omega(C)| + C^{1/2} S(1/2)$$

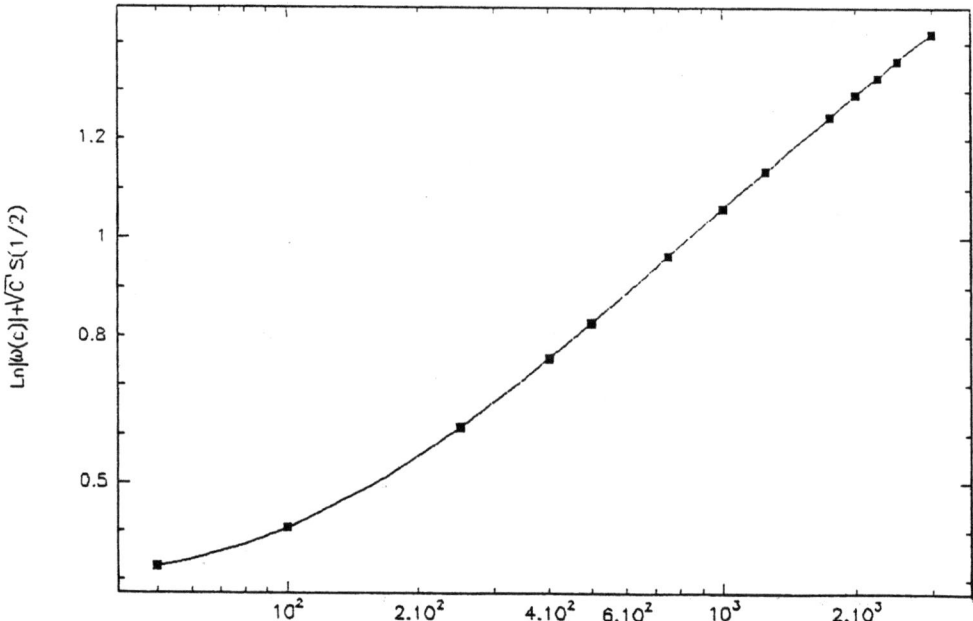

FIGURE 1 Cusp at the tip of the 3d dendrite versus C after removal of the exponential contribution

in terms of ln(C), in order to calculate the numerical exponent τ deduced from our WKB analysis (31).

We have established theoretically $\tau = 2/7 \simeq 0.29$,

we deduce from the slope in figure (1) $\tau \simeq 0.32$ which shows a pretty good agreement between our numerical and analytical analysis.

For these kinds of physical problems, WKB analysis has been used by differents authors to establish the velocity selection directly on the non linear equations or on the approximate equations linearized around a solution found with zero surface tension. It seems that the final results differ only by a numerical constant, in our case A. Let us give an estimate of the ratio between the two constants when either (1) or (36) is considered.

$A_{N.L}/A_L \simeq 1.06$ for 2d $A_{N.L}/A_L \simeq 1.05$ for 3d

So, from a purely numerical point of view,(1) and (36) gives quite the same results, when C is large.

In the intermediate domain, we never find any solution for the needle crystal with isotropic surface tension effects. The function $\omega(C)$ is a monotonic increasing negative function of the C parameter, prohibiting any stationnary solution to the heat diffusion equation, in both geometries studied in this paper.

CONCLUSION

We have presented results of a theoretical study of the needle crystal problem in the low undercooling limit, which is relevant for most experimental situations and for the symmetric model, which is not. Accordingly, there is an obvious need of extending this to the non symmetric situations, in particular to the one sided model (i.e. without diffusion in the solid phase, a fair approximation of the physics, when the growth is controlled by diffusion of impurities). Another major problem left unsolved in the present approach is the sidebranching phenomenon. This violates our basic assumptions, i.e. that the needle crystal keeps a constant shape in a moving frame. The dynamical effects pose a formidable challenge to an analytical approach, although some progress may be expected in the $\Delta \to 0$ limit. It is thinkable (but not certain at all) that these effects could restore in a sense or another the existence of needle crystals, even with isotropic surface tension. This would agree with the fact that the Mullins-Sekerka instability is still here, even without anisotropic surface tension.

ACKNOWLEDGMENTS

We would like to acknowledge useful conversations with G. Caginalp, N. Goldenfeld and H. Müller-Krumbhaar. This research was supported by GRECO 70 "Expérimentation Numérique".

REFERENCES

[1] Nash G. C. and Glicksman M. E., Acta Metal. **22**, 1283 (1974)

[2] Langer J.S, Rev. Mod. Phys. **52**, 1 (1980)

[3] Pelcé P. and Pomeau Y., Studies in App. Math. **74**, 245 (1986)

[4] Langer J.S., Phys.Rev. **A33**, 435 (1986); Caroli B., Caroli C., Langer J.S. and Roulet B.,Phys. Rev. **A33**,442 (1986)

[5] Kruskal M. and Segur H., "Asymptotics Beyond all Orders in a Model of dendritic Crystals" Aero. Res. Ass. of Princeton Tech. Memo. (1985)

[6] Ben Amar M. and B. Moussallam, "Numerical Results on Two Dimensionnal Dendritic solidification", to appear in Physica D

[7] Ben Amar M. and Pomeau Y. Euro. Lett., **2** (4) 307 (1986)

[8] Caginalp G., "The Role of Microscopic Anisotropy in the Macroscopic Behavior of a Phase Boundary", University of Pittsburgh preprint (1986)

[9] Vanden Broeck J.M., Phys. Fluids **26**, 2033 (1983)

[10] Shraiman B., Phys.Rev. lett., **56** 2028 (1986); Hong D.C. and Langer J.S. Phys. Rev. Lett., **56** 2032 (1986); Combescot R., Dombre T., Hakim V., Pumir A. and Pomeau Y., Phys. Rev. Lett. **56** 2036 (1986)

[11] Bender C.M. and Orzag S. A., "Advanced Mathematical Methods For Scientists And Engineers " (McGraw-Hill Co.,new York,N.Y.) 1979

[12] Ben-Jacob E., Goldenfeld N.D., Kotliar G., Langer J.S. Phys. Rev. Lett. **53**, 2110 (1984)

[13] Gradshteyn I.S. and Ryzhik I. M. , "Table of Integrals, Series And Products", (Academic Press) 1980

[14] Kessler D.A., Koplik J. and Levine H., Phys.Rev. **A33**, 3352 (1986); Kessler D.A., Koplik J., Levine H., Proceedings of Nato A.R.W on "Patterns, Defects and Microstructures in Non equilibrium Systems", (Austin, Texas, March 1986)

[15] Caroli B., Caroli C., Roulet B., "On Velocity Selection for 2-d Needle-Crystals in a Fully non Local Model of Solidification", Ecole Normale Superieure Preprint (1986)

[16] Barbieri A., Hong D.C. and Langer J.S. ,"Velocity Selection in the Symmetric model of Dendritic crystal growth", Preprint (1986)

DYNAMICS OF UNSTABLE INTERFACES

Nigel Goldenfeld

Loomis Laboratory of Physics
University of Illinois at Urbana-Champaign
1110 West Green Street
Urbana, Illinois 61801 U.S.A.

1. INTRODUCTION

(a) Qualitative Features of Dendritic Growth

When a solidification front advances into a supercooled melt, its velocity is usually determined by the rate at which the latent heat diffuses away from the front.[1] In this so-called diffusion-limited regime, the interface between the solid and the molten liquid is not simply planar, but instead, it is found to adopt a complex time-dependent shape, known as a dendrite. Under carefully controlled conditions, dendrites exhibit a smooth, approximately parabolic tip, which propagates without apparent change of shape, followed by a train of oscillatory sidebranches.[2] In the laboratory frame, the tip propagates at a steady velocity v_{tip}, while the sidebranches grow away from the main body of the dendrite. As the tip advances, it generates new sidebranches behind it, so that the tip size remains constant. Thus, in the moving frame of the tip, the sidebranches are oscillatory wave-like structures, propagating down the body of the dendrite with a speed $\sim v_{tip}$.

(b) The Mullins-Sekerka Instability

Dendritic crystallization is but one example of the dynamics of an unstable interface. Other examples which have received much attention recently are "viscous fingers", produced when air is pumped into glycerine sandwiched between two closely-spaced plates,[3] electrochemical deposition[4] and diffusion-limited aggregation.[5] These systems share the common feature that the interface evolves through a non-linear coupling to a diffusion or Laplacian field, and accordingly, the planar interface is linearly unstable to long-wavelength perturbations, but linearly stable to short-wavelength perturbations. This feature was first noticed by Mullins and Sekerka,[6] and is a consequence of the fact that a protrusion out of an otherwise planar interface distorts the isotherms in the liquid* so that

---
* For ease of presentation, I shall discuss the situation where there is no thermal diffusion in the solid. No new points of principle emerge when this is included, except a minor enhancement of the interfacial stability. This situation is actually realistic for the case of impurity diffusion.

they become closer together near the protrusion. Thus, the thermal gradient is steeper near the protrusion, heat diffuses away faster, and the protrusion grows outwards faster than the surrounding planar interface. In this way, the planar interface would be unstable to deformations of all wavelengths, were it not for the fact that surface tension stabilizes the interfaces for short wavelengths. This stabilization comes about, because the temperature at the interface, T_s, is reduced below the melting temperature, T_M, if the interface is curved outwards into the liquid, as is the case for our hypothetical protrusion. The amount of this reduction is proportional to both the curvature of the interface, κ, and the surface tension (Gibbs-Thomson effect). A very sharp protrusion, however, will experience such a large excess cooling at its tip, that instead of growing, heat will tend to diffuse to the tip, warming it, and thus causing it to flatten. The length scale at which these two competing effects become equal in magnitude is the stability length, λ_s, given by

$$\lambda_s = 2\pi \sqrt{d_0 \, D/v} \tag{1}$$

where d_0 is the capillary length which is proportional to the surface tension, D is the thermal diffusion coefficient, and v is the velocity of the planar interface in the direction of its normal. The ratio D/v defines the diffusion length, ℓ, over which the temperature profile decays exponentially to its asymptotic value, T_∞, far from the interface. The stability length, λ_s, gives a rough order of magnitude estimate of the characteristic dimensions of a dendrite, such as the tip radius, ρ_{tip}, and the sidebranch spacing.

(c) Formulation

For more precise predictions, it is necessary to take into account the non-linearities which arise from the modified Gibbs-Thomson boundary

$$u_s = \Delta - d_0 \kappa - \beta v_n \tag{2}$$

Here u_s is the dimensionless temperature field $u = (T - T_\infty)/(L/C_p)$ evaluated at the solidification front, $\Delta = (T_M - T_\infty)/(L/C_p)$, v_n is the normal velocity of the interface, κ is the curvature, and L and C_p are the latent heat and specific heat respectively. The coefficient β is often called the kinetic coefficient, and its non-zero value reflects the fact that a moving rough interface is not in thermal equilibrium.[7] The capillary length, $d_0 = \gamma C_p T_M/L^2$, is proportional to the surface tension γ and is typically of the order of 10 Å. Traditionally, dendritic growth has been described by the diffusion equation

$$D\nabla^2 u = \frac{\partial u}{\partial t} \tag{3}$$

away from the interface, the equation of continuity

$$v_n = -[D\nabla u \cdot \hat{n}] \tag{4}$$

at the interface, where the square brackets denote the discontinuity at the interface, and the Gibbs-Thomson boundary condition, equation (2), with $\beta = 0$. The desired solutions of these equations should describe the shape and temporal evolution of the solidification front.

(d) Role of Anisotropy

A note of caution should be sounded regarding equations (2) - (4). When a set of equations is proposed to model a particular phenomenon, it must first be demonstrated that the solutions of the equations actually correspond to the observed phenomenon, before attempting any quantitative comparison between theory and experiment. Until very recently, this was not possible with the traditional formulation ($\beta = 0$) of dendritic growth, owing to the difficulty of solving the equations numerically as well as analytically. The work described in the present article strongly suggests that, in fact, the traditional formulation is deficient, and the solutions of equations (2) - (4) ($\beta = 0$) do not correspond to dendritic growth. The missing ingredient seems to be anisotropy originating in the crystal lattice of the growing solid. This is reflected in a dependence of d_0 and β on the orientation of the solidification front with respect to the lattice. It is clear that anisotropy is important at some level, because of the observed symmetry of snowflakes. What does not seem to have been generally appreciated is that the <u>magnitude</u> of the anisotropy, as opposed to the <u>symmetry</u> of the anisotroy, can have important dynamical consequences. At present, it appears that anisotropy may be important in two possibly related ways: it may influence the steady state solutions and it may influence the dynamics. Firstly, steady state solutions of eqs. (2) - (4) (with $\beta = 0$) do not exist, at small undercoolings, without anisotropy in d_0. It is not yet known whether this result is true away from the limit $\Delta \to 0$. The result is true in both two and three dimensions. This result is significant because the standard approach to understand dendritic growth is to perform linear stability analysis about uniformly propagating steady state solutions (assuming that they exist). Secondly, dynamical computer experiments[7] using a simplified model of solidification, which is valid when $\ell \ll \rho_{tip}$ (the boundary-layer approximation), show that below a critical anisotropy strength, the interface does not evolve through dendritic growth, but forms a branching structure characterized by tip-splitting. That is, the interface generates a sequence of tips which, instead of propagating uniformly without change of shape, undergo successive bifurcation, ultimately generating a branched structure without any of the dendritic attributes summarized in part (a). The results of these computer experiments have been supported by real experiments on a hydrodynamic analogue of solidification,[3] as described below.

2. THE BOUNDARY-LAYER MODEL (BLM)

In this section a simple phenomenological model of solidification in two dimensions is briefly described, and its results summarized. The boundary-layer model[7] (BLM) is important because it is, to date, the only non-local model of interface dynamics which reproduces the qualitative features of dendritic growth, which incorporates memory effects, and which is analytically and numerically tractable. First, an explanation of these terms. "Non-local" refers to the fact that the motion of a given point on the interface is influenced by the motion of other points on the interface, through the coupling to the diffusion field. The BLM sacrifices veracity for simplicity: points on the interface which are close together in space, but which are separated by a large curvilinear distance along the interface are not coupled. This is because in the BLM, diffusion only occurs within a thin boundary-layer at the interface. "Memory" refers to the fact that the motion of a point on the interface at time t is influenced not only by the motion of the other points which make up the interface at time t, but also by the interface at all previous times $t' < t$. This reflects the fact that the dynamics of the interface at time t is determined by the temperature field at time t. However, the temperature field will not, in general, have diffused away to infinity the latent heat generated by the interface at earlier times $t' < t$. Thus the temperature field at time t is determined by the history of the earlier configurations of the interface.

In the BLM, the diffusion length ℓ is assumed to be much smaller than the local radius of curvature of the interface, a situation which corresponds to large undercoolings $\Delta \lesssim 1$. Diffusion of heat is then assumed to occur along the arclength of the interface, with diffusion of heat normal to the interface being accounted for by allowing the diffusion length, ℓ, to vary with time and with the arclength position, s, along the interface. Thus $\ell(s,t)$ is a dynamical variable: $\ell(s,t)$ is large near a depression of the interface, where heat has accumulated, and is small near (e.g.) the growing tip of a dendrite, where the thermal gradient is steepest. The boundary layer approximation is a very attractive one, because it transforms a set of partial differential equations in two space dimensions and one time dimension into a set of partial differential equations in one space dimension and one time dimension. Note that the BLM is a perfectly sensible approximation near the tip and for the first few sidebranches. Once the separation between the sides of adjacent sidebranches is of the order of the boundary-layer thickness, the approximation becomes very poor; such coarsening effects do indeed dominate the behavior away from the tip. We shall simply ignore them for the purposes of understanding how fast the tip moves, and of understanding what controls the tip radius and initial sidebranch spacing.

(a) Equations of the BLM

It is convenient to introduce the heat content of the boundary-layer of the interface, $h = u_s(s,t)\ell(s,t)$. Then h may be assumed to obey

$$\left.\frac{dh}{dt}\right|_n = v_n(1 - u_s) - v_n \kappa h + D \frac{\partial}{\partial s}[\ell(s,t)\frac{\partial u_s}{\partial s}] \tag{5}$$

The first two terms on the right hand side are exact, whilst the last term is an ansatz.[8] It describes thermal diffusion as being driven by temperature gradients along the interface, but other choices are possible for this term.[8] The notation $\left.\frac{d}{dt}\right|_n$ indicates that the derivative is to be taken following a given point on the interface as it moves along its outward normal $\hat{n}$. The curvature and arclength position of this point will then evolve under the exact kinematical equations

$$\left.\frac{\partial \kappa}{\partial t}\right|_n = (\kappa^2 + \frac{\partial^2}{\partial s^2}) v_n \tag{6a}$$

$$\left.\frac{ds}{dt}\right|_n = \int_0^s \kappa v_n \, ds' \tag{6b}$$

Diffusion in the solid is neglected, and the boundary condition (4) becomes

$$v_n \approx D \frac{u_s(s,t)}{\ell(s,t)} \tag{7}$$

Equations (5), (6), (7) and (2) completely specify the BLM.

(b) Numerical Results from the BLM

The BLM equations are simple enough to solve numerically by discretization using a linearized, lowest order, Crank-Nicholson scheme. This is an implicit scheme and is numerically stable; in addition, it is relatively fast, with the number of operations per time step scaling like N, the number of grid points along the interface. Cubic spline interpolation is used at each time step to redistribute the grid points uniformly along the interface. We find that:

(1) The BLM equations do not generate dendritic structures, but rather, tip-splitting structures.

(2) In order for the qualitative features of dendritic growth to be reproduced, anisotropy must be included, either in d_0 or β. For example,

$$\beta = \alpha(1 - \cos 4\theta) \qquad (8)$$

where θ is the angle between $\hat{n}$ and the direction of propagation. The existence of such an anisotropic effect is due to the fact that a dendrite is a single crystal. It is not possible to tell from the early simulations whether or not sidebranching is persistent or transient. Similar conclusions were subsequently obtained with the geometrical model,[9] and with Laplacian interface dynamics.[10]

(3) Recently we have performed a series of very accurate time dependent simulations starting from the exact steady state solutions of the BLM.[11] We restricted ourselves to very small amplitudes, so that the evolution was demonstrably within the linear regime (growth rate proportional to amplitude). We were unable to find any evidence of limit cycle or persistent sidebranching behavior, for strong enough anisotropy $\alpha > a_c$ ($\simeq 0.04$, $\Delta = 0.75$). Below the threshold value of α, we observed strongly unstable behavior, with the growth being faster than exponential. The most unstable mode corresponded to tip-splitting. These observations suggest that in the BLM, steady states are linearly stable for $\alpha > \alpha_c$. If the steady states are indeed linearly stable for large enough anisotropy, then there are only two ways in which persistent sidebranching can occur within the BLM. Firstly, as was first shown by Pieters and Langer,[12] adding a noise source to the model can generate sidebranches. Pieters and Langer allowed the velocity at one point of the tip to undergo random fluctuations, and observed sidebranching to take place. The noise in their simulation was rather strong, but nevertheless their simulation shows that noise may be selectively amplified to generate sidebranches. The sidebranch spacing which they observed was estimated from linear stability analysis. Secondly, the steady states may exhibit a finite amplitude instability to a limit cycle or other time-dependent state. This is currently under investigation.

(c) Analytical Results from the BLM

The analytical results available from the BLM fall into two classes. In the first class are calculations which establish that the BLM does indeed reproduce closely known results from eqs. (2), (3), and (4) (with $\beta = 0$) (hereafter referred to as the free boundary problem FBP). The second class of results concern the steady state solutions of the BLM,[13] and will be mentioned in passing in section 3.

The BLM is intended to describe solidification when the diffusion length is small compared to the local radius of curvature of the interface. This occurs at large undercooling. We have verified explicitly that the BLM does indeed accurately reproduce the known exact results for the FBP, the planar stability spectrum, and the family of steady state parabolic solutions of the FBP, due to Ivantsov[14] which exist when $d_0 = 0$. These latter solutions are completely unstable, but have served, in the past, as the basis for understanding the tip radius and velocity of dendrites.

3. UNIFORMLY TRANSLATING STEADY STATES (NEEDLE CRYSTALS)

Both the FBP and BLM possess a family of parabolic steady state solutions, in the absence of surface tension. These solutions, first discovered by Ivanstov, propagate at some velocity v_{tip} which is not determined uniquely, but which is inversely proportional to the tip radius ρ_{tip}. Although linearly unstable, the resemblance of a dendrite tip to a parabola and the smallness of the capillary length have suggested to various authors that the perturbative inclusion of surface tension into the Ivantsov solution would stabilize these solutions below a certain wavelength, and thus explain the tip radius, tip velocity and sidebranch spacing.[1] Steady state solutions are often referred to as needle crystals. Langer and Muller-Krumbhaar,[15] expanding on an earlier suggestion of Oldfield,[16] proposed that of the family of ostensibly stable needle crystals, which are generated by this procedure, the one with the largest tip radius is selected (the marginal stability hypothesis).

The marginal stability hypothesis is not correct. The same ad hoc procedure which apparently agrees so well with experimental results does not work for the BLM, where all uncertainties and sources of error are absent. The inclusion of surface tension is a singular perturbation, as may be seen from examination of the last term on the RHS of eq. (5). With $\beta = 0$, it reads

$$-Dd_0 \frac{\partial}{\partial s} (\ell \frac{\partial \kappa}{\partial s})$$

showing that d_0 is the prefactor of the highest derivative in the equation. Thus, surface tension cannot be included by naive perturbation theory.

The steady state solutions of the BLM when $d_0 \neq 0$, $\beta \neq 0$ can be calculated numerically,[13] and it is found that instead of there being a continuous family of solutions, there is a discrete set of solutions. The fastest of these needle-crystal solutions has the same tip speed and tip radius as the solutions generated by the time-dependent simulations, the latter being decorated away from the tip by sidebranches (transient or otherwise).

It has also been shown numerically that the FBP does not possess a continuous family of needle crystals.[17-19] When $d_0 \neq 0$, $\beta = 0$, needle crystals only exist when d_0 is anisotropic.

i.e. $d_0 = \tilde{d}_0(1 - \alpha \cos 4\theta)$ \hfill (9)

These important results are beginning to be understood analytically.[20] Using techniques of singular perturbation theory,[21,22] it can be shown that in the limit of $\Delta \rightarrow 0$, the tip of the fastest needle crystal for the two-dimensional symmetrical model (diffusion constant in the liquid = diffusion constant in the solid) is described by

$$v_{tip} \sim \Delta^4 \alpha^{7/4} \qquad (10a)$$

$$\rho_{tip} \sim \Delta^{-2} \alpha^{-7/4} \qquad (10b)$$

Analogous results for the three-dimensional axisymmetrical case[23] have also shown that needle crystals do not exist in the absence of anisotropy in d_0, for $\beta = 0$.

An important point to realize is that the solvability condition, which determines the needle crystals, arises not from a matching of the tail region with the tip, but from the matching of the tip with a point in the complex arc length plane. This sort of matching is common in WKB analysis.†

These calculations raise many questions. Firstly, they are apparently model specific. There is no a priori reason to believe that the results are universal in the sense that the predictions might be insensitive either to the actual form of the anisotropy or to the inclusion of terms other than the surface tension in the Gibbs-Thomson boundary condition. Secondly, since the precise form of the anisotropy is unknown experimentally, it may well be very difficult to check these predictions in practice. Furthermore, the calculations are only possible in situations of special symmetry, such as occur in two dimensions or in the case of axisymmetry. The only reliable check on the validity of the equations is computer simulation, where extraneous effects such as heat loss, convection etc. may be eliminated.

The Ivantsov family of needle crystals is continuous because for an isothermal interface, the problem is scale invariant. (Rescaling lengths by a given amount simply redefines the velocity.) The addition of terms such as $d_0\kappa$ or βv_n to the isothermal boundary condition $u_s = \Delta$ destroys this scale invariance. Does it also automatically imply the non-existence of a continuous family of needle crystals? Surprisingly, the answer seems to be no. Dorsey and Martin[24] have shown that the boundary condition $u_s = \Delta - \beta v_n$ still leads to a continuous family of needle crystals. It seems that the special dynamical significance of surface tension, as reflected in the profound way in which it changes the linear stability spectrum, is also important for the steady states.

4. A HYDRODYNAMIC ANALOGUE FOR SOLIDIFICATION

The numerical results from the boundary-layer model,[7] the geometrical[9] model, and Laplacian interface dynamics[10] suggest that as anisotropy strength is varied at a fixed undercooling, the morphology of the emergent structure can also change. Is this qualitative prediction experimentally verifiable? Since it is rather difficult to vary, in a controlled way, the anisotropy in crystal growth experiments, we turned instead to hydrodynamic experiments.[3,25] A Hele-Shaw cell was constructed by sandwiching a thin layer of glycerine between two plexiglass plates, roughly 0.5 mm apart, and 23 inches in diameter. Air was injected into the center of the top plate for a range of pressures between 50 to 150 mm of Hg.

---

† See the discussion in Landau and Lifschitz "Non-Relativistic Quantum Mechanics", Section 52, (Third Edition), (Pergamon Press, 1981).

The bubble of air does not grow as an expanding circle, but instead undergoes the Mullins-Sekerka instability. In the absence of anisotropy the instabilities grow in a spatially incoherent way, forming a large scale structure by a process of repeated tip splitting.[25] This structure, which is known as the dense branching morphology, is shown in Figure 1. It is very clearly not dendritic. This experiment is probably the most clear cut demonstration that anisotropy of some form or other is essential for dendritic growth.

The equations describing the Hele-Shaw cell follow from the Navier-Stokes equation. Averaging the velocity in the glycerine across the gap of width b yields a two-dimensional velocity field which obeys Darcy's Law:

$$\underline{v} = - \frac{b^2}{12\eta} \underline{\nabla} P \qquad (11)$$

where η is the viscosity of the glycerine. Incompressibility then implies that

$$\nabla^2 P = 0 \qquad (12)$$

in the glycerine, whilst in the air, P is a constant. At the air-glycerine boundary, the pressure drop ΔP is given by

$$\Delta P = d_0 \kappa + \phi(v_n) \qquad (13)$$

where d_0 is the capillary length and $\phi(v_n)$ is a contribution including the effects of the interface motion. In addition to a term $-\beta v_n$ which is presumably present, but may not be numerically significant, $\phi(v_n)$ includes a term proportional to $v_n^{2/3}$ arising from the presence of a thin wetting layer at the plates.[26] It is rather unlikely that this last term is responsible for the absence of dendritic growth. Hence it is concluded that anisotropy is the missing factor.

Anisotropy was included in the Hele-Shaw cell experiments by engraving a regular grid of depth 0.015 in. on the lower plate.[3] In Figure 2 is shown the result. The qualitative features of dendritic growth are indeed reproduced. The anisotropy in this system arises because the plate spacing b is now a periodic function of position. The strength of the anisotropy is represented by the ratio of the plate spacing to the depth of the grooves. It should be noted that when b is function of position, Laplace's equation is no longer valid; there is an additional term $b \underline{\nabla} b \cdot \underline{\nabla} P$, which is small.

Although the Hele-Shaw cell with anisotropy is only an analogue model for solidification, it does indicate some interesting behavior, which is summarized on the morphology diagram sketched in Figure 3. There are two dendritic phases, one due to anisotropy in surface tension,† the other, which appears at higher driving force (and hence higher velocity), due presumably to anisotropy in $\phi(v_n)$. In between there seems to be no steady state regime, but instead tip-splitting is observed. The two dendritic phases have different orientations with respect to the underlying lattice. At low enough driving force, faceted growth is observed, with the motion of kinks on the surface clearly visible. Lastly, at zero, and perhaps very small anisotropy, the dense branching morphology is observed. Similar qualitative results have been observed also in electrochemical deposition,[4] annealing of amorphous alloys,[25] and even in crystal growth experiments on NH_4Br.[27]

---

† I am indebted to Tom Mueller for communicating this result to me prior to publication.

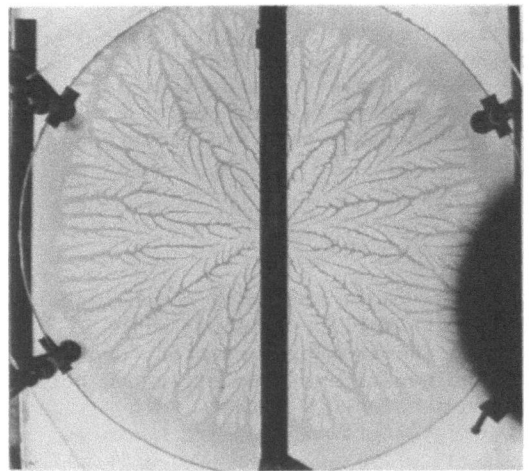 

Fig. 1. The dense branching morphology in a Hele-Shaw cell without anisotropy. The bar across the center of the picture is to maintain constant plate spacing.

Fig. 2. Dendritic growth in a Hele-Shaw cell with anisotropy.

Dendrites can also be generated in the Hele-Shaw experiment by the presence of a small bubble at the tip of one of the fingers, without any externally imposed anisotropy.[28] This effect is presumably an interplay between surface tension and the instabilities of viscous fingers,[11] but is not fully understood. It would be interesting to know if, in a large cell, the Hele-Shaw dendrites propagate in a straight line, or if there is any wandering.

In summary, experiments on the Hele-Shaw cell show that without anisotropy, diffusion-limited dynamics, or its quasi-static Laplacian

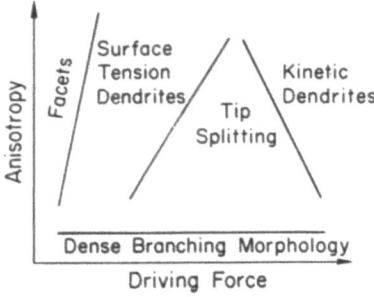

Fig. 3. Morphology diagram for Hele-Shaw cell with anistropy.

limit does not lead to dendritic growth. An important, but largely
unaddressed question is to understand the importance of finite size
effects (in two dimension, the Laplacian Green function depends
logarithmically on the radius of the outer boundary of the system), and
the crossover between the structures observed. A related question is the
role of noise in this system, and the relationship with the structures
grown by diffusion-limited aggregation.

5. SPHERULITIC GROWTH

The discovery in two dimensions of the dense branching morphology as
the outcome of diffusion-limited growth with surface tension, with
velocity dependent modifications to the Gibbs-Thomson condition, and with
sufficiently small anisotropy, raises the question of the existence of a
counterpart to this structure in three dimensions. In fact, such
structures are widely observed during crystal growth, and are known as
spherulites.[29]

Spherulites are polycrystalline aggregates with an approximate radial
symmetry. Close inspection reveals that they comprise a radiating array
of crystalline fibers, which branch at small, non-crystallographic angle,
giving rise to secondary fibers whose crystallographic orientation differs
from that of the primary fibers. Successive generations of fibers branch
repeatedly, apparently at random, to form a space-filling structure with a
typical diameter of microns. Spherulites are commonly formed by minerals
crystallizing from viscous magmas, devitrified glasses, by high polymers
crystallizing from the melt, and by organic compounds crystallizing from
melts with added thickeners. Keith and Padden[30] have pointed out that all
of these materials are effectively alloys, and that as they crystallize, a
minority species is separated from the growing crystal, and forms a thin
boundary layer in front of the advancing solidification front. This
strongly suggests that the growth is diffusion-limited but with a
significant degree of interface control. It is plausible that defects or
crystallization kinetics generate only a small anisotropy, so that the
resultant dynamics is below the threshold for dendritic growth to occur.
This is consistent with both the observed variations in crystallographic
orientation and the non-crystallographic branching. Recognition that
spherulitic structures are due to diffusion-limited growth in the weak
anisotropy regime is important, because it is then possible to make semi-
quantitative predictions about the structure, along the lines of the
analysis leading to equation (1). Preliminary calculations give
modifications to the Keith and Padden theory which are in the appropriate
direction to reconcile theory and experiment.[31]

6. SIDEBRANCHES

The starting point for all attempts to address the question of the
origin of sidebranches are the needle crystal solutions of the equations
of motion. Scenarios for sidebranch generation can be divided into two
classes: noise-driven and deterministic.

(a) Noise-Induced Sidebranching

The premise of the noise-induced scenarios is that, with surface
tension included, the fastest needle crystal is linearly stable. A very
recent calculation[32] by Kessler and Levine, for four-fold anisotropy,
suggests that this is the case for the range of anisotropy strength and
Peclet number (ρ_{tip}/ℓ) which they considered. The dynamical simulations

by Pieters and Langer[12] on the BLM have already been mentioned. In this case, for $\alpha > \alpha_c$, the needle crystals seem to be linearly stable. Time-dependent simulation of the FBP is required to test this scenario there.

The experimental situation is not so clear. At this juncture it seems that there may not be a unique origin for sidebranching. For example, in the recent experiments by Dougherty, Kaplan and Gollub[33] it was found that there was very little correlation between the sidebranches on opposite sides of the tip. This supports the noise-induced sidebranching scenario. On the other hand, the Hele-Shaw experiments of Coudert et. al. seem to show a marked correlation.[28] The noise-induced scenario predicts that for sufficiently small noise, needle crystals should occur. This has not, to my knowledge, been convincingly observed. Reports of needle crystals for the case of ^{4}He crystallization[34] are probably due to the size of the enclosure being of the order of the diffusion length.

(b) Deterministic Mechanisms for Sidebranching

If, for sufficiently small anisotropy, needle crystals are linearly unstable, then a variety of bifurcations and non-linear effects may be invoked to explain sidebranching. Here, I will briefly mention one of these scenarios: solvability-induced sidebranching. A more detailed account of this topic may be found in Ref. 11.

Solvability-induced sidebranching is a mechanism which relies on the discreteness of the family of needle crystals. The basic idea is that one of the unstable eigenmodes of the fastest needle crystal is localized near the tip. An example of such a mode is the tip-splitting mode, which causes the curvature of the tip to be reduced. If this mode, or one like it, grows with a negative amplitude, then the tip of the needle crystal will sharpen rather than flatten. Eventually, non-linear effects may restabilize the tip as a sharper tip. However, the fact that, by hypothesis, the original needle crystal is the sharpest, fastest steady state solution implies that the final state of the system cannot be a steady state. If the tip has restabilized, then sidebranches (chaotic or periodic) must be generated. Unfortunately, this scenario does not seem to occur in the BLM, the only case which we have investigated.

To summarize the main point, there is a good deal of circumstantial evidence that anisotropy must be included in a description of dendritic solidification. Computer experiments and analytical calculations in progress, on solidification models without the boundary-layer approximation will be capable of removing any remaining doubt on this matter. In previous work,[1] it was simply assumed that anisotropy was unimportant. It should be recalled that in the search for the least complicated explanation of a physical phenomenon, it is the neglect of a particular quantity (i.e. anisotropy) which requires justification, not its inclusion. A physical theory should indeed be as simple as possible, but not simpler.

ACKNOWLEDGMENTS

The work reported here was performed in collaboration with E. Ben-Jacob, G. Deutscher, P. Garik, R. Godbey, J. Koplik, G. Kotliar, J. Langer, Y. Lereah, H. Levine, O. Martin, L. Sanders, G. Schön and T. Mueller.

I acknowledge the support of the Materials Research Laboratory through grant NSF-DMR-83-16981-23.

REFERENCES

1. J. S. Langer, *Rev. Mod. Phys.* 52:1 (1980).
2. S. C. Huang and M. E. Glicksman, *Acts. Metall.* 29:701 and 717 (1981).
3. E. Ben-Jacob, R. Godbey, N. Goldenfeld, J. Koplik, H. Levine, T. Mueller, L. M. Sander, *Phys. Rev. Lett.* 55:1315 (1985).
4. Y. Sawada, A. Dougherty, J. P. Gollub, *Phys. Rev. Lett.* 56:1260 (1986).
   D. Grier, E. Ben-Jacob, R. Clarke, L. M. Sander, *Phys. Rev. Lett.* 56:1264 (1986).
5. T. A. Witten and L. M. Sander, *Phys. Rev. B* 27:5686 (1983).
6. W. W. Mullins and R. F. Sekerka, *J. Appl. Phys.* 34:323 (1963).
7. E. Ben-Jacob, N. D. Goldenfeld, J. S. Langer and G. Schön, *Phys. Rev. Lett.* 51:1930 (1983); *Phys. Rev. A* 29:330 (1984).
8. N. D. Goldenfeld (unpublished); W. Van Saarloos and J. D. Weeks, *Phys. Rev. Lett.* 55:1686 (1985).
9. D. Kessler, J. Koplik and H. Levine, *Phys. Rev. A* 30:3161 (1984).
10. D. Kessler, J. Koplik and H. Levine, *Phys. Rev. A* 30:2820 (1984).
11. O. Martin and N. D. Goldenfeld, *Phys. Rev. A* (to appear).
12. R. Pieters and J. S. Langer, *Phys. Rev. Lett.* 56:1948 (1986).
13. E. Ben-Jacob, N. D. Goldenfeld, B. G. Kotliar and J. S. Langer, *Phys. Rev. Lett.* 53:2110 (1984).
14. G. P. Ivantsov, *Dokl. Akad. Nauk.* SSSR 58:567 (1947).
15. J. S. Langer and H. Müller-Krumbhaar, *Acta. Metall.* 26:1681, 1689, 1697 (1977); ibid. 29:145 (1981).
16. W. Oldfield, *Mater. Sci. Eng.* 11:211 (1973).
17. D. Meiron, *Phys. Rev A* 33:2704 (1986).
18. D. Kessler, J. Koplik and H. Levine, *Phys. Rev. A* 33:3352 (1986).
19. M. Ben-Amar and B. Moussallam (*J. de Physique*, to appear).
20. J. S. Langer, *Phys. Rev. A* 33:435 (1986); B. Caroli, C. Caroli, B. Routet, J. S. Langer, *Phys. Rev. A* 33:442 (1986).
21. M. Ben-Amar and Y. Pomeau, *Europhys. Lett.* (to appear).
22. A. Barbieri, D. C. Hong, J. S. Langer, ITP preprint NSF-ITP-86-65.
23. B. Caroli, C. Caroli, C. Misbah, B. Roulet (preprint).
24. A. Dorsey and O. Martin (preprint).
25. E. Ben-Jacob, G. Deutscher, P. Garik, N. D. Goldenfeld and Y. Lereah, *Phys. Rev. Lett.* 57:1903 (1986).
26. C. W. Park and G. M. Homsy, *Phys. Fluids* 28:1583 (1985).
27. S. K. Chan, H. H. Reimer and M. Kahlweit, *J. Cryst. Growth* 32:303 (1976).
28. Y. Coudert. O. Cardoso, D. Dupuy, P. Tavernier, W. Thom (*Europhys. Lett.* to appear).
29. P. H. Geil "Polymer Single Crystals", Wiley (interscience), New York 1963.
30. H. D. Keith and F. J. Padden, Jr., *J. Appl. Phys.* 34:2409 (1963).
31. N. D. Goldenfeld, (in preparation).
32. D. Kessler and H. Levine, (preprint).
33. A. Dougherty, F. Kaplan, J. Gollub, (unpublished).
34. J. P. Franck and J. Jung, *J. Low Temp. Phys.* 64:165 (1986); see also E. Rolley, S. Balibar and F. Gallet (to be published in *Europhysics Letters*) for related work on ^{3}He.

CONVECTIVE AND INTERFACIAL INSTABILITIES DURING SOLIDIFICATION

S. R. Coriell and G. B. McFadden
National Bureau of Standards
Gaithersburg, MD 20899 USA

R. F. Sekerka
Departments of Physics and Mathematics
Carnegie-Mellon University
Pittsburgh, PA 15213 USA

1 Introduction

During solidification and crystal growth from the melt, the crystal-melt interface is subject to morphological instability [1,2,3,4,5,6,7,8]. Under conditions for which unstable interfaces occur, the interface morphology is sometimes cellular, but dendritic (tree-like) growth usually occurs when the degree of instability is large. The solute distribution in the crystal influences the properties of the crystal, and since solid state diffusion is usually very slow, this solute distribution is determined primarily by the solute distribution in the melt at the crystal-melt interface. Thus, the interface morphology and fluid flow in the melt play a central role in determining the properties of the solidified material.

In the following, we shall review the general area of the interaction of morphological instabilities and fluid flow. We first discuss interfacial and convective instabilities during directional solidification and then describe recent experiments in which Glicksman and colleagues demonstrate that there can be very strong interactions between morphological and convective instabilities.

During directional solidification of a single phase solid at constant velocity from a binary liquid melt, there is uniform relative motion of the sample and its thermal environment. Temperature gradients normal to the crystal-melt interface typically of the order of 100 K/cm are used in order to remove the latent heat of fusion; large temperature gradients also help to prevent morphological instability of the crystal-melt interface. Even in vertical growth it is difficult to avoid horizontal temperature gradients since the container walls are not perfectly adiabatic; the release of latent heat and the different thermal conductivities of the container, melt, and crystal give rise to horizontal temperature gradients which drive fluid flow [9]. Even in the absence of horizontal gradients during vertical growth, thermosolutal convection may occur in a binary alloy. Since the equilibrium solute concentration in the melt is different from that in the crystal, the solidification process causes a solute gradient in the melt in front of the crystal-melt interface. For solidification at constant velocity V, the solute gradient is exponential in distance from the interface with a decay distance of D/V, where D is the solute diffusion coefficient in the liquid. The temperature gradient is also exponential with a characteristic distance κ/V, where κ is the thermal diffusivity. However, the thermal diffusivity is usually so large compared to the solute diffusivity that the temperature gradient is essentially constant over the solute distance D/V. For growth vertically upward, the temperature field provides a stabilizing influence on convection for a normal liquid, which expands on heating. The solute field can be either stabilizing or destabilizing depending on whether solute is rejected from or preferentially incorporated in the crystal and on whether it increases or decreases the density of the melt. The convective instability during directional solidification differs from conventional analyses of double diffusive convection [10] in that the solute gradient is exponential rather than linear and the crystal-melt interface is a free boundary. We first discuss the boundary conditions at the crystal-melt interface.

2 Solidification Boundary Conditions

The boundary conditions (see, for example [4,6,8]) at the crystal-melt interface are

$$\boldsymbol{u} \cdot \boldsymbol{t}_1 = \boldsymbol{u} \cdot \boldsymbol{t}_2 = 0 \tag{1}$$

$$\boldsymbol{v} \cdot \boldsymbol{n}(\rho_S - \rho_L) = (\boldsymbol{u} \cdot \boldsymbol{n})\rho_L \tag{2}$$

$$(\boldsymbol{v} \cdot \boldsymbol{n})L_V = (-k_L \nabla T + k_S \nabla T_S) \cdot \boldsymbol{n} \tag{3}$$

$$(\boldsymbol{v} \cdot \boldsymbol{n})(c_S - c) = (\rho_L/\rho_s) D \nabla c \cdot \boldsymbol{n} \tag{4}$$

$$T_S = T = T_M + mc - T_M \Gamma K \tag{5}$$

$$c_S = kc \tag{6}$$

Here, v and u are the local velocities, measured with respect to the crystal, of the crystal-melt interface and the fluid, respectively; t_1, t_2, and n are unit tangent vectors and the unit normal vector to the crystal-melt interface, respectively; ρ_S and ρ_L are densities of crystal and melt, respectively; L_V is the latent heat of fusion per unit volume of solid; T_S and T are the temperature in the crystal and melt, respectively; k_S and k_L are the thermal conductivities of crystal and melt, respectively; c_S and c are the solute concentrations in the crystal and melt, respectively; T_M is the melting point of pure material with a planar interface; m is the change of melting point with solute concentration; Γ is a capillary constant; K is the mean curvature of the crystal-melt interface; and k is the distribution coefficient. The first four boundary conditions and the continuity of temperature follow from conservation laws. The last two boundary conditions assume local equilibrium at the crystal-melt interface; this assumption appears to be a good approximation for most metals at low growth velocities. For faceted materials and for high growth velocities, the last two boundary conditions require modification (see, for example, [11,12,13]). In general, the capillary constant and non-equilibrium effects will depend on crystallographic orientation, and the liquidus slope, m, and the distribution coefficient, k, will depend on concentration. We have also neglected diffusion in the crystal and cross-coupling of the heat and solute fluxes.

3 Morphological Stability in the Absence of Flow

It is well known that, for sufficiently high solute concentrations, a planar crystal-melt interface is morphologically unstable [1] and will develop a cellular or dendritic morphology with concomitant solute segregation (microsegregation). Linear stability analysis shows that the planar interface is stable if

$$G^*/mG_c > S(A,k) \tag{7}$$

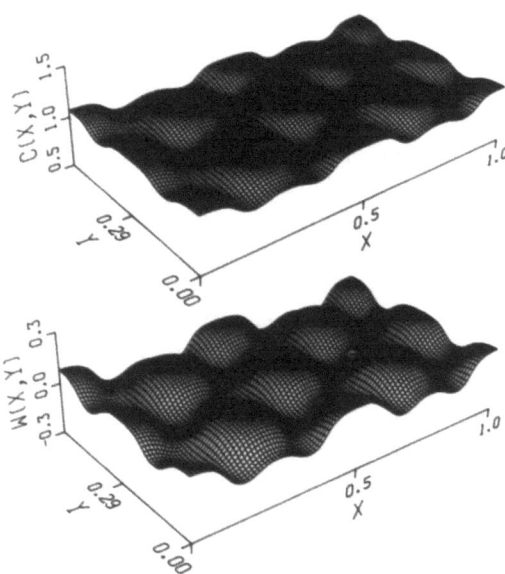

Figure 1 The solute distribution (upper plot) and interface shape (lower plot) during the directional solidification of an aluminum alloy containing chromium. The growth velocity is 0.001 cm/s, the temperature gradient in the liquid is 10 K/cm, the bulk concentration of chromium is 0.275 wt. percent, and the wavelength of the cellular interface is 0.018 cm.

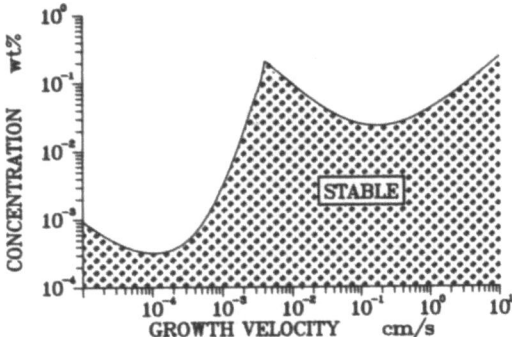

Figure 2 The concentration of tin in lead above which instability occurs during directional solidification vertically upward, with a temperature gradient in the liquid of 200 K/cm. The branch on the right corresponds to morphological modes of instability, and the branch on the left corresponds to convective modes of instability. From [8].

with
$$G^* = \frac{k_S G_S + k_L G_L}{k_S + k_L} \qquad (8)$$

and
$$A = \frac{k^2 T_M \Gamma V}{(k-1) m c_\infty D_L} \qquad (9)$$

Here $G_c = V c_\infty (k-1)/(D_L k)$ is the unperturbed solute gradient at the planar interface in the melt, c_∞ is the solute concentration in the melt far from the interface, and $G_S = (k_L G_L + V L_V)/k_S$ and G_L are unperturbed temperature gradients at the interface in the crystal and melt, respectively. G_L is positive if the temperature increases with distance into the melt. The function $S(A, k)$ depends on the two dimensionless parameters A and k and can be calculated from the roots of a cubic polynomial [14]. For small solidification velocities, $A \ll 1$ and $S(A, k) \approx 1$, so that the approximate stability criterion is $G^*/mG_c > 1$, which is called the modified constitutional supercooling criterion. For large solidification velocities, $A > 1$ and $S = 0$, and the interface is stable provided $G^* > 0$. With local equilibrium at the crystal-melt interface, the onset of instability is non-oscillatory in time and the wavelength of the instability decreases with solidification velocity. For example, for the solidification of aluminum containing copper with a temperature gradient in the melt of 200 K/cm, the wavelength at the onset of instability ranges from 0.06 cm to 0.0002 cm as the solidification velocity varies from 10^{-4} to 10 cm/s [7].

4 Cellular Growth

Recently there has been renewed theoretical interest in the highly nonplanar morphologies that occur during crystal growth and alloy solidification. We will discuss some aspects of cellular growth during directional solidification of a binary alloy in the absence of fluid flow. For interface shapes that do not differ too much from planarity, the free boundary problem may be treated by nonlinear expansion techniques [4,7,15]. Numerical calculations of interface shapes that differ significantly from planarity have also been carried out [16,17,18,19]. An example of recent calculations of a steady state cellular interface and the resulting solute distribution is shown in Fig. 1 [20]. In these calculations, the thermal properties of crystal and melt are assumed to be equal and the latent heat is neglected so that the temperature field is linear. The distribution coefficient k for chromium in aluminum is 1.8 so that the solute distribution $c(x,y)$ at the crystal-melt interface and the interface shape $W(x,y)$ are similar. The computational domain was chosen so that hexagonal symmetry is possible, and, in fact, occurs for the case studied in Fig. 1.

Since cellular spacings are generally very much smaller than sample dimensions, computational domains are used with periodic boundary conditions corresponding to the wavelengths of the anticipated cell spacings. In general, it is possible to find a family of steady state solutions (interface shape, temperature and concentration fields) for a range of wavelengths. The mechanism of wavelength selection is an unsolved fundamental problem.

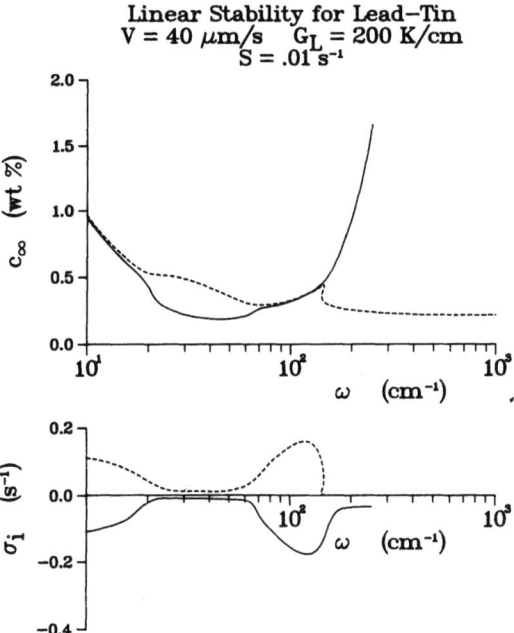

Figure 3 The concentration of tin in lead and the oscillatory frequency σ_i at the onset of instability during directional solidification at 0.004 cm/s as a function of spatial frequency ω of a sinusoidal perturbation for a shear at the interface of 0.01 s^{-1}. The dashed curves correspond to morphological modes and the solid curves to double diffusive modes. From [27].

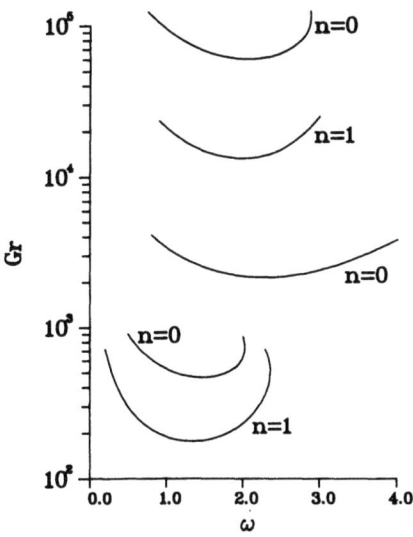

Figure 4 The Grashof number at the onset of instability as a function of the axial wave number ω of a sinusoidal perturbation for Prandtl number of 22.8 and a radius ratio of 0.02. The lowest two curves occur only for a crystal-melt interface, whereas the upper three curves occur for both a crystal-melt interface and a rigid interface. The curves are labeled with the azimuthal wave number n. From [31].

5 Combined Double Diffusive and Morphological Instabilities

The onset of coupled morphological and double diffusive instabilities of the quiescent base state during directional solidification of a binary alloy at constant velocity has been determined by linear stability analysis [21,22,23,24,25,26]. Specific calculations have been carried out for dilute concentrations of tin in lead and ethanol in succinonitrile; for both systems, the solute is rejected at the crystal-melt interface and decreases the density of the melt so that the solute field is destabilizing with respect to convection. The concentration of tin in lead at the onset of instability is shown in Fig. 2 as a function of growth velocity. For small solidification velocities, the onset of instability is via a convective mode, while for large solidification velocities, it is via a morphological mode. The qualitative behavior of the morphological mode can be understood by recognizing that the onset of instability is determined by a balance between the destabilizing solute gradient and the combined stabilizing effect of the positive temperature gradient and the crystal-melt surface tension. The solute gradient is proportional to the interface velocity, and this destabilizing influence increases with velocity. However, at velocities greater than about 1 cm/s, the wavelength at the onset of instability becomes sufficiently small that the stabilizing role of the crystal-melt surface tension becomes important. In other words, for small velocities the temperature gradient is the dominant stabilizing influence while for large velocities the surface tension is dominant. Similarly, the qualitative behavior of the convective mode can be understood by recognizing that the relevant length scale is D/V, that the solute gradient is destabilizing, and that viscosity and the temperature gradient are stabilizing. At low velocities, stability is determined by a balance of the solute gradient and the temperature gradient, and at large velocities by a balance of the solute gradient and viscous forces. The convective mode is obviously a strong function of the gravitational field, whereas the morphological mode is relatively insensitive to gravitational field. Note that very small solute concentrations give rise to convection, for example, at a solidification velocity of 10^{-4} cm/s, the critical concentration is $3.0(10^{-4})$ wt. percent tin. For the lead-tin and succinonitrile-ethanol alloys, the wavelength of the convective instability is usually an order of magnitude greater than that of the morphological instability, and the coupling between them is relatively weak. In general, the onset of instability is non-oscillatory in time; however, near the crossover point where the critical concentrations for convective and morphological instabilities are the same, the onset of instability can be oscillatory [25,26].

The effect of a forced Couette flow, parallel to the horizontal crystal-melt interface, on the onset has also been treated [27]. Such a flow does not affect perturbations with wave vectors perpendicular to the flow. For perturbations with wave vectors parallel to the flow, the onset of morphological instability is somewhat suppressed and double diffusive instability is greatly suppressed. When instabilities occur, they are oscillatory and correspond to travelling waves. For values of the crystal growth velocity for which coupled morphological and convective modes occur, the presence of a forced flow produces sufficient decoupling to allow otherwise degenerate branches to be identified. This behavior is shown in Fig. 3.

Young and Davis [28] have recently studied the effect of buoyancy on morphological stability for alloys with very small distribution coefficients. A weakly nonlinear analysis of steady solutal convection has been carried out [29].

Morphological and convective length scales can differ by orders of magnitude; this has so far precluded numerical calculations of the nonlinear aspects of the coupled problem. Useful results may sometimes be obtained by assuming that the morphological and convective modes are uncoupled.

6 Strong coupling of interfacial and hydrodynamic instabilities

Recent experiments of Glicksman and colleagues, combined with linear stability analysis, provide an example in which this decoupling is not valid [8,30,31,32]. We give a brief discussion of this coupled instability. A long vertical cylindrical sample of high purity succinonitrile is heated by an electrical current passing through a long coaxial heating wire, so that a vertical melt annulus is formed between the coaxial heating wire and the surrounding crystal-melt interface. The outer radius of the crystal is maintained at a constant temperature below the melting point of succinonitrile. This arrangement permits the temperature to decrease monotonically from the melt to the crystal across the crystal-melt interface, a configuration in which the interface would be morphologically stable in the absence of fluid flow.

The thermal gradients in the melt induce buoyancy forces which cause the fluid to flow upward near the heating wire on the axis and downward near the crystal-melt interface. When linear stability analysis is used to calculate the critical Grashof number for instabilities of the axisymmetric flow occurring between two vertical infinite rigid coaxial cylinders held at different temperatures, it is found that for a Prandtl number of 23 (corresponding to succinonitrile), the flow is unstable to an axisymmetric perturbation above a Grashof number of the order of 2000, and the resulting wave speed of this perturbation is comparable to the maximum in the characteristic unperturbed flow velocity.

In contrast, the experimental observations indicate an asymmetric helical instability at a critical Grashof number of about 150 with a wave speed two orders of magnitude less than the unperturbed velocity. Linear stability analysis as shown in Fig. 4 reveals that the instability is due to a coupling between a basic hydrodynamic instability in the buoyant flow and the deformable crystal-melt interface. The crystal-melt interface lowers the critical Grashof number of the analogous rigid-walled system by an order of magnitude; furthermore, the hydrodynamic mode (shear mode) that is actually destabilized by the interface is not the least stable mode (buoyant mode) in the rigid-walled system. The instability is sensitive to the form of the flow field; for example, it does not occur for Couette or Poiseuille flows.

7 Double Diffusive Convection during Directional Solidification

Numerical calculations of the temperature, solute, and fluid flow fields in the melt and the temperature field in the solid have been carried out by using finite differences in a two dimensional, time dependent model that assumes a planar crystal-melt interface and small Prandtl number [33]. The assumption of a planar interface precludes morphological instability, and requires relaxation of the usual solidification boundary conditions; in particular, we do not require that the interface temperature equal the equilibrium temperature. Linear stability calculations have verified the range of valid-

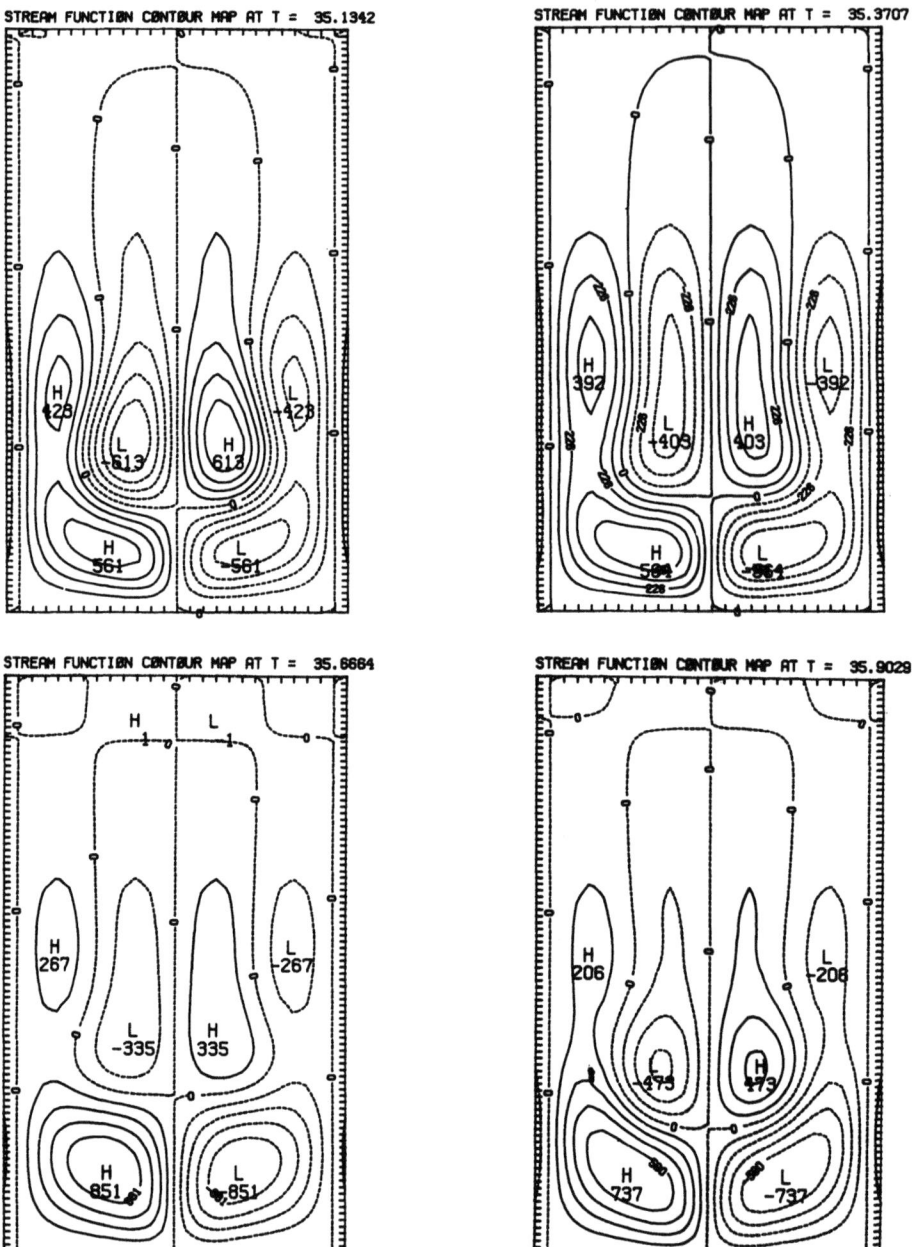

Figure 5 The stream function as a function of position at four different times for the directional solidification of lead containing 0.0015 wt. per cent tin at a growth velocity of 0.0002 cm/s and a temperature gradient in the liquid of 200 K/cm. The tin concentration is 4.3 times the concentration at the onset of convective instability; the flow is time periodic with a period of 880 s. The crystal-melt interface corresponds to the bottom horizontal line; the horizontal and vertical tick marks indicate the mesh used in the calculations.

ity of this approximation with respect to the onset of convective instability. The small Prandtl number of liquid metals and semiconductors, viz., of the order of 0.01, allows the equation for the temperature variation from the base state to be approximated by a linear diffusion equation.

Typical values of the Schmidt number for metals and semiconductors are of the order of 100 and 10, respectively; calculations have been carried out for Schmidt numbers of 81 (corresponding to lead), 10, and unity. The flow is assumed to be periodic in the horizontal direction with a given period. As the solutal Rayleigh number increases from the critical solutal Rayleigh number for the onset of convection, the horizontal wavelength decreases rapidly; a phenomena also indicated by the fastest growing mode of linear stability theory. With the assumption of a planar interface, linear stability calculations for a destabilizing solute field and a stabilizing temperature field (salt finger regime of double diffusive convection [10]) indicate that the onset of convection is not oscillatory in time, and this is in agreement with the finite difference calculations in which steady state convection is found. However, as the solutal Rayleigh number increases, the flow becomes periodic in time. A time periodic flow in the lead-tin system is shown in Fig. 5. The flow velocities are relatively small, i.e., of the order of ten times the crystal growth velocity. However, this is sufficient to cause a redistribution of solute so that the concentration in the solid varies by typically 50 per cent. Steady convection causes segregation transverse to the growth direction, while time-dependent convection causes both transverse and longitudinal segregation.

8 Acknowledgements

The authors are grateful to W. J. Boettinger, R. F. Boisvert, M. E. Glicksman, R. G. Rehm, and R. J. Schaefer for many helpful discussions. This work was conducted with the support of the Microgravity Sciences and Applications Program, National Aeronautics and Space Administration.

References

[1] W. W. Mullins, and R. F. Sekerka, Stability of a Planar Interface during Solidification of a Dilute Binary Alloy, J. Appl. Phys., 34:323 (1964).

[2] R. F. Sekerka, Morphological Stability, in: *Crystal Growth: An Introduction*, P. Hartman, ed., North-Holland, Amsterdam (1973).

[3] R. T. Delves, Theory of Interface Stability, in: *Crystal Growth*, B. R. Pamplin, ed., Pergamon, Oxford (1974).

[4] D. J. Wollkind, A Deterministic Continuum Mechanical Approach to Morphological Stability of the Solid-Liquid Interface, in: *Preparation and Properties of Solid State Materials*, W. R. Wilcox, ed., Vol. 4, Marcel Dekker, New York (1979).

[5] J. S. Langer, Instabilities and Pattern Formation in Crystal Growth, Rev. Mod. Phys. 52:1 (1980).

[6] S. R. Coriell and R. F. Sekerka, Effect of Convective Flow on Morphological Stability, PhysicoChem. Hydrodyn. 2:281 (1981).

[7] S. R. Coriell, G. B. McFadden, and R. F. Sekerka, Cellular Growth during Directional Solidification, Ann. Rev. Mater. Sci. 15:119 (1985).

[8] M. E. Glicksman, S. R. Coriell, and G. B. McFadden, Interaction of Flows with the Crystal-Melt Interface, Ann. Rev. Fluid Mech. 18:307 (1986).

[9] C. J. Chang and R. A. Brown, Radial Segregation Induced by Natural Convection and Melt/Solid Interface Shape in Vertical Bridgman Growth, J. Crystal Growth, 63:343 (1983).

[10] J. S. Turner, *Buoyancy Effects in Fluids*, Cambridge University Press, Cambridge, (1973).

[11] W. J. Boettinger, S. R. Coriell, and R. F. Sekerka, Mechanisms of Microsegregation-free Solidification, Mat. Sci. Eng. 65:27 (1984).

[12] R. L. Parker, Crystal Growth Mechanisms: Energetics, Kinetics, and Transport, Solid State Phys. 25:151 (1970).

[13] B. Caroli, C. Caroli, and B. Roulet, Non-Equilibrium Thermodynamics of the Solidification Problem, J. Crystal Growth 66:575 (1984).

[14] R. F. Sekerka, A Stability Function for Explicit Evaluation of the Mullins-Sekerka Interface Stability Function, J. Appl. Phys. 36:264 (1965).

[15] R. Sriranganathan, D. J. Wollkind, and D. B. Oulton, A Theoretical Investigation of the Development of Interfacial Cells during the Solidification of a Dilute Binary Alloy: Comparison with the Experiments of Morris and Winegard, J. Crystal Growth 62:265 (1983).

[16] L. H. Ungar and R. A. Brown, Cellular Interface Morphologies in Directional Solidification. The One-Sided Model, Phys. Rev. B29:1367 (1984).

[17] G. B. McFadden and S. R. Coriell, Nonplanar Interface Morphologies during Unidirectional Solidification of a Binary Alloy, Physica 12D: 253 (1984).

[18] L. H. Ungar and R. A. Brown, Cellular Interface Morphologies in Directional Solidification. II. The Effect of Grain Boundaries, Phys. Rev. B30:3993 (1984).

[19] L. H. Ungar, M. J. Bennett, and R. A. Brown, Cellular Interface Morphologies in Directional Solidification. III. The Effects of Heat Transfer and Solid Diffusivity, Phys. Rev. B31:5923 (1985).

[20] G. B. McFadden, R. F. Boisvert, and S. R. Coriell, unpublished research

[21] S. R. Coriell, M. R. Cordes, W. J. Boettinger, and R. F. Sekerka, Convective and Interfacial Instabilities during Unidirectional Solidification of a Binary Alloy, J. Crystal Growth, 49:13 (1980).

[22] S. R. Coriell, M. R. Cordes, W. J. Boettinger, and R. F. Sekerka, Effect of Gravity on Coupled Convective and Interfacial Instabilities during Directional Solidification, Adv. Space Res. 1:5 (1981).

[23] D. T. J. Hurle, E. Jakeman, and A. A. Wheeler, Effect of Solutal Convection on the Morphological Stability of a Binary Alloy, J. Crystal Growth, 58:163 (1982).

[24] D. T. J. Hurle, E. Jakeman, and A. A. Wheeler, Hydrodynamic Stability of the Melt during Solidification of a Binary Alloy, Phys. Fluids, 26:624 (1983).

[25] R. J. Schaefer and S. R. Coriell, Convection-Induced Distortion of a Solid-Liquid Interface, Metall. Trans. 15A:2109 (1984).

[26] B. Caroli, C. Caroli, C. Misbah, and B. Roulet, Solutal Convection and Morphological Instability in Directional Solidification of Binary Alloys, J. Phys. (Paris) 46:401 (1985).

[27] S. R. Coriell, G. B. McFadden, R. F. Boisvert, and R. F. Sekerka, Effect of a Forced Couette Flow on Coupled Convective and Morphological Instabilities during Unidirectional Solidification, J. Crystal Growth, 69:15 (1984).

[28] G. W. Young and S. H. Davis, Directional Solidification with Buoyancy in Systems with Small Segregation Coefficient, Technical Report No.8513, Northwestern University, Jan. 1986.

[29] D. R. Jenkins Nonlinear Analysis of Convective and Morphological Instability during Solidification of a Dilute Binary Alloy, PhysicoChem. Hydrodyn. 6:521 (1985).

[30] S. R. Coriell, G. B. McFadden, R. F. Boisvert, M. E. Glicksman, and Q. T. Fang, Coupled Convective Instabilities at Crystal-Melt Interfaces, J. Crystal Growth 66:514 (1984).

[31] G. B. McFadden, S. R. Coriell, R. F. Boisvert, M. E. Glicksman, and Q. T. Fang, Morphological Stability in the Presence of Fluid Flow in the Melt, Metall. Trans. 15A:2117 (1984).

[32] Q. T. Fang, M. E. Glicksman, S. R. Coriell, G. B. McFadden, and R. F. Boisvert, Convective influence on the Stability of a Cylindrical Solid-Liquid Interface, J. Fluid Mech. 151:121 (1985).

[33] G. B. McFadden, R. G. Rehm, S. R. Coriell, W. Chuck, and K. A. Morrish, Thermosolutal Convection during Directional Solidification, Metall. Trans. 15A:2125 (1984).

GENERALIZED GINZBURG-LANDAU EQUATIONS APPLIED TO INSTABILITIES

IN SYSTEMS COUPLING CONVECTION AND SOLIDIFICATION

Thomas Grauer and Hermann Haken

Institut für Theoretische Physik
Universität Stuttgart
D-7000 Stuttgart 80, Federal Republic of Germany

Crystal growth systems having a solid-liquid interface exhibit a variety of instabilities in their spatio-temporal behaviour. Theoretically they are described by nonlinear moving interface problems. We show that an equivalent formulation on fixed domains is possible, but at the expense of getting highly nonlinear boundary conditions. We present an extended version of the method of Generalized Ginzburg-Landau equations, which is appropriate for the weakly nonlinear analysis of this class of problems. Results are given for a solid-liquid two-phase system showing a convective instability of the Rayleigh-Benard type.

1. INTRODUCTION

Systems having a solid-liquid interface are of increasing interest for engineering as well as for theoretical physics. The technological importance for the crystal growth process is quite obvious. On the other hand, theoretical analysis of some of their features gives insight in the behaviour of a much more general class of systems. Especially on the macroscopic level of observation, these systems show phenomena, such as instabilities, cellular or dendritic patterns, which are typical for nonlinear space-time dependent systems. However, there arise particular mathematical difficulties due to the fact that one has to study the dynamics not only of some bulk variables but simultaneously that of the interface.

A recent review (Glicksman et al., 1986) has been devoted to phenomena of interaction between the crystal-melt interface and flow of the liquid phase. The motion of the melt, e.g. as the result of buoyant convection, influences the solidification or melting process and, hence, the location and shape of the interface. The interface, however, acts in various ways. First it determines the fluid motion on account of boundary conditions. Moreover, morphological instability (Mullins and Sekerka, 1964) may cause complicated shapes of the front and consequently steep local gradients near the interface. In some cases, the behaviour of the coupled system differs remarkable from that of the uncoupled one.

In the next chapter of this paper, we present the mathematical formulation of a problem in which Rayleigh-Benard convection is coupled to the dynamics of a crystal-melt interface. A linear stability analysis shows only slight differences to the ordinary Rayleigh-Benard system. However, one can show by means of nonlinear analysis that beyond the instability point there emerges a quite different spatial pattern. The moving interface complicates the nonlinear calculations significantly, in particular direct numerical simulation. It is our impression that there is strong need for weakly nonlinear analysis for these nonlinear moving interface problems.

In the third section we give a version of our method of solution, which is particularly appropriate for this kind of systems for parameter values close to the critical ones. But let us emphasize that this method, which we call the method of Generalized Ginzburg-Landau equations, applies to a much wider class of problems. It is possible to use this method to discuss questions of pattern selection, stability of the resulting states and time dependent solutions. A rather recent point of view is to get information about secondary bifurcations by doing this kind of nonlinear analysis near parameter points where several different types of modes become unstable simultaneously.

We summarize some of the results for the two-phase Rayleigh-Benard system in the fourth section.

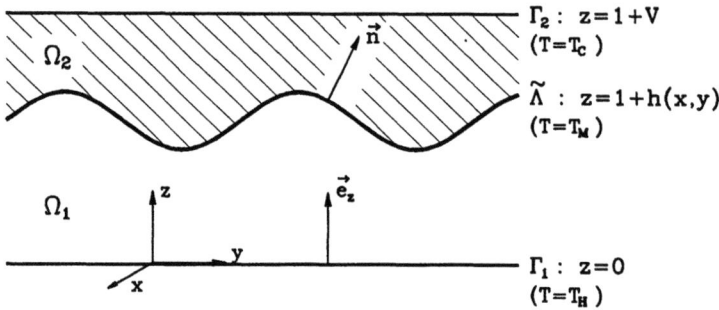

Fig. 1. Schematic illustration of Two-Phase Rayleigh-Benard experiment

2. THE TWO-PHASE RAYLEIGH-BENARD PROBLEM

Within this paragraph we present the type of equations one has to deal with in the context of systems coupling convection to the dynamics of a crystal growth interface. The specific model with horizontal translation symmetry may be called the Two-Phase Rayleigh-Benard problem (TRB). The nonlinear treatment of this is an extension of the work of Schlüter et al. (1965) and has been performed in the limiting case of very weak influence of the solid-liquid interface by Davis et al. (1984). In their work, one can find some experimental results showing the transition from roll-like to hexagonal patterns. For this system we used a somewhat different method, described in the third paragraph, which gives results for the stronger coupling case as well as for transient flows.

Some material enclosed between two rigid plates Γ_1 and Γ_2 is heated from below (Fig. 1). If the melting temperature T_M is kept between the temperatures T_H and T_C of the plates, the substance crystallizes on the upper plate and there is a solidification front Λ separating the two phases. In the case of pure thermal conduction this front remains stationary and flat. But as convection sets in, the interface becomes corrugated. The function $h(x,y,t)$ is a measure for the interfacial deflection and is an essential variable of the system.

We describe the flow of the liquid phase by the Oberbeck-Boussinesq equations. In the crystal there is only heat conduction. We neglect surface tension (Gibbs-Thomson effect), which does not play any role but for curvature radii below 0.1 mm. We further assume equal mass densities and thermal conductivities in both phases. This is not an essential assumption but simplifies some calculations; and this is a good approximation at least for plastic crystals (like succinonitrile or cyclohexane). As mentioned before, there exists a basic quiescent state with flat interface and with a linear temperature profile $T_0(z)$.

Now we present the equations for the deviations from this basic state in a non-dimensional form. In the liquid phase Ω_1 we use the general representation for the solenoidal velocity field $\vec{u}$ by two scalar variables S and R:

$$\vec{u} = \vec{L} S + \vec{M} R , \qquad (2.1)$$

where we defined the operators $\vec{L} = \nabla \times \vec{e}_z$ and $\vec{M} = \nabla \times \vec{L}$.
Then the Navier-Stokes equations read

$$- \partial_t \Delta_2 S - \vec{L} \cdot [(\vec{u} \cdot \nabla)\vec{u}] = - Pr \Delta \Delta_2 S \qquad (2.2)$$

$$\partial_t \Delta \Delta_2 R + \vec{M} \cdot [(\vec{u} \cdot \nabla)\vec{u}] = Pr \Delta^2 \Delta_2 R - Pr\, Ra\, \Delta_2 T \qquad (2.3)$$

$$\partial_t T + (\vec{u} \cdot \nabla)T = - \Delta_2 R + \Delta T , \qquad (2.4)$$

where $\Delta = \partial_x^2 + \partial_y^2 + \partial_z^2$ and $\Delta_2 = \partial_x^2 + \partial_y^2$. In the crystal Ω_2 the heat conduction equation is

$$\partial_t T = \Delta T \qquad (2.5)$$

On the plate Γ_1 we demand fixed temperature condition ($T = 0$) and no-slip condition ($S = R = \partial_z R = 0$) as well as fixed temperature $T = 0$ on Γ_2 ($z = 1 + V$). Now let us turn to the boundary conditions on the interface. The no-slip condition is

$$\vec{u} = 0 \qquad (2.6)$$

The temperature having the fixed value T_M on Λ gives rise to the condition

$$T + T_0 = 0 \qquad (2.7)$$

with our scales. The heat balance on the interface leads to an equation of motion for the interfacial deflection:

$$St\, \partial_t h = \lim_{\varepsilon \to 0} [\, \nabla T |_{z=1+h+\varepsilon} - \nabla T |_{z=1+h-\varepsilon}] \cdot \vec{n} , \qquad (2.8)$$

where $\vec{n} = (-\partial_x h, -\partial_y h, 1)$ is the normal vector on Λ.

Together with this equations we have introduced a set of dimensionless parameters of which the Prandtl number Pr, the Rayleigh number Ra and the Stefan number St are popular ones (see e.g. Normand et al., 1977). An important control parameter is

$$V = \frac{T_M - T_C}{T_H - T_M} \qquad (2.9)$$

which is the ratio of the thicknesses of the crystal with respect to that of the fluid layer in the pure conduction case. The limiting case of $V = 0$ is the ordinary Rayleigh-Benard problem; all results in this limit have to be equal to that of Schlüter et al. (1965).

What makes the analysis of such a system cumbersome is that one has to solve the partial differential equations (2.1) to (2.5) on domains with boundaries which themselves are variable with time. In fact, the dynamics of the moving interface is strongly determined by the values of the bulk variables on this interface. One possibility to overcome this difficulty is the construction of new boundary conditions on a fixed "mean" interface, which replaces the moving boundary problem by a mathematically equivalent fixed boundary value problem. Because of the analycity of the bulk variables in the spatial coordinates, this can be performed by a Taylor expansion around the value $z = 1$. If, for instance, some bulk variable q has a given dependence $g(x,y)$ on the interface Λ we may write

$$g(x,y) = q(z=1+h) = q(1) + \partial_z q|_1 h + \tfrac{1}{2} \partial_z^2 q|_1 h^2 + \ldots \qquad (2.10)$$

thus having derived a boundary condition for the surface $\Lambda = \{ z = 1 \}$. A condition of the form of an infinite series might seem to be not a practical one. But as we are going to show in the next part of this contribution, this procedure is quite useful in the context of weakly nonlinear analysis. Proceeding in this way the conditions (2.6) to (2.8) turn into:

On $\Lambda = \{ z = 1 \}$:

$$\Delta_2 S + h\Delta_2 \partial_z S + \partial_z \vec{u} \cdot \vec{L} h + \tfrac{1}{2} h^2 \Delta_2 \partial_z^2 S + \ldots = 0 \qquad (2.11)$$

$$\Delta_2 R + h\Delta_2 \partial_z R + \tfrac{1}{2}\Delta_2 \partial_z^2 R + \ldots = 0 \qquad (2.12)$$

$$\Delta_2 \partial_z R + h\Delta_2 \partial_z^2 R + \partial_z \vec{u} \cdot \nabla_2 h + \tfrac{1}{2} h^2 \Delta_2 \partial_z^3 R + \ldots = 0 \qquad (2.13)$$

$$T - h + h\partial_z T + \tfrac{1}{2} h^2 \partial_z^2 T + \ldots = 0 \qquad (2.14)$$

$$St \, \partial_t h = \llbracket \partial_z T \rrbracket + h \llbracket \partial_z^2 T \rrbracket - \nabla_2 h \cdot \llbracket \nabla_2 T \rrbracket + \ldots \qquad (2.15)$$

where we use the abbreviations

$$\nabla_2 = (\partial_x, \partial_y, 0)$$

and

$$\llbracket A \rrbracket = \lim_{\varepsilon \to 0} [A|_{z=1+\varepsilon} - A|_{z=1-\varepsilon}]$$

In these equations one can recognize the strong coupling between the bulk variables and the interfacial deflection h.

For simplicity of presentation we introduce a shorthand notation. We combine the state variables of the system by

$$q = q(\vec{x},t) = (S, R, T, h) \quad . \tag{2.16}$$

They obey the nonlinear equations of motion (2.2) to (2.5) and (2.15), the linear parts of which we formally write as

$$M \partial_t q = L_\mu q \quad . \tag{2.17}$$

We define the inner products

$$\langle q_1, M q_2 \rangle = \int_{\Omega_1} (R_1^* \Delta \Delta_2 R_2 - S_1^* \Delta_2 S_2) d^3x$$
$$+ \Pr \mathrm{Ra} \int_\Omega T_1^* T_2 \, d^3x + \Pr \mathrm{St} \int_\Lambda h_1^* h_2 \, d^2x \tag{2.19}$$

and

$$\langle q_1, L_\mu q_2 \rangle = \Pr \int_{\Omega_1} (R_1^* \Delta^2 \Delta_2 R_2 - S_1^* \Delta \Delta_2 S_2) d^3x$$
$$- \Pr \mathrm{Ra} \int_{\Omega_1} (R_1^* \Delta_2 T_2 + T_1^* \Delta_2 R_2) d^3x \tag{2.20}$$
$$+ \Pr \mathrm{Ra} \int_\Omega T_1^* \Delta T_2 \, d^3x + \Pr \mathrm{Ra} \int_\Lambda h_1^* [\partial_z T_2] d^2x \quad .$$

Partial integrations yield the relations

$$\langle q_1, M q_2 \rangle = \langle M q_1, q_2 \rangle$$
$$+ \int_\Lambda \{ \Delta_2 R_1^* \partial_z R_2 - \partial_z R_1^* \Delta_2 R_2 \} d^2x \tag{2.21}$$

and

$$\langle q_1, L_\mu q_2 \rangle = \langle L_\mu q_1, q_2 \rangle$$
$$+ \Pr \int_\Lambda \{ \Delta_2 R_1^* \Delta \partial_z R_2 - \Delta \partial_z R_1^* \Delta_2 R_2 - \Delta_2 \partial_z R_1^* \Delta R_2$$
$$+ \Delta R_1^* \Delta_2 \partial_z R_2 - \Delta_2 S_1^* \partial_z S_2 + \partial_z S_1^* \Delta_2 S_2 \tag{2.22}$$
$$- \mathrm{Ra} \, [(T_1^* - h_1^*) \partial_z T_2] + \mathrm{Ra} \, [(T_2 - h_2) \partial_z T_1^*] \} d^2x \quad .$$

It is now possible to investigate the linear stability of the basic stationary solution. We drop all nonlinear terms in the partial differential equations as well as in the boundary conditions:

$$\lambda M q = L_\mu q \quad \text{in } \Omega \quad ; \quad A_\mu q = 0 \quad \text{on } \Lambda \quad . \tag{2.23}$$

This eigenvalue problem is selfadjoint because of (2.21) and (2.22). The eigenvalues are real and multiple because of rotational symmetry in the horizontal plane. The stationary state of pure conduction remains linearly stable up to some critical Rayleigh number Ra_c, which is slightly decreasing with increasing V (Davis et al., 1984).

3. GENERALIZED GINZBURG-LANDAU EQUATIONS

As we have demonstrated in the preceding part, one can describe a certain class of systems having a moving interface by partial differential equations defined on fixed domains - but at the expense of getting highly nonlinear boundary conditions at the separating surface. Such an expansion technique is of much use if only all the nonlinear terms have rather small values compared to the linear parts. Here we offer a method for solving the remaining nonlinear differential equations on the class of functions determined by the nonlinear boundary conditions. We restrict our analysis to the case that there exists a stationary basic solution which becomes unstable at some critical parameter setting and which bifurcates into some new types of solutions. We are able to discuss questions of stability, mode selection and time dependence of the bifurcating solutions, on condition that their amplitudes remain small.

Our method of solution, introduced by one of us (Haken, 1975), consists in projecting the dynamics of the infinitely dimensional system onto a low dimensional set of ordinary differential equations, which we call the Generalized Ginzburg-Landau equations. The latter can be solved analytically or numerically in order to construct the solutions of the original nonlinear problem. The usual presentation of the method, where the function space of solutions is spanned by the eigenfunctions of the linearized system, does not work in the present case because of the nonlinear boundary conditions. Here we extend a version proposed by Friedrich and Haken (1986).

Presuppositions

We consider a system defined over some spatial domain Ω, bounded by $\partial\Omega$ and separated into two parts by the singular surface Λ. We combine the set of state variables of the system (such as the velocity and temperature fields and the interfacial deflection in the TRB problem) by $q = q(\vec{x},t)$ which is element of some appropriate function space with inner product. In the same way we denote the group of control parameters by μ. Now let us abbreviate the nonlinear partial differential equations on Ω by

$$M \partial_t q = L_\mu q + N^{(2)}(q,q) + N^{(3)}(q,q,q) + \ldots , \qquad (3.1)$$

(where M, L_μ are linear, $N^{(2)}$ bilinear, $N^{(3)}$ trilinear operators) and the linear boundary conditions on $\partial\Omega$ by

$$B q = 0 \qquad (3.2)$$

The nonlinear conditions on Λ are

$$A_\mu q + A^{(2)}(q,q) + A^{(3)}(q,q,q) + \ldots = 0 \qquad (3.3)$$

The linear operator (L_μ, B) and its adjoint define the subspaces

$$S(\Omega) = \{f, B f = 0 \text{ on } \partial\Omega\}$$

and

$$S^+(\Omega) = \{f, B^+ f = 0 \text{ on } \partial\Omega\} \qquad (3.4)$$

We will have to use some kind of "Green's Theorem" for L_μ:

$$\langle f, L_\mu g \rangle = \langle L_\mu^+ f, g \rangle + \int_\Lambda [(A_\mu^+ f)^* \bar{A}_\mu g - (\bar{A}_\mu^+ f)^* A_\mu g] \, d^2x \quad . \tag{3.5}$$

for $f \in S(\Omega)$, $g \in S^+(\Omega)$. The equations have to be scaled in such a way that the stationary basic solution is of the form

$$q^{(0)} = 0 \quad ; \tag{3.6}$$

we assume that this solution is linearly stable in some part of the parameter space.

These assumptions might look artificial. But it can be shown easily that hydrodynamical two-phase instabilities can be formulated in this way - besides a variety of different systems.

Linear Stability Analysis

In order to test the linear stability of the basic solution, we have to solve the following eigenvalue problem:

In Ω: $\quad \lambda_i M \phi_i = L_\mu \phi_i \quad ; \quad \phi_i \in S(\Omega)$. $\tag{3.7}$

On Λ: $\quad A_\mu \phi_i = 0$. $\tag{3.8}$

The stationary solution becomes unstable at some critical parameter value $\mu = \mu_c$. In the vicinity of this critical point we assume to have n "unstable modes"; these are eigenfunctions belonging to eigenvalues with nonnegative real parts, whereas all the remaining eigenvalues have negative real parts and correspond to the "stable" or "slaved modes":

Re $\lambda_i \geq 0 \quad$ for $i = 1, 2, \ldots, n$

Re $\lambda_i < 0 \quad$ for $i = n+1, n+2, \ldots$

Let us add the adjoint eigenvalue problem:

In Ω: $\quad \lambda_i^* M^+ \phi_i^+ = L_\mu^+ \phi_i^+ \quad ; \quad \phi_i^+ \in S^+(\Omega)$. $\tag{3.9}$

On Λ: $\quad A_\mu^+ \phi_i^+ = 0$. $\tag{3.10}$

The adjoint eigenvectors are of importance because we are going to make use of the biorthonormality relation

$$\langle \phi_i^+, M \phi_j \rangle = \delta_{ij} \quad , \tag{3.11}$$

where δ_{ij} denotes Kronecker's symbol.

Order Parameter Equations

It has been shown (Wunderlin and Haken, 1981) that the selforganized dynamics of a system like this in the vicinity of the critical point takes place on a n-dimensional submanifold. In a physical sense this means that the unstable modes are the dominant degrees of freedom whereas the stable modes contribute to the resulting motion in a less important way - they are "slaved" by the unstable modes. This allows us to write for the solution in the weakly nonlinear regime ($\mu \sim \mu_c$)

$$q(\vec{x}, t) = q(\vec{x}; \varepsilon_1(t), \ldots, \varepsilon_n(t)) \tag{3.12}$$

with

$$q(\vec{x};\varepsilon_1=0,\ldots,\varepsilon_n=0) = 0 \qquad (3.13)$$

The (time dependent) local "coordinates" $\varepsilon_1,\ldots,\varepsilon_n$ determine the instantaneous state of the system completely. They are a measure for the deviations from the basic solution and obey some set of first order differential equations, which in the autonomous case is of the form

$$\partial_t \varepsilon_j = f_j(\varepsilon_1,\ldots,\varepsilon_n) \quad , \qquad j = 1,\ldots,n \quad . \qquad (3.14)$$

A power expansion in the ε_j yields

$$q(\vec{x},t) = \sum_{j=1}^{n} \varepsilon_j(t) \, q^{(j)}(\vec{x}) + \frac{1}{2} \sum_{j_1 j_2} \varepsilon_{j_1} \varepsilon_{j_2} \, q^{(j_1 j_2)}(\vec{x}) + \ldots \qquad (3.15)$$

and

$$\partial_t \varepsilon_j = f_j^{(0)} + \sum_k \varepsilon_k f_j^{(k)} + \frac{1}{2} \sum_{k_1 k_2} \varepsilon_{k_1} \varepsilon_{k_2} f_j^{(k_1 k_2)} + \frac{1}{6} \sum_{k_1 k_2 k_3} \varepsilon_{k_1} \varepsilon_{k_2} \varepsilon_{k_3} f_j^{(k_1 k_2 k_3)} + \ldots \qquad (3.16)$$

Because of the analogy to the theory of equilibrium phase transitions we called the ε_j "order parameters" and their equations of motion (3.16) the "Generalized Ginzburg-Landau equations".

It is possible to evaluate the coefficients of the expansion (3.16) as well as the space dependent functions in (3.15) in successive steps through a recursive procedure. Ordinarily one takes into account only the lowest orders of (3.16), so far as they suffice to discuss all qualitative features of the system and to reach the desired quantitative precision.

A Recursive Procedure

Substituting the series (3.15) and (3.16) in the basic equations (3.1), (3.2) and (3.3) and equating the expressions of corresponding powers of ε_j, we get a hierarchy of equations. The equations at zero order imply

$$f_j^{(0)} = 0 \quad . \qquad (3.17)$$

At first order we regain the (unstable part of the) eigenvalue problem:

$$q^{(j)}(\vec{x}) = \phi_j \quad , \qquad f_j^{(k)} = \lambda_j \delta_{jk} \quad . \qquad (3.18)$$

Now let us examine the second order equations in more detail. The partial differential equations defined on Ω read

$$\sum_k f_k^{(j_1 j_2)} M \, q^{(k)} = [\, L_\mu - (\lambda_{j_1} + \lambda_{j_2}) M \,] \, q^{(j_1 j_2)} + N^{(2)}(q^{(j_1)}, q^{(j_2)}) + N^{(2)}(q^{(j_2)}, q^{(j_1)}) \quad , \qquad (3.19)$$

while the boundary conditions on Λ are

$$A_\mu q^{(j_1 j_2)} + A^{(2)}(q^{(j_1)}, q^{(j_2)}) + A^{(2)}(q^{(j_2)}, q^{(j_1)}) = 0 \quad . \qquad (3.20)$$

This is a linear inhomogeneous boundary value problem. In order to get a unique solution we may require the orthogonality of first and second order functions:

$$\langle q^{(i)+}, M\, q^{(j_1 j_2)} \rangle = 0 \qquad (3.21)$$

Multiplying (3.19) by the adjoint eigenfunctions, using (3.5), (3.11) and (3.21) and applying Fredholm's alternative we get the second order coefficients

$$f_i^{(j_1 j_2)} = \langle q^{(i)+}, N^{(2)}(q^{(j_1)}, q^{(j_2)}) + N^{(2)}(q^{(j_2)}, q^{(j_1)}) \rangle$$
$$+ \int_\Lambda (A_\mu^+ q^{(i)+})^* [A^{(2)}(q^{(j_1)}, q^{(j_2)}) + A^{(2)}(q^{(j_2)}, q^{(j_1)})]\, d^2x \quad . \qquad (3.22)$$

After inverting the equations (3.19) together with (3.20) one inserts the second order results into the next order equations and proceeds in the same way. In this context it usually suffices to go up to the third order coefficients.

If one has constructed the Generalized Ginzburg-Landau equations in this way up to the desired accuracy one discusses the dynamical behaviour of this low dimensional system of ordinary differential equations. This allows us not only to calculate transient motions of the physical system but also intrinsically time dependent solutions, if for instance the instability of the basic stationary state leads to periodic, quasi-periodic or chaotic motion.

Let us remark that in the case of multiple eigenvalues because of some symmetry it is possible to use group theoretical methods to determine the vanishing terms of the series (3.16) (see Friedrich and Haken, 1986). For questions of convergence and estimation of rest terms see Haken (1983).

4. RESULTS FOR TWO-PHASE RAYLEIGH-BENARD PROBLEM

For the state variables, we take into account terms up to second order:

$$q(\vec{x},t) = \sum_{j=-3}^{3} \varepsilon_j\, \hat{q}(z)\, e^{i\underline{k}_j \cdot \underline{x}} + \frac{1}{2} \sum_{jj'} \varepsilon_j \varepsilon_{j'}\, \hat{q}^{(jj')}(z)\, e^{i(\underline{k}_j + \underline{k}_{j'}) \cdot \underline{x}} \qquad (4.1)$$

Because we are mainly interested in the stability of roll patterns relative to the competing hexagonal patterns, we perform a six-mode analysis. That means we allow six disturbances with (two-dimensional) wave vectors $\underline{k}_j$, having sixty degree angles, to become unstable. Reality of $q(\vec{x},t)$ implies

$$\varepsilon_{-j} = \varepsilon_j^* \qquad (4.2)$$

The hexagonal symmetry of the problem determines the form of the remaining three Generalized Ginzburg-Landau equations:

$$\partial_t \varepsilon_1 = \lambda\, \varepsilon_1 + A\, \varepsilon_2^* \varepsilon_3^* - \varepsilon_1\, (B|\varepsilon_1|^2 + C|\varepsilon_2|^2 + C|\varepsilon_3|^2) \quad , \qquad (4.3)$$

where one obtains the two other equations by cyclic permutation of the indices. There arise only four independent coefficients λ, A, B and C, which can be computed numerically. Depending on their values, these equations allow for stable roll-like or hexagonal cells (Sattinger, 1977). In order to get transient solutions, we integrate the equations of motion (4.3) and get the time-dependent variables by use of (4.1).

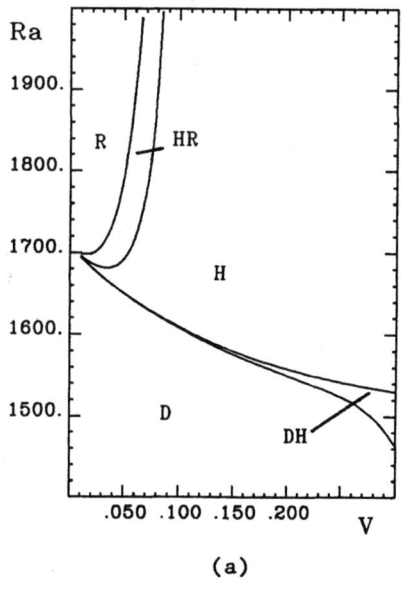
(a)

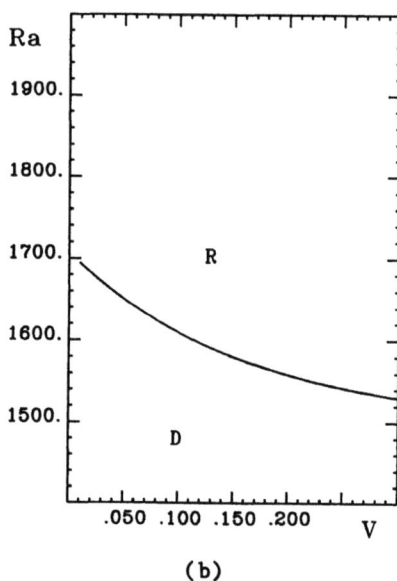
(b)

Fig. 2. Phase diagrams of the TRB system for
(a) cyclohexane (Pr = 17.6, St = $103700/Ra/(V+1)^3$) and
(b) lead (Pr = 0.023, St = $5450/Ra/(V+1)^3$).
The different stability regions are denoted by D : pure heat diffusion; DH : diffusion and hexagonal convection coexist; H : hexagons; HR : hexagons and rolls; R : rolls.

As a consequence of the reflection symmetry about the horizontal midplane, the coefficients A vanishes in the ordinary Rayleigh-Benard case (V = 0). This is the reason for the purely roll-like convection there. This is not the case for V ≠ 0. Here, generally one gets a subcritical bifurcation, which yields hexagonal structure, a jump of the mode amplitudes at the instability point and hysteresis. This already has been reported by Davis et al. We evaluated the phase diagrams for various substances. An error estimate shows to which extend our analysis is of quantitative validity. Fig. 2a and 2b show the results for the data of cyclohexane and for lead. One may generally ascertain that because of the high thermal conductivity of metals and alloys the breaking of the reflection symmetry is strong enough to produce behaviour completely distinct from that of the pure convection system only for very thick solid layers. This is different from the case of poor thermal conductors. Fig. 3. represents a cross section through a stable hexagonal cell, Fig. 4 through roll cells. For the corresponding pattern evolution in horizontally finite geometries see Bestehorn and Haken (1984).

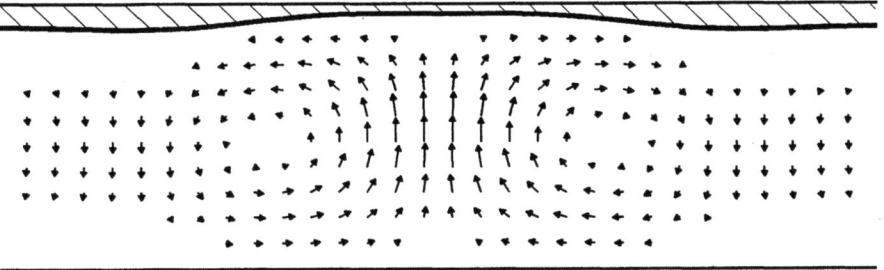

Fig. 3. Cross section through a cyclohexane two-phase layer
(V = 0.08, Ra = 1786) showing hexagonal flow.

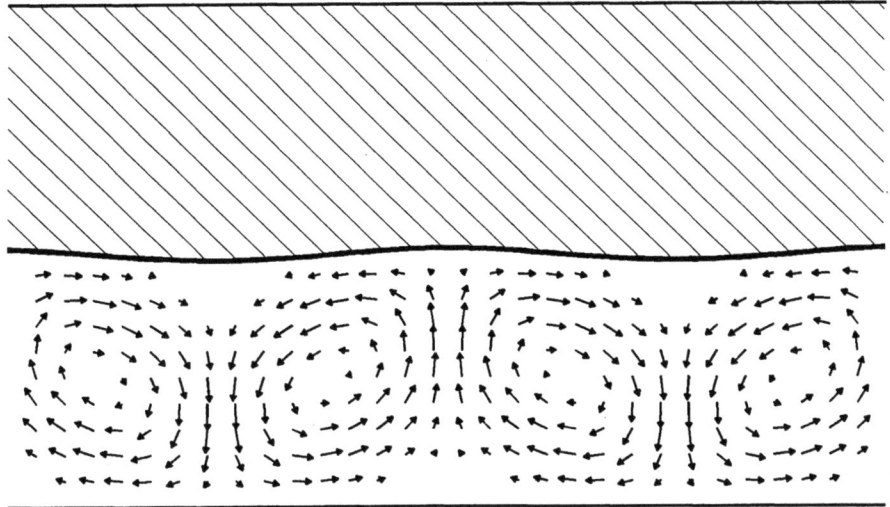

Fig. 4. Cross section through a layer of lead (V = 1, Ra = 1792)
showing roll-like convection.

5. CONCLUSIONS

We think that the method presented above is a very useful tool especially for nonlinear moving interface systems. We do not share the pessimism of some colleagues on the validity of results obtained by weakly nonlinear analysis. Comparison of calculations using our method and direct numerical simulation of hydrodynamical systems (Marx and Haken, 1986) show quantitative agreement in a significantly finite region near instability points. This allows us not only to construct solutions for continuous transitions but also for discontinuous bifurcations as far as the amplitude jumps are not too large. This capability and the possibility of calculating three-dimensional and time dependent solutions are the main advantages of this method.

In a forthcoming paper we will present results for a similar two-phase system showing a helical instability, which has been examined experimentally and by means of a linear theory by Fang et al. (1985). This system shows a more complex temporal behaviour.

It has been demonstrated (Friedrich and Haken, 1986), that hydrodynamic systems may undergo a series of transitions to different complicated time dependent flows in the very vicinity of a critical parameter point if several different modes become unstable simultaneously. At those points, secondary bifurcations and the transition to turbulence become accessible to weakly nonlinear analysis. This gives us the opportunity to investigate the important question how turbulence is altered by the presence of a crystal-melt interface, and how the interaction depends on the solidification parameters.

ACKNOWLEDGMENTS

This work has been supported by the Deutsche Forschungsgemeinschaft. One of us (T.G.) would like to thank the colleagues of the institute, especially R. Friedrich, for very helpful discussions.

REFERENCES

Bestehorn, M., and Haken, H., 1984, Z. Phys. B., 57:329.
Davis, S. H., Müller, U., and Dietsche, C., 1984,
    J. Fluid Mech., 144:133.
Fang, Q. T., Glicksman, M. E., Coriell, S. R., McFadden, G. B., and
    Boisvert, R. F., 1985, J. Fluid Mech., 151:121.
Friedrich, R., and Haken, H., 1986, Phys. Rev. A, 34:2100.
Glicksman, M. E., Coriell, S. R., and McFadden, G. B., 1986,
    Ann. Rev. Fluid Mech., 18:307.
Haken, H., 1975, Z. Phys. B, 21:105.
Haken, H., 1983, "Advanced Synergetics", Springer, Berlin.
Marx, K., and Haken, H., 1986, to be published.
Mullins, W. W., and Sekerka, R. F., 1964, J. Appl. Phys., 35:444.
Normand, C., Pomeau, Y., and Velarde, M. G., 1977,
    Rev. Mod. Phys., 49:581.
Sattinger, D. H., 1977, Arch. Rat. Mech. Anal., 66:31.
Schlüter, A., Lortz, D., and Busse, F., 1965,
    J. Fluid Mech., 23:129.
Wunderlin, A., and Haken, H., 1981, Z. Phys. B, 44:135.

DENDRITIC PATTERN FORMATION

Y. Saito*, G. Goldbeck-Wood and H. Müller-Krumbhaar

Institut für Festkörperforschung der KFA Jülich Postfach 1913, 5170 Jülich, Germany

*Permanent address: Physics Dept., KEIO University Yokohama, Japan

EXTENDED ABSTRACT

We present a numerical analysis of dendritic crystal growth. The growth of dendritic crystals occurs when the relevant transport of heat or matter takes place in the medium ahead of the advancing interface, such that the crystallization front moves into a supercooled or supersaturated region.

A dendrite then is an approximately parabolic needle-crystal with tree-like sidebranches. Under constant experimental conditions it advances at a well-defined velocity, producing sidebranches at an approximately constant rate.

The system can be parameterized by three independent length scales: a) the capillary length d_o being a measure of surface tension and stiffness, b) the critical radius R_c being a measure of the supercooling, c) the diffusion length as a measure of diffusion speed.

Detailed definitions can be found in ref. |1|, a series of precision experiments are reported in ref. |2|. The first approximate solution to this problem |3| ignoring surface-tension leads to a parabolic form of the growing crystal, the Ivantsov-parabola. It gives a relation $V \sim R_o^{-1}$ between the growth rate V and the radius of curvature R_o at the tip of the parabola. Including surface tension in an approximate form it was later argued |1|, that the stability length $\lambda_s = 2\pi \sqrt{\ell d_o}$ plays a crucial role for the velocity selection and sidebranch spacing. More recently it was found |4-6|, that crystalline anisotropy is needed to allow for a discrete set of stationary needle crystals to be solutions of the problem.

The fastest of these needle crystals was conjectured to describe the operating mode of the dendrite, as it is weakly stable against the formation of sidebranches |5,7,8|.

The presently available analytical results, however, are still not conclusive about the sidebranching mode of operation, the usual experimental finding. We have, therefore, performed a numerical analysis using a Green's function technique |9|. The results are summarized as follows. The supercooling should be expressed by the Peclet number $p = R_o/\ell$, which seems to make the result independent of dimensionality. The growth rate then can be scaled for different supercoolings and anisotropies as

$$\sigma = \frac{d_o}{2Dp^2} V \qquad (1)$$

with D being the diffusion coefficient. Then $\sigma = \sigma(\varepsilon)$ is a function of the relative anisotropy ε of the surface tension only (kinetic anisotropy was not considered), whose functional relation is given in |4,10|. The wavelength λ of the sidebranch-spacing, defined as the distance travelled by the dendrite on average between generation of two neighboring branches, scales with the stability length |6,0-8|

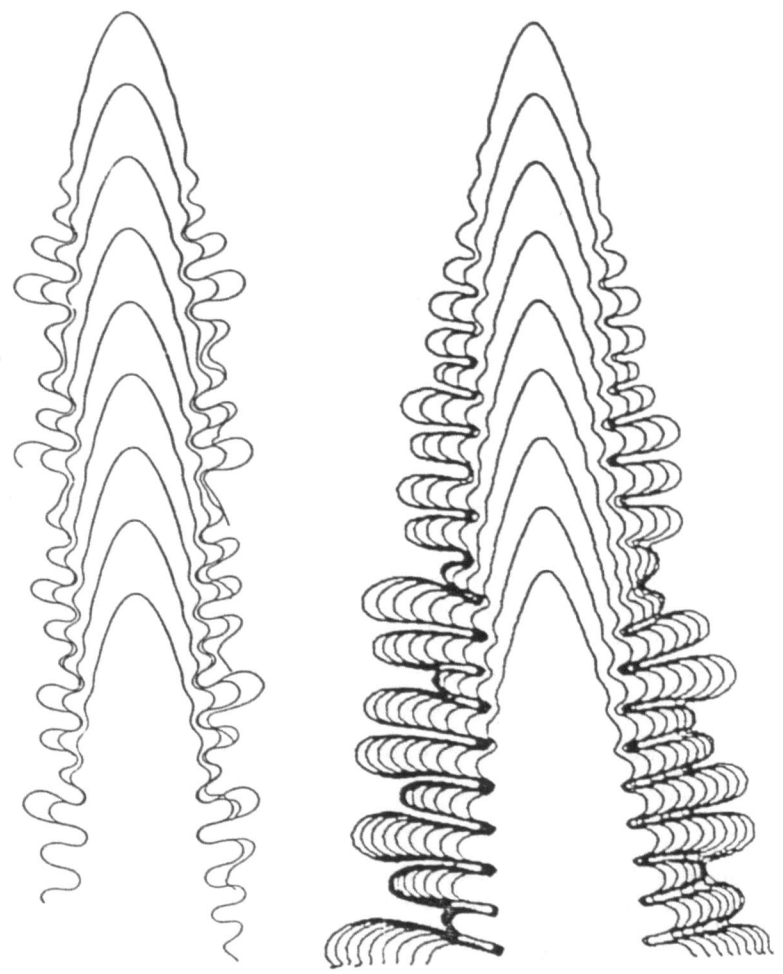

Fig. 1: Computed dendrite (left) |9| compared with experimental dendrite (right) |11|.

$$\lambda \gtrsim \lambda_S \qquad (2)$$

the factor being about 1.3. The overall shape of the primary dendrite averaged over the sidebranches is almost parabolic, the deviation of tip radius from R_o follows the prediction of ref. |4|. A comparison of our simulations |9| with recent image-processing results of experimental dendrites |11| shows striking similitary. (Fig. 1). There are differences as the experiment shows a projection of a three dimensional dendrite where the sidebranches grow narrower than in our two dimensional calculation. Our calculation was furthermore restricted to the tip region (with a smoothing approximation for the tail), while experimentally there is very little known about anisotropy and stiffness of the surface tension, not to speak of kinetic coefficients. Since our previous theory |1|, ignoring the importance of anisotropy, gave the same scaling results for the dependence of V and R on the other parameters, the numerical calculations presented here give the first confirmation for the validity of the more recent results concerning anisotropy.

REFERENCES

|1| J.S. Langer, Rev. Mod. Phys. **52**, 1 (1980); J.S. Langer, H. Müller-Krumbhaar, Acta. Metall. **26**, 1681; 1689, 1697 (1978); H. Müller-Krumbhaar, J.S. Langer, Acta. Metall. **29**, 145 (1981).
|2| M.E. Glicksman, R. Schaefer, J. Ayers, Metall. Trans. A **77**, 1747 (1976); S. Huang, M. Glicksman, Acta. Metall. **29**, 701, 717 (1981); M.E. Glicksman, B. Singh, in "Solidification and Fluid-Dynamics", Proceedings of a workshop at Giesserei-Institut der RWTH-Aachen (Aachen, 1984).
|3| G. Ivantsov, Dokl. Akad. Nauk SSSR **58**, 567 (1947).
|4| M. Ben-Amar, Y. Pomeau, Europhysics Lett. **2**, 307 (1986) and "Theory of the Needle Crystal": These Proceedings.
|5| D.C. Hong, J.S. Langer, Phys. Rev. Lett. **56**, 2032 (1986).
|6| D. Kessler, H. Levine, Phys. rev. A **33**, 2621, 2634 (1986); D. Kessler, H. Levine, "Stability of Dendritic Crystals", Phys. Rev. Lett. (1986).
|7| B. Caroli, C. Caroli, C. Misbah, B. Roulet (preprint).
|8| M.N. Barber, A. Barbieri, J.S. Langer (preprint).
|9| Y. Saito, G. Goldbeck-Wood, H. Müller-Krumbhaar, Phys. Rev. Lett. **58**, 1541 (1987); Proceedings of the EPS conference, Pisa 1987, to appear in Physica Scripta.
|10| C. Misbah, J. de Physique **48**, 1265 (1987).
|11| A. Dougherty, P. Kaplan, J. Gollub, Phys. Rev. Lett. **58**, 1652 (1987).

CELLULAR MORPHOLOGICAL INSTABILITIES DURING DIRECTIONAL SOLIDIFICATION : THIN SAMPLE EXPERIMENTS

S. de Cheveigné, C. Guthmann, and M.M. Lebrun

Groupe de Physique des Solides de l'Ecole Normale Supérieure, Université Paris VII, Tour 23, 2 Pl. Jussieu 75251 Paris-Cedex 05

Metallurgists are familiar with the existence of morphological instabilities[1] which deform the solid-liquid interface of binary alloys during directional solidification. In this situation, samples are pulled at velocity V in a temperature gradient G so that they are progressively solidified. The equilibrium concentration of the solute in the liquid is usually different from that of the solid - often higher, as with the compounds studied here, but sometimes lower. Then an excess or a lack of solute builds up in front of the moving interface and must be evacuated to allow solidification to progress. Above a threshold pulling speed, for given temperature gradient and concentration, the interface becomes unstable and breaks down into a regular, periodic pattern (Fig. 1).

Thin samples of transparent organic compounds, as first suggested by Jackson and Hunt[2] in 1966, allow direct observation, under the microscope, of the dynamics of the solid-liquid interface.

Some organics, in general plastic crystals, present a rough interface which does not facet, and in that respect they behave like metals. The most frequently studied are tetrabromomethane (CBr_4) and succinonitrile $(CH_2CN)_2$.

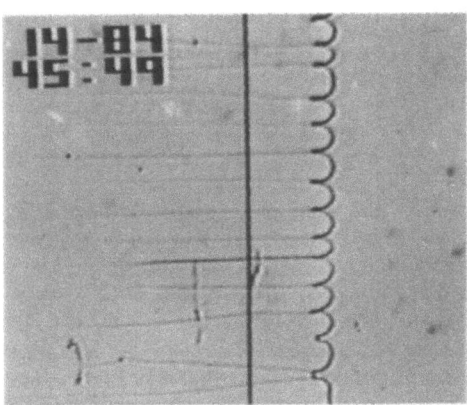

Fig. 1   The solid-liquid interface above threshold ($CBr_4, \lambda \cong 50 \mu m$)

Table 1 - Physico-chemical constants for CBr_4

Molecular weight : 331.65 g
Density 3.42
Latent heat of fusion : 2.41 cal g^{-1} [17]
Solid-liquid interface tension : $\gamma = 1.675 \cdot 10^{-7}$ cal cm^{-2} [18]
Melting point : 93.3 ± .2°C
Specific heat, plastic phase : $c_{pP} = 0.13$ cal K^{-1} g^{-1} [18]
Specific heat, liquid phase : $c_{pL} = 0.11$ cal K^{-1} g^{-1} [18]
Structure : fcc in the plastic phase
Viscosity at 100.7° : 0.0244 g/cm s
Thermal conductivity of solid : 1.1 ± 0.1 W m^{-1} K^{-1} [19]

The work described here was done to gain better knowledge of the behavior of the system at, or near, the threshold. Most of the experiments were performed with CBr_4, but the results were generally checked with succinonitrile, both compounds being used as supplied. Impurity concentrations were therefore fixed, pulling speeds, temperature gradient, and sample thicknesses were varied.

Table 1 gives data concerning CBr_4. We use the compound as provided by Fluka, without further purification. The main impurity is given as Br_2. Differential Scanning Calorimetry (DSC) measurements[3] on the material gave us an impurity concentration $c_\infty = 0.12 \pm 0.02$ mol. %, a liquidus slope m = 2.9 ± 0.2 K/mol.% and a partition coefficient k = 0.16 ± 0.01. A melting temperature of 93.3 ± 0.2°C for the pure compound was calculated. We do not know the diffusion coefficient for Br_2 in CBr_4 but we shall use the value of the threshold pulling speed below to estimate a value of D_L 1.2 ± 0.4 10^{-5} cm^2/s.

Succinonitrile has been extensively characterized both when pure and when mixed with acetone, by Glicksman[4]. We performed DSC measurements and calculations on our material (Koch-light and Eastman) ; they gave c_∞= 0.45±.05 mol. %, m = 2.20 ± 0.05 K/mol.% and k=0.12 ± 0.01. The melting temperature for pure succinonitrile was calculated as 58.1 ± 0.3°C.

The pulling stage is essentially similar to that proposed by Jackson and Hunt[2]. Sample thicknesses are measured to within ± 2-3 μm and are uniform to within ±5 μm. The temperature of the sample is imposed by the glass cell which is a far better conductor of heat than the sample itself. Particular care was taken to avoid residual vertical temperature gradients which strongly affect the experimental results. The experiments are video-taped which allows the pulling speed to be measured precisely (±0.1 μm/s). Wavelengths are averages over at least 10 cells.

We wish to observe the solid-liquid interface in a stationary state and this may take a long time. In figure 2, we have plotted the time it takes the interface to *begin* to deform visibly after a given pulling speed is suddenly applied (the time it takes to reach a steady state is even longer) versus the pulling speed V for CBr_4 ; the points we obtain for succinonitrile fit the same curve, as do most of the response times reported by Somboonsuk et al[5] for changes from one pulling speed to another with a succinonitrile-acetone solution. The final speed is the pertinent parameter since it is at that speed that the concentration profile adjusts.

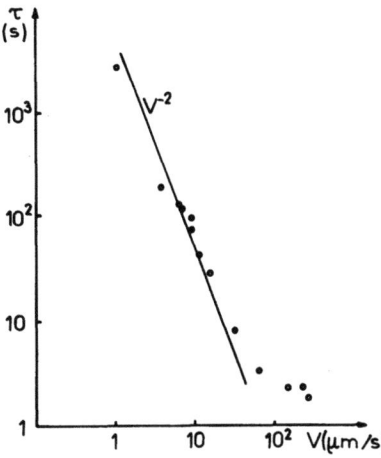

Fig. 2   Response time versus pulling speed.

The response time should in fact contain two terms. The first, a solute diffusion term, is roughly the time it takes to establish the steady state concentration profile after beginning to solidify a sample. This has been estimated by Tiller et al.[6] as :

$$\tau_1 = \frac{D_L}{kv^2}$$

(It is probably greater than the response time which we define just as the time it takes the deformation to begin to appear.) The second term is due to the thermal response of the glass cell and is expected to be preponderant at higher pulling speeds. It is the value at which τ levels off : about 10 s in the present case. The thermal component is negligible at speeds lower than ~ 10 µ/s. At pulling speeds of 10 µ/s, one must wait about 60 s for the system to react, at 1 µ/s about 30 min and extrapolating to 0.1 µ/s, about 16 hours ! Such times make extensive and repeated studies extremely difficult. That is why we prefer to work under conditions (in particular at low concentrations) where critical speeds are above 1 µ/s.

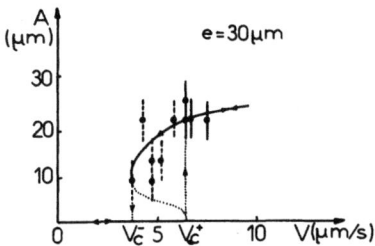

Fig. 3   Amplitude of the periodic deformation versus pulling speed (e=30µm, G=120°/cm). Full segments:increasing speed. Dashed segments : decreasing speed. Curve : interpretation in terms of a subcritical bifurcation.

The bifurcation from a planar to a cellular morphology of the interface is subcritical, i.e. the critical pulling speed V_c^+ for increasing speeds is greater than the critical value V_c^- for decreasing speeds. Fig. 3 shows the amplitude of the deformations of the interface (measured from the cell tip to the point where the cell sides meet) versus the pulling speed. Above V_c^+ a *finite* amplitude deformation sets in. The non-linear analysis of the problem made by Wollkind and Segel[7] and extended by Caroli et al[8], predicts this result. It is of practical consequence for further calculations, since it means that the amplitude of the deformations is never infinitesimally small. In that case, the amplitude equation cannot be used above the threshold.

The linear theory of Mullins and Sekerka[9] predicts a critical pulling speed (V_c^+ in the notation of the previous paragraph) which reduces, at sufficiently low critical temperature gradients G_c, (equal in the liquid and in the solid here since the glass cell imposes the temperature gradient), to what is known as the "constitutional supercooling criterion" (6):

$$\frac{G_c}{V_c} = \frac{C_\infty m(k-1)}{k D_L}$$

(m is the liquidus slope). We used this relation (for relatively thick samples - see explanation below) to estimate the diffusion coefficient as $D_L = 1.2 \pm 0.4 \cdot 10^{-5}$ cm^2/s. This value is quite reasonable and we shall use it for further calculations.

Our experimental values for the wavelength versus the pulling speed are shown on Figure 4, for two different temperature gradients. (The samples considered were thicker than 50 μm to allow comparison with the predicted thresholds - see explanation below). The neutral stability curve deduced from the linear analysis is shown, as well as the value of the fastest growing wavelength[8,9]. According to the linear theory, periodic deformations of wavelengths within the band delimited by the curve will not *a priori* relax, whereas those outside the band will, but no clear selection mechanism within the band has been described theoretically.

Experimentally a wavelength *is* selected, to within a few tens of percent. The threshold value λ_c observed experimentally is 2 to 3 times smaller than the value of the critical wavelength λ_{MS} predicted by the linear analysis (180μ versus 75-80μm for G = 70°/cm, 125μ versus 50-55 μm for G = 120°/cm). When transitory deformations are observed below the threshold pulling speed, λ is greater than λ_c but does not reach λ_{MS}. This is consistent with the existence of a subcritical bifurcation since the prediction of λ_{MS}, made under the hypothesis of infinitesimal deformations, does not hold. We find roughly the same relation between λ_c and λ_{MS} in succcinonitilrile. A ratio 2 has been reported in Pb-Tl by Billia et al[10]. The same ratio can be deduced from the work of Sato et al[11] on Al-Cu (using the value of the surface tension reported by Miyata and Suzuki[12]). One may be observing the second bifurcation, predicted by Ungar and Brown[13], at which the critical wavelength λ_c should jump to $\lambda_{MS}/2$.

Above the threshold, λ varies roughly as $v^{-1/2}$ (we find a functional dependence of $v^{-0.4\pm0.1}$). The $\lambda(V)$ curve does not follow the fastest growing wavelength but is somewhat above it, except close to the threshold. We find λ slightly lower for the lower temperature gradient at a given speed (Fig. 4) but the shift is barely greater than our experimental dispersion. More work, on different systems, is clearly necessary to confirm the experimental picture. We insist on the importance of doing experiments sufficiently slowly : if a pulling speed above the critical value is applied suddenly, a transient pattern of wavelength smaller than the steady state wavelength is observed. This effect, which has been mentioned by various authors[5,14], can be a source of error in measurements of λ, particularly in the case of high response times, i.e. low pulling speeds.

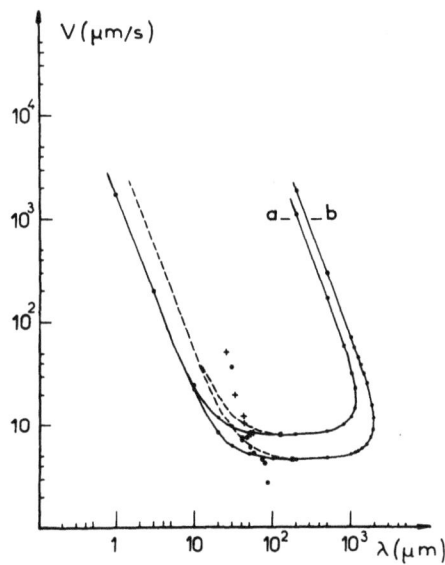

Fig. 4 : Wavelength versus pulling speed for two different temperature gradients. The full line is the calculated neutral stability curve ; the dashed line represents the fastest growing wavelength ; the points are experimental. The shade strip gives the experimental dispersion (e > 50 µm ; (a) and full circles : G = 120°/cm ; (b) and empty circles : G = 70 °/cm).

Sample thicknesses are of the order of tens of micrometers, compared to lengths and widths of several centimeters, and the problem is usually treated theoretically as two dimensional. In reality, sample thickness is an essential parameter : the solid liquid interface becomes less and less stable when the sample thickness decreases. Fig. 5 gives the dependence we observe for the critical pulling speed V_c^+ versus sample thickness, for two temperature gradients (70 and 120°/cm.). We have observed a similar effect with succinonitrile. The curve of V_c^+ versus e

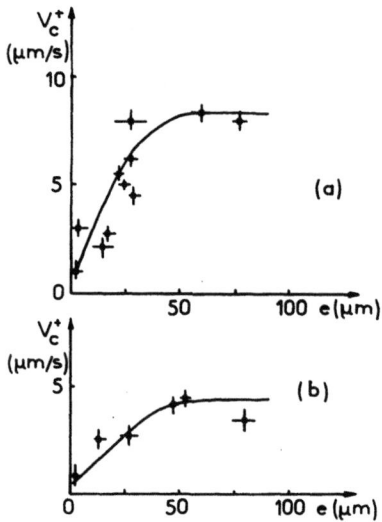

Fig. 5 Critical pulling speed V_c^+ sample thickness for different temperature gradients. a) G = 120°/cm ; b) G = 70°/cm.

flattens off roughly at the pulling speed at which the wavelength, given by Fig.4, equals the sample thickness e.

Convection in the liquid could have been a possible explanation for this effect. We have checked that none takes place, at least on a time scale comparable to that of the phenomena under study, by introducing small (5 μm) particles of silica (of roughly the same density as CBr_4). They were only seen to move in one case where thermocapillary convection was taking place on the liquid-gas interface of a bubble.

The solid-liquid meniscus seems to be the cause of the strong dependence of the critical pulling speed on the sample thickness. The interface is curved to assure mechanical equilibrium at the point of contact with the glass and the contact angle depends on the pulling speed because the solute distribution along the interface varies. The shape of the interface adjusts to assure local thermodynamic equilibrium (Gibbs-Thomson relation). A similar effect may well be responsible for the unusual tip velocity - tip curvature relation observed by Sawada and co-workers[15] for dendritic growth in very thin (5 μm) samples.

Caroli, Caroli and Roulet[16] have calculated the meniscus analytically to the third order and then performed a linear stability analysis of the solid-liquid interface (in a symmetric model). This calculation is valid only for a very weakly curved meniscus, but it shows qualitatively that the effect of the curvature is to decrease the critical pulling speed for thicknesses smaller than the cellular wavelength ; the shift increases as thickness decreases.

Because of this dependence of the critical pulling speed on the sample thickness, one must be cautious in comparing it to the value given by the "constitutional supercooling criterion". The relation is expected to apply to thick samples for which the effect of the meniscus is weak. It is for this reason that we used the value of V_c^+ at which $V_c^+(e)$ levels off to calculate D_L and that we drew Fig. 4 for samples with thicknesses in the plateau region.

The wavelength is apparently not thickness-dependent, except when it becomes smaller than 1/2 to 1/3 of the sample thickness. At that point, several layers of cells are observed and λ is larger than expected according to the relation $\lambda^2 v$ = constant.

We have found the bifurcation from a planar to a cellular interface to be subcritical, as predicted by the non-linear analysis of the problem. The critical wavelength, on the otherhand is 2 to 3 times smaller than predicted by the same theory. Response times are shown to become very long at low pulling speeds and to be a possible source of experimental error. The critical pulling speed is strongly dependent on the sample thickness. This last result should be stressed : the sample cannot be considered two-dimensional.

References

1. For a general review see J.S. Langer, Rev. Mod. Phys. **52** (1980) 1; the special issue of "Materials Science and Engineering" on Solidification Microstructure, vol **65** (1984) illustrates recent metallurgical preoccupations
2. K.A. Jackson and J.D. Hunt, Trans. Met. Soc. AIME **236** (1966) 1929
3. S. de Cheveigné, C. Guthmann, M.M. Lebrun, to appear in Journal de physique
4. M.E. Glicksman, R.J. Shaefer and J.D. Ayers, Metall. Trans. **A7**, (1976) 1747
5. K. Somboonsuk and R. Trivedi, Acta Metall., **33** (1985) 1051
6. W. A. Tiller, K. A. Jackson, J. W. Rutter, B. Chalmers, Acta Metall. **1** (1953) 428
7. D. Wollkind and L. Segel, Phil. Trans. R. Soc. (Lond.) **268** (1970), 351
8. B. Caroli, C. Caroli and B. Roulet, J. Physique **43** 1767 (1982)
9. W.W. Mullins and R.F. Sekerka, J. Appl. Phys. **35** (1964), 444
10. B. Billia, H.Jamgotcian and L.Capella, J. Crystal Growth **66** (1984) 586
11. T. Sato, G. Ohira, J. Crystal Growth **40** (1977) 78
12. Y. Miyata, T. Suzuki, J.-I. Uno, 1799 ; Y. Miyata, T. Suzuki, Metall. Trans **16A** (1985) 1807
13. L.H. Ungar and R.A. Brown, Phys Rev B **29** (1984) 1367 and Phys Rev B **31** (1985) 5931
14. R.M. Sharp and A. Hellawell, J. Crystal Growth **6** (1970) 334-340 and J. Crystal Growth **11** (1971) 77-91
15. H. Honjo, S. Ohta and Y. Sawada, Phys. Rev. Lett., **55** (1985) 841
16. B.Caroli, C.Caroli, B.Roulet,J.Crystal Growth **76** (1986) 31
17. K.J. Frederick and H.J. Hildebrand, J.A.C.S. **61** (1939) 1555
18. W.F. Kaukler and J.W. Rutter, Mat. Sci. Eng. **65** (1984) L1
19. P. Andersson and R.G. Ross, Mol. Phys., **39** (1980) 1379

TRANSPORT PROCESSES DURING DIRECTIONAL SOLIDIFICATION AND CRYSTAL GROWTH :

SCALING AND EXPERIMENTAL STUDY

D. Camel and J.J. Favier

CEA/IRDI/DMG/SEM-Laboratoire d'Etude de la Solidification
Centre d'Etudes Nucléaires, 85 X, 38041 Grenoble Cedex (F)

INTRODUCTION

As it is well known, solute transport during solidification and crystal growth is very easily affected by convection[1]. Thus numerous classes of phenomena may appear depending on the type of couplings occurring between the different involved physical quantities, namely : the temperature field, the concentration field, the interface location and morphology, and the flow field. Moreover, for each class of phenomena, a wide variety of boundary conditions may prevail depending on the particular solidification method under consideration (Czochralski or Bridgman crystal growth, zone melting, casting, welding, surface melting ...).

Powerful numerical methods have been developed which already allowed a much better understanding of various convective effects[2,3]. However, there also appears a need for developing scaling analysis, analytical modelling as well as parametric experimental studies in this field.

This has been tackled in a systematic way to analyse segregation during crystal growth[6,7], morphological stability[8,9], as well as eutectic and dendritic solidification. In the present paper, we particularly consider the simple case where the driving force for convection is external to the solute boundary layer, and the solidification front is planar - which is representative of crystal growth from the doped melt. The methods of order of magnitude analysis are used to define scaling laws for the solute boundary layer extent and concentration in the presence of convection. These laws are introduced into analytical boundary layer models predicting segregation or morphological stability thresholds. At each step, experimental methods designed to check the theoretical predictions are presented. Finally, the extension of this approach to other situations such as eutectic or dendritic solidification is suggested.

PLANAR CRYSTAL GROWTH FROM A DILUTE LIQUID ALLOY

Physical description

The physical system considered is schematically shown on Fig. 1. A doped melt is solidified with a planar growth front by pulling through a gradient furnace at a velocity R. The solute which is rejected builds up in front of the growth interface, thus forming a solute boundary layer, and is transported into the bulk liquid by diffusion and/or convection.

In order to write down the governing equations, the following classical assumptions are made :

The liquid is Newtonian, the Boussinesq approximation is valid, linear phenomenological relations apply with no cross coupling, and transport properties do not depend on temperature and concentration. In addition, the solute content is supposed to be sufficiently low so that density differences are only due to temperature differences. Then, the field equations for mass, momentum, heat and solute in the liquid can be written as follows :

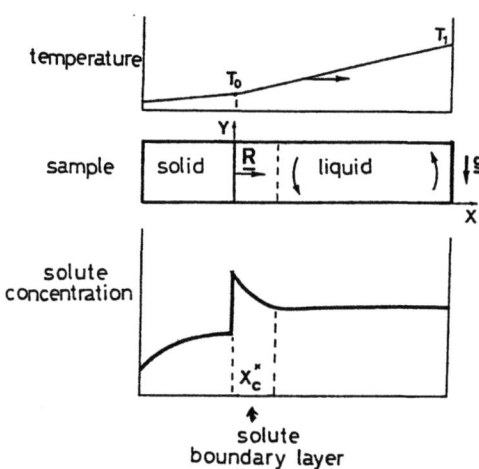

Fig. 1. Schematic representation of the growth of a crystal from the melt.

$$\underline{\nabla} \cdot \underline{V} = 0, \qquad (1\text{-a})$$

$$-\beta_T (T - T_0) \underline{g} - \rho_0^{-1} \underline{\nabla} P_d + \nu \nabla^2 \underline{V} - (\underline{V} \cdot \underline{\nabla}) \underline{V} = 0, \qquad (1\text{-b})$$

$$\alpha \nabla^2 T - \underline{V} \cdot \underline{\nabla} T = 0, \qquad (1\text{-c})$$

$$D \nabla^2 C - \underline{V} \cdot \underline{\nabla} C = 0, \qquad (1\text{-d})$$

where, $\underline{V}$ is the barycentric fluid velocity vector, P_d the dynamic pressure, T the temperature, C the solute concentration, ρ_0 the fluid density at T_0, β_T the thermal expansion coefficient, $\underline{g}$ the gravity vector, ν the kinematic viscosity, α the thermal diffusivity and D the solute diffusion coefficient.

In order to simplify the notation, we will consider a two-dimensional domain and a planar growth interface. X and Y will be the coordinates respectively perpendicular and parallel to the interface (Fig. 1) and, U and V the corresponding components of the velocity vector $\underline{V}$. This is not a fundamental restriction for the scaling analysis to be performed below : Napolitano showed that the mathematical formulation could be generalized to a three dimensional domain and a non-planar interface.

With these notations, the balance equations at the growth interface are written :

$$U(0,Y) = -R \tag{1-e}$$

$$V(0,Y) = 0 \tag{1-f}$$

$$T(0,Y) = T_M \tag{1-g}$$

$$D\left(\frac{\partial C}{\partial X}\right)_{X=0} + R |C(0^+,Y) - C(0^-,Y)| = 0 \tag{1-h}$$

where, R is the growth rate, T_M is the melting point, and $C(0^+,Y)$ and $C(0^-,Y)$ are the solute concentrations at the interface in the liquid and solid respectively. Thermodynamic equilibrium is supposed at the solid-liquid interface :

$$\frac{C(0^-,Y)}{C(0^+,Y)} = k,$$

where k is the equilibrium partition ratio.

Let H be a characteristic dimension, T_0 and T_1 two characteristic temperatures. Introducing the following arbitrary set of non-dimensional quantities :

$$x = \frac{X}{H}, \quad y = \frac{Y}{H}, \quad \underline{v} = \underline{V}\frac{H}{\nu}, \quad \theta = \frac{T - T_0}{T_1 - T_0}, \quad \pi = \frac{H^2}{\rho_0 \nu^2} P_d, \tag{2}$$

and arbitrary non-dimensional concentrations c, eq. (1) are transformed into the following non-dimensional form :

$$\underline{\nabla} \cdot \underline{v} = 0, \tag{3-a}$$

$$-Gr\, \theta \frac{g}{|g|} - \underline{\nabla} + \nabla^2 \underline{v} - (\underline{v} \cdot \underline{\nabla})\underline{v} = 0, \tag{3-b}$$

$$\nabla^2 \theta - Pr\, \underline{v} \cdot \underline{\nabla}\theta = 0, \tag{3-c}$$

$$\nabla^2 c - Sc\, \underline{v} \cdot \underline{\nabla}c = 0, \tag{3-d}$$

and

$$u(o,y) = -Re, \tag{3-e}$$

$$v(o,y) = 0, \tag{3-f}$$

$$\theta(o,y) = \theta_M, \tag{3-g}$$

$$\left(\frac{\partial c}{\partial x}\right)_{x=0} + Pe\, |c(o^+,y) - c(o^-,y)| = 0 \tag{3-h}$$

with

$$Gr = \frac{\beta_T |g| (T_1 - T_0) H^3}{\nu^2}, \quad Pr = \frac{\nu}{\alpha}, \quad Sc = \frac{\nu}{D}, \tag{3-i}$$

$$Re = R\frac{H}{\nu}, \quad Pe = Re\, Sc,$$

Depending on the particular conditions applied along the other boundaries of the system (free surface and/or imposed velocity) two other dimensionless numbers may have to be considered :

$$Re_M = \frac{|\sigma'| \, G \, H^2}{\rho \, \nu^2} \quad , \quad Re_\omega = \frac{\omega H^2}{\nu} \, , \tag{3-j}$$

where σ' is the temperature coefficient of the surface tension, and ω is the rotation rate of crystal or crucible.

Thus, the general problem depends on seven dimensionless parameters (k, Gr, Pr, Sc, Pe, Re_M and Re_ω), the type of geometry of the crucible and the thermal conditions along the crucible wall. However, since we are presently interested in the crystal growth of metals or semiconductors from their impure melt, we will restrict the discussion to the limiting case :

$$Pr \ll 1,$$
$$Sc \gg 1,$$

For such conditions, there is a wide data range where convective solute transport occurs while heat transfer remains mainly conductive. Advantage of the resulting simplification will be taken below.

Scaling of the flow

Let v^* and x^* be the dimensionless bulk flow velocity and boundary layer extent. We first suppose that the solid-liquid interface is at rest (Re = 0). Then, as a consequence of the incompressibility of the melt and the no-slip condition at the interface, characteristic tangential and normal flow velocities at a distance x from the interface are given by :

$$v(x) = \frac{v^*}{x^*} x \qquad \text{if } x < x^* \, ,$$
$$ = v^* \qquad \text{if } x \geq x^* \, , \tag{4-a}$$
$$u(x) = v(x) \cdot x \qquad \text{if } x < 1 \, ,$$
$$ = v^* \qquad \text{if } x \geq 1 \, ,$$

When the growth interface is moving (i.e. Re $\neq$ 0), barycentric velocities relative to this interface are given by :

$$u(x) = \max\left(Re, \, u_1(x)\right) \, , \tag{4-b}$$
$$v(x) = v_1(x) \, ,$$

where $u_1(x)$ and $v_1(x)$ are the velocity components corresponding to the case $Re = 0$.

According to the type of convection prevailing in the melt (forced convection, buoyancy or capillary flow), orders of magnitude of v^* and x^* can be derived from the scaling laws recalled in table 1[4]. However, in order to get sufficiently accurate values for practical situations, estimates of the constants K entering the exact laws are needed. These can be obtained by comparing the previous scaling laws with the results of analytical or numerical calculations for the case of a fluid submitted to buoyancy in enclosures of various geometries[10]. This comparison is given in fig. 2 which shows that the constant should be close to one in the boundary layer regime but much lower than one in the Stokes regime. This can be qualitatively understood by considering that in this last regime there is a strong viscous interaction between the flows along the different boundaries of the enclosure, and thus the length scale available for the flow is not the overall dimension of the enclosure but only a fraction x_0 of it.

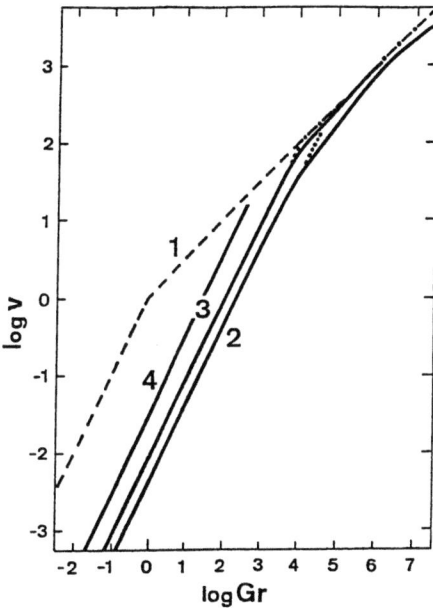

Fig. 2. Plot of the flow velocity v versus Gr
  (1) A-priori scaling laws
  (2-4) Analytical and numerical data for different geometries and aspect ratios A = L/H :
  (2) Square enclosure (A = 1)
  (3) Rectangular enclosure (A > 2)
  (4) Vertical cylinder (A ≃ 4) (with Gr defined from the radius and the maximum horizontal temperature difference).

Table 1. Characteristic length scale x* and flow velocity v* for the different convective modes and parameter ranges.

Convective mode	Parameter range	x*	v*
Buoyancy	$Gr < 1$	1	Gr
	$Gr > 1$	$Gr^{-1/4}$	$Gr^{1/2}$
Capillary flow	$Re_M < 1$	1	Re_M
	$Re_M > 1$	$Re_M^{-1/3}$	$Re_M^{2/3}$
Forced convection	$Re_\omega > 1$	1	Re_ω
	$Re_\omega < 1$	$Re_\omega^{-1/2}$	Re_ω

Then a consistent set of practical characteristic quantities can be defined by the following modified form of the criterion given by Napolitano[5] for the case where buoyancy is the dominant driving force:

$$\max(x^*) \quad ; \quad x^* \leq x_0$$
$$\frac{Gr}{v^*} x^{*2} = 1 ; \qquad (5)$$
$$v^* x^{*2} \leq 1 .$$

The solution of system (5) is given in table 2. The constant x_0 depends on the geometry of the problem, and can be defined by the condition that v^* equals the maximum flow velocity in the Stokes regime, i.e.:

$$x_0 = K^{-\frac{1}{2}}$$

If the upper surface of the melt is free, capillary forces have to be considered as possible driving forces for the flow in addition to buoyancy. Using the same approach, the surface flow velocity will be given by:

$$u = \max\left[\min\left(\frac{Re_M}{k_M}, Re_M^{2/3}\right), \min\left(\frac{Gr}{k_G}, Gr^{1/2}\right)\right] \qquad (6)$$

where $k_M = 4$ and $k_G = 48$ in the case of a planar layer of very large aspect ratio[11].

Table 2. Thermal buoyancy driven flow regimes in an enclosure as predicted from (5): the scaling laws of the momentum boundary layer x^* and the maximum flow velocity v^* are given for the different ranges of the parameter Gr. The constant x_0 depends on the geometry of the enclosure.

Regime	Domain	x^*	v^*
Stokes flow	$Gr < x_0^{-4}$	x_0	$x_0^2 Gr$
Boundary layer flow	$Gr > x_0^{-4}$	$Gr^{-1/4}$	$Gr^{1/2}$

Measurement of surface flow velocities

Bulk flow velocities are very difficult to measure in liquid metals especially in real solidification situations. When free surfaces are present, surface flow velocities can be measured more easily with the help of tracers. Then, a dominant effect of capillary forces on the flow is expected. However, very few studies have been devoted to the analysis of this type of flow in the case of liquid metals.

An experimental method has been developed[11] in order to perform such studies on earth under conditions where buoyancy can be neglected. The method consists of measuring the velocities V of inert particles at the surface of thin horizontal liquid layers (thickness H) subjected to an horizontal thermal gradient G. The method is presently applied to liquid tin. A sketch of the experimental arrangement is given in Fig. 3.

Shallow cavities of depth H, length 2.5 cm, and width 1 cm, are machined in iron plates and wetted with tin under vacuum. These plates are attached at each side on to heating (resp. cooling) blocks the temperature of which are controlled independently. The experiment is performed under high vacuum (< 2.10^{-8} Torr before heating, < 10^{-7} Torr during heating). Various mechanical devices are used in order to remove most of the initial oxide layer, and to adjust the tin quantity. The remaining oxide particles are used as tracers to visualize the flow. Velocity measurements are performed afterwards on video recordings. In the experiments performed up to now, the layer thickness H ranged from 0.3 mm to 4 mm and thermal gradients G from 5 Kcm^{-1} to 40 Kcm^{-1}. The measured dimensionless velocities u have been plotted on Fig. 4 as a function of Re_M.

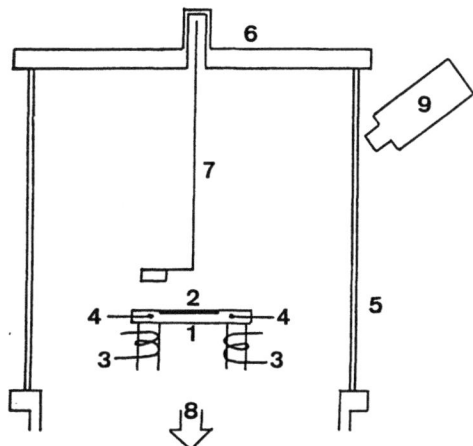

Fig. 3. Sketch of the experimental set-up for Marangoni flow velocity measurements : (1) Fe plate, (2) Sn layer, (3) heating (respectively cooling) blocks, (4) shielded thermocouples, (5) vacuum chamber (pyrex diameter 30 cm), (6) inox plate with magnetic rotation-translation lead-throughs, (7) mechanical devices, (8) ionic pumping, (9) video system.

For $Re_M < 10^3$, a linear law is obtained, which is consistent with a viscous flow regime. Comparatively curve 1 represents the theoretical prediction for pure Sn : It appears that the measured velocities are quite compatible with the σ' value of pure Sn, which indicates that there is no contamination by a surface active element in our experimental conditions.

For $Re_M > 10^3$, the experimental points are systematically lower than curve 1. The contribution of the underlying buoyancy driven flow is not negligible in this range, but should increase the surface velocity rather than decrease it. Thus the observed departure from curve 1 might be due to the occurence of a Marangoni boundary layer flow. (For comparison curve 2 gives the predicted velocities in this regime). Further experiments are needed to confirm this point.

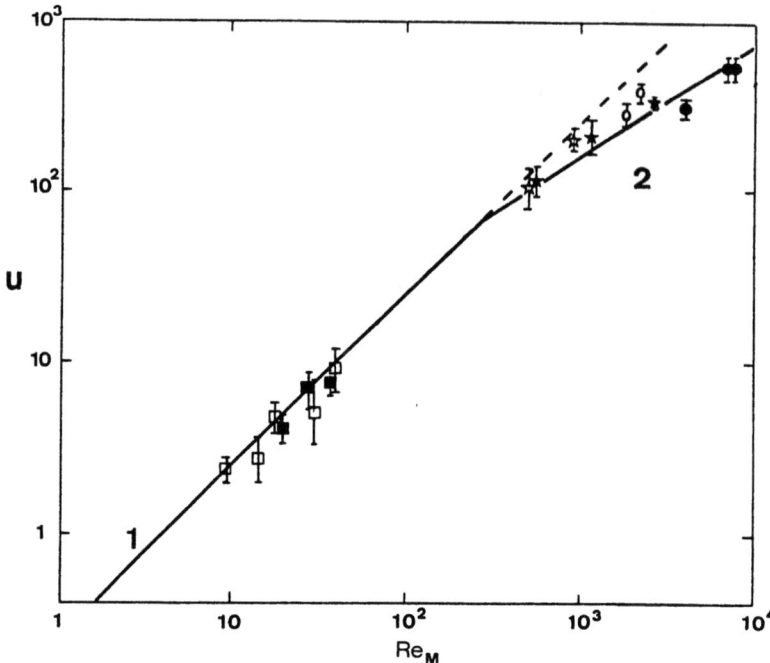

Fig. 4. Plot of the dimensionless surface flow velocity u versus Re_M for pure liquid tin layers :
Experimental values :
(■) H = 0.3 mm G > 0 ; (★) H = 2 mm G > 0 ;
(●) H = 4 mm G > 0 ; (□) H = 0.4 mm G < 0 ;
(☆) H = 2 mm G < 0 ; (○) H = 4 mm G < 0.
(Each bar corresponds to at least 10 measurements).
Theoretical curves :
(1) viscous regime (slope = 1) ;
(2) boundary layer regime (slope = ½).

Similar experiments have been performed with tin-bismuth alloys of various compositions. In the high temperature range (domain 1 of Fig. 5), the motion is in the same direction as for pure tin but slower. In the low temperature range (domain 2 of Fig. 5), the motion direction is reversed, which corresponds to a change of the sign of σ', consistent with the available data on σ for this alloy. When the temperature interval across the sample extends on both sides of the limit between domains 1 and 2, the flow separates in two cells which turn in the opposite way, the separation isotherm being reproducible to within a few degrees for a given composition of the alloy. Thus our experiment allows a precise determination of the temperature at which σ is maximum for a given concentration.

Scaling of solute boundary layer and concentration

Sample of infinite length. In the ideal situation where the sample being solidified is of infinite length, and as long as time dependent convective regimes are not considered, the problem is fully stationary. In the same way as for momentum transport, the solute boundary layer x_c^* can be defined as the scale at which the diffusive and convective contributions are of the same order of magnitude. In order to scale the convective contribution, the normal and tangential components of the flow velocity have to be evaluated within the solute boundary layer[10]. Then, the condition defining x_c^* simply reads :

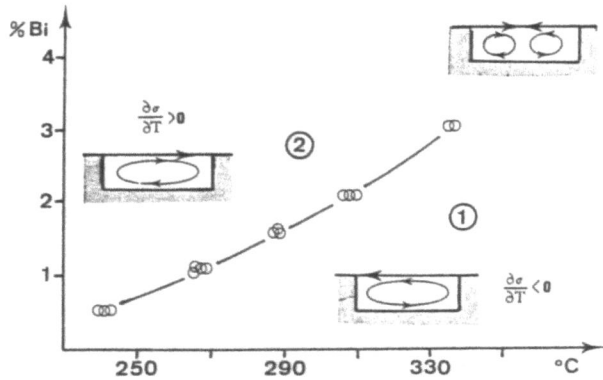

Fig. 5. Flow direction as a function of temperature and Bi content of the Sn layer : For each concentration, there is a separation isotherm above which motion is from hot to cold (domain 1), and below which motion is from cold to hot (domain 2).

$$Sc\ u(x_c^*)\ x_c^* = 1 \tag{7}$$

The solutions of systems (4) to (7) are given in table 3 for the case of buoyancy driven flow. The domains corresponding to the different solute transport regimes are drawn in Fig. 6 for the particular case : $Sc = 10$, $A = 10$, $K = 100$. Regime III, which appears in the case of a sample of finite length to be treated below will not be considered here.

Then, two main regimes have to be distinguished :

(II) When the thickness of the "stagnant" layer is larger than the diffusion length, we recover the diffusive regime ($\Delta = Pe \cdot x_c^* = 1$).

(I) In the opposite case, i.e. when the normal flow velocity in the diffusion layer is larger than the growth rate, there is a convective transport regime. The solute boundary layer is smaller than the diffusion length ($\Delta < 1$) and decreases when convection increases.

Knowing x_c^*, the scale $c(0)$ of the concentration at the growth interface is given by the solute balance at this boundary. Indeed, considering that :

$$\left(\frac{\partial c}{\partial x}\right)_0 \sim \frac{c(0) - c_\infty}{x_c^*},$$

one gets

$$k_{eff} = \frac{k\ c(0)}{c_\infty} \sim \frac{k}{1 - (1-k)\Delta} \tag{8}$$

The radial segregation Δc induced by the convective motion is at most of order $c(0) - c_\infty$. Thus :

$$\frac{\Delta c}{c(0)} \sim (1-k)\Delta. \tag{9-a}$$

This expression is only valid if $x_c^* < 1$.

When $x_c^* > 1$, Δc is only of order $c(0) - c(1)$, and

$$\frac{\Delta c}{c(0)} \sim (1-k)\ Pe. \tag{9-b}$$

Table 3. Solute transport regimes during Bridgman crystal growth as predicted from (4) to (7) : The scaling laws of the solute boundary layer x_c^* are given for the different ranges of the parameters Pe, Gr and Sc, the corresponding domains being drawn on Fig. 6.

Regime	Domain	x_c^*
III	$Pe < A^{-1}$ $x_0^2 \, GrSc < A^{-1}$	A
II	Complement to III and I	Pe^{-1}
I-a	$A^{-1} < x_0^2 \, GrSc < 1$ $Pe < x_0^2 \, GrSc$	$(x_0^2 \, GrSc)^{-1}$
I-b	$1 < x_0^2 \, GrSc < x_0^{-2}$ $Pe < (x_0^2 \, GrSc)^{1/2}$	$(x_0^2 \, Gr \, Sc)^{-1/2}$
I-c	$x_0^{-4} < GrSc < x_0^{-4} \, Sc$ $Pe < (x_0 \, GrSc)^{1/3}$	$(x_0 \, GrSc)^{-1/3}$
I-d	$x_0^{-4} < Gr$ $Pe < Gr^{1/4} \, Sc^{1/3}$	$Gr^{-1/4} \, Sc^{-1/3}$

In the diffusive regime, the radial segregation induced by the residual convective motion will be proportional to the ratio of the flow velocity to the growth rate. Thus, radial segregation is a maximum at the transition between the convective and the diffusive transport regimes.

<u>Sample of finite length</u>. Let us consider a sample of finite length L (aspect ratio A = L/H). There are two reasons for a departure from the steady state conditions described in the previous section :

(1) Solidification starts from a initial liquid of uniform concentration c_0,

(2) The melt length decreases while solidification proceeds.

Then, in the most general case the following steps must be distinguished[6] :

(1) At the beginning of solidification solute builds up in front of the interface during an initial transient which can be subdivided into :
(1-a) a real non stationary stage, during which the solute boundary

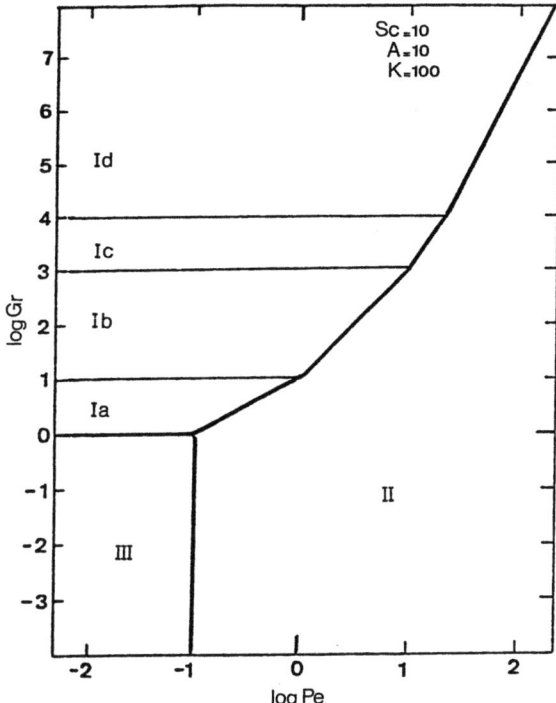

Fig. 6. Domains corresponding to the different solute transport regimes :
  II Diffusive boundary layer regime
  I   Convective boundary layer regime
      Ia : $1 < x^*$ $x^* = x_0$
      Ib : $x_0 < x^* < 1$ $x^* = x_0$
      Ic : $x^* < x_0$ $x^* = x_0$
      Id : $x^* < x_0$ $x^* < x_0$
  III Fast diffusion regime.

layer extends from 0 to its stationary value δ,
(1-b) a pseudo-stationary stage, during which the ratio $c(0)/c_\infty$ rises up to its stationary value.
(2-a) Then, the solute concentration c_∞ in the bulk rises according to the classical Scheil equation.
(2-b) At last, a final transient occurs when the solute boundary layer reaches the end of the melt.

The corresponding time intervals, boundary layer extents δ(t) and concentration scales are summarized in table 4.

Depending on the parameter range, one of the previous steps may prevail. When δ is larger than L (domain III of Fig. 6 and table 3), the liquid is homogenized at any time by diffusion and a Scheil longitudinal segregation law is obtained. In the convective regime (domain I), an identical longitudinal segregation profile is obtained although for a quite different physical reason (the liquid phase now being homogenized by convection). In the diffusive regime (domain II), a sub-regime must be distinguished where there is a longitudinal segregation associated with the initial transient (domain II' corresponding to $A^{-1} < Pe < A^{-1} k^{-1}$). Outside this range, crystals of uniform concentration c_0 are obtained in domain II.

Table 4. Boundary layer extent and concentration scales as a function of time for the solidification of a sample of finite length.

	Time interval	$\delta(t)$	$c(0)/c_\infty$	c_∞/c_0
1-a	$t < \dfrac{\delta^2}{D}$	$(Dt)^{1/2}$	$1 + (\dfrac{R^2}{D} t)^{1/2}$	1
1-b	$\dfrac{\delta^2}{D} < t < \dfrac{\delta^2}{D} \cdot \dfrac{1-k}{1-(1-k)\Delta}$	δ	$1 + \dfrac{Rt}{\delta}$	1
2-a	$\dfrac{\delta^2}{D} \cdot \dfrac{1-k}{1-(1-k)\Delta} < t < \dfrac{L-\delta}{R}$	δ	$\dfrac{1}{1-(1-k)\Delta}$	$(1 - \dfrac{Rt}{L})^{k_{eff}-1}$
2-b	$\dfrac{L-\delta}{R} < t < \dfrac{L}{R}$	$L - Rt$	$\to 1$	$\to \infty$

Segregation curves corresponding to any point within each of the previous domains can be obtained with the help of a stagnant layer model[6], which gives an analytical solution as a function of two parameters only : the dimensionless extent Δ of the stagnant layer and the concentration level at the boundary of this layer.

The scaling laws derived above are no longer valid for concentrated alloys where solutal buoyancy cannot be neglected. For instance, in the case where the solute field is stabilizing, buoyant effects associated with radial thermal gradients are reduced, and thus diffusive transport conditions are more easily obtained[7].

<u>Measurement of the segregation of a dopant in semiconductor crystals grown from the melt</u>

Low concentrations of a non-isoelectric dopant in a semiconductor crystal can be measured very accurately by electrical methods (for instance, 10^{16} atomes/cm^3 Ga in Ge by the spreading resistance method). Several Bridgman crystal growth experiments have been performed on the same Ge (Ga) system by different authors in various thermal conditions, on the ground as well as in microgravity conditions. Witt performed an experiment in Skylab with a 1 cm diameter sample and a low growth rate (8.10^{-4}cm/s)[12]. Favier et al. used a sounding rocket to solidify 0.5 cm diameter samples at higher velocities (3.5 10^{-3}cm/s)[13].

The longitudinal segregation profiles have been fitted with the boundary layer model of[6]. Fig. 7 compares the values obtained for Δ with the scaling laws derived above. The points corresponding to numerical simulations by Polezhaev[14] are also given. For both ground based experiments, which have been performed in a vertical configuration, Gr values are calculated from the maximum horizontal temperature differences in the melt. It can be seen that these residual temperature differences lead to solute mixing. A more recent study[7] shows that with the help of an appropriate design of the furnace and crucible these effects can be reduced but not eliminated. Thus, reducing the g-level is the only way to get purely diffusive transport conditions.

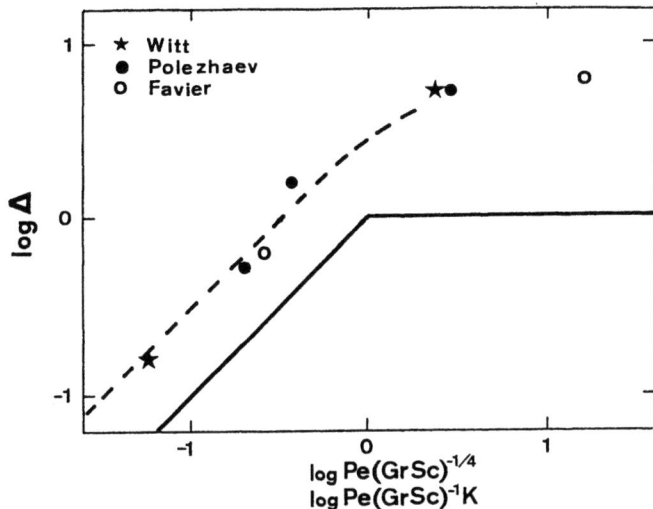

Fig. 7. Scaling laws for the solute boundary layer Δ and points corresponding to ground-based and spatial experiments[12,13] and numerical simulations[14] on the growth of Ga doped Ge crystals from the melt.

The relative effects of convection and interface curvature on the radial segregation Δc/c in these two experiments can be discussed with the help of Fig. 8 and 9. For both cases a-priori estimates of Δc/c derived from above are given as a function of the gravity level, firstly for an ideally planar growth interface, and secondly by taking into account the departure from planarity Δx observed in the experiment. For comparison the vertical bars indicate the experimental radial segregation on the ground and in microgravity. Both experiments show a residual radial segregation in the spatial samples. But, while this segregation can be attributed to residual convection in the case of Witt's experiment which was performed not very far from the convecto-diffusive transition, a dominant effect of interface curvature can be seen in the case of Favier's experiment. These results emphasize the two conditions to be fulfilled in order to get homogeneously doped bulk semiconductor crystals ; namely a low g-level and a low curvature of the crystal-melt interface.

Scaling of the effects of unsteady flows on segregation

Unsteady flows are possible sources of striations in crystals. These may result from intrinsic convective instabilities, or from the application of non constant driving forces. This last case, which is particularly considered here, occurs for instance in the space environment where the residual acceleration varies in direction and intensity with time. Then, in order to analyse the possible effects on segregation, one has to consider three different characteristic times, or inversely, frequencies : ω for the stimulus, ν/H^2 for the flow response, and kR^2/D for the solute boundary layer response. The corresponding non dimensional quantities are : Sr, Sc and Pe^2k, where Sr is the Strouhal number (Sr = $\omega H^2/\nu$). Assuming $Pe^2k < Sc$, three regimes must primarily by distinguished :

(1) if Sr Sc < Pe^2k, the flow and solute fields are any time equivalent to the stationary ones, and the effect of g-jitters on segregation thus obtained is a maximum ;

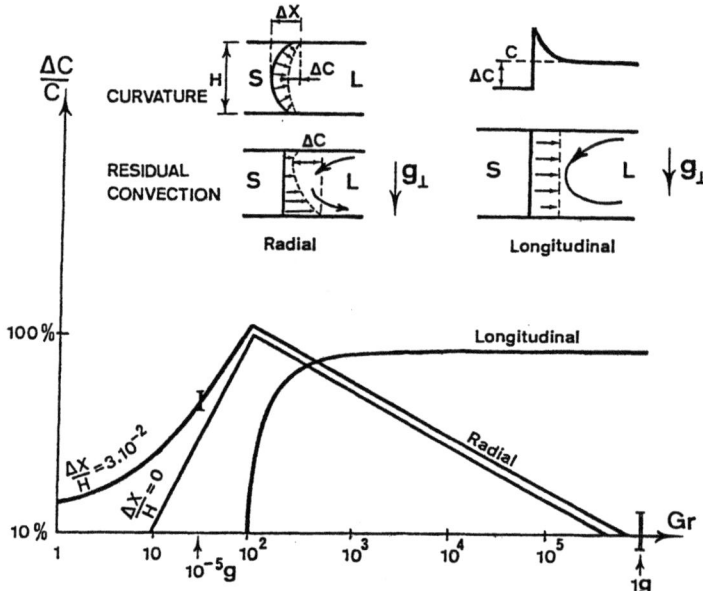

Fig. 8. Effect of g-level and interface curvature on radial segregation : case of Witt's experiment[12].

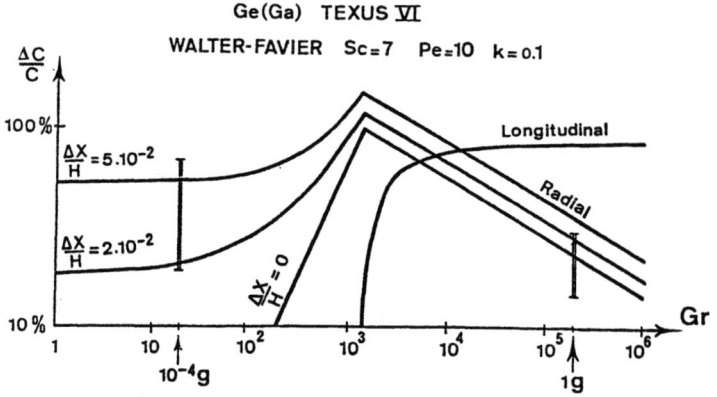

Fig. 9. Effect of g-level and interface curvature on radial segregation : case of Favier's experiment[13].

(2) if $Pe^2 k < Sr\ Sc < Sc$, the flow field is the stationary one but concentration variations are damped ;
(3) if $Sc < Sr\ Sc$, flow velocity variations are also damped.

The corresponding macro- and micro-segregations are drawn on Fig. 10 as a function of Sr, for the case of the typical experiment already considered. Typical concentration profiles along the solidification direction are also shown on Fig. 11. It is clear from these figures that increasing values of ω lead to very different segregation features, and thus to very different properties of the typical "chips" to be taken out of the crystal.

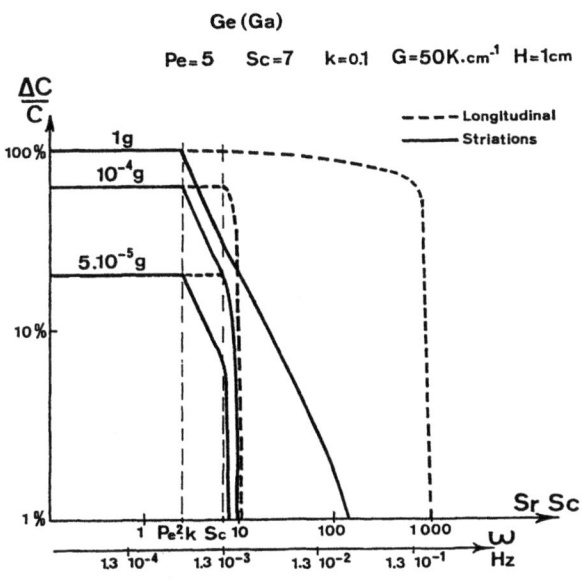

Fig. 10. Effect of g-jitter frequency on macro- and micro-segregation in a crystal grown from the doped melt in a microgravity environment (case of Witt's experiment)[12].

Analysis of morphological instability in the presence of bulk convection

Linear stability analysis has been largely developed in order to study the morphological instability of the growth interface, and so predict the instability threshold as well as the wavelength of the first normal mode. However, specific difficulties are encountered if bulk convection has to be taken into account in the unperturbed reference state, which is generally the case in practice. One way to do so is to perform a linear stability analysis by considering a stagnant layer of dimensionless extent Δ defined by the above described scaling laws. Then, physically realistic results are obtained if the boundary layer is supposed to be deformable instead of rigid [8] : particularly, it is found that for intermediate agitation levels critical wavelengths are reduced while morphological stability is recovered at very high convection levels. This approach can be generalized by considering also the hydrodynamic instability within the solute boundary layer [9]. This type of instability is then found to be very sensitive to the thickness of the solute boundary layer in the unperturbed state.

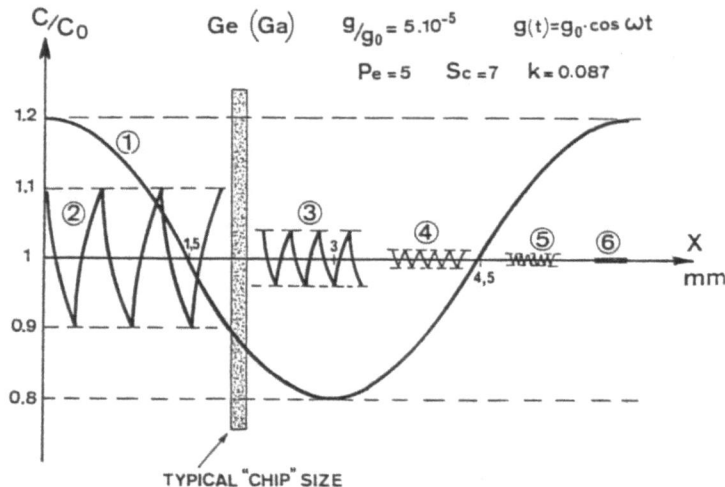

Fig. 11. Typical concentration profiles along the growth direction for increasing g-jitter frequencies.
(1) $SrSc$ = $0.5 < Pe^2 k$ ($\omega = 15 \cdot 10^{-4}$ Hz : EURECA ORBIT PERIOD),
(2) $Pe^2 k < SrSc$ = $5 < Sc$ ($\omega = 1.5 \cdot 10^{-3}$ Hz),
(3) $SrSc$ = $10 > Sc$ ($\omega = 3 \cdot 10^{-3}$ Hz),
(4) $SrSc$ = 16.5 ($\omega = 5 \cdot 10^{-3}$ Hz),
(5) $SrSc$ = 23 ($\omega = 7 \cdot 10^{-3}$ Hz),
(6) $SrSc$ = 25 ($\omega = 8 \cdot 10^{-3}$ Hz).

In-situ Seebeck measurement of the solidification interface temperature

The principle of the method developed in[15] is schematically shown in Fig. 12. A thermoelectric loop is formed between a fixed solid-liquid interface at temperature T_{equ} and a moving one at $T_{equ} - \Delta T$. Cold ends A and B are maintained at the same temperature T_0. Then, the Seebeck voltage between A and B is :

$$E_s = (\eta_s - \eta_1) \Delta T \qquad (10)$$

Due to the low thermoelectric power of solid-liquid metallic interfaces (1.7 µV/°C for tin), much attention has to be paid to the electronic and thermal design of the apparatus in order to ensure sufficiently accurate measurements. With the existing equipment, a sensitivity of a few tens of nanovolts is currently achieved.

When the solidification front is planar, two different contributions to the total undercooling ΔT may appear : a kinetic one if the interfacial attachment is the limiting mechanism, and a chemical one if there is a solute build-up in front of the interface.

Figure 13 shows a typical curve of the Seebeck voltage E_s as a function of time for an horizontal solidification of a 4 mm diameter sample of a dilute alloy showing rapid interfacial kinetics (i.e. rough solid-liquid interfaces). This curve can be directly converted into a $\Delta T(t)$ curve after correction for the offset which appears with the variation of the position of the furnace relative to the sample (dashed line on Fig. 13). The initial increase of the undercooling corresponds to the initial transient during which the solute builds up in front of the interface. Then a quasi-stationary value is maintained, the length solidified being small compared with the total length of the liquid zone. At last, when solidification is stopped, there is a final transient where the liquid is homogenized again, i.e. the constitutional undercooling goes back to zero. Fitting these curves with a boundary layer model gives the value of the parameter Δ for each experiment. Figure 14 compares the results obtained for different alloys and solidification conditions, with the predicted scaling laws.

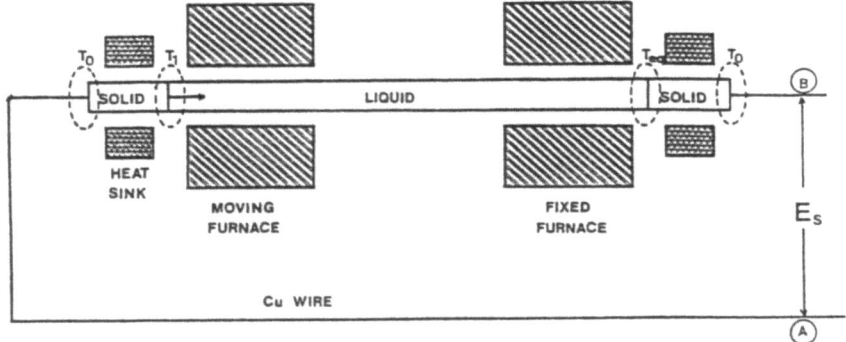

Fig. 12. Sketch of the apparatus for the in-situ Seebeck measurement of the solid-liquid interface temperature during solidification[15].

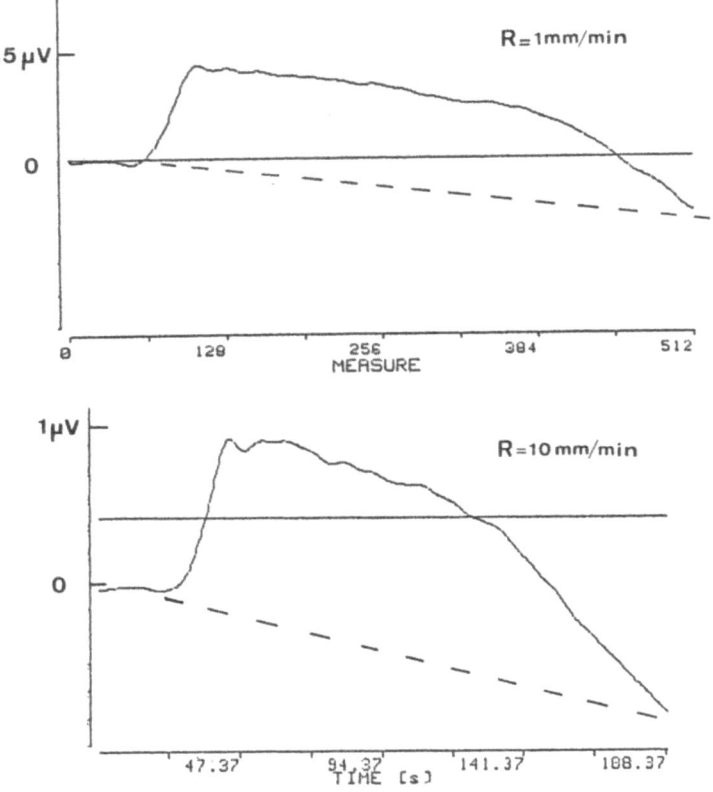

Fig. 13. Typical aspect of the Seebeck signal as a function of time during a solidification run (------ : baseline).

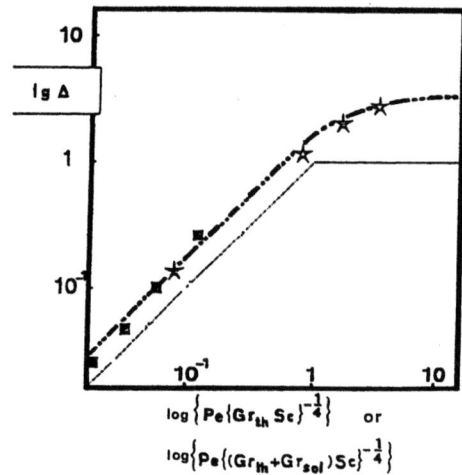

Fig. 14. Dimensionless solute boundary layer Δ derived from the Seebeck signal during different solidifications of tin based alloys.
☆   Sn .1% at Pb as a function of Gr_{th}.
☐   Sn 2% at Bi as a function of $(Gr_{th} + Gr_{sol})$.

The other preliminary results obtained so far illustrate the wide capabilities of the method for performing investigations in areas such as:
- Measurement of interfacial kinetics,
- Real time tracking of morphological thresholds,
- Analysis of the response of the interface to non-stationary stimuli such as produced by unsteady flow regimes in the melt. An example is given in Fig. 15 where frequency spectra are given as a function of the furnace temperature both for the Seebeck signal and the temperature taken at a fixed point of the liquid[16].

NON-PLANAR SOLIDIFICATION OF CONCENTRATED ALLOYS

Eutectic solidification

A-priori estimates of convective effects during eutectic solidification can be made with the help of the approach presented above. As for as bulk convection only is concerned, the flow field in the vicinity of the eutectic front is similar to the one derived above for the case of a planar front. In diffusive transport conditions, the characteristic length scale of the local solute field in front of each lamella is of the order of the interlamellar spacing λ in both the normal and tangential directions. Then, the dominant contribution of convection will be the one associated with the tangential component V of the flow velocity vector. Thus, the condition for the transition from diffusive to convective transport is now written in non-dimensional form as:

$$Sc \times v \left(\frac{\lambda}{H}\right) \times \frac{\lambda}{H} = 1 \qquad (11)$$

or, using (4a)

$$\frac{v^*}{x^*} = \left(\frac{H}{\lambda}\right)^2 \frac{1}{Sc} \qquad (12)$$

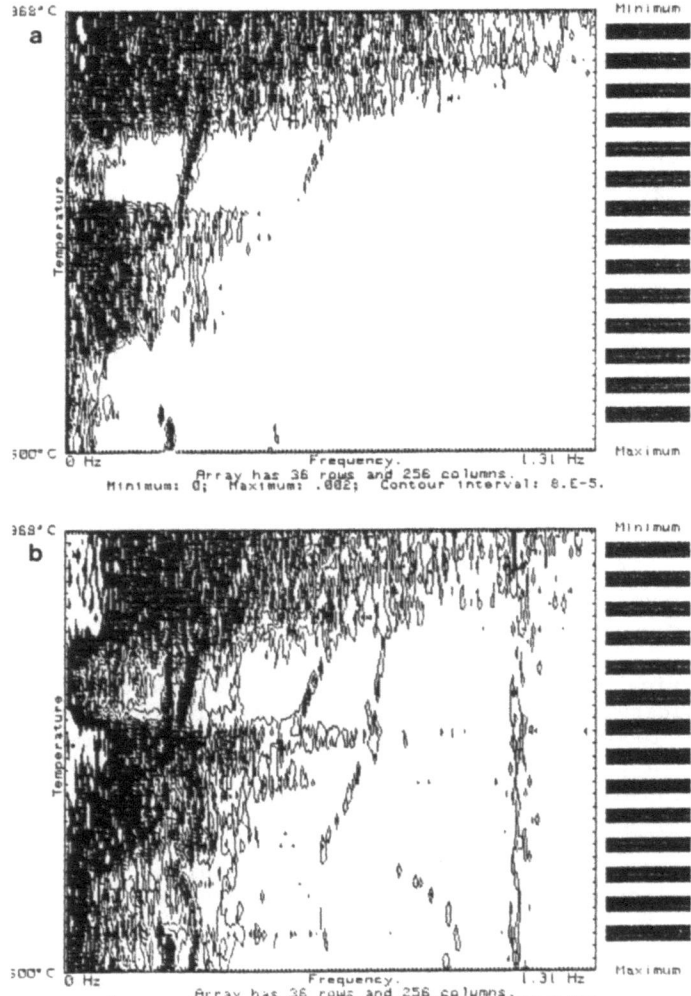

Fig. 15. Frequency spectra at different furnace temperatures for : (a) the temperature measured in a point of the liquid ; (b) the solid-liquid interface temperature as measured by the Seebeck method.

This criterion, which predicts that high agitation levels are needed to perturb eutectic solidification, is in agreement with both numerical simulations and centrifugation experiments[17,18]. Convective effects encountered at lower agitation levels, are in fact due to the average rejection occuring over a length D/R during solidification of a slightly off-eutectic liquid[19].

Dendritic solidification

Physical description. During directional dendritic solidification from the liquid, three different zones can be distinguished in the liquid phase :

1) The bulk liquid where the concentration is uniform, and the only potential driving force for natural convection is the thermal field.

2) The intermediate zone near dendrite tips where a non-unidirectional solute field develops, which can directly couple with the flow.
3) The interdendritic liquid where the concentration can be supposed to be imposed by the local equilibrium between the solid and liquid phases, i.e. is quasi-unidirectional and parallel to the thermal gradient :

$$\frac{dC}{dz} = \frac{G}{m_L} \qquad (13)$$

where m_L is the slope of the liquidus of the equilibrium phase diagram.

In the following, we examine the convective effects under conditions where this last solute gradient is the dominant driving force for convection, which is the most general case due to the generally large values of solutal expansion coefficients β_c as compared with thermal expansion coefficients β_T.

Scaling of interdendritic flow. Natural flows inside the interdendritic liquid are generally accepted to be governed by the generalized Darcy's law :

$$\underline{V} = \frac{K}{\rho \nu} (\underline{\nabla} p + \rho g \beta_c C \underline{\gamma}) \qquad (14)$$

where
$\underline{V}$ is the macroscopic flow rate,
K the permeability.

The physical validity limits of this law can be qualitatively described by considereing the simplified geometry of the classical Hele-Shaw cell. Indeed two-dimensional flow within a porous medium can be modelled by the flow in a liquid layer of thickness h bounded by two plates. In this geometry, the Navier-Stokes equations reduce to a Darcy's law provided that :

- The velocity vector is supposed to be parallel to the plates, i.e. the non-linear inertial terms of the N.S. equations are negligible,
- The buoyant contribution of temperature differences across the layer is neglected.

The convective effect will be maximum is the case of horizontal solidification where the density gradient in the interdendritic liquid is perpendicular to the gravity vector. According to Darcy's law, flow velocity and dynamic pressure then scale as :

$$V_i = A \frac{g \beta_c}{\nu} \frac{G}{m_L} z K$$

$$P_r = \rho g \beta_c \frac{G}{m_L} z^2 \qquad (15)$$

where z is the small dimension of the mushy zone.

Apart from the particular cases of experiments with thin samples or very large solidification intervals, z is a length along the solidification direction. In order to define this length, one has to consider that the permeability K may be strongly non-uniform along this direction. Starting from dendrite tips, K decreases as the liquid fraction f_L decreases. Thus, the deep interdendritic regions, where K is low, may not contribute significantly to the flow, which would then be mainly confined to a region near dendrite tips. Since the determining geometrical factor for the flow rate is the product z K(z), the extent of this region can be determined by considering the variation of this quantity as a function of z. For increasing depth z, z K(z) first increases from zero and then may go through a maximum at z = z* and decrease. z* defines the limit of the zone where the flow is confined, and the corresponding flow rate is obtained by taking $z = z^*$ in (15). In order to estimate z^*, one needs a knowledge of the function $K(z)$, especially away from dendrite tips. If λ is the primary interdendritic spacing, one can generally consider :

$$K = B \, f_L^2 \, \lambda^2 \tag{16}$$

The theoretical analysis of the flow in the case of vertical (stabilizing or destabilizing) solidification is very difficult. Indeed, the flow within the deep interdendritic liquid is then inherently coupled with the flow outside or/and the deformation of the mushy zone through melting or solidification. The main characteristics of the corresponding convective modes can be clarified by performing solidification experiments in different orientations relative to $\vec{g}$ under otherwise identical conditions. In non-transparent systems, where a direct flow visualization cannot be performed, these main characteristics can be derived from the segregations observed a-posteriori[20].

Scaling of segregation. Within the mushy zone, a local solute balance equation can be written at a scale larger than the primary dendritic spacing but lower than the overall dimension of the mushy zone :

$$\frac{\partial f_L}{\partial c_L} = (1 - \frac{V_z}{R}) \frac{f_L}{c_L} \tag{17}$$

and

f_L is the local liquid fraction,
c_L the local concentration of the liquid,
R the solidification rate,
V_z is the component of the flow velocity parallel to the solidification direction, in the laboratory reference frame.

According to (17), the departure of f_L from its value under pure diffusive transport conditions is of the order of the ratio :

$$\Gamma = \frac{V_i}{R} \tag{18}$$

Then, the variation of the local average concentration $\bar{c}$ (liquid + solid) across the mushy zone, and consequently the radial segregation in the solid, are of the order of :

$$\Delta C_r = \frac{G}{m_L} z^* \times \Gamma \tag{19}$$

The induced macroscopic deformation of the dendritic front is given by :

$$\Delta h = \frac{m_L}{G} \Delta C_r = \Gamma \, z^* \tag{20}$$

Criterion for the convecto-diffusive transition. The condition for the transition from a diffusive to a convective transport regime is given by :

$$\Gamma \simeq 1 \tag{21}$$

The scaling law for the primary interdendritic spacing λ in the diffusive regime is :

$$\lambda = 7 \left[D \, \gamma_o \, m_L \, C_\infty \, (1-k) \right]^{1/4} R^{-1/4} \, G^{-1/2} \tag{22}$$

where γ_o is the capillary constant.

Replacing in (21) gives :

$$\Gamma = 1.5 \; 10^{-2} \; \frac{g \; \beta_C (c_\infty - c_E)}{\nu} \; f_E^2 = \left[D \; \gamma_o \; m_L \; c_\infty \; (1-k) \right]^{1/2} \; \frac{R^{-3/2}}{G} \simeq 1 \quad (23)$$

where the constant is adjusted on experimental results on segregation.

Figure 16 shows the corresponding limit in a plane growth rate versus gravity level for a given thermal gradient and alloy concentration, together with the similar limit already given for the case of the planar solidification front occurring at low rates.

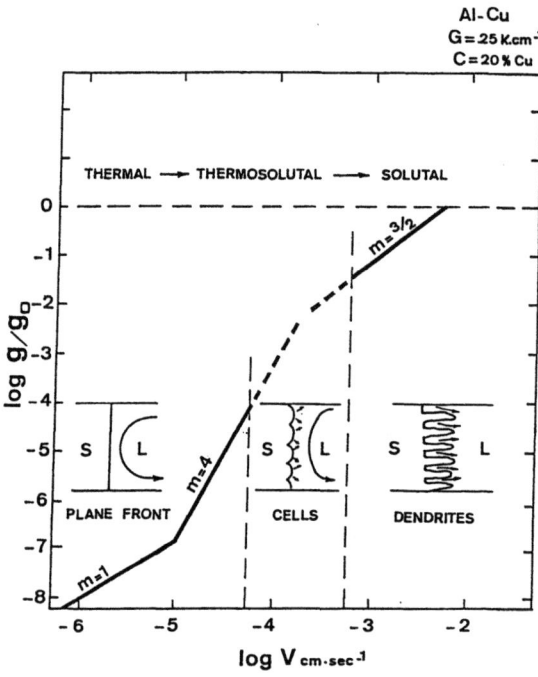

Fig. 16. Maximum allowed stationary g-level to get convection-free solidification in the range of solidification rates covering both planar and dendritic fronts.

CONCLUSIONS

Convective effects during crystal growth from the melt and solidification have been considered both theoretically and experimentally, with a view to derive the scaling laws of the phenomena. Natural convective effects during crystal growth from the doped melt have been particularly emphasized. It has thus been shown that macrosegregation of the dopant can simply be predicted as a function of the physical properties of the system and imposed experimental conditions, i.e. sample dimensions, growth rate, thermal gradients, gravity orientation and level. The generalization of the method to the study of more complex problems such as morphological instability or growth of non-planar fronts under convective conditions has been suggested. The present approach is thus proposed as a useful complement to numerical work.

ACKNOWLEDGEMENTS

Special thanks are given to Dr. A. Rouzaud for the processing of the Seebeck data and other contributions to the reported work, to Mr. P. Tison for carrying out the experiment on surface flow velocities, and to Messrs. J. Coméra, P. Contamin, R. Ginet and G. Marquet for the successful achievement of the experimental method for measuring the temperature of the solidification interface.

The present work was conducted within the framework of the GRAMME agreement between C.N.E.S. and C.E.A. Part of this work was supported by an ESA contract on tolerable g-levels.

REFERENCES

1. S.M. Pimputkar and S. Ostrach, Convective effects in crystals grown from melt, J. Crystal Growth 55 :614 (1981).
2. G.B. Mc Fadden, R.G. Rehm, S.R. Coriell, W. Clark and K.A. Morrish, Thermosolutal convection during directional solidification, Metall. Trans. A15 :2125 (1984).
3. R.A. Brown, Convection and bulk transport, in : "Materials Sciences in Space," B. Feuerbacher, ed. Springer, Berlin (1986).
4. S. Ostrach, Low-gravity fluid flows, Ann. Rev. Fluid Mech. 14:313 (1982).
5. L. Napolitano, Surface and buoyancy driven free convection, Acta Astronaut. 9:199 (1982).
6. J.J. Favier, Macrosegregation : I - Unified analysis during non-steady state solidification ; II - A comparative study of theories, Acta Met. 29:197 and 205 (1981).
7. A. Rouzaud, D. Camel and J.J. Favier, A comparative study of thermal and thermosolutal convective effects in vertical Bridgman crystal growth, J. Crystal Growth 73:149 (1985).
8. J.J. Favier and A. Rouzaud, Morphological stability of the solidification interface under convective conditions, J. Crystal Growth 64:367 (1983).
9. M. Hennenberg, A. Rouzaud, J.J. Favier and D. Camel, Morphological and thermosolutal instabilities inside a deformable solute boundary layer during directional solidification. Part I : Theoretical methods, J. Physique, Submitted.
10. D. Camel and J.J. Favier, Scaling analysis of convective solute transport and segregation in Bridgman crystal growth from the doped melt, J. Physique 47:1001 (1986).
11. D. Camel, P. Tison and J.J. Favier, Marangoni flow regimes in liquid metals, IAF-85-294, 36th IAF Congress, Stockholm (1985).
12. A.F. Witt, H.C. Gatos, M. Lichtensteiger and C.J. Herman, J. Electrochem. Soc. 125:1832 (1978).
13. J.J. Favier and J. de Goër, Analyse de la segrégation du gallium dans des barreaux de germanium solidifiés unidirectionnellement en fusée-sonde, CEA Internal Report, Grenoble (1985).
14. V.I. Polezhaev, K.G. Dubovik, S.A. Nikitin, A.I. Prostomolotov and A.I. Fedyushkin, Convection during crystal growth on earth and in space, J. Crystal Growth 52:465 (1981).
15. J.J. Favier, Etude des cinétiques de cristallisation par application de l'effet thermoélectrique. Analyse de la température d'une interface de solidification, Thesis, Grenoble (1977).
16. A. Rouzaud and J.J. Favier, Influence of various hydrodynamic regimes in a melt on a solidification interface, Submitted to Revue de Physique Appliquée.

17. J.M. Quenisset and R. Naslain, Effect of forced convection on eutectic growth, J. Crystal Growth 54:465 (1981).
18. V. Baskaran and W.R. Wilcox, Influence of convection on lamellar spacing of eutectics, J. Crystal Growth 67:343 (1984).
19. J.J. Favier and J. de Goer , Directional solidification of eutectic alloys, Proceedings of the 5th European Symposium on Material Sciences under Microgravity, Schloss Elmau, 5-7 Nov. 1985 (ESA SP-222).
20. M.D. Dupouy, Convection naturelle en solidification dendritique dirigée d'alliages Al-Cu : Ségrégations et morphologies, Thesis, Grenoble (1986).

MOLECULAR BEAM EPITAXIAL GROWTH AND PROPERTIES OF LAYERED

SEMICONDUCTOR STRUCTURES FOR ADVANCED PHOTONIC AND ELECTRONIC DEVICES

Klaus Ploog

Max-Planck-Institut für Festkörperforschung
D-7000 Stuttgart-80, FR-Germany

ABSTRACT

We demonstrate the unique capability of molecular beam epitaxy (MBE) to synthesize III-V and group-IV semiconductors in well-defined geometrical and spatial arrangements. Based on the shrinkage of the physical size of materials from micron to nanometer dimensions, the concept of band-gap (or wavefunction) engineering in quantum wells and superlattices has been exploited. Quantum well heterostructures made of the materials systems GaAs/Al$_x$Ga$_{1-x}$As, Ga$_{0.47}$In$_{0.53}$As/Al$_{0.48}$In$_{0.52}$As, and GaSb/Al$_x$Ga$_{1-x}$Sb are important for photonic devices covering the wavelength range $0.65 < \lambda > 1.75$ μm and for high-speed electron devices. New electronic phenomena based on quantum confinement and tunneling of carriers in these multiquantum well heterostructures and superlattices have led to the development of novel devices, including quantum well lasers, bistable optical devices, high-speed optical modulators, new avalanche photodiodes and staircase photomultipliers, high electron mobility transistors, hot electron transistors, etc. We will discuss some selected examples of the growth, the physics, and the performance of these devices. In addition, as an example for strained-layer superlattices, we will report on Si/Si$_x$Ge$_{1-x}$ superlattices where the ordering of the electronic bands is strongly affected by the built-in strain. Therefore, depending on the strain distribution, two-dimensional electron or hole gases can be obtained in selectively doped Si/Si$_x$Ge$_{1-x}$ superlattices.

1. INTRODUCTION

Modern photonic and electronic semiconductor devices require thin layers, abrupt junctions, precise dopant control and multiple layers. Many recent achievements in this field have resulted from the application of advanced epitaxial crystal growth techniques to tailor the device structure for optimum performance or for new operating principles. Since nearly two decades the technique of molecular beam epitaxy (MBE) / 1 / has been a good and flexible choice to provide atomic abruptness and smoothness between layers of different lattice-matched and lattice-mismatched crystalline semiconductors at their interfaces or heterojunctions. The abrupt discontinuities in conduction and valence band edges produced by discontinuities in electron affinity across the heterojunction can lead to new electronic phenomena due to the reduced dimensionality in the layered material. The growth of artificially layered periodic semiconductor structures ("superlattices" or "quantum well heterostructures") having dimensions in the order of the electron mean free path and the de Broglie wavelength thus allows the investigation of the 1-dimensional quantization and tunneling of electrons and holes. The first observation of resonant tunneling into quantum wells / 2 / and of optical absorption involving quantized energy levels / 3 / more than a decade ago has opened a wide field of materials science for bandgap engineering in semiconductors, physics of low-dimensional semiconductor systems, and device applications.

After a brief discussion of the growth technique we will present selected examples for the application of molecular beam epitaxy to modify the bulk properties of semiconductors through bandgap engineering by grading the composition and doping both abruptly (on an atomic scale) as well as gradually / 4 / . We will discuss quantum wells made of the lattice-matched material systems AlAs / GaAs, $Al_xIn_{1-x}As$ / $Ga_xIn_{1-x}As$ matched to InP, and AlSb / GaSb, where the ordering of the electronic bands is mainly affected by the discontinuities in electron affinity across the heterojunction. In addition, we will report on Si / Si_xGe_{1-x} strained-layer superlattices where the ordering of the electronic bands is strongly affected by the built-in strain. Finally, the new electronic phenomena based on quantum confinement and tunneling of carriers in these multi quantum well heterostructures and superlattices have led to the development of novel devices, including quantum well lasers, bistable optical devices, high-speed optical modulators, new avalanche photodiodes and staircase photomultipliers, high electron mobility transistors (HEMT), hot electron transistors etc. We will discuss some selected examples of the growth, the physics, and the performance of these devices.

2. KEY ASPECTS OF MBE GROWTH

The unique capability of MBE to create a wide variety of mathematically complex compositional and doping profiles in semiconductors arises from the conceptual simplicity of this growth process / 1 / . In Fig. 1 we show the schematic cross-section of an MBE growth chamber. Basically the MBE process consists of a co-evaporation of the constituent elements and dopants of the epitaxial layer onto a heated crystalline substrate

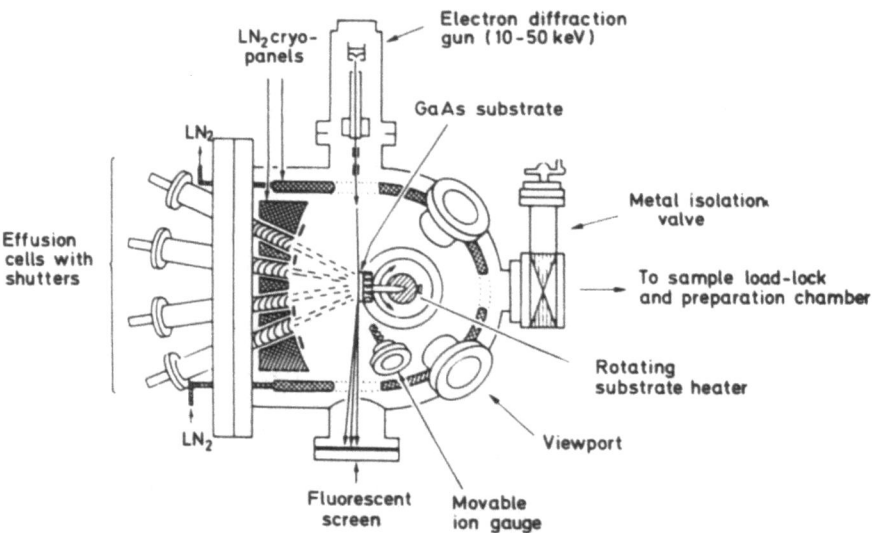

Fig. 1. Schematic cross-section of an MBE growth chamber illustrating the basic evaporation process.

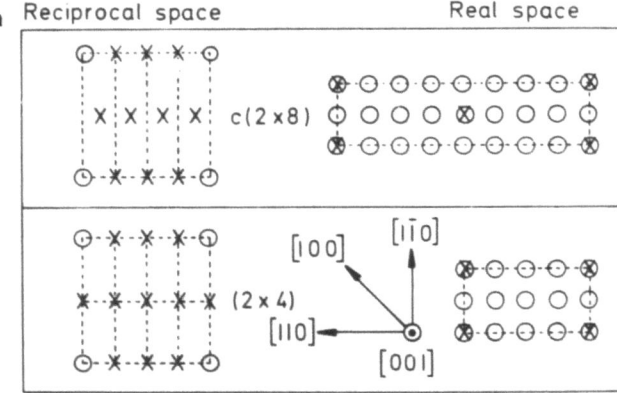
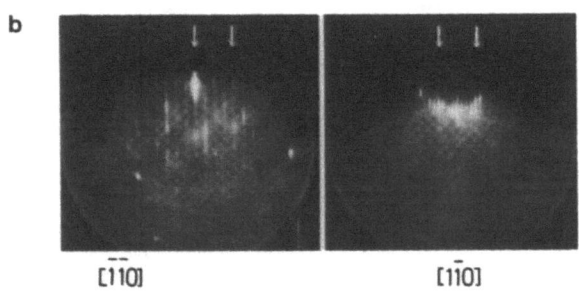

Fig. 2. (a) Real-space and reciprocal-space representation for the c(2x8) and (2x4) surface reconstruction on the growing GaAs (100) face; (b) RHEED patterns of the (2x4) surface reconstruction taken at two orthogonal [110] azimuths.

under ultra-high vacuum (UHV) conditions. The composition of the layer and its doping level depend on the relative arrival rate of the constituent elements which in turn depend on the evaporation rate of the appropriate sources. Accurately controlled temperatures have thus a direct, calculable effect upon the growth process. The growth rate of typically 1 μm/hr is chosen low enough that surface diffusion of the impinging species on the growing surface to the appropriate lattice sites is ensured without incorporating crystalline defects. Simple mechanical shutters in front of the evaporation sources are used to interrupt the beam fluxes to start and stop deposition and doping.

The technique of reflection high energy electron diffraction (RHEED) operated at 10 - 50 KeV in the small glancing angle reflection mode has become an important in-situ method for the study of surface crystallography and kinetics during MBE growth. The diffraction pattern on the fluorescent screen (Fig. 2) contains information from the topmost nanometer of the deposited material that can be related to the topography and structure of the surface / 5 / . Of particular importance is the identification of surface reconstruction, which results from the rehybridization of bonding orbitals of surface atoms in order to lower the free energy of the surface. This reordering of the outermost layers yields a surface symmetry that is modified with respect to that of the bulk lattice. Therefore, additional diffraction streaks due to reconstruction occur at fractional intervals between the bulk diffraction features. In the case of III-V semiconductor surfaces, the form of the reconstruction can be correlated to the surface stoichiometry which is an important growth parameter / 6, 7, 8 /.

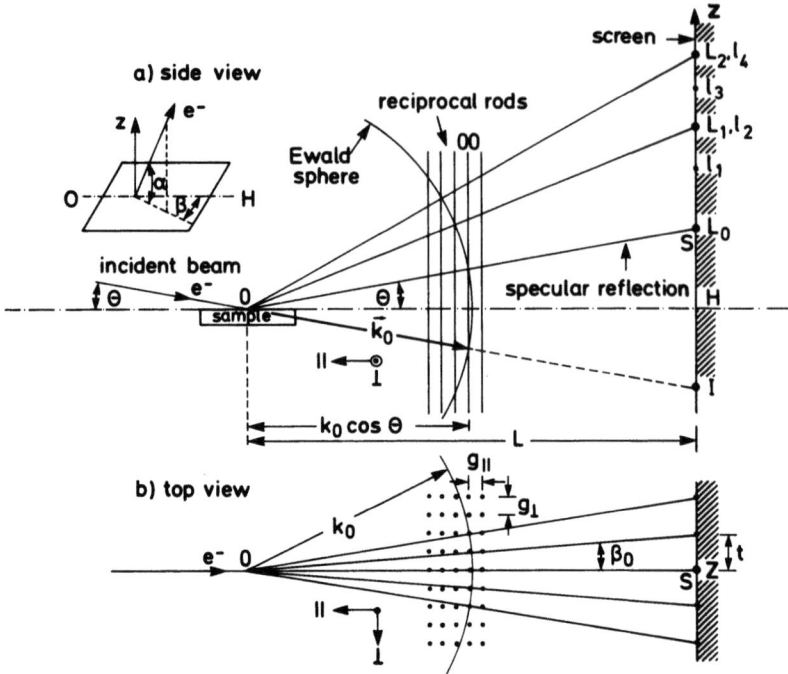

Fig. 3. Ewald construction to illustrate the origin of a typical RHEED pattern during MBE growth (not to scale). (a) Side view, where L_n corresponds to the Laue zones of the ideal surface. The intermediate spots l_n that are present in the case of $g_{||}^j = \frac{1}{2}g_{||}^j$ (g = reciprocal surface vector) are also indicated.
(b) Top view with a projection of the Ewald sphere on a plane parallel to the sample surface / 9 / .

During MBE growth we observe diffraction streaks from atomically flat surfaces instead of spots in most RHEED patterns, as shown in Fig. 2. The main features of a RHEED pattern can be analyzed with the help of the Ewald construction depicted in Fig. 3 / 9, 10 / . The continuous rods of the reciprocal lattice are normal to the surface plane and possess a two-dimensional translational symmetry in that plane. The Ewald sphere intersects all rods contained in the projection of this sphere onto the surface. Diffracted beams are thus observed in the angular directions α and β. From the distance between streaks and the sample-to-screen distance we can determine surface lattice constants at a given electron wavelength. The reconstruction of the surface can be obtained via the measurement of the streak separation under different azimuths, as those shown in Fig. 2. Hernandez-Calderon and Höchst / 9 / have described a more detailed interpretation of RHEED patterns which also allows analysis of intermediate Laue zones.

Several authors / 11, 12, 13 / have recently discovered pronounced periodic intensity oscillations in the specularly reflected and in several diffracted beams in the RHEED pattern during growth of III-V semiconductors and group-V-elements (see Fig. 5 for illustration). The period of oscillation corresponds exactly to the growth rate of a monolayer on the (100) surface (a monolayer GaAs corresponds to a complete layer of Ga plus a complete layer of As). The changing intensity reflects variations in the step density in each layer when growth proceeds. The investigations of the

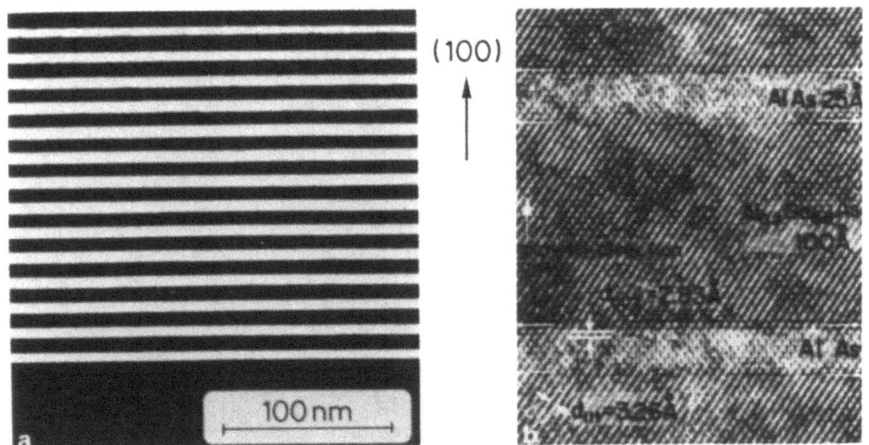

Fig. 4. (a) (100) cross sectional TEM of an AlAs / GaAs superlattice;
(b) High-resolution lattice image / 15 / of a 10 nm $Al_{0.2}Ga_{0.8}As$ / 2.5 nm AlAs superlattice using [110] electron beam incidence.

RHEED oscillations have provided improved understanding of growth dynamics and surface disorder. The intensity oscillations of the specularly reflected beam in the RHEED pattern are now used to monitor the absolute growth rate in-situ and very accurately / 14 / .

Due to the slow growth rate of 1 monolayer/sec during MBE changes in composition and doping can thus be abrupt on an atomic scale. The transmission electron (TEM) micrographs of $Al_xGa_{1-x}As$ superlattices in Fig. 4 demonstrate that this independent and accurate control of the individual beam sources allows the precise fabrication of artificially layered semiconductor structures on an atomic scale. The stoichiometry of most III-V semiconductors during MBE growth is self-regulating as long as excess group-V-element molecules (tetrameric or dimeric) are arriving on the growing surface. The excess group-V-species do not stick on the heated substrate surface, and the growth rate is essentially determined by the arrival rates of the group-III-elements. The simplicity of the MBE process allows composition control from x = 0 to x = 1 with a precision of ± 0.001 and doping control, both n- and p-type, from the 10^{14} cm^{-3} to the 10^{19} cm^{-3} range with a precision of $\pm$ 1%. The accuracy is largely determined by the care with which the growth rate and doping level was previously calibrated in test layers.

High-quality layers require background vacuums in the 10^{-11} Torr range to avoid incorporation of impurities into the growing layers. Therefore, extensive LN_2 cryoshrouds are used around the substrate to achieve locally much less background pressure of condensible species. For MBE growth of III-V semiconductors, the starting materials are evaporated in resistively heated effusion cells made of pyrolytic BN which operate at temperatures up to 1400 °C. In cases where the source material has too low a vapour pressure, as e.g. Si or refractory metals, electron beam evaporators with a special design for constant evaporation rate have to be used.

The preparation of the growth face of the substrate from the polishing stage to the in-situ cleaning stage in the MBE system is of crucial importance for epitaxial growth of heterostructures with high purity and crystal perfection and with accurately controlled interfaces on an atomic scale.

Various cleaning methods have been described for (100) oriented GaAs, the most important substrate material for deposition of III-V semiconductors, which mainly use chemical etching based on a H_2O_2 / H_2SO_2 / H_2O mixture / 1, 4 / . However, the formation of oval defects on MBE grown GaAs and $Al_xGa_{1-x}As$ surfaces, elongated in the $<01\bar{1}>$ direction and with densities ranging from 10^3 to 10^5 cm^{-2}, has long been a severe problem for practical application. We have recently developed a new method for GaAs substrate preparation which effectively reduces the oval-defect density to less than 10^2 cm^{-2} and allows storage of prepared substrates in air under dust-free conditions for several weeks without any degradation. The reproducible preparation of a contamination-free substrate surface was improved as follows / 16 / : The GaAs wafer is first polished with diamond paste to remove saw-cut damage followed by etch-polishing on an abrasive-free lens paper soaked with NaOCl solution leaving a mirror-like finish. Then the wafer is simply placed twice in concentrated H_2SO_4 kept at 300 K and stirred ultrasonically. The slice is then carefully rinsed in water to remove all SO_4^{2-} ions (careful ultrasonic stirring may accelerate SO_4^{2-} removal from the surface) and finally blown dry with filtered N_2 gas. The important final step is heating of the wafer to 300 °C in air under dust-free conditions for about 3 min. During this heating process a stable surface oxide on the (100) GaAs substrate is reproducibly generated to protect the surface from carbon contamination. The passivated wafer is then either soldered with liquid In to a conventional Mo substrate plate or it is fixed to a special Mo substrate holder designed for direct-radiation substrate heating. Both methods give similar results for the subsequent growth of selectively doped n-$Al_xGa_{1-x}As$ / GaAs heterostructures with high-mobility two-dimensional electron gas (2DEG). After transfer to the MBE growth chamber the surface oxide is thermally desorbed by heating the substrate wafer to 550 °C in a flux of arsenic. When the desorption process is finished a clear (2 x 4) surface reconstruction is observed in the reflection high energy electron diffraction (RHEED) pattern.

3. CONTROL OF INTERFACE QUALITY

The heterointerfaces between epitaxial layers of different composition ($Al_xGa_{1-x}As$ / GaAs, $Al_xIn_{1-x}As$ / $Ga_xIn_{1-x}As$, $Al_xGa_{1-x}Sb$ / GaSb, Si / Si_xGe_{1-x}, etc.) are used to confine electrons or holes to two-dimensional (2D) motion. The challenge for the design and growth of materials is to minimize scattering from impurities, alloy clusters or interface irregularities so that the carriers can move freely along the interfaces. For closely lattice-matched systems it should be possible to prepare interfaces by MBE so that compositional changes occur over no more than one monolayer, because MBE growth occurs predominantly in a 2D layer-by-layer growth mode. Direct evidence for this 2D growth mode is provided by the observation of oscillations in the intensity of the RHEED pattern / 17, 18 / , as shown in Fig.5. The intensities of the specularly reflected and of all diffracted beams oscillate, and the period of the oscillations corresponds exactly to the time required to grow a monolayer of GaAs, AlAs, or $Al_xGa_{1-x}As$. To a first approximation we can assume that the oscillation amplitude reaches its maximum when the monolayer is completed (maximum reflection). Although the fundamental principles underlaying the occurrence of these oscillations and the damping of their amplitude are not completely understood, the method is now widely used to monitor and to calibrate absolute growth rates in real time with monolayer resolution.

The oscillatory nature of the RHEED intensities provides direct real-time evidence of compositional effects and growth modes during the formation of heterointerfaces. As for the widely used $Al_xGa_{1-x}As$ / GaAs heterojunctions, the sequence of layer growth is critical for compositional gradients and crystal perfection, which in turn is important for optimizing

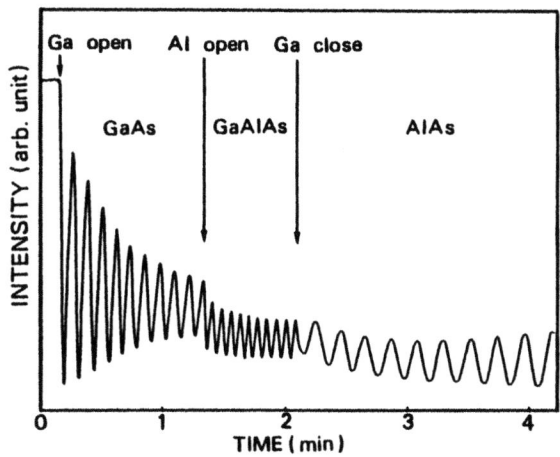

Fig. 5. RHEED intensity oscillation of the specular beam obtained in [100] azimuth on (100) GaAs substrate during continuous growth of a GaAs / $Al_xGa_{1-x}As$ / AlAs heterostructure / 17 / .

2D transport properties. When the Al flux is switched at the maximum of the intensity oscillations, the first period for the growth sequence from ternary alloy to binary compound corresponds neither to the $Al_xGa_{1-x}As$ growth rate nor to the steady-state GaAs rate, but shows some intermediate value (Fig. 5). For the growth sequence from binary compound to ternary alloy or between the two binaries an intermediate period does not exist. A possible explanation for this phenomenon can be found in the relative surface diffusion lengths of the group-III-elements Al and Ga, which were estimated to $\lambda_{Al} \cong 3.5$ nm and $\lambda_{Ga} \cong 20$ nm on (100) surfaces under typical MBE growth conditions / 18 / . These differences in cation diffusion rates have striking consequences on the nature of the interface. While a GaAs layer should be covered by smooth terraces of 20 nm mean length between monolayer steps, those on an $Al_xGa_{1-x}As$ layer would be only 3.5 nm apart. The important result of this qualitative estimate is that the GaAs / $Al_xGa_{1-x}As$ interface is much smoother on an atomic scale than the inverted structure. Direct experimental evidence for this distinct difference in binary-to-ternary layer growth sequence is obtained from inspection of the high-resolution TEM micrograph depicted in Fig. 4b / 15 / . This lattice image of an $Al_{0.2}Ga_{0.8}As$ / AlAs superlattice shows clearly that the heterointerface is abrupt to within one atomic layer only when the ternary alloy is grown on the binary compound but not for the inverse growth sequence.

Since the nature of the heterointerface is critical for optimising excitonic as well as transport properties in quantum wells, various attempts have been made to minimize the interface roughness (or disorder) by modified MBE growth conditions. The most successful modification is probably the method of growth interruption at each interface. Growth interruption allows the small terraces to relax into larger terraces via diffusion of the surface atoms. This reduces the step density and thus simultaneously enhances the RHEED specular beam intensity which can be used for real-time monitoring. The time of closing both the Al and the Ga shutter (while the As shutter is left open) apparently depends on the actual growth condition. Values ranging from a few seconds to several minutes have been reported by different authors / 19, 20 /, in particular for the $Al_xGa_{1-x}As$ / GaAs interface and if the full recovery of the specular beam intensity

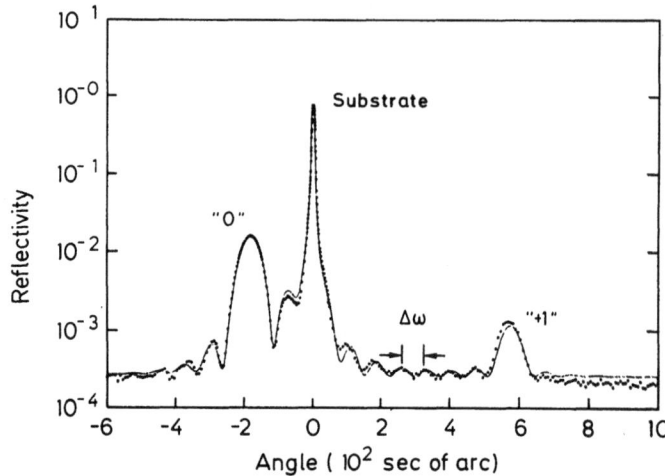

Fig. 6. Experimental (dotted curve) and theoretical (solid line) X-ray diffraction curve of an $(AlAs)_{44}(GaAs)_{46}$ superlattice in the vicinity of the (004) reflection using CuK_{α_1} radiation.

has been allowed for. We have found that in most cases (i.e. growth rate < 1 μm/hr and growth temperature < 650 °C) growth interruption for about 10 sec is sufficient to minimize the interface roughness of the Al-Ga-As system.

High-angle X-ray diffraction could be a powerful technique for investigation of interface disorder effects in superlattices and multi quantum well heterostructures, if a full dynamical analysis of the diffraction curves would be available. We have recently developed a semi-kinematical approach of the dynamical theory of X-ray diffraction to determine the strain profile, the composition, and the interface quality of $Al_xGa_{1-x}As$ / GaAs and of $Al_{0.48}In_{0.52}As$ / $Ga_{0.47}In_{0.53}As$ heterostructures and superlattices /21/. The X-ray reflection curves from the epitaxial layers were recorded with a high-resolution double-crystal X-ray diffractometer in nondispersive (+, -) Bragg arrangement. Since the lattice parameter and scattering factors are subject to one-dimensional (1D) modulation in growth direction, the diffraction patterns consist of satellite reflections located symmetrically around the Bragg reflections, as shown in Fig. 6 for an AlAs / GaAs superlattice. From the positon of the satellite peaks the superlattice periodicity can be deduced. Detailed information about thickness fluctuations of the constituent layers, inhomogeneity of composition, and interface quality can be extracted from the halfwidths and intensities of the satellite peaks. The excellent agreement between experimental and theoretical diffraction curve in Fig. 6 indicates extremely abrupt AlAs / GaAs interfaces to within one monolayer. The existence of interface disorder manifests itself in an increase of the halfwidths and a decrease of the intensities of the satellite peaks, as shown in Fig. 7. During MBE growth of these two $(AlAs)_{42}(GaAs)_{34}$ superlattices, the adjustment of the shutter motion at the transition from AlAs to GaAs and vice versa was changed in the two growth runs. Sample A was grown with growth interruption at each AlAs / GaAs and GaAs / AlAs interface, whereas sample B was grown continuously. While the positions of all the diffraction peaks of sample A coincide with those of sample B, the halfwidths of the satellite

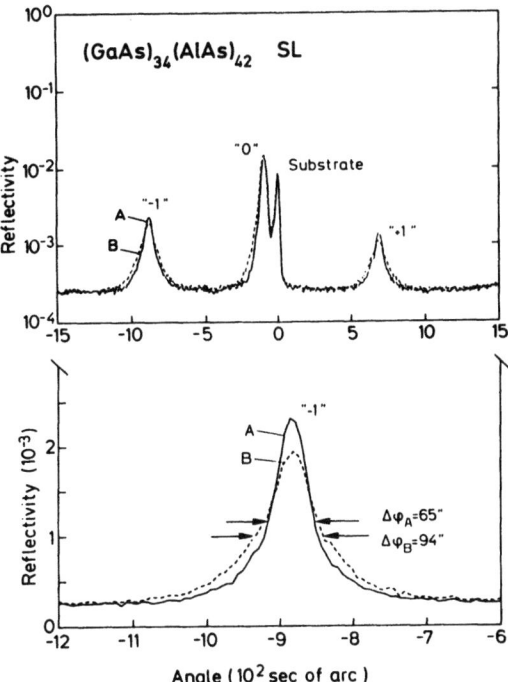

Fig. 7. X-ray diffraction curves (CuK$_{\alpha_1}$) of two (AlAs)$_{42}$(GaAs)$_{34}$ superlattices recorded in the vicinity of the quasi-forbidden (002) reflection. For sample A growth interruption for 10 sec was used at each interface.

peaks from sample A are narrower and their reflected intensities are higher. A growth interruption of 10 sec was sufficient to smooth the growing surface which then provides sharp heterointerfaces. When the heterojunctions are grown continuously, the monolayer roughness of the growth surface leads to a disorder and thus broadening of the interface. In X-ray diffraction this broadening manifests itself as a random variation of the superlattice period of about one lattice constant ($\sim$ 5.6 Å) for sample B.

The quantitative evaluation of the interface quality by X-ray diffraction becomes even more important if the lattice parameters of the epilayers have to be matched to those of the substrate by appropriate choice of the layer composition, as for Al$_{0.48}$In$_{0.52}$As / Ga$_{0.47}$In$_{0.53}$As superlattices lattice-matched to InP substrates. In Fig. 8 we show the X-ray diffraction pattern of such a superlattice with an average lattice mismatch of $\Delta a/a_o$ = 1.6 x 10^{-4} to the InP substrate. The angle separation of $\Delta\phi$ = 911" between the main ("0") superlattice peak and the first ("+1") satellite peak yields a period length of 20.5 nm. The narrow linewidths of the main superlattice peak and of the satellite peaks indicate an excellent homogeneity of the chemical composition throughout the superlattice and fluctuations of the superlattice periodicity of about one monolayer / 22 / .

These few examples indicate that important details of the interface structure can be extracted from high-angle X-ray diffraction if a semi-kinematical treatment of the dynamical theory is applied to analyse the experimental data.

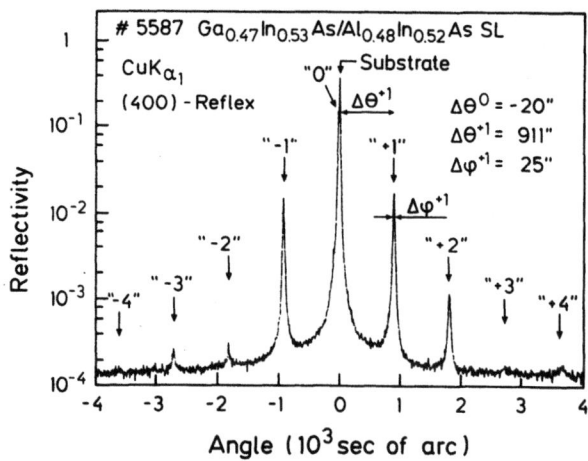

Fig. 8. X-ray diffraction curve of a $Ga_{0.47}In_{0.53}As / Al_{0.48}In_{0.52}As$ superlattice grown lattice-matched to InP [(004) reflection, CuK_{α_1} radiation].

4. CURRENT RESEARCH ACTIVITIES

In this section we will briefly define some selected areas of current research activities on bandgap engineering in semiconductors by MBE growth of artificially layered structures. The first example will emphasize the unique capability of MBE to control monolayer growth, while the other three examples will show that the modification of the electronic bulk properties is mainly due to the quasi-2D or quasi-1D properties of the layered materials. Due to the length restriction the selection will not be exhaustive in this rapidly growing field. In particular, we have to omit all the exciting results associated with the quantum Hall effect /23/.

4.1. Short-Period and Monolayer Superlattices

The recent progress in the control of interface quality using RHEED intensity oscillation and growth interruption has led to the successful growth of very short-period lattice-matched $(AlAs)_n(GaAs)_n$ superlattices / 24 / with n = 1, 2. Growth of monolayer superlattices with n = 1 was achieved by monitoring each deposited monolayer from the RHEED oscillation period, interrupting the group-III-element flux at n = 1 and allowing the RHEED intensity to recover almost to its initial value, and then depositing the next layer. Although the GaAs / InAs system involves a large lattice mismatch of 7%, 2D growth of each layer of GaAs and InAs on both (100) GaAs and (100) InP substrates can be maintained up to n = 2. Beyond two monolayers, InAs tends to grow in a 3D growth mode / 25 / . The monolayer semiconductor superlattices represent a new class of materials which are of great current interest. The well-ordered layer-by-layer arrangement of the group-III-elements on the appropriate lattice sites is expected to modify the band structure strongly as compared to the bulk-type components. While the ternary alloy $Al_{0.5}Ga_{0.5}As$ with random Al and Ga distribution is an indirect semiconductor and when n-type exhibits persistent photoconductivity due to deep donors, the $(AlAs)_1(GaAs)_1$ monolayer superlattice may be a direct-gap material. The $(GaAs)_1(InAs)_1$ monolayer superlattice should exhibit reduced alloy scattering so that even higher mobilities than in

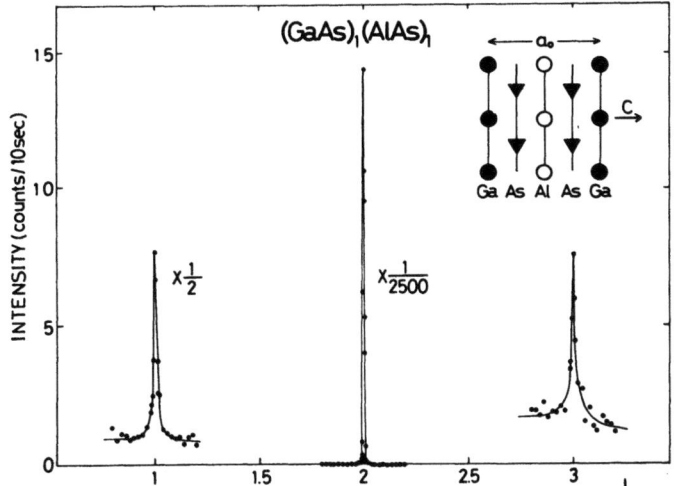

Fig. 9. X-ray diffraction curve of 600-period $(AlAs)_1(GaAs)_1$ monolayer superlattice recorded in the vicinity of the (002) reflection using CuK_{α_1} radiation / 24 / .

the ternary $Ga_{0.47}In_{0.53}As$ alloy can be expected. Although X-ray diffraction (Fig. 9) and Raman scattering / 26 / indicate a periodic monolayer sequence in the $(AlAs)_1 (GaAs)_1$ structures, a thorough examination of the theoretical predictions could not be made, because a perfectly ordered arrangement of Al and Ga (or Ga and In) along the (100) growth direction of the entire epilayer has not yet been accomplished.

Confinement layers composed of short-period superlattices (SPS) $A_n B_n$ with $5 < n < 10$ play an important role for highly improved optical properties of GaAs, GaSb, and $Ga_{0.47}In_{0.53}As$ quantum wells. In Fig. 10 we show schematically the real-space energy bands of a GaAs quantum well with confinement layers composed of either homogeneous ternary $Al_x Ga_{1-x} As$ or of AlAs / GaAs SPS / 27 / . The SPS consist of all-binary AlAs / GaAs for GaAs quantum wells, of all-binary AlSb / GaSb for GaSb quantum wells, and of all-ternary $Al_{0.48}In_{0.52}As$ / $Ga_{0.47}In_{0.53}As$ for $Ga_{0.47}In_{0.53}As$ quantum wells lattice-matched to InP. The effective barrier height for carrier confinement in the quantum wells is adjusted by the appropriate choice of the layer thickness of the lower-gap material in the SPS. The observed improvement of the optical properties of SPS confined quantum wells is due (i) to a removal of substrate defects by SPS layer, (ii) to an amerlioration of the interface between quantum well and barrier, and (iii) to a modification of the dynamics of photoexcited carriers in the SPS barrier. In particular for GaSb and for $Ga_{0.47}In_{0.53}As$ quantum wells we have provided the first direct evidence for intrinsic exciton recombination by application of SPS barriers / 28, 29 / .

4.2. Electric-Field Induced Shift of Excitonic Recombination (Quantum Confined Stark Effect)

Excitons play a significant role for the optical properties of III-V QWH and superlattices / 30 / because of the strongly increased exciton binding energy in these quasi-2D systems. Intrinsic free-exciton recombination dominates the emission spectra, and exciton effects are also pronounced in optical absorption. Well resolved heavy- and light-hole exciton transitions can be observed in absorption as well as in emission due to

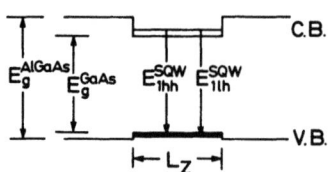

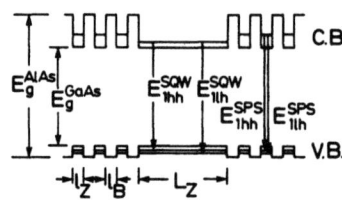

Fig. 10. Schematic real-space energy band diagram of a GaAs quantum well confined either by ternary $Al_xGa_{1-x}As$ layers (top) or by all-binary AlAs / GaAs short-period superlattices (SPS).

the splitting of heavy- and light-hole bands in quantum wells.

An electric field applied perpendicular to the layers of quantum wells significantly changes the luminescence and the optical absorption. Different techniques, including optical absorption, photoluminescence (PL), photocurrent spectroscopy (PCS), and photo- and electroreflectance have been used to study the field-induced changes of the optical properties. Two distinct phenomena have been observed, i.e. a red-shift (Stark shift) of the excitonic recombination in the wells and a simultaneous quenching of the luminescence with increasing field. In Fig. 11 we show the variation of exciton peak energies with electric field observed by PCS on a 9-nm GaAs / $Al_{0.3}Ga_{0.7}As$ multi QWH at 10 K / 31 / . In comparison to the lowest-energy heavy- and light-hole excitons, some of the higher-energy excitonic transitions show only a very small Stark shift. Photocurrent spectra from quantum wells are equivalent to optical absorption spectra. However, PCS measurements do not require any substrate etching process which may cause strain on QWH. Therefore, PCS is a simple and efficient technique to study interband transitions. The application of high electric fields perpendicular to the quantum well allows the clear detection of parity-forbidden ($n \neq n'$) transition peaks owing to the deformation of the well potential. From the intensity of the various forbidden excitonic transitions their oscillator strength can be estimated and compared with theoretical models on valence subband mixing. In addition, the PCS data yield the energy spacing also of higher electron and hole subbands, from which a determination of the band offset can be made.

The early field-induced photoluminescence measurements were performed on rather thin ($\sim$ 5 nm) GaAs quantum wells, where the confinement effects are large. Therefore, only small shifts combined with a strong quenching of the luminescence were observed at increasing field. Only recently, the distinct read-shift of the lowest-exciton emission line has been observed by picosecond luminescence measurements on $Al_xGa_{1-x}As$ / GaAs QWH of larger well width in the quantum-confined Stark (QCS) regime / 32 / . The variation of the PL peak energies as a function of electric field is shown for three different well widths in Fig. 12. The Stark shift of the luminescence is accompanied by a strong enhancement of the recombination lifetime, par-

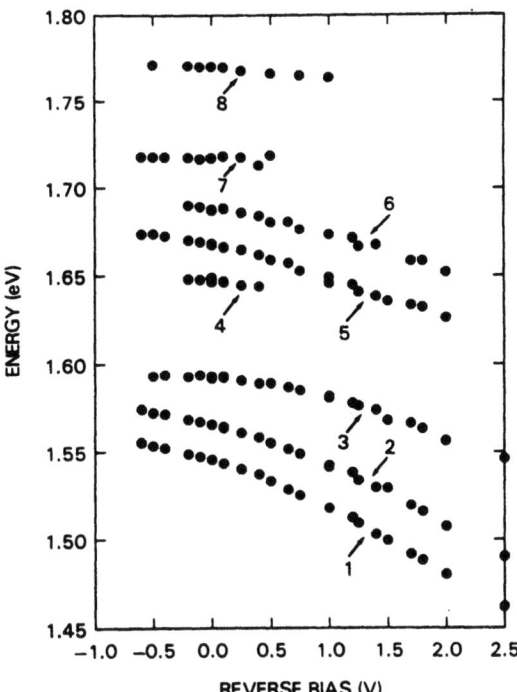

Fig. 11. Variation of exciton peak energy as a function of bias voltage observed by photocurrent spectroscopy on a 9-nm GaAs/Al$_{0.3}$Ga$_{0.7}$As multi QWH. The numbers indicate the various excitonic transitions (1 = lowest heavy-hole exciton, 2 = lowest light-hole exciton).

ticular at well widths beyond 10 nm, due to the spatial separation of electrons and holes by the deformed well potential.

The observed energy shift of the excitonic transitions is caused by shifts of the subband energies by the electric field. Theoretical calculations / 33 / have shown that subbands of lower energy exhibit a larger shift than those of higher energy. The observed luminescence quenching is attributed to a decrease in the overlap of electron and hole wavefunctions, as they are spatially separated by the field, and to carrier tunneling out of the well. Finally, high electric fields result in a large probability for n ≠ n' excitonic transitions which are no longer forbidden due to the strong deformation of the well potential.

The quantum-confined Stark effect is important for application in new photonic devices, such as high-speed optical modulators, self electro-optic effect devices (SEED), and tunable laser sources.

4.3. Strained-Layer Superlattices

Semiconductor superlattices of high quality can be grown from semiconductor materials with lattice mismatches of several percent if the constituent layers are kept sufficiently thin (e.g. < 20 nm). The mismatch in these strained-layer superlattices (SLS) is totally accommodated by coherent elastic strain so that misfit dislocations are not generated at the superlattice interfaces. The absence of misfit dislocations (confirmed by TEM) supports the basic concepts of the Frank-van der Merwe model. The SLS can thus be synthesized from a wide variety of material systems, and they

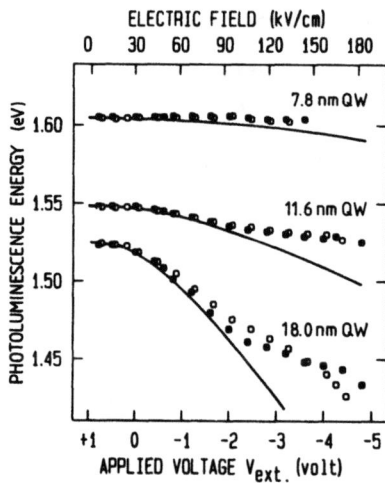

Fig. 12. Variation of photoluminescence peak energy as a function of bias voltage obtained from three GaAs quantum wells of different widths using an excitation energy of 1.675 eV and two different excitation densities. The solid lines are calculated for infinite barrier heights / 32 / .

exhibit several novel electronic properties among which strain-induced band-gap shifts and strain-modified effective masses are of particular importance. The built-in strain in $Ga_{0.8}In_{0.2}As$ / GaAs SLS has been used to remove the degeneracy of the valence band at the Brillouin zone center so that the light-hole band can be preferentially populated / 34 / . Selective doping results in a high-mobility p-type material with the light holes confined to the constituent $Ga_{0.8}In_{0.2}As$ layers.

In selectively n-doped Si / Si_xGe_{1-x} SLS a high-mobility 2D electron gas can be achieved with the electrons confined to the wider-gap Si rather than to the lower-gap Si_xGe_{1-x} layers depending on the built-in strain distribution / 35 / . The distribution of the strain on the constituent superlattice layers can be modified by the appropriate choice of the buffer layer, as indicated in Fig. 13. In Si / $Si_{0.5}Ge_{0.5}$ SLS a $Si_{0.75}Ge_{0.25}$ buffer layer yields a mean lattice spacing between that of Si and of $Si_{0.5}Ge_{0.25}$. In this case both constituent superlattice layers are strained to the same extent but with opposite sign. When Si buffer layers are used, however, the Si superlattice layers remain unstrained while twice the strain exists in the (compressed) $Si_{0.5}Ge_{0.5}$ layers. In Fig. 13 we show that in Si / $Si_{0.5}Ge_{0.5}$ SLS the conduction band of the wider-gap material Si can be lower in energy than that of $Si_{0.5}Ge_{0.5}$ due to the lateral tensile strain in the Si layers and the lateral compressive strain in the $Si_{0.5}Ge_{0.5}$ layers induced by the $Si_{0.75}Ge_{0.25}$ buffer layers. Therefore, in these samples the phenomenon of electron mobility enhancement is only observed when the $Si_{0.5}Ge_{0.5}$ layers are selectively doped and the spatially separated electrons are confined to the Si layers. This effect is indeed observed in the temperature dependence of the Hall electron mobility, as shown in Fig. 14 for samples with different doping positions in selectively doped Si / $Si_{0.5}Ge_{0.5}$ SLS. This intriguing result demonstrates the strong impact of built-in strain on the actual values of bandgap discontinuities in semiconductor QWH and superlattices.

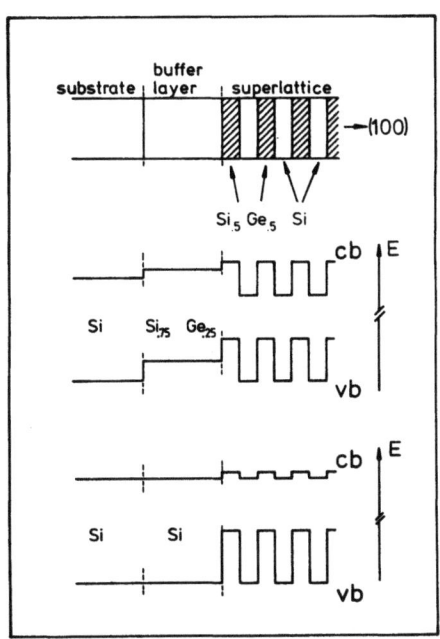

Fig. 13. Schematic layer sequence (top) and real-space energy band diagram of Si / $Si_{0.5}Ge_{0.5}$ strained-layer superlattice grown on either Si or $Si_{0.75}Ge_{0.25}$ buffer layer. cb and vb denote conduction and valence band, respectively / 35 / .

4.4. Vertical Transport (Tunneling)

Originally the hope for new effects through vertical transport / 36 / of carriers, i.e. perpendicular to the heterojunction, due to the formation of extended minibands was the motivation to develop the idea of semiconductor superlattices. Moreover, the resonant transmission (tunneling) of electrons through potential barriers, manifested as negative differential resistance in the current-voltage (I - V) characteristics, provided one of the first evidences for the formation of bound electron levels in semiconductor quantum wells / 2 / . This phenomenon is now of renewed interest because of the dramatic advances in MBE growth control, and negative resistance regions with larger peak-to-valley ratios and even at room temperature have been observed.

In most experiments, resonant tunneling was obtained by applying a voltage to an $Al_xGa_{1-x}As$ / GaAs / $Al_xGa_{1-x}As$ double-carrier structure sandwiched between n^+-GaAs regions (Fig. 15) in order to achieve matching between the Fermi level of the injecting side and the confined quantum states E_1 in the well / 37 / . The resonant transmission of electrons gives rise to a current maximum occurring at the applied voltage of $2E_1$ (the applied voltage is split into two equal voltage drops at each barrier). When the resonance condition is detuned, the I - V characteristics exhibit a negative differential resistance. For fundamental research it is important that these resonant tunneling features provide detailed insight into quantized levels of electrons (holes). The non-linear I - V characteristics may find application for the detection and generation of electromagnetic waves at very high frequencies. High-frequency response with far-infrared lasers has shown response time of less than 10^{-13} sec, in agreement with tunneling times which are given by the uncertainty relation.

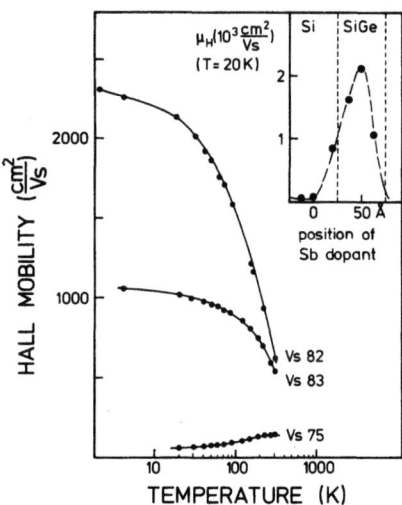

Fig. 14. Hall electron mobility versus temperature obtained from samples with different doping positions in selectively doped Si / $Si_{0.5}Ge_{0.5}$ superlattices (Vs 75: center of Si layer, Vs 82: center of $Si_{0.5}Ge_{0.5}$ layer, Vs 83: edge of $Si_{0.5}Ge_{0.5}$ layer). The inset shows 20 K Hall mobilities observed at different doping positions / 35 / .

Recently, a new $Al_xGa_{1-x}As$ / GaAs tunneling hot electron transistor (RHET) has been demonstrated / 38 / . The electrons are injected from the emitter to the base by resonant tunneling through a quantum well and are ballistically transferred to the collector. The observed peak in the collector current versus base-emitter voltage allows to devise a frequency multiplier or an Exclusive-NOR gate using just one transistor. The field of vertical transport in QWH and superlattices is developing very rapidly. In the SPS clad quantum wells of Sect. 4.1. e.g., the photoexcited carriers moving in conduction and valence minibands of the SPS confinement layers are captured and (optically) detected in the (enlarged) quantum wells which act as probes for vertical transport.

5. DEVICE APPLICATION

At present the most significant impact for MBE to be used for device fabrication comes from high-speed electron devices based on precisely tailored III-V semiconductor heterostructures. The two most prominent devices are the high electron mobility transistor (HEMT) and the heterojunction bipolar transistor (HBT) / 39 / .

Operation of the HEMT device is based on the confinement of a 2D electron gas at a heterointerface, which is achieved by the precise placement of donors in $Al_xGa_{1-x}As$ at a small distance (3 - 10 nm) from the defect-free $Al_xGa_{1-x}As$ / GaAs interface. The device performance is evaluated from the propagation delay measured in ring oscillators. Recently, delay times per stage of 11 psec at 300 K and 8 psec at 77 K have been reported. The doping and interface control and the uniformity of MBE grown material has now made feasible the operation of a 4 K HEMT static random access memory (SRAM), while 16 K HEMT SRAMs are being developed. The heterostructures used for HBT devices typically consist of five $Al_xGa_{1-x}As$ / GaAs layers

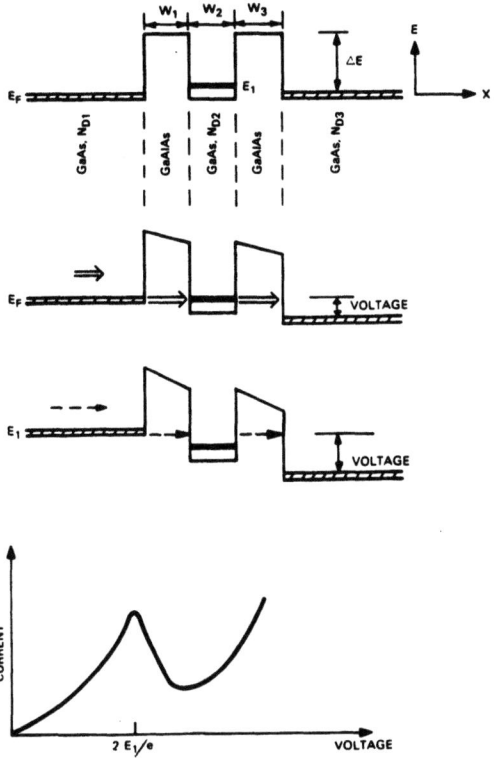

Fig. 15. Schematic real-space energy band diagrams and current-voltage characteristic illustrating resonant tunneling in a double barrier structure / 27 / .

with accurately placed transitions between low and high doping levels. The device performance of HBT has been improved by incorporation of a graded bandgap (i.e. steadily increasing Al content in $Al_xGa_{1-x}As$) in the base region. This improvement resulted in cut-off frequencies above 40 GHz in distrete devices and in propagation delays below 30 psec measured in ring oscillators. Further improvement of the heterostructures used for HBT devices is anticipated, as the performance for optimized devices is projected to 10 psec propagation delay and 100 GHz cut-off frequency.

The most prominent photonic device is undoubtably the heterostructure injection laser / 40 /. The advantages of quantum well lasers are their engineered emission wavelength, their low threshold current density, and their low temperature sensitivity of the threshold current. Recently, threshold current densities below 200 A/cm^2 have been reported for MBE grown $Al_xGa_{1-x}As$ / GaAs graded-index waveguide separate confinement heterostructure (GRIN-SCH) quantum well laser / 41 / . In the long-wavelength Al-Ga-Sb material system the incorporation of a multi QWH in the active region made feasible the first successful room-temperature cw operation of a mesa-stripe laser / 42 / and this was fabricated from MBE grown material.

The signal-to-noise ratio of avalanche photodiodes, which are important for fiber optics communication systems, can be significantly improved if the constituent bulk III-V semiconductor material is replaced by specially designed heterostructures or superlattices. Using the Al-Ga-As system, then the ionization of electrons as compared to holes is favoured, because a larger portion of the energy band discontinuity occurs in the conduction band. This staircase avalanche photodiode / 43 / acts as if it were an electron multiplier tube using dynodes ("solid state photomultiplier").

The first application of the quantum confined Stark effect was in optical modulators / 44 /, where the intrinsic region of a p-i-n diode was replaced by an $Al_xGa_{1-x}As$ / GaAs superlattice. The device exhibited 3 dB modulation depth, and pulse widths down to 131 psec were obtained. Recently, the monolithic integration of a laser diode and an optical waveguide modulator was achieved / 45 /, which required only a low voltage for high-speed modulation.

The same p-i-n structure used as a modulator also operates as a photodiode with nearly unity internal quantum efficiency. The self electro-optic effect devices (SEED) are based on the following feedback mechanism. The photocurrent affects the voltage across the modulator and hence its absorption. An optically bistable SEED operating at 853 nm at room temperature showed switching speeds of 30 nsec with 1.6 mW optical power and a 47 kΩ load / 46 /. A further reduction of the switching power to 500 nW at 100 nsec switching time is expected by optimizing the device structure.

These few examples have shown the importance of MBE grown QWH and superlattices for the development of new device concepts based on bandgap engineering.

6. FUTURE PROSPECTS

Miniaturization in semiconductor devices has resulted in improved speed and lower power consumption. Examination of the limits to further miniaturization requires the fabrication and investigation of semiconductor structures that are atomically engineered in all three dimensions to nanometer design rules. Therefore, in future not only a depth control in z-axis but also a control of the lateral size (x- and y-axis) by a combination with appropriate fine-line lithography techniques with monolayer resolution has to be developed.

ACKNOWLEDGEMENT

This work was sponsored by the Bundesministerium für Forschung und Technologie of the Federal Republic of Germany.

REFERENCES

/ 1 / For a recent extensive review on MBE see: "The technology and physics of molecular beam epitaxy", Ed. E.H.C. Parker (Plenum Press, New York, 1985) ·
/ 2 / L.L. Chang, L. Esaki, and R. Tsu, Appl. Phys. Lett. 24 (1974) 593
/ 3 / R. Dingle, W. Wiegmann, and C.H. Henry, Phys. Rev. Lett. 33 (1974) 827
/ 4 / L.L. Chang and K. Ploog "Molecular Beam Epitaxy and Heterostructures" (Martinus Nijhoff, Dordrecht, 1985) NATO Adv. Sci. Inst. Ser., Vol. E 87
/ 5 / A.Y. Cho, J. Appl. Phys. 42 (1971) 2074
/ 6 / A.Y. Cho, J. Appl. Phys. 47 (1976) 2841
/ 7 / K. Ploog and A. Fischer, Appl. Phys. 13 (1977) 111
/ 8 / J.H. Neave and B.A. Joyce, J. Cryst. Growth 44 (1978) 387
/ 9 / I. Hernandez-Calderon and H. Höchst, Phys. Rev. B 27 (1983) 4961
/ 10 / B.A. Joyce, J.H. Neave, P.J. Dobson, and P.K. Larsen, Phys. Rev. B 29 (1984) 814
/ 11 / C.E.C. Wood, Surf. Sci. 108 (1981) L441

/ 12 / J.H. Neave, B.A. Joyce, P.J. Dobson, and N. Norton, Appl. Phys. A 31 (1983) 1
/ 13 / J.M. van Hove, C.S. Lent, P.R. Pukite, and P.F. Cohen, J. Vac. Sci. Technol. B 1 (1983) 741
/ 14 / T. Sakamoto, H. Funabashi, K. Ohta, T. Nakagawa, N.J. Kawai, and T. Kojima, Jpn. J. Appl. Phys. 23 (1984) L657
/ 15 / Y. Suzuki and H. Okamoto, J. Appl. Phys. 58 (1985) 3456
/ 16 / H.J. Fronius, A. Fischer, and K. Ploog, Jpn. J. Appl. Phys. 25 (1986) L 137
/ 17 / T. Sakamoto, H. Funabashi, K. Ohta, T. Nakagawa, N.J. Kawai, T. Kojima, and K. Bando, Superlattices and Microstructures 1 (1985) 347
/ 18 / B.A. Joyce, P.J. Dobson, J.H. Neave, K. Woodbridge, J. Zhang, P.K. Larsen, and B. Bôlger, Surf. Sci. 168 (1986) 423
/ 19 / M. Tanaka, H. Sakaki, and J. Yoshino, Jpn. J. Appl. Phys. 25 (1986) L 155
/ 20 / F. Voillot, A. Madhukar, J.Y. Kim, P. Chen, N.M. Cho, W.C. Tang, and P.G. Newman, Appl. Phys. Lett. 48 (1986) 1009
/ 21 / L. Tapfer and K. Ploog, Phys. Rev. B 33 (1986) 5565
/ 22 / W. Stolz, L. Tapfer, A. Breitschwerdt, and K. Ploog, Appl. Phys. A 38 (1985) 97
/ 23 / K. von Klitzing, Rev. Mod. Phys. 58 (1986) 519
/ 24 / N. Sano, H. Kato, M. Nakayama, S. Chika, and H. Terauchi, Jpn. J. Appl. Phys. 23 (1984) L 640
/ 25 / R. Katsumi, H. Ohno, T. Takama, and H. Hasegawa, Inst.Phys. Conf. Ser. 79 (1986) 391
/ 26 / M. Nakayama, K. Kubota, H. Kato, S. Chika, and N. Sano, Solid State Commun. 53 (1985) 493
/ 27 / K. Fujiwara, H. Oppolzer, and K. Ploog, Inst. Phys. Conf. Ser. 74 (1985) 351
/ 28 / K. Ploog, Y. Ohmori, H. Okamoto, W. Stolz, and J. Wagner, Appl. Phys. Lett. 47 (1985) 384
/ 29 / J. Wagner, W. Stolz, J. Knecht, and K. Ploog, Solid State Commun. 57 (1986) 781
/ 30 / For a review see: R.C. Miller and D.A. Kleinmann, J. Lumin. 30 (1985) 520
/ 31 / R.T. Collins, K. von Klitzing, and K. Ploog, Phys. Rev. B 33 (1986)
/ 32 / H.J. Polland, L. Schultheis, J. Kuhl, E.O. Göbel, and C.N. Tu, Phys. Rev. Lett. 55 (1985) 2610
/ 33 / J.A. Brum and G. Bastard, Phys. Rev. B 31 (1985) 3893
/ 34 / J.E. Schirber, I.J. Fritz, and L.R. Dawson, Appl. Phys. Lett. 46 (1985) 187
/ 35 / G. Abstreiter, H. Brugger, T. Wolf, H.J. Jorke, and H.J. Herzog, Phys. Rev. Lett. 54 (1985) 2441
/ 36 / L. Esaki and R. Tsu, IBM J. Res. Develop 14 (1970) 61
/ 37 / T.C.L.G. Sollner, W.D. Goodhue, P.E. Tannenwald, C.D. Parker, and D.D. Peck, Appl. Phys. Lett. 43 (1983) 588
/ 38 / N. Yokoyama, K. Imamura, S. Muto, S. Hiyamizu, and H. Nishi, Jpn. J. Appl. Phys. 24 (1985) L 853
/ 39 / For recent reviews on HEMT and HBT technology see: "VLSI Electronics Microstructure Science" Eds. N.G. Einspruch and W.R. Wisseman (Academic Press, Orlando, FL, 1985) Vol. 11
/ 40 / For a review see: W.T. Tsang, IEEE J. Quantum Electron. QE-20 (1984) 1119
/ 41 / O. Wada, T. Sanoda, N. Nobuhara, M. Kuno, M. Makiuchi, and T.Fujii, Inst. Phys. Conf. Ser. 79 (1986) 685
/ 42 / Y. Ohmori, Y. Suzuki, and H. Okamoto, Jpn. J. Appl. Phys. 24 (1985) L 657
/ 43 / F. Capasso, W.T. Tsang, and G.F. Williams, IEEE Trans. Electron Devices ED-30 (1983) 381

/ 44 / T.H. Wood, C.A. Burrus, D.A.B. Miller, D.S. Chemla, T.C. Damen, A.C. Gossard, and W. Wiegmann, IEEE J. Quantum Electron. QE-21 (1985) 117
/ 45 / S. Tarucha and H. Okamoto, Appl. Phys. Lett. 48 (1986) 1
/ 46 / D.A.B.Miller, D.S.Chemla, T.C.Damen, T.H.Wood, C.A.Burrus, A.C.Gossard, and W.Wiegmann, IEEE J. Quantum Electron QE-21 (1985) 1462

SYNCHROTON X-RAY STUDIES OF LIQUID-VAPOR INTERFACES

J. Als-Nielsen

RISØ National Laboratory
DK-4000-Roskilde, Denmark

1. Introduction

The variation of density across the liquid-vapor interface from essentially zero density far out in the vapor phase to a homogeneous density deep in the liquid phase can be determined by X-ray reflectivity measurements[1] at grazing angles θ well beyond the critical angle for total reflection, θ_c. We shall first derive the relation between the density variation, $\rho(z)$, and the reflectivity $R(\theta)$. Next we consider as an experimental example the roughness caused by thermal excitations of capillary waves on simple liquids like water and carbon-tetra-chloride[2]. Then we consider the smectic layering of liquid crystal molecules of the free surface, first in a system with a spontaneous first order phase transition from the isotropic phase to a smectic A phase[3], and afterwards in a system with a second order phase transition from the nematic phase to the smectic A phase[4-6].

2. Reflectivity and Density Profile

We consider in figure 1 a monochromatic X-ray beam (wavelength λ, wavevector $\vec{k}_{in}$) incident on the liquid surface at grazing angle θ and specular reflected to wavevector $\vec{k}_{out}$. Specular reflection means that the wavevector transfer

$$\vec{Q} \equiv \vec{k}_{out} - \vec{k}_{in}, \quad |\vec{Q}| = (4\pi/\lambda)\sin\theta \tag{1}$$

is along the surface normal.

The dimensionless density profile, $\rho(z)$, is normalized to unity for large z where the electron density is ρ_{el}. The gradient of $\rho(z)$ is denoted $\rho'(z)$. In the ideal case of $\rho(z)$ being a step function at z=0, standard derivation from optics obtains the Fresnel reflectivity $R_F(Q)$ as briefly outlined below. In the real case where $\rho(z)$ is a continuous function we shall see that the reflectivity $R(Q)$ is related to $\rho(z)$ by

$$\frac{R(Q)}{R_F(Q)} = \left|\int \rho'(z)\exp[iQz]dz\right|^2 \tag{2}$$

We observe that eq. (2) has the correct limiting behaviour when $\rho(z)$ approaches a step function since $\rho'(z)$ then approaches a delta-function and the Fourier-transform therefore unity.

As far as the Fresnel reflectivity is concerned we recall Snell's law relating the grazing angles θ and θ' for $\theta > \theta_c$, the critical angle of total reflection:

$$\cos\theta/\cos\theta' = n = 1 - \lambda^2 \rho_{el} r_o/2\pi \tag{3}$$

Here n is the index of refraction which for X-rays is given by the second equation in terms of λ, ρ_{el} and r_o, the electron radius or the scattering length of Thompson scattering. As the angles are small eq. (3) can be expanded to

$$\theta^2 = \theta'^2 + \theta_c^2, \quad \theta_c^2 = \lambda^2 \rho_{el} r_o/\pi \tag{4}$$

The reflectivity $R_F(\theta)$ is

$$R_F(\theta) = \left\{\frac{\theta-\theta'}{\theta+\theta'}\right\}^2 \tag{5}$$

For $\theta \gg \theta_c$, $R_F(\theta)$ approaches $(\theta_c/2\theta)^4$ or in terms of wave-vector transfer Q relative to Q_c

$$Q_c = (4\pi/\lambda)\theta_c = 4\pi[\rho_{el}r_o/\pi]^{1/2} \tag{6}$$

$$R_F(Q) \simeq (Q_c/2Q)^4, \quad Q \gg Q_c \tag{7}$$

Anticipating eq. (2) we can estimate that Q must be of the order of the inverse width Δ of $\rho'(z)$ (or the period of $\rho'(z)$ if its oscillating) before an appreciable deviation of $R(Q)$ from $R_F(Q)$ is obtained. The corresponding value of Q/Q_c is

$$Q/Q_c \sim [\Delta \cdot Q_c]^{-1} \simeq [R4\sqrt{\pi(\rho_{el}r_o)^{\frac{1}{2}}}]^{-1} \qquad (8)$$

where we have assumed Δ to be of order the molecular radius R. In a molecule with Z electrons, $\rho_{el} \sim Z/(4/3\pi R^3) \sim Z/(\pi R^3)$, so $\Delta \cdot Q_c \simeq 4(ZRr_o)^{\frac{1}{2}}$ which numerically is typically around 10. The approximate simple expression for $R_F(Q)$ in eq. (7) is therefore useful, and refraction effects can usually be neglected. Absorption, which we have neglected for simplicity in eq. (5), modifies the reflectivity around the critical angle as it rounds the kink at $\theta = \theta_c$, but for $Q \gg Q_c$ the effects are negligible.

The X-ray reflectivity which contains the information about $\rho(z)$ is thus quite low since for $Q \sim 10\, Q_c$, $R_F(Q) \sim 10^{-5}$ and furthermore $R(Q) < R_F(Q)$. That is one reason for using the intense X-ray beams obtainable from synchrotron radiation in this kind of experiments, in particular because the beam height h must be small (0.1 mm) as the "footprint" on the sample, h/θ, must be limited to say 20-30 mm. Another reason is the narrow $\vec{Q}$-resolution which is necessary to distinguish clearly between intensity from specular reflection and intensity due to scattering from the illuminated bulk, which typically has a depth of several microns even for grazing angles of the order of milli-radians.

In order to derive eq. (2) we consider the reflected wave as a superposition of waves reflected from infinitesemal planes at varying depth z implying the phase factor $\exp[iQz]$. In accordance with the order-of-magnitude discussion given above we shall neglect refraction and absorption effects. We only need to know the reflectivity of a thin plate with thickness Δz, cf. fig. 1. The reflected wave is the result of Thompson scattering of the incident photon wave by the individual electrons. The reflected amplitude ΔA_r must be proportional

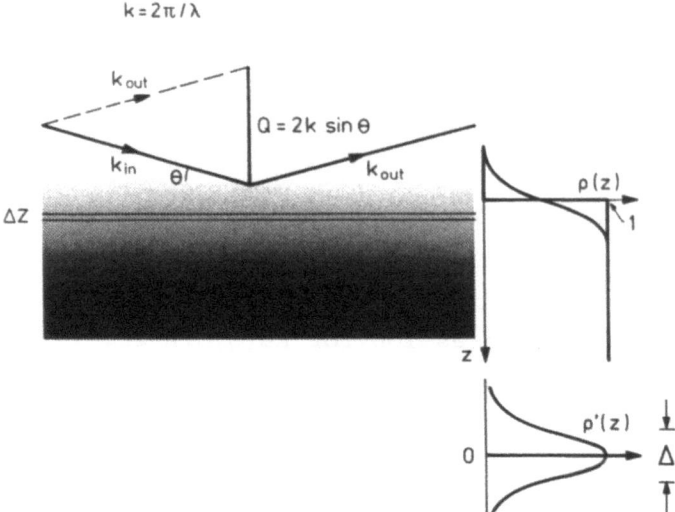

Fig. 1. The density variation across a surface is indicated by shading in the left part and more quantitatively by the function $\rho(z)$ in the right part. The reflectivity versus wavevector-transfer Q is related to the Fourier transform of the gradient of the density, $\rho'(z)$.

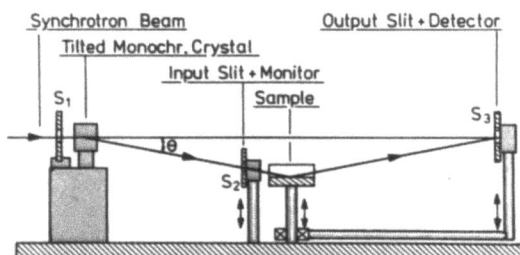

Fig. 2. Side view of the vertical, scattering plane. Beam directions are defined by slits. The monochromatic beam is bent down towards the sample by tilting the monochromator crystal. The incident beam intensity is monitored after slit S_2.

to the incident amplitude A_i, to the scattering length r_o of a single electron and to the number of electrons per unit area perpendicular to the incident beam, $\rho(z) \cdot (\rho_{el} \Delta z / \sin\theta)$. Since $\Delta A_r / A_i$ is dimensionless and the dependence on quantities with length dimensions such as r_o, ρ_{el} and Δz is exhausted by their product the only additional length in the problem, λ, must enter linearly, i.e.

$$\frac{\Delta A_r}{A_i} = c\rho(z)[\lambda r_o \rho_{el}/\sin\theta]\Delta z \tag{9}$$

where c is a dimensionless, complex number. The quantity in the square-bracket is by eq. (6) the same as $Q_c^2/(2Q)^2$. By superposition of all infinitesemal layers and squaring the amplitude ratio to get the reflectivity we find

$$R(Q) = |c|^2 [Q_c^2/(2Q)]^2 |\int \rho(z) \exp[iQz] dz|^2 \tag{10a}$$

$$= |c|^2 [Q_c/(2Q)]^4 |\int \rho'(z) \exp[iQz] dz|^2 \tag{10b}$$

Here equation (10b) follows from eq. (10a) by partial integration. We note that the squared bracket by comparison to eq. (7) is the Fresnel reflectivity $R_F(Q)$ in the present approximation of neglecting refraction effects, i.e. $Q \gg Q_c$. By noting that $R(Q)$ must equal $R_F(Q)$ for a step function $\rho(z)$, or delta function $\rho'(z)$, it follows that $|c|^2 = 1$, and the master formula eq. (2) is obtained within the ab initio assumptions.

3. Experiments

The general experimental set-up is shown in figure 2. A suitable aperture S_1 is placed in the horizontal synchrotron radiation beam incident from the left in figure 2. A monochromatic beam is extracted from the "white" spectrum by Bragg reflection from a monochromator crystal, typically perfect Ge in the (1,1,1) orientation. By tilting the normal to the reflecting planes out of the horizontal plane, the monochromatic beam can be bent down to any glancing angle with the horizontal liquid surface. The

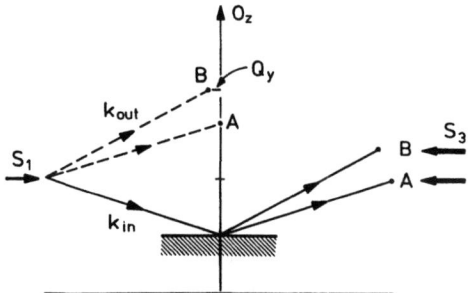

Fig. 3. A transverse wavevector component in the vertical plane may be obtained without rotating the sample by having the incident glancing angle different from the reflected angle.

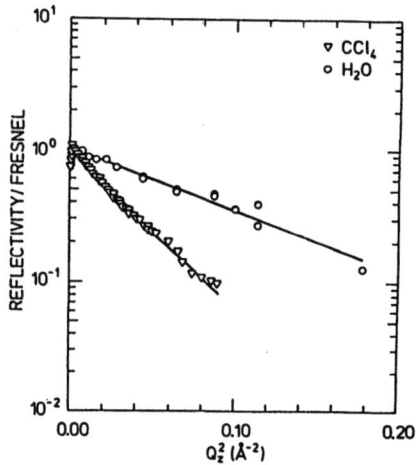

Fig. 4. Reflectivity data versus the squared wavevector transfer for water (open circles) and carbon tetrachloride (triangles). The slope expresses the thermal mean squared displacement of the atoms from the ideal flat surface.

sample is mounted on an elevator so that the liquid surface
always intersect the monochromatic beam. A narrow slit S_2 in
front of the sample defines a suitable footprint of the beam
on the liquid surface, and a beam monitor after S_2 ensures
that intensities measured at different slit heights and at
different currents in the storage ring are all normalized to
the same number of incident photons. A slit S_3 defines the
angular resolution of the reflected beam. This slit together
with the detector, a NaJ scintillation counter, is mounted
on an elevator situated on a spectrometer arm which is
pivoted around a vertical axis through the sample center.
In this way one has enough freedom to obtain an arbitrary
kinematically allowed wavevector transfer $\vec{Q}$ with its three
components: Q_z along the surface normal, Q_y transverse to
Q_z in the vertical plane and Q_x transverse to Q_z in the
horizontal plane, cf. fig. 3.

3a. Thermal Roughness of Simple Liquids[2]

A "quiet" surface of H_2O or CCl_4 was obtained by wetting an
area of approximately 60 mm diameter of a clean glass flat.
The corresponding liquid thickness is a few tenth of a millimeter. Reflectivities of more than 96% was obtained for $\theta < \theta_c$,
and the reflected line shape was essentially identical to the
incident beam line shape. The reflectivity, relative to the
Fresnel reflectivity, turned out to be a Gaussian function of
Q_z as shown in fig. 4, where the logarithm of $R(Q)/R_F(Q)$ is
plotted versus Q^2. The slope of the lines for both H_2O and
CCl_4 reflects essentially the thermal roughness which is
three times bigger for CCl_4 than for H_2O because the surface
tension γ for CCl_4 is only a third of that of H_2O. Thermal
roughness means that the density gradient $\rho'(z)$ is a Gaussian,
$\exp[-z^2/2\langle u^2\rangle]$; the Fourier transform is then also Gaussian,
$\exp[-Q^2\langle u^2\rangle/2]$, and $R(Q)/R_F(Q)$ by eq. (2) equals therefore
$\exp[-Q^2\langle u^2\rangle]$. We shall now express $\langle u^2\rangle$ in terms of the surface
tension γ. To this end consider a sinusoidal perturbation of
a flat square of side length L. The displacement at position

$\vec{r}$ is $u_q \sin(\vec{q}\cdot\vec{r})$. The excess area over the flat surface is $\frac{1}{4}u_q^2 q^2 L^2$ and the associated energy of this mode therefore $\gamma \cdot 1/4 u_q^2 q^2 L^2$. By equipartion follows that the thermal average value of u_q^2, which is denoted $\langle u_q^2 \rangle$, must be $1/2kT/[1/4\gamma q^2 L^2]$. Finally, $\langle u^2 \rangle$ is obtained by summing over all possible $\vec{q}$ values ranging from q_{min} to q_{max}:

$$\langle u^2 \rangle = \frac{1}{2}\sum_{\vec{q}} \langle u_q^2 \rangle = \frac{1}{2}(L/2\pi)^2 \int_{q_{min}}^{q_{max}} \langle u_q^2 \rangle d^2 q \qquad (11a)$$

$$= (2\pi)^{-1}(kT/\gamma)\ln[q_{max}/q_{min}] \qquad (11b)$$

The last equality presumes that the volume element d^2q can be written as $2\pi q dq$. We shall now discuss the wavevector limits. Since the derivation is within the continuum assumption, the smallest wavelength of disturbance must be of the order of a molecular diameter d and $q_{max} \approx 2\pi/d$. The minium value, $\vec{q}_{min}$, is determined by the transverse resolution in $\vec{Q}$ space of the instrumental set up. Since the resolution in the x and y-directions as defined above are different, the assumption behind eq. (11b) is not fulfilled and the integral in (11a) must be evaluated numerically. When this is done one finds $\langle u^2 \rangle^{\frac{1}{2}}_{H_2O} = 2.8$ Å which added in quadrature to a root-mean-square radius of an H_2O molecule of 1.9 Å gives an effective $\langle u^2 \rangle^{\frac{1}{2}}_{eff}$ of about 3.4 Å. The slope for H_2O in figure 3 corresponds to $\langle u^2 \rangle^{\frac{1}{2}}_{eff} = 3.2$ Å. Similar good agreement with no adjustable parameters is found for CCl_4 which has as significantly larger thermal mean squared displacement due to its lower surface tension. We conclude that the thickness of interface of simple liquids far from the critical point is simply determined by the combined effects of molecular size and thermal roughness due to excitation of capillary waves.

3b. Free Surface of Liquid Crystals

In the following two sections we shall review recent experimental studies of the density profile across the surface of liquid crystal materials in different phases. This section provides the minimum background knowledge on liquid crystals. For a general reference, the reader is referred to "The Physics of Liquid Crystals" by de Gennes[7].

Liquid crystals consist of long molecules with a typical length-to-diameter ratio of 5 to 1. In describing the structures, the molecules are considered as rigid rods. The variety of structures or phases is due to the combination of order/disorder between the <u>position</u> of molecules and their <u>orientation</u>. For the present purpose it suffices to recall three liquid phases: The <u>isotropic</u> phases where both position and orientation are disordered as in an ordinary simple liquid, the <u>nematic</u> phase where the position is disordered but all molecules have the same spontaneous average direction, and the <u>smectic A</u> phase where the common orientation is maintained, but in addition the molecules are positoned in layers perpendicular to their long axis with a well-defined repetition distance between layers, but with positional disorder of molecules within the same layer.

Different sequences of transitions between these phases may occur. The high temperature isotropic phase may be followed directly by the smectic A phase, or a nematic phase at intermediate temperatures may intervene between the isotropic and smectic A phases. The transition from the isotropic phase is always discontinuous or first order, whereas the nematic to smectic A phase may be first order or continuous (second order). In the latter case critical fluctuations of short range order smectic A regions in the nematic matrix become more and more pronounced as the nematic to smectic A phase transition temperature is approached.

An example of the molecular structure is the so-called nCB molecules

$$C_nH_{2n+1} -\bigcirc-\bigcirc-C-N$$

$$C-N-\bigcirc-\bigcirc-C_nH_{2n+1}$$

The upper molecule is shown with its aliphatic tail to the left and the polar cyano head to the right. The intramolecular interactions are so strong that this molecule pairs with one of opposite orientation as shown, and the unit rod should be considered as this pair of molecules.

The sequence of phases versus temperatures depends on n, the aliphatic tail length. The phase diagram of nCB is shown in figure 5. In the following we shall describe the smectic layering at the free surface in two distinctly different cases: (i) in the isotropic phase near a strong first order phase transition to the smectic A phase (n = 12), (ii) in the nematic phase near a second order phase transition to the smectic A phase (n = 8).

3c. Quantized Layer Growth at Free Surface of Liquid Crystal in Isotropic Phase

Reflectivity curves at different temperatures for 12 CB are given in fig. 6. Note the logarithmic scale of reflectivity and the wavevector scale relative to Q_0, the reciprocal lattice vector in the smectic A phase. The high temperature data, curve f, are quite similar to the data for H_2O or CCl_4. The root-mean-square displacement, which we prefer here to denote σ_s, is 5.5 Å. As the temperature is lowered towards T_{IA} structure develops in the spectra. The full lines represent a simple model with a set of temperature independent parameter and only one parameter, the number of smectic layers, varying with temperature. Specifically the model is as follows. The density is decomposed into two parts. The first is a step function to liquid density smeared by a Gaussian accounting for thermal roughness etc. The corresponding Fourier transform of gradient density is

$$\Phi_1(Q) = \exp[-Q^2\sigma_s^2/2] \ .$$

This part alone accounts for the curve labelled f in figure 6. The second part is a sinusoidal density wave of amplitude A_0 between z = 0 and z = NL, where L is the layer spacing in the

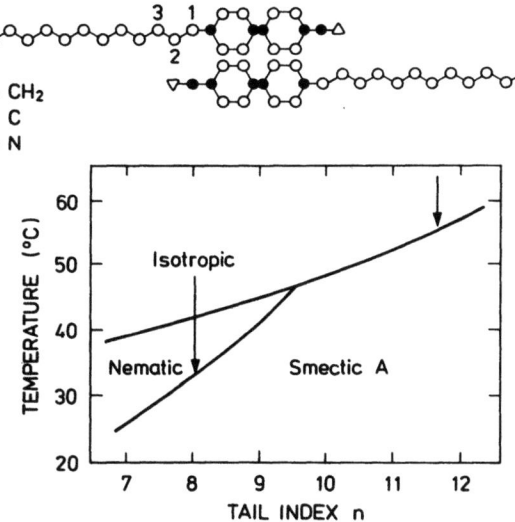

Fig. 5. Phases of nCB at various tail length n and temperature T. Experiments for n = 12 and n = 8 are described in the following.

smectic phase and N an integer number of layers. Also this part is smeared by a Gaussian and the corresponding Fourier transform of gradient density is

$$\Phi_2(Q) = A(x)\exp[-Q^2\sigma_m^2/2]\sin[\pi N]\exp[i\pi N]$$

with $A(x) = 2A_0 x/[(x+1)(x-1)]$
and $x \equiv Q/(2\pi/L) = Q/Q_0$.

In the model is allowed for a phase factor $\exp[iQz_0]$ between the two contributions so the model reflectivity is

$$\frac{R(Q)}{R_F(q)} = |\Phi_1(Q)\exp[iQz_0] + \Phi_2(Q)|^2$$

The full lines in fig. 6 all have the same values of $\sigma_s = 5.5$ Å, $\sigma_m = 4.5$ Å, $z_0/L = -0.35$, $A_0 = 0.12$ whereas N varies from 0 (curve f) to 5 (curve a). The discreteness of layer by layer growth as the temperature approaches T_{IA} is strikingly apparent in figure 7 showing the intensity variation with reduced temperature at $Q/Q_0 = 0.93$.

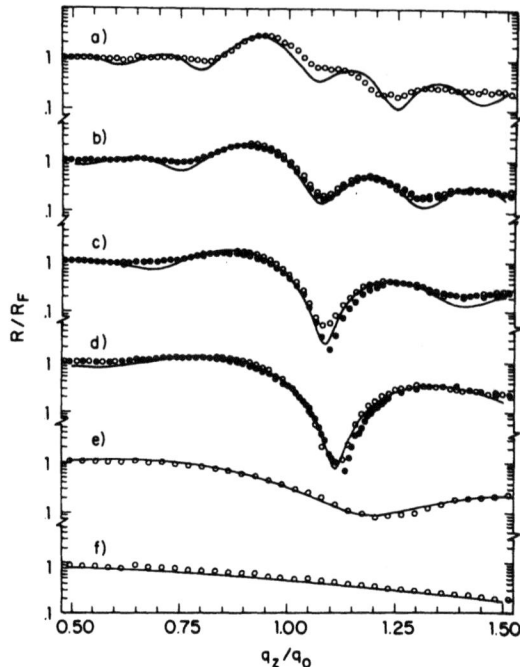

Fig. 6. Reflectivity relative to Fresnel reflectivity versus wavevector at different temperature intervals ΔT above the transition to the smectic A phase. The reflectivity scale is logarithmic and each spectrum is displaced two decades for clarity. The temperature intervals relative to the transition temperature T_{AI} are
a) $t = 3 \times 10^{-5}$
b) $t = 8 \times 10^{-5}$ (open circles), $t = 1.4 \times 10^{-4}$ (closed circles)
c) $t = 3 \times 10^{-4}$ (open circles), $t = 8.3 \times 10^{-4}$ (closed circles)
d) $t = 1.1 \times 10^{-3}$ (open circles), $t = 3 \times 10^{-3}$ (closed circles)
e) $t = 1.9 \times 10^{-2}$
f) $t = 6.1 \times 10^{-2}$
The solid line is for a density model with a sinusoidal modulation terminated after an integral number (a) 5, (b) 4, (c) 3, (d) 2, (e) 1 and (f) 0 of periods.

3d. Layer Growth in Nematic Phase

All it takes in nCB molecules to obtain a nematic phase between the isotropic and smectic A phases is a shortening of the aliphatic tail from $n = 12$ to say $n = 8$. The surface layering is now quite different and so is the reflectivity data. In the

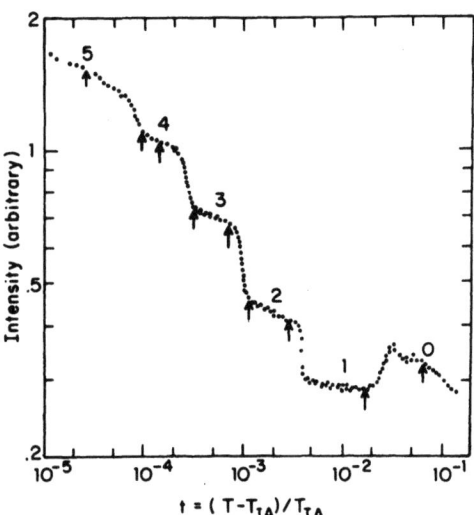

Fig. 7. The reflected intensity from 12CB in the isotropic phase changes in discrete steps as the temperature is lowered towards the transition temperature to the smectic A-phase. The numbers indicate the number of smectic A layers at the surface.

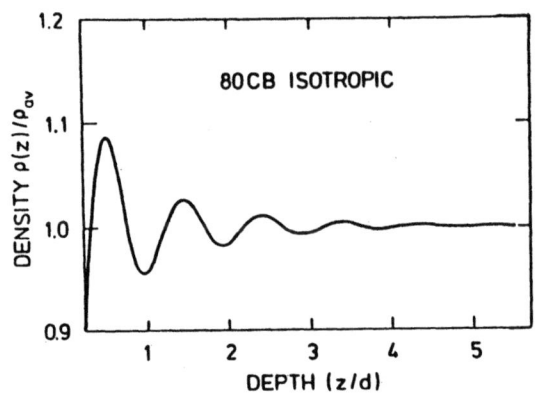

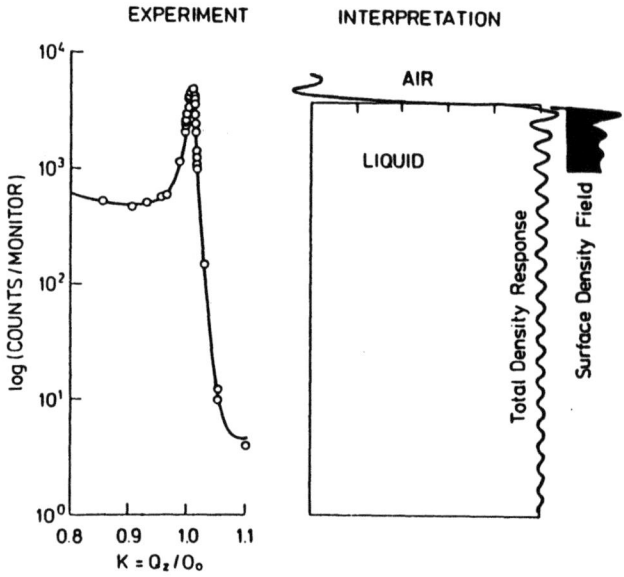

Fig. 8.

Top: Density profile in the isotropic phase slightly above the transiton temperature to the nematic phase of the liquid crystal 8OCB.

Bottom: Reflected intensity versus Q_z/Q_0 in the nematic phase (left) and the corresponding density (right). This can be considered as the response in the nematic phase to smectic layering imposed by the first few top layers.

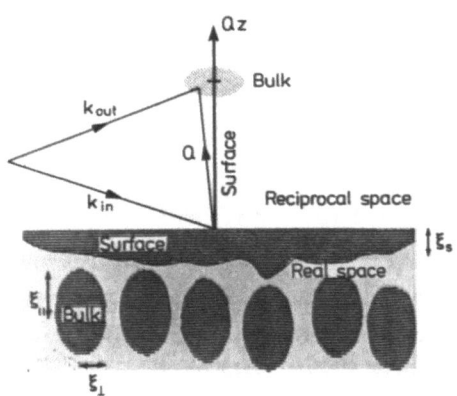

Fig. 9. Smectic layering in the nematic phase. The boundary conditions at the surface impose smectic layering of infinite lateral extent at the surface, decaying exponentially in going into the bulk. Here spontaneous critical fluctuations have correlation ranges $\xi_\parallel$ and $\xi_\perp$ along and perpendicular to the molecular axis, respectively. The top part shows scattering geometries to separate bulk and surface signals.

isotropic phase a few smectic layers are formed on the surface, but now with a decreasing amplitude, as shown in the top of fig. 8, in contrast to 12CB where the model calculation assumed a sinusoidal density variation of constant amplitude up to N periods[8]. The most remarkable difference is however in the nematic phase. The reflected intensity versus Q_z/Q_o is shown in the bottom left part of fig. 8 and the corresponding model density in the right bottom part of fig. 8.

The total picture of smectic layering in the nematic phase is given in fig. 9. The surface layers are of infinite lateral extent and essentially perfect, with an exponentially decaying amplitude into the bulk of penetration depth ξ_s. Deep in the bulk smectic fluctuations in the nematic matrix have one correlation range $\xi_\|$ along the molecular axis and another, finite, correlation range $\xi_\perp$ in the lateral direction. The top part of fig. 9 recapitulates the scattering diagram for separating bulk and surface scattering. The reflected intensity in fig. 8 displays a typical interference lineshape with constructive interference between the ordinary Fresnel wave and the wave scattered from the surface layers for $Q < Q_o$, and destructive interference for $Q > Q_o$. As the temperature approaches the transition temperature T_{NA} to the smectic A phase the correlation ranges $\xi_\|$ and $\xi_\perp$ diverge and also the penetration depth ξ_s increases, because the nematic phase becomes more and more susceptible to the layering imposed by the surface field. Most remarkably, it was found that ξ_s is identical to $\xi_\|$ not only in its temperature dependence but also in its numerical value[4]. The reason may be understood from the Landau theory of phase transitions and, since the argument is quite general, we shall outline it here. The basic quantity in the Landau theory is the order parameter ψ, which in the present case is the smectic density wave, but for a ferromagnet for instance would be the local magnetization. The order parameter varies in space and we shall here only be concerned with the variation along the z-axis. The average value is denoted $\langle\psi\rangle$ and, for a second order phase transition, $\langle\psi\rangle = 0$ for $T > T_c$ but is finite for $T < T_c$. The phenomenological Landau free energy density $f = a\langle\psi\rangle^2 + b\langle\psi\rangle^4$ will lead to such a phase transition if one assumes $a = a_o(T-T_c)$ and $b > 0$. However, spatial fluctuations will cost energy and we therefore add a term of the form $c(\partial\psi/\partial z)^2$ to the free energy density. With this expression for f one readily finds critical fluctuations for $T > T_c$ with a correlation range $\xi_\parallel = (c/a)^{1/2}$, assuming an infinite system. Now we consider the penetration from the surface, requiring a finite value ψ_o of the order parameter at $z = 0$. The order parameter must decay as z increases to reach its bulk average value of zero. On the other hand any spatial change costs energy, cf. the term $c(\partial\psi/\partial z)^2$, so there must be an optimal way for ψ to decay in the sense that the total energy obtained by integrating the energy density becomes minimal, i.e. a typical problem of variational calculus. Explicitly one finds that the optimal decay is exponential with a decay length $\xi_s = (c/a)^{1/2}$, i.e. $\xi_s = \xi_\parallel$.

4. Conclusions

The three examples of X-ray reflectivity data from liquid surfaces illustrate the degree of detail that the density profile across the surface can be determined. A recent example on antiferroelectric surface layering on a liquid crystal material[9] further emphasizes this point. The liquid crystal materials have a rich variety of surface layering which is interesting in itself but in the present context can also be viewed as the testing grounds for development of a new methodology for exploring the liquid-vapor interface in general, and we believe that this method will be widely used in the future.

ACKNOWLEDGEMENTS

The work described in this review has been carried out during the past 5 years in close cooperation with professor P.S. Pershan and his group from Harvard University both as concerns methodology, interpretation and choice of problems. The experiments were carried out at HASYLAB, DESY in Hamburg.

References

1. For a review see J. Als-Nielsen in "Handbook of Synchrotron Radiation", Vol. 3, North Holland (in print).

2. A. Braslau, M. Deutsch, P.S. Pershan, A.H. Weiss, J. Als-Nielsen and J. Bohr, Phys. Rev. Lett. $\underline{54}$, 114 (1985).

3. B. Ocko, A. Braslau, P.S. Pershan, J. Als-Nielsen and M. Deutsch, Phys. Rev. Lett. $\underline{57}$, 94 (1986).

4. J. Als-Nielsen, F. Christensen and P.S. Pershan, Phys. Rev. Lett. $\underline{48}$, 1107 (1982).

5. P.S. Pershan and J. Als-Nielsen, Phys. Rev. Lett. $\underline{52}$, 759 (1984).

6. P.S. Pershan, A. Braslau, A.H. Weiss and J. Als-Nielsen, Phys. Rev. $\underline{xx}$, xxxx (1986).

7. P.G. de Gennes, "The Physics of Liquid Crystals", Clarendon, Oxford, 1974.

8. Footnote: The data actually derive from the material 8OCB, not 8CB. The extra oxygen atom is placed between the aliphatic tail and the first benzene ring of the polar head.

9. E.F. Gramsbergen, W.H. de Jeu and J. Als-Nielsen, J. de Physique $\underline{47}$, 711, 1986.

STATISTICAL MECHANICS OF INTERFACES

B. Widom

Department of Chemistry, Cornell University
Ithaca, New York 14853, U.S.A.

1. Mean-field approximation

The mean-field approximation for interfacial structure may be derived from a form of the potential-distribution theory appropriate to inhomogeneous fluids:

$$\rho(\underline{r}) = \lambda \langle e^{-\Psi/kT} \rangle_{\underline{r}} \qquad (1)$$

where Ψ is the potential measured by a test particle at $\underline{r}$, $\rho(\underline{r})$ is the mean density there, and λ is the uniform activity. Applied to a fluid of attracting hard spheres, with the hard-sphere repulsions treated exactly and the attractions by the mean-field approximation, (1) yields a functional equation for the density profile $\rho(z)$ (density as a function of height z through the liquid-vapor interface):

$$\mu = M[\rho(z)] + \int_{r>b} \phi(r) [\rho(z') - \rho(z)] d\tau, \qquad (2)$$

with $d\tau$ an element of volume at the variable height z', with r the distance of that volume element from a fixed point at the height z, with b the sphere diameter and $\phi(r)$ the potential energy of intermolecular attraction, with $M(\rho)$ the chemical potential of the bulk fluid as a function of density in mean-field approximation (with its van der Waals loops), and with µ the uniform chemical potential of the fluid, obtained from M(ρ) by the equal-areas (Maxwell) construction. For small gradients, (2) reduces to a form analogous to the laws of motion for a particle moving on a line subject to a prescribed potential, and so may be analyzed and solved by the methods of particle dynamics.

The surface tension σ is obtained as the minimum over all possible profiles ρ(z) of a free-energy functional obtained from (2):

$$\sigma = \min_{\rho(z)} \int_{-\infty}^{\infty} \{F[\rho(z)] + \tfrac{1}{2}\rho(z) \int_{r>b} \phi(r)[\rho(z') - \rho(z)] d\tau\} dz, \quad (3)$$

with

$$F(\rho) = \int_{\rho_{g,\ell}}^{\rho} [M(\rho) - \mu] d\rho, \quad (4)$$

where $\rho_{g,\ell}$ is the density of the bulk gas or liquid phase, obtained from M(ρ) by the equal-areas construction.

The mean-field approximation misses the capillary-wave excitations of the interface, and does not describe fluctuations near the critical point entirely accurately, but is nevertheless a very useful theory of interfacial structure and tension. It provides the theoretical framework of much of the current work on interfaces.

2. <u>Ising model (lattice gas)</u>

The ideas of the mean-field theory are illustrated in the Ising model (lattice gas). Neighboring spins interact with energy ±J, or neighboring molecules in the lattice gas with energy ε = 4J. The functional equation (2) for the profile becomes

$$M(\rho) - \mu = c'\varepsilon \Delta^2 \rho(z) \quad (5)$$

where c' is the vertical coördination number (typically 1) and

Δ^2 is the finite-second-difference operator. The solution of (5) gives the profile $\rho(z)$ (where z now indexes discrete lattice planes), and the surface tension follows from the discrete form of (3). The surface tension σ vanishes at the critical point proportionally to $(T_c-T)^{3/2}$, as found by van der Waals.

Neighboring pairs of spins may be interpreted as three species of molecules, ++ ("water"), -- ("oil"), and +- ("amphiphile"). Then the density profiles may be reinterpreted as adsorption profiles for a surfactant (amphiphile) at an oil-water interface: in mean-field approximation $\rho_{+-} \sim \rho_+\rho_- = \rho_+(1-\rho_+)$, which is maximal in the middle of the interface and falls to low values in both bulk phases.

3. Three-phase line; line tension

When three phases α, β, γ are in equilibrium there are three possible interfaces and three tensions, $\sigma_{\alpha\beta}$, $\sigma_{\beta\gamma}$, $\sigma_{\alpha\gamma}$. If $\sigma_{\alpha\gamma}$ is the largest, then either

$$\sigma_{\alpha\gamma} = \sigma_{\alpha\beta} + \sigma_{\beta\gamma} \qquad \text{(Antonow's rule)} \qquad (6)$$

or

$$\sigma_{\alpha\gamma} < \sigma_{\alpha\beta} + \sigma_{\beta\gamma} \qquad \text{(Neumann triangle).} \qquad (7)$$

When (7) holds, the three phases meet in a three-phase line with contact angles that are not 0° or 180°. The grand-canonical free energy Ω exceeds its value $-pV$ in a homogeneous fluid, the excess arising from the three two-phase interfaces and from the three-phase line. The latter contribution, per unit length, is the line tension τ. It may be positive or negative.

In mean-field approximation, τ is given by the constant term in the asymptotic expansion

$$\int_A \Phi \, da \sim (\sigma_{\alpha\beta} + \sigma_{\beta\gamma} + \sigma_{\alpha\gamma}) R + \tau \qquad (R \to \infty), \qquad (8)$$

where R is the radius of a circular cylinder with the three-phase line as axis, A is a cross-section with element of area da, and Φ is the local density of excess Ω that is contributed by the inhomogeneities. This Φ is like the integrand in (3), but now there is in general more than one independently varying density, $\rho_1, \rho_2, \rho_3, \cdots$, and these densities ρ_i now depend not on a single coördinate alone but on the two coördinates perpendicular to the three-phase line. Let ∇ be the associated two-dimensional gradient operator. Then for small gradients Φ takes the form

$$\Phi \simeq F(\rho_1, \rho_2, \cdots) + \tfrac{1}{2} \sum_{i,j} m_{ij} \nabla \rho_i \cdot \nabla \rho_j \qquad (9)$$

with constant coefficients m_{ij} obtained as second moments of the various $\phi_{ij}(r)$.

Kerins and Boiteux [Physica A <u>117</u> (1983) 575] obtained from (8) and (9) the explicit formula

$$\tau = \int [-F(\rho_1, \rho_2, \cdots) + \tfrac{1}{2} \sum_{i,j} m_{ij} \nabla \rho_i \cdot \nabla \rho_j] \, da, \qquad (10)$$

where the integration is now over an infinite cross-section. Significant contributions to the integral in (10) come only from the immediate neighborhood of the three-phase line; the integrand vanishes rapidly with distance from that line. As the point at which the element of area da is located departs from that line toward the interior of any of the three bulk phases, the integrand rapidly vanishes, because both F and the gradients separately do. As the variable point in the integration departs from the line through one of the two-phase interfaces, far from the line the gradients become one-dimensional, confined to the direction perpendicular to the interface. Then, by the dynamical analogy referred to in §1, the integrand in (10) becomes analogous to the Hamiltonian of a dynamical system of constant energy 0 (-F being the potential energy and the gradient terms the kinetic energy); so there is again no contribution to the line tension τ from points far from the three-phase line.

4. Wetting

When (6) holds instead of (7) the β phase spreads at, and is said to wet, the $\alpha\gamma$ interface. The three contact angles are then 180°, 0°, 180°. Sometimes (6) holds over a range of three-phase states, and (7) over another range, with a transition at some point from one to the other. This is the wetting transition. It is (usually) a first-order transition in the structure of the $\alpha\gamma$ interface. When (6) holds, that interface consists of a thick layer of the coexisting bulk β phase, while when (7) holds, the $\alpha\gamma$ interface is ordinarily much thinner and its structure and composition need not be like those of bulk β.

Even when β is not in coexistence with α and γ as a bulk phase, as long as the thermodynamic state of $\alpha + \gamma$ is not too distant from such a triple point, and as long as β, had it been present as a bulk phase, would have wet the $\alpha\gamma$ interface, that interface will again be thick and have a structure and composition like that of β. This is called pre-wetting or premonitory wetting: coexistence with β is anticipated and pre-figured in the αγ interface before it becomes a reality in the bulk. If distance from the triple point is measured by a dimensionless parameter δ, and if the intermolecular forces are of short range, the mean-field approximation (2) implies that the αγ interface thickens proportionally to $\ln(1/\delta)$ as $\delta \to 0$. At the triple point, where β is stable in bulk, the αγ interface is then infinitely thick; i.e., it consists of a macroscopically thick layer of bulk β.

If β is the most dense or least dense of the three phases, and yet wets the αγ interface, most of the β phase will be found above or below the α and γ phases, but a layer of it a few hundred Å thick will remain in equilibrium at the αγ interface. This is seen, for example, in the system studied by Carl Franck and his students at Cornell, in which a binary solution of CH_3NO_2 and CS_2 separates into a phase β rich in CH_3NO_2 and a phase γ rich in CS_2, while they rest on a borosilicate glass surface, which is phase α. The CH_3NO_2-rich phase β wets the αγ interface, so there is a layer of it separating the glass surface and the CS_2-rich phase, even though the CH_3NO_2-rich phase is the lighter, so that most of it is above the CS_2-rich phase. The wetting layer is said to be gravity-thinned; the parameter δ that measures the departure from bulk three-phase coexistence at that depth in the fluid is proportional to $Lg(d_\gamma - d_\beta)$, where L is the height of the γ phase, g the acceleration of gravity, and d_γ and d_β the densities of the γ and β phases. As before, the mean-field approximation, with the assumption of short-range forces, would give a wetting layer of thickness proportional to $\ln(1/\delta)$.

5. Long-range forces

When the forces are of long range, as van der Waals forces are, wetting-layer thicknesses diverge as an inverse power of δ rather than proportionally to $\ln(1/\delta)$, as was emphasized by de Gennes. Take for the density profile of the wetting layer a $\rho(z)$ that jumps discontinuously from the bulk ρ_α to the bulk ρ_β at the $\alpha\beta$ surface of the layer, and then from ρ_β to the bulk ρ_γ at the $\beta\gamma$ surface, with the layer itself of thickness ℓ. For simplicity we may think of common interaction potentials $\phi(r)$ between molecules of all species, and take the ρ_α, ρ_β, ρ_γ to be number densities, with, let us say, $\rho_\gamma < \rho_\beta < \rho_\alpha$. Then this is also the (inverse) order of their energy densities. At the same time we suppose the mass densities d of the phases to be ordered as $d_\beta < d_\gamma < d_\alpha$. Surface forces alone would then favor an ordering of the phases in which β lay between α and γ; but the competition with gravity, which favors the β phase's being on top, then results in the β layer's having only the finite thickness ℓ. The free-energy density f by which β at the $\alpha\gamma$ interface fails to be in equilibrium as a macroscopic phase -- which is just the excess free-energy density of which the earlier parameter δ was a measure -- is then

$$f = (d_\gamma - d_\beta)gL, \qquad (11)$$

with L again the height of the γ phase, as in §4. Then with the assumed step-function $\rho(z)$ described above, which depends parametrically on the presumed thickness ℓ, the integral on the right-hand side of (3) may be evaluated to yield a $\sigma(\ell)$, the layer free energy per unit area as a function of ℓ. If

$\phi(r) \sim -\phi_o/r^n$ as $r \to \infty$, for some $n > 4$, then we find in this way

$$\sigma(\ell) = f\ell + \text{(a constant independent of } \ell\text{)} + (\rho_\beta - \rho_\alpha)(\rho_\beta - \rho_\gamma) j(\ell) \tag{12}$$

with

$$j(\ell) \sim -2\pi\phi_o/(n-2)(n-3)(n-4)\ell^{n-4} \qquad (\ell \to \infty) \tag{13}$$

and with f given by (11). Note that by our assumptions the coefficient of $j(\ell)$ in (12) is negative, while $f > 0$. The equilibrium layer thickness ℓ is that which minimizes $\sigma(\ell)$, and is thus given by

$$\ell_{eq} \sim \left[\frac{2\pi\phi_o(\rho_\alpha - \rho_\beta)(\rho_\beta - \rho_\gamma)}{(n-2)(n-3)f} \right]^{\frac{1}{n-3}} \tag{14}$$

for small f. This replaces the logarithmic dependence of ℓ_{eq} on f which one has with short-range forces. The formula (14), with n=6, is that of de Gennes.

6. Models of microemulsions

Microemulsions (solutions of oil + water + surfactant) may be viewed as having microscopic interfaces -- the surfactant film -- separating oil-coherent from water-coherent regions of the solution. A phenomenological theory of such solutions accounts for the characteristic three-phase equilibrium in which the microemulsion is a middle phase in equilibrium simultaneously with nearly pure oil and nearly pure water; and it accounts also for the ultralow tensions of the interfaces between coexisting phases.

A Hamiltonian model that yields similar results may be

based on the reinterpretation of the Ising model that is described at the end of §2. We assume an interaction energy between amphiphiles that meet with like ends (with their + end or with their − ends) at the same lattice site. The model is then an Ising model with second-neighbor as well as nearest-neighbor interactions (second neighbor defined as separated by two steps on the lattice). In addition, if one distinguishes the interaction between amphiphiles when their + ends meet from that in which their − ends meet, that is equivalent to assuming three-spin interactions in the Ising model; and if we wish to treat unsymmetrical mixtures, in which the configurational chemical potentials of the oil and water differ, then the corresponding Ising model is in an external magnetic field.

In the régime of interaction parameters that is relevant to microemulsions, the nearest-neighbor interactions are ferromagnetic and the second-neighbor ones are antiferromagnetic. That makes the model analogous to the ANNNI model, in which one finds periodically modulated order with periods both commensurate and incommensurate with the underlying lattice. That competition is also what is responsible for producing three-phase equilibrium and ultralow tensions in this model. Both the resulting phase diagram and the quantitative features of the tensions resemble their counterparts in the earlier phenomenological theory of microemulsions referred to above.

The two models, the phenomenological and the microscopic, have two important features in common. First, they distinguish two different homogeneous oil-water solutions, in one of which the mixing occurs at a molecular level, and in the other of which, the microemulsion, the structure is much more complex and includes large oil-coherent and water-coherent regions. Second, the two models agree that the low tensions of the interfaces are of the order of 10^{-3} kT/a^2, where $\underline{a}$ is of molecular size. The small dimensionless coefficient 10^{-3} defines what is meant by "ultralow" tensions in the models.

ACKNOWLEDGEMENTS

This research was supported by the U.S. National Science Foundation and the Cornell University Materials Science Center.

UTILIZATION OF THE SECOND GRADIENT THEORY IN CONTINUUM MECHANICS

TO STUDY THE MOTION AND THERMODYNAMICS OF LIQUID-VAPOR INTERFACES

Henri Gouin

Department of Mathematics and Mechanics
University of Aix-Marseille
Rue Henri Poincaré
13397 Marseille, Cedex 13, France

INTRODUCTION

A thermomechanical model of continuous media based on second gradient theory has been used to study the motions in liquid-vapor interfaces. In the equilibrium state this model is shown to be fundamentally equivalent to molecular theories. Conservative motions in such fluids verify the first integrals that provide Kelvin's circulation theorems and potential equations. The dynamic surface tension of a liquid-vapor interface has been deduced from equations written with a viscosity factor. The result provides and explains the Marangoni effect.

BACKGROUND

Liquid-vapor interfaces are generally represented schematically by a material surface endowed with energy related to Laplace's surface tension. Several studies conducted in the fields of fluid mechanics and thermodynamics represent the interface as a surface separating two media. This surface has its own characteristic behavior and energy properties[1-8]. Detailed theoretical or experimental study shows that when working far from the critical temperature, the capillary layer has a thickness equivalent to a few molecular beams[3,8,9]. For a fluid defined by two parameters such as temperature and density, molecular models such as those used in gas kinetic theory lead to laws of state associated with nonconvex internal energies, e.g., the Van der Waals model[7,9-14]. These models appear advantageous as they provide an even more precise verification of Maxwell's rule applied to isothermal liquid-vapor transition[15]. Nonetheless, they present two disadvantages. First, for densities that lie between the liquid and vapor bulk densities, the pressure may become negative. Simple physical experiments can be used, however, to cause traction that leads to these pressure values[9,12,16]. Second, in the field between vapor and liquid, internal energy cannot be represented by a convex surface associated with the variables density and entropy. This fact seems to contradict the existence of the steady equilibrium state of the matter in this type of region. To overcome these disadvantages, the thermodynamic investigation replaces the nonconvex portion corresponding to internal energy with a plane domain. The fluid can no longer be considered as a continuous medium. The interface is represented by a material surface with a null thickness. In this case the only possible representation of the dynamic behavior of the interface is one of a discontinuous surface, and its essential structure remains unknown. In the equilibrium state it is possible to eliminate the above disadvantages by appropriately modifying the stress tensor of the capillary layer, which is then expressed in an anisotropic form. As a consequence, the energy of the continuous medium must change[10,17,18].

If the interface fluid is to be represented as a continuous medium, how can the stress tensor of the capillary layer be written in a dynamic expression[4]? In recent articles, we have proposed a dynamic theory based on the second gradient: internal capillarity[19-21]. Conceived in the sixties for the static case, this theory shows the advantage of using a three-dimensional approach to capillarity in a continuous media mechanical model[17,22]. The resulting equations of equilibrium provide a satisfactory representation of isothermal liquid-vapor equilibrium states. This approach is not new, and in fact dates back to Van der Waals and Korteweg[23,24]. It was taken up again in 1959 by Cahn and Hilliard in reference to free energy[10], and corresponds to what is known as the Landau-Ginzburg theory[14]. The representation proposed in the present study is based on the notion of internal energy which is more convenient to use when the temperature is not uniform. One of the problems that complicates the study of phase transformation dynamics is the apparent contradiction between Korteweg's classical stress theory[24] and the Clausius-Duhen inequality[25]. Proposals made by Dunn and Serrin[11] and a capillary fluid model[19] rectify this apparent anomaly. In the more general case of internal capillarity, by representing energy in terms of the second gradient[26] and by applying a simple algebraic identity, it is possible to draw a relationship between the energy equation, the motion equation, the mass conservation equation, and entropy. Through deduction, and for a conservative fluid, an additional term that has the dimension of a heat flux may be added to the energy equation. In the case of a viscous fluid, results provide a set of equations that do not modify the Clausius-Duhem inequality, thereby making them compatible with the second thermodynamic principle. For the nonviscous case, classical fluid motions and motions of fluids endowed with internal capillarity reveal a common structure that results in identical thermodynamic forms and potential equations[27-29]. This leads to the same classification of motion[30,31], a generalization of Kelvin's theorem and the Crocco-Vaszonyi equation, and first integrals that keep the same values across the interfaces. By representing internal energy as a function of entropy, density, and the density gradient, using a single constant C for internal capillarity, the resulting equations are thereby identical to those obtained with molecular models in the isothermal case.

For a surface area that is relatively large with respect to the thickness of the capillary layer, surface tension is calculated using integration throughout the interface. Lengthwise it is not constant and depends on the dynamic distributions of density and temperature. These dynamic distributions, based on equations of motion, call upon a Navier-Stokes type viscosity. For interfaces in an isothermal equilibrium, the results are classical[9,10,12,13,18]. The study of motion in the interface without mass transfer introduces the surface tension gradient and the velocity gradients associated with dynamic viscosity. The ensuing Marangoni effect has been interpreted using limit analysis wherein the approximate quantities correspond to the physical dimensions of the interface. (For the case where mass flow across the interface is not null, a general dynamic form of Laplace's equation is given in ref. 32). The method presented herein is completely different from classical calculations based on balance equations established for both sides of a discontinuous surface, which take into account density variations only as a difference across the interface[33]. In the particular case of isothermal liquid-vapor equilibria, an invariant integral of motions compatible with the interface coincides with a generalization of the rule advanced by Maxwell, associated with phase transitions[15].

The model of a viscous fluid endowed with internal capillarity is therefore substantiated by consequences verified in both the equilibrium state and in non-isothermal motion. This model provides a better understanding of the behavior of liquid-vapor interfaces in motion. It gives at least a partial answer to the question: "is the fluid at the interface rigid or moving?"[34], and proposes a theory that takes into account the stress tensor and dynamics in the essential structure of the interface[4]. The resulting laws of behavior are not the thermodynamic laws governing classical fluids since they include an anisotropic stress tensor in the momentum equation and an additional heat flux term in the energy equation.

1. EQUATIONS OF MOTION FOR A FLUID ENDOWED WITH INTERNAL CAPILLARITY

1.1 Case of a Conservative Fluid

Second gradient theory[26], conceptually more straightforward than Laplace's theory, can be used to construct a theory of capillarity. In the present text the only addition is an internal mass energy ε that is a function of density ρ and entropy s, as well as grad ρ. The internal mass energy characterizes both the compressibility and capillarity properties of the fluid, independent of the bodies with which it is in contact. For an isotropic fluid, it is assumed that[17]:

$$\varepsilon = f(\rho, s, \beta) \tag{1}$$

where
$$\beta = (\text{grad } \rho)^2$$

The equation of motion for a conservative fluid is written:

$$\rho \, \Gamma = \text{div } \sigma - \rho \, \text{grad } \Omega \tag{2}$$

where Γ denotes the acceleration vector, Ω the extraneous force potential, and σ is the general stress tensor:

$$\sigma = -p\, I - C\, (\text{grad } \rho)(\text{grad } \rho)^t \tag{3a}$$

or
$$\sigma = -p\, I + C\, \{(\text{grad } \rho)^2 \, I - (\text{grad } \rho)(\text{grad } \rho)^t\} \tag{3b}$$

$$\sigma_{ij} = -p\, \delta_{ij} - C\, \rho_{,i}\, \rho_{,j} \qquad i,j \in \{1,2,3\}$$

or
$$\sigma_{ij} = -p\, \delta_{ij} + C\, \rho_{,k}\, \rho_{,k}\, \delta_{ij} - C\, \rho_{,i}\, \rho_{,j}$$

where
$$C = 2\, \rho\, \varepsilon'_\beta \tag{4}$$

and
$$p = \rho^2\, \varepsilon'_\rho - \rho\, \text{div}(C\, \text{grad}\, \rho) \tag{5a}$$

or
$$\wp = \rho^2\, \varepsilon'_\rho - \rho\, \text{div}(C\, \text{grad}\, \rho) - C\, (\text{grad}\, \rho)^2 \tag{5b}$$

(Depending on the calculations, either one of the forms of the stress tensor associated with p and $\wp$ may be used). It should be noted that:

$$\theta = \varepsilon'_s \tag{6}$$

is the Kelvin temperature expressed as a function of ρ, s and β.

Demonstration. The equation of motion is obtained in the clearest manner by using the virtual work principle[26,35]. Virtual displacement, denoted δx, is that which has been defined by J. Serrin in ref. 35 (page 145). For a fluid endowed with internal capillarity, the virtual work principle (see ref. 35, IV, Section 14) is stated as follows. The motion of a fluid is such that

$$\delta \int_V \rho (\epsilon + \Omega) \, dv - \int_{\partial V} \rho \Gamma \, \delta x \, dv = 0$$

where V is an arbitrary material volume. The variation of entropy (ref. 35, page 148) must satisfy the condition $\delta s = 0$,

while taking into account that

$$\delta \frac{\partial \rho}{\partial x} = \frac{\partial \delta \rho}{\partial x} - \frac{\partial \rho}{\partial x} \frac{\partial \delta x}{\partial x}$$

which implies

$$\delta \beta = 2 \left(\frac{\partial \delta \rho}{\partial x} - \frac{\partial \rho}{\partial x} \frac{\partial \delta x}{\partial x} \right) \text{grad } \rho ,$$

and

$$\delta \epsilon = \epsilon'_\rho \, \delta \rho + \epsilon'_s \, \delta s + \epsilon'_\beta \, \delta \beta$$

as well as equations (14.5) and (14.6) of the cited reference. For virtual displacement where δx is null at the edge of V, (integrations by parts being taken, the integrals calculated on the edge of V having a null contribution), we obtain:

$$\int_V (\rho \Gamma - \text{div } \sigma + \rho \, \text{grad } \Omega) \, \delta x \, dv = 0$$

where σ is expressed by eq. (3). Classical methods of the calculus of variations then lead to eq. (2). It appears that a single term, C, accounts for the effects of the second gradient in the equation of motion. C, like ϵ, depends on ρ, s, and β. In a study of surface tension based on gas kinetic theory, Rocard[9] obtained the same expresssion (3) for the stress tensor, but with C constant. If C is constant (see ref. 36), specific energy ϵ is written

$$\epsilon(\rho, s, \beta) = \alpha(\rho, s) + \frac{C}{2\rho} \beta$$

i.e., the second gradient term $C\beta/2\rho$ is simply added to the energy $\alpha(\rho, s)$ of the classical compressible fluid. Pressure for this fluid is written $P = \rho^2 \alpha'_\rho$, and temperature $\Theta = \alpha'_s$.

This gives $p = P - C(\frac{3\beta}{2} + \rho \Delta \rho)$; $p = P - C(\frac{\beta}{2} + \rho \Delta \rho)$; $\theta = \Theta$

For P, Rocard uses either Van der Waal's pressure equation

$$P = \rho \frac{R\Theta}{1-b\rho} - a\rho^2$$

or other laws, of which he provides a comparison. It should be noted that if C is constant, this implies that $\theta = \Theta$, and that there exists a relationship between θ, ρ and s that is independent of the second gradient terms.

1.2 Case of a Viscous Fluid

If the fluid is also endowed with viscosity, the equation of motion includes not only the stress tensor σ, but also the classical stress tensor σ_v, due to viscosity:

$$\sigma_v = \lambda \, \text{tr}(D) + 2\mu D \quad (7)$$

where D is the deformation tensor, a symmetrical gradient of the velocity field:

$$D = \frac{1}{2} \left\{ \frac{\partial V}{\partial x} + \left(\frac{\partial V}{\partial x} \right)^t \right\}$$

It would of course be coherent in second gradient theory to add to viscosity those terms accounting for the influence of higher order derivatives of this velocity field. This has not been done in the present case. Equation (2) is written by adding the virtual work of the forces of viscosity:

$$\rho \Gamma = \text{div}(\sigma + \sigma_v) - \rho \, \text{grad} \, \Omega \qquad (8)$$

2. ENERGY EQUATION FOR A VISCOUS FLUID ENDOWED WITH INTERNAL CAPILLARITY

Let e be total volumic energy: $e = \rho(\frac{1}{2}v^2 + \varepsilon + \Omega)$,

q the heat flux vector, r radiant heating, and $h = \varepsilon + \frac{p}{\rho}$

Let $M = \rho \Gamma - \text{div}(\sigma + \sigma_v) + \rho \, \text{grad} \, \Omega$

$B = \dot{\rho} + \rho \, \text{div} \, V$

$S = \rho \theta \dot{s} + \text{div} \, q - r - \text{tr}(\sigma_v D)$

$E = \frac{\partial e}{\partial t} + \text{div}[(e - \sigma - \sigma_v)V] - \text{div}(C \, \dot{\rho} \, \text{grad} \, \rho) + \text{div} \, q - r - \rho \frac{\partial \Omega}{\partial t}$

Theorem. For internal energy written as in (1) and for any motion in the fluid, the relation

$$E - M \cdot V - (\frac{1}{2}v^2 + h + \Omega)B - S = 0 \qquad (9)$$

is an identity.

Demonstration. In the first member of (9), the dissipative terms q, r and σ_v cancel out. The same is true for the extraneous force potential and the inertia terms. After having replaced σ, p, and θ by their respective values (3)-(6), it remains to be proved that the terms from internal energy ε also cancel out. These terms include the following:

a) in E: $\frac{\partial}{\partial t}(\rho \varepsilon) + \text{div}(\rho \varepsilon V) - \text{div}(\sigma V) - \text{div}(C \, \dot{\rho} \, \text{grad} \, \rho)$

i.e., $\rho \dot{\varepsilon} + \varepsilon(\dot{\rho} + \rho \, \text{div} \, V) - \text{div}(\sigma)V - \text{tr}(\sigma \frac{\partial V}{\partial x}) - \text{div}(C \, \dot{\rho} \, \text{grad} \, \rho)$

or, finally $\rho(\varepsilon'_s \dot{s} + \varepsilon'_\rho \dot{\rho} + \varepsilon'_\beta \dot{\beta}) + \varepsilon(\dot{\rho} + \rho \, \text{div} \, V) - \text{div}(\sigma)V$

$+ p \, \text{div} \, V + C \, (\text{grad} \, \rho)^t \frac{\partial V}{\partial x} \text{grad} \, \rho - \text{div}(C \, \dot{\rho} \, \text{grad} \, \rho)$

b) in M . V: $-\text{div}(\sigma)V$

c) in $(\frac{1}{2}v^2 + h + \Omega)B$: $(\varepsilon + \frac{p}{\rho})(\dot{\rho} + \rho \, \text{div} \, V)$

d) in S: $\rho \varepsilon'_s \dot{s}$

This leaves the following in the first member of (9):

$$\rho \varepsilon'_\rho \dot{\rho} + \frac{C}{2} \dot{\beta} - \frac{p}{\rho} \dot{\rho} + C(\text{grad} \, \rho)^t \frac{\partial V}{\partial x} \text{grad} \, \rho - \text{div}(C \, \dot{\rho} \, \text{grad} \, \rho)$$

Finally, replacing p by its expression (5), we obtain

$$C (\text{grad } \rho)^t \frac{d}{dt} \text{grad } \rho + \dot{\rho} \text{ div}(C \text{ grad } \rho) + C (\text{grad } \rho)^t \frac{\partial V}{\partial x} \text{grad } \rho - \text{div} (C \dot{\rho} \text{ grad } \rho)$$

which is identically null.

<u>Corollary 1</u>. In motion of a conservative fluid endowed with internal capillarity, conservation of entropy on the trajectories is equivalent to:

$$\frac{\partial e}{\partial t} + \text{div}[(e - \sigma) V] - \text{div}(C \dot{\rho} \text{ grad } \rho) - \rho \frac{\partial \Omega}{\partial t} = 0 \qquad (10)$$

Eq. (10) is naturally designated the energy conservation equation. This relation is a direct result of (9) where $\sigma_v = 0$, div $q - r = 0$ and $M = 0$, $B = 0$, $S = 0$. This brings us to add the additional term div ($C \dot{\rho}$ grad ρ) to the energy equation. The vector $C \dot{\rho}$ grad ρ has the dimension of a heat flux vector and occurs even in the conservative case.

<u>Corollary 2</u>. In motion of a viscous fluid endowed with internal capillarity, the energy equation

$$\frac{\partial e}{\partial t} + \text{div}[(e - \sigma - \sigma_v) V] - \text{div}(C \dot{\rho} \text{grad} \rho) + \text{div } q - r - \rho \frac{\partial \Omega}{\partial t} = 0 \qquad (11)$$

is equivalent to the entropy equation

$$\rho \theta \dot{s} + \text{div } q - r - \text{tr}(\sigma_v D) = 0 \qquad (12)$$

Equation (12) corresponds to the classical version of the entropy variation expressed by the function of dissipation of stress due to viscosity:

$$\Psi = \text{tr}(\sigma_v D)$$

3. THE PLANCK INEQUALITY AND THE CLAUSIUS-DUHEM INEQUALITY

For any motion in a viscous fluid endowed with capillarity, it has been assumed that $\Psi \geq 0$. Equation (12) implies that

$$\rho \theta \dot{s} + \text{div } q - r \geq 0 \qquad (13)$$

This represents the Planck inequality[37]. Supposing that the Fourier principle is expressed in a very general manner:

$$q \cdot \text{grad } \theta \leq 0 \qquad (14)$$

the Clausius-Duhem inequality can be deduced directly:

$$\rho \dot{s} + \text{div} (\frac{q}{\theta}) - \frac{r}{\theta} \geq 0$$

The present analysis reveals that the second thermodynamic principle leads to the existence of a heat flux vector within a fluid in motion if the fluid is endowed with internal capillarity. This heat flux vector adds an additional term to the classical equation of energy. It is present even if the fluid is not viscous and the motion is conservative. Generalization allows these results to be applied to non-fluid continuous media which display an internal energy that contains second gradient terms.

4. TRANSFORMATION OF MOTION EQUATIONS FOR A FLUID ENDOWED WITH INTERNAL CAPILLARITY

This procedure consists of writing equation (2) in other forms.

4.1 Special Case where C is Constant

Let us note

$$\omega = \Omega - C \Delta\rho \qquad (15)$$

Eq. (2) may then be written

$$\rho \Gamma + \text{grad } P + \rho \text{ grad } \omega = 0 \qquad (16)$$

This is a formal representation of a perfect fluid where P, for example, is the pressure of Van der Waals. All capillarity terms are grouped together in ω. Verification is achieved directly: eqs. (3) and (5) give

$$\sigma_{ij} = -P \delta_{ij} + C\{(\tfrac{1}{2} \rho_{,k} \rho_{,k} + \rho\rho_{,kk}) \delta_{ij} - \rho_{,i} \rho_{,j}\}$$

thus

$$\sigma_{ij,j} = -P_{,i} + C \rho \rho_{,ijj}$$

i.e.,

$$\text{div } \sigma = -\text{grad } P + C \rho \text{ grad}(\Delta\rho)$$

It may be observed that for a viscous fluid, (16) is written:

$$\rho \Gamma + \text{grad } P + \rho \text{ grad}(\Omega - C \Delta\rho) - \text{div } \sigma_v = 0 \qquad (17)$$

4.2 Thermodynamic Form of the Equation of Motion

In general, and not only when C is constant, the equation of motion (2) may be written in the following form:

$$\Gamma = \theta \text{ grad } s - \text{grad}(h + \Omega) \qquad (18)$$

which in the present case will be designated as the thermodynamic form of the equation of motion. Eq. (18) is widely recognized (see ref. 35, p.171) when capillarity does not exist ($\varepsilon = \alpha(\rho,s)$, $\varepsilon'_\beta = 0$); h is then the enthalpy. It is significant to observe that this equation remains valid even in the present case. Verification consists simply of writing out the second member of eq. (18). Let us note

$$A = \theta \text{ grad } s - \text{grad } h \qquad \text{i.e.,} \qquad A_i = \theta s_{,i} - h_{,i}$$

According to (6), this gives

$$A_i = \varepsilon'_s s_{,i} - \partial_i (\varepsilon + \tfrac{P}{\rho})$$

Thus

$$A_i = \varepsilon'_s s_{,i} - \{\varepsilon'_s s_{,i} + \varepsilon'_\rho \rho_{,i} + \varepsilon'_\beta \partial_i(\rho_{,k} \rho_{,k})\}$$

$$- \tfrac{P}{\rho^2} \rho_{,i} - \tfrac{1}{\rho} P_{,i}$$

By replacing p by its value in (5), A_i is found to be identical to $\tfrac{1}{\rho} \sigma_{ij,j}$ given in (3).
This results in $A = \tfrac{1}{\rho} \text{div } \sigma$.

5. GENERALIZED KELVIN THEOREMS

Let J be the circulation of the velocity vector on a closed fluid curve C convected by the stream

$$J = \oint_C v^t \, dx$$

and

$$\frac{dJ}{dt} = \oint_C \Gamma^t \, dx \qquad \text{(see ref. 35).}$$

Based on eq. (18), it is possible to deduce $\oint_C \Gamma^t \, dx = \oint_C \theta \, ds$ and consequently the following theorems which are valid for fluids endowed with internal capillarity.

<u>Theorem 1</u>. Velocity circulation on a closed, isentropic fluid curve is constant.

<u>Corollary</u>. In a homentropic flow[27], velocity circulation on a closed fluid curve is constant.

<u>Theorem 2</u>. Velocity circulation on a closed, isothermal fluid curve is constant.

<u>Corollary</u>. In an isothermal flow, velocity circulation on a closed fluid curve is constant.

6. POTENTIAL EQUATIONS FOR CONSERVATIVE FLUIDS

The results obtained in refs. 27-30 may be applied to conservative motions in fluids endowed with internal capillarity. To any motion within these fluids there corresponds scalar potentials φ, ψ, τ and χ whose evolutions verify

$$\left| \begin{array}{l} \dot{\varphi} = \frac{1}{2} v^2 - h - \Omega \\ \dot{\psi} = \theta \\ \dot{s} = 0 \end{array} \right. \qquad \left| \begin{array}{l} \dot{\tau} = 0 \\ \dot{\chi} = 0 \end{array} \right. \qquad (19)$$

Fluid velocity is given by

$$V = \operatorname{grad} \varphi + \psi \operatorname{grad} s + \tau \operatorname{grad} \chi \qquad (20)$$

The following also holds:

$$\frac{\partial \rho}{\partial t} + \operatorname{div} \rho V = 0 \qquad (21)$$

7. CLASSIFICATION OF THE MOTIONS

The preceding equations may be used to classify motions in the same manner as in the case of non-capillary, conservative, perfect fluids[27,30].

7.1 <u>Homentropic Motions</u>

Throughout the fluid s is constant. Eq. (20) is written: $V = \operatorname{grad} \varphi + \tau \operatorname{grad} \chi$. This leads to Cauchy's theorem[35]:

$$\frac{d}{dt} \left(\frac{\operatorname{rot} V}{\rho} \right) = \frac{\partial V}{\partial x} \frac{\operatorname{rot} V}{\rho}$$

7.2 Oligotropic Motions

Eq. (20) is written $V = \text{grad } \varphi + \psi \text{ grad } s$. These flows verify the relation $\text{rot } V \cdot \text{grad } s = 0$. Surfaces with equal entropy are eddy surfaces. Velocity circulation on a closed, isentropic fluid curve is null.

Remark. Given the form of equations (16) and (18), the results in ref. 31 may also be generalized.

8. GENERALIZED CROCCO-VAZSONYI EQUATION

The energy equation (10) may be written in an equivalent manner as:

$$\frac{\partial e}{\partial t} + \text{div}[(e+p)V] - \text{div}(C \frac{\partial \rho}{\partial t} \text{grad } \rho) - \rho \frac{\partial \Omega}{\partial t} = 0 \tag{22}$$

Observing that $\quad e + p = \rho(\frac{1}{2}V^2 + h + \Omega)$

and noting $\quad H = \frac{1}{2}V^2 + h + \Omega$

Eq. (22) may be written:

$$\frac{\partial \rho H}{\partial t} + \text{div } \rho H V = \rho \frac{\partial \Omega}{\partial t} + \frac{\partial p}{\partial t} + \text{div}(C \frac{\partial \rho}{\partial t} \text{grad } \rho) \tag{23}$$

The value of the first member in (23) is $\rho \dot{H} + H (\dot{\rho} + \rho \text{ div } V)$. Therefore, in virtue of mass conservation, the final form of the energy equation is written

$$\rho \dot{H} = \rho \frac{\partial \Omega}{\partial t} + \frac{\partial p}{\partial t} + \text{div}(C \frac{\partial \rho}{\partial t} \text{grad } \rho) \tag{24}$$

In parallel, by taking into account the following identity

$$\Gamma = \frac{\partial V}{\partial t} + \text{rot } V \wedge V + \text{grad}(\frac{1}{2}V^2)$$

eq. (18) may be written

$$\frac{\partial V}{\partial t} + \text{rot } V \wedge V = \theta \text{ grad } s - \text{grad } H \tag{25}$$

If the motion is a steady flow, then eq. (25) makes it possible to conclude that H is constant along the stream lines ($\dot{H} = 0$), and eq. (23) shows that the partial energy flux ρHV has a divergence that is null. Eq. (25) also serves as a basis for establishing the Crocco-Vazsonyi equation generalized to fluids in stationary motion, endowed with internal capillarity:

$$\text{rot } V \wedge V = \theta \text{ grad } s - \text{grad } H,$$

where, as for a perfect fluid, s and H are constant on each stream line.

Remark. The equations (19) show that τ and χ, like entropy, are scalars that remain constant on each trajectory, and therefore across the interfaces. They represent first integrals of the motion. It is obviously possible to find others; the Kelvin integrals, for instance,

$$J = \int_C \tau (\text{grad } \chi)^t \, dx = \int_C \tau \, d\chi$$

are constant for any closed or non-closed fluid curve C. Due to Noether's theorem, it is known that any law of conservation can be represented by an invariant group. It has been shown[38] that the law of conservation expressed by the Kelvin theorems associated with isentropic fluid curves corresponds to the group of permutations consisting of particles of equal entropy. It is clear that this group keeps the equations of motion invariant for both a classical perfect fluid and a fluid endowed with internal capillarity, and the same is also true for any perfect fluid with an internal energy that depends on ρ, s, and their gradients of any order. It is even tempting to define a general perfect fluid by identifying it with this invariant group or, consequently, with a continuous medium whose motions verify the Kelvin theorems.

9. DYNAMIC SURFACE TENSION OF A LIQUID-VAPOR INTERFACE

Far from the neighborhood of the critical temperature, the thickness of a liquid-vapor interface is very small[9]. Outside the capillary layer, density and its spatial derivatives have regular variations. The density in each phase is reached at points located within the immediate vicinity of this layer. In the present case bubbles and drops the size of a few molecular beams have not been examined. Surfaces of equal density that materialize the interface are considered to be parallel surfaces, and are used to define a system of orthogonal coordinates[39]. Notations are those used in ref. 40. The subscript 3 refers to a direction normal to surfaces of equal density in the ascending order of density. e_3 denotes the unit vector of this index. The mass flow through the interface is assumed to be null and the surfaces of equal density in the capillary layer are material surfaces. In the capillary layer

$$\text{div } V = 0 \qquad (26)$$

By neglecting extraneous forces (which simplifies calculations), eq. (17), representing motions compatible with the interface, is written:

$$\rho \Gamma_3 + \frac{1}{h_3} \frac{\partial P}{\partial x_3} = C \rho \frac{1}{h_3} \frac{\partial}{\partial x_3} \Delta\rho + 2 \mu \text{ (div D)}_3 \qquad (27)$$

$$\rho \Gamma_{tg} + \text{grad}_{tg} P = C \rho \text{ grad}_{tg} \Delta\rho + 2 \mu \text{ (div D)}_{tg} \qquad (28)$$

(The subscript tg denotes the component tangent to the surfaces of equal density). Molecular models indicate that the viscosity coefficient μ is independent of density, but does depend on temperature[9,16]. The different linear sizes of the interface should be taken into consideration. The capillary layer is measured in Angstroems and the surface curvature radii are in non-molecular dimensions. The relations obtained are the result of a limit analysis wherein the parameter related to the thickness of the capillary layer approaches zero. The subscripts ν and ℓ designate, respectively, the vapor and liquid bulks. Assuming that Γ_3 is bounded, integration of eq. (27) on the third coordinate line gives

$$P - P_\nu = C \rho \Delta\rho - C \int_{x_3^\nu}^{x_3} \Delta\rho \frac{\partial \rho}{\partial x_3} dx_3 + 2 \mu \int_{x_3^\nu}^{x_3} \text{(div D)}_3 h_3 dx_3 \qquad (29)$$

The partial derivatives of velocity with respect to coordinates x_1 and x_2 are assumed to be bounded. Taking into account (26), the last term in eq. (29) is negligible. Furthermore,

$$\Delta\rho = - \frac{2}{R_m} \frac{1}{h_3} \frac{\partial \rho}{\partial x_3} + \frac{1}{h_3} (\frac{1}{h_3} \frac{\partial \rho}{\partial x_3})_{,3} \qquad (30)$$

where R_m represents the mean curvature of the surfaces of equal density in the capillary layer, oriented by e_3[26]. This gives

$$P - P_\nu = C \{ \rho \Delta\rho - \frac{1}{2} (\text{grad } \rho)^2 \} + 2 \frac{C}{R_m} \int_{x_3^\nu}^{x_3} (\text{grad } \rho)^2 h_3 dx_3 \qquad (31)$$

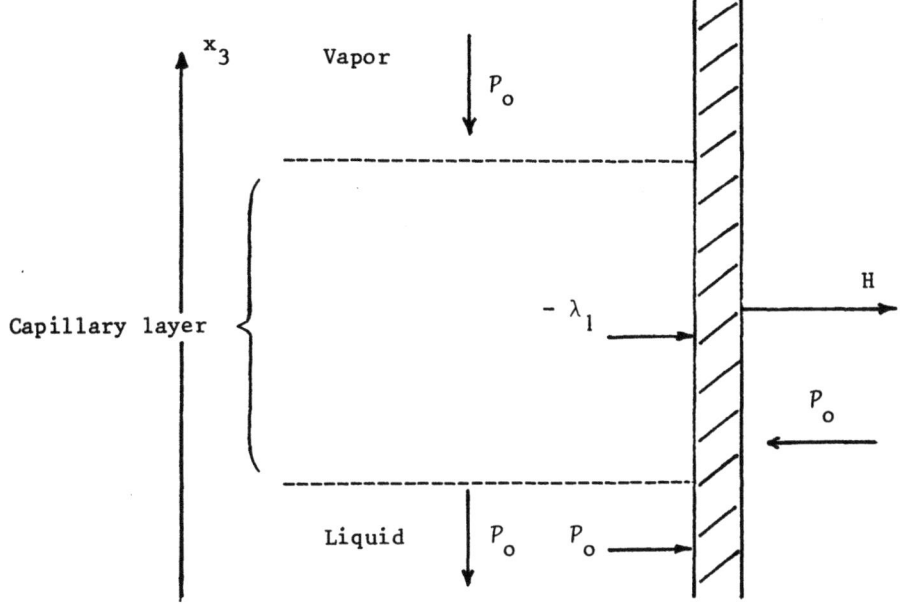

Figure 1.

Noting $dn = h_3 \, dx_3$, one obtains

$$P_\ell - P_v = 2 \frac{H}{R_m} \qquad \text{where} \qquad H = C \int_{n_v}^{n_\ell} (\text{grad } \rho)^2 \, dn \qquad (32)$$

This expresses Laplace's equation for interfaces in motion. H is interpreted as the fluid surface tension.

Remark: Interpretation of the surface tension of a plane interface in equilibrium by internal capillarity. The eigenvalues of the internal capillarity stress tensor are deduced from (3): $\lambda_1 = -p + C \, (\text{grad } \rho)^2$ is the eigenvalue associated with the plane perpendicular to grad ρ; $\lambda_2 = -p$ is the eigenvalue associated with the direction grad ρ. In the system of curvilinear coordinates joined at the interface, the stress tensor is written:

$$\sigma = \begin{vmatrix} \lambda_1 & 0 & 0 \\ 0 & \lambda_1 & 0 \\ 0 & 0 & \lambda_2 \end{vmatrix}$$

The equilibrium equation for the plane interface is drawn from (2), and by neglecting extraneous forces, implies:

$$\lambda_2 = -P_o$$

where P_o denotes the pressure common to the liquid and vapor bulks. Force per unit length along the edge of the interface (see Figure 1), is written:

$$F = \int_{x_3^v}^{x_3^\ell} \lambda_1 \, h_3 \, dx_3 = -P_o \, e + \int_{x_3^v}^{x_3^\ell} C \, (\text{grad } \rho)^2 \, h_3 \, dx_3$$

where e denotes the capillary thickness layer. $P_0 e$ is obviously negligible. Let us note

$$H = \int_{x_3^v}^{x_3^\ell} C (\text{grad } \rho)^2 h_3 \, dx_3$$

Force per unit length is H; this represents the surface tension of the plane interface in equilibrium.

10. PRACTICAL CALCULATION OF SURFACE TENSION

Eq. (27) is written:

$$\Gamma_3 + \frac{1}{h_3} \frac{1}{\rho} \frac{\partial P}{\partial x_3} = C \frac{1}{h_3} \frac{\partial}{\partial x_3} \Delta\rho + \frac{2\mu}{\rho} (\text{div } D)_3$$

Integration on the third coordinate line, the viscosity term becoming negligible, gives

$$C \Delta\rho = \frac{\partial}{\partial \rho} \left(\rho \int_{\rho_v}^{\rho} \frac{P - P_v}{\rho^2} \, d\rho \right) \quad (33)$$

Note that $\frac{\partial P}{\partial \Theta} \frac{\partial \Theta}{\partial x_3}$ is negligible with respect to $\frac{\partial P}{\partial \rho} \frac{\partial \rho}{\partial x_3}$.

Eq. (30) being given, performing integration on the third coordinate line once again, we obtain

$$\frac{C}{2} (\text{grad } \rho)^2 = \frac{2C}{R_m} \int_{x_3^v}^{x_3} (\text{grad } \rho)^2 h_3 \, dx_3 + \rho \int_{\rho_v}^{\rho} \frac{P - P_v}{\rho^2} \, d\rho \quad (34)$$

By denoting x_3^i the third coordinate of a surface of equal density, $\rho = \rho_i$ such that for $x_3 \in [x_3^v, x_3^i]$, $\frac{2C}{R_m} \int_{x_3^v}^{x_3} (\text{grad } \rho)^2 h_3 \, dx_3$ is negligible with respect to $\frac{C}{2} (\text{grad } \rho)^2$. In general, $\rho_i = \frac{1}{2} (\rho_v + \rho_\ell)$ satisfies this condition. For $\rho \in [\rho_v, \rho_i]$ eq. (34) is written:

$$\frac{C}{2} (\text{grad } \rho)^2 = \rho \int_{\rho_v}^{\rho} \frac{P - P_v}{\rho^2} \, d\rho \quad (35)$$

A similar result is obtained for $\rho \in [\rho_i, \rho_\ell]$. This leads to

$$H = \sqrt{2C} \left(\int_{\rho_v}^{\rho_i} \sqrt{u \int_{\rho_v}^{u} \frac{P - P_v}{\rho^2} \, d\rho} \, du + \int_{\rho_i}^{\rho_\ell} \sqrt{u \int_{\rho_\ell}^{u} \frac{P - P_\ell}{\rho^2} \, d\rho} \, du \right) \quad (36)$$

This relation expresses the surface tension of a liquid-vapor interface animated by motions compatible with the interface. P is not only a function of density, but also temperature, which varies according to the location of the point within the interface. Viscosity has explicitly disappeared from the H expression. The expression given by Rocard[9] is found in the case of a plane interface in isothermal equilibrium. The value of the internal capillarity constant can therefore be calculated numerically on the basis of experimental values of H and known expressions of P. By reinjecting the value of C in eq. (36), it is thus possible to calculate surface tension for any dynamic temperature distribution.

Remark. Equation (33) is written:

$$C \Delta \rho = \frac{P}{\rho} - \frac{P_v}{\rho_v} + \int_{\rho_v}^{\rho} \frac{P}{\rho^2} d\rho$$

We deduce

$$\int_{\rho_v}^{\rho_\ell} \frac{P}{\rho^2} d\rho = \frac{P_v}{\rho_v} - \frac{P_\ell}{\rho_\ell} \qquad (37)$$

which represents an invariant integral corresponding to motions compatible with the interface. This gives the equation (4-11) in ref. 15 demonstrated for the specific case of an interface in isothermal equilibrium. The case of the plane interface results in Maxwell's theorem, known as Maxwell's equal area rule[9,15].

11. MARANGONI EFFECT IN LIQUID-VAPOR INTERFACES

Let us note

$$H_j(x_3) = \int_{x_3^j}^{x_3} (\text{grad } \rho)^2 h_3 dx_3 \quad , \quad j \in \{v, \ell\}$$

Using a calculation similar to that described in section 10, for example with $x_3 \in [x_3^v, x_3^\ell]$, the result gives

$$H_v(x_3) = \sqrt{2C} \int_{\rho_v}^{\rho} \sqrt{u \int_{\rho_v}^{u} \frac{P - P_v}{r^2} dr} \, du$$

By replacing the terms in eq. (28) with those in eq. (31), it is possible to obtain

$$\rho \, \Gamma_{tg} = \frac{C}{2} \text{grad}_{tg} (\text{grad } \rho)^2 - \text{grad}_{tg} (\frac{2}{R_m} H_j(x_3)) - \text{grad}_{tg} P_j + 2 \mu (\text{div } D)_{tg} \qquad (38)$$

Eq. (26) and the hypotheses applied to the partial derivatives of velocity imply

$$\int_{x_3^v}^{x_3^\ell} (\text{div } D)_{tg} h_3 dx_3 = [D e_3]_{x_3^v}^{x_3^\ell}$$

Integration of eq. (38) gives

$$\frac{C}{2} \int_{x_3^v}^{x_3^\ell} (\text{grad}_{tg} (\text{grad } \rho)^2 h_3 dx_3 + 2 [\mu D e_3]_{x_3^v}^{x_3^\ell} = 0$$

Taking into consideration that

$$\frac{C}{2} \int_{x_3^\nu}^{x_3^\ell} (\text{grad}_{tg} (\text{grad } \rho)^2) h_3 \, dx_3 = \text{grad}_{tg} \{ C \int_{x_3^\nu}^{x_3^\ell} (\text{grad } \rho)^2 h_3 \, dx_3$$

it is possible to conclude

$$\text{grad}_{tg} H + 2 [\mu D e_3]_\nu^\ell = 0 \qquad (39)$$

Supposing that the viscosity stress tensor of the vapor bulk is negligible, the following holds:

$$\text{grad}_{tg} H + 2 \mu D^\ell e_3 = 0 \qquad (40)$$

The Marangoni condition is usually presented in this form in free boundary problems[41]. In the limit case where the viscosity coefficient is null (as for superfluid helium), the problem must be posed in another manner: the momentum associated with the interface can no longer be neglected and other physical effects must be taken into account[5]. It may be observed that there exist other phenomenological presentations of the Marangoni effect (see for example ref. 42), but to our knowledge, they all consider the interface as a discontinuous surface of the fluid medium. The calculations performed in the present text do not call upon the use of any linear approximations, and only take into account the various physical quantities distinctive of a liquid-vapor interface while working far from the critical temperature. The use of second gradient theory in the representation of interfaces has, of course, been considered by several authors. In refs. 11 and 43-45, their source is found to be in free energy that is more directly useable in isothermal flows. Ref. 32 discusses interfaces that display material transfer and provides additional demonstrations of the Marangoni effect.

REFERENCES

1. R. Defay, I. Prigogine and A. Sanfeld, Surface Thermodynamics, J. Colloid Interf. Sci., 58:498, (1977).
2. A. Sanfeld, Thermodynamics of Surfaces, in: "Physical Chemistry, vol. 1", Academic Press, New York, (1971).
3. G. Emschwiller, "Chimie Physique", Presses Universitaires de France, Paris, (1961).
4. S. Davis, Rupture of Thin Liquid Films in: "Waves on Fluid Interfaces", R. E. Meyer, ed., Academic Press, New York, (1983).
5. L. Landau and E. Lifchitz, "Fluid Mechanics", Mir, Moscow, (1958).
6. A. K. Sen and S. H. Davis, Steady Thermocapillarity Flows in Two-Dimensional Slots, J. Fluid Mech., 121:163, (1982).
7. S. Ono and S. Kondo, Molecular Theory of Surface Tension in Liquid, in: "Structure of Liquids", S. Flügge, ed., Encyclopedia of Physics, X, Springer-Verlag, Berlin, (1960).
8. V. Levitch, "Physicochemical Hydrodynamics", Prentice-Hall, Englewood Cliffs, New Jersey, (1962).
9. Y. Rocard, "Thermodynamique", Masson, Paris, (1952).
10. J. W. Cahn and J. E. Hilliard, Free Energy of a Nonuniform System. III. Nucleation in a Two-Component Incompressible Fluid, J. Chem. Phys., 31:688, (1959).
11. J. E. Dunn and J. Serrin, On the Thermomechanics of Interstitial Working, Arch. Ration. Mech. Anal., 88:95, (1985).
12. A. Brin, "Contribution à l'étude de la couche capillaire et de la Pression Osmotique", Thesis, Paris, (1956).
13. J. S. Rowlinson and B. Widom, "Molecular Theory of Capillarity", Clarendon Press, Oxford, (1982).
14. P. C. Hohenberg and B. I. Halperin, Theory of Dynamic Critical Phenomena, Reviews of Modern Physics, 49:435, (1977).

15. E. C. Aifantis and J. B. Serrin, The Mechanical Theory of Fluid Interfaces and Maxwell's Rule, J. Colloid Interf. Sci., 96:517, (1983).
16. G. Bruhat, "Cours de Physique Générale - Thermodynamique", Masson, Paris, (1968).
17. P. Casal, La théorie du second gradient et la capillarité, C. R. Acad. Sc. Paris, 274:1571, (1972).
18. V. Bongiorno, L. E. Scriven, and H. T. Davis, Molecular Theory of Fluid Interfaces, J. Colloid Interf. Sci., 57:462, (1976).
19. P. Casal and H. Gouin, Connection between the Energy Equation and the Motion Equation in Korteweg Theory of Capillarity, C. R. Acad. Sc. Paris, 300:231, (1985).
20. P. Casal and H. Gouin, Kelvin's Theorems and Potential Equation in Korteweg's Theory of Capillarity, C. R. Acad. Sc. Paris, 300:301, (1985).
21. H. Gouin, Dynamical Surface Tension and Marangoni Effect for Liquid-Vapor Interfaces in Internal Capillarity Theory, C. R. Acad. Sc. Paris, 303:5, (1986).
22. P. Casal, Capillarité interne en Mécanique des Milieux Continus, C. R. Acad. Sc. Paris, 256:3820, (1963).
23. J. D. Van der Waals, Thermodynamique de la Capillarité dans l'hypothèse d'une variation continue de densité, Arch. Néerlandaises, XXVIII:121, (1894).
24. D. J. Korteweg, Sur la forme que prennent les équations du mouvement des fluides si l'on tient compte des forces capillaires, Arch. Néerlandaises, II,VI:1, (1901).
25. M. E. Gurtin, Thermodynamics and the Possibility of Spatial Interaction in Elastic Materials, Arch. Ration. Mech. Anal., 19:339, (1965).
26. P. Germain, La méthode des puissances virtuelles en mécanique des milieux continus, J. Mécanique, 12:235, (1973).
27. P. Casal, Principes variationnels en fluide compressible et en magnétodynamique des fluides, J. Mécanique, 5:149, (1966).
28. R. L. Seliger and G. B. Whitham, Variational Principles in Continuum Mechanics, Proc. Roy. Soc., A.305:1, (1968).
29. P. Casal, La théorie du second gradient et la capillarité, C. R. Acad. Sc. Paris, A.274:806, (1972).
30. H. Gouin, "Contribution à une étude géométrique et variationnelle des milieux continus", Thesis, University of Aix-Marseille, (1978).
31. H. Gouin, Examples of Non-Conservative Perfect Fluid Motions, J. mécanique, 20:273, (1981).
32. H. Gouin, Interpretation of Motions in Liquid-Vapor Interfaces through a Model in Continuum Mechanics, to appear in J. Méc. Th. Appl.
33. M. Ishi, "Thermo-Fluid Dynamic Theory of Two-Phase Flow", Eyrolles, Paris, (1975).
34. G. Birkhoff, Numerical Fluid Dynamics, SIAM Review, 25:31, (1983).
35. J. Serrin, Mathematical Principles of Classical Fluid Dynamics, in: "Fluid Dynamics 1", S. Flügge, ed., Encyclopedia of Physics, VIII/1, Springer-Verlag, Berlin, (1959).
36. J. Pratz, "Contribution à la théorie du second gradient pour les milieux isotropes", Thesis, University of Aix-Marseille, (1981).
37. C. Truesdell, "Rational Thermodynamics", p. 30, Mac Graw Hill, New York, (1969).
38. H. Gouin, Noether Theorem in Fluid Mechanics, Mech. Res. Comm. 3:151, (1976).
39. G. Valiron, "Equations fonctionnelles. Applications", Masson, Paris, (1950).
40. P. Germain, "Mécanique des milieux continus", p. 334-340, Masson, Paris, (1962).
41. A. K. Sen and S. H. Davis, Steady Thermocapillarity Flow in Two-dimensional Slots, J. Fluid Mech., 121:163, (1982).
42. L. E. Scriven, Dynamics of a Fluid Interface. Equation of Motion for Newtonian Surface Fluids, Chem. Engng. Sci., 12:98, (1960).

43. M. Slemrod, Admissibility Criteria for Propagating Phase Boundaries in a Van der Waals Fluid, <u>Arch. Ration. Mech. Anal.</u>, 81:301, (1983).
44. M. Slemrod, Dynamic Phase Transitions in a Van der Waals Fluid, <u>J. Diff. Equ.</u>, 52:1, (1984).
45. J. Dunn, Interstitial Working and a Nonclassical Continuum Thermodynamics <u>in</u> "New Perspectives in Thermodynamics", ed. J. Serrin, Springer-Verlag, Berlin, (1986).

THE SURFACE TENSION OF IONIC LIQUIDS

M. Baus*
Chimie-Physique II**, C. P. 231
Université Libre de Bruxelles
B-1050 Brussels
Belgium

1. INTRODUCTION

Ionic liquids usually form electric double layers, i. e., a local separation between the negative and positive charges, in regions of spatial non-uniformity occuring close to electrodes or to interfaces. The presence of an interfacial electric double layer, at say, the liquid-vapour interface of a molten salt, will modify the expression of the surface tension of that interface. The corresponding modification of the surface tension cannot beobtained from the standard expressions for ordinary liquids, as encountered in textbooks[1, since the latter inmediately lead to divergencies in the presence of the long-ranged Coulomb forces set up by this double layer.

The first principles determination of the microscopic expression of the surface tension is certainly one of the oldest examples where statistical

---
* Chercheur Qualifié du F. N. R. S.
** Association Euratom-Etat Belge

mechanics has been used successfully in the context of interfacial problems. This progress is associated with the works of Kirkwood and Buff[2], Yvon[3], Triezenberg and Zwanzig[4] and more recently of Schofield and Henderson[5]. The particular case of ionic liquids has been considered by Evans and Sluckin[6], Senatore and Tosi[7] and by Baus and Tejero[8].

In the present lecture we will review the three traditional routes to the surface tension, with a particular emphasis on the case of ionic liquids, and conclude with a brief discussion of their equivalence.

2. THE THERMODYNAMIC DEFINITION

In thermodynamics the surface tension, σ, is defined as:

$$\sigma = (\delta F/\delta A)|_{T,V,\{N_\alpha\}} = (\delta\Omega/\delta A)|_{T,V,\{\mu_\alpha\}} \quad (1)$$

and hence related to the infinitesimal change of the Helmholtz free energy, $F = F(T,V,A,\{N_\alpha\})$, or the grand potential, $\Omega = \Omega(T,V,A,\{\mu_\alpha\})$, resulting from an infinitesimal change of the interfacial area, A, at constant temperature, T, volume, V, and number of particles, N_α or chemical potential, μ_α, of the different species α ($\alpha = 1, ..., s$). In this way the surface pressure, $-\sigma$, is viewed as the surface equivalent of the bulk pressure $P = -(\delta F/\delta V)|_{T,\{N_\alpha\}}$. The difficulty with this definition however is that we do not know the dependence of, say F, on the interfacial area A explicitly. This problem can be circumvented by adapting the Bogoliulov-Born-Green scaling trick for the bulk pressure to the case of the surface tension, as first done by Buff[9]. To this end let us consider a two-phase system enclosed in a rectangular vessel with edges of lengths L_x, L_y and L_z, oriented along the cartesian x ,y,z axes. The cross sectional area perpendicular to the x-axis being $A_x = L_y L_z$ and similarly for $A_y = L_x L_z$ and $A_z = L_y L_x$ while $V = L_x L_y L_z$ is the volume. Let us assume the two phases to be separated by a planar interface which we take perpendicular to the z-axis so that A_z is identical to the interfacial area, $A = A_z$. When the vessel is rescaled at constant volume, for ex. as $L_x(\lambda) = \lambda L_x, L_y(\lambda) = L_y, L_z(\lambda) = \lambda^{-1} L_z$ so that $V(\lambda) = V$ and $A(\lambda) = A_z(\lambda) = \lambda A_z = \lambda A$, we can consider $F(\lambda) = F(T,V,A(\lambda),\{N\alpha\})$ and rewrite eq. (1) as:

$$\sigma = (\delta F(\lambda)/\delta A(\lambda))|_{\lambda=1} = \left\{(\delta F(\lambda)/\delta\lambda)/(\delta A(\lambda)/\delta\lambda)\right\}_{\lambda=1} = (1/A)(\delta F(\lambda)/\delta\lambda)|_{\lambda=1} \quad (2)$$

where we have omitted to indicate the constant variables $(T,V,\{N_\alpha\})$. The rescaling of the vessel $(L_i \to L_i(\lambda)\ ;\ i = x,y,z)$, which is a global transformation, can be brought about at the local level by a corresponding change of coordinates, $\mathbf{r} \to \mathbf{r}(\lambda) = (\lambda x, y, \lambda^{-1} z)$. In a classical particle system, such a change of coordinates affects only the interaction potential and we can hence compute the r.h.s. of eq. (2) in two steps. First we consider an arbitrary change in the interaction potential, $\phi_{\alpha_1\alpha_2}(\mathbf{r}_1,\mathbf{r}_2)$ between two particles of species α_1 and α_2:

$$\sigma = (1/A)\Big\{\sum_{\alpha_1\alpha_2}\int d\mathbf{r}_1(\lambda)\int d\mathbf{r}_2(\lambda)\big(\delta F(\lambda)/\delta\phi_{\alpha_1\alpha_2}(\mathbf{r}_1(\lambda),\mathbf{r}_2(\lambda))\big)$$

$$\big(\partial\phi_{\alpha_1\alpha_2}(\mathbf{r}_1(\lambda),\mathbf{r}_2(\lambda))/\partial\lambda\big)\Big\}_{\lambda=1} \quad (3)$$

or, taking into account that when F is viewed as a functional of the pair potentials $\phi_{\alpha_1\alpha_2}(\mathbf{r}_1,\mathbf{r}_2)$ we have[10]:

$$\delta F/\delta\phi_{\alpha_1\alpha_2}(\mathbf{r}_1,\mathbf{r}_2) = (1/2)\rho_{\alpha_1\alpha_2}(\mathbf{r}_1,\mathbf{r}_2) \quad (4)$$

where $\rho_{\alpha_1\alpha_2}(\mathbf{r}_1,\mathbf{r}_2)$ is the average pair density of particles of species α_1, and α_2. Next we consider the particular coordinate transformation introduced above and compute the last factor in the r.h.s. of eq.(3) as:

$$\left(\delta\phi_{\alpha1\alpha2}(|r_1(\lambda)-r_2(\lambda)|)/\delta\lambda\right)\big|_{\lambda=1} = \left((x_1-x_2)\nabla_{x1-x2} - (z_1-z_2)\nabla_{z1-z2}\right)\phi_{\alpha1\alpha2}$$

$$-\left\{\left((x_1-x_2)^2-(z_1-z_2)^2/|r_1-r_2|\right)\right\}\phi'_{\alpha1\alpha2}(|r_1-r_2|)$$

(5)

where we have assumed central pair potentials, $\phi_{\alpha1\alpha2}(r_1,r_2) \ldots \phi_{\alpha1\alpha2}(|r_1-r_2|)$ and introduced the abbreviation $\phi'(r) = d\phi/d|r|$. Combining eqs. (3) to (5) we obtain:

$$\sigma = (1/2A)\left\{\sum_{\alpha1\alpha2}\int_v dr_1 \int_v dr_2 \rho_{\alpha1\alpha2}(r_1,r_2)\left\{\left((x_1-x_2)^2-(z_1-z_2)^2/|r_1-r_2|\right)\right\}\right\}\phi'_{\alpha1\alpha2}(|r_1-r_2|)$$

(6)

where all quantities refer now to the undistorted situation $(\lambda=1)$. In the thermodynamic limit of an infinite system we have moreover for a planar interface perpendicular to the z-axis:

$$\rho_{\alpha1\alpha2}(r_1,r_2) = \rho_{\alpha1\alpha2}(|R_1 - R_2|; z_1, z_2) \qquad (7)$$

where $r = (R,z)$ with $R=(x,z)$. Combining eqs. (6) and (7) we obtain finally:

$$\sigma = (1/4) \sum_{\alpha1\alpha2} \int dz_1 \int dz_2 \int d^2(R_1-R_2)\, \rho_{\alpha1\alpha2}(|R_1-R_2|;z_1,z_2)$$

$$\left\{\left((x_1-x_2)^2-(z_1-z_2)^2/|r_1-r_2|\right)\right\}\phi'_{\alpha1\alpha2}(|r_1-r_2|) \qquad (8)$$

which is a completely microscopic expression for the surface tension σ in terms of the pair potentials $\phi_{\alpha1\alpha2}$ and the pair densities $\rho_{\alpha1\alpha2}$ of a multicomponent system.

When the distance $|r_1-r_2|$ is much larger than the range of the potentials, $\phi_{\alpha1\alpha2}(|r_1-r_2|)$, the pair density factorizes into one-particle densities, $\rho_{\alpha1\alpha2}(r_1,r_2) \to \rho_{\alpha1}(r_1)\,\rho_{\alpha2}(r_2)$, which, in the bulk phases, are constants so that the convergence properties of the integrand of eq. (8) are controlled by the decay properties of the potential. When some of the components are charged, introducing the long-ranged Coulomb interaction potential, the r.h.s. of eq. (8) is not well behaved and we have to return to eq. (6) before the Coulomb divergence can be cured. An expression equivalent to eq. (8) but applicable also to ionic liquids can then be obtained as follows. Let us separate the interaction potential into a short ranged part, $V_{\alpha 1 \alpha 2}$, and a Coulomb part as:

$$\phi_{\alpha 1 \alpha 2}(|r_1-r_2|) = V_{\alpha 1 \alpha 2}(|r_1-r_2|) + e_{\alpha 1} e_{\alpha 2} (|r_1-r_2|) \tag{9}$$

with e_α the electric charge of the α species and $V(|r|) = 1/|r|$ the Coulomb potential. Let us now return to the finite volume expression (6) and before going to the thermodynamic limit we add and substract the asymptotic value of pair density, $\rho_{\alpha 1 \alpha 2}$, by introducing the so-called truncated pair density, $\rho_{\alpha 1 \alpha 2}{}^T$, according to:

$$\rho_{\alpha 1 \alpha 2}{}^T (r_1-r_2) = \rho_{\alpha 1 \alpha 2}(r_1-r_2) - \rho_{\alpha 1}(r) \rho_{\alpha 2}(r_2) \tag{10}$$

Using eqs. (9) and (10) we can separate off all Coulomb contributions to (6) as follows:

$$\sigma = (1/2A)\Big\{\sum_{\alpha 1 \alpha 2}\int_v dr_1 \int_v dr_2 \rho_{\alpha 1 \alpha 2}(r_1,r_2)\Big\{\big((x_1-x_2)^2-(z_1-z_2)^2/|r_1-r_2|\big)\Big\}V'_{\alpha 1 \alpha 2}(|r_1 r_2|)$$

$$+ (1/2A)\Big\{\int_v dr_1 \int_v dr_2 \rho_{ee}{}^T(r_1,r_2)\Big\{\big((x_1-x_2)^2-(z_1-z_2)^2/|r_1-r_2|\big)\Big\}V'(|r_1-r_2|)$$

$$+(1/2A)\Big\{\int_v dr_1 \int_v dr_2 \rho_e(r_1)\rho_e(r_2)\Big\{(x_1-x_2)\nabla_{x1-x2}-(z_1-z_2)\nabla_{z1-z2}\Big\}V(|r_1-r_2|)$$

$$\tag{11}$$

where we have introduced the average electric charge density, $\rho_e(r) = \sum_\alpha e_\alpha \rho_\alpha(r)$, the pair charge density, $\rho_{ee}(r_1,r_2) = \sum_{\alpha 1 \alpha 2} e_{\alpha 1} e_{\alpha 2} \rho_{\alpha 1 \alpha 2}(r_1,r_2)$ and the truncated pair charge density, $\rho_{ee}{}^T(r_1,r_2) = \rho_{ee}(r_1,r_2) - \rho_e(r_1)\rho_e(r_2)$ which is short-ranged so that the second term of eq. (11) is well behaved even if $V(r)$ is long ranged, e.g., $1/r$. In order to compute the infinite volume limit of the third term of (11) we use the following partial Fourier series representation of the Coulomb potential:

$$V(|\mathbf{r}|) = (1/A) \sum_q (2\pi/q) e^{-q|z|} e^{i\mathbf{q} \cdot \mathbf{R}} \tag{12}$$

where as before $\mathbf{r} = (\mathbf{R}, z)$. The final result reads then[8]:

$$\sigma = (1/4) \sum_{\alpha_1 \alpha_2} \int dz_1 \int dz_2 \int d^2(\mathbf{R}_1 - \mathbf{R}_2) \rho_{\alpha_1 \alpha_2}(|\mathbf{R}_1 - \mathbf{R}_2|; z_1, z_2)$$

$$\left\{ \left((\mathbf{R}_1 - \mathbf{R}_2)^2 - 2(z_1 - z_2)^2 / |\mathbf{r}_1 - \mathbf{r}_2| \right) \right\} V'_{\alpha_1 \alpha_2}(|\mathbf{r}_1 - \mathbf{r}_2|)$$

$$- (1/4) \int dz_1 \int dz_2 \int d^2(\mathbf{R}_1 - \mathbf{R}_2) \rho_{ee}^T(|\mathbf{R}_1 - \mathbf{R}_2|; z_1, z_2)$$

$$\left\{ \left((\mathbf{R}_1 - \mathbf{R}_2)^2 - 2(z_1 - z_2)^2 / |\mathbf{r}_1 - \mathbf{r}_2|^3 \right) \right\}$$

$$+ 2\pi \int dz_1 \int dz_2 \, \rho_e(z_1) |z_1 - z_2| \rho_e(z_2) \tag{13}$$

where the first term is well behaved because $V_{\alpha_1 \alpha_2}(|\mathbf{r}_1 - \mathbf{r}_2|)$ is short-ranged, the second term because $\rho_{ee}^T(|\mathbf{r}_1 - \mathbf{r}_2|)$ is short-ranged, while the last term equals minus the finite electrostatic energy of the interfacial electric double layer. Indeed, in the present geometry Poisson's equation reduces to:

$$\nabla_z^2 \phi(z) = -4\pi \rho_e(z) \tag{14}$$

which, except for the factor 4π which reminds us that the original system is three-dimensional, appears as a one-dimensional Poisson equation for which the solution which remains finite at $z=0$ can be written:

$$\phi(z) = \int dz' \, \{-2\pi |z - z'|\} \rho_e(z') \tag{15}$$

whereas the last term of eq. (13) can then be rewritten as:

$$(1/4\pi) \int dz (E(z))^2 = \int dz \rho_e(z) \phi(z) = -2\pi \int dz \int dz' \rho_e(z) |z - z'| \rho_e(z') \tag{16}$$

where $E(z) = -\nabla_z \phi(z)$ is the electric field corresponding to the electric potential $\phi(z)$, which in turn, results from the electric charge density $\rho_e(z)$. The final expression of the surface tension of ionic liquids, eq.(13), depends thus both on the three-dimensional Coulomb potential, $1/r$, characterizing the interactions between the particles, and on the one-dimensional Coulomb potential, $-2\pi |z|$, characterizing the interactions between the charged sheets building up the interfacial electric double layer. Expression (13), first proposed by Baus and Tejero[8], appears as a fairly natural generalisation to ionic mixtures of expression (8) first obtained by Buff for neutral mixtures [9].

3. THE MECHANICAL DEFINITION

An alternative route to the surface tension of a planar interface defines the latter as the surface excess value of the tangential stresses, i.e.

$$\sigma = \int dz \, (\sigma_{xx}(z) - \sigma_{zz}(z)) \tag{17}$$

where z denotes again the coordinate perpendicular to the planar interface, while within the present geometry, $\sigma_{xx}(z) = \sigma_{yy}(z)$ denotes the tangential stresses and $\sigma_{zz}(z)$ the normal stresses, $\sigma_{ij}(z)$ being the cartesian components of the stress tensor, i.e. minus the pressure tensor. When the interface is stabilized by an infinitesimal external field we have, $\sum_i \nabla_i \sigma_{ij}(z) = 0 = \nabla_z \sigma_{zz}$, so that the normal stresses are constant and eq. (17) effectively defines σ as a surface excess value (over the constant bulk value σ_{zz}). This mechanical definition of Kirkwood and Buff[2] can be related to the thermodynamic definition of eq. (1) by considering the underlying stress-strain relation. Proceeding again in two steps we first consider a general strain change and compute eq. (2) as:

$$\sigma = (1/A) \left\{ \sum \int dr(\lambda) \, ((\delta F/\delta u_{ij}(r(\lambda)))((\delta u_{ij}/\delta\lambda)(r(\lambda))) \right\}_{\lambda=1} \tag{18}$$

Using the general stress-strain relation:

$$(\delta F/\delta u_{ij}(\mathbf{r})) = \sigma_{ij}(\mathbf{r}) \qquad (19)$$

we finally compute the last factor of eq. (18) by considering a particular strain which produces an infinitesimal change in the interfacial area A. To this end we again consider the above change of coordinates, $\mathbf{r} \to \mathbf{r}(\lambda) = (\lambda x, y, \lambda^{-1}z)$, which in the language of elasticity theory corresponds to a local displacement vector $\mathbf{u}_\lambda(\mathbf{r})$, $\mathbf{r}(\lambda) = \mathbf{r} + \mathbf{u}_\lambda(\mathbf{r})$, of the form $\{\epsilon x, 0, -\epsilon z\}$ where we have put, $\lambda = 1 + \epsilon$, and for an infinitesimal transformation, $\lambda^{-1} \approx 1 - \epsilon$. The corresponding strain, $u_{ij}(\mathbf{r}) = (1/2)(\nabla_i u_j + \nabla_j u_i)$, is diagonal ($u_{xx} = \epsilon$, $u_{yy} = 0$, $u_{zz} = -\epsilon$), produces no volume change (e.g. $\Sigma_j u_{jj} = 0$) and transforms eqs. (18 - 19) into:

$$\sigma = (1/A)\int d\mathbf{r} \{\sigma_{xx}(\mathbf{r}) - \sigma_{zz}(\mathbf{r})\} \qquad (20)$$

which in the present geometry, $\sigma_{ij}(\mathbf{r}) = \sigma_{ij}(z)$, reduces to eq. (17). A microscopic expression for σ can then again be obtained by combining eq. (17) with the microscopic expression for the stress tensor $\sigma_{ij}(\mathbf{r})$. From the momentum balance equation one obtains the divergence of the stress tensor as:

$$\nabla \cdot \sigma'(\mathbf{r}) = -\sum_{\alpha,\alpha'} \int d\mathbf{r}' \rho_{ee}(\mathbf{r},\mathbf{r}') \nabla \phi_{\alpha\alpha'}(|\mathbf{r}_1-\mathbf{r}_2|) \qquad (21)$$

where $\sigma'(\mathbf{r})$ designs only the potential part of $\sigma = \sigma^k + \sigma'$, since the kinetic part, $\sigma^k(\mathbf{r}) = I k_B T \Sigma_\alpha \rho_\alpha(\mathbf{r})$, being diagonal, does not contribute to σ. Proceeding, inmediately to the case of the ionic liquids we use eqs. (9 - 10) to transform eq. (21) into:

$$\nabla \cdot \sigma'(\mathbf{r}) = -\sum_{\alpha,\alpha'} \int d\mathbf{r}' \rho_{\alpha,\alpha'}(\mathbf{r},\mathbf{r}') \nabla V_{\alpha\alpha'}(|\mathbf{r}_1-\mathbf{r}_2|)$$
$$- \int d\mathbf{r}' \rho_{ee}^T(\mathbf{r},\mathbf{r}') \nabla V(|\mathbf{r}_1-\mathbf{r}_2|)$$
$$- \rho_e(\mathbf{r})\int d\mathbf{r}' \rho_e(\mathbf{r}) \nabla V(|\mathbf{r}_1-\mathbf{r}_2|) \qquad (22)$$

Introducing the average electric potential, $\phi(\mathbf{r})$, and electric field, $\mathbf{E}(\mathbf{r})$, defined as usual by:

$$\mathbf{E}(\mathbf{r}) = -\nabla \phi(\mathbf{r}) = -\nabla \int d\mathbf{r}' \rho_e(\mathbf{r}') V(|\mathbf{r}_1 - \mathbf{r}_2|) \qquad (23)$$

together with the corresponding Maxwell stress tensor, $\sigma^M(\mathbf{r})$:

$$\nabla \cdot \sigma^M(\mathbf{r}) = \rho_e(\mathbf{r}) \mathbf{E}(\mathbf{r}); \quad \sigma^M(\mathbf{r}) = (1/4\pi) \{ \mathbf{E}(\mathbf{r})\mathbf{E}(\mathbf{r}) - (1/2) E^2(\mathbf{r}) \mathbf{I} \} \qquad (24)$$

where we have used the fact that here we have $\nabla \cdot \mathbf{E} = 0$ and $\nabla \cdot \mathbf{E} = 4\pi \rho_e(\mathbf{r})$. The total stress tensor, $\sigma(\mathbf{r})$, can now be written as the sum of three contributions, the kinetic part $\sigma^K(\mathbf{r})$, the Maxwell stresses $\sigma^M(\mathbf{r})$ and a remainder which we will call the truncated stresses, $\sigma^T(\mathbf{r})$, since it involves the truncated pair charge density according to:

$$\nabla \cdot \sigma^T(\mathbf{r}) = (-1/2) \sum_{\alpha,\alpha'} \int d\mathbf{r}' \{ \rho_{\alpha,\alpha'}(\mathbf{r}-\mathbf{r}',\mathbf{r}) - \rho_{\alpha,\alpha'}(\mathbf{r},\mathbf{r}+\mathbf{r}') \} \nabla \cdot V_{\alpha\alpha'}(|\mathbf{r}'|)$$

$$- (1/2) \int d\mathbf{r}' \{ \rho_{ee}^T(\mathbf{r}-\mathbf{r}',\mathbf{r}) - \rho_{ee}^T(\mathbf{r},\mathbf{r}+\mathbf{r}') \} \nabla \cdot V(|\mathbf{r}'|) \qquad (25)$$

where we have symmetrized the original expression (21) so as to allow us to extract one gradient according to:

$$\rho_{\alpha,\alpha'}(\mathbf{r}-\mathbf{r}',\mathbf{r}) - \rho_{\alpha,\alpha'}(\mathbf{r},\mathbf{r}+\mathbf{r}') = -\nabla \cdot \{\mathbf{r}' \, \bar{\rho}_{\alpha,\alpha'}(\mathbf{r},\mathbf{r}+\mathbf{r}')\} \qquad (26)$$

$$\bar{\rho}_{\alpha,\alpha'}(\mathbf{r},\mathbf{r}+\mathbf{r}') = \int_0^1 d\mu \, \rho_{\alpha,\alpha'}(\mathbf{r}-\mu\mathbf{r}', \mathbf{r}+(1-\mu)\mathbf{r}') \qquad (27)$$

where $\bar{\rho}_{\alpha,\alpha'}$ is an auxiliary pair density first introduced by Irving and Kirkwood [11]. Returning to the surface tension, we can write finally σ in the present geometry as:

$$\sigma = \int dz \{ \sigma^T_{xx}(z) - \sigma^T_{zz}(z) \} + \int dz \{ \sigma^M_{xx}(z) - \sigma^M_{zz}(z) \} \qquad (28)$$

$$\sigma = (1/2) \int dz \int dz' \int d\mathbf{R}' \, \bar{\rho}_{\alpha,\alpha'}(|\mathbf{R}'|;z,z') \{ x' \nabla_{x'} - z' \nabla_{z'} \} V_{\alpha,\alpha'}(|\mathbf{r}'|)$$

$$+ (1/2) \int dz \int dz' \int d\mathbf{R}' \, \bar{\rho}_{ee}^T(|\mathbf{R}'|;z,z') \{ x' \nabla_{x'} - z' \nabla_{z'} \} V_{\alpha,\alpha'}(|\mathbf{r}'|)$$

$$- (1/4\pi) \int dz \, (E_z(z))^2 \qquad (29)$$

From eq.(28) it is seen that one cannot, as originally proposed by Frenkel [12, simply add the Maxwell stresses to the total stresses but, as is seen from eq. (28) and (29), only to the truncated stresses. Furthermore, the mechanical formula (29) and the thermodynamic formula (13) will become equivalent if one performs a change of variables from $z - \mu z'$ to z in eq. (29) and permutes the z and μ integrations. Since the integrand is not a decreasing function of z these operations are not completely trivial except if one performs the **R'** integration first and assumes (or proofs) that one is then left over with a decreasing function of z.

4. FLUCTUATION THEORY

It is well known that the compressibility can be obtained from two complementary routes: via the equation of state (thermodynamics) or via the static structure factor (fluctuation theory). The same is true also for the surface tension where the fluctuation route was initiated by Yvon [3] and Triezenberg and Zwanzig [4].

Let us consider again a local distorsion of the form, $\mathbf{r} \to \mathbf{r} + \mathbf{u}(\mathbf{r})$ and assume that $u_x(\mathbf{r}) = 0 = u_y(\mathbf{r})$ so that the density profiles $\rho_\alpha(z)$ are shifted by the distorsion into $\rho_\alpha(z+u_z(\mathbf{r}))$, for each α. If we want the volume change to vanish, $\nabla \cdot \mathbf{u} = \nabla_z u_z = 0$, then $u_z(\mathbf{r})$ can only depend on $\mathbf{R} = (x,y)$; $u_z = u_z(\mathbf{R})$.

At each point **R**, the Gibbs dividing surface (say, z=0) is now locally shifted by the amount $u_z(\mathbf{R})$ and this then produces a change δA, in the interfacial area δA equal to:

$$\delta A = \int_A d\mathbf{R} \{(1 + |\nabla u_z|^2)^{1/2} - 1\} = (1/2) \int_A d\mathbf{R} |\nabla u_z(\mathbf{R})|^2 \qquad (30)$$

where we assumed ∇u_z to be small. The latter can be achieved automatically by choosing a long-wavelength periodic disturbance, $u_z(\mathbf{R}) \sim u_z \exp i\mathbf{q}.\mathbf{R}$, in which case eq. (30) becomes:

$$\delta A = (A/2) q^2 |u_z|^2 \qquad (31)$$

so that, indeed, $\delta A \to 0$ with $q \to 0$.

The cange in density, $\delta\rho_\alpha(\mathbf{R},z) = \rho_\alpha(z+u_z) - \rho_\alpha(z)$, produces also a change in free energy, which, since, according to eq. (31), δA is second order in u_z, has to be computed to second order in this density change. We thus have:

$$\delta F = \sum_\alpha \int dz\, (\delta F/\delta\rho_\alpha(\mathbf{r}))\Big|_{\delta\rho=0} \cdot \delta\rho_\alpha(\mathbf{r}) + (1/2)\sum_{\alpha\alpha'} \int d\mathbf{r}\int d\mathbf{r}'\, \delta\rho_\alpha(\mathbf{r})\,\delta\rho_{\alpha'}(\mathbf{r}')$$

$$\left(\delta^2 F/\delta\rho_\alpha(\mathbf{r})\,\delta\rho_{\alpha'}(\mathbf{r}')\right)\Big|_{\delta\rho=0} + \ldots \qquad (32)$$

where the subscript $\delta\rho=0$ indicates the undistorted system. For the latter we have moreover:

$$(\delta F/\delta\rho_\alpha) = \mu_\alpha - \phi_{\text{ext},\alpha}(\mathbf{r}) \qquad (33)$$

$$\beta\left(\delta^2 F/\delta\rho_\alpha(\mathbf{r})\,\delta\rho_{\alpha'}(\mathbf{r}')\right) = \delta_{\alpha\alpha'}\left(\delta(\mathbf{r}-\mathbf{r}')/\rho_\alpha(\mathbf{r})\right) - c_{\alpha\alpha'}(\mathbf{r},\mathbf{r}') \equiv \beta K_{\alpha\alpha'}(\mathbf{r},\mathbf{r}') \qquad (34)$$

where, μ_α is the chemical potential of species $\alpha, \beta = 1/k_B T$ the inverse temperature, $c_{\alpha\alpha'}(\mathbf{r},\mathbf{r}')$ the partial direct correlation functions and $\phi_{\text{ext},\alpha}(\mathbf{r})$ an external (gravitation) potential which we take infinitesimally small. The first term in the r.h.s. of eq. (32) drops out because the $\rho_\alpha(z)$ are assumed to be the equilibrium profiles for which the μ_α in eq. (33) are constants while there is no density change ($\int d\mathbf{r}\,\delta\rho_\alpha(\mathbf{r}) = 0$) and $\phi_{\text{ext},\alpha} = 0$. The surface tension becomes the combining eqs. (1) and (31-32);

$$\sigma = \delta F/\delta A$$
$$= \lim_{q\to 0} (1/(q^2|u_z|^2 A)) \sum_{\alpha,\alpha'} \int d\mathbf{R}\int dz \int d\mathbf{R}'\int dz'\, \delta\rho_\alpha(\mathbf{R},z)\delta\rho_{\alpha'}(\mathbf{R}',z') K_{\alpha\alpha'}(|\mathbf{R}-\mathbf{R}'|; z,z') \qquad (35)$$

Expanding $\delta\rho_\alpha(R,z) = \rho_\alpha(z+u_z) - \rho_\alpha(z) = u_z(R)\rho_\alpha'(z) + \sigma(u_z^2)$, and using $u_z(R) \sim u_z \exp iq.R$, we obtain from eq. (35):

$$\sigma = \lim_{q \to 0} (1/q^2) \sum_{\alpha,\alpha'} \int dz \int dz' \, \rho_\alpha'(z) K_{\alpha\alpha'}(|q|; z,z') \rho_{\alpha'}'(z') \qquad (36)$$

where $K_{\alpha,\alpha'}(|q|)$ is the Fourier transform of $K_{\alpha,\alpha'}(|R-R'|)$:

$$K_{\alpha,\alpha'}(|q|; z, z') = \int d(R - R') \, e^{iq(R - R')} K_{\alpha,\alpha'}(|R-R'|; z, z') \qquad (37)$$

and $\rho_\alpha'(z) = \nabla_z \rho_\alpha(z)$. From eq. (33) there also results that:

$$\nabla (\mu_\alpha - \phi_{ext,\alpha}(r)) = \nabla (\delta F/\delta\rho_\alpha(r)) = \sum_{\alpha'} \int dz' \, K_{\alpha\alpha'}(r,r') \nabla' \rho_{\alpha'}(r) \qquad (38)$$

which is the so-called Lovett-Mou-Buff-Wertheim equation [13]. In the present case eq. (38) reduces to

$$\lim_{q \to 0} \sum_{\alpha'} \int dz' \, K_{\alpha\alpha'}(|q|; z,z') \rho_{\alpha'}'(z') = 0 \qquad (39)$$

Let us also expand $K_{\alpha\alpha'}(q; z,z')$, exploiting the rotational invariance, as:

$$K_{\alpha\alpha'}(q; z, z') = K_{\alpha\alpha'}(q=0; z, z') + q^2 K_{\alpha\alpha'}^{(2)}(z, z') + \sigma(q^4) \qquad (40)$$

Substracting eq. (39) from eq. (36) and using eq. (40) we obtain:

$$\sigma = \sum_{\alpha\alpha'} \int dz \int dz' \, \rho_\alpha'(z) K_{\alpha\alpha'}^{(2)}(z, z') \rho_{\alpha'}'(z') \qquad (41)$$

and using eqs. (37) and (40) we also have:

$$K_{\alpha\alpha'}^{(2)}(z, z') = (1/2) \int dR \, (iqR/q)^2 K_{\alpha\alpha'}(|R|; z, z')$$

$$= -(1/4) \int dR \, |R|^2 K_{\alpha\alpha'}(|R|; z, z')$$

$$= (1/4\beta) \int dR \, |R|^2 c_{\alpha\alpha'}(|R|; z, z') \qquad (42)$$

where we took into account that the delta function of eq. (34) does not contribute to eq. (42). From eqs. (41 - 42) we obtain finally:

$$\beta\sigma = (1/4) \sum_{\alpha\alpha'} \int dz \int dz' \int dR |R|^2 \, c_{\alpha\alpha'}(|R|; z, z') \, \rho'_\alpha(z) \, \rho'_{\alpha'}(z') \tag{43}$$

which is the so called Triezenberg-Zwanzig formula. For the particular case of ionic liquids we again introduce a truncated quantity by splitting off explicitly the Coulomb terms:

$$c_{\alpha\alpha'}(r, r') = c_{\alpha\alpha'}{}^T(r, r') - \beta(e_\alpha e_{\alpha'} / |r - r'|) \tag{44}$$

$$c_{\alpha\alpha'}(q; z,z') = c_{\alpha\alpha'}{}^T(q; z, z') - \beta \, e_\alpha e_{\alpha'} \, V(q; z, z') \tag{45}$$

where according to eq. (12):

$$V(q; z) = (2\pi/q) \, e^{-q|z|} \; ; \; (V_z^2 - q^2) \, V(q; z) = -4\pi\delta(z) \tag{46}$$

As above, the truncated direct correlation function of eq. (44), $c_{\alpha\alpha'}{}^T(r, r')$, can be considered as short ranged and the above treatment applicable but the long-ranged Coulomb term of eq. (44) needs a separate treatment since, for instance, eq. (40) is no longer valid when eqs. (45 - 46) are used. Returning a few steps backwards we can write for the ionic liquids, using eq. (45) in eq. (36):

$$\beta\sigma = (1/4) \sum_{\alpha\alpha'} \int dz \int dz' \int dR |R|^2 \, c_{\alpha\alpha'}{}^T(|R|; z, z') \, \rho'_\alpha(z) \, \rho'_{\alpha'}(z')$$

$$+ \lim_{q \to 0} (\beta/q^2) \int dz \, \rho'_e(z) \left\{ \int dz' \, V(q; z-z') \, \rho'_e(z') - \phi'(z) \right\} \tag{47}$$

where, as above, $\rho'_e(z)$ is the charge density and $\phi(z)$ the corresponding electric potential of eq. (15). To obtain eq. (47) we have used eq. (39) which becomes here:

$$\lim_{q\to 0} \sum_\alpha \int dz' \, K_{\alpha\alpha'}(q; z,z') \, \rho'_{\alpha'}(z') - \sum_{\alpha'} \int dz' \, K_{\alpha\alpha'}^T(q=0; z,z') \rho'_{\alpha'}(z') + \beta e_\alpha \phi'(z) \quad (48)$$

The small-q limit in eq. (47) can now be taken without difficulty provided the following three conditions are satisfied. First we have:

$$\int dz \, \rho_e(z) = 0 \quad (49)$$

which is the condition of overall electroneutrality, next:

$$\int dz \, \rho'_e(z) = 0 \quad (50)$$

which is the condition of electroneutrality of the bulk phases ($\rho_e(z=\pm\infty)=0$), and finally:

$$\int dz' \, |z-z'| \, \rho'_e(z') = \nabla_z \int dz' \, |z-z'| \, \rho_e(z') \quad (51)$$

where the integration by parts in eq. (51) requires $\rho_e(z)$ to vanish as $|z| \to \infty$ more rapidly than $1/|z|$. When these conditions are satisfied eq. (47) can be transformed into the final result:

$$\beta\sigma = (1/4) \sum_{\alpha\alpha'} \int dz \int dz' \int d\mathbf{R} |\mathbf{R}|^2 \, c_{\alpha\alpha'}^T(|\mathbf{R}|; z, z') \, \rho'_\alpha(z) \, \rho'_{\alpha'}(z')$$

$$- \beta \int dz \, \rho_e(z) \, \phi(z)\} \quad (52)$$

which was first obtained by Evans and Sluckin[6] and Senatore and Tosi[7] in a somewhat less rigorous manner and in the present form by Baus and Tejero[8].

5. THE EQUIVALENCE PROBLEM

We have now arrived at three different expressions for the surface tension of ionic liquids: eqs. (13), (29) and (52). If we drop the electric contributions we recover the expressions applicable to a mixture of neutral particles, namely eq. (8), (21) and (43). In each case we have computed the surface tension by considering the change in free energy (δF) produced by a volume preserving distortion producing an infinitesimal change in the interfacial area (δA). It is however not at all obvious whether the final expressions we arrived at are stricktly equivalent. The explicit proof that these expressions are indeed equivalent was first given by Schofield[14] for the neutral fluid case and later extended to the case of ionic fluids by Baus and Tejero[8]. This proof is however very technical and will not be repeated here[8,14] especially since a simplified proof appears to be underway[15].

REFERENCES

1. J. S. Rowlinson and B. Widom, *"Molecular Theory of Capilarity"*, Clarendon, Oxford (1982).

2. J. G. Kirkwood and F. P. Buff, *J. Chem. Phys.*, **17**, 338 (1949)

3. J. Yvon in Proc. IUPAC Symposium on Thermodynamics (Brussels 1948).

4. D. G. Triezenberg and R. Zwanzig, *Phys. Rev. Lett.*, **28**, 1183 (1972)

5. Schofield and J. R. Henderson, *Proc. R. Soc. Lond.* **A379**, 231 (1982)

6. R. Evans and T. J. Sluckin, *Molec. Phys*, **40**, 413 (1980).

7. G. Senatore and M. P. Tosi, *Nuovo Cim.*, **56B**, 109 (1980).

8. M. Baus and C. F. Tejero, Molec. Phys., **47**, 1211 (1982).

9. F. P. Buff, *J. Chem. Phys.*, **23**, 419 (1955).

10. R. Evans, *Adv. Phys.* **28**, 143 (1979).

11. J. H. Irving and J. G. Kirkwood, *J. Chem. Phys.* **18**, 817 (1950).

12. J. Frenkel, p. 362 in *"Kinetic Theory of Liquids",* Dover, New York (1946)

13. M. Baus, *Molec. Phys.* **51**, 211 (1984).

14. M. H. Waldor and D. E. Wolf, *J. Chem. Phys.* (submitted June 1986).

LANDAU THEORY OF WETTING TRANSITIONS

E.H. Hauge

Institutt for teoretisk fysikk
Universitetet i Trondheim, NTH
N 7034 Trondheim-NTH, Norway

1. INTRODUCTION

An interface between two coexisting bulk phases costs free energy[1]. The macroscopic surface tension is nothing but this excess free energy per unit area. Suppose now that <u>three</u> bulk phases coexist. A natural question is then: Are all three interfaces stable against the intension of a macroscopic layer of the third phase ? If every interfacial tension is less than the sum of the other two, the answer is yes. One could express this by saying that none of the bulk phases <u>wet</u> the interface between the other two. (There might be a wetting layer of microscopic thickness. In that case one could reasonably call the interface <u>partially wet</u>.) On the other hand, if the triangular inequality does not hold for the three tensions, one of the interfaces will be <u>completely wet</u> by the third phase. In 1977 Cahn predicted[2] that for suitably chosen systems there should be a (interfacial) phase transition between states of partial and complete wetting at a temperature below bulk criticality. This prediction has since been verified experimentally[3].

A large amount of work has now been done on various aspects of the wetting transition. In this introductory exposition, we shall not attempt a review of the field, but instead approach wetting from one particular angle. In fact, we shall adopt the simplest theoretical framework possible, the phenomenological one of a Landau, or Landau-Ginzburg theory[4]. We use "Landau-Ginzburg" in a generic sense here. In the particular context of a free interface, this theory was first formulated by van der Waals[5] in 1894. It was already introduced in this school by B. Widom (this volume), and here we shall exploit it further in the context of wetting. Other

aspects of wetting than those highlighted here are discussed in this
volume in the contributions of D. E. Sullivan, M. M. Telo da Gama, P.
Tarazona and R. Evans. The reader interested in a more complete picture
is referred to a number of recent reviews[6].

2. ONE-COMPONENT LANDAU-GINZBURG THEORY[7,8]

For simplicity, let one of the three coexisting bulk phases be an
inert <u>wall</u>, and the two remaining ones be gas (g) and liquid (ℓ),
respectively. With gas occupying the bulk, the question is then whether
the wall is sufficiently attractive that a macroscopically thick layer
of liquid forms on it. We shall discuss this question on the basis of
the postulate that the total surface free energy per unit area, σ, can
be written as a Landau-Ginzburg functional of the density profile, $\rho(x)$,
where x is the distance from the flat wall.

$$\sigma[\rho(x)] = \int_0^\infty dx \left\{ \tfrac{1}{2}\left(\frac{d\rho}{dx}\right)^2 + f(\rho(x)) - f(\rho_g) \right\} - $$
$$ - H(\delta\rho(0) - \delta\rho_\ell) + \tfrac{1}{2} c \left(\delta\rho(0)^2 - \delta\rho_\ell^2\right) \tag{2.1}$$

Here $f(\rho)$ is the Helmholtz free energy per unit volume, from which the
bulk part, $f(\rho_g)$, has been subtracted. The term $\tfrac{1}{2}(d\rho/dx)^2$ represents
the free energy cost of inhomogeneities, and the unit of energy has been
judiciously chosen to make the prefactor (which is actually proportional
to the tension of a free gas-liquid interface[1]) equal to $\tfrac{1}{2}$. The surface
terms have been added in the Landau spirit: They are the two first terms
(allowed by the symmetry) in an expansion in the order parameter. For the
gas-liquid transition, that order parameter is $\delta\rho = \rho - \rho_c$, the deviation
from the critical density. The quantity H represents a net attractive
surface field of short range (on the scale of the bulk correlation length).
The constant c could be called a "surface temperature". Its molecular
origin is complex and will be left unspecified here.

Basic to a free energy functional of the form (2.1) is the assumption that the gradients, $d\rho/dx$, are small. I.e., the thickness of interfacial layers, which is generally of the order of the bulk correlation
length, must be considerably larger than the lengths associated with the
microscopic Hamiltonian. A large correlation length implies a state in
the vicinity of bulk criticality. One should therefore expect (2.1) to be
a reliable starting point for a discussion of the wetting transition only
if that transition occurs reasonably close to the bulk critical tempera-

ture, T_c. When the wetting transition takes place at temperatures considerably below T_c, a more detailed theory is necessary. The simplest candidate is the mean field theory, which (with fluids) should be of the van der Waals type[1,9,10]. Already at the level of mean field theory, one must distinguish between different kinds of basic interactions, whether they are algebraic (for example, of the Lennard-Jones type) or "short-range" (for example, exponential). This distinction turns out to be important for wetting exponents[11], but it is ignored by the Landau-Ginzburg approximation (2.1). Crossover between various regimes should therefore be expected.

Finally, one should worry about renormalization corrections, and more so the closer to bulk criticality. For short range forces they are very interesting[12], since the upper critical dimensionality, d_c, for that wetting problem is 3. Algebraic forces[11] should give mean field exponents since $d_c < 3$! From a practical point of view, however, the evidence[13] is that renormalization effects on wetting exponents can probably be ignored entirely in $d=3$, even with short range forces.

Worries and limitations aside, from now on we shall simply accept (2.1) and study its consequences. The Landau-Ginzburg functional has the form of a classical Lagrangian for a particle of unit mass, moving in the conservative potential, $V(\rho) = -f(\rho) + f(\rho_g)$, when ρ is interpreted as "position" and x is relabeled "time". The mechanics of this particle is given by

$$\tfrac{1}{2}\left(\frac{d\rho}{dx}\right)^2 + V(\rho) = E = 0 \qquad (2.2)$$

with initial condition from the surface terms in (2.1)

$$\frac{d\rho(0)}{dx} = c\delta\rho(0) - H \qquad (2.3)$$

and final condition, $\rho(\infty) = \rho_g$.

From this mechanical interpretation, a qualitative discussion of the wetting transition inherent in (2.1) is immediate. In Fig. 1 is shown

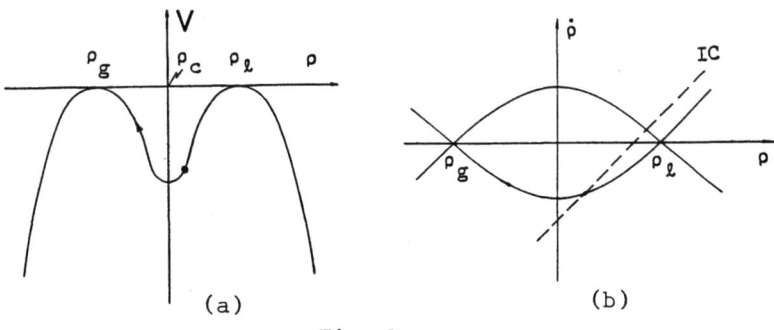

Fig. 1

(a) the potential $V(\rho)$, and (b) the corresponding trajectory in phase space $(\rho,\dot{\rho})$, [with $\dot{\rho}\equiv d\rho/dx$] for which the total energy vanishes, $E=0$. The dashed line (IC) with slope c in Fig. 1b corresponds to the initial condition (2.3). [In the Landau spirit, we shall ignore weak temperature dependences in the phenomenological surface parameters c and H, and treat them, near the wetting transition, as constants.] When c exceeds the slope of the $E=0$ trajectory at $\rho=\rho_\ell$, i.e., when $c > \sqrt{-V''(\rho_\ell)}$, the wetting transition is continuous. Fig. 2 demonstrates this: As the temperature ($T<T_c$) increases, the $E=0$ trajectory in the $(\rho,\dot{\rho})$ plane shrinks, to $\rho_g = \rho_\ell = \rho_c$ when $T=T_c$. The continuous transition occurs at a temperature for which $c\delta\rho_\ell(T_W)=H$. Since $\delta\rho_\ell$ is bounded both from below and from above, it is clear that a wetting transition only exists for intermediate fields,

$$0<H<c(\delta\rho_\ell)_{max} \tag{2.4}$$

With H below/above this range, the wall is never/always completely wet, respectively.

Our Landau-Ginzburg approach also allows a first-order wetting transition. This happens when c is sufficiently small, as demonstrated in Fig. 3. Exactly at which point one jumps from a partially to a completely wet state, must be determined by a free energy calculation, and cannot be read off from Fig. 3.

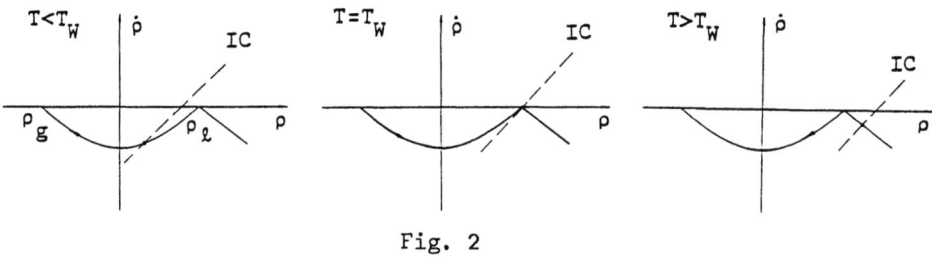

Fig. 2

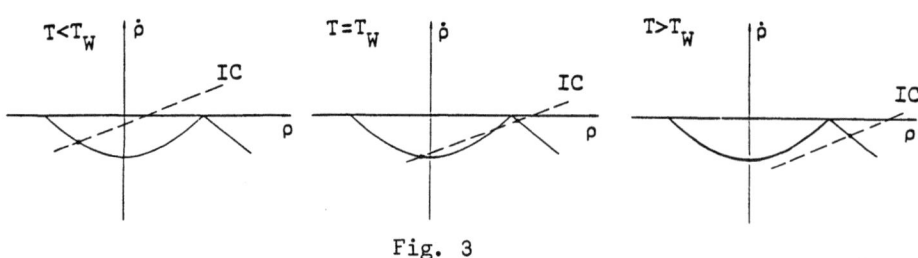

Fig. 3

3. LANDAU THEORY

The above Landau-Ginzburg picture is not, in a strict sense, a Landau theory of the wetting transition. If one is exclusively interested in that transition, the starting point (2.1) is too rich from a purist Landau point of view. In the Landau spirit, one should start by identifying the <u>order parameter</u>, not for the bulk system, but appropriate to the wetting transition itself. A simple choice is the thickness, h, of the wetting layer. This choice is convenient because the excess surface density is proportional to h for large h. <u>Fluctuations</u> in $h=h(y,z)$ are therefore, in direct analogy with the bulk case, closely related to a susceptibility, through a surface fluctuation theorem. It is unusual, of course, that in the disordered state $(T>T_W)$ the order parameter is infinite rather than zero, but this only counts as a minor nuisance.

In order to effect the contraction of the description from (2.1) down to a Landau theory proper, we postulate that the density profile of a <u>free</u> interface, $\rho_0(x-h)$, located at $x=h$, can be substituted for $\rho(x)$ in (2.1). This approximation is excellent for $x \sim h$, when h is large. The fact that it is dubious close to the wall, turns out not to be of qualitative significance[8,14].

The form of $\rho_0(x-h)$ follows from (2.1) itself, when the wall is moved to $x=-\infty$. With the standard quartic approximation for $f(\rho)$, $\rho_0(x-h)$ has the form of a hyperbolic tangent[1], interpolating between ρ_ℓ and ρ_g over a bulk correlation length, λ^{-1}, where $\lambda = \sqrt{-V''(\rho_\ell)}$. When $\rho_0(x-h)$ is inserted into (2.1), now with the wall back at $x=0$, one finds[8]

$$\Delta\sigma(h) \equiv \sigma(h) - \sigma_0 \simeq (H - c\delta\rho_\ell)\alpha e^{-\lambda h} + \tfrac{1}{2}(c-\lambda)\alpha^2 e^{-2\lambda h} + \ldots \qquad (3.1)$$

Here σ_0 is the tension of a free interface, and $\alpha = 2\delta\rho_\ell$. When $c>\lambda$, we recognize our previous result: There is a continuous wetting transition as the coefficient of the leading term changes sign at $T=T_W$, where $c\delta\rho_\ell(T_W) = H$. On the other hand, $c<\lambda$ will produce a negative second term which (assuming higher terms to be positive) will cause the wetting transition to be first order. This argument parallels the standard one that a cubic term, which would be the second one in a Landau expansion of the free energy of a bulk magnetic system, is responsible for the corresponding transition going first order[4].

From a theorist's standpoint, a continuous wetting transition is more exciting than a first-order one, since in the former case, exponents describing how various quantities diverge or vanish as $T \to T_W^-$ can be defined and calculated. From this point of view it is unfortunate that algebraic forces favor first-order wetting transitions[11,15]. Coexisting magnetic or structural phases may therefore be more likely candidates for the observation of continuous wetting transitions than are fluid phases[16].

Be that as it may, we shall now investigate, on the basis of (3.1) with $c > \lambda$, how the equilibrium thickness of the wetting layer, h_m, diverges as $T \to T_W^-$. Also, we want to determine how the correlation length ξ diverges as $T \to T_W^-$, where ξ is associated with the interface fluctuations. These fluctuations are described by the correlation function $\langle \delta h(\vec{0}) \delta h(\vec{r}_\perp) \rangle$; $\vec{r}_\perp = (y,z)$. The answers[8] to these questions follow readily from (3.1). The equilibrium condition $\partial \Delta\sigma / \partial h = 0$ gives h_m as

$$\lambda h_m \sim \ln(c\delta\rho_\ell(T) - H)^{-1} \sim \ln(T_W - T)^{-1} \qquad (3.2)$$

Furthermore, the surface fluctuation theorem together with (3.1) gives

$$(\partial^2 \Delta\sigma / \partial h^2)_{h = h_m} \sim \xi^{-2} \sim (T_W - T)^2 \qquad (3.3)$$

from which the exponent ν, defined by $\xi \sim (T_W - T)^{-\nu}$, follows as $\nu = 1$.

4. TWO-COMPONENT THEORY

When the bulk order parameter has more components than one, the formal generalization of (2.1) is straightforward[1,17,14]: The scalar density $\rho(x)$ is simply replaced by the vector field $\vec{\rho}(x)$. Suppose now that the three coexisting phases are the wall and two bulk phases corresponding to two maxima of equal height in the "mechanical" potential $V(\vec{\rho})$. Let us, for convenience, use the same terminology as before, "liquid" and "gas". Now, however, $\vec{\rho}_\ell$ and $\vec{\rho}_g$ are points in a two-dimensional "position" space, (ρ_1, ρ_2). Equipotential lines in Fig. 4 illustrate the situation. The

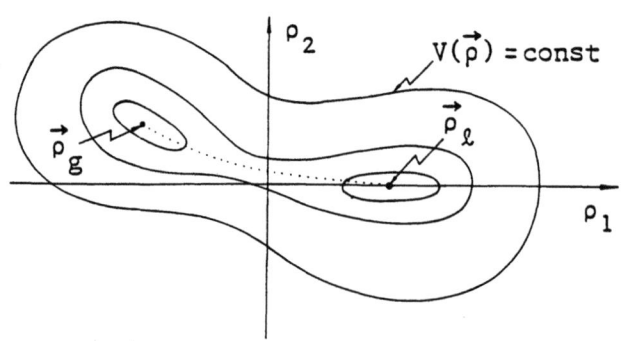

Fig. 4

dotted line from $\vec{\rho}_\ell$ to $\vec{\rho}_g$ indicates the trajectory corresponding to a free interface between these two bulk phases. For convenience, the coordinate system has been oriented parallel to the major axis of the hill near $\vec{\rho}_\ell$.

With an attractive wall field $\vec{H}=(H_1,H_2)$ (i.e., with $H_1>0$ in Fig.4), the question is again whether the wall will be wet by "liquid", in spite of the bulk being filled with "gas". The properties of the corresponding (continuous or weakly first-order) wetting transition can be calculated by the same technique as in Sec. 3: Insert $\vec{\rho}_0(x-h)$ with large h into the vector version of (2.1) to find $\Delta\sigma(h)$. There is a difficulty, however: In contrast to the one-component case, the free interface, described by the hill-to-hill excursion dotted in Fig. 4, is not known explicitly. Another look at (3.1) nevertheless gives hope: The difference $\Delta\sigma(h)$ has only been used to second order. The same holds true for more components: One really needs the details of the motion only in the region close to $\vec{\rho}_\ell$, where the equations of motion can be linearized. One has to worry a little about how to formulate the condition that the particle is properly aimed so as to end up on top of the $\vec{\rho}_g$ hill. But this problem is solved[14] by the observation that, as the particle leaves the $\vec{\rho}_\ell$ neighborhood, it must already be exponentially close to the trajectory of the free interface.

As a result one finds[14]

$$\Delta\sigma(h) \simeq (H_1 - c\delta\rho_{\ell 1})\alpha_1 e^{-\lambda_1 h} - H_2\alpha_2 e^{-\lambda_2 h} + \\ + \tfrac{1}{2}(c-\lambda_1)\alpha_1^2 e^{-2\lambda_1 h} + \tfrac{1}{2}(c-\lambda_2)\alpha_2^2 e^{-2\lambda_2 h} + \ldots \quad (4.1)$$

Here $\lambda_i = \sqrt{-\partial^2 V(\vec{\rho})/\partial\rho_i^2}$, (i=1,2), are the curvatures at the liquid hill, and α_1, α_2 are treated as constants. We restrict ourselves to models where $c > \lambda_i$. Assume first that $2\lambda_1 < \lambda_2$. In that case, the two leading terms in (4.1) are of precisely the same form as those in (3.1), and the same conclusions follow. If, on the other hand, $\lambda_1 < \lambda_2 < 2\lambda_1$, interesting new possibilities arise. If $H_2 > 0$, the wetting transition is first order, even for $c > \lambda_1$. If $H_2 < 0$, the transition is again continuous, unless higher terms conspire to drive it first-order. Assuming this not to be the case, a calculation like in (3.3) gives the correlation length exponent as

$$\nu = \frac{\lambda_2}{2(\lambda_2-\lambda_1)} \quad (4.2)$$

where $\lambda_i = \lambda_i(T_W)$. Since the curvatures are continuous functions of temperature, and since T_W depends on the basic parameters of the problem, this wetting exponent is non-universal and continuously varying here. The growth of the wetting layer is still (universally) logarithmic, however. Finally, when $\lambda_2 < \lambda_1$, there is no wetting transition with a $V(\vec{\rho})$ of the type shown in Fig. 4.

5. CONCLUSIONS

We have shown above that the simple Landau-Ginzburg theory, with its Landau contraction, gives a remarkably rich picture of the wetting transition. Continuous and first-order transitions are incorporated and, with more components than one, the exponents may be continuous functions of the parameters. Needless to say, this phenomenological theory is not the last word on the problem. More basic approaches can, in addition to providing a theoretical underpinning to the parameters in the Landau-Ginzburg functional, add important new elements[11,12,15,18]. In fact, when the wetting transition takes place far removed from bulk criticality, the problem becomes even more sensitive to details. We refer the interested reader to the literature.

ACKNOWLEDGEMENTS

What I know about the wetting transition, I have learned in active and pleasant collaboration with Michael Schick, Trond Aukrust, Ding E-Jiang, Kåre Olaussen and Jarek Piasecki. I am grateful to them all. This project has been supported in part by Statoil through the VISTA program.

REFERENCES

1. As an excellent basic reference, see J. S. Rowlinson and B. Widom, The Molecular Theory of Capillarity (Clarendon, Oxford, 1982).
2. J. W. Cahn, J. Chem. Phys. 66, 3667 (1977); C. Ebner and W. F. Saam, Phys. Rev. Lett. 38, 1486 (1977).
3. The transition was first seen by M. R. Moldover and J. W. Cahn, Science 207, 1073 (1980); since then a number of groups have studied the wetting transition experimentally in various contexts. For references, see the review by D. E. Sullivan and M. M. Telo da Gama, Ref. 6.
4. L. M. Landau and E. M. Lifshitz: Course of Theoretical Physics, Vols. 5 and 9: E. M. Lifshitz and L. P. Pitaevskii: Statistical Physics, Part 1 (3rd ed.) & 2 (Pergamon, Oxford, 1980).
5. J. D. van der Waals, Z. Phys. Chem. 13, 657 (1894); English translation in J. Stat. Phys. 20, 197 (1979).

6. Early work is reviewed in R. Pandit, M. Schick and M. Wortis, Phys. Rev. B 26, 5112 (1982); For an introductory survey, see E. H. Hauge in *Fundamental problems in Statistical Mechanics VI*, E. G. D. Cohen, ed. (North-Holland, Amsterdam, 1985); Statics and dynamics are reviewed by P. G. de Gennes, Rev. Mod. Phys. 57, 827 (1985); The most comprehensive review is that by D. E. Sullivan and M. M. Telo da Gama, in *Fluid Interfacial Phenomena*, C. A. Croxton, ed. (Wiley, New York, 1986).
7. R. Pandit and M. Wortis, Phys. Rev. B 25, 3236 (1982).
8. E. Brézin, B. I. Halperin and S. Leibler, J. Physique 44, 75 (1983).
9. P. C. Hemmer and J. L. Lebowitz in *Phase Transitions and Critical Phenomena*, Vol. 5B, C. Domb and M. S. Green, eds. (Academic, New York, 1976) pp 107-203.
10. D. E. Sullivan, Phys. Rev. A 25, 1669 (1982).
11. R. Lipowsky, Phys. Rev. Lett. 52, 1429 (1984); S. Dietrich and M. Schick, Phys. Rev. B 31, 4718 (1985) and 33, 4952 (1986); C. Ebner, W. F. Saam and A. K. Sen, Phys. Rev. B 31, 6134; D.M. Kroll, R. Lipowsky and R. K. P. Zia, Phys. Rev. B 32, 1862 (1985).
12. E. Brézin, B. I. Halperin and S. Leibler, Phys. Rev. Lett. 50, 1387 (1983); R. Lipowsky, D. M. Kroll and R. K. P. Zia, Phys. Rev. B 27, 4499 (1983); E. H. Hauge and K. Olaussen, Phys. Rev. B 32, 4766 (1985).
13. K. Binder, D. P. Landau and D. M. Kroll, Phys. Rev. Lett. 56, 2272 (1986).
14. E. H. Hauge, Phys. Rev. B 33, 3322 (1986).
15. D. M. Kroll and T. F. Meister, Phys. Rev. 31, 392 (1985); P. Tarazona, U. Martini Bettolo Marconi and R. Evans, to be published.
16. D. M. Kroll and G. Gompper, to be published.
17. G. Forgacs, H. Orland and M. Schick, Phys. Rev. B 33, 95 (1986).
18. T. Aukrust and E. H. Hauge, Phys. Rev. Lett. 54, 1814 (1985).

THE DYNAMICS OF WETTING

A.M. Cazabat and M.A. Cohen Stuart[*]

Collège de France, Physique de la matière condensée
11 Place M. Berthelot, 75231 Paris Cedex 05

INTRODUCTION

The problem to describe how a liquid front advances or recedes on a solid surface has already some hystory in physics. It has been the subject of many theoretical and experimental studies [1-8]; of these, the treatment by De Gennes and Joanny [7] deserves attention because it has put much emphasis on the role of the precursor film in the dynamics of complete wetting.

The aim of the present work is to check some of the predictions for complete wetting of smooth surfaces. We have also studied rough surfaces, for which data are lacking; most of the available literature deals with the case of partial wetting [9-11].

These latter studies [9-11] prompted us to carry out a few spreading experiments in non wetting situations, where specific instabilities of the contact line occur. We concluded that the mechanism which drives these instabilities has much in common with the progression of a precursor film in the case of complete wetting.

In the following, we shall first recall and discuss briefly the theoretical predictions for the dynamics of complete wetting on smooth surfaces. Next, we compare them with experiments. The experiments on rough surfaces are then presented and analyzed in terms of a very simple model. Finally, the non wetting case and its characteristic instabilities are described.

---

[*] Permanent address: Agricultural University, Department for Physical and Colloid Chemistry, De Dreijen 6, 6703 BC Wageningen, The Netherlands.

COMPLETE WETTING OF SMOOTH SURFACES

Theoretical predictions

Let us consider a drop of a non volatile liquid (dry spreading) which spreads on a smooth horizontal surface.

During the spreading, the volume Ω of the drop is conserved. The rate of spreading is characterized by the time dependence of the radius $R(t)$ of the wetted spot, or by the dynamic contact angle $\Theta(t)$.

In the case of complete wetting, the spreading parameter S

$$S = \gamma_{SG} - \gamma - \gamma_{LS}$$

is positive. Her, γ_{SG} is the interfacial tension between the solid and the gas (air, as the liquid is non volatile), γ the surface tension of the liquid, γ_{LS} the interfacial tension between solid and liquid.

The driving forces for spreading are capillarity and gravity forces, they are balanced by viscous forces.
Per unit length of contact line, the capillary forces are

$$F_c = S + \gamma(1-\cos\Theta) \approx S + \frac{1}{2}\gamma\phi^2$$

According to De Gennes [7], S is exactly balanced by friction forces in a very thin precusor film which always precedes the droplet in the case of complete wetting. The dynamics of the macroscopic droplet (the "cap") is then independent of S. Two different situations may occur:

(i) For small drops, $\frac{1}{2}\gamma\Theta^2$ is larger than the gravity force. The spreading is driven mainly by capillarity. This domain has been extensively studied by Tanner [8]. Here

$$R(t) \sim \Omega^{3/10} \left(\frac{\gamma t}{\eta}\right)^{1/10},$$

η being the viscosity of the liquid.

For larger drops, gravity forces are dominant [5]. Hence

$$R(t) \sim \Omega^{3/8} \left(\frac{\rho g t}{\eta}\right)^{1/8};$$

ρ is the liquid density, g the gravitational acceleration

The transition between these situations is expected to occur for

$$R(t) = cst \sim \left(\frac{\gamma}{\rho g}\right)^{\frac{1}{2}}$$

Discussion

A very important result of the theoretical analysis[7] is that the dynamics of the cap does not depend on the spreading parameter S, a point which has often been overlooked[12].

However, the treatment of the precursor film is based on rather questionable assumptions.

First, the movement of the precursor is treated by macroscopic hydrodynamics. From the explicit expressions given in ref[7], one can see immediately that its thickness is usually in the molecular range[1]. (An example is the case of pure Van der Waals forces where the spreading parameter and the disjoining pressure are just proportional). Ellipsometric measurements (1, 13, 14) of precursor thickness confirm this point. All this suggests that molecular diffusion plays a role in the spreading process and that the precursor probably behaves as a normal molecular film.

A second point is that the dynamics of the precursor is derived for a one-dimensional model (a liquid wedge moving with constant velocity U). For a drop the result cannot be used immediately, especially at times so long that the variation of precursor length l, $\frac{dl}{dt}$, is larger than the edge velocity $\frac{dR}{dt}$.

However, it is highly probable that the cancellation of S in the precursor film is correctly predicted [15]. So, even if the predictions for the precursor dynamics are rather questionable, we expect the macroscipic cap to follow the predicted power laws.

Experimental results

We have studied the spreading of silicone oils (PDMS) on hydrophilic or hydrophobic glass surfaces. The drop volume varied between 0.35 and 40 µl, the maximum drop thickness was always larger than 40 µm. With these values, capillarity and gravity ranges are successively observed during spreading, and the transition between them can be studied. The dependences of R on time t, volume Ω and viscosity η (1 pa.s $>\eta>$0.02 pa.s) have been checked and found in perfect agreement with expected power laws in both ranges[16]. On the contrary, the cross-over between them is not observed for R, but rather around a given value of $R(\frac{dR}{dt})^{-2/3}$, which is independent of the viscosity (see figures 1 and 2) but which does depend on the spreading parameter S[16].

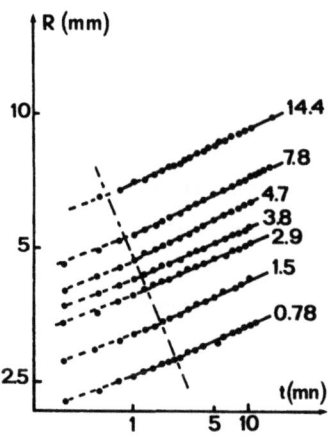

Figure 1 Double logarithmic plot of wetted spot radius R versus time t. Dashed line: slope 1/10. Full line: slope 1/8. The numbers on the curves give the drop volume in µl. The transaction between the two domains is tentitatively modicated. The liquid viscosity is 0.02 Pa.s

Conclusion

As expected, the theoretical power laws are well obeyed both in the capillarity and in the gravity ranges. In these limiting cases, the spreading parameter actually does not appear in the cap dynamics.

It influences the crossover between these ranges, but more refined theories would be needed to account for this effect.

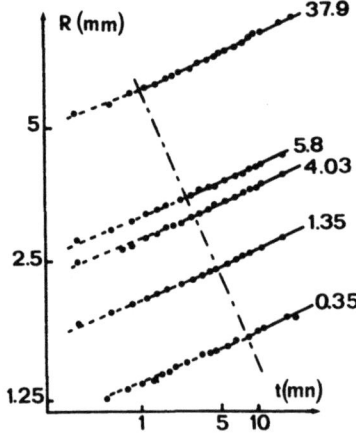

Figure 2 Same as (1) but for η = 1 Pa.s. The transition does not depend on the liquid viscosity.

Contact angle hysteresis on rough surfaces has been extensively studied by Mason and coworkers[9-11]. The roughness was due to parallel, radial or spiral grooves or produced by bead blasting.

Our work concerns the dynamics of complete wetting on rough surfaces, obtained by sand blasting (for large roughnesses) or by using different grades of abrasive powder. They were prepared and characterized at the workshop of the Ecole Supérieure d'Optique (Orsay-France).

Experimental observations

We have studied the spreading of silicone drops (PDMS) on hydrophilic rough glass surfaces. The maximum values h and the root mean squares Δh deduced from the scanning profiles of the glass samples are given in the table.

Four successive stages are observed during the spreading. First, the drop spreads as a whole (I). After some time, a "foot" of liquid finds "navigable channels"[11] into the roughness and precedes a "cap", which spreads with unchanged velocity and acts as a reservoir, feeding the foot (II). When this reservoir is consumed, the drop again spreads as a whole (III). A slowing down of the spreading occurs when the drop thickness becomes much smaller than the roughness (IV).

During the first stage, the power laws for smooth surfaces (1/10, then 1/8) are observed. During the second stage, the edge of the drop does not obey a universal power law. Indeed, the velocity of the edge results from two contributions: the motion of the cap and the growth of the foot. Each of them obeys a simple power law. The radius Rc of the cap increases as $t^{1/8}$. The length ΔR of the foot is found to be

$$\Delta R(t) = (A\, t/\eta)^{\frac{1}{2}} = (Dt)^{\frac{1}{2}}$$

where A depends on the sample roughness. The values of A can be found in the table. Data showing the proportionality between D and η^{-1} are summarized in Figure 3.

TABLE 1

Sample	h μm	Δh μm	$A = \eta D$ $\times 10^{11}$ N
I	50	12	–
III	29	6.5	89
V	18	3.6	40
VI	4.7	0.71	8
XIV	3.4	0.48	4.2
XVI	2.8	0.49	1.65
XII	2.4	0.37	1.1
IX	2	0.33	0.5

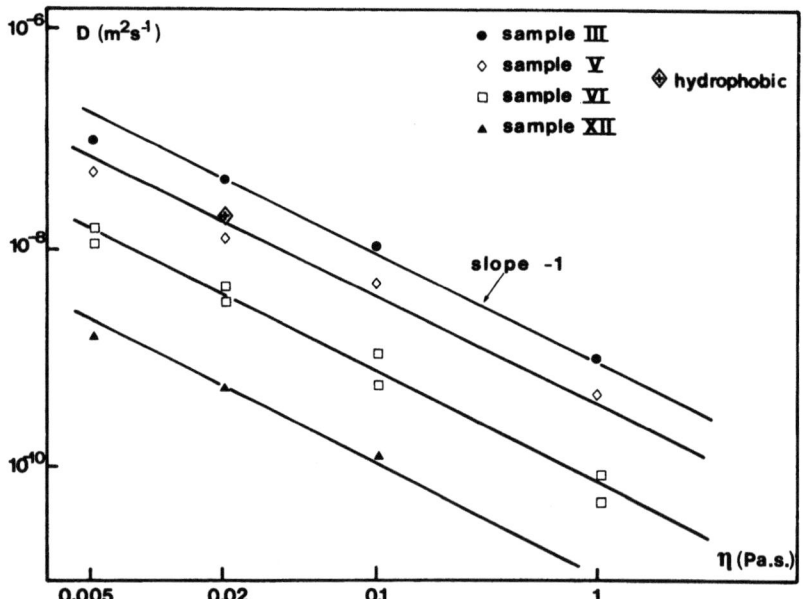

Figure 3 Log-Log plot of D versus η for various samples. All the glass surfaces are hydrophobic. The sample V has been made hydrophobic (cross) to check that the spreading parameter has no influence here, which is actually the case.

During the third stage, the drop radius increases as $t^{1/4}$. During the last stage, we found $R \sim t^x$ ($0.16 < x < 0.17$ - Perhaps $x = 1/6$).

In order to account for these observations we will now propose a very crude and naive model which is based on simplified descriptions of the physical processes driving the successive stages of spreading.

Tentative explanation

As for smooth surfaces, the dynamics of spreading is the outcome of the balance between capillarity and gravity as driving forces, and viscous forces. Now, the dynamic contact angle is a local variable which can be formally written as

$$\theta^2(t) = \theta_S^2(t) + \overline{\delta^2(t)}$$

θ_S being the contact angle on a smooth surface. $\overline{\delta^2(t)}$ is a positive mean contribution of the roughness, which is expected to be <u>independent of time</u> when the thickness of the drop is larger than the roughness.

We suppose that this is the case for the three first stages of spreading.

Stage 1: capillarity $\gamma\theta_S^2$ and gravity are much larger than $\gamma\overline{\delta^2}$. This is a "smooth regime", with the power laws corresponding to smooth surfaces.

Stage 2: cap + foot with $\Omega_{cap} \gg \Omega_{foot}$. Now $\gamma\overline{\delta^2}$ is the dominant term. The cap spreads on a smooth surface and its volume does not change significantly: the power law for the cap is unchanged.

The foot spreads under a constant driving force $\gamma\overline{\delta^2}$ and is connected to a reservoir. If e is the thickness of the foot, which is of the order of the maximum roughness h, one finds

$$\Delta R \sim \left(\frac{et}{\eta}\right)^{1/2} \sim \left(\frac{ht}{\eta}\right)^{1/2} .$$

This is equivalent to the Washburn equation[17] which governs imbition phenomena[18].

A new prediction of our simple model is that $A \sim h$, at least if the roughness profiles of various samples are homothetic, which is approximately true. So we expect A and h to be proportional. Such a trend is actually observed (see table).

We note that the same results are obtained by using a more rigorous model of triangular ducts proposed by Lenormand and Zarcone[19].

Stage 3: flat drop. $\gamma\overline{\delta^2}$ is still the dominant term. The drop spreads as a whole under a constant driving force. Taking into account the conservation of the volume we get

$$R \sim t^{1/4}$$

in good agreement with experiment. (Note that the same power law is obtained for smooth surfaces when the cancellation of S in the precursor film is not assumed[12].)

The last stage shows a $\underline{\text{slowing}}$ down of the spreading, which we attribute to a decreasing value of $\overline{\delta^2}$. The drop thickness is well below the roughness h, and the random character of the surface obviously plays a role. We have no quantitative predictions for this range. A model of partial filling of fractal structures proposed by De Gennes[20] gives $R \sim t^{1/5}$. Obviously, a better description of the statistical properties of the roughness profiles is needed.

In the next section we will dicuss a particular instability phenomenon observed in non wetting situations (on smooth surfaces). Since we consider our observation as evidence of molecular surface diffusion, thus should be relevant to the discussion of precursor dynamics in the case of complete wetting (see section 1).

INSTABILITIES IN NON WETTING SITUATIONS

Dynamic studies can also be performed in non wetting situations, by swelling a drop[4][9-11] or immersing a solid plane into a liquid.

Very often, instabilities occur during the motion of the contact line. Since these were first reported by Haines[21] we will refer to them as "Haines jumps"[3][22].

Haines jumps are due to traces of surface active impurities present in the liquid, which seem to diffuse on the dry solid plane because they appear ahead of the contact line[22][23]. Below, we shall describe the dynamics of the unstabilities and how we extract information on the molecular diffusion of these impurities.

Experimental observations

A glass slide is immersed slowly with constant velocity U ($10^{-2} \mu s^{-1}$ < U < $10^2 \mu s^{-1}$) in the liquid. α is the angle of the plane with the horizontal and can be varied between 0° and 50°. Liquids we used were Trioctylamine (TOA - Advancing contact angle $\theta_a \approx 30°$, receding contact angle $\theta_a \approx 20°$) and hexadecane (HD - $\theta_r \approx 20°$, $\theta_r \approx 10°$).

The advantage of a plane geometry is that one avoids complicated coupling effects observed for closed lines[22]. Moreover, dipping in under $\alpha \approx \theta$

ensures a minimal deformation of the liquid surface, which facilitates both theoretical analysis and experimental observation (optical imaging).

With some precautions, the jumps become well reproducible. They are periodic waves, propagating at constant velocity and without deformation, along the contact line. They provide a mechanism for the contact line to pass from one contact angle θ_1 to a smaller contact angle θ_2[23].

An example of such a wave is shown on the photograph.

The difference $\theta_1-\theta_2$ is reproducible and of the order of 10°, which is just $\theta a-\theta r$ for the two liquids we studied. The details of the experiment and of the analysis can be found elsewhere[23].
Here, we shall only discuss the effect of dipping velocity (u), on the number n of jumps observed for a given immersed length of the glass plane (15 mm). The results can be found in figure 4.

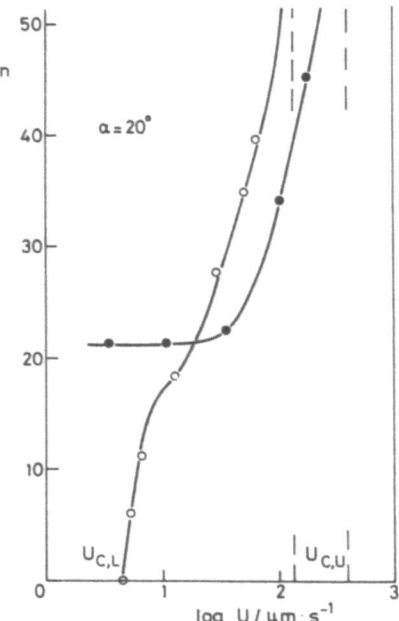

Figure 4 Number of jumps, n, versus Log u ; u is the velocity of the slide.

We distinguish roughly three domains. For intermediate velocities, $U \sim 10$ μm s^{-1} n is fairly constant (HD), but increases steeply up to a high velocity threshold U_{cu} where the number of "nucleations" is so large that propagation is no longer possible; the contact line has a "noisy" appearance. A low velocity threshold U_{cl} is also present (TOA), although not seen for HD (longer experiments would be required, calling for a better control of the atmosphere).

Interpretation

The origin of the wave-like instabilities seems to be the presence of surface active molecules, just ahead of the contact line, locally increasing the contact angle.

If the immersion velocity is very low $U < U_{cl}$, the advancing line will find a surface uniformly covered by these molecules, no unstabilities are expected in this case. The contact angle has a given value θo.
For increasing velocity, the contact line will "see" a gradient of surface properties which may be able to block it during some time, up to an angle $\theta 1$, after which it must jump over the barrier and relax. The liquid is then in contact with clean surface, with a contact angle $\theta 2$. As a result of adsorption, a new barrier builds up, θ increases again, and so on. At very high velocities, one expects the waves to disappear and the contact angle to stay at the value $\theta 2$. In fact, we observe the disparition of the waves, but rather by mutual annihilation. This effect is probably purely hydrodynamical and not related to the rate of molecular diffusion. (The size of the waves does not change.) So we expect $\theta 1 \sim \theta o$, and $\theta 1 - \theta 2$ constant and comparable to the contact angle hysteresis, which is actually the case.

The interesting result here is the lower threshold velocity, which is related to the molecular diffusion rate of impurities[22][23]. It is very high for TOA (4.4 μs^{-1}), much lower for HD ($\sim 10^{-2}$μs^{-1}, an order of magnitude which was already found by Bayramli and co.[22]). These values are in the range of velocities observed for the precursor film.

CONCLUSION

On the basis of the experiments discussed in this paper we believe that theoretical treatments and experimental studies of the precursor film must take "dry" surface diffusion into account, both for the "pure" film itself, and for impurities which can be present in the liquid. In addition, it should be stressed that very small amounts of surface active impurities can change drastically the spreading process[23]. We can learn from this, that in wetting studies both macroscopic and microscopic phenomena must be considered. This seems a complicated situation. Nevertheless, we found that simple models, which allow to identify the basic physical processes can give quite satisfactory descriptions.

ACKNOWLEDGEMENTS

It is a pleasure to thank the workshop of the Ecole Supérieure d'Optique, where the rough surfaces were prepared and characterized, M.A. Cuedeau for silane grafting and A. Bouillault who asisted with most of the experiments.

We also benefitted much from discussions with P.G. de Gennes, P. Fromherz and K. Mysels. We gratefully acknowledge N. Churaev, R.J. Good, T.G.M. van de Ven and F. van Voorst Vader for providing us with bibliographic references.

REFERENCES

1. B.B. Derjaguin, Zh.Fis.Khim., 14, 157 (1940).
   B.V. Derjaguin, V.V. Karasev and I.A. Lavygin, Spec. Disc. Faraday Soc. (1970), no. 1, p. 98.
   B.V. Derjaguin, Z.M. Zorin, N.V. Churaev, V.A. Shishin in Wetting Spreading and Adhesion (Padday - Ac. Press 1977), p. 201.
2. J.W. Cahn, J. Chem. Phys. 66, 3667 (1977).
3. E.B. Dussan V., Ann. Rev. Fluid Mech., 371 (1979)
4. A.W. Neumann, R.J. Good, J. Coll. Interface Sci. 38, 341 (1972)
   J.D. Eick, R.J. Good, A.W. Neumann, J. Coll. Interface Sci. 53, 235 (1975).
   R.J. Good, E.D. Kotsidas, J. Coll. Interface Sci. 66, 360 (1978).
5. J. Lopez, C.A. Miller, E. Ruckenstein, J. Coll. Interface Sci. 56, 460 (1976).
6. G.F. Teletzke, Ph.D. Thesis, Univ. Minnesota, (1983) and references therein.
7. P.G. de Gennes, rev. Mod. Phys. 57, 827 (1985).
8. L.H. Tanner, J. Phys. D 12, 1473 (1979).
9. J.F. Oliver, S.G. Mason, "The fundamental properties of paper related to its uses", Cambridge (1973).

E. Bayramli, T.G.M. van de Ven, S.G. Mason, Can. J. Chem. $\underline{59}$, 1954, 1962 (1981).

10. J.F. Oliver, S.G. Mason, J. Coll. Interface Sci. $\underline{60}$, 480 (1977).
11. J.F. Oliver, S.G. Mason, J. Materials Sci. $\underline{15}$, 431 (1980).
12. B.D. Summ, I.V. Gorionov, Chimiya (Russ) (1976).
13. W.D. Bascom, R.L. Cottington, C.R. Singleterry, Adv. Chem. $\underline{43}$, Contact Angle, Wettability and Adhesion, 355 (1964).
14. D. Ausserré, A.M. Picard, L. Leger, submitted for publication (1986).
15. P.G. de Gennes, private communication.
16. A.M. Cazabat, M.A. Cohen Stuart, J. Phys. Chem. (in press) (1986).
17. E.D. Washburn, Phys. Rev. $\underline{17}$, 374 (1921).
18. A. Okagawa, S.G. Mason, in 'Fibre-water interactions in paper-making', 581, Oxford (1977).
19. R. Lenormand, C. Zarcone, 59th Ann. Tech. Congr. and Exhibition - Soc. Petroleum Eng., Houston (1984).
20. P.G. de Gennes in 'Physics of disordered materials', D. Adler, H. Fritzsche, S.R. Ovshinsky eds., Plenum (1985).
21. W.G. Haines, J. Agric. Sci. $\underline{20}$, 97 (1930).
22. E. Bayramli, T.G.M. van de Ven, s.G. Mason, Colloids Surfaces $\underline{3}$, 131 (1981).
23. M.A. Cohen Stuart, A.M. Cazabat, Colloid Polymer J. (in press) (1987).

SPREADING OF LIQUIDS ON SOLID SURFACES

L. Leger

Laboratoire de Physique de la Matiére Condensée
Collège de France - 11 place M. Berthelot
75231 Paris Cédex 05. France

INTRODUCTION

The spreading of a liquid on a solid conditions a large number of practical situations (metal anticorrosive coating, textile dying, paints, plant nutrition and treatments, lubrication, water imbibition of porous rocks, enhanced oil recovery....) . In spite of their technical importance, these processes are not yet fully understood and indeed present a kind of challenge for both theorists and experimentalists: experimentally, all these processes are very sensitive to contamination by impurities which may adsorb preferentially on the solid interface and are difficult to control as the presence of the liquid does not allow the use of the sophisticated characterisation techniques developed for the solid-vacuum interfaces. Moreover, practical systems are often rather complicated, the liquid sometimes is a polymer or an emulsion while the solid can be finely divided (suspensions, porous media, fibers) implying additional drastic curvature effects . Theoretically, the problem of the moving contact line (3 phases: solid-liquid-gas boundary) is highly singular.

Since the pionnering work of Young[1], Laplace[2] and, more recently, Zisman[3], Fowkes[4] and Padday[5], the role of the spreading parameter $S = \gamma_{SG} - \gamma_{SL} - \gamma$ (with γ_{SG}, γ_{SL} and γ, respectively the solid-gas, solid-liquid, liquid-gas interfacial tensions) has been recognized: it controls the thermodynamic wettability of the surface. If S is negative, a liquid drop deposited on the solid adopts an equilibrium shape corresponding to a finite contact angle θ_e, defined by the Young relation $\cos \theta_e = (\gamma_{SL} - \gamma_{SG})/\gamma$. If S is positive, spontaneous spreading occurs, and the equilibrium state corresponds to a complete coverage of the solid by a thin liquid film, replacing the costly solid-gas interface by the less energetic solid-liquid and liquid-gas interfaces S is a measure of the interfacial energy gained per unit area during the spreading (see fig.1).

However, S is not the only quantity which plays a role, and it is important to know which parameters condition both the final equilibrium state of the system (final shape of the drop) and the kinetics of the spreading process. We shall devote the present discussion to the description of those two last points following the recent theory by de Gennes [6] and Joanny [7], which emphasizes the role of long range forces in the thicknenning of the wetting film (parts I and II for the statics and the dynamics, respectively). Comparison with recent experimental work to test the validity of these concepts will be presented in part III. Incomplete, partial wetting will not be discussed in these notes.

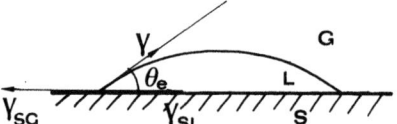

Fig.1. Schematic representation of the equilibrium of a small liquid drop placed on a horizontal solid substrate (partial wetting case).

I. WETTING: STATIC ASPECTS

1. Role of long range forces: the disjoining pressure

Let us consider the two situations schematically presented on fig.2 For a given liquid and a given substrate, we want to compare the energies of two flat films, one thin (a) and the other thick (b). With the usual thermodynamic definition, γ_{SL} represents the energy necessary to create one unit area of interface between two semiinfinite liquid and solid phases. If all intramolecular interactions are short-ranged the energy of the film is only the energy necessary to create the two interfaces, $E = \gamma_{SL} + \gamma$, whatever the film thickness. On the contrary, if the film thickness becomes comparable or smaller than the range of the interactions, the film energy is different from the energy of the two interfaces as the liquid can no longer be considered as a semi-infinite medium. One has then to write

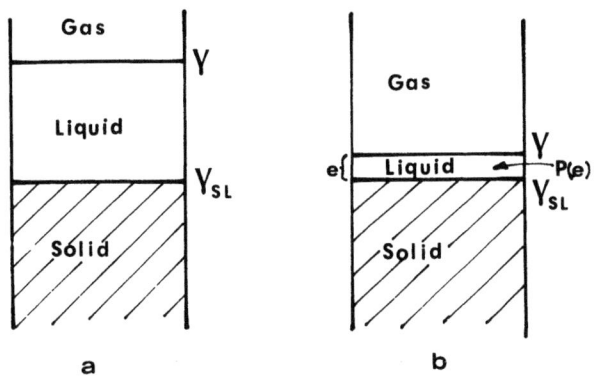

Fig.2. Comparison of the energy necessary to create a liquid film of constant thickness on top of a solid when the thickness of the liquid is large(a)or small(b)relative to the interaction force range.

$$E(\zeta) = \gamma_{SL} + \gamma + P(\zeta), \quad \text{with } \lim_{\zeta \to \infty} P(\zeta) = 0 \quad (1)$$

Usually, $P(\zeta)$ is a positive decreasing function of the film thickness ζ. The formation of a thin film needs to break more intermolecular cohesive interactions than the formation of a thick film, and thus costs more energy. $P(\zeta)$ is related to the disjoining pressure introduced by Deryagin[8], $\Pi(\zeta)$, by

$$\Pi = -dP/d\zeta, \quad \text{or} \quad P(\zeta) = \int_{\zeta}^{\infty} \Pi(\zeta') d\zeta' \quad (2)$$

The explicit thickness dependence of $\Pi(\zeta)$ depends on the nature of the cohesive forces, and has been discussed by Teletzke[9]. We shall only recall here the result for two important cases.

a) <u>Van der Waals forces</u>

The Van der Waals interaction potential[10] can be integrated to evaluate $P(\zeta)$, and one gets

$$P(\zeta) = \begin{cases} \dfrac{A_{SL} - A_{LL}}{12\pi\zeta^2} & \zeta < \lambda_o \\[2ex] \dfrac{B_{SL} - B_{LL}}{3\zeta^3} & \zeta > \lambda_o \end{cases} \quad (3)$$

where the wave-length λ_o, usually in the UV range, separates the length scale range of respectively non retarded ($\zeta < \lambda_o$) and retardated Van der Waals interactions. The Hamaker constants A_{SL} and A_{LL} are related to the static polarizabilities of the solid and the liquid α_S, α_L, and usually are in the range 10^{-20}, 10^{-21} Joules. In the case of complete wetting, $\alpha_S > \alpha_L$, and the effective Hamaker constant $A = A_{SL} - A_{LL}$ is usually positive. The corresponding disjoining pressure is

$$\Pi(\zeta) = \begin{cases} \dfrac{A}{6\Pi\zeta^3} & \zeta < \lambda_o \\ \dfrac{B}{\zeta^4} & \zeta > \lambda_o \end{cases} \quad (4)$$

b) <u>Double layer forces</u>

If the liquid is an electrolyte solution (Debye screening length K_D^{-1}) it has a large dielectric constant compared to the vapour, and the electric field has to vanish at the liquid gas interface. The Poisson-Boltzmann equation can be solved by electric images technics and, again, the potential integrated to yield Π [11].

In the large ζ limit ($K_D \zeta > 1$), one gets:

$$\Pi(\zeta) = C \exp - 2K_D \zeta$$
$$C = 64 n k_B T \tanh(\psi_o e / k_B T) \quad (5)$$

with ψ_o the electric potential at the solid surface, n the charge density, e the electron charge, and $k_B T$ the usual thermal energy. We have assumed a monovalent electrolyte.

2. Equilibrium shape of the wetting film

With a positive spreading parameter, we gain interfacial energy by intercalating liquid between the solid and the gas. We can guess that the final state of spreading will correspond to a thin rather flat film. The free energy of such a film with a uniform thickness e, covering an area A of the solid is

$$f = -AS + A P(e) \quad (6)$$

The first term tends to make the film thinner, while the second one tends to increase e. It is tempting to balance those two terms and minimize f in order to get the equilibrium thickness of the film e. However, before doing that minimization, one has to know what are the additional constraints imposed on the system. If the liquid is non volatile, its volume $\Omega = A$ e has clearly to stay constant. If the liquid is volatile, in order to reach a true equilibrium, one has to saturate the vapour phase, and the liquid volume may not remain constant. We shall treat these two cases separately.

a) Case of a volatile liquid

Let us assume that, to be sure that the vapour phase is saturated, the system is enclosed in a box containing a reservoir of the liquid, as schematically presented on fig. 3. There is then no limitation in the area of the solid which can be covered by the liquid, and there will not appear any contact line. If H is the difference in height between the reservoir free surface and the solid surface, the equilibrium thickness of the film can be obtained by minimazing the free energy with only disjoining pressure and gravitational terms, and one gets

$$\Pi(e) = \rho g (H + 2) \qquad (7)$$

(ρ is the fluid density and g the gravitational acceleration).

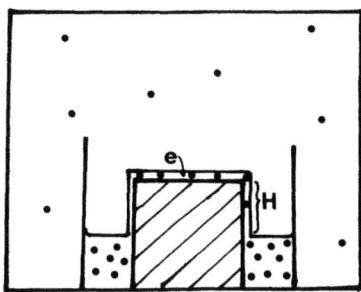

Fig.3. Schematic experiment in which a thin film of a volatile liquid can be deposited on top of a solid at equilibrium.

Equation (7) has been discussed at length by Deryagin and Dzyaloshinky[8,12], and describes well Rollin films of or normal fluids[13-15]. For Van der Waals interactions, eq. (7) leads to

$$e = \left(\frac{A}{6 \Pi \rho g H} \right)^{1/3} \qquad (8)$$

If H is macroscopic, (1cm), e falls in the 10^2Å range, while e should increase, and enter the range of retarded Van der Waals interactions (8) has then to be modified) for $H \to 0$.

In all cases, the equilibrium thickness of the wetting film formed by a volatile fluid is governed by an external parameter (H in the above experiment).

b) Case of a non-volatile liquid

We are back to eq.(6) which has to be minimized with the constraint $\Omega = A e$ = cte. Introducing a Lagrange multiplier, this leads to an im-implicit equation for e^6

$$S = e \Pi(e) + P(e) \tag{9}$$

If the explicit form of the disjoining pressure is known, one can deduce e from (9). For Van der Waals non-retarded interactions, one gets

$$e = \left(\frac{A}{8\Pi S}\right)^{1/2} = a \left(\frac{3\gamma}{2S}\right)^{1/2} \tag{10}$$

where the prefactor $a = (A/12 \Pi \gamma)^{1/2}$ is of the order of the molecular size. When electrostatic interactions dominate one gets:

$$e = \frac{1}{2 K_D} \ln \frac{C}{S K_D} \tag{11}$$

i.e. which strongly depends on the salinity of the solution.

Of course, the above presentation is oversimplified and neglects the effects associated with the curvature of the liquid-gas interface, i.e. the edges of the wetting film. The complete free energy functionnal which has to be minimized at constant volume is more complicated than (6). For a one dimensionnal system (invariant in the y-direction) the full free energy is

$$f = f_0 + \int dx \left\{ - S + \frac{\gamma}{2} \left(\frac{d\zeta}{dx}\right)^2 + P(\zeta) + G(\zeta) \right\} \tag{12}$$

It takes into account possible curvature and gravity terms ($r/2 (d\zeta/dx)^2$ and $G(\zeta)$ respectively).

The minimistion of (12) has been performed by Joanny and de Gennes[6,7], and yields the complete equilibrium shape of the drop, which looks like a "pancake", almost flat everywhere, with a thickness given by (9) except close to the edge where its thickness goes to zero over distances comparable to e. Such a "pancake" is schematically presented in fig. 4, for Van der Waals interactions. The edge profile is then a parabola, with e given by (10).

The important conclusion is that for small values of S, the "pancake" thickness may become much larger than the molecular size, a, while for large S values, e a. This is in agreement and explains the practical Cooper-Nuttal[16] rule that thin films form on high-energy surfaces.

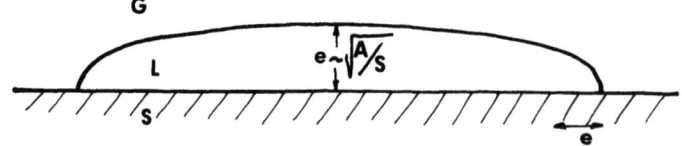

Fig.4. Equilibrium final spreading step of a droplet of non volatile liquid.

II. THE DYNAMICS OF SPREADING

For many practical situations, the kinetics of the film formation is essential . For example, a liquid spontaneously forming a perfect film in more than one week is of no use in the coating of textile fibers which come out, when extrudated, at a speed of several meters per second.

An a priori simple but usual situation is the spreading of one isolated liquid droplet deposited on a horizontal substrate. One has, in here again, to distinguish between volatile and non-volatile liquids, as schematically presented in fig. 5. if the liquid is volatile, a very efficient transport of molecules of the liquid from the drop to the surface can take place through the vapour phase. Molecules of the liquid can thus recondense on the solid to form a thin film with a thickness fixed by an external parameter analogous to H (see I, 2a). The remaining drop then spreads over a "solid" of identical chemical structure, and S vanishes. On the contrary, if the liquid is non-volatile, the only way of forming the "pancake" is by liquid flow and displacement of the contact line. We shall restrict our attention to the latter case.

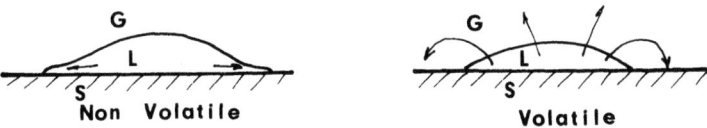

Fig.5. Comparison of spreading process of a non volatile liquid(flow) and a volatile one(transport via vapor phase and recondensation)

1. Experimental results

The experimental situation has been recently reviewed by Marmur[17]. Essentially, two types of information are available in order to understand the spreading kinetics of a drop.

a) Macroscopic observations

Several optical techniques have been used to characterize the temporal evolution of a small drop deposited on a horizontal solid substrate. From pictures or direct observation, the size of the drop can be measured as a function of time; from interferometry, its shape can be characterized, while the determination of the focal length of the equivalent lens yields the apparent contact angle θ_a.

All observations permit a description of the drop as schematically presented in fig. 6:

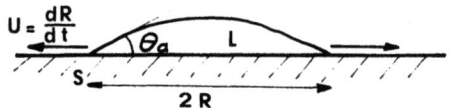

Fig.6. Apparent behavior of a drop during spreading (macroscopic scale)

- Close to the contact line, the shape is that of a simple edge, characterized by an apparent contact angle θ_a which decreases with time.

- The contact line moves with a velocity V, also decreasing with time.

- If the drop is smaller than the capillary length $l_c \simeq (\gamma/\rho g)^{1/2}$, gravity effects can be neglected, and the whole shape of the drop is that of a spherical cap. Its maximum thickness is $h \sim \frac{1}{2} R \theta_a$ for small enough θ_a. For a non-volatile liquid the drop volume is a constant,

$$R^3 \theta_a = 2 h R^2 = \text{cte} \qquad (13)$$

Indeed, experiments performed with a large variety of liquid and substrates seem to indicate power-law variations for both the size and the apparent contact angle:

$$R(t) \sim t^{0.1 \pm 0.03}$$
$$\theta_a(\theta) \sim t^{-0.3 \pm 0.05} \qquad (14)$$

These exponents agree with (13), and appear rather independent of the spreading parameter S [18,19,20,21]. Moreover, Tanner[27], using liquids of different viscosities, has established that $\theta_a^3 \sim \eta/\gamma \, U$, again independently of the spreading parameter.

These results represent a puzzling paradox: one can think of the total dissipation during the spreading as equal to the product of a force by a flux. The obvious force acting on the contact line is the unbalanced Laplace force, due to the fact that the apparent contact angle θ_a does not correspond to the equilibrium of the three tensions acting on the line, while the flux is related to the line velocity U, i.e.

$$T \overset{\circ}{\Sigma} = F \cdot U = [S + \gamma(1 - \cos \theta_a)] U \qquad (15)$$
$$\sim S U + \frac{1}{2} \gamma \theta_a^2 \, U$$
for small θ_a

As S is positive and eventually large, while θ_a goes to zero, the dominant term in the dissipation is proportional to S and the whole spreading kinetics shouls strongly depengly depend on S.

b) **Microscopic observations**

All the above mentionned experiments have been performed using optical techniques, with a thickness resolution comparable to a wavelength of light. Some authors, however, have reported observations on shorter length scales. In some cases, an "invisible" precursor film extending in front of the visible macroscopic contact line, has been detected through indirect manifestations of the presence of the liquid: dust particules deposited on the solid started to move before entering in contact with the drop edge [23], a gas supersaturated with water vapour blown on to the surface gave vapour blowing pattern, i.e. condensation on the substrate which stopped before reaching the contact line [23-24], and, by ellipsometric measurements, Bascon et al [25] have indeed detected a thin liquid film ahead the drop . Other curious effects have also been reported, such as spreading kinetics depending on the proximity of the edges of the substrate or of other droplets [26].

These results can be interpreted by assuming that a spreading drop allways develops a thin precursor film ahead of the apparent macroscopic contact line. However many authors rather conclude that the presence of such a film is due to impurity effects, vapour condensation in the case of volatile liquids, or surface roughness, and is not inherently related to the spreading process.

2. de Gennes and Joanny's description

Assuming the existence of a precursor film, Joanny and de Gennes have proposed a thourough description of the spreading kinetics of a drop evolving towards the final pancake, which allows to understand the apparent autonomy of the spreading parameter S [6,7]. We shall only present here qualitatively the main steps of their theory.

The basic idea is to assume that the dissipation can be decomposed into three parts:

$$T \, \dot{\Sigma} = T \, (\dot{\Sigma}_{macro} + \dot{\Sigma}_{precursor} + \dot{\Sigma}_{line}) \qquad (16)$$

each contribution in (16) comes from the corresponding part of the drop schematically presented in fig 7. Comparing (15) and (16), and assuming a negligible contribution of the line, one gets

$$T \, \dot{\Sigma}_{macro} + T \, \dot{\Sigma}_{precursor} = \frac{1}{2} \gamma \theta_a^2 \, U + S \, U \qquad (17)$$

The basic hypothesis of the de Gennes and Joanny's approach is to identify each term of the sum as

$$\begin{aligned} T \, \dot{\Sigma}_{macro} &= \frac{1}{2} \gamma \theta_a^2 \, U \\ T \, \dot{\Sigma}_{precursor} &= S \, U \end{aligned} \qquad (18)$$

Then, computing the dissipation directly in each region and assuming Poiseuille type flows, one can deduce the evolution law of $\theta_a(t)$ and the shape of the precursor film.

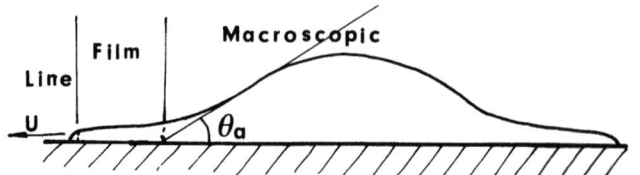

Fig.7. Parts of a drop spreading on a solid(semimicroscopic scale)

a) <u>Dissipation in the macroscopic drop</u>

For well advanced spreading, the apparent contact angle θ_a is small and close to the edge of the macroscopic part of the drop, the shape is that of a simple wedge. The spreading implys flow in that wedge, that we assume to be of Poiseuille type, with a velocity u_x in the x direction parallel to the surface (for a simple one-dimensional system), and boundary conditions $u_x(z = 0)$, $\partial u_x/\partial z = 0$ at $z = \zeta$.

The average velocity $U = (1/3) \int u(z)dz$ is such that

$$u_x(z) = \frac{3U}{2\zeta^2}(-z^2 + 2\zeta z) \qquad (19)$$

and the dissipation, for a thin slab dx,

$$\int_0^z dz\, \eta \left(\frac{du}{dz}\right)^2 = \frac{3\eta U^2}{\zeta} \qquad (20)$$

Then the total dissipation in the macroscopic part of the drop is

$$T \overset{\circ}{\Sigma} = \int_{x_{min}}^{x_{max}} \frac{3\eta U^2}{\zeta} d|x| = \frac{3\eta U^2}{\theta_a^2} \ln\left|\frac{x_{max}}{x_{min}}\right| \qquad (21)$$

as $\zeta \sim \theta_a x$, with $x_{max} \sim R$, and where $x_{min} = \zeta_{min}/\theta_a$ is related to the cut-off thickness ζ_{min} below which one enters the precursor film region.

Assuming a term of order unity, and comparing (21) and (18), one obtains Tanner's law

$$\theta_a^3 \sim \frac{\eta}{\gamma} U \qquad (22)$$

which appears, indeed, independent of the spreading parameter S, except for the prefactor which depends weakly (ln) on the cut-off x_{min}.

b) **Dissipation in the precursor film**

Here the problem is more complicated as 1) we do not know, a priori, the shape of the film, and 2) the thicknesses are now small, and long range forces have to be taken into account.

Taking again the one-dimensional model, schematically presented in fig. (8), the local pressure can be written

$$P(z) = P_G - \gamma \frac{d^2 \zeta}{dx^2} - \Pi(\zeta) + \Pi(\zeta) \qquad (22)$$

where P_G is the pressure in the gas, $\gamma(d^2\zeta/dx^2)$ the Laplace pressure term taking into account the curvature of the liquid-gas interface, and the two last terms come from the long-range forces. If we have local hydrostatic equilibrium, the vertical force on a small volume of fluid vanishes and

$$-\frac{\partial p}{\partial z} + \frac{\partial \Pi}{\partial z} = 0 \qquad (23)$$

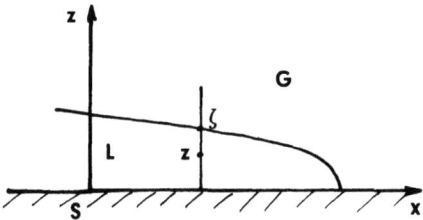

Fig.8. Parameters defining the precursor film for a one-dimensional system (uniform in the y-direction)

The current parrallel to the surface, in the lubrication approximation then simplifies to

$$J_s = \frac{\zeta^3}{3\eta} \left[-\frac{\partial p}{\partial x} \right]$$

In the frame of the apparent macroscopic contact line, for steady solutions the total flux vanishes and $J_s = U \zeta$

Thus we can derive the differential equation of the pressure film profile:

$$\frac{3\eta U}{\zeta^2} = \frac{d}{dx} \left[-\gamma \frac{d^2\zeta}{dx^2} - \Pi(\zeta) \right] \qquad (24)$$

Eq. (24) has been numerically studied in detail by Hervet and de Gennes[27] in the case of Van der Waals interactions. Their main results are summarized in fig. 9., for a well developped film for which curvature effects are negligible. The film has a hyperbolic shape

$$\zeta(x) = \frac{a}{\omega(x_2 - x)} \tag{25}$$

again with $a = A/6\Pi\gamma$ or order of a molecular size, and $\omega = U\eta/\gamma$ the capillary number.

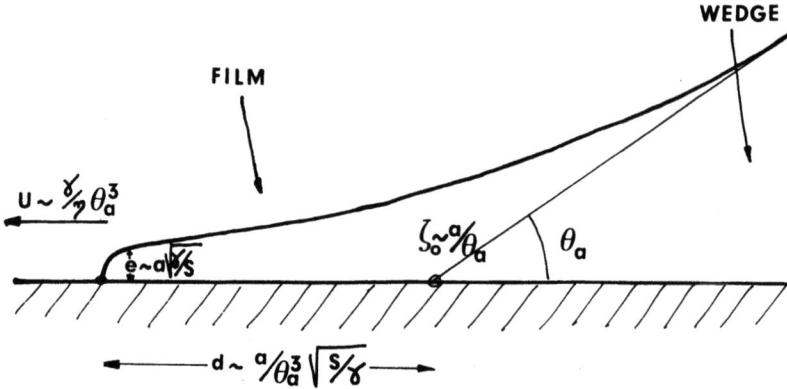

Fig.9. Characteristics of a macroscopic wedge, of the precursor film and of the cross-over zone between both as predicted by de Gennes and Joanny for a Van der Waals liquid.

The cross-over towards the macroscopic wedge takes place for $\zeta_o \sim a/\theta_a$, and has a width $r_o \sim a/\theta_a^2$

The cross-over towards the tiple line region takes place for $\zeta \sim e \sim a (2\gamma/3S)^{\frac{1}{2}}$, the equilibrium thickness of the pancake.

The film extension is thus

$$d = \frac{a^2}{\omega e(S)} \sim \frac{a}{\omega}\left(\frac{S}{\gamma}\right)^{\frac{1}{2}} \sim \frac{a}{\theta_a^3}\left(\frac{3}{\gamma}\right)^{\frac{1}{2}} \tag{26}$$

The whole film profile is thus controlled by the spreading parameter S, the effective Hamaker constant included in a, and the apparent contact angle θ_a.

With the above profile, Eq. (25), one can compute the total dissipation in the precursor film:

$$T \Sigma \text{ precursor} = \int_{-d}^{0} \frac{3 \eta U^2}{\zeta} dx = \frac{3 \gamma a^2 U^2}{2 e^2(S)} = S U \qquad (27)$$

Eq. (27) shows the coherence of the whole picture.

Let us finish with a possible explanation of the paradox of the S-independent macroscopic spreading kinetics:

- The surface energy S gained during the spreading in eliminated dissipation in the precursor film

- The remaining small term $\frac{1}{2} \gamma \theta_a^2$ (work of the Laplace pressure force) is dissipated in the macroscopic part of the drop, and yields Tanner's law $\theta_a^3 = \frac{\eta}{\gamma} U$.

III. EXPERIMENTAL TESTS

In order to establish without ambiguity the existence of a precursor film and then test the above predictions a series of experiments on model-systems have been under taken in our laboratory, in collaboration with D. AUSSERRE and A.M. PICARD.

To ensure that the liquid was a little volatile as possible, we have used high molecular weigh polydimethyl siloxane (PDMS). It is a linear flexible polymer, liquid at room temperature (glass transition temperature $T_g \sim 120$ °C). Three fractions of narrow molecular weight distribution have been kindly given to us by C. Strazielle from the Institut C. Sadron in Strasbourg. Their characteristics are listed in Table I.

Table I. Characteristics of the PDMS sample used.

M_w	M_w/M_n	η_{cp}
79.000	1.15	4.10^3
160.000	1.11	$3.3.10^4$
280.000	1.19	$1.8.10^5$

The surface tension is molecular weight independent, and about 23 dynes/cm.

For the solid substrate we have used silicon wafers, i.e., discs of silicon monocrystals, covered by a 20 Å thick oxyde later. The residual surface roughness was better than 10 Å. Two surface states were achieved: ultraclean bare wafers lead to a high energy surface, i.e., to large positive S values; wafers grafted with octadecyl trichlorosilane lead to a low energy surface, of critical surface tension $\gamma_c \sim 24$ dynes/cm[28,29]. In the latter case, for our liquid with $\gamma \sim 23$ dynes/cm, S is positive, and small (about 1 dyne/cm).

Small liquid drops were deposited on the surfaces inmediatly after either cleaning or grafting, and each wafer was then enclosed in a sealed box to prevent further contamination of the system during the spreading process which is long owing to the high viscosity of the liquid used. The boxes were equiped with a glass window in order to allow for optical investigation of the drops.

Several types of measurements have been conducted:

1. The spreading kinetics of the macroscopic part of the drop has been characterized through simultaneous size and apparent contact measurements[30]. The size was deduced from direct microscope observations, while the apparent contact angle was measured by a recently developed technique using light reflection on the whole drop[29].

Typical results for both θ_a and R as a function of time are summarized in Fig. 10 for different drop volumes, different molecular weights, and the two differential values of S. In all cases power laws are observed over more than three time decades, with exponents which appear totally insensitive to the above mentioned parameters. The drop volumes have all been chosen small enough to ensure that even at the latest stage of spreading studied the drop radius remains smaller than the capillary length $l_c \sim (\gamma/\rho g)^{1/2} \sim 2.5$ mm for our system. Thus all data in Fig. 10 correspond to the capillary regime where gravity is negligible compared to capillarity and where the drops are spherical caps.

These data confirm previous results[19,20,21,22], and lead to

$$R \sim t^{0.105 \pm 0.01}$$

$$\theta_a \sim t^{-0.303 \pm 0.012}$$

independently of S, and in good agreement with Tanner's law $\theta_a^3 \sim \dfrac{dR}{dt}$

2. The precursor film has been visualized by reflectometry in polarized light under a microscope[30], and its profile has been characterized by ellipsometry[31].

Both techniques rely on the same principle: when polarized light is reflected by a metal surface covered by a thin dilectric film, the reflected intensity depends on the orientation of the polarization with respect to the plane of incidence. Any polarization can be decomposed in two vectors $\vec{n}_\parallel$ and $\vec{n}_\perp$, respectively in and normal to the plane of incidence. For a beam polarized parallel to $\vec{n}_\parallel$, the complex reflection coefficient is $r_\parallel$, while it is $r_\perp$ for a beam polarized normal to the plane of incidence. The ellipsometric coefficient ψ, and Δ are defined by

$$r_\parallel = r_\perp \, tg\, \psi\, e^{i\Delta}.$$

r, ψ and Δ depend on the optical characteristics of the substrate (complex index of refraction) and of the film (index of refraction and thickness) on the angle of incidence θ, and are given by Frensel's laws[32].

When observing the sample through a microscope in relfected polarized light, the collected intensity coming from one point in the object plane is a mean value of the intensities over all θ in the aperture cone of the microscope and over all directions of the plane of incidence with respect to the incident polarization. Such a double average can be numerically performed [30], for either a bare wafer (Silicon + a 20Å oxyde layer of index of refraction 1.45) and give I_{wafer}, or a wafer covered with a polydimethy-siloxane film of thickness e (index of refraction n = 1.43) giving I_{Film}. The contrast thus evaluated by $\dfrac{I_{Film} - I_{wafer}}{I_{Film} + I_{wafer}}$ is reported as a function of the liquid film thickness in Fig.11. Going from thick to thin films, the contrast increases and passes through a maximum for film thicknesses close to 400Å, before decreasing slowly with decrasing thicknes. This behaviour contrasts with what happens in reflected natural light where the presence of the dielectric film decreases the reflection coefficient and thus leads to a reverse much weaker constrast. Using such a method of visualization, that we have called ellipsocntrast[30], we have been able to clearly establish the existence of precursor films slowly developping ahead of the macroscopic part of all the drops we have studied. Such a film, as observed through the microscope is reported in Fig.12, for a drop deposited on a silanated wafer (small S value). For long enough spreading times, the precursor films can extend as far as 1.5 mm from the macroscopic drop edge. Moreover, the aspect of the precursor films appears very sensitive to the spreading parameter. All films corresponding to high S valve appear, when observed through the microscope, very uniform, and the contrast slowly dies out at large distances, not allowing for a determination of the location of the film edge. On the contrary, the films deposited on silnated wafers appear non uniform, and strong thickness fluctuations are clearly visible in Fig. 12. At the same time, the edge of the film can be located at the beginning of the development of the film. It seems that the thickness fluctuations of the film are driven by inhomogeneities in the surface properties of the silanated wafers, invible before the film deposition, that are revealed by the film.

In order to gain more quantitative information on the film profiles, we have performed spatially resolved ellipsometry, using and ellipsometer built by Erman and Theeten[33], in the Laboratoire d'Electro-

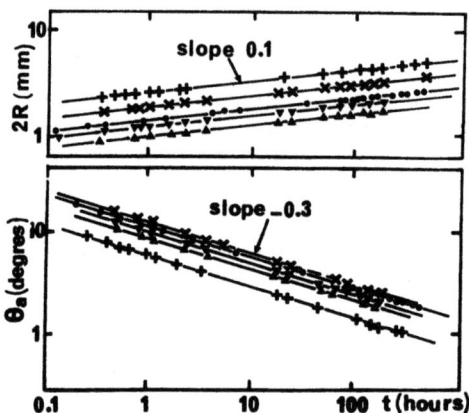

Fig.10. Macroscopic spreading of small PSMS drops deposited on horizontal silicon wafer for various drop volumes, Ω, with different molecular weights and the two surface states, large and small S. Power laws appear for both the size R and the apparent contact angle, with exponent insensitive to the above mentioned parameters.

o large S , M_W = 1.6 10^5 , Ω = 6.5 $10^{-5} cm^3$
X small S , M_W = 1.6 10^5 , Ω = 1.8 $10^{-4} cm^3$
∇ small S , M_W = 1.6 10^5 , Ω = 3.5 $10^{-5} cm^3$
△ small S , M_W = 1.6 10^5 , Ω = 1.5 $10^{-5} cm^3$
+ small S , M_W = 7.9 10^4 , Ω = 2.1 $10^{-4} cm^3$

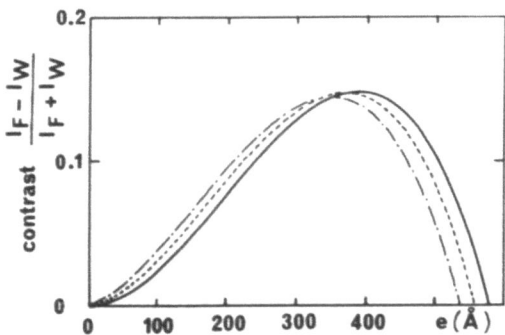

Fig.11. Estimates of the contrast in reflected polarized light as a function of the PDMS film thickness. The silicon wafer(complex index of refraction n=4-i 0.03) is assumed to be covered with an oxyde layer of thickness 0 Å(solid line), 20 Å (dotted line) and 40 Å (broken-dotted line). The index of refraction of the PDMS is assumed to be the same as the oxyde layer, n=1.43. The angle of incidence is fixed at 20°.

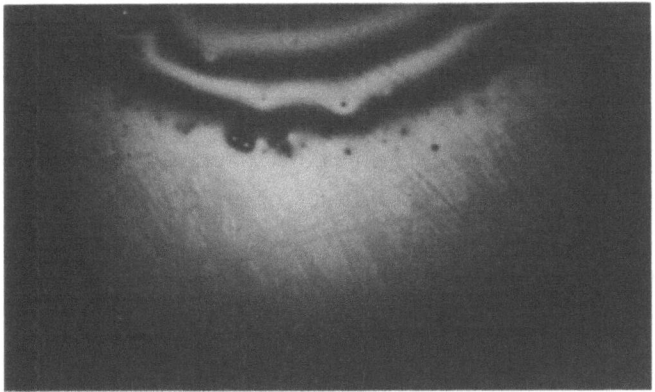

Fig.12. Visualization of a precursor film surrounding a small PDMS drop(M_w=79,000) deposited on silonated wafer. Using reflected polarized light fringes are ordinary equal thickness fringes visible in macroscopic part of drop. Ahead of last fringe, the bright area corresponds to a precursor film of thickness between 100 and 200 Å. The full view diameter is 1800μm.

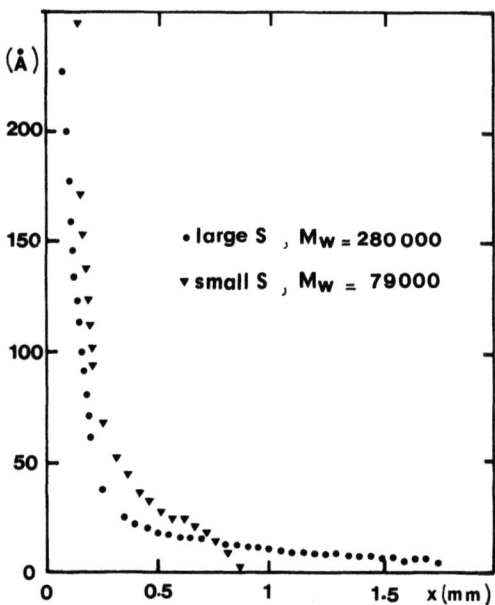

Fig.13. Preliminary results obtained in spatially resolved ellipsometry(spatial resolution 10x20 m) for the precursor film profile in the two surface states achieved with two different molecular weights.

nique et de Physique Appliquée (Limeil-Brevannes, France) with a spatial resolution of 10 x 20 µm. Rather preliminary results are reported in Fig.13 for two drops spreading at the same time, on wafers of large and small S respectively. The origin of the x axis is an arbitrary point inside the macroscopic drop, close to the macroscopic contact line. The two precursor films are clearly visible. Their thicknesses chose to the edge of the film are in qualitative agrement with $e \sim a\sqrt{3\gamma/2S}$, if one takes $a \sim 4.35$ Å, i.e., about a molecular size. The film thickness is clearly larger for small S than for large S, while the film extension goen in the opposite way.

These results are in rather good qualitative agreement with de Gennes and Joanny's predictions. More systematic and accurate measurements are presently in progress and should permit a test of the exact form of the profile, along with the characteristics of the cross-over towards the macroscopic region.

CONCLUSION

From experiments on model systems (non-volatile liquid drops deposited on well characterized smooth horizontal surfaces), we can deduce that a spreading drop effectively displays several regions:

- For the macroscopic part, the spreading kinetics follows scaling laws characterized by exponents which are independent of the spreading parameter S, and agree with Tanner's laws:

$$R(t) \sim t^{0.1}$$

$$\theta_a(t) \sim t^{-0.3}$$

$$R^3 \theta_a = cte$$

$$\theta_a^3 \sim \frac{dR}{dT}$$

- For the precursor film, our experiments clearly demonstrate its existence, even in the case of non-volatile, pure liquids. From both direct visualization through a microscope in polarized reflected light and ellipsometric measurements, we establish that its structure depends on S and is in good qualitative agreement with de Gennes-Joanny's predictions: the film is thicker for smaller and more developped for larger S or smaller θ_a.

A more detailed analysis is presently in progress in order to establish the whole film profile and test quantitatively de Gennes's predictions. Such an analysis is not easy as the independent determination of both S and the Hamaker constants entering into play is a classically difficult problem in interfacial phenomena. Moreover, for small values of S the fluctuations in the film thickness may complicate the picture. Lastly, the liquid used is a high molecular weight polymer and specific polymeric features could also enter in the dynamics. However, the qualitative agrement yet obtained appears valerable in itself. Long range forces indeed play an important role in the spreading of liquids on solid substrates and are responsible in the existence of the precursor film. Systematic studies of such precursor films could thus provide a way of investigating such forces.

REFERENCES

1. T. Young, Philos, trans. R. Soc. London $\underline{95}$, 65 (1805).
2. P. S. Laplace, "Mecánique Céleste" Suplement. X Livre Coursier, París ,(1806).
3. W. Zisman, in "Contact angle, Wettability and adhesion", edited by F. M. Fowkes, Advances in Chemistry Series, $\underline{43}$, 1, American Chemical Society, Washington D.C, (1964).
4. F. M. Fowkes, Ed. Advances in Chemistry Series , $\underline{43}$, (1964).
5. J. F. Padday, Ed. Wetting, Spreading and Adhesion, Academic, New York, (1978).
6. P. G. de Gennes, Rev. Mod. Phys. $\underline{57}$, 827, (1985).
7. J. F. Joanny, Thèse Université París VI. (1985)
   J. F. Joaany, P.G. de Gennes, C.R. Acad. Sci. París, $\underline{299\ II}$,279 (1984).
8. B. Deryagin, Zh. Fiz. Khim., $\underline{14}$, 137, (1940).
9. G. H. Teletzke, H.T. Davis., L.E. Scriven., Journ. Colloid Interface Sci. (1984).
10. J. Lyklema, in "Study of weak molecular forces", ed. Pontifical Academy of Sciences , North Holland, Amsterdam, and Wiley- Interscience, New York, (1967).
11. I. Langmuir., J. Chem. Phys. $\underline{6}$, 893, (1938).
12. I. E. Dzyaloshinskii., E.M. Lifshitz., L.P. Pitaevskii., Adv. Phys. $\underline{10}$, 165, (1961).
13. D. V. Brewer., in "The Physics of Liquid and Solid Helium" edited by K. Benneman and J. Ketterson, Wiley-New York, 537, (1978).
14. B. Deryagin, Z. Zorin, N. Churaev, V. Shishin, in ref. 5
15. M. Moldover, J. Schmidt., J. Chem. Phys. $\underline{79}$, 379, (1983).
16. W. Cooper, W. Muttal, J. Agr. Sci. $\underline{7}$, 219, (1915).
17. A. Marmur, Adv. Colloid Interface Sci. $\underline{19}$, 75, (1983).
18. H. Schonhorn, H.L. Frisch, T.K. Kwei, J. App. Phys. $\underline{37}$, 4967 (1966).
19. V. A. Ogarev, T.N. Timonina, V.V. Arslanov, A.A. Trapeznikov., J. Adhesion $\underline{6}$, 337 (1974).
20. G. G. Sawicki, in ref. 5.
21. M. D. Lelah, A. Marmur., J. Colloid Interface Sci. $\underline{82}$ (1981).
22. L. Tanner, J. Phys. D. $\underline{12}$, 1473 (1979).
23. W. Hardy, Philos. Mag. $\underline{38}$, 49 (1919).
24. D. Bangham, Z. Saweris., Trans. Farad. Society, $\underline{34}$, 554, (1938)
25. W. Bascon, R. Cottington, C. Singleterry, in ref. 4.
26. A. Marmur, M. D. Lelah., J. Colloid Interface Sci. $\underline{78}$, 262, (1980).
27. H. Hervet, P.G. de Gennes, C.R. Acad. Sci. París $\underline{299\ II}$, 499,(1984).
28. E. G. Shafrin, W. A. Zisman, J. Colloid Interface Sci., $\underline{7}$, 166,(1952).
29. C. Allain, D. Ausserre, F. Rondelez, J. Colloid Interface Sci. $\underline{107}$, 5, (1985).
30. D. Ausserre, A.M. Picard, L. Leger, Phys. Rev. Letters. to be published (1986).
31. L. Leger, D. Ausserre, A.M. Picard, M. Erman, to be published.
32. M. Born, E. Wolf., Principle of Optics, Pergamon Press Oxford. pp. 38-61, (1980).
33. M. Erman, J.B. Theeten, J. Applied Phys., to be published

THE EFFECTS OF POLYMERS ON WETTING BY NONVOLATILE LIQUIDS

A. Halperin

Department of Physical Chemistry and The Fritz Haber
Research Center for Molecular Dynamics, The Hebrew
University, Jerusalem 91904, Israel

INTRODUCTION

Polymers may modify the wettability of a solid surface. A variety of situations may be envisaged: (1) Spreading at dilute[1] and semidilute[2] polymer solution on a non adsorbing surface. (2) Spreading of a pure solvent on a surface coated by grafted or adorbed polymers[3]. These effects may be of importance to adhesion, permeation of porous media, etc. From the polymer science point of view, thin wetting films provide an interesting confinement medium for macromolecules. Here we discuss the effects of linear, flexible and uncharged polymers on thin wetting films of nonvolatile solvents. We do so using a continuun theory of wetting proposed by Joanny and de Gennes[4,5].

A liquid placed on a planar solid surface may form a drop (finite contact angle, $\theta > 0$) or spread into a wetting film ($\theta=0$). The final outcome is determined by the spreading coefficient[5,6,7]

$$S = f_d - f_w \qquad (I-1)$$

where f_d and f_w are respectively the free energies per unit area of the dry and the wet surface. For $S < 0$ the liquid forms a drop ($\theta > 0$), while $S = 0$ leads to complete wetting ($\theta = 0$). It should be noted that S is defined for macroscopic liquid layers. in a true equilibrium, obtained for mutual saturation of the phases, S is never greater than zero[6]. However, this limitation does not apply to constrained equilibria. Specifically, low vapor pressure liquids may be considered as nonvolatile ("dry") on the experimentally relevant time scale[2,8], and in such cases it is meaningful to consider $S > 0$ as well as $S \leq 0$.

The wetting film formed by a 2dry" solvent for $S > 0$ was considere by Joanny and de Gennes[4,5,7]. the equilibrium thickness of the film (h_0) is given by the balance between the 'spreading force' ($-S$) and long range van der Waals forces which may favor thick films.

These forces may be written as

$$P_{vw}(h) = \frac{1}{2} \gamma \frac{a^2}{h}$$ (I-2)

where γ is the liquid-vapour surface tension and h the thickness of the film. γa^2 is proportional to the Hamaker constant which fixes a to be an atomic dimension. the free energy of the film is

$$f - f_0 = \int [-S + P_{vw}(h) + \frac{1}{2}\gamma(\nabla h)^2 + \mu h] d^2 r \;,$$ (I-3)

where μ is a Lagrange multiplier assuring the constant volume constraint. Two characteristic length scales emerge. First, the equilibrium film thicknees h_0,

$$h_0 = a(3\gamma/2S)^{1/2} \;.$$ (I-4)

For most practical situations with $S > 0$, h_0 is less than about 100 Å. The second is healing length[1], λ, characterizing the deformation of the film due to a localized perturbation

$$\lambda = h_0^2 / a \;.$$ (I-5)

For $h_0 \sim 10^2$ Å, $\lambda \sim 10^3$–10^4 Å.

Our immediate expectation of interesting polymer effects follows from the comparison of h_0 and the Flory radius[9] R_F

$$R_F \approx N^{3/5} a$$

where N is the number of monomers in the chain and a is the monomer size. R_F measures the size of a macromolecule in a good solvent in the bulk, and because R_F can easily exceed h_0 we may expect strong polymer-film coupling.

WETTING A NON ADSORBING SURFACE BY POLYMER SOLUTIONS

Monodispersed polymers consisting of N monomers are dissolved in a good nonvolatile solvent. We now discuss the spreading of the polymer solution on a flat, nonadsorbing, solid surface. We assume the pure solvent and the surface are such that S is small but positive ($h_0 \sim 10^2$ Å, $\lambda \sim 10^3$–10^4 Å), and $R_F \gg h_0$. We consider first, following de gennes, the spreading of a semidilute polymer solution[2]. for such a system the coils interpenetrate. As the separation between the coils is then much smaller than λ, we may assume the film surface to be flat. The total free energy of the film consists of two terms due tothe pure solvent film and to the polymers. For a flat film ("pancake") of area A, the solvent contribution may be approximated as[5,7]

$$f_s = f_0 + A(P_{vw}(h) - S) \tag{II-1}$$

keeping in mind the constant volume constraint $V = Ah$.

Both the solid-liquid and liquid-vapor interfaces are nonadsorbing in our case. Accordingly, the polymer concentration is lower in the depletion layers[9] is roughly ξ_i, where

$$\xi_i \sim a \phi_i^{-3/4} \tag{II-2}$$

is the polymer correlation length in the interior of the film and ϕ_i the monomer volume fraction in the interior. As a result f_w increases and S decreases. The polymer free energy per unit volume increases with ϕ_i as[9]

$$F \approx \frac{kT}{\xi_i^3} = \frac{kT}{a^3} \phi_i^{2.25} \quad . \tag{II-3}$$

Because of the depletion layers and the constant volume constraint, ϕ_i increases with spreading. This leads to a polymer driven long range interaction which supplements $P_{vw}(h)$, and tends to thicken the film. To make these observations more precise, we approximate the polymer concentration profile as a "square well" i.e., we assume no monomers are present in the depletion layers and a constant monomer volume fraction ϕ_i in the interior. Monomer conservation imposes

$$\phi_i(h - 2\xi_i) = \phi_b h \tag{II-4}$$

where ϕ_b is the monomer volume fraction in the bulk. We confine ourselves to the case of thick films $h_0 \gg \xi_b \sim a\phi_b^{3/4}$. In this case the polymer contribution to the film free energy may be written as

$$f_p \approx \frac{kT}{\xi_i^3} A(h - 2\xi_i) \approx \lambda_1 \frac{kT}{\xi_b^3} V + \lambda_2 \frac{kT}{\xi_b^2} A + \lambda_3 \frac{kT}{\xi_b h} A \tag{II-5}$$

where $A(h - 2\xi_i)$ is the effective volume of the polymer solution and the λ_i's are positive numerical constants. to obtain the right hand side of (II-5) we expand f_p in powers of ξ_i/h. The first term in (II-5) represents the bulk free energy, the second the correction due to the increase in surface tension. the third term accounts for the long range interactions due to the polymers.

The full free energy of the film is then given by

$$f = f_p + f_s = f_0 + \lambda_1 \frac{kT}{\xi_b^3} V - S - \lambda_2 \frac{kT}{\xi_b^3} A + P_{vw}(h) \lambda_3 \frac{kT}{\xi_b} \frac{1}{h} A \tag{II-6}$$

Both S and $P(h)$ are modified because of the polymer presence. S is lowered and may even become negative. In the latter case the will retract into a droplet. A lower, but still positive S, leads to a thicker film. The polymer term supplementing $P_{vw}(h)$ also tends to thicken the film. Thus, for a semidilute wetting film the curvature

743

term in the free energy density $\frac{1}{2} \gamma (\nabla h)^2$ may be disregarded, and the presence of polymers leads to global modification of $P_{vw}(h)$ and S, both favoring thicker films.

Let us now consider the opposite case: at wetting by a dilute polymer[1] solution, such that the separation between the coils exceeds λ. Now we can no longer ignore polymer induced deformations of the film envelope. For polymers such that $R_F \sim N^{3/5} a \gg h_0$ we expect each polymer to produce a "bump" in the film envelope. In such a case the curvature term, $\frac{1}{2}\gamma(\nabla h)^2$ is important. The effects of the polymers are now local: $P_{vw}(h)$ is modified locally and not globally as in the semidilute case. We expect no global change in the surface tension and thus no change in S.

We first sketch the approach to the "single bump" problem. For a weak local perturbation, the "bump" envelope $(h = h_0 + \delta h)$ is given by[1]

$$\frac{\delta}{h_0} \approx \psi_0 \, K_0(\rho) \tag{II-7}$$

where $\rho = \sqrt{3}r/\lambda$ and K_0 is a zero order bessel function of imaginary argument. Asymptotically $K_0 \to (\pi/2\rho)^{1/2} e^{-\rho}$ for $\rho \to \infty$, while near the origin K_0 diverges as $-\ln \rho$. Because $\delta h / h_0$ is slowly varying for small arguments it may be approximately taken to be a constant over the region occupied by the polymer, i.e. the "bump" envelope is assumed to be flat near the origin, with a maximal height h_m (ψ_0). In effect we assume that the polymer coil is confined between parallel plates. This assumption is justified if the transverse radius of the confined coil, R_{2F}, is much smaller than the healing length λ. R_{2F} is given by[10]

$$R_{2F} \approx N^{3/4} \frac{a}{h_m}^{1/4} a . \tag{II-8}$$

For $N = 10^3$ and $h_0 = 10^2 \text{Å}$, $R_{2F} \ll \lambda$ as required even though $R_F > h_0$. The confinement free energy of a coil between two plates is[10]

$$\frac{f_p}{kT} \approx N \frac{a}{h_m}^{5/3} . \tag{II-9}$$

Now we may write the free energy of a film distored by a single polymer coil as

$$f = f_0 + \pi\gamma \, h_0^2 \int_0^\infty [\psi^2 + (\nabla\psi)^2] \rho d\rho + f_p(\psi_0) \tag{II-10}$$

where $\psi = \delta h / h_0$. The integral in (II-10) is obtained from (I-3) for weak localized perturbation. Both the integral and p depend on the yet to be determined ψ_0. By minimizing (II-10) with respect to ψ_0 we obtain ψ_0, h_m and f.

For distances of order 2λ and less we expect attractive interaction between "bumps". At such distances the envelopes of the two "bumps" overlap. Two results follow: (1) h_m increases bringing about a relief of the polymer confinement; (2) The curvature of the overlapping "bumps" is lowered. Both effects lower the free energy of the interacting "bumps", thus leading to mutual attraction. As each "bump" is associated with a polymer coil, our reasoning suggests polymer-polymer long range attractive interaction, which is mediated by the polymer induced surface distortion. The attractive interaction is replaced by strong steric repulsion when the two coils begin to overlap. The net "bump-bump" interaction may be approximated as

$$u(r) = \begin{cases} \infty & r \leqslant 2R_{2F} \\ f_p(h_m + h(r)) - f_p(h_m) \approx -\frac{5}{3} NkT \left(\frac{a}{h_0}\right)^{5/3} \frac{\delta h \, r}{h_0} & r \geqslant 2R_{2F} \end{cases}$$ (II-11)

In terms of $u(r)$, the two dimensional second virial coefficient, V, between polymer coils ("bumps") is

$$V = \int [1 - e^{-u(r)/kT}] d^2 r \qquad (II-12)$$

The free energy F, governing the plase behavior of the system may be written as follows

$$\frac{F - F_0}{kT} = \phi_s \ln \phi_s + \frac{V}{R_{2F}^2} \phi_s^2 + \text{terms linear in } \phi_s. \qquad (II-13)$$

For a negative second virial coefficient we expect a phase separation whenever $|V|/R_{2F}^2 > \phi_s^{-1}$, where ϕ_s is the average fraction of the film area occupied by chains. The biphasic domain spans the range $1 > \phi_s > (h_0/R_{2F})^{2/3} \gamma a^2/2kT$. In this domain the film separates into a thick polymer-rich film and a thin polymer-poor film. The thickness of the polymer-poor film is approximately $h = h_0[1 + (h_0/R_{2F})^{2/3} \, 2kT/\gamma a^2]$, while the thickness of the polymer-poor film is roughly h_0. For $h_0 \approx 10^2 \, \text{Å}$ the thickness of the polymer-rich film may reach 1.5 h_0.

In a film of dilue polymer solution in a good nonvolatile solvent with $S > 0$ ($h_0 \approx 10^2 \, \text{Å}$), polymer-induced film distortions lead to long range attractive interactions between the polymers. This interaction can cause phase separation into a thick polymer-rich film and a thin polymer-poor film. The phase separation occurs in a good solvent and it is not associated with polymer collapse.

WETTING OF POLYMER COATED SURFACES BY PURE NONVOLATILE SOLVENTS[3]

A continuons polymer coating of a planar solid surface modifies it wettability. Both and the net long range interaction $P_t(h)$ increase. usually, the change in S is the more significant, leading to thinner films ($< h_0$). A comparison with the wetting of a nonadsorbing bare surface by a semidilute polymer solution (section II) may be instructive. In this last case, for $h_0 \gg a$, S decreases while $P_t(h)$ increases leading to thicker films ($> h_0$). The reason for the different signs of Δ in the two cases is a follows: For bare nonadsorbing surfaces f_w increases due to the presence of depletion layers. For coated f_w is lowered because of the mixing free energy of the polymer coating. The precise effects of the polymer coating differ with the coating method. We distinguish between adsorption[11] and grafting[12]. Densely grafted polymers are distinguished by a flat concentration profile along the normal to the surface. Adsorbed polymers exhibit a self-similsr structure ($\phi \sim z^{-4/3}$, where z is the distance along the normal to the surface). Here we will discuss grafted surfaces.

Because the grafted layer is mostly uniform and rather dense, the Flory mixing free energy provides an appropiate description of it once we discard the translational term, wich is suppressed by the grafting. the free energy per lattice site is then

$$\frac{F_{mix}}{kT} = (1-\phi)\ln(1-\phi) + \chi\phi(1-\phi) \approx \phi(-1+\chi) + \frac{1}{2}\phi^2(1-2\chi) \quad (III-1)$$

where χ is the Flory interaction parameter. To obtain S we now write

$$d = d^{bulk} + \gamma_{pv} + \gamma_{sp}(1-\sigma) \quad (III-2)$$

$$d_w = d^{bulk} + \Gamma F_{mix} + \gamma + \gamma_{sl}(1-\sigma)$$

where γ_{sp} γ_{pv} γ and γ_{sl} are respectively the solid-polymer, polymer-vapour, liquid-vapour and the solid-liquid surface tensions. Γ, the total number of monomers per unit area is given by N/D^2 where D is the average distance between grafted sites. The graft density is given by $\sigma = (a/D)^2$, and the factor $(1-\sigma)$ in (III-2) accounts for the reduction in the solid fluid contacts due to the grafted sites. Using equations (I-1) and (III-1) we obtain

$$S = S_0 + kT\Gamma\left[(1 - \chi) - \frac{1}{2}\phi(1 - 2\chi) + ...\right] \qquad (III-3)$$

$$S_0 = \gamma_{pv} - \gamma + (\gamma_{sp} - \gamma_{sl})(1 - \sigma)$$

$kT\Gamma$ is rather large and in many cases we way may find $S \approx kT\Gamma$. Let us note again that S is defined for a macroscopic wetting layer forwhich long range interactions play no role.

Consider now the wetting film of a good nonvolatile solvent on a densely grafted surface such that $S \approx kT\Gamma$. Our starting point is equation (II-1), once we augment $P_{vw}(h)$ with a term, $P_p(h)$, accounting polymer effects, and denote $P_t(h) = P_{vw}(h) + P_p(h)$. Minimizing the corrected (II-2) with respect to subject to the constant volume constraint $V = Ah$, we obtain

$$S = h\pi_t(h) + P_t(h) \qquad (III-4)$$

where $\pi_t(h)$, the disjoining pressure, is defined by $\pi = -dP/dh$. Equation (III-4) determines the equilibrium thickness of the wetting layer. First, however, we must obtain the explict form of $P_p(h)$ and $\pi_p(h)$.

A rough form of π_p and P_p may be obtained by using a Flory-type argument. We start by writing the free energy of a single chain as a sum of entropic and elastic terms

$$\frac{1}{kT}F_p = \frac{hD^2}{a^3}(1 - \phi)\ln(1 - \phi) + \frac{1}{2}\frac{h^2}{R_0^2}. \qquad (III-5)$$

Here we identify the thickness of the wetting layer with that of the grafted layer. hD^2/a^3 is the number of lattice sites per grafted chain and $R_0 = N^{\frac{1}{2}}$ is the radius of an ideal chain. As in (III-1), the translational term is discarded because of the grafting. To obtain π_p we use

$$\pi_p = -\frac{\partial F_p}{\partial h} \qquad (III-6)$$

keeping in mind the constraint

$$\phi = \frac{Na^3}{hD^2} = N\sigma\frac{a}{h}. \qquad (III-7)$$

We eventually find

$$\phi \approx 0.9 \qquad (III-8)$$
$$h \approx N\sigma a.$$

Thus, the wetting film on a grafted layer is much thinner in comparison to a grafted layer immersed in a solvent bath $(\approx N\sigma^{1/3}a)^{11}$, or to a wetting film at a pure solvent on the bare surface.

Adsorbed or grafted polymers may be considered as laterally immobile. It is then possible to consider the wetting of a non uniformly coated surface. The spreading of a wetting film on such a surface may present interesting behavior[3]. Consider for example a surface consisting of a bare circular patch surrounded by a coated surface. Imagine an adiabatically growing film whose center coincides with that of the bare patch. When the contact line overtakes the patch boundary the wetting film is transformed into a "two level film". Simultaneously, the film thickness at the central zone undergoes an abrupt change.

SUMMARY

Polymers may affect wetting films in a wide variety of ways. Take a thick ($\sim 10^2 \overset{\circ}{A}$) film of a pure nonvolatile solvent on a bare surface as a reference system. For the cases considered, polymers may thicken the film, thin it, or cause it to phase separate into a "two level film" (thick, polymer rich film and a thin, polymer poor film). The final outcome depends on the method used to introduce the polymers into the system. Two factors play an important role:
(1) The initial polymer area fraction, i.e. whether the separation between the coils is much larger than the healing length, λ.
(2) The lateral mobility of the coil is whether the polymers are attached to the surface.

When the separation between the coils is much smaller than λ, the film envelope is essentially flat. The film thickness is then determined by the balance between S and the net long range interaction $P_t(h)$. In the three cases considered, the polymers reinforce $P_t(h)$ so as to favor thick films. The effect on S is the source of differences between the cases. For immobile polymers (grafted or absorbed) S increases because of the dissolution free energy of the polymer coating. On the other hand, mobile polymers on a nonadsorbing surface lower S because of the depletion layers. Thus, in the first case we expect thinner films while in the second thicker films are expected.

Dilute mobile coils (where the initial coil-coil separation is larger than λ) may destabilize the film, leading to the formation of a "two level film". This is due to polymer induced film deformations leading to long range attractive interactions between coils. Note that the resulting interactions mediate between different coils but not between monomers belonging to a single chain. Accordingly, the resulting phase separation is not associated with polymer collapse.

ACKNOWLEDGEMENTS

The autor is grateful for instructive discussions with S. Alexander, M. Boudoussier, F.Brochard, P.G. de Gennes and P. Pincus. The Fritz Haber Research Center is supported by the Minerva Gesellschaft für die Forschung, mbH, Munich, FRG.

REFERENCES

(1) A. Halperin, P. Pincus, and S. Alexander, J. Physique Lett 46:L-543 (1985).

(2) a) P.G. de Gennes, C.R. Hebd. Sean.Acad. Sci. 300II:29 (1985).

b) M. Boudoussier, to be published.

(3) a) P.G. de Gennes, lecture at the College de france (1985).

b) A. Halperin, and P.G. de Gennes, J. Physique 47:1243 (1986).

(4) J.F. Joanny, and P.G. de Gennes, C.R. Hebd. sean. Acad. Sci. 299II:279 (1984).

(5) P.G. de Gennes, rev. Mod. Phys. 57:827 (1985).

(6) J.S. Rowlinson, and B. Widom, "Molecular Theory of Capillarity", Oxford University Press, Oxford (1982).

(7) L. Leger, this proceedings.

(8) W. Zisman, in: "Contact Angle, Wettability and Adhesion", F.M. Fowkes, ed., American Chemical Society, Washington D.C. (1964).

(9) P.G. de Gennes, "Scaling Concepts in Polymer Physics", Cornell University Press, Ithaca (1979).

(10) M. Daoud, and P.G. de Gennes, J. Physique 38:85 (1976).

(11) P.G. de Gennes, Macromolecules 15:492 (1982).

(12) a) S. Alexander, J. Physique 38:983 (1977).

b) P.G. de Gennes, Macromolecules 13:1075 (1980).

PHASE EQUILIBRIA OF FLUIDS IN NARROW PORES

R. Evans and U. Marini Bettolo Marconi

H.H. Wills Physics Laboratory
University of Bristol, Bristol BS8 1TL, U.K.

Despite its practical importance the nature of the phase equilibria of fluids confined in narrow pores remains poorly understood. Much experimental effort is focussed on gas adsorption measurements in materials such as vycor glass or carbon powders. Such measurements can, in principle, yield important information concerning the character of the confined fluid. However, the theory of adsorption in mesoporous solids is still in its infancy. While many of the difficulties in real materials are associated with the complex connectivity of the pores and the fact that these are not of a uniform size, even the idealized case of a single, infinitely long pore has received relatively little attention from theoreticians. Recently we have embarked upon a detailed study of phase equilibria, adsorption and related phenomena of simple fluids confined in slit-like and cylindrical capillaries which act as models of a single pore. Our approach is based on a mean-field, density functional theory for inhomogeneous fluids that treats the walls of the capillary as exerting an external potential on the fluid molecules. This type of approach has already proved successful for a variety of other interfacial problems, providing much physical insight into wetting transitions, contact angles and adsorption isotherms for both solid-fluid and fluid-fluid interfaces.

The lecture summarizes some of our results for capillary condensation, the phenomenon whereby an undersaturated 'gas' condenses to a dense 'liquid' in a pore. Corrections to the macroscopic Kelvin equation for the condensation pressure have been ascertained for both complete and partial wetting. For certain models complete capillary phase diagrams have been obtained in terms of the variables μ (chemical potential) T (temperature) and H (pore size). The diagrams exhibit rich structure, characterized by a condensation surface terminating in a line of capillary critical points. In some cases this surface intersects a prewetting surface, on which thick and thin liquid films coexist, in a line of triple points. The implications of our results for adsorption in real porous media and for measurements of the forces between solid mica surfaces immersed in fluids are discussed.

Details of the theory, calculations and results are presented in a series of publications listed below. References 2 and 4 contain brief introductions to the subject.

REFERENCES

1. R. Evans and U. Marini Bettolo Marconi, 'The Role of Wetting Films in Capillary Condensation and Rise: Influence of Long Ranged Forces', Chem.Phys.Lett., **114**, 415 (1985).
2. R. Evans, U. Marini Bettolo Marconi and P. Tarazona, 'Fluids in Narrow Pores: Adsorption, Capillary Condensation and Critical Points', J.Chem.Phys., **84**, 2376 (1986)
3. R. Evans and U. Marini Bettolo Marconi, 'Capillary Condensation versus Pre-wetting', Phys.Rev.A., **32**, 3817 (1985)
4. R. Evans, U. Marini Bettolo Marconi and P. Tarazona, 'Capillary Condensation and Adsorption in Cylindrical and Slit-like Pores', Faraday Symp. of Chemical Society No. 20 (to appear)
5. E. Bruno, U. Marini Bettolo Marconi and R. Evans, 'Phase Transitions in a Confined Lattice Gas: Pre-Wetting and Capillary Condensation', Physica A (to appear)
6. P. Tarazona, U. Marini Bettolo Marconi and R. Evans, 'Phase Equilibria of Fluid Interfaces and Confined Fluids: Non-local versus Local Density Functionals', Molec.Phys. (to appear)
7. R. Evans and U. Marini Bettolo Marconi, 'Phase Equilibria and Solvation Forces for Fluids Confined between Parallel Walls', J.Chem.Phys, (submitted)

ORIENTATIONAL ORDER AND SURFACE PHASE TRANSITIONS

M.M. Telo da Gama

Departamento de Física e CFMC/INIC
Faculdade de Ciências da Universidade de Lisboa
1700 Lisboa, Portugal

In this talk we review surface induced orientational order in liquid crystals and in mixtures containing one amphiphilic component. We discuss, in particular, the possibility of orientational surface phase transitions in these systems and we establish the connection between the latter and the wetting transitions that occur in simpler systems[1]. Finally we present some results of calculations based on a microscopic mean field theory for model liquid crystals[2-5] and ternary oil-water-surfactant systems[5-8].

1. SURFACE INDUCED ORIENTATIONAL ORDER

Fluids of anisotropic molecules range in complexity from almost spherically symmetric fluids of diatomic molecules such as N_2 to liquid crystals and oil-water-surfactant mixtures. While the former are isotropic over the whole liquid range, liquid crystals[9] and surfactant mixtures[10-12] exhibit a rich variety of bulk liquid phases some of which are orientationally ordered. Examples of the latter are the nematic and smectic phases of liquid crystalline materials and the lamellar phases of sufactant mixtures. In these ordered phases (mesophases) the rotational invariance of the isotropic liquid is spontaneously broken, i.e. the molecules are preferentially aligned along a direction in space. (In addition in the smectic and lamellar phases translational invarieance is broken along one spatial direction and the systems exhibit one-dimensional positional order). In the absence of external fields or boundaries the ordered phases are degenerate since the free energy is independent of the direction of alignment of the molecules. This degeneracy is usually broken in systems with two (or more) coexisting phases e.g. in systems with a liquid-vapour or a liquid-liquid interface.

Let us digress for a while and consider a simple molecular fluid, say N_2. In the bulk, the N_2 molecules are randomly oriented. At a liquid--vapour interface, however, the translational symmetry is broken and this induces a preferred orientation of the molecules in the interfacial region, since the rotational and translational degrees of freedom in the partition function are usually coupled. For a simple molecular fluid the rotational symmetry is broken only in the direction of the density gradient, i.e. the system exhibits uniaxial symmetry (see fig.1). For a system of rod-like molecules such as N_2 the preferred molecular orientation is perpendicular to the interface on the liquid side and parallel on the vapour side of the interfacial region.[13] For N_2 this preference is weak i.e. only a few % of the molecules are aligned[13] (see fig.1).

At the nematic-vapour or nematic-isotropic interfaces on the other hand, the degree of alignment can be very large (more than 50%). Furthermore, the bulk nematic phase is also ordered and the direction of alignment in the bulk is determined by the alignment of the molecules in the interfacial region, ie. the degeneracy of the bulk phase is broken. In the simplest case the molecules are perpendicular to the interface which has then uniaxial symmetry. This occurs, for example, for the liquid crystal 5CB[14] (see fig.2).

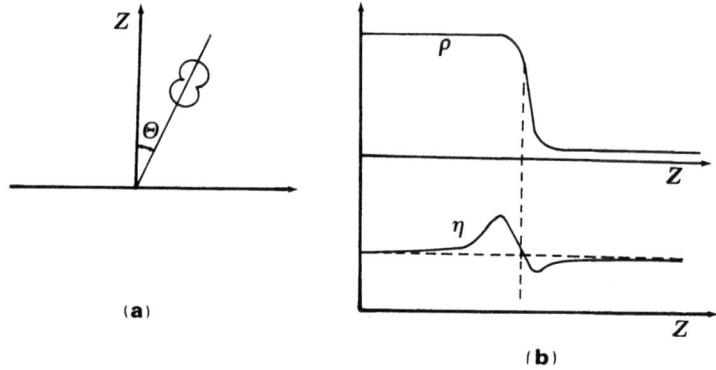

Fig.1 Schematic density (ρ) and orientational order parameter (η) profiles at the liquid-vapour interface of a molecular fluid of rod-like molecules such as N_2. The molecules in the interfacial region are preferentially aligned along a direction that makes an angle θ (a) with the normal. For N_2, θ is zero in the liquid side (positive η) and $\pi/2$ in the vapour side (negative η) of the interface (b). The orientational order parameter η is the weighted angular average of the orientational distribution function $f(z,\theta)$ (see text) which gives the fraction of molecules at height z with orientation θ.

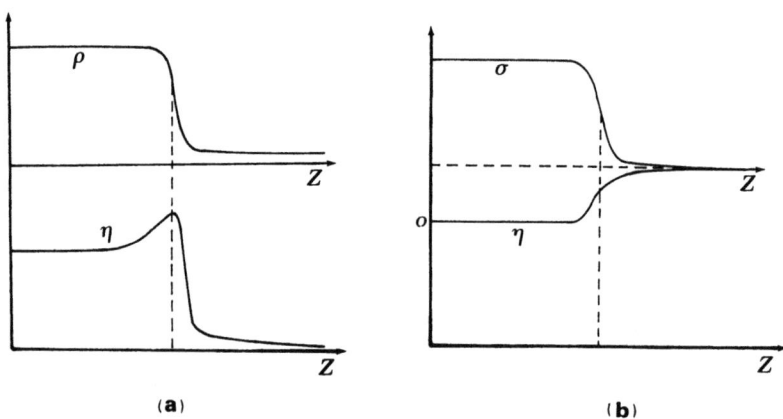

Fig.2 Schematic density (ρ) an orientational order parameter (η and σ) profiles at the nematic-vapour interface of a liquid crystal. In (a) the preferential alignment is perpendicular while in (b) it is parallel to the interface. Consequentll in (b) two order parameters (η and σ) are required to describe the orientation of the system. At the interface in (a) there is excess surface order, while in (b) there is not.

In general the molecules may be aligned at an angle $\theta \neq 0$ (the tilt angle) to the normal. Indeed, this seems to be always the case at the nematic-isotropic interface.[15] In such systems the interface is biaxial and we need more than one order parameter to describe its orientation[2] (see fig.2).

Surface induced order plays also an important role in oil and water surfactant mixtures. At low surfactant compositions these are isotropic liquid mixtures, but they exhibit strong orientational ordering at the liquid-vapour (eg. water-air) or at the liquid-liquid (eg. water-oil) interface. At these interfaces the surfactant molecules are strongly adsorbed and are oriented with their polar heads into the water rich phase (see fig.3). The interface is rotationally invariant in the xy plane and the orientational order parameter is odd (i.e. it changes sign) with respect to an inversion through the origin. (This should be contrasted with the liquid-crystal case where the order parameters are invariant under inversions). The odd parity of the surfactant orientational order parameter reflects the amphiphibic nature of the surfactant molecules and distinguishes them from the liquid crystals mentioned above.

Both liquid crystals and surfactant mixtures undergo a variety of phase transitions in the liquid range (eg. nematic-isotropic) when the thermodynamic fields (eg. temperature) are varied. Typical phase diagrams for a liquid crystalline material are shown in fig.4. In the next section we discuss what happens to the interface discussed in this section when one of the bulk phases undergoes a phase transition, ie. under conditions of three phase equilibrium.

2. INTERFACIAL PROPERTIES UNDER CONDITIONS OF THREE PHASE EQUILIBRIUM

Let us start with a liquid crystal in the isotropic phase at the vapour pressure. At temperatures much higher than the transition temperature to the ordered nematic or smectic phase at the vapour pressure (ie. T_{NI} in fig.4a or T_{SI} in fig.4b) the interface may look somewhat like that in fig 1.

However as T_{NI} or T_{SI} is approached from above the interfacial order may increase since at that temperature the ordered phase becomes a stable bulk

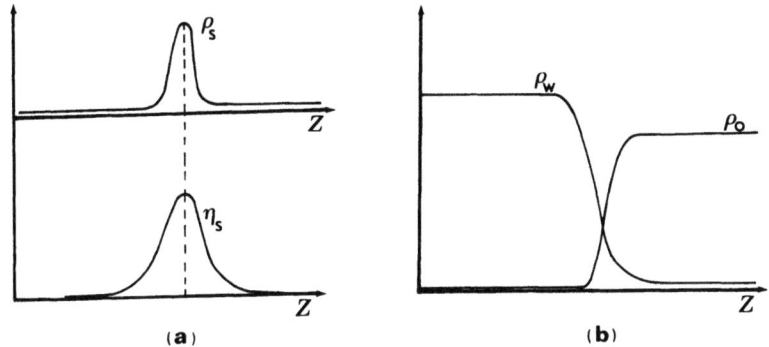

Fig.3 Schematic density (ρ_s, ρ_w, ρ_o) and orientational order parameter (η_s) profiles at the oil-water interface of a ternary surfactant mixture. The surfactant molecules are strongly adsorbed and oriented at the interface (a) while the bulk phases are almost pure water and oil (b).

phase and it may wet or partially wet the isotropic-vapour interface[1]. If complete wetting is obtained at T_{NI} or T_{SI} the ordered layer at the interface will reach macroscopic dimensions and the interface undergoes a triple point wetting transition [1]. This has indeed been observed for 5CB[16] (at the nematic-isotropic-vapour triple point) while partial wetting was observed at the liquid-vapour interface of 12CB[17] (at the Smectic A-isotropic-vapour triple point).

More generally at the triple point T_{OI} one ordered phase (smectic or nematic) coexists with the isotropic liquid and with vapour. At that temperature we can define three different surface tensions (one for each pair of coexisting phases) and the stability of the interface requires that[1,2]

$$|\Delta\sigma| = |\sigma_{OV} - \sigma_{IV}| \leq \sigma_{OI} \qquad (1)$$

ie. that the difference between the two largest tensions, which in this case are those of the liquid-vapour interfaces σ_{OV} and σ_{IV}, be less than the tension of the liquid-liquid interface σ_{OI}.

Two distinct situations can arise: (i) $\sigma_{OV} > \sigma_{IV}$ and (ii) $\sigma_{OV} < \sigma_{IV}$. When (ii) holds, the inequality (1) becomes,

$$\sigma_{IV} \leq \sigma_{OV} + \sigma_{OI} \qquad (2)$$

and the isotropic-vapour interface is wet (when $\sigma_{IV} = \sigma_{OV} + \sigma_{OI}$) or partially wet (when $\sigma_{IV} < \sigma_{OV} + \sigma_{OI}$) by the ordered phase. The first case corresponds to the liquid-vapour interface of 5CB mentioned above where the ordered layer diverges as T_{OI} is approached from above while the second corresponds to the situation found for 12CB where the thickness of the ordered layer remains finite at T_{OI}. On the other hand when (i) holds the inequality (1) can be written,

$$\sigma_{OV} \leq \sigma_{IV} + \sigma_{OI} \qquad (3)$$

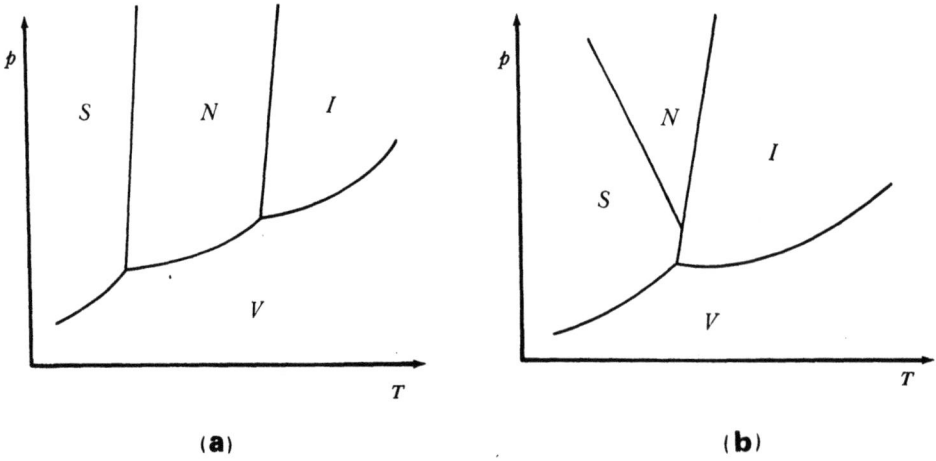

Fig.4 Typical bulk fluid diagrams of liquid crystalline materials, exhibiting smectic (S), nematic (N), isotropic liquid (I) and vapour (V) phases. In (a) there are NIV and SNV triple points, while in (b) the triple points are SNI and SIV.

and the smectic or nematic vapour interface is wet (=) or partially wet (<) by the isotropic liquid. Some liquid crystals are known to have $\sigma_{OV} > \sigma_{IV}$ and at least for model calculations complete wetting by the isotropic phase can also occur[2,18].

We note, in passing, that eq.(1) sets an upper bound for the surface tension gap $\Delta\sigma$ at the transition temperature[2]. The ternary surfactant mixtures considered in section 1 may also exhibit a line of three liquid phase equilibria for overall surfactant compositions of the order of a few percent. The nature of the third liquid phase which is in equilibrium with the water and oil rich phases, and the structure of the oil-water interface on the triple line are the subject of intense experimental and theoretical work[10-12, 19-21]. For systems with good surfactants the microemulsion middle phase (rich in oil and water) does not wet the oil-water interface[19] while it seems to do so in systems with weaker surfactants[20]. Some of these effects are reproduced in model calculations of ternary mixtures with an amphiphibic component[21,5-8].

3. RESULTS FOR MODEL LIQUID CRYSTALS AND SURFACTANT MIXTURES

In this section we summarize very briefly some of the results of our calculations[2-8]. In the last few years we have developed and used a microscopic mean field theory for non-uniform molecular fluids[2] to calculate the interfacial properties of very simple models of nematic liquid crystals[2-5] and of ternary surfactant mixtures. The approximate grand potential free energy is the minimum of the functional[2]

$$\Omega[\{\rho_i\}] = \int d\underline{r} f_0(\underline{r}) + 1/2 \sum_{ij} \int\int d\underline{r} d\underline{r}' d\omega d\omega' \rho_i(\underline{r},\omega)\rho_j(\underline{r}',\omega') \times$$
$$\phi_{ij}(\underline{r},\underline{r}',\omega,\omega') - \sum_i \mu_i \int d\underline{r} d\omega \rho_i(\underline{r},\omega) \quad (4)$$

where $\{\rho_i\}$ is the set of density orientational profiles, μ_i is the bulk chemical potential of species i and ϕ_{ij} the long-ranged part of the intermolecular potential between species i and j. The short ranged repulsive interactions are treated in a local approximation and $f_0(\underline{r})$ is the free energy density of a mixture interacting with those potentials evaluated at the local centre of mass densities, $\rho_i(\underline{r}) = \int d\omega \rho_i(r,\omega)$ and with orientation $f_i(r,\omega) = \rho_i(r,\omega)/\rho_i(\underline{r})$.

The interfacial properties-ie. the structure and thermodynamics of the systems characterized by the potentials ϕ_{ij} are obtained by minimizing Ω with respect to $\{\rho_i(\underline{r},\omega)\}$. (see refs. 2 and 5 for details)
In the simplest case a nematic liquid crystal was modelled by the potential[2]

$$\phi(\underline{r},\omega_1,\omega_2) = -A(\sigma/r)^6 \{1 + B/A\, P_2(\omega_1,\omega_2)\}, \quad r > \sigma \quad (5)$$

where A and B are positive energy parameters, σ is a molecular length scale and P_2 is the second order Legendre polynomial. The bulk phase diagram of this fluid has a NIV triple point (cf with fig. 4a) and at T_{NI} the NV interface is wet by the isotropic phase (see fig.5) in agreement with a recent computer simulation[18] for a very similar model. (see ref 2 for details)
A ternary surfactant mixture (components 1,2,3) was also modelled by the isotropic potentials,

$$\phi_{ij}(r) = -(\varepsilon_0 + \delta_{ij}\varepsilon_i)\,(\sigma/r)^6 \quad i,j = 1,2,3, \quad r > \sigma \quad (6)$$

where ε_0 and ε_i are energy parameters and δ_{ij} is the Kronecker delta and by additional anisotropic interactions between species 3 (the surfactant) and species 1 and 2,

$$\phi'_{i3}(\underline{r},\omega_3) = (-1)^i \varepsilon'(\sigma/r)^6 P_1(\underline{r},\omega_3) \quad i = 1,2, \quad r > \sigma \qquad (7)$$

where ε' is the strength of the anisotropic interactions and P_1 is the first order Legendre polynomial. The bulk phase diagrams of these mixtures are extremely complicated[5-8]. Here it suffices to say that there are regions of two liquid (oil-water) phase equilibrium which, at constant temperature and pressure, end at a triple point as the surfactant compostion is increased.

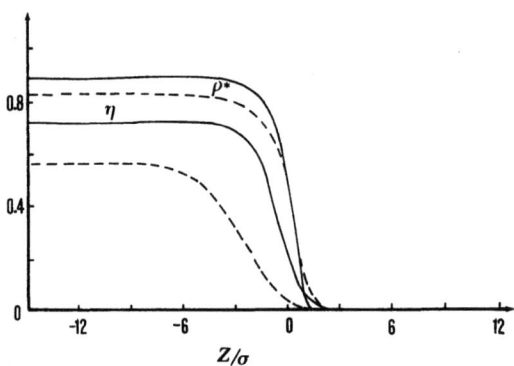

Fig.5 Density $\rho^* = \rho\sigma^3$ and orientational order parameter η profiles at the nematic vapour interface for a system with B/A = 0.3 at $T \ll T_{NI}$ (full line) and $T \lesssim T_{NI}$ (dashed line). As $T \Rightarrow T_{NI}$ a disordered liquid layer appears at the interface. At $T = T_{NI}$ the NV interface is wet by the isotropic liquid phase.

Wetting of the oil-water interface by this third (middle) phase is favored by weak anisotropy of the surfactant molecules and high temperatures in agreement with experiment (see fig.6) (for details see refs 5-8)

In conclusion, although a lot of work still remains to be done, we can begin to understand some of the rich interfacial behaviour exhibited by complex fluids such as liquid crystals and surfactant mixtures.

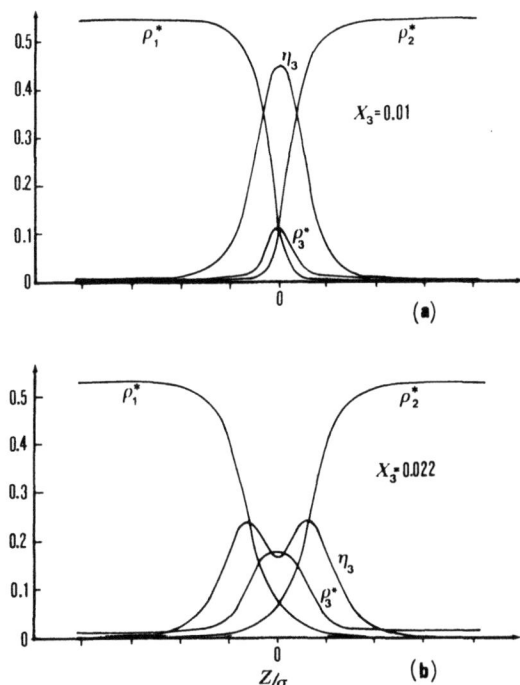

Fig.6 Density ρ^* and orientational order parameter η_3 profiles at the liquid 1-liquid 2 interface of a ternary surfactant mixture at fixed temperature and pressure for two bulk compositions of the surfactant, $x_3 = 0.01$ (a) and $x_3 = 0.022$ (the triple point composition) (b). As x_3 increases the concentration of 3 at the interface increases and the orientational order decreases. In (b) the system is close to a wetting transition and an incipient layer of the middle phase in the interfacial region can already be seen.

References

1. D.E. Sullivan and M.M. Telo da Gama, in Fluid Interfacial Phenomena, ed. C.A. Croxton, Ch 2 (Wiley, London 1986)
2. M.M. Telo da Gama, Mol. Phys., $\underline{52}$, 585 (1984)
3. M.M. Telo da Gama, Mol. Phys., $\underline{52}$, 611 (1984)
4. J.H. Thurtell, M.M. Telo da Gama and K.E. Gubbins Mol. Phys., $\underline{54}$, 321 (1985)
5. M.M. Telo da Gama and J.H. Thurtell, Faraday Symp. $\underline{20}$,000 (1985)
6. M.M. Telo da Gama and K.E. Gubbins, Proceedings of the 4th International Chemical Engineering Conference, Coimbra, Portugal (1985)
7. M.M. Telo da Gama and K.E. Gubbins, Mol. Phys., $\underline{58}$,000 (1986)
8. M.M. Telo da Gama, Langmuir, in press.
9. D.L. Johnson, J. Chim. Phys., $\underline{80}$ 45 (1983), see also M. Papoular, Sol. State Com., $\underline{7}$, 1961, (1969)
10. A.M. Bellocq, J. Biais, P. Botherel, B. Clin, G. Fourche, P. Lalanne, B. Lemaire, B. Lemanceau and D. Roux, Adv. Colloid Interf. Sci, $\underline{20}$, 167 (1984)
11. Micellization, Solubilization and Microemulsions, ed. K.L. Mittal (Plenum, New York 1984)
12. Surfactants in Solution, eds. K. Mittal and B. Lindman (Plenum, New York 1977)
13. K.E. Gubbins in Fluid Interfacial Phenomena ed. C.A. Croxton, Ch 10, (Wiley, London 1986)
14. M.G. Gannon and T.E. Faber, Phil. Mag. A, $\underline{37}$ 117 (1978)
15. P. Chiarelli, S. Faetti and L. Fronzoni, J.Physique, $\underline{44}$, 1061 (1983); Lett. Nuovo Cimento, $\underline{36}$, 60 (1983); Phys. Lett; $\underline{101A}$, 31 (1984); S. Faetti and V.Palleschi, J. Physique Lett., $\underline{45}$, L313 (1984); Phys. Rev. A, $\underline{30}$, 3241 (1984)
16. D. Beaglehole, Mol. Cryst. Liq. Cryst., $\underline{89}$, 319 (1982)
17. B.M. Ocko, A. Braslau, P.S. Pershan, J. Als-Nielsen and M. Deutch, Phys. Rev. Lett., $\underline{57}$, 94 (1986)
18. G.R. Luckhurst, T.J. Sluckin and H.B. Zewdie, preprint
19. A. Pouchelon, J. Meunier, D. Langevin, D. Chatenay and A.M. Cazabat, Chem. Phys Lett, $\underline{76}$, 277 (1980); A. Pouchelon, D. Chatenay, J. Meunier and D. Langevin, J. Colloid Interf. Sci, $\underline{82}$, 418 (1981)
20. A.M.Bellocq, D. Bourbon and B. Lemanceau, J. Disp. Sci. Tech. 2, 27 (1981); J.C. Lang, P.K. Lim and B. Widom, J. Phys. Chem. $\underline{20}$ 1719 (1976)
21. B. Widom, J. Phys. Chem, $\underline{88}$ 6508 (1984); J. Chem Phys. $\underline{81}$ 1030 (1984); C. Borzi, R. Lipowsky and B. Widom, Faraday Symp. $\underline{20}$,000 (1985); B.Widom, J. Chem. Phys., $\underline{84}$ 6943 (1986)

WETTING AND ORIENTATIONAL TRANSITIONS

IN NEMATIC LIQUID CRYSTALS

D.E. Sullivan and A.K. Sen[*]

Department of Physics and Guelph-Waterloo
Program for Graduate Work in Physics
University of Guelph
Guelph, Ontario N1G 2W1

INTRODUCTION

This paper discusses some recent developments in the theory of the interfacial properties of nematic liquid crystals. Our focus will be on the relation between orientational ordering and wetting at a nematic-wall interface. We shall base our presentation on the Landau-de Gennes[1,2] theory for non-uniform nematics. Earlier applications of that theory to nematic interfacial structure and wetting, which comprise an important background to the present discussion, have been reviewed recently by Sluckin and Poniewierski[2]. An alternative approach to the same topic is provided by the molecular mean-field theory described by Telo da Gama at this symposium (see also refs. 3 and 4). Due to its explicit treatment of microscopic interactions and inclusion of coupling between orientational and density order parameters, the latter is a richer and basically more realistic approach. The much simpler structure of Landau-de Gennes theory, which renders it more susceptible to analytic treatment, justifies this account. A more detailed presentation of various technical aspects of this work will be published elsewhere[5].

STRUCTURE OF THE NEMATIC-WALL INTERFACE

It may be worthwhile to first clarify a few concepts pertaining to the interfacial behavior of nematics. As is well known, a <u>bulk</u> nematic phase is characterized by spontaneous orientational ordering of elongated molecules, such that on average the molecular axes are aligned in a direction n_b, the "director". The degree of alignment is traditionally indicated by an order parameter η_b defined as $\eta_b = <P_2(e \cdot n_b)>$, where P_2 is the second Legendre polynomial and the brackets denote a thermal average over all orientations of the molecular axis e. In a completely uniform phase, one for which n_b and η_b are constants, and in the absence of external fields coupling to n_b, the free energy is invariant under rotations of the director.

---

[*] Present address: Dept. of Mechanical and Aerospace Engineering, North Carolina State University, Raleigh, N.C. 27695

Consider now the interface between the nematic and, say, a solid substrate, which we shall idealize as a smooth ("unrubbed") planar wall. The following discussion is unchanged in essence if the "substrate" is replaced by an immiscible liquid phase or vapor. The structure of the interfacial region will normally differ from that in the bulk liquid over some characteristic scale ξ, typically on the order of several molecular lengths. Over this region, one can generally expect both the local mean direction of alignment n(z) and degree of ordering, as well as the number density, to vary with height z above the wall, as illustrated schematically in Fig. 1. In contrast to the bulk nematic, where the probability distribution of molecular axes is cylindrically symmetric about the director n_b and the single orientational order parameter η_b suffices, the interface is characterized by <u>two</u> distinct axes of ordering, namely n(z) and the interface normal vector k. If these do not coincide, the distribution of molecular axes cannot be cylindrically symmetric about either axis and the interface becomes <u>biaxial</u>[2-6]. The appropriate generalization of η_b is the tensor defined by

$$Q(z) = \frac{1}{2} \langle (3ee - I) \rangle , \qquad (1)$$

which, by tracelessness and symmetry, can always be represented as

$$Q(z) = \frac{\eta(z)}{2} [3n(z)n(z) - I] + \frac{\sqrt{3}}{2} \mu(z)[\ell(z)\ell(z) - m(z)m(z)] , \qquad (2)$$

where I is the unit tensor and the axes n, ℓ, m form a local orthonormal triad. In general, Q(z) is specified by five variables, namely $\eta(z)$, $\mu(z)$, and the direction cosines of its axes relative to the laboratory frame. With the assumptions that the order parameters vary only in the normal direction z and the director n(z) remains always in one plane, say the xz plane, only a single angular variable $\psi(z) = \cos^{-1}[k \cdot n(z)]$ is required.

Under the convention that n(z) is the direction of maximum orientational ordering, it follows from (1) and (2) that the order parameters must satisfy $\eta(z) \geq |\mu(z)|/\sqrt{3}$. We see that $\mu(z)$ measures the non-equivalence of two ordering axes perpendicular to n(z), hence the degree of biaxiality. This necessarily vanishes in the bulk liquid at $z \gg \xi$, while n(z) and $\eta(z)$ approach the bulk values n_b and η_b, respectively.

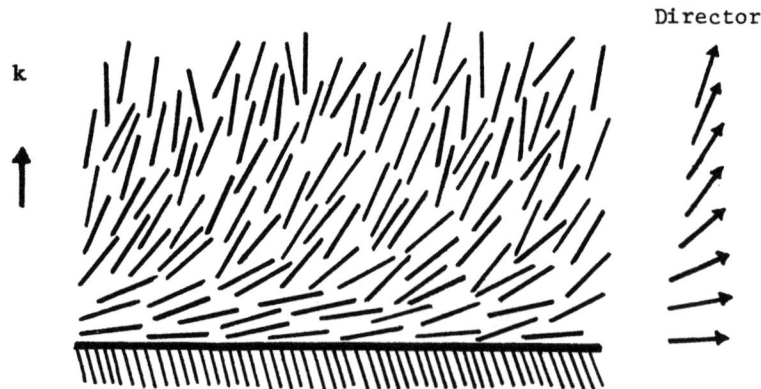

Fig. 1 Schematic configuration of molecules in a nematic liquid near a substrate.

The picture in Fig. 1 has been drawn for some arbitrary fixed orientation of the bulk director n_b, making an angle ψ_b with the k axis. One could imagine varying ψ_b, e.g., by momentarily applying an external field which couples to n_b. It is naturally expected that such a change in ψ_b changes the whole microscopic structure of the interfacial region, i.e., altering the form of the profiles $\eta(z)$, $\mu(z)$, and $\psi(z)$. Those changes will be reflected in a modified value of the excess free energy per unit area associated with the interface, denoted σ. (Recall that due to rotational invariance, there will be no change in the bulk free energy of the liquid.) Hence we can regard $\sigma = \sigma(\psi_b)$ to be a function of the bulk "tilt angle" ψ_b. Now, in the absence of any constraint on n_b due to external fields, or other walls, the system will spontaneously seek that physical state which minimizes $\sigma(\psi_b)$, thus determining the equilibrium values of the interfacial free energy and bulk alignment.

The alignment induced by nematic interfaces has, of course, been the object of many experimental[7,8] and theoretical[2-4,9] studies, although our level of understanding of the phenomenon remains low. Let us emphasize that it is the <u>bulk</u> tilt angle ψ_b which usually has been deduced from measurements. As indicated in Fig. 1, the mean orientation in the interface can in principle deviate from the limiting bulk value, but the microscopic extent of that region yields a negligible contribution when the usual optical probes are employed. Many studies have also been devoted to the angle dependence of the free energy $\sigma(\psi_b)$, the so-called "anchoring energy" in liquid-crystal parlance, by applying suitable constraints to the bulk director. In most such cases, n_b is not actually constant but rather varies spatially on some macroscopic length scale, which can be taken into account by standard elastic continuum theory[10]. Due to the large separation in length scales, it is entirely reasonable to suppose that $\sigma(\psi_b)$ in those circumstances can be identified with the interfacial free energy in the absence of any bulk spatial variation, at some appropriate "matching" value of ψ_b, but this deserves some further clarification.

LANDAU-de GENNES EQUATIONS

The Landau-de Gennes theory is based on the following expression for the interfacial free energy as a <u>functional</u> of the order-parameter tensor $Q(z)$ (restricted to one-dimensional spatial variation):[2,11,12]

$$\sigma = \int_0^\infty dz \left\{ f_G[\dot{Q}(z)] + f_L[Q(z)] - f_L[Q_b] \right\} + f_s[Q(0)] , \qquad (3)$$

where $\dot{Q}(z) \equiv dQ/dz$ and Q_b is the order-parameter tensor in the bulk liquid. The various quantities appearing in (3) are discussed below.

The function $f_L(Q)$ is the free-energy density of a <u>uniform</u> liquid with given order parameter Q, usually represented by the small-Q expansion

$$f_L(Q) = A \, TrQ^2 - B \, TrQ^3 + C[TrQ^2]^2 , \qquad (4)$$

where B and C are constants while A is linearly related to temperature. As is well known, the cubic term in $f_L(Q)$ accounts for the first-order character of the bulk nematic-isotropic transition. The symbol Tr in (4) denotes the trace operation and thus ensures the rotational invariance of the bulk free energy. When Q and its components η and μ, cf.(2), are expressed in units of $B/(6C)$, and f_L in units of $(B^2/24)^2/C^3$, (4) becomes

$$f_L = t(\eta^2+\mu^2) - 2\eta(\eta^2-3\mu^2) + (\eta^2+\mu^2)^2 , \qquad (5)$$

where $t = 24AC/B^2$ is a dimensionless temperature. This expression is identical to that describing the free-energy density of the 3-state Potts model in Landau approximation[13]. From the analysis of that model, one finds that f_L displays 3-fold symmetry in the (η,μ)-plane, with equivalent local minima representing the ordered (i.e., nematic) phase at points $(\eta_c,0)$ and $(-\eta_c/2, \pm \sqrt{3}\eta_c/2)$ for all $t \leq 9/8$, where

$$\eta_c = \frac{3}{4} [1 + \sqrt{1-8t/9}] \quad . \tag{6}$$

A local minimum representing the isotropic phase occurs at $(\eta,\mu) = (0,0)$ for all $t \geq 0$. The first-order transition, where the nematic and isotropic free energies become equal, is at $t = 1$.

Due to the continuous rotational symmetry of the model, only one of the three nematic minima of f_L has any physical significance. Under our earlier convention that n corresponds to the direction of maximum ordering, and the ensuing inequality $\eta \geq |\mu|/\sqrt{3}$, the relevant minimum is that at $(\eta,\mu) = (\eta_c,0)$. As anticipated, the stable nematic state in bulk is uniaxial.

The function $f_G(\dot{Q})$ gives the contribution to the local free-energy density due to gradients of Q. To quadratic order in Q, the most general expression for f_G allowed by symmetry is

$$f_G(\dot{Q}) = \frac{L_1}{2} \text{Tr}\dot{Q}^2 + \frac{L_2}{2} k\cdot\dot{Q}^2\cdot k \tag{7}$$

(which generalizes straightforwardly[1] when Q has gradients in arbitrary directions). This expression depends on two "elastic constants" L_1 and L_2, which can be related[14] to the Frank elastic constants K_i characterizing long-wavelength director distortions in a bulk nematic. Equn. (7) predicts the equality of the splay (K_1) and bend (K_3) elastic constants, contrary to both observation and theory, which is due to the neglect of order parameters beyond those describing averages of second-order spherical harmonics, cf.(1).[15] As discussed in ref. (5), these considerations suggest that "relevant" values of the elastic constants obey $L_2 > L_1 > 0$, with L_2/L_1 occurring in the range 2 to 4.

The remaining term $f_s[Q(0)]$ in (3) is a specific surface energy which represents several possible effects -- direct interactions between the substrate and adjacent liquid molecules, "broken bonds", i.e., reduction in the number of nearest-neighbor pairs for a molecule in contact with the substrate, and substrate-mediated "surface enhancement" of the interaction between nearest-neighbor pairs -- all of which are familiar from other contexts, e.g., at surfaces in Ising models.[16] An additional consideration arises in the nematic case, concerning the rotational symmetry of $f_s[Q(0)]$. As mentioned earlier, we assume the substrate surface to be smooth (i.e., without "grooves"), which implies that it does not single out a specific tangential ordering axis. This is not inconsistent with the occurrence of spontaneous nematic alignment along an axis having a definite tangential projection, such as shown in Fig. 1. The assumption of tangentially isotropic substrate symmetry means that $f_s(Q)$ must be invariant to rotations about the surface normal vector k. To lowest "relevant" (quadratic) order, a phenomenological model for $f_s(Q)$ with these properties is[2]

$$f_s(Q) = c_1 k\cdot Q\cdot k + c_2 \text{Tr}Q^2 + c_3 (k\cdot Q\cdot k)^2 + c_4 k\cdot Q^2\cdot k \quad . \tag{8}$$

An approximate derivation of (8) from a molecular mean-field theory[3,4] has been described in ref. (5). Due to the various possible effects represented by the constants $c_1 - c_4$, these can be considered as essentially independent of the other material constants appearing in (4) and (7).

One can evaluate $f_G(\dot{Q})$ and $f_S(Q)$ using the principle-axis representation of Q in (2). Although that is useful for exhibiting some aspects of the angular dependence of the free energy[5], the Landau-de Gennes theory is more readily analyzed when Q is expressed in terms of its components in a lab-fixed frame. Letting i and j(= m) denote unit vectors along the x and y axes, respectively (recall that the x-direction is <u>defined</u> to coincide with the projection, if any, of the director onto the surface plane), we now have

$$Q(z) = \frac{\eta_s(z)}{2}(3kk - I) + \frac{\sqrt{3}}{2}\mu_s(z)(ii - jj)$$

$$+ \frac{\sqrt{3}}{2}\nu_s(z)(ik + ki) .\qquad(9)$$

On comparing (9) with (2), the order parameters η_s, μ_s, and ν_s can be related to the previous variables η, μ, and ψ. For example, it follows that the tilt angle is given by

$$\tan 2\psi(z) = \frac{2\nu_s(z)}{\sqrt{3}\eta_s(z) - \mu_s(z)} .\qquad(10)$$

Using (9) in (7) and (8), we find that f_G and f_S have the diagonal forms:

$$f_G(\dot{Q}) = \frac{1}{2}\left\{\left(\frac{d\eta_s}{dz}\right)^2 + L\left(\frac{d\mu_s}{dz}\right)^2 + M\left(\frac{d\nu_s}{dz}\right)^2\right\} ,\qquad(11a)$$

$$L = \frac{L_1}{L_1 + 2L_2/3} , \quad M = \frac{L_1 + L_2/2}{L_1 + 2L_2/3} ,\qquad(11b)$$

$$f_S(Q) = c_1\eta_s + c_{2\eta}\eta_s^2 + c_{2\mu}\mu_s^2 + c_{2\nu}\nu_s^2 ,\qquad(12)$$

where we have applied the same scaling to Q and f_G as described before (5), while f_S and distance z are expressed in units of $\xi_0(B^2/24)^2/C^3$ and $\xi_0 = [24C(L_1+2L_2/3)/B^2]^{1/2}$, respectively. The coefficients c_1, $c_{2\eta}$, $c_{2\mu}$, and $c_{2\nu}$ in (12) are certain dimensionless combinations of those in (8).[5] This diagonalization of f_G and f_S is achieved at the price of formally making f_L a function of all three order parameters η_s, μ_s, and ν_s, instead of just two variables as in (5), but that outweighs the increased complexity of f_G, in particular, when the latter is expressed in terms of the principle-axis parameters.

Note that the linear term in (12) contains only the order parameter η_s, which measures (cf.(9)) the degree of orientational order along the normal vector k. This reflects the fact that the direct substrate-fluid interactions (which enter only via the coefficient c_1[5]) do not impose a preferred tangential ordering axis.

Functionally minimizing the free energy σ yields coupled Euler-Lagrange equations for determining the order parameters $\eta_s(z)$, $\mu_s(z)$, and $\nu_s(z)$. Denoting these by $\phi_i(z)$, $i = 1,2,3$, respectively, we have

$$\tilde{L}_i \frac{d^2\phi_i(z)}{dz^2} = \frac{\partial f_L}{\partial \phi_i(z)}, \quad z \geq 0 \tag{13}$$

and

$$\tilde{L}_i \frac{d\phi_i(z)}{dz}\bigg|_{z=0} = \frac{\partial f_s}{\partial \phi_i(0)}, \tag{14}$$

where $\tilde{L}_i = 1, L, M$ for $i = 1, 2, 3$, respectively.

In an extended treatment including fluctuation effects, the functional (3) would be regarded as the model Hamiltonian rather than a thermodynamic potential. The tensor Q would become a fluctuating field with spatial gradients in all directions, thus requiring consideration of its full 5-component character[17] (as well as a generalization of (3) to include integration over the additional spatial dimensions). We shall not discuss such a treatment here.

ORIENTATIONAL WETTING

Up to the present time, with one exception, there have been no studies of the model outlined in the previous section including variations in all three order parameters η_s, μ_s, and ν_s (or η, μ, and ψ). The exception is a paper by Marcus[18], who described a numerical solution of the theory for the interface between coexisting isotropic (I) and nematic (N) phases at temperature $t = 1$, in the absence of a wall. Taking an elastic constant ratio of $L_2/L_1 = 2$, Marcus found the interfacial free energy σ_{IN} to be minimized for a state of constant director parallel to the interface, i.e., $\psi(z) = \pi/2$ for all z. This finding agrees with an earlier approximate analysis of the theory by de Gennes[1], who predicted that the equilibrium alignment at the I-N interface would be $\psi = \pi/2$ for $L_2 > 0$ and $\psi = 0$ (perpendicular or "homeotropic") for $-(3/2)L_1 < L_2 < 0$. As mentioned earlier, we expect $L_2 > 0$ in realistic cases.

It has been shown that the "bare" wall energy f_s given by (8) could favor any possible orientation $0 \leq \psi \leq \pi/2$, depending on the values of the coefficients $c_1 - c_4$[2,5]. The true equilibrium alignment for a nematic liquid near a substrate, however, must be determined by minimizing the total free energy σ, and there could be competing tendencies associated with different terms in (3). This is clearly apparent for, but is not limited to, situations involving wetting of the substrate by either the isotropic or nematic phase when these are in coexistence.

Let us denote by α the bulk phase far from the wall. When the thermodynamic state is such that α coexists with a phase denoted β, but the latter is not present in any macroscopic amount, it is nonetheless possible that the wall may favor formation of an adjacent layer resembling the β phase. This is indicated schematically in fig. 2a, where the thickness of the β-layer is denoted ℓ. When that thickness is infinite (in practice, some finite but nearly macroscopic value), the β-layer completely wets the substrate-α interface. Otherwise, the β-layer only partially wets that interface or may be absent altogether. The definition of the thickness ℓ is usually made by reference to suitable inflections in the order-parameter profiles. This is illustrated in Fig. 2b by the $\eta_s(z)$ profile in the hypothetical case of near-complete wetting by a homeotropic nematic layer when the bulk phase is the isotropic liquid. A picture similar to the latter in which ℓ is finite would also apply to a system slightly removed from coexistence, but which becomes completely wet <u>at</u> coexistence, exhibiting what Professor Widom has called "premonitory" wetting. In the present context, where N-I coexistence occurs at only the single temperature t = 1, we can imagine changing the wetting properties of a system, i.e., causing a wetting transition[16], by varying material parameters such as the coefficients in f_s.

Let us restrict attention to cases where the tilt angle is constrained to be constant, either $\psi = 0$ or $\pi/2$, which has so far been the situation in actual calculations. We shall also take $L_2 > 0$, so the free N-I interface favors planar (parallel) alignment. We can classify the possible wetting states according to whether the substrate is preferentially wet by the nematic (β = N, α = I) or isotropic (β = I, α = N) phase. The nematic in turn can be in either homeotropic ($N_\parallel$) or planar ($N_\perp$) alignment. This leads to the four possibilities listed in Table 1. For example, the first entry corresponds to wetting by a homeotropic nematic layer, expected to occur on a substrate which strongly favors perpendicular molecular orientations. If that wetting is complete, however, then in reality the director would undergo a distortion across the layer[19] in order to allow the occurrence of planar alignment at the N-I interface existing far from the substrate. This has not been recognized in previous calculations bearing on this situation[11,12]. In contrast, there is little reason to expect director distortion in a nematic wetting film on a substrate favoring planar alignment.

The last two cases in Table 1 correspond to wetting by the isotropic phase, which one intuitively expects to find on substrates with rather weak aligning forces. As discussed earlier, the alignment in the bulk nematic is not arbitrary and one must compare the free energies of those two cases to determine which one is more stable (a restricted version of varying the bulk alignment over all possible tilt angles, $0 \leq \psi_b \leq \pi/2$). It turns out that <u>complete</u> wetting by I when the bulk phase is $N_\perp$ is not possible. This can be shown by application of Antonow's rule[20]. If we suppose that the isotropic phase <u>does</u> completely wet the interface between the substrate (or wall, denoted by W) and bulk $N_\perp$, then at equilibrium the free energy $\sigma_{WN_\perp}$ of that interface would obey

$$\sigma_{WN_\perp} = \sigma_{WI} + \sigma_{IN_\perp} . \tag{15}$$

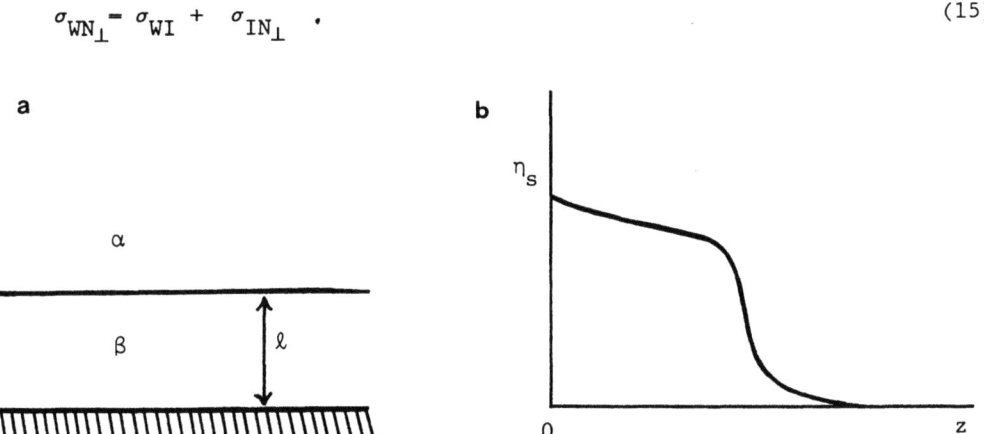

Fig. 2 (a) A wetting layer of phase β between the substrate and bulk phase α; (b) Schematic profile of the order parameter η_s for near-complete wetting by a homeotropic nematic layer.

Table 1. The four possible wetting configurations at nematic-isotropic coexistence in presence of a substrate, when the nematic alignment is either homeotropic or planar.

α	β
I	$N_\perp$
I	$N_\parallel$
$N_\perp$	I
$N_\parallel$	I

But, as already discussed, the free energies of I-N interfaces satisfy $\sigma_{IN_\perp} > \sigma_{IN_\parallel}$. Hence $\sigma_{WN_\perp}$ in (15) exceeds the sum $\sigma_{WI} + \sigma_{IN_\parallel}$, which would equal the free energy $\sigma_{IN_\parallel}$ for complete isotropic wetting of the interface between the wall and bulk $N_\parallel$ phase. Therefore, according to Landau-de Gennes theory, complete wetting by the isotropic phase is stable only when the bulk nematic alignment is planar. Of course, that complete-wetting configuration may in turn be unstable relative to partial-wetting states.

The features described above can be demonstrated more explicitly by considering a model recently introduced by Sluckin and Poniewierski[21]. This is based on taking the elastic constant $L_1 = 0$. While somewhat unrealistic, it has the great virtue of enabling one to exactly solve the Landau-de Gennes theory in circumstances which include biaxial planar alignment. Note that when the alignment is restricted to be either planar or homeotropic, as we shall continue to maintain, the order parameter $\nu_s(z)$ vanishes identically, cf.(10). One can show that the function $f_L(\eta_s, \mu_s, \nu_s=0) = f_L(\eta_s, \mu_s)$ is then given by the same expression as in (5). When $L \propto L_1 = 0$, the Euler-Lagrange equation in (13) for $\mu_s(z)$ reduces to $\partial f_L/\partial \mu_s = 0$, which yields a relation between $\mu_s(z)$ and $\eta_s(z)$:

$$\mu_s = [(\eta_+ - \eta_s)(\eta_s - \eta_-)]^{1/2}, \quad \eta_- < \eta_s < \eta_+ ,$$
$$= 0, \text{ otherwise} , \quad (16)$$

where

$$-\eta_\pm = \frac{3}{2}[1 \mp \sqrt{1-2t/9}] . \quad (17)$$

Here, non-zero $\mu_s(z)$ signifies (cf.(9)) biaxial symmetry-breaking due to planar alignment. (The only exception to biaxiality is when the order parameters equal the values appropriate to a <u>bulk</u> planar nematic, satisfying $\mu_s = -\sqrt{3}\eta_s$ with $\eta_s = -\eta_c/2$, which are necessarily consistent with (16) and (17).) Vanishing $\mu_s(z)$ describes either a state of homeotropic alignment, when $\eta_s(z) > 0$, or a random, disordered configuration of molecules with mean orientation parallel to the interface, when $\eta_s(z) < 0$. When we substitute (16) into the appropriate Euler-Lagrange equation (13) for $\eta_s(z)$, that becomes an effective one-component equation which, together with the corresponding initial condition in (14) for $\eta_s(0)$, can be analyzed by well-known analytic methods[16]. Consistency between the relation (16) and the initial condition in (14) for $\mu_s(0)$ strictly requires setting $c_{2\mu} = 0$, but it can be shown that the results are unchanged when $c_{2\mu} \neq 0$ by considering the <u>limit</u> $L_1 \to 0$.

Some of the results[5] of this model are given in Fig. 3, which shows the surface "phase diagram" at bulk N-I coexistence in terms of the wall parameters c_1 and $c_{2\eta}$. The broad domains at sufficiently large positive and negative values of c_1, which we recall measures the strength of the wall-fluid interaction, correspond to complete wetting (CW) by planar and homeotropic nematic films, respectively. When $c_{2\eta} \geq -0.21$, the complete-wetting states are separated by solid lines from a domain in which the nematic partially wets the interface. These lines are light or bold according to whether the associated wetting transitions are first- or second-order, respectively. The separate wetting lines meet at a triple point and, for more negative values of $c_{2\eta}$, corresponding to strong "surface enhancement" of the pair interactions, are replaced by a single direct line of first-order transitions between planar and homeotropic CW states. For the reasons discussed earlier, however, we msut expect all transitions to complete wetting by $N_\perp$ to be modified when the constraint of constant tilt angle is relaxed.

The dashed line running diagonally through the figure is the locus of (first-order) transitions in the surface-induced bulk orientation, such that the stable bulk nematic alignment is planar above that line and homeotropic below. In the region of negative $c_{2\eta}$, that line lies slightly above the line of $N_\perp$ - $N_\parallel$ wetting transitions, an effect which can be explained by an Antonow's rule argument similar to that used earlier. Over most of its extent at positive $c_{2\eta}$, the orientational transition line is also a locus of wetting transitions to complete wetting by the isotropic phase. The latter state is otherwise bounded by the solid lines at or near to $c_1 = 0$ and thus occurs only when the stable bulk nematic alignment is planar, in accord with our earlier argument.

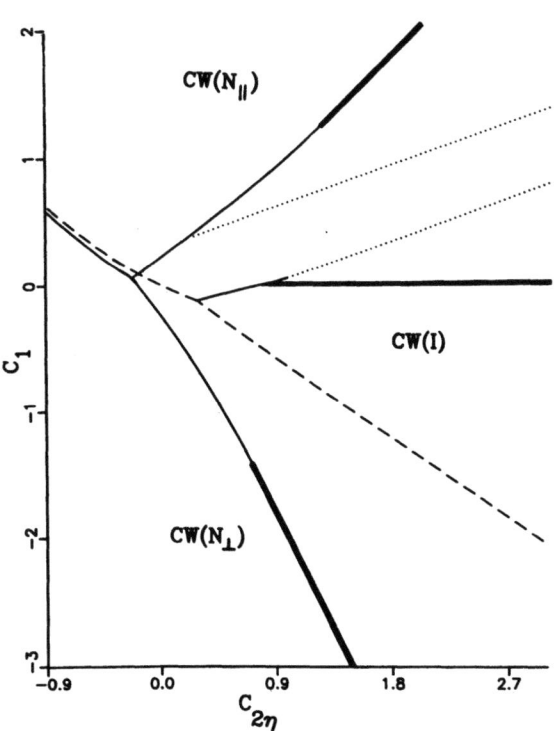

Fig. 3 Surface phase diagram of the Sluckin-Poniewierski model at bulk N-I coexistence.

The two dotted lines in the figure signify continuous symmetry-breaking transitions[21] between "random planar" and biaxial partial-wetting states. The upper line applies when the bulk phase is isotropic, such that the adsorbed nematic layer is biaxial above that line and uniaxial below. The lower dotted line holds when the bulk phase is N∥, now such that the layer immediately adjacent to the wall is uniaxial below that line and biaxial above.

THE PARABOLIC APPROXIMATION

Except in the Sluckin-Poniewierski limit[21] $L_1 = 0$, it is generally not possible to obtain analytic solutions of the two-component Landau-de Gennes theory for biaxial surface phases in planar alignment. We have used[5] a so-called "parabolic approximation" to solve the theory for arbitrary values of L_1. This method has recently been developed by Lipowsky and co-workers[22,23] to examine interfacial phenomena in some related multi-component Landau models. The method is also closely related, though not strictly identical, to that described by E.H. Hauge in his lecture at this symposium for treating continuous wetting transitions[24]. It can be generalized to deal with the three-component Landau-de Gennes theory for arbitrary director alignment, and work on that aspect is currently in progress.

The gist of this approach is to replace the function $f_L(\eta_s,\mu_s)$ by the composite of the quadratic functions obtained by expanding the exact f_L about each of its minima. These minima or "fixed points" are the same as those described below (5) using the principle-axis variables η and μ, except that there is now a significance attached to the distinct nematic minima. The one at $(\eta_s,\mu_s) = (\eta_c,0)$ describes a homeotropic nematic, while that at $(\eta_s,\mu_s) = (-\eta_c/2, \sqrt{3}\eta_c/2)$ pertains to a planar nematic. (When generalized to allow arbitrary tilt angles, so that $\nu_s \neq 0$, the nematic fixed points lie on a <u>continuous curve</u> in the three-dimensional (η_s,μ_s,ν_s)-space.) Fig. 4 shows the location of the nematic and isotropic fixed points in the (η_s,μ_s)-plane. The dash-dot lines in that figure represent the matching loci at $t = 1$ where the quadratic expansions of $f_L(\eta_s,\mu_s)$

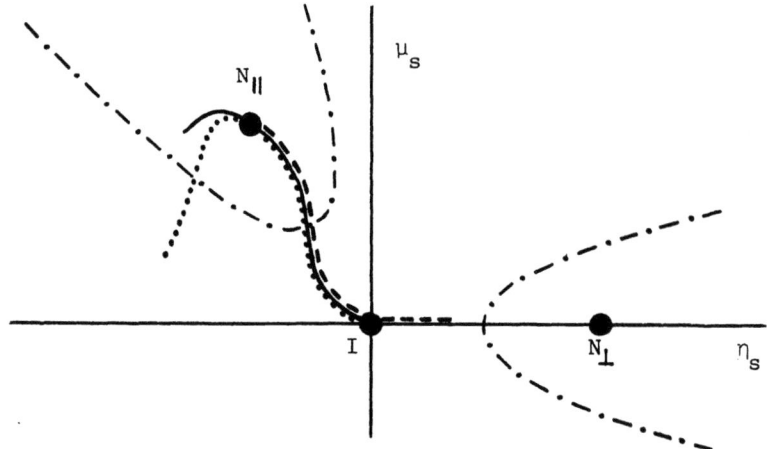

Fig. 4 Fixed points, matching contours, and representative complete-wetting trajectories in parabolic approximation. The trajectories share a common path between the N and I fixed points, describing the free I-N interface.

about its nematic and isotropic minima have equal values. Within each of the regions separated by those curves, $f_L(\eta_s,\mu_s)$ is replaced by the corresponding parabolic approximant and the solutions of the associated Euler-Lagrange equations (13) for $\eta_s(z)$ and $\mu_s(z)$ are simply exponential functions or linear combinations of exponentials. The coefficients of those exponentials are determined from the initial conditions (14) and by requiring the order parameters to be smooth functions of z whenever the "trajectory" in the (η_s,μ_s)-plane crosses a matching locus in Fig. 4. Further mathematical details as well as limitations of this approach are discussed in ref. 5.

A schematic trajectory that corresponds to complete wetting by the planar nematic is indicated by the solid curve in Fig. 4. Here the initial point $[\eta_s(0),\mu_s(0)]$ is located within the region encompassing the $N_{\parallel}$ fixed point. The trajectory passes through that fixed point, then subsequently crosses into the isotropic region and ends at the fixed point of the latter, which in this case represents the bulk ($z\to\infty$) phase. A partial-wetting trajectory would be one which does not traverse the nematic fixed point. In practice,[5] the equilibrium trajectory under any given conditions is found by minimizing the interfacial free energy σ with respect to a variable ℓ, equivalent to the wetting-layer thickness, here defined as the distance z where matching of the order-parameter profiles occurs. A complication arises in that it is often necessary, in order to obtain stable states of complete wetting by $N_{\parallel}$, to consider "reentrant" trajectories such as indicated by the dotted curve in Fig. 4. That entails solving the Euler-Lagrange equations with matching conditions imposed at two different distances. For complete wetting at coexistence, one can avoid such a double-matching calculation by invoking Antonow's rule and thus treating separately the two segments of the trajectory from $[\eta_s(0),\mu_s(0)]$ to the $N_{\parallel}$ fixed point and from the latter to the I fixed point, each of which involves only a single-matching calculation.

Our main application of the method outlined above has been to study the limits of complete wetting by the planar nematic phase. Representative results are shown by the phase diagram in Fig. 5, corresponding to a fixed ratio $L_2/L_1 = 10$. While that is considerably larger than what we believe to be physically relevant values of L_2/L_1, the behavior indicated is characteristic of all smaller values of that ratio as well as a range of large values. The figure exhibits the surface of wetting transitions in the three-dimensional (c_1, $c_{2\eta}$, $c_{2\mu}$) - parameter space, where the various solid lines are slices through that surface at different fixed values of $c_{2\eta}$. The region enclosed by the surface corresponds to complete wetting by $N_{\parallel}$, while outside there is only partial wetting. Hence, starting at a sufficiently low value of c_1 for some fixed values of $c_{2\eta}$ and $c_{2\mu}$ below the apex of the surface, there is first encountered a transition to complete wetting on increasing c_1, and subsequently a <u>dewetting</u> transition with further increase in c_1. The domain of partial wetting is thus "reentrant" as a function of the substrate strength c_1. For $c_{2\mu}$ greater than a small positive value which increases slowly with L_2/L_1, complete wetting does not occur for any value c_1. As $c_{2\mu}$ approaches the negative value ($-\sqrt{L/2}$), the width of the complete-wetting domain increases and the dewetting limit recedes to infinite values of c_1, although simultaneously the free energy exhibits a negative divergence. The latter may signal a "surface-enhanced" transition of strong first-order character, inadequately treated by the parabolic approximation, but this interpretation is not firm.

The wetting and dewetting transitions across most of the surface in Fig. 5 are found to be first order; continous wetting transitions occur in the section shown ruled. The latter is actually accompanied by a surface of <u>first-order</u> symmetry-breaking transitions (not displayed) which extends outside the wetting surface and joins it along the dashed seam in the figure.

The qualitative behavior shown in Fig. 5 is found for all values of L_2/L_1 in the range $0 \leq L_2/L_1 \leq 15.771...$ When L_2/L_1 exceeds the upper limit of this range, the dewetting side of the surface in Fig. 5 disappears, as likewise does the limiting upper value of $c_{2\mu}$. Once complete wetting by $N_\parallel$ has been attained for some value of c_1, it remains stable for all larger values of that parameter. This behavior is, of course, consistent with that given by exact solution of the Sluckin-Poniewierski model, which corresponds to the limit $L_2/L_1 \to \infty$. One can give a heuristic argument, based on the competition between various terms in the expression (3) for the free energy σ, which suggests that the existence of a limiting value of L_2/L_1, below which complete wetting cannot occur for arbitrarily large c_1, should be of general validity and is not an artefact of the parabolic approximation.[5]

We have also used this method to study complete wetting by the isotropic phase as well as transitions in the surface-induced alignment of the bulk nematic phase. Now, it is found that the qualitative behavior under conditions of bulk N-I coexistence does not differ significantly from that given by the Sluckin-Poniewierski model (cf. Fig.3), as was anticipated by our arguments in the previous section. An important difference, however, arises when we consider temperatures $t \leq 1$, i.e., slightly below that of bulk coexistence. A typical trajectory which describes complete wetting by the isotropic phase <u>at</u> coexistence with the bulk nematic in planar alignment is indicated by the dashed curve in Fig. 4. That trajectory begins at a point on the positive η_s-axis, representing a weakly homeotropic and uniaxial layer near the wall, subsequently traverses the isotropic fixed point, leading to the intrusion of an infinite isotropic wetting layer, and is finally completed by a path to the bulk $N_\parallel$ fixed point which oscillates about the line $\mu_s = -\sqrt{3}\eta_s$, describing the biaxial I-$N_\parallel$ interface. The trajectory to $N_\parallel$ maintains that same form at temperatures $t \leq 1$ in the Sluckin-Poniewierski model (cf. (16) and (17)). In particular, it remains strictly uniaxial for positive η_s and

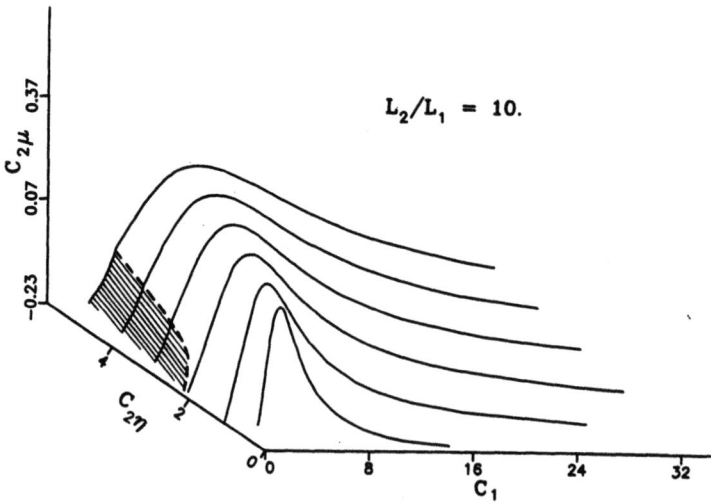

Fig. 5 Surface of wetting and dewetting transitions by planar nematic films in parabolic approximation.

crosses the isotropic fixed point; since the latter is only metastable at t < 1, this trajectory produces "premonitory" rather than complete wetting by the isotropic phase. When L_2/L_1 is finite, we find a different behavior at $t \leq 1$ using the parabolic approximation. Now the trajectory at positive values of η_S is shifted slightly above the η_S-axis and does not quite traverse the isotropic fixed point. This behavior is a bit unphysical, since it implies that the nematic fluid near the substrate is aligned homeotropically but is weakly biaxial. (Note that a similar result is obtained in the Sluckin-Poniewierski limit at temperature t < 0.). While this feature might be due to a failure of the parabolic approximation, the difficulty may be resolved on allowing the director to vary continuously with z. Support for this idea is suggested by measurements[25] of director tilt angles on SiO substrates, which exhibit complete or near-complete wetting by the isotropic phase[26]. These experiments indicate a continuous variation in bulk tilt angle on approaching N-I coexistence from lower temperatures, tending toward a near-planar alignment at coexistence.

CONCLUDING REMARKS

We have discussed, in the context of Landau-de Gennes theory, the structure of a liquid crystal/wall interface. The necessity to account for multiple position-dependent order parameters due to the presence of biaxiality and varying tilt angle has been emphasized. With the restriction to either homeotropic or planar alignment, an explicit picture of the possible surface phases that can occur has been derived in the Sluckin-Poniewierski[21] limit $L_1 = 0$, showing the connection between wetting behavior and orientational alignment. We have noted that in situations where the wall-liquid interactions favor complete wetting by a homeotropic nematic layer, it is expected that the director alignment should actually vary across the layer in order to accomodate the planar alignment favored by elastic forces at the nematic-isotropic interface.

As was first pointed out in ref.(19), director distortion should produce a contribution varying asymptotically as $1/\ell$ in the free energy $\sigma(\ell)$ as a function of wetting-layer thickness ℓ. This is essentially the same effect as one characterizing Bloch domain walls in a Heisenberg ferromagnet[27]. The inverse-ℓ dependence is also similar to that caused by elastic strain in a solid wettting film[28], but differs in that its contribution to $\sigma(\ell)$ is <u>repulsive</u>, i.e., favors separation of the two interfaces bounding the wetting layer. One may speculate about whether analogous effects operate in other systems characterized by a continuous rotational symmetry, e.g., wetting films of superfluid helium.[29]

In the present theory, director distortion should not generally be of concern for nematic wetting films on a substrate favoring planar alignment. On the other hand, our calculations based on the parabolic approximation indicate that complete wetting by such films should be of rather limited occurrence for realistic values of L_2/L_1. That would appear to be consistent with experimental results (summarized in ref. (2)), since no definite cases of complete nematic wetting on substrates with tangentially isotropic boundary conditions have yet been found. Let us note that the "reentrant" nature of wetting as a function of the substrate strength c_1 seen in Fig. 3 is similar to that found both experimentally[30] and theoretically[28] for solid wetting films, where it is usually attributed to the strain-induced effect mentioned above.

As we have noted earlier, the Landau-de Gennes theory predicts the stable alignment at the nematic-isotropic interface to be either homeotropic or planar, depending on the sign of the elastic constant L_2. In reality, most such interfaces appear to be characterized by an <u>oblique</u>

tilt angle. Possible reasons for this discrepancy include the neglect of fluctuations as well as of higher-level (e.g. "P_4") orientational order parameters in the Landau-de Gennes theory.

REFERENCES

1. P.G. de Gennes, Mol. Cryst. Liq. Cryst. 12:193 (1971).
2. T.J. Sluckin and A. Poniewierski, in: "Fluid Interfacial Phenomena", C.A. Croxton, Ed., Wiley, Chichester (1986).
3. M.M. Telo da Gama, Mol. Phys. 52:585, 611 (1984).
4. J.H. Thurtell, M.M. Telo da Gama, and K.E. Gubbins, Mol. Phys. 54:321 (1985).
5. A.K. Sen and D.E. Sullivan, Phys. Rev. A (to be published).
6. T.C. Lubensky, Phys. Rev. A 2:2497 (1970).
7. J. Cognard, Mol. Cryst. Liq. Cryst. 78, Supp. 1:1.
8. S. Faetti and V. Palleschi, Phys. Rev. A 30:3241 (1984).
9. J.D. Parsons, Phys. Rev. Lett. 41:877 (1978); Mol. Phys. 42:951 (1980).
10. P.G. de Gennes, "The Physics of Liquid Crystals", Clarendon, Oxford (1974).
11. D.W. Allender, G.L. Henderson, and D.L. Johnson, Phys. Rev. A 24:1086 (1981); P. Sheng, Phys. Rev. A 26:1610 (1982).
12. A. Poniewierski and T.J. Sluckin, Mol. Cryst. Liq. Cryst. 111:373 (1984); 126:143 (1985).
13. J.P. Straley and M.E. Fisher, J. Phys. A 6:1310 (1973).
14. "Introduction to Liquid Crystals," E.B. Priestley, P.J. Wojtowicz, and P. Sheng, Eds., Plenum, New York (1974), Ch.10.
15. J.P. Straley, Phys. Rev. A 8:2181 (1973); W.M. Gelbart and A. Ben-Shaul, J. Chem. Phys. 77:916 (1982).
16. D.E. Sullivan and M.M. Telo da Gama, in: "Fluid Interfacial Phenomena," C.A. Croxton, Ed., Wiley, Chichester (1986).
17. V.L. Pokrovskii and E.I. Kats, Sov. Phys. JETP 46:405 (1977).
18. M.A. Marcus, Mol. Cryst. Liq. Cryst. 100:253 (1983).
19. E. Perez, J.E. Proust, and L. Ter-Minassian-Saraga, Coll. Poly. Sci. 256:784 (1978).
20. J.S. Rowlinson and B. Widom, "Molecular Theory of Capillarity", Clarendon, Oxford (1982).
21. T.J. Sluckin and A. Poniewierski, Phys. Rev. Lett. 55:2907 (1985).
22. M. Nilges, dipom thesis, University of Munich (1984); R. Lipowsky and M. Nilges (unpublished).
23. C. Borzi, R. Lipowsky, and B. Widom, J. Chem. Soc. Faraday Symp. 20 (1985).
24. E.H. Hauge, Phys. Rev. B 33:3322 (1986).
25. H.A. Van Sprang and R.G. Aartsen, J. Appl. Phys. 56:251 (1984).
26. H. Yokoyama, S. Kobayashi, and H. Kamei, Mol. Cryst. Liq. Cryst. 99:39 (1983).
27. M. Wortis, in: "Fundamental Problems in Statistical Mechanics VI", E.G.D. Cohen, Ed., North Holland, New York (1985).
28. D.A. Huse, Phys. Rev. B 29:6985 (1984); F.T. Gittes and M. Schick, Phys. Rev. B 30:209 (1984.
29. P. Taborek and L. Senator, Phys. Rev. Lett. 57:218 (1986).
30. M. Bienfait, J.L. Seguin, J. Suzanne, E. Lerner, J. Krim, and J.G. Dash, Phys. Rev. B 29:983 (1984).

A MODEL FOR CAPILLARY CRYSTALLIZATION

G. Navascués and P. Tarazona

Departamento de Física del Estado Sólido (UAM)
Instituto de Física del Estado Sólido (CSIC)
Universidad Autónoma de Madrid, 28049 Madrid
Spain

ABSTRACT

We analyse the phase diagram of a lattice gas model with both condensation and order-disorder phase transitions, when the system is confined between two walls. The gas-liquid transition is shifted into the, so called, capillary condensation. The crystallization, both from the gas and from the liquid, is also shifted from the bulk values, but the ordered structure is frustrated or enhanced depending on its commensuration with the walls separation, H. This produces a strong oscillatory dependence of the phase diagram with H. Even the macroscopic Kelvin equation, valid for the capillary condensation at large H, has to be corrected to describe the capillary crystallization and the gas-liquid-solid triple point.

INTRODUCTION

The gas-liquid condensation in a confined system, as a porous media, takes place before the bulk pressure p reaches the saturation value p_0. This phenomenon called capillary condensation has been known for long time, however only recently a microscopic basis has been given[1]. Here using a microscopic model we extend the analysis to obtain the first insight of what can be called capillary crystallization and capillary triple point, i.e. we study the changes of the full phase diagram including crystallization and triple point when the system is confined. As it is usually done[1-3] we confine the system between parallel walls. When the distance H between walls is small enough to freeze the degrees of freedom associated with the normal direction to the walls we obtain a two-dimensional (2D) system. In this 2D limit the particles of the system will condensate at temperatures $T<T_c^{(2D)}$. In the opposite limit ($H \to \infty$) the condensation takes place at $T<T_c^{(3D)}$; then if $T_c^{(2D)}<T<T_c^{(3D)}$ the capillary condensation isotherm, $p(H,T)$, which describes the condensation transition with H at a given temperature, will end in a critical point before the 2D limit is reached[4].

Below the triple point temperature, $T_T^{(3D)}$, a gas condensates to a crystalline solid instead of a disordered liquid. For real fluids the ratio $T_T^{(3D)}/T_c^{(3D)}$ is about the same that the theoretical predictions for $T_c^{(2D)}/T_c^{(3D)}$, so that for the argument given above it is important to consider how the confinement affects not only to the condensation but also to the crystallization and the triple point coexistence. New experimental techniques for normal fluids[5] and colloidal particles[6] give evidence of strong layering and fully ordered phases in capillaries a few particle diameters width, therefore the capillary crystallization already deserves a theoretical attention.

We first propose a simple model with a bulk phase diagram which imitates a real phase diagram, i.e. it includes gas, liquid and solid phases, a condensation critical point and a gas-liquid-solid triple point. The model, a lattice gas of a binary mixture, is however much simpler than any real system. We analyse the results when the lattice is confined between parallel walls. We discuss the changes of phase diagram as a function of the distance between walls H. We find not only the known[1-3] capillary critical points for the gas-liquid condensation but also for the gas-solid and the liquid-solid crystallization and even the possible existence of a series of critical points which appears when the system structure near each wall interferes wich each other. We also develop a macroscopic approach for the capillary crystallization but, as it happens with the Kelvin equation for the capillary condensation, this approach is unable to describe the richness of the phase diagram for small H. We shall also see that the macroscopic approach for the capillary crystallization though it is a straightforward extension of the Kelvin equation has quite interesting consequences.

MICROSCOPIC MODEL AND MACROSCOPIC THEORY

Consider a lattice gas model of a binary mixture where each site may be occupied by a particle of species A or B, or or else be empty. We assume nearest neighbours interactions with symmetric coupling constants, α for AB pairs and $\alpha + \beta$ for AA and BB pairs. In the mean field approximation (MFA) the equilibrium values of the density, $\rho_i^{(A)}$ and $\rho_i^{(B)}$, at any lattice i-site, at given temperature T and chemical potentials $\mu^{(A)}$ and $\mu^{(B)}$, are obtained by minimizing the grand potential energy:

$$\Omega[\{\rho_i(\nu)\}_{\nu=A,B}]_{i=1,N} = kT \sum_{i=1}^{N} [\rho_i(A) \lg \rho_i(A) + \rho_i(B) \log \rho_i(B) + (1 - \rho_i(A) - \rho_i(B))$$
$$\lg (1 - \rho_i(A) - \rho_i(B))] + (1/2) \sum_{i,j} [(\alpha + \beta)(\rho_i(A)\rho_j(B) + \rho_i(B)\rho_j(A))$$
$$+ 2\alpha \rho_i(A)\rho_j(B)] + \sum_{i=1}^{N} \sum_{\nu=A,B}^{(n-n)} (v_i(\nu) - \mu(\nu)) \rho_i(\nu) \quad (1)$$

where V_i^ν and μ^ν are the external potential and the chemical potentials of species ν. All over this work we take $\mu^{(A)} = \mu^{(B)} = \mu$, so that the perfect symmetry between both species may only be broken by the external potentials.

Figure 1 shows the phase diagram of the above model in the absence of external potentials: for negative $\alpha + \beta/2$ there is a first order condensation at $T<T_c=-(\alpha+\beta/2)$ from a low density 'gas' to a high density 'liquid', both with equal concentration of species A and B. If β is negative and at given high values of the total density $\rho(=\rho^{(A)}+\rho^{(B)})$ there is also a liquid-liquid segregation with an A-rich phase coexisting with its symmetric B-rich phase (shadow region in Fig.1). There is a triple line where the two segregated phases coexist with the 'liquid' phase at $T>T_T$ and with the 'gas' phase at $T<T_T$. The temperature T_T is a quadruple point where the 'gas', 'liquid' and the two segregated phases coexist. The triple line ends in a tricritical point at $T_{Tc}=-7\beta/9$.

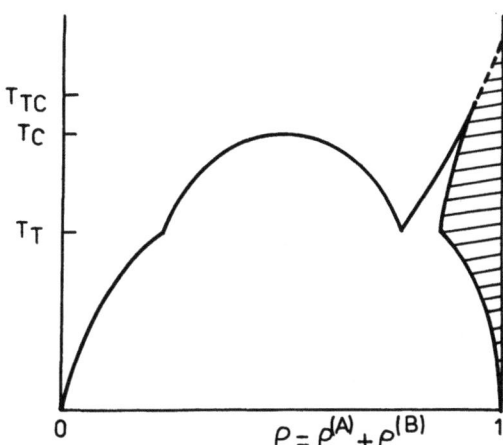

Figure 1.- Bulk phase diagram (outline) of the proposed model (see text).

For $\beta>0$, the attraction is stronger between molecules of different species, so that there is no segregation, but using the equivalent of the ferromagnetic-antiferromagnetic symmetry in the Ising model, it is easy to show that the system presents an ordering phase transition at high values of the total density. This is a first order transition from an ordered phase, with a sublattice rich in species A and the complementary sublattice rich in species B, to the symmetric disordered phase. It is straightforward to show that the phase diagram with $\beta>0$ (ordering) maps into the case $\beta<0$

(segregation) by the transformation $\alpha = \alpha + \beta$ and $\beta = -\beta$ which preserves the mean coupling constant $\alpha + \beta/2 = \alpha + \beta/2$, so that the condensation is not affected. Therefore Fig.1 can properly be used to describe the ordering case phase diagram where the triple line corresponds to the coexistence of two ordered phases with the liquid one at $T>T_T$ and with the gas one at $T<T_T$. Now the tricritical temperature is $T_{Tc} = 7\beta/9$.

Inside the MFA there is no qualitative effect of the dimensionality, so that for the sake of simplicity we restrict ourselves to a one dimensional lattice.

As we are going to use our binary mixture model to analyse the capillary behaviour of a simple substance with a phase diagram including gas, liquid and solid phases, it is worth stablishing the proper comparison and limits of this real phase diagram with that of Fig.1. As in the case of the simple substance we have a unique chemical potential variable, namely $\mu = \mu^{(A)} = \mu^{(B)}$. At the same time we only need the total density $\rho^{(A)} = \rho^{(B)} = \rho/2$ for the fluid phases and $\rho_{odd}^{(A)} = \rho_{even}^{(B)} = \rho x$ and $\rho_{odd}^{(B)} = \rho_{even}^{(A)} = \rho(1-x)$, where the concentration variable x is a function of T and ρ; for the fluid phases $x=1/2$. Therefore, taken the details of the mixture inside a black box, our ρ-T phase diagram includes a low density gas like phase, a high density liquid like phase and a high density ordered phase which is twice degenerated and will be called 'solid'. One of the degenerated phases has high concentration of species A in a sublattice (say in the odd sites) and high concentration of species B in the complementary sublattice (even sites); the other degenerated phase is its symmetric. Each one of these degenerated phases may be obtained from each other not only by interchanging species A and B (as in segregation case), but also by shifting the lattice by one lattice space.

In a real space diagram, a crystalline solid phase is in fact composed by infinite number of microscopically different phases, obtained by arbitrary shifts or rotations of the crystal lattice. The presence of this degenerate phases is trivial for bulk systems; it becomes important for a crystal close to a single wall, because then the degeneracy is broken by the boundary condition and one, or a few, states will be selected. When the system is confined between two walls this aspect becomes crucial, because depending on whether the phase selected by both independent boundaries is the same or not, it would be an enhancement or frustration of the ordering. This is the aspect we are trying to reproduce in our model with the two degenerate configurations representing the solid and which may be obtained one from the other by a lattice traslation.

We chose $\alpha = -3$ and $\beta = 1.5$ which locate the triple point at $T_T = 0.9618\, T_c$ and the tricritical one at $T_{Tc} = 1.037\, T_c$. The confining walls are modelled by the external potential

$$V_i^\nu = \infty \quad \text{for } i<1 \text{ or } i>H, \quad \nu = A, B$$

$$V_i^\nu = 0 \quad \text{for } 1<i<H, \quad \nu = A, B \qquad (2)$$

$$V_1^{(A)} = V_H^{(A)} = -V_1^{(B)} = -V_H^{(B)} = -10\, kT_c$$

This mimics two symmetric hard walls separated by H lattice spaces, with a strong first layer atraction (repulsion) for species A (B). A single wall will produce a large concentration of A in the first layer, which will produce a large concentration of B in the second and so on; then there would be a local ordering produced by the wall. If the bulk phase far from the wall is fluid, the oscillatory concentration structure will decay out of the wall, but if it is a solid bulk, the external potential will select the configuration with high A concentration in the first layer. There would also be in this case an enhancement of the oscillating structure near the wall, but the main effect of the wall will be to discard one of the degenerated structures in favour of the other. The strength V_H^v of the first layer attraction/repulsion is such that the surface phase diagram presents perfect wetting of the liquid and the solid in the gas-wall interface and perfect wetting of the solid in the wall-liquid interface, i.e.

$$\gamma_{WG} = \gamma_{WL} + \gamma_{LG} \quad (T>T_T), \qquad (3)$$

$$\gamma_{WG} = \gamma_{WS} + \gamma_{SG} \quad (T<T_T) \qquad (4)$$

and

$$\gamma_{WL} = \gamma_{WS} + \gamma_{SL} \quad (T>T_T) \qquad (5)$$

where γ_{ij} are the interfacial tensions of the (ij) interface (W=wall, S=solid, L=liquid, G=gas).

When the system is confined between the two walls the grand potential of the gas and the liquid configurations are given respectively by

$$\Omega_G = -pH + \omega_G(H) \qquad (6)$$

$$\Omega_L = -p_L^+ H + \omega_L(H) \qquad (7)$$

where p and p_L^+ are the bulk pressure of the gas and of the metastable liquid configurations respectively at a given T and μ. ω is the capillary excess of the grand potential which is determined by the two inferfacial contributions, i.e.

$$\omega_G(H) = 2\gamma_{WG}(H) \qquad (8)$$

$$\omega_L(H) = 2\gamma_{WL}(H). \qquad (9)$$

The capillary condensation occurs when $\Omega_G = \Omega_L$, or when the pressure satisfies

$$p-p_L^+ = -\frac{\Delta\Omega_{LG}(H)}{H} \qquad (10)$$

where $\Delta\Omega_{LG} = 2(\gamma_{WL}(H) - \gamma_{WG}(H))$. In the $H \to \infty$ limit

$$p - p_L^+ = (p - p_0)(\rho_G - \rho_L)/\rho_G \qquad (11)$$

where ρ_G and ρ_L are the coexisting bulk densities at T. From (10) and (11) the Kelvin equation is obtained

$$\Delta p = p - p_0 = \frac{\rho_G \, \Delta\Omega_{LG}}{(\rho_L - \rho_G)H} \qquad (12)$$

where $\Delta\Omega_{LG}$ is evaluated in the limit $H = \infty$; this macroscopic expression predicts a presaturation proportional to H^{-1}. Each wall generates an oscillating structure which decays towards the centre of the capillary; when H is small enough these structures interfere each other positively or negatively depending on if H is odd or even respectively, decreasing or increasing the free energy of the system. The result is a larger capillary shift of Δp for odd H than for even H instead the monotonous prediction of Kelvin equation (12). In a continuous description the capillary condensation isotherm should have an oscillating structure for small H^7. For large H this effect decreases and the smooth dependence predicted by (12) is recovered.

The above analysis when applied to a solid configuration gives place to a different capillary behaviour. For odd H the ordered structure induced by the walls is the same, so the capillary is filled up with the ordered phase whose structure is enhanced near each wall; in this case the capillary excess of grand potential is

$$\omega_s = 2\gamma_{ws} \quad \text{(odd H)} \qquad (13)$$

For even H the walls select ordered structures which are incompatible with each other, then if the wall-particle interaction is weak enough, the ordered structure selected by one of the walls extends over the capillary with an enhancing ordering near this wall and a damping ordering near the other. The corresponding capillary excess is

$$\omega_s = \gamma_{ws} + \gamma_{ws}' \quad \text{(small } V_H\text{, even H)} \qquad (14)$$

where γ_{ws} and γ_{ws}' are the interfacial tensions of the two asymmetric interfaces. In the case of a strong wall-particle interaction, as it is our case, the ordered structures induced by the walls prevails from each wall to a region in the middle of the capillary where they interfere negatively producing a solid-solid interface. If we call γ_{ss} the free energy of this frustration, then

$$\omega_s = 2\gamma_{ws} + \gamma_{ss} \quad \text{(strong } V_H\text{, even H)} \qquad (15)$$

Doing a similar analysis to that done to obtain the Kelvin equation we would find for the capillary crystallization from the liquid:

$$\Delta p = \frac{\rho_L \, \Delta\Omega_{SL}}{(\rho_S - \rho_L)H} \qquad (16)$$

where ρ_L and ρ_S are the bulk coexistence densities and $\Delta\Omega_{SL} = \omega_S - \omega_L$. An analogous results is obtained for the crystallization from the gas phase. As ω_S depends on whether H takes odd or even values, so it happens to the capillary pressure shift (16); we have then two Kelvin equations for capillary crystallization one for odd H and other for even H no matter how large H is. This contrasts with the unique expression (12) for the capillary condensation. A continuous description of the capillary crystallization would predict an oscillating structure for the capillary isotherm which extends to large wall separations.

In our specific case (strong V_H) there is perfect wetting in different configurations (see eqs.(3), (4) and (5)), then the Kelvin equation for crystallization reduces to

$$\Delta p = -\frac{\rho_L}{\rho_S - \rho_L} \, 2\gamma_{SL} \qquad \text{(odd H)} \qquad (17)$$

and

$$\Delta p = -\frac{\rho_L}{\rho_S - \rho_L} (2\gamma_{SL} - \gamma_{SS}) \qquad \text{(even H)} \qquad (18)$$

and similar expressions for crystallization from the gas. It could happen, as it does for our chosen interaction parameters, that a liquid wetting film would appear in the mismatch of the solid-solid interface for even H; the resulting structure is similar to that of the liquid with two solid wetting layers at the walls. Then the crystallization from the liquid goes smoothly without sharp transition for even H but it is still a first orden transition for odd H. In this case $\gamma_{SS} = 2\gamma_{SL}$ and $\Delta p = 0$ for even H. In a continuous description this effect should split the oscillating structure of the capillary crystallization (from the liquid) isotherm into a series of segments. Each segment would end in two critical points and they should be centred at wall separations where the wall induced structures are compatible. These consequences of the wetting regime should be observed no matter how large H is. Of course as H increases, though this effects exist, they are damped and it could happen that the critical points of two sucessive segments colapse in a unique critical point. These kind of questions can only be answered with a continuous model. The macroscopic approach can also be done for the capillary temperature shift of the triple point. The condition of phase coexistence $\Omega_S = \Omega_L = \Omega_G$ in the H 0 limit gives

$$\Delta T = T_T(H) - T_T(\infty) = -(1/H)\left[\frac{\Sigma(\omega_\alpha \rho_\beta - \omega_\beta \rho_\alpha)}{\Sigma(\rho_\alpha - \rho_\beta)(dp/dT)_{\alpha\beta}}\right] T_T(\infty) \qquad (19)$$

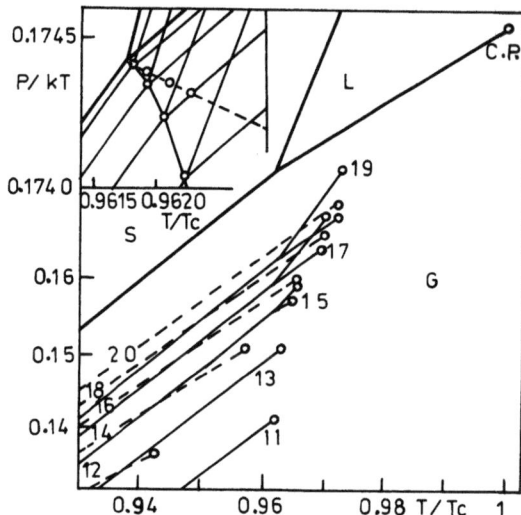

Figure 2.- p-T phase diagram of the present lattice gas model for bulk (thick line), for capillaries with odd H (continuous line) and for capillaries with even H (dashed line). H is the capillary size. Triple point shift for large capillaries is shown in the inset: dashed line: macroscopic prediction; continuous line: microscopic calculation. The marked points correspond to H=99,149,299,499 and 999.

where $\alpha, \beta = S, L, G$ and $(dp/dT)_{\alpha\beta}$ is the slope of the $\alpha\beta$ coexistence line in the bulk phase diagram. Here again the differences of ω_s with H are transmited to $T_T(H)$. In our wetting regime where the solid-liquid has been smoothed for even H the triple point dissapears while ΔT increases with H^{-1}. In a continuous description and in a non-wetting regime it is expected that ΔT oscillates with H with a period of the order of the layer width. The oscillation should remain no matter how large H is, though it should be damped with H. The inset of Fig.2 shows the macroscopic prediction of the triple point shift for odd H. This is a relative large shift which is determined basically by γ_{SG} and γ_{LG}. For $\gamma_{SG} \simeq 2\gamma_{LG}$ the macroscopic analysis predicts at given T a relative greater pressure shift for the capillary crystallization (from gas) than for the capillary condensation. This makes the crossing of the capillary solid-gas and liquid-gas coexistence lines to move towards high temperatures. However, the macroscopic analysis begins to fail, in our model, for H<1000. This is a

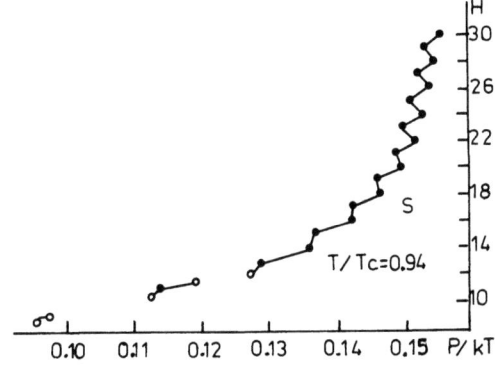

Figure 3.- Capillary coexistence at T=0.94 Tc. Vertical asymptote corresponds to the bulk solid-gas transition.

direct consequence of the presence of wetting layers which preclude the use of the microscopic analysis for small capillaries[8].

NUMERICAL RESULTS

Fig.2 also shows the results of the microscopic calculations of our model for H<20, and for few large values of H (inset). The increasing pressure shift Δp with H^{-1} is observed not only for the condensation but also for crystallization as expected; there are clear asymmetries between the coexistence lines corresponding to odd values and those corresponding to even values of H. The most evident corresponds to the lack of the solid-liquid coexistence line for even H as a consequence of the wetting regime, which at the same time makes the triple point to disappear. As expected the condensation critical point recedes towards lower temperatures as H decreases; we also find (for odd H) a critical point where the solid-liquid coexistence line ends whose behaviour with H is similar to that of the critical condensation. As bulk phase diagrams of real systems have no crystallization critical points the question if the existence or not of these points is an artefact which arises from our model. We believe that the confinement can be an explanation for these points when H is small enough. In any case, most probably, the quantitative receding towards low temperatures of the solid-liquid critical point shown in figure 2 is affected by the existence of the tricritical point in the bulk solid-liquid coexistence line of our model. The crystallization and condensation critical points collapse into a unique solid-fluid critical point for H<15, producing the disappearance of the triple point. For small H the solid-fluid critical point moves towards lower temperatures much faster for even H than for odd H. The macroscopic-microscopic predictions are compared in the inset where the agreement is only found for H>1000.

In Fig.3 and 4 we present capillary isotherms including gas, liquid and solid phases. The first one corresponds to T=0.94 T_c slightly below T_T(H=∞), so that the only bulk phase transition is the solid-gas one which is marked as the vertical asymptote. This transition is shifted inside the capillary, and for high H the shift is well approximated by the modified Kelvin equations. The frustation of the solid for even H produces the oscillatory structure. For narrower capillaries this effect is stronger and from N=12 to 7 the phase transition disappears for even H while it exists for odd H.

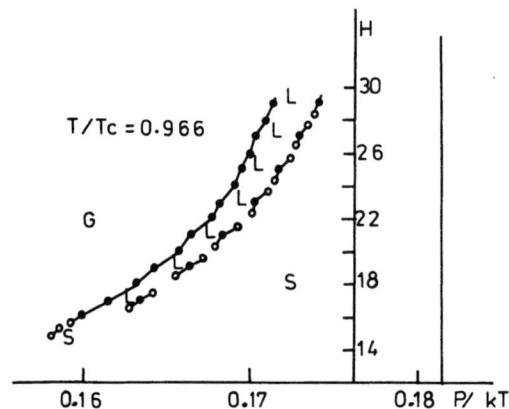

Figure 4.- Capillary coexistence at T=0.966 T_c. Vertical asymptotes correspond to the bulk gas-liquid and liquid-solid transitions.

Figure 4 corresponds to T=.966 T_c, above the bulk triple point; it shows two vertical asymptotes related to the bulk gas-liquid condensation and to the bulk liquid-solid crystallization. The most remarkable feature of the capillary phase diagram is the presence of a series of pockets of liquid between the gas and the solid states, for odd values of H, while for even H (in our case) the transition from solid to liquid is always smooth, as explained above. If a different kind of walls were used, avoiding the complete wetting of the wall-liquid interface by the solid phase, one should expect that for large enough H the solid-liquid transition in the capillary will appear, approaching the bulk value asymtotically as predicted by (16). In this case the pockets of liquid will merge in a single liquid region.

ACKNOWLEDGEMENT

The authors thank the British Council and Ministerio de Educación y Ciencia of Spain for the "Acciones Integradas" grant which allowed us helpful discussions with Dr. R. Evans and Dr. U. Marini Bettolo Marconi. We also wish to thank Prof. K. Gubbins for helpful discussions. The work has been suported by the Comisión Asesora de Investigación y Técnica of Spain and the Spain-USA Joint Committee for Scientific and Technical Cooperation (CCB8504023).

References

1. R. Evans and P. Tarazona, Phys. Rev. Lett., **52**, 557 (1984).
2. H. Nakanishi and M.E. Fisher, J. Chem. Phys., **78**, 3279 (1983).
3. R. Evans, U. Marini Bettolo Marconi and P. Tarazona, J. Chem. Phys., **84**, 2376 (1986).
4. See the discussion in the Faraday Society Symposium No.20 (1986). In the opposite limit, for very large H, the capillary critical point is better understood in terms of the shift of the bulk critical point and may be analysed with finite size scaling methods, see for example H.N. Barber in Phase Transition and Critical Phenomena, v.18, p.145, ed. C. Domb and I.L. Lebowitz, Academic Press (1983).
5. R.A. Horn and I.N. Israelachvili, Chem. Phys. Lett., **71**, 192 (1980).
6. B. Pansu, P. Pieranski and L. Strzelecki, J. Physique, **44**, 531, (1983); P. Pieranski, Contemp. Phys., **24**, 25 (1983).
7. This effect has already been observed with a continuous model by R. Evans, U.M.B. Marconi and P. Tarazona (To be published).
8. R. Evans and U. Marini Bettolo Marconi, Chem. Phys. Lett., **144**, 415 (1985).

THE LIQUID-SOLID TWO-PHASE COEXISTENCE

M. Baus*
Chimie-Physique II**, C. P. 231
Université Libre de Bruxelles
B-1050 Brussels
Belgium

1. INTRODUCTION

Phase transitions are macroscopic phenomena which one should in principle be able to describe by equilibrium statistical mechanics. Progress has however been very slow in this field over the past decades. This is particularly true for the liquid-solid transition which is nevertheless a very general property of matter. During the present decade some progress has been achieved in the theoretical study of the freezing of simple model systems such as the hard sphere system. Analysis of the experimental [1] and computer simulation[2] studies makes it moreover plausible that the freezing of more realistic systems is monitored by the freezing of some underlying hard sphere system so that the theoretical study of the liquid-solid coexistence of more realistic systems may soon also become accessible to equilibrium statistical mechanics. Considerable progress has been realized in recent years in the theory of freezing by

---

* Chercheur Qualifié du F. N. R. S.
** Association Euratom-Etat Belge.

reformulating the pioneering (but unsuccessful) work of Kirkwood and Monroe into the more modern language of the density functional theory[3]. The original work of Kirkwood and Monroe[4] was formulated on the basis of the Born-Green-Yvon hierarchy which in retrospect is not a good starting point since this hierarchy depends explicitly on the interaction potential whereas freezing is known[2] to be largely independent of the details of the potential. The recent success of the density functional theory of freezing rests partly on the fact that it is a direct correlation function based theory in which the potential does not appear explicitly. The density functional theory gives moreover easy access to the free energy, a central quantity for the study of phase transitions, whereas this is not the case if one starts from the ordinary correlation functions.

2. DENSITY FUNCTIONAL THEORY

In the density functional theory the central quantity is the Helmholtz free energy, F, which is viewed as a functional of the one-particle density, $\rho(r)$, a fact which is indicated by a square bracket, $F = F[\rho]$ The free energy density functional can be split into three terms, $F = F_{conf} + F_{corr} + F_{ext}$. The first term:

$$\beta F_{conf}[\rho] = \int_v \delta r \, \rho(r) \, (\ln (\lambda^3 \rho(r)) - 1) \tag{1}$$

describes the configurational part of the free energy in terms of the one-particle configuration or density, $\rho(r)$, and thermal de Broglie wavelength λ ($\lambda = k/2\pi m k_B T)^{1/2}$ for particles of mass m and inverse temperature $\beta = 1/k_B T$. In the appropriate limit eq. (1) reduces to the ideal gas free energy. The second term, F_{corr}, corresponds to the correlational contribution to the free energy. This part is not known explicitly but is directly related to the direct correlation function, $C(\mathbf{r},\mathbf{r}';[\rho])$, through the relation[3]:

$$\beta \frac{\delta^2 F_{corr}[\rho]}{\delta\rho(\mathbf{r}) \delta\rho(\mathbf{r}')} = -C(\mathbf{r},\mathbf{r}';[\rho]) \qquad (2)$$

which can be integrated functionally in density space along a linear trajectory running between the actual density and the density of some referencece state (subscript R) as[3]:

$$\beta (F_{corr}[\rho] - F_{corr}[\rho_R]) = -\int_V d\mathbf{r} \int_V d\mathbf{r}' \int_0^1 \delta\lambda \, C(\mathbf{r},\mathbf{r}';[\rho_R + \lambda\Delta\rho]) \times \Delta\rho(\mathbf{r}) \Delta\rho(\mathbf{r}') \qquad (3)$$

where $\Delta\rho(\mathbf{r}) = \rho(\mathbf{r}) - \rho_R(\mathbf{r})$. Finally, there also is the contribution from the external field, $\phi_{ext}(\mathbf{r})$:

$$\beta F_{ext}[\mathbf{r}] = \int_V d\mathbf{r} \, \rho(\mathbf{r}) \phi_{ext}(\mathbf{r}) \qquad (4)$$

which allows us to fix the solid with respect to the given volume V but which, in the thermodynamic limit of an infinite system, plays no further role provided the intensive quantities, such as the average density ρ:

$$\rho = (1/V) \int_V d\mathbf{r} \, \rho(\mathbf{r}) \qquad (5)$$

remain finite. From the free energy, F, one can obtain the grand potential, Ω, by a Legendre transformation:

$$\Omega[\rho] = F[\rho] - \int d\mathbf{r} \, \rho(\mathbf{r}) (\delta F[\rho]/\delta\rho(\mathbf{r})) \qquad (6)$$

and hence the pressure, p, from $\Omega = F - \mu \tilde{N} = -pV$ where $\tilde{N} = \int d\mathbf{r} \, \rho(\mathbf{r})$ and $\mu = \delta F[\rho]/\delta\rho(\mathbf{r})$ is the chemical potential. In this way we obtain easy acces to the conditions of two phase coexistence, namely the equality of the chemical potentials, $\mu_S = \mu_F$, and of the pressures $p_S = p_F$ of the solid (S) and fluid (F) phases of equal temperature ($T_S = T_F$).

3. LIQUID-PHASE BASED APPROXIMATIONS

The density functional theory provides us with a convenient framework in which the equations governing the two-phase liquid-solid coexistence conditions can be compactly formulated. To proceed it is however necessary to introduce approximations. It was the original idea of Kirkwood and Monroe [4] to suggest that one could approximate the unknown pair correlations of the solid phase by those of the much better known liquid phase while the anisotropy of the solid would then be retained only at the one-body level. The idea behind this approximation being that the liquid-phase correlations are similar to those of the angular averaged solid. Basically, the direct correlations of the anisotropic solid phase will be described here in terms of the direct correlations of some isotropic reference liquid and the different theories of freezing differ, between other things, mainly in the way in which this approximation is carried out. A first group of authors[5] has considered the reference liquid to be the coexisting liquid itself. Since the density gap between the solid and the coexisting liquid is not small these authors usually have to include higher order terms in the expansion of the free energy of the solid around that of the coexisting liquid. A second group of authors[6] has described the free energy of the solid by that of the coexisting liquid but evaluated in terms of a coarse-grained solid density. Here the whole burden of the theory amounts to properly determine the weighting function which is used to coarse-grain the solid density. A final group of authors[7] has optimized the free energy of the solid with respect to the density of the reference liquid by considering this density as a variational parameter or by scaling the correlations of the reference liquid with those of the solid. Finally, a more technical distinction concerns the approaches based on a reciprocal lattice Fourier expansion of the one particle density and those using a real space expansion of the density in a set of Gaussian peaks.

Whereas in reality there are always departures from gaussianity, the latter are usually small. The Fourier expansion can, in principle, cope with these departures but, in practice, this method requires a very large number of Fourier components to be determined in order to describe the strongly localized density peaks of the solid and usually one runs into convergence problems.

4. THE FREEZING OF HARD SPHERES

As an illustration of the results which can be obtained by the above methods we consider the freezing of a system of hard spheres of diameter σ. This infinitely steep repulsive potential only mimics the repulsive part of the more realistic potentials but this is enough for freezing which is due to a competition between the excluded volume effect (F_{corr}) and the configurational entropy contribution (F_{conf}) to the free energy (F). The hard sphere transition has also been thoroughly studied by computer simulations[2] and offers hence a good testing ground for the theoretical approaches to freezing. Recently, the simulation results have been corroborated by a laboratory experimet[8] preformed with concentrated suspensions of spherical colloidal particles.

For the theoretical results on the freezing of hard spheres according to the above ideas we will follow our recent results[9]. The fluid of hard spheres (there is no liquid-gas transition for this system) can be described fairly accurately by the Percus-Yevick equations for the correlation functions. Since the latter can be solved analytically [7,9] for the direct correlation function we can, when using the Gaussian trial functions for the one-particle density, obtain an explicit expression for the free energy. Minimizing this free energy with respect to the widths of the Gaussians we obtain the

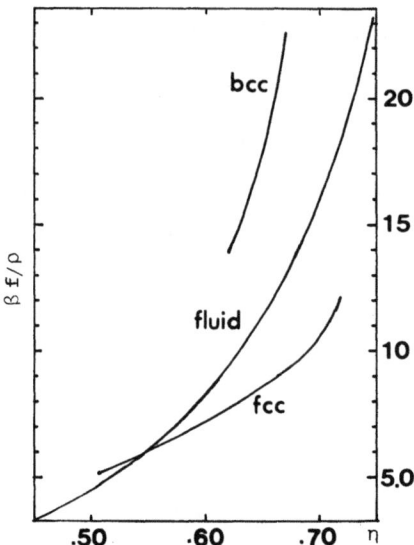

Fig. 1. The reduced free energy per particle, $\beta f/\rho$ (f is the free energy density and $\beta = 1/k_B T$), versus the packing fraction, $\eta = (\pi/6)\sigma^3\rho$, for three phases (fluid, bcc and fcc solids) of a system of hard spheres of diameter σ and average number density ρ.

results[9] shown in Figure 1. It is seen there that the body centered cubic hard sphere solid is metastable relative to the hard sphere fluid whereas the face centered cubic solid is stable at high density and metastable at low density. In each case the solid is stable (relative to particle localisation) only in a finite density domain (the temperature scales out for hard spheres) extending from a lower bifurcation point, below which the solid is mechanically unstable, up to the density of crystal close packing.

The pressure-density phase diagram of the hard sphere fluid and fcc solid is compared to the corresponding result of the computer simulations[2] in Figure 2. It is seen that the pressure is slightly overestimated by the theory in both the fluid and the solid phase as are the coexisting densities.

Some of the details of the hard sphere transition can also be found in Table 1.

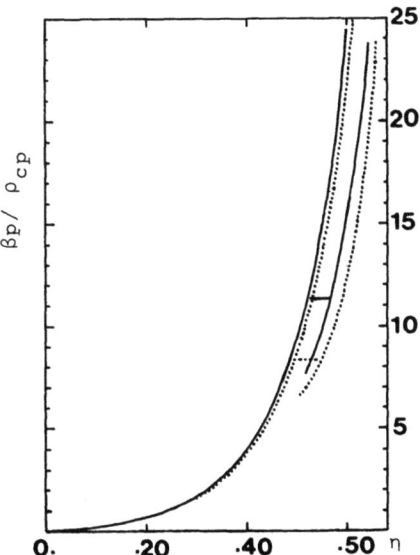

Fig. 2. The reduced pressure, bp/ρ_{cp} (ρ_{cp} is the density of fcc close packing), versus the packing fraction, η, for the stable and metastable fluid and fcc-solid phases of a system of hard spheres. The full lines correspond to the theoretical results[9] and the dotted lines to the results of the computer simulations[2].

Table 1. Characteristics of the hard sphere transition

	η(fluid)	η(solid)	βp/ρ (coex)	$(s_F - s_S)/k$
Theory	0.567	0.520	16.2	1.36
Computer simulations[2]	0.545	0.494	12.4	1.16

s - entropy per particle

4. CONCLUSIONS

From the above results it appears clearly that at least the simplest of all possible freezing transitions, the hard sphere transition, is correctly described by the above density functional theory. Some progress on more complicated systems has also been made recently[10]. Several studies are now considering the explicit inclusion within this theory of a description of the liquid-solid interface[11]. Although not everything is understood and a systematic approximation scheme has still to be devised the progress realized is considerable.

REFERENCES

1. S. M. Stishov, *Sov. Phys. Usp.* **17**, 625 (1975)

2. D. Frenkel and J. P. McTague, *Ann. Rev. Phys. Chem.*, **31**, 491 (1980)

3. R. Evans, *Adv. Phys.*, **28**, 143 (1979)

4. J. G. Kirkwood and E. Monroe, *J. Chem. Phys.*, **9**, 514 (1941)

5. T. V. Ramakrishnan and M. Yussouff, *Phys. Rev.*, **B19**, 2775 (1979)
   V. N. Ryzhov and E. E. Tareyeva, *Theor. Math. Phys*, **48**, 835 (1981)
   N. H. March and M. P. Tosi, *Phys. Chem. Liq*, **11**, 79 (1981)
   A. D. J. Haymet and D. W. Oxtoby, *J. Chem. Phys.*, **74**, 2559 (1981)
   G. L. Jones and U. Mohanty, *Molec. Phys.*, **54**, 1241 (1985)

6. P. Tarazona, *Molec. Phys*, **52**, 81 (1984)
   W. A. Curtin and N. W. Ashcroft, *Phys. Rev*, **A32**, 2909 (1985)

7. M. Baus and J. L. Colot, *Molec. Phys*, **55**, 653 (1985)
   J. L. Barrat, M. Baus and J. P. Hansen, *Phys. Rev. Lett*, **56**, 1063 (1986)
   J. P. Stoessel and P. B. Wolynes, *J. Chem. Phys* (to appear)
   F. Igloi and J. Hafner, *J. of Phys. C* (to appear)

8. P. N. Pusey and W. Van Megen, *Nature*, **320**, 340 (1986)

9. M. Baus and J. L. Colot, *J. of Phys* C, **18**, L365 (1985)
   J. L. Colot and M. Baus, *Molec. Phys*, **56**, 807 (1985)
   J. L. Colot and M. Baus and H. Xu, *Molec. Phys*, **57**, 809 (1986)

10. C. Marshall, B. B. Laird and A. D. J. Haymet, *Chem. Phys. Let.*, **122**, 320 (1985)
    W. A. Curtin and N. W. Ashcroft, *Phys. Rev. Lett*, **56**, 2775 (1986)

11. D. W. Oxtoby and A. D. J. Haymet, *J. Chem. Phys*, **76**, 6262 (1982)
    S. M. Moore and H. J. Ravéché, to be published

COEXISTENCE OF TWO PHASES:

LONG-RUN MOLECULAR DYNAMICS COMPUTER SIMULATIONS

J.J. Morales[1], F. Cuadros[1] and L.F. Rull[2]

(1) Dpto. Física. Facultad de Ciencias, 06071 Badajoz
(2) Dpto. Física Teórica. Facultad de Física, 41012 Sevilla

In traditional DM experiments, the number of atoms in a volume is fixed and the total energy conserved as the dynamics of the systems evolves in time. The time average of any property is an approximate measure of the microcanonical ensemble average of that property MD (EVN).

For the same thermodynamic constraints, averaging over ensemble space and averaging over time are identical if the system is ergodic.

The MD method allows one to obtain not only the static but also the dynamics quantities. This is one advantage over the MC method. However, a disadvantage of the MD method is that the conditions of the simulations are not the same as those normally encountered in experiments; for example, constant temperature, constant pressure or (TPN) conditions. In this regard, Andersen's (1) introduction of the constant pressure MD method represented a significant breakthrough. To perform MD (HPN) simulations, we can keep the pressure constant by the scaling the dimension of the cell at each time step, and to perform MD (TVN) simulations we can keep the kinetic energy constant by scaling the velocities at each time step. The extension of the MD method to treat ensembles other than the traditional microcanonical ensemble has attracted considerable attention. Thus Abraham and Koch (2), using a special computational technique MD (TPN) inspired by the MC (TPN) ensemble, investigated melting in a two dimensional Lennard-Jones systems.

Toxvaerd (3) made a critical study of this technique calculating the isothermal compressibility from the fluctuation of volume. He established that the scaling procedure was incorrect and cannot be used to obtain definitive statements about the nature of the plase transition. The reason was because the calculations of Abraham and Koch are obtained from the Laplace transform of the canonical partition function, and this transformation has to be performed for constant extensive state variables except for the volume.

In an analysis of the ensembles used in MD simulations, Lado (4) found that, in the classical limit, for the interacting systems with hard-core interaction, the MC and MD calculations without corrections should yield the same equation of state.

Nosé (5) has made an exhaustive study of the ensembles in MD to reproduce both the canonical and the isothermal-isobaric probability densities in phase space. The physical system of interest consists of N particles, with f degrees of freedom, to which an external macroscopic variable and its conjugate momentum are added, permitting the total energy of the system, by introduction of an additional degree of freedom s, to fluctuate in a way similar to Haile and Gupta's results (6) with a thermal bath. This method is purely dynamical and, in the extended system of the particles and the coordinate s, the total Hamiltonian is conserved and all the equations of motion are solvable without introducing any stochastic process.

Recently, Hoover (7) modified Nosé's equations and developed a slightly different set of equations, free of time scaling s, by introducing two variables which act as thermodynamic coefficients of friction with Gaussian distributions. This procedure is called the Nose-Hoover method and is coming to be applied in computer simulations for the study of fluid and solid (8).

In this chapter we present some results of long runs in two dimensions with the traditional MD technique (EVN), i.e., without any scaling procedures (pressure and/or temperature), because we think it is incorrect to perform simulations using the constant pressure and/or constant temperature methods in the melting zone as there are problems related with the fluctuations in space and time independent of the nature of the phase transition (first or second order). In Sect. I, we shall describe the details of the systems and the algorithms used in our computer simulations. In Sect. II, we show the problems of the fluctuations close to and far from the melting zone with a system of N=256 particles. Our results with a system of N=576 particles compared with those obtained by Barker et al (9) with MC (TVN) at a point within the melting zone are shown in Sect. III. Finally, we have made a study of the correlations in the data sequences obtained in our simulations using the method suggested by Smith and Wells (10).

I. SYSTEMS AND ALGORITHMS

We used a two dimensional systems of N=256 and 576 particles in a rectangular cell with the ratio $\overline{\sqrt{3}}/2$ between the edge lengths. The initial position of the particles formed a triangular lattice and their interaction was through a truncated Lennard-Jones potential:

$$u(r) = 4\{(\sigma/r)^{12} - (\sigma/r)^6\}, \quad r < 2.5\, r_m$$
$$u(r) = 0, \quad r > 2.5\, r_m \quad (1)$$

where r_m is the distance of the minimum of the potential ($r_m = 2^{1/6}\sigma$).

This potential is strongly repulsive at short distances with a weak long range atraction. The rapid variation in time for the short range force (11) makes the differential equations stiff (12) and, near melting, the high energy collisions might be a trigger of the phase transition by producing the artificial fluctuations needed to reach the new state. Therefore we used two different algorithms to solve numerically the Newtonian differential equations in order to see the influence they would have on the macroscopic properties: the Verlet (13) algorithm (third order predictor) and Toxvaerd (14) algorithm (a fifth order predictor).

The Verlet algorithm is defined by

$$\vec{r}_i(t+h) = 2\vec{r}_i(t) - \vec{r}_i(t-h) + h^2 \vec{a}_i(t) + \theta(h^4) \quad (2)$$

$$\vec{v}_i(t+h) = (\vec{r}_i(t+h) - \vec{r}_i(t))/2h + \theta(h^2) \quad (3)$$

with a time step $h = 0.005\, (m\sigma^2/\epsilon)^{1/2} \approx 10^{-14}$ s.

The Toxvaerd algorithm is defined by

$$\vec{r}_i(t+h) = 2\vec{r}_i(t) - \vec{r}_i(t-h) + h^2 \vec{a}_i(t) + \frac{1}{12} h^4 \vec{a}''_i(t) + \theta(h^6) \quad (4)$$

$$\vec{v}_i(t+h) = (\vec{r}_i(t+h) - \vec{r}_i(t))/2h - \frac{1}{6} h^2 \vec{r}'''_i(t) + \theta(h^4) \quad (5)$$

where the prime means derivative with respect to time, and:

$$\vec{a}''_i(t) = \sum_{j \neq i}^{N} \{[B_{ij}(\vec{r}_i \vec{r}''_j + \vec{r}'^2_{ij}) + C_{ij}(\vec{r}_{ij} \vec{r}'_{ij})^2] \vec{r}_{ij} + 2B_{ij}(\vec{r}_{ij} \vec{r}'_{ij}) \vec{r}'_{ij} + A_{ij} \vec{r}''_{ij}\} \quad (6)$$

and

$$\vec{r}'''_i(t) = \sum_{j \neq i}^{N} \{A_{ij} \vec{r}'_{ij} + B_{ij}(\vec{r}_{ij} \vec{r}'_{ij}) \vec{r}_{ij}\} \quad (7)$$

where

$$A_{ij} = -\frac{1}{r_{ij}} \frac{du(r_{ij})}{dr_{ij}}, \quad B_{ij} = \frac{1}{r_{ij}} \frac{dA_{ij}}{dr_{ij}}, \quad C_{ij} = \frac{1}{r_{ij}} \frac{dB_{ij}}{dr_{ij}} \quad (8)$$

We calculated $\vec{a}_i''(t)$ and $\vec{r}_i'''(t)$ for distances lying between σ and r_m, where the pair potential is strongly repulsive.

The first algorithm has the advantage that as it is symmetric in time, there will always be conservation of energy, even in zones where there are large fluctuations. Toxvaerd's algorithm, on the other hand, gives a greater accuracy in the position of the particles when they are close together, but has the disadvantage of not being symmetric in time, because the determination of $\vec{r}_i'''(t)$ and $\vec{a}_i''(t)$ implies knowing not only $\vec{a}_i(t)$ but also the velocities $\vec{r}_i'(t)$ which are calculated using a third order predictor (14).

$$h\vec{r}_i'(t) = \vec{r}_i(t) - \vec{r}_i(t-h) + (1/6)h^2 \{ 2\vec{a}_i(t) + \vec{a}_i(t-h) \} + \theta(h^4) \quad (9)$$

Although a small drift in the energy exists, this problem can be solved by rescaling the velocities of the particles through the heat capacity (15).

The thermodynamic variables (T,p,U) are calculated during the free evolution of the systems (without any constraints) as time averages of the associated microscopic variables and c_v and $(\partial p/\partial T)$ from the theory of fluctuations (16) in the microcanonical ensemble:

$$c_v^{-1} = 1 - N <\delta T^2>/T^2 \quad (10)$$

and

$$\left(\frac{\partial p}{\partial T}\right)_\rho = -N\rho <\delta (\frac{p}{\rho} - T)\delta T> c_v/T^2 \quad (11)$$

where the brackets represent time averages, and $\delta x = x(t) - <x(t)>$

II. THE PROBLEM OF THE FLUCTUATIONS

To study the problem of the fluctuations we have chosen a system of N=256 particles at the densities $\rho\sigma^2=0.7937$ and $\rho\sigma^2=0.9048$.

These choices correspond respectively to the fluid state, far for the melting zone, and the solid state very close to the melting zone. This latter zone was located thermodynamically by Toxvaerd (17) who calculated the points where fluid and solid have equal chemical potentials corresponding to the temperatura $kT/\epsilon=1.00$, the difference in our case being that we have performed long runs (40 000 time steps for the liquid and 80 000 time steps for de solid) and we used Toxvaerd's own algorithm to integrate the equations of motion while he used the Verlet algorithm.

Both systems evolved in exactly the same way. After every ten time steps (updating the table of the nearest neighbors), the velocities were renormalized to give the reduced temperatura $kT/\varepsilon = 1.00$ (ε: minimum of the Lennard-Jones potential). The rescaling procedure was performed during the first 8 000 time steps. After that the evolution of the system was free, without any scaling procedure, the thermodynamic properties being then obtained as time averages for each 800 time steps. The virial pressure was obtained for the truncated Lennard-Jones potential, not for a full Lennard-Jones potential.

The time evolution in the liquid system for temperature and pressure shows the "regular" fluctuation typical of the liquid system far from the melting zone, and the system is in equilibrium once free evolution commences (18). But this is not the case for the system close to the melting zone. In Figure 1, we show the results of the simulations for reduced temperature (top) and reduced pressure (bottom) versus time for the solid. Each point in these figures corresponds to an average over 800 time steps. One can see that equilibrium is reached after 20 000 time steps, and that the temperature spontaneously drops twice (arrows) and consequently the pressure rises as would be expected in the microcanonical ensemble (EVN) in free evolution.

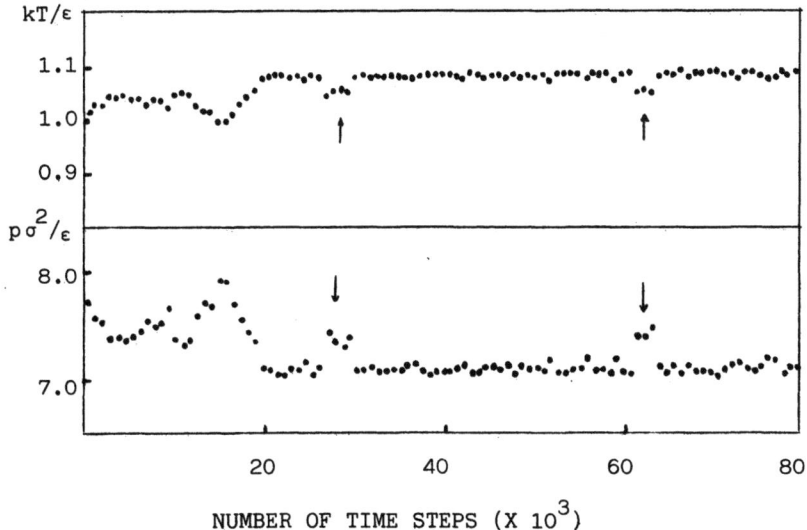

FIGURE 1. Reduced temperature (top) and reduced pressure (bottom), versus time, for a system of N=256 particles at the density $\rho\sigma^2 = 0.9048$.

Toxvaerd (17) posed the question of whether the equilibrium state of the system could be obtained in a time interval of 10 000 time steps. Our answer is no because in the melting zone (or close to it) the system spontaneously goes back and forth between two different points of state. In our calculations the uncertainty in the temperature and pressure in the solid is four times greater than in the liquid as a consequence of the system's melting and freezing.

It seems evident that far from a transition in liquid or solid systems, computer simulations using the (TVN) or (TPN) methods are reliable because the fluctuations are small. However fluctuations play an important role in the melting zone, so to obtain a definitive statement about the nature of the phase transition one should use MD simulations in free evolution, without any scaling procedure, i.e. without temperature and/or pressure constraints. These constraints imply that the system is not isolated but is in contact with an energy reservoir, the energy is transferred by a generalized force (6), and the spontaneous fluctuations in the system are erased.

III. RESULTS OF MC(TVN) AND MD(EVN) CALCULATIONS

In this section our MD results will be compared with MC results for the same state points in the relevant parts of the phase diagram.

Tha phase diagram of a two dimensional system has been studied by several authors (9,17). Barker et al (9) carried out MC simulations for a system of N=256 particles for certain temperatures. Their MC(TVN) results agree with perturbation theory along the liquid branch, and with the self-consistent cell model along the solid branch. In the melting zone they used the isothermic-isobaric MC(TPN) method, and found that the results were the same as those obtained with MC(TVN). To calculate the melting pressure they chose a point P in the two phase region, with density $\rho\sigma^2$=0.84 and temperature kT/ϵ =0.7. The particle trajectories at that point clearly show the coexistence of solid and liquid regions, and the pressure obtained is the expected melting pressure.

Our results have been obtained with a two dimensional system of N=576 particles at three characteristic densities along the same isotherm kT/ϵ=0.7. The first point is in the liquid region ($\rho\sigma^2$=0.75), the second in the solid ($\rho\sigma^2$=0.90), both far from the melting zone, and the third is in the melting region ($\rho\sigma^2$=0.84), the point P in Ref. (9). The point P was reached by means of density scaling from both the liquid and solid state.

The particles were constrained to rescale their velocities (updating the table of the nearest neighbor) every ten time steps up to a total of

4 000, so that the systems adjust themselves to the isotherm $kT/\epsilon = 0.7$. After that, the systems were allowed to evolve freely for 16-28 000 time steps, and then instantaneously rescaled to the density of the point P. In this process, the liquid (solid) system has warmed up (cooled down) on passing to the melting zone and therefore the temperature has to be adjusted as before over 4 000 time steps. From this point on, the systems were left to evolve freely over long runs (105-140 000 time steps).

Table 1 shows results from both the MC(TVN) simulations of Barker et al (9) at $kT/\epsilon = 0.7$ for pressure and total energy per particle (we have included the kinetic contribution to the energy) and our MD(EVN) simulations using Verlet and Toxvaerd algorithms at the same state points. At the melting density the parentheses show whether the results have been scaled from the liquid (L) or from the solid (S). The last column shows the total number Nt of MD time steps. In our results we have added the asymptotic contribution to the pressure and energy per particle (17), setting $g(r)=1$ for all densities.

The pressures with the Verlet algorithm for the liquid and with the Toxvaerd algorithm for liquid and solid are the same as those obtained by Barker et al (9), while the pressure for the solid with the Verlet algorithm is a little higher.

At the point P, in the melting zone, there appear slight differences between the algorithms. While the two pressures obtained with the Toxvaerd algorithm are the same (within the deviation of the mean temperatures and the calculation errors), the Verlet algorithm gives higher values when starting from the solid phase, and those resulting from the liquid phase are lower than those of Barker et al. Thus the Verlet algorithm gives a sharper difference between the two pressures. These differences could be explained in the following way: while in the Toxvaerd algorithm the positions and velocities of the particles are calculated up to fifth and fourth order in h respectively in the repulsive part of the potential, in the Verlet algorithm the positions are calculated up to third order in h by means of $\vec{a}(t)$ obtained from the interparticle potential and the velocities are calculated to first order in h with a simple midpoint evaluation.

The fluctuations behave the same with both algorithms, being two or three times greater in the solid than in the liquid, while the fluctuations are increased eight to ten times when the point P is scaled from the solid, and three to five times from the liquid. These fluctuations are typical of melting: the systems have both solid and liquid zones, as Barker et al (9) concluded from the particle trajectories at the same density, and we agree with them that the pressure at the point P provides us with a "direct coexistence" estimate of the melting pressure $p\sigma^2/\epsilon=2.53$.

Table 1. MonteCarlo results of Barker et al, and our results with the Verlet and Toxvaerd algorithms. At the density 0.84 the parentheses show whether the results were scaled from the liquid (L) or from the solid (S).

	$\rho\sigma^2$	kT/ϵ	$p\sigma^2/\epsilon$	$U/N\epsilon$	$Nt \times 10^3$
Barker et al's results MC(TVN)	0.75 0.90 0.84	0.7 0.7 0.7	1.20 3.60 2.53	-1.508 -2.053 -1.813	---
Our MD(EVN) results with the Verlet algorithm	0.75 0.90 0.84(L) 0.84(S)	0.683±0.002 0.702±0.001 0.702±0.010 0.680±0.009	1.16±0.02 3.70±0.01 2.63±0.08 2.34±0.08	-1.4578 -1.9559 -1.7112 -1.7789	16 20 140 120
Our MD(EVN) results with the Toxvaerd algorithm	0.75 0.90 0.84(L) 0.84(S)	0.074±0.003 0.694±0.001 0.707±0.009 0.693±0.010	1.25±0.03 3.64±0.01 2.64±0.08 2.52±0.08	-1.4296±0.0001 -1.9707±0.0007 -1.7162±0.0004 -1.7434±0.0002	28 20 115 105

The total energy per particle is qualitatively the same, although our results are slightly higher than those of Barker et al in all cases. This could be due to the ensemble differences and to the asymptotic contribution approximation. The main difference between the energies obtained with the two algorithms is the small drift with the Toxvaerd algorithm. It is of the order of 10^{-4} and the scaling of the velocities of the particles was not necessary because the drift produces a temperature variation of 10^{-7}, smaller than the precision of the calculations.

There is a similarity of behaviour between c_v and $\partial p/\partial T$ results with both algorithms. When the transition is reached from the liquid there exists an appreciable variation in these two values, but if it is from the solid there is no variation in the mean values. The dynamic time evolution of these magnitudes is qualitatively the same. In Figure 2, we show, as an example, the time evolution for c_v (top) and $\partial p/\partial T$ (bottom) during the free evolution of the solid system with the Verlet algorithm before the density scaling, at 20 000 time steps, and after that up to the total 140 000 time steps. One sees that the behaviour of $\partial p/\partial T$ is the same as c_v but sharper. That means that fluctuations in the kinetic energy (c_v) predominate over those in the pressure, and determine the characteristic behaviour of $\partial p/\partial T$. The value of $\partial p/\partial T$ is about four or five times greater than c_v and the uncertainty is ten times greater in all cases.

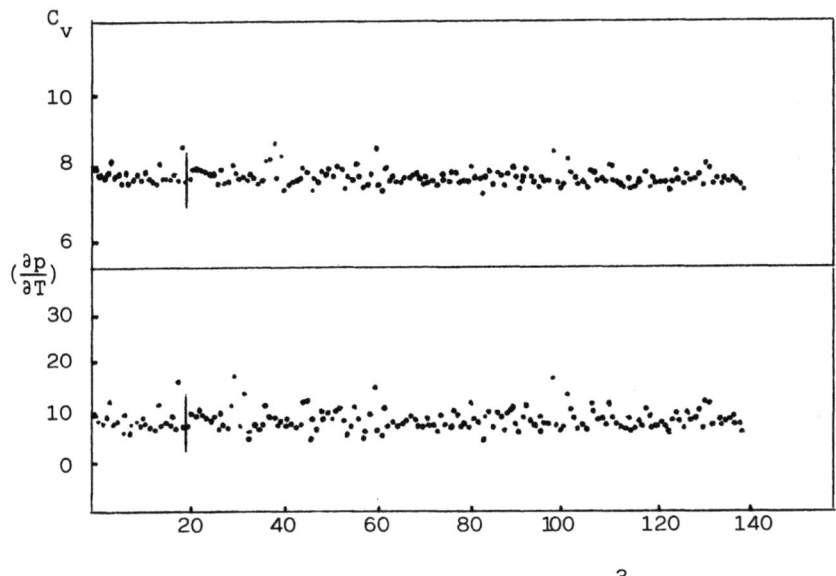

FIGURE 2. Heat capacity (top) and ∂p/∂T (bottom) versus time for a system of N=576 particles in the solid (first part) and melting using Verlet's algorithm.

In general, there is a considerable increase in fluctuations of the thermodynamic variables in the melting region as a consequence of the coexistence of the two phases, and one observes no real differences in using one or the other algorithm.

Averaging over space in the canonical ensemble with the MC method gives results in agreement with averaging over time in the microcanonical ensemble with the MD method, as would be expected in an ergodic system.

IV. STUDY OF CORRELATIONS IN THE DATA SEQUENCES

The general problem in the calculations of averages from computer simulation is the statistical error in the resulting serially correlated sequences.

In order to see if the averages were long enough, we studied the correlations using the method put forward by Smith and Wells (10). As a measure of the correlation in the data sequences, they propose the autocorrelation function $\hat{r}_k$, defined by

$$\hat{r}_k = \frac{\sum_{k=1}^{N}(Z_i-\bar{Z})(Z_{i-k}-\bar{Z})}{\sum_{i=1}^{N}(Z_i-\bar{Z})^2} \tag{12}$$

The $\{\hat{r}_k\}$ decay roughly geometrically as k increases.

805

The order of the correlation in the data sequences is calculated through the variance of the mean $\{Var(\bar{Z})\}$. Zeroth order is the value that would be obtained if the data were assumed to be uncorrelated, and first order is that calculated on the assumption that the series is a first order autoregressive process. The ratio between them is called the sampling ratio γ and is defined by

$$\gamma = \frac{\{Var(\bar{Z})\}_1}{\{Var(\bar{Z})\}_0} = \frac{1+\hat{r}_1}{1-\hat{r}_1} \qquad (13)$$

The values of γ that arises from any given data sequence is the factor by which the number of configurations sampled must be increased in order to obtain the same precision that would result from randomly distributed data points. If the data points were independent trials, γ would be unity.

Tables 2a and 2b show the correlation of the results from the simulation for temperature and pressure respectively using Verlet's algorithm. Tables 3a and 3b show the same for Toxvaerd's algorithm. In these tables, N is the number of sub-averages analysed and m is the number of time steps per average, so that mxN is the total number of steps taken in the free evolution of the systems ($\simeq N_t$ of Table 1).

For the cases of temperature and pressure of the solid and liquid systems with either algorithm, one sees that the results obtained for the variables $\hat{r}_1-\hat{r}_4$ are values which fluctuate positive and negative very close to zero, and with a value of γ near to unity. These results are therefore considered to be uncorrelated. This was to be expected from Table 1 in which the fluctuations are small in solid and liquid states far enough away from the phase transition.

This is not so for the case of the systems near the melting zone. Only the first five configurations have been represented (m=800, 1 600, 2 400, 3 200, and 4 000), up to fourth order in $\hat{r}_k$. In general, within a given configuration one observes a fall-off of the correlation with the increase in the order in k, and this same behaviour is seen when all configurations for a given value of $\hat{r}_k$ are taken into account. These features lead one to deduce that the correlations get drastically smaller with the increase in the number of steps per average and with the order in k.

The value of γ decreases with the increase in the number of time steps per average, tending to the value unity. In all cases in the melting zone, the values of γ with the Toxvaerd algorithm are greater than

Table 2a. Correlations of the temperature using the Verlet algorithm. The symbols are defined in the text.

	Liquid	Melting (from liquid)				
m	800	800	1600	2400	3200	4000
N	20	170	85	56	42	34
$\hat{r}_1$	.1658	.6747	.5182	.4714	.4077	.2662
$\hat{r}_2$	.0445	.4251	.2554	.1834	.0529	-.0216
$\hat{r}_3$	-.3148	.2962	.1819	-.1045	-.0517	-.1003
$\hat{r}_4$	-.0792	.2100	-.0058	-.0727	-.1578	-.0648
γ	1.40	5.15	3.15	2.78	2.38	1.73

	Solid	Melting (from solid)				
m	800	800	1600	2400	3200	4000
N	25	150	75	50	37	30
$\hat{r}_1$	-.1061	.6791	.4695	.2608	.0802	.0279
$\hat{r}_2$	-.2738	.3857	.0206	-.1404	-.2188	-.2418
$\hat{r}_3$	-.1569	.1559	-.0952	-.2413	-.0681	.1916
$\hat{r}_4$	-.0554	.0134	-.1700	-.0583	.1671	.0489
γ	.81	5.23	2.77	1.70	1.17	1.06

Table 2b. Correlations of the pressure using the Verlet algorithm. The symbols are defined in the text.

	Liquid	Melting (from liquid)				
m	800	800	1600	2400	3200	4000
N	20	173	86	57	43	34
$\hat{r}_1$	-.2966	.6591	.5510	.5138	.4514	.4112
$\hat{r}_2$	-.0731	.3923	.3122	.2465	.0434	-.1381
$\hat{r}_3$	-.0414	.3108	.2018	-.0901	-.0589	-.1674
$\hat{r}_4$	-.0125	.2260	.0050	-.1070	-.1433	-.0824
γ	.54	4.87	3.45	3.11	2.65	2.40

	Solid	Melting (from solid)				
m	800	800	1600	2400	3200	4000
N	25	150	75	50	37	30
$\hat{r}_1$	.1584	.6665	.4551	.2652	.1431	.0468
$\hat{r}_2$	-.2372	.3758	.0561	-.1388	-.2853	-.2444
$\hat{r}_3$	-.0723	.1639	-.1136	-.2287	.0257	.3125
$\hat{r}_4$	.1316	.0376	-.1783	-.0039	.2862	.0691
γ	1.38	5.00	2.67	1.72	1.33	1.10

Table 3a. The same as Table 2a but using the Toxvaerd algorithm.

	Liquid	Melting (from liquid)				
m	800	800	1600	2400	3200	4000
N	35	143	71	47	35	28
$\hat{r}_1$	-.2829	.6906	.6250	.6457	.5189	.5087
$\hat{r}_2$	.1553	.4848	.4241	.3073	.1601	-.0323
$\hat{r}_3$	-.0793	.4417	.2767	.0377	-.1155	-.1839
$\hat{r}_4$	.0928	.3994	.1341	-.1713	-.0953	.1469
γ	.56	5.46	4.33	4.64	3.16	3.07

	Solid	Melting (from solid)				
m	800	800	1600	2400	3200	4000
N	25	129	64	43	32	25
$\hat{r}_1$	-.0653	.7463	.6284	.5821	.4356	.3558
$\hat{r}_2$	-.0091	.5519	.3632	.2119	.0637	-.0551
$\hat{r}_3$	.0112	.4297	.1540	-.0150	-.1664	-.3446
$\hat{r}_4$	-.0923	.3673	-.0024	-.1619	-.2877	-.3049
γ	.88	6.88	4.38	3.79	2.54	2.10

Table 3b. The same as Table 2b but using the Toxvaerd algorithm.

	Liquid	Melting (from liquid)				
m	800	800	1600	2400	3200	4000
N	35	143	71	47	35	28
$\hat{r}_1$	-.1589	.7200	.6633	.6408	.5141	.4912
$\hat{r}_2$	-.1328	.5433	.4389	.2863	.1783	-.0162
$\hat{r}_3$	.0525	.4779	.2770	.0478	-.1087	-.1102
$\hat{r}_4$	-.2245	.4139	.1180	-.1600	.0051	.2773
γ	.73	6.14	4.94	4.56	3.11	2.93

	Solid	Melting (from solid)				
m	800	800	1600	2400	3200	4000
N	25	129	64	43	32	25
$\hat{r}_1$	-.0863	.7194	.5642	.5461	.3905	.3399
$\hat{r}_2$	-.1601	.4833	.3119	.1699	.0733	-.0426
$\hat{r}_3$	-.0828	.3690	.1232	-.0115	-.1358	-.1751
$\hat{r}_4$	-.0595	.3136	-.0045	-.1176	-.1448	-.1771
γ	.84	6.13	3.59	3.41	2.28	2.03

those obtained with the Verlet algorithm. This means that the Toxvaerd algorithm retains the correlations better than the Verlet algorithm, and in most cases for m=4 000, there continues to be a certain correlation between the results of the subaverages.

Even though the behaviour of T and p in the melting zone is qualitatively the same, one notices a more pronounced fall-off in the correlation when the rescaling was from the solid than when from the liquid side.

With respect to the results for c_v and $(\partial p/\partial T)$, the calculated values of $\hat{r}_1 - \hat{r}_4$ for the five values of m are all (positive or negative) very close to zero, giving values of $\gamma \simeq 1$ for the liquid, solid, and melting zone. The absence of correlations for these cases is because fluctuations in the values of c_v and $(\partial p/\partial T)$ are too high (greater than 0.1) and the resulting data sequence is uncorrelated.

We conclude that long observation times for temperature and pressure are only required where two phases coexist, because the fluctuations are intermediate in size (~0.01) and there is an appreciable correlation between the subaverages.

References

1) H.C. Andersen, J. Chem. Phys. $\underline{72}$, 2384 ((1980)
2) F.F. Abraham and S. Koch, Phys. Rev. B $\underline{29}$, 2824 (1984)
3) S. Toxvaerd, Phys. Rev. B $\underline{29}$, 2821 (1984)
4) F. Lado, J. Chem. Phys. $\underline{75}$, 5461 (1981)
5) S. Nosé, Mol. Phys. $\underline{52}$, 255 (1984),
6) J.M. Haile and S. Gupta, J. Chem. Phys. $\underline{79}$, 3067 (1983)
7) W.G. Hoover, Phys. Rev. A $\underline{31}$, 1695 (1985)
8) J.J. Morales, S. Toxvaerd and L.F. Rull, Phys. Rev. A, $\underline{34}$, 1495 (1986)
9) J.A. Barker, D. Henderson and F.F. Abraham, Physica 106A, 226 (1981)
10) E.B. Smith and B.H. Wells, Mol. Phys. $\underline{53}$, 701 (1984)
11) W.B. Street, D.J. Tildesley and G. Saville, Mol. Phys. $\underline{35}$, 639 (1978)
12) G.W. Gear, "Numerical initial value problems in ordinary differential equations". Chap. 11. Prentice Hall, Englewood Cliffs. NJ (1971)
13) L. Verlet, Phys. Rev. $\underline{159}$, 98 (1967)
14) S. Toxvaerd, J. Comp. Phys. $\underline{47}$, 444 (1982)
15) S. Toxvaerd, J. Comp. Phys. $\underline{52}$, 214 (1983)
16) J.L. Lebowitz, J.K. Percus and L. Verlet, Phys. Rev. $\underline{153}$, 250 (1967)
17) S. Toxvaerd, Phys. Rev. A $\underline{24}$, 2735 (1981)
18) L.F. Rull, J.J. Morales and F. Cuadros, Phys. Rev. B $\underline{32}$, 6050 (1985)

PHASE SEPARATION OF BINARY FLUIDS NEAR A CRITICAL POINT

UNDER MICROGRAVITY

F. Perrot, P. Guenoun and D. Beysens
Service de Physique du Solide et de Résonance Magnétique
CEN-Saclay, 91191 Gif-sur-Yvette Cedex, France

I INTRODUCTION

The experiments which are reported are the first steps of a general study of the phase separation of fluid mixtures under microgravity conditions.

Phase separation occurs when a homogeneous system is put in a thermodynamically non stable state. This separation is a very non equilibrium and non stationnary process which drives the system to the final equilibrium state, where it is formed of two well-defined phases. Classically two phase separation processes can be distinguished; nucleation or spinodal decomposition[1]. Note that most of the applications of phase separation are in metallurgy[2].

Mixtures of organic fluids are very suitable for studying phase separation; their critical temperature is near room temperature, they are transparent and the phase separation time scale is well adapted to experiments.

Phase separation processes are strongly affected by the earth's gravitational acceleration (g) through convection due to the density difference between the two phases. This prevents any valuable study of the late stages being made. More subtle effects[3], connected to capillary fluctuations, are also expected.

The convection due to gravity can be suppressed, in principle, if the density difference of the two phases is made nearly zero, by using isodensity mixtures, and/or if the gravitational acceleration itself is reduced, by performing experiments in a microgravity environment. Such an environment is provided by sounding rockets for which the microgravity period is 6 minutes, a period which is precisely that necessary to study the spinodal decomposition in the critical region of a binary fluid.

This paper is organized as follows: Section II describes the basic ideas of phase separation in binary fluids, Section III deals with the ground-based experiments in isodensity mixtures, Section IV reports the microgravity experiments and Section V compares both types of experiments.

II PHASE SEPARATION IN A BINARY FLUID

1) Phase diagram

Binary fluids are formed of two liquids A and B which are miscible in all proportions for temperatures T larger than the critical temperature T_c. Below this temperature, both liquids are partially miscible. A typical phase diagram is shown on Figure 1 in the diagram temperature T, concentration c of one component (say A). Below the coexistence curve the system separates into two phases, one rich in A, the other rich in B. The point (c_c, T_c) is the critical point; in its vicinity the thermodynamic properties of binary fluids have been extensively studied[4]. We thus report in the following only the basic behaviors.

a) Critical properties

Considering the static properties, the binary fluid belongs to the same universality class as the 3-dimensional Ising model. The order parameter is the difference $c-c_c$, which takes two values below T_c. With the reduced temperature

$$\epsilon = \frac{T-T_c}{T_c}, \qquad (1)$$

the coexistence curve is described by

$$c^+ - c^- = B_c(-\epsilon)^\beta \qquad (2)$$

where $\beta = 0.325$ is an universal exponent and B_c is system dependent.

The density difference between the two phases $(\Delta\rho)$ is related to the concentration of both phases and can be expressed approximatively by

$$\Delta\rho = (\Delta\rho)_0 (-\epsilon)^\beta \qquad (3)$$

the correlation length ξ^- measures the range of the order parameter correlation in the 2-phase region:

$$\xi^- = \xi_0^- (-\epsilon)^{-\nu} \qquad (4)$$

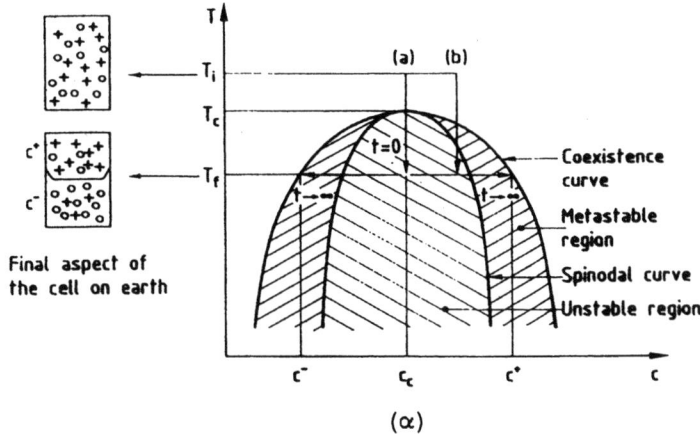

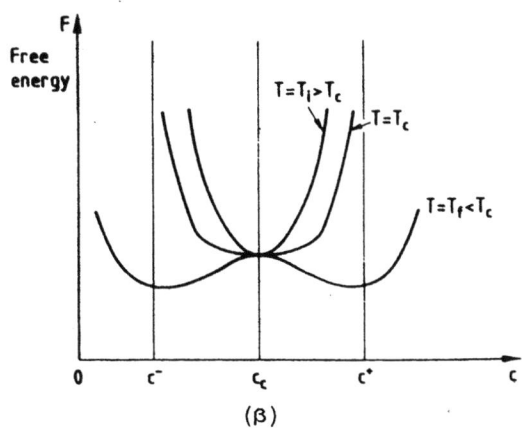

Figure 1. Phase separation of a binary fluid: (α) phase diagram. The spinodal curve classically separates the nonstable region under the coexistence curve in two regions according to the sign of the osmotic compressibility which is related to the curvature of the free energy; (β) free energy at different temperatures.

where $\nu = 0.63$ is an universal exponent, and ξ_0^- is an amplitude.

The value ξ_0^- is related to the value ξ_0^+ (above T_c) by the universal relation $\xi_0^+ \simeq 2\xi_0^-$. Below T_c, the correlation length ξ^- is the same in both phases.

the surface tension σ between the two liquid phases is given by

$$\sigma = \sigma_0 (-\epsilon)^\mu \qquad \text{with} \qquad \mu = 2\nu , \qquad (5)$$

and typically[5]

$$\frac{k_B T_c}{\sigma_0} \sim 2.6 \left(\xi_0^+\right)^2$$

It is not so simple to describe the dynamics owing to the velocity fluctuations which strongly influence the transport coefficients. The time relaxation τ_ξ of the order parameter fluctuations can be interpreted as the diffusion time of Brownian fluid particles of size ξ in a liquid of same viscosity η_s (Stokes Einstein law)[4,6]

$$\tau^- = \frac{6\pi \eta_s}{k_B T} \left(\xi^-\right)^3 \qquad (6)$$

where k_B is the Boltzmann constant. The parameter η_s is the viscosity of the fluid mixture which is weakly divergent near the critical point.

The constants B_c, $(\Delta\rho)_0$, ξ_0^- are system dependent (see values in Table I for the systems investigated here).

b) **Phase separation**

The non-stable region, which occurs below the coexistence curve, can be further separated in two regions[1] depending on the sign of the osmotic compressibility χ:

- The metastable region, where χ is positive: the decay of the metastable state takes place by nucleation.

Table I. Values of the critical amplitudes for the studied mixtures

System	c_c (mass fraction)	T_c (°C)	ξ_0^- (A°)[a]	B_c	$(\Delta\rho)_0$ (g/cm)^3
C*-M	0.7307	42.130	1.80	0.713	0.15[b]
C+C*-M (c' = 3%)	0.7046	44.745	1.63	0.754	3×10^{-5}[c]

a) From ξ_0^+ and using $\xi_0^+/\xi_0^- \simeq 2$.
b) From the pure component assuming the volume additivity[10].
c) From the Laplace radius[10].

- The unstable region, where the compressibility χ is negative. Here the system is thermodynamically unstable and phase separation takes place by spinodal decomposition. We will only be concerned with this process in the following.

2) <u>Spinodal decomposition process</u>

Let us consider a thermal quench in a binary mixture, from a temperature T_i in the homogeneous phase, to a temperature T_f corresponding to an unstable state, inside the coexistence curve (see quench (a), figure 1). The system phase separates.

Generally, different regimes can be defined:

a) <u>Linear regime</u>[7] $\left(t/\tau^- \lesssim 1\right)$

Just after the quench, the compressibility is negative and the fluctuations of concentration are growing; the growing rate however depends on the fluctuation wavevector (figure 2b). Indeed, fluctuations of small wavelength require a large amount of energy to grow, because of the concentration gradients, and fluctuations of large wavelength cannot grow rapidly, because they need an important mass transfer. A characteristic wavelength λ_m thus appears in the system whose order of magnitude is the correlation length ξ^-, and the fluid appears to be formed of interconnected clusters of typical size λ_m. This regime has been seen in computer simulations[8] and in experiments with polymers or microemulsions[9]. Such interconnected structures have been observed also in isodensity mixtures[10], but not in the linear growth regime.

b) <u>"Diffusion" growth</u> $\left(1 \lesssim t/\tau^- \lesssim 10\right)$

The growth of the fluctuations at the characteristic wavelength λ_m is limited by the equilibrium concentration c^+ and c^- as the time increases. This means that the local equilibrium is reached at the scale λ_m (figure 2c). Due to non-linearities the characteristic length λ_m increases with time (figure 2d) with a variation law[11]:

$$\lambda_m \sim t^a \quad \text{with} \quad a \simeq 0.2 - 0.33$$

In this cross-over region, the transition regions between the domains are still of order of the correlation length ξ^-, while the size of the domains increases; this corresponds to the formation of well-defined interfaces.

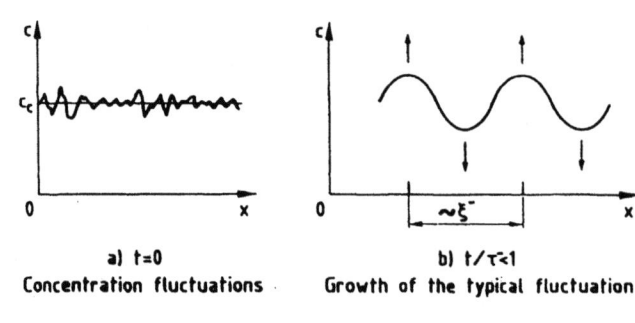

a) t=0
Concentration fluctuations

b) t/τ⁻≲1
Growth of the typical fluctuation

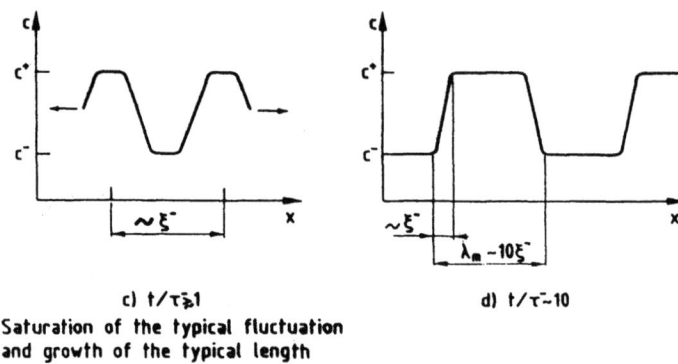

c) t/τ⁻≳1
Saturation of the typical fluctuation
and growth of the typical length

d) t/τ⁻~10

Figure 2. Phase separation by spinodal decomposition: early stages.

c) <u>Hydrodynamic regime</u>

With well-defined interfaces, instabilities due to surface tension can take place. Deformations of the interfaces induce a flow which increases in turn this deformation, leading to a necking down of the tube-like domains and their break up. The typical behavior of the characteristic length λ_m is linear in time[12]:

$$\lambda_m = K \frac{\sigma}{\eta_s} t \quad \text{with} \quad K \simeq 10^{-2} \quad (7)$$

This region is generally called "late stages".

In these three regimes, the behavior of the characteristic length λ_m can be scaled using the variables $q_m \xi^-$ and t/τ^-, where $q_m = \frac{2\pi}{\lambda_m}$ and ξ^- and τ^- are respectively the correlation length and the relaxation time of the concentration fluctuations at the final temperature T_f. As an example, a typical curve from Furukawa[13] is plotted in figure 3.

d) Remarks on the growth processes[14]

When the final state of the system is in the metastable region, nucleation of the minority phase occurs and spherical droplets of this phase can develop. The connectivity of these droplets is low and the interfaces always have the same sharp profile during the growth. The growth law of a droplet of radius R is

$$R \sim t^{1/3} \qquad (8)$$

which is very slow compared to the hydrodynamical evolution of the spinodal decomposition (late stage). The morphology of the pattern in the early stages of nucleation is therefore very different when compared to spinodal decomposition, where the structures are highly interconnected.

Such interconnected structures can be obtained only in systems where the volumes of each phase are nearly identical. An order of magnitude of the frontier between connectivity and non connectivity can be obtained by comparing with the percolation limit, where the volume ratio percolation threshold is 0.2^{12}.

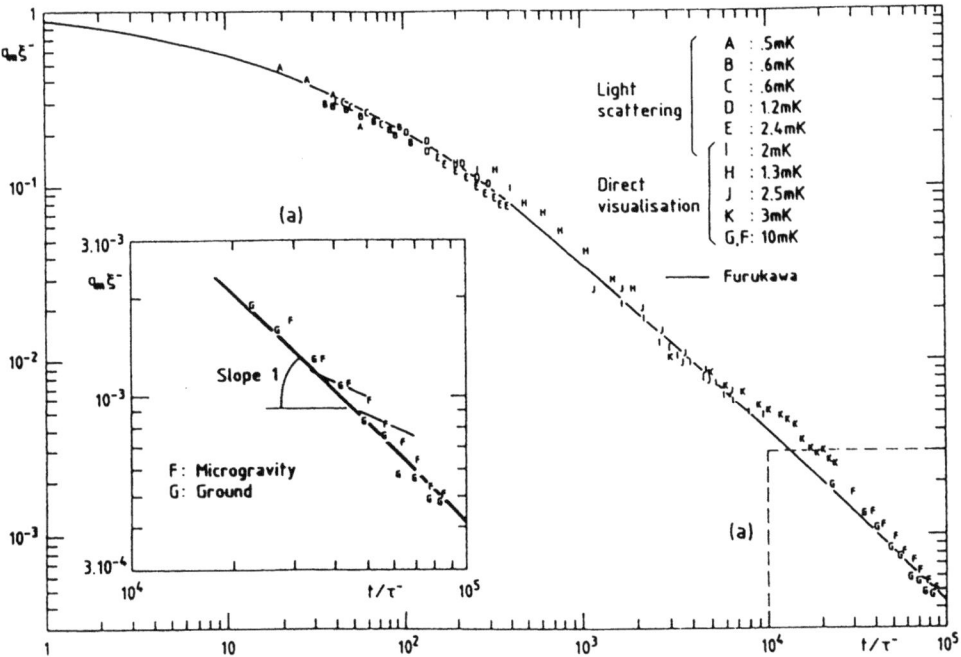

Figure 3. Evolution of the caracteristic wavevector (q_m) versus time (t) in reduced variables (see text).

3) Gravity effects

In earth based experiments, gravity accelerates the phase separation by inducing convections: the denser phase sinks down and the lighter rises up. In the process of spinodal decomposition, these gravity effects become important in the late stages, preventing any valuable observation being made. In particular, the final state of the system (equilibrium state) is due to gravity, the denser phase being at the bottom of the cell and the interface between the two phases being horizontal (see figure 1).

The other constraints of the system are the walls of the container which act on the fluids according to the wetting properties. One of the phases generally completely wets the container in the vicinity of T_c[15].

Let us estimate the time domain where gravity effects become dominant; according to Siggia[12], the capillary length l_c measures the balance, at equilibrium, between the surface forces due to surface tension and the volume forces due to gravity

$$l_c = \left(\frac{\sigma}{g\Delta\rho}\right)^{1/2} , \qquad (9)$$

where σ is the surface tension between the two liquid phases and $\Delta\rho$ the density difference between them. To study the cross-over between the hydrodynamic region and the gravity-driven region, we can estimate the different contributions in the Navier-Stokes equation, at the scale of domains of size λ_m:

$$\lambda_m^3 \rho \left[\frac{\partial \vec{v}}{\partial t} + (\vec{v}\cdot\vec{\nabla})\vec{v}\right] = \left(-\vec{\nabla}p + \eta_s \Delta\vec{v} + \Delta\rho\vec{g}\right)\lambda_m^3 \qquad (10)$$

where ρ represents the mean density of the two phases. The pressure term is related to the surface tension via the Laplace law:

$$\vec{\nabla}p \simeq \frac{\sigma}{\lambda_m^2} . \qquad (11)$$

If we consider a steady state with a small velocity, the only contributions to (10) are in the right hand side where we can note that the first two terms are of the order of magnitude of λ_m and that the last one is of the order of λ_m^3. Then, the third term is negligible compared to the two first terms when

$$\lambda_m^2 < \frac{\sigma}{g\Delta\rho} \qquad \text{or} \qquad \lambda_m < l_c$$

and the order of magnitude of the velocity is given by:

$$\eta \Delta \vec{v} \sim \vec{\nabla} p \quad \text{or} \quad v \sim \frac{\sigma}{\eta_s} \qquad (12)$$

which is comparable to the expression (7).

When λ_m is greater than the capillary length l_c, the third term due to buoyancy becomes predominant; it corresponds to an accelerated motion and the convection term $(\vec{v}.\vec{\nabla})\vec{v}$ can no longer be neglected. The typical velocity in the system is then given by:

$$\rho(\vec{v}.\vec{\nabla})\vec{v} \sim \Delta\rho\vec{g}$$

or

$$v \sim \sqrt{\frac{\Delta\rho}{\rho} g \lambda_m} \sim \frac{\sqrt{(\sigma/\rho)\lambda_m}}{l_c} \qquad (13)$$

This expression is not valid if l_c is of the order of the size of the sample.

We have illustrated this point by the study of the evolution of a non-isodensity system of deuterated cyclohexane-methanol at critical concentration, after a thermal quench from $T_i = T_c + 5$ mK to $T_f = T_c - 10$mK.

The aspect of the cell (2 mm thickness, 2 cm diameter) is shown at different times after the quench in figure 4. At time t = 1 mn after the quench, we see clearly the convections which are well developped at t = 2mn. A simple calculation shows that convections should take place at times longer than 15 s, with velocities of order 0,2 cm.s^{-1}, which agrees well with the observation.

In order to study the late stages of the spinodal decomposition, it is therefore necessary to suppress the gravitational convections. The typical parameter being $\Delta\rho g$, it appears that we have to lower either $\Delta\rho$ using isodensity mixtures and/or g by performing experiments in a microgravity environment.

III EARTH'S BASED EXPERIMENTS IN A ISODENSITY SYSTEM

1) The system

We first consider the cyclohexane (C)-methanol (M) binary fluid whose densities are nearly matched with $\rho_C \lesssim \rho_M$. The deuterated

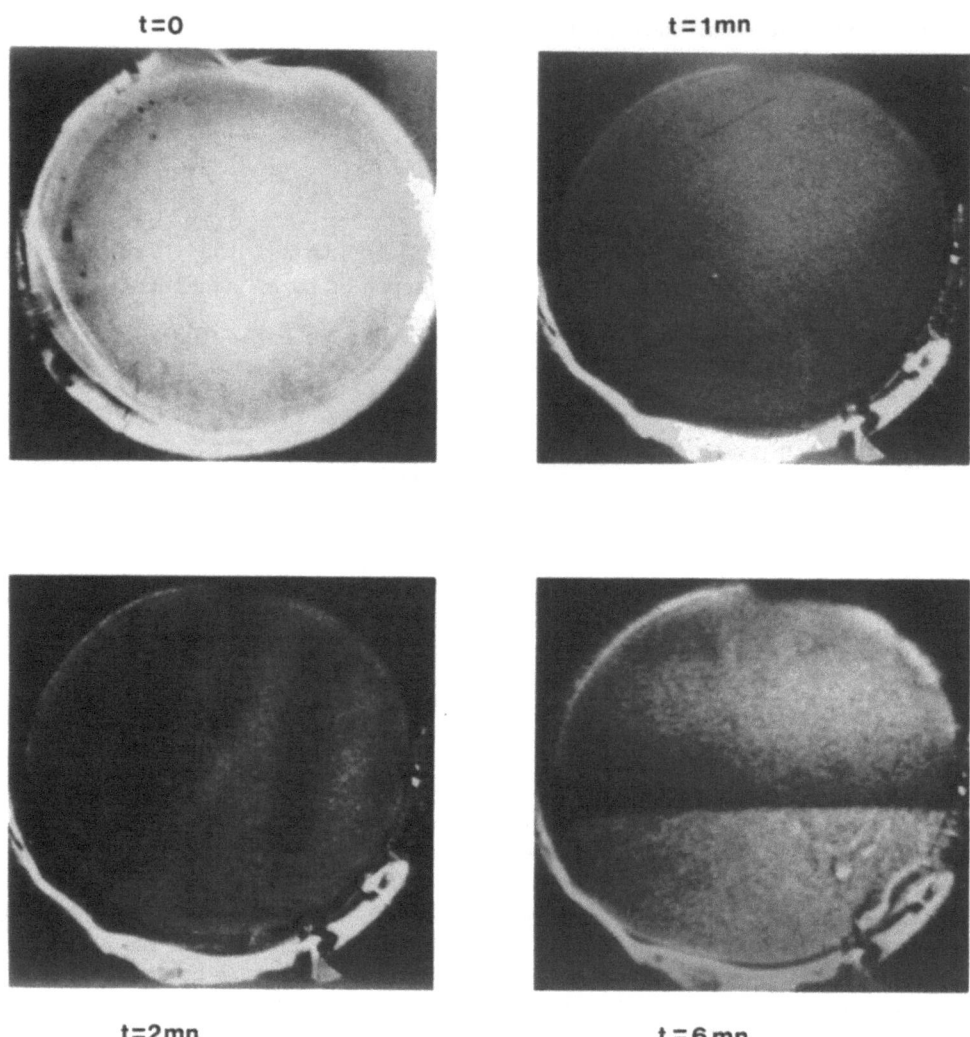

Figure 4. Aspect of a sample of the deuterated cyclohexane methanol mixture on earth at the time t after a critical quench of 10 mK below T_c.

cyclohexane (C^*) exhibits a density ρ_{c^*} such that

$$\rho_c \lesssim \rho_M < \rho_{c^*} \qquad (14)$$

Thus, by adding a small amount of deuterated cyclohexane, we can match the densities of both phases. The matching of densities is obtained for the mass fraction $c' = \dfrac{[C^*]}{[C]+[C^*]}$ such that[10]

$$c' = 3\% \pm 1\%$$

The uncertainty is due to the presence of uncontrolled impurities (mainly water).

Typically, one obtains

for the (C-M) system[5]: $(\Delta \rho)_0 = 3.10^{-3}$ g/cm^{-3}

for the $(C+C^*-M)$ system[10]: $(\Delta \rho)_0 = 3.10^{-5}$ g/cm^3

The isodensity system being strictly speaking a ternary mixture, we had to test its critical properties (critical exponents, universal ratios of critical amplitude, universality class). We have demonstrated that the isodensity system behaves in the critical region as a real binary fluid[10]. Critical amplitudes of this mixture are given in Table I.

2) **Experimental study of the phase separation**

a) **Experimental set-up**

The experimental set-up is the same as that one used in the microgravity experiments (figure 5).

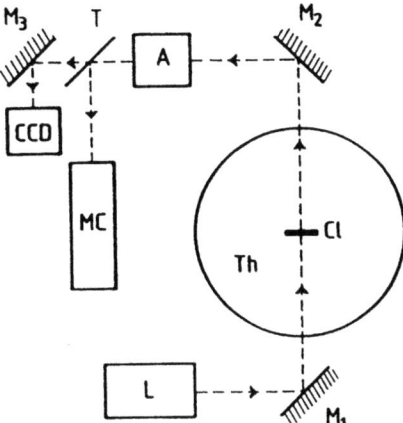

Figure 5. Experimental set up for studying the phase separation processes by direct visualization: L: white lamp; M_1, M_2, M_3: mirrors; Th: thermostat; Cl: cell; A: achromat; T: semitransparent mirror; CCD: CCD camera; MC: 16 mm film camera.

The thermostat has two stages, each one is regulated using a PID regulator. An extra passive shield is used to maintain the system at atmospheric pressure. The thermal regulation is better than 1 mK.

The optical cell is made of quartz with a 2 mm thickness and a 2 cm diameter. An ultrasonic device allows the system to be homogenized.

The visualization system depends on the scale of the observation:

- for $\lambda_m < 100 \lambda$ (λ wavelength of the light), laser light scattering at small angles is used. The observed pattern is a ring whose radius is a measurement of the inverse of λ_m [16,17]. This ring collapses with time. In this case, the measured intensity $I(\vec{q},t)$ ($\vec{q}$ transfer wavevector) is proportionnal to the structure factor $S(\vec{q},t)$ defined as the Fourier transform of the order parameter correlation function.

- if $\lambda_m > 100 \lambda$, the structures can be observed directly. The observation volume is constituted of a sample slide close to the exit wall of the cell[18].

b) **Results and discussion**

Figure 6 shows typical structures obtained in a critical isodensity mixture for different times, after a thermal quench from $T_i = T_c + 5$ mK to $T_f = T_c - 10$ mK. At the temperature T_f, the density difference between the two phases was estimated to be $\Delta\rho = 10^{-6}$ g/cm^3.

The video pictures are digitized using a computer (figure 7a). We then perform the Fourier transform of the intensity $I(x,y,t)$ of the picture and we calculate the function $F(\vec{q},t)$ defined by

$$F(\vec{q},t) = \left| \int\int I(x,y,t) \, e^{i(q_x x + q_y y)} \, dxdy \right|^2 \qquad (15)$$

where q_x and q_y are the $\vec{q}$ components in the observation plane.

In the figure 7b, we have drawn the function $F(\vec{q},t)$ corresponding to the picture 7a. The function $F(\vec{q},t)$ is a ring whose radius q_m corresponds to the characteristic length λ_m of the structures $\left(\lambda_m = \dfrac{2\pi}{q_m}\right)$.

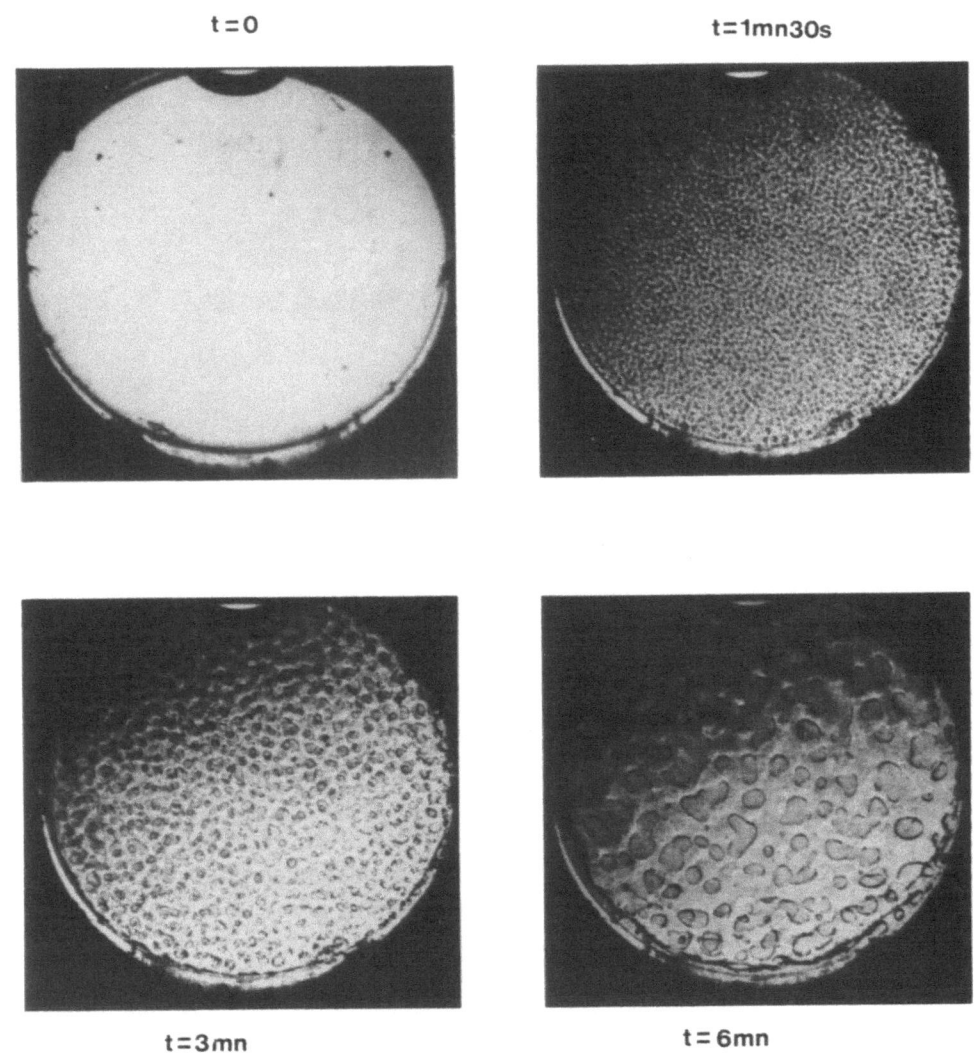

Figure 6. Aspect of a sample of the C+C*-M isodensity mixture on earth at the time t after a critical quench of 10 mK below T_c.

The function $F(\vec{q},t)$ is seen to behave as the structure factor $S(\vec{q},t)$ obtained by laser light scattering[18].

We have plotted in figure 3, the reduced characteristic wavevector $(q_m \xi^-)$ versus the reduced time (t/τ^-), together with laser light

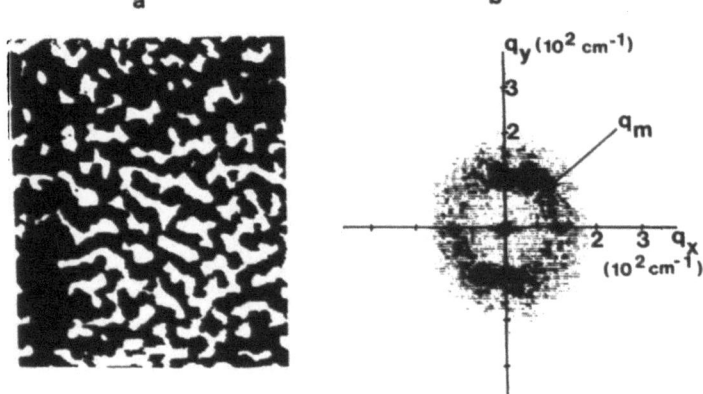

Figure 7. (a) digitized picture of a sample; (b) corresponding function $F(\vec{q},t)$ as defined by (eq. 15). The amplitude of this function is proportional to the point density.

scattering experiments. Both techniques give data in good agreement with each other and with the Furukawa function.

The shape of the function $F(q,t)$ can be scaled as $\tilde{S}(\tilde{q},t)$

$$\tilde{S}(\tilde{q},t) = \frac{q_m^3 F\left(\frac{q}{q_m},t\right)}{\int_{q_{min}}^{q_{max}} F(q,t)\, q^2 dq} \qquad (16)$$

where

$$\tilde{q} = \frac{q}{q_m}$$

The experimental results are reported in figure 8 and compared with laser light scattering results[18] and the function given by Furukawa[19].

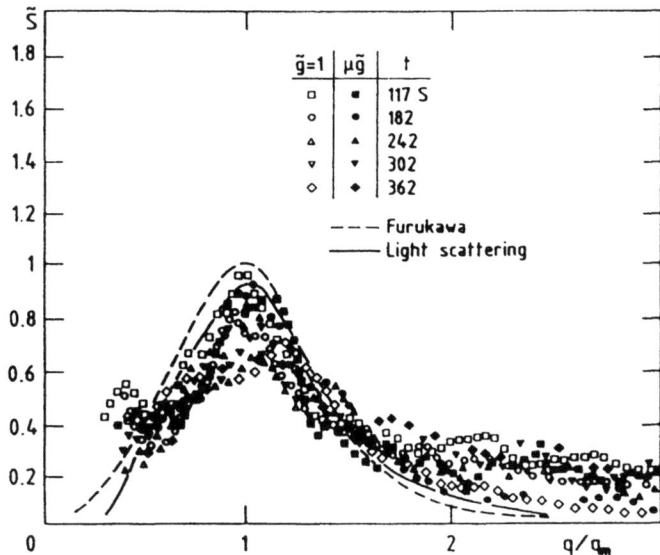

Figure 8. Scaled function $\tilde{S}(\tilde{q},t)$ as defined by (eq. 16) obtained with the isodensity mixture at different times t after the quench; $\tilde{g} = 1$: earth based experiment (see quench G in figure 3); $\mu\tilde{g}$: microgravity experiments (see quench F in figure 3).

IV MICROGRAVITY EXPERIMENTS

1) Experimental set-up

The microgravity period was obtained during the flight of a sounding rocket (TEXUS, ESA program). The rocket includes 4 parts; from the bottom to the top, one distinguishes:

- Two stages of propulsion.

- A telemetry module which links the experiments to the ground during the flight and contains the µg stabilization system.

- The payload which is composed of five experiments. The module that we used contained a thermostat, an optical system and electronics in a cylinder of roughly 40 cm diameter cylinder, 50 cm height and 45 kg weight.

- A top section which contains the parachute.

The flight is parabolic (maximum altitude 300 km) and consists of three periods:

- Lift off and ascent period (70 s).

- Microgravity period (360 s).

- Reentry and recovery of the payload and of the telemetry module. The total time of the flight is 15 mn.

In the telemetry module, accelerometers in the three directions allow measurements of the microgravity level to be made. A typical value is $|g| \lesssim 2.10^{-4}\ g_0 = 0.2$ CGS with no preferential direction, g_0 being the earth gravity.

During the count-down, the critical temperature was checked and the sample set at the temperature $T_i = T_c + 5$ mK. An automatic thermal step, quenching the cell at the temperature $T_f = T_c - 10$ mK was performed at the beginning of the microgravity period. During all the flight, the temperature was regulated within a precision better than 1 mK and a direct visualization of the cell was obtained using a real time video line.

We present here two experiments in microgravity:

2) <u>Texus 11 experiments (April 1985)</u>

The sample was a critical mixture of deuterated cyclohexane and methanol. At T_f

$$|g|\ \Delta\rho \lesssim 10^{-3}\ \text{CGS}$$

This value was the same as the one obtained on earth with the isodensity mixture (cf. III- 2-b).

The criticality of the mixture has been tested by turbidity: at 1 mK above T_c, the measured turbidity is the same as the critical value to within a precision of 1%.

The aspect of the cell is shown in figure 9 at different times after the quench. Clearly phase separation occurs, but no macroscopic structures are present. The bubble in the pictures is the vapor phase.

We consider two possible explanations to account for this absence of macroscopic pattern:

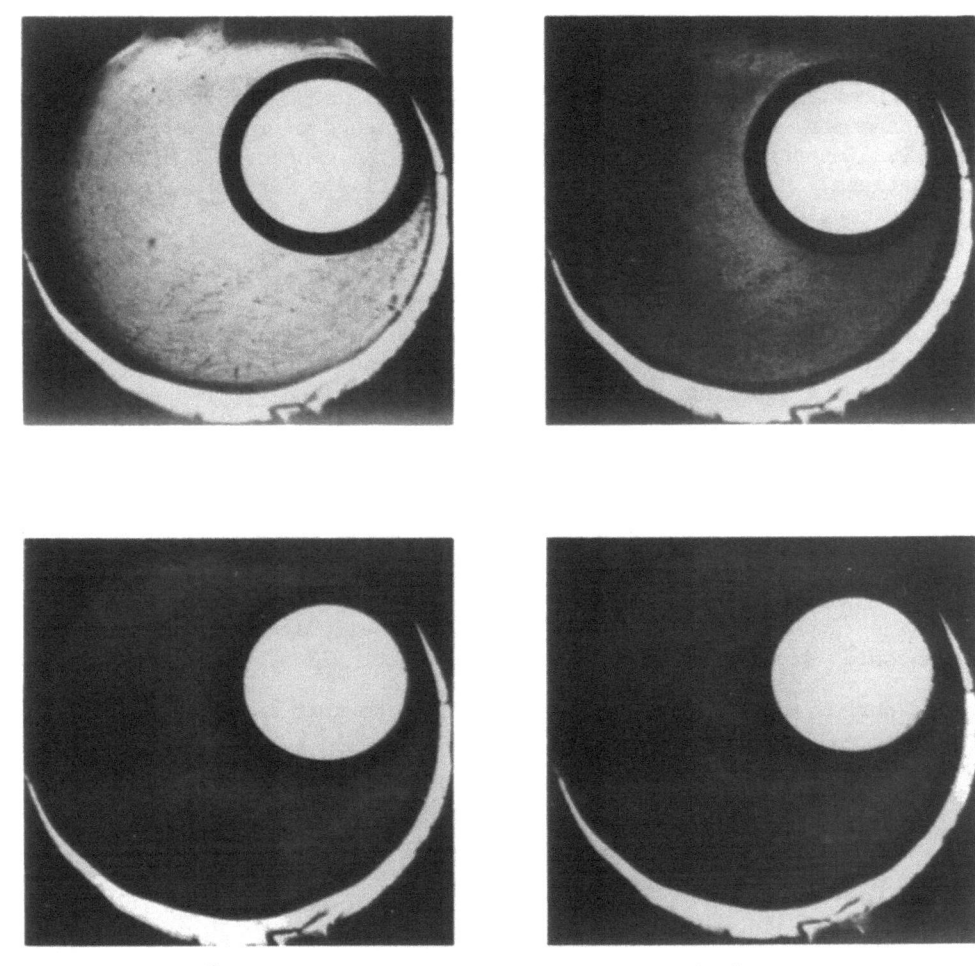

Figure 9. Aspect of a sample of the C^*-M mixture during the microgravity period at the time t after a critical quench of 10 mK below T_c.

- The sample was only partially mixed and the stirring by the vapor phase during the lift-off could have probably induced some concentration gradients.

Using the percolation threshold 0.2 (see II-2-d) for the limit of the metastable region, a simple calculation shows that this threshold at $T_f = T_c - 10$ mK corresponds to a variation of concentration $|c_c-c|/c_c \sim 2\%$. We have checked experimentally this point using the isodensity mixture.

Therefore, if the quench is not performed at criticality, the time scale of the pattern growth would have been much longer than a spinodal evolution, according to the expression (8).

- Problems in the interfaces formation: some predictions[3] have been made for infinite systems: in the absence of external fields, as gravity, interfaces could not form due to the enhancement of capillary fluctuations. However the influence of the finite size of the sample is known to considerably reduce these fluctuations.

To discriminate between these two possible explanations, we have performed a new microgravity experiment using the isodensity mixture, in which the appearance of a spinodal pattern is a very sensitive test of the critical concentration. Moreover the cell has been redesigned to have an homogeneous stirring of the sample and a minimized vapor phase.

3) Texus 13 experiments (April 1986)

The sample is an isodensity mixture at critical concentration. On earth, macroscopic spinodal structures have been already observed in this system (see figure 6).

The aspect of the cell in microgravity is shown in figure 10. Macroscopic structures do appear. Using part III, we can compare the scaled curve $\left[\left(q_m \xi^-\right) \text{ versus } \left(t/\xi^-\right)\right]$ on earth to that in microgravity (see figure 3a) and also the scaled factor $\tilde{S}(\tilde{q}, t)$ (equation 16) at different times in both cases (figure 8). Quantitatively the results are seen to be identical within the experimental errors.

Nevertheless, by comparing at the same time, figure 10 under microgravity to figure 6 on earth, we can observe a difference in the morphology of the droplets. One can remark that the capillary length is quite different:

- On earth: l_c = 0.24 cm.

- In microgravity: l_c = 38 cm, which is much larger than the cell dimension.

We can observe that the percolated structures exhibit a rapid time evolution (typically $\sim t$). In these structures small droplets remain whose evolution is much slower than the structures. On earth, using the isodensity mixture, the separated phases appear to be granular-like in the final stage. This seems due to the absence of connectivity of these clusters, which can no longer grow by capillary or gravity instabilities. Their growth is thus limited only by diffusion which is a much slower process.

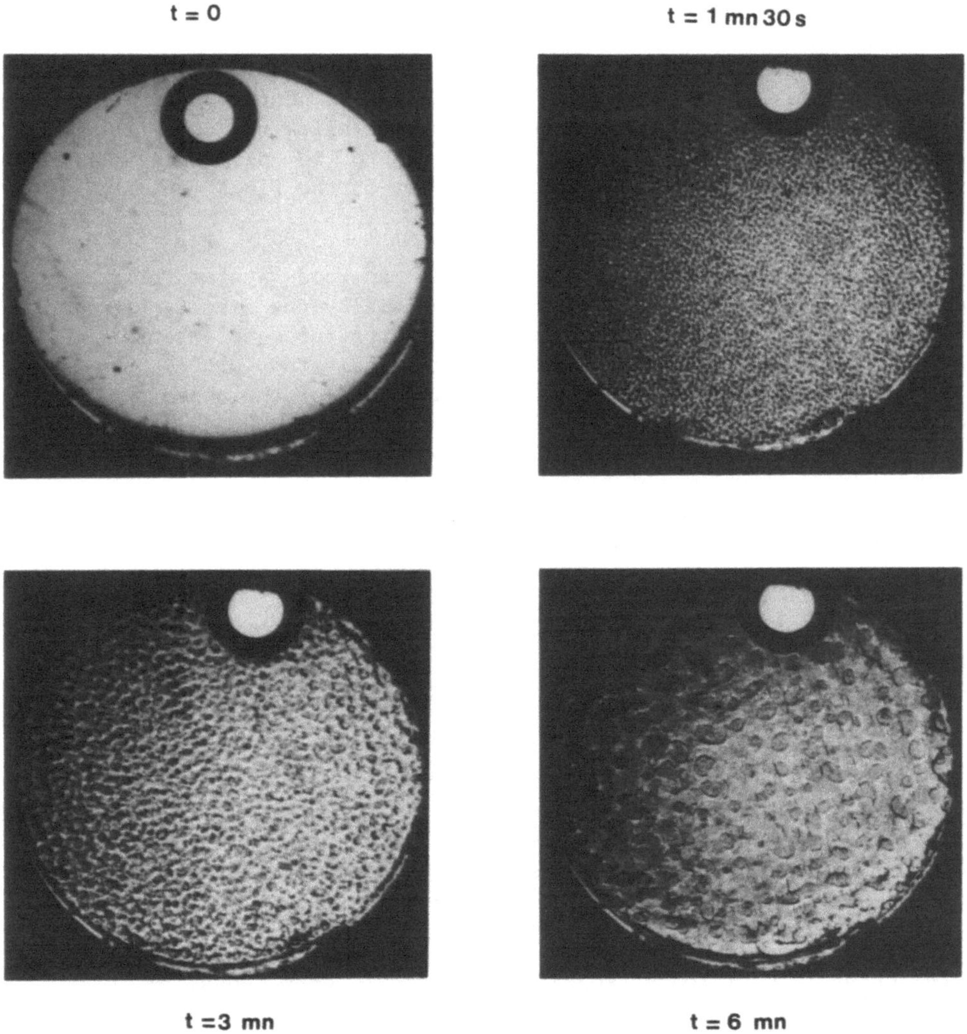

t = 0   t = 1 mn 30 s

t = 3 mn   t = 6 mn

Figure 10. Aspect of a sample of the C+C*-M isodensity mixture during the microgravity period at the time t after a critical quench of 10 mK below T_c.

V CONCLUSIONS AND PERSPECTIVES

The last spinodal decomposition experiment under microgravity demonstrates that the typical wavelength of the phase separation pattern increases with time (hydrodynamic region), which means that the formation of interfaces is not seriously affected by microgravity. Moreover this characteristic length exhibits a quite similar behavior on earth and under microgravity.

Experiments under microgravity with a time longer than 6 mn would be very useful to study the evolution of the morphology of the droplets and the final equilibrium stage. In the absence of gravity, wetting forces and finite size effects will play a predominant role to determine this equilibrium stage. These experiments will allow the comparison with earth experiments on isodensity mixtures to be performed, and the influence of the residual gravitational flows on the final equilibrium to be evaluated.

The isodensity mixture is however a good system for creating microgravity-like environment on earth. Especially the direct observation of the sample allows more information on the patterns to be obtained (in particular their morphology) than the statistical study provided by laser light scattering. A systematic investigation of the phase diagram is presently underway with such isodensity systems.

ACKNOWLEDGMENTS: A CNES grant is gratefully acknowledged. We thank ESA for having supported the space experiments.

REFERENCES

1. A recent review is:
   J.D. Gunton, M. San Miguel and P.S. Sahni in "Phase transition and critical phenomena" (edited by C. Domb, J.L. Lebowitz, Academic Press, 1983) p. 267.
2. See, e.g., "Material Sciences in Space" edited by B. Feuerbacher, H. Hamacher and R.J. Naumann (Springer-Verlag, 1986).
3. M. Robert, Phys. Rev. Lett. $\underline{54}$, 444 (1985).
4. See e.g. A. Kumar, H.R. Krishnamurthy and E.S.R. Gopal, Phys. Rep. $\underline{98}$, 57 (1983).
5. H. Chaar, M.R. Moldover and J.W. Schmidt, J. Chem. Phys. $\underline{85}$, 418 (1986).
6. K. Hamano, S. Teshigawara, T. Koyama and N. Kuwahara, Phys. Rev. $\underline{A33}$, 485 (1986) and refs therein.
7. J.W. Cahn, J. Chem. Phys. $\underline{42}$, 93 (1965).
8. F.F. Abraham, Phys. Rep. $\underline{53}$, 93 (1979).
9. T. Izumitani and T. Hashimoto, J. Chem. Phys. $\underline{83}$, 3694 (1985)
   D. Roux, Preprint.
10. C. Houessou, P. Guenoun, R. Gastaud, F. Perrot and D. Beysens, Phys. Rev. $\underline{A32}$, 1818 (1985).
11. J.S. Langer, M. Baron and M.D. Miller, Phys. Rev. $\underline{A11}$, 1417 (1975).
    K. Kawasaki and T. Ohta, Prog. Theor. Phys. $\underline{59}$, 362 (1978).
    K. Binder and S. Stauffer, Phys. Rev. Lett. $\underline{33}$, 1006 (1974).
12. E.D. Siggia, Phys. Rev. $\underline{A20}$, 595 (1979).
13. H. Furukawa, Adv. Phys. $\underline{34}$, 703 (1985).
14. C.M.F. Jantzen and H. Herman in "Phase diagrams: Material Science and Technology" vol 5 (A.A. Alper ed., Academic Press, 1978) p. 127.
15. J.W. Cahn, J. Chem. Phys. $\underline{66}$, 3667 (1977), for a recent review, see

"Fundamental problems in statistical mechanics VI" (E.G.D. Cohen, ed., Elsevice, 1985).
16. W.I. Goldburg, C.H. Shaw, J.S. Huang and M.S. Pilant, J. Chem. Phys. 68, 484 (1978).
Y.C. Chou and W.I. Goldburg, Phys. Rev. A20, 2105 (1979); Phys. Rev. A23, 858 (1981).
17. N.C. Wong and C.M. Knobler, J. Chem. Phys. 69, 725 (1978); Phys. Rev. A24, 3205 (1981).
18. P. Guenoun, R. Gastaud, D. Beysens and F. Perrot, in preparation.
19. H. Furukawa, Physica 123A, 497 (1984).
20. D. Beysens, P. Guenoun and F. Perrot, in preparation.

INFLUENCE OF SODIUM SULFATE ON THE PHASE DIAGRAM AND RHEOLOGICAL BEHAVIOUR FOR A SODIUM DODECYLBENZENESULFONATE/POLYOXYETHYLENE FATTY ALCOHOL/WATER SYSTEM

José Muñoz, Críspulo Gallegos, Vicente Flores and José M. Pérez

Departamento de Química Técnica
Facultad de Química, Universidad de Sevilla
41012 Sevilla, Spain

INTRODUCTION

For years a growing interest in properties of surfactant solutions has been seen[1]. In addition to isotropic solutions and crystals, several different liquid crystalline phases can appear[2].

Surprisingly, at the present time, relatively little work has been done on the rheological properties of surfactant solutions, especially on liquid crystal rheology[3]. These macroscopic properties are very important in order to distinguish the liquid crystalline state[4] as well as in many practical applications of surfactants.

In most applications surfactant mixtures rather than pure species are used[5]. Modern formulations for use in the laundry process contain a combination of nonionic and anionic surfactants, a necessity in order to obtain optimum removal of soil from different kinds of fabric[6]. Salinity and hardness tolerances can be increased in ionic surfactant solutions by the addition of nonionic surfactants[7].

Taking these factors into account, a study on the rheological behaviour of aqueous systems containing a polyoxyethylene fatty alcohol (AEO-8) and a sodium dodecylbenzenesulfonate, with different sodium sulfate contents has been carried out.

EXPERIMENTAL

A nonionic surfactant, polyoxyethylene (EO:8) fatty alcohol (C12-C14) and an ionic surfactant, sodium dodecylbenzenesulfonate, "LAS-Na", (containing 4 or 16% by weight of sodium sulfate) have been used.

The phase diagrams were determined within a temperature range from 278 to 333 K. The maximum concentration of surfactant was 40% by weight.

Phase separation was verified by means of a "Hettich" thermostatic centrifuge, model "Rotanta K".

The liquid crystalline regions were detected by using a "Carl Zeiss" polarising microscope with a thermostatic stage and a camera, which permits photomicrographs to be taken.

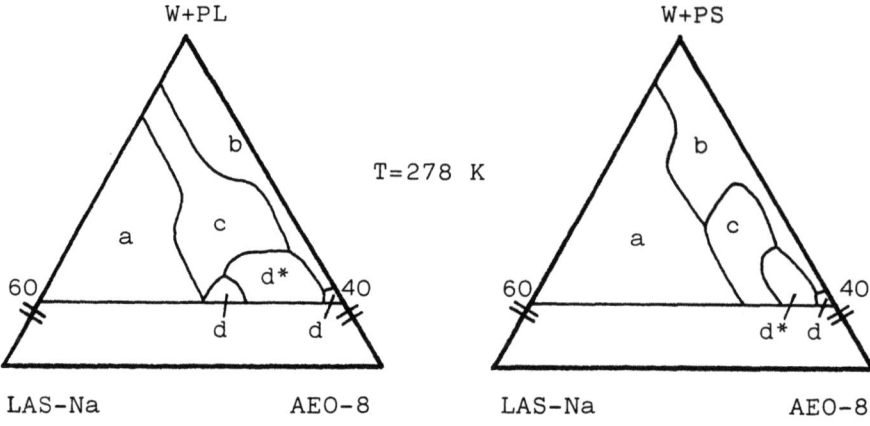
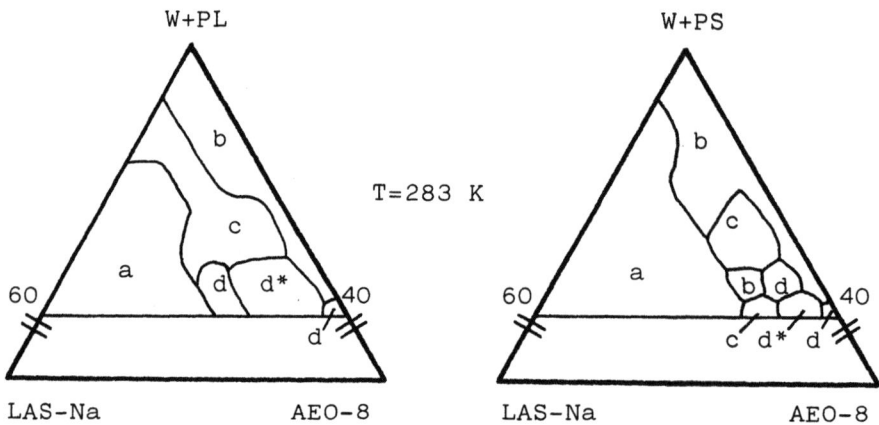
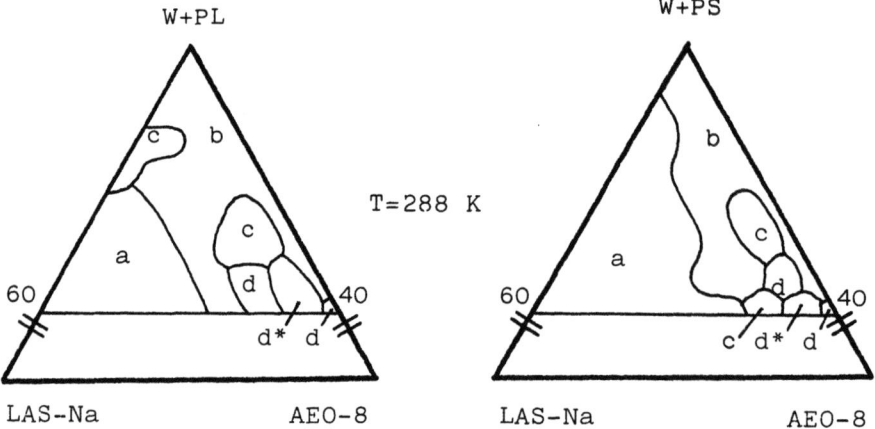

Fig. 1. Phase diagrams at different temperatures. a) Biphasic systems. b) Globular aggregates. c) Rod-shaped aggregates. d) Hexagonal mesophase: pseudoplastic behaviour; d*) Hexagonal mesophase: plastic behaviour.

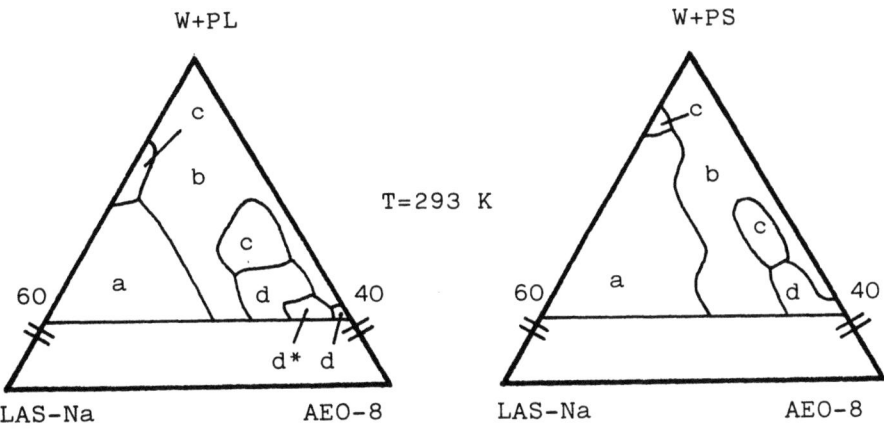

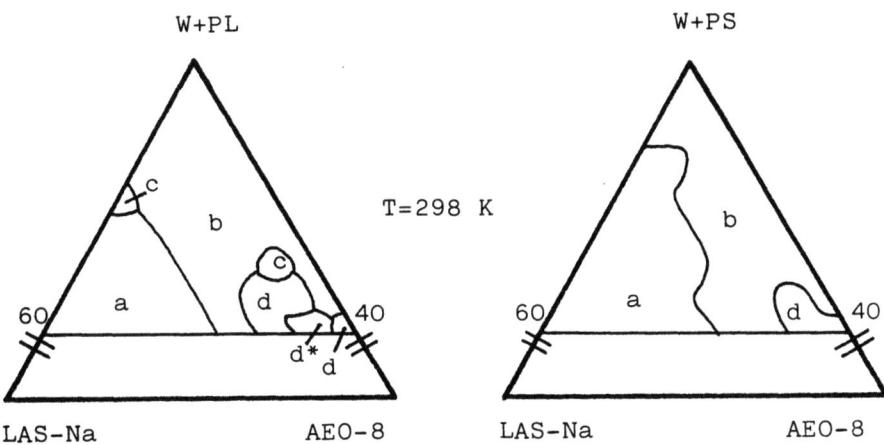

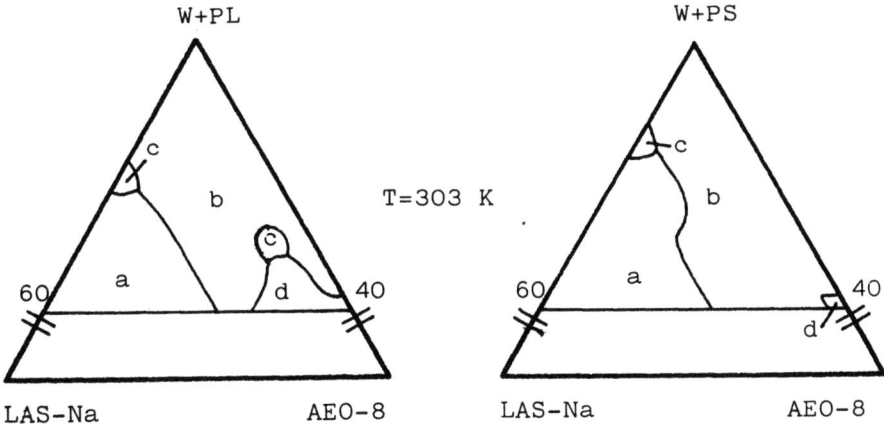

Fig. 2. Phase diagrams at different temperatures. a) Biphasic systems. b) Globular aggregates. c) Rod-shaped aggregates. d) Hexagonal mesophase: pseudoplastic behaviour. d*) Hexagonal mesophase: plastic behaviour.

The rheological measurements were taken by using a rotational viscometer "Haake", model "Rotovisco RV-3". "NV", "SV-I" and "SV-II" sensor systems were used. The shear rate at the surface of the inner cylinder was derived from a series solution to the integral equation of the concentric cylinder viscometer. This solution can be used for plastic behaviours[8].

RESULTS AND DISCUSSION

In Figures 1 and 2, phase diagrams of the above mentioned systems are shown. Their notations are, respectively, "PL" or "PS" for systems containing "LAS-Na" with 4 or 16% of sodium sulfate. The temperature range was varied between 278 and 303 K. Different areas can be observed in the phase diagrams:

a.- A biphasic area, above all, at high "LAS-Na" concentrations and low temperatures. This is a consequence of the high critical micellar temperatures of the anionic surfactant[9].

b.- An isotropic micellar solution region exhibiting a Newtonian rheological behaviour, which can be attributed to the occurrence of globular micelles. This area appears at high temperatures and low "LAS-Na" concentrations.

c.- An isotropic micellar solution region exhibiting a non-Newtonian rheological behaviour, which fits the Ostwald-de Waele model[10]:

$$\tau = K \dot{\gamma}^n$$

where, τ = shear stress; $\dot{\gamma}$ = shear rate. This behaviour is attributed to the appearance of rod-shaped aggregates, at low temperatures and intermediate concentrations of both "LAS-Na" and "AEO-8".

d.- A hexagonal liquid crystalline area exhibiting a plastic rheological behaviour, that fits the Herschel-Bulkley model[11]:

$$\tau - \tau_o = K' \dot{\gamma}^{n'}$$

where, τ_o = yield point. However, a pseudoplastic behaviour, fitting the Ostwald-de Waele model, has been predominantly found at temperatures close to the melting point of the mesophase. This region appears at high concentrations of "AEO-8", low "LAS-Na" concentrations and temperatures below 303 K.

An increase in the sodium sulfate concentration provokes the disappearance of non-Newtonian behaviours as well as the appearance of a larger biphasic area.

From a quantitative point of view, the influence of the sodium sulfate concentration on dynamic viscosities and rheological parameters depends on the nonionic surfactant concentration.

In Figures 3 and 4 dynamic viscosity values are plotted against temperature. At low concentrations of anionic surfactant two opposite effects which depend upon the nonionic concentration can be observed.

At nonionic surfactant concentrations below 25%, Figure 3, a dramatic increase in the dynamic viscosity, and the occurrence of secondary micellar aggregates can be detected (AEO-8:15%). If the concentration of sodium sul-

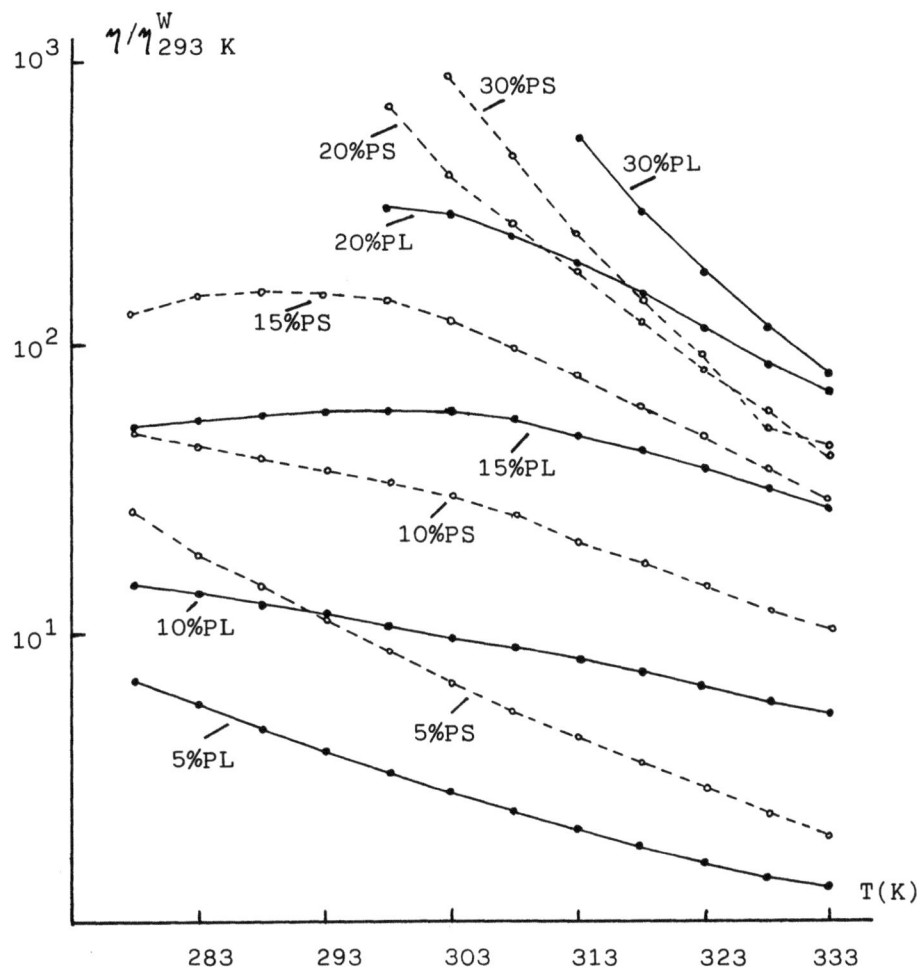

Fig. 3. Relative viscosities for systems containing 5% of sodium dodecylbenzenesulfonate.

fate is raised, this secondary aggregation occurs at lower temperatures[12].

Above 25% of nonionic concentration, on the contrary, the viscosity values become similar; and, finally, a significant decrease can be observed.

This last effect becomes more important as the "LAS-Na" concentration increases.

As far as the Ostwald-de Waele and Herschel-Bulkley rheological parameters are concerned, systems containing 5% of anionic surfactant can be studied in function of:

a) Power law exponents. These are plotted against temperature in Figure 5-A. There are two ranges of values. The first one, comprised between 0.78 and 0.99, is due to the occurrence of rod-shaped micelles. The values of the second range are lower than those of the first due to the existence of a hexagonal mesophase. The systems with a high content of sodium sulfate show the lowest exponent values.

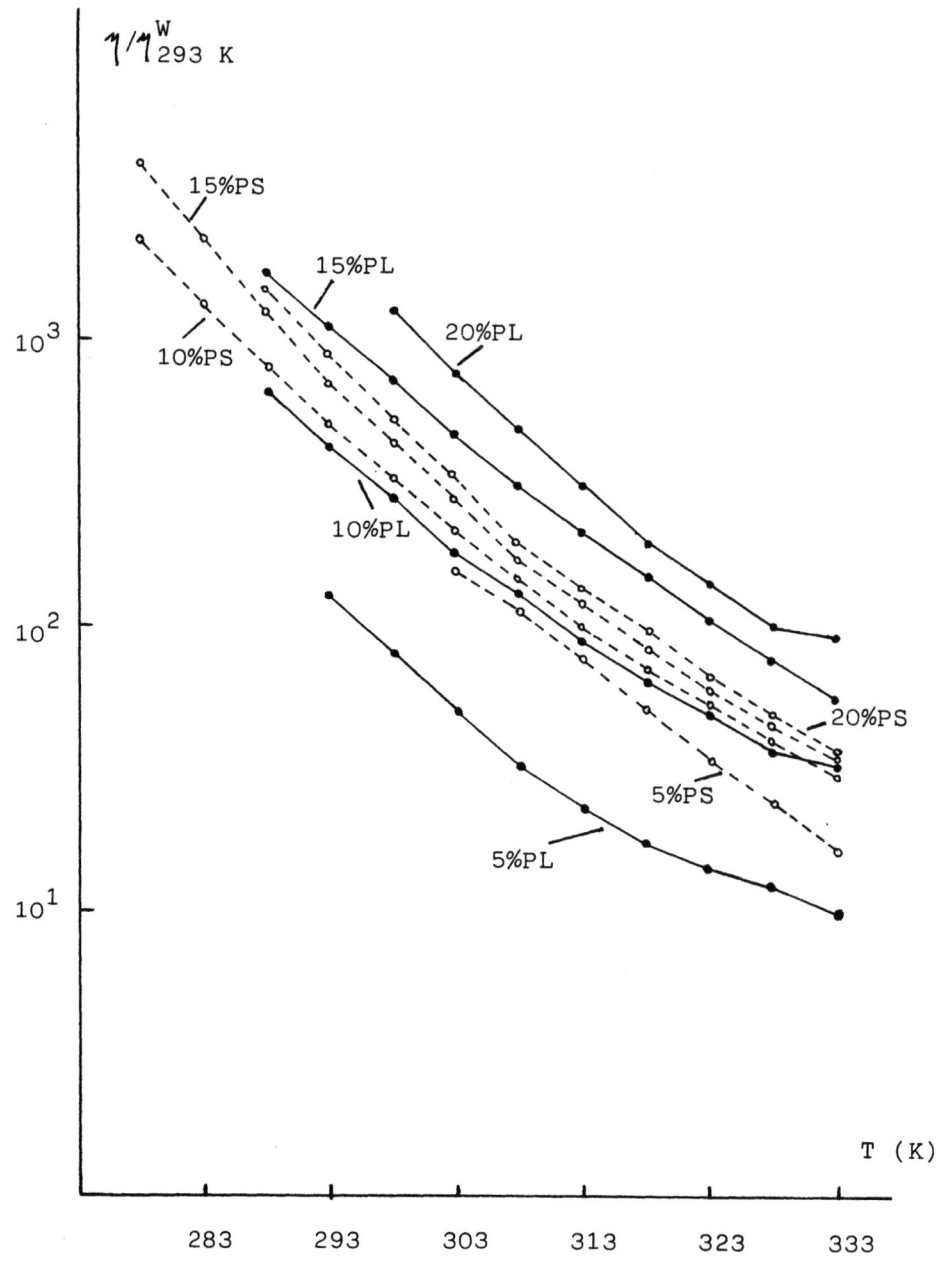

Fig. 4. Relative viscosities for systems containing 10% of sodium dodecylbenzenesulfonate.

b) Apparent viscosity values, at 25 and 200 s^{-1}, for systems within the rod-shaped micellar region. These are plotted against temperature in Figure 5-B. An increase in apparent viscosities can be seen in those systems which contain the highest concentration of sodium sulfate.

c) Differential viscosity values, at 0.1 and 1.0 s^{-1}, within the hexagonal mesophase area, These values decrease as the sodium sulfate concentration increases (Figure 5-C).

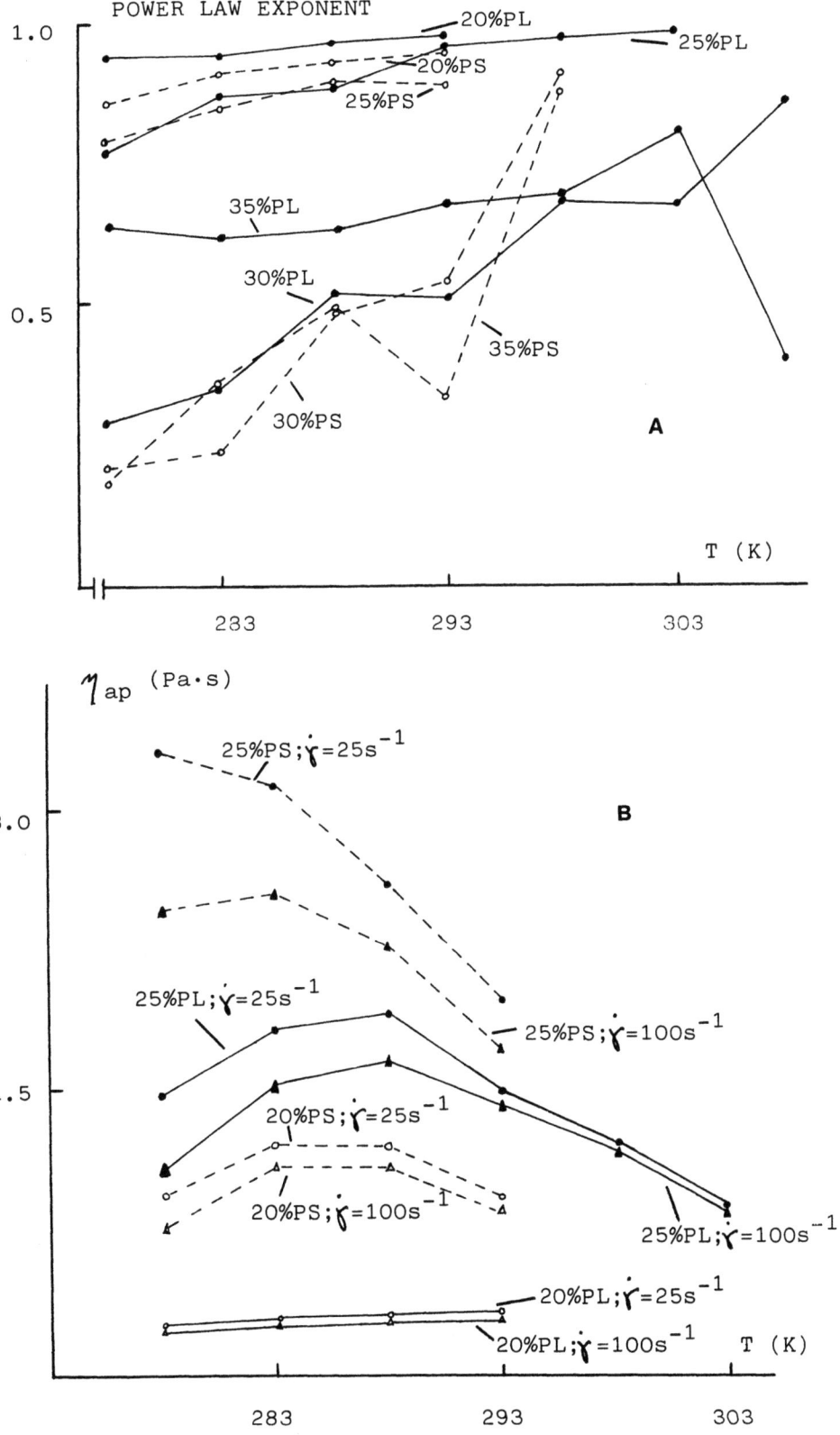

Fig. 5. Cont.

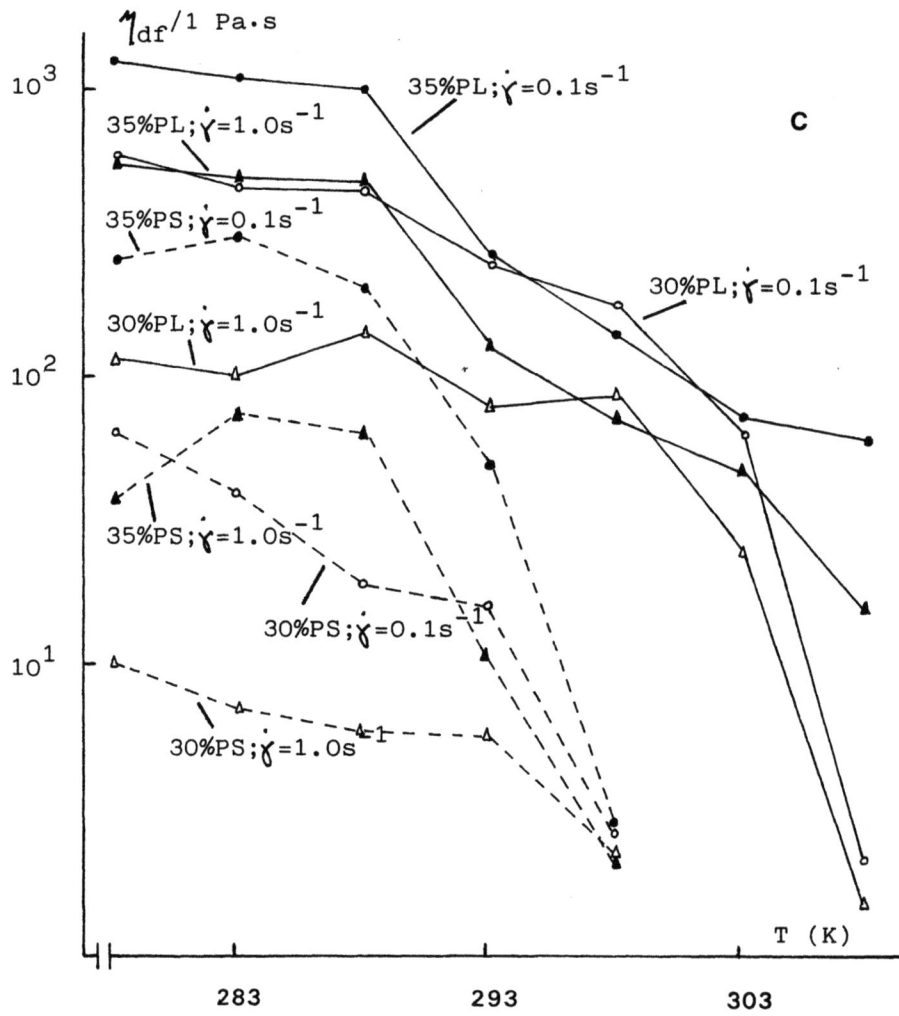

Fig. 5. Systems containing 5% of anionic surfactant. A) Power law exponents. B) Apparent viscosities rod-shaped aggregates. C) Differential viscosities hexagonal area.

To summarize, an increase in the sulfate concentration favours rod-shaped micelles; while, on the other hand, it tends to produce the disappearance of the liquid crystalline structure.

As the "LAS-Na" concentration and the sodium sulfate content increase, both the rod-shaped micelles and the hexagonal mesophase show a tendency to disappear. This gives rise to a spectacular increase in the power law exponents as well as to a significant decrease in consistency.

All these results can be interpreted on the basis of several considerations:

The presence of "LAS-Na" favours the occurrence of rod-shaped micelles in systems with water and nonionic surfactant. This tendency becomes more and more important as the anionic concentration increases. These bulky aggregates lead to an ordered structure, as a consequence of the interaction between the aggregates. This results in non-Newtonian rheological behaviours and viscoelasticity[13].

The addition of sodium sulfate exerts a salting-out effect giving rise to biphasic systems, and an increase in micellar size that induces the appearance of an ordered structure.

However, the addition of salts also provokes the opposite effect. It reduces the repulsion between the aggregates and, therefore, favours the destruction of the structure.

Thus, depending on surfactant and sodium sulfate concentrations one of the above mentioned effects will predominate.

REFERENCES

1. H. Hoffmann, H. Löbl, H. Rehage and I. Wunderlich, Rheology of surfactant Solutions, Tenside Detergents 22:290 (1985).
2. G. J. T. Tiddy and M. F. Walsh, Lyotropic Liquid Crystals, in: "Aggregation Processes in Solution", E. Wyn-Jones and J. Gormally, ed., Elsevier, New York (1983).
3. H. A. Barnes, Detergents, in: "Rheometry. Industrial Applications", K. Walters, ed., Wiley, New York (1980).
4. T. Asada and S. Onogi, Rheological and Rheo-Optical Studies of Polymer Liquid Crystals, Polymer Engineering Reviews 3:323 (1983).
5. I. W. Osborne-Lee, R. S. Schechter, W. H. Wade, and Y. Barakat, A New Theory and New Results for Mixed Nonionic-Anionic Micelles, Journal of Colloid and Interface Science 108:60 (1985).
6. H. Sagitani and S. E. Friberg, Phase Diagrams and Association Structures of Combined Surfactants-Water Systems, Bull. Chem. Soc. Jpn. 56:31 (1983).
7. K. L. Stellner and J. F. Scameborn, Surfactant Precipitation in Aqueous Solutions Containing Mixtures of Anionic and Nonionic Surfactants, JAOCS 63:566 (1986).
8. T. M. T. Yang and I. M. Krieger, Comparison of Methods for Calculating Shear Rates in Coaxial Viscometers, Journal of Rheology 22: 413 (1978).
9. A. Helenius and K. Simons, Solubilization of Membranes by Detergents, Biochimica et Biophysica Acta 415:29 (1975).
10. R. Darby, "Viscoelastic Fluids", Marcel Dekker, New York (1976).
11. Y. R. Chen, Rheological Properties of Sieved Beef-Cattle Manure Slurry: Rheological Model and Effects of Temperature and Solids Concentration, Agricultural Wastes 15:17 (1986).
12. C. Tanford, Y. Nozaki, and M. F. Rohde, Size and Shape of Globular Micelles Formed in Aqueous Solution by n-Alkyl Polyoxyethylene Ethers, The Journal of Physical Chemistry 81:1555 (1977).
13. J. Ulmius, H. Wennerström, L. B. Johansson, G. Lindblom, and S. Gravsholt, Viscoelasticity in Surfactant Solutions. Characteristics of the Micellar Aggregates and the Formation of Periodic Colloidal Structures, The Journal of Physical Chemistry 83:2232 (1979).

STATISTICAL THERMODYNAMICS OF PURE AND MIXED AMPHIPHILIC AGGREGATES

I. Szleifer and A. Ben-Shaul

Department of Physical Chemistry, and
The Fritz Haber Molecular Dynamics Research Center
The Hebrew University, Jerusalem, 91904 Israel

W.M. Gelbart

Department of Chemistry and Biochemistry
University of California, Los Angeles
Los Angeles, California 90024, USA

INTRODUCTION

In this paper we review a mean field ('single chain') statistical thermodynamic theory for molecular organization of pure and mixed amphiphilic (micellar) aggregates of different geometries[1]. In the last few years many studies of these topics have been reported[2], including large scale computer simulations[3], mean field (single chain) theories[1,2,4] as well as simple phenomenological models[5]. Computer simulations provide the most detailed results; so far however, they are limited to very few systems – primarily because of the computational difficulties involved. On the other hand, the mean field models, although approximate, are easily applied to any system; and as was shown elsewhere (for bilayers[1a,4a]) yield very good agreement with molecular dynamics results for conformational and thermodynamic properties. Furthermore, the mean field theory presented below provides explicit expressions for the probability distribution function (pdf) of chain conformations, it is based on very few assumptions and involves no adjustable parameters.

The discussion below centers on two main issues: the conformational properties of amphiphile chains in pure and mixed aggregates, and the thermodynamic stability of such aggregates as a function of composition and geometry. By conformational properties we refer for example to bond order parameter profiles of chains packed in different micellar geometries, spatial distribution of chain segments and the effects of surface roughness on these properties. In the thermodynamic analysis we will compare chains' contributions to the free energy with those arising from the "opposing forces" operating at the micellar interface, with particular emphasis on the interplay between the two in determining the preferred aggregation geometry. For mixed aggregates we will also consider the role of short/long chain composition in determining the preferred micellar geometry.

THE pdf OF CHAIN CONFORMATIONS, PACKING CONSTRAINT AND THERMODYNAMICS

In accordance with experimental observations, we assume that the interior of an amphiphilic aggregate (e.g. spherical or cylindrical micelles, bilayers, etc.) is liquid like. More explicitly it is assumed that the chain segment (monomer) density, is constant and liquid-like throughout the hydrophobic core, except for possible small density fluctuations near the aggregate's interface. A quantitative characterization of the (monomer) density profile in an aggregate of arbitrary geometry is conveniently achieved by dividing the interior of the aggregate into L imaginary layers, which are parallel (or concentric) to the interface, see Fig. 1. Then, the average volume available, per molecule in layer i, for the three basic geometries treated here, is $m_i \propto (L - i + 1)^d - (L - i)^d \sim (L - i)^{d-1}$ where $d = 1,2,3$ for bilayers, cylinders and spheres, respectively, and i is the layer's number starting from the interface.

The chain packing constraints are

$$\sum_b P(b) \phi_i(b) = <\phi_i> = \rho_i m_i \qquad i=1,\ldots,L \qquad (1)$$

where $\phi_i(b)$ is the volume occupied by the chain, in conformation b, in layer i (which is proportional to the number of monomers there). ρ_i is the relative density in layer i; $0 \leq \rho_i \leq 1$, with $\rho_i = 1$ representing the liquid hydrocarbon density. Density fluctuations near the interface are mimicked by taking a lower density in, say, the first two imaginary layers, i.e., $\rho_1 < \rho_2 < \rho_3 = \ldots = \rho_L = 1$, (typically the layer width is $\sim 1 - 2\text{Å}$).

Eq. 1 applies to pure aggregates, i.e. those comprising one type of chains. The appropriate packing constraints for binary aggregates are

$$x_A <\phi_i>_A + x_B <\phi_i>_B = \rho_i m_i \qquad i=1,\ldots,L \qquad (2)$$

where x_A (x_B) is the mole-fraction of A (B) chains, and $<\phi_i>_A$ is the average volume taken up by an A chain in layer i, etc.

The pdf is derived by minimizing the free energy functional of the system, (see below), subject to the packing constraints. For pure aggregates this free energy is:

$$\mu_t^o = E_t - TS_t = \sum_b P(b) \epsilon(b) + kT \sum_b P(b) \ln P(b) \qquad (3)$$

whereas for mixed, binary, aggregates the appropriate function is

$$\mu_t^o = x_A \mu_{t,A}^o + x_B \mu_{t,B}^o + kT (x_A \ln x_B + x_B \ln x_B) \qquad (4)$$

where $\mu_{t,K}^o$, (K = A,B), is the free energy of the molecule of type K and the last term accounts for the mixing entropy. (The subscrip 't' is used here for 'tails'; the surface (head-group) contributions will be denoted by 's', see below).

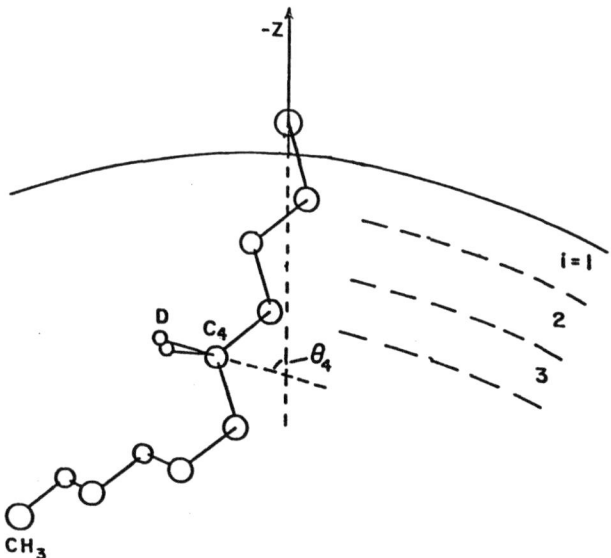

Fig. 1. A schematic representation of an hydrocarbon chain in a micellar aggregate. The interior of the micelle is divided into imaginary parallel (or concentric) to the interface. θ_i is the normal to the interface.   is the angle between the C_i - D(H) bond and the normal to the interface (shown θ_4 as an example).

The resulting pdf's are:

For pure aggregates

$$P(b) = \frac{1}{z} \exp[-\beta\varepsilon(b) - \beta\sum_i \Pi_i \phi_i(b)] \tag{5}$$

For mixed aggregates

$$P_K(b) = \frac{1}{z_K} \exp[-\beta\varepsilon_K(b) - \beta\sum_i \Pi_i \phi_{i,K}(b)], \quad K = A, B \tag{6}$$

Here $\varepsilon_K(b)$, $(K = A,B)$, is the internal energy ('trans gauche') of the conformation. The Π_i's are the lateral pressures exerted on the central molecule, in layer i, by the neighboring molecules, and are determined by substitution of (5) or (6) to (1) or (2), respectively. Note that the Π_i's are the same for both A and B chains in the mixed aggregates. Note also that while for pure aggregates $<\phi_i>$ is an 'input parameter' (via the constraints), this is not the case for mixed aggregates. In the latter case only the weighted sum (2) is given, while $<\phi_i>_K$ (K = A,B) are determined by the resulting pdf's.

Once the P(b)'s are known, i.e., after determining the Π_i's, it is possible to calculate any conformational property of the chain. Similarly, thermodynamic properties such as entropy, internal energy and free energy can be calculated; e.g., the free energy is given by substituting (5) or (6) into (3) or (4), respectively.

Eqs. 3 and 4 describe the chain contribution to the free energy. Another important contribution to the free energy arises from the 'opposing forces' acting at the hydrocarbon-water interface[5]. These are: (i) The 'hydrophobic effect' which tends to minimize the water-hydrocarbon contact area. This is often modeled by a γa term, where γ is the oil-water surface tension and a is the average area per head-group. (ii) The repulsion between head-groups, which tends to maximize the area per molecule, and is most simply represented in the form c/a, where c is a parameter depending on the specific nature of the head-groups and the composition of the aqueous solution. This implies a surface free energy of the form

$$\mu_s^o = \gamma a + c/a = 2\gamma a_o + \gamma a (1 - a_o/a)^2 \tag{7}$$

where $a_o = (c/\gamma)^{1/2}$ is the area per head-group which minimize μ_s^o.

In the next section we will show that both contributions to the free energy are important in determining the relative stability of aggregates of different geometries. (At variance with the standard treatments[5] which determine the optimal geometries using only the surface terms; i.e., assuming that the hydrophobic core is a liquid hydrocarbon droplet whose free energy is independent of the core's geometry).

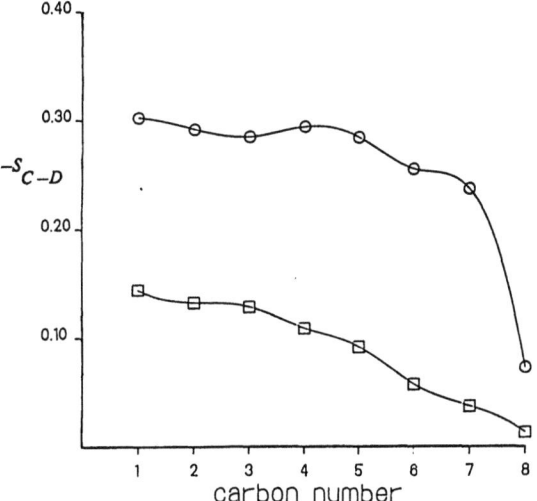

Fig. 2. Bond order parameter profiles of a $-(CH_2)_7-CH_3$ chain in a planar bilayer. $-a = 25\text{Å}^2$ $-a = 32.4\text{Å}^2$

REPRESENTATIVE RESULTS AND DISCUSSION

All the calculations reported here are for chains represented by the rotational isomeric state model[7]. Details concerning the computational procedure can be found elsewhere[1a].

a) Pure aggregates

A quantity which measures the extent of chain alignment due to the dense packing in micelles and bilayers, is the bond order parameter profile defined by

$$S_{C_i-D} = \langle 3/2 \cos^2 \theta_i - 1/2 \rangle \tag{8}$$

where θ_i is the angle between the $C_i - D$ (i.e. selectively labelled $C_i - H$) bond and the normal to the interface. The brackets denote averaging over all possible conformations. (When all bond angles are equally probable $S_{C-D} = 0$. For an all-trans chain perpendicular to the interface $S_{C-D} = 0.5$). These parameters can be determined experimentally by NMR (deuterium quadrupole splitting or ^{13}C relaxation time) measurements[6].

Fig. 2 shows bond order parameter profiles for a 8-carbon chain packed in a bilayer with two different areas per head-group. The differences in the two profiles are due to the fact that for $a = 25 \text{Å}^2$ the chains are very tightly packed, (note that for a bilayer the minimum area per head-group is $\sim 21 \text{Å}^2$, i.e. when all chains are in the all-trans conformation), while for $a = 32.4 \text{Å}^2$ the degree of order is lower.

Fig. 3 shows bond order parameter profiles for 8-carbon chains in a spherical micelle. One of the curves ( ) represents the experimental results of Walderhaug et al.[6a]. the other three are theoretical calculations based on Eqs. (1), (5) and (8), and correspond to different density profiles as shown in the inset. Qualitatively, the order parameters increase with the extent of surface roughnees (as modeled by the gradient $\rho_1 < \rho_2 < \rho_3 = \ldots = \rho_L$). The best agreement with experiment is obtained for $\rho_1 = 0.30$, $\rho_2 = 0.74$, $\rho_{i \geq 3} = 1$, suggesting a roughness layer of $\sim 3.5 \text{Å}$. It is interesting to note that a qualitatively similar conclusion has been arrived at by recent molecular dynamics calculations[3d].

Fig. 4 shows a similar set of data for cylindrical micelle. Again, one of the curves is experimental, Klason et al.[6d]. In this case the results for a 'compact micelle' ($\rho_i = 1$ for all i) show quite agreement with experiment, which further improves upon alloving a small degree of roughness ($\rho_1 = 0.09$, $\rho_2 = 0.92$).

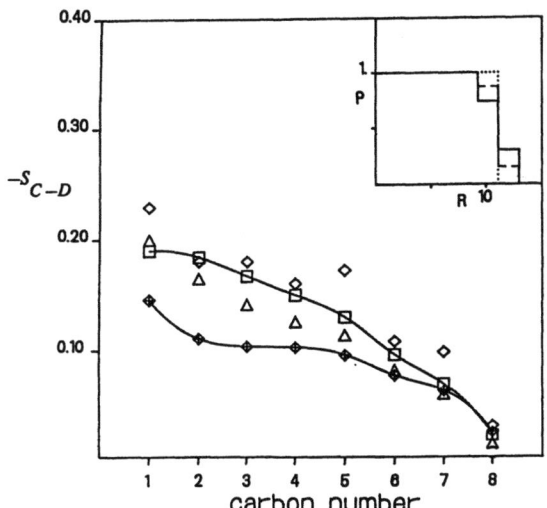

Fig. 3. Bond order parameter profiles of a $-(CH_2)_7-CH_3$ chain in a spherical micelle. experimental results, from ref 8a. compact micelle ($\rho_1, \rho_2, \rho_{i \geq 3}$) = (0,1,1), dotted line in inset. (0.15,0.87,1), dashed line in inset. − (0.30,0.74,1), solid line in inset.

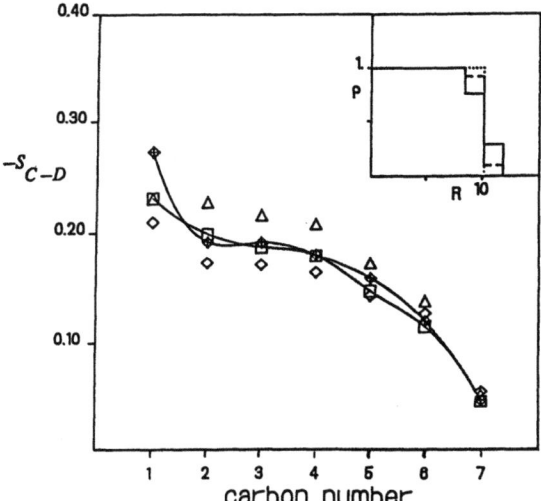

Fig. 4. Bond order parameter profiles of a $-(CH_2)_6-CH_3$ chain in a cylindrical micelle. experimental results, from ref 8d. compact micelle ($\rho_1, \rho_2, \rho_{i \geq 3}$) = (0,1,1), dotted line in inset. − (0.094, 0.92,1), dashed line in inset. (0.28,0.76,1), solid line in inset.

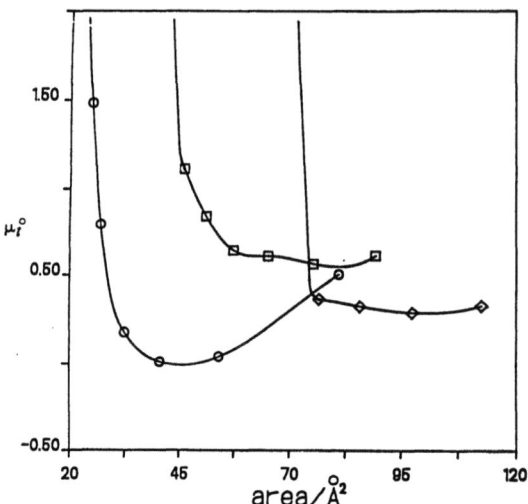

Fig. 5. Tail's free energy as a function of the average area per head-group, for a 8-carbon chain, measured relative to the free energy of a free chain a bilayer, see text. 'compact' bilayer. spherical micelle $(\rho_1, \rho_2, \rho_{i \geq 3}) = (0.30, 0.79, 1)$.

cylindrical micelle $(0.13, 0.89, 1)$.

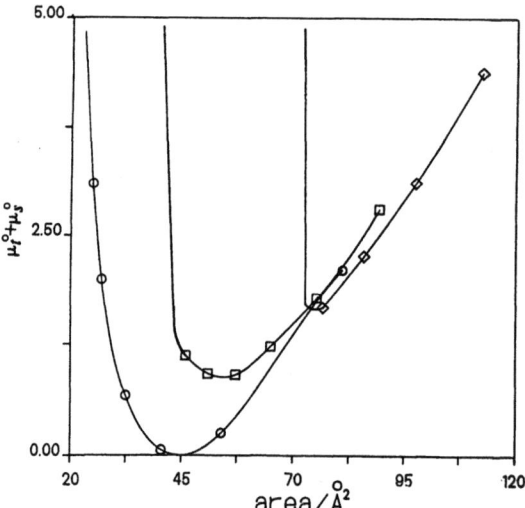

Fig. 6. Total free energy of a 8-carbon chain as a function of the average area per head-group. $a_0 = 45 Å^2$, $\gamma = 0.1\ kT/Å^2$. bilayer, spherical micelle, cylindrical micelle.

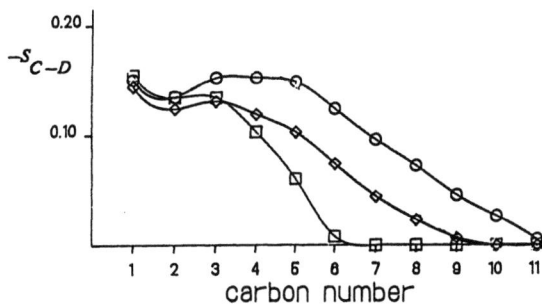

Fig. 7. **a)** Bond order parameter profiles of the long chain (11 carbons) in mixed planar bilayer. — $x_A = 1.0$, $a = 32.4 \text{Å}^2$, — $x_A = 2/3$, $a = 33.75 \text{Å}^2$, — $x_A = 0.2$, $a = 32.4 \text{Å}^2$. (x_A is the mole fraction of long chains).

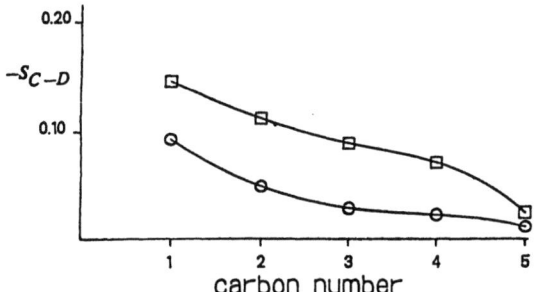

Fig. 7. **b)** Bond order parameter profiles of the short chain (5 carbons) in mixed planar bilayers. — $x_A = 2/3$; — $x_A = 0.2$. (x_A is the mole fraction of long chains).

Fig. 5 shows the chain contribution to the free energy, μ_t^0, as a function of the average area per head-group, for the three basic geometries. The free energy is measured relative to that of a 'free chain' in a bilayer. (The free chain is a chain subject to the same boundary conditions, but no packing constraints due to neighboring chains. Clearly, its free energy is lower than that of all packed chains). The calculations reveal that the differences in free energy between different geometries are on the order of $0.1-1.0\ kT$. Note for example the $\sim 1.5\ kT$ difference corresponding to molecules packed with average areas per head-group of 25Å^2 and 36Å^2. This is due to the very tight packing of the chains in the first case which is also reflected in the conformational properties - see Fig. 2. Such differences clearly confirm the stament made earlier, that chain packing in microenviroments with different geometrical characteristics involve non-negligible free energy differences.

Fig. 6 shows the total, chains + heads, (core + surface) free energy permolecule, assuming that $a_0 = 45\text{Å}^2$ (see eq. 7). for this value of a_0 the chains can pack either into cylindrical micelles or planar bilayers. Simple considerations, based on the notion of the 'opposing forces' and the hydrocarbon droplet assumption[5], would suggest that cylindrical micelles will be formed - due to their higher translational ('mixing') entropy. On the other hand our results suggest that this will only be the case if the translational entropy of a solution of cylindrical micelles will exceed the $\sim 1 kT$ difference between μ_{bil}^0 and μ_{cyl}^0 ($\mu^0 = \mu_t^0 + \mu_s^0$).

b) **Mixed aggregates**

All the results in this section are for mixed aggregates of 11-carbon and 5-carbon chains[1b] in different proportions.

The bond order parameter profiles for bilayers at different composition but similar average area per head-group are shown in Fig. 7. Fig. 7a is for the long chains and Fig. 7b for the short ones. the effects of the presence of short chains on the alignment of the long ones is mainly apparent in the 'second half' of the long chain. This portion of the chains carries the burden of filling up the inner volume (around the midplane) of the bilayer. This implies increased lateral placements of the last few bonds and consequently reduced order parameters. The effect increases with the fraction of short chains. For the short chains we observe that the order parameters increase as their proportion in the aggregate increases. This may be attributed to the fact that by stretching themselves a little, the short chains allow more lateral (conformational) freedom to their long neighbors.

The tail contribution to the free energy of mixed aggregates, calculated using eq. 4, is shown in Fig. 8. More precisely, the figure shows - for each aggregation geometry and composition - the minimal free energy of chain packing is concerned, the bilayer is always the most stable geometry. That is, for every composition the minimum in μ_t^0 as a function of a is always lowest for the bilayer. The fact that spheres appear more stable than cylinders is a direct consequence of their higher surface roughness; for 'compact' aggregates cylinders are more stable than spheres[1a].

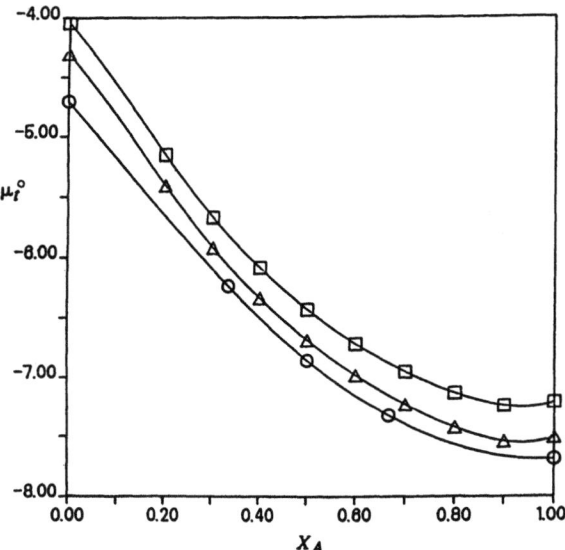

Fig. 8. Tail's free enrgy, calculated using eq. 4, as a function of the long chain mole fraction, in mixed aggregates, planar bilayer, spherical micelle, cylindrical micelle.

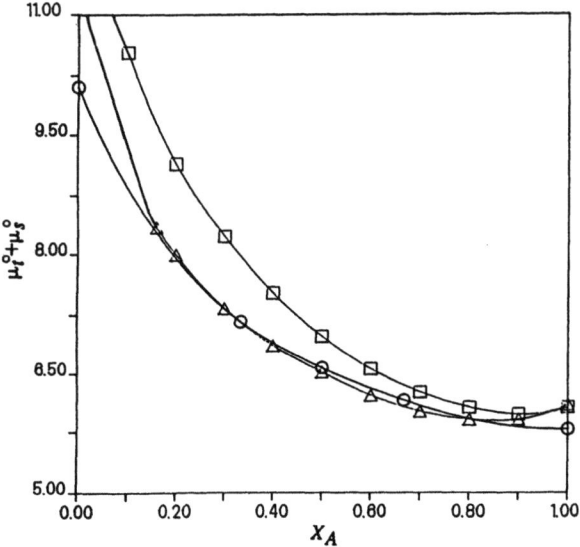

Fig. 9. Total free energy as a function of the long chain mole fraction, in mixed aggregates, $a_0 = 66 \text{Å}^2$; $\gamma = 0.1 \, kT / \text{Å}^2$. Symbols as in fig. 8.

In Fig. 9 we plot the total, tail+head, free energy per amphiphile as a function of composition, (again minimizing with respect to a for each geometry). Note that μ_s^0 does not depend on geometry, but depends strongly on the assumed a_0. In Fig. 9 the value chosen for a_0 is 66 Å^2. It should be noted that the value of a wich minimizes $\mu_t^0 + \mu_s^0$ is, in general, different from the value of a corresponding to the separate minima of μ_t^0 and μ_s^0. Fig. 9 indicates that when the total free energy is minimized the free energy hierarchy between different geometries may depend on composition, e.g. for the above choice of a_0, spherical micelles appear as (or even slightly more) stable as bilayers in the region $x_A \sim 0.3$-0.8. (Translation entropy of small spherical micelles, which is not accounted for by $\mu_t^0 + \mu_s^0$, will further favor these small aggregates).

CONCLUDING REMARKS

The conformational and thermodynamic properties of amphiphile molecules in micelles and bilayers are well described by a mean field theory.

Surface roughness appears more important in small, curved, aggregates. (It was found[1a] that a 'compact' bilayer shows excellent agreement with molecular dynamics simulations and also with experiment). The differences in free energy between different geometries due to chain packing are of the same order of magnitude as those associated with the surface ('opposing') forces. Thus, when treating shape and size distributions in micelles, both contributions should be considered.

The conformational properties of chains comprising mixed aggregates differ from those in pure aggregates. The areas per head group allowed for amphiphiles in mixed aggregates are generally smaller compared to those in pure aggregates. Upon combining the head and tail contributions to the aggregate's free energy we conclude that relative stabilities of different geometries may vary with composition.

ACKNOWLEDGMENT

The financial support of the US-Israel binational foundation is gratefully acknowledged. The Fritz Haber Molecular Dynamics Research Center is supported by the Minerva Gesellschaft fur die Forschung. Munich FRG.

REFERENCES

(1) a) I. Szleifer, A. Ben-Shaul and W. M. Gelbart. J. Chem. Phys. (in press), b) I. Szleifer, A. Ben-Shaul and W.M. Gelbart (submitted), c) A. Ben-Shaul, I. Szleifer and W.M. Gelbart. J. Chem. Phys. **88**, 3597 (1985). d) I. Szleifer, A. Ben-Shaul and W.M. Gelbart. J. Chem. Phys. **88**, 3612 (1985).

(2) For a recent review see: A. Ben-Shauld and W.M. Gelbart, Ann. Rev. Phys. Chem. **36**, 179 (1985).

(3) a) P. van der Ploeg and H.J.C. Berendsen. Mol. Phys **49**, 233 (1983). b) B. Owenson and L.R. Pratt. J. Phys. Chem. **88**, 2905 (1984), **88**, 6048 (1984). c) J.M. Haile and J.P. O'Connell, J. Phys. Chem **88**, 6363 (1984). d) B. Jonsson, O. Edholm and O. Teleman, J. Chem. Phys. **85** 2259 (1986).

(4) a) D.W.R. Gruen. J. Phys. Chem. **89**, 146 (1985), **89**, 153 (1985), and references therein. b) K.A. Dill and R.S. Cantor. Macromolecules **17**, 380 (1984), and references therein.

(5) a) J.N. Israelachvili, in "Physics of the Amphiphiles: Micelles, Vesicles and Microemulsions". V. Degiorgio and M. Corti eds. (North Holland, Amsterdam, 1985) p. 24, and references therein, b) C. Tanford, "The Hydrophobic Effect". 2nd ed. (Wiley-Interscience, New York, 1980).

(6) a) H. Walderhaug, O. Soderman and P. Stilbs. J. Phys. Chem. **88**, 1655 (1984). b) H. Wennerstrom, B. Lindman, O. Soderman, T. Drakenberg and J.B. Rosenholm. J. Am. Chem. Soc. **101**, 6860 (1979). c) J. Chavrolin, J. Chim. Physique **80**, 15 (1983). d) T. Klason and U. Henriksson."Solution Behaviour of Surfactants: Theoretical and Applied Aspects". K.L. Mittal and E.J. Fendler, Eds. **1**, 417 (1982).

(7) a) P.J. Flory, "Statistical Mechanics of Chain Molecules". (Wiley-Interscience, New York, 1969).

STRUCTURAL STUDIES IN LANGMUIR MONOLAYERS BY FLUORESCENCE MICROSCOPY : A NEW APPROACH TO THE PHASE DIAGRAM OF N-PENTADECANOIC ACID AT THE AIR-WATER INTERFACE

F. Rondelez [*,+], J.F. Baret, and K.A. Suresh [¦]

Etudes et Fabrication Dowell Schlumberger
BP 90, 42003 Saint Etienne, France

C.M. Knobler

UCLA, Department of chemistry, Los Angeles, CA 90024 USA

¦ Permanent address : Raman Research Institute
  Bangalore 560080, India

* Also at Collège de France, Laboratoire de Physique de la
  Matière Condensée, 11 place Berthelot, 75231 Paris Cedex France

+ Author to whom all correspondence should be sent

ABSTRACT

Fluorescence microscopy (sometimes also called epifluorescence), when coupled with low-light level camera detection, is sensitive enough to observe the structure of monomolecular films a few Å thick deposited on solid or liquid substrates. We have applied this technique to the study of Langmuir films of pentadecanoic acid mixed with small concentrations of fluorescent probes (0.1-1% by mole) and spread at the air-water interface at 20°C. The surface density can be continuously changed by a movable barrier and the corresponding surface pressure is simultaneously measured by the Wilhelmy hanging plate method. A wide variety of structures is observed : uniform dark or bright backgrounds, spherical islands, irregularly-shaped objects, white foams and dark foams. All these structures can be related to the various states of the two-dimensional layer. At very low densities, probably higher than 2000 $Å^2$ per molecule, only the gaseous phase exists. However, as the monolayer is gradually compressed, the liquid expanded-gas two-phase region is entered. The transition is first-order, as evidenced by the coexistence of liquid and gaseous phases. At densities between 70 and 45 $Å^2$ per molecule, the monolayer is in the liquid expanded state and looks homogeneous. Below 45 $Å^2$ and down to 22 $Å^2$, a coexistence between liquid expanded and liquid condensed domains is observed. As the molecular surface density is gradually increased by compression, the denser phase becomes prominent over the other until the entire monolayer is converted to the liquid condensed state. The reverse situation is observed on expansion. These observations definitely ascertain that the liquid expanded-liquid condensed transition

in pentadecanoic acid is first-order. Here we clear up a long-standing
problem, since some authors have suggested that the transition could
actually be second-order. The main argument was the imperfect flatness of
the plateau in the pressure isotherm in the coexistence region. Our
experiments indicate that electrostatic interactions, which are not
included in the usual theories, are certainly present and may explain why
the compressibility of the monolayer remains finite in the transition
region. Our observations agree with independent experiments on monolayers
of phospholipids by Möhwald and Lösche, and McConnell, Tamm and Weis.
Finally, the boundaries of the liquid expanded-liquid condensed and liquid
expanded-gas coexistence curves can be located accurately using the
technique of epifluorescence microscopy. The values observed for
pentadecanoic acid suggest that some of the literature data, which are all
derived from surface pressure isotherms measurements, may be in large error
especially at the liquid expanded-gas transition.

INTRODUCTION

The phase changes occurring in insoluble films of amphiphilic
molecules at the air/water interface have been extensively studied by
surface manometry (1). This word, coined by Pethica (2), describes the
measurement of the surface pressure exerted by the monomolecular film onto
a movable barrier. A common variation of the same technique is to derive
the interfacial tension from the pulling force exerted by the liquid
surface onto a partially immersed platinum blade. These relatively
unsophisticated mechanical methods have produced a surprisingly large
wealth of experimental data and several phase changes have been identified
upon compression of the monolayer.

Fig 1 shows a typical pressure-area diagram for two-dimensional phases
of pentadecanoic acid. At sufficiently high areas per molecule, or low
surface pressures, the film exists in a gaseous state (called G in fig 1)
and its behavior is very close to that of an ideal two-dimensional gas. As
the area is decreased however, attractive intermolecular forces will tend
to condense the film in the so-called liquid expanded state (3). The L_E/G
transition is first-order and shows up in the surface pressure isotherm as
a plateau region of infinite compressibility (points A to B in fig 1). As
soon as the transition is completed, the surface pressure re-increases ;
the monolayer compressibility is then fairly low, as is normal for a liquid
phase. However the compressibility increases again sharply when the area
per molecule reaches a value A_t which is a decreasing function of
temperature (point C in fig 1). This slope change in the isotherm has been
interpreted as indicative of a transition from a liquid-expanded to a
liquid-condensed monolayer state. In the L_E/L_C transition region (points C
to D in fig 1), the slope of the isotherm approaches but generally never
attains zero value. This had led certain authors to contest that the liquid
expanded-liquid condensed transition was of the first-order type (4). On
the other hand, several arguments have been proposed to explain the
non-zero slope in the transition region while retaining the first-order
character : limited cooperativity of the transforming units (5), impurity
effects (2,6,7) long-range electrostatic forces (8). Finally as the area
per molecule approaches the molecular cross-section in the plane of the
interface, the pressure increases sharply and the monolayer becomes almost
incompressible. Surface pressure isotherms show a small kink when the
liquid condensed phase converts into a solid phase (5).

The temperature dependence of the various physical states of the
monolayer is also shown in fig 1. For both liquid expanded-gas and liquid
expanded-liquid condensed transitions, there is a critical temperature,

called respectively T_{c1} and T_{c2}, above which the phase transition disappears. These critical temperatures are generally thought to be of the classical van der Waals type (4). There is however a notable exception (5) in T_{c2} where it has been ascribed to a tricritical point (i.e. that the transition is first-order below T_{c2} and second-order above). It is quite astonishing that such fundamental questions concerning the various thermodynamic states of monolayers of insoluble films are still unraveled after almost sixty years of intense experimental activity. The main reason behind that rather unsettling situation is the lack of a palette of experimental tools which would allow to probe monolayer properties other

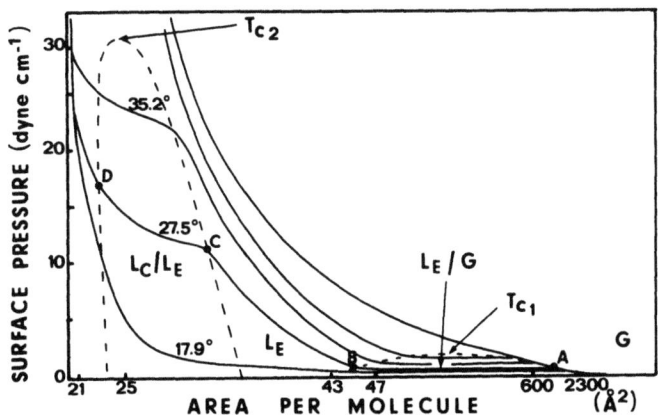

Fig 1. Typical pressure area diagram for monolayers of fatty acids at the air-water interface. The example shown here is for n-pentadecanoic acid. The curves have been taken from Harkins, Young and Boyd, J. Chem. Phys. <u>8</u>, 954 (1940). The two-dimensional phases are gaseous (G), liquid expanded (L_E) and liquid condensed (L_C). Solid phases are also known to exist at low area per molecule but they have not been indicated here. The dotted lines indicate the liquid expanded-liquid condensed (L_C/L_E) and the liquid expanded-gas (L_E/G) coexistence regions. In both cases there is a critical temperature, T_{c2} and T_{c1} respectively, above which the two-phase region disappears and there exists only a single phase.

than surface pressure. For instance the direct optical observation of coexisting liquid expanded and liquid condensed domains in the plateau region would be a very convincing argument in favor of a first-order transition, despite the fact that the compressibility is still finite. Similarly the detection of the liquid expanded-gas transition from surface pressure data requires to push the measuring technique to its sensitivity limit (9). As a consequence there is large disagreement in the literature on the boundaries of the coexistence curve for a given compound, not to mention the value of the critical temperature T_{c1} nor the values of the critical exponents for the compressibility and the density difference between the liquid and the gaseous phases. (9-11)

In this respect, the observation by von Tscharner and McConnell of microscopic images of phospholipids monolayers doped with a small amount of fluorescent dye has truly been a breakthrough (12). Using an epifluorescence microscope, they were able to detect well-defined structures, owing to the strong partitioning of the dye into the more fluid phase. This first experiment was not exempt of drawbacks however. In particular the images could only be observed after transfer of the monolayer onto a solid substrate (alkylated glass slides) and no photographs of the images could be taken. These difficulties were later removed by the use of a very low light level camera tube for the optical detection and also by eliminating the streaming normally present in monolayers at the air-water interface. Images of monolayers taken in-situ have been reported almost simultaneously by Peters and Beck (13), Lösche, Sackmann and Möhwald (14) and McConnell, Tamm and Weis (15) on various phospholipids doped with 1-10% of carbocyanine and NBD-phosphatidylcholine or ethanol amine dyes. The lateral resolution was about 2 μm ; the most striking result was the observation of liquid-condensed (also called solid by some authors) domains of size 10 - 100 μm in equilibrium with a fluid continuum in the liquid expanded-liquid condensed coexistence region. Under certain conditions, periodic arrays could form (14,15), demonstrating the existence of long-range forces, probably electrostatic in origin, and which had been neglected in all previous theories of the phase transitions. In addition, non-equilibrium snow-flake structures were also observed following a sudden step compression (6,8,15). Recently Miller, Knoll and Möhwald (16) have demonstrated that these metastable structures are self-similar and characterized by a fractal Hausdorff dimension d = 1.5 (17), thus connecting these biological monolayers with the most recent solid-state theories of crystallization and aggregation (18). Up to now, most of the recent experimental effort has been devoted to the liquid expanded-liquid condensed phase transition. The liquid expanded-gas phase change has been comparatively much less studied, apart from a sketchy description in the original paper of Lösche et al. (14)

Here we present a complete scan across the phase diagram for the much studied pentadecanoic acid at the air-water interface. This is obtained by varying the area per molecule between 20 and 2000 Å² while keeping the temperature fixed at 20°C. Following a brief description of the experimental set-up, we will present photographs of the structures observed upon compression and expansion cycles. We will successively discuss three situations i) compression from the liquid expanded phase through the liquid expanded-liquid condensed coexistence region, ii) re-expansion from the liquid condensed phase towards the liquid expanded phase, iii) expansion from the liquid expanded phase into the liquid expanded-gas coexistence region. Whenever possible, we will compare our findings for n-pentadecanoic fatty acid monolayers with those previously obtained for phospholipids. Lastly, we will show how our qualitative experiments allow to seriously question the data of the literature for the liquid expanded-gas transition region.

EXPERIMENTAL

Pentadecanoic acid ($CH_3(CH_2)_{13}COOH$) was selected because its monolayer properties have been extensively studied in the past. It was purchased from Sigma Chemicals at a purity of 99% and was generally used as received. In some instances, it was further purified by multiple recristallizations (6 times) from saturated chloroform solutions at room temperature followed by slow cooling to 7°C. Colourless flakes with white tint were obtained which were much purer than the starting materials according to our ^{1}H NMR analysis. The use of the purified crystals however did not seem to change our optical observations. On the other hand, precise measurements of the surface pressure isotherms, which could have shown some differences according to Pethica et al (2), were not attempted. At any rate the use of highly purified starting materials was rendered superfluitous by the necessity of adding finite amounts of a fluorescent dye in order to perform the epifluorescence measurements. 4 -(hexadecylamino) - 7 - nitrobenz - 2 - oxa - 1,3 - diazole (in the following we will use NBD- hexadecylamine for short) was obtained commercially from Molecular Probes Inc (Junction City, Oregon). This non-ionic lipid probe is highly fluorescent in non-polar environments but has a very low quantum yield in water. We have checked that it readily forms fluorescent monolayers at the air-water interface when excited at 470 nm. The emission band is centered around 530 nm. Since NBD-hexadecylamine has a hydrocarbon chain length very close to that of pentadecanoic acid, it can be assumed that the two compounds will not phase-separate. Typical concentrations of dye used in the experiments were between 0.1 and 1%.

Pentadecanoic acid and NBD-hexadecylamine were dissolved in spectro-grade chloroform (Merck) at a very low concentration (10^{-3} g/cm³). Chloroform was chosen as the solvent because it is highly volatile and a good spreading solvent. Hexane is also very popular when dealing with long-chain fatty acids (2, 19) but its use has been recently questioned by Bois et al (20). A drop of the pentadecanoic acid-chloroform solution was deposited at the air-water interface using an Agla microsyringue or calibrated micropipettes (2 - 10 µl). After allowing some time for solvent evaporation, the monolayer was compressed and brought into the liquid expanded phase. The monolayer was then allowed to equilibrate thermally in this homogeneous state, for at least 10 minutes, before starting the expansion-compression cycles.

The subphase was distilled water passed through a Millipore Q - TM filtering system. It was adjusted to pH 2 by addition of dilute hydrochloric acid. The carboxylic group of the polar head does not dissociate at low pH and therefore there is little chance for the fatty acids in the monolayer to react with the metallic impurities, particularly divalent ions, inevitably present in the substrate.

The trough containing the monolayer was made entirely of teflon and had small dimensions 7 x 4 cm² so it was very easy to clean with chromic acid. It was temperature-controlled to 0.2°C by circulating water from a thermostat into a heat exchanger made of stainless steel and immersed in the subphase. A thin tin-oxide coated glass was used simultaneously as a protective cover for the monolayer, reducing surface streaming from ambient air currents, and also as a heater. Its temperature was independently controlled by an electrical rheostat. Temperature readings were taken in the subphase, close to the monolayer, with a Platinum resistance thermometer. The trough was equipped with a moving barrier, also made of teflon. Its motion was controlled by an actuator and a digital optical encoder to an accuracy of 0.1 µm. This allowed a precise knowledge of the various barrier positions. The corresponding molecular densities were calculated by dividing the surface of the trough by the number of molecules deposited onto the interface.

The whole mechanical assembly was attached onto the stage of a Polyvar-Met metallurgical reflection microscope (Reichert-Jung, Austria). Contrary to Lösche and Möhwald (21), we did not use an inverted system with a microscope objective of the water-immersion type to look at the monolayer from below. The objective used had a power of x20 and a numerical aperture in air of 0.40 for a working distance of 5 mm. A 200 W Hg lamp (HBO super pressure) was used for illumination of the monolayer. A beam-splitter with a dichroïc mirror and suitable narrow band filters allowed to excite the monolayer at 455-490 nm, well into the absorption band of the dye, and to observe it at 525-540 nm, again well into the emission band of the dye. The monolayer was imaged onto the photocathode of a sensitive Vidicon camera (model LH 4036, Thomson-CSF, France), the electrical signal of which was sent to a TV monitor and to a 3/4" video-recorder (model NV 9210, Panasonic, Japan) for later image analysis. Photographs of the monolayer structures were taken directly from the monitor screen using a Minolta X-700 camera and Ilford HP-5 film with a sensitivity of 400 ASA extended to 1600 ASA by developer adjustment.

RESULTS

There is common agreement among the many experimentalists who have studied pentadecanoic acid monolayers that the liquid expanded phase should be observed in the range of 35 to 65 $Å^2$ per molecule (2,7,9,10,22,23). Our epifluorescence observations confirm this point since, at surface densities between 45 and 70 $Å^2$ per molecule, homogeneous fluorescence is detected throughout the monolayer which appears uniformly white. As far as possible, we have always selected this phase as the initial state of the monolayer before either expanding or compressing.

1) Compression into the liquid expanded-liquid condensed coexistence phase

Fig 2 to 7 show a sucession of photographs taken as the monolayer is gradually compressed from 50 $Å^2$ down to 20 $Å^2$ per molecule. Compression rate is approximately 1 $Å^2$ per molecule per minute. At 45 $Å^2$, we can observe an abrupt transition in the field of view of the microscope, viz., the homogeneous all-white liquid expanded phase changes to a coexistence phase containing a number of dark droplets. We believe that these droplets are two-dimensional islands of the denser liquid condensed phase in equilibrium with the continuous liquid expanded phase. Their shape is spherical, evidencing that line tension is the dominant mechanism and tries to minimize the perimeter to surface ratio of the growing objects.

Fig 2 shows the aspect of the monolayer at 41 $Å^2$ per molecule. The scale is indicated by the bar which corresponds to 120 µm. The size of the islands is fairly uniform and is centered around 35 ± 9 µm. Fig 3 is taken at 39 $Å^2$ per molecule. A careful observation of the photograph shows that the number of islands has remained constant. On the other hand, the average size has markedly increased to 48 ± 11 µm. Using peculiarities of the structure such as defects (there is one good example in the upper left of the image) or relative positions between neighboring droplets, it is possible to follow the growth of an individual island and therefore to accurately monitor the growth rate. Going from 41 to 39 $Å^2$ per molecule increases the island size by 30%. Fig 4 taken at 34 $Å^2$ per molecule merely confirm our previous observation. The average island size is larger, of the order of 57 ± 14 µm, and the growth rate for a given droplet is 80% between fig 2 and 4.

Photographs taken at still lower areas per molecule allow to observe two additional phenomena which are particularly striking. Fig 5, 6 and 7 were taken at an area per molecule of respectively 32, 28 and 25 $Å^2$. They should not be compared however with figures 2 to 4 because they correspond to different experiments. Fig 5 evidences the fact that, as the dark

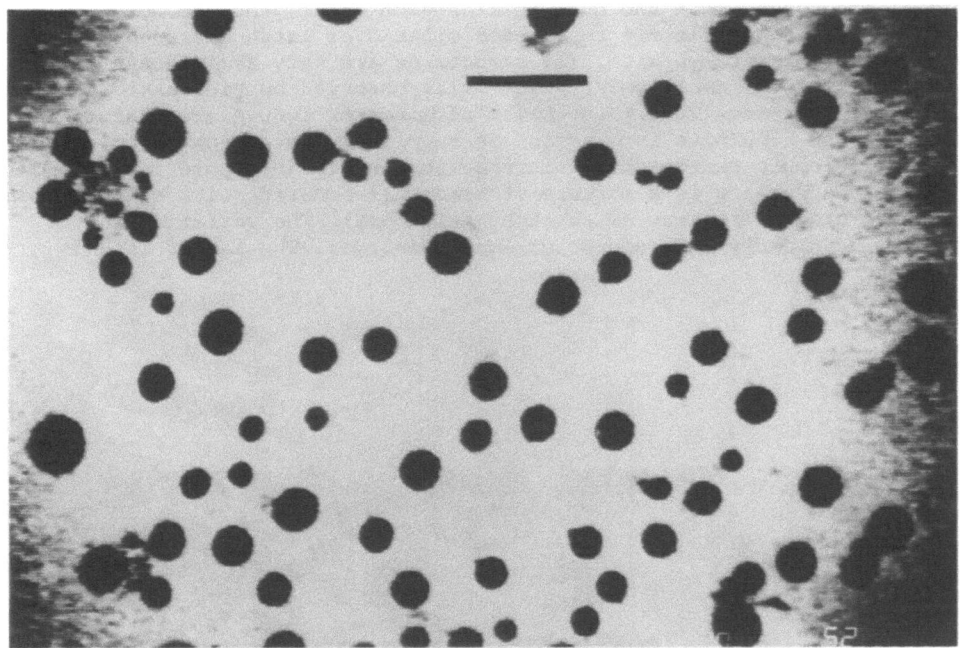

Fig 2. Fluorescence photograph of the liquid expanded-liquid condensed coexistence region for pentadecanoic acid monolayers. Image was taken while compressing from the liquid expanded phase. Surface density is 41 Å^2 per molecule. The black dots are liquid condensed domains nucleating into the continuous liquid expanded phase which appears as a white background. T = 20°C. The bar length is 120 μm.

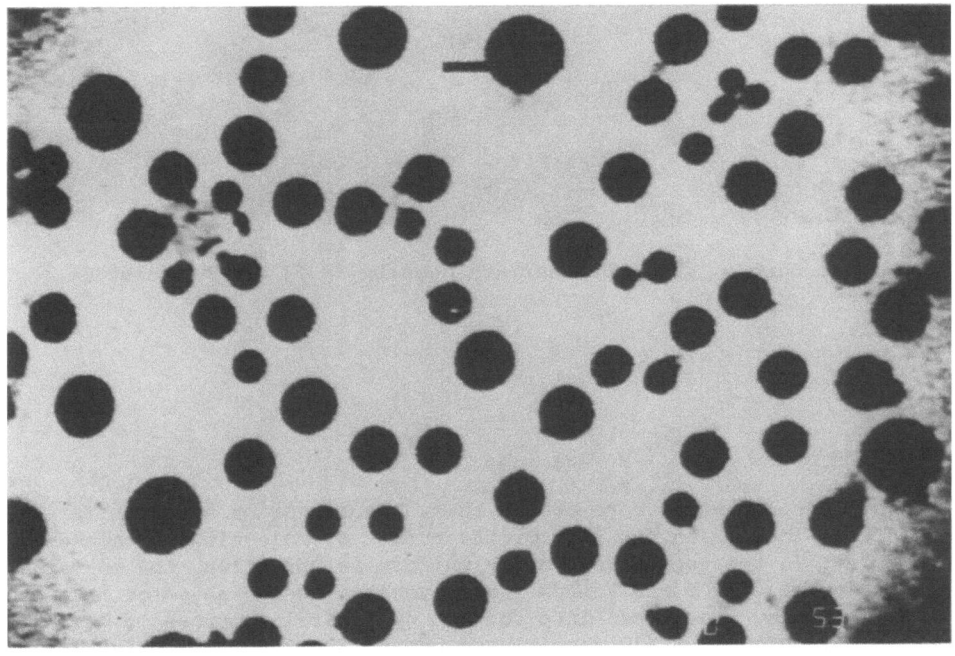

Fig 3. Same as fig 2 but surface density is 39 Å^2 per molecule

islands are pushed more and more towards each other, they do not want to merge. Rather, they clearly repel each other over large distances, roughly comparable to their own size. The structures are very reminiscent of the Wigner crystals of small polystyrene balls observed by Pieranski at the air-water interface (24). These balls of diameter 1000 Å are charged and therefore the repulsive interaction is supplied by electrostatics, either through Coulombic charge-charge interaction or dipole-dipole repulsion. In our case the pattern is a mixture of hexagonal-ordered (with six neighbors) and cubic-ordered regions (with eight neighbors). The variation in the island size clearly is a source of local disorder. The larger islands

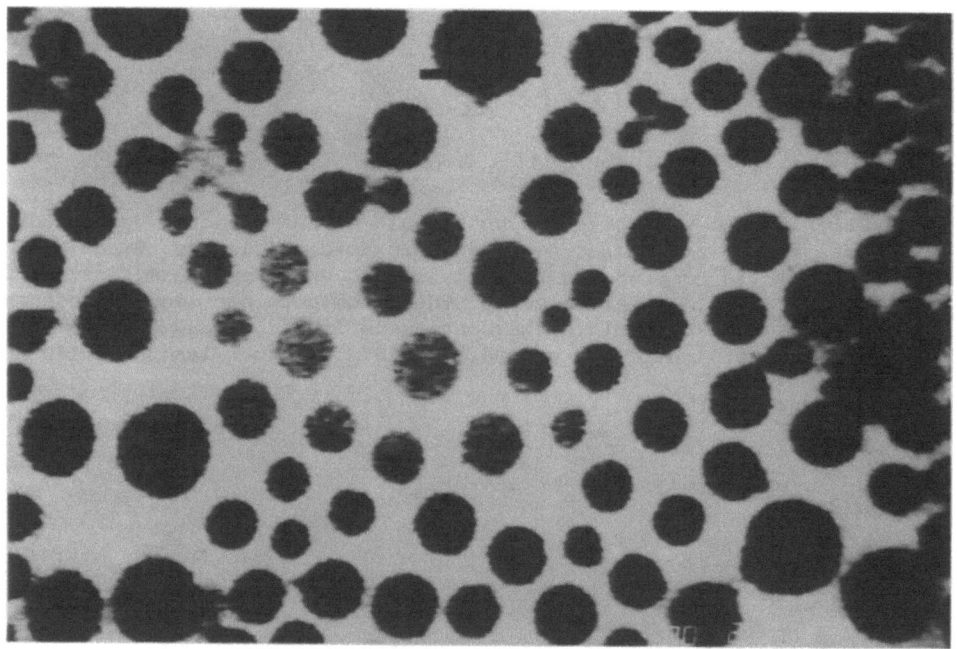

Fig 4. Same as fig 2 but surface density is 34 Å² per molecule

induce a wider depletion zone around them. The repulsion strength is therefore an increasing function of the number of molecules contained in one droplet. In the regions of fairly uniform island sizes, the pattern is generally cubic. Fig 6 and 7 show the distorsion in the shape of the islands when they are forced into contact by lateral compression. At an area per molecule of 28 Å², the islands become facetted. The orientation of the facets is such that the distance to the neighboring objects is maximized. It is evident that the system chooses this way in order to

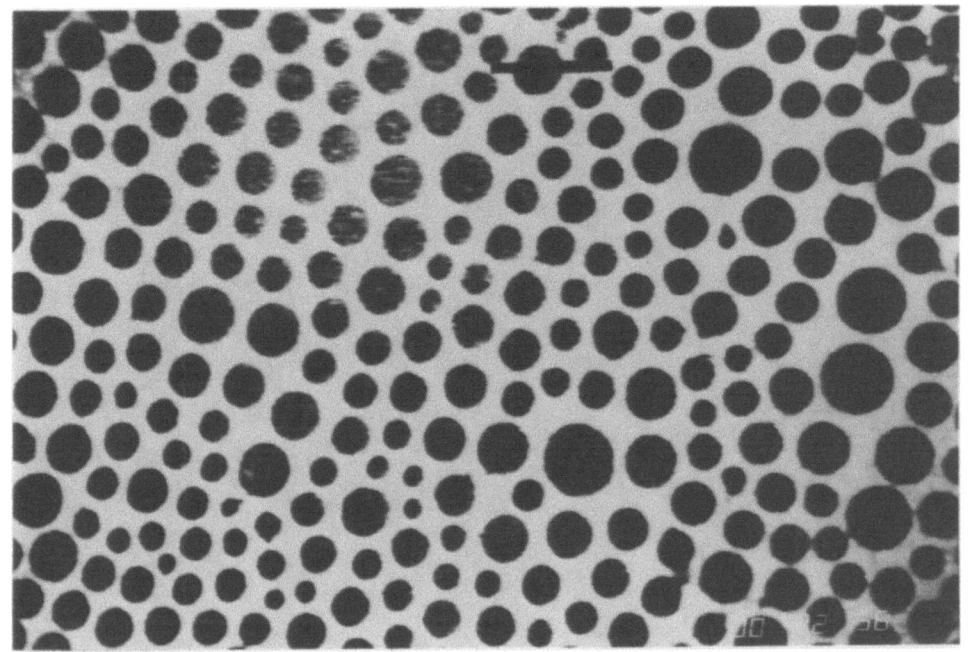

Fig 5. Same as fig 2 but surface density is 32 $Å^2$ per molecule. Also this is a different region of the monolayer compared to the previous photographs

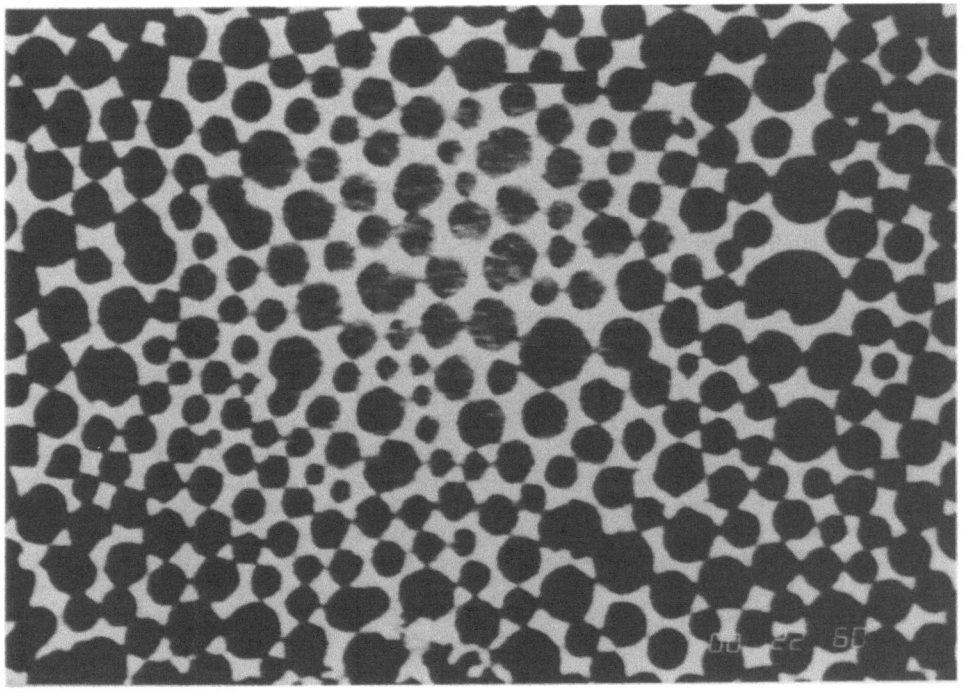

Fig 6. Same as fig 5 but surface density is 28 $Å^2$ per molecule

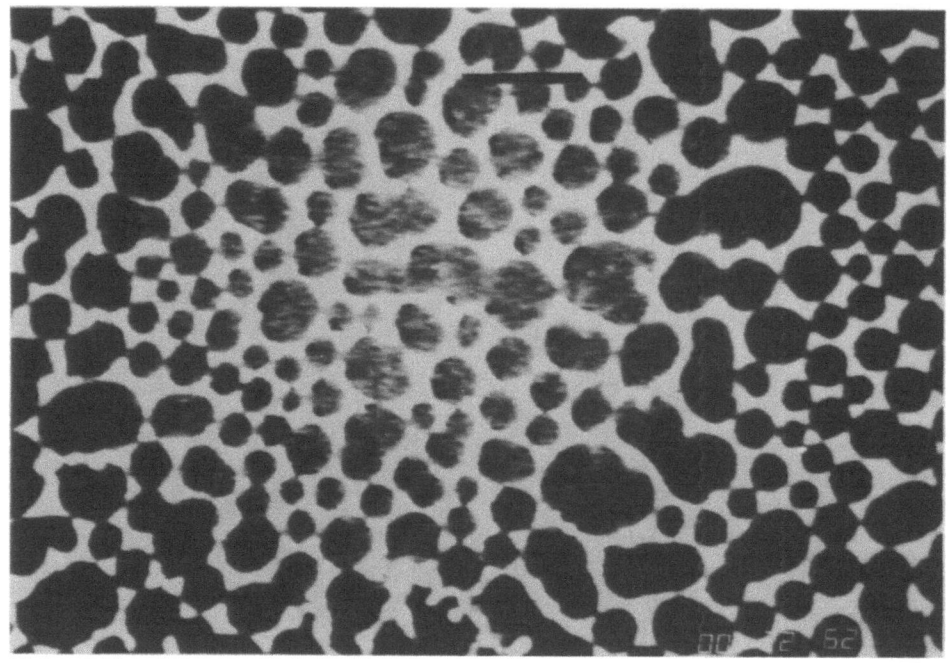

Fig 7. Same as fig 5 but surface density is 25 Å² per molecule

reduce the electrostatic repulsion. Two neighboring islands are only in close proximity by their angular points. An extremely peculiar feature is that these angular points are connected, forming a continuous pathway across the field of view of the photograph. It seems that this corresponds to the first stage of the merging mechanism. In fig 7, about 25% of the initial islands have merged, generally by group of two or three at the most. The overall shape is highly non-symmetrical and is reminiscent of the shape of the individual islands before merging. There is clearly little positional rearrangement inside an island, which must therefore be extremely viscous. It seems that we are dealing with plastic deformations. This collapsing process can be pursued until the whole field of view becomes black. At this point, the transition to the liquid condensed state is complete. In our experiments this was observed at an area of 22 Å² per molecule.

2) Expansion from the liquid condensed state

Expanding the monolayer from the homogeneous liquid condensed state, one expects to observe the growth of droplets of the liquid expanded phase as the liquid condensed-liquid expanded coexistence region is entered. This is indeed what can be seen in fig 8 to 12 taken at respectively 23 - 26 - 30 - 35 - 40 Å² per molecule. White droplets appear on a dark background and at first sight these photographs seems to be a mere negative of the previous series. There are important differences however. First, the size distribution of the white droplets is extremely large. In fig 8 we can observe variation in sizes all the way from 8 to 100 µm. The largest droplet of the photograph has the interesting feature that it still contains a small fraction of liquid condensed phase in its interior. Secondly, the number of white droplets does not stay constant upon monolayer expansion. Rather it markedly decreases as can be seen by comparing fig 8, 9 and 10. This is due to the fact that two droplets of liquid expanded phase easily merge as soon as they touch. There seems to be no repulsive force as in the case of the liquid condensed droplets. The larger droplets grow very rapidly at the expense of the smaller ones. Their sizes largely exceed a few hundreds of μm in fig 10 and 11. Fig 12 shows a small fraction of a bubble which is even in the millimeter range. This is clearly a third difference with the growth of liquid condensed droplets which have a maximum size of 120 to 240 μm. Lastly, the overall shape of the islands always tend to be spherical and this indicates considerable rearrangement after merging. The interior of the white droplets is therefore highly fluid. Eventually, the field of view becomes uniformly white when the homogeneous liquid expanded state is reached, corresponding to an area per molecule of 45 Å^2 at the temperature of the experiment which was 20°C.

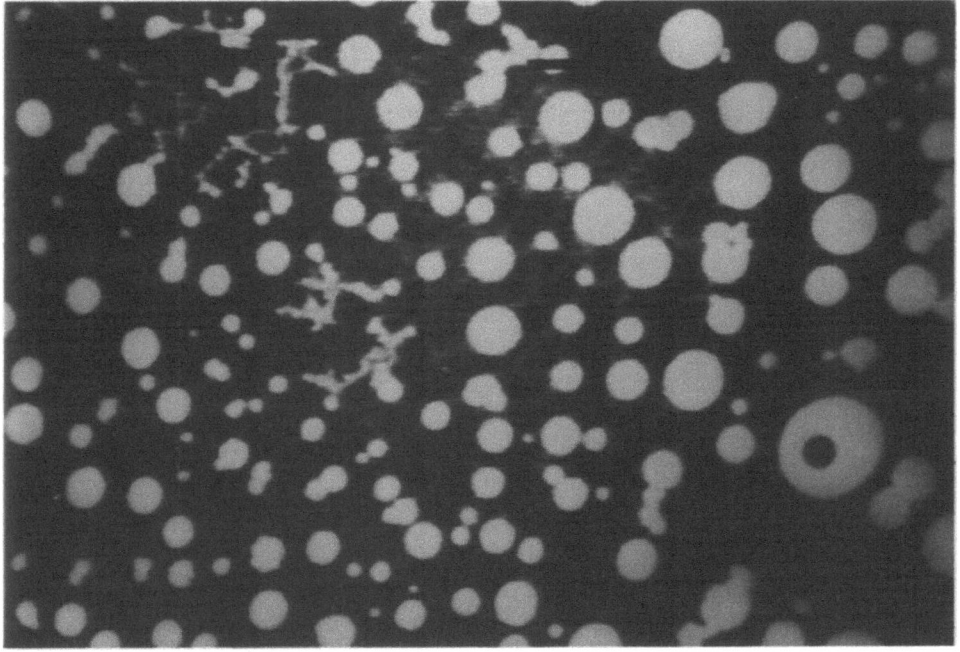

Fig 8. Fluorescence photograph of the liquid expanded-liquid condensed coexistence region for pentadecanoic acid monolayers. Image was taken while expanding from the liquid condensed phase. Surface density is 23 Å^2 per molecule. The white droplets are liquid expanded domains nucleating into the continuous liquid condensed phase which appears as a dark background. T = 20°C. The bar length is 120 μm.

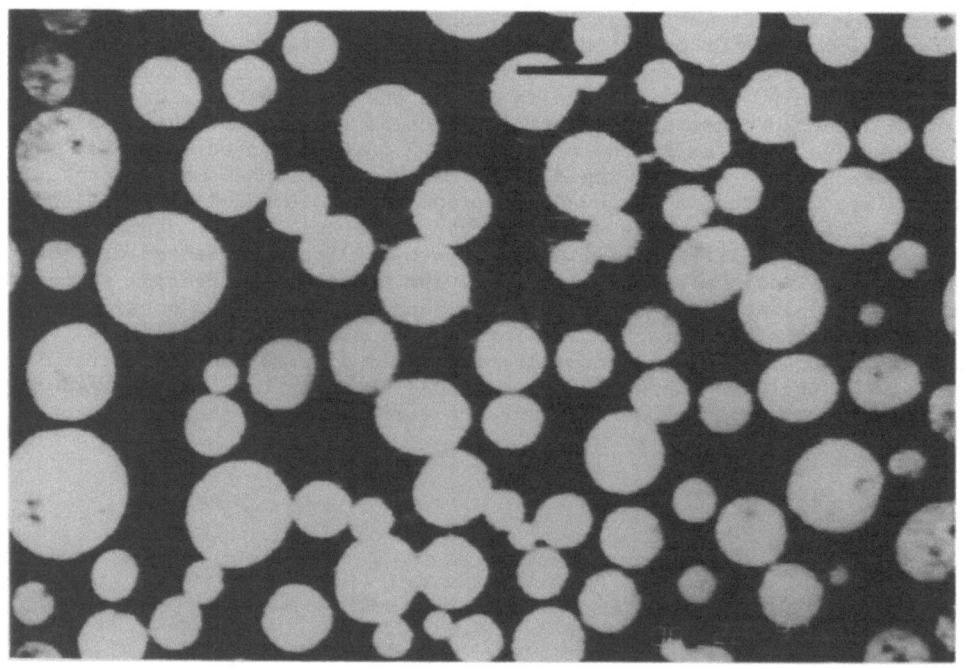

Fig 9. Same as fig 8 but surface density is 26 Å² per molecule

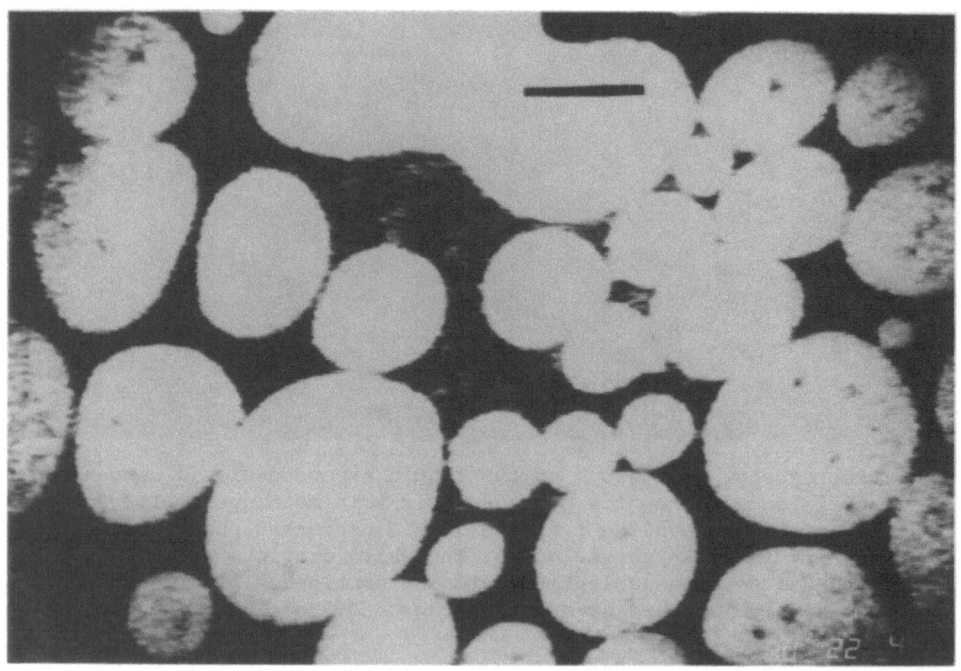

Fig 10. Same as fig 8 but surface density is 30 Å² per molecule

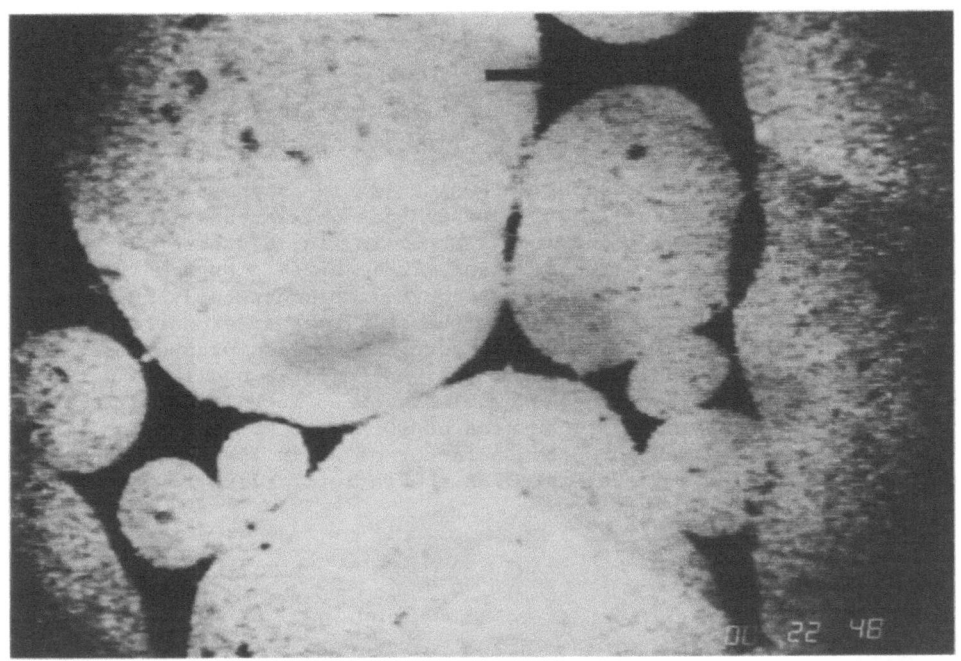

Fig 11. Same as fig 8 but surface density is 35 Å² per molecule

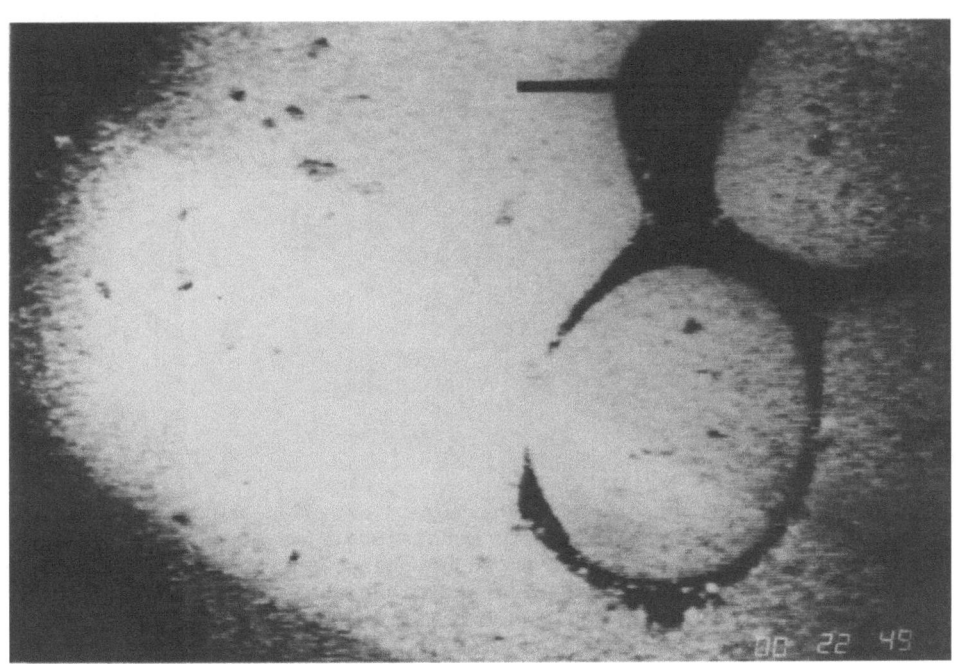

Fig 12. Same as fig 8 but surface density is 40 Å² per molecule

3) Expansion from the liquid expanded phase into the liquid expanded-gas coexistence region

For this particular experiment, one has to work at extremely low surface densities. As a consequence, the molecular areas given in fig 13 to 19 are only approximate.

Starting from the homogeneous all-white liquid expanded phase, gas bubbles are nucleated as the liquid expanded-gas coexistence line is crossed at about 70 Å² per molecule. As shown in fig 13, these gas bubbles appear as black domains of circular shape. Their size is extremely small, typically 1 µm or less. They also appear quite inhomogeneously in the field of view. This may be an artefact due to the fairly rapid expansion rate, of the order of 10 Å² per minute, and therefore the surface pressure may not have equilibrated throughout the monolayer. As the area available per molecule is kept expanding, the gas bubbles grow to much larger sizes, as seen in fig 14, corresponding to an area of 90 Å² per molecule. The size distribution is very wide, more than a factor of 20 in the photograph. The larger bubbles grow at a faster rate. As the ratio of black to white

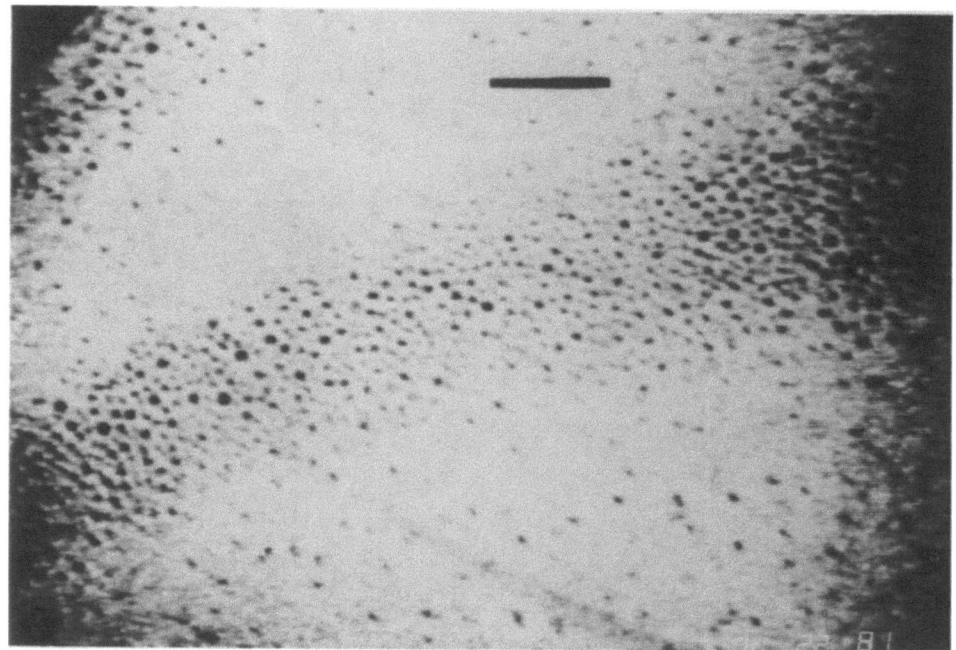

Fig 13. Fluorescence photograph of the liquid expanded-gas coexistence region for pentadecanoic acid monolayers. Image was taken while expanding from the liquid expanded phase. Surface density is 70 Å² per molecule. The dark droplets are gaseous bubbles nucleating into the continuous liquid expanded phase which appears as a white background. T = 20°C The bar length is 120 µm.

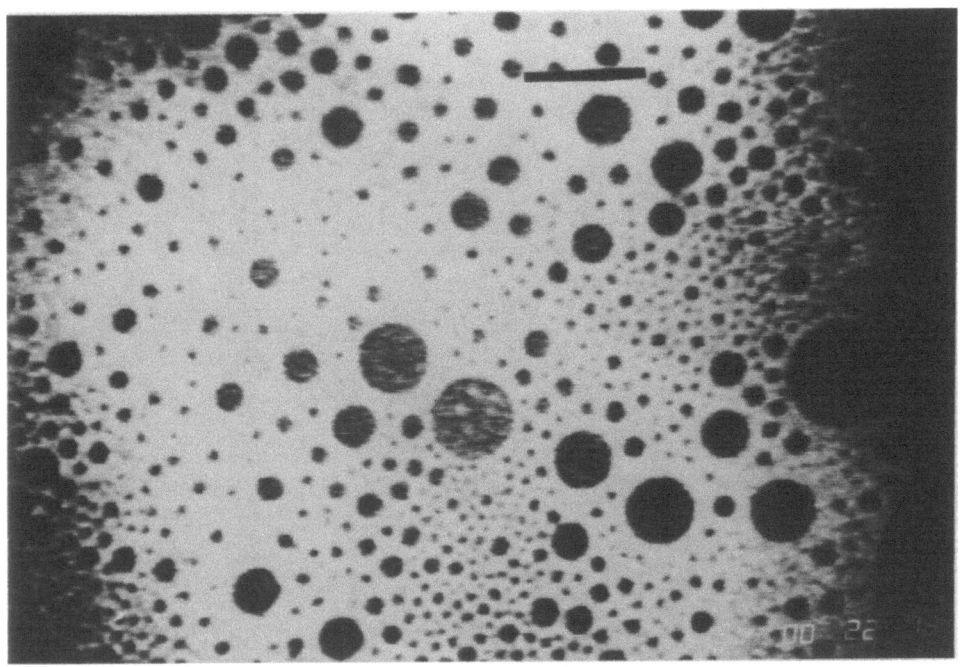

Fig 14. Same as fig 13 but surface density is 90 Å² per molecule. The "Swiss cheese" structure is clearly discernable.

regions increases, the monolayer takes a structure reminiscent of Swiss cheese with isolated dark circles of all sizes in a uniformly white background.

At still larger areas per molecule the Swiss cheese structure gradually evolves into a foam. The white regions have become tenuous, interconnected, straight lines. They are connected at vertices of coordination number 3. Vertices with higher coordination number, 4 for instance, have also been observed but they appear to be very unstable (some can be seen on the lower left side of fig 15). This two-dimensional foam is observed over a wide range of surface densities, typically from 200 up to 600 Å² per molecule. At a given surface density, the structure is metastable and the average cell size grows continuously as the result of two mechanisms typical of soap films, namely, i) fusing of neighboring cells by the disappearance of an edge ii) shrinkage of the smallest cells into vertices (25). Fig 15 to 17 show the progressive growth of the foam cells up to average sizes of several hundreds of μm, at a fixed surface density of 400 Å² per molecule. The time period is of order of tens of minutes, which prove the foam to be reasonably stable. One reason for this good stability may be the absence of gravity effects at the air-water interface which is horizontal by definition. Drainage mechanisms which are important in ordinary three-dimensional foams become irrelevant here.

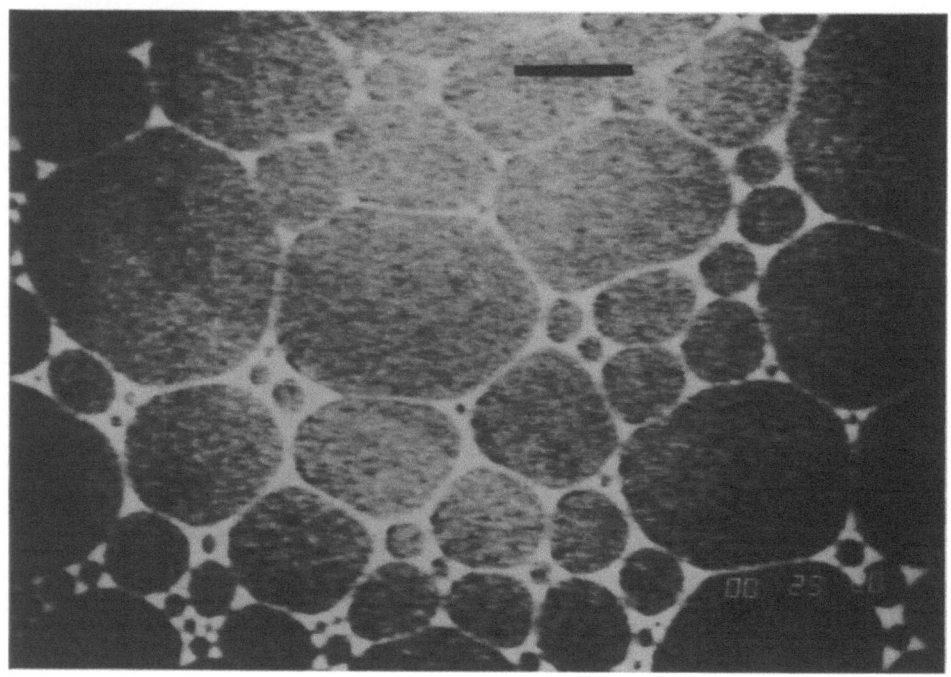

Fig 15. Fluorescence photograph of the liquid expanded-gas coexistence region at a surface density of 400 Å^2 per molecule. The continuous liquid expanded phase of fig 14 has evolved into a series of interconnected, tenuous, white lines surrounding dark gaseous regions. This cellular structure is reminiscent of a foam.

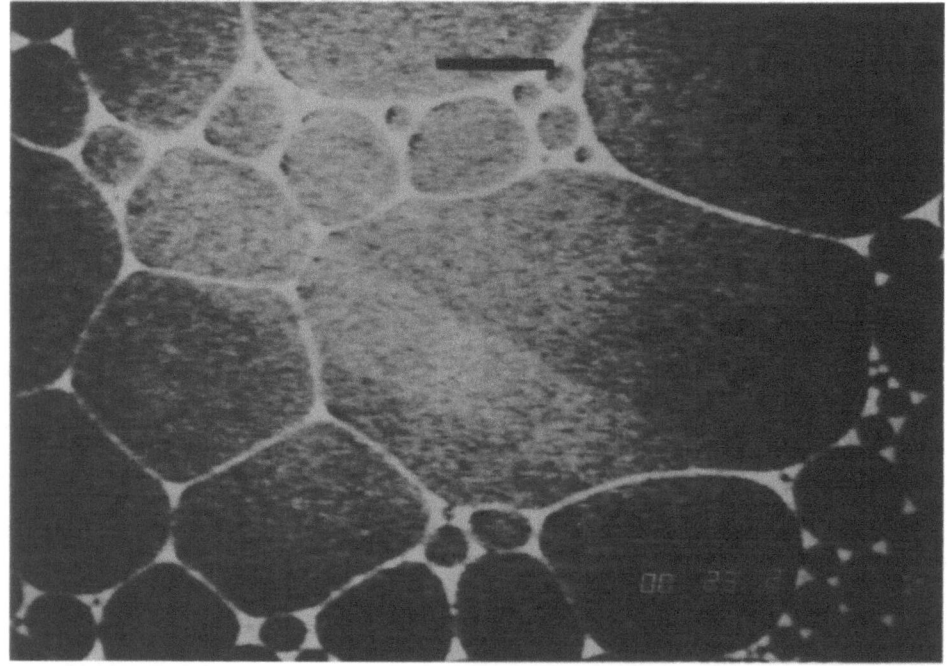

Fig 16. Same as fig 14 but photograph is taken 5 minutes later

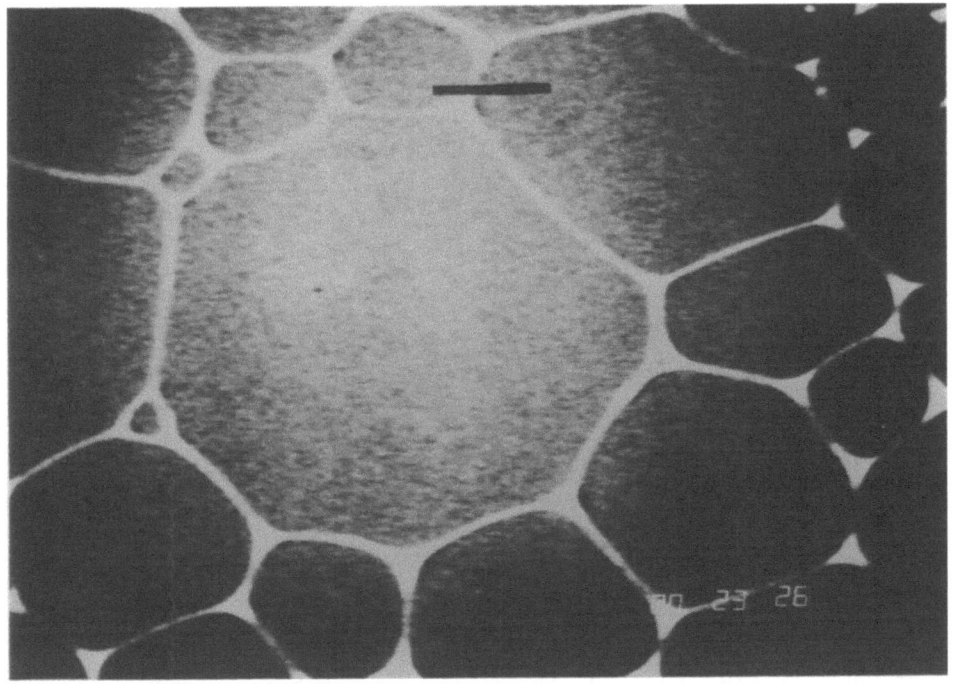
Fig 17. Same as fig 15 but photograph is taken 10 minutes later.

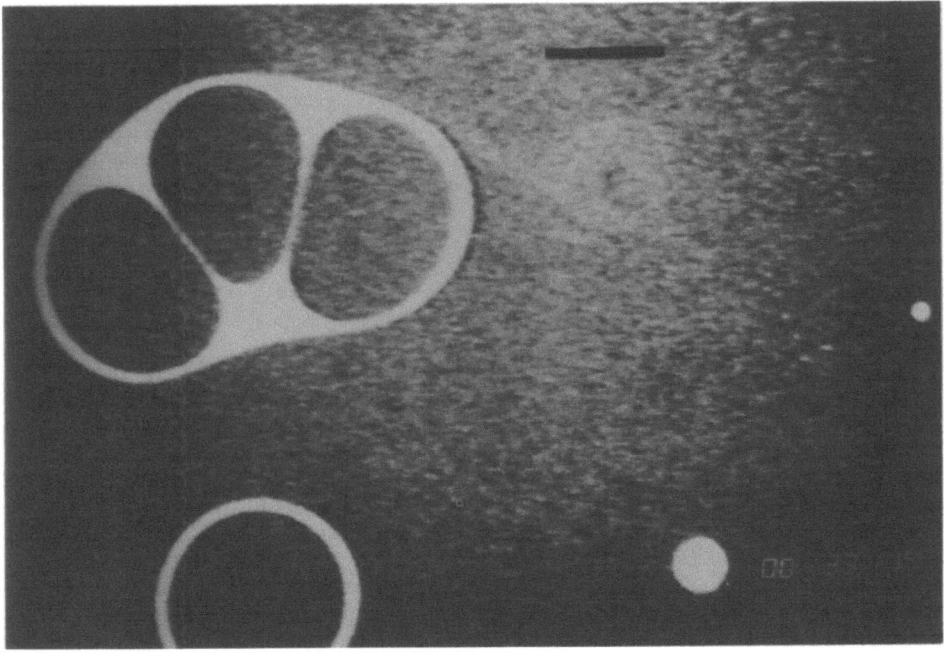
Fig 18. Same as fig 15 but surface density is 800 Å^2 per molecule
The cell structure has become discontinuous. The white circular droplets are remnants of previous cells which have collapsed.

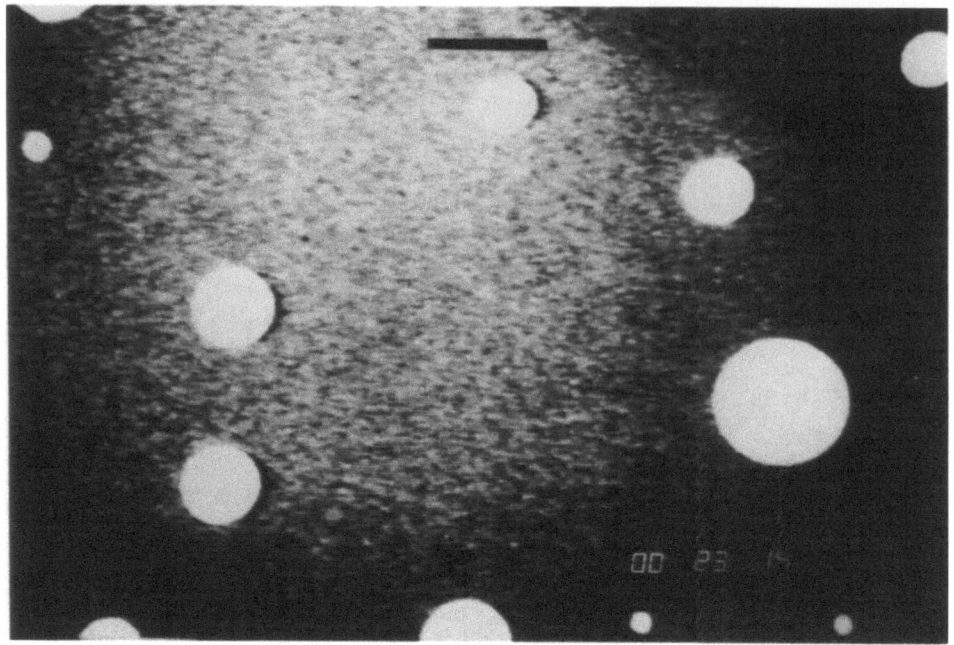

Fig 19. Same as fig 18 but all the cells have now been converted to white circular droplets. These dots correspond to liquid expanded domains floating in a continuous gaseous phase which appears as a dark background.

Non-interconnected foam cells are observed at larger areas. Fig 18, which corresponds to an area per molecule of 800 Å^2, shows a three-cell foam together with a one-cell bubble in which a 19 µm-thick liquid expanded layer surrounds a circular gas domain. Such structures are highly unstable and they collapse into tiny spherical droplets of the liquid expanded phase such as the ones on the right hand side of fig 18. Other examples are shown in fig 19 for a surface density of 800 Å^2 per molecule. Experiments performed at 1500 Å^2 still show isolated droplets of the liquid expanded phase. Therefore, we have to admit that the right-hand side of the liquid expanded-gas coexistence curve has not yet been reached even at these extremely low surface densities. We surmise that the minimum area for an homogeneous gas phase may exceed 2000 Å^2 per molecule.

DISCUSSION

We have first to explain why these structures can be observed by epifluorescence microscopy. The reason actually differs according to whether one is in the liquid condensed-liquid expanded or liquid expanded-gas coexistence regions. In the former case, droplets of the liquid condensed phase appear black because they are depleted in fluorescent molecules relative to the liquid expanded phase. It should be remembered that the liquid condensed domains have an average density roughly twice as large as the liquid expanded phase. The molecules are therefore much more tightly packed, with their polar heads in close proximity and their aliphatic tails oriented vertically in an all-trans conformation. The fluorescent NBD- hexadecylamine dye, with its bulky polar head, does not easily penetrate into the liquid condensed phase. On the contrary, the steric restrictions are much less severe in the liquid expanded phase. Therefore, the dye partitions preferentially into this latter phase. Such an explanation has also been proposed by Möhwald et al (14) and McConnell et al (15) in their monolayer studies of phospholipids labelled with NBD fluorescent groups.

The same reason obviously cannot explain why the structures of the liquid expanded-gas coexistence region are observable by epifluorescence microscopy. In the above, we have assumed that the gaseous domains appear black in a white liquid expanded background. We believe that this is due to the differences in the molecular orientation of the fluorescence probe in the two phases. In the liquid expanded phase, the molecular state is much more disorganized than in the liquid condensed phase. However, on the average, the orientation of the hydrocarbon chains is still perpendicular to the air-water interface. This is no longer the case in the gaseous phase where the average area per molecule largely exceeds the molecular dimensions and where the molecules exhibit little steric constraint on each other. Langmuir (26) was the first to suggest that fatty acid chains in the two dimensional gas phase were more or less parallel to the subphase. As mentioned earlier, the lipid dye used in this experiment is highly fluorescent only in an oil phase but has a very low quantum yield in water. Therefore if there is reorientation of the molecule between the liquid expanded and the gas phase and if this reorientation brings the fluorescent NBD head group in intimate contact with water, the natural fluorescence of the dye will be quenched and the gas phase will appear dark. Unpublished data by Moore and Knobler on a similar dye dissolved either in an organic phase (carbon tetrachloride) or in an oil in water microemulsion show that the fluorescence is decreased by two orders of magnitude when there is water present in the system.

The observations of the monolayer structures by epifluorescence microscopy allow to locate the limits of the coexistence regions with great accuracy. If the compression or expansion rates are slow enough, the monolayer states are at thermodynamic equilibrium and the areas per molecule corresponding to the onset of two coexisting phases can be determined to $\pm$ 0.5 $Å^2$. For n-pentadecanoic acid at 20°C, we estimate that i) the homogeneous liquid expanded phase is observed between 45 and 70 $Å^2$ per molecule ii) the homogeneous liquid condensed phase is entered below 22 $Å^2$ per molecule iii) the homogeneous gaseous phase is about 2000 $Å^2$ per molecule. If we compare these values with the literature data, we find both agreements and discrepancies.

There is a consensus to say that the liquid expanded to liquid condensed phase change is completed at 20-21 $Å^2$ per molecule, independent of temperature (2,7,22,23). This is in good agreement with our value of 22 $Å^2$. The slight discrepancy may come from the fact that the measurements derived from the shape of the surface pressure isotherms are always hampered by the rounding-off of the plateau region on the low area per molecule side.

The literature data for the onset of the liquid expanded phase are more scattered. The exact value seems to be highly dependent on the spreading procedures for the formation of the monolayer and on the impurities contents. It is also a strong function of temperature. The published values at 20°C are in the range of 34.5 $Å^2$ (7), 39 $Å^2$ (22) and 42 $Å^2$ (23) per molecule . The present experimental data of 45 $Å^2$ per molecule is not in disagreement with those values, although it is clearly in the high range. It is interesting to note that the largest values seem to be obtained for the less pure compounds. In our experiments, the pentadecanoic acid has been doped voluntarily with a fluorescent dye. Considering our high measuring accuracy, it would be interesting to investigate systematically if the onset of the liquid expanded phase is shifted to lower areas per molecule as the amount of dye is decreased.

The determinations of the boundaries for the liquid expanded-gas coexistence region which can be found in the literature are highly controversial. In table 1, we have listed the values published by three different groups (9, 10,11) at 20°C. A_{max} is the largest area per molecule at which the monolayer is still in the homogeneous liquid expanded phase. A_{min} is the lowest area per molecule where the gas phase exists. These data have been derived from surface pressure measurements, by estimating the points of deviation of the isotherms from a flat plateau. We have also indicated the value of the pressure in the plateau region. Note that the pressures to be measured are extremely low and are thus prone to errors, despite the ingeniosity and care of the various experimentalists.

Table 1. Surface manometry data for the liquid expanded-gas coexistence region of n-pentadecanoic acid monolayers at 20°C

Reference	A_{max} (Liquid expanded) Å²/Molecule	A_{min} (Gas phase) Å²/molecule	$\pi_{plateau}$ 10^{-3} dyne/cm
Hawkins and Benedek (1974)	87	750	90
Kim and Cannell (1976)	84	666	132
Pallas & Pethica (1986)	41.5	1500	132

Table 1 shows that there is little agreement between the three groups. Hawkins and Benedek (10) roughly agree with the boundary limits of Kim and Cannell (9) but disagree on the plateau surface pressure value by 50%. Pallas and Pethica (11) disagree with the other teams for the boundary limits and seem to have observed a much wider coexistence region. On the other hand their pressure value is identical to that of Kim and Cannell(9). Our data of 70 Å² per molecule for A_{max} and our rough estimate of 2000 Å² for A_{min} differ from everybody else's results. We are confident however that, in the limits of our method, A_{max} for the onset of the liquid expanded-gas coexistence region is larger than 42 Å² per molecule, since at this very same value we observe the beginning of the liquid expanded-liquid condensed two-phase region. Our estimation of 2000 Å² for A_{min} needs confirmation. It is not inconsistent with Pallas and Pethica's result and also with a much earlier determination by Harkins, Young and Boyd (22). We cannot be absolutely positive at this stage that the small bright circles observed in fig 19 are not dye-rich regions which are phase-separated from the pure pentadecanoic monolayer. However, this seems rather improbable since we have checked that monolayers of pure dye are also in the gaseous phase at comparable surface densities.

Observations of monolayers of fluorescent-labelled fatty acids yield richer information than the mere delimitation of two-phase coexistence regions. For instance, the unambiguous evidence for the coexistence of separated liquid condensed islands into a continuous liquid expanded phase proves that the liquid expanded-liquid condensed phase transition is first-order. This has been a subject of long-standing dispute. Advocates of the order-disorder theories basically predict a first-order phase change and explain the transition by a collective hydrocarbon-chain disordering (27). Contenders of the orientation models consider the molecule to be a rigid entity and expect second-order phase transitions if one admits the possibility of anisotropic molecular orientations in the plane of the monolayer (28). On the other hand, the theoreticians of the first group were at loss to explain why the slope of the plateau region was generally measured to be non-zero. The direct optical observation of long-range repulsive forces between the growing liquid condensed domains easily explains why the compressibility is not infinite in the coexistence region. So far, all models only considered attractive forces of the van der Waals type. A posteriori arguments for the non-flatness of the plateau in the surface pressure isotherm had therefore to bear on difficult to prove effects such as finite cooperativity of the transition (5) and/or impurity effects (6,7). The possibility of electrostatic repulsions has first been introduced by Möhwald et al (8) and McConnell et al (15) in their epifluorescence studies of phospholipids monolayers.

Some information on the nature of the liquid condensed phase can be derived from the way two dark islands merge when the lateral pressure externally applied exceeds the electrostatic repulsion. If the phase was fluid, the final shape would be circular, as imposed by minimization of the interfacial energy between the liquid condensed domains and the liquid expanded continuous phase. The fact that one merely observes a juxtaposition of the initial objects indicate that the phase is at best plastic. Mass diffusion measurements of fluorescent impurities by the technique of Fluorescence Photobleaching Recovery (13,15) have shown that the diffusion coefficient in the liquid condensed phase is three orders of magnitude smaller than in the liquid expanded phase (13). Similarly, electron diffraction experiments on liquid condensed monolayers of phospholipids transferred onto microscope grids have given evidence that the liquid condensed phase either is crystalline with hexagonal symmetry (8) or at least possesses long range positional ordering (29). Our observations of liquid condensed domain with spherical, instead of facetted, shape is an indication that these islands cannot however be single crystals.

The existence of a two-dimensional foam structure in the liquid expanded-gas coexistence region is somewhat unexpected. In three dimensions, formation of a foam between a liquid and a gas requires a third component which is surface-active and acts as a stabilizer (30). It is not clear if this is the dye which is acting here to reduce the line tension. Experiments on monolayers containing intrinsically-fluorescent lipids or fatty acids should help to clarify this point. The time evolution of a similar foam has already been discussed in one of our earlier papers (31) in the case of monolayers containing 1% of stearic acid molecules labelled at the 12 position with a NBD fluorescent group. Since we find again the same qualitative features on a different system and with a different dye, there is at least some evidence that the formation of 2-dimensional foams may be a general phenomenon. This will be a subject of further investigation.

CONCLUSION

We have demonstrated that epifluorescence microscopy allows to observe the structural properties of fatty acid monolayers with spatial resolution of the order of a few μm. Since these monolayers are only a few Angstrom thick, they can be considered as model two-dimensional systems. In that respect, the direct optical observation of a wide variety of phases : liquid condensed (sometimes called solid), liquid expanded, gas, foams, etc.., opens up new areas of research. Physico-chemical hydrodynamic experiments on two-dimensional fluid droplets in slow viscous flow, calculated long ago by the theoreticians (32), become feasible. Two-dimensional solid growth can also be studied, together with the instabilities of the solidification front (33). One of the peculiarities of the air-water interface is that it is a zero-gravity system. Therefore there is also hope to observe macroscopic spinodal decomposition structures, and wetting layers, during phase separation processes close to the two-dimensional critical points (34).

REFERENCES

1. G.L. Gaines, "Insoluble monolayers at liquid-gas interfaces", Interscience Publishers, Wiley, New York (1966).
2. S.R. Middleton, M. Iwahashi, N.R. Pallas and B.A. Pethica "Absolute surface manometry : thermodynamic fixed points for air-water monolayers of pentadecanoic acid at 25°C", Proc. R. Soc. Lond. A396, 143 (1984).
3. The terminology used here is that of N.K. Adam, a pioneer in this field. See for example N.K. Adam, "The Physics and Chemistry of Surfaces", 3rd edition, Oxford University Press, Oxford, England (1941) or the original articles by N.K. Adam and G. Jessop, "The properties and molecular structure of thin films : part II, III, VII, VIII", Proc. R. Soc. London A101, 452, 516 (1922)
      ibidem A110, 423 (1926)
      ibidem A112, 362 (1926)
4. J.F. Baret, "Phase transitions in two-dimensional amphiphilic systems", Prog. Surface and Membrane Sci. 14, 291 (1981).
   G.M. Bell, L.L. Combs and L.J. Dunne, "Theory of cooperative phenomena in lipid systems", Chem. Rev. 81, 15 (1981).
5. O. Albrecht, H. Gruler and E. Sackmann, "Polymorphism of phospholipid monolayers", J. Physique (Paris) 39, 301 (1978)
6. M. Lösche and H. Möhwald, "Impurity-controlled phase transitions of phospholipid monolayers", Eur. Biophys J. 11, 35 (1984).
7. N.R. Pallas and B.A. Pethica, "Liquid-expanded to liquid-condensed transitions in lipid monolayers at the air/water interface", Langmuir 1, 509 (1985). The values at 20°C given in this paper are estimations obtained from extrapolation of the data published at 30°C and 25°C.
8. A. Fischer, M. Lösche, H. Möhwald and E. Sackmann, "On the nature of the lipid monolayer phase transition", J. Physique Lett. 45, L-785 (1984).
9. M.W. Kim and D.S. Cannell, "Experimental study of a two dimensional gas-liquid phase transition", Phys. Rev. A 13, 411 (1976), "Surface potential of a two dimensional film of pentadecanoic acid in the coexistence region", Phys. Rev. A 14, 1299 (1976).
10. G.A. Hawkins and G.B. Benedek, "Measurements of the equation of state of a two-dimensional gas near its critical point" Phys. Rev. Lett. 32, 524 (1974).
11. N.R. Pallas and B.A. Pethica, "The liquid-vapour transition in monolayers of n-pentadecanoic acid at the air-water interface", preprint (1986) and N.R. Pallas, Ph. D. Dissertation, Clarkson University, Potsdam, New York (1983).
12. V. von Tscharner and H.M. McConnell, "Physical properties of lipid monolayers on alkylated planar glass surfaces", Biophys. J. 36, 409, 421 (1981).
13. R. Peters and K. Beck, "Translational diffusion in phospholipid monolayers measured by fluorescence microphotolysis", Proc. Natl. Acad. Sci. (USA) 80, 7183 (1983).

14. M. Lösche, E. Sackmann and H. Möhwald, "A fluorescence microscope study concerning the phase diagram of phospholipids", Ber. Bunsenges Phys. Chem. 87, 848 (1983).
15. H.M. McConnell, L.K. Tamm and R.M. Weis, "Periodic structures in lipid monolayer phase transitions", Proc. Natl. Acad. Sci. (USA) 81, 3249 (1984).
16. A. Miller, W. Knoll and H. Möhwald, "Fractal growth of crystalline phospholipid domains in monomolecular layers", Phys. Rev. Lett. 56, 2633 (1986).
17. B.B. Mandelbrot, "The fractal geometry of nature", Freeman, San Francisco (1982).
18. J.S. Langer, "Instabilities and pattern formation in crystal growth", Rev. Mod. Phys. 52, 1 (1980).
19. See ref.1, p 32
20. A.G. Bois, M.G. Ivanova and I.I. Panaiotov, "Solvent influence on the liquid expanded-liquid condensed transition of pentadecanoic acid monolayers", preprint submitted for publication (1986).
21. M. Lösche and H. Möhwald, "Fluorescence microscope to observe dynamical processes in monomolecular layers at the air/water interface", Rev. Sci. Instrum. 55, 1968 (1984).
22. W.D. Harkins, T.F. Young and E. Boyd, "The thermodynamic of films : energy and entropy of extension and spreading of insoluble monolayers", J. Chem. Phys. 8, 954 (1940).
23. Th. Rasing, Y.R. Shen, M.W. Kim and S. Grubb, "Observation of molecular reorientation at a two-dimensional liquid phase transition", Phys. Rev. Lett. 55, 2903 (1985).
24. P. Pieranski, "Two dimensional interfacial colloïdal crystals", Phys. Rev. Lett. 45, 569 (1980).
25. D. Weaire and N. Rivier, "Soap, cells and statistics - Random pattern in two dimensions", Contemp. Phys. 25, 59 (1984).
26. I. Langmuir, "Oil lenses on water and the nature of monomolecular expanded films", J. Chem. Phys. 1, 756 (1933).
27. J.F. Nagle, "Theory of the main lipid bilayer phase transition", Ann. Rev. Phys. Chem. 31, 157 (1980).
28. J.P. Legré, J.L. Firpo and G. Albinet, "Mean-field simulation in a renormalization - group procedure and application to amphiphilic monolayers", Phys. Rev. A31, 1703 (1985)
J.P. Legré, G. Albinet, J.L. Firpo and A.M.S. Tremblay, "Liquid expanded-liquid condensed phase transition in amphiphilic monolayers : a renormalization group approach to chiral-symmetry breaking of hydrocarbon-chain defects", Phys. Rev. A30, 2730 (1984).
29. A. Fischer and E. Sackmann, "Electron microscopy and diffraction study of phospholipid monolayers transferred from water to solid substrates" J. Physique (Paris) 45, 517 (1984).
30. J.J. Bikerman, "Foams", Springer-Verlag, Berlin (1973).
31. B. Moore, C.M. Knobler, D. Broseta and F. Rondelez, "Studies of phase transitions in Langmuir monolayers by fluorescence microscopy", Faraday Symp. Chem. Soc. 20, 000 (1985).
32. S. Richardson, "Two-dimensional bubbles in slow viscous flows" J. Fluid Mech. 33, 476 (1968).
J.D. Buckmaster and J.E. Flayerty, "The bursting of two-dimensional drops in slow viscous flow", J. FLuid Mech. 60, 625 (1973).
33. See for instance, in the Proceedings of this Conference, the article of N. Goldenfeld, "Pattern formation in dentritic solidification", for a discussion of the theoretical aspects and experiments on Hele-Shaw cells, an hydrodynamic analogue of solidification. See also the article of S. de Cheveigné, C. Guthmann and M.M. Lebrun, "Cellular instabilities in crystal growth : thin samples experiments", for the description of an experiment of directional solidification under well-controlled conditions.

34. This type of experiments has recently been reported for neutrally-buoyant three-dimensional binary fluids by C. Houessou, P. Guenoun, R. Gastaud, F. Perrot and D. Beysens, "Critical behavior of the binary fluids cyclohexane-methanol, deuterated cyclohexane-methanol and of their isodensity mixture : Application to microgravity simulations and wetting phenomena", Phys. Rev. A **32**, 1818 (1985). See also F. Perrot, P. Guenoun and D. Beysens, "Phase separation of binary fluids near its critical point in microgravity", Proceedings of this Conference.

ESR STUDY OF Cu(II) IN POLYACRYLAMIDE NETWORKS

Shulamith Schlick

Department of Chemistry
University of Detroit
Detroit, MI 48221

Introduction

Surface interactions between water and polymer networks have a profound effect on the water structure. The properties of water in these, and other, heterogeneous systems are sensitive to the size of the network pores and have been described by the two-phase model which assumes partition of the water between the "bulk" and the "bound" water phases[1,2]. Evidence for this partition has been obtained in numerous proton NMR studies[3-6] and also in ESR studies of paramagnetic probes in zeolites[7], silica gels[8] and in water containing polymers[9].

Although this model has been successful for interpretation of results in a variety of systems, recent studies indicate that it represents an oversimplification[1]. For instance dilatometry, specific conductivity and differential scanning calorimetry results obtained on networks swollen by water can be better described if three types of water are assumed: bound, interfacial and bulk water[10,11]. While the properties of bulk water and to some extent of bound water are reasonably well understood, the interfacial water phase is poorly defined, probably due to the fact that it might represent a wide range of structures which depend on and are sensitive to the local environment and to the specific interactions in the system studied.

Metal ions in heterogeneous systems are expected to be solvated by the different types of water, leading to a partition of the guest between the various phases present. If the guest is paramagnetic, its solvation can be studied using the technique of Electron Spin Resonance (ESR) and used to derive information on the network-solvent interactions.

In this report we analyze the binding of hydrated Cu(II) in crosslinked polyacrylamide (PAA) networks, based on ESR spectra measured at X-band (9 GHz) and S-band (2.4 GHz), in the temperature range 77 K to 300 K. It will be shown that the results obtained in this study can best be rationalized in terms of cation solvation by water whose properties vary gradually, as a function of the distance from the polymer network. Cu(II) bound to the polymer through one nitrogen ligand was observed in networks with pore size smaller than 0.7 nm. No evidence was observed for the presence of measurable amounts of bulk water in pores of diameter in the range studied.

Experimental

Crosslinked polyacrylamide gels were obtained by free radical polymerization of acrylamide in the presence of the tetrafunctional monomer N,N'-methylene-bisacrylamide (BIS) at 288 K. Gels with pore sizes in the range 0.6 to 6 nm were prepared by varying the total monomer concentration in the polymerization mixture. The procedure has been described in detail[12,13]. Normal copper is a mixture of magnetic isotopes ^{63}Cu and ^{65}Cu in a ratio of ~2:1. In this study some magnetically dilute samples of Cu(II) in PAA gels were prepared using isotopically enriched ^{63}Cu in order to obtain maximum spectral resolution. ^{63}Cu (98%) as CuO was purchased from the Oak Ridge National Laboratory and was reacted with a stoichiometric amount of H_2SO_4 in order to obtain the $CuSO_4$ used for doping of the gels. The copper concentration in the gels varied between 100 and 500 acrylamide units for each cation, depending on the water content of the gels.

ESR spectra at X-band in the temperature range 77-300 K were measured with Varian E-9 and Bruker ER 200D-SRC spectrometers operating at 9.3 GHz and 9.7 GHz, respectively. Spectra at 77 K were taken in a liquid nitrogen Dewar inserted in the ESR cavity. Above this temperature the Bruker Variable Temperature unit ER 4111 VT was used. The S-band spectra at 2.4 GHz and 120 K were measured at the National Biomedical ESR Center in Milwaukee, Wisconsin, using a spectrometer equipped with a loop gap resonator cavity[14]. The X-band spectra were checked at 120 K and found identical to those obtained at 77 K. The absolute value of the magnetic field was measured using a Bruker ER 035M NMR Gaussmeter. Calibration of g-values was based on DPPH (g=2.0036) and Cr(III) in MgO (g=1.9800). The scan was calibrated by using a ^{55}Mn(II)-doped MgO single crystal. A width of 433.5 for the total separation of the ^{55}Mn sextet was used.

Spectra were calculated by a Burroughs 6800 mainframe computer at the University of Detroit and plotted on a Hewlett-Packard 7470A digital plotter.

Results

ESR spectra of Cu(II) are easier to interpret, because of narrower lines, at low temperatures. We have measured ESR spectra at 77 K in gels quickly frozen by insertion in liquid nitrogen and assumed that in this process Cu(II) retains its hydration shell. This assumption is justified by results in systems which are extremely sensitive to changes in temperature, such as micelles and reaction centers in photosynthetic bacteria[15,16].

In this study spectra at 77 K will be used to define the rigid complexes, while results at 298 K will be interpreted in order to deduce the effect of the cavity in terms of dynamics and averaging of ESR parameters.

Typical spectra at 77 K for hydrated Cu(II) in PAA gels crosslinked by BIS are shown in Figure 1 for total monomer concentrations of 4.0, 50.0 and 60.0%. All spectra are identical in the perpendicular region, with $g_\perp$ = 2.081 and $A_\perp$ = 18 x 10^{-4}cm^{-1}. In the parallel orientation, spectra are similar for the two lower monomer concentrations, with $g_\parallel$ = 2.398, $A_\parallel$ = 135 x 10^4cm^{-1}; the linewidths however decrease as % M increases. The widths at half maximum intensity for the m_I = -3/2 transition are 33G and 25G, for total monomer concentrations of 4.0 and 50.0% M, respectively.

The most striking result in Figure 1 is the presence of two sets of hyperfine parameters in the parallel orientation for a total monomer concentration of 60%, Figure 1C. These sets are assigned to hydrated Cu(II) in two sites which differ in their values for $g_\parallel$ and $A_\parallel$, as indicated by arrows.

For site I, $g_\parallel^I$ = 2.403 and $A_\parallel^I$ = 135 x 10^{-4}cm^{-1}. These values are identical to ESR parameters of hydrated Cu(II) in gels with large pore sizes, Figures 1A and 1B.

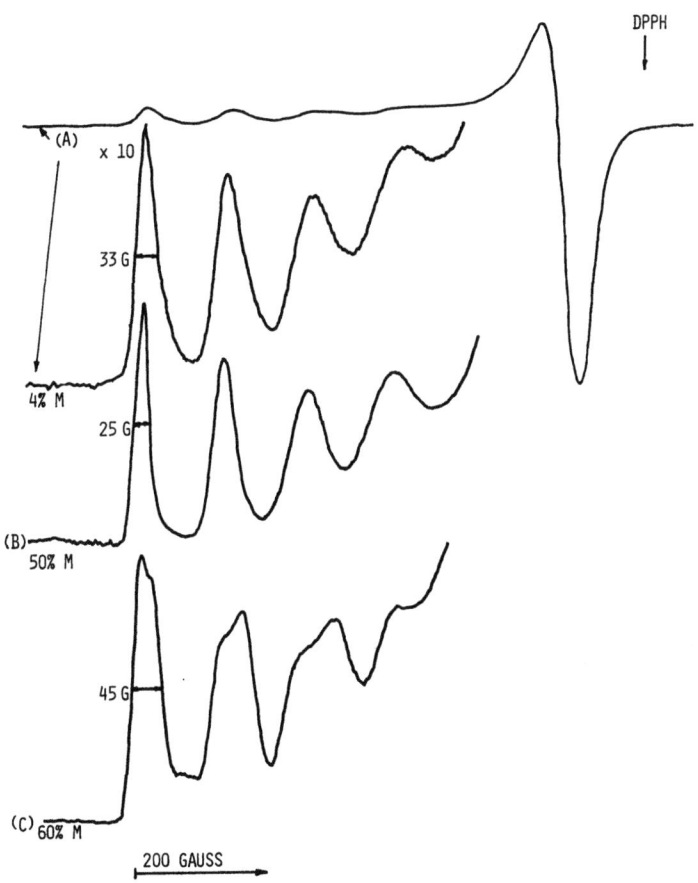

Figure 1 ESR spectra of Cu(II) at X-band in chemically crosslinked PAA gels at 77 K, prepared from the indicated values of the total monomer concentration.

For site II, $g_{\parallel}^{II}$ = 2.368 and $A_{\parallel}^{II}$ = 150 x 10^{-4}cm^{-1}.

It is important to note that the gels prepared with 50.0% M and 60.0 M have the same swelling ratio and therefore the same <u>average</u> cavity size. The difference in the ESR spectra of these two gels, Figures 1B and 1C, must be due to a different <u>distribution</u> of cavity sizes. Because one site for Cu(II) is observed for the gel prepared from 50.0% M, we must assume an homogeneous distribution of cavity sizes in this case. The immediate conclusion is that the gel prepared from 60.0% M, because of the two sites observed by ESR, contains some larger and some smaller pores, i.e. a bimodal distribution.

The values of $g_{\parallel}$, $g_{\perp}$, $A_{\parallel}$ and $A_{\perp}$ for Cu^{2+} in site I are typical of the completely hydrated ion. In this type of complexation, the values of $A_{\parallel}$ and $g_{\parallel}$ are very sensitive to the nature of the ligands. The effect of ligand on $A_{\parallel}$ and $g_{\parallel}$ as well as the expected change in these parameters when one or more of the oxygen ligands are replaced by nitrogen are presented in the Peisach-Blumberg (PB) plots[17]. For $g_{\parallel}$ = 2.368 and $A_{\parallel}$ = 150 x 10^{-4}cm^{-1}, as detected in site II of the chemically crosslinked gels, the PB plots strongly suggest replacement of an oxygen ligand by one nitrogen ligand. In PAA this ligand is from the pendant amide group $-CONH_2$. Inspection of the PB plots indicates that more than one nitrogen ligand is virtually impossible for above values of $g_{\parallel}$ and $A_{\parallel}$.

From the relative intensities of sites I and II in Figure 1C and the network swelling ratio of the network, it was possible to determine that the pore sizes corresponding to the bimodal distribution are 0.6 nm and 0.12 nm[12].

We will now analyze ESR spectra of Cu(II) in pores larger than 0.7 nm with Cu(II) enriched in ^{63}Cu as the paramagnetic probe.

ESR spectra of Cu(II) in crosslinked PAA at X-band and 77 K for pore diameters of 1.3 and 4.0 nm are shown in Figure 2. All spectra are identical in the perpendicular region, with $g_{\perp}$ = 2.0081 and $A_{\perp}$ =~10 x 10^{-4} cm^{-1}. $A_{\perp}$ was evaluated by dividing the width of the perpendicular signal by three; $g_{\perp}$ was taken at the magnetic field where the perpendicular signal crosses the baseline. The low field quartet of lines are centered at $g_{\parallel}$ = 2.403. The hyperfine splitting $A_{\parallel}$ = 135 x 10^{-4} cm^{-1} is due to the ^{63}Cu nucleus (I=3/2). In the parallel orientation at a given pore size we observe a gradual increase in the linewidths corresponding to m_I values of -3/2, -1/2, 1/2 and 3/2. For a given m_I value, the linewidth increases with pore size. The linewidth at half maximum intensity of the m_I = -3/2 and m_I = -1/2 transitions as a function of pore diameter can be measured directly from the ESR spectra and are given in Table I.

In many systems ESR linewidths of Cu(II) depend on the microwave frequency. It was shown in some early papers that lines are broader at Q-band (35 GHz) and narrower at S-band, compared with those at X-band. This was our initial motivation for measuring ESR at S-band, in order to interpret the experimental variation of the linewidth with pore size and m_I values.

ESR spectra of Cu(II) in crosslinked PAA at S-band and 120 K, for pore diameters of .7 nm and 3.2 nm are shown in Figure 3. Similar to the X-band spectra is the increase in linewidth for a given value of m_I as the pore diameter of the network increases. The variation of the linewidth in one type of gel as a function of the m_I value is however markedly different at S-band, compared to that observed at X-band: the linewidth corresponding to m_I = 1/2 is the narrowest line observed in this system. The linewidths corresponding to the m_I = -3/2 and -1/2 transitions at S-band were measured directly from the spectra and are included in Table I, together with the widths measured at X-band.

The results presented in Figures 1, 2 and 3 are typical of Cu(II) complexes with tetragonal symmetry and can be interpreted in terms of an axial spin hamiltonian. Solution of the spin hamiltonian to second order gives the magnetic field for the allowed transitions.

Table I

Variation of the experimental linewidth ΔH at X-band and S-band, the residual width ΔH^R, the width of the $g_{\parallel}$ ($\delta g_{\parallel}$) and the $A_{\parallel}$ ($\delta \hat{A}_{\parallel}$) distributions as a function of pore diameter of the network.

Pore Diameter (nm)	ΔH(Gauss) S-band		ΔH(Gauss) X-band		ΔH^R(Gauss)	$\delta g_{\parallel}$	$\delta A_{\parallel}$ (cm^{-1}) $\times 10^4$	$\delta A_{\parallel}/\delta g_{\parallel}$ (cm^{-1})
m_I	-3/2	-1/2	-3/2	-1/2				
.7	22.3	11.3	19.2	35.6	10.9	0.0390	22.7	0.0582
1.0	23.1	11.6	19.4	36.4	11.2	0.0400	23.5	0.0587
1.3	24.2	11.8	20.8	38.0	11.3	0.0420	24.7	0.0588
1.7	24.8	12.4	21.2	39.4	12.0	0.0432	25.3	0.0586
2.3	25.3	12.7	21.6	40.0	12.2	0.0439	25.7	0.0585
3.2	26.0	14.3	22.0	40.6	14.0	0.0440	25.7	0.0584
4.0	26.4	15.4	23.6	43.2	15.2	0.0465	26.2	0.0563
5.8	26.7	15.7	25.2	44.4	15.3	0.0474	26.4	0.0557

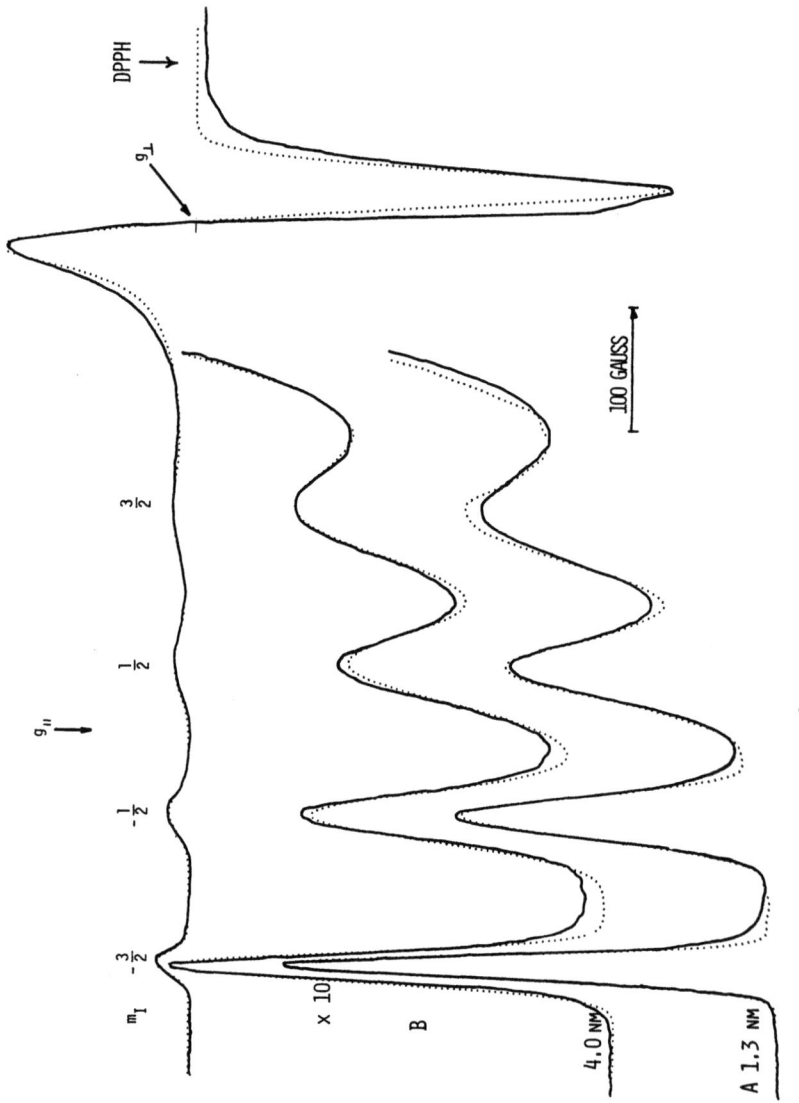

Figure 2 ESR spectra of ^{63}Cu(II) at X-band and 77 K, in chemically crosslinked polyacrylamide gels with pore diameters of 1.3 nm (A) and 4.0 nm (B). Solid lines are experimental spectra; dotted lines are spectra calculated using the appropriate values of ΔH_L^R, $\delta A_{//}$, and $\delta g_{//}$ given in Table I and with $g_{//} = 2.403$, $g_\perp = 2.080$, $A_{//} = 0.0134$ cm^{-1}, $A_\perp = 0.0009$ cm^{-1} and $\Delta H_L^R = 30.0$ Gauss.

$$H(m_I,\Theta) = \frac{h\nu}{g\beta} - \frac{Km_I}{g\beta} + \frac{A_\perp^2(A_\|^2 + K^2)\left[I(I+1) - m_I^2\right]}{4g\beta h\nu K^2} -$$

$$\frac{g_\|^2 g_\perp^2 (A_\|^2 - A_\perp^2)^2 \cos^2\Theta \sin^2\Theta m_I^2}{2g^5 \beta h\nu K^2} \quad (1)$$

In this expression Θ is the angle between the magnetic field and the symmetry axis while g and K are defined below.

$$g^2 = g_\|^2 \cos^2\Theta + g_\perp^2 \sin^2\Theta \quad (2)$$
$$K^2 g^2 = g_\|^2 A_\|^2 \cos^2\Theta + g_\perp^2 A_\perp^2 \sin^2\Theta \quad (3)$$

Eqs. 1, 2 and 3 can be used to simulate a powder spectrum, taking into consideration the correction for the transition probability necessary in a field-swept ESR spectrum[18]. In most cases an orientation dependent linewidth is assumed in generating the spectrum:

$$\Delta H^2 = \Delta H_\|^2 \cos^2\Theta + \Delta H_\perp^2 \sin^2\Theta \quad (4)$$

The experimentally observed variation of the linewidths with m_I and with the microwave frequency presented in Figures 2 and 3 cannot be reproduced by eqs. 1-4. An m_I variation similar to that presented here has been observed before for Cu(II) in frozen glasses and treated phenomenologically[19], by assuming an explicit m_I dependence of the linewidth ΔH, or by generating a powder spectrum through superposition of spectra with different values of the g tensor and hyperfine coupling constants[20-22].

A model proposed by Froncisz and Hyde[23] has satisfactorily explained both the m_I and the microwave frequency dependence of the linewidths. The model is based on the assumption that small site fluctuations of molecular bonding parameters affect the range of values for $g_\|$ and $A_\|$. According to this model, the linewidths measured in the parallel orientation can be described by the distribution parameters $\delta g_\|$ and $\delta A_\|$ and are frequency dependent. We will interpret our results in terms of this theory.

The linewidth at half maximum intensity $\Delta H_\|$ is composed of the residual width $\Delta H_\|^R$ and a contribution from the distribution δH, as shown in eq. 5.

$$\Delta H_\|^2 = (\Delta H_\|^R)^2 + (\delta H)^2 \quad (5)$$

The width ΔH due to the distribution depends on the m_I value, on the microwave frequency ν and on the distribution parameters $\delta g_\|$ and $\delta A_\|$.

$$(\delta H)^2 = (m_I \delta A_\|)^2 + \left(\frac{h\nu \delta g_\|}{g_\|^2 \beta}\right)^2 + \frac{2\varepsilon m_I h\nu}{g_\|^2 \beta} \delta g_\| \delta A_\| \quad (6)$$

In eq. 6 ε is a parameter which indicates the extent of correlation between $\delta g_\|$ and $\delta A_\|$. If $\varepsilon=1$, these distributions are "perfectly correlated", in the sense that all complexes studied have the same ratio $\delta A_\|/\delta g_\|$.

The last term in the expression for the linewidth, eq. 6, depends on the microwave frequency and can be either positive or negative. For negative values of m_I, there is one value of m_I which, due to cancellation of terms in eq. 6, results in the narrowest line observed. In the system we studied, the narrowest line is observed at ν=2.4 GHz and m_I=-1/2, as seen in Figure 3.

The four parameters which are involved in eqs. 5 and 6, $\delta A_{\parallel}$, $\delta g_{\parallel}$, $\Delta H_{\parallel}^{R}$ and ε, have been calculated from the experimental values of the linewidths corresponding to m_I=-3/2 and -1/2 at X- and S-bands, using a least squares fitting program. The values obtained are given in Table I and the variation of the calculated $\delta g_{\parallel}$ and $\delta A_{\parallel}$ values with the pore diameter is plotted in Figure 4.

In all networks studied we found ε=0.97±0.02; we can therefore assume that in all complexes studies the variations in $g_{\parallel}$ and $A_{\parallel}$ values are perfectly correlated.

In order to check further the correctness of this procedure, we used the deduced values for the distribution parameters and the residual linewidth to simulate the experimental spectra at X-band and S-band, using an expression which specifically includes the linewidths dependence on the distribution parameters. An orientation dependent linewidth was used, eq. 4, with $\Delta H_{\perp} = \Delta H_{\perp}^{R}$. The effect of the $g_{\parallel}$ and $A_{\parallel}$ distributions on the linewidth was included only for the parallel orientation[23].

The results of the simulations are shown in Figures 2 and 3, superimposed on the experimental results. The agreement between calculated and experimental spectra is very good. Numerous simulations were performed in order to assess the effect of the various parameters. The results indicate that the simulated spectra are very sensitive to the choice of the distribution parameters and to the values of the residual widths, $\Delta H_{\parallel}^{R}$ and $\Delta H_{\perp}^{R}$. Given the limited possibilities of measuring ESR spectra at S-band, we believe that computer simulations are a viable alternative.

The variation of the spectra above 77 K up to 310 K was measured for pore diameters of 0.7, 1.7, 4.0 and 5.8 nm. The temperature variation is very similar for all the networks studied. Typical results are shown in Figure 5 for 0.7 nm pores. The most important effect of the increase in temperature is the appearance of a signal centered at a field corresponding to g_{iso} = 2.1910 at ~245 K. The intensity of this signal increases with temperature and indicates the isotropic quartet due to Cu^{2+} splitting in a hydrated complex in which the tumbling is averaging the g- and hyperfine coupling anisotropy. In networks with pore diameters of 5.8 nm the isotropic signal is observed even at 225 K. The best resolution of the isotropic quartet is observed in all networks studied at 275 K.

Discussion

The results obtained will be interpreted in order to obtain details on the immediate environment of the paramagnetic probe and used to define the structure of water and the range of interactions between the solvent and the network.

The concept of bound or non-freezable water appears often in studies of systems which are heterogeneous on the molecular level. The term applies to water which is different from bulk water because its properties show a continuous variation as it warms up from low temperatures through the melting point of 273 K. From a large number of ESR studies of various paramagnetic probes it becomes clear that in bulk water at low temperatures, around 77 K, the spectral resolution is lost because of ice formation and aggregation of the paramagnetic probes which results in dipolar broadening. In such systems a broad (~300 G) line is the only signal observed[24,25]. Therefore, in many cases various salts are added in order to obtain a glass in which ice formation and ionic aggregation are prevented.

For Cu(II) in water adsorbed on silica gels it was reported[24] that in

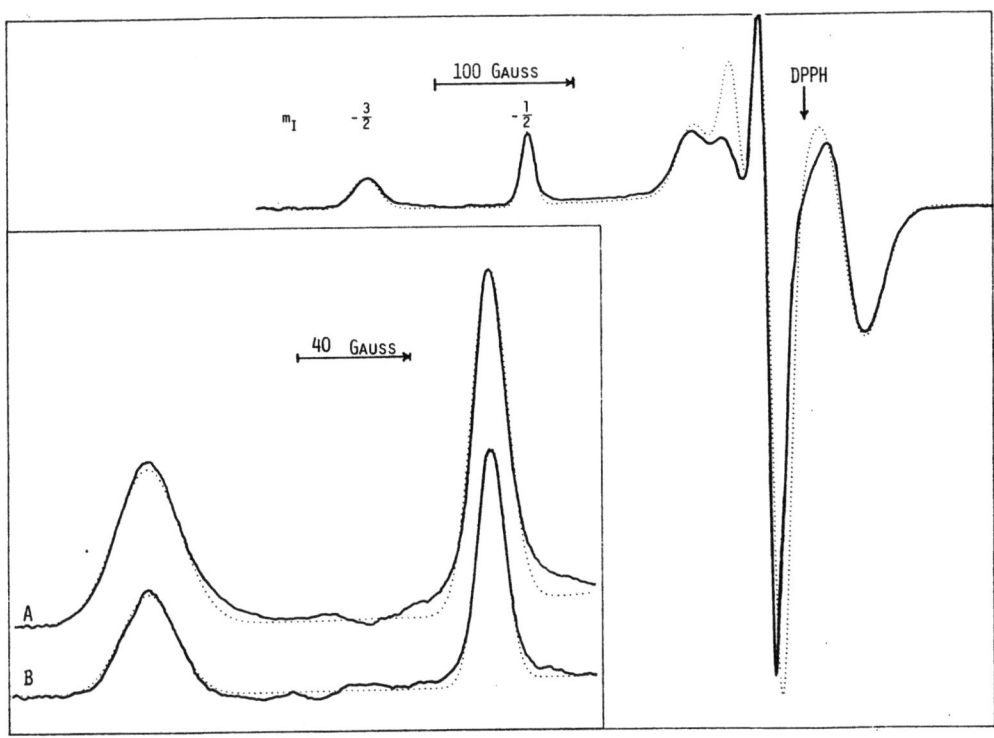

Figure 3 ESR spectra of ^{63}Cu(II) at S-band (2.4 GHz) and 120 K in chemically crosslinked polyacrylamide gels with pore diameter of 3.2 nm. The inset shows the two low field lines of the parallel quartet for pore diameter of 3.2 nm (A) and 0.7 nm (B). Solid lines are experimental spectra; dotted lines are spectra calculated using the appropriate values of ΔH^R, $\delta A_\parallel$ and $\delta g_\parallel$ given in Table I, and with $g_\parallel = 2.403$, $g_\perp = 2.080$, $A_\parallel = 0.0133$ cm^{-1}, $A_\perp = 0.0005$ cm^{-1} and $\Delta H^R_\perp = 16.5$ Gauss.

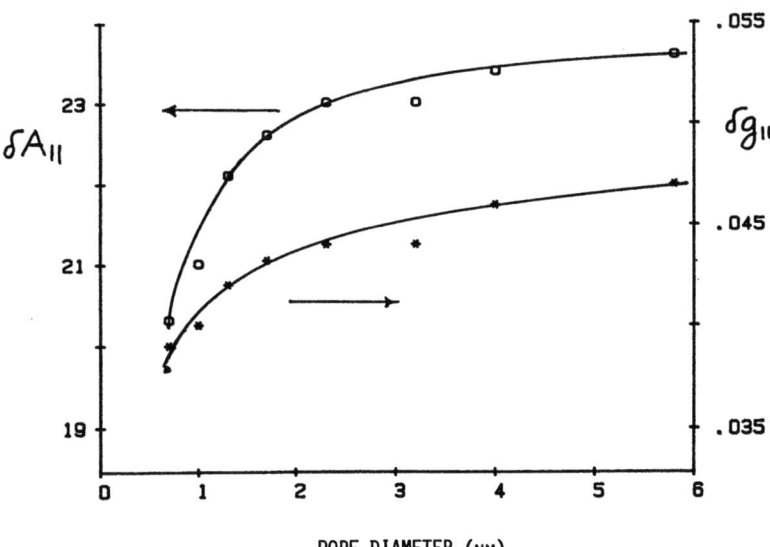

Figure 4 Values of $\delta A_\parallel$ and $\delta g_\parallel$ for Cu(II) in polyacrylamide gels as a function of pore diameter, deduced from experimental spectra at X-band, 77 K and at S-band, 120 K.

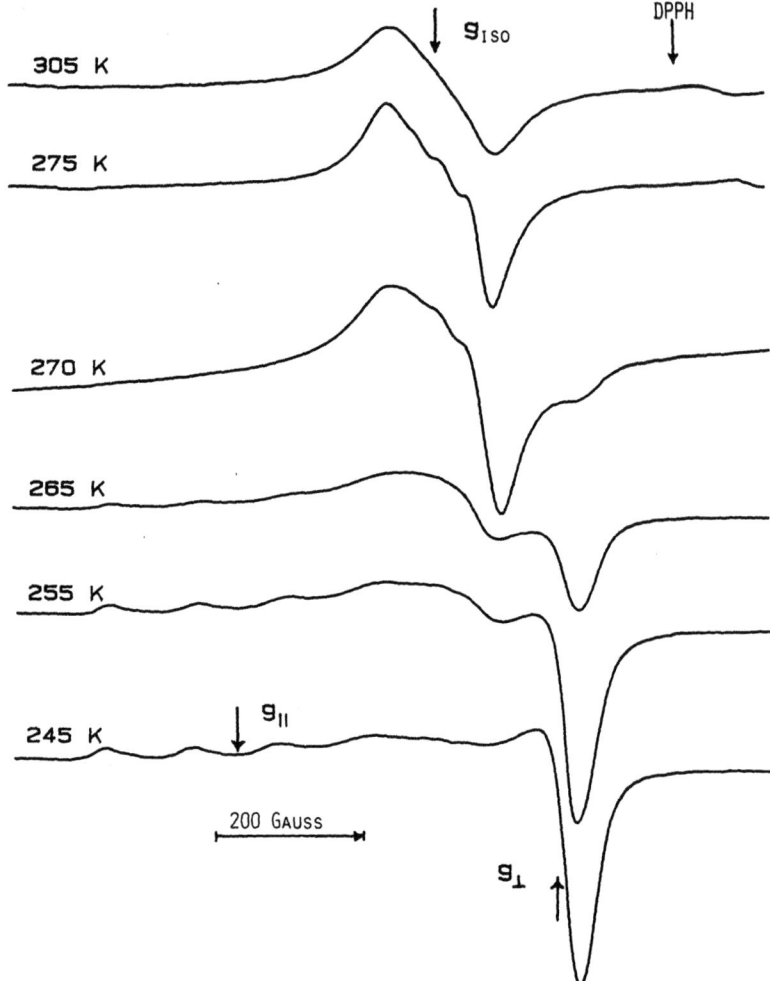

Figure 5 ESR spectra of Cu(II) at X-band in chemically crosslinked polyacrylamide gels with pore diameter of 0.7 nm as a function of temperature. The appearance of the quartet centered at g_{iso} is visible above 245 K.

gels with small pores isolated hydrated ions are detected at 77 K; in pores larger than ~4 nm a broad signal is superimposed on the spectrum of isolated ions. The appearance of the broad signal indicates aggregation of cations and the presence of bulk or freezable water.

Examination of the ESR spectra shown Figures 2 and 3, shows no indication of the dipolar broadened line at 77 K in the networks studied even when the samples were cooled to 77 K from ambient temperature during more than four hours. We particularly checked the S-band spectra for this line. We expect the dipolar broadening to be the same at the two frequencies but much more conspicuous at S-band because the spectrum from isolated ions is spread over a smaller range of magnetic fields at this microwave frequency. The absence of the broad line indicates that in all the networks measured Cu(II) hydrated by freezable, or bulk, water <u>is not detected</u>.

The conclusions from these ESR studies can be compared with the results of an NMR study of proton spin lattice relaxation in crosslinked polyacrylamide gels[6]. This NMR study report various <u>small</u> (up to 6% of the total) amounts of bound water as a function of the length of the network between the crosslinking points which was taken as a measure of the pore size. Most of the gels studied[6] contained a high concentration of the crosslinker BIS and the resulting pore size was not calibrated. It has been shown in a study of gel filtration of model proteins that at high levels of BIS the pore size increases whereas a decrease in the pore size is expected if the results obtained at low levels of BIS are extrapolated[26]. A bimodal distribution of pore diameters is also a definite possibility. In networks prepared with high levels of BIS we have detected by ESR a bimodal distribution of pore size using the vanadyl ion, VO^{2+}, as a paramagnetic probe[27].

The large amount of bulk water reported in ref. 6 might therefore be due to the presence of very large pores which are swollen by large quantities of water.

In our system, because the amount of bulk water is negligible, we have a simpler problem and can consider only bound water within the picture of a two-phase system or bound and interfacial water if the three-phase model is adopted.

Assuming as a first approximation that the pores are spherical and that the effective diameter of the water molecule is 0.03 nm[24], we can roughly estimate that the number of water layers in the pores studied ranges from 1, in the networks with pore diameters of .7 nm, to 10 in the largest pores studied. The absence of bulk water in these networks might indicate that the effect of the polymer interaction is felt through at least ten layers of water molecules. This conclusion might seem to be in contradiction with previous results which indicate that strong modification of the water structure is evident in the first two to three surface layers and probably not beyond about six layers[24]. The contradiction can be rationalized in two ways.

First, the strength of the polymer-water interaction must diminish at larger distances from the network, and is clearly reflected in the distribution of the $g_{\|}$ and $A_{\|}$ values. As has been stated before, this distribution is due to local variation in the values of the in plane and to a smaller degree of the out-of-plane bonding parameters between Cu(II) and the oxygen ligands[23]. In the smallest pores studied most of the water is in the first two to three layers and is strongly affected by the proximity of the network, so that fluctuations in the bonding parameters are limited, if not impossible. The result is a relatively narrow distribution in the values of $g_{\|}$ and $A_{\|}$ and narrow observed lines. As the pores increase the ligands are less affected by the interaction with the network and wider fluctuations in the bonding parameters are observed. The deduced values of the distribution parameters increase significantly up to pore diameters of ~2 nm and seem to approach a plateau for pores of diameter ~3 nm, where approximately five to six layers of water exist in the pores. In order to visualize the distribution of water in the various layers, we have indicated in Figure 6 the cumulative fraction of water molecules in the first six layers, as a function of pore diameter. As seen from Figure 6, the number of layers in the largest pore is ~10 but the fraction of molecules in the first 6 layers is .94, indicating that the increase in pore diameter does not increase proportionately the number of molecules of water distant to the network. The conclusion is that even if hydrated Cu(II) in layers remote from the network might be able to distort, their number is not large enough to contribute to the total signal intensity observed.

An alternative explanation for the plateau observed in the distribution parameters $\delta A_{\parallel}$ and $\delta g_{\parallel}$ is suggested by the deduced values of ΔH^R as a function of pore size.

In Table I we observe that the residual width changes very slightly up to and including networks with pore diameters of 2.3 nm but increases above pore diameters of 3.2 nm. We checked the source of the broadening by measuring ESR spectra of gels doped with a 1/1 mixture of Zn(II) and Cu(II) ions, for pore diameters of 0.7 and 5.8 nm. We chose to monitor the effect of this substitution by measuring the linewidth of the signal corresponding to the $m_I=-3/2$ parallel transition at X-band, which can be measured most accurately from experimental spectra. The Zn(II) substitution had no effect on the linewidth in networks with pore diameters of 0.7 nm. In the largest pores, however, the linewidth is reduced by 3 G. The increase of the residual linewidth with increase in pore diameters can therefore be assigned to dipolar broadening. Assuming that the distribution parameters do not change on Cu(II) substitution by Zn(II), we can calculate from eq. 6 the residual linewidth excluding dipolar broadening, $\Delta H^R(0)$. The results is $\Delta H^R_{\parallel}(0)=10$ G. This value is, within experimental error, identical to the residual linewidth deduced in networks with small pores and is considered the limiting value of the linewidth in all pores, in the absence of dipolar broadening. The distance d between the cations in the large pores can be calculated by observing that the dipolar contribution $\delta(\Delta H)$ to the residual linewidth for large pores is 5.3 G. Using the expression for the second moment of Gaussian lines, we can write[28]

$$\delta(\Delta H) = 2\left[\frac{3}{5} g^2 \beta^2 S(S+1)\right]^{\frac{1}{2}} d^{-3} \qquad (7)$$

If the isotropic value of g for the hydrated cation in eq. 7 above is used, we obtain the relation between the interion distance d (in nm) and the observed broadening.

$$\delta(\Delta H) = 27.193 \, d^{-3} \qquad (8)$$

We deduce that the intercation distance is 1.8 nm.

This value of d suggests that in the large pores, for example in pores with diameter of 5.8 nm, the cations are approximatelyy at 2.0 nm from the network, not at 2.9 nm if only one cation per pore is measured. The presence of two cations per pore reduces even more the number of ions which are expected to be solvated by water removed from the polymer network. Te result is that even though the pores increase in size, the distance between the cation and the network decreases.

It is possible that both explanations are valid and contribute to the observed results.

The gradual variation in the distribution parameters can be compared to the gradual change observed in the rate of proton transfer from the fluorescent probe pyranine in reversed micelles as a function of the size of the water pool[29]. The reaction rate reaches a limiting value for water pools of diameter ~4 nm. In addition, a study of electron solvation in reversed micelles by pulse radiolysis indicates that both the width and the

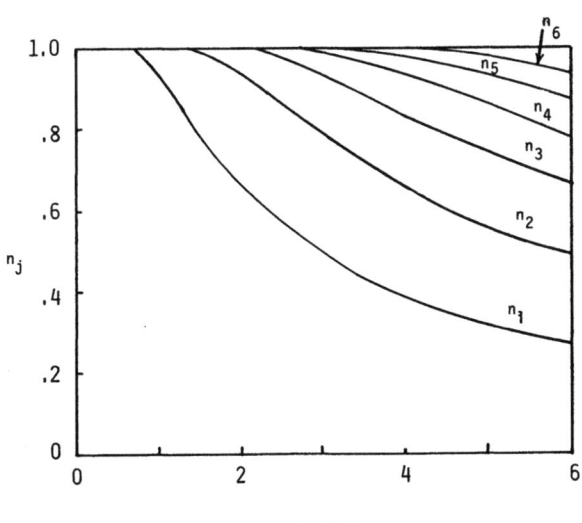

Figure 6 Cumulative fraction of water molecules n_j as a function of pore diameter. We have used $n_j = \sum_{i=1}^{j} n_i$ and n_i is the fraction of water molecules in the ith layer.

wavelength at maximum absorption vary gradually with the size of the water pool and reach limiting values in large water pools, typically of diameter 6 nm[30]. While the probes and the experimental measurements are quite different in these, and our, studies, we believe that the picture is consistent and points to an entire range of water properties, depending on the distance from network in crosslinking polymers or from the micelle boundary in reversed micelles. The picture of two or even three water phases in these systems seems to be an oversimplification indeed.

The variation of the spectra as a function of temperature in the range 77 K-310 K support the idea of the gradual change in the properties of water as the distance to the network changes. The isotropic signal indicates rapid tumbling on the ESR timescale, with a correlation time of 1×10^{-8}s. The width of the signal decreases and its intensity increases gradually above 235 K indicating a change in the mobility of the hydration water. In bulk water a sudden change is observed in the mobility around the melting point of 273 K. The fact that in the gels studied the isotropic signal is observed well below the melting point of ice indicates that the water is not "bulk" but rather bound or interfacial. At the temperatures where the isotropic signal was observed, an anisotropic signal, from hydrated Cu complexes with slower tumbling rates, is also observed. This might indicate hydration of the cation by water bound to the network (anisotropic signal) and by interfacial water (isotropic signal). The relative intensity of the two signals changes with temperature, indicating a temperature-dependent equilibrium. The fact that in networks with pore diameters of 5.8 nm the isotropic signal is detected at even lower temperatures is very likely indicative of the larger amount of interfacial water in these networks.

Conclusion

This report indicates that studies of polymer networks using paramagnetic probes is capable of providing information on the polymer-water interactions as well as on cation binding to the polymer network.

Additional studies using paramagnetic probes in reversible polyacrylamide gels are in progress. Preliminary results indicate significant differences between the linear and the crosslinked gels in terms of cation binding and linear network density.

Acknowledgements: This research was supported by a grant from the Research Corporation and by an NSF equipment grant DMR-8501362 for the purchase of the ESR spectrometer. Support of the ESR measurements at S-band was provided by a grant NIH-RR-01008 to the National Biomedical ESR Center. We thank Drs. J.S. Hyde and C. Felix for their assistance with these measurements.

References
1. J.H. Fendler, Science 223:888 (1984) and references therein.
2. I.D. Kuntz, in "Magnetic Resonance in Colloid and Interface Science", H.A. Resing and C.G. Wade, eds.; ACS Symposium Series, 1976, Vol 34.
3. R.W. Creekmore, and C.N. Reilley, Anal. Chem. 42:570, 725 (1970).
4. J. Fripiat and M. Letellier, J. Magn. Reson. 57:279 (1984).
5. D. Doskocilova, B. Schneider and J. Jakes, J. Magn. Reson. 29:79 (1978).
6. S. Katayama and S. Fujiwara, J. Am. Chem. Soc. 101:4485 (1978).
7. Y.B. Taarit and C. Naccache, Chem. Phys. Lett. 11:11 (1971).
8. G. Martini, M. Bindi, M.F. Ottaviani and M. Romanelli, J. Colloid Interface Sci. 108:140 (1985).
9. K. Hamada, T. Iijima and R. McGregor, Polymer J. 17:1245 (1985).

10. H.B. Lee, M.S. Jhon and J.D. Andrade, J. Colloid Interface Science, 51:225 (1975).
11. K.F. O'Driscoll and D.G. Mercer in "Contemporary Topics in Polymer Science", M. Shen, ed., Plenum Press, New York, (1979), Vol. 3, p. 319.
12. G.C. Rex and S. Schlick, J. Phys. Chem. 89:3598 (1985).
13. G.C. Rex and S. Schlick, Amer. Chem. Soc. Polym. Div. Preprints 27:339 (1986).
14. W. Froncisz and J.S. Hyde, J. Magn. Reson. 47:515 (1982).
15. P.A. Narayana, A.S.W. Li and L. Kevan, J. Am. Chem. Soc. 103:3603 (1981).
16. J.R. Norris and J.J. Katz in "The Photosynthetic Bacteria", R.K. Clayton and W.R. Sistrom, eds., Plenum Press, New York (1978), Chapter 21.
17. J. Peisach and W.E. Blumberg, Arch. Biochem. Biophys. 165:691 (1974).
18. R. Aasa and T. Vanngard, J. Magn. Reson. 19:308 (1975).
19. L.D. Bogomolova, V.A. Jachin, V.N. Lazukin, T.K. Pavlushkina and V.A. Shmuckler, J. Non-Cryst. Sol. 28:375 (1978).
20. D.L. Griscom, E.J. Friebele and G.H. Siegel, Jr., Solid State Commun. 15:479 (1974).
21. P.C. Taylor and P.J. Bray, J. Magn. Reson. 2:305 (1970).
22. D.L. Griscom, J. Non-Cryst. Sol. 64:229 (1984).
23. W. Froncisz and J.S. Hyde, J. Chem. Phys. 73:3123 (1980).
24. V. Bassetti, L. Burlamacchi and G. Martini, J. Am. Chem. Soc. 101:5471 (1979).
25. P.A. Narayana, A.S.W. Li and L. Kevan, J. Phys. Chem. 86:3 (1982).
26. J.S. Fawcett and C.J.O.R. Morris, Sep. Sci. 1:9 (1966).
27. G.C. Rex and S. Schlick, unpublished results.
28. A. Abragam, "Principles of Nuclear Magnetism", Oxford University Press, Oxford (1983), Chapter IV, p.106.
29. E. Bardez, B.T. Goguillo, E. Keh and B. Valeur, J. Phys. Chem. 88:1909 (1984).
30. M.P. Pileni, B. Hickel, C. Ferradini and J. Pucheault, Chem. Phys. Lett. 92:308 (1982).

ELECTRIC POTENTIAL AND CHARGE DENSITY PROFILES IN INHOMOGENEOUS INTERFACIAL REGIONS

Vishnampet S. Vaidhyanathan

Department of Biophysics, School of Medicine
State University of New York at Buffalo
Buffalo, New York, 14214 U.S.A.

INTRODUCTION

This paper is concerned with certain fundamental questions raised by our studies (Vaidhyanathan, 1985, 1986), regarding the relations between concentration distributions of charged species and the electric potential profile in inhomogeneous interfacial regions. The existance of a dielectric profile and the finite nature of the extent of the inhomogeneous region are examined. One of the unsolved problem of major interest, in interfacial surface phenomena, is the relation existing between surface charge density and the value of the electric potential at the surface, when a surface containing fixed charges is in contact with an electrolyte of known composition. An expression, known as Gouy equation, is available in literature (Grahame,1947) for this relation, when the electrolyte contains only univalent ions. The validity of this equation is also restricted to the situation when the positional dependence of dielectric coefficient can be ignored. A general equation, valid for the case of multivalent many ion system, such as one encounters in biology, is not available. A critical examination of the classical theory of double layers, (Verwey and Overbeek, 1945), suggests that the assumption of constant dielectric coefficient should be justified. In biophysical systems of interest, the concentrations of ions are of the order of unimolar. Therefore, the results of strong electrolyte theory, whose validity is limited to extremely dilute solutions, are not evidently applicable to these systems, without corrections.

In recent years, the study of the inhomogeneous interfacial region has received impetus with the techniques of statistical mechanics. A review of these advances are presented by Carnie and Torrie (1984). These can be classified as HNC/mean spherical approximation (Nielaba and

Forstmann, 1985), Modified Poisson Boltzmann equation (Outhwaite et al, 1980) and the Born-Green-Yvon integral equation method (Blum et al, 1983). One may conclude that for low ion concentrations, all three approaches predict correct deviations from modified Gouy-Chapman theory. However, at higher inhomogeneities, induced by greater surface charges, all the statistical theories fail qualitatively. In our opinion, a major deficiency of these statistical approaches is the assumption of a constant dielectric background. It is possibly appropriate to point out at this point, that one of the main results obtained from our analysis, is the nonvalidity of the assumption of constant dielectric coefficient.

The main questions to which we seek answers in this analysis are, 1. the existance of a dielectric profile in the interfacial region, near a charged surface; 2. the finite nature of the magnitude of extent of inhomogeneous region, and 3. the nonvalidity of the Nernst expression relating the concentration distribution of ions of electrolyte and the difference in electric potential at two locations, say, x_1 and x_2, in interfacial regions at equilibrium. It should be stated that the Nernst potential expression and the Nernst-Planck equations of electrodiffusion form the main basis of theoretical investigations of bioelectric phenomena in the field of biophysics.

The salient result of our investigations is that, in the interfacial region, in any plane parallelepiped layer of the electrolyte solution, the repulsive forces between like charges play a dominant role, in determining the local concentration distribution of ionic species. Therefore, to a first order of approximation, the Poisson-Boltzmann equation needs to be modified with the presence of a term, denoting the local deviation from microscopic electroneutrality. This charge density term, is denoted in our analysis as $[H/4\pi]Y(x)$ in this paper. $Y(x)$ equals $-(4\pi e)\sum Z_\sigma C_\sigma(x)$, where e denotes the protonic charge, Z_σ and $C_\sigma(x)$ denote respectively the signed valence charge number and concentrations of ions of kind σ, at location x, in the interfacial region. H represents a molecular integral, denoting the charge independent ion-ion interaction contribution to the electrostatic free energy of the electrolyte system. The nonvanishing nature of ion-ion interaction energy contribution to the chemical potential of an ion in interfacial region, requires the existance of a position dependent dielectric profile.

The value of the dielectric coefficient at the interface, denoted by $\epsilon(o)$, is given exactly by the Gauss theorem,

$$\int_0 Y(x)\,dx = -\epsilon(o)\phi'(o)$$

$$Y(x) = \nabla \cdot \epsilon(x) \nabla \phi(x); \quad \phi'(x) = [d\phi/dx] \qquad (1)$$

In equation (1), $\phi(x)$ is the electric potential felt by a unit charge placed at x. The position variable x is defined as normal to the plane of the interface. The inhomogeneous region is assumed to extend from x = 0 to x = d. It is our thesis, that both charge density profile and electric potential profile should exhibit an extremum value in the interfacial region.

Another significant conclusion that we obtain, which is at drastic disagreement with the classical picture of the interfacial region, is that the electric potential profile and charge density profile must be schematically similar and opposite in sign with the values of Y(x). The classical picture of Gouy-Chapman theory requires that, in the interfacial region, the ratio $[\phi(x)/Y(x)]$ should be positive definite for all values of x. Recall that the classical theory correctly requires that the surface charge density S and the surface potential should have the same sign. This is expressed by the approximate relation, valid for the 1-1 electrolyte system with the assumption that positional dependence of dielectric coefficient can be ignored, (Kruyt, 1952) viz.,

$$S = [\epsilon \varkappa / 4\pi] \phi(o) \quad (2a)$$

$$\varkappa^2 = (4\pi e^2 / \epsilon(d) kT) \sum_\sigma z_\sigma^2 c_\sigma(d) \quad (2b)$$

where $\varkappa$ is the Debye-Huckel parameter of strong electrolyte theory. k is the Boltzmann constant and T is temperature in Kelvin scale. $\epsilon(d)$ is the value of dielectric coefficient of electrolyte solution in the homogeneous region, far from the interface. The inconsistancy between the two conclusions is evident from the definition of Y(x). From equation (1) it follows that Y(o) must be necessarily positive, if the term $\sum_\sigma Z_\sigma C_\sigma(o)$, is negative, surface has excess of negative ions, and the potential at surface $\phi(o)$ must therefore be negative (!).

THE CLASSICAL THEORY OF THE DOUBLE LAYER

The classical Gouy-Chapman theory is based on the assumed validity of the limiting expression for the electrochemical potential μ_σ, of an ion of kind σ,

$$\mu_\sigma(x) = \mu_\sigma^*(T,P) + kT \ln C_\sigma(x) + Z_\sigma e \phi(x) \quad (3)$$

where μ_σ^* is the composition independent part of the chemical potential of species σ, in its standard state, and P denotes the pressure. Since at equilibrium, the chemical potential is independent of position, equation (3) leads to the Nernst expression, relating distribution of ions at two locations with the difference in electric potential at these two locations, viz.,

$$Z_\sigma \beta^{Nernst} = \ln\{C_\sigma(x_1)/C_\sigma(x_2)\} = (e/kT)[\emptyset(x_2) - \emptyset(x_1)] \tag{4}$$

where $\emptyset(x)$ is the value of the electric potential at x. Equation (4), together with the Poisson's equation yields the Poisson-Boltzmann equation. Equation (3) stipulates that when the surface is negatively charged, there occurs an accumulation of positive ions of the electrolyte, near the surface such that the concentration profile of positive ions, decrease monotonically from its extremum value to its bulk value in the homogeneous part of the solution. The electric potential is presumed to increase monotonically from some finite value $\emptyset(o)$ at the interface to the value of zero in bulk, $[\emptyset(d) = 0]$, in an asymptotic manner. This conclusion that $C^+(x)$ decreases while $\emptyset(x)$ increases in the interfacial region is valid when equation (3) is valid at equilibrium. One assumes that the distribution of ions in solution is governed, by Boltzmann relation, expressing that at places of positive potential, the negative ions are concentrated and positive ions are repelled. In obtaining the solution of Poisson-Boltzmann equation, one neglects the positional dependence of the dielectric coefficient. In this case, the classical theory requires that the ratios, $[\emptyset(x)/Y(x)]$ and $[\emptyset'(x)/Y'(x)]$ should be positive definite for all values of x, in the interfacial region. $[\emptyset'(x) = [d\emptyset/dx]; Y'(x) = [dY/dx]]$. Evidently this conjecture is due to the assumption that the dielectric coefficient, $\epsilon(x)$ is not a function of x. It should be noted that the values of concentration distribution of ions of electrolyte solution in regions very close to the interface, (hence of the charge density) are never seriously discussed in classical theory !

The equilibrium Boltzmann relation, relates concentration distribution of a specified molecular species at two different locations with the difference in their energies. Replacement of this difference in their energies with the difference in the electric potential at these two locations, for charged species, should be justified (Booth, 1953). It is evident that the surface should contain negative charges to evolve a negative potential, in order that the concentrations of the positive ions exceed the concentrations of the negative ions in regions close to the surface. This view indicates the existance of an extremum value in the charge density profile. If this picture is close to reality, and the dielectric coefficient is position independent, then the Debye-Huckel kind of relation,

$$[d^2\emptyset/dx^2] = \chi^2 \emptyset(x) \tag{5}$$

will require the existance of an extremum value in the electric potential profile also.

The electric potential specified in equation (3), describes essentially, the potential due to the negative charges fixed on the surface, and ignores any contribution to electric potential $\phi(x)$, arising from other ions present in the solution. The excess positive ions, which are in closer proximity to an arbitrary location x, in the interfacial region, evidently contributes positive potentials. Thus, apriori, the magnitude and sign of the electric potential at location x, that a unit charge detects when placed at x, is arbitary and is subject to question and speculation.

If ion distribution in interfacial region is similar to the case of a parallel plate condenser, then the potential and charge density at all locations must have similar signs. If this is valid, then the ratio $[\phi(x)/Y(x)]$ should be negative definite ! When this contention that the charge density and the electric potential at every location must have similar signs, is valid, the dielectric profile cannot be a constant independent of x, since

$$Y(x) = \epsilon'(x)\phi'(x) + \epsilon(x)\phi''(x) = -(4\pi e)\sum_\sigma Z_\sigma C_\sigma(x)$$

$$\phi''(x) = [d^2\phi/dx^2] \qquad (6)$$

The validity of equations of the kind (5) require that $\phi(x)$ and $\phi''(x)$ must have similar signs. One usually assumes that the difficulties of obtaining the solutions of the nonlinear Poisson-Boltzmann and Nernst-Planck equations are somewhat reduced by invocation of the assumption, that positional dependence of the dielectric coefficient can be ignored. The dielectric coefficient is a function of many variables, such as composition, temperature and electric field (Frohlich, 1949). The electric fields that one encounters in biophysical systems, are of the order of 10^5 (volts/cm), or higher. Therefore, the variation of ϵ near the interface cannot be ignored. The dielectric coefficient of lipids, with which most biological membranes are composed of, is of the order 2, and the dielectric coefficient of the aqueous electrolyte is of the order 80 at 20°C. Unless there exists, a mathematical discontinuity a continuous variation of dielectric coefficient with x, at or near the interface must exist. (Booth,1951; Luzar et al,1985).

The validity of equation (3) is restricted to extremely dilute solutions. The contribution to chemical potential, arising from intermolecular interactions has been neglected. It should be recalled that the logarithmic dependence on the concentration arises from the entropy of mixing terms of ideal solution expression, in which the mole fractions have been replaced by concentrations for extremely dilute solution. The contribution from electric potential to this expression for the

chemical potential is phenomenologically added for charged
species. Thus, for the electrolyte systems, with significant
amount of solute electrolyte concentrations, such as one encounters in biophysical systems, the validity of equations (3) and
(4) are highly questionable. In strong electrolyte systems,
the presence of unlike kinds of charges in closer vicinity,
introduces some order, resulting in decrease of entropy terms.

It is generally assumed, that the deficiencies of the
Nernst expression are corrected, when one replaces the concentration terms of equations (3) and (4) by the activity terms.
The variation of the chemical potential with concentrations
of a specified species is a fundamental, not yet solved problem
of physical chemistry (McMillan and Mayer, 1945; Kirkwood and
Buff, 1951). In spite of being highly useful in experimental
physical chemistry, one must acknowledge that the concept of
activity coefficients are merely an empirical measure of
our ignorance factors. In attempting to improve the expression
for chemical potential for charged species, equation (3), one
must include additional terms arising from ion-ion interaction
and ion-dipole interaction energy terms to expressions for the
free energy and chemical potential expression denoted in equation (3). This is extremely important for concentrated solutions such as one encounters in biology.

If equation (4) is indeed valid for interfacial regions,
then for a value of the surface potential of say about 100 mV,
the concentrations of univalent ions can be computed to equal
about 54.6 times their concentrations in homogeneous bulk
regions. For divalent ions, the concentrations at the interface, $C^{++}(o)$ will equal 2981 $C^{++}(d)$. Obviously, these
computed concentrations near the interface are physically
unrealistic, for ions to be accommadated at or near the surface.
It is evident that such crowding will result in significant
amount of repulsion between (like) charges, resulting in much
lower concentrations than predicted by equation (4).

The assumption that the dielectric coefficient can be
regarded as constant, can be shown easily to be nonvalid,
within the context of the classical theory itself. If one
has $\epsilon \neq \epsilon(x)$, then $Y(x)$ must be proportional to $\emptyset''(x)$.
One can assume for example, the profile of $\beta(x)$ in the interfacial region, and can compute precisely the values of $\emptyset(x)$,
$\emptyset'(x)$, $\emptyset''(x)$, $C_\sigma(x)$ and $Y(x)$ for every value of x, using
equations (4) and (6). One may verify, that the computed
values of $Y(x)$ are never proportional to the computed values
of $\emptyset''(x)$. Also, one may convince oneself, that the computed
values of the integrals of $Y(x)$, of equation (1), for any
assumed profile of $\beta(x)$, resulting in the calculated value
for the dielectric coefficient at interface, $\epsilon(o)$, [since $\emptyset'(o)$
is known]. never equals the value $\epsilon(d)$, of the dielectric
coefficient of aqueous electrolyte solution in bulk.

EFFECTS OF INTRODUCTION OF ION-ION INTERACTION TERMS

The limiting expression (3) for the electrochemical potential of ionic species, σ, can be modified to include the contribution to free energy arising from interparticle interaction terms, as

$$\mu_\sigma(x) = \mu_\sigma^*(T,P) + Z_\sigma e\, \phi(x) + kT \ln C_\sigma(x) +$$
$$+ \sum_\eta C_\eta(x) H_{\sigma\eta} + \sum_j C_j(x) H_{\sigma j} \qquad (7)$$

$$\mu_j(x) = \mu_j^*(t,P) + kT \ln C_j(x) + \sum_k C_k(x) H_{jk} +$$
$$+ \sum_\sigma C_\sigma(x) H_{j\sigma} \qquad (8)$$

In equations (7) and (8), the greek subscripts denote an ionic species, and the roman subscripts are utilized to indicate a nonionic, uncharged species present in the system. $H_{\sigma\eta}$ are molecular integrals, over the potential energy of interactions and the probability functions of molecular interactions over the whole available volume of space. $H_{\sigma j}$ and H_{kj} are also similar molecular integrals, representing respectively the ion-neutral molecular interaction contributions and the dipole-dipole interactions (also dispersion) contributions to free energy. Thus, $C_\eta H_{\sigma\eta}$ represents the integral contribution of all ions of kind η, present in the system to the chemical potential of an ion of kind σ. The introduction of an additional assumption, that the pair potential energy of interactions can be expressed as, $Z_\sigma Z_\eta e^2$ times a charge independent function of position variable, (mutual distance of separation), for ionic species, enables one write that $H_{\sigma\eta} = Z_\sigma Z_\eta e^2 H$ and $H_{\sigma j} = Z_\sigma e\, H^*$. Probability functions are positive definite. Equation (7) can now be expressed as

$$\mu_\sigma(x) = \mu_\sigma^*(T,P) + kT \ln C_\sigma(x) + (Z_\sigma e) F(x)$$
$$F(x) = \phi(x) - (H/4\pi) Y(x) + H^* \sum_j C_j(x) \qquad (9)$$

In obtaining equation (9) Poisson's equation (6) has been utilized. Therefore, the introduction of ion-ion interaction energy term contribution to free energy, results in the presence of the charge density term, $[-Y(x)]$, in addition to the electrical potential $[\phi(x)]$ term, the expression for the chemical potentials. Neglecting the H^* term for the sake of brevity, one has that, in interfacial regions, the concentration distribution of ions are determined by the expression,

$$Z_\sigma \beta(x) = \ln \{C_\sigma(x)/C_\sigma(d)\}$$
$$= (e/kT)[\phi(d) - \phi(x)] + [He/4\pi kT]\, Y(x)$$
$$Y(d) = 0 = \phi(d) \qquad (10)$$

When equations of the kind (7) and (8) are valid, it follows that Poisson-Boltzmann and Nernst-Planck equations (Vaidhyanathan, 1979) should contain terms representing local deviation from electroneutrality. One should in addition refrain from invoking the assumption of position independent dielectric coefficient, to obtain realistic answers.

It is our contention that H can be regarded as position independent, since it represents a definite integral. One can therefore approximate that

$$-[4\pi/H] = K(d) = \varkappa^2(d)\epsilon(d)$$

$$K(x) = \varkappa^2(x)\epsilon(x) = (4\pi e^2/kT)\sum_{\sigma}C_{\sigma}(x)Z_{\sigma}^2 \qquad (11)$$

Equation (11) has been shown to be valid, (Vaidhyanathan, 1982), for the case of a symmetrical ion system, when a dielectric profile exists. The validity of equation (11) for the general case of multi-ion unsymmetrical electrolyte system is assured, when an additional boundary condition that $\emptyset'''(d) = 0$, is valid. When equations (11) and (9) are valid, one has

$$\beta(x) = -[e/kT]\emptyset(x) + \left\{[\sum_{\sigma}Z_{\sigma}C_{\sigma}(x)]/[\sum_{\sigma}Z_{\sigma}^2 C_{\sigma}(d)]\right\} \qquad (12)$$

Equation (12) can be written as

$$\sinh\beta(x) - \beta(x) = (e/kT)\emptyset(x) : \text{1-1 ion system} \qquad (13a)$$

$$3[e/kT]\emptyset(x) = e^{2\beta} - e^{-\beta} - 3\beta(x) : \text{2-1 ion system} \qquad (13b)$$

$$12[e/kT]\emptyset(x) = 2e^{2\beta} + 3e^{\beta} - 5e^{-\beta} - 12\beta(x): \text{ for a 3-ion system, with } [C^{++}(d)/C^-(d)] = 0.2 \qquad (13c)$$

The results of equations (11), (12) and (13) state that in the interfacial regions, $\beta(x)$ and $\emptyset(x)$ must have the same signs, opposite to that of $Y(x)$. This conclusion is consistant with the picture of a parallel plate condenser of the electric double layer in ionic solutions, near a surface containing fixed charges. At equilibrium, the validity of equation (9) requires only that the profiles of $C^+(x)$ and $F(x)$ must be schematically opposite. The relations between β and $\emptyset$ that must exist in interfacial region, as given by equations (13) are presented in Figure 1. One should notice that β is no longer proportional to $\emptyset$. β approaches an asymptotic value for large values of $\emptyset$. This approach to an asymptotic value is accelerated by the presence of divalent ions in the system. This may have significance in the explanation of role of ions like calcium in biological system. In our opinion, the results of equations (12) and (13) present the concentration distribution of ions in interfacial region, more realistically than the distribution given by Nernst expression (4).

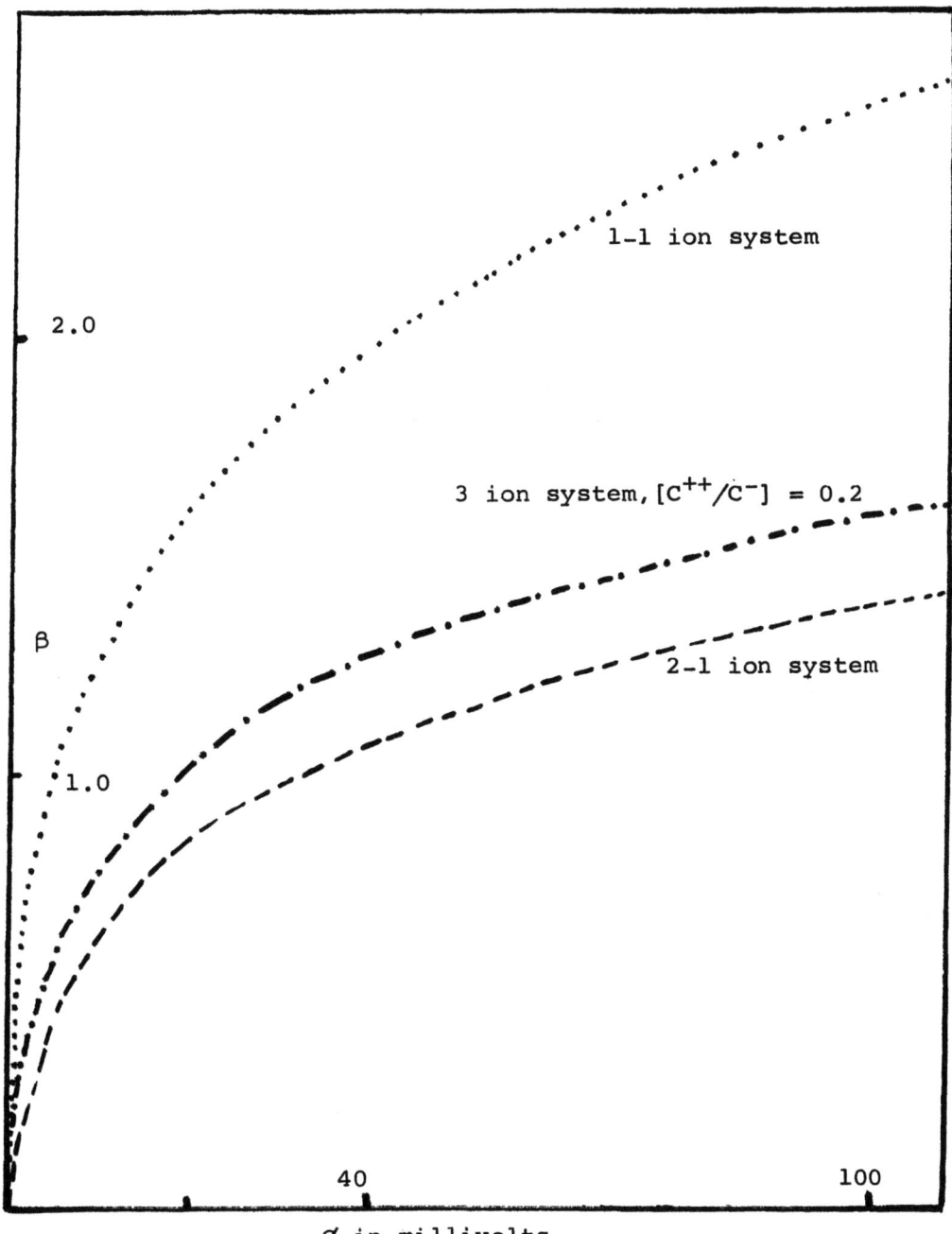

FIGURE 1. The relation between β and ϕ, when $\phi(d) = 0$, and surface contains fixed negative charges, for the three electrolyte systems.

When equation (9) is valid for every kind of ions in the system, one obtains the relations (14a) and (14b), which are exact.

$$(4\pi kT) \, K(x) A'(x) = Y(x) Y'(x) \tag{14a}$$

$$K(x) \phi'(x) = Y'(x) \left\{ 1 + (H/4\pi) K(x) \right\} \tag{14b}$$

$$A(x) = \sum_\sigma C_\sigma(x) \; ; \; A'(x) = [dA/dx] \tag{14c}$$

Equation (14a) is obtained, by summing equation (9) over all ionic species present in the system. Equation (14a) is the generalized version of Maxwell's Osmotic balance equation. When H* terms and H terms of equation (9) are neglected, and the dielectric coefficient is regarded as position independent, one obtains by integration, the famous familiar result, (Agin, 1971; Lakshminarayaniah, 1969),

$$\left\{ [8\pi kT] / \epsilon(d) \right\} [A(o) - A(x)] = \phi'(o)^2 - \phi'(x)^2 \tag{15}$$

Equations (14) and (15) indicate the essential nonvalidity of famous constant field approximations, often assumed in membrane biophysics. The equation (14b) is obtained by multiplication by charge, $Z_\sigma e$, of each equation of the kind (9) for species σ, and then summing over all ionic species present in the system. Poisson's equation has been reutilized. It is easy to verify that for most electrolyte system of any composition, even with classical expressions, (3) and (4), that $K(x_1)$ is always greater than $K(x_2)$ in interfacial region, provided that x_2 is always greater than x_1. [The exception to above statement occurs for unsymmetrical electrolyte system, when β is negative. For a 2-1 ion system, when β assumes values in the range, $-1 < \beta < 0$, such an exception occurs. For three ion system, such as one given in equation (13c), exception to above statement occurs for much smaller range of negative values of β.] Therefore, in the interfacial region, $K(x)$ is always larger than $K(d)$ [Vaidhyanathan, 1982]. If equation (11) is valid, than it necessarily follows that the ratio, $[\phi'(x)/Y'(x)]$ should be negative definite in interfacial regions.

If the surface at interface, contains fixed negative charges, above results indicate that $Y(x)$ will be negative over major extent of the interfacial inhomogeneous region. Thus, the integral of $Y(x)$, indicated by Gauss equation (1), will be negative. Since the value of dielectric coefficient at $x = 0$, $\epsilon(o)$ is positive definite, it follows that the gradient of electric potential at interface, $\phi'(o)$ should alwo be positive definite. Since β will be positive, in regions, where $Y(x)$ will be negative, $[Y(o)$ should be positive definite], the electric potential and charge density profiles should exhibit extremum values. Such extremum values, in charge density profiles, a feature never shown by Stern-Gouy-

Chapman theory, is obtained also in the paper of D'Aguanno et al,(1987), suing a local HNC/HNC approximation for the 2-2 Restrictive Primitive Model electrolytes, for certain values of bulk electrolyte concentrations.

The electric potential profile and the negative of charge density profile, $Y(x)$, consistant with the results of our analysis is presented in Figure 2. In the same figure, for comparison, the profiles resulting from classical concepts are also presented. When concentrations of salts in aqueous solution is of the order of unimolar, the value of the magnitude of extent of inhomogeneous interfacial region, d, can be estimated to equal about 20×10^{-8} cm. The magnitude of this inhomogeneous region increases by a factor of $[10]^{1/2}$, for every tenfold decrease in concentration. The product $\varkappa$ d, is approximately a constant, in interfacial region, though not necessarily equal to unity. Computed values of the integrals of equation (1), for the three electrolytes of Figure 1, or equations (13), indicate that the values are not significantly different from each other, (Vaidhyanathan, 1985, 1986). The computed value of the dielectric coefficient at interface, $\epsilon(o)$, surprisingly turns out to have a value very close to unity!

RESOLUTION OF THE CONFLICT AND ESTIMATION OF H AND d

It is evident, that the conclusions obtained with the use of the equations (9) and (13) are drastically different from the conclusions of the classical theory. Recent advances such as Modified Poisson-Boltzmann equation, Mean Spherical Approximation, or the Restricted Primitive Model, do not show such abnormal discrepancy from classical theory. It is also evident that the conclusions presented so far in this paper, is plausible and reasonable. In our opinion, the recent statistical mechanical advances, do not show such startling divergent conclusions, due to the fact that these theories still assume that the medium is a dielectric continuum, with a constant dielectric coefficient. The results of a Nernst-Planck analog approach for electrodiffusion problems of the Membrane Biophysics, (Vaidhyanathan, 1979) indicate the fruitfulness of Taylor expansion approach in dealing with the non-linear differential equations of biophysics. Specifically, one obtains the result that the electric potential profile is a polynomial in position variable, and that it assumes a third order function in x, in membranes, when the flux and gradient of all noncharged species vanish. Similar expansion in Taylor series, and truncation of the corresponding expressions for the dielectric profile with the leading three terms and truncation of the expression for electric potential profile with the leading seven terms, lead to an evidently approximate relation,

$$K(d)d^2 = 42 \Delta\epsilon \; ; \; \Delta\epsilon = \epsilon(d) - \epsilon(o). \qquad (16)$$

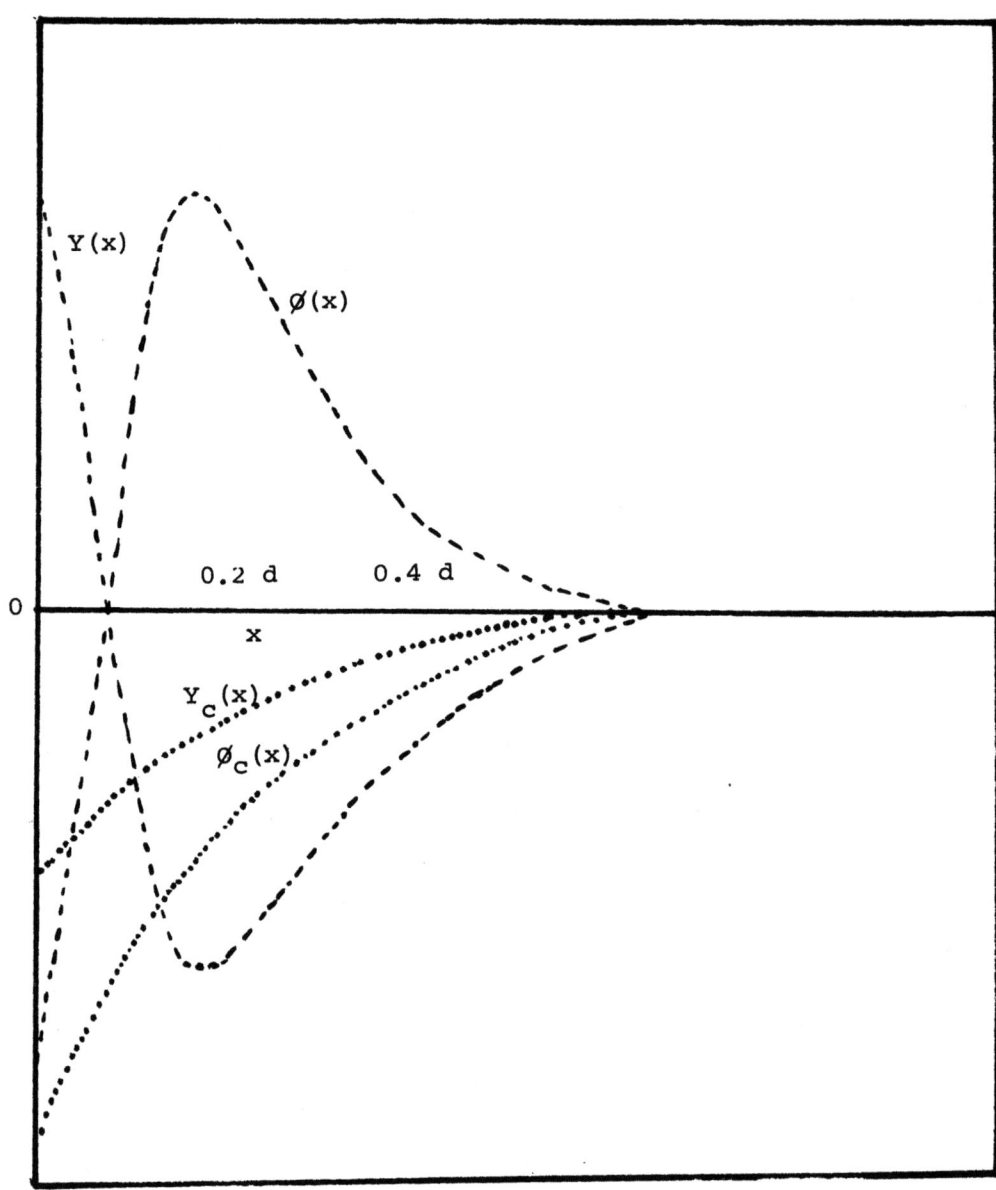

FIGURE 2. Schematic representations of electric potential profiles, $\emptyset(x)$, and negative of charge density profiles, $Y(x)$, given by the results of this paper, denoted by (-----) and corresponding profiles given by classical theory, denoted by (.....). Scales in y-axis are arbitrary.

Though approximate, equation (16) indicates that the magnitude of the extent of inhomogeneous interfacial region, d, is finite, when $\Delta\epsilon$ is nonzero, and that it becomes indefinite, when one assumes that the dielectric coefficient is a constant. In the same limit of approximation, one obtains an expression for the symmetrical electrolyte system, relating β with dielectric coefficients,

$$1 - \cosh \beta = [7\Delta\epsilon [10 \epsilon(o) - \Delta\epsilon]]/[10 \epsilon(d)^2] \quad (17)$$

Equation (17) indicates the dependence of β on the dielectric profile or vice versa.

Since almost all analysis of the electrokinetic studies in biological systems are based on the assumed validity of the Nernst equation (4) and transport studies based on analysis of solutions of Nernst-Planck equation for electrodiffusion, it is extremely important to decide between the validity of our conclusions, and the divergent implications of the classical Gouy-Chapman theory (Verwey & Overbeek, 1945). Unfortunately, all experimental determinations involve either the implicit assumption of the validity of equations (3) and (4), or based on the experiments carried out in locations where local electroneutrality is preserved. Thus, it is our opinion that at this stage, an experimental determination about the validity of either of these two opposite conclusions cannot be performed.

However, if one has knowledge of the dielectric profile and the electric potential profile, then one can compute the charge density profile and the concentration profiles of ions in interfacial regions. This will enable one to determine the signs of various quantities involved. Assuming that the profiles in question are continuous and analytic, one can expand these in Taylor series, truncating at convenient number of terms. For example, if the dielectric profile can be represented as,

$$\epsilon(x) = \sum_{i=0}^{m} \epsilon_i x^i \quad ; \quad (i!) \; \epsilon_i = [d^i \epsilon/dx^i]_{x=0} \quad (18)$$

where m is the order of the highest power of x, retained. When m = 2, and the boundary conditions that $\epsilon'(d) = 0$, $\epsilon''(d) = 0$, are valid, one has the results, $\epsilon'(o) = [2\Delta\epsilon/d]$, and $\epsilon''(o) = 2 \epsilon_2 = -[2\Delta\epsilon/d^2]$. Similarly, when m = 3, one obtains the results, $\epsilon'(o) = [3\Delta\epsilon/d]$ and $\epsilon''(o) = -[6\Delta\epsilon/d^2]$. In this manner, the plausible dielectric profile is specified. One may expand the electric potential profile in the interfacial region in a like manner.

$$\phi(x) = \sum_{i=0}^{n} \phi_i x^i \quad ; \quad (i!)\phi_i = [d^i\phi/dx^i]_{x=0} \quad (19)$$

Again utilizing plausible boundary conditions that all order derivatives of the electric potential profile, vanish at x = d, one can evaluate the values of Taylor coefficients retained. These values for various values of n, where (n+1) terms of equation (19) are retained are listed in Table 1.

TABLE 1. Computed values of various order derivatives of the electric potential using equation (19).

	n = 4	n = 5	n = 6
$\phi(o)$	$\phi_4 d^4$	$-\phi_5 d^5$	$\phi_6 d^6$
$\phi'(o)$	$-4\phi_4 d^3$	$5\phi_5 d^4$	$-6\phi_6 d^5$
$\phi''(o)$	$12\phi_4 d^2$	$-20\phi_5 d^3$	$30\phi_6 d^4$
$\phi'''(o)$	$-24\phi_4 d$	$60\phi_5 d^2$	$-120\phi_6 d^3$

TABLE 2. Computed values of $Y(o)$ and $Y'(o)$, assigining different values for m and n of equations (18) and (19).

m	n	$Y(o)$	$Y'(o)$	$[Y(o)/\phi(o)]$	$[Y'(o)/\phi'(o)]$
2	4	-6.8	93.6	-6.8	-23.4
2	5	8.0	-84	-8.0	-16.8
2	6	-9.0	60	-9.0	-10.0
3	4	-10.8	93.6	-10.8	-23.4
3	5	13.0	-144	-13.0	-28.8
3	6	-15.0	204	-15.0	-36.0

The values of $Y(o)$ are in $[\phi_n d^{n-2} \Delta\epsilon]$

The values of $[Y(o)/\phi(o)]$ and $[Y'(o)/\phi'(o)]$ are of the dimension $[\Delta\epsilon/d^2]$.

The computed values of $Y(o)$, $Y'(o)$ and the ratios, $[Y(o)/\emptyset(o)]$ and $[Y'(o)/\emptyset'(o)]$ that one obtains, from equations (18) and (19), without assigning values for d, or $\emptyset(o)$, or bulk electrolyte concentrations are presented in Table 2. In computation of the values of $Y(o)$ and $Y'(o)$, it is assumed that $\in(o)$ equals $0.1\Delta\in$. It should be noted that all computed values of the ratios, $[\emptyset(o)/Y(o)]$ and $[\emptyset'(o)/Y'(o)]$ presented in Table 2, are negative definite. These results are in direct disagreement with the implications of equation (3) of classical theory.

Consider the case of a 1-1 electrolyte with concentrations in bulk homogeneous region, of values, $C^+(d) = C^-(d)$ equal to 5×10^{-5} [moles/cm^3], at 20^0C. The value of $K(d)$ can be computed as 43.159×10^{14} [cm^{-2}]. If one assigns the values of 1, 2 and 3 respectively to $\beta(o)$, one can compute the values of $K(o)$ as 66.598, 162.372 and 434.51 [$\times 10^{14}$] respectively. From equation (14b), the values of $(H/4\pi)$ are given by the relations,

$$(H/4\pi) = [\emptyset'(o)/Y'(o)] - [K(o)]^{-1} \quad (20)$$

If the value of d equals 2×10^{-7} cm, and $\in(o) = 0.1\Delta\in$, from the values listed in Table 2, when $m = 2$ and $n = 4$, one obtains

$$- (H/4\pi) = 0.8498 \times 10^{-16} \text{ cm}^2, \text{ if } \beta(o) = 2. \quad (21)$$

The corresponding value of $[K(d)]^{-1}$ equals in this case, 2.317×10^{-16}. The computed negative value of $(H/4\pi)$ and its magnitude being of the same order as the value of $[K(d)]^{-1}$ is extremely gratifying. These calculations indicate the plausible validity of results presented in equation (11). If equation (11) is valid, when $\beta(o) = 1$, one obtains the result that

$$[d^2/\Delta\in] = 10.08187 \times 10^{-16}, \text{ when } m = 2 \text{ and } n = 4.$$
$$\quad (22)$$
$$\text{or, } [d^2/\Delta\in] = 29.3566 \times 10^{-16}, \text{ when } m = 3 \text{ and } n = 6.$$

Thus, one computes that the extent of inhomogeneous region from surface, when electrolyte concentration equals 0.5 molar, d = 37.336 angstroms, when $\beta(o) = 1$, $m = 2$ and $n = 4$. d equals 46.31 angstroms, when $\beta(o) = 1$, $m = 3$ and $n = 6$. It is interesting to note that the approximate relation presented in equation (16) yields the value that d = 8 angstroms, for 1-1 electrolyte system, when concentrations are of the order of unimolar.

When $\beta(o)$ equals 2, $m = 3$ and $n = 6$, one obtains the result, d equals 66.89 angstroms. The dependence of d on $\beta(o)$ indicate that d is also a function of the surface electric potential, $\emptyset(o)$. Therefore the contention that $\varkappa$ d is a constant is only an approximate relation.

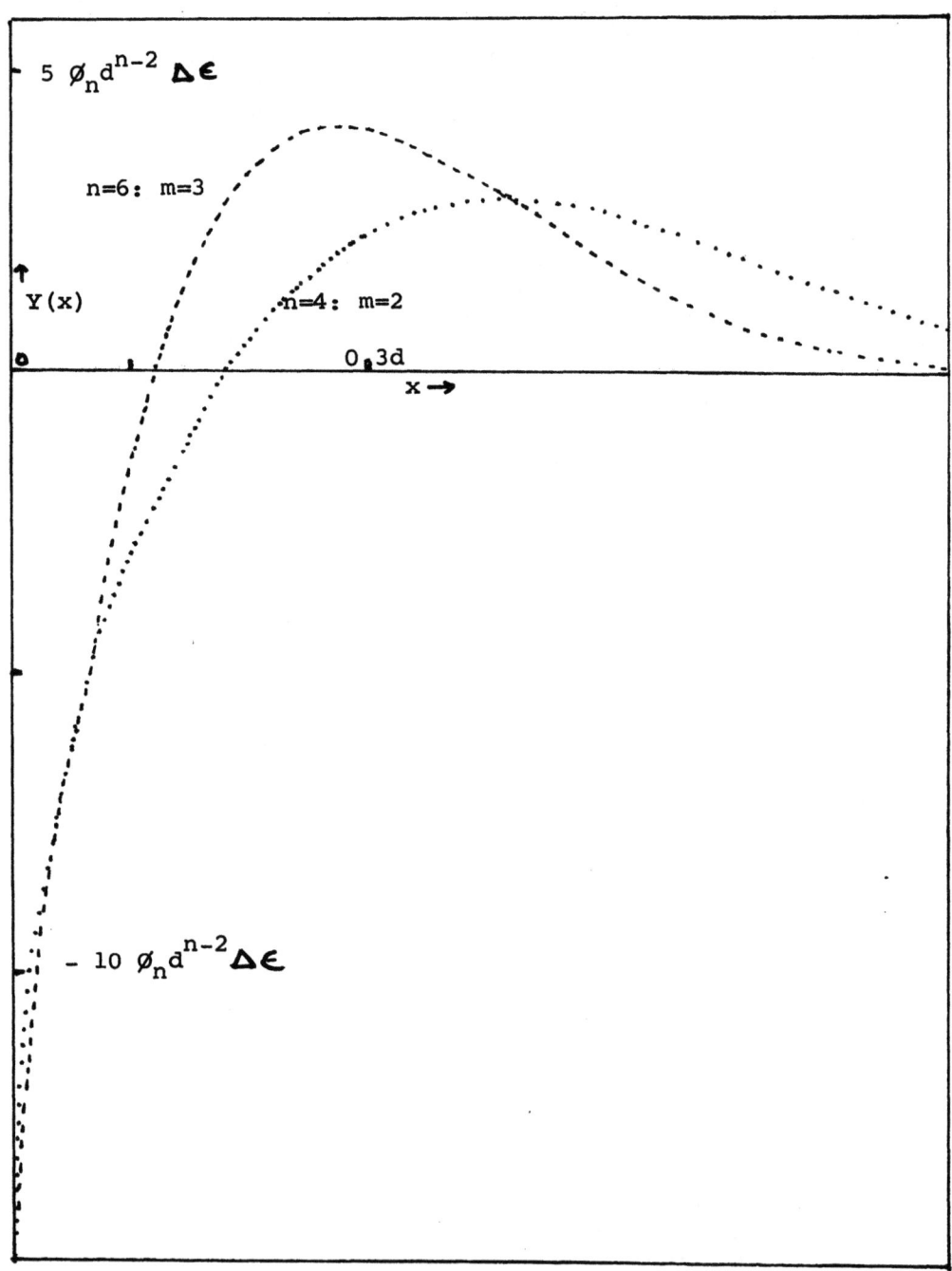

FIGURE 3. Computed values of negative of charge density profiles, [- Y(x)], using equations (6), (18) and (19), with assigned values for m and n. $\phi(0) = \phi_n d^n$.

DISCUSSION

The inclusion of the contribution to chemical potential of an ion, from ion-ion interaction energy terms, (equation 9) leads to a number of significant conclusions about ion distributions in interfacial regions. These are, 1. in the interfacial region, ion distributions, $Z_\sigma \beta(x)$, are determined by the magnitude and sign of local charge density, in addition to difference in electric potential at these locations. 2. The electric potential profile and concentration profiles of positively charged ions are schematically similar. This is due to significant contribution arising from repulsion between similarly charged ions in interfacial region. 3. Both charge density profile and electric potential profile should exhibit extremum values in interfacial region, as depicted in Figure 2. 4. A dielectric profile must exist, to ensure the negative nature of $[\phi(x)/Y(x)]$. 5. The magnitude of the extent of the inhomogeneous region is finite and is determined by the system, in a self-consistant manner. Two other plausible conclusions that one may obtain on the basis of these studies are, viz., 1. the value of dielectric coefficient at the interface is very close to unity, and that d, $\beta(o)$ and $\phi(o)$ are interrelated in a yet undetermined manner.

The electric potential profile presented in Figure 2, is obtained by retention of the leading ten terms of equation (19),(Vaidhyanathan, 1986), where the location at which the extremum value in $\phi(x)$ occurs was arbitrarily assumed. When the leading five terms of equation (19) are retained, one has $\phi_0 = \phi_4 d^4$; $\phi_1 = -4\phi_4 d^3$; $\phi_2 = 6\phi_4 d^2$ and $\phi_3 = -4\phi_4 d$, when one utilized the boundary conditions that $\phi'(d) = \phi''(d) = \phi'''(d) = \phi(d) = 0$. Assuming that the dielectric profile in the system is given by retention of the leading three terms of equation (18), the charge density profile can be computed. The calculated charge density profile $[-Y(x)]$, thus obtained is presented in Figure 3. Though the electric potential profile obtained by neglect of ϕ_5 and higher order terms is monotonic, the charge density profile calculated exhibits an extremum value. Similarly, when one retains the leading seven terms of equation (19), and assumes in addition that $[d^5\phi/dx^5]_{x=d}$ equals zero, such that $0/5 = -6\phi_6 d$, one obtains, $\phi_4 = 15\phi_6 d$ $\phi_3 = -20\phi_6 d^3$, $\phi_2 = 15\phi_6 d^4$ and $\phi_1 = -6\phi_6 d^5$. The electric potential profile computed in this case also is monotonic. Assuming m = 3, the dielectric profile computed using equation (18) can be utilized to compute the charge density profile. The resultant is presented also in Figure 3, which again contains an extremum value. Though, one is tempted to conclude that these two charge density profiles of Figure 3, indicate the essential validity of classical theory, in regions where deviation from electroneutrality is small, it is our opinion that the strict requirement of the validity of Gauss relation (1) prohibits the validity of such conclusions.

REFERENCES

Agin, D., (1971) in 'Foundations of Math. Biology' Rosen, ed. Academic Press.
Booth, F., (1951) J. Chem. Physics.,$\underline{19}$, 39, 1327.
Blum, L., Hernando, J., and Lebowitz, J.L., J. Phys. Chem., (1983)$\underline{87}$, 2895.
Carnie, S.L., and Torrie, G.M., Adv. Chem. Phys., (1984) $\underline{56}$, 141.
D'Aguanno, B., Nielaba. P., Alts, T., and Forstmann, F., (1987) Proc. NATO-ASI on PCH-Interfacial Phenomena, La Rabida, Spain, this volume, Plenum Publications.
Frohlich, H., (1949) 'Theory of Dielectrics' Oxford University Press.
Grahame, D.C., (1947) Chemical Reviews $\underline{41}$, 441.
Kirkwood, J.G., and Buff, F.P., (1951) J. Chem. Phys., $\underline{19}$, 774.
Kruyt, H.R., (1952) 'Colloid Science' Elsevier Publ. Co. N.Y..
Lakshminarayaniah, N., (1969) 'Transport Phenomena in Membranes'.
Luzar, A., Svetina, S., and Zeks, B., (1985) J. Chem. Phys., $\underline{82}$, 5146.
McMillan, W.G., and Mayer, J.E., (1945) J. Chem. Phys., $\underline{13}$, 276.
Nielaba, P., and Forstmann, F., (1985) Chem. Phys. Letters, $\underline{117}$, 46.
Outhwaite, C.W., Bhuiyan, L.B., and Levine, S., (1980) J. Chem. Soc. Fraday Trans.,II, $\underline{76}$, 1388.
Vaidhyanathan, V.S., (1979) Bull. Math. Biology, $\underline{41}$, 365.
Vaidhyanathan, V.S., (1982) J. Biol. Phys., $\underline{10}$, 153, 167.
Vaidhyanathan, V.S., (1985) Studia Biophysica, $\underline{110}$, 29.
Vaidhyanathan, V.S., (1986) in 'Electrical Double Layers in Biology', M. Blank, editor, Plenum Press. p.31.
Verwey, E.J.W., and Overbeek, Th.G., (1945) 'Theory of Stability of Lyophobic Colloids' Elsevier Publishing Co. N.Y..

PACKING AND PAIRING OF IONS NEAR A CHARGED ELECTRODE

T. Alts, B. D'Aguanno, P. Nielaba, and F. Forstmann

Institut für Theorie der Kondensierten Materie
Freie Universität Berlin, Arnimallee 14
D-1000 Berlin 33

ABSTRACT

Local HNC/MSA and local HNC/HNC approximations are developed and applied to symmetric 1:1, 2:2 and asymmetric 2:1 electrolytes in front of a charged electrode. The results are compared with Monte Carlo data and other theories such as MGC, HNC/HNC, HNC/MSA, BGY and MPB5. A remarkable improvement of density profiles and of diffuse layer potentials is obtained. Packing and pairing of ions, charge inversion, non-monotonous profiles of electrostatic potential at higher surface charges and related effects are predicted in accordance with Monte Carlo simulations.

The local approximation is related to the density functional approach for inhomogeneous fluids. Its success can be traced back to an approximately right treatment of the ion-ion direct correlation functions within the non-neutral double layer.

1. INTRODUCTION

The electrical double layer at an electrolyte solution/electrode interface is one of the great problems in electrochemistry. It became in the last few years the object of intense theoretical interest [1]. Different approximation schemes within the statistical mechanics of inhomogeneous liquids have been developed to study the distributions of ions near a charged surface [2-15]. The difficulties due to extreme density gradients are such, that the electrolyte model is reduced in most cases to the "(Restricted) Primitive Model", (RPM), an ensemble of equal size hard spheres with centered point charges in a homogeneous background of dielectric constant ε, and that the electrode is approximated by a non-polarizable hard wall with constant surface charge density ω.

Only very few attempts to include the dipoles of solvent molecules in the statistical calculations of the double layer have been published [16-19]. The contribution of a polarizable metal electrode to the double layer capacitance has been studied with a jellium model for the metal electrons [20-22].

We concentrate in the following on the most simple electrolyte model, the RPM. This is mainly done to test statistical approximation schemes for the ionic contributions to the structure, potential and capacitance of the highly inhomogeneous double layer. The treatment of a more realistic electrolyte as a mixture of dipolar solvent molecules and ions requires the solution of the RPM-problem first.

The earliest treatment of the double layer go back to the pioneering works of Gouy /23/, Chapman /24/ and Stern /25/. In this so-called Modified Gouy-Chapman (MGC) approach the ionic size is only introduced by defining a distance of closest approach to the charged surface. The ion distributions and the mean electrostatic potential are calculated from the Poisson Boltzmann equation for charged point particles. Recent approaches are instead based on the integral equation theories of liquid state physics. They all account for the finite ion size and for the correlations between the ions. These two aspects originate, for certain values of the electrolyte bulk density and the surface charge density, an oszillatory behaviour in the density profile near the surface, a feature never shown by the MGC-theory.

The integral equation approaches can be classified into three broad groups: The Hypernetted Chain (HNC) /2-4,8-11,15/, the Born-Green-Yvon (BGY) /12-14/, and the Modified Poisson-Boltzmann (MPB) /5-7/ based theories. All these theories are different approximations to the exact equations of statistical mechanics of liquids and they distinguish mainly by different assumptions about the particle-particle correlations within the inhomogeneous electrostatic field of the double layer. The accuracy of the approximations are tested against the essentially exact results of Monte Carlo (MC) simulations /26-28/, which serve as the "experimental" background for the RPM electrolytes.

For 1:1 electrolytes the agreement between the previous theories and MC-data is good only for regions of small surface charge density but it even deteriorates gradually when decreasing the electrolyte bulk density from 2 M to 0.01 M. For the 2:2 and 2:1 electrolytes no theory so far was able to yield reasonable ion distributions /29/ and consequently these systems appear to be a severe test for any theoretical approach.

In what follows we concentrate only on HNC-based theories. The HNC theory divides in two branches, HNC/MSA and HNC/HNC, according to the approximation scheme used to describe the particle-particle correlations inside the electrolyte solution. The wall-particle correlations are described in both cases by the HNC closures of the otherwise exact Ornstein-Zernike (OZ) equations for non uniform systems.

The HNC-theories for the double layer start from the integral equations for the ion-density profiles /3,4,30/; $\underline{r} = (0,0,z)$:

$$\rho_\alpha(z) = \rho_\alpha^{bulk} \exp\{-\frac{V_\alpha(z)}{kT} + \sum_\beta \int d^3r'[\rho_\beta(\underline{r}') - \rho_\beta^{bulk}]c_{\alpha\beta}(\underline{r},\underline{r}')\} \qquad (1.1)$$

in which z is the distance from the electrode and $V_\alpha(z)$ the external potential energy of an ion α due to the charged surface. Equ. (1.1) can be solved iteratively, when the direct ion-ion correlation functions $c_{\alpha\beta}(\underline{r},\underline{r}')$ are known. The $c_{\alpha\beta}$ are required throughout the strongly inhomogeneous region near the electrode surface.

Previously the ion-ion correlations have been approximated by those of the neutral bulk electrolyte of density $\rho^B = \{\rho_\alpha^{bulk}\}$:

$$c_{\alpha\beta}(\underline{r},\underline{r}') \simeq c_{\alpha\beta}(|\underline{r}-\underline{r}'|;\rho^B). \qquad (1.2)$$

In homogeneous systems, the $c_{\alpha\beta}$ can be evaluated analytically in the mean spherical approximation (MSA) and numerically in the HNC approximation. In general, the HNC approximation gives the better structure /31,32/ in homogeneous systems. Therefore the expectation is, that the HNC/HNC must provide a better description of the double layer than the HNC/MSA. However, we know that for both 1:1 and 2:2 electrolytes this is usually not the case (see Fig. 2 and 5 of Ref. /29/). At elevated surface charge densities both procedures fail to describe the packing of counterions in a second layer /1/ for 1:1 electrolytes. In the 2:2 case the disagreement with MC-density profiles has been found also for low charge densities /29/. For the asymmetric 2:1 electrolytes with equal ion size but unequal ion charges non-monotonic ion profiles and corresponding charge inversions are not properly reproduced. Moreover, at higher surface charges no solutions have been obtained for the HNC/HNC equations.

All these shortcomings can be traced back to the approximation of Equ. (1.2). An idea to improve it is provided by introducing a "local density" concept /8-11/. The particle-particle direct correlations are taken from homogeneous systems of the composition $\bar{\rho}(z) := \{\bar{\rho}_\alpha(z)\}$ as found locally for the electrode in the non-neutral double-layer region, where there are nearly only ions of one kind and the ionic surroundings is strongly different from that of the neutral bulk system /9,33/. Then, the essential step in the local HNC/MSA and the local HNC/HNC approximations is:

$$c_{\alpha\beta}(\underline{r},\underline{r}') \simeq c_{\alpha\beta}(|\underline{r}-\underline{r}'|; \bar{\rho}(z)). \tag{1.3}$$

The densities $\bar{\rho}_+(z)$ and $\bar{\rho}_-(z)$ are evaluated by averaging the density profiles $\rho_\alpha(z)$ around position z. A similar idea has recently been used in /34,35/ to describe the freezing transition of hard spheres, which is another example of strongly varying density.

In this article we investigate the symmetric 1:1, 2:2 and the asymmetric 2:1 electrolytes using the local HNC/MSA and the local HNC/HNC approximations. All the unwanted features (poor agreement with MC-data of the diffuse layer potential, inferior description of ionic density profiles near the electrode and break down in the solution of the HNC/HNC approximations) are eliminated. It turns out, that for the weakly correlated 1:1 electrolytes the local HNC/MSA and the local HNC/HNC approximations yield practically the same results; we therefore use the simpler and analytically known particle-particle correlations in MSA-closure. However, for the stronger correlated 2:2 and 2:1 electrolytes the better description of the particle-particle correlations by the HNC-closure is necessary.

The structure of the article is as follows: In Sections 2 and 3 we derive a generalisation of (1.1) from density functional theory for inhomogeneous mixtures and discuss closure conditions. In Section 4 these results are specialized to RPM-electrolytes in front of a plane uniformly charged electrode. The local approximation and its motivation are given in Section 5. A short discussion of the numerical iterative procedure and accuracy can be found in Section 6. Finally, in Section 7, we discuss the results of solving the local HNC/MSA and the local HNC/HNC equations and compare them to MC data and to the results of other approximate theories. The conclusions are given in Section 8.

It can be stated, that the local HNC/MSA and the local HNC/HNC formulation yields an essential improvement in the understanding of inhomogeneous fluids, of ionic structure and electrostatics of the double layer.

We mention that among the very recent theories there are two /13,14,15/ which are also relatively successful in describing the ionic contributions to the double layer problem. The "local concept" in these theories is used indirectly either by adding a bridge contribution to Equ. (1.1) and otherwise using particle-particle correlations of the neutral bulk electrolyte /15/, or by using an ansatz for the particle-particle correlations in the strongly inhomogeneous region near the charged electrode which is determined by satisfying the local electroneutrality condition and eliminating unphysical values of the total correlation functions in the neighbourhood of the electrode surface /13,14/.

2. SOME EXACT RESULTS OF DENSITY FUNCTIONAL THEORY FOR AN IONIC SOLUTION NEAR A CHARGED ELECTRODE

A general starting point for the treatment of inhomogeneous liquids is the density functional theory due to Mermin /36/ and Saam & Ebner /37/. Since the extension to long range Coulomb forces needs some care, we include a few steps of derivation.

We consider a grand canonical ensemble, a mixture of charged particles of different kinds α ($\alpha = 1, 2, \ldots, \nu$) at fixed volume and temperature T with external potential energies $V_\alpha(\underline{r})$ and chemical potentials μ_α. Given

$$U_\alpha(\underline{r}) := V_\alpha(\underline{r}) - \mu_\alpha , \quad (2.1)$$

which we henceforth call external potentials, there exists a functional of the particle density $[\rho] := \{\rho_\alpha(\underline{r})\}$:

$$\Omega_U[\rho] = \sum_\alpha \int d^3r \, \rho_\alpha(\underline{r}) U_\alpha(\underline{r}) + \quad (2.2)$$

$$+ kT \sum_\alpha \int d^3r \, \rho_\alpha(\underline{r}) [\ln(\Lambda_\alpha^3 \rho_\alpha(\underline{r})) - 1] - \Phi[\rho] ,$$

which takes its minimum value for those equilibrium densities that belong to the fixed $U_\alpha(\underline{r})$. $\Omega_U[\rho]$ given for the equilibrium density is the grand free energy of the system. The second integral in (2.2) is the ideal and $-\Phi[\rho]$ the interaction contribution to the free energy. $\Lambda_\alpha := h(2\pi m_\alpha kT)^{-1/2}$ denotes the de Broglie wavelength of particles α with mass m_α (h = Planck's constant, k = Boltzmann's constant).

The equilibrium densities satisfy the necessary conditions for the minimum of $\Omega_U[\rho]$:

$$\frac{\delta \Omega_U[\rho]}{\delta \rho_\alpha(\underline{r})} = 0 = kT \ln[\Lambda_\alpha^3 \rho_\alpha(\underline{r})] + U_\alpha(\underline{r}) - kT c_\alpha(\underline{r};[\rho]) , \quad (2.3)$$

where

$$- kT c_\alpha(\underline{r};[\rho]) := - \frac{\delta \Phi[\rho]}{\delta \rho_\alpha(\underline{r})} \quad (2.4)$$

is the potential of mean force due to particle interactions; it is a function of $\underline{r}$ and a functional of $[\rho]$. The second functional derivatives of $\Phi[\rho]$ define the Ornstein-Zernike direct correlation functions

$$c_{\alpha\beta}(\underline{r},\underline{r}';[\rho]) := \frac{1}{kT} \frac{\delta^2 \Phi[\rho]}{\delta \rho_\alpha(\underline{r}) \delta \rho_\beta(\underline{r}')} . \quad (2.5)$$

Knowing the direct correlation functions $c_{\alpha\beta}$, the interaction contribution $\Phi[\rho]$ of the free energy can be obtained by functional integration. Since

$\Phi[\rho]$ is a unique functional of $[\rho]$ we may choose a linear density path

$$\rho_\alpha(\underset{\sim}{r};\lambda) := \rho_\alpha^i(\underset{\sim}{r}) + \lambda[\rho_\alpha(\underset{\sim}{r}) - \rho_\alpha^i(\underset{\sim}{r})], \qquad 0 \le \lambda \le 1 \qquad (2.6)$$

between initial equilibrium densities $\rho_\alpha^i(\underset{\sim}{r})$, belonging to external potentials $U_\alpha^i(\underset{\sim}{r})$ and satisfying the equilibrium conditions; $[\rho^i] := \{\rho_\alpha^i(\underset{\sim}{r})\}$,

$$0 = kT \ln[\Lambda_\alpha^3 \rho_\alpha^i(\underset{\sim}{r})] + U_\alpha^i(\underset{\sim}{r}) - kT c_\alpha(\underset{\sim}{r};[\rho^i]), \qquad (2.7)$$

and the final equilibrium densities $\rho_\alpha(\underset{\sim}{r}) = \rho_\alpha(\underset{\sim}{r},1)$. Functional integration of (2.4) and (2.5) yields then; $[\rho^\lambda] := \{\rho_\alpha(\underset{\sim}{r};\lambda)\}$:

$$c_\alpha(\underset{\sim}{r};[\rho]) = c_\alpha(\underset{\sim}{r};[\rho^i]) + \sum_\beta \int d^3 r' [\rho_\beta(\underset{\sim}{r}') - \rho_\beta^i(\underset{\sim}{r}')] \int_0^1 d\lambda\, c_{\alpha\beta}(\underset{\sim}{r},\underset{\sim}{r}';[\rho^\lambda]) \qquad (2.8)$$

and

$$\begin{aligned}\Phi[\rho] = \Phi[\rho^i] &+ kT \sum_\alpha \int d^3 r [\rho_\alpha(\underset{\sim}{r}) - \rho_\alpha^i(\underset{\sim}{r})] c_\alpha(\underset{\sim}{r};[\rho^i]) \\ &+ kT \sum_{\alpha,\beta} \iint d^3 r d^3 r' [\rho_\alpha(\underset{\sim}{r}) - \rho_\alpha^i(\underset{\sim}{r})][\rho_\beta(\underset{\sim}{r}') - \rho_\beta^i(\underset{\sim}{r}')] \times \\ &\times \int_0^1 d\lambda \int_0^\lambda d\lambda' c_{\alpha\beta}(\underset{\sim}{r},\underset{\sim}{r}';[\rho^{\lambda'}]).\end{aligned} \qquad (2.9)$$

With (2.7) we can eliminate $c_\alpha(\underset{\sim}{r};[\rho^i])$ from (2.9) and can thus rewrite (2.2) as follows

$$\begin{aligned}\Omega_U[\rho] = \Omega_U i[\rho^i] &+ \sum_\alpha \int d^3 r \rho_\alpha(\underset{\sim}{r})[U_\alpha(\underset{\sim}{r}) - U_\alpha^i(\underset{\sim}{r})] + \\ &+ kT \sum_\alpha \int d^3 r \rho_\alpha(\underset{\sim}{r}) \ln\left[\frac{\rho_\alpha(\underset{\sim}{r})}{\rho_\alpha^i(\underset{\sim}{r})}\right] - kT \sum_\alpha \int d^3 r [\rho_\alpha(\underset{\sim}{r}) - \rho_\alpha^i(\underset{\sim}{r})] - \\ &- kT \sum_{\alpha,\beta} \iint d^3 r d^3 r' [\rho_\alpha(\underset{\sim}{r}) - \rho_\alpha^i(\underset{\sim}{r})][\rho_\beta(\underset{\sim}{r}') - \rho_\beta^i(\underset{\sim}{r}')] \times \\ &\times \int_0^1 d\lambda \int_0^\lambda d\lambda' c_{\alpha\beta}(\underset{\sim}{r},\underset{\sim}{r}';[\rho^{\lambda'}]).\end{aligned} \qquad (2.10)$$

The second term on the r.h.s. contains the difference of the external potentials, since $\Omega_U i[\rho^i]$ is the grand free energy in the initial equilibrium state. For a one component system Saam & Ebner /37/ presented a slightly different form of (2.10) with $\Omega_U[\rho^i]$ instead of the initial grand free energy. Finally, eliminating $c_\alpha(\underset{\sim}{r};[\rho])$ from (2.3) with (2.8) and using (2.7) the equations for the equilibrium densities $\rho_\alpha(\underset{\sim}{r})$ belonging to external potentials $U_\alpha(\underset{\sim}{r})$ can be rewritten in the explicit form:

$$\ln\left[\frac{\rho_\alpha(\underset{\sim}{r})}{\rho_\alpha^i(\underset{\sim}{r})}\right] = -\frac{1}{kT}[U_\alpha(\underset{\sim}{r}) - U_\alpha^i(\underset{\sim}{r})] + \sum_\beta \int d^3 r' [\rho_\beta(\underset{\sim}{r}') - \rho_\beta^i(\underset{\sim}{r}')] \int_0^1 d\lambda\, c_{\alpha\beta}(\underset{\sim}{r},\underset{\sim}{r}';[\rho^\lambda]). \qquad (2.11)$$

With long range Coulomb interactions the most important contribution on the r.h.s. of (2.11) is the mean electrostatic potential $\psi(\underline{r})$. This can be separated from the other contributions by the following two steps:
i.) The direct correlation functions are split into short range parts and Coulomb contributions

$$c_{\alpha\beta}(\underline{r},\underline{r}';[\rho^\lambda]) = -\frac{1}{kT}\frac{Z_\alpha Z_\beta e^2}{\epsilon|\underline{r}-\underline{r}'|} + c_{\alpha\beta}^{SR}(\underline{r},\underline{r}';[\rho^\lambda]). \qquad (2.12)$$

Z_α is the valence (including the sign) of ion α and e denotes the absolute value of the electron charge. ii.) The external potential energy $V_\alpha(\underline{r})$ is decomposed into a Coulomb contribution $V_\alpha^c(\underline{r})$ due to the surface charge density $\omega(\underline{s})$ at location $\underline{s}$ on the surface of the charged electrode and a short range part $V_\alpha^{SR}(\underline{r})$. Accordingly, $U_\alpha(\underline{r})$ is decomposed as follows

$$U_\alpha(\underline{r}) = V_\alpha^c(\underline{r}) + V_\alpha^{SR}(\underline{r}) - \mu_\alpha = \frac{Z_\alpha e}{\epsilon}\int_S d^2s\,\frac{\omega(\underline{s})}{|\underline{r}-\underline{s}|} + W_\alpha(\underline{r}). \qquad (2.13)$$

All Coulomb interactions are devided by ϵ, the dielectric constant of a background, simulating the solvent of the RPM electrolyte. With the total charge density

$$q(\underline{r}) = \sum_\alpha Z_\alpha e\,\rho_\alpha(\underline{r}) \qquad (2.14)$$

and the mean potential

$$\psi(\underline{r}) = \psi_q(\underline{r}) + \psi_\omega(\underline{r}) = \frac{1}{\epsilon}\int d^3r'\,\frac{q(\underline{r}')}{|\underline{r}-\underline{r}'|} + \frac{1}{\epsilon}\int_S d^2s\,\frac{\omega(\underline{s})}{|\underline{r}-\underline{s}|} \qquad (2.15)$$

the equilibrium condition (2.11) is finally transformed into

$$\ln\left[\frac{\rho_\alpha(\underline{r})}{\rho_\alpha^i(\underline{r})}\right] = -\frac{Z_\alpha e}{kT}[\psi(\underline{r}) - \psi^i(\underline{r})] - \frac{1}{kT}[W_\alpha(\underline{r}) - W_\alpha^i(\underline{r})]$$
$$+ \sum_\beta \int d^3r'\,[\rho_\beta(\underline{r}') - \rho_\beta^i(\underline{r}')]\int_0^1 d\lambda\, c_{\alpha\beta}^{SR}(\underline{r},\underline{r}';[\rho^\lambda]). \qquad (2.16)$$

Equation (2.10) yields correspondingly

$$\Omega_U[\rho] = \Omega_U^i[\rho^i] + \sum_\alpha \int d^3r\,\rho_\alpha(\underline{r})[W_\alpha(\underline{r}) - W_\alpha^i(\underline{r})] +$$
$$+ \int d^3r\,q(\underline{r})[\psi_\omega(\underline{r}) - \psi_\omega^i(\underline{r})] + \frac{1}{2}\int d^3r\,[q(\underline{r}) - q^i(\underline{r})][\psi_q(\underline{r}) - \psi_q^i(\underline{r})]$$
$$+ kT\sum_\alpha \int d^3r\,\rho_\alpha(\underline{r})\ln\left[\frac{\rho_\alpha(\underline{r})}{\rho_\alpha^i(\underline{r})}\right] - kT\sum_\alpha \int d^3r\,[\rho_\alpha(\underline{r}) - \rho_\alpha^i(\underline{r})]$$
$$- kT\sum_{\alpha,\beta}\iint d^3r\,d^3r'\,[\rho_\alpha(\underline{r}) - \rho_\alpha^i(\underline{r})][\rho_\beta(\underline{r}') - \rho_\beta^i(\underline{r}')] \times \qquad (2.17)$$
$$\times \int_0^1 d\lambda \int_0^\lambda d\lambda'\,c_{\alpha\beta}^{SR}(\underline{r},\underline{r}';[\rho^{\lambda'}]).$$

The exact integral equations (2.16) allow the calculation of the ionic density profiles $\rho_\alpha(r)$, provided the short range direct correlation functions $c_{\alpha\beta}^{SR}(r,r';[\rho^\lambda])$ are known for all $0 \leq \lambda \leq 1$.

3. CORRELATION FUNCTIONS FOR THE IONIC SOLUTION WITHIN THE DOUBLE LAYER AND CLOSURE CONDITIONS

Another exact result of density functional theory of liquid mixtures are the Ornstein-Zernike-Equations, which in the inhomogeneous region of the double layer take the form

$$h_{\alpha\beta}(r,r';[\rho^\lambda]) - c_{\alpha\beta}(r,r';[\rho^\lambda]) =$$

$$= \sum_\gamma \int d^3 r'' \, h_{\alpha\gamma}(r,r'';[\rho^\lambda]) \rho_\gamma(r'';\lambda) \, c_{\gamma\beta}(r'',r';[\rho^\lambda]) \, . \tag{3.1}$$

$h_{\alpha\beta}(r,r';[\rho^\lambda])$ are the total correlation functions between the ions. They are functions of position r (of ion α) and position r' (of ion β) and functionals of the ionic density distribution. A derivation of (3.1) may be found e.g. in /38/.

In order to have a solvable set of equations, (3.1) must be completed by <u>closure conditions</u>. We take the HNC-approximation (hypernetted chain)

$$1 + h_{\alpha\beta}(r,r';[\rho^\lambda]) = \tag{3.2}$$

$$= \exp\left\{ -\frac{1}{kT} V_{\alpha\beta}(r,r') + h_{\alpha\beta}(r,r';[\rho^\lambda]) - c_{\alpha\beta}(r,r';[\rho^\lambda]) \right\}$$

or the MSA-approximation (mean spherical), a linearisation of (3.2) for hard core potentials:

$$c_{\alpha\beta}(r,r';[\rho^\lambda]) = -\frac{1}{kT} V_{\alpha\beta}(r,r') \quad \text{for} \quad |r - r'| > \frac{1}{2}(\sigma_\alpha + \sigma_\beta) \,,$$
$$h_{\alpha\beta}(r,r';[\rho^\lambda]) = -1 \quad \text{for} \quad |r - r'| < \frac{1}{2}(\sigma_\alpha + \sigma_\beta) \,. \tag{3.3}$$

$V_{\alpha\beta}(r,r')$ is the potential energy for the ion-ion interaction. It contains a short range $V_{\alpha\beta}^{SR}(r,r')$ and a Coulomb $V_{\alpha\beta}^{C}(r,r')$ contribution and simplifies for the PM-electrolyte to

$$V_{\alpha\beta}(r,r') = V_{\alpha\beta}^{SR}(|r-r'|) + \frac{Z_\alpha Z_\beta e^2}{\epsilon |r-r'|} \tag{3.4}$$

with

$$V_{\alpha\beta}^{SR}(|r-r'|) = \begin{cases} \infty & \text{for } |r-r'| < \frac{1}{2}(\sigma_\alpha + \sigma_\beta) \,, \\ 0 & \text{for } |r-r'| > \frac{1}{2}(\sigma_\alpha + \sigma_\beta) \,, \end{cases} \tag{3.5}$$

where σ_α is the diameter of the hard spherical ion α with charge $Z_\alpha e$.

The equs. (2.11) [or (2.16)] for the ion densities and (3.1), with (3.2) or (3.3) for the ion-ion correlations form a closed and coupled set of integral equations for known density distributions $\rho_\alpha^i(\underline{r})$ of the chosen initial equilibrium state, but no one so far has solved these. In a homogeneous and isotropic ionic solution, however, the densities $[\rho^\lambda] = \{\bar{\rho}_\alpha\}$ are spatially constant and the correlation functions depend only on the distance $|\underline{r}-\underline{r}'|$ of the ions. In this case the densities uncouple from the OZ-equations (3.1) and the correlations $h_{\alpha\beta}(|\underline{r}-\underline{r}'|;\{\bar{\rho}_\gamma\})$ and $c_{\alpha\beta}(|\underline{r}-\underline{r}'|;\{\bar{\rho}_\gamma\})$ become functions of the constant densities. For any arbitrary choice of $\bar{\rho}_\alpha$, in general for electrically non-neutral compositions, the OZ-equations together with their closure conditions can then be solved. For MSA-closure an analytic solution is known /33/, whereas HNC-closure requires numerical solution. These observations allow the introduction of a <u>local concept</u> whose physical content will be substantiated in the sequel.

4. RPM-ELECTROLYTE IN FRONT OF A PLANE ELECTRODE

Equations (2.16) and (2.17) are exact and general. We pass now to an infinite plane hard wall at $z = 0$ with a constant surface charge density ω and no induced polarisation, i.e. there are no image forces caused by the interface. The ions have hard cores of equal diameter σ (RPM).

The initial state $[\rho^i]$ in (2.16) shall now be identified with the constant densities $\{\rho_\alpha^B\}$ of the neutral bulk electrolyte. This requires setting $v_\alpha^{SR,i}(\underline{r}) \equiv 0$, $q^i(\underline{r}) \equiv 0$ and $\omega^i(\underline{r}) \equiv 0$ for the initially homogeneous state. For the translationally invariant double layer, (2.16) simplifies then to ; $\underline{r} = (o,o,z)$:

$$\ln\left[\frac{\rho_\alpha(z)}{\rho_\alpha^B}\right] = -\frac{Z_\alpha e}{kT}\psi(z) + \frac{1}{kT}(\mu_\alpha - \mu_\alpha^B) +$$

$$+ \sum_\beta \int d^3r'[\rho_\beta(z') - \rho_\beta^B]\int_0^1 d\lambda\, c_{\alpha\beta}^{SR}(\underline{r},\underline{r}';[\rho^\lambda]), \quad z \geq \frac{\sigma}{2} ;$$

$$\rho_\alpha(z) \equiv 0, \quad z < \frac{\sigma}{2} ; \qquad (4.1)$$

where $c_{\alpha\beta}^{SR}(\underline{r},\underline{r}';[\rho^\lambda]) = c_{\alpha\beta}^{SR}(z,z',R';[\rho^\lambda])$ with $R' = \sqrt{x'^2 + y'^2}$ and integration must be performed over all $\underline{r}'$-space.

In passing to the infinite plane surface each integral in the mean potential (2.15) is divergent. Only the exact compensation of the surface charge density by the electrolyte screening charge per unit area (overall charge neutrality)

$$\omega + \int_0^\infty dz\, q(z) = 0 \qquad (4.2)$$

leads to a finite potential

$$\psi(z) = \bar{\psi}(z) + \psi_B = \begin{cases} \dfrac{4\pi}{\varepsilon}\int_z^\infty dz' \cdot (z-z')q(z') + \psi_B, & z \geq 0, \\[6pt] -\psi_B, & z \leq 0; \end{cases} \qquad (4.3)$$

$$\psi_B = \frac{2\pi}{\varepsilon} \int_0^\infty dz' \cdot z' q(z') \quad . \tag{4.4}$$

Remarkably, the potential in the bulk of the electrolyte does not go to zero. It takes a constant value ψ_B which depends on the external potential (on ω). In order to reach in (4.1) the bulk densities $\{\rho_\alpha^B\}$ as $z \to \infty$ we must therefore require

$$\mu_\alpha = \mu_\alpha^B + Z_\alpha e \psi_B := \eta_\alpha^B \quad . \tag{4.5}$$

The η_α^B are the electrochemical potentials in the bulk /39/. According to (2.3) for zero external potential the chemical potentials μ_α^B in the bulk are functions of T and $\{\rho_\alpha^B\}$:

$$\mu_\alpha^B(T,\{\rho_\gamma^B\}) = kT \ln(\Lambda_\alpha^3 \rho_\alpha^B) - kT\, c_\alpha(T,\{\rho_\gamma^B\}). \tag{4.6}$$

Consequently, charging the wall at fixed bulk densities and temperature makes it necessary to change the electrochemical potentials. This requires contacting the system with particle baths of changing chemical potentials μ_α.

Inserting (4.5) and (4.3) into (4.1) yield the final equations for the equilibrium density profiles in front of a plane electrode, $\underline{r} = (o,o,z)$:

$$\frac{\rho_\alpha(z)}{\rho_\alpha^B} = \exp\left\{-\frac{1}{kT}\left[Z_\alpha e\,\bar\psi(z) - kT \sum_\beta \int d^3 r'[\rho_\beta(z') - \rho_\beta^B] \int_0^1 d\lambda\, c_{\alpha\beta}^{SR}(\underline{r},\underline{r}';[\rho^\lambda])\right]\right\},$$
$$z \geq \frac{\sigma}{2} \; ;$$
$$\rho_\alpha(z) \equiv 0, \quad z < \frac{\sigma}{2} \quad . \tag{4.7}$$

Here $\bar\psi(z)$ satisfies the one-dimensional Poisson-equation with the boundary conditions $\left.\frac{\partial\bar\psi(z)}{\partial z}\right|_{z=0} = -\frac{4\pi}{\varepsilon}\omega$, $\lim_{z\to\infty}\bar\psi(z) = 0$, such that overall charge neutrality is fulfilled. Integration is to be performed over all $\underline{r}'$-space.

The limit to an infinite planar electrode has previously been reached by blowing up a spherical particle of the system /2-4/. By this concept one introduces in principle another double layer at infinity which assures that the potential value at infinity stays at zero. Grimson & Rickayzen /40/ have introduced a Lagrangean parameter by which they forced the electrostatic potential to zero in the bulk. Our approach shows that the electrochemical potentials must be adjusted in order that the bulk densities can be fixed independent of the surface charge density ω. In a similar treatment of Coulomb systems due to Evans & Sluckin /41/ between uncharged planar liquid/vapour interfaces it was not necessary to distinguish between chemical and electrochemical potentials.

The variational functional (2.17) diverges when we pass from the finite system to the infinite one in front of the plane electrode. Instead we introduce the functional $w_U[\rho]: = \lim_{B\to\infty}(\Omega_U[\rho]/S)$ per unit area of the infinitely extended plane electrode. Proper assortion of long range terms yields then still a divergent contribution (independent of $[\rho]$) of the form:

$$-\frac{1}{2\varepsilon S}\int_S d^2 s\,\omega \int_S d^2 s'\,\frac{\omega}{|\underline{s}-\underline{s}'|}\xrightarrow[S\to\infty]{} -\frac{1}{2}\omega\psi_\omega(0) = -\frac{2\pi}{\varepsilon}\lim_{R\to\infty}\int_0^R dr \quad . \tag{4.8}$$

Grimson & Rickayzen /40/ proposed to add the interaction energy of the charges on the electrode (divided by ε) to the functional (2.17). This leads to an exact cancellation of (4.10) and to the result, $\mathbf{r} = (0,0,z)$:

$$w_U[\rho] = w_{UB}[\rho^B] - \sum_\alpha \int_{-\infty}^\infty dz \rho_\alpha(z)(\mu_\alpha - V_\alpha^{SR}(z) - \mu_\alpha^B) + \frac{1}{2}\int_{\sigma/2}^\infty dz q(z) \bar\psi(z) + \frac{1}{2}\omega\bar\psi(0)$$

$$+ kT \sum_\alpha \int_{-\infty}^\infty dz \left\{\rho_\alpha(z) \ln\left[\frac{\rho_\alpha(z)}{\rho_\alpha^B}\right] - [\rho_\alpha(z) - \rho_\alpha^B]\right\}$$

$$- kT \sum_{\alpha,\beta} \int_{-\infty}^\infty dz \int d^3r'[\rho_\alpha(z) - \rho_\alpha^B][\rho_\beta(z') - \rho_\beta^B] \times \qquad (4.9)$$

$$\times \int_0^1 d\lambda \int_0^\lambda d\lambda' c_{\alpha\beta}^{SR}(\underline{r},\underline{r}';[\rho^{\lambda'}]) .$$

(4.9) is the exact one-dimensional formulation of the variational density functional in front of an infinite uniformly charged electrode. Though (4.9) still contains divergent contributions, minimizing $w_U[\rho]$ with respect to $\rho_\alpha(z)$ for fixed $\rho_\alpha^B(z)$, μ_α and ω, T leads with (4.5) to the equilibrium relations (4.7).

5. THE LOCAL HNC/MSA- AND THE LOCAL HNC/HNC-APPROXIMATION

As already mentioned, the double layer equations (4.7) uncouple from the OZ-equations (3.1) and the closure conditions (3.2) or (3.3) for the ion-ion correlation functions, when the ion densities in (3.1) are fixed. Previously constant bulk densities were choosen in (3.1) leading to (1.2) with the shortcomings mentioned in the introduction. The main reason is that the correlations of the bulk liquid, where one ion is surrounded by a shell of ions of the opposite charge, are transported into the inhomogeneous region of the double layer near the charged wall, where a counterion is almost completely surrounded by ions of the same charge. This fact changes drastically the ion-ion correlation functions near the electrode compared to those in the bulk.

Using the ion-ion correlation functions $c_{\alpha\beta}(|\underline{r}-\underline{r}'|;\{\rho_\alpha^B\})$ in (4.7) instead of the correct ones yields the standard HNC/MSA- or the standard HNC/HNC equations for ion densities in the double layer, and the notation /MSA or /HNC is used to indicate the closure conditions for the OZ-equations. These names stem historically from the cluster expansion method, by which the standard equations have been derived first. We use these names also in connection which our local concept for the ionic correlation functions in the inhomogeneous double layer and denote the corresponding equs. (4.7) then the local HNC/MSA- or local HNC/HNC-approximations for the ion densities.

The correct cylinder symmetric correlation functions $c_{\alpha\beta}^{SR}(z,z',R;[\rho^\lambda])$ in (4.7) are presently unknown; therefore the λ-integration cannot be performed. However, an improvement over using the bulk correlations can be achieved by using correlation functions $c_{\alpha\beta}^{SR}(|\underline{r}-\underline{r}'|;\{\bar\rho_\gamma(z)\})$ of homogeneous systems with non-neutral electrolyte composition $\{\bar\rho_\gamma(z)\}$ as found locally in front of the charged electrode. Hence, we replace

$$\int_0^1 d\lambda c_{\alpha\beta}^{SR}(\underline{r},\underline{r}';[\rho^\lambda]) \simeq c_{\alpha\beta}^{SR}(|\underline{r}-\underline{r}'|;\{\bar\rho_\gamma(z)\}) . \qquad (5.1)$$

It is immediately clear, that $\bar{\rho}_\alpha(z)$, the composition at z, cannot be identified with the probability densities $\rho_\alpha(z)$ for finding a number of ionic centers of sort α per unit volume around z. These can take values above 100 M (Mol per liter) at the electrode, whereas close packing of spherical ions with equal diameter is already reached for a 30.59 M system for our choice of ion diameter σ = 4.25 Å. Instead of the probability densities we define local particle densities $\hat{\rho}_\alpha(z)$ by averaging over a layer thickness of one ion diameter

$$\hat{\rho}_\alpha(z) = \begin{cases} \dfrac{1}{\sigma} \int\limits_{z-\sigma/2}^{z+\sigma/2} dx\, \rho_\alpha(x) & \text{for } z \geq \dfrac{\sigma}{2}, \\[2ex] \dfrac{1}{z+\sigma/2} \int\limits_{0}^{z+\sigma/2} dx\, \rho_\alpha(x) & \text{for } z \leq \dfrac{\sigma}{2}. \end{cases} \qquad (5.2)$$

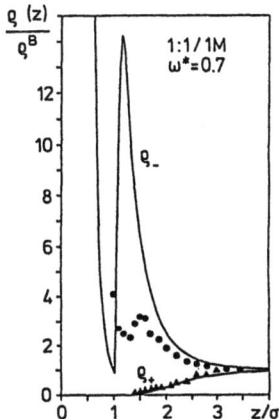

Fig.1. Ion density $\rho_\alpha(z)$ as calculated with the local concept from the particle densities $\hat{\rho}_\alpha(z)$.
●●▲▲ MC data from /26/

This gives continuous profiles for the local particle densities with reasonable contact values at the electrode well below the value of close packing, and the profiles reach for z→∞ the bulk densities ρ_α^B.

Identifying the local composition $\bar{\rho}_\alpha(z)$ with the local particle densities $\hat{\rho}_\alpha(z)$ we can solve the local HNC/MSA- or HNC/HNC-equations for the probability densities $\rho_\alpha(z)$. For 1:1 electrolytes at high electrode charge $\omega^* = \omega\sigma^2/e = 0.7$ ($\omega = 0.62$ C m^{-2}), for instance, the result is drawn in Fig. 1. It yields a layering of counterions in a second layer. The standard

HNC/MSA- or HNC/HNC theories using the bulk correlations do not predict this layering (Fig. 2). Hence, the appearance of counterion packing in a second layer must be considered as the great success of the local concept, although the layering is largely exaggerated in comparison to the MC results.

The reason for this exaggeration is due to the identification of the local composition $\bar{\rho}_\alpha(z)$ with the local particle densities $\hat{\rho}(z)$ corresponding to the final density profiles $\rho_\alpha(z)$. This leads to too strong effective correlations $c_{\alpha\beta}^{SR}(|\underline{r}-\underline{r}'|;\{\hat{\rho}_\gamma(z)\})$ and obviously to too much structure in $\hat{\rho}_\alpha(z)$. The effective correlations $c_{\alpha\beta}^{SR}(|\underline{r}-\underline{r}'|;\{\bar{\rho}_\gamma(z)\})$ in (5.1), however, correspond to summation of all correlation functionals $c_{\alpha\beta}^{SR}(\underline{r},\underline{r}';[\rho^\lambda])$ over the complete linear density paths $\rho_\alpha(x;\lambda) = \rho_\alpha^B + \lambda[\rho_\alpha(x) - \rho_\alpha^B]$ between the bulk densities ρ_α^B and the final densities $\rho_\alpha(x)$ about some neighbourhood $N(z)$, x $N(z)$, of the distance z from the electrode. The local composition $\bar{\rho}_\alpha(z)$, as a measure of the effective correlations, must therefore be identified with some intermediate density profile with less structure as the particle density $\hat{\rho}_\alpha(z)$. The local composition $\bar{\rho}_\alpha(z)$ may thus be defined by smoothening $\hat{\rho}_\alpha(x)$ about some neighbourhood x $N(z)$:

$$\bar{\rho}_\alpha(z) = \begin{cases} \frac{1}{2\Delta} \int_{z-\Delta}^{z+\Delta} dx\, \hat{\rho}_\alpha(x) & \text{for } z \geq \Delta, \\[2ex] \frac{1}{z+\Delta} \int_0^{z+\Delta} dx\, \hat{\rho}_\alpha(x) & \text{for } 0 \leq z \leq \Delta. \end{cases} \quad (5.3)$$

This simple definition avoids intermediate density profiles $\rho_\alpha(z;\lambda)$, whose calculation is not possible at present due to limitations of computer time and space. Moreover, it yields continuous profiles $\bar{\rho}_\alpha(z)$ for the local composition and it accounts in the right way for the effective correlations, that are needed in the convolution integral in (4.7) for the short range contribution of the potential of mean force, c.p. /9/.

The density averaging range Δ is a parameter of the theory. It may depend on the input parameters ρ_α^B, T, ε, σ, ω of the double-layer problem. By dimensional analysis of the basic equations together with the boundary conditions it can be proven for all two-component RPM-mixtures, that the dimensionless parameter Δ^*: $= \Delta/\sigma$ can depend at most on the dimensionless plasma energy E_p^*: $= 4\pi e^2/(\varepsilon kT\sigma)$, on the dimensionless electrode charge $\omega^* = \omega\sigma^2/e$ and on the dimensionless Debye-Hückel screening length L^*: $= L/\sigma$, where

$$L = \left[\frac{4\pi e^2}{\varepsilon kT} \sum_\alpha z_\alpha^2 \rho_\alpha^B\right]^{-1/2}. \quad (5.4)$$

On physical grounds we expect the density averaging range Δ^* to be in the order of magnitude of the Debye-Hückel screening length L^*; hence

$$\Delta^* = A(E_p^*,\omega^*,L^*) \cdot L^* \quad (5.5)$$

with an unknown function $A(E_p^*,\omega^*,L^*)$. This is obtained by a fit of the diffuse layer potential $\bar{\psi}(\sigma/2)$ to the corresponding MC-value for the highest available surface charge. It turns out that Δ^* is independent of the surface charge.

We have found the following values:

1:1 electrolytes: $\Delta^* = 1.0$ (2M), $\Delta^* = 1.5$ (1M), $\Delta^* = 6.5$ (0.1M)
2:1 electrolytes: $\Delta^* = 2.0$ (0.5M), $\Delta^* = 5.0$ (0.05M)
2:2 electrolytes: $\Delta^* = 1.5$ (0.5M), $\Delta^* = 4.0$ (0.05M) (5.6)

for $\sigma = 4.25$ Å, $\varepsilon = 78.5$, T = 298 K.
These scale approximately linear with L^*. Consequently $A = A(E_p^*)$ is at most a <u>universal</u> function of the plasma energy for all two-component RPM-electrolytes. For constant plasma energy then one single fit suffices to determine A. Using $\Delta^* = 5.0$ for a 2:1/0.05M-electrolyte (with $\sigma = 4.25$ Å, $\varepsilon = 78.5$, T = 298 K) we obtain the value

$$A = 2.8 \quad \text{for } E_p^* = 21.11 \ . \tag{5.7}$$

This empirical scaling law, (5.5) and (5.7), makes the local approximation scheme powerful, because it allows the prediction of Δ for other RPM-electrolytes at different electrolyte concentrations ρ_α^B, different dielectric constant ε, different temperature T and different ionic diameter σ, provided $E_p^* = 21.11 = $ const. MC-simulations for different E_p^* are needed to determine $A(E_p^*)$.

6. THE NUMERICAL PROCEDURE

Inserting (5.1) into (4.7) the integration parallel to the wall can be performed, such that

$$\rho_\alpha(z) = \rho_\alpha^B \exp\left\{-\frac{Z_\alpha e}{kT}\psi(z) + \sum_\beta \int_{-\infty}^{\infty} dz' C_{\alpha\beta}(|z-z'|; \{\bar{\rho}_\gamma(z)\}) \cdot [\rho_\beta(z') - \rho_\beta^B]\right\}, \quad z \geq \frac{\sigma}{2}$$

$$\rho_\alpha(z) \equiv 0, \quad z < \frac{\sigma}{2} ; \tag{6.1}$$

where with $\underline{r} = (0,0,z)$

$$C_{\alpha\beta}(|z-z'|; \{\bar{\rho}_\gamma(z)\}) = \int\int_{-\infty}^{\infty} dx' dy' c_{\alpha\beta}^{SR}(|\underline{r}-\underline{r}'|; \{\bar{\rho}_\gamma(z)\})$$

$$= 2\pi \int_{|z-z'|}^{\infty} dt \cdot t \, c_{\alpha\beta}^{SR}(t; \{\bar{\rho}_\gamma(z)\}) . \tag{6.2}$$

We solve (6.1) iteratively for two-component 1:1-, 2:1- and 2:2 electrolytes.

For the numerical solution of the local HNC/MSA- and the local HNC/HNC equations we proceed as follows. We define a grid of densities

$$\bar{\rho}_+ = \rho_+^1, \rho_+^2, \ldots, \rho_+^n$$

$$\bar{\rho}_- = \rho_-^1, \rho_-^2, \ldots, \rho_-^m \tag{6.3}$$

Chosen values for the density grid lie between zero and well above the largest particle densities that appear in the calculations.

We solve the OZ-equation together with MSA- or HNC-closure for homogeneous but non-neutral systems for the pairs of densities in the grid. With MSA-closure we use the analytically known solution of Parrinello & Tosi /33/;

details in /9/. With HNC-closure the solution is obtained numerically by iteration using the mixing procedure of Ng /43/; details in /11/. The outputs are short range direct correlation functions $c_{\alpha\beta}^{SR}(r;\rho_+^k,\rho_-^l)$ from which we construct the integrated versions $C_{\alpha\beta}(|z-z'|;\rho_+^k,\rho_-^l)$ according to (6.2). We perform the calculation of $C_{\alpha\beta}(|z-z'|;\rho_+^k,\rho_-^l)$ once and for all and store the results in a matrix indexed by the density grid.

Now, the iteration procedure to solve Eqs. (6.1) starts and we proceed as follows: We begin solving the MGC-eqations to get the input densities $\rho_\alpha(z)$. We evaluate $\bar\rho_\alpha(z)$ from (5.2) and (5.3). For each z, we select $C_{\alpha\beta}(|z-z'|;\bar\rho_+(z),\bar\rho_-(z))$ by linear interpolation between the grid points of the stored matrix $C_{\alpha\beta}(|z-z'|;\rho_+^k,\rho_-^l)$ and determine the convolution integrals in (6.1). The mean potential $\bar\psi(z)$ is evaluated from the one-dimensional Poisson-equation

$$\frac{\partial^2 \bar\psi(z)}{\partial z^2} = -\frac{4\pi}{\epsilon} \sum_\alpha Z_\alpha e \rho_\alpha(z) \tag{6.4}$$

by a Runge-Kutta method satisfying the boundary conditions

$$\left.\frac{\partial \bar\psi(z)}{\partial z}\right|_{z=0} = -\frac{4\pi}{\epsilon}\omega, \quad \lim_{z\to\infty} \bar\psi(z) = 0. \tag{6.5}$$

This is added to the convolution integrals in (6.1). New values of $\rho_\alpha(z)$ are obtained from (6.1) and they become the input for the next iteration step.

This algorithm is quickly performed and after around ten iterations we reach convergence of:

$$\left\{\frac{1}{10L}\sum_\alpha \int_0^{10L} dz \left|\frac{\rho_\alpha^{out}(z)}{\rho_\alpha^B} - \frac{\rho_\alpha^{in}(z)}{\rho_\alpha^B}\right|^2\right\}^{1/2} < 10^{-5}$$

in which L is the Debye-Hückel screening length and 10L is the maximum distance from the wall up to which all functions have been calculated.

The calculations for the weakly correlated 1:1 electrolytes have been performed with the simpler MSA-closure (control with HNC-closure for the 1M case gave practically the same results). The stronger correlated 2:1- and 2:2-electrolytes, however, required HNC-closure.

The total charge neutrality (4.2) is satisfied within 2 %. The contact value relation /4/

$$\sum_\alpha [\rho_\alpha(\frac{\sigma}{2}) - \rho_\alpha^B] = \frac{2\pi\omega^2}{kT\epsilon} - 2\pi\sum_{\alpha\beta}\rho_\alpha^B\rho_\beta^B \int_0^\infty dt \cdot t^2 c_{\alpha\beta}^{SR}(t;\rho_+^B,\rho_-^B)$$

has for the calculated contact densities $\rho_\alpha(\frac{\sigma}{2})$ a deviation less than 5 %.

7. RESULTS

We apply the local method to calculate several aspects of the double layer structure for 1:1-, 2:1- and 2:2-electrolytes in the RPM. The distinct differences between the symmetric 1:1- and 2:2-electrolytes can be shown best by a direct comparison, whereas the different characteristics of the asymmetric 2:1-electrolyte are discussed separately.

7.1 Packing and pairing of ions in the double layer

In Fig. 2 are shown the density profiles $\rho_\alpha(z)$ for a 1:1/1M-electrolyte at high electrode charge $\omega^* = \omega\sigma^2/e = 0.7$ ($\omega = 0.62$ C m^{-2}) and for a 2:2/0.5M-electrolyte for intermediate charge $\omega^* = 0.24$ ($\omega = 0.21$ C m^{-2}). Due to weak ion-ion correlations of the monovalent ions in the 1:1 electrolyte the coions are repelled from the charged wall, whereas the attracted counterions form a second layer as a result of competition between Coulomb attraction to the wall and Coulomb- and hard core repulsion among themselves. The situation is different for the 2:2 electrolytes. The strongly

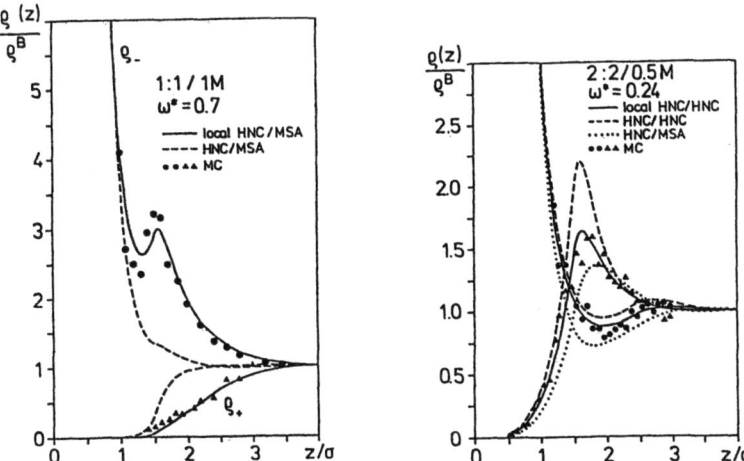

Fig. 2. Packing and pairing of ions. Density profiles $\rho_\alpha(z)$ in front of a charged electrode at $z = 0$. Weak correlations in the 1:1-electrolyte favour counterion packing in a second layer, whereas for the stronger correlated 2:2-electrolytes ion-pair formation results in a coion-layer close to the electrode. MC-data from /26,29/.

correlated divalent ions have due to Coulomb attraction between ions of different charge a stronger tendency to form pairs. This pairing is not destroyed in the competition between wall-attraction of counterions and wall-repulsion of coins. As a result a first layer of coins is formed around the position $z = 1.5\sigma$ of the second particle layer. The tendency of layer formation is reduced when the electrode charge is decreased, Fig. 3.

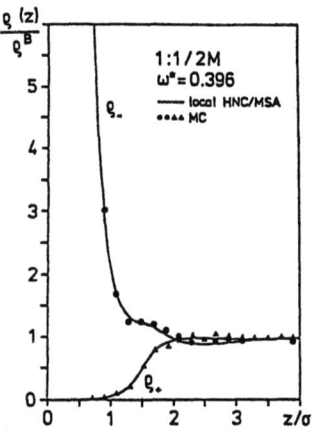
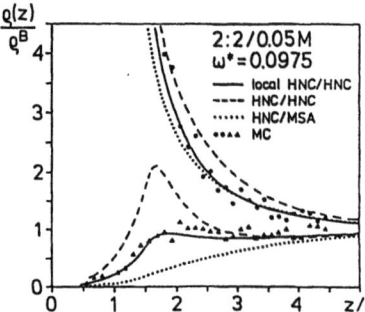

Fig. 3. Reduced tendency to packing and pairing at smaller surface charge. MC-data from /26,29/.

The results of the <u>local</u> HNC/MSA-calculations for the 1:1-electrolytes and of the <u>local</u> HNC/HNC-calculations for the 2:2-electrolytes are in very good agreement with MC simulations. The reason for this is the use of local correlation functions (details in /9/), whose dependence on the local composition provides approximately the right ionic surroundings in the inhomogeneous region in front of the electrode. The standard HNC/MSA- and HNC/HNC-calculations fail to describe the right ionic structure near the wall.

7.2 Charge dependence of the mean electrostatic potential

In Fig. 4 the diffuse layer potential $\psi^*(\sigma/2)$ is drawn versus the reduced surface charge ω^* of the electrode:

$$\psi^* \left(\frac{\sigma}{2}\right) = e \, (kT)^{-1} \, \bar{\psi} \left(\frac{\sigma}{2}\right), \quad \omega^* = \omega \frac{\sigma^2}{e} \, . \tag{7.1}$$

We have evaluated $\psi^*(\sigma/2)$, as function of ω^* for different bulk concentrations and compared the results to several previous calculations and to MC data. The strong increase of the potential with increasing surface charge for the 1:1-electrolytes has its origin in the counterion-layer formation, which increases the dipole moment of charge distribution in front of the electrode, c.p. Equs. (4.3), (2.14) and Figs. 3,2 (l.h.s.). For 2:2-electrolytes the diffuse layer potential behaves differently in the ω^* range considered. After an increase with charging the wall it passes through a maximum and decreases at intermediate electrode charge. This behaviour has its origin in the formation of a coion-layer with charging, which neutralizes the charge distribution already for small distances from the wall and decreases its dipole moment, c.p. Figs. 3,2 (r.h.s.). Further increase of the surface charge ω^* produces a steep increase of the diffuse layer potential, since then a second counterion layer develops, see /10/.

For the 1:1-electrolyte the local HNC/MSA- and the local HNC/HNC-calculations give approximately the same results and do well reproduce the available MC-data. For 2:2-electrolytes only the local HNC/HNC-calculations are in agreement with MC-results. The standard approximations like HNC/MSA, HNC/HNC and MGC fail, see Fig. 4.

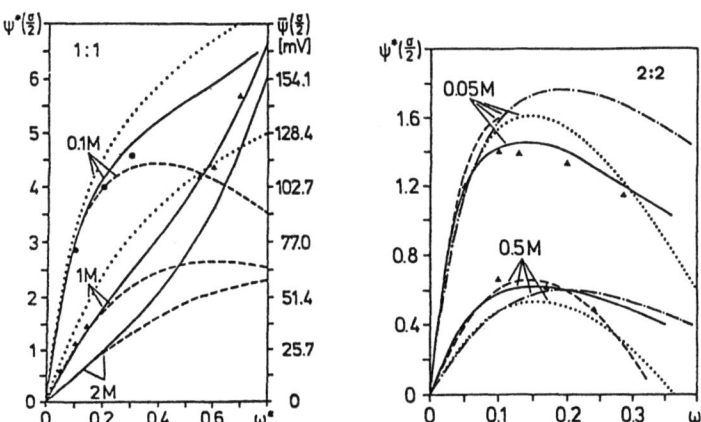

Fig. 4. Diffuse layer potential $\psi^*(\sigma/2)$ versus reduced electrode charge ω^* for 1:1- and 2:2-electrolytes at different bulk concentrations. Parameters $\sigma = 4.25$ Å, $\varepsilon = 78.5$, $T = 298$ K.

Left: —— local HNC/MSA
--- HNC/MSA
··· MGC
●●▲▲ MC data from /26/

Right: —— local HNC/HNC
-·- local HNC/MSA
--- HNC/HNC
··· HNC/MSA
▲▲▲ MC data from /29/

7.3 Charge dependence of the double layer capacitance

The differential capacitance C (per unit area) of the electrical double layer at a charged electrode is often split into an inner part C_I and a diffuse layer contribution C_D:

$$C^{-1} = \frac{\partial \bar{\psi}(0)}{\partial \omega} = \frac{\partial \bar{\psi}(\sigma/2)}{\partial \omega} + \frac{4\pi}{\epsilon}\frac{\sigma}{2} = C_D^{-1} + C_I^{-1} \ . \tag{7.2}$$

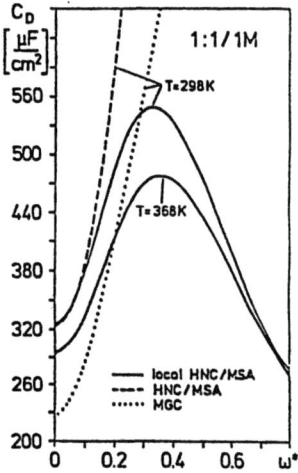

Fig. 5. Surface-charge dependency of diffuse layer capacitance

We plot in Fig. 5 the diffuse layer capacitance C_D for 1:1-electrolytes. This is the inverse of the derivative of the curves in Fig. 4 (l.h.s.). It is more sensitive to the surface charge ω^* than C itself. Important is the maximum in C_D which is also seen in MC-data and is related to the onset of a second layer formation. All previous theories gave capacitances increasing infinitely. Though the RPM is admittedly a rough model, the maximum in the diffuse layer capacitance may be of relevance to measurements of C. The measured capacitances usually level off at higher charges and sometimes go through maxima and decrease. The explanation was generally sought in properties of the inner Helmholtz layer (of C_I) because former theories always predicted a monotonous strong increase of C_D. Our calculation reveals that structure in the diffuse layer can produce nontrivial dependence of C_D on ω.

Fig. 5 shows in addition the temperature dependence of C_D. The maximum decreases and shifts slightly by raising the temperature. Since there are no MC-data available, we have calculated Δ from (5.4) - (5.6), leaving ε unchanged, and used $\Delta(T = 368 \text{ K}) = 1.67 \sigma$.

7.4 Change of ionic bulk concentrations

With increasing bulk concentration the Debye-Hückel length decreases, the screening charge is getting condensed to very few layers and more structure in the ion cloud near the wall is the consequence. For instance, the charge inversion in the double layer of a 1:1/2M-system at $\omega^* = 0.396$, see Fig. 3 (l.h.s.) and /44,13,14/, contributes to the low surface potential $\psi^*(\sigma/2)$, see Fig. 4 (l.h.s.); the agreement of our local scheme with MC-data is good and it agrees also in this low charge regime with MPB5 results /44/. At higher surface charge no MC results are available for the 1:1/0.1M/2M-systems. We provide in Fig. 6 ion densities for these systems at the same charge $\omega^* = 0.7$ as was studied in Fig. 2 for the 1M case. Though the densities in the first layer are rather similar (e.g. $\bar{\rho}_{max}(\bar{\sigma}/2) = 111$M for 0.1M, 103M for 1M, 105M for 2M) there is no second layer maximum in the extended double layer of the 0.1M system, the layering is developed for 1M and is enhanced for the 2M case, where also

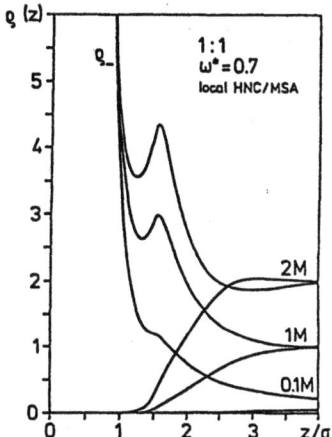

Fig. 6. Density profiles $\rho_\alpha(z)$ at different bulk concentrations

charge inversion develops. This means, that with increasing bulk density the dipole moment of the charge distribution q(z) in front of the wall is reduced and the diffuse layer potential is diminished, see Fig. 4 (l.h.s). It would be desirable to check these predictions by MC simulations.

7.5 The asymmetric 2:1 electrolyte

The asymmetric 2:1 electrolytes show both effects, ion packing when the counterions are monovalent, and ion pairing when the counterions are divalent. The diffuse layer potentials develop correspondingly. In Fig. 7 we show a summary of local HNC/HNC calculations, MC simulations and other approximate theories. More details may be found in /11/.

The monovalent ions are chosen to carry one negative elementary charge, the divalent ions are then doubly positively charged. For a positively charged electrode the monovalent ions are thus the counterions, at high surface charge they form a second layer (packing). The diffuse layer shows

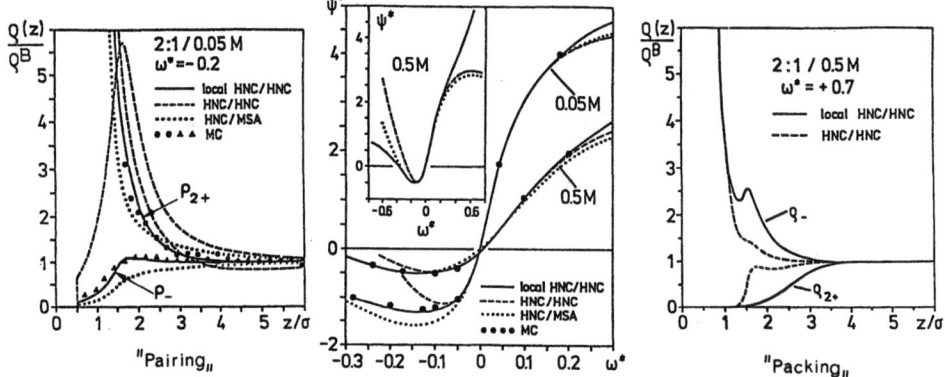

Fig. 7. Density profiles $\rho_\alpha(z)$ and diffuse layer potentials $\psi*(\sigma/2, \omega*)$ for 2:1-RPM-electrolytes. MC-data from /28,45/

a steep increase with surface charge, as for the 1:1 electrolyte. However, if the electrode is negatively charged, the divalent ions are the counterions; a reduced pairing effect builds then up with increasing absolute values of the surface charge. The absolute values of the diffuse layer potential increase with charging up to a certain charge and decrease with further charging, similar as for the 2:2 electrolyte. So, the 2:1 electrolytes embrace the features of both symmetric 1:1 and 2:2 cases.

At zero surface charge the local HNC/HNC as well as standard HNC/HNC give a non-zero potential of zero charge (pzc), Fig. 7. This effect is very small: $\psi*(\sigma/2) = -2 \cdot 10^{-2}$ for a bulk density $\rho_{2+}^B = 0.05M$ and develops to $\psi*(\sigma/2) = -7 \cdot 10^{-2}$ for $\rho_{2+}^B = 0.5M$ close to MC-data /45/. For unequal ion sizes one gets much larger pzc /46/ but it is interesting, that the <u>charge asymmetry</u> produces a non-zero pzc at all. The less careful treatment by standard HNC/MSA misses this effect. The pzc is negative, this fact is caused by a dipole layer with the negative end towards the wall. Why is it more than twice as probable to find negative ions near the wall than positive ions? We believe, that due to the tendency of ions to form locally neutral conformations, it is favourable that the doubly charged positive ion has its singly charged negative ions preferably at opposite sides. Such ion triplets near the wall will touch the electrode primarily with a negative ion. This angular correlation in the ion triplet is obviously roughly represented even in the two particle correlation functions which contain the information, that it is more probable for the negative ions to stay at distances related to opposite sides of the positive ion.

8. CONCLUSIONS

Density functional theory for inhomogeneous charged liquids leads to the idea, that the standard Ornstein-Zernike HNC/MSA- or HNC/HNC-integral equation methods can be improved by employing ion-ion correlations of homogeneous electrolytes with non-neutral compositions as found in the double-layer region in front of the electrode. We have developed these <u>local HNC/MSA-</u> or <u>local HNC/HNC</u> concepts and applied them to 1:1, 2:2-, and 2:1-electrolytes with equal size hard core ion-mixtures in a continuous background of constant dielectric constant, the RPM.

The calculated results on double-layer structure, diffuse layer potential and differential capacitance agree very well with Monte-Carlo simulations and are superior to all previous calculations, especially at high electrode charge. The reason is the better treatment of local correlations near the wall.

Comparing these local concepts among each other we have found, that the more elaborate HNC-approximation is necessary for a proper treatment of the strongly correlated 2:2- and 2:1-electrolytes. However, the weakly correlated 1:1-electrolytes allow also application of the analytically available MSA-approximation for the determination of local direct correlation functions.

Our local scheme is based on the concept of local coarse grained ionic densities, which represent the local electrolyte composition. This concept contains a parameter Δ, the density averaging range, which is determined by fitting of one value of the diffuse layer potential to its corresponding MC value. As an empirical law we find, that Δ scales linearly with the Debye-Hückel screening length. Consequently, only one single fit is required to cover all RPM-electrolytes for different ionic sizes and valences, different bulk concentrations, different temperatures and different dielectric constants. Thus, this local scheme is a powerful and quick scheme for the evaluation of the double layer structure, potential and related features.

We recognize the argument, that with the fitted averaging range we are bound to obtain better results than previous theories. Our defense are the many detailed improvements. Nevertheless, we are trying to overcome this single adjustment to MC-data by arguments of thermostatic consistency.

<u>Acknowledgment</u>

We acknowledge the financial support by the Deutsche Forschungsgemeinschaft under Sonderforschungsbereich 6.

References

/ 1/ S.L. Carnie and G.M. Torrie: Adv. Chem. Phys. 56, 141 (1984)
/ 2/ D. Henderson, F.F. Abraham and J.A. Barker: Mol. Phys. 31, 1291 (1976)
/ 3/ D. Henderson and L. Blum: J. Chem. Phys. 69, 5441 (1978)
/ 4/ S.L. Carnie, D.Y.C. Chan, D.J. Mitchell and B.W. Ninham: J. Chem. Phys. 74, 1472 (1981)
/ 5/ S. Levine and C.W. Outhwaite: J. Chem. Soc. Faraday Trans. II 74, 1670 (1978)
/ 6/ C.W. Outhwaite, L.B. Bhuiyan and S. Levine: J. Chem. Soc. Faraday Trans. II 76, 1388 (1980)
/ 7/ S. Levine, C.W. Outhwaite and L.B. Bhuiyan: J. Electroanal. Chem. 150, 291 (1981)
/ 8/ P. Nielaba and F. Forstmann: Chem. Phys. Lett. 117, 46 (1985)
/ 9/ T. Alts, P. Nielaba, B. D'Aguanno and F. Forstmann: Chem. Phys., in press
/10/ B. D'Aguanno, P. Nielaba, T. Alts and F. Forstmann: J. Chem. Phys., 85, 3476 (1986)
/11/ P. Nielaba, T. Alts, B. D'Aguanno and F. Forstmann: Phys. Rev. A34, 1505 (1986)
/12/ T.L. Croxton and D.A. McQuarrie: Mol. Phys. 42, 141 (1981)
/13/ C. Caccamo, G. Pizzimenti and L. Blum: Phys. Chem. Liq. 14, 311 (1985)
/14/ C. Caccamo, G. Pizzimenti and L. Blum: J. Chem. Phys. 84, 3327 (1986)
/15/ P. Ballone, G. Pastore and M.P. Tosi: J. Chem. Phys. 85, 2943 (1986)
/16/ S.L. Carnie and D.Y.C. Chan: J. Chem. Phys. 73, 2949 (1980)
/17/ L. Blum and D. Henderson: J. Chem. Phys. 74, 1902 (1981)
/18/ J.P. Badiali: Mol. Phys. 55, 939 (1985)
/19/ J.P. Badiali, M.L. Rosinberg and V. Russier: Mol. Phys. 56, 105 (1985)
/20/ J.P. Badiali, M.L. Rosinberg, F. Vericat and L. Blum: J. Electroanal. Chem. 158, 253 (1983)
/21/ W. Schmickler and D. Henderson: J. Chem. Phys. 80, 3381 (1984)
/22/ V.I. Feldman, A.A. Kornyshev and M.B. Partenskii: Solid State Commun. 53, 157 (1985)
/23/ G. Gouy: J. Phys. Radium 9, 457 (1910)
/24/ D.L. Chapman: Phil. Magazine 25, 475 (1913)
/25/ O. Stern: Z. Elektrochem. 30, 508 (1924)
/26/ G.M. Torrie and J.P. Valleau: J. Chem. Phys. 73, 5807 (1980)
/27/ G.M. Torrie, J.P. Valleau and G.N. Patey: J. Chem. Phys. 76, 4615 (1982)
/28/ G.M. Torrie, J.P. Valleau and C.W. Outhwaite: J. Chem. Phys. 81, 6296 (1984)
/29/ S.L. Carnie: Mol. Phys. 54, 509 (1985)
/30/ D. Henderson, L. Blum and W.R. Smith: Chem. Phys. Lett. 63, 381 (1979)
/31/ J.P. Valleau and L.K. Cohen: J. Chem. Phys. 72, 5935 (1980)
/32/ P.J. Rosski, J.B. Dudowicz, B.L. Tembe and H.L. Friedman: J. Chem. Phys. 73, 3372 (1980)
/33/ M. Parrinello and M.P. Tosi: Chem. Phys. Lett. 64, 579 (1979)
/34/ P. Tarazona: Phys. Rev. A31, 2672 (1985)
/35/ W.A. Curtin and N.W. Ashcroft: Phys. Rev. A32, 2909 (1985)
/36/ N.D. Mermin: Phys. Rev. 137, 1441 (1965)
/37/ W.F. Saam and C. Ebner: Phys. Rev. A15, 2566 (1977)
/38/ R. Balescu: "Equilibrium and Nonequilibrium Statistical Mechanics", John Wiley & Sons 1975
/39/ R. Haase: "Elektrochemie I - Thermodynamik electrochemischer Systeme", Grundzüge der Physikalischen Chemie in Einzeldarstellungen Bd. V (Ed. R. Haase), Dr. Dietrich Steinkopff Verlag 1972
/40/ M.J. Grimson and G. Rickayzen: Mol. Phys. 45, 221 (1982)
/41/ R. Evans and T.J. Sluckin: Mol. Phys. 40, 413 (1980)
/42/ J.R. Henderson and F. van Swol: Mol. Phys. 51, 991 (1984)
/43/ K.C. Ng: J. Chem. Phys. 61, 2680 (1974)

/44/ C.W. Outhwaite and L.B. Bhuiyan:
J. Chem. Soc. Faraday Trans. II **79**, 707 (1983)
/45/ G.M. Torrie and J.P. Valleau:
J. Phys. Chem. **86**, 3251 (1982)
/46/ U. Marini Bettolo Marconi, J. Wiechen and F. Forstmann:
Chem. Phys. Lett. **107**, 609 (1984)

PROPAGATION OF SOUND WAVES IN THE PRESENCE OF WEAK DENSITY INHOMOGENEITY:
APPLICATION TO INTERFACIAL PHENOMENA AND PHASE TRANSITIONS IN FLUIDS

J. Maza, F. Miguelez, A. Veira and F. Vidal

Laboratorio de Física de Materiales, Facultad de Física
Universidad de Santiago de Compostela
Spain

I. INTRODUCTION

Continuous or discontinuous phase transitions, interfacial phenomena, and in general any critical behavior of real systems, are always affected by the so-called "rounding" effects[1]. These effects, which mitigate and round the possible sharp variations or the critical divergences of the system, are caused, in particular, by the presence of impurities, strains, nonequilibrium situations, finite-sample effects or inhomogeneities. In the case of continuous phase transitions (as, for instance, the lambda transitions) these effects may be considered as nonintrinsic, because they will be absent in an ideal case. However, in other situations, as for instance in the case of interfacial phenomena or in the presence of discontinuous phase transitions, the inhomogeneities are intrinsic, and the associated rounding effects will be present even in an ideal sample.

Let us illustrate with an example how these rounding effects arise. Consider a fluid in the presence of the Earth's gravity. The gravity field will cause the pressure in the fluid to vary with height. As a consequence, each pressure-dependent parameter of the fluid is subject to a gradient. This is the case, for instance, for density, which is coupled to pressure through the compressibility of the system. In general, the coupling coefficients remain finite, and in a sample of reasonable depth (a few cm or less) the corresponding gradients are negligible. However, in the presence of critical phenomena, the coupling coefficients may suffer very sharp variations or even critical divergences. In such cases the spatial inhomogeneities associated with the gravitational field may be very important. So, any bulk measurement on this inhomogeneous medium will result in a kind of average ("rounding") over the local magnitudes[1,2].

Even near critical situations, these rounding effects are, in general, relatively small. However, mainly due to the very accurate theoretical and experimental results availables now, these nonintrinsic rounding mechanisms are one of the basic problems encountered at the present time in the study of critical phenomena. In addition, in general it is very difficult to separate or to estimate the magnitude of these effects, mainly in the case of time-dependent or transport critical parameters for which intrinsic rounding mechanisms, associated precisely with the time-dependent behavior, may also be present. In fact, in spite of its increasing importance, until now there exist very few quantitative studies of these nonintrinsic rounding contributions in any system. Recently[3,4], we have presented an analysis of the local and non-local rounding effects on (first) sound waves propagating near the lambda transition in liquid 4He. In that case, the rounding mechanism was due to the gravity-induced inhomogeneity in a sample of finite height, and a numerical procedure to separate the gravity contribution from the dispersion effect was proposed. In this paper, some of these results are generalized and extended to the case of the sound propagating in any system presenting weak inhomogeneity. Our work here is centered on the relations between the local, average and measured velocities. These relations play a crucial role in the understanding of the rounding effects on sound. In addition to its formal interest, the propagation of sound is one of the most useful probes for static and dynamic critical phenomena. So, the knowledge of these relations is of very general interest. Let us indicate, finally, that as the results presented here are valid in all cases with uniform or nonuniform weak inhomogeneity, they may easily be applied to continuous phase transitions as well as to discontinuous critical phenomena as, for instance, interfacial phenomena.

II. FRAMEWORK AND APPROXIMATIONS

This paper is centered on the case of standing sound waves propagating in a resonant cavity. This is so because most of the high-precision experiments on sound propagation in any system are performed with this method. However, some considerations about the pulse propagation technique will be also summarized here. In what concerns the standing waves propagation, we are going to consider two different experimental situations : the sound waves propagating either in the same direction as that of the inhomogeneity gradient or normally to it. These two pictures, which need two different treatments, correspond, for instance, to the sound propagating either parallely or normally to the gravity acceleration in a fluid submitted to the Earth's gravity.

In both situations, the effects of the inhomogeneity on the sound velocity (the sound attenuation in almost insensitive to weak inhomogeneity) will be obtained by calculating the eigenfrequencies of the corresponding wave equation of the medium. In each case, the wave equation will be resolved by using a perturbative technique similar, for instance, to that already used by Hohenberg and Barmatz in the analysis of the sound propagating near critical points in fluids subject to a parallel (to the sound propagation direction) gravity field[2,5]. To some extent, our work here systematize these results and also extend them to the very interesting situations in which the sound propagates normally to the inhomogeneity gradient.

In order to simplify the calculations, we suppose an "unidimensional" inhomogeneity, i.e., that the local properties (and, in particular, the density) of the sample change only in one direction, that we choose as z direction. Also we choose, arbitrarily, as z = 0 the position in the sample of lowest density. Note that, then, in the example of a fluid submitted to the Earth's gravity, this point will coincide with the top of the sample. Naturally, this choice of z = 0 should not have any influence on the final results. The total "height" of the sample in this z direction is noted H. We shall assume, also, the sample in thermodynamic equilibrium, characterized by a temperature T (independent of z, indeed) and a pressure P(z).

In this framework, our main problem is then to find the relation between the local velocity u(z) and the experimental velocity <u>. As we are going to see, both velocities can only be reasonably defined, in our case, through the wave equation (note that the experimentally accessible magnitude is just a resonant-frequency). Finally, to make possible the use of the perturbation approach to solve the wave equation, we must suppose that the inhomogeneity is weak, i.e.

$$\frac{u(z) - u(0)}{u(0)} \ll 1, \qquad (1)$$

for any z. In other words, we restrict our treatment to the case where $u(z)$ does not vary substantially in the sample.

III. SOUND PROPAGATION PARALLEL TO THE INHOMOGENEITY GRADIENT

We shall begin with the wave equation

$$\frac{d^2\Phi(z,t)}{dz^2} - \frac{1}{u^2(z)} \frac{d^2\Phi(z,t)}{dt^2} = 0, \qquad (2)$$

where $\Phi(z,t)$ is the wave function. We have written the unidimensional wave equation since in the usual experimental arrangement the sound is excited in wave-plane modes. The pertinent contour conditions are $d\Phi/dz = 0$ for $z = 0$ and $z = H$. Since we are interested in the eigenmodes of the cavity, which is assumed cylindrical, we write

$$\Phi(z,t) = f(z)\, e^{i\omega t}, \qquad (3)$$

where ω is the angular frequency. Our basic equation is then

$$\frac{d^2 f(z)}{dz^2} - \frac{\omega^2}{u^2(z)} f(z) = 0. \qquad (4)$$

It is useful to remember that for homogeneous media, i.e., u constant, the eigenfunctions consistent with the contour conditions are

$$f_o(z) = C \cos \frac{\pi p z}{H}, \qquad (5)$$

where C is an arbitrary constant and p is an integer positive number labeling the mode. The corresponding eigenfrequencies are

$$\omega_o = \frac{\pi p}{H} u_o. \qquad (6)$$

Note that we use the subscript o for the parameters of a homogeneous medium.

For an inhomogeneous medium, we define as in Ref.2 the "average velocity" by using an expression similar to that of Eq.(6), i.e.:

$$\omega = \frac{\pi p}{H} \langle u \rangle, \qquad (6')$$

where ω is now the measured resonant frequency in the inhomogeneous sample. At this point, we remember that our fundamental problem is now to find the relation between the velocity for a homogeneous sample and $\langle u \rangle$, this last defined by Eq.(6'). For the velocity of the homogeneous system we may choose, for instance, $u_o = u(o)$. It is obvious that the answer to this problem requires calculating the eigenfrequencies in Eq.(4). For weak inhomogeneous systems, this can be done by using a perturbative technique. Considering Eq.(1), we introduce

$$\delta(z) \equiv u(z) - u(o),$$

so,

$$\frac{\omega^2}{u^2(z)} \approx \frac{\omega^2}{u(o)} - \frac{2\omega_0^2}{u(o)^3} \delta(z), \qquad (7)$$

and Eq.(4) reduces to

$$[\frac{d^2}{dz^2} + \frac{2\omega_0^2}{u(o)^3} \delta(z)] f(z) = \frac{\omega^2}{u(o)^2} f(z). \qquad (8)$$

This perturbative spectral equation in the weak perturbation $[2 \omega_0^2/u(o)^3] \delta(z)$ posseses, as can be easily found, the new eigenvalues

$$\frac{\omega^2}{u(o)^2} = \frac{\omega_0^2}{u(o)^2} + \frac{4\omega_0^2}{u(o)H} \int_0^H \delta(z) \cos^2 \frac{\pi p z}{H} dz, \qquad (9)$$

where for the old eigenfunctions in Eq.(5) to be normalized a value $C = (2/H)^{\frac{1}{2}}$ has been taken. Introducing now the average velocity, Eq.(9) results in

$$<u> = u(o) + \frac{2}{H} \int_0^H \delta(z) \cos^2 \frac{\pi p z}{H} dz, \qquad (10)$$

valid in first order in $\delta(z)$. Making use of the defining equation for δ, we finally obtain,

$$<u> = \int_0^H u(z) [\frac{2}{H} \cos^2 \frac{\pi p z}{H}] dz. \qquad (11)$$

We see then that the experimental velocity is an average of the local velocity weighted by a modedependent "weight factor" verifying

$$\int_0^H (2/H)(\cos^2 \pi p z/H) dz = 1.$$

Note that the result is, as it should be, independent of the choice for $u(o)$. An example of the weight factor, which shows its influence on the sound propagation, has been plotted in Fig. 1. The dotted-dashed line

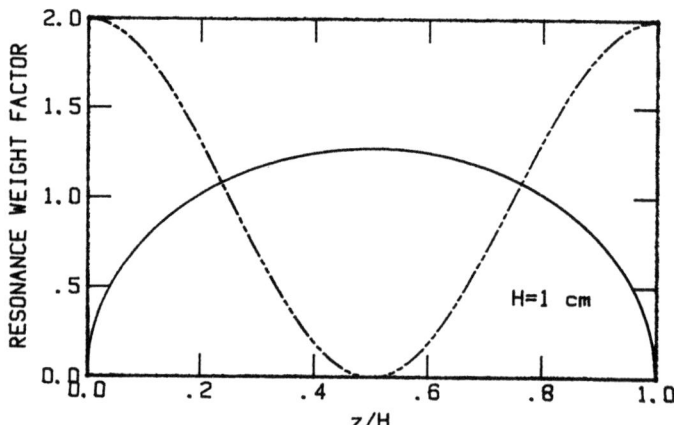

Fig.1. Examples of the "weight factor" acting on the local velocity to obtain the average velocity. For details see text.

corresponds to the case of a sound wave (the fundamental mode) propagating parallely to the inhomogeneity gradient. Note that in this case the weight factor favors the local velocity around the z extremes. We stress also that this behavior of the weight factor is independent of the uniformity or not of the inhomogeneity, provided that the condition given by Eq.(q) is satisfied. The solid line case will be commented on in part IV.

IV. SOUND PROPAGATION NORMAL TO THE INHOMOGENEITY GRADIENT

In this Section, the physical situation under study is that of sound propagation in a cylindrical cavity in which the medium presents an inhomogeneity perpendicular to the longitudinal axis. We shall use for geometrical specification cylindrical coordinates : ρ radial coordinate, ϕ azimuthal angle and y axial coordinate (so chosen to not be confused with z, in this paper standing for the coordinate along the inhomogeneity gradient). The length of the cylinder will be L and its radius R.

Since for this arrangement it is not possible the existence of only y-dependent modes, we begin with the complete wave equation for harmonic functions

$$\nabla^2 \Phi - \frac{\omega^2}{u^2(z)} \Phi = 0, \qquad (12)$$

where Φ is the wave function, which must satisfy the contour conditions $\nabla \Phi = 0$ for $\rho = R$ and for $y = 0, L$.

It is again useful to remember the basic solutions for a homogeneous medium for which $u(z) = u(0)$. The eigenfunctions are

$$f_0 = C' J_m \left(\frac{\pi \alpha_{mn}}{R} \rho \right) \cos m\phi \cos \frac{\pi p}{L} y, \qquad (13)$$

where C' is an arbitrary constant and m,n and p are nonnegative integers labeling the mode. J_m is the Bessel function of order m and α_{mn} discrete numbers satisfying

$$\frac{d J_m}{d\rho} \left(\frac{\pi \alpha_{mn}}{R} \rho \right) = 0, \qquad (14)$$

for $\rho = R$. The eigenfrequencies, in turn, can be expressed as

$$\omega_0^2 = u(0)^2 \left[\left(\frac{\pi \alpha_{mn}}{R} \right)^2 + \left(\frac{\pi p}{L} \right)^2 \right]. \qquad (15)$$

Following the same reasoning as in the preceeding Section, we begin by defining the experimental velocity <u> for inhomogeneous media as

$$\omega^2 = <u>^2 \left[\left(\frac{\pi \alpha_{mn}}{R} \right)^2 + \left(\frac{\pi p}{L} \right)^2 \right], \qquad (16)$$

ω being the experimental resonant frequency.

In order to relate <u> to u(0), we must, as before, first relate ω to ω_0. This can be done by combining Eq.(7) and Eq.(12) to obtain

$$[\nabla^2 + \frac{2\omega_0^2}{u(o)^3}\delta(z)]\phi = \frac{\omega^2}{u(o)^2}\Phi. \tag{17}$$

To obtain the eigenvalues of this equation, the constant C' that normalizes the old eigenfunctions in Eq.(13) must be determined. In this respect, we have that the integral

$$\int_0^L \int_0^{2\pi} \int_0^R J_m^2(\frac{\pi\alpha_{mn}}{R}\rho)\cos^2 m\phi \cos^2 \frac{\pi p y}{L} \rho d\rho d\phi dy$$

is equal to

$$\frac{R^2}{2}(1 - \frac{m^2}{\pi^2 \alpha_{mn}^2}) J_m^2(\pi\alpha_{mn}) \pi \frac{L}{2},$$

for $m \neq 0$ and $n \neq 0$, and

$$\pi R^2 \frac{L}{2},$$

for $m = 0$ and $n = 0$. These formulas fix the value of C' to be used in what follows.

With these preambles, the new eigenvalues can now be immediately evaluated, in analogy with Eq.(9), to give

$$\frac{\omega^2}{u(o)^2} = \frac{\omega_0^2}{u(o)^2} + \int_0^R \int_0^{2\pi} \int_0^L C'^2 J_m^2(\frac{\pi\alpha_{mn}}{R}\rho)\cos^2 m\phi (\cos^2 \frac{\pi p}{L} y)\frac{2\omega_0^2}{u(o)^3} \times$$

$$\delta(z) \rho d\rho d\phi dy. \tag{18}$$

Notice that for modes in plane waves, i.e., $m = n = 0$, which are the usually excited in practice, the above expression simplifyes to

$$\frac{\omega^2}{u(o)^2} = \frac{\omega_0^2}{u(o)^2} + \frac{1}{\pi R^2} \int_0^R \int_0^{2\pi} \frac{2\omega_0^2}{u(o)^3}\delta(z) \rho d\rho d\phi, \tag{19}$$

that in terms of the average velocity leads to

$$<u> = u(o)[1 + \frac{1}{\pi R^2}\int_0^R \int_0^{2\pi} \frac{\delta(z)}{u(o)} \rho d\rho d\phi], \tag{20}$$

valid in first order in $\delta(z)$. By using the definition of $\delta(z)$, we obtain the simpler relation

$$<u> = \frac{1}{\pi R^2}\int_0^R \int_0^{2\pi} u(z) \rho d\rho d\phi. \tag{21}$$

We conclude, then, that for plane-wave modes the measured velocity is simply the average of the local velocity over the cross section of the cylinder. However, such a result can be taken to a closer analogy to Eq.(11) by writing the surface element in Eq.(21) in terms of z. In this way, we obtain

$$\langle u \rangle = \int_0^H u(z) \frac{8}{\pi H^2} [z(H-z)]^{\frac{1}{2}} dz. \qquad (22)$$

The weight factor in this equation, which corresponds to a sound propagation perpendicular to the inhomogeneity gradient, has also been plotted in Fig.1. Note that

$$\int_0^H \frac{8}{\pi H^2} [z(H-z)]^{\frac{1}{2}} dz = 1,$$

so the weight factor is properly normalized. An example of this weight factor for a "horizontal" cavity (solid line in Fig.1 for H=1 cm) shows that in contrast with the "vertical" case, here the local velocity values around mid-z are outweighted.

A first conclusion here is that for sound velocity waves propagating normally to the weak inhomogeneity, also, as in the parallel case, the measured velocity appears as a weighted average of the local velocity. So, in both cases, the inhomogeneity will produce a rounding effect. The precise consequences of this rounding will depend not only on the detailed inhomogeneity profile but also on the experimental technique implemented. However, the results presented here are general and may easily be applied to each particular case.

V. SUMMARY

In this paper we have established the relations between the local and the measured velocity of sound waves propagating in a weak inhomogeneous medium. These results permit inhomogeneity correction to be made in a self-consistent way in different experimental situations. The general relations here obtained are independent of both the cause and type of inhomogeneity. Also, they are independent of the spatial distribution of the inhomogeneity, provided that the weak inhomogeneity condition, expressed in Eq.1, is satisfied. This last conclusion is very important because it is then possible to directly apply our results to sound propagating through interfaces, provided that

$$\frac{u_I - u_{II}}{u_{I,II}} \ll 1,$$

where u_I and u_{II} are the sound velocity in, respectively, medium I and II. A case particularly interesting, and for which the associated rounding effects on sound may be important, is when sound waves propagate through a liquid-gas interface. The corresponding density discontinuity may be allowed for directly by using the results of Section III and IV.

It is worth to stress here, too, that the foregoing results apply even when sound is simultaneously affected by other rounding mechanisms (in addition to that caused by inhomogeneity) as, for instance, velocity dispersion. Besides, this treatment has revealed the influence of each particular experimental arrangement on the rounding effect. These two points are illustrated in Fig.2 and Fig.3 for the specific case of (first) sound propagating near the lambda normal-superfluid transition in liquid He. The rounding inhomogeneity is produced in this case by the Earth's gravity field in a sample of finite height. In Fig.2 (a) and (b) we consider the limiting case where the "rounded" velocity is the thermodynamic zero-frequency one, i.e., we neglect the "intrinsic" rounding effect associated with (critical or not) dispersion. In both figures, the solid line represents the local velocity, which in this case (at a pressure of 0.05 bar) undergoes a minimum of 21730 cm/s at the lambda transition (out of range in the

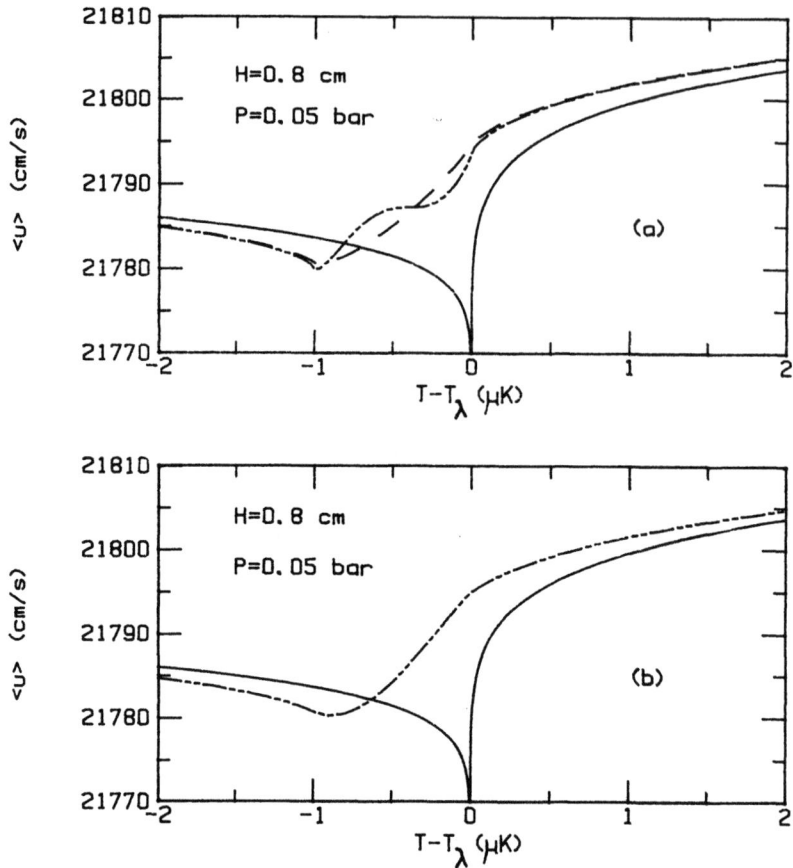

Fig.2. Examples of the rounding of the sound velocity due to the gravity induced inhomogeneity in a sample of finite height. Fig.2(a) applies to a vertical cavity and 2(b) to a horizontal cavity. These examples correspond to the specific case of (first) sound propagating near the lambda normal-superfluid transition in liquid 4He. In both figures, the solid line is the local velocity, whereas the dashed line is the average velocity for a fly-time technique and the dotted-dashed lines correspond to the average velocity in the case of a resonant-frequency method (for the fundamental mode) in a cavity of 0.8 cm length. For other details see text.

figures). Fig.2(a) applies to a vertical cavity, and the dotted-dashed line correspond to the rounded velocity of the fundamental mode in a vertical cavity of 0.8cm in height. As can be seen, both the minimum and the absolute value of the velocity is strongly modified by the gravity-induced inhomogeneity. Other physical aspects of this particular case may be seen in Ref.4. However, we must stress here that although some local details (say, for temperature intervals less than 0.5 μK) of this curve have probably not physical significance, the general shape of this curve depends on the harmonic used. In contrast, this dependence of the shape of the rounded velocity with the resonant mode disappears, and this is an striking result, in the case of sound propagating in horizontal cavity. This situation is illustrated in Fig.2(b).

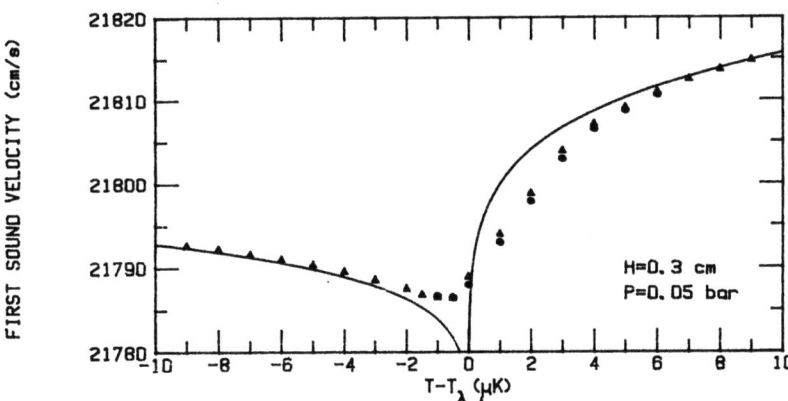

Fig.3. An example of the relation between the local and the measured velocity in an inhomogeneous sample, but in presence of critical dispersion, an intrinsic rounding effect. This example corresponds to first sound standing waves in liquid helium near T_λ resonating in a horizontal cavity of 0.3 cm in height. So, the sound propagates normally to the inhomogeneity gradient. The solid line is again the zero-frequency thermodynamic velocity and without gravity inhomogeneity. The triangles represent the real velocity data as measured by Thomlison and Pobell[6]. So, these data are affected by both dispersion and gravity-induced inhomogeneity effects. When the results of Section IV are applied to suppress gravity influence, we obtain the round points. The minimum of u local is out of range in the Figure.

The other main point noted before, namely the influence of the presence of other rounding mechanisms, is illustrated also for the case of (first) sound near T_λ in Fig.3. Here the solid line is again the zero-frequency velocity, whereas triangles are the experimental velocity data of Buchal and Pobell[6]. In spite of its very low frequency (2.3 KHz), the fact that these data are "already" affected by ("rounding") dispersion, here an intrinsic effect, highly reduces the influence of the gravitational inhomogeneity. When we "suppress" this last effect we obtain the round points in Fig.3. The inhomogeneity influence is just the difference between the round and the triangle points values. Note also that our treatment seems to reveal a small experimental anomaly, the local velocity points affected by dispersion being lower than the thermodynamic velocity in the range o-4 μK.

Finally, let's just indicate the relation between u and <u> in the case of the pulse propagation technique. In this method, the measured parameter is the (average) flytime T of the pulse throughout the known distance H. In this case, the average velocity can be defined as $<u>_{FT} \equiv H/T$. In this expression, T may easily be related to the local velocity $u(z)=dz/dt$ (here t is the time) to obtain

$$<u>_{FT} = [\frac{1}{H}\int_0^H \frac{dz}{u(z)}]^{-1}.$$

The behavior of the average velocity for this technique can be compared, for a particular case, with that for a resonance technique in Fig. 2(a).

REFERENCES

1. See, for instance, M.R. Moldover, J.V. Sengers and R.W. Gammon, and R.J. Hocken, Rev.Mod.Phys.$\underline{51}$,79 (1979).

2. P.C. Hohenberg and M. Barmatz, Phys.Rev.A$\underline{6}$,289 (1972).

3. J. Maza and F. Vidal, Phys.Lett.A$\underline{117}$,417 (1986);and in Proceedings of the Ultrasonics International $\overline{85}$, King College, London, 1985,edited by Z. Novak (Butterworth, G.B. 1985), p.661.

4. F. Vidal and J. Maza, Phys.Rev.B$\underline{34}$,   (1986).

5. M. Barmatz and P.C. Hohenberg, Phys.Rev.Lett.$\underline{24}$,1225 (1970).

6. W.C. Thomlinson and F. Pobell, Phys.Rev.Lett.$\underline{5}$,283 (1983).

FLOW AND TRANSPORT IN SYSTEMS WITH SHAPE CHANGE: MASS TRANSFER IN ELECTROCHEMICAL SYSTEMS

Robert L. Sani, Adele P. Peskin and Miriam K. Maslanik

Department of Chemical Engineering and
Cooperative Institute for Research in Environmental Sciences
University of Colorado
Boulder, Colorado 80309

INTRODUCTION

There are many important technological and engineering science applications in which a free and/or moving fluid and/or solid interface play a dominant role. Such systems encompass general static and dynamic applications encountered, for example, in capillarity, crystal growth, electrochemical plating and and corrosion, coating and polymer technology, separation processes, metal and glass forming processes and many other areas of engineering and science. Specifically, the equilibrium shape and stability of menisci between pairs of immiscible fluids in containers; coating flows in which a viscous fluid is deposited on a rigid, or flexible, substrate as commonly encountered in the manufacturing of photographic films and plate glass or the coating of paper; solidification processes which are commonly encountered in the material science of crystal growth, e.g., in open boat configuration, Czochralski or floating zone methods; extrusion of liquid from nozzles as encountered in the continuous production of fibers or curtain coating operations; domain shape changes associated with electrodeposition and chemical etching processes. A quantitative description and solution of such problems has been allusive even in the simplest cases because of firstly, the inherent nonlinearities in the continuum equations and secondly, heat and/or mass (and sometimes concomitant heat) transport which often play a decisive role in the process.

While all the applications mentioned above are important technologically electrodeposition and etching systems are singled out herein for investigation. These processes are extremely important to the electronics industry in the manufacture of printed circuit boards. Additional insight gleaned from numerical simulation and physical experiments on prototype systems would lead to increased fundamental understanding in these process, the potential for increased production rates from existing on-line processes as well as insight into scale-up and operation of *new generation techniques* currently being developed. Moreover, the numerical techniques described and applied to electrochemical problems in subsequent sections can be utilized in modeling the other physical systems with minor modifications.

BRIEF DESCRIPTION OF ELECTROCHEMICAL PHYSICAL SYSTEMS

Two steps in the manufacturing of many printed circuit boards is the selective chemical etch of a polymer masked surface and electrodeposition onto exposed metallic surface. The latter is accomplished by the passage of current between electrodes. The latter creates a potential field which is significantly influenced by cell geometry, choice of operating conditions, and all materials. Closely related to the potential field are the many pathways along which the current flows from one electrode to another, known as the "current distribution." Predicting the current distribution requires, in addition to the potential field, a knowledge of mass transfer processes near electrode surfaces, and of interfacial processes at the surface itself. Commercial processes are invariably limited by mass transfer processes, or by ohmic resistance, rather than by surface reaction limitations. As a consequence, engineering design of commercial processes requires careful attention to cell geometry and to fluid flow patterns.

The central problem is to predict the *rate of reaction at a surface of complex and changing shape*. The challenges which make this problem interesting are that fluid flow patterns in cavities are complex, and that the course of the cavity shape evolution depends critically upon accurate prediction of local transport processes. The present state of the art cannot predict easily these aspects of behavior, and therefore requires substantially improved methods. Particularly important are the electrical potential field, fluid flow patterns, and mass transfer conditions, all of which depend strongly upon the shape of the surface.

Although several decades ago, electrochemical engineers began modeling these processes by making several simplifications in the problem significant progress was limited due to the inherent nonlinearities in the process. Anodic leveling of surfaces was simulated for simple cylindrical geometries and equipotential surfaces using perturbation techniques (see Fitz-Gerald, et al. (1969), Rasmussen and Christiansen (1976)). Finite difference techniques made it possible to examine more complex geometries in several electrochemical machining applications for which equipotential surface approximations were useful (see Forsyth and Rasmussen (1979)). Irregular surfaces with non-linear potential distributions along the electrodes have only recently been attempted with both finite difference and finite element methods. (See, for example, Alkire, Berg and Sani (1978), Prentice and Tobias (1982), Sautebin and Landolt (1982), Riggs, Muller, and Tobias, (1981), Prentice and Tobias (1982).) Because of the numerical difficulties of these electrochemical problems, there appear to be no *general* numerical algorithm or modeling surface shape changes in such systems.

The objective of the research proposed herein is to perform some numerical experiments on electrodeposition and chemical etching in geometries typical of those encountered in the fabrication of printed circuit boards.

BASIC EQUATIONS AND GEOMETRIES

The continuum mathematical characterization of the electrochemical systems of interest herein are generated by coupling the appropriate form of Maxwell's equations with a momentum and specie balance (Newman (1973)). Neglecting free charge effects, assuming constant physical properties and isothermal conditions (assumptions which can be relaxed if necessary) and employing the Boussinesq approximation (Mihaljan (1962)) in these cases in which buoyance effects are important leads to the following dimensionless equations:

$$Re\left[\frac{\partial \mathbf{u}}{\partial t} + \mathbf{u} \cdot \nabla \mathbf{u}\right] = \nabla \cdot \tau - \nabla p + \mathbf{b}(C)$$

$$\nabla \cdot \mathbf{u} = 0$$

$$Sc\left[\frac{\partial C}{\partial t} + \mathbf{u} \cdot \nabla C\right] = \nabla^2 C$$

$$\nabla^2 \phi = 0$$

Where

$\mathbf{u}$ = fluid velocity
τ = viscous shear stress
C = reactant concentration,
p = pressure,
ϕ = electric field
Re = Reynolds number,
Sc = Schmidt number.
$\mathbf{b}(C)$ = gravitational body force term

In order to complete the mathematical formulation an appropriate set of initial and boundary conditions must be specified. The usual conditions of a specified magnitude or appropriate flux, for the field variables ($\mathbf{u}$, p, C, ϕ) are usually adequate for those portions of the domain's boundary which are fixed and passive. The moving and active portions of the boundary require special consideration. Take, for example, a typical case of electrodeposition onto a cathodic portion of the boundary is one in which the electrochemical reaction rate can be represented by a dimensionless form of the Butler-Volmer rate equation including concentration overpotential effects (Prentice and Tobias (1982)):

$$j = \xi_\phi \left[C^{\gamma-\alpha_a} \exp(\alpha_a \phi) - C^{\gamma+\alpha_c} \exp(-\alpha_c \phi) \right]$$

where ξ_ϕ is a dimensionless exchange current parameter which is a ratio of electrokinetic effects to electrical conduction, α_a and α_c are kinetic parameters. For example, utilizing this expression for the current appropriate boundary conditions for the potential and concentration fields at the cathode undergoing shape change are:

$$\mathbf{n} \cdot \nabla \phi = -j$$

$$\mathbf{n} \cdot \nabla C = -(\xi_c/\xi_\phi) j$$

$$\frac{\partial h}{\partial t} = \beta_\phi j$$

where the last equation represents a *Faraday balance*

In many cases encountered in practice the diffusion dominated (negligible convective transport) limit, the Stokes limit, i.e., Re = 0, is a valid approximation *except, for example, in the new generation technology currently being developed which focuses on a utilization of forced convective mass transfer. Typical geometries are illustrated in Fig. 1*; in the etching problems the ϕ-field need not be determined. The appropriate boundary conditions are the usual ones of specified value or zero flux at passive, non-moving boundaries while Faraday's law coupled with specified electrochemical kinetics are employed at active electrode surfaces.

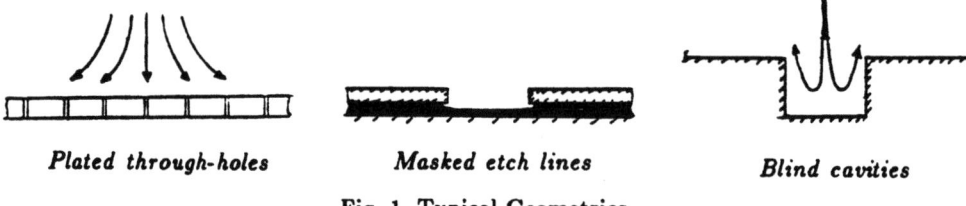

Plated through-holes   *Masked etch lines*   *Blind cavities*

Fig. 1. Typical Geometries

Plated through-holes, as shown in Fig. 1, again involve penetration of the hole by the external flow or the process can be limited by a diffusive transfer process. The latter severely limits the plating speed if a uniform layer is desired. For the three geometries of interest, the main theoretical and experimental topics are as follows:

a. For the single through-hole, the effect on hydrodynamics of the entrance shape can have a pronounced effect on the plating process. Sharp corners may be expected to generate a stable eddy near the entrance, and thus to promote locally poor mass transfer conditions.

b. For the multiple-hole/single jet, the variation of mass transfer rate in different holes is important.

c. For the multiple-hole/multiple jet, the variation of flow rate through different holes will effect the mass transfer and hence the plating process. The effect of jet interactions, which are not present in part (b), should be to make the mass transfer rates more uniform hole to hole.

d. In the diffusion dominated (negligible convective transport) limit the effect of geometry and transport and kinetic parameters on the shape change is of interest.

Blind cavities involve penetration of the cavity by the external flow and, in most cases, formation of eddies in the recessed corners.

Masked etch lines involve, in addition to the foregoing effects, undercutting of the surface mask owing to shape change. The presence of fluid eddies in undercut regions will act to decrease etching rates in these areas and therefore to enhance directional etching. Thus the important features to be studied is the tradeoff between flow rate, etch rate, and directionality of the etch process.

The investigation of these systems must proceed both from a physical experiment and numerical modeling standpoint. The two must be developed concurrently because experimental observation and measurement of certain transport properties and features such as the mass transfer and concomitant shape change are necessary to guide and corroborate modeling efforts. The size and complexity of some of the systems of industrial interest limit the ability of physical experimentalists to make quantitative measurements. Here again numerical, or analytical, models can be used to *scale down* experimental observations made on larger systems as well as to assess the effect of different operating conditions in a cost-effect manner. The development of one such model is discussed in the next section and some preliminary results for a mass diffusion-ohmic loss limited regime are presented in a subsequent section.

NUMERICAL ALGORITHM: GALERKIN FINITE ELEMENT TECHNIQUE

While numerical simulation in certain area of computational Physics is to a state in which many of the commercially available algorithms can be used as a "black box" by reference to user's manuals, this is generally not true in the electrochemical problems of interest herein, i.e., those involving convective and diffusive mass transport and shape change. A basic knowledge of numerical methods as well as of the physics of the problem is essential. The simulation of these convective transport and/or moving domain problems can easily lead to spurious numerical oscillations in the solutions which must first be recognized and then dealt with in a proper fashion, such as, for example, selective mesh refinement, etc.

Finite difference methods (FDM) have played an essential role in the development of methods in computational physics (see, for example, Roach (1972)). However, within the last ten years descretization methods such as Galerkin's technique using local polynomial basis functions, herein referred to as Galerkin-finite element method (GFEM) have become serious competitors in many cases. The GFEM method has inherently built into it the properties of being appropriate to irregularly shaped domains and allowing the modeler to grade his "mesh" to allow for any known local structure in the solution to the problem. These properties plus the accuracy of the GFEM method make it an attractive method for the simulation of electrodeposition or etching problems where the geometries are oftentimes irregular due to the initial configuration or the shape change induced by the process.

The algorithm which is to be in this research is built around a GFEM incorporating an adaptive mesh capability. The algorithm solves the "weak form" of the appropriate mathematical model of the electrochemical system. The domain shape is than determined as a function of time, either by using explicit, or implicit, time integration techniques. The movement of the domain is determined by solving an interface material balance equation and the position of the interface which is interpolated in a natural fashion using finite element basis functions in advanced each time step. The mesh adaptation scheme to be used in the algorithm will be a modified version of that used by Kistler and Scriven (1982) and Engelman and Sani (1983) and will not only allow the movement of "boundary nodes" and other "nodes" along prescribed lines in space and the lines themselves to be tilted automatically in order to maintain a "reasonable" mesh of elements with predetermined aspect ratios; but also will allow the *injection* of elements into the domain at certain times, a feature *necessary* in, for example, etching and corrosion simulations. The latter guarantees that the accuracy of the technique is maintained as the domain shape enlarges and possibly undergoes a shape change which is difficult to track by sliding nodes along tilting lines; such is the case, for example, in etching problems which exhibit undercutting. For cost-effectiveness the algorithm will also allow only those parts of the computational domain specifically tagged to be dynamically altered during the simulation. A typical mesh illustrating a free surface with an adjoining moving node region above a fixed reference line with a fixed node region below is illustrated in Fig. 2.

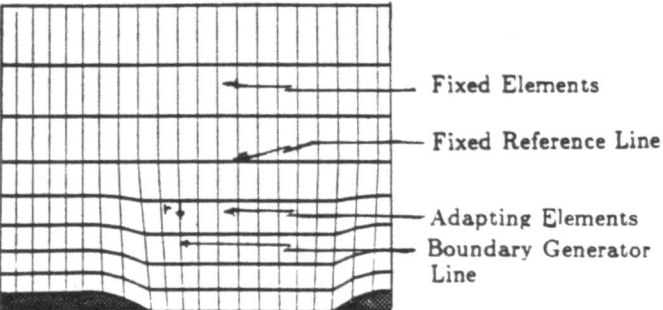

Fig. 2. Adaptive Mesh Techniques

The total domain can be made up of patches of such subdomains. The GFEM and adaptive mesh technique which will be modified in this study are discussed in more detail in the manuscript included in Appendix A and below we focus on some of the special features to be incorporated in this research and more background material on mesh adaptation.

<u>Finite Element Algorithm</u>

The finite element algorithm developed during the course of this work was constructed in a "tree and branch" structure so that not only a wide range of applications in etching and deposition could be accommodated but also modifications in the basic continuum balance equations or boundary conditions could be effected easily. The latter feature facilitates the interaction between numerical and physical experiments since oftentimes physical experiments suggest changes in the continuum model or its discrete analog.

Hybrid elements as the eleven noded 2D element and its associated 5 noded 1D element displayed in Fig. 3 will be incorporated and utilized as the "front tracking" elements in certain cases in which additional boundary resolution is desirable; the latter is efficient since the resolution is localized to those regions requiring it.

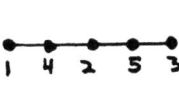 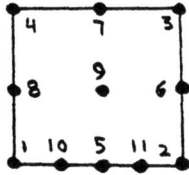

Fig. 3 Example of Hybrid Elements

Initially only variable step size, implicit, first order (backward Enter) and second order (trapezoid) time integration schemes will be implemented but some effort will be focused on variable step explicit integration schemes which may be more cost effective for certain type problems. Finally, the implementation of these techniques to 3D problems which, although formally straight forward because of the *primitive variable* formulation utilized herein, requires a significant effort in the development of efficient mesh adaptation and solution strategies although the availability of large mainframe computers with large solid state disks have somewhat relieved this problem.

Adaptive Mesh Technique

Efficient mesh adaptation is important for solving these problems, since moving elements or remeshing can be computationally expensive. There are three general classes of methods for tracking moving boundaries. (See Crank (1984) for a general discussion and selected examples.) The first class involves the use of a fixed mesh. A second class of moving boundary problems is solved by making a coordinate transformation to a frame of reference in which the moving boundary remains stationary. A third class involves moving the mesh itself along with the boundary. The position of the surface is solved for directly in the governing set of equations. Of course these techniques can be implemented in both a finite difference or a finite element setting. In the area of electrochemical modeling electrochemical machining problems provided the initial modeling impetus. Fitz-Gerald, McGeough and McMarsh (1969) and Rasmussen and Christiansen (1976) used perturbation techniques to study shape change during electromaching and later Christiansen and Rasmussen (1976) used a boundary integral technique to determine the potential field coupled to a finite difference technique to calculate the new anode surface. Forsythe and Rasmussen (1979) and Riggs, Muller and Tobias (1981) studied electromaching using a finite difference method. Prentice and Tobias (1982) also used such a method to study electrodeposition in a V-notch as well as a sinusoidal surface but had to develop a method to prevent what appeared to be numerical instabilities from propagating on the anodic boundary.

Finite element techniques were used to model several geometries by Sautebin et al., who investigated the rate of leveling of triangular surfaces under primary, secondary, and tertiary current distribution conditions (Sautebin, Froidevaux, and Landolt (1980), Sautebin and Landolt (1982)). Alkire, Bergh, and Sani (1978) studied the electroplating of a single metal strip using a finite element technique. System equations were solved for the potential field, and the mesh was moved according to the current distribution found at each step.

Finite element techniques have also been used in other related areas in which moving and/or free boundaries are encountered. Ettouney and Brown (1983) present a very detailed study of an application to a phase change problem and Silliman and Scriven (1980), Saito and Scriven (1980) and Kistler and Scriven (1982) have presented a comprehensive development and application of finite element methods with concomitant mesh adaptation to the modeling of coating flow problems. The mesh adaptation techniques developed in the latter studies form the foundation of the algorithm described by Engelman and Sani (1983) (see also Appendix A) and the one to be developed herein for electrochemical problems. That is the positions of the surface nodes are solved for directly in the governing equations and then a portion of the computational domain is remeshed at each timestep. Each surface node is the endpoint of a line segment containing other nodes in the mesh, while the other endpoint of the line segment remains fixed. Each segment is represented parametrically by a new variable, r_i, where i designates a particular surface node. The position of each node along a line segment is given by:

$$x = \alpha_x r + \beta_x$$
$$y = \alpha_y r + \beta_y$$

where $r=0$ at the fixed end and $r=1$ at the free surface before any motion occurs. The line segment can tilt in any direction; that is, α_x, α_y, β_x, and β_y are all functions of the angle of the line segment. Each moving boundary equation determines a new r_i value for the location of node i. After an r_i value is found, all nodes along the line segment are moved so that elements maintain their relative sizes along the line segment (Kistler and Scriven (1982)). For efficiency remeshing need not occur over the whole domain (Engelman and Sani (1983)).

In this scheme surface coordinates are approximated by basis functions (Γ_j):

$$x = \sum_j x_j \Gamma_j = \sum_j \left(\alpha_{x_j} r_j + \beta_{x_j} \right) \Gamma_j ,$$

$$y = \sum_j y_j \Gamma_j = \sum_j \left(\alpha_{y_j} r_j + \beta_{y_j} \right) \Gamma_j .$$

Surface growth h, is described in terms of the r values at each node on the surface. Electrode growth occurs normal to, for example, the cathodic surface. At a given surface node, h is the projection of the normal onto its parameterized line. A surface node i, with r_i value equal to r grows with time $\left(\dfrac{\partial h}{\partial t} \right)$ as:

$$\sum_j \Gamma_j \left(\alpha_{x_j} n_x + \alpha_{y_j} n_y \right) \frac{\partial r_j}{\partial t}$$

where $\mathbf{n} = \left(n_x, n_y\right)$, the normal vector.

It is convenient to transform to a frame of reference that moves along a specified line segment relative to the velocity of the moving boundary, in order to specify the potential at each node's new location.

The above represents a brief description of current generation techniques applied to the electrochemical problems of interest herein. An additional *new development* which will be incorporated in this study is the ability to *inject* elements at selected times and at selected points in the domain. The time and position will be determined such that solution accuracy is maintained during the simulation. In contrast to many former studies alluded to above the cases to be studied here are all transient phenomena and in particular, the simulation of a typical etching scenario is one in which the computation domain may increase substantionally and the ability to inject new elements into the computational domain is needed to maintain accuracy. Without this capability the simulation must be stopped periodically, remeshed and restarted; this process can easily lead to artificial transients being observed in the restart process due to the necessity of interpolating field variables between meshes.

EXAMPLES OF PRELIMINARY NUMERICAL EXPERIMENTS IN THE DIFFUSION DOMINATED LIMIT

The preliminary results displayed herein are all limited to cases in which convective transport can be assumed to be negligible. Since the dimension of the systems of interest in many electroplating or etching applications in the electronics industry is small (trenches, or holes, 3-10 mils wide and 10-70 mils deep or lines from 1-10 mils wide), the systems can be within the viscous sublayer and experience a locally plane parallel flow, or locally a very feeble flow. Thus, one extreme of the *operation window* is reasonably assumed to be a mass diffusion-ohmic loss limited one. The systems illustrated here are:

(1) Electrodeposition onto stripes of differing width on a flat surface.

(2) The initial stage of noticeable interface instability.

These systems represent typical elements and corresponding plating scenarios which are experienced in, for example, the fabrication of printed circuit boards. In this manner the electroplating and etching process can be investigated from a very fundamental standpoint and the results can be used to suggest appropriate operating windows or improvements in the actual process. For example, as illustrated below, even the results of the simplified model presented here exhibit some shape changes as well as interface instabilities oftentimes observed industrially. However, by numerical experiments we have been able to quantify these effects and characterize their occurrence by delimiting an *occurrence window* in an appropriate parameter space.

The first case considered is that of a series of alternating thick-thin stripes shown in Fig. 4 where the computation domain shown is obtained by utilizing symmetry. The temporal evolution of the electroplated interface is displayed in Fig. 5. The current density and hence plating is larger on the smaller cathode-----a result verified by experimental observation (results courtesy of Rohlev, (1985)). This growth rate difference is sensitive to operating conditions and a factor of *four* can be realized under certain operating conditions. Fig. 6 illustrates the automatic adaptation of the mesh during the simulation. Fig. 7 illustrates the effect of the polarization parameter, ξ,

$$\xi \alpha \frac{\text{surface electrochemical kinetic effects}}{\text{electrical conduction effects}}$$

on the cathode shape change at a time when the *horizontal growth of the cathode is identical in all cases*. It is apparent that increasing the polarization factor causes increased horizontal growth rates and nonuniformity in shape. Both effects and their dependence on system parameters are important in technological applications. The results of such numerical experiments can be used to suggest *operating windows* for such processes.

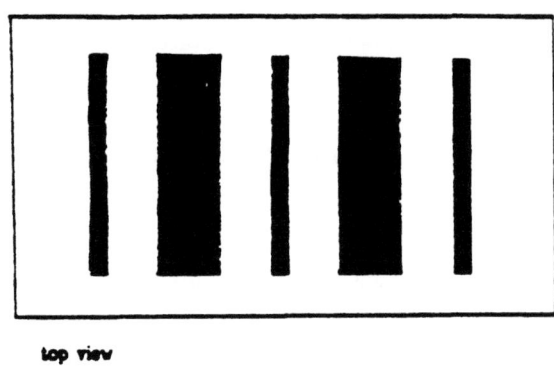

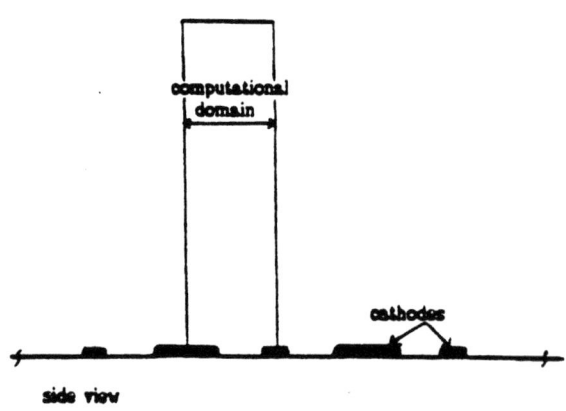

Fig. 4. Strip Electroplating Example

As the final example electrodeposition of a layer into the bottom of a trench was considered for a case in which the system was *very* close to its limiting current configuration. The physical process of electrodeposition can be shown to be a *physically unstable* process, i.e., in general the interface has a tendency to exhibit nodule or dendrite growth which is usually undesirable. Fortunately, in most electroplating scenarios the instabilities (nodules or dendrites) grow too slowly to be a problem. However, under certain conditions, for example, near limiting current in the example being considered here, their

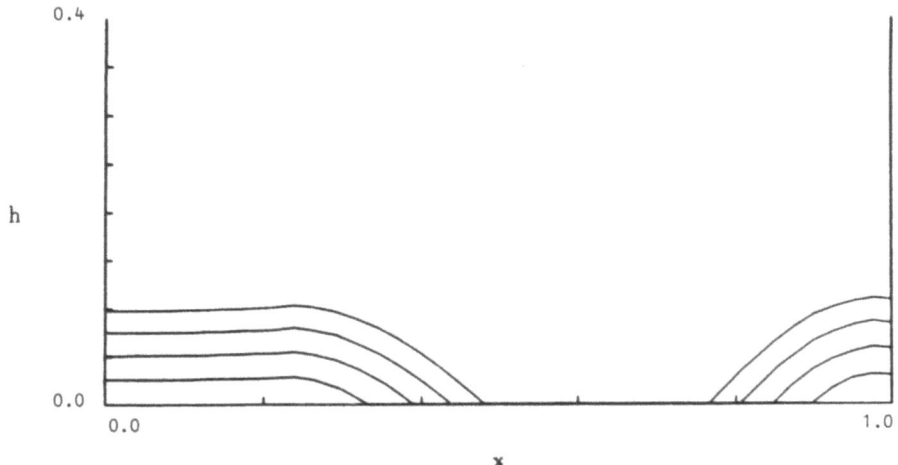

Fig. 5. Time History of Electroplating

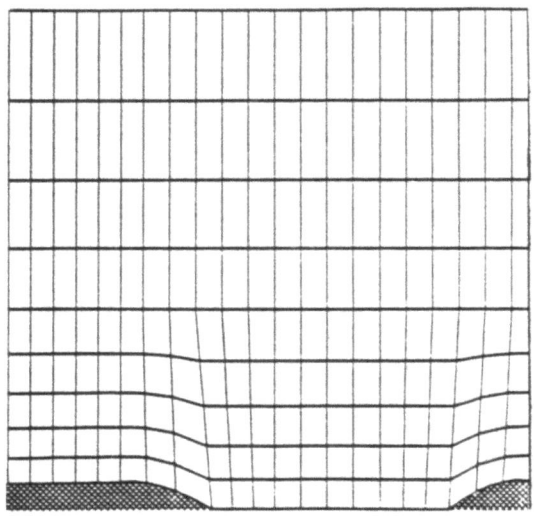

Fig. 6. Mesh At Timestep 8

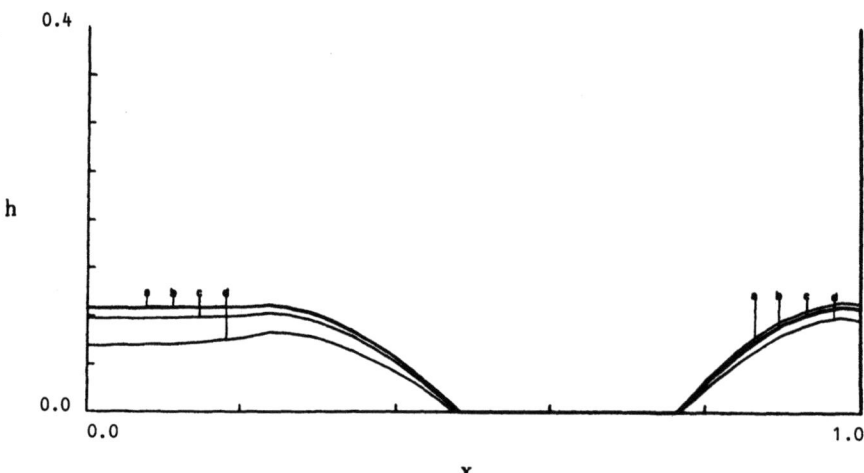

Fig. 7. Effect of Polarization Parameter
a-d: 0.1, 0.5, 1.0, 5.0

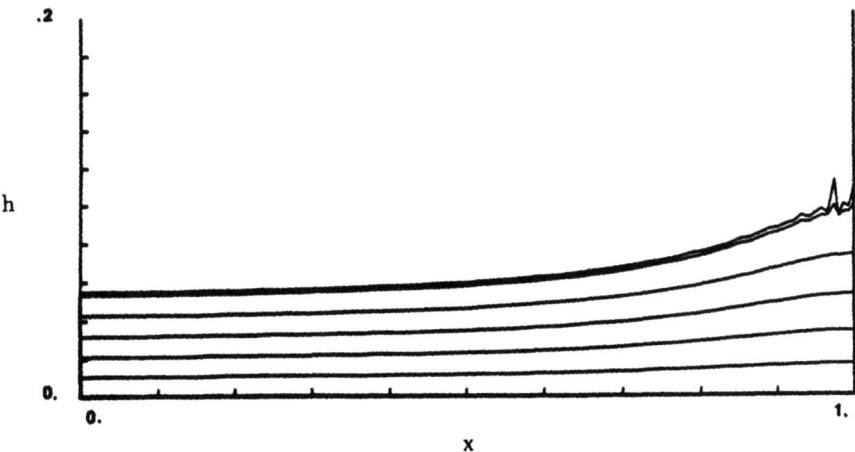

Fig. 8. Time History of Electroplating Exhibiting A Surface Instability

growth rate is sufficiently large to cause a noticeable surface imperfection. Such a surface imperfection is apparent in Fig. 8 near the righthand wall of the trench. Since the imperfections are very small scale and the mesh was *too coarse* to capture it they appear as oscillations in the numerical results. A further refinement of the mesh would be necessary to track such a phenomenon after its initiation. However, from the numerical experiment it is evident that firstly, the initiation of an *observable* surface instability is modeled and also it is evident, as one might intuitively expect, that it is first observed on the fastest growing portion of the surface.

CONCLUSIONS

The examples presented herein represent preliminary results of numerical experiments performed with a new adaptive-mesh Galerkin finite element algorithm which is still in a developmental stage. These examples illustrate the prediction of shape changes in diffusion limited systems, i.e., no flow systems or systems in which convective effects are negligible demonstrate feasibility of the numerical and mesh adaption schemes. Such complications as concomitant convective and shape change effects can also be addressed ---- an essential feature in many cases. For example, the electrodeposition examples in the previous section would in general lead to asymmetric profiles with the inclusion of convective effects. Current techniques of, for example, air sparging in plating tanks in which the printed circuit boards are vertically hung oftentimes lead to severe mass transfer limitations when plating into very *small* holes, or pits and consequently, new technology in this area utilizes convective transport. There are numerous other areas, for example the modeling of pit corrosion (Alkire, Reiser and Sani (1984) or liquid and vapor phase crystal growing techniques, in which the inclusion of convective transport and shape change effects are essential to the basic understanding and possible improvement of the technique.

ACKNOWLEDGMENTS

The authors would like to acknowledge support from the U.S. Army Research Office (DAAG29-83-K-01112) and the National Science Foundation (CBT-8513570).

REFERENCES

Alkire, R.C., Bergh, T., Sani, R.L., 1978, Predicting electrode shape change with use of finite element methods, *J. Elec. Soc.*, *125:1981*.

Alkire, R.C., Reiser, D.B. and Sani, R.L., 1984, Effect of fluid flow on removal of dissolution products from small cavities, *J. Elec. Soc.*, *131:2795*.

Crank, J., 1984, "Free and Moving Boundary Problems", Clarendon Press, Oxford, UK.

Engelman, M.S. and Sani, R.L., 1983, FEM simulation of incompressible fluid flows with a free/moving surface, *in*: "Numerical Methods in Laminar and Turbulent Flow," Taylor, ed., Peneridge Press, Swansea, UK.

Fitz-Gerald, J.M., McGeough, T.A., 1969, Mathematical theory of electrochemical machining 1. Anodic smoothing, *J. Int. Math. Applica.*, 5:387.

Fitz-Gerald, J.M., McGeough, T.A., McMarsh, L., 1969, Mathematical theory of electrochemical machining 2. Anodic Shaping, *J. Int. Math. Applica.*, 5:409.

Forsyth, P., Rasmussen, H., 1979, Solution of time dependent electrochemical machining problems by a co-ordinate transformation, *J. Int. Math. Applica.*, 24:411.

Kistler, S. and Scriven, L.E., 1982, Coating Flow Theory by Finite Element and Asymtotic Analysis of the Navier-Stokes System. *Fourth International Symposium on Finite Element Method in Flow Problems*, Tokyo, 26-29 July.

Mihalijan, J.M., 1962, A rigorous exposition of the Boussinesq approximations applicable to a thin layer of fluid, *Astrophys. J.*, 136:1126.

Newman, J.S., 1973, "Electrochemical Systems," Prentice-Hall, Inc., Englewood Cliffs.

Prentice, G.A., Tobias, C.W., 1982, Simulation of electrode profiles, *J. Elec. Soc.*, 129:78.

Rasmussen, H., Christiansen, S., 1976, A perturbation solution for a two-dimensional annular electrochemical machining problem, *J. Int. Math. Applica.*, 18:149.

Riggs, J.B., Muller, R.H., Tobias, C.W., 1981, Prediction of work piece geometry in electrochemical cavity sinking, *Electrochimica Acta*, 26:961.

Roache, P., 1972, "Computational Fluid Dynamics," Hermosa Publishers, Albuquerque.

Rohlev, S., 1985, Hewlett-Packard, Loveland, Co. Division.

Sautebin, R., Froidevaux, H., Landolt, D., 1980, Theoretical and experimental modeling of surface leveling in ECM under primary current distribution conditions, *J. Elec. Soc.*, 27:1096.

Sautebin, R., Landolt, D., 1982, Anodic leveling under secondary and tertiary current distribution condition, *J. Elec. Soc.*, 129:946.

Silliman, W.J., Scriven., 1980, Separating flow near a static contact line: Slip at a wall and shape of a free surface, *J. Comp. Phys.*, 34:387.

NUMERICAL SIMULATIONS OF BUOYANCY DRIVEN FLOWS IN
CYLINDERS AND CAVITIES FOR VAPOUR CRYSTAL GROWTH

P. Bontoux, F. Elie, C. Smutek, G.P. Extremet,
A. Randriamampianina, E. Crespo*, H. Branger, and
B. Roux

I.M.F.M., rue Honnorat, Marseille, France

INTRODUCTION

The paper reports about numerical works carried out by the authors at I.M.F.M. in the group of Numerical Fluid Mechanics. The common objectives of these works are related to the analysis of fluid dynamics in closed tube methods employed in crystal growth from vapours. Several numerical methods are considered and the solutions are discussed with respect to the results obtained with different approaches of the problem as the asymptotic theories, the stability analyses and the experiments. Some groups were associated in relation with these approaches and also the development of the numerical methods, finite difference and spectral methods. They are the groups of Prof. G. De Vahl Davis at University of New South Wales-Australia, Prof. F. Rosenberger at University of Utah-U.S.A. (with Dr. G.H. Schiroky, B.L. Markham, A.C. Hurford..), Prof. R. Sani at University of Colorado-Boulder-U.S.A. (with G. Hardin), Prof. M.G. Velarde at U.N.E.D.-Madrid-Spain (with E. Crespo), Dr. R. Peyret at Université de Nice-France (with Dr. J. Ouazzani and J.M. Lacroix).

The closed ampoule vapour transport techniques are used for the preparation of single crystal and very useful for research in crystal growth (however costly for production where open tube methods can be generally preferred). The ampoule is usually cylindrical and heated differentially inside a furnace which produces the thermal conditions for vaporization at the source and condensation at the sink (see Omaly et al., 1981). The process involves very complex physical mechanisms, i.e. the sublimation of a material at the hot source, the transport of the volatile component A across the ampoule through other components B and the crystal growth at the cold end of the tube (see Rosenberger and co-workers, 1979 to 1984). The convection is the principal fluid mechanism and the transport can also be enhanced by tilting the tube with respect to the gravity field (Pamplin, 1980). Depending on the values of the

*permanent address: Dpto. Fis. Fund., U.N.E.D., Madrid, Spain

physical parameters (Rayleigh numbers...), the aspect ratio and the inclination angle, many complex flows can develop in the ampoule which can be highly or moderately three-dimensional or simply two-dimensional as in axisymmetric situations. The understanding of convection in the nutrient by predictions of velocity scales and flow patterns is very important for the researchers because it affects the growth behaviour of the crystal (Carruthers, 1977, Olson and Rosenberger, 1979, Ostrach and co-workers, 1979, 1981, 1982).

The mathematical model is given by the Navier-Stokes equations with the Boussinesq approximation. The three-dimensional simulation of thermal convection is made with a finite-difference method. Both hermitian finite differences and spectral approximations were considered for the two-dimensional simulation of physical vapour transport with a simple analytical law or the Fick's law to modelize the mass flux at the interfaces. The characteristics of the numerical methods are shortly given. The typical flows obtained in horizontal, vertical and inclined cylinders are presented. An assessment of the validity of the two-dimensional approximation is discussed with respect to the experimental, analytical and numerical results in the symmetry plane of a horizontal cylinder. The fluid dynamics of crystal growth from vapours is studied when the thermal and solutal Rayleigh numbers vary.

PHYSICAL MODELS AND GEOMETRIES

Crystal growth by physical vapour transport (PVT) in a closed tube is experimentally simple. It consists in the vaporization of a material A at a source and its condensation to a crystal generally held at a "colder" temperature than the source (Greenwell et al., 1981, Markham et al., 1981). The transport of the gaseous component A through (inert) components B across the ampoule is governed by the competition between the advection-diffusion driven by the interfacial mass flux and the convection generated by the density gradients in the changes of temperature and concentration.

The cylindrical geometry is depicted in Fig.1 with the associate rectangular cavity model. The reference frames are given for both geometries with the velocity components $\underline{u}$, $\underline{v}$, $\underline{w}$ defined in the radial ($\underline{r}$ or $\underline{x}$), azimuthal (ϕ) and axial ($\underline{z}$) directions, respectively. The axial length is L, R is the radius and D is the height (D=2R). The aspect ratio is a=L/2R=L/D. The enclosure is filled with a binary mixture of gaseous components A and B (molecular weights M_A and M_B). At the source and the crystal, the solid S consists of A only. The interfaces (of length D) are flat, stationary and held at constant different temperatures (hot at the source T_2 ($\underline{z}$=L), cold at the sink T_1 ($\underline{z}$=0)) and concentrations ($S_{A2} \geq S_{A1}$). The growth rate is assumed to be limited by the mass transfer of A in the vapour phase. The flux of the (inert) component B vanishes at the interfaces (active walls). The passive side walls are impermeable to both species.

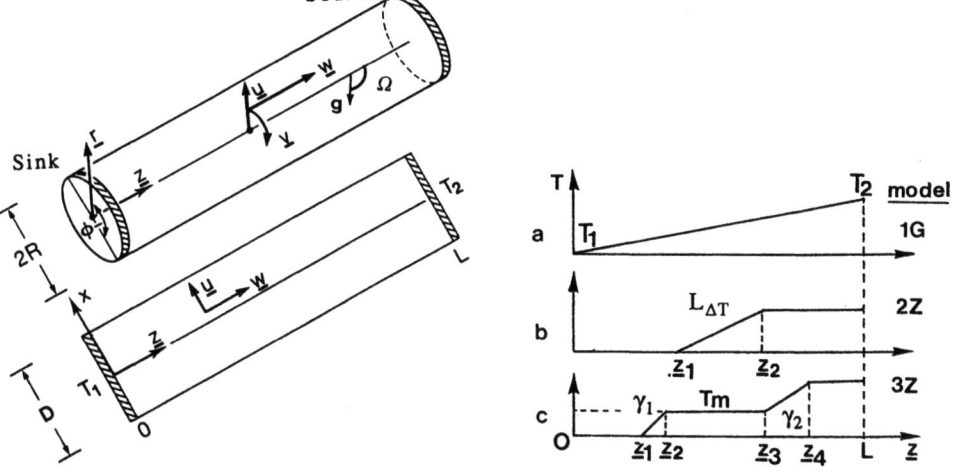

Fig.1. Differentially heated cylinder and cavity at inclination with the gravity vector: definition sketch for dimensions, position coordinates and velocity components.

Fig.2. Temperature profiles along the (axial) walls: 1-G model (a), 2-Z model (b) and 3-Z model (c).

Depending on the applications, various kinds of temperature profiles can be produced at the axial walls by the furnace in order to enhance the growth process (Omaly et al., 1981, 1983, Launay, 1982, Brisson, 1981, Extrémet et al., 1986). They are presented in Fig.2. The simplest profile is linear and corresponds to the one-gradient model (abbreviated 1-G). In the general case one or more gradient can be inserted between isothermal zones (multizone models, with two and three zones abbreviated 2-Z and 3-Z). The conditions at the side walls are expressed as:

$$T = T_1 + \Delta T \, \Theta(z) \qquad (1)$$

where $\Theta(z) = z/L$ with $\Delta T = T_2 - T_1$, when $0 \leq z \leq L$ for the 1-G model (Fig.2a); $\Theta(z)=0$ when $0 \leq z \leq z_1$ and $\Theta(z)=1$ when $z_2 \leq z \leq L$ for the 2-Z model (Fig.2b); $\Theta(z)=0$ when $0 \leq z \leq z_1$, $\Theta(z)=\Theta_m=(T_m-T_1)/\Delta T$ when $z_2 \leq z \leq z_3$, $\Theta(z)=1$ when $z_4 \leq z \leq L$ for the 3-Z model (Fig.2c); in the 2-Z and 3-Z models the constant gradient zones (of widths $\gamma_i = (L_{\Delta T})_i/L$) are inserted between the isothermal zones and $T_1 \leq T_m \leq T_2$.

The axis of the ampoule will not be always perpendicular to the gravity field, that is in microgravity environment but also on earth in the applications when the inclination is used to enhance convection (Pamplin, 1980). As defined in Fig.1, the inclination angle, Ω, is referred to the vertical. The inclination $\Omega=0°$ will correspond to the vertical cavity heated from below.

MATHEMATICAL MODELS

The governing system is given by the Navier Stokes, energy and species equations with the Boussinesq approximation and the linear Fick's law.

The Boussinesq Approximation and the Linear Fick's Law

The density is linearly related to the variation of temperature, T, and mass fraction, $S_A = \rho_A/\rho$, of component A ($S_A + S_B = 1$) with the Boussinesq approximation (Joseph, 1976) as follows :

$$\rho = \rho_0 \{ 1 - \beta (T-T_0) - \alpha_A (S_A - S_{A0}) \} \qquad (2)$$

where β is the thermal expansion factor ($1/T_0$ for ideal gas), and $\alpha_A = M_0(M_B - M_A)/M_A M_B$ is the solutal expansion factor expressed in terms of the molecular weights (subscript 0 refers to average quantities).

The diffusive flux of the components with respect to the mass average velocity, $V = S_A V_A + S_B V_B$, is given by the linear Fick's law as (see Rosenberger, 1979):

$$S_A(V_A - V) = -D_{AB} \nabla S_A \qquad (3)$$

where D_{AB} is the binary diffusivity.

The Two-Dimensional Model

The governing equations are written using the dimensionless vorticity, ζ, streamfunction, ψ, temperature, $\theta = 2(T-T_0)/(T_2-T_1)$, and mass fraction, $S = 2(S_A - S_{A0})/(S_{A2} - S_{A1})$, as :

$$\partial \zeta / \partial t + \eta_i V \cdot \nabla \zeta = \eta_d \Delta \zeta + \eta_{tb} g_0 \cdot \nabla \theta + \eta_{sb} g_0 \cdot \nabla S \qquad (4)$$

$$\partial \theta / \partial t + \lambda_i V \cdot \nabla \theta = \lambda_d \Delta \theta \qquad (5)$$

$$\partial S / \partial t + \sigma_i V \cdot \nabla S = \sigma_d \Delta S \qquad (6)$$

$$\Delta \psi = \zeta \qquad (7)$$

where ∇ and Δ are the gradient and Laplacian operators, $V = (u = \partial \psi / \partial z, w = -\partial \psi / \partial x)$ and g_0 is the unity gravity vector. Depending on the choice of the scaling factors for the velocities and space variables, the coefficients η, λ, σ are expressed in terms of the physical parameters (De Vahl Davis, 1986): the Prandtl number, $Pr = \nu/\kappa$, the Lewis number, $Le = \kappa/D_{AB}$, or the Schmidt number, $Sc = \nu/D_{AB}$, the thermal and solutal Rayleigh numbers, $Ra_T = \beta g \Delta T (H)^3/\nu \kappa$ and $Ra_S = \alpha_A g \Delta S_A (H)^3/\nu D_{AB}$, or the Grashof numbers, Gr=Ra/Pr, where $\Delta S_A = S_{A2} - S_{A1}$ and H=D/2 (or R in what will concern the cylindrical cavity). With κ/H (κ is the thermal diffusivity) as scaling factor for the velocity, the dimensionless coefficients in system (4-7) are $\eta_i = \lambda_i = \sigma_i = 1$, $\eta_d = Pr$, $\lambda_d = 1$, $\sigma_d = 1/Le$, $\eta_{tb} = Ra_T Pr$ and $\eta_{sb} = Ra_S Pr/Le$. The common scaling for the space variables in the following results will be based on the mid-height D/2 of the cavity (respectively R for the cylinder).

The mass fraction at the side walls is governed by the no-flux condition for both species, $\partial S/\partial x = 0$. For the velocity the usual no-slip and no-permeability conditions are used (Jhaveri et al., 1981, 1982).

Two models are considered to analyse the effect of the interfacial mass flux of component A at the active walls:

- one is derived directly from the Fick's law (3) (see Rosenberger and co-workers, 1981 to 1984); the w-profile is then:

$$w_{interface} = (\partial S/\partial z) \, \xi(x) / Le \, (E - S) \qquad (8)$$

where E is a dimensionless number characterizing the mass fractions at the source and the sink (and then related to the molecular weights M_A and M_B and partial pressure conditions at sublimation and deposition, see Elie, 1984):

$$E = 2 \, (1 - S_{A0}) / \Delta S_A . \qquad (9)$$

The transport of components A and B is computed from the mass average solution using a second number F, defined as $F + E = 2/\Delta S_A$. The function $\xi(x)$ is artificially introduced in order to regularize the solution near the corners. Its constant is also adjusted at the sink in order to ensure the mass conservation of species A (see Ouazzani, 1984, Elie, 1984).

- an analytical model is also considered as a perturbation of the convective flow and following the quartic w-velocity profile (Extrémet et al., 1986)

$$w_{interface} = v_d \, q \, [\, x \, (x-1) \,]^2 \qquad (10)$$

where q is related to the maximal velocity generated by buoyancy in the core (see following relation (14)) and v_d is a proportionality factor suitably chosen to correspond to realistic conditions.

The Three-Dimensional Model

The governing system is considered with the velocity (V)-vorticity ($\zeta = \nabla \times V$) formulation where V(u,v,w) (see definitions in Fig.1). With R and κ/R as scaling factors for the coordinates and the velocities and with θ as dimensionless temperature, the governing equations for thermal convection are

$$\partial \zeta / \partial t = \nabla \times (V \times \zeta) - 0.5\, Ra_T Pr\, \nabla \times (\theta\, g_0) - Pr\, \nabla \times (\nabla \times \zeta) \qquad (11)$$

$$\partial \theta / \partial t = -\nabla \cdot (V \cdot \theta) + \nabla^2 \theta \qquad (12)$$

$$\nabla^2 V = -\nabla \times \zeta + \nabla (\nabla \cdot V) \approx -\nabla \times \zeta \qquad (13)$$

The thermal boundary conditions correspond to the 1-G model. In addition to the usual no-slip conditions, the conditions on the vorticity at the rigid walls are derived as : $\zeta_r = -\partial v / \partial z$, $\zeta_\phi = \partial u / \partial z$, $\zeta_z = 0$ at the end walls and $\zeta_r = 0$, $\zeta_\phi = -\partial w / \partial r$, $\zeta_z = \partial v / \partial r$ at the side walls.

ANALYTICAL SOLUTIONS

The basic flow in long horizontal cavities or cylinders filled with a mono-component fluid corresponds to two counter-flows coming from the hot zone in the upper part, and from the cold zone in the lower part. Analytical approximations for the axial velocity profile are available for the core of rectangular cavities and cylinders when they are differentially heated using a 1-G temperature model (see Bontoux et al., 1986). These approximations are based on temperature gradient parameters derived from experiments and analyses. At low Ra_T, the horizontal temperature gradient is constant and generates the main buoyancy forces, which corresponds to the core driven regime, denoted CDR (see Cormack et al., 1974, Imberger, 1974). At larger Ra_T, the main buoyancy forces are located in the end regions (boundary-layer driven regime, denoted BLDR). In these cases the functional laws available from Cormack et al. (1974), Bejan and Tien (1978) and Hart (1983) can be used to predict the flow profiles. The analysis was also recently extended for cavities in the case of multizone models by Extrémet et al. (1986).

1-G Thermal Model for Cavities and Cylinders

Parallel flow solutions were proposed by Klosse and Ullersma (1973), Birikh (1966), Hart (1972), and Bejan and Tien (1978). Elaborate approximations, including some interaction with the end regions and allowing for secondary flows in cavities, were given by Cormack et al. (1974), Bejan and Tien (1978) and Shih (1981) (see also Ostrach et al., 1980, 1982). These core solutions are still limited to the S-shaped profile for the horizontal velocity component. Further improvements, which remove this limit for higher Ra_T conditions, are due to Tichy and Gadgil (1982). For cylinders, Schiroky and Rosenberger (1984) proposed a third-order power series in Ra_T which well predicts the shift of the maximum in the core velocity profile towards the wall for the low Ra_T-range of the BLDR.

The analytical expressions for the core flow (horizontal velocity) depend on the axial dimensionless temperature gradient $k_1=0.5(\partial\theta/\partial z)$. They are expressed as follows in terms of $k_1 Ra_T$:

- for cavities:
$$w^{2D} = (k_1 Ra_T)(x^2-1)x/6 \tag{14}$$

- for cylinders:
$$w^{3D} = (k_1 Ra_T)\cos\phi\,(r^2-1)r/8 \tag{15a}$$

$$u^{3D} = -(k_1 Ra_T)^2 \cos 2\phi\,(2r^6-15r^4+24r^2-11)r/184320 \tag{15b}$$

$$v^{3D} = -(k_1 Ra_T)^2 \sin 2\phi\,(8r^6-45r^4+48r^2-11)r/184320 \tag{15c}$$

In the CDR the value of k_1 is $(2a)^{-1}$ and the flow is mainly axial. For the CDR up to the beginning of the BLDR, Cormack et al. (1974) derived an expression for k_1 in terms of Ra_T^2 and a. For the fully developed BLDR, Bejan and Tien (1978) proposed a $Ra_T^{-3/5}$ dependence. More recently, Hart (1983) derived from Cormack et al's works an expression for k_1 that is valid for the whole CDR and BLDR range. The expression is implicit but, for large values of Ra_T, one recovers the explicit relation

$$k_1 = (1024Q)^{-1/3} Ra_T^{-2/3} \tag{16}$$

which is independent of the aspect ratio a (Q is a constant parameter), as Bejan and Tien's relation. For cylinders Schiroky (1982) proposed an extension of the end-integral method used by Bejan and Tien (1978) for the prediction of k_1 (see also in Bontoux et al., 1986). When Ra_T is increased, three-dimensional flow structures superimpose to the basic flow as described by relations (15b-c).

<u>Multizone Thermal Models for Cavities</u>

With a multizone temperature profile, the analysis was developed by Extrémet et al. (1986) for the conduction regime (CDR). The temperature gradient in the middle of the ith temperature-gradient zone is not only connected to the aspect ratio a as in the 1-G model but also depends on the reduced width of the temperature gradient γ_i. The functional laws for k_1 are empirically determined from computations (see Fig.3) when the wall gradient is inserted between two isothermal regions (2-Z model) as

$$2k_1 = 1 \quad \text{at} \quad a\gamma_i < 1 \tag{17a}$$

$$2a\gamma_i k_1 = 1 \quad \text{at} \quad a\gamma_i > 1 \tag{17b}$$

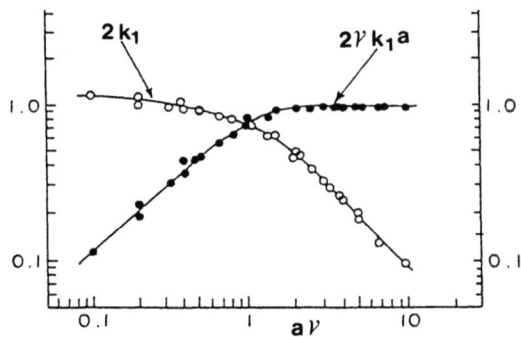

Fig.3. Variation of the temperature gradient k_1 in 2-Z models with $10 \geq a \geq 1$ and $1 \geq \gamma \geq 0.05$ for conduction regime. From Extrémet et al., 1986.

The maxima of the velocity profile (14) can be expressed by:

$$|w^{2D}_{max}| = A/9\sqrt{3} \tag{18}$$

with $A = Ra_T \, k_1 \, \beta_i \, \delta_i$ and where β_i is a discrepancy factor determined from comparisons with computed solutions (Extrémet et al., 1986). β_i was found to vary between 1 and 0.4 for a wide range of aspect ratios, a, and reduced widths of the temperature gradient, γ_i. The factor δ_i is 1 for the 2-Z model and is proportional to the local reduced temperature step, as $\Theta_m = (T_m - T_1)/\Delta T$ and $(1 - \Theta_m)$ for the 3-Z model (see Extrémet et al., 1986).

<u>Binary Mixture</u>

The parallel flow solution is extended to the case of thermal and solutal convection, as follows, where $k^S{}_1$ is related to the axial concentration gradient, $k^S{}_1 = 0.5(\partial S/\partial z)$:

$$w^{2D} = (k_1 Ra_T \beta_i \delta_i + k^S{}_1 Ra_S/Le)(x^2 - 1)x/6 \tag{19}$$

The motion corresponds again to two counterflows, but which direction depends on the sign of $(k_1 Ra_T \beta_i \delta_i + k^S{}_1 Ra_S/Le)$. For the 1-G model and at low Ra_T, the horizontal gradients of temperature and concentration are constant in the core. Both these gradients are then nearly $(2a)^{-1}$. At the large Ra_T, the functional laws available from Bejan and Tien (1978) and Hart (1983) can be used as a first approximation for a binary fluid. Also, the solution (19) was discussed with respect to computed solutions by Extrémet et al. (1986) for multizone models.

NUMERICAL SOLUTIONS

The solution of governing systems (4-7) and (11-13) was made by using different numerical methods. Classical finite differences known for their "robustness" were used for the computation of the complex three-dimensional flow patterns in cylinders. In cavities filled with a binary mixture higher-order accurate methods were considered as the hermitian finite-difference method and the spectral method based on Chebyshev-polynomial expansion of the variables. Their characteristics are briefly recalled with the details corresponding to the present applications.

Classical Finite-Difference Methods

The approximation is local and based on the classical explicit relations between the derivatives and the variables over three discretizing points. With a constant step size in space, Δz, the relations are second-order accurate and written as:

$$(\partial f/\partial z)_k = (f_{k+1} - f_{k-1})/2\Delta z \tag{20a}$$

$$(\partial^2 f/\partial z^2)_k = (f_{k+1} - 2f_k + f_{k-1})/\Delta z^2 \tag{20b}$$

The details of the method developed for the three-dimensional convection problem are reported in Leong and De Vahl Davis (1979), Leong (1983) and Smutek et al. (1983, 1985). The method is based on centered finite differences with an uniform mesh composed of $L \times M \times N$ discretizing points in r-, ϕ- and z-directions as $r_i = (i-1/2)\Delta r$ with i=1 to L, $\phi_j = (j-1)\Delta \phi$ with j=1 to M and $z_k = (k-1)\Delta z$ with k=1 to N, where $\Delta r = 1/(L-1/2)$, $\Delta \phi = 2\pi/M$, $\Delta z = 2a/(N-1)$. No mesh points are located on the axis to avoid singularity problems (De Vahl Davis, 1979).

The continuity equation, $\nabla \cdot V = 0$, is not automatically satisfied. Then, it is used to check the accuracy resulting from a given mesh size (see Bontoux et al., 1986) The solution of Poisson velocity equations (13) is made with the Fourier serie direct method (Le Bail, 1972) and using the FFT algorithm (Cooley and Tukey, 1965). The advancement in time is based on Samarskii-Andreyev ADI scheme (Mallinson and De Vahl Davis, 1973, Samuels and Churchill, 1967, Peaceman and Rachford, 1955).

The convergence of the solution towards a steady state with a 9x32x33 mesh can be obtained with less than 100 iterations up to $Ra_T \approx 4000$ and starting from initial conditions derived from relation (15) at a=5 and Pr=0.73. After the transition to the boundary layer driven regime (BLDR), convergence is generally obtained by incrementing Ra_T. For the most severe conditions at $Ra_T = 18,720$ and when the axis is tilted from the horizontal to nearly the vertical ($90° \geq \Omega \geq 20°$) the convergence is achieved after about 500 to 3500 iterations using a 9x32x65-mesh and varying progressively Ω by steps of 10° to 15°. The computation cost per iteration is about 0.32, 0.60 and 1.19 sec on CRAY1/S with 9x32x33, 65 and 129-meshes, respectively.

Hermitian Finite-Difference Methods

The method considers implicit fourth-order accurate relations between the derivatives and the variable at three discretizing points (Krause and co-workers, 1972-76, Hirsh, 1975). The derivatives are considered as additional unknowns of the problem but can be eliminated in the case of linear solvers as in Poisson's equation (Adam, 1977). The implicit relations are:

$$(\partial f/\partial z)_{k+1} + 4(\partial f/\partial z)_k + (\partial f/\partial z)_{k-1} = 3(f_{k+1} - f_{k-1})/\Delta z \qquad (21a)$$

$$(\partial^2 f/\partial z^2)_{k+1} + 10(\partial^2 f/\partial z^2)_k + (\partial^2 f/\partial z^2)_{k-1} = 12(f_{k+1} - 2f_k + f_{k-1})/\Delta z^2 \qquad (21b)$$

The method was applied to the two-dimensional governing system, using for the closure of the system additional relations at the boundaries as the Padé approximant (Roux et al., 1978, Bontoux et al., 1978, Peyret and Taylor, 1982). With the vorticity and streamfunction formulation a third-order accurate relation is considered for the vorticity at the walls. A mixed method was used here and based on classical finite differences for the transport equations combined with the hermitian method for Poisson's equation and Hirsh's relation for the wall vorticity (Loc and Daube, 1978, Roux et al.,1980). The integration is made with an A.D.I. false-transient method (see Peaceman and Rashford, 1955, Mallinson and De Vahl Davis, 1973). The false-transient method is used which introduces relaxation factor to improve the convergence process.

For a converged solution obtained after 240 iterations with 21x81 discretizing points the computations take 3.15 seconds on CRAY1/S computer.

Spectral Methods

The spectral methods are based on the method of the weighted residuals and on trial-function expansions of the variables. Their use has become very popular as numerical methods after pioneering works by Orszag (1971) (see Gottlieb and Orszag, 1977) and the development of vector computers. The approximation gives highly accurate derivations in the spectral plane. The method is based on powerful vectorized algorithms as Fast Fourier Transforms (Temperton, 1983) and Poisson-Helmholtz solvers (Haidvogel and Zang, 1979, Bondet de la Bernardie, 1980).

In the case of Chebyshev-polynomial expansions the variable and its derivatives are expressed as:

$$f \approx \sum_{n=0}^{N} a_n \, T_n(x) \tag{22a}$$

$$(\partial f/\partial z) \approx \sum_{n=0}^{N} a_n^{(1)} \, T_n(x) \tag{22b}$$

$$\text{with } a_n^{(1)} = (2/c_n) \sum_{\substack{p=n+1 \\ p+n \text{ odd}}}^{N} p \, a_p$$

$$(\partial^2 f/\partial z^2) \approx \sum_{n=0}^{N} a_n^{(2)} \, T_n(x) \tag{22c}$$

$$\text{with } a_n^{(2)} = (1/c_n) \sum_{\substack{p=n+2 \\ p+n \text{ even}}}^{N} p(p^2-n^2) \, a_p$$

The solution of system (4-7) is made with the Tau-Chebyshev spectral method using Chebyshev polynomial expansions both in x- and z-directions (see Bondet de la Bernardie, 1980, Elie et al., 1983, Randriamampianina, 1984) as

$$(\zeta, \theta, \psi, S) \approx \sum_{n=0}^{N} \sum_{m=0}^{M} (a,b,c,d)_{nm} \, T_n(x) \, T_m(z) \tag{23}$$

It results in a set of ordinary differential equations in the spectral space for the transport equations as:

$$\frac{da_{nm}}{dt} + \eta_i (e_{nm}^{10} + e_{nm}^{01}) - \eta_d (a_{nm}^{20} + a_{nm}^{02}) - \eta_{tb} \, b^*_{nm} - \eta_{sb} \, d^*_{nm} = 0 \tag{24}$$

and for the Poisson equation as

$$c_{nm}^{20} + c_{nm}^{02} = a_{nm}, \tag{25}$$

with $0 \le n \le N-2$ and $0 \le m \le M-2$ and where b^*_{nm} and d^*_{nm} are expressed similarly as $b^*_{nm} \sim \cos\Omega \, b^{10}_{nm} + \sin\Omega \, b^{01}_{nm}$. The superscripts ij for the variables a, b, c and d refer to the components of the ith x- and j^{th} z-derivatives of ζ, θ, ψ and S (diffusion and source terms). The components e_{ij}, h_{ij} and g_{ij} correspond to the non-linear convective terms. The closure of the system with the conventional boundary conditions was detailed elsewhere (Bontoux et al., 1981, Elie, 1984, Randriamampianina, 1984).

Among the various FFT algorithms available for the evaluation of the non-linear terms (Singleton-IMSL, CRAY-SCILIB library, Lhomme et al., 1982, Temperton, 1983), the algorithm of Temperton is chosen as it saves computation time on vector computers and works for any numbers of Fourier components such that $2^p 3^q 5^r$. Poisson equation (25) is solved using the Matrix Diagonalization Technique (Haidvogel and Zang, 1979, Bondet de la Bernardie, 1980).

Various schemes can be used for the integration in time (Adams-Bashforth, Adams-Bashforth-Crank-Nicolson schemes, Runge-Kutta, LSODA methods). The multistep LSODA method (Livermore Solver for Ordinary Differential Equations with Automatic method switching between non-stiff and stiff problems, ODEPACK) utilizes two predictor-corrector schemes adapted for non-stiff and stiff problems, respectively the Adams-Moulton scheme (AM) and the Backward-Differentiation Formula (BDF) (for details see Hindmarsh, 1976, 1982, Petzold, 1980).

With the LSODA method, no crucial stability problem was encountered as both the time step and order of integration are adapted automatically. The cost of LSODA is approximatively the same as the explicit AB method at small (NxM). When (NxM)>400 the iteration cost (as the memory required) increases strongly due to the internal generation of the jacobian which varies as 2(NxM) (see Randriamampianina et al., 1985).

RESULTS

The results bring some insights about three points of interest concerning the prediction of vapour transport in closed tubes.

- The first point is related to the understanding of the complex three-dimensional flows which develop in cylinders when the buoyancy force is governed by the temperature differences and when the inclination varies. The finite difference solutions are analysed with respect to experimental and theoretical results when they exist.

- The second point concerns the validity of two-dimensional solutions to predict significantly the main features of the flow patterns. Two-dimensional finite difference and spectral solutions are analysed with respect to the various results in a horizontal cylinder.

- The last point is related to the analysis of the fluid dynamics of physical vapour transport in a two-dimensional model with various axial wall temperature models. The analysis is made for a binary mixture in some horizontal and vertical situations.

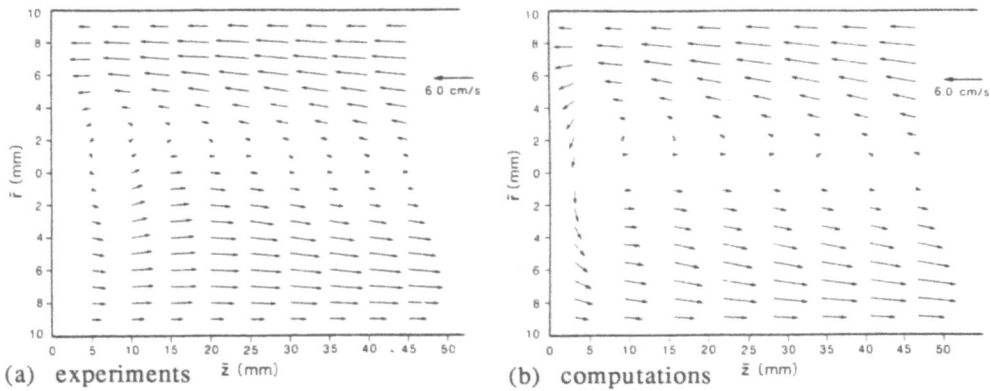

Fig.4. Velocity fields in vertical midplane ($\phi=0°,180°$) for $Ra_T=18,720$, a=5 (L=10cm, R=1cm) and $\Omega=90°$: (a) experiments, (b) computations. From Smutek et al., 1985.

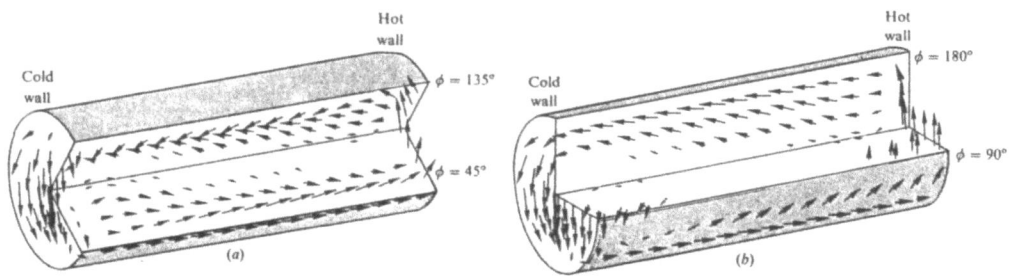

Fig.5. Three-dimensional velocity field (magnitude larger than 20% of the maximum) at various azimuths: (a) $\phi=45°,135°$ and (b) $\phi=90°,180°$; $Ra_T=18,720$, a=5. From Bontoux et al., 1986.

Three-Dimensional Convection in Tilted Cylinders

Convection regimes in horizontal cylinders. The core driven regime (CDR) is characterized by a parallel flow in the core and the w(r)-profile is S-shaped as predicted by relation (15a) (see Smutek et al., 1985, also in following Fig.18 for $Ra_T=660$). At $Ra_T \approx 18,700$ which is after the transition to the BLDR for a=5 (L=10cm, R=1cm, see Bontoux et al., 1986), the flow is inclined with respect to the axis (see Fig.4). The maximum w-velocity in the core has increased and shifted towards the walls (see following Fig.18 for $Ra_T=18,720$). As shown in Fig.5 secondary vorticies develop near the end walls and boundary layers expand along the lateral walls (see Schiroky and Rosenberger, 1984, Smutek et al., 1985). In Fig.5 the fully three-dimensional velocity patterns are given at four regularly spaced azimuths : $\phi=\pi/4$, $3\pi/4$, $\pi/2$ and π and the velocity vectors are plotted only when they exceed some 10% of their maximum. In the regions hidden by the lateral (shaded) surfaces, they are only plotted at the last row in the r-direction. The circular section at the left of the cylinder corresponds to the cross-section at the plane nearest the cold end wall. The cross-flow structure superimposed on the main counterflows in the entire cavity is demonstrated in Fig.6 at various

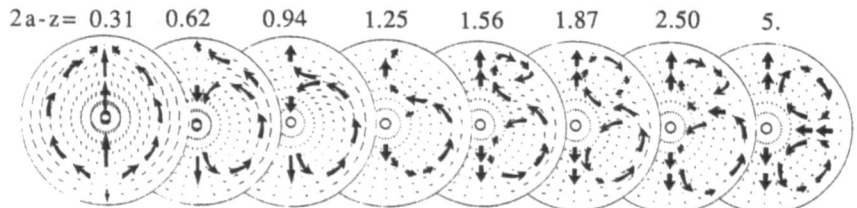

Fig.6. Flow structures in the (r,φ)-plane at various vertical cross-sections between the hot wall (z=2a) and the centre (z=a) for $Ra_T=18,720$ and a=5. From Smutek et al., 1983.

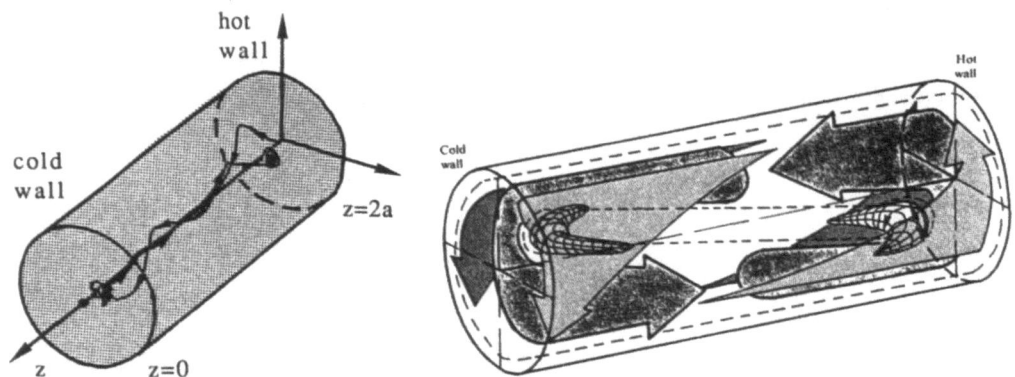

Fig.7. Track of a particle released near the axis in the hot end region for $Ra_T=18,720$ and a=5.

Fig.8. Schematic flow pattern in a horizontal cylinder in the BLDR. Primary, secondary flows. From Bontoux et al., 1986

cross-sections (r,φ) between the hot wall and the centre. The track of a fluid particle released at a point near the axis of the cylinder in the hot region is displayed in Fig.7. It reveals the occurence of both secondary vortices which set near the hot and cold end walls, and azimuthal transport in the axial direction (refer to the helicoidal flow structures emphasized in Fig.6). A sketch of the complete flow pattern is presented in Fig.8 at $Ra_T \approx 18,700$ (see Bontoux et al., 1986).

Effect of the inclination Ω. Experimental results by Rosenberger and co-workers are available in an a=5 cylinder (L=10cm, R=1cm) when the axis is tilted at an angle of $20° \leq \Omega \leq 150°$ with the gravity (see Schiroky and Rosenberger, 1984, also in Bontoux et al., 1986) at $Ra_T=3580$ and 18,720 which correspond to supercritical values in vertical cylinders heated from below ($Ra_{Tc} \approx 431a$ when $\Omega=0°$ and a>>1). The computations were carried out for severe flow patterns at $Ra_T=18,720$ and $20° \leq \Omega \leq 90°$ (heating from below) with a 9x32x65-mesh.

The computed and experimental w-results in the core (z=5) are plotted vs. radius in Fig.9a. The comparison shows fairly good agreement (less than a 10% difference on the maximum of w when $30° \leq \Omega \leq 60°$). Between $\Omega=90°$ and $30°$, the w-profile changes from a Z-shape in BLDR at $\Omega=90°$ to a S-shape (again, as in CDR at $\Omega=90°$) at $\Omega=30°$. The

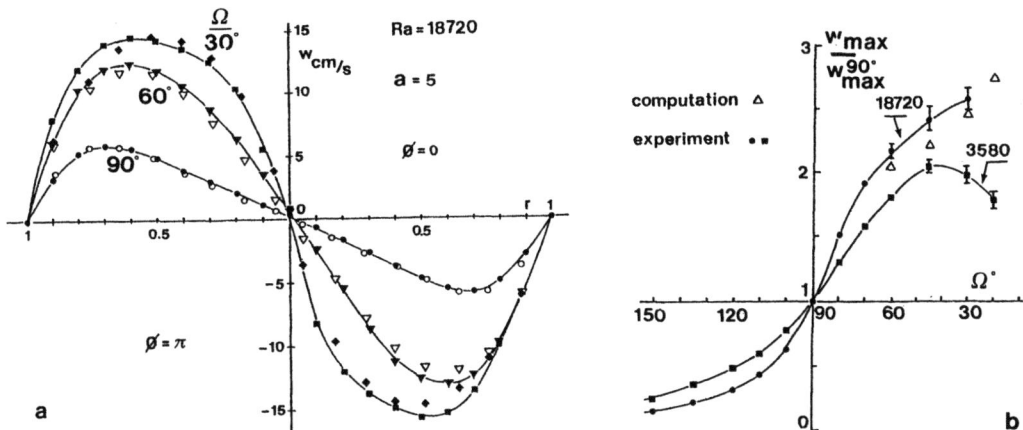

Fig.9. Effect of inclination Ω: (a) axial velocity (w) profiles vs radius at various Ω, computed ($\bigcirc, \triangledown, \blacklozenge$) and experimental ($\bullet, \blacktriangledown, \blacksquare$) results; (b) maximal core w-velocity vs Ω, computed ($\triangle$) and experimental ($\bullet, \blacksquare$) results. From Hurford et al., 1982/84, Bontoux et al., 1986.

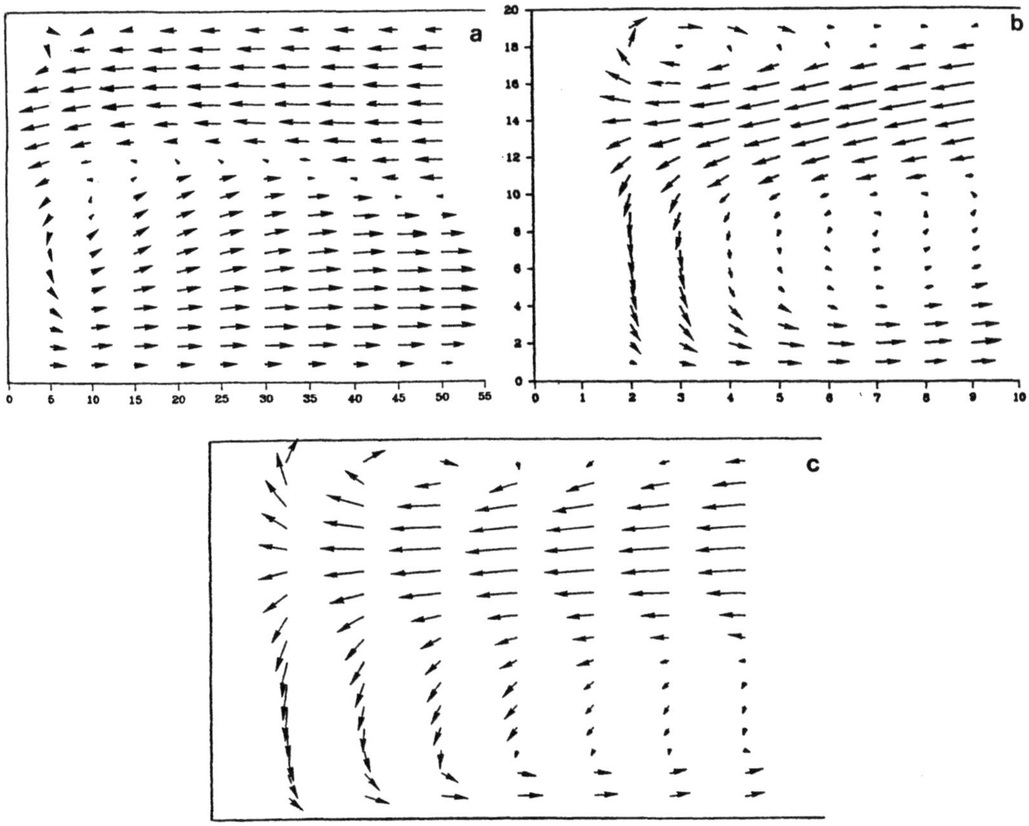

Fig.10. Velocity fields in the vertical midplane ($\phi=0°,180°$) at an inclination angle $\Omega=30°$ for $Ra_T=18,720$ and a=5. Experiments (a,b) and computations (c) over $0 \leq z \leq a$ (a) and $0 \leq z \leq a/5$ (b,c) from the cold end. (Refer to Fig.4.) From Hurford et al., 1982/84, Bontoux et al., 1986.

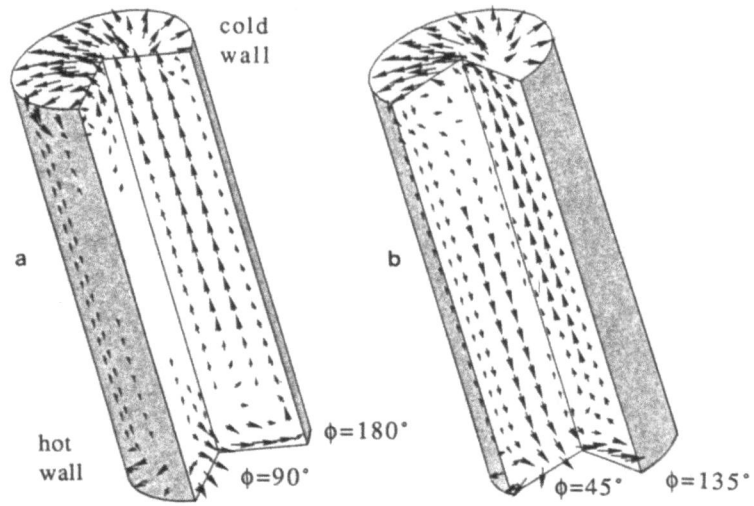

Fig.11. Three-dimensional velocity field (magnitude larger than 20% of the maximum) at various azimuths: (a) $\phi=90°,180°$ and (b) $\phi=45°,135°$; $Ra_T=18,720$, a=5, $\Omega=30°$. From Bontoux et al., 1986.

variation vs. Ω of the maxima of w is displayed in Fig.9a. Experimental results exhibit a dependence on both Ω and Ra_T. At $Ra_T=3580$ the variation of w is well represented with $w_{max}(\Omega)/w_{max}(90°) \approx 1-\sin(2\Omega-\pi)$ up to $\Omega=20°$. The experiments reveal a slight asymmetry at $\Omega \leq 45°$ when $Ra_T=3580$ and $\Omega \leq 60°$ when $Ra_T=18,720$. Also, the numerical results for $Ra_T=18,720$ are shown in Fig.9b to slightly underpredict the experimental variation.

The velocity fields in the vertical symmetry plane and in the vicinity of the cold end wall are shown in Fig.10a-c. The basic flow coming from the hot end wall separates into two parts in plane $\phi=0-\pi$ at the junction with the cold wall : - the main part is turned downward by the cold wall and at a certain distance from the endwall the flow becomes parallel again to the axis ; - the second part is driven into a secondary counter-rotating vortex which develops in azimuth. The agreement is obvious between experiments (Fig.10b) and computations (Fig.10c) in both magnitude and direction; the discrepancies are small and mainly concern to the size and location of the vortices in the vertical plan .

The fully three-dimensional velocity field is given in Fig.11 at $\Omega=30°$ (same details as for $\Omega=90°$ in Fig.5). The velocity patterns show the occurence of a weak flow everywhere in the plane ($\phi=\pi/2, 3\pi/2$) except near the end walls where boundary layers and vortices develop. The secondary co-rotating vortex which exists at the cold end for $\Omega=90°$, has disappeared at $\Omega=60°$. The counter-rotating vortex (see Fig.10b-c) is much smaller and expands in azimuth in a way which is similar to the

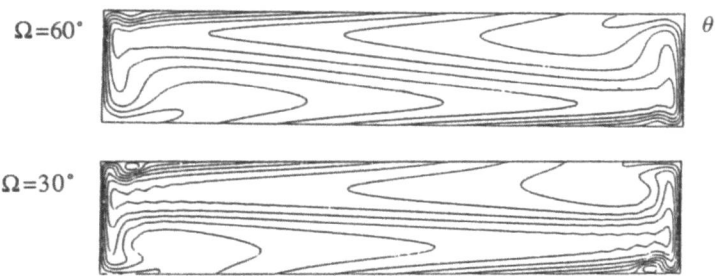

Fig.12. Isotherm patterns at $\Omega=30°$ and $60°$ for $Ra_T=18,720$ and a=5 (isotherms spaced by $\Delta T/10$). From Bontoux et al., 1986.

vertical case (Smutek, 1984). Except in the end regions the flow is mainly parallel to the axis as shown in Fig.11 for $\phi=\pi/4$, $3\pi/4$ and π.

The isotherm patterns displayed in Fig.12 exhibit distortions close to the end walls and in the core which result from the increase of transport at $\Omega<90°$ (see following Fig. 19 for $\Omega=90°$). The isotherms also reveal the onset of the secondary counter-rotating vortices shown in Fig.10 at $\Omega<60°$.

Axisymmetric and asymmetric regimes in vertical cylinders $(\Omega=0°)$. The first transition from rest to steady convection in a vertical cylinder heated from below is predicted by linear stability analyses (see Charlson and Sani, 1971, Gershuni and Zhukhovitski, 1976, Buell and Catton, 1983) to be dominated by the axisymmetric mode in flat cylinder ($a<a_c$) and by the antisymmetric mode in long cylinder ($a>a_c$). The study of these regimes is detailed elsewhere (see Smutek, 1984, Bontoux et al., 1986), however, some points are reported here.

Axisymmetric flows were computed at small a=0.5 and at $a=0.625 \approx a_c$. For this last value of aspect ratio a, both axisymmetric and asymmetric patterns were obtained depending on the initial disturbances imposed to the fluid at rest. Also, as shown in Fig.13, the axisymmetric solutions can involve either an upstream or a downstream at the centre.

Attention was focused on the asymmetric solution in an a=1 cylinder above critical Ra_{Tc}. At $Ra_T/Ra_{Tc} \approx 5.3$ secondary counter-rotating vortices are superimposed on the basic flow in the symmetry plane $(\phi=0,\pi)$ and develop in azimuth into a four-eddy pattern in the plane $(\phi=\pi/2,3\pi/2)$ as shown in the two-dimensional velocity fields displayed in Fig.14 at $Ra_T=6250$. The three-dimensional velocity field is displayed in Fig.15 which emphasizes the azimuthal expansion of the plane rolls shown in Fig.14. A schematic pattern is proposed in Fig.16. The occurence of such patterns is confirmed by solutions obtained with insulated side walls (see Müller et al., 1984).

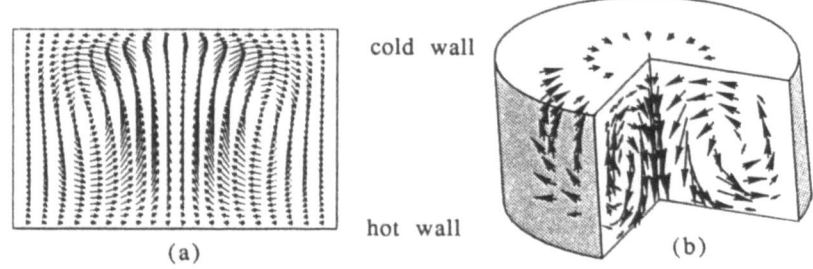

Fig.13. Axisymmetric velocity field at a=0.625, $\Omega=0°$ and $Ra_T=2000$: (a) ascending and (b) descending flow in the centre. From Smutek, 1984, and Bontoux et al., 1986.

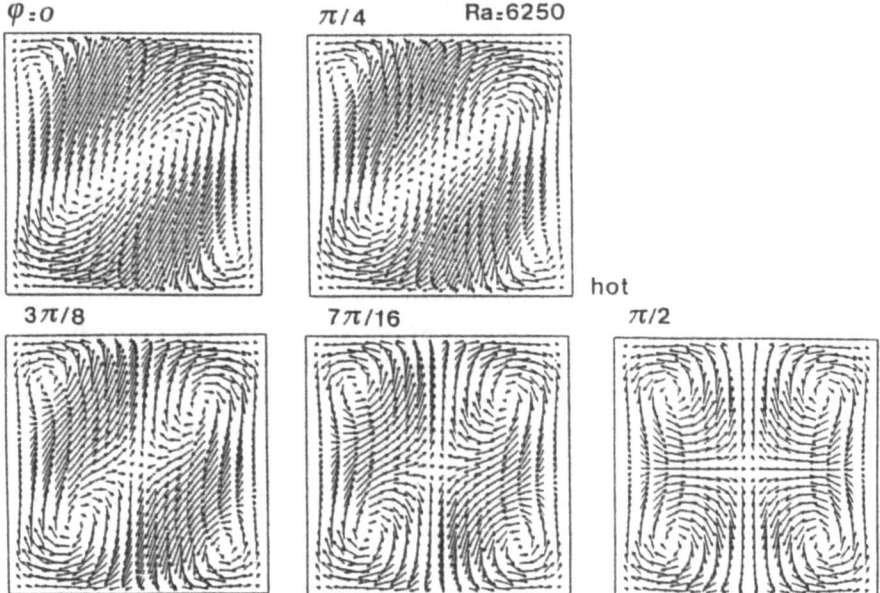

Fig.14. Asymmetric solution at a=1, $\Omega=0°$ and $Ra_T=6250$. Velocity fields in various planes (r,z). From Smutek, 1984.

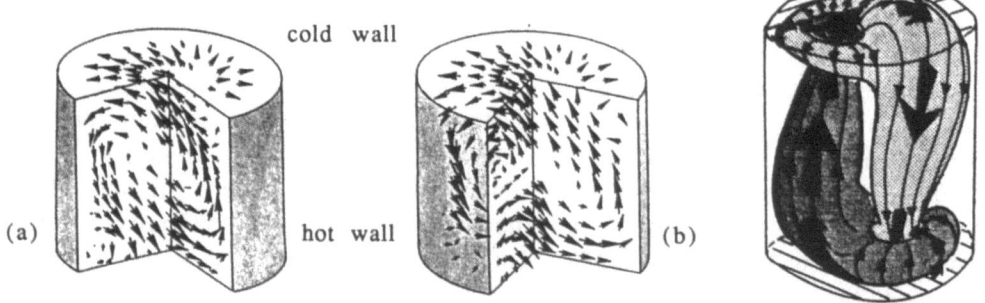

Fig.15. Three-dimensional velocity field (magnitude larger than 20% of the maximum) at various azimuths: (a) $\phi=45°,135°$ and (b) $\phi=90°,180°$; $Ra_T=6250$, a=1, $\Omega=0°$. From Bontoux et al., 1986.

Fig.16. Schematic flow pattern for the asymmetric regime in a vertical cylinder heated from below. From Bontoux et al., 1986.

Comparisons of Two- and Three-Dimensional Solutions in a Horizontal Cylinder

Analytical solutions for the maximum values of the horizontal w-velocity component can be derived from (14) and (15a) respectively as

$$w^{2D}_{max} = k_1 Ra_T / 9\sqrt{3} \qquad (26)$$

and

$$w^{3D}_{max} = k_1 Ra_T / 12\sqrt{3} \qquad (27)$$

These solutions are plotted for a=5 with the conduction condition $k_1=(2a)^{-1}$ in Fig.17 together with various experimental and numerical results. The numerical results are obtained with the three-dimensional (3-D) finite difference solutions (FD3D), the two-dimensional (2-D) hermitian finite difference (HFD2D) and Tau-Chebyshev spectral (TC2D) solutions. In addition, Fig.17 contains plots of (27) with evaluations of k_1 obtained for the BLDR and also with the third-order expansion term in ($Ra_T k_1$) proposed by Schiroky and Rosenberger (1984) (see Bontoux et al., 1986). The analysis of these curves results in the following observations:

(i) The experimental and 3-D numerical results agree well over the range covered by the numerical data (74<Ra_T<30,000). The 2-D results show significant deviations from these 3-D results not only, as expected, in the CDR but also in the BLDR.

(ii) At low Ra_T (in the CDR) the 3-D analytical solution (27) yields good predictions. The 2-D solution (26), though correctly reflecting the linear dependence of w_{max} on Ra_T, overestimates by 1/3 with respect to the cylinder.

(iii) At high Ra_T (i.e. in the BLDR) the experimental and 3-D numerical results for the core velocity vary as $Ra_T^{1/2}$, parallel to Gill (1966)'s relation for boundary-layer. The velocity in the cylinder is about 1.5 times larger than the 2-D numerical solution, which however shows a realistic Ra_T-dependence. The velocities obtained from the analytical solutions derived by Schiroky (1982) (see Bontoux et al., 1986) are too low and their Ra_T-dependence is nearly $Ra_T^{0.4}$.

(iv) If we define a critical Ra_{Tc} for the transition between the CDR and BLDR, then we find a factor of about 3 for a=5, between the 2-D ($Ra_{Tc}^{2D} \sim 2000$) and 3-D ($Ra_{Tc}^{3D} \sim 6000$) behaviours.

The above points are further illustrated by w(r)-profiles for two Ra values in the CDR and BLDR, and by streamline and isotherm patterns compared to 3-D velocity fields and isotherm patterns. In Fig.18a one sees that in the CDR ($Ra_T=660$) the 3-D analytical results agree well with the experimental and numerical values. The 2-D numerical values lie too

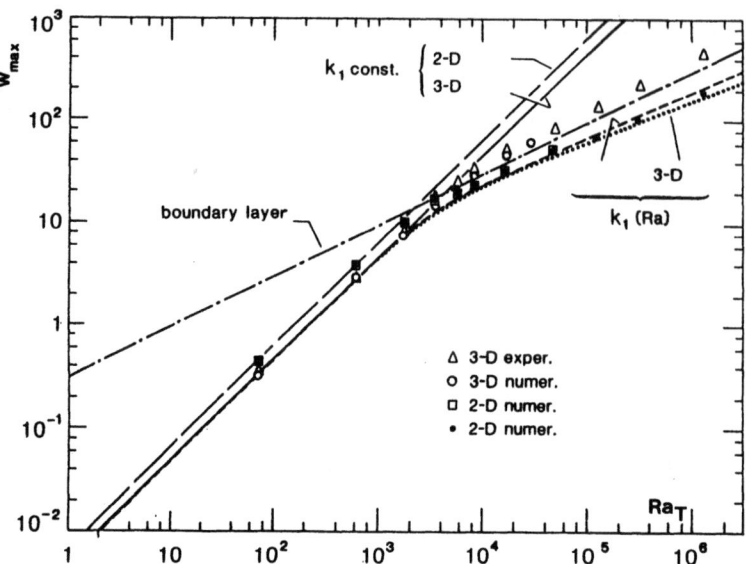

Fig.17. Maximum value of dimensionless axial velocity (core velocity) vs Ra_T for $\Omega = 90°$: comparison of experimental, analytical and numerical results.

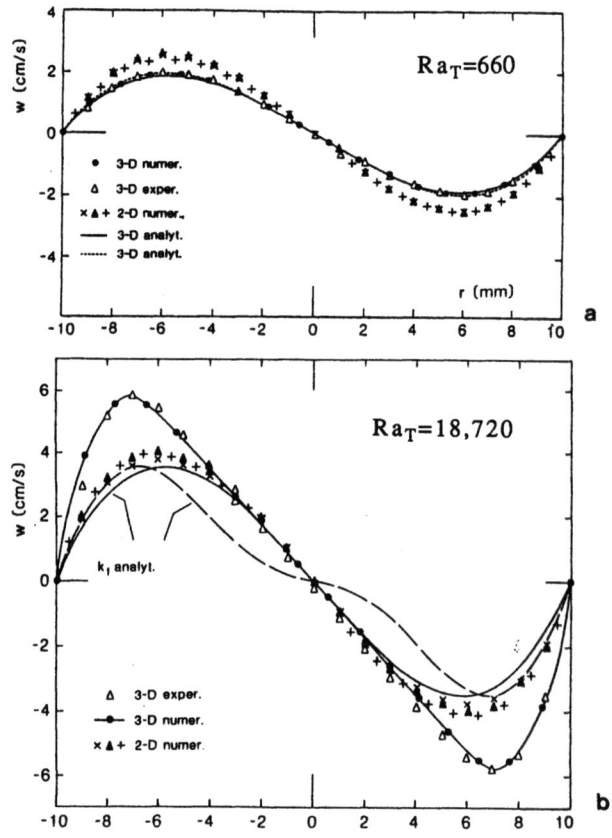

Fig.18. Core velocity (w) profiles for: (a) $Ra_T = 660$ and (b) $Ra_T = 18,720$. $\Omega = 90°$, a=5. From Bontoux et al., 1986.

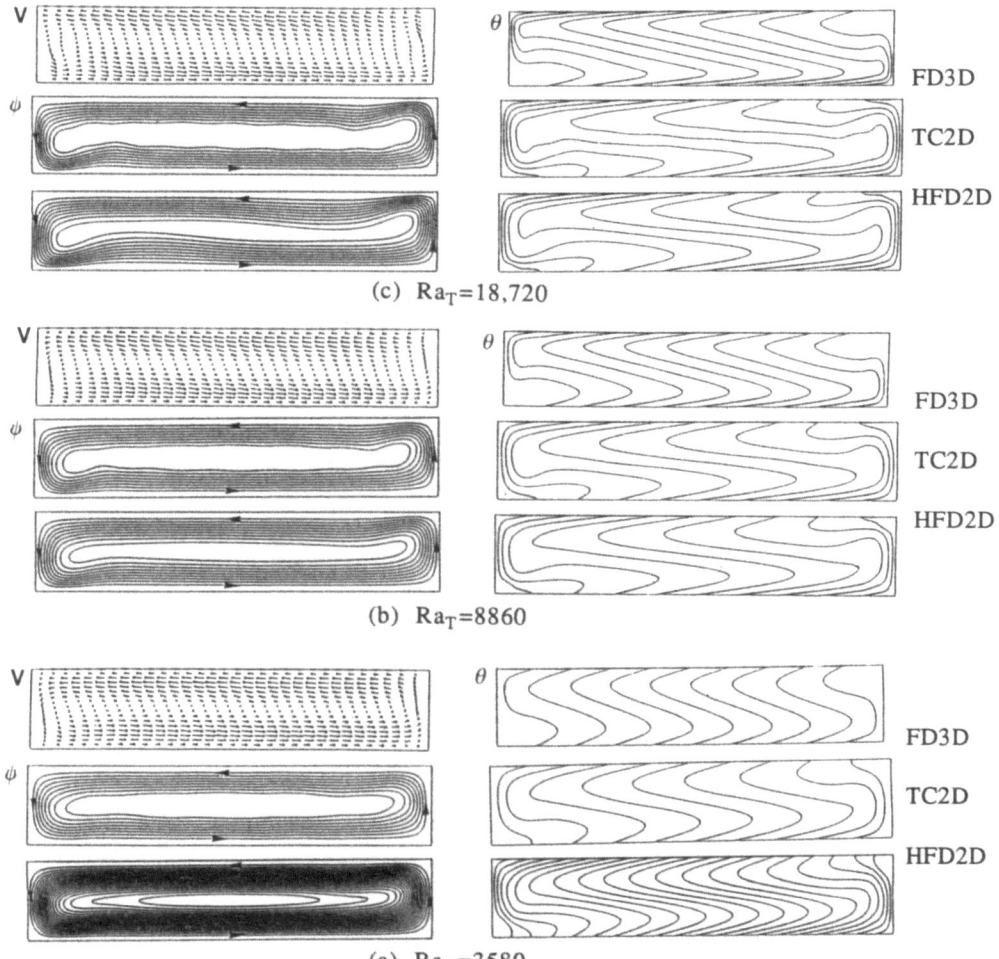

Fig.19. Computed (FD3D) velocity fields and isotherms in the vertical midplane of a cylinder. Computed (HFD2D and TC2D) streamlines and isotherms in a cavity. a=5 and $\Omega=90°$. Comparison for various Ra_T: (a) $Ra_T=3580$, (b) $Ra_T=8860$ and (c) $Ra_T=18,720$.

high. At $Ra_T=18,720$ (Fig.18b), the 2-D numerical results and 3-D analytical velocity profiles differ strongly from each other and from experiments. There are also distinctions into "S-shapes" and "Z-shapes". The evolution of the Z-shape from the S-shape occurs slower for 2-D than for 3-D solutions. One should point out the good agreement between experiment and 3-D numerical solutions.

The streamline and isotherm patterns are presented with the velocity fields computed in the vertical plane of the cylinder in Fig.19 near and above the transition to the BLDR. The agreement is rather good

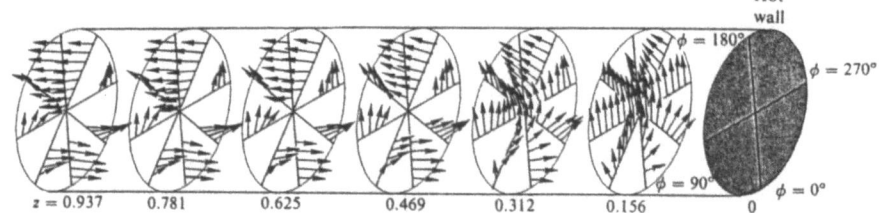

Fig.20. Three-dimensional velocity field (magnitude larger than 20% of the maximum) over a distance of a/5 from the hot wall. Ra_T=18,720, a=5. From Bontoux et al., 1986.

between the (21x41) finite difference solutions and the (12x27) Tau-Chebyshev solutions. The streamlines in the cavity remain nearly parallel to the horizontal walls up to Ra_T=18,720 while the velocity fields are already inclined at Ra_T=8860 in the cylinder. Also, the onset of secondary vortices near the vertical walls is revealed in both 2-D and 3-D computations but they arise differently near the cold wall : below the axis in the cavity, and above in the vertical plane of the cylinder. The distortion of the isotherms near the end wall is more important in the cavity than in the cylinder. This reflects the three-dimensional transport. As suggested by the experimental results of Schiroky and Rosenberger (1984) and observed in the 3-D numerical simulation by Smutek et al. (1985) and Bontoux et al. (1986), the U-turn of the flow occurs there over more than 180° in azimuth as displayed in Fig.20.

Simulation of Physical Vapour Transport in Cavities

The numerical solution of the governing system (4-7) is considered for PVT with both spectral Tau-Chebyshev method (TC2D) and mixed $O(h^2, h^4)$ finite-difference method (HFD2D) detailed in previous sections. Chebyshev-polynomial expansions are well adapted when the gradient zones are located near the boundaries. They are used with the 1-G thermal model (Fig.2a) and the linear Fick's law (8) at the interfaces. For a=4 the computations are carried out with no more than 17x17 Chebyshev components when $Ra_T \leq 6.10^3$ and $|Ra_S| \leq 2.10^3$ (Elie et al., 1986). For the multizone thermal models, 2-Z and 3-Z, the mixed finite differences are used with a 21x81 uniform mesh (Extrémet et al., 1986). Also, the simple analytical model (10) is taken at the interfaces.

Fick's Law and 1-G Thermal Model. Negative solutal Rayleigh numbers correspond to binary mixtures where the heavier component ($M_A \geq M_B$) is the active one (A) in PVT process ($S_{A2} \geq S_{A1}$). Then, solutal and thermal buoyancy terms act in opposite directions giving rise to more complex flow patterns than with $Ra_S > 0$. In Fick's model (8) the values for E (and F) are derived from relation (9) with $1 < S_{A0} < 0.5$ and $\Delta S_A < 0.5$. The computations are carried out for steady solutions in cavities with $1 \leq a \leq 4$ at horizontal ($\Omega = 90°$) and vertical inclinations (with heating from below, $\Omega = 0°$, and from above, $\Omega = 180°$).

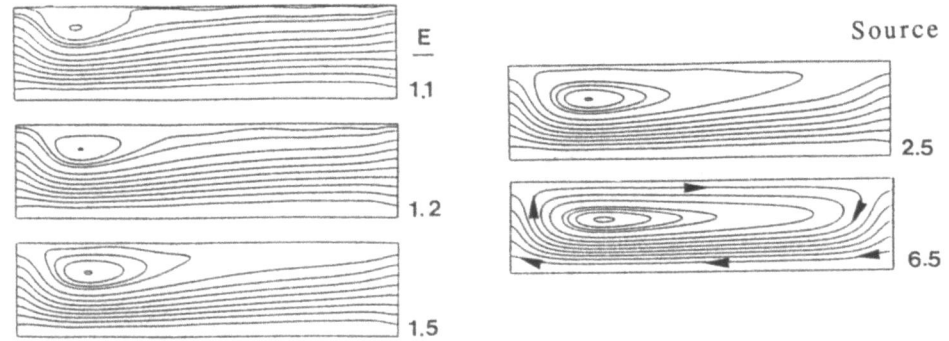

Fig.21. Streamlines for various E at $Ra_S=-12.5$, $Ra_T \approx 0$, $a=4$. (Elie, 1984)

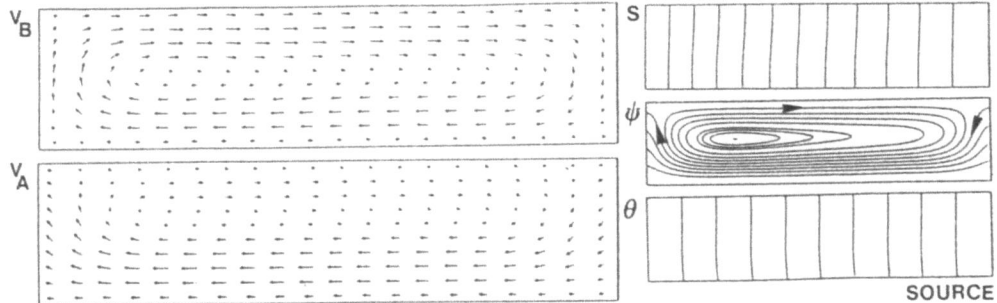

Fig.22. Streamline, isotherm, iso-concentration patterns and velocity fields (A-B components) for $E=10$ at $Ra_S=-12.5$, $Ra_T \approx 0$, $a=4$.

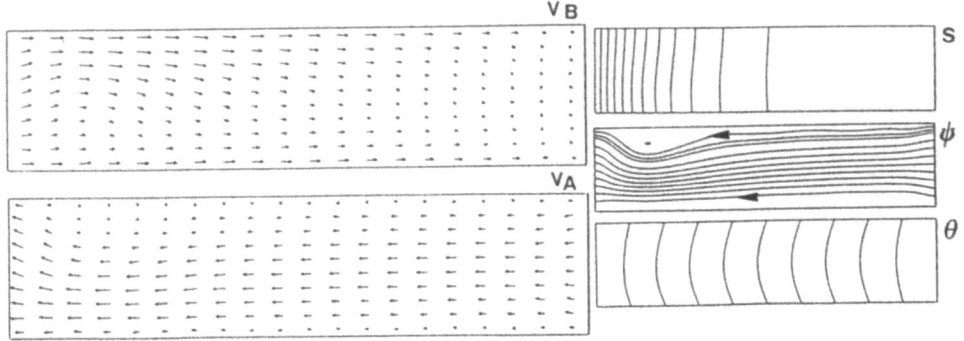

Fig.23. Streamline, isotherm, iso-concentration patterns and velocity fields (A-B components) for $E=1.05$ at $Ra_S=-12.5$, $Ra_T \approx 0$, $a=4$. From Elie, 1984, also Elie et al., 1986.

(i)Horizontal cavities ($\Omega = 90°$). In horizontal cavities the temperature and concentration gradients between the source and the sink are perpendicular to the gravity vector. As above mentioned a basic flow, then, exists even at very low Ra and can be descending at the (hot) source depending on the density gradient due to **dominant** temperature or concentration change. Also the mass flux of component A superimposes to thermal and solutal convection an advective diffusive motion through the cavity. A variety of such complex flows is studied with Pr=Le=1 at a=4.

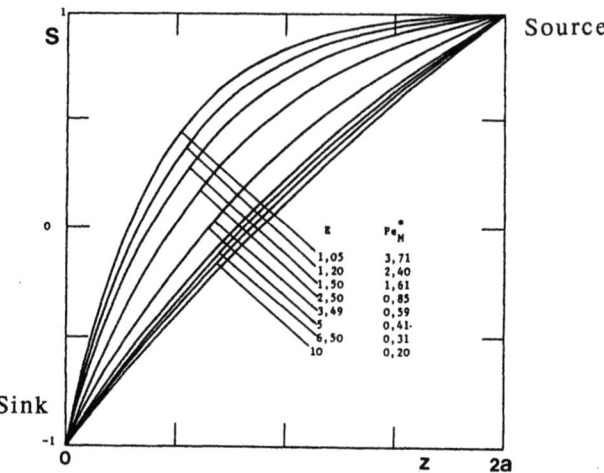

Fig.24. Computed concentration profiles along the axis for various E (related to a Peclet number) at $Ra_S=-12.5$, $Ra_T \approx 0$, a=4. From Elie, 1984. (Refer also to Greenwell et al., 1981.)

When $M_A > M_B$ the streamline pattern (at $Ra_S=-12.5$ and $Ra_T \approx 0$ in Fig.21) is strongly affected by an increase of the interfacial mass flux at the source (as E varies from 10 to 1.05). The basic cell is descending at the source and ascending at the sink. The mass flux of A organizes into a forced flow which develops from the source to the sink below the solutal cell. When the flux rate increases, the cell is progressively damped towards the sink. Typical velocity fields are given in Fig.22 and 23 with streamline, isotherm and iso-concentration patterns, at small (E=10) and large (E=1.05) flux rates and for F=10. When E=10 (Fig.22) the motion of B is mainly driven by the solutal cell, while component A is dominantly advected in the lower part of the cavity by the net flow between the interfaces. When E=1.05 (Fig.23) the advection from the source dominates. The concentration varies then strongly near the sink (see the concentration profiles at various E in Fig.24). The mass average and A-component velocity fields are very similar and the motion of B occurs in the regions where A-velocities are the lowest, i.e. in the solutal cell and near the horizontal walls.

When $M_A \approx M_B$ ($Ra_S \approx 0$) the basic cell is driven by the temperature changes. The basic cell is ascending at the hot source and descending at the cold sink (see Fig.25). Then, the net flow develops in the upper part of the cavity and gets more and more confined as Ra_T is increased. At $Ra_T=5860$ the streamlines suggest the onset of a secondary cell near the sink (crystal). (See in Fig.19 for a mono-component gas). Informative patterns and velocity fields are presented in Fig.26 to 28 for $Ra_T=19.5$, 586 and 1950. The flow of A is mainly driven by the net flow. The transport of B is associated with the thermal cell. Strong gradients develop at $Ra_T=1950$ (Fig.28) near the vertical walls.

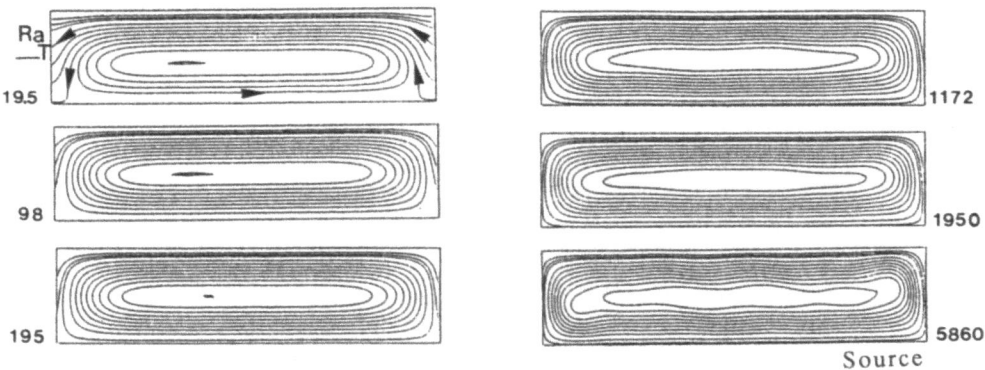

Fig.25. Streamline patterns when Ra_T varies at $Ra_S \approx 0$, $E=5.27$, $a=4$.

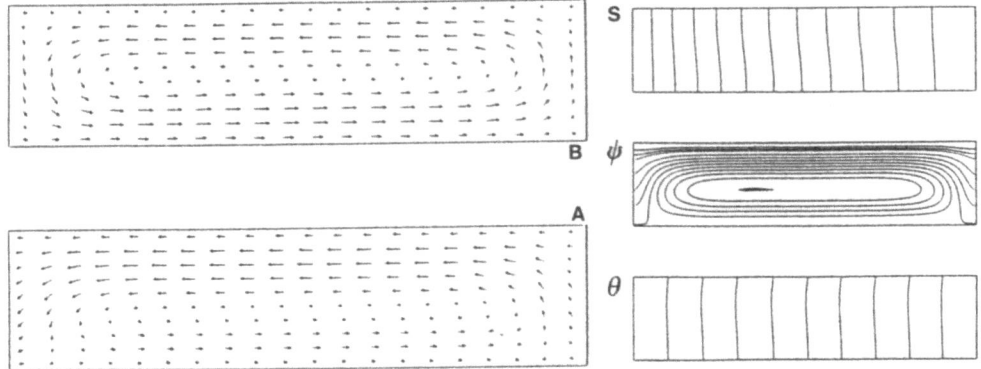

Fig.26. Streamline, isotherm, iso-concentration patterns and velocity fields (A-B components) for $E=5.27$ at $Ra_T=19.5$, $Ra_S \approx 0$, $a=4$.

When $Ra_T \approx |Ra_S|/Le$ there is a direct competition between the thermal and solutal buoyancy forces. At a large flux rate ($E=1.20$) the results are displayed in Fig.29 for $Ra_T=11.7$ and $Ra_S=-12.5$. The net flow dominates but the opposite effects of concentration and temperature gradients are still revealed by slight distortions in the streamline pattern. In the following section the competition will be emphasized for a multizone thermal model.

(ii) Vertical cavities heated from below ($\Omega=0°$). In a cylindrical enclosure filled with monocomponent gases, a steady antisymmetric instability is predicted to onset at $a=1$ by both stability analyses and computations (see above sections). Computed 2-D solutions involving a net flow ($E=2.3$) are found, however, to be axisymmetric at $Ra_T=3800$ and $Ra_S \approx 0$ with $Pr=1$ and $Le=0.114$ (Fig.30) in agreement with previous results by Markham et al. (1980, 1981). The flow is driven upward along the axis and two cells develop on both sides due to temperature gradients.

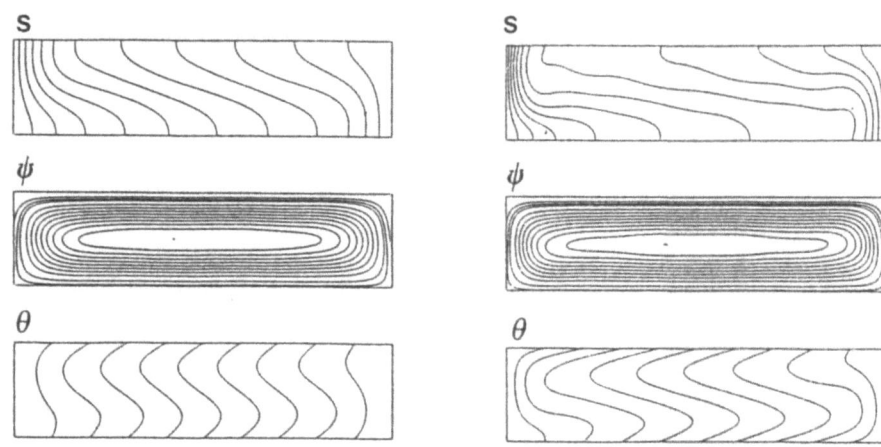

Fig.27. Streamline, isotherm, iso-concentration patterns for E=5.27 at $Ra_T=586$, $Ra_S \approx 0$, a=4.

Fig.28. Streamline, isotherm, iso-concentration patterns for E=5.27 at $Ra_T=1950$, $Ra_S \approx 0$, a=4.

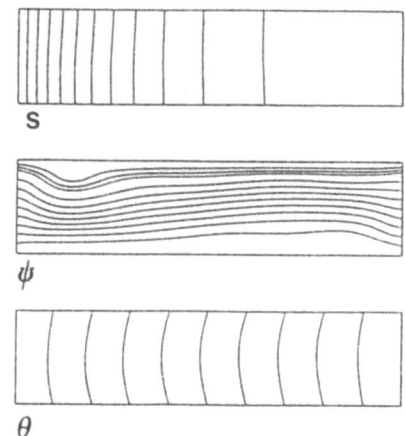

Fig.29. Streamline, isotherm, iso-concentration patterns for E=1.20 at $Ra_T=11.7$, $Ra_S=-12.5$, a=4.

(iii) Vertical cavities heated from above ($\Omega = 180°$). The computations are made again for a=1 with $Ra_T=500$, $Ra_S=-1660$, Pr=1, Le=0.18, E=2.3 ($S_{A0}=0.99$ and $\Delta S_A=0.009$). Due to dominant advection diffusion an axisymmetric downward motion develops along the axis (Fig.31) with two small side cells near the vertical walls and driven by the density gradients. When a=4 with $Ra_T=541$ and $Ra_S=-1475$ with Pr=1, Le=0.18 and same E=2.3 (Fig.32) the flow pattern is noticeably different. The fluid does not flow longer downward along the axis but now along the side walls with two recirculating cells in between symmetrically to the axis.

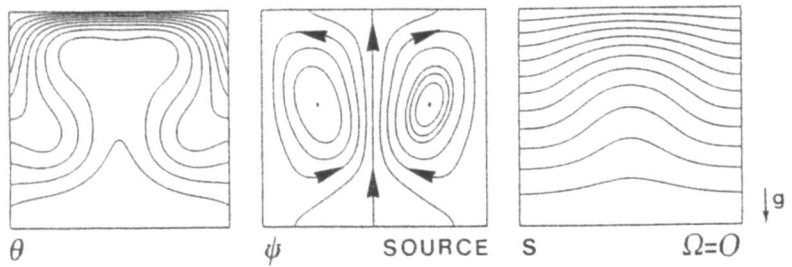

Fig.30. Streamline, isotherm, iso-concentration patterns for E=2.3 at $Ra_T=3800$, $Ra_S \approx 0$, Le=0.114, Pr=1, a=1 and $\Omega=0°$.

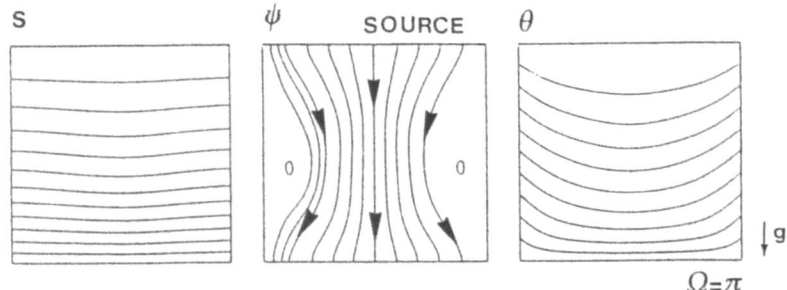

Fig.31. Streamline, isotherm, iso-concentration patterns for E=2.3 at $Ra_T=500$, $Ra_S=-1660$, Le=0.18, Pr=1, a=1 and $\Omega=180°$.

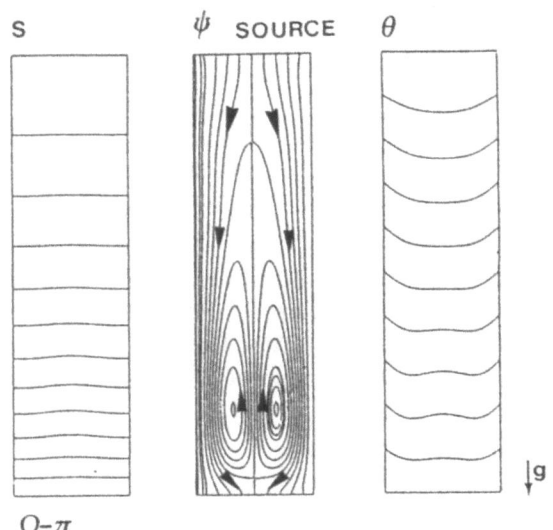

Fig.32. Streamline, isotherm, iso-concentration patterns for E=2.3 at $Ra_T=541$, $Ra_S=-1475$, Le=0.18, Pr=1, a=4 and $\Omega=180°$.

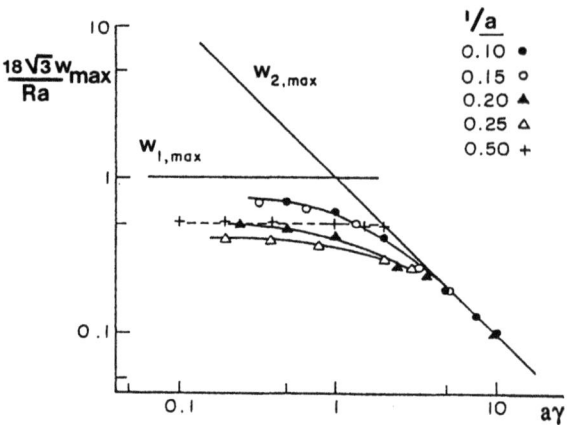

Fig.33. Variation of the maximum axial velocity with $a\gamma$ in the conduction regime for various computed values of a and the analytical laws (26) and (17). From Extrémet et al., 1986.

Multizone Thermal Models and Horizontal Cavities ($\Omega = 0°$). The analytical solution (26) with k_1 derived from (17a-b) have been compared in Fig.33 to the maximum velocities computed for 2-Z thermal models (Fig.2b) at various a and γ. The agreement is fairly good as the computed solutions shows a ($a\gamma$)-dependence when $a\gamma > 1$ and are independent on $a\gamma$ when $a\gamma < 1$. When $a\gamma < 1$ the computed values, however, stay substantially below the prediction by about 60% at most (refer to discrepancy factor β in (18)). Also, the analytical solution (19) is shown elsewhere to predict correctly the order of magnitude (at least) of the velocities for 3-Z thermal models (Fig.2c) and binary mixtures (see Extrémet et al., 1986).

With a 3-Z thermal model more than with a 1-G model, completely different flow patterns can be expected if $Ra_T \gg |Ra_S|/Le$ (thermal convection dominated regime), $Ra_T \ll |Ra_S|/Le$ (solutal convection dominated regime) or $Ra_T \approx |Ra_S|/Le$ (mixed regime). For the mixed regime the buoyancy terms have coefficients of comparable magnitudes everywhere and the density gradients determines locally the resulting ascending or descending motion. The flow patterns also depend on the strength of interfacial fluxes modelized with relation (10) and characterized by v_d.

(i) Without interfacial flux ($v_d = 0$) for $a = 4$ and $\Theta_m = (T_m - T_1)/\Delta T = 0.75$, the flow in the mixed regime ($Ra_S \approx -40$, $Ra_T \approx 19.5$, $Le \approx 1.4$) is composed of two counter-rotating cells. (See Fig.34 and compare with the solution for $Ra_S \approx 0$). The larger one, extending over 5/6 of the cavity, is driven by dominant solutal buoyancy. The motion is clockwise and the effect of the small temperature gradient (of width γ_2) which drives opposite buoyancy forces, is revealed by the streamline pattern.

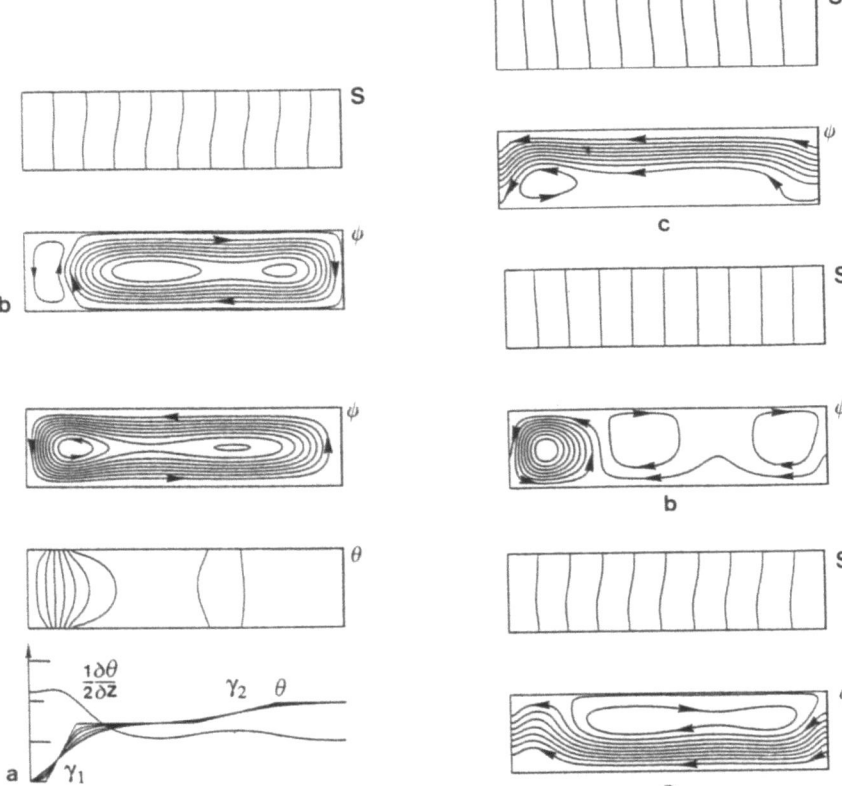

Fig.34. Streamlines, isotherms, iso-concentrations and axial temperature profiles for a=4, $Ra_T=19.5$, $Le \approx 1.4$, and: (a) $Ra_S=0$, (b) $Ra_S=-40$. From Extrémet et al., 1986.

Fig.35. Streamlines, iso-concentrations for a=4, $Ra_T=19.5$, $Le \approx 1.4$ and $|v_d|=2$ at various Ra_S: (a) $Ra_S=-40$, (b) $Ra_S=-21$, (c) $Ra_S=-3$. From Extrémet et al., 1986.

(ii) With an interfacial flux ($v_d \neq 0$) the modifications of the streamline pattern are displayed in Fig.35. The forced flow entering the cavity from the source, develops under the "solutal cell" then above the "thermal cell" (Fig.35a). In Fig.35b-c we show the flow patterns when Ra_S decreases. The forced flow mostly runs in the lower part of the domain when $Ra_T<|Ra_S|/Le$ at $Ra_S \approx -40$ (except in the close vicinity of the sink, see Fig.35a). When $Ra_T>|Ra_S|/Le$ ($Ra_S \approx -3$) it moves entirely in the upper part of the cavity (Fig.35c). In this case, the secondary cells in the bottom of the cavity are anti-clockwise and mostly driven by the thermal buoyancy force in the two gradient-zones. In Fig.35b, the thermal and solutal buoyancy balances as $Ra_T \approx |Ra_S|/Le$ ($Ra_S \approx -21$). Solutal buoyancy still dominates near the source but the basic "solutal" cell is, then, cut into two co-rotating cells by the second thermal gradient (γ_2). Near the sink, the "thermal" cell corresponding to the first gradient (γ_1) extends over the entire height. The forced flow visualized by one streamline, is shown to slide between these three cells.

CONCLUSIONS

The paper is an amalgam of various (published and un-published) numerical works which common purpose is the understanding of the fluid dynamics and the heat and mass transfer for vapour crystal growth applications in enclosures. The experimental studies carried out in the group of Prof. F. Rosenberger were used for the validation of the three-dimensional numerical solutions in cylinders. The results concern the following points:

(i) The complex flow patterns were analysed for various inclinations with respect to the gravity vector. The transition from conduction dominated regime (core driven regime, CDR) to boundary layer driven regime (BLDR) is characterized for a horizontal cylinder. A good agreement is obtained between computations and experiments in long tilted cylinders for the BLDR, and between computations and stability analyses at the onset of convection in vertical cylinders (at moderate aspect ratios).

(ii) The limitations of the 2-D model to predict the main features of the flow were studied for a mono-component gas and aspect ratio a=5. The velocity scales can be under- or over-estimated by 30 to 50% depending on Ra_T. The Ra_T for the transition between CDR and BLDR for the core velocity is underestimated with the 2-D solutions by a factor of 3 ($Ra_{T_c}{}^{2D} \approx 2000$, $Ra_{T_c}{}^{3D} \approx 6000$).

(iii) The analysis of a vapour growth process (PVT) was made with 2-D solutions and using mass-flux models at the interfaces. Both analytical approximations and direct simulations were considered for one-gradient and multizone (wall) temperature profiles. Very complex flow patterns were emphasized for various Ra_T and Ra_S. Agreement was obtained also with other numerical results obtained with different methods.

Experiments are in process in the laboratory of F. Rosenberger and data should be soon available for comparison with numerical solutions. Some effects connected with the presence of a seed located (for cristallization) at the side walls of a horizontal ampoule were discussed by Extrémet et al. (1986). Also, 3-D solutions were computed for multizone devices (Extrémet, 1986). A recent modelling has been used in the interface conditions which introduce a coupling between convection in the fluid and surface kinetics (see Zappoli and Elie, 1984, and Zappoli, 1986).

Acknowledgments: The authors want to gratefully acknowledge Dr. S.S. Leong for use of code CYL3D and Dr. J.M. Lacroix for participation in the elaboration of graphics softwares. They wish to thank Prof. R. Cadoret and Dr. J.C. Launay for focusing their attention on the fluid dynamics problems involved in their growth experiments. Collaboration with Dr. B. Zappoli is also acknowledged. Financial supports by the D.R.E.T. (G.6), the C.N.E.S. (Division "Matériau") and the C.N.R.S. (GRECO-64). Computing centers: C.C.V.R. (CRAY1/S), C.N.U.S.C. and P.A.-Saint Charles (K. Dang Quoc, G. Barisone, D. Pardini, J.L. Zubillaga).

REFERENCES

Adam Y., 1977, J. Comput. Phys., 24, 10-22.
Bejan A., Tien C.L.,1978, Int. J. Heat Mass Transfer, 21, 701.
Bejan A., Tien C.L., 1978, J. Heat Transfer, 100, 641-647.
Birikh R.V., 1966, J. Appl. Math. & Tech. Phys., 30, 432.
Bondet de la Bernardie B., 1980, Doct. Thesis, Univ. Aix-Marseille III.
Bontoux P., Forestier B., Roux B.,1978, J. Méc. Appl., 2, 3, 291-316.
Bontoux P., Bondet de la Bernardie B., Roux B.,1981, in Numer. Meth. Coupled Prob., Ed. Hinton, Bettess, Lewis, Pineridge, 1018-1030.
Bontoux P., Smutek C., Roux B., Lacroix J.M., 1986, J. Fluid Mech.,169, 211-227.
Bontoux P., Roux B., Schiroky G.H., Markham B.L., Rosenberger F., 1986, Int. J. Heat Mass Transfer, 29, 227-240.
Bontoux P., Smutek C., Roux B., Extrémet G.P., Schiroky G.H., Hurford A.C., Rosenberger F.,1986, in Numer. Meth. Non-Linear Prob., 3, Ed. Taylor, Owen, Hinton, Damjanic, Pineridge, 1102-1115.
Bontoux P., Smutek C., Randriamampianina A., Roux B., Extrémet G.P., Hurford A.C., Rosenberger F., 1986, in Advances in Space Research, Ed. Legros, Pergamon.
Bontoux P., Smutek C., Roux B., Hardin G., Sani R., 1986, Three-Dimensional Buoyancy Driven Flows in Cylindrical Cavities with Differentially Heated End Walls. Part II : Vertical Cylinders (submitted to J. Fluid Mech.).
Brisson P., 1981, Ing.-Doct. Thesis, Univ. Clermont 2, France.
Buell J.C., Catton I.,1983, J. Heat Transfer, 105, 2, 255-260.
Carruthers J.R., 1977, in Prep. and Prop. of Solid State Materials, 3, Ed. Wilcox and Lefever, Dekker.
Charlson G.S., Sani R.,1971, Int. J. Heat Mass Transfer, 14, 2157-2160.
Cormack D.E., Leal L.G., Imberger J.,1974, J. Fluid Mech. ,65 , 209.
Cormack D.E., Leal L.G., Seinfeld J.H.,1974, J. Fluid Mech. ,65 , 231.
Cooley J., Tukey J.W.,1965, Math. Comp., 19, 297.
De Vahl Davis G.,1979, Numer. Heat Transfer, 2, 261.
De Vahl Davis G.,1986, Proc. 8th Int. Heat Transfer Conf., ASME,101-109.
Elie F., Chikhaoui A., Randriamampianina A., Bontoux P., Roux B.,1983, Notes Numer. Fluid Mech.,7, Vieweg, 57.
Elie F.,1984, Doct. Thesis, Univ. Aix-Marseille II, France.
Elie F., Randriamampianina A., Bontoux P., Extrémet G.P., Roux B., 1986, in Numer. Meth. Non-Linear Prob., 3, Ed. Taylor, Owen, Hinton, Damjanic, Pineridge, 738-753.
Extrémet G.P., Bontoux P., Roux B.,1986, Int. J. Heat Fluid Flow (in print)
Extrémet G.P., Roux B., Bontoux P., Elie F.,1986, J. Crystal Growth(in print)
Extrémet G.P., Bontoux P., Roux B., 1986, in Advances in Space Research, Ed. Legros, Pergamon.
Extrémet G.P., 1986, "Numerical Modellings for Multizone Physical Vapour Transport. Two- and Three-Dimensional Solutions", to be published in Materials Sciences under Micro-Gravity, ESA Scien. & Tech. Publications.

Gershuni G.Z., Zhukhovitskii E.M.,1976, "Convective Stability of Incompressible Fluids", Keters/Wiley.
Gill A.E., 1966, J. Fluid Mech., 26, 3, 515-536.
Gottlieb D., Orszag S.A., 1977, CBMS-NSF Conf. Series Applied Math.
Greenwell D.W., Markham B.L., Rosenberger F.,1981, J. Crystal Growth, 51, 413.
Haidvogel D.B., Zang T.,1979, J. Comp. Phys., 30, 167.
Hart J. E.,1972, J. Atmos. Sci.,29, 687-697.
Hart J. E.,1983, Int. J. Heat and Mass Transfer, 26, 1069-1074.
Hindmarsh A.C.,LLNL Rep.,UCID-30130 ,1976 & UCRL-88007,1982
Hirsh R.S.,1975, J. Comp. Phys.,10, 90-109.
Hurford A.C., Schiroky G.H., Rosenberger F., 1982/84-Reports, Univ. Utah, Salt Lake City, USA.
Imberger J., 1974, J. Fluid Mech., 65, 247.
Jhaveri B.S., Markham B.L., Rosenberger F., 1981, Chem. Eng. Comm., 13, 65.
Jhaveri B.S., Rosenberger F., 1982, J. Crystal Growth, 57, 57.
Joseph D.D.,1976,"Stability of fluid motions II", Springer Tracts in Natural Philosophy, 28.
Klosse K., Ullersma P., 1973, J. Crystal Growth, 18, 167-174.
Krause E., 1972, V.K.I. Lecture Series on Numer. Meth. Fluid Dynamics, Rhode-St-Genèse, Belgium.
Krause E.,1974, Numerical solutions of the Navier-Stokes equations", L.N. I.C. for Mech. Sc., Udine, Italy.
Krause E., Hirschel E.H., Kordulla W., 1976, Computers and Fluids, 4, 2, 77-92.
Launay J.C., 1982, J. Crystal Growth,60, 185-190
Le Bail R.C.,1972, J. Comput. Phys., 9 , 440 ,
Leong S.S., De Vahl Davis G., 1979, Numer. Meth. Thermal Prob., Ed. Lewis & Morgan, Pineridge, 287
Leong S.S.,1983, PhD thesis, Univ. of NSW, Kensington, Australia.
Lhomme B., Morgenstern J., Quandalle P.,1982, Euromech159, Nice.
Loc T.P., Daube O., 1978, J. Mécanique,17, 5, 651-678.
Mallinson G.D., De Vahl Davis G., 1973, J. Comput. Phys.,12, 436.
Markham B.L., Rosenberger F., 1980, Chem. Eng. Comm., 5, 287.
Markham B.L., Greenwell D.W., Rosenberger F., 1981, J. Crystal Growth, 51, 426.
Markham B.L., Rosenberger F., 1984, J. Crystal Growth, 67, 241.
Müller G., Neumann G., Weber W., 1984, J. Crystal Growth, 70, 78.
Ouazzani J., Peyret R., 1983, Notes Numer. Fluid Mech.,7, Vieweg, 57.
Ouazzani J., 1984, Doct. Thesis, Univ. Nice, France.
Olson J.M., Rosenberger F., 1979, J. Fluid Mech., 92, 4, 631-642.
Omaly J., Robert M., Cadoret R., 1981, Mat. Res. Bull.,16, 1261.
Omaly J., Robert M., Brisson P., Cadoret R., 1983, Nuclear Inst. and Meth., North-Holland, 213, 19-26.
Orszag S.A.,1971, J. Fluid Mech.,149, 75.
Ostrach S., Loca R.R., Kumar A.,1980, Natural Convection in Enclosures, Ed. Catton & Torrance, ASME, HTD-8, 1.

Ostrach S.,1982, Ann. Rev. Fluid Mech., 14, 313-345.
Pamplin B., 1980, Int. Series Sciences Solid State, 16, Ed. Pamplin, Pergamon.
Peaceman D.W., Rachford H.H., 1955, J. Soc. Ind. Appl. Math., 3, 28.
Petzold L.R. ,1980, Sandia Nal. Lab. Rep. SAND80-8230.
Peyret R., Taylor T.D., 1982,"Computational Methods for Fluid Flow", Springer-Verlag.
Pimputkar M., Ostrach S., 1981, J. Crystal Growth, 55, 614-646.
Randriamampianina A.,1984, Doct. Thesis, Univ. Aix-Marseille II, France.
Randriamampianina A., Bontoux P., Roux B., Argoul P., 1985, Notes Numer. Fluid Mech. ,Vieweg ,13, 302.
Rosenberger F.,1979, Phys. Chem. Hydrodyn., 1, 1.
Rosenberger F.,1979, Fundamentals of Crystal Growth I, Solid State Sci., 5, Springer.
Rosenberger F., 1982, Convective Transport and Instability Phenomena, Ed. Zierep & Oertel ,Verlag-Braun , 469.
Roux B., Grondin J.C., Bontoux P., Gilly B., 1978, Numer. Heat Transfer, 1, 331-349.
Roux B., Bontoux P., Ta Phuoc Loc, Daube O., 1980, in L.N. in Math. 771, Springer, 450-468.
Samuels M.R., Churchill S.W.,1967, A.I.Ch.E. J., 13, 77.
Shih T.S.,1981, Int. J. Heat Mass Transf., 24, 1295-1303.
Schiroky G.H., 1982, Ph.D. Thesis, Univ. Utah, Salt Lake City, USA.
Schiroky G.H., Rosenberger, F.,1984., Int. J. Heat Mass Transf., 27, 587
Smutek C., Bontoux P., Roux B., Schiroky G.H., Hurford A., Rosenberger F.,1985, Numer. Heat Transfer, 8, 613 ,
Smutek C.,1984, Doct. thesis, Univ. Nice, France.
Smutek C., Roux B., Bontoux P., De Vahl Davis G.,1983, Notes Numer. Fluid Mech., Vieweg ,7, 338,
Solan A., Ostrach S.,1979, Preparation and Properties of Solid State Materials, 4, 63-110.
Temperton C.,1983, J. Comp. Phys., 52, 81.
Tichy J., Gadgil A., 1982, J. Heat Transfer, 104, 103.
Zappoli B., Elie F.,1984, "On Boundary Conditions for Hydrodynamic Equations in Reduced Gravity", 35th Int. Astronautical Federation Conf., Lausanne, Switzerland.
Zappoli B., 1986, "On the Interaction between Convection and Surface Reactions in Rectangular Horizontal Enclosures", (submitted to J. Crystal Growth).

ONSET OF OSCILLATORY CONVECTION IN HORIZONTAL LAYERS OF LOW-PRANDTL-NUMBER MELTS

Hamda Ben Hadid, Bernard Roux, Anthony Randriamampianina, Emilia Crespo* and Patrick Bontoux

Institut de Mécanique des Fluides , 1 Rue Honnorat, F-13003 Marseille
* U.N.E.D. Dept. Fís. Fund. , Apartado 50:487 , E-28.080 Madrid

INTRODUCTION

This study is devoted to horizontal layers of low-Prandtl-number, Pr, fluids subjected to buoyancy forces in a long rectangular cavity which vertical endwalls are maintained at different temperatures . Our main motivation is to study temperature fluctuations occurring during the growth of metals and semi-conductor crystals (like GaAs) in horizontal-boats (e.g. by Bridgman technique).

As shown in a survey paper given by Pimputkar and Ostrach (1981) a number of experiments in open horizontal boats containing molten metals (low Pr fluids) have shown that beyond a certain critical value of the Grashof number thermal oscillations occur (Cole and Winegard (1964) ; Hurle (1967) ; Utech et al. (1967) ; Carruthers and Winegard (1967) and Carruthers (1968)). Such oscillations in liquid metals cause undesirable impurities striations in melt-grown crystal.

Pimputkar and Ostrach (1981) pointed out that these oscillations were initially attributed to a variety of cyclic phenomena (Ueda (1961); Gatos et al.(1961)) but that it is now fairly conclusively established that one of the major causes of crystal striations is stable and periodic temperature fluctuations in the melt (Jakeman and Hurle (1972)). Experiments without solidification , carried out by Hurle et al. (1974) for pure molten Ga in open cavity, exhibit oscillations of few periods by second.

A tentative explanation of these oscillations has been done by Gill (1974) who developed a linear stability analysis which gives some qualitative agreement with the experiments of Hurle et al. (1974), but which doesn't contain the length L as a parameter and thus cannot confirm the strong L-dependency shown in experiments. More recently, Hart (1983a) proposed an other stability analysis, involving three-dimensional disturbances, but these results mainly applied to large cavities , due to the basic (steady state) solution used (i.e. the one cell Hadley circulation). As pointed out by Simpkins and Chen (1983), the numerical simulation already done by Cormack et al. (1974) and more recently by Hart (1983b) shows that this Hadley circulation is not appropriate for the domain of values of Gr at which oscillations occur. An other approach, in which the basic steady flow is more appropriate, has been proposed by Winters et al. (1986) in the case of an aspect ratio (height/length),Az=4.

The numerical study of the onset of oscillatory regimes in low-Pr fluids received an increasing interest these last five years. Two-dimensional simulations of buoyancy-driven flow (Crochet et al. (1983), Roux et al. (1984) and Winters et al. (1986)) show the existence of oscillatory regimes in rectangular open cavity. But, the agreement with the Hurle et al. (1974) experiments, concerning the threshold value of Gr for the onset of oscillations and the frequency of these oscillations, is not good enough to prove that the modelisation used is completely relevant . Three-dimensional computations, carried out by Crochet et al. (1986), show a decrease of about 40% of the threshold value, compared to the two-dimensional model ; but this value is again far from the experimental values . All these simulations do not allow us to conclude what is the exact cause of the oscillations.

Thermal oscillations have also been found in channel boats (Utech and Early,1967; Favier et al. 1986) and several authors mention other possible causes of oscillations as : (i) the competition between horizontal and vertical (adverse) temperature gradients (see Birikh et al. (1969) and Bhattacharyya and Nadoor (1976)) , (ii) the thermocapillary (Marangoni) shear on the upper open boundary (see the papers by Birikh (1966) and Davis et al. (1983)), (iii) the upper boundary deformation , (iv) the dissymmetry due to the strong temperature dependence on the transport properties, (v) impurities effects (double diffusion, Soret, ...).

The aim of the present paper is to contribute to review all these various possible causes of thermal oscillations in order to propose a more relevant modelisation (equations and boundary conditions) of the experiments . Unsteady two-dimensional Navier-Stokes equations are numerically solved by using a finite difference technique to analyze the behaviour of a low-Pr fluid layer in a rectangular cavity , the upper surface of which can be either rigid, free or subjected to a shear due to thermocapillary effect . This study is done for a few values of Pr including Pr=0, which corresponds to the case where the temperature field would be frozen .

POSITION OF THE PROBLEM

Experimental Configuration   (Hurle et al.,1974)

As in Hurle et al. (1974) experiments, we consider a parallelepipedic cavity having two differentially heated vertical endwalls, at temperatures T1 and T2, and filled up with a low-Prandtl-number fluid (see Fig.1). A buoyancy-driven flow occurs as soon as T1≠ T2, and becomes oscillatory beyond a certain critical value, $Gr_{c,osc}$, of the Grashof number defined as $Gr = g \beta G_h H^4/\nu^2$, where the cavity height, H , is taken as reference length and G_h = ΔT/L, with ΔT= T2- T1. L denotes the length of the cavity and Az its aspect ratio : Az = L/H .

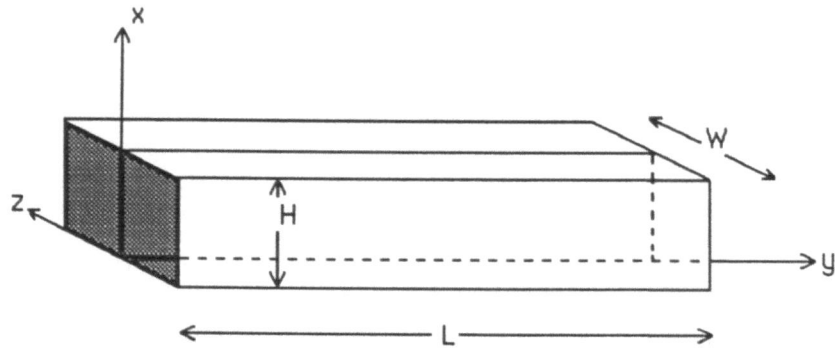

Fig.1   Geometry of the problem

Two-Dimensional Governing Equations

As far as the modelisation is concerned, we assume that the velocity is small enough to consider the flow as laminar . In addition , the fluid is assumed to be Newtonian and quasi-Incompressible (Boussinesq approximation) .

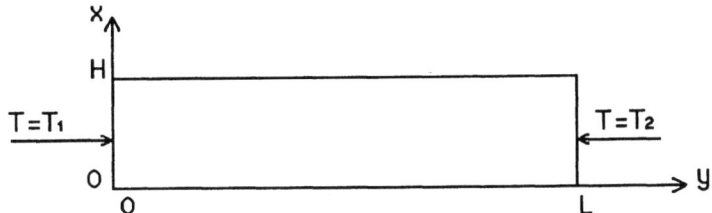

Fig. 2   Two-dimensional model

With the notation proposed by De Vahl Davis (1986), the equations can be written in the streamfunction and vorticity formulation, ψ and Ω, as ;

$$\Omega_t + V_l [u \Omega_x + v \Omega_y] = V_d [\Omega_{xx} + \Omega_{yy}] + V_b \theta_y ; \quad (1)$$
$$\psi_{xx} + \psi_{yy} + \Omega = 0 ; \quad (2)$$
with : $u = \psi_y$; $v = - \psi_x$ and $\Omega = v_x - u_y$. (3)

The transport equation for energy is

$$\theta_t + T_l [u \theta_x + v \theta_y] = T_d [\theta_{xx} + \theta_{yy}] ; \quad (4)$$

where : $\theta = (T-T_1)/ T_{ref}$,   with $T_{ref} = \Delta T / Az$.

Boundary Conditions

- on the rigid walls (x=0 , y=0 , y=Az) $u = v = 0$; (5)

- on the upper boundary (x=1) $u = 0$ and
  - either   v=0   (rigid boundary; noted **R-R** case) , (6a)
  - or $\partial v/\partial x=0$ (stress-free surface; noted **R-F** case), (6b)
  - or $\partial v/\partial x = f (\partial \sigma/\partial T)$ (Marangoni; noted **M** case) ; (6c)

- on the vertical walls $\theta_{(x,0)} = \theta_1 = 0$; $\theta_{(x, L)} = \theta_2 = Az$; (7)

- on the horizontal boundaries ; either $\theta_x = 0$ (adiabatic condition) (8a)
  - or $\theta = y$ (perfectly conducting condition) . (8b)

Dimensionless Parameters

Taking H^2/ν for time scale , we obtain: $V_d= 1$; $T_d = 1/ Pr$; $V_l = T_l = v_{ref}$ H / ν and $V_b = - g \beta H^2 T_{ref} /(\nu v_{ref})$. Thus, depending on the values of v_{ref}, different types of nondimensionalization can be considered :

A1 - For small values of Gr (viscous ~ buoyancy), after Hart (1983) we could take
$$v_{ref} = (\nu /H) Gr \quad (9a)$$
and thus : $V_l = T_l = Gr$ and $V_b = -1$

A2 - For larger values of Gr (inertia ~ buoyancy ; typically : Gr $\geq 10^4$) it is better to use (cf. Ostrach, 1976)
$$v_{ref} = (\nu /H) Gr^{0.5} \quad (9b)$$
and thus : $V_l = T_l = Gr^{0.5}$ and $V_b = - Gr^{0.5}$

A3 - In the case of strong thermocapillary (Marangoni) effect , but assuming that unsteadiness does not depend on Ma (i.e. keeping $t_{ref} = H^2/\nu$), we can choose

$$v_{ref} = (\nu/H)\ Ma/Pr \quad (\text{where } Ma = -\partial\sigma/\partial T\ G_h\ H^2/(\rho\nu\kappa)) \tag{9c}$$

and thus: $\quad V_l = T_l = Ma/Pr \quad$ and $\quad V_b = -GrPr/Ma$

Note that in this case, the condition (6c) writes: $\partial v/\partial x = -\partial\theta/\partial y$

Limiting Case Pr = 0

In the limiting case Pr=0 (finite ν and infinite κ), equation (4) simplifies to $\theta_{xx} + \theta_{yy} = 0$; the solution of which, taking into account (7) and (8), is :

$$\theta = y \tag{10}$$

The rest of the system corresponds to the classical Navier-Stokes equations, with a constant source term ($\theta_y = 1$):

$$\Omega_t + V_l [u\,\Omega_x + v\,\Omega_y] = V_d[\Omega_{xx} + \Omega_{yy}] + V_b\,\theta_y\ ; \tag{11}$$

$$\psi_{xx} + \psi_{yy} + \Omega = 0. \tag{12}$$

Infinitely Long Cavity, Az ∞ (Hadley Circulation)

In the case of infinitely long cavity (finite H and infinite L), Gr goes to zero; thus the nondimensionalization A1 applies, with $v_{ref} = \nu\,Gr/H$. In that case the equations (1) and (2) admit a particular one-dimensional steady solution (referred by Hart (1972) as the Hadley circulation) such that u=0 and v=v(x) (see Birikh (1966a, 1966b) and Hart (1972, 1983a)). Accounting for the mass flux conservation in a vertical plane and the condition (5), the longitudinal velocity writes :

- for condition (6a): $\quad v_{R-R}(x) = -(V_b/V_d)\,\theta_y\,(2x^2 - 3x + 1)\,x/12\ ; \tag{13}$

- for condition (6b): $\quad v_{R-F}(x) = -(V_b/V_d)\,\theta_y\,(8x^2 - 15x + 6)\,x/48\ ; \tag{14}$

- for condition (6c): $\quad v_M(x) = -(V_b/V_d)\,\theta_y\,(8x^2 - 15x + 6)\,x/48\ - (3x-2)\,x/4. \tag{15}$

A vertical temperature profile, T(x), can be associated to v(x), such that $\theta(x,y) = T(x) + y$. T(x) is derived from (4) as $T_{xx} = T_l/T_d\,v\,\theta_y$. As an example, for the R-F case (14) we have

$$T(x) = -(V_b T_l/V_d T_d)\,(\theta_y)^2\,[x^3(24x^2 - 75x + 60)/2880 + Bx + C]\ ; \tag{16}$$

The values of the constants B and C are given in Annexe 1; together with the analytical expressions of T(x) for the other sets of boundary conditions (6) and (8).

ANALYSIS OF THE PREVIOUS STUDIES (MELTS WITHOUT SOLIDIFICATION)

We recall the main results of the previously mentioned studies concerning the oscillatory regimes in horizontal layers subjected to a mainly horizontal temperature gradient, without any solidification process (i.e. without any phase change, or moving front, or morphological instability).

Experimental results

Conditions for the Onset of Temperature Oscillations

The available experimental conditions observed for the onset of temperature oscillations have been summarized by Carruthers (1968). They are interpreted in terms of $Gr_{c,osc}$ and Az in Tab.1.

The results in Tab.1 can be interpreted as showing a flow stabilization when Az diminishes. In addition the relatively low threshold value obtained by Utech and Flemings (1967) could be connected to some adverse temperature gradient on the top surface, as no insulation at all was employed at this surface in order to permit direct observation. Hurle et al. (1974) give the critical value of the Rayleigh number for the onset of oscillations for different L at H =1.2cm and for different H at L=3cm, at Pr=0.02. We summarize all these results in one curve giving critical values of the Grashof number, $Gr_{c,osc}$, vs Az (Fig. 3a). This curve seems to show the existence of an asymptotic limit, not far from $Gr=3.10^4$, when Az > 5. In addition, it exhibits a strong stabilization effect when Az diminishes. This can probably be connected, as in the Bénard case, to an increasing effect of viscous shear forces exerced by the wetted walls, compared to volume forces. Hurle et al. (1974) also give the value of oscillations frequency near the threshold, f_c, in terms of L and H. A correlation of these results shows that f_c increases with Az (Fig. 3b) and $Gr_{c,osc}$ (Fig. 3c).

Table 1. Conditions observed for the onset of temperature oscillations in open boats

Authors	H(cm)	L(cm)	Az	G_h (°C/cm)	$H^4 G_h$	$Gr_{c,osc}$
Cole & Winegard (1964)	0.5	15.2	30.4	27.0	1.68	26 600
(liquid Sn ; Pr=0.015)	0.6	15.2	25.3	17.0	2.20	35 000
	0.8	15.2	19.0	7.0	2.87	45 500
	1.0	15.2	15.2	3.0	3.00	47 500
Utech & Flemings (1967) (liquid Sn ; Pr=0.015)	0.94	14	15	1.1	0.86	13 600
Hurle (1967)	1.2	4	3.33	5.0	10.4	166 400
(liquid Ga ; Pr=0.020)	1.2	2.6	2.22	7.5	15.6	250 000

In Hurle et al. (1974) experiment, fluid conductivity is close to k= 8 10^{-2} cal/ (cm.sec.°C), i.e. higher than the wall conductivity (kw = 3.5-5.5 10^{-3} cal/(cm.sec.°C)) and the air conductivity, making both the upper and lower horizontal surfaces nearly adiabatic .

Behaviour of Temperature Oscillations after their Onset

The amplitude and frequency of the fluctuations have been shown by Cole and Winegard (1964) and by Utech and Flemings (1967) to increase (quite strongly) with the horizontal temperature gradient, G_h . For a liquid tin layer of 0.94 cm, Utech and Flemings (1967) found a frequency of a few cycles per second near the threshold but of thousand cycles per second when G_h is ten times this threshold value (turbulent regime ?) .

Flow Pattern

The flow pattern in an open horizontal boat containing molten tin has been studied (using a thermocouple delay technique) by Utech et al. (1967a) at a gradient sufficiently steep to cause temperature fluctuations. Superimposed on the generally steady flow along the bottom and top of the boat is a pattern of co-rotative cells (six), for an aspect ratio Az=8. The oscillations could correspond to fluctuations of this multiple-transverse-rolls structure (which could represent a first (steady) bifurcated regime after the one-cell Hadley circulation, according to the linear stability theory discussed hereafter). Utech et al. (1967a) also carried out experiments in liquid sodium chloride, a transparent molten salt with a value of Pr=0.13 (approaching the one for molten metals). They observed temperature fluctuations for Az=8 at G_h = 30 °C / cm .

Temperature Oscillations in Closed Cavities

Experiments have been done by Utech and Early (1967) in a 1.3 cm tube, filled up with liquid tin. Oscillations are observed for a length of liquid zone of 3.8 cm (Az=3) and for G_h=14°C/cm, corresponding to Gr=6.3 10^5; while the flow is steady at Gr=8.10^5 for L=2.1cm (i.e. Az=1.6).

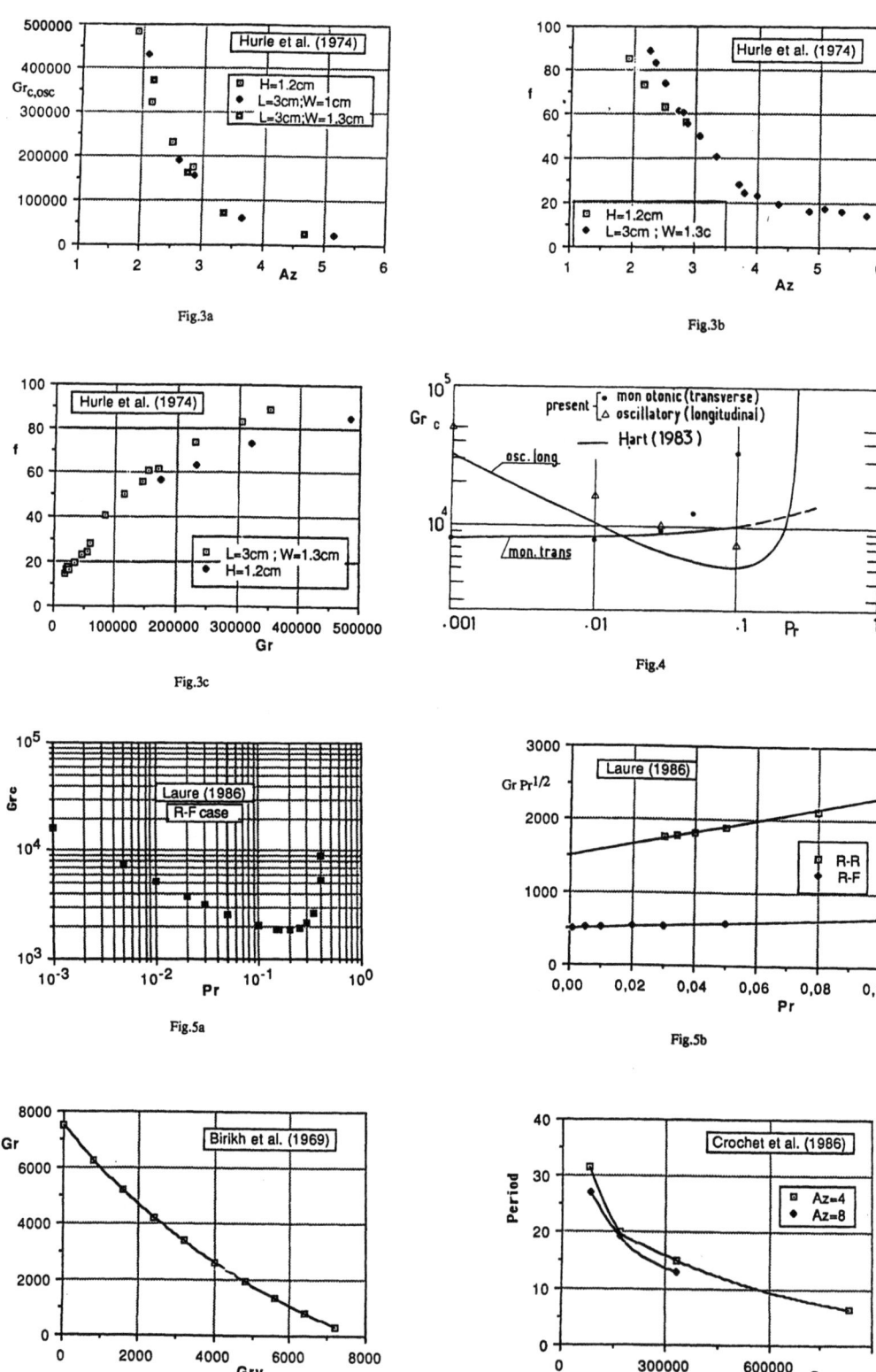

Fig.3a

Fig.3b

Fig.3c

Fig.4

Fig.5a

Fig.5b

Fig.6

Fig.7

Flow Velocity and Flow Pattern in Closed Cavities

Flow velocity and flow pattern of liquid tin in a long channel boats have been determined by MacAulay and Weinberg (1973) using radioactive tracer techniques. The results obtained in a square aluminium channel of H=0.64 cm inside width, with melts length of 28.8 , 37.5 and 48.5 cm show a linear dependency of the velocity in terms of the horizontal temperature gradient. We can remark that such experiments correspond to large Az (respectively 45, 59 and 76) and relatively small Gr (less than 4.7 10^4). We will see further that such a behaviour corresponds to the" conducting" regime. No oscillations are detected in such conditions .

Properties Temperature Dependency

MacAulay and Weinberg (1973) also mentioned the effect of the properties variation with the temperature. The velocity is shown to increase from 0.37 to .47 cm/sec when the average melt temperature increases from 300 to 400 °C; this can be connected to the corresponding diminution of the values of Pr (from 0.011 to 0.009). Transport properties for Ga, GaAs, Sn and Si are given in Annexe 3, while the temperature dependence of transport properties of gallium is given in Annexe 4.

Non-Monotonic Horizontal Temperature Gradient in Closed Cavities

Stable oscillations have been shown by Favier et al. (1986) to appear in a long horizontal cylindrical channel (of diameter 0.4 cm), filled up with liquid tin and subjected to a non-uniform horizontal temperature gradient, generating four contra-rotative cells. These oscillations occur when the temperature gradient G_h, close to the solid/liquid interface, exceeds 130°C/cm, which corresponds to Gr=4.3 10^4. Dimensionless frequency is close to 30. In these experiments oscillatory regimes are initiated by imposing regular perturbations at the solid/liquid interface. The authors point out the high dependence of fluid motion selection on the boundary conditions. Nevertheless they exhibit the existence of some chaotic zones separated by quasi-monochromatic ones, in a relatively short range of moderate values of Gr, typically 4.3 10^4 < Gr< 10^5.

Multi-Directional Heat Fluxes

Several authors point out the influence of multi-directional heat fluxes. In addition to the imposed horizontal temperature gradient, G_h, lateral heat flow can exist as a result of a badly insulated apparatus or just as the main heating system of the melt (Hurle,1967; Carruthers,1968 ; Hurle et al., 1974). Vertical temperature gradients, G_v, can also be present; such gradients are positive when induced by the bulk flow (due to thermal convection generated by G_h as shown by Utech and Early, 1967) and negative (adverse) when induced by a bad insulation of longitudinal surfaces (MacAulay and Weinberg, 1972 and Carruthers,1977).

Hurle and Jakeman (1973) reported on natural oscillations in water-methanol layer (Pr=7) heated from below, but subjected to a slight imposed horizontal gradient (G_h = G_v /60) leading to a steady convective motion in the pre-oscillatory regime. They wanted to prove that the success in earlier experiments in generating stable finite amplitude oscillations was connected with a slight G_h caused by non-uniformities in the apparatus. Hurle (1967) also mentioned an other kind of non-uniformity, namely off-axis horizontal temperature gradient, that could favorize the onset of stable oscillations .

Thermocapillary (Marangoni) Effect

Only a few results about thermocapillary effect are available in the literature. In a private communication Hurle (1986) points out that in his previous experiments with molten gallium (Hurle et al. 1974) the effect of removing the oxide from the surface of the gallium and of establishing a thick stable oxide were looked into. Both situations gave rise to oscillatory behaviour and Hurle (1986), therefore, does not believe that Marangoni convection is significant to explain oscillations. This point of view seems to agree with recent experiments presented by Camel et al. (1986) for a long open boat (L=25cm, H varying from .04 cm to .4 cm) which only show steady-state results up to a Reynolds-Marangoni number, Re_M = Ma/Pr, equal to 10^4.

Theoretical and Numerical Results

Stability of Hadley Circulation (Linear Stability)

In an attempt to explain the oscillations observed by Hurle et al. (1974), Gill (1974) developed a stability theory based on the basic flow (Hadley circulation) described by Hart (1972). As this basic flow does not contain Az as a parameter, Hart (1983a) and Simpkins and Chen (1983) already pointed out that this model can only be valid for large aspect ratio cavities. The physical mechanism for the oscillations discussed by Gill (1974) is the following one: for an inviscid fluid of infinite conductivity the particle displacement and temperature perturbations remain in phase so that the fluid oscillates as a damped internal gravity wave. For large conductivity (small but non-zero Pr), diffusion is no longer instantaneous and the particle reaches its minimum temperature a little after it reaches its maximum elevation. This will lead to growth of perturbations (as an overstable oscillation) if the resulting phase shift provides a restoring force which is sufficient to overcome the effects of viscous damping. These perturbations, which grow primarily as longitudinal rolls (with axes parallel to G_h), occur once Gr exceeds a critical value. In his study which is limited to conducting horizontal boundaries, Gill (1974) obtained the following criteria for the onset of oscillations: $Gr_{c,osc}$=1030/ Pr in R-R case and $Gr_{c,osc}$= 240/ Pr in R-F case. Thus the threshold is lowered by a factor of four when bounding the layer with rigid walls. For both kinds of boundary conditions (R-R and R-F conducting), the wavelength in transverse (z) direction and the angular frequency, at the threshold, are respectively equal to λ_z = 3.7 H and ω_c=22, independently of Pr, for small Pr.

Hart (1983a) remarks that Gill theory in essence concerns free conductors. He discusses the importance of boundary conditions on the stability properties of the Hadley flow. For rigid horizontal adiabatic walls he shows that in the range .015 $\leq$ Pr $\leq$ 0.27, the first perturbations to grow are overstable (oscillatory) longitudinal rolls with axes perpendicular to the endwalls, but with very large cross-stream wavelengths of about 9 to 15 layer depths (depending on Pr). An experiment with mercury (Pr=0.026) in a large cavity (H=1.2cm, L=W=21cm) reasonably confirms this prediction, oscillations being observed at Gr=9300 with an angular frequency ω=54, instead of Gr=7100 and ω =36.5 . Hart (1983a) also gives an argument to show that, in the region of constant v_x , ω could be roughly expressed as $\omega^3 = k\ Gr^2\ v_x$. Thus, for low Gr (conducting regime) where v_x is proportional to Gr, ω itself would be proportional to Gr. In the case of an open cavity, the threshold is lowered by a factor of 5 to 8; oscillations could appear for Gr as low as Gr=2000. For both kinds of boundary conditions (R-R and R-F adiabatic) the angular frequency is almost independent of Pr, in the range $10^{-3} \leq$ Pr $\leq 5.10^{-2}$; with ω_c= 32 in R-F conditions and ω_c= 37 in R-R conditions. But, when extending the method to supercritical regimes, ω is shown to rapidly increase with Gr, while the most rapidly growing wavenumber, k, increases. For R-R adiabatic case, ω =36.5 for k_c=0.7 at Gr=7.1 10^3 , while ω is close to 108 for k=1.1 at Gr= 2.10^4 .

The major difference between Hart's results and the previous ones is that oscillatory instabilities predicted by his theory should occur at large wavelengths in z-direction, λ_z (9 to 15 layer depths), and thus could probably be observed only in parallelepipedic cavities with very large z-extension. In his experiment with mercury (Pr=0.026) in large cavity (Az=Ay=17.5), Hart (1983a) observed oscillatory regime occurring with λ_z=8 and ω =54 at Gr=9.3 10^3 . This experiment confirms the major predictions of the theory; i.e. the dominant instability corresponds to longitudinal rolls and takes its energy from the basic lateral temperature gradient of the Hadley circulation .

Roux et al. (1984) repeated this last theoretical study in the case of rigid adiabatic horizontal walls, by using an accurate spectral technique (based on Chebyshev expansions) to solve the linear perturbations equations given by Hart (1983a). They found a discrepancy for the threshold values, probably due to round-off errors in Hart's calculations, as they were performed on a 61/2 digit machine. Roux et al. (1984) predict overstable (oscillatory) longitudinal rolls after Pr = 0.034 (instead of Pr = 0.015 in Hart's solutions); see Fig. 4.

Laure (1986) confirmed these last results and extended this last work developing a non linear stability analysis and computing the bifurcated solutions beyond the thresholds given by the linear stability theory. In the case of rigid horizontal walls, he observed either steady transverse rolls (with axis parallel to the endwalls) for $0 \leq Pr \leq 0.034$ or time-periodic (travelling waves) for $0.034 \leq Pr \leq 0.2$. In this last case the critical $Gr_{c,osc}$ reaches a minimum (close to 7,300) for Pr close to 0.1. For open cavities (R-F adiabatic), Laure (1986) found two forms of time-periodic solutions, either travelling waves propagating in the (y,z)-plane for $0 \leq Pr \leq 0.38$, or almost stationary ones (in fact slowly propagating in y direction), for $0.38 \leq Pr \leq 0.41$. In this case, the $Gr_{c,osc}$ curve given in Fig. 5a for the range $0 \leq Pr \leq 0.41$ presents a minimum at Gr= 1885 for Pr=0.155. At Pr=0.02, the period of oscillations would approximatively be of 60 sec. in R-R adiabatic case and of 120 sec. in R-F adiabatic case. Plotting the results in terms of $Gr_{c,osc} Pr^{1/2}$, in Fig. 5b, shows that this quantity behaves almost linearly in Pr in the range $0 < Pr < 0.1$, and that the bounding of upper surface is stabilizing and increases $Gr_{c,osc}$ by a factor of 3 to 4.

The value of angular frequency at the threshold is found, once again, to be almost independent of Pr, for small Pr, in agreement with Hart's results; for R-R adiabatic case it varies of less than 1% around $\omega_c = 36$ (in excellent agreement with Hart's results), in the range $0.034 \leq Pr \leq 0.06$; while for R-F adiabatic case it varies of less than 1% around $\omega_c = 15.6$ (instead of $\omega_c = 32$ in Hart's results), in the range $0.001 \leq Pr \leq 0.1$. An other important feature, that agrees with Hart's result is the very low value of the wavenumber in y-direction (which corresponds to instability in the form of rolls having their axes parallel to y-direction) and the relatively large value of the wavelength in z-direction (λ_z / H equals 9 for R-R adiabatic case, and varies from 14 at Pr=0.03 to 23 at Pr=0.01 for R-F adiabatic case). Hart's experiments with mercury (Pr=0.026) in large cavity (Az=Ay=17.5) give the onset of oscillations at Gr=9.3 10^3, while the theory overestimates this threshold at a Gr close to 10^4.

Stability of Semi-Infinite Layer with Two-Dimensional Heat Flow

Stability of semi-infinite layers with two-dimensional (horizontal + vertical) heat flow has been studied in various situations. A large review of the stability of these layers (which corresponds to slightly different plane-parallel convective motions) has been made by Carruthers (1977) for low, moderate and large Pr. We will note the main previous results given in the literature for low Pr fluids.

For an infinite fluid layer heated on one side and tilted of an angle α with the vertical (characterized by a cubic profile velocity and a linear temperature distribution), Gershuni and Zhukhovitskii (1969) found that even a small deviation of the layer from the horizontal position ($\alpha=-90°$) can give rise to intensive convective motion. In this case, even at angle close to -90° (i.e. in the Rayleigh-Bénard region) the instability is of hydrodynamic nature and related to the development of transverse rolls (normal to the basic flow direction). The enhanced destabilization of flows at low Pr has been theoretically confirmed by Korpela (1974) by using the same kind of linear stability analysis. Korpela (1974) points out that this behaviour occurs as soon as Pr is less than 0.024. Below this value, the product (Pr tan α) becomes the significant parameter.

Birikh et al. (1969) study the stability with a vertical (upward or downward) temperature gradient, of a steady convective motion generated between two semi-infinite vertical plates, differentially heated. In the absence of vertical gradient, the flow becomes unstable at

Pr=0.2 for a critical Gr of 7520 (here $Gr=g \beta G_h H^4 / \nu^2$ and H is the horizontal distance between the plates). In the case of bottom heating, increase in G_v is stabilizing untill Ra = π^4 where $Ra = g \beta G_v H^4 / \nu\kappa$, and destabilizing after. The dependence of the critical value of Gr (noted Gr_m) vs Gr_v (defined as Ra/Pr) is shown in the Fig. 6. For $Ra > \pi^4$, it exists a domain of value of Gr and Ra in which oscillatory perturbations are not damped. For top heating (negative Ra) a stabilizing vertical stratification occurs.

Liang and Acrivos (1970) considered the stability of natural convection in a thin, horizontal layer slightly tilted with respect to the horizontal and heated from below. It is shown for a large range of values of Pr ($0.1 \leq Pr \leq 10^4$) that the critical Ra has the same value as for an exactly horizontal layer, however, the predicted motion, rather than being indeterminate, is one of longitudinal rolls (having axes parallel to the mean flow).

An other case of combined vertical and horizontal temperature gradients has been considered by Weber (1973) who studies the effect of an horizontal temperature gradient imposed on an horizontal layer heated from below. Contrary to Liang and Acrivos (1970) results, this study shows an increase in critical Ra (stabilization) resulting from the shear flow generated by a non zero G_h. In addition, the predicted motion (stationary) can be with transverse rolls (having axes perpendicular to the mean flow) or with longitudinal rolls (with axes parallel to the mean flow) depending on whether Pr is less or larger than 5.1. It seems, nevertheless, that the preferred mode could be longitudinal rolls again when Pr is less than 0.03. But the method which is based on a perturbation technique using G_h as a small perturbation quantity, seems not to be quite appropriate for the very low values of Pr considered herein.

Sweet et al. (1977) proposed an extension of previous Hart (1972) and Gill (1974) theories, including an adverse vertical temperature gradient. They develop their study in the case of free conducting horizontal walls where the eigenvalue problem can be handled analytically, with respect to perturbations of the form: $e^{\omega t} \sin(l_y y) \sin(l_x x)$. In that case the critical value of Ra for the onset of stationary instability is: $Ra_c = 27 \pi^4/4 = 657.5$ (compared to 1708 for rigid conducting walls). In the limit of small Pr and large K, where $K = 0.52 \, G_h/G_v$, they found the critical value of Ra for the onset of oscillatory instability as: $Ra_{c,osc} = [2/(3-\gamma^2)]^{1/2} (1+\gamma^2)^2 \pi^4 /\gamma K$, where $\gamma = l_y/l_x$. That expression presents a minimum for $\gamma = 0.536$ (i.e. for $\lambda_y / \lambda_x = 1.86$, if λ_y and λ_x are wavelengths in y and x directions), and then writes as $Ra_{c,osc} = 2.655 \, \pi^4 /K$. That threshold can be re-written, in the limit of strong K (i.e. small G_v) in terms of the usual Gr, as $Gr_{c,osc} = 500/$ Pr. Surprisingly enough this new expression no longer depends on G_v. It seems that this theory trivialy recovers the Gill's one (without G_v); but there is a not yet explained difference (by a factor of two) with the value of the threshold proposed by Gill (1974), i.e. $Gr_{c,osc} = 240/$ Pr. Nevertheless, at Pr=0.02 for instance, Sweet et al. theory would predict oscillations, for two free conducting horizontal surfaces, at $Gr = 2.5 \, 10^4$, for any small G_v. The value of the frequency at this threshold, always in the limit of small Pr and large K, can be written as $\omega H^2/\nu = (4.89/Pr)^{1/2} \pi^2$ (i.e. $\omega H^2/\nu = 154.3$ at Pr=0.02, representing periods of 2 sec., for H=1cm and $\nu = 3. \, 10^{-3}$).

Numerical Two-Dimensional Solution in Closed Rectangular Cavities

A lot of papers have been devoted to the numerical solution of the Boussinesq equations in the case of a rectangular cavity, differentially heated and filled up with a low-Prandtl-number fluid. For cavities with small aspect ratios at Pr=0.013, Stewart and Weinberg (1971) have shown that the flow is steady and unicellular, the streamlines becoming more and more circular, despite the rectangular shape of the enclosure, when Gr is increased from 2.10^3 to 2.10^5. Consequently, as mentioned by Carruthers, boundary layer separation and secondary vortex generation is expected in the corners rather than inside the unicellular flow. Nikitin et al. (1979 and 1981) reported unicellular stationary flow in the range $0 < Gr < 10^7$, for a closed rectangular cavity (at $Az=4$), filled up with liquid germanium doped with impurities ($Pr=0.016$). For the conditions $Pr.Gr << 10^3$ they found that the velocity profile in median cross-section is in good agreement with the Hadley solution.

Hart (1983b) studied the end influence length in long cavity ($Az \leq 0.1$), in situations where multi-rolls can occur. He suggests to correlate his results obtained in the ranges $10^2 < a\,Gr < 10^5$ and $10^{-3} < Pr < 10^{-1}$, using a parameter, a, which matches core (Hadley) flow and end region flow through the convected heat flux $Q = \iint u\,T\,dxdy$. This parameter is given by the following implicit relationship $a = 1 - (2\,a^2\,Gr^2\,Pr^2\,Q/Az)/(1 + 2\,a^2\,Gr^2\,Pr^2\,Q/Az)$. In the domain of values of Gr and Pr, we have $Q = 1.76$ (almost independently of Gr and Pr). Unless $Gr^2\,Pr^2/Az$ is greater than 10^5, we have $a=1$. Compared to the critical value $Gr_c = 7980$ given by the linear stability theory (Hart, 1983a), the numerical results exhibit an imperfect bifurcation, on the form of cat's eye in the corner, at $Gr=6900$; these cat's eye weaken as Pr is increased. Multicell structure, affecting all the cavity, is exhibited at $aGr = 8,400$ with several weak vortices; for $aGr=11,900$ several strong vortices are present, the number of which regularly increases with Az. No computation has been done for higher Gr, allowing to oscillations regime.

Numerical two-dimensional simulation in open rectangular cavities

Crochet et al. (1983 and 1986) studied the generation of periodic flow in open horizontal crucibles of molten gallium arsenide under the action of an horizontal temperature gradient. They used steady-state finite element, time-dependent finite difference and time-dependent finite element algorithms which all lead to identical results, and they found that the loss of convergence of the steady-state finite element method coincides with the onset of oscillations detected by the time-dependent methods. The computations carried out for $Pr=0.015$ show steady two-cells flow at $Az=4$, in both conducting or adiabatic upper free surface, untill $Gr= 79,000$, that seems to represent approximately the end of a damped oscillatory regime (the period of the oscillations being close to 32 sec. at $Gr= 79,000$). When the temperature gradient is increased, the oscillatory behaviour detected in the generation of vortices is undamped; at $Gr=166,700$ periods of respectively 17.5 and 20 sec. are found in the adiabatic and temperature-imposed conditions. At $Pr=0.05$, for $Az=4$ and $Gr=166,700$ and for temperature-imposed conditions oscillations occur with periods of 19 sec. (against 20 sec. at $Pr=0.015$); the isotherms appear to be much more distorted at the higher value of Pr. Additional computations at $Az=8$, show that an increase of the aspect ratio for a given Gr slightly decreases the period of the oscillations. This dependence of the period on Gr, for $Az=4$ and $Az=8$ at $Pr=0.015$, is shown in Fig. 7. Note that when representing the evolution of the period in terms of the Grashof number, Gr.Az, used by Crochet et al. (1986), the effect of Az is to increase the period (see Fig.12 of their paper). They remark that it is more suitable to employ the definition of Gr used herein (instead of Gr.Az).

Cartage et al. (1985) repeated the computations done by Crochet et al. (1983) for $Az=4$, and claimed that the agreement was good. In addition, they give results for a rectangular closed cavity filled up with tin, for $Pr=5.64\ 10^{-3}$. For their $Ra=500$ ($Gr=2.2\ 10^4$?) they found steady-state solution; at $Ra=1000$, the time history is not given for sufficient large time to conclude if the result is with damped oscillations or not damped ones.

Numerical Three-Dimensional Solution

Dupont et al. (1986) extends, in a companion paper, the two-dimensional study made by Crochet et al. (1986). In order to see the three-dimensional effect on Bridgman growth process, they consider parallelepipedic cavity where the width is twice the depth and the length four times this depth. Using a steady-state finite difference algorithm, they found a converged solution at $Pr=0.069$ up to $Gr=1.25\ 10^5$ (against $Gr=1.77\ 10^5$ with a corresponding two-dimensional model). The Author's conclusion is that oscillations would appear for (about 40%) lower Gr, when using a three-dimensional model. In addition they found that the 3D-flow structure which is unicellular (but presenting a recirculation region near the cold endwall), is quite different from the 2D one that is multicellular after $Gr=10^4$.

This last result is in agreement with the results given by Roux et al. (1984) for a 3D flow simulation in a confined horizontal cylinder subjected to an axial (horizontal) temperature gradient, at $Pr=0$ and $Gr=10^5$ (here Gr is based on the diameter of the cylinder).

Dupont et al. (1986) also reported numerical results for a cylindrical crucible made of half a quartz tube (with a free upper surface and a linear temperature distribution along the generators), in which again they observed that the side walls inhibit the generation of multiple vortices and that the flow presents a recirculation region near the cold endwall. A converged solution is observed up to $Gr=1.38\ 10^5$.

Viskanta et al. (1986) developed a numerical method for the solution of the steady-state three-dimensional Boussinesq equations in a parallelepipedic domain. They want to discuss the question of two-dimensional vs three-dimensional modelling of natural convection in side-heated enclosures filled up with a low-Prandtl-number fluid (liquid gallium: $Pr=0.02$). They considered two types of enclosures (with well insulated lateral and horizontal walls) for which they also do experiments, namely: L=6.35cm, H=6.35cm and W=4.44cm (i.e. Az=1, Ay=0.7) and L=8.88cm, H=4.44cm and W=3.81cm (i.e. Az=2, Ay=0.86). They find steady-state results, up to $Gr=5.10^7$, where their numerical results exhibit complicated multiple longitudinal flow in the cavity ; the longitudinal temperature distributions in the vertical symmetry plane, both from experiments and comptutations, indicate a strong convective effect on the thermal field (the numerical solution even presents a sign change of T_y in the middle of the cavity).

Computations have been done by Bullister et al. (1986) in the case of an horizontal closed cavity of large extent (Ay=Az=17.5), for Pr=0.026; a situation corresponding to Hart (1983b) experiments. The authors found a steady-state solution at $Gr=7.1\ 10^3$, which agrees in the central part of the cavity with the asymptotic solution, for infinite Az (see expression A1.1). This solution well agrees with our stability prediction which gives $Gr_{c,st}$ close to 9.10^3, for Pr=0.026, a value that also agrees with the threshold experimentally obtained by Hart (1983a). For $Gr=1.8\ 10^4$, the solution computed by Bullister et al. (1986) presents in the vertical symmetry plane the trace of (five) transverse rolls ; that agrees with our stability predictions given in Fig.10, which predict that transverse modes are the most unstable untill Pr=0.034 (instead of Pr=0.015 in Hart's results). The time history reported by Bullister et al. (1986) shows that their numerical solution is probably oscillatory with a single frequency, but this solution is not well enough established to give an accurate evaluation of f.

Stability of Two-dimensional Solutions Using Bifurcation Theory

An interesting recent work has been done by Winters et al. (1984 and 1986) who developed a theory to determine Hopf bifurcation in the steady-state (Boussinesq) equations for buoyancy-driven flows. They discretize these Boussinesq equations using a finite element method and write the resulting equations as $f(x, \lambda, \alpha) = 0$, where f is a smooth non-linear function, x is the vector solution, λ the bifurcation parameter (here Gr) and α the control parameters (here Pr, Az, ...). It should be stressed that this method predicts the exact point of instability, whereas transient algorithms must estimate the point of onset from a series of calculations at different parameter values and they must distinguish true oscillations from transient oscillations which would lead to a steady-state after sufficient time. The Authors consider a two-dimensional rectangular container with Az=4, filled with liquid gallium-arsenide (Pr=0.015) to make comparisons with Crochet et al. (1983) results. The main conclusions reported by Winters et al. (1984 and 1986) and by Winters (1986) are:

(i) for a given grid size, $Gr_{c,osc}$ is predicted much less accurately than the flow velocities and temperature (a coarser grid overestimates by 40% $Gr_{c,osc}$ but still gives a correct steady flow).

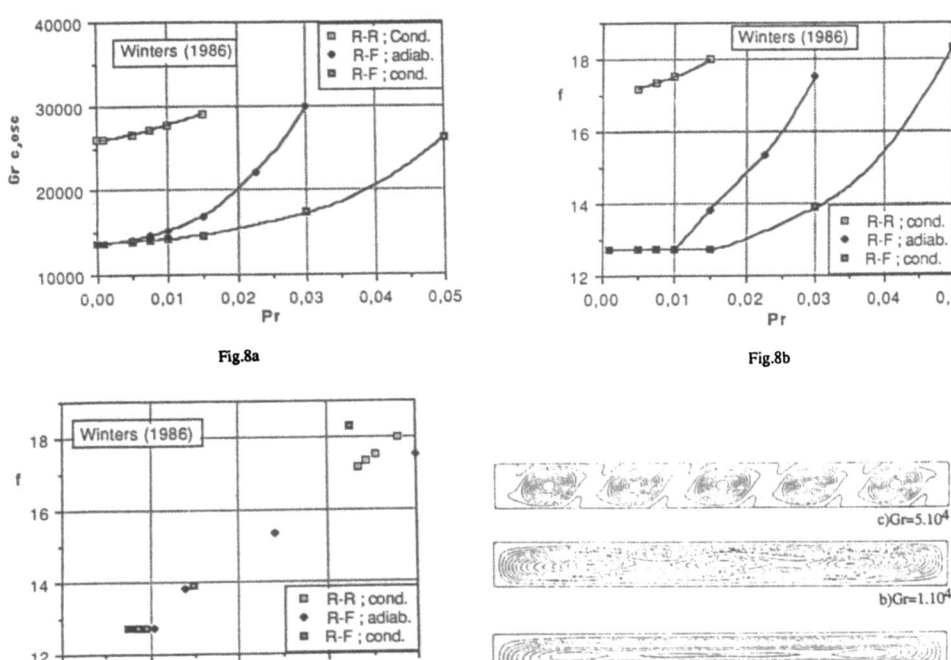

Fig.8a

Fig.8b

Fig.8c

Fig. 9   Multicellular structures, collaboration with Nikitin and Polezhaev (1984)

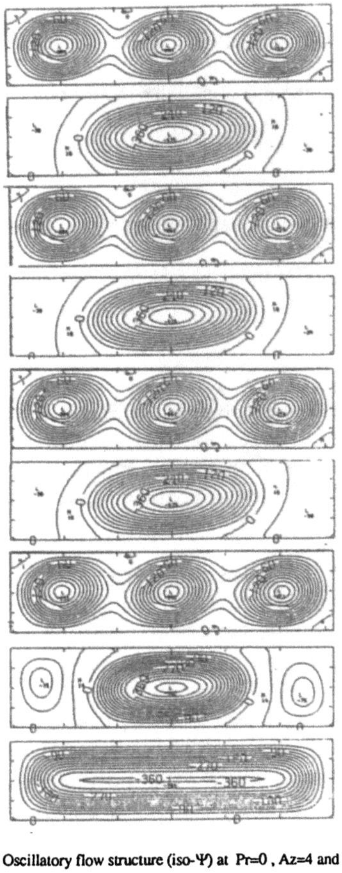

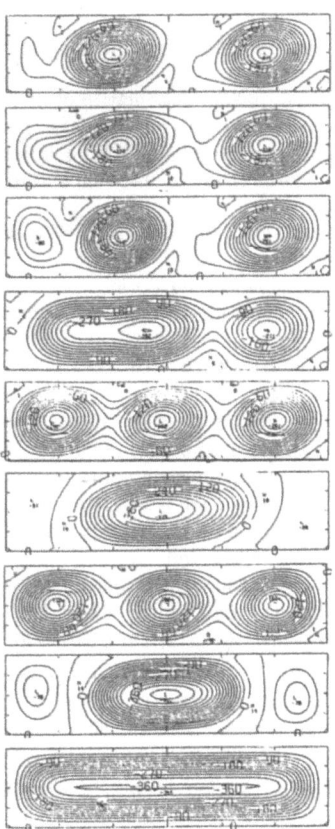

Fig.10a   Oscillatory flow structure (iso-Ψ) at Pr=0, Az=4 and Gr=5.10^4
IT= 0, 71, 151, 267, 349, 465, 547, 663, 745

Fig.10b   Transient flow structure (iso-Ψ) at Pr=0,
Az=4 and Gr=5.10^4 (with a perturbation factor of 1.5)
IT= 0, 71, 151, 267, 349, 411, 482, 512, 560

(ii) $Gr_{c,osc}$ tends to a value of $1.38\ 10^4$ in the limit of Pr=0 and increases from $1.48\ 10^4$ to $2.63\ 10^4$ as Pr changes from 0.015 to 0.05, for the R-F conducting case.

(iii) for a fixed Pr, $Gr_{c,osc}$ strongly increases when reducing Az (from $1.48\ 10^4$ to $7.75\ 10^4$ when Az changes from 4 to 2.4, at Pr=0.015) ; this strong variation agrees with Hurle's results in Fig.3a.

(iv) bounding the upper surface makes $Gr_{c,osc}$ twice higher ($2.82\ 10^4$ at Pr=0.015).

Changing the thermal conditions on this surface, from conducting to insulating, has a similar inhibiting effect on the onset of oscillatory convection. The dependency of $Gr_{c,osc}$ and f vs Pr, has been given by Winters (1986) for various boundary conditions (see Fig. 8a and 8b). Owing to the similarity of the $Gr_{c,osc}$ and f curves vs Pr, we also plotted the variation of f vs $Gr_{c,osc}$, in Fig. 8c.

NUMERICAL FINITE DIFFERENCE METHOD

The numerical technique for the solution of the system (1)-(8) is an extension to purely unsteady equations of the "false-transient" algorithm detailed in previous papers by Roux et al. (1978 and 1979). A relaxation process is used in order to improve the convergence rate of the iterative process imposed by the non-linearities and the coupling of the governing equations. The main features are the following:

(i)   a second order accurate central differenciation for spatial derivatives

(ii) a fourth order accurate compact Hermitian method for equation (2)

(iii) a second order accurate approximation of the boundary condition for the vorticity

(iv) an alternating direction implicit (ADI) method for solving the finite difference form of the transport equations (1) and (4); this method includes a compatibility condition for variables on boundaries at the intermediate time level (Fairweather and Mitchell ,1971),and an iterative process at each time step.

(v)   internal iterations are needed to adjust the values of ψ, at each iteration.

This technique is detailed in a paper by Ben Hadid & Roux (1986a) in which the efficiency and the accuracy of the technique are discussed.

Long Confined Cavity (Az=10) at Pr=0

The code has been used, in its pseudo-unsteady form (without internal iterations on ψ), in a comparative numerical study with Nikitin & Polezhaev (1984), at Az=10 and Pr=0, to study the onset of monotonic instabilities and the bifurcation to steady multicells structure already observed for moderate values of Pr (Pr < 0.1) and for adiabatic horizontal walls by Hart (1983b), and in a vertical cell (an equivalent problem at Pr=0) by Lee & Korpela (1983) at Pr=0 and Jones (1982) at Pr=0.035.

The results obtained with our code, at Az=10 and Pr=0, are given as a sample of shear flow instability (like the one exhibited by Gershuni et al. 1969) generated on the boundary of two opposite flows (Fig. 9). At Gr=7000, below the critical value, Gr_c=7932, given by Korpela et al. (1973), this solution presents a one cell motion except small single cat's eyes near each endwall. At Gr=10^4, a smooth transition can be observed, through the streamline (iso-Ψ) patterns, to a stable multicell solution which has been previously found by Lee and Korpela (1983) and by Hart (1983b) who present this as an example of imperfect bifurcation. Here the correct solution is with three stationary cells. At Gr= $5\ 10^4$, the observed solution corresponds to five equally strong and regularly spaced stationary cells. This pattern is quite different from the one corresponding to the beginning of the shear flow instability (Gr=10^4), with completely separated cells occupying all the interval between the two horizontal planes. We also experienced in the comparative work with Nikitin & Polezhaev (1983) that when the computation is made with a too coarse mesh size (98 x9 at Az=10) a wrong but "converged" four-cells solution is obtained at Gr=10^4. Furthermore, the four-cells solution was maintained at Gr= 5.10^4 when using this solution (with interpolation) as initial condition for a new calculation with a finer mesh. The final solution (which is considered as wrong or not observable) is very tricky as it corresponds to a y-wavelength that would better agree with the critical wavelength given by the linear stability theory carried out by Korpela et al. (1974) for Az=∞. In fact, this is an hidden effect of the endwalls which play an important role, even for Az=10.

Moderately Long Confined Cavity (Az=4) at Pr=0

Our code was used, in its pseudo-unsteady form (Roux et al.,1984) to confirm not yet published results obtained by Gresho & Sani (1984) with a finite element method carried out for fully transient Boussinesq equations at Pr=0. One of the interesting features observed by these authors was that the solution for Az=4 and Gr=5.10^4 presents very stable oscillations (the pattern changing alternatively into one and three cells) during several ten cycles, then presented a sudden transient state, and finally converged to a stationary two-cell solution. Gresho & Sani observed, in addition, that during the intermediate transient phase the centro-symmetry of the solution was lost.

We repeated these computations in two ways and we analyzed the time-depend behaviour of the flow structure through the iso-Ψ curves drawn for some characteristic times (IT.ΔT), corresponding to successive maximum and minimum values of the maximum absolute value of Ψ, Ψ_{max}, which generally occurs at the centre (for closed cavity and centro-symmetric boundary conditions). In the first way, the computation was started with an initial condition derived from asymptotic solution (13), then a stable oscillatory behaviour was rapidly obtained, like in the beginning of Gresho & Sani's solution (Fig. 10a). In a second way, the initial condition was disturbed asymmetrically, by multiplying the initial value of Ψ at a certain place (for x= 6 Δx and y = 18 Δy) by a given factor. For a factor of 1.25, the stable oscillatory solution is broken after only three cycles (for 61 x 21 grid meshes), giving rise to a stationary two-cells solution. For a factor of 1.5, the transition to a two cells solution is again more rapid (Fig.10b).

We obtained the same kind of results with the new version (fully transient) of our code at Gr=5.10^4, i.e. stable oscillations at the beginning and stationary two-cells solution after. But probably due to the better accuracy of the new results (91 x 31 grid meshes) the transition to the stationary solution occurs after a larger number of cycles (five when using a perturbation factor of 1.25).

A sample of a steady flow structure for smaller Gr is given on Fig.11a, for Gr= 2.10^4. It presents a strong central vortex and two small symmetrical co-rotative vortices. The corresponding damping of oscillations of Ψ_{max}, which characterizes the strength of the central vortex, is clearly shown on Fig.11b.

Moderately Long Confined Cavity for Small but Finite Pr

(Weakly Stable) Oscillatory Regimes (at Az=4)

For Pr = 0.035, computations have been carried out for increasing values of Gr. For Gr=10^4 the solution rapidly converges (498 time steps) to a stationary one-cell solution. For Gr= 1.5 10^4 the solution converges again to a stationary one-cell solution, but with marked damped oscillations. At Gr=2.10^4 the solution presents a stable oscillatory solution, without any damping during eight cycles, even when a perturbation with a factor of 1.25 was used. For 5 10^4 we repeated the comparison made at Pr=0, without perturbation and for perturbations with factors of 1.25 and 1.5. In the first case we found again stable oscillatory solution with

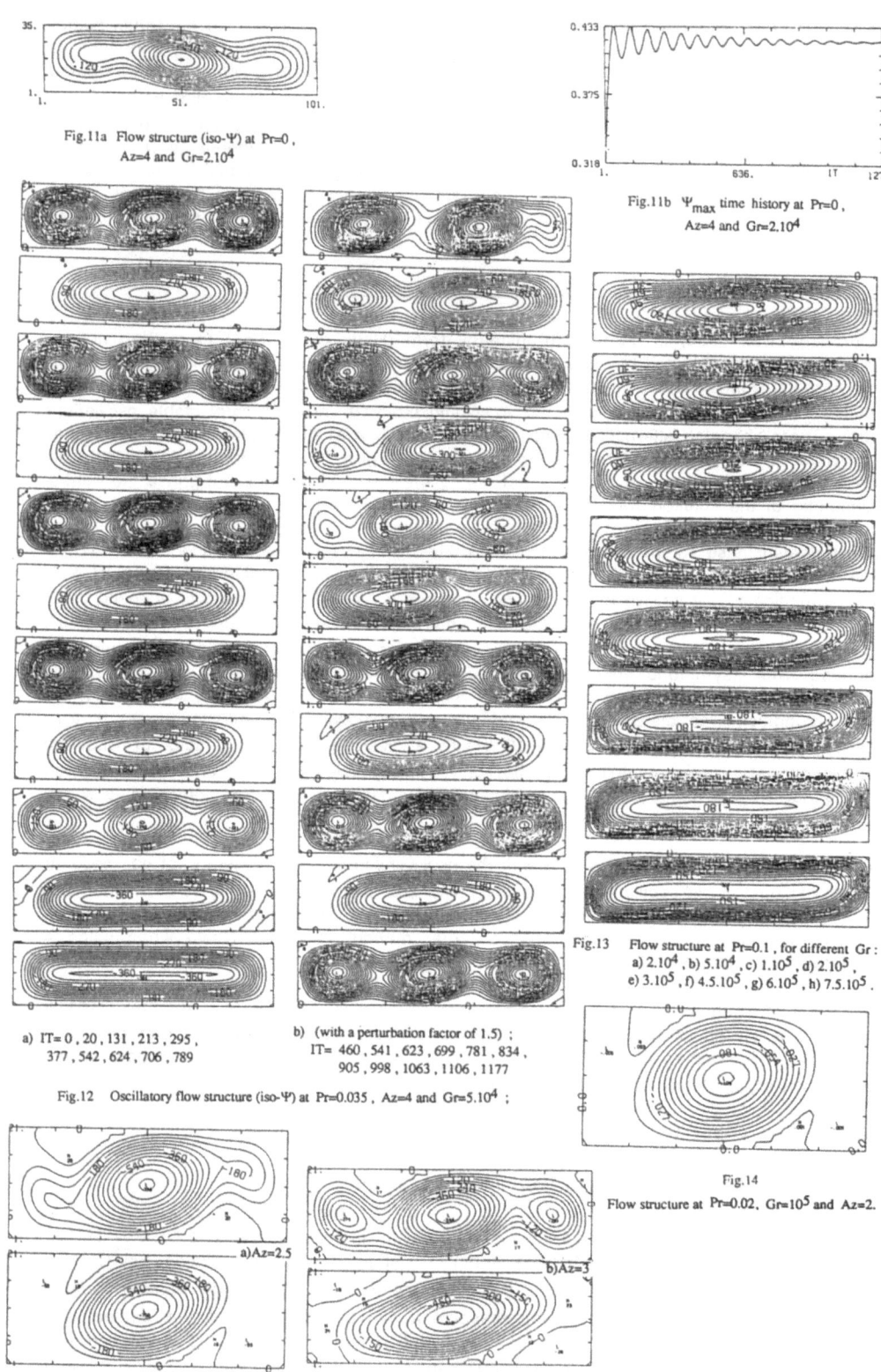

Fig.11a Flow structure (iso-Ψ) at Pr=0, Az=4 and Gr=2.10^4

Fig.11b Ψ_{max} time history at Pr=0, Az=4 and Gr=2.10^4

a) IT= 0, 20, 131, 213, 295, 377, 542, 624, 706, 789

b) (with a perturbation factor of 1.5); IT= 460, 541, 623, 699, 781, 834, 905, 998, 1063, 1106, 1177

Fig.12 Oscillatory flow structure (iso-Ψ) at Pr=0.035, Az=4 and Gr=5.10^4;

Fig.13 Flow structure at Pr=0.1, for different Gr: a) 2.10^4, b) 5.10^4, c) 1.10^5, d) 2.10^5, e) 3.10^5, f) $4.5.10^5$, g) 6.10^5, h) $7.5.10^5$.

Fig.14 Flow structure at Pr=0.02, Gr=10^5 and Az=2.

Fig.15 Spurious flow structures at Pr=0.02 and Gr=10^5.

alternatively one and three cells (Fig. 12a). The damping effect of initial perturbations seems not to be so strong than for Pr=0; here the structure change appears only after six cycles, for the factor of 1.25, and after five cycles for a factor of 1.5 (Fig.12b).

For Pr = 0.070 the solution at Gr= 10^4 converges again more rapidly (386 time steps) to a stationary single-cell solution which is also reached at 2. 10^4 but after damped oscillations. Additional results obtained for increasing Gr show that the solution is again much more stable than for the lower Pr. For Gr= 8.10^5 the solution converges (173 time steps from the solution obtained at 7.5 10^5) to a single stationary cell. This "stabilizing" effect of Pr, when this parameter is increased in the range 0 ≤Pr≤ 0.1, is consistent with the results of marginal stability for transverse modes shown in Figs 4 and 8a. It corresponds to the increase of viscous effect with Pr. We can observe in Fig.13 that the flow structure at Pr=0.1 is quite different than for smaller Pr (Pr ≤ 0.035); only single-cell solutions are exhibited in a wide range of Gr, 2.10^4≤ Gr ≤ 7.5 10^5. This kind of flow pattern is more like the one observed for moderate-Prandtl-number fluids by Wirtz & Tseng(1980) for long water-filled cavities (Pr=6.8), by Lee & Sernas (1980) for air-filled cavities (Pr=0.7) , by Ostrach et al. (1980) for oils (Pr=10^3-2.10^3) and for air, and by Tichy & Gadgil (1982) at Pr=1. In all that cases the fluids are flowing all along the geometry of the cavity (instead of only recirculate in the central region). This is also consistent with our neutral stability curve, presented in Fig.4 , which exhibits a strong increase , near Pr= 0.1 , of the critical value of Gr for the onset of y-periodic stationary modes (even if this curve is strictly valid when Az goes to infinity). Thus it can be questionable to compare results obtained for Pr< 0.1 to those for Pr > 0.1.

More recent computations made in a comparative study with Pulicani & Peyret (1986) who develop a spectral method, seem to confirm that the stable oscillations obtained at Gr= 5. 10^4 are maintained for a time much more longer if the number of degree of freedom (discretizing points or coefficients of Chebyshev expansion) is increased, before to reach the(probably much more stable) stationary two-cells solution. Thus, these stable oscillations can probably be interpreted as a possible, but weakly stable, solution of the problem . Note that such stable oscillations are only observed for specific values of Az, Pr and Gr. They were not found at Gr=5.10^4 for high Az, neither for Pr≥ 0.1.

Aspect Ratio Effect (2≤Az≤3) and Spurious Oscillatory Regimes

The flow structure for aspect ratios corresponding to the ones used in Hurle et al. (1974) experiments has been analyzed for Pr=0.02. For Az=2, a steady one-cell solution has been found which presents a strong vortex concentrated in the central region of the cavity, at Gr= 10^5 (Fig.14). The false transient code was giving oscillations which are not "broken" by an initial asymmetrical perturbation. Some of them have been proved to be "spurious" solutions, probably induced by the right hand side of equation (2), Ψ_t, which is used in the false transient technique. This spurious behaviour can occur when equation (2) is solved without internal iteration, i.e. without reducing Ψ_t to zero; in that case, Ψ_t plays the role of a forcing term and can generate spurious oscillatory behaviours. Examples of such regimes are given for respectively Az=2.5 and Az=3, at Gr=10^5 and Pr=0.02 in Figs.15a and 15b (Figs.15a1 and 15b1 correspond to an instant where Ψ_{max} is maximum, while Figs.15a2 and 15b2 correspond to an instant where Ψ_{max} is minimum).

Flow Characteristics in Stationary Regimes

It is interesting to analyze the evolution of Ψ_{max} and of some other flow characteristics such as the maximum of the longitudinal velocity V_{max}, and the longitudinal temperature gradient at the center, $(\partial\theta/\partial y)_m$ as a function of Gr, in the pre-oscillatory regime. As in a previous study about buoyancy-driven convection in an air-filled cavity by Roux et al. (1978), we can define a "conducting" regime, as the one in which the isotherms are only slightly (less than 10% variation of $\partial\theta/\partial y$) distorted by the bulk flow (i.e. here: 0.9 < $(\partial\theta/\partial y)_m$ < 1). For Pr=0.02 the end of such a conducting regime occurs for Gr close to 10^4 (Fig. 16a) . In that

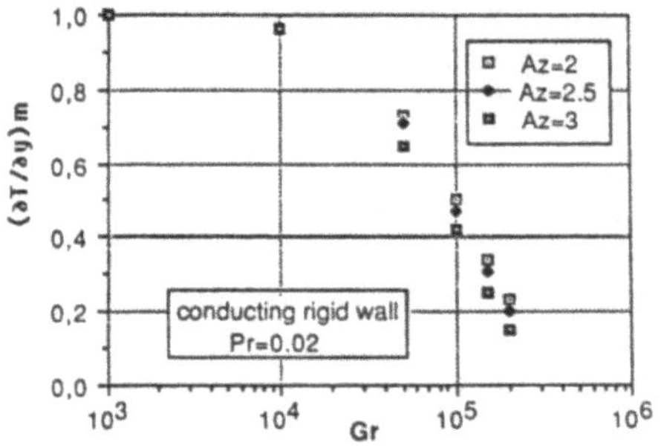

Fig.16a (∂T/∂y)m vs log(Gr) for different Az at Pr=0.02

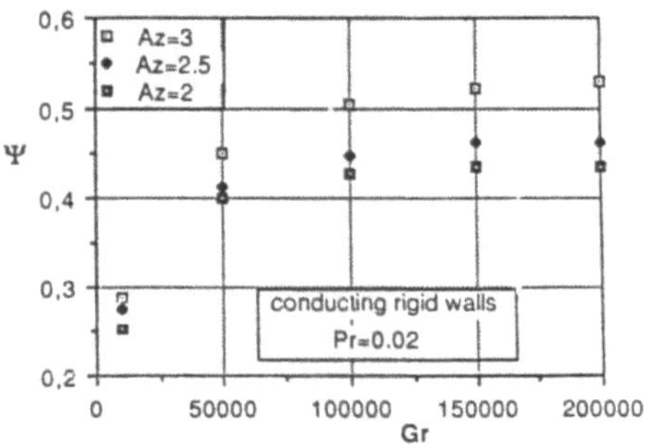

Fig.16b Ψ_{max} vs Gr for different Az at Pr=0.02

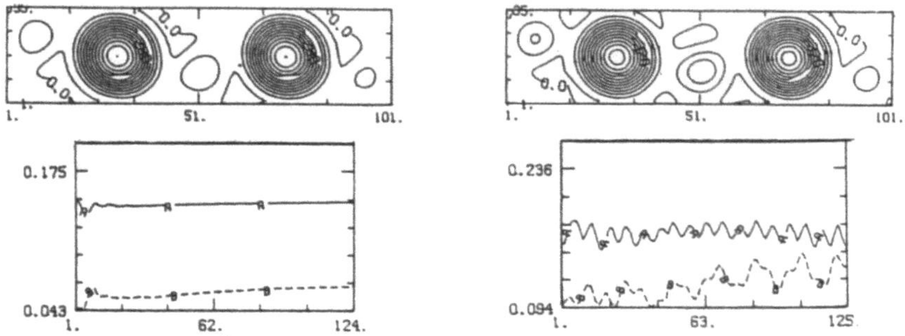

Gr=5.10^5

Gr=10^6

Fig.17 Flow structure (iso-Ψ) at Pr=0 , Az=4 ,

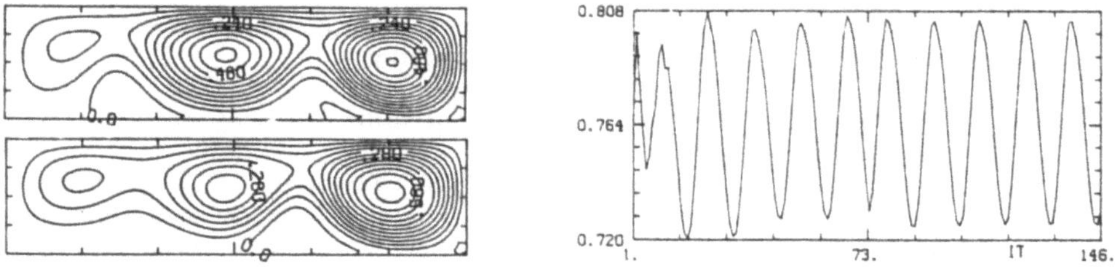

a)time evolution of the iso-Ψ pattern

b) Ψ_{max} time history .

Fig.18 Flow structure for a stress-free surface at Pr=0.015 .

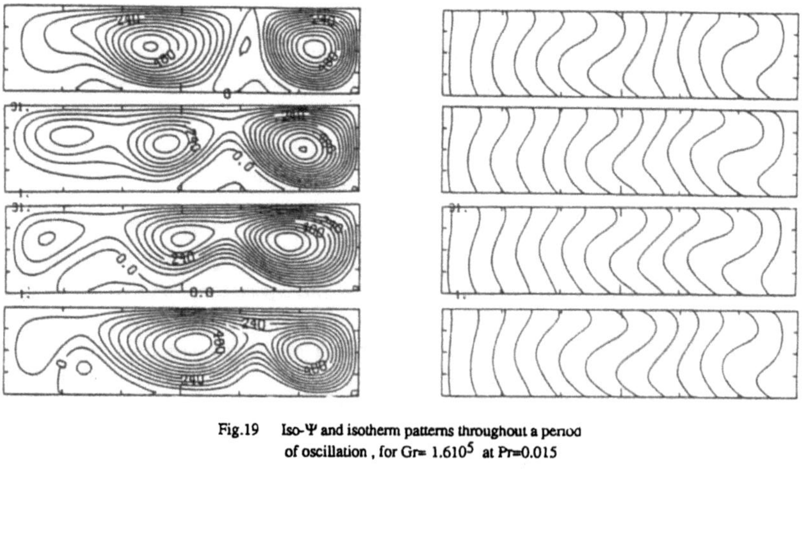

Fig.19 Iso-Ψ and isotherm patterns throughout a period of oscillation , for Gr= $1.6 10^5$ at Pr=0.015

conducting regime, the flow velocity increases almost linearly with Gr, as in (9a), due to a balance between viscous and buoyancy forces. For higher Gr the velocity behaves proportionally to $Gr^{1/2}$, as in (9b), due to a balance between inertia and buoyancy forces. For still higher Gr, a vertical temperature stratification occurs and again slows down the increase of the flow velocity. This can be shown in the Fig.16b, for Pr=0.02 and Az=4, through the variation of Ψ_{max}, which is nondimensionalized by $Gr^{1/2}$.

Flow Structure at High Gr (for Az=4 at Pr=0)

Results obtained with the fully transient version of our code are presented in Figs. 17a and 17b, for respectively Gr=5. 10^5 and 10^6, at Pr=0. By comparison with the final flow structure presented in Fig. 10b for Gr=5. 10^4, the structure at high Gr corresponds again to a two-cells solution, but the form of the cells is now almost circular. The solution at Gr= 10^6 seems to present small undamped fluctuations.

Stress-Free Upper Surface (R-F case) at Az=4

The fully-transient version of our finite-difference code has been recently used to study the case where the upper surface of the cavity is free (Ben Hadid and Roux ,1986b). Oscillation regimes are mainly characterized through the time history of Ψ_{max}. When solutions present Ψ_{max} fluctuations, they are tracked for several periods, in order to observe their damping or stabilization. In fact, it is always difficult to accurately determine the exact threshold of the oscillatory regime, as the rate of damping slows down (and the amplitude of fluctuations diminishes) when this threshold is approached. The present method had to be used very carefully; if internal iterations were stopped without requiring enough accuracy, spurious stable oscillations were present. To perform very accurate computations, the mesh size would have to be adjusted and refined for increasing Gr, and the test accuracy (for stopping iterations) correspondingly changed. For convenience, such sophisticated procedure has not been used; most of the computations have been done for a fixed mesh grid (31x61), except for highest Gr, for which a finer (37x85) grid has been used.

Limiting Case Pr=0

Table 2 Flow characteristics as a function of Gr, R-F case, at Pr=0

Gr Ψ_{max}	Ψ	-fluct.	f
5.0 10^3	0.43	-	-
1.0 10^4	0.56	-	-
1.5 10^4	0.63	-	-
2.0 10^4	0.67	-	-
2.5 10^4	0.70*	1.1%	20
3.0 10^4	0.72*	15.5%	22
4.0 10^4	0.74*	15.4%	27
5.0 10^4	0.78*	16.0%	31
8.0 10^4	0.82*	15.1%	43

* oscillatory mean value

Preliminary results obtained with a finite difference technique for Pr=0 when using the A2-nondimensionalization are shown in Tab.2. They suggest that the A2-nondimensionalization is suitable for all the test cases considered herein. Small-amplitude oscillations are present (with Ψ_{max} fluctuations of 1.1%) for Gr= 2.510^4, i.e. for Gr above the threshold value (Gr=1.37510^4) obtained by Winters (1986) and given in Fig. 8a.

For higher Gr the flow still oscillates with a single frequency that increases with Gr. Ψ_{max} fluctuations increase suddenly (15.5% at Gr= 3.10^4) after the threshold and then remain nearly constant.

Pr=0.015 , Conducting Horizontal Surfaces

At $Gr= 2.5 \cdot 10^4$, small-amplitude fluctuations (2.7%) of Ψ_{max} have been observed during several ten periods. The flow characteristics are given in Tab.3 for several values of Gr ranging from $1.0 \cdot 10^4$ to $8.0 \cdot 10^4$. At $Gr= 8.10^4$ the flow still oscillates with a frequency f=41.7. The time evolution of the iso-Ψ pattern, for $Gr= 8.10^4$, is presented in Fig.18a; the fluctuations of Ψ_{max} reach 10%, as shown through the time history of Ψ_{max}, in Fig.18b. This Gr corresponds to the case Ra=4750, presented by Crochet et al.(1983) in their Fig.4 as the beginning of oscillations; thus, the present code determines the onset of oscillations for a Gr substantially smaller. May be, that difference results from a coarse grid mesh in Crochet et al. calculations, according to the remark done by Winters et al.(1986) who mention that their own computations overestimate $Gr_{c,osc}$ by 40%, using a similar grid. Nevertheless a nice agreement is observed for the frequency (f=40) found by Crochet et al.

Table 3 Flow characteristics as a function of Gr , R-F case , at Pr=0.015

Gr	Ψ_{max}	Ψ-fluct.	f
$1.0 \cdot 10^4$	0.56	-	-
$2.0 \cdot 10^4$	0.66	-	-
$2.5 \cdot 10^4$	0.69*	2.7%	19
$3.0 \cdot 10^4$	0.70*	9.2%	20
$6.0 \cdot 10^4$	0.76*	8.4%	37
$8.0 \cdot 10^4$	0.765*	10,%	42

* oscillatory mean value

At $Gr= 1.6 \cdot 10^5$, where the oscillations are well established also in Crochet et al. results (see the results presented in their Fig.6, at $Ra= 10^5$), the comparison is good . A sequence of a few instantaneous flow structures (four instants evenly distributed over a period) presented in Fig.19 , shows the same basic mechanism as described by Crochet et al. (1983) who claim that "the oscillations are caused by the rising and diving of two (contra-rotative) eddies which try to rise the free surface but are unable to do so simultaneously".

Pr=0.030 , Conducting Horizontal Surfaces

For a slightly higher value of Pr, a different behaviour has been found in the sense that oscillations are observed with very small-amplitude fluctuations (e.g. 2.5% of Ψ_{max}) at $Gr=10^5$, but these oscillations desappear for slightly higher Gr (i.e. $Gr \geq 1.2 \cdot 10^5$) (see Figs. 20a and 20b, for respectively $Gr= 10^5$ and $Gr=1.4 \cdot 10^5$).

Table 4 Flow characteristics as a function of Gr , R-F case , at Pr=0.03

Gr	Ψ_{max}	Ψ-fluct.	f
$3.0 \cdot 10^4$	0.56	-	18**
$6.0 \cdot 10^4$	0.71	1.5% (?)	35
$8.0 \cdot 10^4$	0.71*	0.7%	42
$1.0 \cdot 10^5$	0.70*	2.5%	48
$1.1 \cdot 10^5$	0.70*	1.8%	47
$1.2 \cdot 10^5$	0.70	-	46**
$1.4 \cdot 10^5$	0.70	-	35**
$1.6 \cdot 10^5$	0.69	0	0

* oscillatory mean value
** damped oscillations

Flow characteristics given in Tab.4 show that the frequency reaches a maximum at $Gr= 10^5$ and decreases for higher Gr. At $Gr=6.10^4$ the damping of Ψ_{max} fluctuations is very slow,

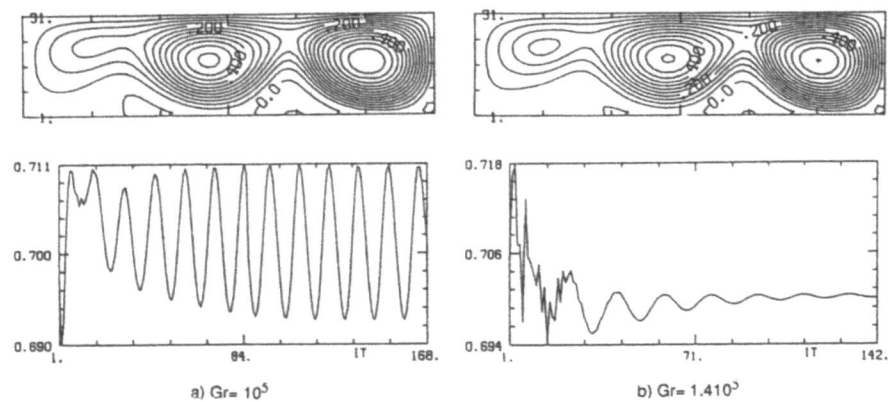

a) Gr= 10^5 b) Gr= 1.410^5

Fig.20 Flow structure for a stress-free surface at Pr=0.030

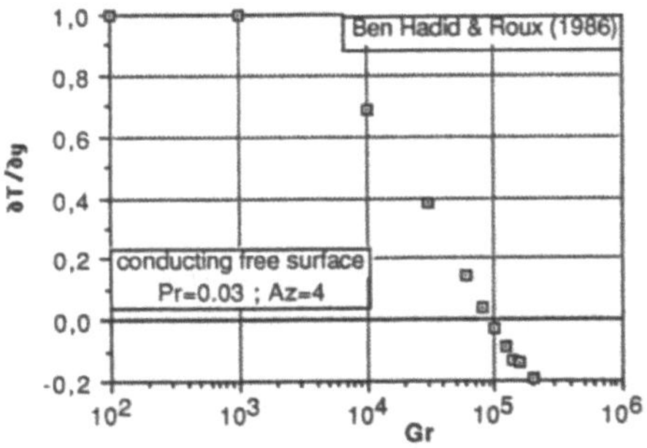

Fig.21

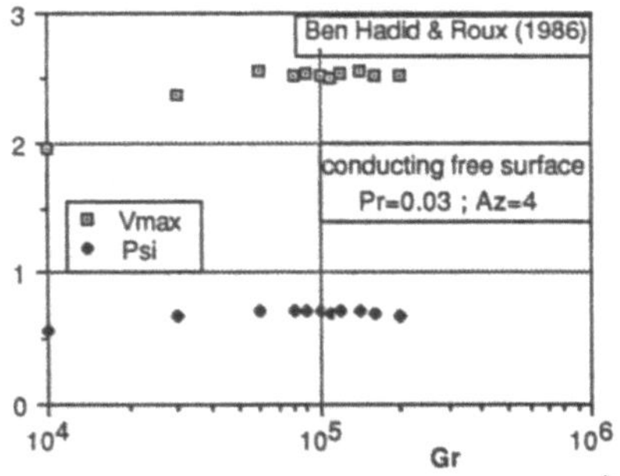

Fig.22

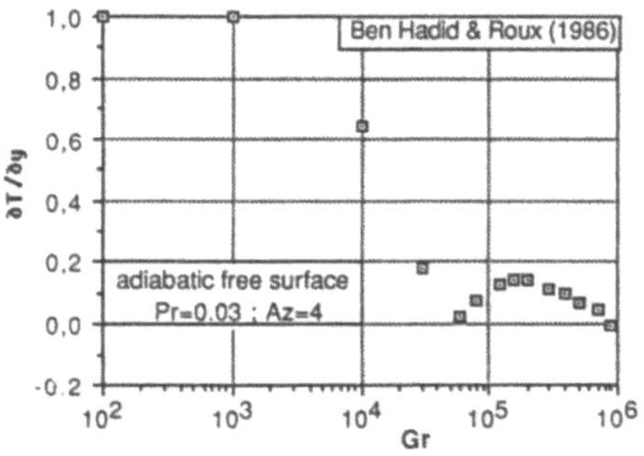

Fig.23

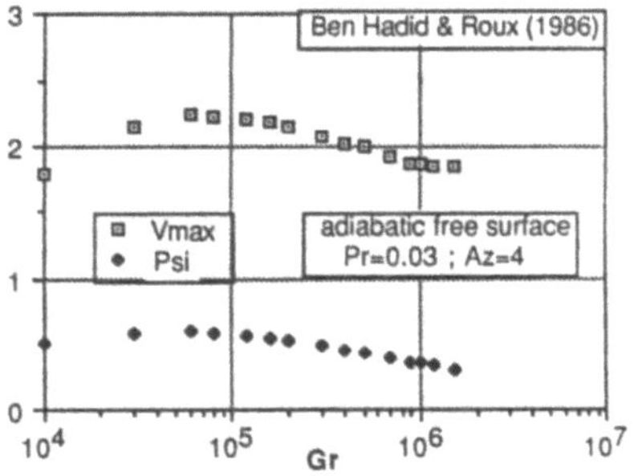

Fig.24

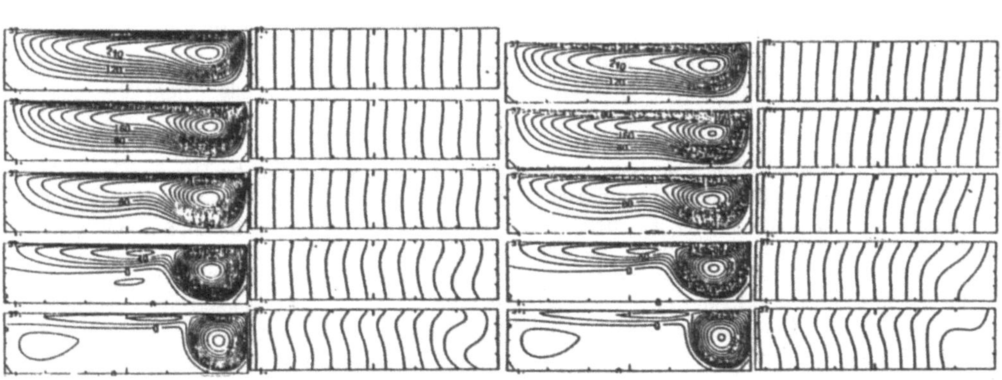

Fig.25a Flow pattern in the conducting case
Ma = 5, 10, 20, 100, 500

Fig.25b Isotherm pattern

Fig.26a Flow pattern in the adiabatic case
Ma = 5, 10, 20, 100, 500

Fig.26b Isotherm pattern

from 3% to 1.5% after 200 time steps corresponding to about 10 periods (and about 12mns on a Cray-1S computer). This behaviour presents some similarity with the one observed for closed cavity at $Gr=10^4$ for Az=4; it involves some stabilizing mechanism which is not well understood at this time. One possible explanation is that after a destabilization range due to a strong enough adverse vertical temperature gradient created near the upper surface (by the bulk flow), a re-stabilization occurs due to the flow in the boundary layer adjacent to this upper surface, the strength of which increases as $Gr^{1/2}$. This last mechanism is evoked as it presents some similarities with to the one described for the stability of thermally stratified Poiseuille flow by Gage & Reid (1968), Artiz-Cohen and Velarde (1981) and by Platten & Legros (1979) for confined geometries. A thermally stratified layer (heated from below) is increasingly stabilized when the speed of the Poiseuille flow is increased; in addition, it was shown by Platten & Legros (1979) that for narrow cavities (W/H of order one), the flow structure is with transverse rolls for low speed and changes into longitudinal rolls for a critical speed, the value of which increases when W/H diminishes and when Pr increases. In any way, it is important to note that in some situations the oscillations could appear for moderate Gr and disappear for higher Gr. In such circonstances, the use of stability theory predictions to interpret experimental results has to be very carefully done as these experimental results could correspond to a "next" bifurcated region. Main flow characteristics are given for increasing Gr, for $(\partial\theta/\partial y)_m$ as a function of log(Gr) in Fig.21 and for Ψ_{max} and V_{max} as a function of Gr, in the range $5\ 10^3 \leq Gr \leq 1.4\ 10^5$, in Fig.22.

Pr=0.030 . Adiabatic Horizontal Surfaces

For adiabatic horizontal surfaces, the stability of the flow against oscillations is strongly increased . In fact, at Pr=0.03 , we didn't find any oscillation for all the range $10^4 \leq Gr \leq 1.5\ 10^6$, thus far away from the threshold predicted by Winter's (1986) bifurcation theory, $Gr_{c,osc}= 1.1\ 10^5$ (Fig.8a) . Nevertheless, the present numerical data confirm an interesting result of this theory, by showing that an adiabatic condition is more stabilizing than a conducting one (this effect being the opposite of the one for the usual Benard problem).

The evolution of $(\partial\theta/\partial y)_m$ is given in Fig.23 and the variation of Ψ_{max} and V_{max}, in Fig.24, as a function of log(Gr), in the range $5 10^3 \leq Gr \leq 1.5\ 10^6$. As for the other situations previously mentioned, i.e. R-R conducting (Fig.16b) and R-F conducting (Tabs.3-4 and Fig. 22), Ψ_{max} and V_{max} rapidly reach a maximum close to $Gr=10^4$. Here, the considered domain of Gr is larger enough to observe a diminution of Ψ_{max} and V_{max}, after $Gr=2.10^4$ probably due to a positive thermal stratification.

Marangoni Effect on the Upper Boundary

Transient Marangoni-convection computations have been done by Polezhaev et al. (1981), at Pr=1 and Az=2 , for Ma equals to 10^2, 10^3 and 10^4. In all cases, the solution converges to a steady-state and presents a single-cell structure. The centre of this cell approaches the cold wall and tends to go down , creating temperature inhomogeneity when Ma increases.

Computations have been done by Wilke & Loser (1983) for $NaNO_3$ and Si melts, in small-Az rectangular cavities . For Az=1 and Ma=7, the flow starts with a single-cell structure and converges into a steady-state structure, with two contra-rotative cells near the colder vertical wall.

Marangoni-convection has also been considered in a separated paper by Ben Hadid & Roux (1986c), for both adiabatic and conducting boundary conditions and for a wide range of Ma, starting from a value of one, i.e. below the threshold value given by Smith & Davis (1983), for $Pr \leq 0.1$. The computations have been done in the case of no buoyancy effect (g=0, or Gr= 0) in order to uncouple the buoyancy and thermocapillary effects, and in order to be close to the domain of application of the theory of Smith and Davis, who studied the stability of a one-dimensional flow, like the one given by Birikh (1966) for large Az (expression (15) but with Gr=0), associated to a linear temperature distribution in y-direction. The linear stability theory applied to this basic flow by Smith & Davis (1983) exhibits oscillatory instability that takes the form of two hydrothermal waves (with the angles +ø and -ø with respect to negative y-axis) that propagate obliquely to the direction of the surface flow; for low Pr, the mechanism of instability involves a transfer of energy from the horizontal temperature gradients of the basic state to the perturbations through horizontal convection. In the adiabatic case (Biot number equal to zero), the critical Marangoni number, Ma_c, for the onset of unsteady regime is close to 2, for $Pr=10^{-3}$, and behaves as $Pr^{1/2}$ when Pr goes to zero . The Authors state in addition that an increase of the Biot number, Bi, always results in a more stable system (Bi =0 minimizes Ma_c). For Pr=0.025 (where Ma_c =10), we would have ø close to 90°, and that corresponds to instability in form of nearly longitudinal roll (with axes nearly in the direction of G_h). At the threshold, the wavenumber, k_c, is close to 0.38 and the angular frequency (nondimensionalized by t_{ref} = Ma/Pr. H^2/ν) is close to 0.028, i.e. with our notations ω_c=11.2.

Conducting Horizontal Surfaces

Computations have been done for a wide range of Ma; 1. ≤ Ma ≤ 750. They don't show any evidence of oscillatory regime. The flow structure evolution can be seen in Fig.25a for selected values of Ma (5 ; 10 ; 20 ; 100 ; 500) that exhibit the formation of a concentrated vortex near the cold wall, the strength of which increases with Ma. After Ma=20, a contra-rotative cell occurs in the lower part of the layer and wets the hot endwall . The strength of this secondary motion increases when Ma is further increased. The corresponding isotherm patterns are shown in Fig.25b; these isotherms appear to be only slightly affected by the bulk flow, except for the highest Ma value (Ma=500), in the region of the strong primary vortex near the cold wall. The evolutions of Ψ_{max} and V_{max} are plotted in Fig.27a as a function of Ma ; they present an asymptotic behaviour for high Ma.

Adiabatic Horizontal Surfaces

For adiabatic horizontal wall the flow pattern shown in Fig.26a is quite similar to the one presented in Fig.25a for rigid walls. Similarly the isotherms patterns, in Fig. 26b, present a little difference with respect to the ones presented in Fig.25b, except near the cold wall , in the region of the strong vortex. The evolutions of Ψ_{max} and V_{max} , plotted in Fig.27b, as a function of Ma also present an asymptotic behaviour, as for the conducting case in Fig.27a . The conducting or adiabatic nature of the horizontal boundaries doesn't play an important role in the range Ma < 750.

The v-velocity profile across the layer at y=Az/2, has been compared in Figs.28 with the theoretical one given by (15) at Gr=0. For Ma=1 and Az=4, the Fig.28a shows an excellent agreement with the theory valid for infinitely large Az. Note that the theoretical profile is independent of Ma and Pr, with the velocity reference given in 9(c), proportional to Ma/Pr. The v-profile, at y=Az/2, rapidly differs from the theoretical one, when Ma is increased (Fig. 28b). In addition the value of v at x=1, increasingly differs from the theoretical value: 0.25. For Ma=20, the v-profiles at y=0.5.Az and y=0.88.Az, are compared to the theoretical one in Fig.29.

The evolution of the velocity at the surface (x=1) along the y-direction, is presented in Fig.30, on the interval 0.5 ≤ y/Az ≤ 1. The curves in this last figure present a maximum closer to the cold endwall, the (nondimensionalized) value of which decreases, showing an increasing endwall effect (near the cold wall) when Ma is increased; a larger value of Az would be necessary to reach an asymptotic value (independent of Az) at x=1. This asymptotic value is 0.25 for moderate Ma/Pr, but would decrease , when the boundary layer regime is reached, as shown experimentally by Camel et al. (1986). We can say in conclusion that the flow pattern , for moderate Az, rapidly differs from the theoretical one and becomes strongly y-dependent.

DISCUSSION AND CONCLUSION

We presented a review of several studies devoted to the problem of spontaneous oscilllations

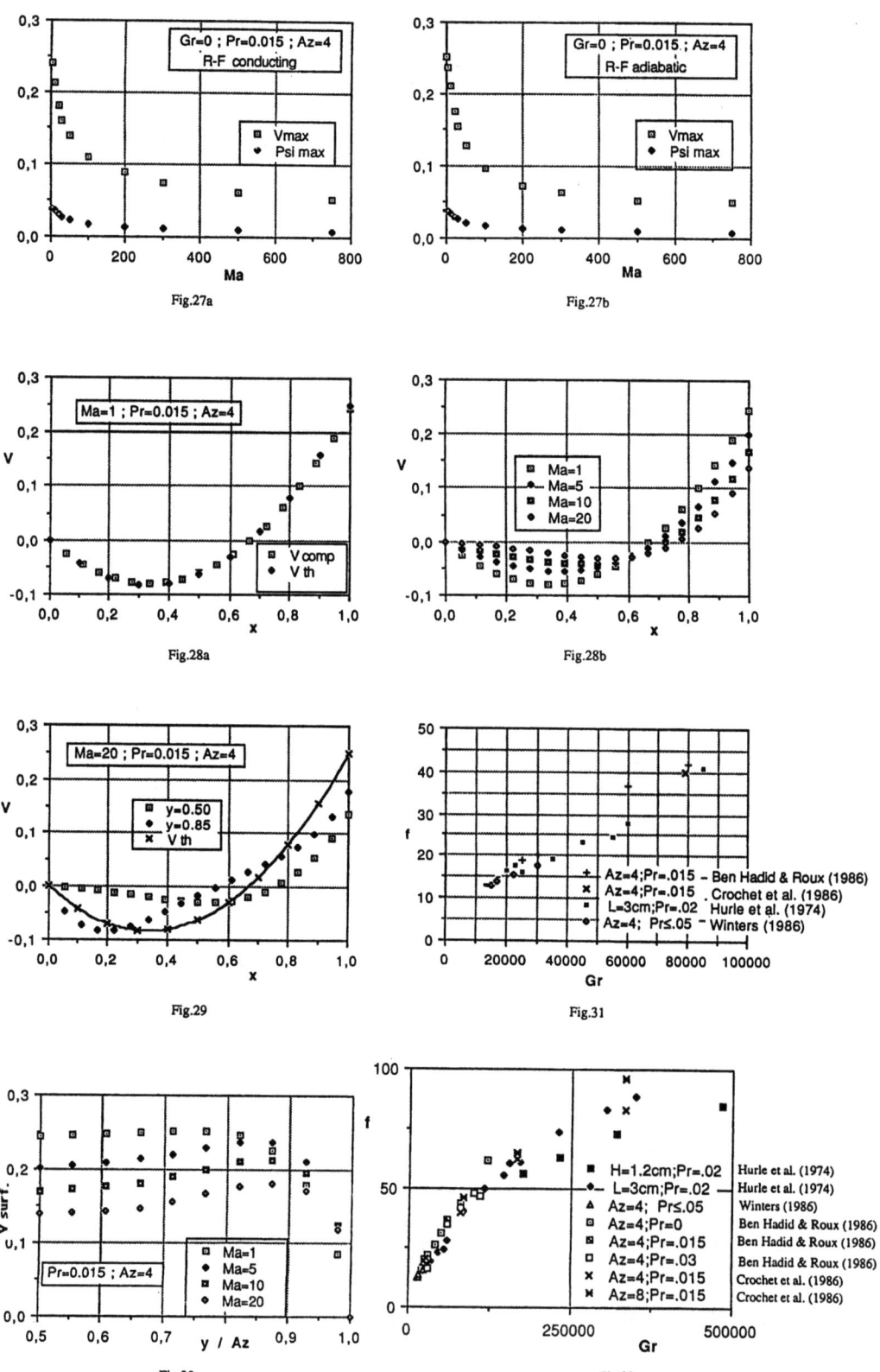

Fig.27a

Fig.27b

Fig.28a

Fig.28b

Fig.29

Fig.31

Fig.30

Fig.32

occurring in horizontal layers of low-Prandtl-number fluids subjected to an horizontal temperature gradient, G_h. Such situations can appear in melts during unidirectional crystallization in a boat, by a Bridgman technique for example. Following Hurle et al. (1967) we assumed that the origin of oscillations is not connected to the crystallization process itself (i.e., we don't deal with interfacial instability, or phase change, or moving front, ...).

Thus we mainly considered, as in Hurle et al. (1974) experiments, a parallelepipedic cavity and tried to handle with all the results given in the literature (from experiments, stability theories or direct numerical simulations), about the mechanism of oscillations. New computed results in rectangular cavities have been presented in order to study the influence of a large variety of boundary conditions; considering the cases where the upper horizontal surface is either rigid (R-R case) or stress-free (R-F case) or subjected to Marangoni forces (M case). The horizontal boundaries have been considered as either adiabatic or conducting. One of the results of these numerical simulations is to show that flow structure seems to become substantially different when $Pr \geq 0.1$, thus limiting the notion of small Pr to $Pr < 0.1$.

Most of the theoretical studies concern the stability of a one-dimensional flow (depending only in the vertical direction) like the Hadley circulation (Gill,1974; Hart, 1983; Roux et al.,1984 and Laure,1986) or the flow generated by a constant Marangoni force (Smith and Davis,1983). Different oscillatory regimes have been found to occur as the first bifurcation of this basic one-dimensional flow. All these regimes correspond to longitudinal rolls, with their axes in the direction of the imposed temperature gradient, G_h. But the (transverse) wavelength, λ_z/H, of these rolls strongly depends on the theories; it is equal to 3.7 in Gill theory for R-R and R-F conducting cases, equal to 9 to 16 in Hart theory for R-R and R-F adiabatic cases, and takes again higher values in Roux et al. (1984) results for R-R conducting and in those by Laure for R-R and R-F adiabatic cases. In general, the values of the frequency at the threshold are almost independent of the Prandtl number, in the range of small Pr. But this frequency would be strongly increasing for $Gr > Gr_{c,osc}$. All theories predict a decrease of $Gr_{c,osc}$ when Pr increases. In Marangoni case, the value of λ_z/H given by Smith and Davis is close to $2.6\,Pr^{1/2}$. In any way, due to the one-dimensional nature of the basic flow, these theories can only be applied for cavities having very large extension in both horizontal dimensions. They can't definitely permit to interpret the experiments in narrow boats, as the ones usually employed in Bridgman technique, in which the lateral confinement doesn't permit structures like longitudinal-rolls to occur. We have also shown, in the Marangoni case, that the confinement in y-direction also strongly affects the structure of the basic state flow, that is supposed to be y-invariant in the stability analysis. All the stability results are summarized in Tab. 5

Table 5 Stability theories for large cavities

Authors	Basic state	Bound. condit	Critical condit.	Pr	ω_c	λ_z/H
Gill (1974)	Hadley	R-R cond.	$Gr_c = 1030/Pr$	any	22	3.7
		R-F cond.	$Gr_c = 240/Pr$	any	22	3.7
Hart (1983)	Hadley	R-R adiab.	$Gr_c = 1130\,Pr^{-1/2}$	$.015 \leq Pr < .1$	37	9
		R-F adiab.	$Gr_c = 350\,Pr^{-1/2}$	$.010 \leq Pr < .1$	32	16
Laure (1986)	Hadley	R-R adiab.	$Gr_c = 1820\,Pr^{-1/2}$	$.034 \leq Pr < .1$	36	9
		R-F adiab.	$Gr_c = 550\,Pr^{-1/2}$	$.010 \leq Pr < .1$	15.6	14-23
Smith and Davis (1983)	Marangoni	R-F adiab.	$Ma_c = 632\,Pr^{1/2}$		11.2	$2.6\,Pr^{-1/2}$

For the narrow boats, the three-dimensional computations done by Dupont et al. (1986) almost suggest that the flow structure is not far, in the symmetry plane, from the one obtained with a two-dimensional model by Crochet et al. (1986). Computations carried out for a two-dimensional model don't permit at least with the boundary conditions used up to now to give a clear conclusion about the relevance of the model to explain the occurrence of oscillatory regime.

One of the main conclusion of the two-dimensional simulations is that oscillatory flow regimes are difficult to obtain in the R-F adiabatic case, in the range of Gr (typically Gr $\leq$ 10^6) that correspond to situations where the experiments exhibit such oscillatory regimes (4.10^4 to 10^5, in Hurle et al. (1974)).

In agreement to Winters (1986) stability analysis, the most unstable situations , for buoyancy-driven flow and constant transport properties , is the R-F conducting case. A possible explanation of the de-stabilizing role of conducting horizontal surfaces is the existence of some adverse vertical temperature gradient that do not exist in the corresponding adiabatic situations . In addition , one can remark that if , at a first view, the adiabatic conditions would seem more appropriate to model experiments in open boats, it always exists in these experiments heat losses that induce such adverse temperature gradient.

Even in the R-F insulated case, the oscillatory regime has been found only for relatively small Pr, and for a narrow range of Gr. After a short domain of Gr in which clearly the flow exhibits oscillations ($6.10^4 \leq Gr \leq 1.210^5$ for Pr=.03) or "tendency" to oscillate (slowly damped oscillations) for higher Pr, the flow is re-stabilized (see mainly Tab. 4). This stabilization behaviour has not been previously mentioned in the literature; it is perhaps to be connected to the behaviour exhibited recently by Favier et al. (1986) in their experiments for a long cylindrical channel which present limited range of Gr in which stable oscillations are maintained . But, it is impossible at this moment to firmly conclude about the relevance of such re-stabilization behaviour which could be due to a positive thermal stratification in the bulk or to interaction between the mechanism of instability due an adverse thermal stratification at the upper surface and the increasing inertia forces generated in the boundary layer adjacent to this upper surface . This last mechanism is evoked as it presents some similarities with to the one described for the stability of thermally stratified Poiseuille flow by Gage & Reid (1968) and by Platten & Legros (1979) for confined geometries.

Infact similar oscillations or, more precisely similar tendency to oscillate, have also been observed, in the R-R conducting case at Pr=0, in a narrow range of Gr ($2.5\ 10^4 \leq Gr \leq 5.10^4$); but at Pr=0 the temperature field is "frozen", thus the stabilization mechanism, attributed to positive thermal stratification for Pr> 0, is not uniquely related to that.

For R-F adiabatic case , the computations have only been done for Pr=.03 , they don't exhibit oscillatory regime up to Gr=$1.5\ 10^6$. But we can't conclude, at the present time , if oscillatory regime exists for smaller Pr . Computations have been done by Crochet et al. (1986) that exhibit oscillations at Pr=0.015 , but for mixed conditions (adiabatic upper surface and conducting lower surface).

Substantial and disappointing differences are observed between experiments, computations and stability theories concerning $Gr_{c,osc}$; namely, at Az=4 for R-F case, this threshold value is found as $1.7\ 10^4$ by Winter's bifurcation theory for adiabatic upper surface , as $2.5\ 10^4$ by the present computations and as $7.9\ 10^4$ in Crochet et al. results ; all for Pr=0.015 . While Hurle et al. (1974) experiments, for Pr close to .020, gives $Gr_{c,osc}$ close to $4.\ 10^4$. Despite these differences, an interesting general result, shown in Figs. 31-32 , is the very nice agreement found for the values of the frequency, f, when correlated with Gr, between experiments (Hurle et al.,1974), two-dimensional simulations (Crochet et al.,1986 and Ben Hadid & Roux,1986b) and bifurcation theory (Winters,1986). In that correlation we considered not only the threshold values (as for Hurle's experiments) but values corresponding to sub-critical (damped oscillations) or supercritical (stable oscillations) situations. This dependence between f and Gr, appears only slightly affected by the variations of Az and Pr, in the ranges $2. \leq Az \leq 8$ and $.015 \leq Pr \leq .07$, mainly for low Gr (Fig.31).

The question of the relevance of modelisation and numerical simulation to analyze the oscillatory regime for low-Prandtl-number fluids in open boat, like in Hurle et al. (1974) experiments, is not a trivial problem and is still open. To address this problem, a GAMM-Workshop devoted to "transient convection in low-Prandtl-number fluids", is now being prepared. Additional experiments also would be very useful in this important area of fluid mechanics.

Acknowledgments
The authors wish to thank Mme C. Guibaud for improving the translation of their manuscript. They gladly acknowledge the contribution of several well-known searchers, mainly Profs. D.T.J. Hurle, S. Ostrach, V.I. Polezhaev, R. Sani, G. De Vahl Davis and K.H. Winters, and the Group of J.J. Favier who contributed to this study through enlighting discussions and private communications. They also acknowledge the Centre National d' Etudes Spatiales (Division Matériaux), for financial support and the Centre de Calcul Vectoriel pour la Recherche for providing access to its Cray-1S computer.

REFERENCES

Artiz-Cohen J.A. and M.G. Velarde (1981). Natural versus forced convection in the two-component Bénard problem: New theoretical results. J. Non-equilib. Therm., 6, 159-164.

Balasubramanian and Ostrach S. (1984). Physico-Chemical-Hydrodynamics, 5, 3-18

Ben Hadid H. and Roux B. (1987a). Numerical simulation of time-dependent natural convection in horizontal layers. To be published.

Ben Hadid H. and Roux B. (1987b). Oscillatory buoyancy-driven flow in horirontal liquid-metal layer. 6th European Symposium Materials Sciences in Microgravity conditions, Bordeaux, France ,2-5 Dec.

Ben Hadid H. and Roux B. (1987c). Numerical simulation of Marangoni convection in horizontal layers. To be published.

Bhattacharyya S.P. and Nadoor S. (1976). Stability of thermal Convection between non-uniformly heated plates. Appl. Sci. Res., 32, 555-570.

Birikh R.V. (1966a). J. Appl. Math. Mech. (P.M.M.), 30, 356-361

Birikh R.V. (1966b). Thermocapillary convection in a horizontal layer of liquid, J. Appl. Mech. Tech. Phys.,7 , 43.

Birikh R.V., Gershuni G.Z., Zhukhovitskii E.M. and Rudakov R.N. (1969) Stability of the steady convective motion of a fluid with a longitudinal temperature gradient. J. Appl. Math. Mech. (P.M.M.), 33, 937

Bontoux P., Roux B., Schirocky G.H., Markham B.L. and Rosenberger F. (1986). Convection in the vertical midplane of a cylinder - comparison of two-dimensional approximations with three-dimensional results. Int. J. Heat Mass Transfer, 29, 227-240.

Brenier B., Roux B. and Bontoux P. (1986). Comparaison des méthodes Tau-Chebyshev et Galerkin dans l'étude de stabilité des mouvements de convection naturelle. J. Mécanique Théorique et Appliquée, 5, 95 -119.

Bullister E.T., Cartage T., Deville M. and Patera A.T. (1986). Spectral simulation of thermal convection in complex geometries. Tenth Int. Conf. Num. Meth. in Fluid Dyn., Beijing, China, 23-27 June.

Camel D., Tison P. and Favier J.J. (1986). Marangoni Convection in Liquid Metal Layers. Sixth European Symp. Material Sc. under Microg. conditions. FP-20, ESA SP-256, 87.

Carruthers J.R. (1968). Thermal Convection in Horizontal Crystal Growth. J. Crystal Growth ,2,18

Carruthers J.R. (1977). Thermal Convection Instabilities Relevant to Crystal Growth from Liquids. Preparation and Properties of Solid State Materials, Vol.3, Ed. Wilcox and Lefever (M. Dekker, New York)

Carruthers J.R. and Winegard W.C. (1967). Thermal Convection and Solute Segregation during Horizontal Melting and Solidification. J. Phys. Chem. Solid, Supplement n°1, 645-649.

Cartage T., Demaret P. and Deville M. (1985). Chebyshev Spectral and Pseudo-Spectral Solutions of the Navier-Stokes equations. Lect. Notes in Phys., 218, 127-132.

Cole G.S. and Winegard W.C. (1964). Thermal Convection During Horizontal Solidification of Pure Metals and Alloys. J. Institute. Metals, 93, 153-164.

Cormack D.E., Leal L.G. and Imberger J. (1974). Natural Convection in a Shallow Cavity with Differentially Heated Endwalls, J. Fluid Mech., 65, 209-230

Crochet M.J., Geyling F.T. and Van Schaftingen J.J. (1983). Numerical Simulation of the Horizontal Bridgman Growth of a Gallium Arsenide Crystal. J. Crystal Growth, 65, 166-172

Crochet M.J., Geyling F.T. and Van Schaftingen J.J. (1986). Numerical Simulation of the horizontal Bridgman growth. Part I: two-dimensional flow. Submitted for publication.

Dupont S., Marchal J.M., Crochet M.J. and Geyling F.T. (1986). Numerical Simulation of the horizontal Bridgman growth. Part II: Three-dimensional flow. Submitted for publication.

Fairweather G. and Mitchell A.R. (1967). SIAM J. Num. Anal., 4, 2.

Favier J.J., Rouzaud A. and Comera J. (1986). Influence of Various Hydrodynamic Regimes in Melts on Solidification Interface. submitted to Revue de Physique Appliquée

Gage K.S. and Reid W.H.(1968) .The stability of thermally stratified plane Poiseuille flow. J. Fluid Mech.,33, 21-32.

Gershuni G.Z. and Zhukhovitskii E.M. (1969). Stability of plane-parallel convective motion with respect to spatial perturbations. J. Appl. Math. Mech. (P.M.M.), 33, 855-860.

Gill A.E. (1974). A Theory of Thermal Oscillations in Liquid Metals, J. Fluid Mech., 64, 577-588

Gresho Ph. and Sani R. (1984), private communication

Hart J.E. (1972). Stability of thin non-rotating Hadley circulations. J. Atm. Sc., 29, 687-697

Hart J. E. (1983a). A Note on the Stability of Low Prandtl Number Hadley Circulation. J. Fluid Mech., 132, 271.

Hart J. (1983b). Low Prandtl Number Convection between Differentially Heated Endwalls, Int. J. Heat Mass Transfer, 26, 1069

Hurle D.T.J. (1967). Thermo-Hydrodynamic Oscillations in Liquid Metals: The Cause of ImpuritiesStriations in Melt-Grown Crystals. J. Phys. Chem. Solid, Supplement n°1, 659-663.

Hurle D.T.J. and Jakeman E. (1973). Natural Oscillations in Heated Fluid Layers. Physics Lett.,43A, 127-129

Hurle D.T.J., Jakeman E. and Johnson C.P. (1974). Convective Temperature Oscillations in Molten Gallium, J. Fluid Mech, 64, 565-576

Jakeman E. and Hurle D.T.J. (1972). Rev. Phys. Technol., 3, 3.

Jones I.P. (1982). Low Prandtl number free convection in a vertical slot. Harwell Report AERE-R 10416.

Jordan (1985). J. Crystal Growth, 71, 551-558.

Korpela S.A., Gozum D. and Baxi C.B. (1973). Int. J. Heat Mass Transfer, 16, 1683-1690.

Korpela S.A. (1974). A study on the effect of Prandtl number on the stability of the conduction regime of natural convection in an inclined slot. Int. J. Heat Mass Transfer, 17, 215-222.

Laure P. (1986). Etude des mouvements de convection dans une cavité rectangulaire avec un gradient de température horizontal. Submitted to J. Mécanique Théorique et Appliquée.

Lee E.I. and Sernas V. (1980). Numerical study of heat transfer in rectangular air closures of aspect ratio less than one. A.S.M.E Paper, 80 /HT-43

Lee Y. and Korpela S.A. (1983). Multicellular convection in a vertical slot. J. fluid Mech. ,126, 91-121

Liang S.F. and Acrivos A. (1970). Stability of buoyancy-driven convection in a tilted slot. Int. J. Heat Mass Transfer, 13, 449-458.

Mueller A. and Wilhelm M. (1964). Z. Naturforch., 19a, 254.

Nikitin S.A. and Polezhaev V.I. (1980). Mathematical simulation of impurity distribution in space processing experiments with semiconductors. XXIIIrd COSPAR, ISC G.2.3. ,Budapest, Hungary.

Nikitin S.A., Polezhaev V.I. and Fedyushkin A.I. (1981). Mathematical simulation of impurity distribution in crystals prepared under microgravity conditions. J. Crystal Growth, 52, 471-477.

Nikitin S.A. and Polezhaev V.I. (1983), private communication

Ostrach S. (1976), Proc. Second European Symposium on Material Sciences in Space, ESA-SP-114.

Ostrach S., Loka R.R. and Kumar A. (1980). Natural convection in low aspect-ratio rectangular enclosures. *Natural Convection in Enclosures* -Torrance Ed., A.S.M.E. Heat Transfer Division, Vol.8, New-York, 1-10.

Pamplin B.R. and Bolt G.H. (1976). J. Phys. (Appl. Phys.), 9, 145.

Patterson J. and Imberger J. (1980). Unsteady natural convection in a rectangular cavity. J. Fluid Mech., 100, 65-86.

Pimputkar S.M. and Ostrach S. (1981). Convective Effects in Crystals Grown from Melts, J. Crystal Growth, 55, 614-646

Platten J.K. and Legros J.C. (1979). Stabilité de la convection forcée non isotherme. Mémoire Acad. Royale de Belgique.

Polezhaev V.I. (1979). Convective processes at low gravity . Proc. 3rd European Symp. on Material Sciences in Space . ESA-SP-142 , 25-31.

Polezhaev V.I., Dubovik K.G. , Nikitin S.A. , Prostomolotov A.I. and Fedyushkin A.I. (1981) . Convection during crystal growth on Earth and in Space. J. Crystal Growth , 52, 465-470 .

Pulicani J.P. and Peyret R. (1986). private communication

Roux B. ,Grondin J.C. , Bontoux P. and Gilly B. (1978). On a high-order accurate method for the numerical study of natural convection in a vertical square cavity. Numerical Heat Transfert1, 331-349

Roux B. , Bontoux P. , Loc T.P. and Daube O. (1979) . Optimisation of Hermitian methods for N.S. equations in vorticity and streamfunction formulation. Lect. Notes in Math., Ed. SpringerVerlag, 771, 450-468 .

Roux B. , Bontoux P. and Henry D. (1984) . Numerical and Theorical Study of Different Regimes occuring in Horizontal Fluid Layers, Differentially Heated. Lect. Notes in Physics, Ed. SpringerVerlag, 230, 202-21.

Simpkins P.G. and Chen K.S. (1983). Natural Convection in Horizontal containers with Applications to Crystal Growth , AT&T Bell Lab. Report.

Smith M.K. and Davis S.H. (1983) . Instabilities of dynamic thermocapillary liquid layers - Part.1 Convective instabilities . J. Fluid Mech. , 132 , 119-144 .

Smithells (1976). Metals Reference Book . Butterworth Ed. , 5th Edition, London, p. 940 .

Stewart M.J. and Weinberg F. (1971) . J. Crystal Growth , 12 , 217-228 .

Sweet D. , Jakeman E. and Hurle D.T.J. (1977) . Free convection in the presence of both vertical and orizontal temperature gradients. Phys. of Fluids , 20 , 1412-1415 .

Tichy J. and Gadgil A. (1982). High Rayleigh number laminar convection in low aspect ratio enclosures with adiabatic horizontal walls and differentially heated vertical walls. J. Heat Transfer, 104, 103-110 .

Utech H.P. and Early S.G. (1967a) . On the Presence of Thermal Convection in the Kinetics Experiments of Rigney and Blakely . Acta Metallurgica, 15, 1238-1239 .

Utech H.P., Brower W.S. and Early S.G. (1967b) . Thermal Convection and Crystal Growth in Horizontal Boats ,in : *Crystal Growth* , Ed. Peiser (Pergamon, Oxford), 201-205.

Utech H.P. and Flemings M.C. (1967) . Thermal Convection in Metal-Crystal Growth : Effect of a Magnetic Field . J. Phys. Chem. Solid, Supplement n°1, 651-658 .

De Vahl Davis G. (1986). Finite difference methods for natural and mixed convection in enclosures. *Heat Transfer*, Ed. Hemisphere , Washington, 101-109 .

Viskanta R. , Kim D.M. and Gau C. (1986) .Three-dimensional natural convection heat transfer of a liquid metal in a cavity. Int. J. Heat Mass Transfer , 29 , 475-485. Weber J.E. (1973) . On thermal convection between non-uniformly heated planes .Int. J. Heat Mass Transfer, 16,961-970

Wilke H. and Loser W. (1983). Numerical calculation of Marangoni convection in a rectangular open boat. Crystal Res. & Technol. , 18, 825-833

Winters K,H. , Clife K.A. and Jackson C.P. (1984) . A review of extended systems for finding critical points in coupled problems. *Numerical Methods for transient and coupled problems*. Ed. Pineridge Press, Swansea, U.K.

Winters K.H. and Clife K.A. (1986a). The onset of Convection in a bounded fluid with a free surface.Harwell Report HL86/1335. Submitted to J. Comp. Physics

Winters K.H. , Clife K.A. and Jackson C.P. (1986b) . The Prediction of Instabilities using Bifurcation Theory. Harwell Report HL86/1147. To appear in *Transient and Coupled System* , Ed. John Wiley, Chichester, U.K.

Wirtz R.A. and Tseng W.F. (1980). Natural convection across tilted rectangular enclosures of small aspect ratio. *Natural Convection in Enclosures* -Torrance Ed., A.S.M.E. Heat Transfer Division, Vol.8, New-York , 47-54 .

ANNEXE 1. TEMPERATURE PROFILE ASSOCIATED TO THE HADLEY SOLUTION

From (4), a temperature profile is associated to $v(x)$ by $T_{xx} = T_i/T_d \, v \cdot \theta_y$.

In the case (13) we have $T_{xx} = -CB(2x^2 - 3x + 1)x/12$ where $CB = (V_b T_i / V_d T_d)(\theta_y)^2$ so,

for adiabatic conditions (8a) on both the horizontal walls, $B = 0$ and $\int_0^1 T(x)\,dx = 0$,

$$T_{R-R}(x) = -CB\,[x^3(6x^2 - 15x + 10) - 1]/720 ; \tag{A1-1}$$

for perfectly conducting conditions (8b) on both the horizontal walls,

$$T_{R-R}(x) = -CB\,[x^3(6x^2 - 15x + 10) - 1]/720. \tag{A1-2}$$

In the case (14), we have $T_{xx} = -CB(8x^2 - 15x + 6)x/48$; so,

for adiabatic conditions (8a),

$$T_{R-F}(x) = -CB\,[x^3(24x^2 - 75x + 60) - 4]/2880 ; \tag{A1-3}$$

for perfectly conducting conditions (8b),

$$T_{R-F}(x) = -CB\,[x^3(8x^2 - 25x + 20) - 3x]/960 ; \tag{A1-4}$$

for mixed conditions (8a) at $x=1$ and (8b) at $x=0$,

$$T_{R-F}(x) = -CB\,[x^3(8x^2 - 25x + 20)]/960. \tag{A1-5}$$

In the case (15), $T_{xx} = -CB(8x^2 - 15x + 6)x/48 - CM\,[3x^2 - 2x]/4$, where $CM = T_i/T_d)(\theta_y)$ and so,

for conditions (8a),

$$T_M(x) = -CB\,[x^3(24x^2 - 75x + 60) - 4]/2880 - CM\,(3x^4 - 4x^3 + 0.4)/48 ; \tag{A1-6}$$

for conditions (8b),

$$T_M(x) = -CB\,[x^3(8x^2 - 25x + 20) - 3x]/960 - CM\,(3x^4 - 4x^3 + x)/48 ; \tag{A1-7}$$

for mixed conditions,

$$T_M(x) = -CB\,[x^3(8x^2 - 25x + 20)]/960 - CM\,(3x^4 - 4x^3)/48. \tag{A1-8}$$

ANNEXE 2. ANALYTICAL TWO-DIMENSIONAL CONTINUATION OF THE HADLEY SOLUTION

To insure a better evaluation of the initial velocity components near the endwalls of the cavity, we use an analytical y-continuation (satisfying (5a)) of the streamfunction: $\Psi = \varepsilon(y)\,\psi(x)$ with $\varepsilon(y) = \tanh[C\,y^2(Az - y)^2]$. The constant C permits to match the 1D solution at a distance (from vertical walls) equal to Az_{ref} ($y = Az_{ref}$ and $Az - Az_{ref}$);

$$C = 3.27 / [Az_{ref}(Az - Az_{ref})]^2.$$

Setting $\varepsilon(y) = \tanh p$ we have :

$$u = \psi\,\varepsilon_y = \psi\,\varepsilon_p\,p_y ; \tag{A2-1}$$

$$v = -\varepsilon\,\psi_x ; \tag{A2-2}$$

$$\Omega = \varepsilon\,\psi_{xx} + \psi\,\varepsilon_{yy} = \varepsilon\,\psi_{xx} + \psi\,[\varepsilon_p\,p_{yy} + p_y^2\,\varepsilon_{pp}]. \tag{A2-3}$$

ANNEXE 3. TRANSPORT PROPERTIES OF Ga, GaAs, Sn AND Si AT MELTING TEMPERATURE

Table A3   Transport properties of liquid metals (Ga, GaAs, Sn and Si)

	Smithells (Ga)	Hurle (Ga)**	Smithells (Sn)	Jordan (GaAs)	Ostrach (Si)
melting temperature (°K)	303		505	1511	1412*
density (g cm^{-3})	6.09	6.10	7.00	5.72	2.52
volume exp.coef., α (°K^{-1})	1.01 10^{-4}	1.26 10^{-4}	.875 10^{-4}	1.87 10^{-4}	2. 10^{-4}
viscosity, μ, (g cm^{-1}sec^{-1})	0.0204	0.0170	0.0185	0.0279	0.01
specific heat (J g^{-1} °K^{-1})	0.398	0.401	0.250	0.434	0.930
thermal conductivity (W cm^{-1} °K^{-1})		0.255	0.34	0.300	0.178
Cinematic viscosity, ν, (cm^2 sec^{-1})	3.35 10^{-3}	2.78 10^{-3}	2.64 10^{-3}	4.87 10^{-3}	3.97 10^{-3}
diffusivity, κ (cm^2 sec^{-1})	1.05 10^{-1}	1.39 10^{-1}	1.71 10^{-1}	0.717 10^{-1}	1.28 10^{-1}
Prandtl number, ν/κ,	**0.032**	**0.020**	**0.015**	**0.068**	**0.031**
Surface tension, σ dyn cm^{-1})	718		544		865*
$\partial\sigma / \partial T$	- 0.1		-0.07		-0.43* -0.13*

* (Smithells, 1976)
** at some meantemperature

ANNEXE 4. TEMPERATURE DEPENDENCE OF TRANSPORT PROPERTIES OF GALLIUM

The temperature dependence of ρ, μ and σ, is given by Smithells (1976) as : $\rho = \rho_m$ [1 + 1.01 10^{-4} (T- T$_m$)] ; μ= 4.359 10^{-3} exp (4. 103 / 8.3144 T)   g / sec. cm and σ=718 [1 + 1.39 10^{-4} (T- T$_m$)] dyn / c. The temperature dependence of the density does not play a significative role on the value of Pr. The temperature dependence of the conductivity, k, is given by Smithells as :

T (°C)	T (°K)	k (W cm^{-1} °K^{-1})
30	303	0.255
100	373	0.300
200	473	0.350
300	573	0.392

We have approximately : $k = 0.255 + 0.045 (T-T_m) / 70$, for $0 < T - T_m < 70$; where the melting temperature, T_m, is equal to 303 °K. Thus, the temperature dependence of the Prandtl number, **Pr**, can be deduced as follows, for $T \approx T_m$:

$$Pr_m/Pr = \exp[1.588 (T-T_m) / T] \cdot [1 + 2.52 \cdot 10^{-3} (T-T_m)] \approx 1 + 7.76 \cdot 10^{-3} (T-T_m)$$

Table A4 Temperature dependence of transport properties of liquid gallium

T (°K)	T-T_m	μ_m / μ	k / k_m	$(\nu k)_m / \nu k$	$(\nu_m / \nu)^2$	Pr_m/Pr	Pr
303	0	1.	1.	1.	1.	1.	0.032
308	5	1.026	1.013	1.01	1.053	1.04	0.031
313	10	1.052	1.025	1.03	1.107	1.08	0.030
323	20	1.103	1.050	1.05	1.217	1.16	0.028
333	30	1.154	1.076	1.07	1.332	1.24	0.026
343	40	1.203	1.101	1.09	1.447	1.32	0.025
353	50	1.252	1.126	1.11	1.568	1.41	0.023

THE KURAMOTO-SIVASHINSKY EQUATION: SPATIO-TEMPORAL CHAOS AND INTERMITTENCIES FOR A DYNAMICAL SYSTEM

Basil Nicolaenko

Center for Nonlinear Studies and Theoretical Division
Los Alamos National Laboratory
Los Alamos, New Mexico 87545, U.S.A.

I. INTRODUCTION

During the last decade we have seen a number of major developments which show that the long-time behavior of solutions of a very large class of partial differential equations (PDEs) possess a striking resemblance to the behavior of solution of finite dimensional dynamical systems, or ordinary differential equations (ODEs). The first of these advances was the discovery (by a number of researchers) that a dissipative PDE has a compact, maximal attractor X with finite Hausdorff and fractal dimensions. More recently [6, 12-15] it was shown that some of these PDEs possess a finite dimensional inertial manifold, i.e., an <u>invariant</u> manifold that contains the attractor X. For the later equation, the connection with ODEs is no longer a mere resemblance, instead it has become a striking reality! The reason for this is that where one restricts the PDE to the inertial manifold one obtains an ODE, which we call an <u>inertial form</u> for the given PDE. Since an inertial manifold contains the universal attractor, this means that the long-time behavior of solutions of a PDE with an inertial manifold is <u>completely</u> determined by the inertial form.

There is, indeed, a class of phenomena of incipient turbulence which can be modeled by scalar field PDEs undistinguishable in practice from dynamical systems: weak turbulence on interfaces between complex flows, upon which appear well localized patterns and structures. Such are quasiplanar flame fronts; thin viscous fluid films flowing over inclined planes; and even, under some conditions, dendritic phase change fronts in binary alloy mixtures [30-31, 38, 40-42]. In such physical contexts, the onset of destabilization of the simplest laminar regime is heralded by the cohesive organization of cells and patterns (often hexagonal) on the moving and buckling front or interface.

Many such interfaces with localized turbulence, including flames, can be modeled by the simple Kuramoto-Sivashinsky (K-S) PDE [33-36, 38]. This equation accurately accounts for the thermo-diffusive and convective mechanics of flow-field coupling across an interface before turbulence breaks away from the interface and reaches deeply into the fluid.

In one space dimension, the K-S equation modeling a small perturbation u(x,t) of a metastable planar front or interface is

$$u_t + \nu u_{xxxx} + u_{xx} + \frac{1}{2}(u_x)^2 = 0 \ , \ (x,t)\epsilon R^1 \times R_+,$$
$$u(x,0) = u_o(x) \ , \ u(x+L) = u(x,t)$$

(1.1)

Here the subscripts indicate partial differentiation, ν is a positive fourth-order viscosity and u_o is L-periodic; L being the size of a typical pattern scale. The natural bifurcation parameter is the renormalized dimensionless parameter $\tilde{L} = L/(2\pi\sqrt{\nu})$. $[\tilde{L}]$ is also the number of linearly unstable Fourier modes, where the symbol [ ] designates the integer part of a real number.

In a previous work our computer simulations of the K-S equation [28] demonstrated an uncanny, low-dimensional behavior for the values of the bifurcation parameter up to $\tilde{L}$ = 3.67 (L = 23.1). A low dimensional structure does also underline an example of onset of chaos at $\tilde{L} = 5.42$ (L = 34.05). In general, the attracting solution manifolds undergo a complex bifurcation sequence including multimodel fixed points, invariant tori, traveling wave trains and homoclinic orbits. Moreover, amidst lengthy and complex chaotic time series, puzzling intermittencies do occur at random: these are protracted oscillations in a small neighborhood of some metastable state. We conjectured in [28] that such a behavior shadows a perturbed homoclinic orbit and betrays a hidden underlying low-dimensional dynamical system.

The long time behavior of such smooth dissipative differential systems is characterized by the presence of an universal attractor X toward which all trajectories converge. The structure of X may be very complicated even in the case of simple ordinary differential equations: X may be a fractal on parafractal set. In the case of dissipative partial differential equations, although the phase space (i.e. the function space) is an infinite dimensional space, X has finite fractal dimension (see [2-5] [41-42]). However, the already possibly complex nature of X is in this case further complicated by the infinite number of degrees of freedom of the ambient space. In the case the dissipative system admits an inertial manifold the attractor is dynamically embedded in a finite dimensional manifold and thus becomes the attractor of a dynamical system with a finite number of degrees of freedom.

Since [28], it has indeed been demonstrated that the K-S equations are rigorously equivalent to a finite-dimensional dynamical system. The approach (introduced in [12-15]) consists in constructing a finite dimensional Lifschitz manifold $\sum$ (called <u>inertial</u>) in the phase space of the PDE such that:

(i) $\sum$ is invariant and has compact support; that is if $(S(t,\cdot))_{t\geq 0}$ is the nonlinear semigroup associated with the initial value problem for the equations, then $S(t,\sum)$ is contained in $\sum$ for all $t \geq 0$.

(ii) All solutions converge exponentially to $\sum$. In particular, the universal attractor, X, is included in $\sum$ and the dissipative system reduces on $\sum$ to a finite system (called an <u>inertial ODE</u>).

(iii) asymptotic completeness holds: for every initial value for the full K-S equation, there exists some initial point on the inertial manifold $\sum$ decreases exponentially to zero [6].

The last point does fully establish the equivalence between the PDE and the inertial ODE on $\sum$. Concretely, given a chaotic trajectory for the exact PDE, we can find a finite dimensional chaotic trajectory for the inertial ODE, such that the two trajectories converge exponentially.

The existence of such an inertial manifold has been demonstrated [6, 14, 15, 37] for the K-S equation with Neumann boundary conditions. In similar results hold for a PDE model of 2-D weak turbulence in Kolmogorov shear flows [8].

Hence, weak turbulence on interfaces modelled by the K-S equation is strictly equivalent to chaos for a finite inertial dynamical system. Still, weak spatio-temporal turbulence involves complex mechanisms within the bifurcations of the inertial manifold. To unravel these, we must obtain a clear picture of those few nonlinear states (spatial structures) which form a reduced nonlinear coordinates basis for the manifold. A nonlinear representation of the inertial manifold must be constructed, based on reduced coordinates patches, with the goal of establishing reduced, low-dimensional, inertial normal forms for the inertial ODEs, valid for some range of the bifurcation parameter. These inertial normal forms control the global vector field bifurcations into weak turbulence and ultimately account for the universality of transition to chaos in infinite dimensional systems.

In this conference presentation, we propose to give a partial survey of the questions raised above, in the context of the K-S equation. Rather than another catalogue of bifurcations, we establish that successive transitions to chaos and intermittent relaminarizations are ruled by the stable and unstable manifolds of a small number of nonlinear states. This reduced representative sample changes along with the bifurcation parameter; its dimensionality is much smaller than the rough estimates for the dimension of the inertial manifold obtained in [6, 14-15]. In part III, we evidence such low dimensional canonical vector field bifurcations for regimes within $23.1 \leq L \leq 34.05$ ($3.67 \leq \tilde{L} \leq 5.42$, where $[\tilde{L}]$ is the number of unstable modes); for such regimes, the fractal dimension of chaos does not exceed 5. We have systematically searched for classical dynamical systems bifurcations and for multiple basins of attractions for the K-S model. We used a general PDE solver code developed by J. M. Hyman at LANL [29, 30]. The interactions of multiple basins through their fractalized boundaries have been evidenced. Intermittencies in turbulent time series are one of the key mechanisms in bridging the gap between PDEs and dynamical systems. They enable us to track the unstable manifolds of key hyperbolic points. These intermittencies are random time windows where dynamics remain highly oscillatory yet are confined in a relatively small neighborhood of some metastable point or circle. Such critical states are the natural candidates for local nonlinear coordinates of the inertial manifold.

In part IV, we survey an intrinsic geometric construction of inertial manifolds, inspired by the exponentially fast lock-up of dynamical trajectories onto such manifolds. This surveys joint work with P. Constantine, C. Foias and R. Temam [6].

II. OVERVIEW OF COMPUTATIONAL SIMULATIONS AND THEORETICAL RESULTS

We have normalized the K-S equation to an interval of length 2π; set the damping parameter to the original value derived by Sivashinsky, $\nu = 4$, and introduced the bifurcation parameter $\alpha = 4\tilde{L}^2 = L^2/4\pi^2$. The equation can now be written as

$$u_t + 4u_{xxxx} + \alpha[u_{xx} + \frac{1}{2}(u_x)^2] = 0 , \quad 0 \leq x \leq 2\pi,$$
$$u(x + 2\pi, t) = u(x, t) , \quad u(x, 0) = u_o(x)$$
(2.1)

This equation is equivalent to Eq. (1.1) with a different time scaling.

The mean value of the solution to Eq. (2.1)

$$m(t) = \frac{1}{2\pi} \int_0^{2\pi} u(x, t) dx$$
(2.2)

satisfies the drift equation

$$\dot{m}(t) = \frac{-\alpha}{4\pi} \int_0^{2\pi} (u_x)^2 dx.$$
(2.3)

To normalize this drift to zero, we numerically solved the equation for

$$v(x, t) = u(x, t) - m(t).$$
(2.4)

That is, the drift-free K-S equation is

$$v_t + 4v_{xxxx} + \alpha[v_{xx} + \frac{1}{2}(v_x)^2] + \dot{m}(t) = 0.$$
(2.5)

we have scanned the domain $54 \leq \alpha \leq 320$, i.e., $3.67 < r \leq 8.95, 25.8 < L < 56.2$. Many previous investigations have solved Eq. (1.1) on an interval $[0,L]$, with viscosity $\nu = 1$ [31]. Our rescaled time t in (2.1) and the usual unnormalized time t_{Phys} are related through:

$$t_{Phys} = 4\tilde{L}^4 t = \alpha^2 t/4;$$
(2.6)

the rescaling grows quadratically in α. The bifurcations diagrams (Fig. 1, 2, 3) were obtained by scanning in α and varying initial data.

In our computer experiments, we found that high precision was necessary because of the extreme sensitivity of the simulations to numerical accuracy. Nonconverged numerical solutions of Eqs. (1.1) and (2.1) can occur in regimes we are interested in if the time integration errors are greater than 10^{-6} per unit time-step. In fact, small effects of the order of 10^{-6} in the energy for some sensitive Fourier modes critically impact on the nonlinear dynamics. To alleviate this, we use in our calculations high-precision, pseudospectral approximation to the spatial derivatives [9, 10] and a variable time-step, variable-order integration method in time to keep the solution errors between 10^{-8} and 10^{-10} per unit time [29, 30]. The PDE solver was an all purpose code developed by J. M. Hyman at LANL [29, 30].

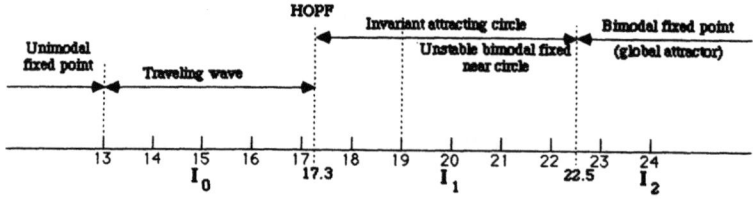

Fig. 1. The stable solution manifolds have a simple structure when α is small.

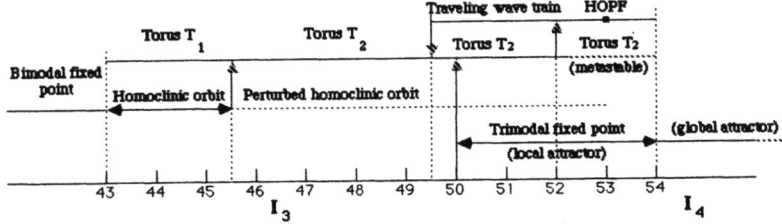

Fig. 2. As the bifurcation parameter increases, complex homoclinic and quasiperiodic behavior sets in (number of unstable modes between 3.2 and 3.6).

A typical example of the extreme numerical sensitivity of the numerical solutions to the K-S equation is the disappearance of homoclinic orbits if the precision is too low. The hyperbolic fixed points degenerate into stable fixed points with a numerically artificial basin of attraction the size of the error control. Because of the artificially stable fixed point, our numerical results of the K-S equation differ from some of the previously published simulations that relied on second-order schemes with only modest control over time integration errors.

We systematically tracked the domains of stability of each attractor with respect to the bifurcation parameter by varying α and reinitializing v(x,0) to the final solution from the previous run with a different α. Many problems were recalculated several times with different grid resolutions and time truncation error criteria to ensure that the numerical solutions were converged within an acceptable accuracy.

A remarkable feature of the K-S equations is the alternating sequence of intervals in α containing laminar behavior (some fixed point is ultimately attracting) with intervals of persistent oscillatory and/or chaotic behavior. Let $I_j = [\alpha_j, \alpha_{j+1}]$ be the jth interval. Then $I_o = [0, \alpha_1]$, where at α_1 comes the first Hopf bifurcation; a classical pitchfork steady-state bifurcation occurs at $\alpha = 4 < \alpha_1$. For j even, I_j is characterized by the ultimate decay to a globally attracting fixed point $\tilde{u}_q(x), q = (j/2)+1, j \geq 2$. These fixed points have most of their energy concentrated in a cos qx mode. The higher harmonics appear with exponentially decreasing energy and the fixed point has a lacunary Fourier expansion:

$$\tilde{u}_q(x) = a_{1q}\cos qx + \epsilon a_{2q}\cos 2qx \\ + \epsilon^2 a_{3q}\cos 3qx + \cdots + \epsilon^{n-1} a_{nq}\cos nqx + \cdots, \quad (2.7)$$

where $q = j/2 + 1$. Numerically, we have found that a_{1q} is O(1) and $\epsilon \cong 10^{-1}$. We call these sinks associated with I_j, j even, <u>cellular states</u>. When the Fourier expansion (2.7) of a cellular state is dominated by cos qx as we call it a <u>q-modal cellular state</u>.

These relaminarization intervals I_j, j even, are consistent with experiments at small and moderate Reynolds numbers [40]. Moreover, as j and α increase, the ultimate decay follows long periods of transient chaos. Transient chaos is observed in the K-S equations beginning in the interval I_4, provided enough modes are excited in the initial data. Moreover, as α increases, the mean lifetime of transient chaos increases exponentially in $\tilde{L}$: this growth makes transient chaotic intervals undistinguishable in practice from chaotic intervals in the strongly chaotic regimes (say, when the fractal dimension of the universal attractor, X, for the flow is large, $dim_F(X) \geq 10$).

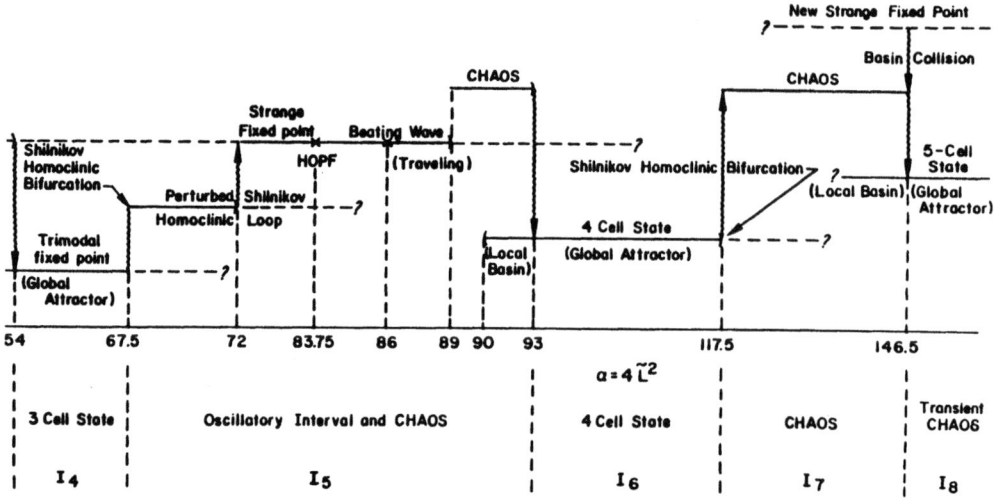

Fig. 3. A low-dimensional vector field skeleton underlies these weakly chaotic regimes (number of unstable modes between 3.6 and 6).

When j is odd, the intervals I_j have persistent oscillatory and/or chaotic behavior. For moderate values of α (say, up to I_7), the quasiperiodic and/or chaotic behavior reflects a competition between the previous $(j+1)/2$ cellular state, dominated by the $cos((j+1/2)x)$ mode, and the $(j+3)/2$ cellular state, dominated by $cos((j+3/2)x)$. This competition creates a complex interplay between temporal chaos and spatial coherence. In some sense, the (low-dimensional) temporal chaos corresponds to adjustment from one (low-dimensional) space pattern to the next one. Unfortunately, this simple picture is not borne by our computations at strongly chaotic regimes ($dim_F(X) > 6$) where a zoo of strange hyperbolic fixed points appear in intermittencies. Their strangeness resides in that they are not cellular in the sense of (2.7) and possess a broad energy spectrum band covering all the unstable modes up to $[\tilde{L}]$.

Finally, the best current estimate [15] on the dimension of $\sum$, the inertial manifold for Eq. (2.1) is:

$$dim(\sum) \leq c\alpha^{1.75} , \qquad (2.8)$$

which is still too large when compared to the upper estimate of the fractal dimension of the universal attractor X obtained in [36]:

$$d_f(X) \leq c\alpha^{0.75} . \qquad (2.9)$$

The severe numerical sensitivity of the K-S equation demonstrates that the dynamics of the inertial ODE and the bifurcations of the inertial manifold are very sensitive to the accuracy of numerical algorithms. Conventional PDE algorithms do not carry over.

III. LOW DIMENSIONAL CHAOS FOR THE KURAMOTO-SIVASHINSKY EQUATION

In this section we describe the behavior of the solutions to the K-S equation for parameter values in the intervals $I_4, I_5, I_6, 54 < \alpha \leq 117.5$, that is $3.67 < \tilde{L} \leq 5.42, 23.1 < L \leq 34.05$. The windows I_1 (oscillatory, $17.3 \leq \alpha \leq 22.5$) I_2 (2-cell state globally attracting, $22.5 < \alpha < 43$) and I_3 (quasiperiodicity, $43 \leq \alpha < 54$) were investigated in [28]; results are summarized in Fig. 1 and Fig. 2. Our preliminary catalogue for the intermediate values of α, $54 \leq \alpha \leq 117.5$, is presented in Fig. 3. It contains a sequence of "laminar" intervals and intervals with complex oscillatory behavior:

I_4	=	$54 < \alpha < 67.5$:	a 3-cellular state global attractor
I_5	=	$67.5 < \alpha < 93$:	complex oscillatory behavior
I_6	=	$93 < \alpha < 117.5$:	a 4-cellular state global attractor
I_7	=	$117.5 < \alpha < 146.5$	chaos

Within these intervals, we evidence canonical vector field bifurcations leading to quasi-periodic motion and chaos, and systematically explore multiple basins of attraction, in a low-dimensional situation. The mechanisms which we pin down are truly generic for both onset of chaos and relaminarization crises in regimes of "strong" chaos (see part IV), and more representative than the bifurcations in $I_1 - I_3$ studied in [28]. The classical homoclinic loop bifurcations from a saddle point observed around $\alpha \sim 22.5$ and $\alpha \sim 43$ do not reoccur; repeated onsets of oscillatory, and/or chaotic regimes are in fact triggered by perturbed homoclinic loops bursting from spiral hyperbolic points. T_1 tori (invariant circles) are usually metastable. Strange fixed points are the rule rather than the exception, spanning the range from $\alpha = 49.5$ through $\alpha=93$. A travelling beating wave obsreved in Fig. 1b from $\alpha=49.5$ to $\alpha=54$ is a true harbinger of such a strange (two-humped) fixed point. Also, there is a wealth of reverse bifurcation, and attractors which alternatively destabilize and restabilize again at some larger α! Last, the crisis of chaos observed at $\alpha=93$ is likely triggered by the two-humped strange fixed point sitting on the basin boundary of the chaotic attractor and shadowing the turbulent time series through multiple intermittencies. Such low-dimensional mechanisms pervade the strongly chaotic regimes of part IV.

In the discussion below, the "energy" is the integral of $(u_x)^2$ and the "energy in mode k" is the modulus of the kth Fourier coefficient.

The trimodal cellular state $\tilde{u}_3(\alpha)$ is a global attractor in I_4 until it bifurcates at $\alpha=67.5$. The bifurcation is neither of Hopf type, nor through a classical homoclinic loop. This is explored in Figs. 4-7, where $u_o = \tilde{u}_3(67.5) = 2.95 cos 3x + 0.44 cos 6x + \cdots$. Figure 4 heralds two regimes: at roughly periodic intervals the orbit bursts away on the unstable manifold of $\tilde{u}_3$ and puffs into a spiked intermittency at a much lower energy level; then it spirals back around the hyperbolic point $\tilde{u}_3$. Figure 5 confirms that the energy in the first mode is low during the small oscillations around the spiral hyperbolic point $\tilde{u}_3$; the bursts have a much higher level in the first mode. Small amounts of energy trigger the bursts around the loop. The energy in the third mode, Fig. 6, is the mirror image of Fig. 5. It oscillates in a small neighborhood of 2.9, before bursting away from $\tilde{u}_3$ into sharp spikes at much lower levels. The energy in mode 6 (Fig. 7) is substantial in the vicinity of $\tilde{u}_3$, at a level of 0.4. It clearly shows two different scales in the dynamics of the orbit next to $\tilde{u}_3$; first very high frequency, small amplitude oscillations around $\tilde{u}_3$, followed by slower spiraling around the trimodal point. This bifurcation has many of the characteristics of a perturbed Shilnikov homoclinic loop [25]. This is a homoclinic loop associated with a spiral hyperbolic point and persists until $\alpha=72$.

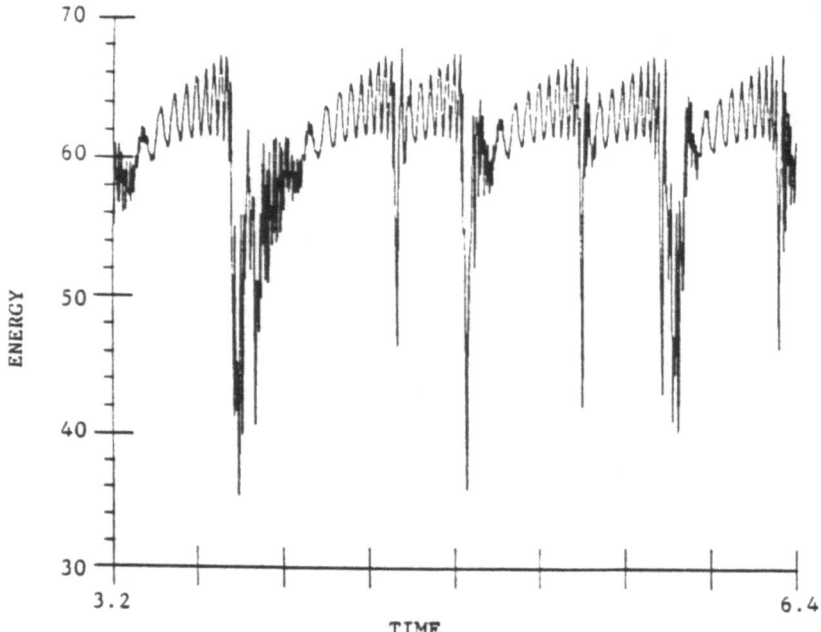

Fig. 4. The energy has near periodic bursts on the Shilnikov homoclinic loop and then spirals around the hyperbolic point ($\alpha = 67.5$).

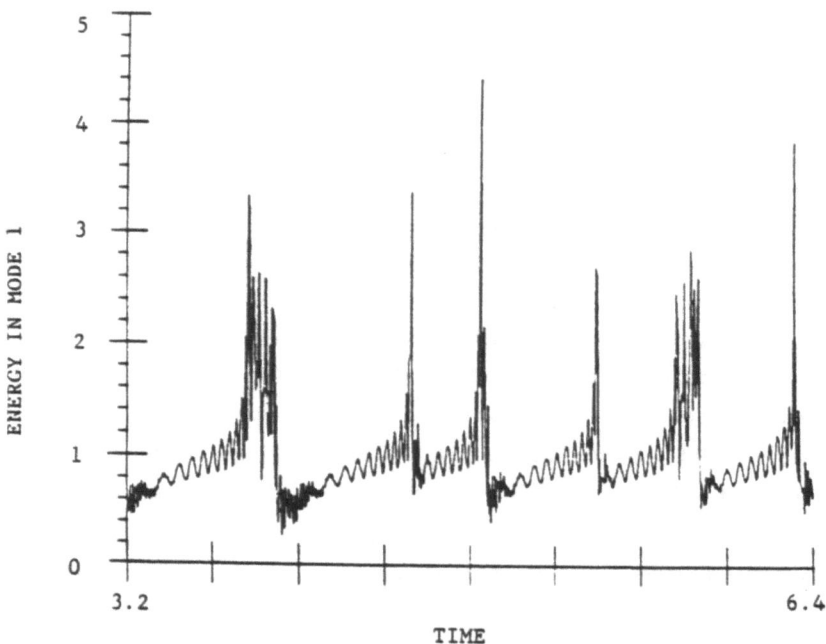

Fig. 5. The energy in the first mode is low during the small oscillations near the trimodal cellular hyperbolic point ($\alpha = 67.5$).

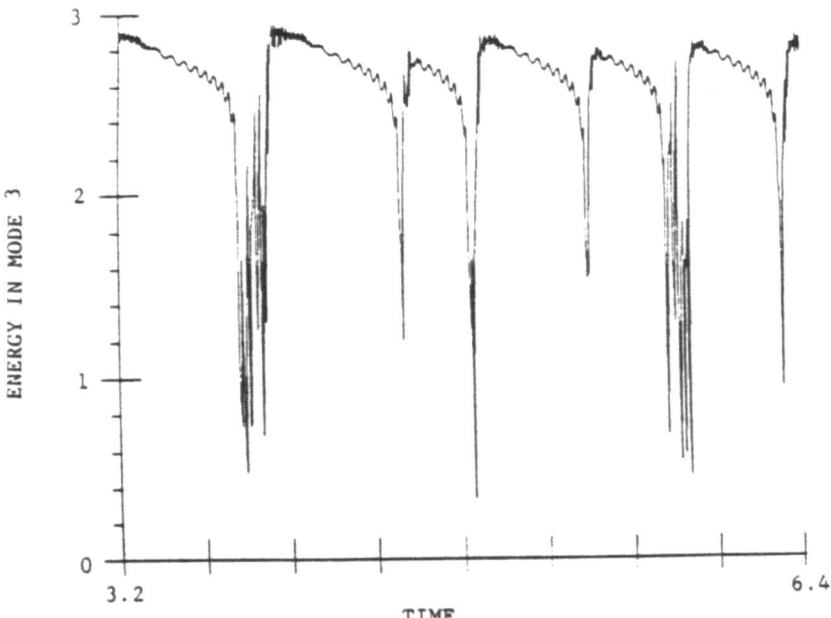

Fig. 6. The energy in the third mode is high in the neighborhood of the spiral hyperbolic point, whose components are only harmonics of three ($\alpha = 67.5$).

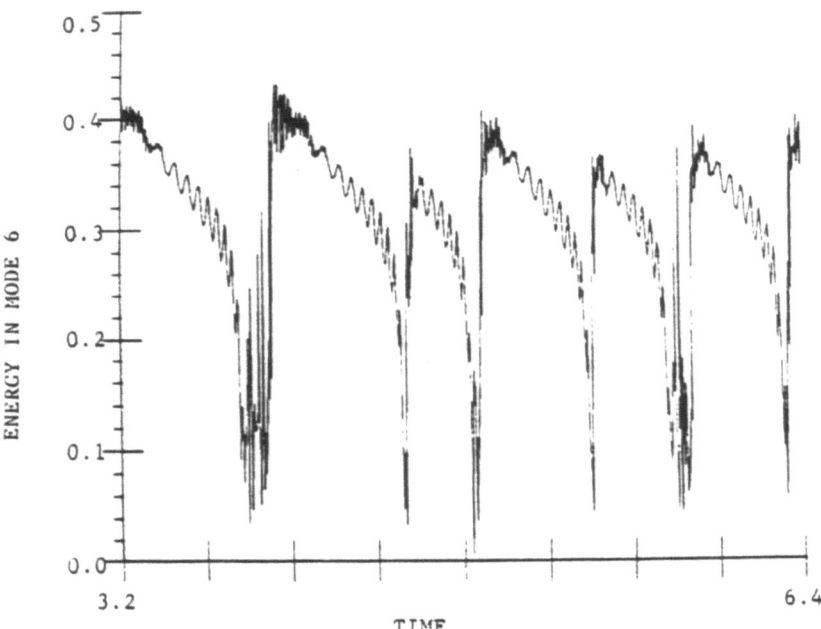

Fig. 7. The energy in the sixth mode is substantial next to $\tilde{\mu}_3$; high frequency oscillations are clearly distinguished close to this cellular point ($\alpha = 67.5$).

The Shilnikov loop is quickly deformed into a homoclinic tangle, as evidenced in Figs. 8-10 ($\alpha=68$, initial conditions continued from $\alpha=67.5$). The duration of the chaotic excursions is now comparable to the transit times in the vicinity of $\bar{u}_3$, any semblance of periodicity is lost and spiked bursts occur at random times (Fig. 8). The energy in mode one (Fig. 9) demonstrates that the high frequency, small oscillations around $\bar{u}_3$ prevail upon the spiraling time-scale dynamics; a computer movie "zoom" onto such time intervals reveals transient dynamics hardly distinguishable from those on a metastable circle (Torus T_1). Energy in mode 3 (Fig. 10) confirms the picture of an homoclinic tangle between the stable and unstable manifolds of the spiral hyperbolic point $\bar{u}_3$, with chaotic time series interrupted by random intermittencies around $\bar{u}_3$. The above generic picture will permeate the onset of chaos at $\alpha=117.5$.

At $\alpha=72$, a strange fixed point $u^*(\alpha)$ suddenly becomes a global attractor. It is not related to any cellular state; its Fourier expansion is rather flat, with energy present in all first six modes ($\tilde{L} = 4.24$). u^* has a typical profile with two humps, a large one and a small one (Fig. 11). An entirely similar two-humped structure has been observed as a traveling and beating wave, from $\alpha=49.5$ to $\alpha=54$; the strange fixed point $u^*(\alpha)$ has indeed undergone a reverse bifurcation back to stability! It persists as a global sink, until it undergoes some kind of Hopf bifurcation at $\alpha=83.75$. As the contour levels show in Fig. 12 ($\alpha=84.25$, initial data by continuation), the rapid oscillations are strictly localized in space, on the top of the higher hump. Such a spatio-temporal localization is a forerunner of spatially concentrated zones of turbulence. This peculiar example of spatial complexity does not seem to be ruled by a standard Hopf mechanism. At $\alpha=86$, the localized oscillating pattern bifurcates into a travelling beating wave. The contour levels plot (Fig. 13, $\alpha=87$, initial data by continuation) manifest fast oscillations still localized on the higher hump. At $\alpha=89$, the picture reverts to chaotic behavior, as if the "horseshoe" attractor observed from $68 \leq \alpha < 72$ had undergone a basin boundary crisis. The interval I_5 ends at $\alpha=93$, where the 4-modal cellular state

$$\bar{u}_4 = 2.94 cos 4x + 0.28 cos 8x + \cdots \tag{3.1}$$

mutates into a global sink. The apparent crisis of chaos at $\alpha=93$ is further complicated by the fact $\bar{u}_4$ has a limited, albeit small basin of attraction for $90.5 < \alpha \leq 93$. This suggests basin boundary crisis [19-24]. Chaotic time series both prior to and at the crisis exhibit multiple intermittencies around some hyperbolic point, which is obligatory non cellular (since $\bar{u}_4$ is a local sink). This is illustrated at $\alpha=91$, with initial data $u_o = \sum_{j=1}^{4}(cos jx + sin jx)$, Figs. 14-17. In Fig. 14, energy in mode one goes through two broad intermittencies; these are characterized by small amplitude, high frequency oscillations at an average level of 4. Within the intermittent windows, energy in mode 2 (Fig. 15), at an average level of 2.5, is comparable to mode 1. The (average) Fourier energy spectrum at the intermittencies is significantly comparable to that of the strange point $u^*(\alpha)$. Movies unmistakedly betray the two-humped structure. Energy in mode 3 (Fig. 16) confirms high frequency, small scale oscillations whenever the orbit wanders close to the stable manifold of u^*; as if the boundary of the basin for the chaotic attractor were vested with multiple fingers close to

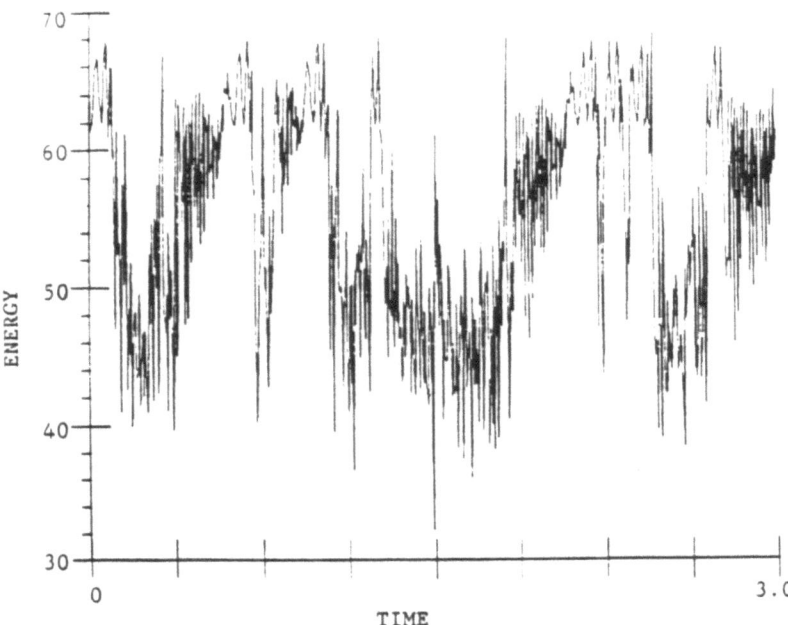

Fig. 8. For the perturbed Shilnikov loops, the orbit goes into chaotic puffs array from the trimodal point ($\alpha = 68$).

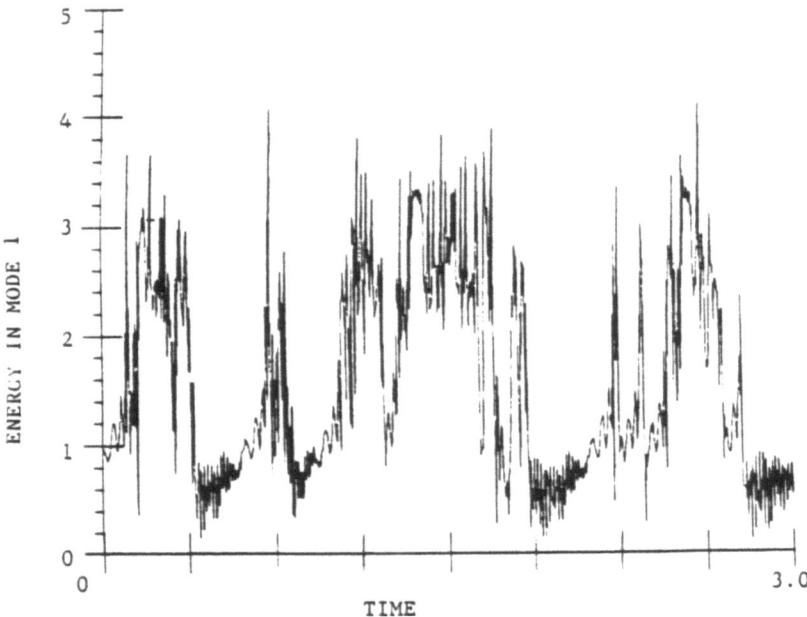

Fig. 9. The energy in the first mode is high clearing the chaotic excursions; their duration is comparable to the stagnation time near $\tilde{\mu}_3$ ($\alpha = 68$).

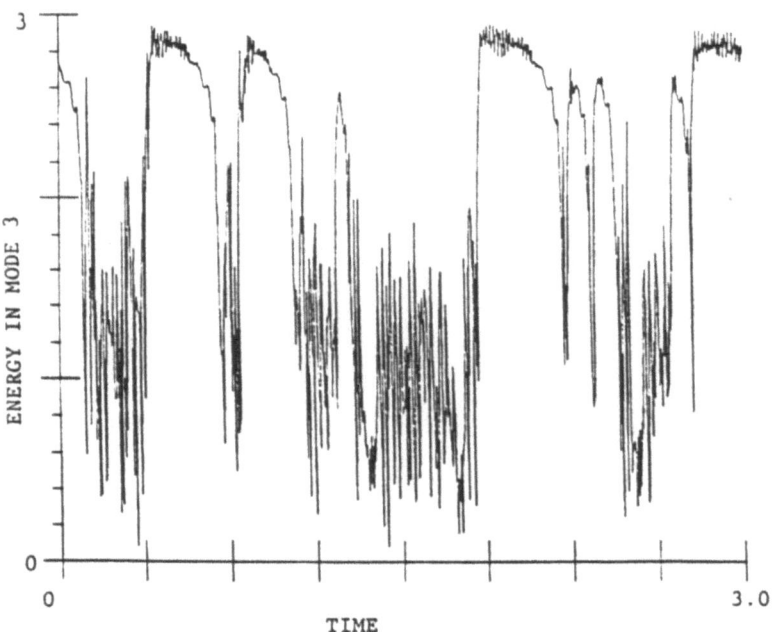

Fig. 10. The energy in Mode 3 clearly shows the homoclinic tangle between the stable and unstable manifolds of $\tilde{\mu}_3$ ($\alpha = 68$).

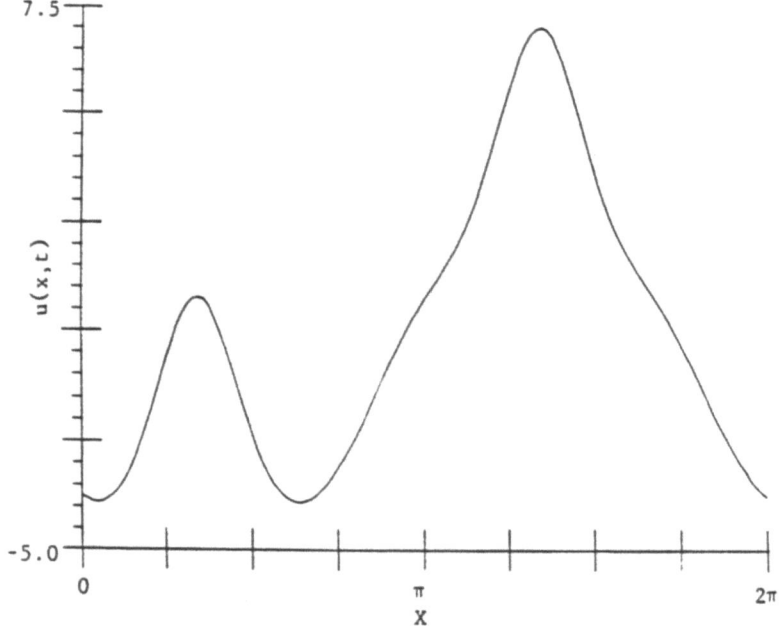

Fig. 11. This strange fixed point becomes a global attractor at $\alpha = 72$. It has energy in all first seven Fourier Modes ($\alpha = 72$).

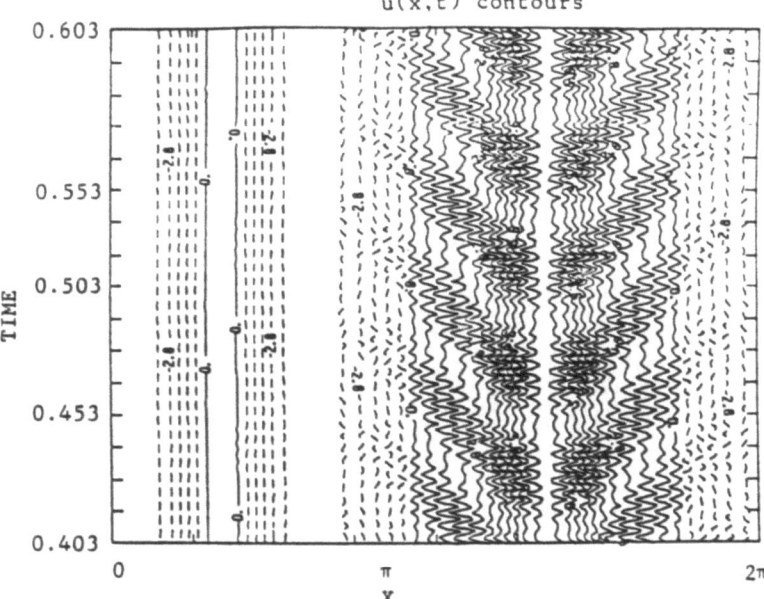

Fig. 12. The strange fixed point undergoes Hopf bifurcation. The oscillations are localized on the tip of the higher hump ($\alpha = 84.25$).

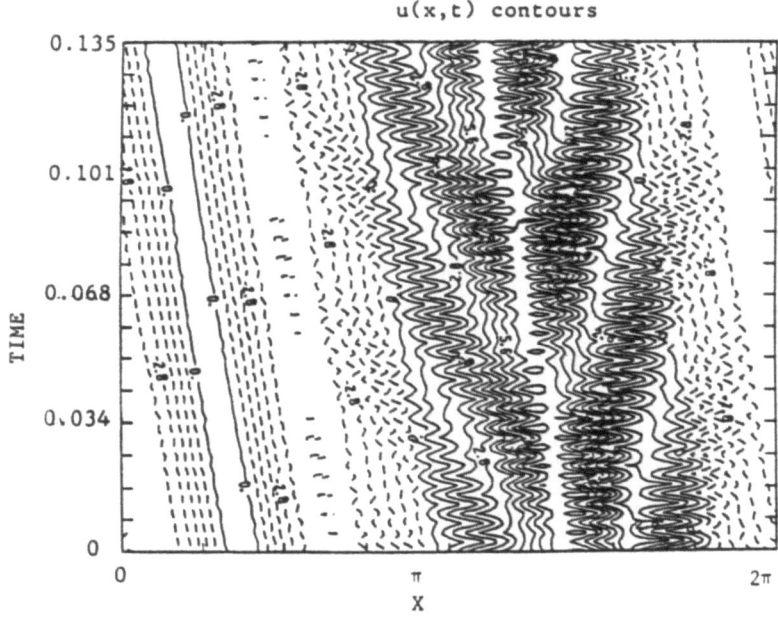

Fig. 13. The strange fixed point has further gone into a traveling beating wave. The beating is still localized on the higher hump ($\alpha = 87$).

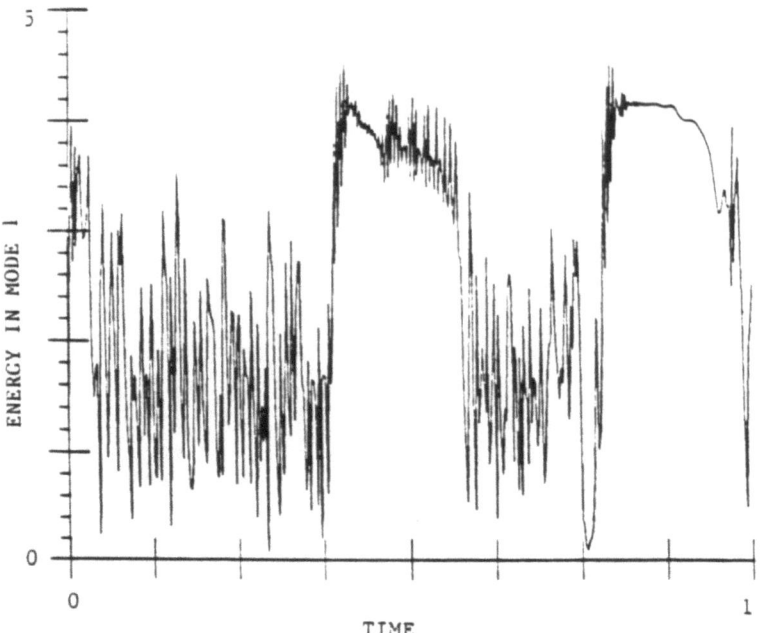

Fig. 14. Crisis of chaos looms. Intermittencies are observed, with small amplitude, high frequency oscillations ($\alpha = 91$).

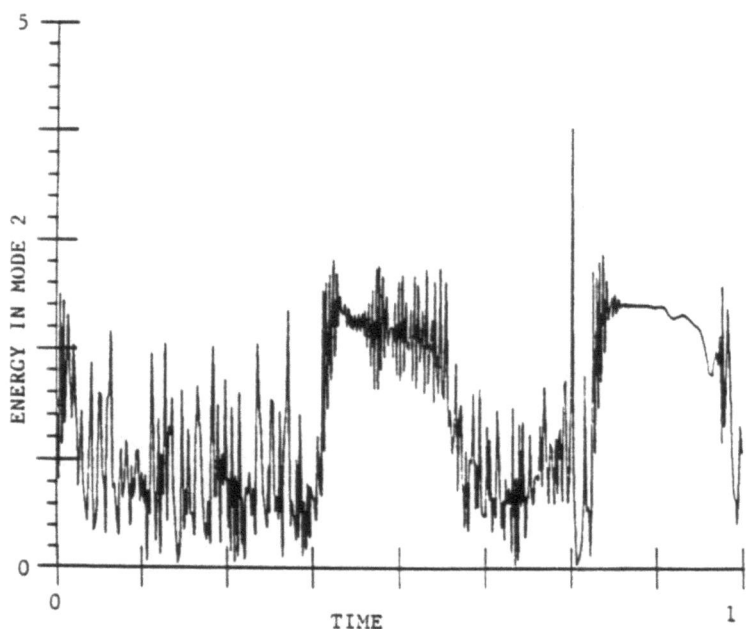

Fig. 15. Within the intermittent windows, energy in Modes 2 and 1 are comparable. They betray the strange, two-humped fixed point ($\alpha = 91$).

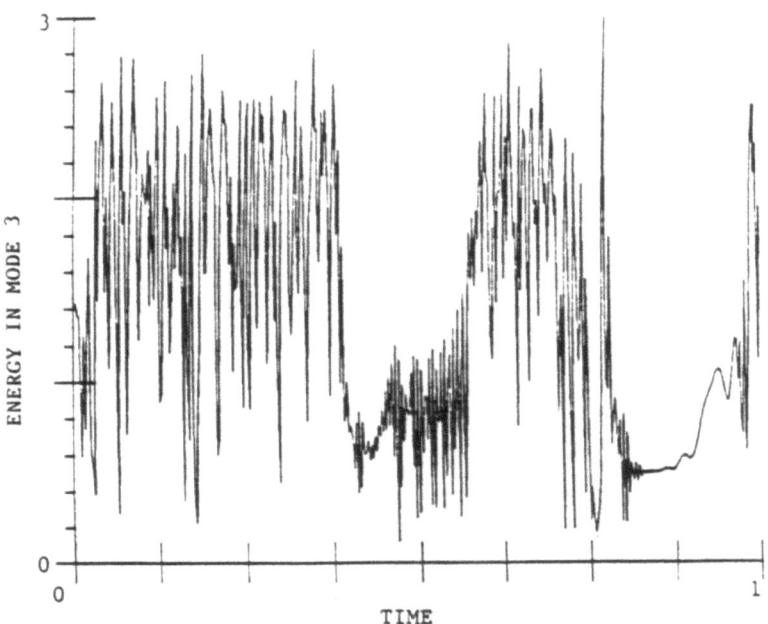

Fig. 16. Energy in Mode 3 is small when the orbit wanders close to the stable manifold of the strange hyperbolic point ($\alpha = 91$).

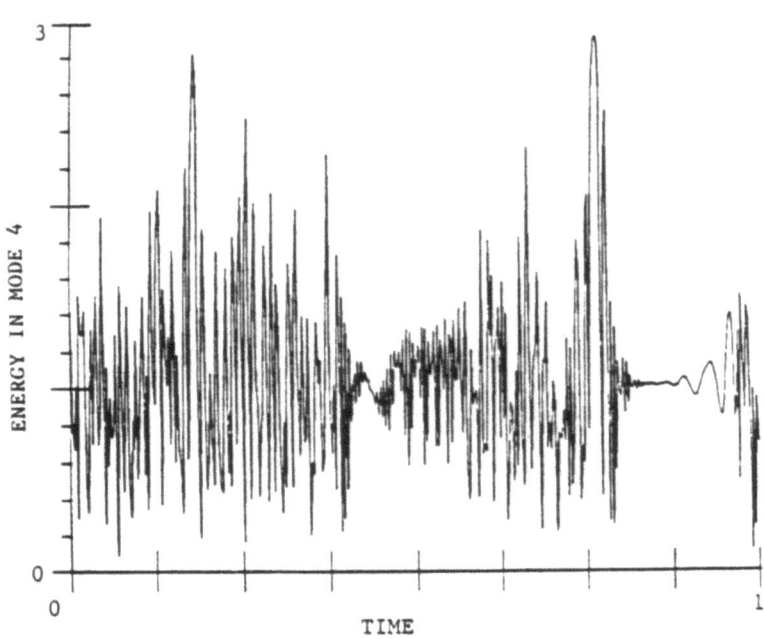

Fig. 17. Energy in Mode 4 confirms that the two-humped strange fixed point sits on the boundary between the basins of resp. the chaotic attractor and the quadrimodal state $\tilde{\mu}_4$ ($\alpha = 91$).

that stable manifold. Energy in mode 4 (Fig. 17) confirms that u* sits on the boundary delineating the basins of $\tilde{u}_4$ and is responsible for this basin boundary crisis [19-24].

The interval I_4 of global stability for $\tilde{u}_4$ ends at $\alpha=117.5$. In [28] we suspected some homoclinic skeleton to underline the onset of chaos at $\alpha=117.5$. We can now give a much more precise microscopy of this bifurcation. At $\alpha=117.5$, we took for initial data

$$\tilde{u}_4 + 0.1 sinx = 0.1 sinx + 2.995 cos4x + 0.43 cos8x + \cdots. \tag{3.2}$$

The time series in Figs. 18-20 are remarkably akin to, albeit more chaotic than those for the perturbed Shilnikov loop in Figs. 8-11; just replace $\tilde{u}_3$ by $\tilde{u}_4$. The energy (Fig. 18) undergoes very high frequency oscillations within intermittencies close to the quadrimodal state, before exploding into chaos with an higher average energy. The energy in mode 2 (Fig. 19) dips at very small levels in the vicinity of $\tilde{u}_4$. The energy in mode 4 (Fig. 20) clearly shows the orbit nearly locking onto some metastable torus, around 2.9 before bursting into homoclinic tangles. For the energy in mode 8 (Fig. 21), the intermittencies center at 0.4; this confirms the picture of a perturbed Shilnikov tangle around $\tilde{u}_4$, as a mechanism for onset of chaos. As computed by Manneville [35], the Lyapunov dimension of chaos is slightly larger than 5 in this case.

The bifurcations of the K-S equation, unravelled in this part, occur on low-dimensional inertial manifolds. Multiple forward and reverse bifurcations of several fixed points are entangled in a web of Tori, together with "strange" hyperbolic points. For these regimes, we conjecture that it may be possible to construct a simple reduced normal form for the ODEs on the inertial manifold using the unstable manifolds of $\tilde{u}_3(\alpha), (\tilde{u}_4(\alpha)$ and the two-humped "strange" fixed point $u^*(\alpha)$.

IV. A GEOMETRIC CONSTRUCTION OF THE INERTIAL MANIFOLDS

The K-S equations possess inertial manifolds $\overline{\sum}$. These are positively invariant regular objects toward which all solutions tend at (at least) a uniform exponential rate. Let H be the Hilbert phase space (usually a Sobolev space) and let $S(t)u_0$) denote the trajectory (solution of the system) starting at $t = 0$ from u_0. By an inertial manifold for $S(t)$ we mean a set $\overline{\sum}$ satisfying

$$\overline{\sum} \text{ is a finite dimensional Lipschitz manifold} \tag{4.1}$$

$$S(t)\overline{\sum} \subset \overline{\sum} \text{ for } t \geq 0 \tag{4.2}$$

There exists a constant k such that for every $u_0 \epsilon H$, there exists $t_0 \geq 0$ (4.3)

(uniformly for u_0 in bounded sets) such that, for $t \geq t_0$

$$dist(S(t)u_0, \overline{\sum})$$

$$\leq dist(S(t_0)u_0, \overline{\sum}) exp(-Kt)$$

We shall present here a geometric method of constructing $\overline{\sum}$ for a class of dissipative systems large enough to contain the one dimensional Kuramoto-Sivashinski and one and two dimensional parabolic reaction diffusion equations. Full details will be found in a forthcoming paper by P. Constantine, C. Foias, R. Temam and B. Nichols [6]. The K-S

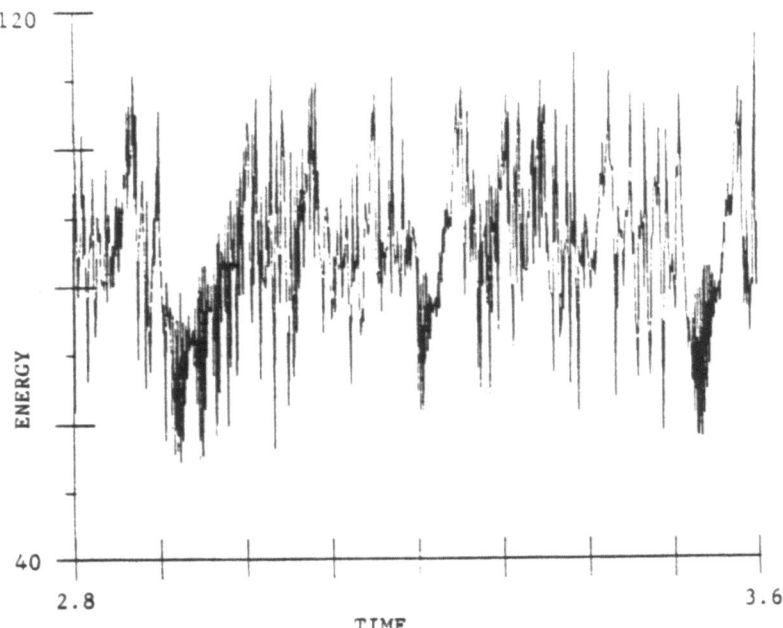

Fig. 18. The energy undergoes high frequency oscillations close to the quadrimodal cellular state, before going into chaotic excursions ($\alpha = 117.5$).

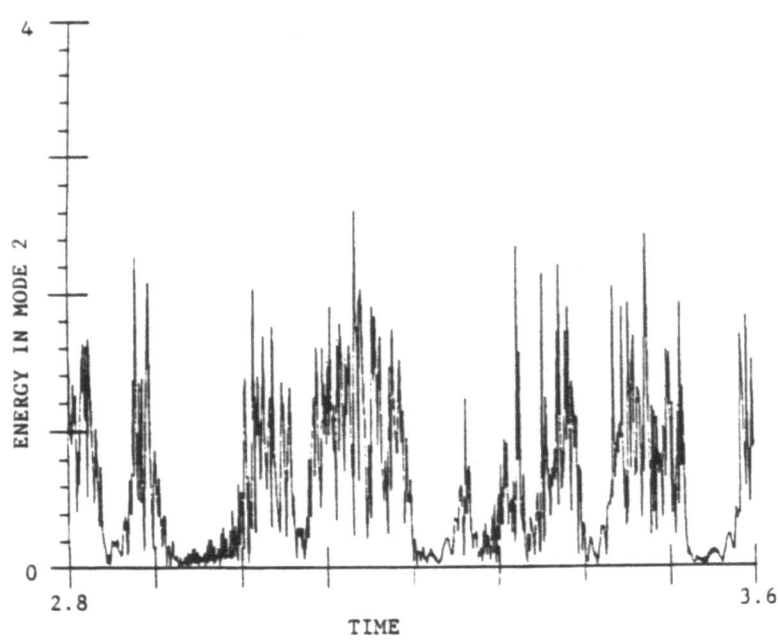

Fig. 19. The energy in Mode 2 is very small in the neighborhood of $\tilde{\mu}_4$ ($\alpha = 117.5$).

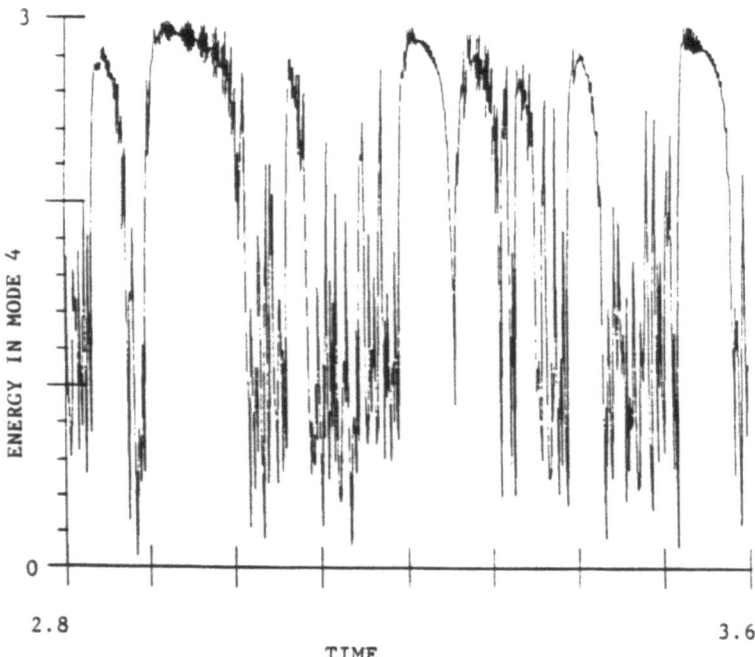

Fig. 20. The energy in Mode 4 clearly shows the trajectory spiraling around the cellular state, before bursting into homoclinic tangles ($\alpha = 117.5$).

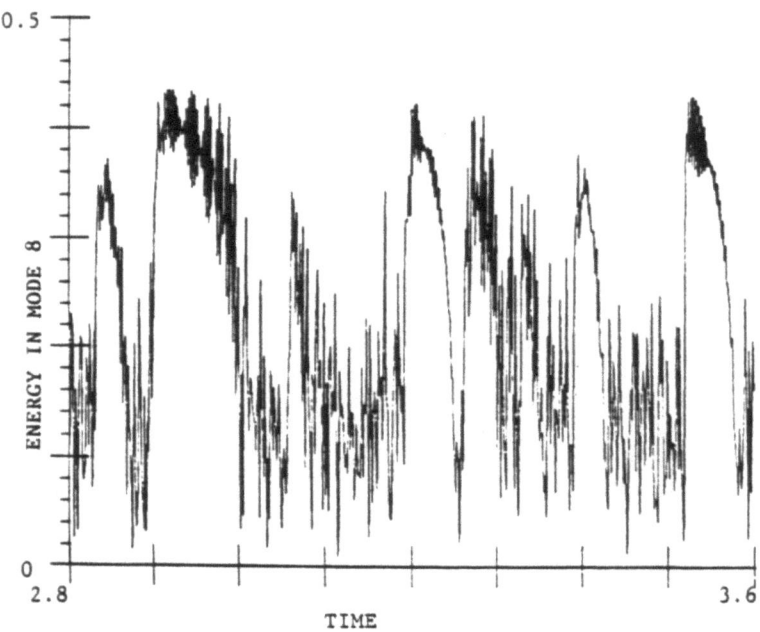

Fig. 21. The energy in Mode 8 confirms the picture of a perturbed Shilnikov loop around $\tilde{\mu}_4$, as a mechanism for onset of chaos ($\alpha = 117.5$).

equation can be restated abstractly as:

$$\frac{du}{dt} + N(u) = 0 \text{ with} \tag{4.4}$$

$$N(u) = Au + R(u) \tag{4.5}$$

where A is a positive selfadjoint operator and $R(u)$ is a lower order nonlinear nonhomogeneous term. We denote by $(\Lambda_j)_j$ the increasing sequence of distinct eigenvalue of A and J_y (λ_j) the nondecreasing sequence of eigenvalues counted with their multiplicities. The linearized around $u(t)$ of $N(u)$, will be denoted by $A(t)$

$$L(t)v = \frac{\partial R}{\partial u}(u(t))v \ , \ A(t) = A + L(t) \tag{4.6}$$

The key idea is to use the transport properties of finite dimensional contact elements. By a finite dimensional contact element we mean a pair (u_0, P_0) with $u_o \in H$ and P_0 a finite dimensional projector (orthogonal projection operator) in H. One regards P_0 as the projector on the tangent space at u_0 to an infinitesimal surface passing through u_0. The transport under $S(t)$ of this surface induces the transport of (u_0, P_0) according to

$$u(t) = S(t)u_0 \tag{4.7}$$

$$\frac{d}{dt}P(t) + (I - P(t))A(t)P(t) + P(t)A(t)^*(I - P(t)) = 0 \tag{4.8}$$

$$P(0) = P_0 \tag{4.9}$$

where $A(t)$ is the linearized (0.6) and $A(t)^*$ is the adjoint in H. For any N dimensional contact element (u, P) we introduce the quantities

$$\Lambda(u) = Max\{(Ag, g) | |g| = 1 \ , \ Pg = g \ , \ g \in D(A)\} \tag{4.10}$$

$$\lambda(u) = Min\{(Ag, g) | |g| = 1 \ , \ Pg = 0 \ , \ g \in D(A)\} \tag{4.11}$$

where $(,)$ and $||$ denote the scalar product and the norm in H; $D(A)$ is the domain of A. It follows from the minimax and mimimin theorem that $\Lambda(u) \geq \lambda_N$, $\lambda(u) \leq \lambda_{N+1}$. These two quantities measure the position of the linear space $ker(I - P)$ relative to the fixed orthonormal system of coordinates formed with the eigenvectors (w_j) of $(Aw_j = \lambda_j w_j)$. We assume that $L(t)$ satisfy bounds of the type

$$|L(t)v|^2 \leq K_2|v|^2 + K_2|Av^{1/4}|^2 + K_3|A_v^{1/2}|^2 \tag{4.12}$$

$$|L(t)^*v|^2 \leq K_1|v|^2 + K_2|Av^{1/4}|^2 + K_3|A_v^{1/2}|^2 \tag{4.13}$$

reflecting the fact that $R(u)$ is assumed to be of lower order (half the number of derivatives at most) than A. We derive under these assumptions differential inequalities for the transported quantities $\lambda(t) = \lambda(P(t))$ $\Lambda = \Lambda(P(t))$. If the linear diffusion operator A has gaps in the spectrum which are large with respect to constant K_1, K_2, K_3 more precisely if

$$(\Lambda_{m+1} - \Lambda_m)^2 > K_1 + K_2 \left(\frac{\Lambda_m + \Lambda_{m+1}}{2}\right)^{1/2} + K_3 \frac{\Lambda_m + \Lambda_{m+1}}{2} \tag{4.14}$$

for some m, then we can deduce the powerful spectral blocking proposition:

Theorem 4.1 (Spectral blocking property).

Let $\lambda(t) = \lambda(P(t))$ $\Lambda(t) = \Lambda(P(t))$ be defined in (4.10), (4.11) for $P(t)$ solving (4.7)-(4.9), then

$$\text{if for some } t_0 \geq 0 \ \Lambda(t_0) < \frac{\Lambda_m + \Lambda_{m+1}}{2} \tag{a}$$

for some m satisfying (4.14) then $\Lambda(t) < \frac{\Lambda_m + \Lambda_{m+1}}{2}$ for all $t \geq t_0$,

$$\text{if for some } t_0 \geq 0, \ \lambda(t_0) > \frac{\Lambda_m + \Lambda_{m+1}}{2} \tag{b}$$

for some (possibly different) m satisfying (4.14) then $\lambda(t) > \frac{\Lambda_m + \Lambda_{m+1}}{2}$ for all $t \geq t_0$. Thus $\lambda(t)(resp \Lambda(t))$ cannot cross large gaps in the spectrum of A from the right (resp left).

We note here that although a condition of the type $\lambda(t_0) > \frac{\Lambda_m + \Lambda_{m+1}}{2}$ can be realized only if the dimension N of $P(t_0)$ is large enough ($\lambda_{N+1} > \frac{\Lambda_m + \Lambda_{m+1}}{2}$) conditions of the type $\Lambda(t_0) < \frac{\Lambda_m + \Lambda_{m+1}}{2}$ do not impose restrictions on the dimension of $P(t_0)$ provided the set of m's for which (4.14) is valid is not founded. In particular the blocking of $\Lambda(t)$ in the $N = 1$ case has important consequences. Let us denote by P_m the spectral projector of A on the span of $w_1, \cdots, w_m$. Let us consider the cone in H

$$c = \{\omega \epsilon H \| |(I - P_n)\omega| \leq \frac{1}{3}|P_n\omega|\} \tag{4.15}$$

We prove the strong squeezing properties.

Theorem 4.2 Let n be large enough. Let $\omega(t)$ be a solution of

$$\begin{aligned} \frac{d\omega}{dt} + A(t)\omega &= 0, \\ \omega(0) &= \omega_0, \end{aligned} \tag{4.16}$$

the linearized equation around $S(t)u_0 = u(t)$. If for some $t_0 \geq 0$ $\omega(t_0)$ belongs to C, then for all $t \geq 0$ $\omega(t)$ belongs to C. Moreover, the following alternative holds:

$$|\omega(t)| \leq |\omega(0)|exp(-Kt) \text{ for all } t \geq 0 \tag{a}$$

or there exists a finite

$$t_0 > 0 \text{ such that the irregularity in (a) holds for } t \leq t_0 \text{ and for } t \geq t_0 \omega(t) \text{ belongs to C}. \tag{b}$$

The precise condition on the size of n is given in [8], but essentially the requirement is that $\lambda_n > 5(\Lambda_m + \Lambda_{m+1})$ for some n satisfying the gap condition (4.14) Theorem 4.2 is a direct consequence of Theorem 4.1 for $N = 1$. Using a slight modification of Theorem 4.1. We obtain, also,

Theorem 4.3 (Strong squeezing property) let n be large enough (same conditions as in Theorem 4.2). Let $\omega(t) = S(t)u_0 - S(t)u$ be the difference of two solutions. Then the conclusions of Theorem 4.2 hold for $\omega(t)$. The strong squeezing property was established for the Kuramoto-Sivashinski equation in [15]. The consequences of this property regarding the universal attractor are studied further. We prove

Theorem 4.4. If n is large enough to insure the validity of Theorem 4.3 then the projector P_n is injective when restricted to the universal attractor X and its inverse is Lipschitz.

More precisely
$$|(1-P_n)(x-y)| \le \frac{1}{3}|\Gamma_n(x-y)|$$
for every x,y in X.

Theorem (4.4) follows easily from Theorem 4.3 but is an important fact. It was known that because X has finite fractal dimension, there are many projectors that are injective on X; however, P_n is an important explicit one.

Denoting $C_{n,X} = x\epsilon X\{u \in H \| (1-P_n)(u-X)| \le \frac{1}{3}|P_n(u-x)|\}$ we deduce from Theorem 4.3 that $S(t)C_{n,X} \subset C_{n,X}$ if n is large enough, that $X \subset C_{n,X}$ (Theorem 4.4) and that as long as a solution $S(t)u_0$ remains in the complement of $C_{n,X}$, its distance to X decreases exponentially. Finally, we conclude by showing that the complement of a large ball in H is included in $C_{n,X}$. For a further consequence of strong squeezing we consider a smooth N dimensional positively invariant surface. We assume that it is "blocked" in the sense that $\lambda(u) > \frac{\lambda_N + \lambda_{N+1}}{2}$ for $u \in \Sigma$ and $\lambda(u) = \lambda(P(u))$ with $P(u)$ the projector on the tangent space at u to Σ. We show that under these assumptions, as long as the distance from some solution $S(t)u_0$ to Σ in attained on Σ, it must decay exponentially (at an explicit uniform rate).

We now proceed to describe the initial data for our construction. They form the smooth oriented boundary Γ of an bounded, open, connected set D included in $P_N H$. N is chosen sufficiently large such that $\lambda_{N+1} - \lambda_N > 0$ is a gap satisfying (4.14) and such that $\lambda_{N+1} > 5(\lambda_m + \lambda_{m+1})$ with a satisfying also (4.14). We denote at each net by $P(u)$ the projection on the space $N(u)R + T_n(\Gamma)$ where $T_u(\Gamma)$ is the tangent space at u to Γ; we design by $\nu(u)$ the outward unit normal to Γ and we set $\lambda(u) = \lambda(P(u)), \Lambda(u) = \Lambda(P(u))$. Then the properties of Γ are

$$\Lambda(u) < \frac{\lambda_N + \lambda_{N+1}}{2} \text{ for any } u\epsilon\Gamma \qquad (I)$$

$$\lambda(u) > \frac{\lambda_N + \lambda_{N+1}}{2} \text{ for any } u\epsilon\Gamma \qquad (II)$$

$$(N(u), \nu(u)) > 0 \text{ for any } u\epsilon\Gamma \qquad (III)$$

$$\Gamma \subset C_{N,X} \qquad (IV)$$

$$\text{For any } u \in \Gamma, \ N(u)\mathbf{R} + T_u(\Gamma) \subset C \qquad (V)$$

Properties (I) and (II) assert that the initial surface Γ is "blocked". Property (III) shows that $\frac{d}{dt}S(t, u_0)\big|_{t=0}$, at any $u_0 \in \Gamma$ points toward the interior of D. In applications Γ is usually a simple explicit set: a large sphere for the Kuramoto-Sivashinski equation.

Ultimately, we use the spectral blocking, strong squeezing and volume decay properties in order to construct starting from Γ the inertial manifolds. We denote by Σ the integral manifold having Γ as initial data:

$$\Sigma = \bigcup_{t>0} S(t)\Gamma \qquad (4.17)$$

We establish first using (I) and the spectral blocking property the fact that projection P_N at any point of Σ is a regular map (has invertible Jacobian). From the results in Theorem 4.4 and condition (IV) for Γ it follows that $\Sigma \subset C_{N,X}$. Since we may take Γ to lie far away from X; it follows that $\Sigma \cap X = \phi$ and thus, $P_N \Sigma \cap P_N X$ is void. We show that the closure of $P_N \Sigma$, $\overline{P_N \Sigma}$ is included in the union of the disjoint sets $P_N X \cup \Gamma \cup P_N \Sigma$. We use next the isoperimetric inequality and the exponential decay of surfaces of dimension larger or equal to $N-1$ to show that $\overline{P_N \Sigma} \supset D$. From the backward uniqueness theorem for solutions of an equation and (III) we deduce that for P in a neighborhood of Γ in D the fiber $P_N^{-1}\{p\} \cap \Sigma$ consists of a single point. Since P_N is regular at Σ and since $P_N \Sigma$ is connected we deduce that P_N restricted to Σ is injective. It follows that $D = P_N \Sigma \cup P_N X \cup \Gamma$ and we can define on D the inverse Φ of P_N, $\Phi : D \to E$, $\Phi\big|_\Gamma$ = {identity}. We show, using the strong squeezing property and (V) (Theorem 4.2) that

$$|(1 - P_N)(\Phi(r_1) - \Phi(r_2))| \leq \frac{1}{3}|P_N(\Phi(r_1) - \Phi(r_2))| \qquad (4.18)$$

for any r_1, r_2 in D. Finally we show using (II) and the spectral blocking that for any u_0, $dist(S(t)u_0, \overline{\Sigma})$ decreases exponentially. We conclude that $\overline{\Sigma}$ is an inertial manifold satisfying, beside properties (4.1), (4.2), (4.3), and (4.19) $\overline{\Sigma}$ is the graph of an explicit Lipschitz map (4.20) $\overline{\Sigma}$ is the closure of a smooth manifold (21) the N-dimensional volume of $\overline{\Sigma}$ is finite.

This concludes the outline of the construction of the integral inertial manifold. Needless to say, it lends itself to fast and robust numerical algorithms. Full details may be found in [8].

V. CONCLUSION

A low dimensional vector field skeleton underpins "strong" chaos for the K-S models of turbulent interfaces. Heretofore it was unsuspected, because of the extreme numerical sensitivity of chaos in dissipative PDEs. Indeed, low precision methods of integration based on second order schemes [31] are adequate to compute tables of Lyapunov exponents; they wash out the subtle architecture mirrored by repeated bifurcations and intermittencies. High precision, high speed, parallel codes on future parallel architecture machines shall play a crucial role in definitely bridging the gap between strong dynamical chaos and fully developed turbulence. From a theoretical view point, the fact that a small, yet exotic zoo of hyperbolic points and Tori generates strong chaos supports the current analytic work initiated [6] by C. Foias, P. Constantin, R. Temam and B. Nichols. We presently aim at constructing optimal inertial normal forms for the dynamical vector fields on the inertial manifolds at different α-regimes. We suspect that the zoo of dynamically relevant strange fixed points will be enhanced by specimens with a Cantor-like structure in space (this has been proven by Michelson [45], for $\alpha = \infty$, for K-S). Hence, spatial chaos would intermingle with temporal chaos as the bifurcation parameter is increased to another order of magnitude.

ACKNOWLEDGMENTS

The author was supported by the US Department of Energy, Office of Scientific Computing under contract KC-07-01-01-0 and the US Department of Energy under contract W-7405-ENG-36. Different parts of this survey article focus on joint research with P. Constantine, C. Foias, J. M. Hyman and R. Temam. Travel support for this research is acknowledged from NATO Grant 85-0509.

REFERENCES

1. P. Clavin, "Dynamical Behavior of Premixed Flame Fronts in Laminar and Turbulent Flows," Prog. Energy. Combust. Sci. 11 (1985), 1-59.
2. P. Constantine and C. Foias, Comm. Pure Appl. Math. 38 (1985), 1-27.

3. P. Constantine, C. Foias and R. Temam, Memoirs AMS 53 (1985) 314, vii+.67 pp.
4. P. Constantine, C. Foias, O. P. Manley and R. Temam, C. R. Acad. Sci. Paris I, 297 (1983), 599-602.
5. P. Constantine, C. Foias, O. P. Manley and R. Temam, J. Fluid Mech. 150 (1985) 427-440.
6. P. Constantine, C. Foias, B. Nicolaenko and R. Temam, "Integral Manifolds and Inertial Manifolds for Dissipative PDEs (submitted).
7. J. D. Farmer, E. Jen, A. Brandstäter, J. Swift, H. L. Swinney, A. Wolff and J. P. Crutchfield, "Low Dimensional Chaos in a Hydrodynamic System," Phys. Rev. Lett. 51, 16 (1983) 1442-1445. See also J. P. Gollub and H. L. Swinney, Phys. Rev. Lett. 35 (1975) 927.
8. C. Foias, B. Nicolaenko and R. Temam, "Asymptotic Study of an Equation of G. I. Sivashinsky for Two Dimensional Turbulence of the Kolmogorov Flow," to appear, Proc. Paris Acad. Sci.
9. C. Foias and R. Temam, C. R. Acad. Sci. Paris, I, 295 (1982) 239-241.
10. C. Foias and R. Temam, C. R. Acad. Sci. Paris, I, 295 (1982) 523-525.
11. C. Foias and R. Temam, Mathematics of Computation 43 (1984), 117-133.
12. C. Foias, G. R. Sell and R. Temam, C. R. Acad. Sci. Paris, I, 301 (1985) 139-141.
13. C. Foias, G. R. Sell and R. Temam, "Inertial Manifolds for Dissipative PDEs (submitted).
14. C. Foias, B. Nicolaenko, G. R. Sell and R. Temam, C. R. Acad. Sci. Paris, I, 301 (1985) 285-288.
15. C. Foias, B. Nicolaenko, G. R. Sell and R. Temam, "Inertial Manifolds and an Estimate of Their Dimension for the Kuramoto-Sivashinsky Equation (submitted).
16. P. L. Garcia-Ybarra and M. G. Velarde, Phys. Fluids 30 (1987) 1649.
17. P. L. Garcia-Ybarra, J. L. Castillo and M. G. Velarde, Phys. Lett. A 122 (1987) 107; Phys. Fluids 30 (1987) 2655.
18. M. Golubitsky and D. G. Schaeffer, "Singularities and Groups in Bifurcation Theory," Springer-Verlag, New York (1985).
19. C. Grebogi, E. Ott and Y. A. Yorke, "Crisis, Sudden Changes in Chaotic Attractors and Transient Chaos," Physica 7D (1983) 181-200.
20. C. Grebogi, E. Ott and Y. A. Yorke, "Fractal Basin Boundaries, Long-Lived Chaotic Transients and Unstable-Unstable Pair Bifurcation," Phys. Rev. Lett. 50, 13 (1983) 935-938.
21. C. Grebogi, S. W. McDonald, E. Ott and Y. A. Yorke, "Final State Sensitivity: An Obstruction to Predictability," Phys. Lett. 99A, 9 (1983) 415-418.
22. C. Grebogi, S. W. McDonald, E. Ott and Y. A. Yorke, "Structure and Crises of Fractal Basin Boundaries," Phys. Lett. 107A, 2 (1985) 51-54.
23. C. Grebogi, E. Ott and Y. A. Yorke, "Super Persistent Chaotic Transients," Ergod. Th. and Dynam. Sys. 5 (1985) 341-372.
24. C. Grebogi, S. W. McDonald, E. Ott and Y. A. Yorke, "Fractal Basin Boundaries," Physica 17D (1985) 125-153.
25. J. Guckenheimer and P. H. Holmes, "Nonlinear Oscillations, Dynamical Systems and Bifurcation of Vector Fields," Springer-Verlag (1985).

26. J. Guckenheimer, "Strange Attractors in Fluids: Another View," Annual Review of Fluid Mechanics (1986).
27. J. Guckenheimer, private communication.
28. J. M. Hyman and B. Nicolaenko, "The Kuramoto-Sivashinsky Equation: a Bridge Between PDEs and Dynamical Systems," Physica 18D (1986), 113-126.
29. J. M. Hyman, "Numerical Methods for Nonlinear Differential Equations," Nonlinear Problems: Present and Future, A. R. Bishop, D. K. Campbell and B. Nicolaenko, Eds., North-Holland Publ. Co., (1982) 91-107.
30. J. M. Hyman and M. Naughton, "Adaptive Static Rezoning Methods," Lectures in Applied Math. 22, Part I (1985) 321-343.
31. P. Manneville, "Liapunov Exponents for the Kuramoto-Sivashinsky Model," Proc. Conf. on Macroscopic Modeling of Turbulent Flows, U. Frisch, Ed., Springer-Verlag, Lecture Notes in Physics 230 (1985), 319-326.
32. B. Nicolaenko and S. Zaleski, "Crisis of Chaos for Models of Interface Turbulence," to appear in Physica D.
33. B. Nicolaenko and B. Scheurer, "Remarks on the Kuramoto-Sivashinsky Equation," Physica 12D (1984) 331-395.
34. B. Nicolaenko, B. Scheurer, and R. Temam, "Quelques proprietes des attracteurs pourl'equation de Kuramoto-Sivashinsky," C. R. Acad. Sci. Paris 298 (1984) 23-25.
35. B. Nicolaenko, B. Scheurer, and R. Temam, "Attractors for the Kuramoto-Sivashinsky Equations," Physica 16D (1985) 155-183.
36. B. Nicolaenko, B. Scheurer, and R. Temam, "Attractors for the Kuramoto-Sivashinsky Equations," AMS-SIAM Lectures in Applied Mathematics 23, 2 (1986) 149-170.
37. B. Nicolaenko, B. Scheurer, and R. Temam, "Attractors for Classes of Nonlinear Evolution of Partial Differential Equations," in preparation.
38. G. I. Sivashinsky and A. Novick-Cohen, "Interfacial Instabilities in Dilute Binary Mixtures Change of Phase, to appear in Physica D.
39. Z. S. She, U. Frisch and O. Thual, "Homogenization and Visco-Elasticity of Turbulence," Proc. Conf. on Macroscopic Modeling of Turbulent Flows, U. Frisch, Ed., Springer-Verlag Lecture Notes in Physics 230 (1985) 1-13.
40. K. Shreenivasan, "Transition and Turbulence in Fluid Flows and Low-Dimensional Chaos," Frontiers in Fluid Mechanics, S. H. Davis and J. L. Lumley, Eds., Springer-Verlag (1985) 41-67.
41. R. Temam, "Navier Stokes Equations and Nonlinear Functional Analysis," SIAM, Philadelphia, 1983.
42. R. Temam, "Infinite Dimensional Dynamical Systems of Fluid Mechanics," AMS-Summer Res. Institute "Nonlinear Funct. Anal. Appl." (Berkeley, 1983).
43. M. G. Velarde and C. Normand, Sci. Amer. 243 (1980) 92.
44. M. G. Velarde and J. L. Castillo, "Convective Transport and Instability Phenomena, J. Zierep and H. Oertel, Jr., Eds., Braun-Verlag, Karlsruhe (1982).
45. D. M. Michelson, "Steady-States for the Kuramoto-Sivashinsky Equation," preprint.

SPATIAL COHERENCE AND TEMPORAL CHAOS IN HYDRODYNAMIC INSTABILITIES

S. Ciliberto

Istituto Nazionale di Ottica
Largo E. Fermi 6, 50125 Firenze, Italy

I - INTRODUCTION

In the last decade many experiments have demonstrated that the transition to chaos is a low dimensional phenomenon even in hydrodynamic instabilities governed by an infinite number of degrees of freedom /1/.

However in fluid systems the physical origin of the chaotic behavior is not very well understood. Mathematical models that incorporate the correct dynamics and allow a prediction of the behavior as a function of the control parameter are not generally available. Very often the experimental observations can be just correlated with the behavior of simple maps, as in the case, for example, of the Feigenbaum cascade /2/, and quasi periodicity /3/.

Besides another problem remains open: is low dimensional chaos a precursor of fully developed turbulence, where the fluid flow exibits chaotic behavior both in space and time? To give new insight to this problem there is nowadays a growing interest in the study of the relatioship between spatial order and temporal chaos.

For example it has been observed in numerical studies of certain partial differential equations (P.D.E.) /4,5/, in theoretical studies of chemical reactions /6/ that coherent spatial structures coexist with temporal chaos. From an experimental point of view the behavior of spatial patterns have been quantitatively analysed in time-dependent chaotic regimes only in few experiments /7-10/. We describe here two of these experiments in which spatial patterns have been studied in time dependent regimes, either periodic or chaotic.

In Section II we report experiments on surface wave instabilities /9/, where the competition between two spatial patterns produces time dependent behavior and chaos. The results of this experiment are in good agreement with a low dimensional model obtained from Navier-Stokes equations /11/.

In Section III we describe experiments on time dependent behavior /10, 12/ of a horizontal fluid layer, heated from below, that is Rayleigh-Benard convection (R-B). We show that time dependent regimes are characterized by the presence of either travelling waves or localized oscillations. These spatio-temporal regimes turn out to be similar to those observed in numerical simulations /4/ of Kuramoto-Shivanshisky (K-S) /13/ and Kuramoto-Velarde (K-V) /14/ partial differential equations.

II - SURFACE WAVE INSTABILITIES

The system of interest is a cylindrical fluid layer in a container that is subjected to a small vertical oscillation of amplitude A and frequency f_0. It is well known that, if the driving amplitude exceeds a critical value $A_c(f_0)$, which is a function of frequency, the free surface develops a pattern of standing waves. The surface deformation $S(r,\theta,t)$ can then be written as a superposition of normal modes:

$$S(r, \theta, t) = \sum_{l,m} a_{lm}(t) \, J_l(k_{lm} r) \cos l\theta,$$

where J_l are Bessel functions of order l and the allowed wave numbers k_{lm} are determined by the boundary condition that the derivative $J'_{lm}(k_{lm} R) = 0$, where R is the radius of the cylinder. The modes may be labeled by the indices l (giving the number of angular maxima) and m (related to the number of nodal circles). The mode amplitude $a(t)$ develops an instability when the corresponding eigenfrequency (given by the dispersion law for capillary gravity waves) is approximately in resonance with half the driving frequency f_0 and A exceeds $A_c(f_0)$.

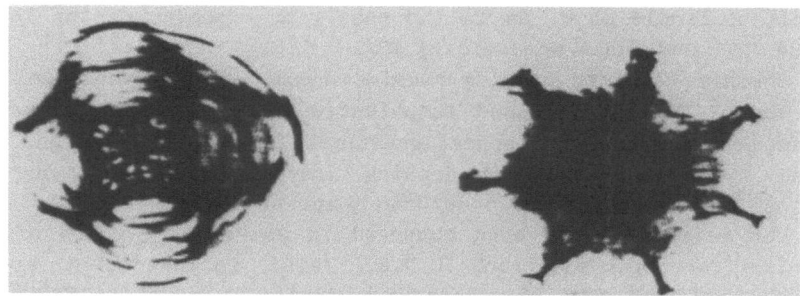

Figure 1. Optical intensity patterns for the (4,3) and (7,2) modes. The first index gives the number of angular maxima.

This parametric instability leads to standing waves in which the mode amplitude oscillates at $f_0/2$. To take into account the possibility of a further slow modulation of the mode amplitudes, which, in fact, occurs due to mode competition, we write each amplitude in terms of fast oscillations at $f_0/2$ and slow envelopes $C_l(t)$ and $B_l(t)$:

$$a_l(t) = C_l(t)\cos(\pi f_0 t) + B_l(t)\sin(\pi f_0 t).$$

We omit the second subscript because, in practice, only a single value of m is significant for a given value of l. In our experiment the working fluid was water of depth of 1 cm, and the radius of the tank was 6.35 cm. Examples of stable patterns involving a single mode (and possibly harmonics) are shown in Fig. 1 for the (7,2) and (4,3) modes. The index l is obvious from the simmetry while m was determined by matching the frequency to the known dispersion law. The white areas correspond to surface depressions (tipically 0.5 mm) and the black ones to surface elevations. The driving amplitude A was about 1.1. A_C and the frequency were at the minimum of the stability curve in each case. Both pictures have been obtained with a focusing technique.

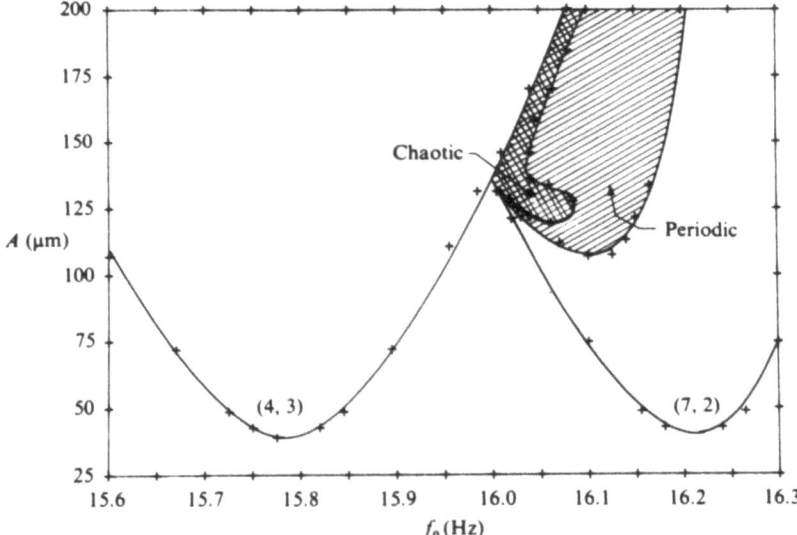

Figure 2. Phase diagram as a function of driving amplitude A and frequency f_0. The crosses are experimentally determined points on the stability boundaries. Stable patterns occur in the regions labeled (4,3) and (7,2). Slow periodic and chaotic oscillations involving competition between these modes occur in the shaded regions.

The behavior of the system as a function of A and f_0 is shown in Fig. 2, where a small part of the phase diagram is reported. Below the parabolic stability boundaries, the surface is essentially flat. Above the stability boundaries, the fluid surface oscillates at half the driving frequency in a single stable mode, C_1 and B_1 are constant as function of times. The shaded areas are regions of mode competition, in which the surface can be described as a superposition of the (4,3) and (7,2) modes with amplitudes having a slowly varying envelope in addition to the fast oscillation at $f_0/2$. They oscillate periodically or chaotically at a mean frequency that is two order of magnitude smaller than f_0.

Our experimental apparatus, described in /9/, allows us to study a fixed linear combination of the slow coefficients $C_1(t)$ and $B_1(t)$, which we denote by $a_l^o(t)$. In Fig. 3 is shown the time dependence of a_4^o and a_7^o.

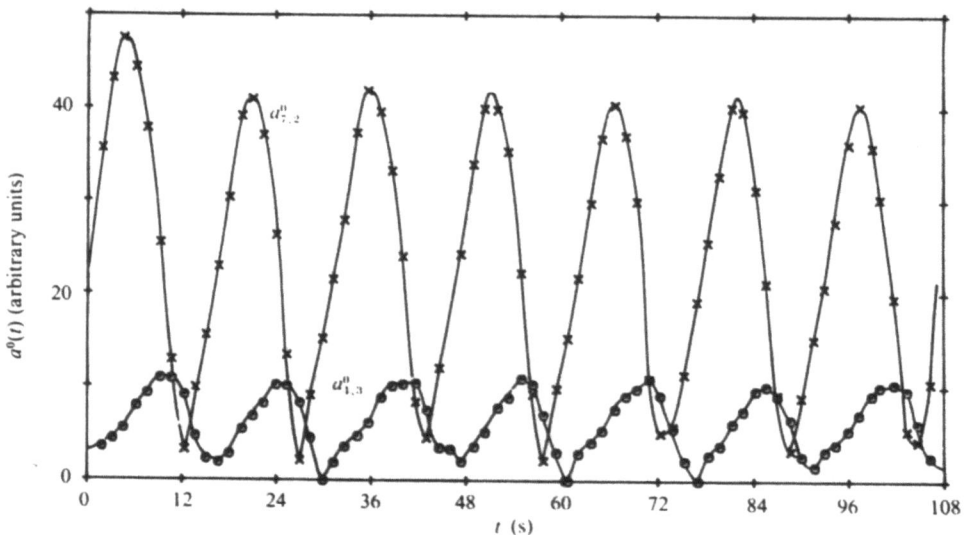

Figure 3. The slowly varying amplitudes a_4^o and a_7^o ocillate periodically.

The slow oscillation resulting from mode competition is periodic in this case and a_4^o leads a_7^o by about 90°. This phase relationship is significant it implies that the mode (7,2) pump (4,3). The dynamic of the slow oscillation was explored by varying A and f_0 separately inside of the interaction region. In Fig. 4 time series and corresponding power spectra are shown for three different driving amplitudes but fixed driving frequency of 16.05 Hz.

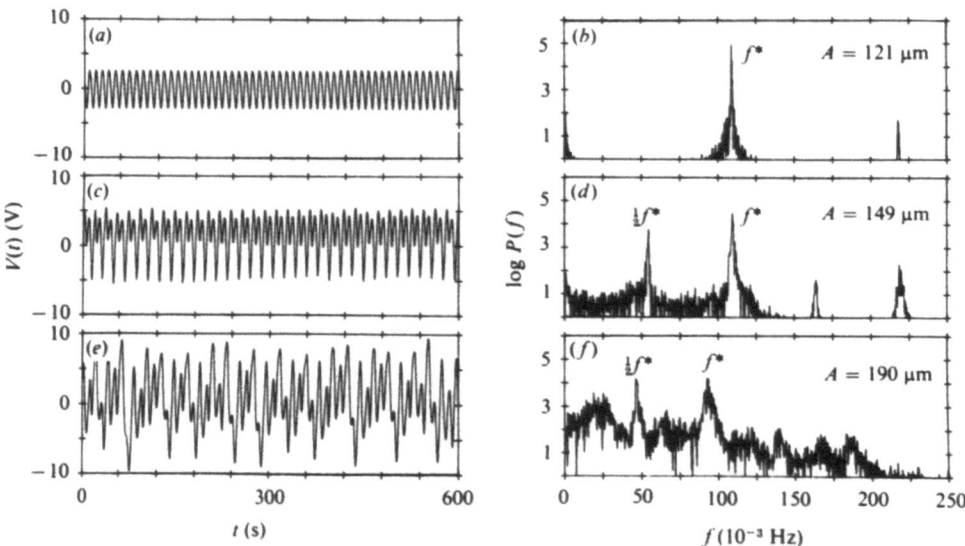

Figure 4. The transition from periodic to chaotic oscillation. Time series and corresponding power spectra of the slow oscillation are shown for f_0=16.05 Hz and three different driving amplitudes. Broad-band noise is associated with the appearance of a subharmonic f*/2 of the dominant oscillation.

As the driving amplitude is increased, a chaotic state with a broad power spectrum is obtained. We characterize the chaotic behavior quantitatively by computing from the experimental data the correlation dimension ν of the attractor and a lower bound K_2 for the Kolmogorov entropy K. When the oscillation is periodic (A=121 μm), we find $\nu = 1.0 \pm 0.04$ and $K_2 = (0.01 \pm 0.01)$ sec^{-1}. On the other hand when the slow oscillation in chaotic (A=190 μm), $\nu = 2.22 \pm 0.04$ and $K_2 = (0.1 \pm 0.01)$s^{-1}. These measurements clearly demonstrate that the attractor has a low (and fractional) dimension and that there is at least one positive Lyapunov exponent.

This result may seem in contrast with the fact that time resolved spatial Fourier spectra show the presence of many other modes with l=3,8,11,14 besides the mode (7,2) and (4,3). However this has been theoretically explained by Meron and Procaccia /11/. They start from Navier-Stokes equations with suitable boundary conditions. They can prove rigorously, using center-manifold and normal form theories that the dynamics is governed by the modes (7,2) and (4,3) and all the other modes are slaved by these two. Finally they obtained a system of four coupled ordinary differential equations for the amplitude of the mode (7,2) and (4,3). The phase diagram of this system is in very good agreement with the experimental one. Time dependent behavior either chaotic or periodic is indeed reproduced at the intersection of the stability curves of the two modes.

This experiment and the associated theory have shown how the drastic reduction from a large to a very small number of degrees of freedom occurs in practice. They have also shown that temporal chaotic behavior is produced by the interaction of spatial modes and that spatial order can be preserved in time-dependent aperiodic regimes. This result was made possible by a time resolved analysis of the spatial patterns which allows a more direct comparison with the theory.

In other instabilities, such as R-B convection a time resolved analysis of spatial patterns in temporal chaotic regimes has been carried out only in few experiments and we will show in the next paragraph that also in this case the study of spatial patterns is very useful to relate the observed behavior with that of a realistic model.

III - RAYLEIGH-BENARD CONVECTION

We remind very briefly the properties of thermal convection in a fluid layer heated from below, that is Rayleigh-Benard instability /16/. When the temperature difference ΔT between the two horizontal plates, confining the fluid exceeds a critical value ΔT_c convection begins and the fluid motion forms a periodic structure, a set of parallel rolls, with a wave number $q \simeq /d$, where d is the depth of the layer. The most relevant parameters are the Rayleigh number $R = \alpha g \Delta T d^3 / \nu\kappa$ and the Prandtl number $P = \nu/\kappa$. Here α, g, ν, κ are respectively the volumetric expansion coefficient, the acceleration of gravity, the kinematic viscosity and the heat diffusion constant. It has been shown that for an infinitely extended horizontal layer the critical Rayleigh number at which convection sets in is $R_c = 1708$.

Increasing R above R_c another threshold R_T is reached where the fluid motion becomes time-dependent. The value of R_T and the behavior of the fluid strongly depends on P and on the aspect ratio Γ, that is the ratio between the horizontal length and depth of the layer.

Rayleigh-Benard convection has been widely used to study the transition from a regular to a chaotic motion. Nevertheless spatial patterns in R-B have been studied just near the threshold of the instability in large aspect-ratio cells and a good agreement with theories has been found /15/. On the contrary, convective patterns in time dependent states has been investigated just in a few experiments /7-8/ leaving open the question of the role that the spatial degrees of freedom play in the transition into these regimes.

Thus to have a better insight into the mechanisms leading to chaos in thermal convection and to allow a more direct comparison with numerical models we have experimentally /10/12/ studied the evolution of the temperature field in time-dependent regimes of R-B convection.

In our set up the fluid layer has horizontal size l_x=4 cm, l_y=1 cm and height d=1 cm. The x and y axes of the coordinate reference frame are respectively perpendicular and parallel to the rolls axis, (Fig. 5). The z axis is the vertical one.

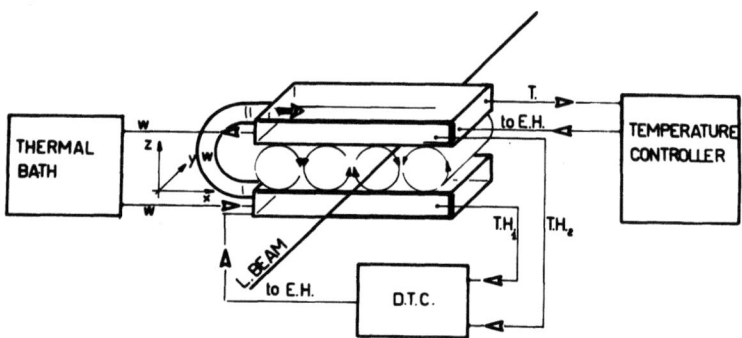

Figure 5. Schematic diagram of the cell: T,TH_1, TH_2 thermistors, W water circulation, DTC temperature difference control, EH electric heater.

The fluid is silicone oil with Prandtl number 30. The bottom and top plates are made of copper and the long term stability of the temperature difference is better than 4 m°C. This stability has been obtained with three independent temperature regulators. The first one is a water bath with a stability of about 0.05°C. The water circulates in the bottom and top plates where a electrical resistor is inserted in each of them. The two heating resistors are connected with other two stages of regulation. One stabilizes the temperature of the upper plate, the other controls temperature difference between the two plates. The last regulator is connected to a micro-computer that allows a complete automatization of the measurement.

The lateral walls of the cell are made of glass to permit visual observation.

The detection system consists of a laser beam that crosses the silicone oil perpendicular to the (x,z)-plane and is deflected by the thermal gradient inside the fluid. The laser beam sweeps the (x,z)-plane and we can measure the temperature gradient averaged along y in 1024 points of the (x,z)-plane by a method described elsewhere /17. The 1024 points are in an array with 16 rows of 64 points each. Precisely, for each position of the impinging beam, the unperturbed zero gradient is measured by a position-sensitive detector and recorded in a computer. Later upon application of temperature gradients we can measure the horizontal-and vertical-shift components, respectively proportional to the horizontal and vertical refractive index gradients $\partial n/\partial x$ and $\partial n/\partial z$ averaged along the y-axis, that is along the optical path of the laser beam. From these gradients one infers the temperature through the relation $\frac{\partial}{\partial} = (\partial n/\partial T \cdot \partial T/\partial x)$ and similar for z. The temperature field is then easily recovered by numerical integration of the two recorded gradients. The sweeping time is fast compared to the time scales of the phenomena under study. Therefore, by this method we can study the time evolution of the temperature field.

We perform the experiment in the following way. We start from zero temperature difference between the two plates and then we increase the temperature of the bottom plate till the maximum allowed in our apparatus corresponding to about $R=400\ R_c$. The steps in which the temperature has been increased are separated by sufficient amount of time to allow the system to relax to a stable state. This type of run has been repeated several times to check the dependence of the found regimes on the way in which the control parameter has been varied.

Analysing the fluid behavior as a function of $r=R/R$ we find a stable four roll structure at $r \leqslant 80$. Above this threshold the regimes of the system are outlined in Table I.

TABLE I

Interval	r	regime	Spatial structure
I1	80-90	TD *	R4 + LO
I2	90-95	S	R4
I3	95-130	TD	R4 + LO+TW
I4	130-150	S	R4
I5	150-182	TD	R4 + LO
I6	182-186	SHO	R4 + TW
I7	186-200	**	**
I8	200-300	TD	R4 + LO+TW

TW = Travelling waves
TD = Time dependent
S = Stationary
LO = Localized oscillations
SHO= Shilnikov type homoclinic orbit
R4 = 4 rolls
* The interval I1 is not observed in all of the runs
** the interval I7 presents a stationary regime in some runs and localized oscillations of very small amplitude in others.

From a run to another the interval initial positions are reproducible within 10%, whereas the length of the interval does not change sensitively. Instead the behavior of the system in the time dependent regimes can be different from a run to another. For example, in the interval I5 we can find other subintervals of periodic, biperiodic and chaotic behavior, but their existence is related to the speed with which the temperature gradient is increased and to the history of the system.

So we focus just on the general features that we always observe. In particular we see in Table I that the time-dependent regimes are associated with two different spatial patterns, one characterized by localized oscillations, the other by travelling waves. We will describe first how the temperature field has been reconstructed and then we will discuss more specifically the spatial patterns indicated in Table I.

To reconstruct the time-dependent component of the temperature field we proceed as follows. We first record the stationary temperature field $\tilde{T}(x,z)$ that is obtained by time-averaging the instantaneous temperature field $T(x,z,t)$. An example of $\tilde{T}(x,z)$ at r=82.66 in a periodic regime is reported in Fig.6(a). On the vertical axis of the figure is reported the temperature measured in terms of the temperature difference ΔT between the two plates. The lengths of the laser sweeps L_x =3.6 cm, L_z =0.8 cm in the x,z directions, respectively, are slightly smaller than the corresponding dimensions of the layer to avoid deflections of the light from the cell walls. A sequence of temperature field fluctuations $T_d^{\cdot}(x,z,t)=\tilde{T}(x,z)-T(x,z,t)$ is reported in Fig 7. The sequence has been taken ar r=83.66 where a periodic regime with frequency $f_1 \approx$ 22mHz was present.

The fields shown in Fig. 7(a) and 7(c) are recorded after almost a period of oscillation. Of course, the coincidence of the two images is not perfect because the sampling time was not an exact submultiple of the oscillation period. The oscillation amplitude depends on the space coordinates and reaches its maximum, that is about 1/10 of the stationary pattern amplitude, near the bottom plate at z=0.3 cm. Averaging numerically $T(x,z,t)$ along x (we remind that $T(x,z,t)$ is already averaged along the optical path y) we obtain the dependence on the z coordinate of the temperature field $T(z,t)$ averaged in a horizontal plane. The result obtained from the field recorded at r=82.66 is shown in Fig.6(b) for four different times.

Because $L_z <$ 1 cm the stable boundary layers are not completely seen in Fig.6(b) but it is still possible to measure their depth $\bar{\delta}$ that is about 2mm and then to compute the frequency of oscillation by with formula $f = K/\delta^2$ where we recall that k is the thermal diffusivity of the fluid. We find that f is about 28 mHz that is almost the value of f_1. Furthermore, the average field reported in Fig.6(b) has a time evolution quite similar to that obtained from numerical simulation /18/. Nevertheless there is a large asymmetry in Fig. 6 between the upper part of the field and the lower one that is not present in the result of the numerical simulation. This difference can be related to the non-Boussinesquian character of the fluid in our experiment. A deviation from the Boussinesquian approximation used in Ref. /18/ may be due to temperature difference between the two plates, which is five degrees.

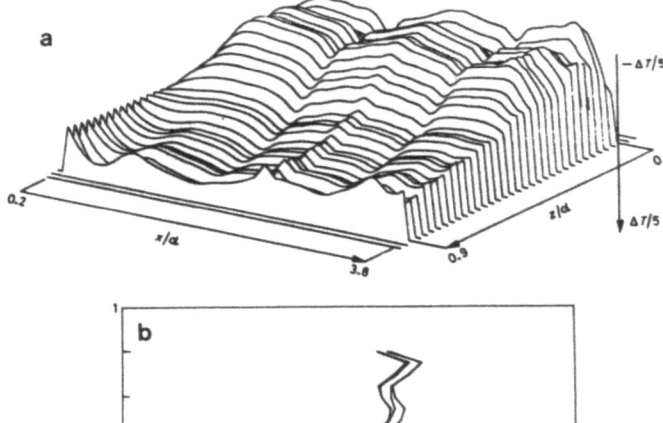
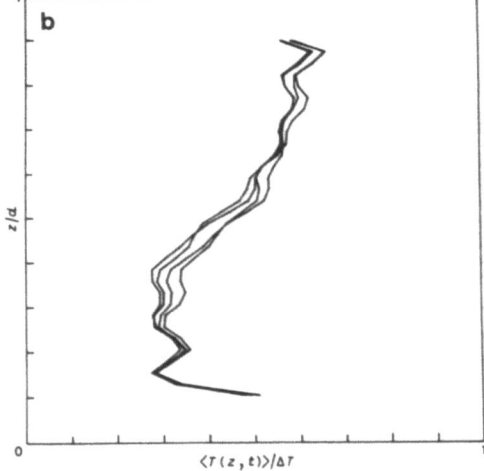

Figure 6.
a) Temperature field in the (x,z) plane averaged along the y axis. The field here reported is time averaged at r=82.66 where an oscillatory regime is present. The vertical axis of the figure is the temperature amplitude measured in terms of the temperature difference between the two plates.
b) Temperature distribution averaged in a horizontal plane as a function of z. The temperature has been recorded at four different times at r=82.66.

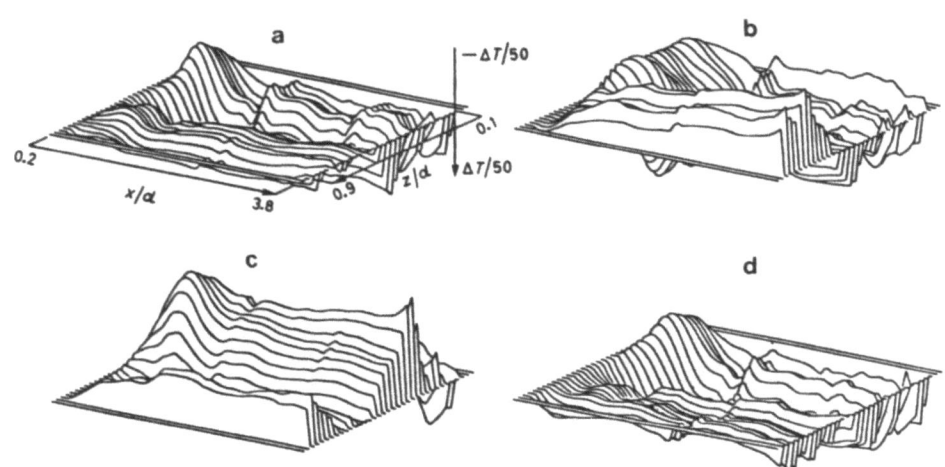

Figure 7. Evolution of the time dependent component of the temperature field T_d in the (x,z) plane at r=82.66 in the periodic regime. The records a),b),c),d) are respectively taken at 0,10,20,45 sec. The scales reported in a) are the same for all of others recordings. This sequence shows that the period of oscillation is about 45 sec.

In biperiodic and chaotic regimes the number of reconstructed temperature fields that must be recorded to have a resonable precision in fractal dimension calculation and in the Fourier transform, is very large (at least 4096). Therefore the amount of stored data would be excesive for our computer (about 8M byte for each time-series). After verifying that the dynamics does not depend sensitively on the z coordinate, except for the amplitude of T , we focus our attention just on the evolution of the horizontal component of the gradient $u(x,t) = \partial T / \partial x$ measured at a fixed z. The horizontal gradient is infact a direct result of the measurement and furthermore it does not contain the amplitude of the stationary gradient imposed between the two plates.

In what follows $w(x,t) = u(x,t) - \bar{u}(x)$ and $\bar{u}(x)$ is the time-average of $u(x,t)$. The "energy" $E(t)$ is the spatial average of $w^2(x,t)$.

Besides, the study of u just in one direction allows a more direct comparison with numerical simulations done in one-dimensional partial differential equations (P.D.E.).

As an example we show in Figure 8 the evolution of w as a function of x and time t. We see here that the oscillations are localized in space both in the periodic regime Fig. 8(a) at r = 83 and in the chaotic one, Fig. 8b) at r = 87.5. The localization can be also quantitatively measured by making the Fourier spectrum $S(f,x)$ of time series $w(x,t)$ recorded in different position of the cell. The spectrum $S(f,x)$, with $\bar{x} = 1.5$ cm is

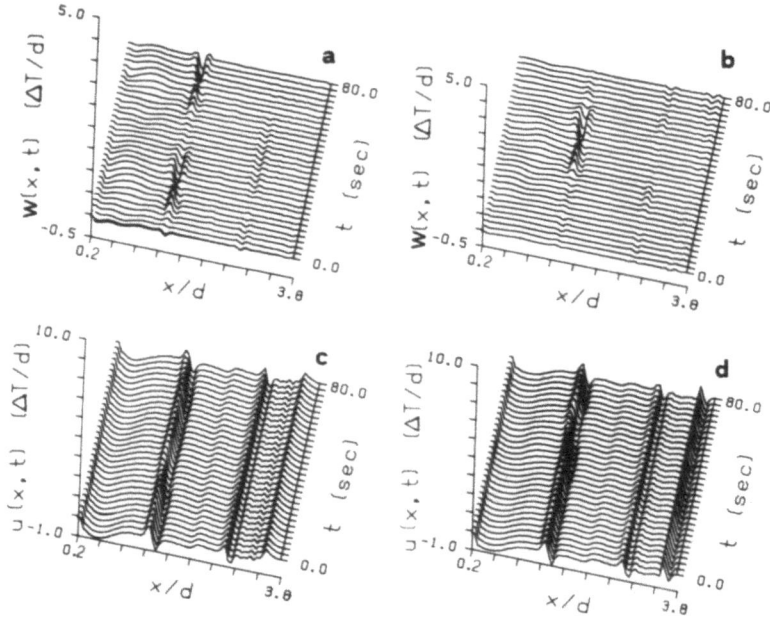

Figure 8. Evolution of the time-dependent component of the horizontal temperature gradient w (x,t) recorded at r = 83.5 (a) and r = 87.5 (b). The corresponding horizzontal gradients u (x,t) are instead reported in (c) and (d), respectively.

shown in Fig. 9a) at r=86.5 biperiodic regime and Fig. 9(b), chaotic regime at z=87.5. The amplitude of $S(f,x)$ at f_1 and f_2 as a function of x at r=86.5 and at r=87.5 is shown in Figure 10(a-b) respectively.

The amplitude of $S(f,x)$ changes three orders of magnitude by moving the measuring point just 4 mm. We see the high degree of localization of the oscillations. We also observe in Fig. 10(b) that the maximum amplitude of the two frequencies tends to become equal at the onset of chaos.

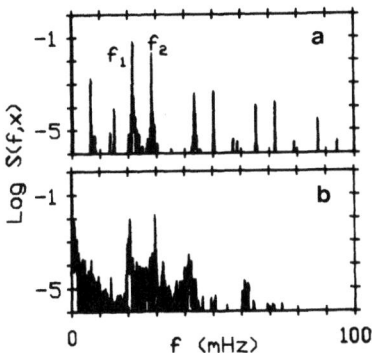

Figure 9. Fourier Spectrum $S(f,x)$ of the time series of $u(x,t)$ recorded in the point $\bar{x}=1.5$ at r=86.49 biperiodic regime a) and at r=87.49 b) chaotic regime.

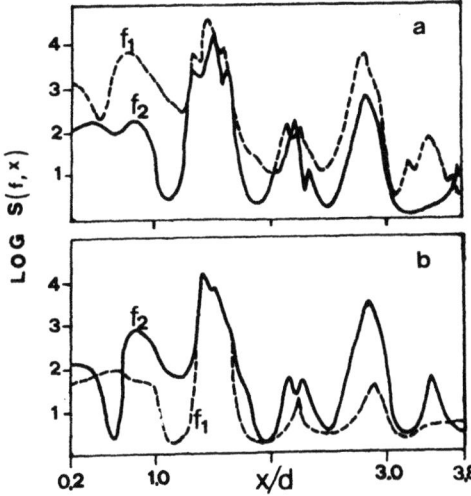

Figure 10. Amplitude $S(f,x)$ at frequencies f_1 and f_2, as a function of the x coordinate in the biperiodic regime at r=86.49 and in the chaotic regimes (r=87.5). The vertical. dashed lines show the limits of the horizontal laser sweep.

This spatio-temporal regime with localized oscillations is not the only one that we observe. Increasing R we find other windows of time dependent regimes that we associate with travelling waves.

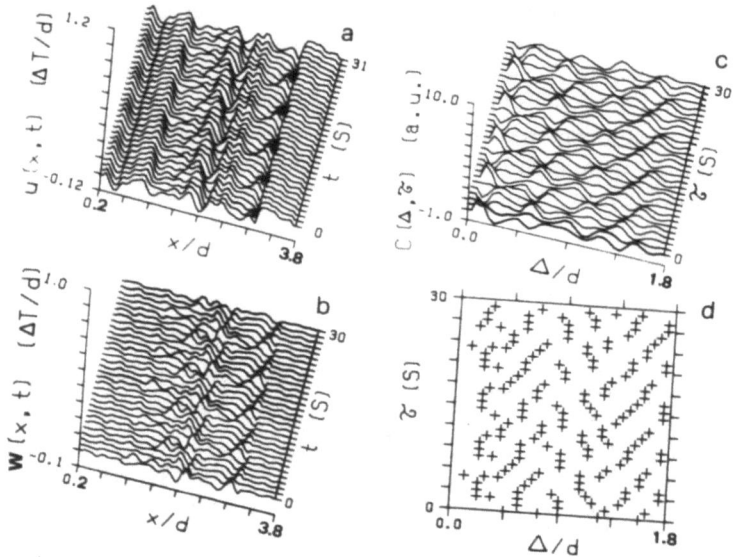

Figure 11. Travelling waves. Time evolution of $u(x,t)$ a) and $w(x,t)$ b) recorded at r=230. $C(\Delta,\tau)$ is reported in c). In d) the time evolution of the position of the maxima of C (crosses) is shown.

As an example we report in Fig.11(b) the evolution of $w(x,t)$ at r=230 where a biperiodic regime is present. We see that there are waves starting in the center of the cell that propagate towards the sides of the cell. This travelling structure is more evident in Fig. 11(c) where the spatio temporal correlation function $C(\Delta,\tau) = \int\int w(x+\Delta, t+\tau) w(x,t) dx\, dt$ is reported. We see that the extrema (Fig. 11d) of C propagates with a velocity of about 0.06 cm/sec. This velocity is consistent with the velocity scale constructed with $\nu/d = 0.03$ cm/sec for our fluid. $w(x,t)$ the time evolution of the maxima of $C(\Delta,\tau)$ measured in the chaotic regimes at r=268 are reported in Fig. 12. We see that the spatial behavior does not change sensitively when the system is driven from a periodic to a chaotic time dependent regime.

The transition between localized oscillations and travelling waves occurs at $r \simeq 180$ with a regime that has an evolution like that shown in Fig. 13(a). This evolution is characterized by the presence of quasi laminar oscillations that are interrupted by very large bursts. This regime is produced by a sort of competition between two spatial structures, one associated with the laminar period, the other with the fast transient This is shown in Figure 13(b) where the temporal evolution of $u(x,t)$, averaged over 4 periods of the fast oscillation of Figure 13(a), is reported. We see that during the fast transient of Figure 13(a), the time averaged structure of the convective motion shifts in an appreciable way the position of the rolls boundaries (where $\frac{\partial T}{\partial x} = 0$).

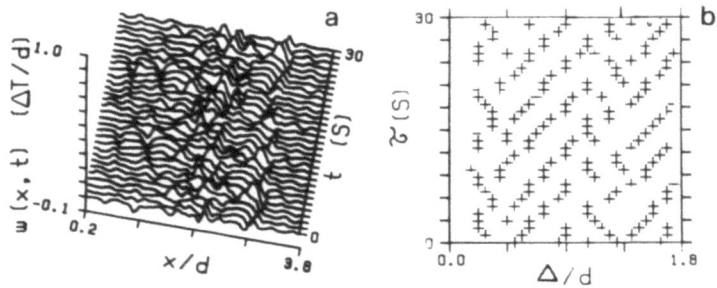

Figure 12. Travelling waves. a) time evolution at w (x,t) recorded at r=268; b) time evolution of the maxima positions of the spatio temporal correlation functions C (Δ,τ).

This change corresponds to a switch of energy (the energy in a mode is the amplitude of the spatial Fourier spectrum) between the odd and even modes of the spatial Fourier transform of w (x,t). The period of time T_0 between two bursts diverges with the following law $T_0 = 850(r-r_0)^{-0.1}$ sec when the bifurcation point for this regime $r_0 = 182.5$ is approached.

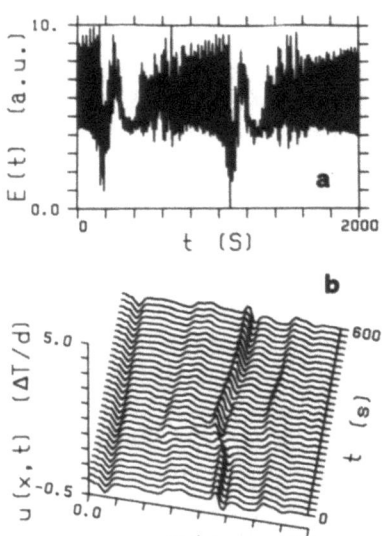

Figure 13. Transition between localized oscillations and travelling waves.
a) temporal evolution of the energy.
b) temporal evolution of the spatial structure u(x,t) averaged over 4 periods of the fast oscillation.

It is important to note that the presence of two different time dependent spatial patterns, one characterized by localized states and the other by travelling waves, is also found on the numerical simulations /4/ of Kuramoto-Shivanshinsky /11/ and Kuramoto-Velarde /12/ equations. In these systems it has also been observed that the transition between these two different behaviors takes place via a Shilnikov type homoclinic bifurcation that has the same properties that we have observed in our experiment, in the time dependent regime at $r \simeq 180$ corresponding to the transition between localized oscillations and travelling waves.

III - CONCLUSION

We have described two experiments on hydrodynamic instabilities where coherent spatial states coexist with temporal chaos. In the surface waves case the experimental results are in very good agreement with theory whereas in Rayleigh-Benard convection there is not yet a realistic model with which to compare the results. Nevertheless the fact that the behavior of this instability is very similar to that observed in the numerical simulation of (K-S) and (K-V) equations, suggests that a partial differential equation of this type can account for many of the observations made in R-B convection.

Much more experimental and theoretical work is still necessary to understand the relationship between spatial patterns and temporal chaos.

The experiments here described show that techniques that allow to resolve in time spatial patterns produced by the fluid motion can be very useful to give relevant information in this problem.

REFERENCES

/1/ A recent review of all the concepts of chaotic dynamics is: P. Bergé, Y. Pomeau, C. Vidal, *Order within Chaos*, J. Wiley, New York, 1986.
/2/ M. Giglio, S. Musazzi, E. Perini, Phys. Rev. Lett. $\underline{53}$, (1984) 240.
/3/ M.H. Jensen, L.P. Kadanoff, A. Libchaber, I. Procaccia, J. Stavans, Phys. Rev. Lett. $\underline{55}$ (1985) 2798.
/4/ J.M. Hyman, B. Nicolaenko, Los Alamos Report UR-86-1388. J.M. Hyman, B. Nicolaenko, Physica $\underline{18D}$ (1986) 13.
/5/ A.R. Bishop, K. Fesser, P.S. Lomdahl, W.C. Kerr, M.B. Williams, S.E. Trullinger, Phys. Rev. Lett. $\underline{50}$ (1983) 1095.
/6/ Y. Oono, M. Kohmoto, Phys. Rev. Lett. $\underline{55}$ (1985) 2927.
/7/ J. Gollub, S.V. Benson, J. Fluid Mech. $\underline{100}$ (1980) 449.
/8/ M. Dubois, P. Bergé, Phys. Lett. $\underline{93}$ (1983) 365.
/9/ S. Ciliberto, J.P. Gollub, Phys. Rev. Lett. $\underline{53}$ (1984) 922; S. Ciliberto, J.P. Gollub, J. Fluid Mech. $\underline{158}$ (1985) 381.
/10/ S. Ciliberto, F. Simonelli, Europhysics Lett. $\underline{2}$ (1986) 285.
/11/ E. Meron, I. Procaccia, Phys. Rev. Lett. $\underline{56}$ (1986) 1323.
/12/ S. Ciliberto, M.A. Rubio, Phys. Rev. Lett. $\underline{58}$ (1987) 2652.
/13/ Y. Kuramoto, Y. Tsuzuky, Progr. Theor. Phys. $\underline{55}$ (1976) 356.
/14/ P.L. Garcia-Ybarra, J.L. Castillo, M.G. Velarde, Phys. Lett. $\underline{A122}$ (1987) 107.
/15/ J.P. Gollub, A.R. Carriar, Phys. Rev. A $\underline{26}$ (1982) 270; A. Pocheau, V. Croquette, P. Le Gal, Phys. Rev. Lett. $\underline{55}$ (1985) 1094; M.S. Heutmaker, P.N. Fraenkel, J.P. Gollub, Phys. Rev. Lett. $\underline{54}$ (1985) 1369; G. Ahlers, D.S. Cannell, V. Steinberg, Phys. Rev. Lett. $\underline{54}$ (1985) 1373.
/16/ C. Normand, Y. Pomeau, M.G. Velarde, Rev. Mod. Phys. $\underline{49}$ (1977) 581.
/17/ S. Ciliberto, F. Francihi, F. Simonelli, Opt. Comm. $\underline{54}$ (1985) 251.

INTERFACES IN ELECTROPHORESIS, TWO-PHASE MODELS, AND EXCITABLE MEDIA

Paul C. Fife

Mathematics Department
Univ. of Arizona
Tucson, Arizona 85721, USA

INTRODUCTION

Three examples of interfaces will be discussed: all of them somewhat away from the mainstream of the conference. They are modelled by "interior transition layers" in solutions of partial differential equations of reaction-diffusion-advection type. It is there fore no surprise that two of the examples are involved with the transport of chemical species in solution; the interface then is formed by the action of competition between a diffusive process, which acts to smooth out the interface, and a force (chemical reaction or applied electric field) which tends to sharpen it. The third example, though not involving chemicals, does have a diffusive type process, related to a Gibbs free energy, and also an aggregative force (a double-well potential wich attempts to force states into one or another of two preferred ones). Interfaces resulting from the balance of diffusive and aggregative forces like this occur in other contexts as well (especially in biology and geophysics). It is hoped therefore that this exposition will serve to be illustrative and also suggestive of phenomena well beyond the confines the specific examples discussed here.

ELECTROPHORESIS

Description of the process

This term refers to a group of techniques used to separate mixtures of charged particles in solution by means of an imposed electric field. These techniques are used extensively in the separation of proteins and other biological materials[1]. The compounds of interest in solution are generally capable of reacting either as an acid or as a base (amphoteric compounds); in the ionization process the species may acquire either positive or negative charges. When a solution containing amphoteric compounds is exposed to an electric field, the movement of ions and uncharged species occurs simultaneosuly with dissociation-recombination reactions. The resulting transport is highly specific, so that a mixture of different compounds can be separated on the basis of different transport properties.

The dissociation-recombination reactions occur on such shorter time scales than that of transport by diffusion or electromigration, and it is often assumed that those reactions are everywhere in equilibrium. Under this assumption, the model for electrophoresis simplifies, taking the form of nonlinear advection-diffusion equations coupled with nonlinear algebraic relations characterizing the chemical equilibria. Moreover, the experimental setup is often such that variation occurs in only one spatial dimension (along a column). A mathematical model for electrophoresis with these simplifications (which we assume here) was given in[2].

There are several different electrophoretic techniques which are described by the same equations, but which differ in the boundary and/or initial conditions. For example if the ends of the column are impermeable to the amphoteric compounds, the process will reach steady state and the various species will focus in neighborhoods of their respective isoelectric focusing. The initial concentrations of compounds in the isoelectric focusing process are usually uniform.

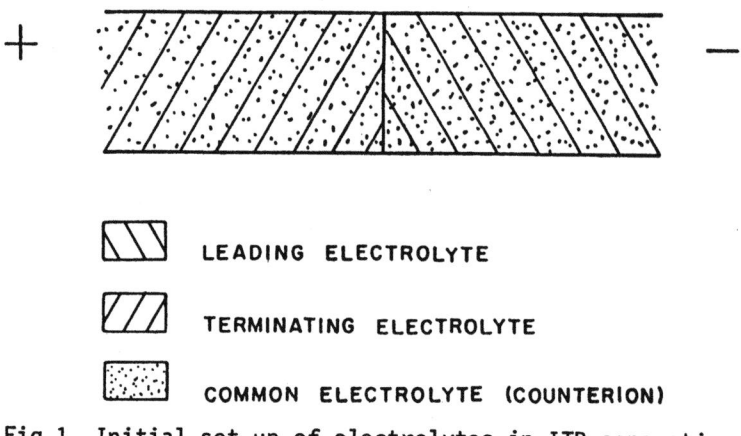

Fig.1. Initial set-up of electrolytes in ITP separation

In this article, however, we deal with another extensively used separation technique, called isotachophoresis (ITP). The usual setup for ITP consists of a reaction column connected at both ends to reservoirs of buffer electrolytes. The reservoirs are large enough that the processes in the column have a negligible effect on the composition of the buffers. This assures a fixed concentration of components at the column ends. The column is filled with specially prepared solutions (Fig.1), called common, leading, and terminating electrolytes, and a sample containing a mixture of species is then inserted at the interfaces between the leading and terminating solutions. Such nonuniform initial conditions are characteristic for the ITP process. After the sample insertion, an electric field is applied. Following a transient period, a set of continguous zones is formed (Fig.2) with all zones traveling along the column with the same constant velocity. This particular feature of the process gives rise to its name. Each zone contains only one of the (positively charged, say) species. The sample has therefore become unmixed, and interfaces have formed between the zones. My object here is to describe the structure of these interfaces mathematically.

Modeling these traveling waves, as well as other aspects of electrophoresis, uses the concept of mobility. An ion will move with velocity proportional to the product of its charge and the electric fiel. The proportionality constant is defined to be its mobility Ω. The mobilities (M^2/v.sec) of the positive ions A^+, B^+, and D^+ in the three zones depicted in Fig.2 are Ω_1, Ω_2, and Ω_4, and they satisfy the inequalities $\Omega_4 < \Omega_1 < \Omega_2$. Within a given zone, the composition is approximately uniform, changes being mainly concentrated in transitional regions (interfaces) between the zones. The composition increments from one zone to the next are governed by well known regulating functions, first

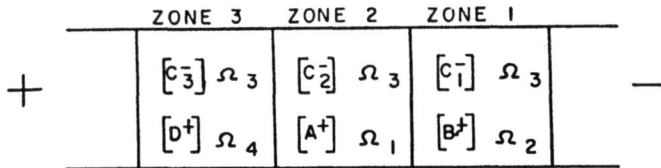

Fig. 2. Traveling zones in ITP steady state

derived by Kohlrausch[3] for strong (fully ionized) electrolytes. The regulating relation between zones 1 and 2 is

$$\frac{\hat{n}_1}{\hat{n}_2} = \frac{\Omega_1 (\Omega_2 + \Omega_3)}{\Omega_2 (\Omega_1 + \Omega_3)}$$

where $\hat{n}_1$ and $\hat{n}_2$ are the species concentrations in zones 1 and 2 respectively, and Ω_3 is the mobility of the common, negatively charged, species called counterion. This regulating relation is readily derived from the governing equations, assuming experimentally observable constant velocity of movement and uniform composition away from the transition regions.

There are several variations on the idea of regulating functions in the literature; they determine the concentrations of species inside the various zones and the traveling velocity. They were derived for various systems like multicomponent configurations involving zones traveling in both directions[4], simple weak acids and bases[5] and simple ampholytes[2].

The analysis of the transition regions is much more complicated and is usually performed with the aid of a digital computer used to generate numerical approximations to the solutions of governing equations.

In this paper we consider a prototypical transition region separating two positive ions with different mobilities; a common negative counterion is also present. The pype of solutions we obtain are expected to describe typical transition regions in more complex isotacthophoretic waves with a greater number of ions; but we idealize by assuming that only the ions mentioned are present.

The Mathematical Model

The basic object of study is then a mixture of charged species in solution, subject to an imposed electric fiel. As indicated above, we will shortly specialize to the case when the number of species is only 3, but for now denote it by m. The concentrations of the m species (in moles/volume) are denoted by the vector $\hat{n} = (\hat{n}_1, \ldots, \hat{n}_m)$. The basic assumption is that each species has a characteristic mobility constant Ω_i such that the velocity of its ions, when acted upon by a potential gradient, is given by

$$v_i = \Omega_i z_i \frac{\partial \phi}{\partial \hat{x}}, \qquad (1.1)$$

where z_i is its charge, in units of the absolute value of the electron charge, and ϕ is the potential. The species also diffuse, with diffusivities D_i. The flux f_i of the ith species is then given by

$$f_i = v_i \hat{n}_i - D_i \frac{\partial \hat{n}_i}{\partial \hat{x}}.$$

From this and (1.1), assuming strong electrolytes, we may write the species balance equations for a one-dimensional configuration in the form

$$\frac{\partial \hat{n}_i}{\partial t} = \frac{\partial}{\partial \hat{x}} \left[z_i \Omega_i \hat{n}_i \frac{\partial \phi}{\partial \hat{x}} + D_i \frac{\partial \hat{n}_i}{\partial \hat{x}} \right], \quad i = 1, \ldots, m. \qquad (1.2)$$

There is an additional equation, since ϕ must satisfy Poisson's equation

$$\hat{\varepsilon} \frac{\partial^2 \phi}{\partial \hat{x}^2} = -e \Sigma z_k \hat{n}_k, \qquad (1.3)$$

where e is the molar charge and $\hat{\varepsilon}$ the permittivity of the solvent.

We assume the Einstein relation $D_i = (RT/e) \Omega_i$. For the isotachophoresis problem, it is appropriate to look for traveling wave solutions, and so we introduce the traveling wave coordinate $x = \hat{x} - Ut$, U being the velocity. Let $\hat{E} = \partial \phi / \partial x$ be the elctric field. Written in the traveling wave coordinate, (1.2) becomes an ordinary differential equation which can be integrated once to yield

$$\frac{RT}{e} \Omega_i \frac{d \hat{n}_i}{d x} = \Omega_i z_i \hat{n}_i E - U \hat{n}_i + C_i, \qquad (1.4)$$

The C_i being integration constants. Also (1.3) is transformed in the obvious manner. These equations are to be solved under appropriate boundary conditions at $\pm\infty$. In general there may also be integral conditions, but not in the special case which we shall now consider.

Suppose that $m = 3$, there being two positive ions ($\hat{n}_1$ and $\hat{n}_2$) with different mobilities $\Omega_1 < \Omega_2$, and one negative ion ($\hat{n}_3$). Thus $z = (1, 1-1)$. We look for traveling wave solutions which serve to separate the two positive ions, the one with higher mobility ($\hat{n}_2$) being ahead of the other. In other words,

$$\hat{n}_1(+\infty) = \hat{n}_2(-\infty) = 0 \tag{1.5}$$

The remaining boundary conditions, the velocity U, and the constants C_i are parameters in the problem which should either be prescribed or deduced from the equations. Still another important parameter is the (constant) electric current through the medium, which is given in terms of the other parameters by

$$I = e \sum z_i (f_i + U \hat{n}_i)$$

In view of (1.3) (the right side vanishes at $\pm\infty$) and the fact that the diffusion term in f_i also vanishes there, the expression for the current simplifies when evaluated at :

$$I = e \sum (z_i)^2 \Omega_i \hat{n}_i E_{\pm\infty} \tag{1.6}$$

where $E_\pm \equiv \hat{E}(\pm\infty)$.

It turns out to be physically reasonable to specify only the concentration ν of the trailing positive ion at $-\infty$ ($\hat{n}_1(-\infty) \equiv \nu$) and the current I; all other parameters, as we shall see, can then be obtained a priori in a unique manner. First of all, if we set $\mu = \hat{n}_2(+\infty)$, the vanishing of the right side of (1.3) at $\pm\infty$, together with (1.5) implies

$$\hat{n}_1 = \hat{n}_3 = \nu \text{ at } -\infty; \quad \hat{n}_2 = \hat{n}_3 = \mu \text{ at } +\infty. \tag{1.7}$$

We use the following dimensionless variables and parameters:

$$x = \frac{e E_-}{RT} \bar{x},$$

$$n_i = \hat{n}_i / \nu,$$

$$E = \hat{E} / E_-,$$

$$b_i = \frac{\Omega_1}{\Omega_i}, \quad i = 2, 3$$

$$\theta = \nu / \mu,$$

$$\varepsilon = \hat{\varepsilon} E_-^2 / RT\mu.$$

The inequality involving mobilities translates into

$$b_2 < 1. \tag{1.8}$$

Introducing this notation into (1.4), (1.3) and evaluating the C_i from the conditions at both $+\infty$ and $-\infty$ yields the following system, dots denoting differentiation with respect to x:

$$\dot{n}_1 = n_1 E - n_1,$$
$$\dot{n}_2 = n_2 E - b_2 n_2,$$
$$\dot{n}_3 = n_3 E - b_3 n_3 + \theta(1 + b_3) \tag{1.9}$$
$$\epsilon \dot{E} = n_1 + n_2 - n_3.$$

As boundary conditions, we have:

$$\text{At } x = -\infty : n = (\theta, 0, \theta), \; E = 1; \tag{1.10a}$$
$$\text{at } x = +\infty : n = (0, 1, 1), \; E = b_2. \tag{1.10b}$$

Finally, all the remaining parameters $E\pm$, μ, U, θ are obtained in terms of the Ω_i, ν, and I by the relations

$$I = e(\Omega_1 + \Omega_3) \nu E_- = e(\Omega_2 + \Omega_2) \mu E_+,$$
$$E_+ = b_2 E_-,$$
$$U = \Omega_1 E_-,$$
$$\theta = \nu/\mu.$$

The first relation here implies

$$\theta(1 + b_3) = (b_2 + b_3). \tag{1.11}$$

For $\epsilon = 0$ (the electroneutrality assumption), the problem can be solved explicitly[2]. The electrically neutral solution $\bar{n}$, $\bar{E}$ is given by the formula

$$\bar{n}_1(x) = \kappa \theta \, e^{-\kappa x}(1 + e^{\alpha x})^{-\frac{1}{2}} \int_{-\infty}^{x} e^{\kappa t}(1 + e^{\alpha t})^{-\frac{1}{2}} dt,$$
$$\bar{n}_2 = \bar{n}_1 \, e^{(1 - b_2)x}$$
$$\bar{n}_3 = \bar{n}_1 (1 + e^{(1 - b_2)x}),$$
$$\bar{E} = 1 + \frac{1}{\bar{n}_1} \frac{d\bar{n}_1}{dx}.$$

In [6], the problem was solved for small positive ϵ. More precisely, it was proved that there exists a unique (except for shifts in the traveling wave coordinate) solution of (1.9, 1.10), and it is approximated to within $O(\epsilon)$ by the electrically neutral solution:

$$|n(x) - \bar{n}(x)| \leq C\epsilon \; ; \quad |E(x) - \bar{E}(x)| \leq C\epsilon$$

uniformly in x for some constant C independent of . Similar inequalities hold for the derivatives of the solution up to any order.

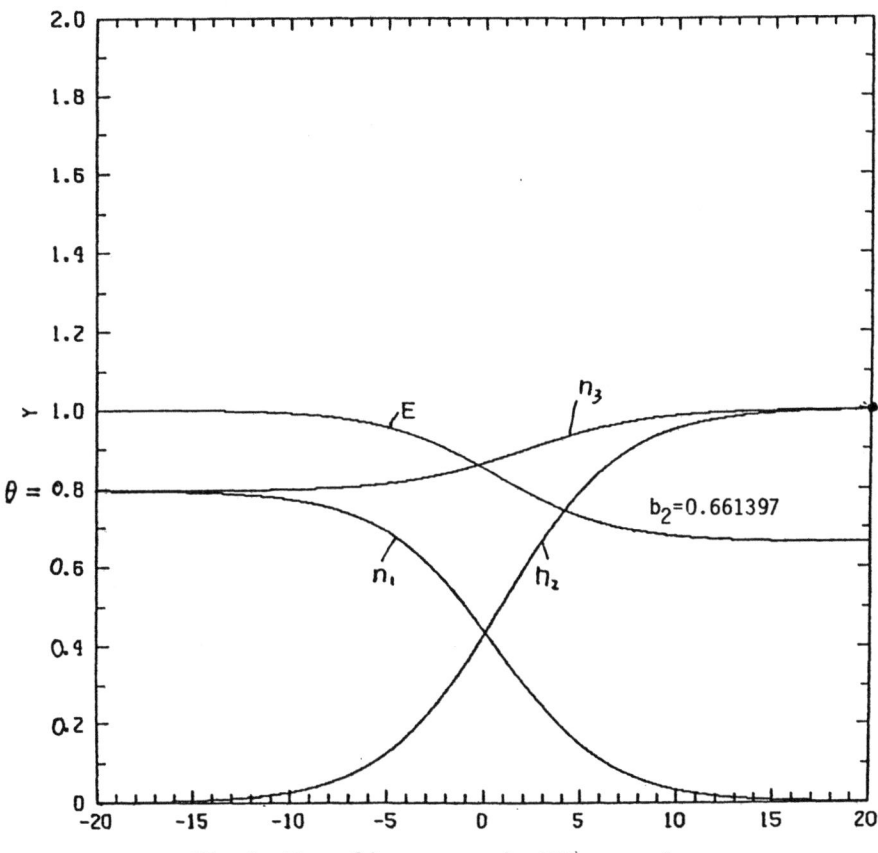

Fig.3. Traveling waves in ITP. $\epsilon = 0$

The profile in the transition zone has the monotonicity properties shown in Fig. 3, the result of a computation for small ϵ. These monotonicity properties, in fact, were proved to hold [6]. Finally, an analytic treatment of the transient problem, involving the process by which the initial mixed sample becomes unmixed, is under way.

PHASE INTERFACES

Consider a material which may be in either of two phases, e.g. liquid or solid, in a region Ω of space. We consider mathematical models for the statics and dynamics of such a continuum, including the motion, shape, and internal structure of the interface Γ (t) between the two phases.

The simplest such model is the classical Stefan formulation, in which the interface is consider to be infinitesimally thin, the temperature u(t,x), at the interface Γ(t) between solid and liquid, is zero, the temperature u(t,x) satisfies the usual heat diffusion equation in each of the two phases, and an energy balance condition is imposed on Γ . The latter relates the normal speed of Γ , the heat flux there, and the latent heat of the material.

The Stefan model does not accomodate various other physical realities which may be relevant. One approach to remedy this situation is through refinements of the Stefan model based on a Landau-Ginzburg free energy functional [7,8,9], where in addition to the temperature, an order parameter is also envisaged as a field quantity. These refinements are known as phase field models. The free energy functionals can be derived in a formal way from discrete lattice models with assumed interactions among the nodes (e.g. [10,11]). The resulting field equations can then be analyzed. In so doing, it is possible to connect lattice concepts such as interaction length and anisotropy in lattice interactions with properties of the continuum model which are left out of the Stefan formulation[10]. The latter include (1) a (generalized) Gibbs-Thompson relation among the equilibrium temperature at the interface, its curvature, and its normal velocity [12]; (2) anisotropy in growth velocities[11]; (3) finiteness of the interface thickness; (4) certain supercooling and superheating effects (the above papers and [13]); and (5) interaction of the interface with the boundary of the vessel containing the material [14].

Thus the phase field models automatically incorporate these phenomena, which otherwise may have to be added to existing models in an ad hoc manner. Moreover, successful numerical methods for these models have been developed and used [15].

The field equations assume the form

$$\tau \phi_t = \xi^\circ L \phi + f(u,\phi), \qquad (2.1)$$

$$u_t + \frac{1}{2} l \phi_t = K \Delta u, \qquad (2.2)$$

where u is temperature, ϕ is the order parameter, f is a functional sigmoidal in ϕ for each u (for example , $f = \frac{1}{2} (\phi - \phi^3) + 2 u$), L is an elliptic operator, and ξ (an interaction length figuring into the basic lattice model) and τ (relaxation time) are small paremeters.

Summary of recent results

G. Caginalp and I have obtained several mathematical results about the phase field models; they will be briefly described.

In [16], the equilibrium problem was considered in the isotropic case. Given a temperature field, it was proved that under certain conditions a stationary solution exists for which the order parameter has a internal layer, representing the phase interface. The small thickness of the layer is linked to the interaction length, a basic small parameter in the problem. It was shown in that paper that the equilibrium interfacial shape satisfies the Gibbs-Thomposn relation.

In [10,11], the effect of retaining higher order interaction terms in the continuum model was studied. More and more complicated anisotropic effects are possible when the order of the retained interaction terms is increased. The inner problem at the interface involved solving the equation.

$$L_o \phi = f(\phi) \tag{2.3}$$

on the entire real line, with ϕ approaching finite limits at $\pm\infty$ (zeroes of f). Here f is an old function with positive and negative zeroes. L_o is an even order differential operator with constant coefficients, whose order depends on the order of the interaction terms included. Very little is known about this boundary value problem.

In [14], it was shown how to incorporate interaction with the boundary into the model. With certain types of interaction, it was shown that the essential effect was to prescribe an angle-of-contact condition between the internal layer and the boundary. (This is a condition which has been traditionally used and derived from other considerations.) It was conjectured that this was also true for much more general types of interaction. Finally, the effect of the boundary interactions on the stability of the equilibrium configuration was examined.

In [12], a complete formal asymptotic analysis of the dynamics of the interface was given; among other things, it clarified the dynamic Gibbs--Thompson relation and its dependence on the other parameters of the problem. The main arguments in this last paper will now be given.

Phase field interface dynamics

We deal with the moving internal layer problem for the phase field equation (2.1).(2.2), in which we take simplest model, namely when $L = \Delta$ (the Laplacian).

Let the location of the interface, defined as where $\phi = 0$, be given by

$$r(x,y,t) = 0 \tag{2.4}$$

In a neighborhood of Γ, in fact, we define $r(x,y,t)$ to be $\pm$ the distance from (x,y) to Γ (+ on one side; -on the other, so that r is a smooth function). It follows that in that neighborhood,

$$\nabla r \equiv 1; \text{ on } \Gamma, \quad \Delta r = \kappa, \tag{2.5}$$

where ∇ and Δ refer only to the spacial variable x an y, and κ is the curvature of Γ. The function $s(x,y,t)$ is now defined so that (s,r) is a local coordinate system near Γ which is orthogonal and such that on Γ, s measures arclength from some fixed point, depending smoothly on t.

We shall apply a formal asymptotic procedure to solutions of these equations. In particular, there will be an outer and an inner expansion. The outer one is as follows:

$$u = u(x,y,t,\xi) = u^0(x,y,t) + \xi u^1(x,y,t) + \ldots, \quad (2.6a)$$

$$\phi = \text{(similar)} \quad (2.6b)$$

The position of the interface will itself depend on ξ, and so we write

$$r(x,y,t,\xi) = r^0(x,y,t) + \xi r^1(x,y,t) + \ldots, \quad (2.6c)$$

with a similar expansion for $s(x,y,t,\xi)$. The terms on the right of (2.6a,b) may be discontinuous at $r = 0$; otherwise they are smooth.

The inner expansion proceeds by defining $z \equiv r/\xi$ and thinking of u and ϕ as depending in a regular manner on the variables $z, s,$ and t near Γ:

$$u = U(z,s,t,\xi) = U^0(z,s,t) + \xi U^1(z,s,t) + \ldots \quad (2.7a)$$

$$\phi = \Phi(z,s,t,\xi) = \text{(similar)}. \quad (2.7b)$$

By definition of r, we require

$$\Phi(0,s,t,\xi) \equiv 0 \quad (2.8)$$

There are standard matching conditions between the inner and outer expansions:

$$U^0(\pm\infty,t) = u^0(Y^0\pm,t), \quad (2.9a)$$

$$U^1(z,t) \simeq u^1(Y^0\pm,t) + z u^0_x(Y^0\pm,t) \quad (z \to \pm\infty) \quad (2.9b)$$

$$U^2(z,t) \simeq (u^2(Y^0\pm,t) + Y^1(t)u^1_x + Y^2 u^0_x + \frac{1}{2}(Y^1)^2 u^0_{xx}) +$$
$$z(u^1_x + u^0_{xx} Y^1) + \frac{1}{2} z^2 u^0_{xx} \quad (z \to \pm\infty). \quad (2.9c)$$

Here the argument of the u^i and their derivatives are as in (2.9b), and that of the Y's is t.

Outer expansion. Setting (2.6) into (2.1) and (2.2) and equating coefficients of corresponding powers of ξ, we obtain a sequence of outer problems:

$O(2.1)$:

$$u^0_t + \frac{l}{2}\phi^0_t = K\Delta u^0 \quad (r \neq 0), \quad (2.10a)$$

$$f(\phi^0) + 2u^0 = 0; \quad (2.10b)$$

$O(\xi)$:

$$u^1_t + \frac{l}{2}\phi^1_t = K\Delta u^1, \quad (2.11a)$$

$$f'(\phi^0)\phi^1 + 2u^1 = 0; \quad (2.11b)$$

$O(\xi^2)$:

$$u^2_t + \frac{l}{2}\phi^2_t = K\Delta u^2, \qquad (2.12a)$$

$$f'(\phi^0)\phi^2 + 2u^2 = -\frac{1}{2}f''(\phi^0)(\phi^1)^2 + \alpha\phi^0_t - L\phi^0, \qquad (2.12b)$$

etc.

It will be assumed that (2.10b) can be solved for ϕ^0 as a function of u^0, for u^0 in the range of interest, and for ϕ^0 in a neighborhood of +1; this is then substituted into (2.10a) to obtain an equation for u^0 alone, which should hold on the liquid side of the layer, which we arbitrarily specify to be the side where $r > 0$. Similarly, it is assumed that (2.10b) can be solved with ϕ^0 in a neighborhood of -1; this will generate an equation for u^0 valid for $r < 0$. The higher order equations (2.11), (2.12), etc. likewise reduce to equations for u^k alone. To complete the solution, interface conditions at $r = 0$ will be derived later.

Inner expansion. A standard calculation with use made of (2.5) and the orthogonality of the coordinate system (r,s) shows that in terms of that system,

$$\Delta u = u_{rr} + u_{ss}(s_x^2 + s_y^2) + u_r\Delta r + u_s\Delta s.$$

In terms of z, s, and t, the equations (2.1), (2.2) become

$$KU_{zz} + \gamma(-r_t U_z - \frac{l}{2}r_t\Phi_z + K\Delta r\, U_z) +$$

$$+ \xi^2(U_t + U_s s_t + \frac{l}{2}\Phi_t + \frac{l}{2}\Phi_s s_t + K(U_{ss}\Delta s + U_s|\nabla s|^2)) = 0, \qquad (2.13)$$

$$\Phi_{zz} + f(\Phi) + 2U + \alpha r_t\xi\Phi_z +$$

$$+ \xi\Phi_z\Delta r + \xi^2(\Phi_{ss}|\nabla s|^2 + \Phi_s\Delta s) = 0 \qquad (2.14)$$

We shall now proceed to examine the various orders of approximation of (2.13) and (2.14) obtained by substituting (2.7) therein.

<u>$O(1)$.</u>

$$U^0_{zz} = 0 \qquad (2.15a)$$

$$\Phi^0_{zz} + f(\Phi^0) + 2U^0 = 0 \qquad (2.15b)$$

We want bounded solutions, so from (2.15a), $U^0 = $ const, and the existence of a bounded solution of (2.15b) with distinct limits at $\pm\infty$ requires $U^0 \equiv 0$. From this and matching condition (2.9a) we obtain

$$U^0 \equiv u^0 \big|_{r=0} = 0, \qquad (2.16a)$$

and

$$\phi^0 = \phi^0(z) \equiv \psi(z), \qquad (2.16b)$$

$\psi(z)$ being the unique solution of

$$\psi'' + f(\psi) = 0; \quad \psi(\pm\infty) = \pm 1, \quad \psi(0) = 0. \qquad (2.17)$$

Now (2.9a), applied to ϕ, yields

$$\phi^0 \big|_{r=0_\pm} = \pm 1,$$

which was clear anyway from (2.10b) and (2.16a).

$\underline{O(\xi)}$

$$KU^1_{zz} = r^0_t U^0_z + \frac{l}{2} r^0_t \phi^0_z - K\Delta r^0 U^0_z = \frac{l}{2} r^0_t \psi'(z), \qquad (2.18)$$

the latter because of (2.16). Integrating, we obtain

$$KU^1_z = -\frac{l}{2} r^0_t \psi(z) + c_1(s,t). \qquad (2.19)$$

But (2.9b) applies, and we let $z \to \pm\infty$ in (2.19) to obtain

$$ku^0_r \big|_{r=0_\pm} = \mp \frac{l}{2} r^0_t + c_1(s,t). \qquad (2.20)$$

Subtracting this equation with the $-$ sign from the same equation with the $+$ sign yields

$$K[u^0_r]_{r=0} = -l r^0_t, \qquad (2.21)$$

which is the Stefan condition for the lowest order approximation. If appropriate initial conditions are given, then (2.10) and (2.21), with those initial conditions, may be solved to give a unique value for $u^0(x,y,t)$, $r^0(x,y,t)$, and $\phi^0(x,y,t)$.

Once u^0 and r^0 are found this way, they can be used in (2.20) to determine $c_1(s,t)$. We now proceed toward the determination of U^1. Integrate (2.19) to obtain

$$KU^1(z,s,t) = -\frac{l}{2} r^0_t \int_0^z \psi(z')dz' + c_1(s,t)z + c_2(s,t)$$

$$= \frac{l}{2} r^0_t \int_0^z (\text{sgn } z' - \psi(z'))dz' - \frac{l}{2} r^0_t |z| + c_1(s,t)z + c_2(s,t). \qquad (2.22)$$

Now let $z \to \infty$ and use (2.9b) to obtain (2.20) again and also

$$ku^1|_{r=0\pm} = c_2(s,t) + \frac{l}{2} r_t^0 \int_0^{\pm\infty} (\text{sgn } z - \psi(z))dz. \tag{2.23}$$

Substraction yields

$$k[u^1|_{r=0}] = \frac{l}{2} r_t^0 \int_0^{\pm\infty} (\text{sgn } z - \psi(z))dz. \tag{2.24}$$

The final determination of u^1 and c_2 (hence U^1) will be done later.

The $O(\xi)$ terms in (2.14) are

$$\phi^1_{zz} + f'(\phi^0)\phi^1 + 2U^1 = -\alpha r_t^0 \psi'(z) - \Delta r \psi'(z).$$

Let $\Delta' \equiv (\partial_z)^2 + f'(\psi(z))$; then this becomes

$$\Delta' \phi^1 = -2U^1 - \alpha r_t^0 \psi'(z) - \Delta r^0 \psi'(z). \tag{2.25}$$

As an operator on $L_0(-\infty, +\infty)$, Δ' has an eigenvalue at the origin, with eigenfunction $\psi'(z)$. Moreover, it will be simple, so that the solvability condition for $\Delta'\phi = g \in L_2$ is orthogonality of g to this same eigenfunction, which we denote by ψ' for short. We know from (2.9a), however, that it can be expected that ϕ^1 may be unbounded (growing linearly in z at $\pm\infty$). Nevertheless the solvability condition for (2.25) remains the same:

$$2\int U^1 \psi' dz + (\alpha r_t^0 + \Delta r^0) \int (\psi')^2 dz = 0. \tag{2.26}$$

Now the substitution of (2.22) into (2.26) gives the function $c_2(y,t)$ uniquely, hence U^1, since everything else in (2.22) is known. Specifically,

$$c_2 = v^0 \left[-\frac{1}{4} \int_{-\infty}^{\infty} \Psi(z) \psi(z) dz - \frac{1}{4}\alpha KA \right] - \frac{1}{4} Kk^0 A; \tag{2.27}$$

where

$$A \equiv \int_{-\infty}^{\infty} (\psi')^2 dz.$$

From this and (2.23), one calculates that

$$u^1|_{\Gamma\pm} = v^0 \left[\frac{l}{2K} B - \frac{1}{4}\alpha A \right] - \frac{1}{4} Ak^0, \tag{2.28}$$

where

$$B = -\int_{-\infty}^{\infty} (\psi(z) - \text{sgn}(z))(\psi(z) + \frac{1}{2}\text{sgn}(z))dz,$$

These latter two functions provide the needed boundary conditions at the interface to be used with (2.11) in the unique determination of $u^1(x,y,t)$ on both sides of the interface. Of course initial data are also needed; if the initial temperature is given as not depending on ξ, then u^1 will simply have zero initial values.

Since U^1 is now known, the function ϕ^1 can be found uniquely from (2.25), the condition (from (2.8)) that it vanish when $z = 0$, and that it satisfy the required behavior at $\pm\infty$. The latter, obtained from (2.9b) applied to Φ, turns out to be the same as one obtains directly from (2.25), letting $z \to \pm\infty$, $\Phi_{zz} \to 0$, and using the known asymptotic properties of U^1. The orthogonality condition having been imposed, we obtain a unique solution Φ^1.

We now have the functions u, ϕ, U, Φ to orders 0 and 1, and r to Higher approximations to any desired order may be obtained by continuing this type of analysis.

Law of motion of the interface. We wish to draw attention to (2.28), which is the analog of the stationary Gibbs Thompson, but for a moving interface. It is linear relation among the temperature $u = \xi u^1$ on Γ, the latter's normal velocity v, and its curvature k^0. Such a relation has been used before, but here it follows in a systematic way from the field equations themselves, and the coefficients in this linear relation are given explicitly in terms of them.

Higher order phase field models may in principle be analyzed in the same way as above, although mathematical difficulties arise in solving (2.15b).

EXCITABLE MEDIA

Roughly speaking, an excitable medium is a biological or chemical medium which can exist, at each spatial point, in three states: excitable, excited, and refractory. A second requirement is that there is a local dynamics which governs the changes of state. If the right kind of stimulus is applied to a unit while it is in its excitable state, it will change to the excited state; but then it automatically becomes refractory after a period of time, and after another one, reverts to being unexcited. During the refractory state, which is not an excited state, it cannot be excited. Thirdly, there is communication between nearby sites in the medium in the sense that if one site is excited, then it may act to excite other states which are near enough to it.

This describes, in more or less reasonable terms, a number of actual media, of which one is the Belousov-Zhabotinski (BZ) reagent under certain common conditions. One characteristic of excitable media which is often observed, both in the laboratory and in computer simulations, is the presence of spiral-shaped rotating structures called "rotors" (see[18] for the earliest account of them, at least in the West). Excitable media can exist in many forms, and can be modeled mathematically in many ways, but stable and persistent rotors seem to be a common feature of most any model[19,30]. Other structures such as expanding ring patterns and planar traveling pulses and periodic waves are also commonplace, and were even discovered before rotors.

In the case of the BZ reagent, it turns out that chemical modeling leads to the "Oregonator" skeleton [20] which (after application of a quasi-steady state relation[21] can be adequately described by a pair of partial differential equations of "propagator-controller" (PC) type[22,23,24]. It further turns out that the relevant solutions of these equations have an interfacial structure, in that they have thin layers representing interfaces between adjacent excited and nonexcited (excitable or refractory) states. These interfaces characteristically move into the unexcited state if it is excitable, and recede from the unexcited state if it is refractory. Their laws of motion can easily be derived from an asymptotic analysis of the governing equations [24,25,26,27].

In view of this, it appears logical to try to understand the origin and persistence of rotors, at least in the context of the BZ reagent, in terms of the PC system. This in fact can be done, and was reported in [22,23,27]. The main lines of the argument will be outlined here.

The PC system can be written in the form

$$u_t = \varepsilon \Delta u + \varepsilon^{-1} f(u,v), \quad (3.1)$$

$$v_t = \varepsilon \Delta v + g(u,v), \quad (3.2)$$

(ε is a small positive parameter) where the nullcline { $f(u,v) = 0$ } in the u,v plane has a sigmoidal (more properly, N-shaped) configuration and the g-nullcline intersects the other one near one of its "knees" (being N-shaped, it has two of them). Looking for traveling wave solutions with an internal layer of characteristic thickness ε, one finds with the aid of formal asymptotics that such layers do exist, with u but not v undergoing a sharp transition there, provided v at the interface lies in a certain range of values, and provided the velocity of the interface is a specific function of v there. For a special value $v = v^*$, this velocity is zero.

Moreover, asymptotics can be used to construct pulse-like solutions, representing unidirectional plane waves in the form of a band of excited state entering a region of excitable state and leaving behind another region of refractory state (which later becomes excitable). If one now perturbs this traveling band by mixing the chemicals in a segment of it, so that the segment is effectively deleted, the two ends of the remaining undeleted parts of the band begin twisting and curling. Interestingly, this action can be readily understood and almost quantified on the basis of the local laws of motion of the interfaces.

This is the birth of a pair of spirals, rotating in opposite directions. The twisting must continue, one can reasonably argue, until a fully developed rotor, rotating at a constant angular velocity, is present. The mathematical problem now is to model and analyze the fully developed rotor. This can be dome, again on the basis of formal asymptotics, by (1) recognizing the characteristic space and time scales of the rotor, (2) rescaling in the corresponding way, and (3) examining the resulting partial differential equations to leading order in the small parameter appearing there (which happens to be $\varepsilon^{1/3}$

The mathematical result is a free boundary problem, in which the free boundary, which will be spiral-shaped, represents the layer between the excited region and one or the other of the two other unexcited regions. The characteristic spacing between branches of the spiral is of the order $O(\varepsilon^{2/3})$, the characteristic thickness of the interface $O(\varepsilon)$, and the characteristic period of rotation $O(\epsilon^{1/3})$. the free boundary problem involves a linear elliptic partial differential equation for the rescaled control variable v in each of the regions (excited and unexcited) and a nonlinear interface condition relating, at each point of the interface, the value of v at that point to the angular velocity, the slope of the interface, the value of v at that point to the angular velocity, the slope of the interface there, and its curvature.

A slightly different scaling was used in a subsequent paper by Keener and Tyson[28]. They observed also that if one makes the ad hoc assumption (which seems to be borne out by observations) that v is constant on the interface except very near its center, then the problem reduces to an ordinary differential equation and one can characterize what should be the unique angular velocity.

Finally, it should be noted that if the two diffusion terms in (3.1), (3.2) are not taken to be the same, but rather of different orders of magnitude, then the nature of the limiting problem changes. In fact if the ratio of the diffusivity of u to that of v is large, then one obtains a problem like that of [28], without having to make their ad hoc assumption. On the other hand if that ratio is very small, then one obtains a free boundary problem with different interface condition: the effect of the curvature is no longer present. This case was derived and studied in a bounded region by Sultan and Ortoleva[29]. One does obtain the expected spiral sturctures in any case, further attesting to their ubiquitous nature.

REFERENCES

1. Z. Deyl, ed., Electrophoresis: A Survey of Techniques and Applications, Elsevier, Amsterdam, 1979.
2. D. A. Saville and O. A. Palusinksi, Theory of electrophoretic separations, Part 1 and Part e, AICHE Journal, 32, Nº 2 (1986).
3. F. Kohlrausch, Ueber Concentrations-Verschiebungen durch Electrolyse im Inneren von Losungen and Losungsgemischen, Ann. Phys., Leipzig, 62, 209 (1897).
4. V. P. Dole, A theory of moving boundary systems formed by strong electrolytes, J. Amer. Chem. Soc. 67, 119 (1945).
5. R. A. Alberty, Moving boundary systems formed by weak electrolytes: Theory of simple systems formed by weak acids and bases, J. Amer. Chem. Soc., 72, 361 (1950).
6. P. C. Fife, O. A. Palusinski and Y. Su, Electrophoretic traveling waves, submitted.
7. J. S. Langer, Theory of the condensation point, Annals of Physics 41, 108-157 (1967).
8. P. C. Hohenberg and B.I. Halperin, Theory of dynamic critical phenomena, Reviews of Modern Physics 49, 435-430 (1977).
9. G. Caginalp, An analysis of a phase field model of a free boundary, Arch. Rat. Mech. Analysis 92, 205-245 (1986).
10. G. Caginalp and P. C. Fife, Phase field methods for interfacial boundaries, Phys. Rev. B, 33, 7792-7794 (1986).
11. G. Caginalp and P.C. Fife, Higher order phase field models and details anisotropy, Phys. Rev. B, to appear (1986).
12. G. Caginalp and P.C. Fife, Dynamics of layered interfaces arising from phase boundaries, in preparation.
13. P. C. Fife and H. Geng, Solidification waves in the phase-field model, in preparation.

14. G. Caginalp and P.C. Fife, Phase field models of free boundary problems: exterior boundaries, higher order equations, and anisotropy, Proc. NATO Workshop on Partially Solidified Systems, D. Loper, ed., to appear (1986).
15. M. Mimura et al, Videotape of computer simulations of instabilities and dendritic growth, using phase field model (1986).
16. G. Caginalp and P.C. Fife, Eliptic problems involving phase boundaries satisfying a curvature condition, submitted.
17. G. Caginalp and P.C. Fife, Eliptic problems with layers representing phase interfaces, Proc. Conf. on Nonlinear Diffusion Equations, Rome, to appear.
18. A. T. Winfree, Spiral waves of chemical activity. Sci. $\underline{175}$, 634-636 (1972).
19. A. T. Winfree, Stably rotating patterns of reaction and diffusion. Pp. 1-51 in Theoretical Chemistry Vol. 4 (H. Eyring and D. Henderson, eds.). Academic Press, New York (1978).
20. R. J. Field, E. Körös and R.M. Noyes, Oscillations in chemical systems. II. Thorough analysis of temporal oscillation in the bromate-cerium-malonic acid system. J. Amer. Chem. Soc. $\underline{94}$, 8649-8664 (1972).
21. J. Tyson, On scaling and reducing the Field-Körös-Noyes mechanism of the Belousov-Zhabotinskii reaction. J. Phys. Chem. $\underline{86}$, 3006-3012 (1982).
22. P. C. Fife, Propagator-controller systems and chemical patterns, Nonequilibrium Dynamics in Chemical Systems, A. Pacault and C. Vidal, eds., Springer-Verlag, pp. 76-88 (1984).
23. P. C. Fife. Understanding the patterns in the BZ reagent, J. Stat, Physics $\underline{39}$, 687-703 (1985).
24. J. Tyson and P.C. Fife, Target patterns in a realistic model of the Belousov-Zhabotinskii reaction. J. Chem Phys. $\underline{73}$, 2224-2237 (1980).
25. L. A. Ostrovskii and V.G. Yahno, The formation of pulses in an excitable medium. Biofizika $\underline{20}$, 489-493 (1975).
26. P. C. Fife, Pattern formation in reacting and diffusing systems. J. Chem. Phys. $\underline{64}$, 854-864 (1976).
27. P. C. Fife, Current topics in reaction-diffusion systems. Pp. 371-412 in Nonequilibrium Cooperative Phenomena in Physics and Related Fields, M.G. Velarde, ed., Plenum Pub. Corp., New York (1984).
28. J. P. Keener and J.J. Tyson, Spiral waves in the Belousov-Zhabotinski reaction, Physica D, to appear.
29. R. Sultan and P. Ortoleva, Rotating waves in reaction-diffusion systems with folded slow manifolds, J. Chem. Phys., to appear.
30. Y. Kuramoto, Chemical Oscillations, Waves, and Turbulence. Springer Series in Synergetics 19, Springer-Verlag, New York (1984).

FIELD INDUCED PATTERN FORMATION IN NEMATIC LIQUID CRYSTALS

F. Sagués[+] and M. San Miguel[*]

[+]Departament de Química Física, Univ. de Barcelona
Diagonal 647, Barcelona 08028, Spain

[*]Departament de Física, Univ. de les Illes Balears
Palma de Mallorca, Spain

INTRODUCTION

The Fréedericksz transition in the nematic phase of a liquid crystal[1] appears as the result of a competition between elastic and magnetic torques when a magnetic field is applied to a sample in which boundary conditions fix the director field at the plates containing the sample. There is now broad experimental evidence[2-6] that this transition is accompanied by a transient pattern which appears when a large enough magnetic field is switched-on. The pattern has a characteristic periodicity and dissapears in the final stationary state. This phenomenon has been observed in thermotropics and lyotropics and in the twist and splay geometries. In general one finds a stripe pattern with a one dimensional periodicity, but two dimensional patterns have also been observed[6].

The mechanism responsible for the emergence of the pattern is the coupling of the director field and the velocity flow. This gives rise to an effective wake number-dependent viscosity such that the most unstable mode turns out to be not the homogeneous one. This mode characterizes the periodicity of a well developed pattern. It has been pointed out[2-3] that this is analogous to what happens in the classical Cahn-Hilliard theory of spinodal decomposition[7]. In this case phase separation dynamics is also dominated by a most unstable mode $q \neq 0$. It is however well know[7-8] that the Cahn-Hilliard theory of spinodal decomposition is unsatisfactory at least in two points. First, a proper description of phase separation has to include thermal fluctuations self-consistently. Second, the time domain of validity of such linear theory is too short to be accesible to experimentation for Ising like systems with short-range interactions. These two issues have been examined in the context of the dynamics of the Fréedericksz transition for low magnetic fields for which hydrodynamic coupling can be neglected and no pattern is formed[9-10]. The analysis is based on a Langevin dynamical model[11]. The selfconsistent inclusion of thermal fluctuations allows a calculation of an onset time associated with the decay of the unstable state. Mathematically this is defined as a mean first passage time (MFPT). For a typical sample of MBBA it is of the order of 200 sec.[9-11]. A nonlinear calculation of the structure factor associated with transient fluctuations of the orientation of the director field shows that the first stage of evolution, up to the MFPT, is well described by a linearized theory. These results indicate that the initial stages of development of the transient pattern should be properly described by a linearized calculation of the structure factor when hydrodynamic couplings are included. Such a calculation is presented here. Through this calculation we describe the time evolution associated with the emergence of a characteristic periodicity giving predictions for the time scales of appearance and formation of the pattern[12].

DYNAMICAL EQUATIONS

Stochastic nematodynamics is defined by a set of coupled nonlinear Langevin equations for the director and velocity fields[12]. These equations are of the general type of time-dependent Ginzburg Landau models. The stochastic forces satisfy fluctuation-dissipation relations with nonlinear dissipative terms in a such a way that the stationary distribution for director and velocity is P_{st} = Ne - βF, where F is the appropriate free energy. We choose a twist geometry in which the nematic sample is contained between two plates perpendicular to the z-axiz separated a distance d. The director $\vec{n}(\vec{r})$ is initially aligned along the x axis and the magnetic fied is directed along the y axis. We assume homogeneity in the y direction. We also assume that macroscopic flow only exists in the y direction $V_y(x,z)$. Setting $n_x(x,z) = \cos\phi(x,z)$ and $n_y(x,z) = \sin\phi(x,z)$, the general stochastic nematodynamic equations become in this geometry and in a minimal coupling approximation[12].

$$d_t \begin{pmatrix} \phi \\ V_y \end{pmatrix} = \begin{pmatrix} -\frac{1}{\gamma_1} & \frac{1}{2\rho}(1+\lambda)\partial_x \\ \frac{1}{2\rho}(1+\lambda)\partial_x & -\nu_2 \partial_z^2 \nu_3 \partial_x^2 \end{pmatrix} \begin{pmatrix} \frac{\delta F}{\delta \phi} \\ \frac{\delta F}{\delta V_y} \end{pmatrix} + \begin{pmatrix} \xi \\ \partial_x \Omega_{yx} \partial_z \Omega_{yz} \end{pmatrix}$$

(1)

ρ is the mass-density, $\lambda = -\gamma_2/\gamma_1$, and γ_1, γ_2, ν_2 and ν are viscosity coefficients[1]. The free energy F is given by[1]

$$F = \int d\vec{r} \left\{ \frac{K_1}{2}(\vec{\nabla}\cdot\vec{n})^2 + \frac{K_2}{2}(\vec{n}\cdot\vec{\nabla}\times\vec{n})^2 \frac{K_3}{2}(\vec{n}\times(\vec{\nabla}\times\vec{n}))^2 \right\} - \int \frac{1}{2}\chi_a(\vec{n}\cdot\vec{H})^2 + \frac{1}{2}\rho \int d\vec{r} V^2$$

(2)

The stochastic forces λ, Ω_{yx} and Ω_{yz} are Gaussian white noise and satisfy the following fluctuation-dissipation relations

$$<\xi(\vec{r},t)\xi(\vec{r}',t')> = \frac{2 K_B T}{\gamma L} \delta(x-x') \delta(z-z') \delta(t-t')$$ (3)

$$<\Omega_{yx}(\vec{r},t)\Omega_{yx}(\vec{r}',t')> = 2\frac{K_B T}{L}\nu_3 \delta(x-x') \delta(z-z') \delta(t-t')$$ (4)

$$<\Omega_{yx}(\vec{r},t)\Omega_{yx}(\vec{r}',t')> = 2\frac{K_B T}{L}\nu_2 \delta(x-x') \delta(z-z') \delta(t-t')$$ (5)

where L is the typical sample dimension in the y direction.

The diagonal terms in the matrix in (1) give a self-adjoint operator which implies a purely dissipative dynamics. The equation coming from diagonal terms are decoupled. The equation for ϕ is the one studied in the absence of hydrodynamic coupling[9-11] and the equation for V_y is nothing but the Navier-Stokes equation. The coupling between ϕ and V_y wich is responsible for pattern formation is introduced by the nondiagonal terms in (1). These nondiagonal terms form and anti-adjoint operator implying non-dissipative dynamics.

The Fourier modes $\theta_{m,q_x}(t)$ of the director angle ϕ are defined by

$$\phi(x,z) = \sum_m \sum_{q_x} \theta_{m,q_x}(t) \cos(2m+1)\frac{\pi z}{d} e^{iq_x x} \tag{6}$$

Neglecting inertia ($\rho d_t V_y = 0$) and in a linearized approximation one can obtain from (1)-(5) the following equation for $\theta_{m,q_x}(t)$

$$\partial_t \theta_{m,q_x}(t) = \frac{1}{\bar{\gamma}}(\chi_a H^2 - K_2(2m+1)^2 \frac{\pi^2}{d^2} - K_3 q_x^2)\theta_{mq_x}(t) + W_{m,q_x}(t) \tag{7}$$

$\bar{\gamma}_1$, is an effective viscosity

$$\bar{\gamma}_1 = \gamma_1 - \frac{\alpha_2}{\eta_c + \eta_a Q^{-2}} \quad ; \quad Q = q_x/(2m+1)\pi/d \tag{8}$$

Where α_2, η_c and η_a are Meiscowitz coefficients[1]. In this approximation the whole effect of the hydrodynamic coupling is included in the effective viscosity $\bar{\gamma}_1$. The consistency of the approximation leading to (7) is seen in the fact that for the random force $W_{m,q_x}(t)$ one obtains a fluctuation-dissipation relation involving $\bar{\gamma}_1$.

$$\langle W_{m,q_x}(t) W^*_{n,q'_x}(t')\rangle = 2\frac{2K_B T}{\bar{\gamma}_1 V}\delta_{m,n}\delta_{q_x,q'_x}\delta(t-t') \tag{9}$$

STRUCTURE FACTOR

The transient dynamics of the system when the magnetic field is suitched on can be described solving (7) for the structure factor $C_{q_x,m}(t) = \theta_{m,q_x}(t)\theta_{m,-q_x}(t)$. The range of stability of the different modes is the same than in the absence of hydrodynamic coupling[11]: A mode m becomes unstable for $H > (K_2(2m+1)^2\pi^2/\chi_a d^2)^{\frac{1}{2}}$, so that only $m = 0$ is unstable for $H < 3H_c$, $H_c = (K_2\pi^2/\chi_a d^2)^{\frac{1}{2}}$. For a given unstable mode m there is a range of unstable q_x modes given by $\chi_a H^2 - K_2((2m+1)\pi/d)^2 - K_3 q_x^2 > 0$. Also as in the case without hydrodynamic coupling the most unstable mode is m = 0. However, the wave number dependence of the viscosity implies now that the q_x mode of fastest growth is different from zero when

$$\left(\frac{H}{H_c}\right)^2 > 1 + \frac{K_3 \eta_a \gamma_1}{K_2 \alpha_2^2} \tag{10}$$

In these conditions the system does not respond homogeneously byt a pattern appears. A detailed description of the emergence of the pattern is given by the time evolution of the structure factor shown in Fig.1. The initial condition is taken as the solution of (7) for the equilibrium structure factor for an initial field $H_i < H_c$. The structure factor,

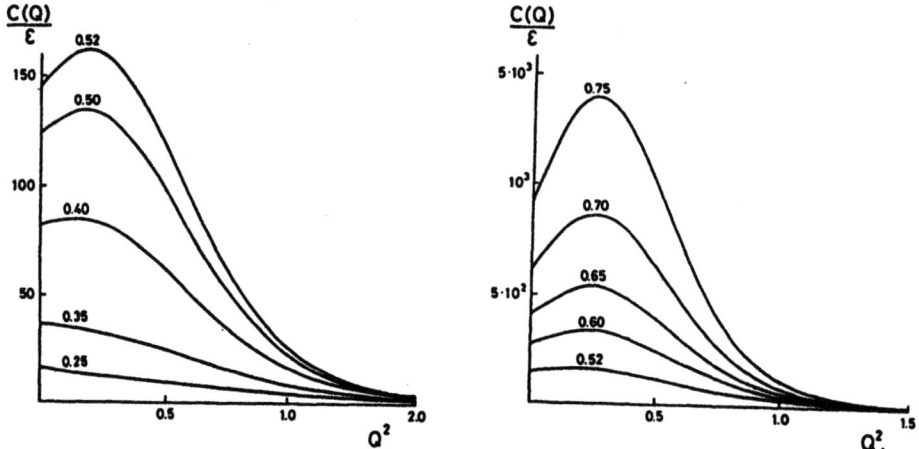

Fig.1. Structure factor vs wavenumber at different times. Parameter values correspond to MBBA at room temperature. $(H_i/H_c)^2=0.5$, $(H/H_c)^2= 5, d=10^{-2}$ cm, $V=10^{-2}$ cm^3, $\varepsilon=2k_B$ T/Vk $2(\pi/d)^2$. Time is measured in units of $\tau_0 = \gamma_1 / \chi_a H_c^2$ about 10 sec.

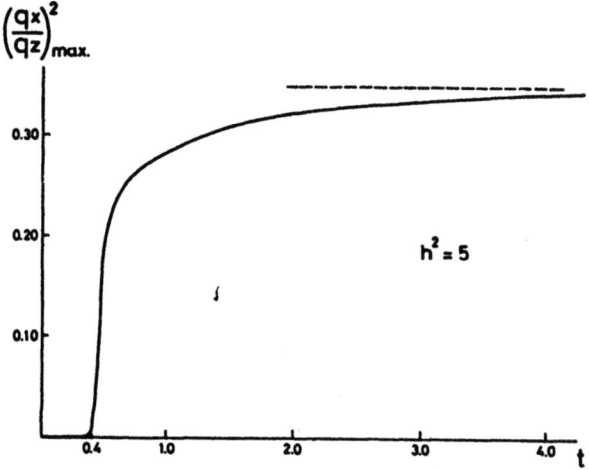

Fig.2. Maximum of the structure factor depicted in Fig.1. vs time. (time scale as in Fig.1.)

initially peaked at $Q = 0$, develops a well defined maximum as time goes on. The wavenumber Q_{max} for which the maximum occurs tends asymptotically to the mode of fastest growth characterizing the pattern periodicity in earlier theories [2-5]. This is more clearly seen in Fig.2 where the time dependence of Q_{max} is shown. It shows a well defined time $t \simeq 0.4 \simeq 4$ sec. at which Q_{max} becomes sharply nonzero. This time is associated with the appearance of the pattern. Fig.2 also shows a subsequent stage of evolution associated with the formation of the pattern. In this stage Q_{max} grows smoothly during a time which is an order of magnitude larger than the time of appearance of the pattern. The mode of fastest growth is only reached at long times which are beyond the range of validity of the theory estimated by a MFPT calculation to be of the order of $t \simeq 2.5 \simeq 25$ sec. The reduction of the MFPT by an order of magnitude with respect to the case without hydrodynamic coupling is mainly due to the larger applied magnetic field.

ACKNOWLEDGMENTS

Partial financial support from Comisión Asesora de Investigación Científica y Técnica (CAYCIT, Spain) and from the US-Spanish Cooperative Research Program is acknowledged.

REFERENCES

1. P. G. de Gennes, "The Physics of Liquid Crystals," Clarendon Press, Oxford (1975).
2. E. Guyon, R. Meyer and J. Salán, Mol. Cryst. Liq. Cryst. 54:261 (1979).
3. F. Longerg, S. Fraden, A.J. Hurd and R.B. Meyer, Phys. Rev. lett. 52: 1903 (1984).
4. Y. W. Hui, M.R. Kuzma, M. San Miguel, M.M. Labes, J.Chem.Phys. 83:288 (1985).
5. A. J. Hurd, S. Fraden, F. Longerg and R.B. Meyer, J. Physique 46:905 (1985).
6. M. R. Kuzma, Phys. Rev. Lett., 57: 349 (1986).
7. J. D. Gunton, M. San Miguel and P.S. Sahni in: "Phase Transitions and Critical Phenomena", vol. 8, ed. C. Domb and J.L. Lebowitz, Academic Press, N.Y. (1983).
8. M. San Miguel in: "Stochastic Processes Applied to Physics", ed. L. Pesquera and M.A. Rodriguez, World Scientific, Singapore (1985).
9. M. San Miguel and F. Sagués in: "Recent Developments in Nonequilibrium Thermodynamics: Fluids and Related Topics", Lecture Notes in Physics vol. 253, ed. J. Casas, D. Jou and M. Rubí, Springer, Berlin (1986).
10. F. Sagués and M. San Miguel, Phys. Rev. A 33:2769 (1986).
11. F. Sagués and M. San Miguel, Phys. Rev. A 32:1843 (1985). (see the appendix).
12. M. San Miguel and F. Sagués, preprint.

PRODUCTION OF OZONE BY HIGH FREQUENCY SURFACE DISCHARGE ON A HIGH PURITY

ALUMINA CERAMIC DIELECTRICS

Senichi Masuda

Department of Electrical Engineering
Faculty of Engineering, University of Tokyo
7-3-1, Hongo, Bunkyo-ku, Tokyo, Japan 113

INTRODUCTION

Ozone has a very broad potential area of applications because of its inherent large advantages: a very strong oxidizing capacity next to fluorine, no hazardous compounds to be left after reaction, etc. But, there are three problems to be solved before general use of ozone: a too low energy efficiency of ozone generation (60 gO_3/kWh for air; 120 gO_3/kWh for oxygen compared to the theoretical value 1200 gO_3/kWh) making ozone excessively expensive, too low volume yield of ozonizers (ca. 0.4 gO_3/1h) making them too large in size, and difficulty of using a high frequency voltage for driving ozonizers (below 1000 Hz) that makes the power supply excessively large in size. The latter two probles are especially important when medium and small sized applications of ozone are considered. These problems could be solved by using high frequency surface discharge in combination with high purity alumina ceramic (92 %) for ozone generation.

CONSTRUCTION OF NOVEL OZONIZER

Figures 1 and 2 shows schematically the construction of the present ozonizer consisting of a ceramic cylinder. A number of parallel discharge electrodes(tungsten) are attached on its inner surface, and facing to a film like induction electrode (tungsten) across a high purity alumina dielectrics layer with 0.5 mm thickness. A high frequency high voltage is applied therebetween to produce surface corona discharges from the discharge electrodes to extend along the inner surface. Either dry air or oxygen is introduced into inside of the cylinder and flows tangentially in a gap between the cylinder and the glass tube inside. Ozone is produced by the plasma chemical process taking place inside the surface discharge. The outer surface of the ceramic cylinder is colled by water so as to avoid an excessive temperature rise of the reaction zone and keep the outlet gas temperature below a critical value (52 °C) beyond which ozone generation is detriorated by thermal decomposition.

Unique features of this ozonizer are: (1) use of a thin alumina dialectrics between the electrodes which allows the use of a substantially lower voltage (5 - 10 kV_{pp}), (2) use of alumina cylinder also as the construction member which allows a substantially increased heat conduction from the reaction zone and the operation at an increased gas pressure up to 6 kg/cm^2,

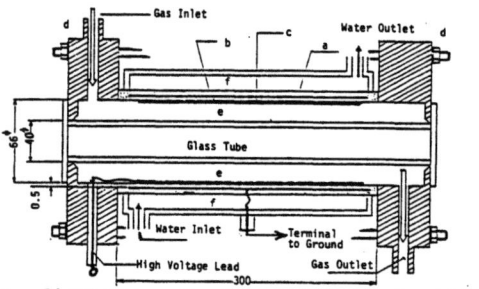
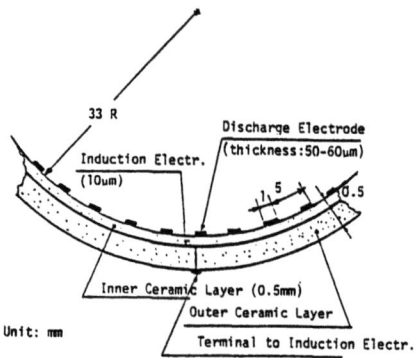

a: discharge electrodes, b: inducttion electrode, c: inner ceramic layer, d: insulator heads, e: gas passage way, f: cooling water jacket

Figure 1 Cross-Sectional View of Ceramic Ozonizer

Figure 2 Longitudinal View of Ceramic Ozonizer

(3) use of surface discharge, easy to cool and having a free space above in which ozone is not decomposed by discharge, and (4) use of high frequecy, which is possible by (2) and (3), enabling both ozonizer and its power supply very small in size. All these features contribute to opening the way of using ozone in many medium and small sized applications.

EXPERIMENTAL RESULTS OF OZONE GENERATION

The author confirmed by experiments that ozone generation rate rises proportinally with increasing frequency and voltage provided the concurrent increase in heat generation be removed by sufficient heat conduction so as to keep the gas temperature in the reaction zone constant. However, there is a limit in cooling ability and a critical temperature is reached beyond which the ozone generation drops darstically. Figure 3 shows one of such examples, indicating the ozone concentration, C (vol. ppm), to take maximum at a certain voltage depending upon the frequency. The conventional ozonizer using the silent glow discharge in a gas space between two parallel electrodes with an insulating sheet inserted has an inherent difficulty in its construction to enhance cooling of reaction zone.

Figure 4 indicates the ozone generation rate, G (g/h), of the present ozonizer being originally independent of gas flow rate up to a certain voltage level, beyond which the effect of excessive temperature rise appears. Of course, the heating is resulted by gaseous discharge and dielectric loss of the insulating layer. The higher gas flow rate produces a higher peak of G at a higher voltage V_p because of its cooling effect.

Figure 5 indicates the ozone concentration, C, being originally proportional to V_p, as described above, provided the temperature rise is hampered by using an increased gas pressure and thereby hampering heat generation. This unique effect of increased gas pressure appears only in the present ozonizer, but not in the conventional type ozonizer.

The performance of an ozonizer is to be expressed in terms of ozone concentration, C, ozone generation rate, G, and energy yield of ozone generation, Y_o (gO_3/kWh), plotted in a three coordinate sysyem. Figures 6 and 7 illustrates the typical performance of the present ozonizer, which show a very broad varion range in C, G, and Y_o, depending upon the selection of operating parameters: voltage, V_p; frequency, f; gas pressure, P_g; gas flow rate Q; gas and water temperatures at the inlet, $T_g(IN)$ and $T_w(IN)$. For the purpose of comparison the same data of a conventional type ozonizer having the same outer cylinder dimension with 1.5 mm electrode gap and 0.5 mm die-

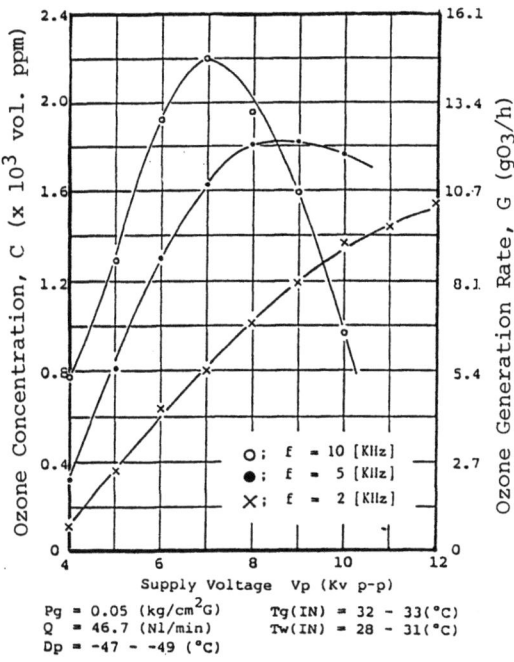

(gas: dry air)

Figure 3 Ozone Concentration and Ozone Generation Rate as a Function of Voltage (f varied)

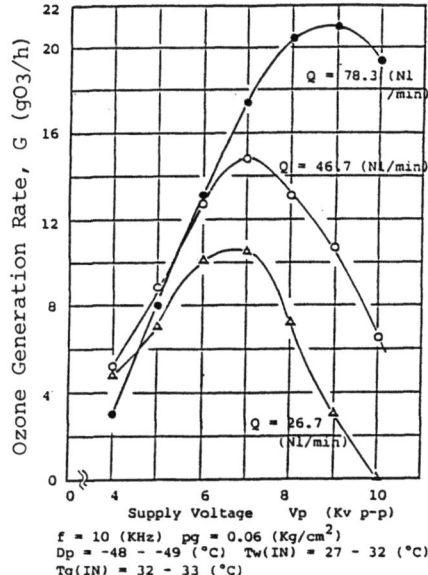

(gas: dry air)

Figure 4 Ozone Generation Rate as a Fuction of Voltage (Q varied)

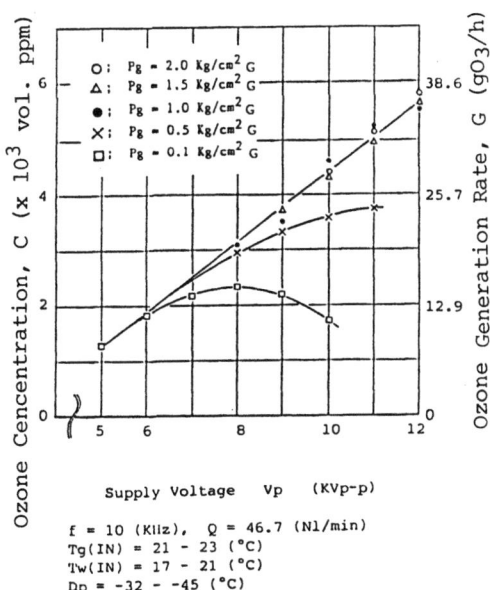

(gas: dry air)

Figure 5 Ozone Concentration and Generation Rate as a Function of Vp (Pg varied)

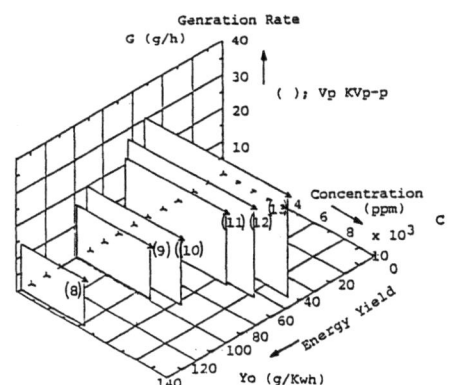

(gas: dry air)

Figure 6 Ozone Generation Performance of Ceramic Ozonizer - (I)

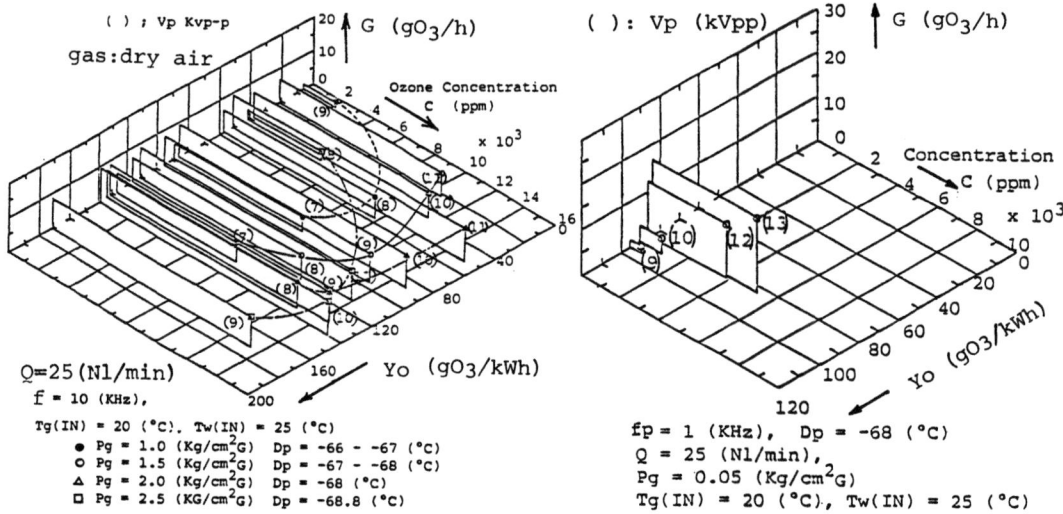

Figure 7 Ozone Generation Performance of Ceramic Ozonizer - (II)

Figure 8 Ozone Generation Performance of Conventional Ozonizer

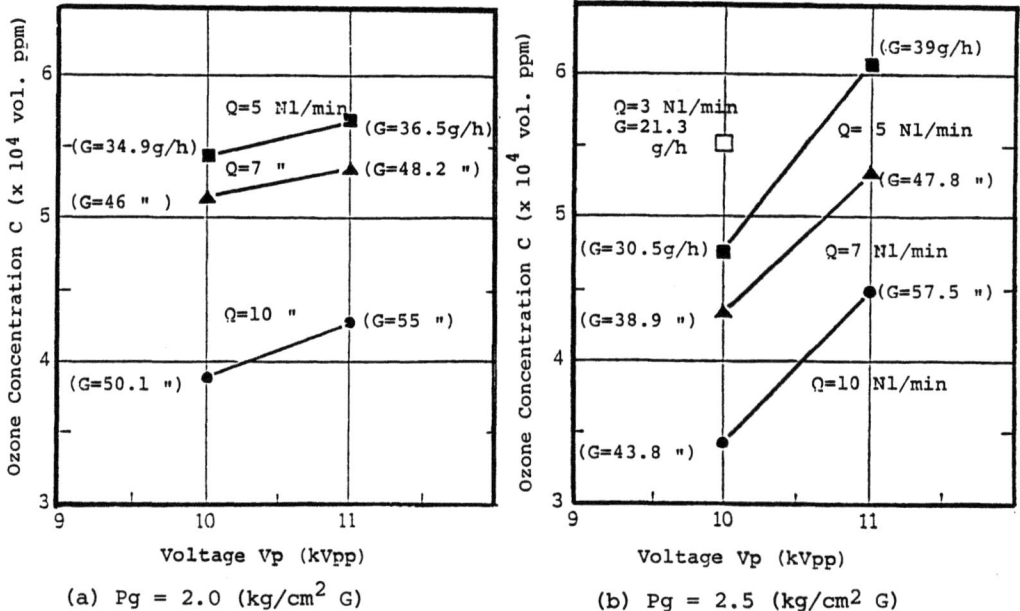

(a) $Pg = 2.0$ (kg/cm^2 G)

(b) $Pg = 2.5$ (kg/cm^2 G)

Figure 9 Ozone Concentration as a Function of Voltage for Oxygen Input

lectric layer thickness. Thus, it is evident that the present ozonizer provides a wide freedom of maximizing either one or two of G, C, and Yo by selecting proper operating parameters so as to meet with indivisual application requirements.

Figure 9 shows examples of the ozone generating performance of the present ozonizer in the case when oxygen is used at an increased pressure. A very high ozone concentration, beyond 50,000 (vol. ppm) or even up to 100,000 (vol. ppm) can be easily obtained by the present ozonizer. This also is a unique feature of the present ozonizer, enabling its application in special applications where a very high ozone concentration is needed, such as ozone ashing of photo-resist film in LSI fabrication and an instantaneous disinfection of containers and bottles in pharnaceutical and medical industries.

CONCLUSIONS

The use of high frequency surface discharge in combination with the 92 % high purity alumina ceramic dielectrics as an inter-electrode sheet and construction material provide a number of unique features that enables the ozone applications in madium and small scale areas. The conclusions derived so far are as follows:

(1) Use of high frequency, up to say 10 kHz, is possible, leading to a substantial decrease in size of the ozonizer itself as well as its power supply.

(2) A very wide freedom exists in maximizing either one or two of the performance variables, C, G, and Yo, by a proper selection of operating parameters. As a result the energy efficiency as high as $Y_o = 170$ (gO_3/kWh) can be achieved with the cost of decreased G value.

(3) Ozone concentration and ozone generation rate rise linearly with increasing frequency or voltage, independent of any other parameters such as gas flow rate, gas pressure, etc., so far as the temperature of the reaction zone be kept unaltered by sufficient cooling. However, the ozone concentration and generation rate reach maximum at certain levels of the frequency and voltage at which the temperature exceeds a critical level (ca. 52 °C) to cause thermal decomposition of ozone. Increase in gas flow rate produces a higher magnitude of maximum to occur because of its increased cooling effect. The increased gas pressure hampers the generation of heat by loss. This pressure effect does not appear in the conventional ozonizers.

(4) An excessive increase in gas pressure, say beyond 2 (kg/cm^2 G) detriorates the surface discharge and thereby hampers the ozone generating performance.

(5) By using oxygen at an increased pressure and with using water-cooling an ozone concentration as high as 50,000 to 100,000 (vol. ppm) can be easily achieved.

The surface discharge of the present ozonizer produces an instantaneous jump in its inter-electrode capacity in each half cycle of the high frequency to produce a "jumping phenomenon" of current beyond a certain threshold voltage, a phenomenon specific to "non-linear circuit". A special power supply has been developed to overcome this difficulty. As a result a broad use of this unique ozonizer is being expected.

REFERENCES

1. Masuda, S. and Kiss, E., Ceramic-Made Electric Curtain Devices and Their Applications, Invited Talk at Int. Conf. on Industrial Electrostatics (May 17-18, 1984 in Budapest, Hungary).
2. Masuda, S. and Kiss, E., Investigation on Ceramic-Made Ozonizer of High Frequency Surface Discharge Type, Proc. 1983 Ann. Conf. Inst. Electrostatics Japan (Japanese; Oct. 1983 in Fukuoka).
3. Masuda, S., Akutsu, K., Kuroda, M., Awatsu, Y. and Shibuya, Y., A Ceramic-Based Ozonizer Using High Frequency Surface Discharge, Record of IEEE/IAS 1985 Annual Conference, pp. 1353-1358 (Oct., 1985 in Tronto, Canada)

NATO-ASI & EPS CONFERENCE-JULY 1-15,1986-LA RABIDA(HUELVA)SPAIN-PARTICIPANTS

PARTICIPANTS

ACRIVOS, PROF. A.	Dpt. Chemical Engineering, Stanford University, Stanford, California 94305 - U.S.A.
ADLER, DR. M. M.	Lab. D'Aerothermique-CNRS, 4 Route Des Gardes, 92190 Meudon - France.
AKDENIZ, DR. Z.	Itu Fen-Edebiyat Fak./Muhen., Bilim. Bol., Itu Kampusu, Maslak-Istambul - Turkey.
ALMEIDA, DR. B.	Centro Quimica Estructural, Instituto Sup. Tecnico, 1096 Lisboa Codex - Portugal.
ALS-NIELSEN, DR. J.	Risø National Laboratory, DK-4000 Roskilde-Denmark.
ALTS, PROF. T.	I. T. Kondensierten Materie, Freie Univ. Berlin19, Arnimallee 14 - R. R. Germany.
ANTORANZ, DR. J. C.	Dpto. Fisica Fundamental Uned, Apdo. Correos 60.141, 28.071, Madrid - Spain.
AVNIR, PROF. D.	Dept. Organic Chemistry, Hebrew University, Jerusalem 91904 - Israel.
BALL, MR. P. C.	H. H. Wills Phys. Lab., Bristol University, Bristol BS8 1TL - U. K.
BARRERO, PROF. A.	E.S.I. Industriales, Mecanica Fluidos, Avda. Reina Mercedes S/N, 41012 Sevilla -Spain.
BARTHES-BIESEL, PROF. D.	Dep. Genie Biologique, Univ. Tech. Compiegne, 60206 Compiegne Cedex - France.
BAUS, DR. M.	Chimie Physique II, ULB, Campus Plaine - Boulevard Triomphe, 1050 Brussels - Belgium.
BEN AMAR, DR. M.	Phys. Solides-ENS, 24 Rue Lhomond, 75231 Paris Cedex 05 - France.
BOCHACHEVSKY, DR. I.O.	Los Alamos Nat. Lab., P.O. Box 1663 - M.S. F600, Los Alamos, NM 87545 - U.S.A.
BONTOUX, DR. P.	I.M.F.M., 1, Rue Honnorat, 13003 Marseille - France.

BOUDOUSSIER, MR. M.	Lab. Phys. Matiere Condensee, 11, Place Marcelin-Berthelot, 75231 Paris Cedex 05 - France.
BRENNER, PROF. H.	Dept. Chemical Engineering, MIT Room 66-566, Cambridge, Massachusetts 02139 - U.S.A.
BRUNO, DR. E.	Ist. Fisica Teorica, Univ. Di Messina, 98100 Messina - Italy.
CAGINALP, PROF. G.	Dept. Mathematics, Pittsburgh Univ., Pittsburgh, PA 15260 - U.S.A.
CASTELLANOS, PROF. A.	Dpto. Electricidad - F. Fisica, Univ. Sevilla, Avda. Teina Mercedes S/N, 41012 Sevilla - Spain.
CASTILLO, DR. J. L.	Dpto. Fisica Fundamental Uned, Apdo. Correos 60.141, 28.071 Madrid - Spain.
CERISIER, PROF. P.	Systemes Energetiques, University of Provence - Saint Jerome, 13397 Marseille Cedex 13 - France.
CHACON, DR. E.	Fac. Ciencias C-XII, Univ. Autonoma Madrid, Cantoblanco (Madrid) - Spain
CHANG, PROF. J. -S.	Dept. Engineering Physics, McMadter University, Hamilton Ontario - Canada L8S 4M1
CHU, MR. X. -L.	Physics Dept., Hust-Huazhong Univ. Science and Tech., Wuhan Hubei - The People's Republic of China.
CILIBERTO, DR. S.	Istituto Nazionale Di Ottica, Largo E. Fermi, 6, 50125 Firenze - Italy.
CLAVIN, PROF. P.	Dept. Combustion, Universite Provence, Saint Jerome, 13397 Marseille Cedex 13 - France.
CONDE, MR. L.	Instituto Plasticos - CSIC, C/Juan De La Cierva, S/N, 28006 Madrid - Spain.
CORIELL, DR. S. R.	Metallurgy Division A153, 233, National Bureau of Standards, Gaithersburg, MD 20899 - U.S.A.
COURVILLE, MR. M. P.	Lab. Reactivite Solides, Faculte Des Sciences Mirande, BP-138, 21004 Dijon Cedex - France.
CRESPO, DR. E.	Depto. Fisica Fundamental Uned, Apdo. Correos 60141, 28.071 Madrid - Spain.
CUADROS-BLAZQUEZ, DR.E.	Fac Ciencias, Dpto. Termodinamica, Universidad de Extremadura, 06071 Badajoz - Spain.
DAVIS, PROF. S. H.	Dept. Engineering Sciences, Northwestern University, Evanston, Illinois 60201 - U.S.A.

DE BRUIJN, IR. R. A.	Unilever Research Laboratory, P.O. Box 114, 3130 Ac Vlaardingen - The Netherlands.
DE CHEVEIGNE, DR. S.	E.N.S., Groupe Physique Des Solides, 2, Place Jussieu-Tour 23, 75251 Paris Cedex 05 - France.
DE LA TORRE, MR. M.	Depto. Fisica Fundamental Uned., Apdo. Correos 60141, 28.071 Madrid - Spain.
DIEZ-SANZ, MR. F.	Depto. Quimica Tecnica, Univ. Oviedo, Oviedo - Spain.
DIJHSTRA, IR. H. A.	Mathematical Inst. R. U. G., P.O. Box 800, 9700 Av Groningen - The Netherlands.
DOU, MR. R.	Physics Department, University of Oslo, Oslo - Norway.
EVANS, PROF. R.	H.H. Wills Physics Lab., University of Bristol, Bristol, BS8 1TL - U.K.
FAVIER, PROF. J. J.	C.E.N. Grenoble, Bab. Etudes Solidification, 85 X, 38041 Grenoble Cedex - France.
FERNANDEZ-CRUZ, MR. R.	Dpto. Fisica Fundamental Uned, Apado. Correos 60.141, 28.071 Madrid - Spain.
FERNANDEZ-SAN JUAN, MR. M. A.	E.T.S. Arquitectura, Ciudad Universitaria, 28040, Madrid - Spain.
FIFE, PROF. P. C.	Dept. of Mathematics, Building 89, University of Arizona, Tucson, Arizona 85721 - U.S.A.
FRANKOWICZ, DR. M.	Univ. Jagellonian, Fac. Chemistry, UL. Karasia 3, 30060 Krakow - Poland.
GALLEGOS- MONTES, DR. C.	Dpto. Quimica Tecnic, Univ. Sevilla, C/ Tramontana S/N 41012 Sevilla - Spain.
GAMBI, DR. C. M. C.	Dpto. Fisica - Univ. Firenze, Largo E. Fermi, 2, 50125 Firenze - Italy.
GAÑAN-CALVO, MR. A. M.	E.S.I. Industriales, Mecanica Fluidos, Avda. Reina Mercedes S/N, 41012 Sevilla - Spain.
GARCIA-SANZ, DR. J.	Dpto. Fisica Fundamental Uned, Apdo. Correos 60141, 28.071 Madrid - Spain.
GARCIA-YBARRA, DR. P.L.	Dpto. Fisica Fundamental Uned, Apdo. Correos 60141, 28.071 Madrid - Spain.
GLEASON, DR. K. J.	Dept. of Chemical Engineering, University of Colorado, Campus Box 424, Boulder, Colorado 80309-0424 - U.S.A.
GOLDENFELD, PROF. N.	Physics Dept., Univ. of Illinois, 1110 West Green St., Urbana, Illinois 61801 - U.S.A.
GOLIA, DR. C.	Istituto U. Nobile, Univ. Napoli, Ple. V. Tecchio, 8, 80125 Napoli - Italy.
GOUIN, PROF. H.	Fac. Sciences, Univ. Aix-Marseille, Rue Henri Poincare, 13397 Marseille Cedex 13 - France.

GRAUER, MR. T.	I. Theoretische Physik I, Pfaffenwaldring 57, 7000 Suttgart 80 - F. R. Germany.
GUERRERO-CONEJO, Mr. A. F.	Dpto. Quimica Tecnica, Univ. Sevilla, Tramontana S/N, 41012, Sevilla - Spain.
GUTHMANN, PROF. C.	E.N.S., Groupe Physique Des Solides, 2, Place Jussieu-Tour 23, 75251 Paris Cedex 05 - France.
HALPERIN, DR. A.	Dpt. Phys. Chemistry - MDRC, Hebrew Univ. Jerusalem, 91904, Jerusalem - Israel.
HAUGE, PROF. E. H.	Institut for Teoretisk Pysikk, Universitetet I Trondheim, 7034 Trondheim - Norway.
HERNANDEZ, MR. E.	Dpto. Fisica Teorica, Univ. Central Barcelona, Diagonal 647, 08028 Barcelona - Spain.
HIGUERA, DR. F.	E.T.S.I. Aeronauticos, Ciudad Universitaria, 28040 Madrid - Spain.
HOLLY, PROF. F. J.	Dry Eye Institute, P.O. Box 98069, Lubbock, Texas 79499 - U.S.A.
IVANOV, PROF. I. B.	Faculty of Chemistry, University of Sofia, 1126 Sofia - Bulgaria.
JIMENEZ-FERNANDEZ, DR. J.	E.T.S. Arquitectura, Ciudad Universitaria, 28040 Madrid - Spain.
KAGAN, MR. M.	Dpt. Organic Chemistry Hebrew Univ. Jerusalem, Jerusalem 91904 - Israel.
KASSEMI, DR. S. A.	Case Western Reserve Univ., Cleveland, Ohio 44106 - U.S.A.
KOSCHMIEDER, PROF. E. L.	Atmospheric Sciences Group, University of Texas, Austin, Texas 78712 - U.S.A.
LANGEVIN, DR. D.	Lab. Spectro. Hertzienne Ens, 24, Rue Lhomond, 75231 Paris Cedex 05 - France.
LEBON, PROF. G.	Dept. Mecanique Thermo., Univ. Liege, B5 Sart Tilman, 4000 Liege - Belgium.
LEGER, DR. L.	Phys. Mateire Condensee, College De France, 75231 France Cedex 05 - France
LEGROS, PROF. J. C.	Chemical Physics Dept. CP. 165, Univ. Libre Bruxelles, 50 Ave, F. D. Roosevelt, 1050 Bruxelles - Belgium.
LETAMENDIA, DR. L.	Centre De Physique Moleculaire Optique et Hertzienne, Univ. De Bordeaux I, 33405 Talence - France.
LEVICH, PROF. B. G.	Institute Applied Chemical Physics, City College - Steinman 202, Convent Ave, 138 St., New York, N.Y. 10031 - U.S.A.
LIMBOURG, DR. M. C.	Chemical Physics Dept. CP 165, Univ. Libre Bruxelles, 50 Ave F. D. Roosevelt, 1050 Bruxelles - Belgium.

LIÑAN, PROF. A.	E.T.S.I. Aeronauticos, Ciudad Universitaria, 28.040 Madrid - Spain.
MARTINEZ-HERRANZ, DR. I.	E.T.S. Ingenieros Aeronauticos, Ciudad Universitaria, 28.040 Madrid - Spain.
MASLANIK, MS. M. K.	Colorado Univ., Boulder, Cires - Campus Box 449, Boulder, Colorado 80309 - U.S.A.
MASUDA, PROF. S.	Dept. Electrical Engineering, University of Tokyo, 7-3-1-Hongo, Bunkyo--ku, Tokyo 113 - Japan.
MATSUI, PROF. G.	Institute of Engineering Mechanics, University of Tsukuba, Sakura, Ibaraki 305 - Japan.
McCLUSKEY, DR. F.	L.E.M.S. - C.N.R.S., B. P. 166 X, 38042 Grenoble Cedex - France.
McGIVERN, MR. R. C.	Dpt. Pure Applied Physics, Queen University of Belfast BT7 1NN - U.K.
MESSIER, MS. A. M.	Lab. Phys. Matiere Condensee - C.F., 11, Place M. Bethelot, 75231 Paris Cedex 05 - France.
MULLER, DR. S. C.	Max Planck I, Ernahrungsphysiologie, Rheinlanddamm 201, 46 00 Dortmund 1 - F. R. Germany.
MULLER, PROF. U.	Kernforschungszentrum Karlsruhe, Postfach 36400 - Ins. Reaktor, 7500 Karlsruhe 1 - F. R. Germany.
MUÑOZ-GARCIA, MR. J.	Dpto. Quimica Tecnica, Univ. Sevilla, Tramontana S/N, 41012 Sevilla - Spain.
NICHOLS, PROF. B.	Los Alamos Nat. Lab., CNLS, M. S. B-258, Los Alamos, New Mexico 87545 - U.S.A.
OSTRACH, PROF. S.	Case Institute Technology, Western Reserve Univ., Cleveland, Ohio 44106 - U.S.A.
PAREDES-GOMEZ, MR. V.	Dpto. Termodinamica, Facultad De Ciencias, Universidad De Extremadura, 06071 Badajoz - Spain.
PEREZ, MR. A. T.	Dpto. Electricidad - F. Fisica, Univ. Sevilla, Avda. Reina Mercedes, 41012 Sevilla - Spain.
PEREZ-VILLAR, PROF. V.	Facultad Ciencias Fisicas, Univ. Santiago Avda. Ciencia, Santiago De Compostela - Spain.
PERROT, DR. F.	Resonance Magnetique, Cen Saclay, 91191 Gif Sur Yvette Cedex B.P. 2 - France.
PLATTEN, PROF. J. K.	Chimie Phys. Thermodynamique, Universite Mons, Av. Maistriau, 21, 7000 Mons - Belgium.

PLOOG, PROF. K.	Max Planck Institutt, Heisenbergstrasse 1, Postfach 800665, 7000 Stuttgart 80 - F. R. Germany.
POMEAU, DR. Y.	Cen Saclay, Orme Des Merisiers, 91190 Gif-Sur-Yvette Cedex B.P. 2 - France.
REBELO, DR. M. J. R.	Cecul, R. Escola Politecnica 58, 1200 Lisboa - Portugal.
RIVAS, DR. D.	Dpt. Mech. & Aerospace Eng., Case Inst. Tech., Cleveland, Ohio 44106 - U.S.A.
RODRIGUEZ-LUIS, MR. A. J.	Dpt. Electricidad, Fac. Fisica, Univ. Sevilla Avda. Reina Mercedes 41012 Sevilla - Spain.
RODRIGUEZ, DR. J. R.	Dpto. Termologia, Fac. Fisica Univ. Santiago, Avda. Ciencia, Santiago De Compostela - Spain.
RONDELEZ, DR. F.	Dowell Schlumberger B. P. 90, 42003 Saint Etienne Cedex 1 - France.
ROUX, DR. B.	I.M.F.M., 1, Rue Honnorat, 13003 Marseille - France.
RUBIO, DR. M. A.	E.T.S. Arquitectura, Ciudad Universitaria, 28040 Madrid - Spain.
RULL, PROF. L. F.	Dpto. Electricidad, Fac. Fisica, Univ. Sevilla, Avda. Reina Mercedes S/N, 41012 Sevilla - Spain.
SANFELD, PROF. A.	Chimie Physique II - ULB, Campus Plaine - Boulevard, Triomphe (CP 231), 1050 Brussels - Belgium.
SANI, PROF. R. L.	C.I.R.E.S., Campus Box 449, University of Colorado, Boulder, Colorado 80309-0449 - U.S.A.
SARMA, DR. G. S. R.	DFVLR-Inst. Theoret., Stromungsmechanik, Bunsenstrasse 10, 3400 Gottingen - F. R. Germany.
SAVILLE, PROF. D. A.	Dpt. Chemical Eng., Princeton Univ., Princeton, New Jersey 08540 - U.S.A.
SCHLICK, PROF. S.	Chemistry Dept., Detroit University, 401 W. McNichols Road, Detroit, Michigan 48221 - U.S.A.
SCHMITZ, MR. G.	Giesserei-Institutt, Rhein.-Westf. Tech. Hochschule, Intzestrasse 5, 5100 Aachen - F. R. Germany.
SCHWABE, PROF. D.	I. Physikalisches Institutt, Heinrich-Buff-Ring 16, 6300 Giessen - F. R. Germany.
SEKERKA, PROF. R. F.	College Science, Carnegie-Mellon Univ., 115 Scaife Hall, Pittsburgh, Pennsylvania 15213 - U.S.A.

SHAIN, MR. P.	Dept. Chemical Eng., California Univ. Berkeley, Berkeley, California 94720 - U.S.A.
STALIDIS, PROF. G. A.	Lab. Gen. Inorg. Chem. Tech. (114), 54006 Thessaloniki - Greece.
STEEN, DR. P.	School Chemical Eng., Olin Hall, Cornell University, Ithaca, NY 14853 - U.S.A.
STEINCHEN, DR. A.	Chimie Physique II-ULB, Campus Plaine, Boul Triomphe (CP. 231), 1050 Brussels - Belgium.
SULLIVAN, DR. D. E.	Dept. of Physics, University of Guelph, Guelph, Ontario N1G 2W1 - Canada.
SZLEIFER, MR. I.	Dept. Physical Chemistry, Hebrew Univ. Jerusalem. Givat Ram, Jerusalem 91904 - Israel.
TABELING, DR. P.	Group Phys. Solides - ENS, 24 Rue Lhomond, 75005 Paris - France.
TARAZONA, DR. P.	Fac. Ciencias C XII, Univ. Autonoma Madrid, Cantoblanco (Madrid) - Spain.
TEJERO, DR. C. F.	Dpto. Termologia - F. Fisicas, Univ. Complutense Madrid, 28040 Madrid - Spain.
TELO DA GAMA, DR. M.M.	Dpto. Fisica - Fac. Ciencias, Univ. Lisboa, Rua Ernesto Vasconcelos, C1 4 - Portugal.
TYVAND, DR. P. A.	Dept. Physics, Agricultural Univ. Norway, 1432 AAS - NLH - Norway.
VAIDYANATHAN, DR. V.S.	Medecine School, State Univ. New York, 118 Carry Hall, Buffalo, N.Y. 14214 - U.S.A.
VELARDE, PROF. M. G.	Dpto. Fisica Fundamental Uned., Apdo. Correos 60.141, 28071 Madrid - Spain.
VILCHEZ, PROF. J. C.	Centro Asociado Uned, Sanlucar De Barrameda, 1, 21001 Huelva - Spain.
VON DEN BRINCKEN, MR. P.	Giesserei-Institutt, Rhein-Westf. Tech. Hochschule Intzestrasse 5, 5100 Aachen - F. R. Germany.
WIDOM, PROF. B.	Dpt. Chemistry, Baker Lab., Cornell Univ., Ithaca, N.Y. 14853 - U.S.A.
ZAPPOLI, DR. B. C.	CNES-PFM/ Germe, 18, Av. Edouard Belin, 31055 Toulouse Cedex - France.
ZUÑIGA, DR. I.	Dpto. Fisica Fundamental Uned, Apdo. Correos 60141, 28.071 Madrid - Spain.

AUTHOR INDEX

Acrivos	1	D'Aguano	915
Adler	19	Davis	227
Als-Nielsen	639	De Cheveigné	587
Alts	915	D'Humières	71
Avnir	417	Dijstra	337
Baret	857	Earnshaw	163
Barrero	53	Elie	963
Barthes-Biesel	97	Evans	751
Baus	683, 787	Extremet	963
Ben Amar	527		
Ben Hadid	997		
Ben-Shaul	843	Favier	595
Bertrand	387	Fernández-Nóvoa	411
Beysens	811	Fife	1067
Bontoux	963, 997	Flores	833
Borghi	469	Forstmann	915
Branger	963		
		Gallegos	833
Camel	595	Gañán	53
Cazabat	709	García-Ybarra	179
Cerisier	199	Gelbart	843
Chang	115, 125, 135	Goldbeck-Wood	583
Chen	343	Goldenfeld	547
Chu	183, 343	Gouin	667
Ciliberto	1053	Grauer	571
Clavin	435	Guenon	811
Cohen-Stuart	709	Gillon	387
Coriell	559	Guthmann	587
Courville	387		
Crespo	963, 997		
Cuadros	797		

Haken	571
Halperin	741
Hauge	699
Hess	423
Holly	401
Jacquin	19
Kagan	417
Knobler	857
Koschmieder	189
Kumazawa	87
Lallemand	71
Lallemant	387
Lamprecht	291
Langevin	147
Lebon	253
Lebrun	587
Leger	721
Legros	209
Libchaber	515
Limbourg-Fontaine	209
López-Quintela	411
Mahé	19
Marconi	751
Martínez	25
Maslanik	949
Masuda	1091
Matsui	87
Maza	939
McFadden	559
McGivern	163
Meister	97
Meunier	147
Miguelez	939
Morala	125
Morales	797
Moulai-Mostefa	97
Müller	423

Muller-Krumbhaar	583
Muñoz	833
Nakache	359
Navascués	775
Nicolaenko	1029
Nielaba	915
Occelli	199
Pantaloni	199
Pérez	833
Perrot	811
Peskin	949
Petré	209
Platten	311
Plesser	423
Ploog	619
Pomeau	487, 527
Quibén	411
Raharimalala	359
Randriamampianina	963, 997
Rondelez	857
Roux	963, 997
Rull	797
Said	469
Saito	583
Saqués	1085
Sanfeld	367
Sani	949
San Miguel	1085
Sarma	271
Scharman	291
Schlick	881
Schwabe	291
Searby	71

Sekerka	559
Sen	761
Smutek	963
Steinchen-Sanfeld	367, 387
Sullivan	761
Suresh	857
Szleifer	843
Tabeling	515
Tarazona	775
Telo da Gama	753
Tyvand	173
Vaidhyanathan	897
Veira	939
Velarde	179, 183, 343
Vidal	939
Vignes-Adler	19
Villers	311
Widom	657
Yamashita	87
Zocchi	515

SUBJECT INDEX

Adaptive mesh technique, 955
Antonow's rule, 659

Bénard convection, 181, 189, 196, 199, 200, 204, 213, 272, 343, 391
Biot number, 273, 319, 338
Bond number, 26, 43, 183, 211, 273, 395
Born-Green-Yvon Eqs.,
Bubbles, 71, 82, 87, 90, 997, 133
Butler-Volmer Eq., 951

Capillary/Crispation number, 1, 26, 97, 179, 256
Capillary crystallization, 775
Capillary waves, 122, 147, 155, 163, 179, 183, 273
Cellular automata, 472
Chaos, 1029
Chemical waves, 423, 426
Combustion, 435, 443
Contact lenses, 407
Convection and chemical reaction, 417, 423, 426
Creeping flow, 3
Critical phenomena, 811
Crocco-Vazzonyi Eq., 675

Crystal growth, 39, 45, 454, 547
(*see also* Microgravity)
and convection, 283, 291, 559, 571, 616, 963
dendritic growth, 547, 587, 613
directional solidification, 595
floating zone, 39
MBE growth, 619
vapour, 963
zone refining, 39

de Gennes theory
liquid crystals, 761
wetting, 729
Disjoining pressure, 722
Double-diffusion convection, 348, 565
Drops, 1, 19, 730
breakdown, 1
creeping flow, 3
shape, 53
slender-body theory, 6
stability, 1

Elasticity modulus/number, 186, 343, 369, 372
Electrical double layer, 382, 683, 724, 899, 915
Electrochemistry, 949
Electrodeposition, 957

Electro-magneto-capillarity, 115, 120, 155
Electroplating, 959
Ellipsometry, 158, 738
Energy stability theory, 234
Epifluorescence microscopy, 877
ESR, 881
Eutectic solidification, 612

Finite-difference method, 1010
Finite-element algorithm, 954
Flames, 436
    stability, 441, 461, 469
Floating zone, 39, 4 2, 46, 300
Fluorescence microscopy, 857, 877
Fredericksz transition, 1085

g-jitter, 609
Grashof number, 211, 565, 1006
Gravity waves, 173, 179 (*see also* Capillary waves)

Hadley circulation, 1004
Hamaker constant, 733
Hard spheres, 791
HBT, 634
Hele-Shaw cell, 554
HEMT, 634
Hickman number, 395
Hydrothermal instability, 239

Infrared thermography, 393
Interfaces
    canthoaxis, 32
    configurations, 27
    shape, 31
    stability, 31
Intermittencies, 1029, 1053
Ising model (lattice gas), 658
Ivantsov parabola, 509

Kelvin-Helmholtz instability, 76
Kuramoto-Sivashinsky Eq., 1029, 1054
Kuramoto-Velarde Eq., 1054

Landau-Ginzburg theory, 571, 700
Langmuir law, 184, 346, 857
Lattice gas, 71, 78, 658
Lewis number, 435, 457
Light scattering technique, 154, 163
Line tension, 659
Liquid bridge, 33, 47, 54
    configurations, 35, 4 4, 46, 54
    solidification radius, 49
    stability, 33
Liquid crystals, 647, 757, 761, 1085
Liquid-solid coexistence, 787
Lovett-Mou-Buff-Wertheim Eq., 694

Marangoni convection, 181, 185, 199, 227, 231, 253, 273, 291, 295, 311, 337, 1003 (*see also* Bénard convection)
    and chemical reaction, 381, 417
    and evaporation, 387
Marangoni number, 179, 183, 211, 230, 254, 273, 303, 319, 338, 343, 362, 395, 986
Maxwell-Caltaneo fluid, 259
Micelle, 848
Microemulsions, 411, 664
Microgravity, 26, 196, 216, 299, 421, 609, 811
    D1 mission, 220
    parabolic flights, 217
    Texas flights, 216, 825
Mullins-Sekerka instability, 547, 559, 571

Needle crystal, 463, 505, 527, 552
Neumann triangle, 659
Non-Boussinesquian effects, 256, 273
Nusselt number, 136

Ozone generation, 1092

Peclet number, 607
Phase diagram, 40, 833, 947
Prandtl number, 179, 230, 565, 997, 1006

Rayleigh-Bénard convection, 572, 1057
Rayleigh number, 195, 273, 338, 343, 392, 988, 1006
Reaction-diffusion, 1081
Reynolds number, 461, 951
RHEED, 621

Saffman-Taylor fingers, 488, 516, 527
Seebeck effect, 610
Schmidt number, 183, 338, 567, 951
Slaving principle, 577
Solidification front, 452
Solutal convection, 343, 359 (*see also* Double-diffusion convection)
Spectrophotometry, 425
Spinodal decomposition, 815
Spreading, 721, 727
Strouhal number, 607
Surface phase transitions, 753
Surface tension minimum, 211
Surfactants, 148, 168
Surface viscosities, 370

Taylor number, 273
Tear film, 401
　formation, 403
　rupture, 404
Thermocapillary convection, (*see* Bénard convection and Marangoni convection)
Two-Phase flow, 125, 135
Two-Phase models, 1068

Ultrasonic diagnosis, 126
　pulse echo,

Van der Waals theory, 657, 723
Vapor recoil instability, 398

Wetting, 661, 699, 709, 722, 761
　precursor film, 731, 741
　transitions,

Zeldovich number, 439

MIX
Papier aus verantwortungsvollen Quellen
Paper from responsible sources
FSC® C105338

If you have any concerns about our products,
you can contact us on
ProductSafety@springernature.com

In case Publisher is established outside the EU,
the EU authorized representative is:
**Springer Nature Customer Service Center GmbH
Europaplatz 3, 69115 Heidelberg, Germany**

Printed by Libri Plureos GmbH
in Hamburg, Germany